PROCESS CONTROL

Designing Processes and Control Systems for Dynamic Performance

McGraw-Hill Chemical Engineering Series

Building the Literature of a Profession

Fifteen prominent chemical engineers first met in New York more than 60 years ago to plan a continuing literature for their rapidly growing profession. From industry came such pioneer practitioners as Leo H. Baekeland, Arthur D. Little, Charles L. Reese, John V. N. Dorr, M. C. Whitaker, and R. S. McBride. From the universities came such eminent educators as William H. Walker, Alfred H. White, D. D. Jackson, J. H. James, Warren K. Lewis, and Harry A. Curtis. H. C. Parmelee, then editor of *Chemical and Metallurgical Engineering,* served as chairman and was joined subsequently by S. D. Kirkpatrick as consulting editor.

After several meetings, this committee submitted its report to the McGraw-Hill Book Company in September 1925. In the report were detailed specifications for a correlated series of more than a dozen texts and reference books, which have since become the McGraw-Hill Series in Chemical Engineering and which became the cornerstone of the chemical engineering curriculum.

From this beginning there has evolved a series of texts surpassing by far the scope and longevity envisioned by the founding Editorial Board. The McGraw-Hill Series in Chemical Engineering stands as a unique historical record of the development of chemical engineering education and practice. In the series one finds the milestones of the subject's evolution: industrial chemistry, stoichiometry, unit operations and processes, thermodynamics, kinetics, and transfer operations.

Chemical engineering is a dynamic profession, and its literature continues to evolve. McGraw-Hill, with its editor, B. J. Clark, and its consulting editors, remains committed to a publishing policy that will serve, and indeed lead, the needs of the chemical engineering profession during the years to come.

The Series

Bailey and Ollis: *Biochemical Engineering Fundamentals*
Bennett and Myers: *Momentum, Heat, and Mass Transfer*
Brodkey and Hershey: *Transport Phenomena: A Unified Approach*
Carberry: *Chemical and Catalytic Reaction Engineering*
Constantinides: *Applied Numerical Methods with Personal Computers*
Coughanowr: *Process Systems Analysis and Control*
de Nevers: *Air Pollution Control Engineering*
de Nevers: *Fluid Mechanics for Chemical Engineers*
Douglas: *Conceptual Design of Chemical Processes*
Edgar and Himmelblau: *Optimization of Chemical Processes*
Gates, Katzer, and Schuit: *Chemistry of Catalytic Processes*
Holland: *Fundamentals of Multicomponent Distillation*
Katz and Lee: *Natural Gas Engineering: Production and Storage*
King: *Separation Processes*
Lee: *Fundamentals of Microelectronics Processing*
Luyben: *Process Modeling, Simulation, and Control for Chemical Engineers*
McCabe, Smith, and Harriott: *Unit Operations of Chemical Engineering*
Marlin: *Process Control: Designing Processes and Control Systems for Dynamic Performance*
Middleman and Hochberg: *Process Engineering Analysis in Semiconductor Device Fabrication*
Perry and Chilton (Editors): *Perry's Chemical Engineers' Handbook*
Peters: *Elementary Chemical Engineering*
Peters and Timmerhaus: *Plant Design and Economics for Chemical Engineers*
Reid, Prausnitz, and Poling: *Properties of Gases and Liquids*
Smith: *Chemical Engineering Kinetics*
Smith and Van Ness: *Introduction to Chemical Engineering Thermodynamics*
Treybal: *Mass Transfer Operations*
Valle-Riestra: *Project Evaluation in the Chemical Process Industries*
Wentz: *Hazardous Waste Management*

PROCESS CONTROL
Designing Processes and Control Systems for Dynamic Performance

Thomas E. Marlin
McMaster University

McGraw-Hill, Inc.
New York St. Louis San Francisco Auckland Bogotá Caracas Lisbon
London Madrid Mexico City Milan Montreal New Delhi
San Juan Singapore Sydney Tokyo Toronto

This book was set in Times Roman by Publication Services, Inc.
The editors were B. J. Clark and John M. Morriss;
the production supervisor was Kathryn Porzio.
The cover was designed by Initial Graphic Systems, Inc.
Project supervision was done by Publication Services, Inc.
R. R. Donnelley & Sons Company was printer and binder.

PROCESS CONTROL
Designing Processes and Control Systems for Dynamic Performance

This book is printed on acid-free paper.

1 2 3 4 5 6 7 8 9 0 DOC DOC 9 0 9 8 7 6 5

ISBN 0-07-040491-7

Library of Congress Cataloging-in-Publication Data

Marlin, T. E.
Process control: designing processes and control systems for dynamic performance / Thomas Marlin.
p. cm.—(McGraw-Hill chemical engineering series)
Includes index.
ISBN 0-07-040491-7
1. Chemical process control. I. Series.
TP155.75.M365 1995
660′.2815—dc20 94-45967

ABOUT THE AUTHOR

Thomas E. Marlin is Professor of Chemical Engineering at McMaster University, Hamilton, Ontario, Canada, where he holds the NSERC Industrial Research Chair in Process Control. He received his Ph.D. degree in Chemical Engineering from the University of Massachusetts in 1972, after which he applied advanced process control in industry for 15 years. In 1987, he served as the Visiting Fellow for a Warren Centre Study of process automation in Australia that identified potential benefits and technology needs in a wide range of industries. He is currently Director of the McMaster Advanced Control Consortium, a group of university personnel and industrial companies collaborating on automation research for the process industries. His research interests include real-time advanced control and optimization. In addition to teaching university undergraduate and graduate courses, Dr. Marlin consults widely and teaches numerous industrial short courses on control technology, control benefits analysis, and plant operations optimization.

CONTENTS

PREFACE

Automation via feedback principles is not new. Early application of automatic control principles appeared in antiquity; widespread use of automation began in the nineteenth century, when machinery was becoming the dominant method for manufacturing goods. Great advances have been made in theory and practice, so that automation is now used in systems as commonplace as room heating and automotive speed control and as exciting as the navigation of interplanetary exploration and telecommunications. The great change over the recent years is the integral—at times essential—role of automation in our daily lives and industrial systems.

Process control is a subdiscipline of automatic control that involves the selection and tailoring of methods for the efficient operation of chemical processes. Proper application of process control can improve the safety and profitability of a process while maintaining consistently high product quality. The automation of selected functions has relieved plant personnel of tedious, routine tasks, providing them with time and data to monitor and supervise operations. Essentially every chemical engineer designing or operating plants is involved with, and requires some expertise in, process control. This book provides an introduction to process control with emphasis on topics that are of use to the general chemical engineer as well as the specialist.

GOALS OF THIS BOOK

Since many process control books already exist, it is worthwhile commenting on the goals of this book. Most university textbooks present essentially the same principles; in contrast, books written for practitioners tend to concentrate on heuristics and case studies. The intent of this book is to present the full complement of fundamental principles with clear ties to application and with guidelines on their reduction to practice. The presentation is based on four basic tenets.

Fundamentals

First, engineers must master the fundamental control technology, because there is no shortcut set of heuristics that can serve them through their careers. (Topics that can be shortcut are by definition not fundamental.) Since these fundamentals must be presented with mathematical rigor, the mathematical tools—basically modelling and differential equations—are reviewed in the earlier chapters. The automation principles associated with stability, feedback, controllability, and so forth are introduced as the book progresses. Finally, it may be worth emphasizing that these principles were selected because they provide the *simplest* approaches for solving meaningful problems.

Practice

It is not necessary or efficient to start from scratch every time a problem is encountered; similar situations can be analyzed to develop guidelines for a defined set of applications. Also, the fundamental concepts can be best reinforced and enriched through the presentation of good engineering practice. With this perspective, important design guidelines and enhancements are presented as logical conclusions and extensions to the basic principles. The coverage of implementation issues includes pitfalls with the straightforward "textbook" approaches, along with modifications for practical application. The examples of translating theory to practice are intended to encourage the readers to begin to develop their own heuristics as they study the material and prepare for engineering practice.

Complexity

The presentation in this book follows the guideline, "Everything should be made as simple as possible, and no simpler." Naturally, many issues are easily resolved using straightforward analysis methods. However, the engineer must understand the complexity of automating a system, because closed-form mathematical solutions do not exist at the present time for feedback control performance. This situation is encountered in addressing the latter topics of multivariable control and control design.

Design

Design is a capstone topic that enables engineers to specify, build, and operate equipment that satisfies predetermined goals. Currently, closed-form solutions do not exist for this activity; thus, a comprehensive design method for managing the numerous interlocking design tasks is presented, along with a step-by-step approach to guide the engineer through problem definition, preliminary analysis of degrees of freedom and controllability, and selecting process and control structures. Many guidelines, checklists, and examples aid the student in making well-directed initial decisions and refining them through iterations to achieve the design goals.

The book begins with a brief introduction to the goals of process control in Part I, which emphasizes the close relationship between the process and its automation.

Parts II and III present the fundamentals of mathematical modelling, stability analysis, and analytical and numerical solutions of differential equations, whereas the engineering practice is addressed in the tuning, applications issues, and digital implementation chapters. Part IV presents enhancements such as cascade, feedforward, and adaptive tuning in a manner that reinforces the fundamental role of feedback, builds on its strengths, and compensates for its limitations. The level of complexity increases significantly in Part V on multivariable control. Analysis methods are given that provide a firm fundamental understanding and are useful in making many decisions, but it is recognized that the final control system and tuning may require some trial and error via simulation or plant trials. All parts provide intermediate practical design results, and Part VI addresses the topic of design from a rather broad perspective that chooses control calculations, sensors, final elements, control structure, and even the process itself to achieve good dynamic performance.

THE READERS

Hopefully, readers with different backgrounds will find value in this treatment of process control. A few comments are now addressed to the three categories of likely readers of this book: university students, instructors, and practitioners. (Perhaps the comments will be of interest to all readers.)

Students

Process control can be one of the most interesting and enjoyable courses in the curriculum. In this course you begin to apply the skills built in fluid mechanics, heat transfer, thermodynamics, mass transfer, and reactor design to process systems and begin to consider how they should be designed and operated in realistic situations: when goals change and disturbances occur. This presentation emphasizes the central role of the process in the performance of control systems. Process control is not (merely) the study of algorithms, because the process dynamic response plays an integral part of every process control system. Therefore, quantitative understanding using dynamic models is introduced early and used throughout the book. To help the student, realistic process systems are used as examples. Several example processes are used repeatedly in the book, and the student can refer to Appendix G for a guide to these key example processes.

In this topic the student may notice a subtle difference in emphasis from other courses. First, process control is often concerned with operating plants in which process equipment has been built and can be modified only at considerable cost. Thus, the proper answer to the question, "How can the exchanger outlet temperature be raised to 56°C?" is not, "Increase the heat transfer area"; rather, the heating medium flow rate should be modified. Thus, the variables available for change are different in the plant operations situation. Also, process control must operate over a wide range of conditions in which the process behavior will change; thus, the engineer must design controls for good performance with an imperfect knowledge of the plant. Deciding operation policies for imperfectly known, nonlinear processes is challenging but provides an excellent opportunity to apply skills from previous courses, while building expertise in process control.

Instructors

This book is intended to provide a basis of fundamentals and practice to enable the instructor to formulate a course meeting the needs and integrated with the curricula of most departments. It is expected that each instructor will modify the offering to provide his or her special insights, perhaps placing more emphasis on instrumentation, mathematical analysis, or a special process type such as pulp and paper, batch processing, or polymer processing. The fundamental topics have been selected to enable subsequent study of many processes, and the organization of the last three parts of the book allows the selection of material most suited for the particular course.

This book offers considerably greater coverage of single-loop enhancements, multivariable control, and control design than most other texts. This extension required some reductions in other areas. In particular, detailed coverage of z-transforms has been excluded from this text. The book presents analytical methods for continuous systems and, where necessary, provides the discrete formulation of real-time algorithms for digital implementation. Thus, the student is able to program the PID algorithm with anti–reset windup in single-loop, cascade, feedforward, gain scheduling, decoupling, and other structures, as well as single-loop IMC and multivariable DMC algorithms. Also, the student will be able to simulate control systems numerically and to fit dynamic models using the least squares method. This is quite sufficient at the undergraduate level.

After reviewing the table of contents, the instructor may ask where the hot topic of robustness is addressed. The answer is, "Everywhere." Modelling, starting with the first worked example, addresses the effects of imperfect knowledge and linearization on the accuracy of our predictions of dynamic behavior. This analysis continues throughout the book, since new models are developed in all sections (never stop modelling!). Also, control performance is defined to include the behavior of all important variables (controlled *and* manipulated) for variations in process dynamics and significant measurement noise. Engineers trained in this manner are sensitive to the real needs of feedback algorithms and open to more advanced analysis using higher-level mathematical tools.

The material in this course certainly exceeds that necessary for a single-semester course. Most instructors will want to cover Parts I through III along with selected topics from the remainder of the book in a first course. A second-semester course can be built on the multivariable and design material, along with some nonlinear simulations of such chemical processes as binary distillation. Finally, some of the topics in this book should also be helpful in other courses. In particular, topics in Parts IV through VI could be integrated into the process design course. Certainly, sensor selection, manipulated-variable selection, and inferential monitor selection are appropriate topics in process design. In addition, the analysis of operating windows, degrees of freedom, and controllability is facilitated by the flowsheeting programs used in a design course.

The calculations associated with process control analysis can sometimes be performed by hand for simple systems. However, these calculations—stability analysis, frequency response, and dynamic responses—are complex and time-consuming for most realistic systems; they must be performed by computer. To facilitate process control education, a complementary Software Laboratory and Workbook are available for use at universities. With this software, students can repeat over 100 textbook examples and the Workbook suggests many extensions to investigate additional issues. The Software and Workbook can be obtained by contacting a McGraw-Hill representative or the author.

Practitioners

This book should be useful to practitioners who are building their skills in process control. Fundamental theoretical concepts are reduced to practice throughout: in empirical modelling; controller tuning; filtering; designing enhancements such as cascade and feedforward; modifying methods for special situations such as level control; selecting loop pairings; and control system design. The development of practical correlations, design rules, and guidelines is explained so that the engineer understands the basis, correct application, and limitations of each. This book includes more modelling and analysis, such as frequency response, than is typical for practitioners. It is my strong belief that these topics are essential for good engineering practice; in fact, one industrial challenge problem, recently defined by Tennessee Eastman, specifies control performance not only as the variability of selected variables but also as the frequency range of the variability (Downs, J., and E. Vogel, "A Plant-wide Industrial Process Control Problem," *Comp. Chem. Eng., 17,* 245–255, 1993). Hopefully, the emphasis on the interpretation of the amplitude ratios (e.g., in Chapter 13) will convince the reader of the merit in learning and using these concepts.

ACKNOWLEDGMENTS

I am pleased to acknowledge the helpful comments and suggestions of many colleagues. The following people reviewed drafts of this book and used it in teaching undergraduate courses: Yaman Arkun, Ali Cinar, Will Cluett, Andy Hrymak, Tom McAvoy, Jim McLellan, Michel Perrier, Andrew Ogden-Swift, and Russ Rhinehart. They provided many corrections and suggestions, which improved the final product. Naturally, any remaining errors are my responsibility.

Also, many students have used early drafts of the book and have, through their questions, indicated where the explanations fell short of complete clarity.

The Natural Science and Engineering Research Council of Canada provides the opportunity for experienced practitioners to return to university teaching and research through their Industrial Research Chair program. Without this program, I would not have enjoyed the stimulating environment at McMaster University with, among others, Cam Crowe, Andy Hrymak, John MacGregor, and Paul Taylor, nor would I have written this book.

Finally, I would like to acknowledge the great assistance provided me by two mentors. Professor Tom McAvoy has always set high standards of rigor in investigating meaningful engineering problems; this project suffered greatly when he decided not to participate in this book in order to serve in a demanding administrative position. Dr. Nino Fanlo, one of the best practitioners of process control, gently reminded me that good control theory must work in the plant. I can only hope that this book passes on some of the benefits from collaboration with these fine engineers.

FEEDBACK

Feedback—using the output or results to determine an input—is the basic concept in process control, but it also applies to a good textbook! I would appreciate comments and suggestions from readers and can assure you that they will be considered seriously.

Thomas E. Marlin

PART I

INTRODUCTION

There is an old adage, "If you do not know where you are going, any path will do." In other words, a good knowledge of the goal is essential before one addresses the details of a task. Engineers should keep this adage in mind when studying a new, complex topic, because they can easily become too involved in the details and lose track of the purpose of learning the topic. Process control requires the application of previously studied principles in mathematics and the engineering sciences when learning the new concepts associated with the dynamic operation of process systems. Thus, the topic offers many interesting details, and the challenge to the reader is to learn these details as essential elements leading to the ultimate goal. Process control is introduced in this first, brief part of the book so that the reader will understand the overall goal of process automation and appreciate the need for the technical rigor of the subsequent parts.

The study of process control introduces a new perspective to the mastery of process systems: *dynamic operation.* Prior engineering courses in the typical curriculum concentrate on steady-state process behavior, which simplifies early study of processes and provides a basis for establishing proper equipment sizes and determining the best constant operating conditions. However, no process operates at a steady state (with all time derivatives exactly zero), because essentially all external variables, such as feed composition or cooling medium temperature, change. Thus, the process design must include systems that respond to external disturbances and maintain the process operation in a safe region that yields high-quality products in a profitable manner. The emphasis on good operation, achieved through proper plant design and automation, requires a thorough knowledge of the dynamic operation, which is introduced in this part and covered thoroughly in Part II.

In addition, the study of process control introduces a major new concept: *feedback control.* This concept is central to most automation systems that monitor a process and adjust some variables to maintain the system at (or near) desired conditions. Feedback is one of the topics studied and employed by engineers of most subdisciplines, and chemical engineers apply these principles to heat exchangers, mass transfer equipment, chemical reactors, and so forth. Feedback control is introduced in this part and covered in detail in Part III.

Finally, the coverage of these topics in this part is qualitative, because it precedes the introduction of mathematical tools. This qualitative presentation is not a shortcoming; rather, the direct and uncomplicated presentation provides a clear and concise discussion of some central ideas in the book. The reader is advised to return to Part I to clarify the goals before beginning each new part of the book.

CHAPTER 1

INTRODUCTION TO PROCESS CONTROL

1.1 INTRODUCTION

When observing a chemical process in a plant or laboratory, one sees flows surging from vessel to vessel, liquids bubbling and boiling, viscous material being extruded, and all key measurements changing continuously, sometimes with small fluctuations and other times in response to major changes. The conclusion immediately drawn is that the world is dynamic! This simple and obvious statement provides the key reason for process control. Only with an understanding of transient behavior of physical systems can engineers design processes that perform well in the dynamic world. In their early training, engineering students learn a great deal about steady-state physical systems, which is natural, because steady-state systems are somewhat easier to understand and provide appropriate learning examples. However, the practicing engineer should have a mastery of dynamic physical systems as well. This book provides the basic information and engineering methods needed to analyze and design plants that function well in a dynamic world.

Control engineering is an engineering science that is used in many engineering disciplines—for example, chemical, electrical, and mechanical engineering—and it is applied to a wide range of physical systems from electrical circuits to guided missiles to robots. The field of *process control* encompasses the basic principles most useful when applied to the physicochemical systems often encountered by chemical engineers, such as chemical reactors, heat exchangers, and mass transfer equipment.

Since the principles covered in this book are basic to most tasks performed by chemical engineers, control engineering is not a narrow specialty but an essential topic for all chemical engineers. For example, plant designers must consider the dynamic operation of all equipment, because the plant will never operate at steady state (with time derivatives exactly equal to zero). Engineers charged with operating plants must ensure that the proper response is made to the ever-occurring disturbances so that operation is safe and profitable. Finally, engineers performing experiments must control their equipment to obtain the conditions prescribed by their experimental designs. In summary, the task of engineers is to design, construct, and operate a physical system to behave in a desired manner, and an essential element of this activity is sustained maintenance of the system at the desired conditions—which is process control engineering.

As you might expect, process control engineering involves a vast body of material, including mathematical analysis and engineering practice. However, before we can begin learning the specific principles and calculations, we must understand the goals of process control and how it complements other aspects of chemical engineering. This chapter introduces these issues by addressing the following questions:

- What does a control system do?
- Why is control necessary?
- Why is control possible?
- How is control done?
- Where is control implemented?
- What does control engineering "engineer"?
- How is process control documented?
- What are some sample control strategies?

1.2 WHAT DOES A CONTROL SYSTEM DO?

First, we will discuss two examples of control systems encountered in everyday life. Then, we will discuss the features of these systems that are common to most control systems and are generalized in definitions of the terms *control* and *feedback control.*

The first example of a control system is a person driving an automobile, as shown in Figure 1.1. The driver must have a goal or objective; normally, this would be to stay in a specific lane. First, the driver must determine the location of the automobile, which she does by using her eyes to see the position of the automobile on the road. Then, the driver must determine or calculate the change required to maintain the automobile at its desired position on the road. Finally, the driver must change the position of the steering wheel by the amount calculated to bring about the necessary correction. By continuously performing these three functions, the driver can maintain the automobile very close to its desired position as disturbances like bumps and curves in the road are encountered.

The second example is the simple heating system shown in Figure 1.2. The house, in a cold climate, can be maintained near a desired temperature by circulating hot water through a heat exchanger. The temperature in the room is determined by a thermostat, which compares the measured value of the room temperature to a desired range, say 18 to 22°C. If the temperature is below 18°C, the furnace and pump are turned on, and if the temperature is above 22°C, the furnace and pump are turned off. If the temperature is between 18 and 22°C, the furnace and pump statuses remain

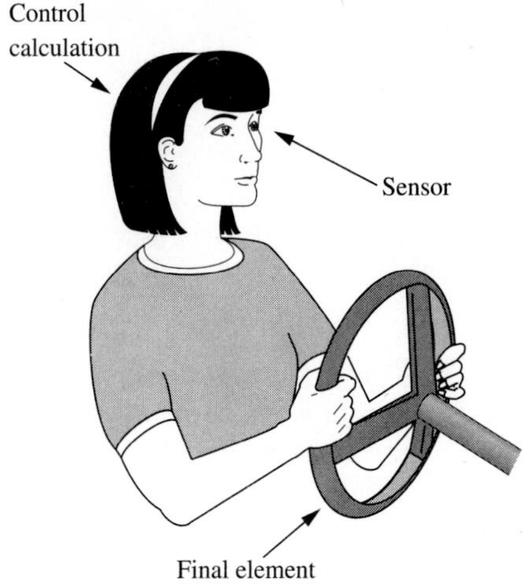

FIGURE 1.1
Example of feedback control for steering an automobile.

unchanged. A typical temperature history in a house in given in Figure 1.3, which shows how the temperature slowly drifts between the upper and lower limits. It also exceeds the limits, because the furnace and heat exchanger cannot respond immediately. This approach is termed "on/off" control and can be used when precise control at the desired value is not required. We will cover better control methods, which can maintain important variables much closer to their desired values, later in this book.

Now that we have briefly analyzed two control systems, we shall identify some common features. The first is that each uses a specific value (or range) as a desired value for the controlled variable. When we cover control calculations in Part III, we will use the term *set point* for the desired value. Second, the conditions of the system are measured; that is, all control systems use sensors to measure the physical variables that are to be maintained near their desired values. Third, each system has a control calculation, or *algorithm,* which uses the measured and the desired values to determine a correction to the process operation. The control calculation for the room heater is very simple (on/off), whereas the calculation used by the driver may be very complex. Finally, the results of the calculation are implemented by adjusting some item of equipment in the system, which is termed the *final control element,* such as

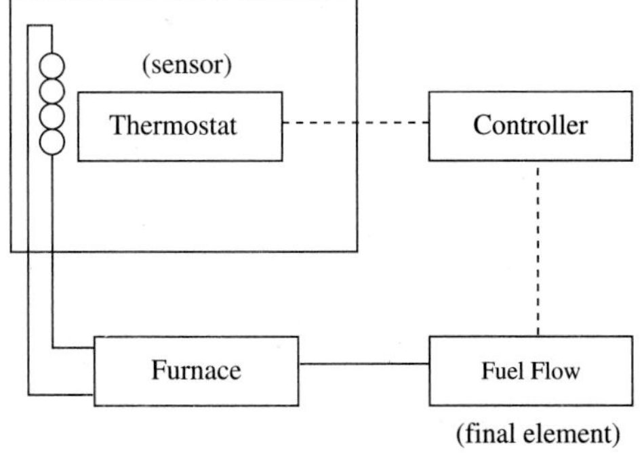

FIGURE 1.2
Example of feedback control for controlling room temperature.

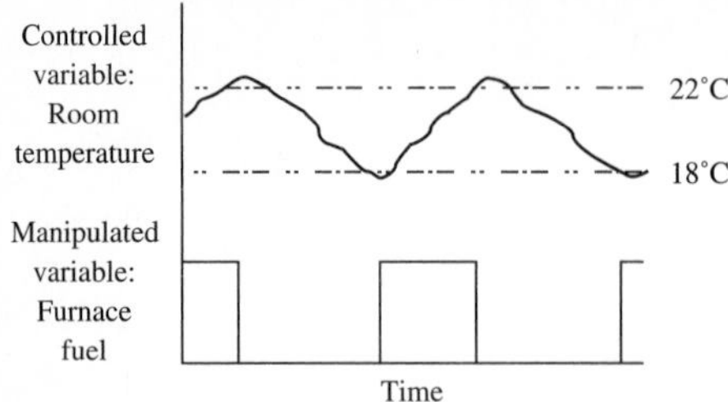

FIGURE 1.3
Typical dynamic response of the room temperature when controlled by on/off feedback control.

the steering wheel or the furnace and pump switches. These key features are shown schematically in Figure 1.4, which can be used to represent many control systems.

Now that we have discussed some common control systems and identified key features, we shall define the term *control.* The dictionary provides the definition for the verb *control* as "to exercise directing influence." We will use a similar definition that is adapted to our purposes. The following definition suits the two physical examples and the schematic representation in Figure 1.4.

> **Control** (verb): To maintain desired conditions in a physical system by adjusting selected variables in the system.

The control examples have an additional feature that is extremely important. This is *feedback,* which is defined as follows:

> **Feedback control** makes use of an output of a system to influence an input to the same system.

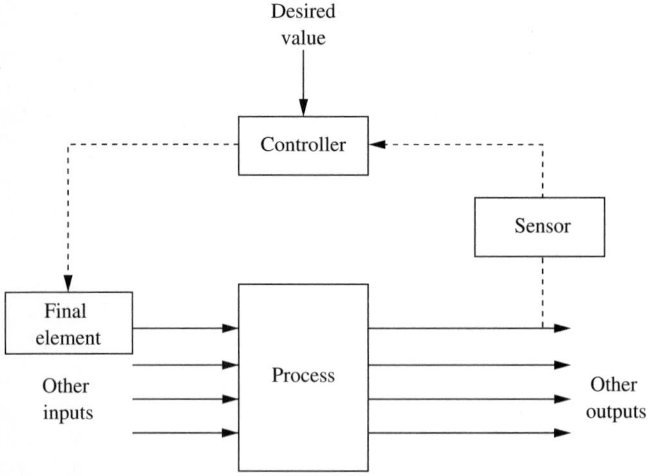

FIGURE 1.4
Schematic diagram of a general feedback control system showing the sensor, control calculation based on a desired value, and final element.

For example, the temperature of the room is used, through the thermostat on/off decision, to influence the hot water flow to the exchanger. When feedback is employed to *reduce* the magnitude of the difference between the actual and desired values, it is termed "negative feedback." Unless stated otherwise, we will always be discussing negative feedback and will not use the modifier *negative.* In the social sciences and general vernacular, the phrase "negative feedback" indicates an undesirable change, because most people do not enjoy receiving a signal that tells them to correct an error or to reverse a tendency to deviate from the desired condition. Most people would rather receive "positive feedback," a signal telling them to continue a tendency to approach the desired condition. This difference in terminology is unfortunate; we will use the terminology for automatic control, with "negative" indicating a change that tends to approach the desired value, throughout this book without exception.

The importance of feedback in control systems can be seen by considering the alternative without feedback. For example, an alternative approach for achieving the desired room temperature would set the hot water flow based on the measured outside temperature and a model of the heat loss of the house. (This type of predictive approach, termed *feedforward,* will be encountered later in the book, where its use in combination with feedback will be explained.) The strategy without feedback would not maintain the room near the desired value if the model had errors—as it always would. Some causes of model error might be changes in external wind velocity and direction or inflows of air through open windows. On the other hand, feedback control can continually manipulate the final element to achieve the desired value. Thus, feedback provides the powerful feature of enabling a control system to maintain the measured value near its desired value *without requiring an exact plant model.*

Before we complete this section, the terms *input* and *output* are clarified. When used in discussing control systems, they do not necessarily refer to material moving into and out of the system. Here, the term *input* refers to a variable that *causes* an output. In the steering example, the input is the steering wheel position, and the output is the position of the automobile. In the room heating example, the input is the fuel to the furnace, and the output is the room temperature. It is essential to recognize that the input causes the output and that this relationship cannot be inverted. The causal relationship inherent in the physical process forces us to select the input as the manipulated variable and the output as the measured variable. Numerous examples with selections of controlled and manipulated variables are presented in subsequent chapters.

Therefore, the answer to the first question about the function of control is, "A feedback control system maintains specific variables near their desired values by applying the four basic features shown in Figure 1.4." Understanding and designing feedback control systems is a major emphasis of this book.

1.3 WHY IS CONTROL NECESSARY?

A natural second question involves the need for control. There are two major reasons for control, which are discussed with respect to the simple stirred-tank heat exchanger shown in Figure 1.5. The process fluid flows into the tank from a pipe

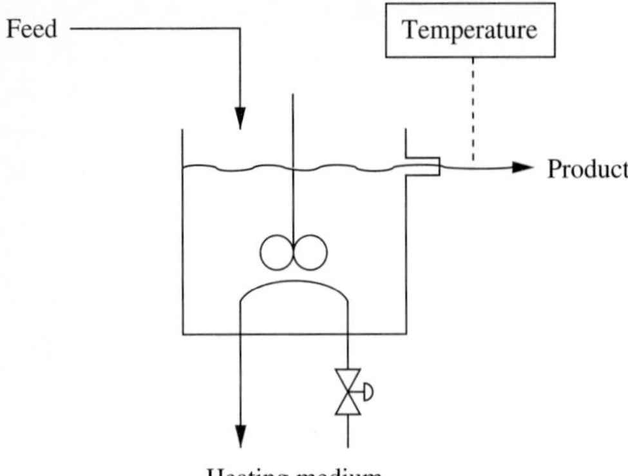

FIGURE 1.5
Schematic drawing of a stirred-tank heating process.

and flows out of the tank by overflow. Thus, the volume of the tank is constant. The heating fluid flow can be changed by adjusting the opening of the valve in the heating medium line. The temperature in the tank is to be controlled.

The first reason for control is to maintain the temperature at its desired value when *disturbances* occur. Some typical disturbances for this process occur in the following variables: inlet process fluid flow rate and temperature, heating fluid temperature, and pressure of the heating fluid upstream of the valve. As an exercise, you should determine how the valve should be adjusted (opened or closed) in response to an increase in each of these disturbance variables.

The second reason for control is to respond to changes in the *desired value.* For example, if the desired temperature in the stirred-tank heat exchanger is increased, the heating valve percent opening would be increased. The desired values are based on a thorough analysis of the plant operation and objectives. This analysis is discussed in Chapter 2, where the main issues are arranged in seven categories:

1. Safety
2. Environmental protection
3. Equipment protection
4. Smooth plant operation
5. Product quality
6. Profit optimization
7. Monitoring and diagnosis

These issues are translated to values of variables—temperatures, pressures, flows, and so forth—which are to be controlled.

1.4 WHY IS CONTROL POSSIBLE?

If the plant equipment is not designed properly, control will perform poorly or be impossible; therefore, the control and dynamic operation is an important factor in plant design. Based on the key features of feedback control shown in Figure 1.4, the plant design must include adequate sensors of plant output variables and appropriate final control elements. The sensors must respond rapidly so that the control action can be taken in *real time.* Sensors using various physical principles are available for the basic process variables (flow, temperature, pressure, and level), compositions (e.g., mole fraction) and physical properties (e.g., density, viscosity, heat of combustion). Many of these sensors are inserted into the process equipment, with a shield protecting them from corrosive effects of the streams. Others require a sample to be taken periodically from the process; note that this sampling can be automated so that a new sensor result is available at frequent intervals. The final control elements in chemical processes are usually valves that affect fluid flows, but they could be other manipulated variables, such as power to an electric motor or speed of a conveyor belt.

Another important consideration is the capacity of the process equipment. The equipment must have a large enough maximum capacity to respond to all expected disturbances and changes in desired values. For the stirred-tank heat exchanger, the maximum duty, as influenced by temperature, area, and heating medium flow rate, must be large enough to maintain the tank temperature for all anticipated disturbances. This highest heat duty corresponds to the the highest outlet temperature, the highest process fluid flow, the lowest inlet fluid temperature, and the highest heat loss to the environment. Each process must be analyzed to ensure that adequate capacity exists. Further discussion of this topic appears in the next two chapters.

Therefore, the answer to why control is possible is that we anticipate the expected changes in plant variables and provide adequate equipment when the plant is designed. The adequate equipment design for control must be calculated based on *expected changes;* merely adding a percentage extra capacity, say 20 percent, to equipment sizing is not correct. In some cases, this would result in waste; in other cases, the equipment capacity would not be adequate. If this analysis is not done properly or changes outside the assumptions occur, achieving acceptable plant operation through manipulating final control elements may not be possible.

1.5 HOW IS CONTROL DONE?

As we have seen in the automobile driving example, feedback control by human actions is possible. In some cases, this approach is appropriate, but the continuous, repetitious actions are tedious for a person. In addition, some control calculations are too complex or must be implemented too rapidly to be performed by a person. Therefore, most feedback control is automated, which requires that the key functions of sensing, calculating, and manipulating be performed by equipment and that each element communicate with other elements in the control system. Currently, most automatic control is implemented using electronic equipment, which uses levels of current or voltage to represent values to be communicated. As would be expected, many of the computing and some of the communication functions are being performed increasingly often with digital technology. In some cases control systems use pneumatic, hydraulic, or mechanical mechanisms to calculate and communicate; in these systems, the signals are represented by pressure or physical position. A typical process plant will have examples of each type of instrumentation and communication.

Since an essential aspect of process control is instrumentation, this book introduces some common sensors and valves, but proper selection of this equipment for plant design requires reference to one of the handbooks in this area for additional details. Readers are encouraged to be aware of and use the general references listed at the end of this chapter.

Obviously, the other key element of process control is a device to perform the calculations. For much of the history of process plants (up to the 1960s), control calculations were performed by analog computation. Analog computing devices are implemented by building a physical system, such as an electrical or mechanical system, that obeys the same equations as the desired control calculation. As you can imagine, this calculation approach was inflexible. In addition, complex calculations were not possible. However, some feedback control is still implemented in this manner, for reasons of cost and reliability in demanding plant conditions.

With the advent of low-cost digital computers, most of the control calculations and essentially all of the complex calculations are being performed by digital computers. Most of the principles presented in this book can be implemented in either analog or digital devices. When covering basic principles in this book, we will not distinguish between analog and digital computing unless necessary, because the distinction between analog and digital is not usually important as long as the digital computer can perform its discrete calculations quickly. Special aspects of digital control are introduced in Chapter 11. In all chapters after Chapter 11, the control principles are presented along with special aspects of either analog or digital implementation; thus, both modes of performing calculations are covered in an integrated manner.

For the purposes of this book, the answer to the question "How is control done?" is simply, "Automatically, using instrumentation and computation that perform all features of feedback control without requiring (but allowing) human intervention."

1.6 WHERE IS CONTROL IMPLEMENTED?

Chemical plants are physically large and complex. The people responsible for operating the plant on a minute-to-minute basis must have information from much of the plant available to them at a central location. The most common arrangement of control equipment to accommodate this need is shown in Figure 1.6. Naturally, the sensors and valves are located in the process. Signals, usually electronic, communicate with the control room, where all information is displayed to the operating personnel and where control calculations are performed. Distances between the process and central control room range from a few hundred feet to a mile or more. Some control is performed many miles from the process; for example, a remote oil well can have no human present and would rely on remote automation for proper operation.

In the control room, an individual is responsible for monitoring and operating a section of a large, complex plant, containing up to 100 controlled variables and 400 other measured variables. Generally, the plant never operates on "automatic pilot"; a person is always present to perform tasks not automated, to optimize operations, and to intervene in case an unusual or dangerous situation occurs, such as an

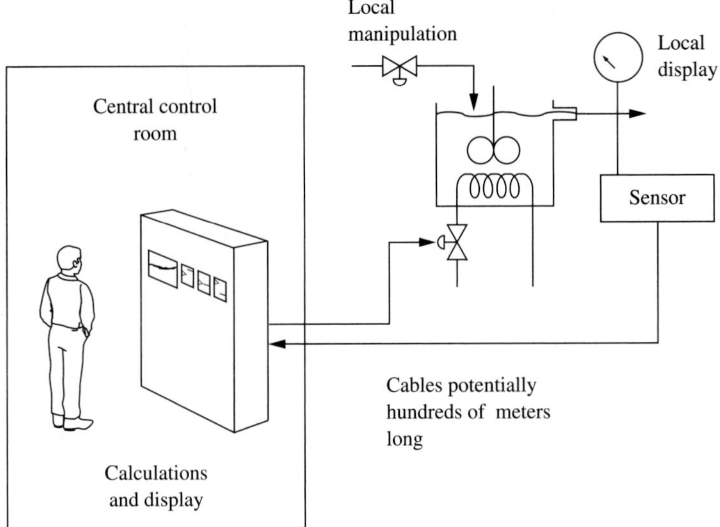

FIGURE 1.6
Schematic representation of a typical control system showing both local and centralized control equipment.

equipment failure. Naturally, other people are present at the process equipment, usually referred to as "in the field," to monitor the equipment and to perform functions requiring manual intervention, such as backwashing filters. Thus, well-automated chemical plants involve considerable interaction between people and control calculations.

Other control configurations are possible and are used when appropriate. For example, small panels with instrumentation can be placed near a critical piece of process equipment when the operator needs to have access to the control system while introducing some process adjustments. This arrangement would not prevent the remainder of the plant from being controlled from a central facility. Also, many sensors provide a visual display of the measured value, which can be seen by the local operator, as well as a signal transmitted to the central control room. Thus, the local operator can determine the operating conditions of a unit, but the individual local displays are distributed about the plant, not collected in a single place for the local operator.

The short answer to the location question is

1. Sensors, local indicators, and valves are in the process.
2. Displays of all plant variables and control calculations are in a centralized facility.

It is worth noting that increased use of digital computing makes the distribution of the control calculation to the sensor locations practical; however, all controllers would be connected to a computing network that would function like a single computer for the purposes of the material in this book.

1.7 WHAT DOES CONTROL ENGINEERING "ENGINEER"?

What can engineers do so that plants can be maintained reliably and safely near desired values? Most of the engineering decisions are introduced in the following five topics.

Process Design

A key factor in engineering is the design of the process so that it can be controlled well. We noted in the room heating example that the temperature exceeded the maximum and minimum values because the furnace and heat exchanger were not able to respond rapidly enough. Thus, a more "responsive" plant would be easier to control. By *responsive* we mean that the controlled variable responds quickly to adjustments in the manipulated variable. Also, a plant that is susceptible to few disturbances would be easier to control. Reducing the frequency and magnitude of disturbances could be achieved by many means; a simple example is placing a large mixing tank before a unit so that feed composition upsets are attenuated by the averaging effects of the tank. Many more approaches to designing responsive processes with few disturbances are covered in the book.

Measurements

Naturally, a key decision is the selection and location of sensors, because one can control only what is measured! The engineer should select sensors that measure important variables rapidly and with sufficient accuracy. A rapid sensor response is important because the feedback correction cannot be determined until the measurement is available; thus, the fast sensor is part of a responsive plant. The accuracy of the sensor depends on the physical principle upon which the sensor is based and the level of noise accompanying the "true" measurement. In this book, we will concentrate on the process analysis related to variable selection and to determining response time and accuracy needs. Details of a few common sensors are also presented as needed in exercises; a full review of sensor technology and commercial equipment is available in the references at the end of this chapter.

Final Elements

The engineer must provide *handles*—manipulated variables that can be adjusted by the control calculation. For example, if there were no valve on the heating fluid in Figure 1.5, it would not be possible to control the process fluid outlet temperature. The control performance of a process can often be improved by adding extra flexibility through bypass lines, which enable the control system to direct flow through or around heat exchangers and other units to achieve the desired product properties. Again, this book concentrates on the process analysis related to final element location. We will typically be considering control valves as the final elements, with the percentage opening of these valves determined by a signal sent to the valve from a controller. Specific details about the best final element to regulate flow of various fluids—liquids, steam, slurries, and so forth—are provided by references noted at the end of this chapter. These references also present other final elements, such as motor speed, that are used in the process industries.

Control Structure

The engineer must decide some very basic issues in designing a control system. For example, which valve should be manipulated to control which measurement? As an everyday example, one could adjust either the hot or cold water valve opening to control the temperature of water in a shower. Naturally, the correct structure, or *pairing,* would require a causal relationship between the final element and the controlled variable. Many other issues must be considered in complex chemical processes, such as favorable dynamic responses and reducing interaction among controllers. These topics are presented in later chapters, after a sound basis of understanding in dynamics and feedback control principles has been built.

Control Calculations

After the variables and control structure have been selected, equation(s) are chosen that use the measurement and desired values in calculating the manipulated variable. As we shall learn, only a few equations are sufficient to provide good control for many types of plants. After the control equations' structure is defined, parameters that appear in the equations are adjusted to achieve the desired control performance for the particular process.

It is important to recognize that good control engineering always requires a thorough understanding of the process. For example, the rider of a bicycle uses techniques different from those used by the driver of an automobile. Process understanding is provided to chemical engineers in courses on, among others, fluid mechanics, heat transfer, mass transfer, thermodynamics, and chemical reactor design and is reinforced and extended in this book, which covers modelling required for control engineering.

1.8 HOW IS PROCESS CONTROL DOCUMENTED?

As with all activities in chemical engineering, the results are documented in many forms. The most common are equipment specifications and sizing, operating manuals, and technical documentation of plant experiments and control equations. In addition, control engineering makes extensive use of drawings that concisely and unequivocally represent many design decisions. These drawings are used for many purposes, including designing plants, purchasing equipment, and reviewing operations, and safety procedures. Therefore, many people use them, and to avoid misunderstandings standard symbols have been developed by the Instrument Society of America for use throughout the world. We shall adhere to a reduced version of this excellent standard in this book because of its simplicity and wide application.

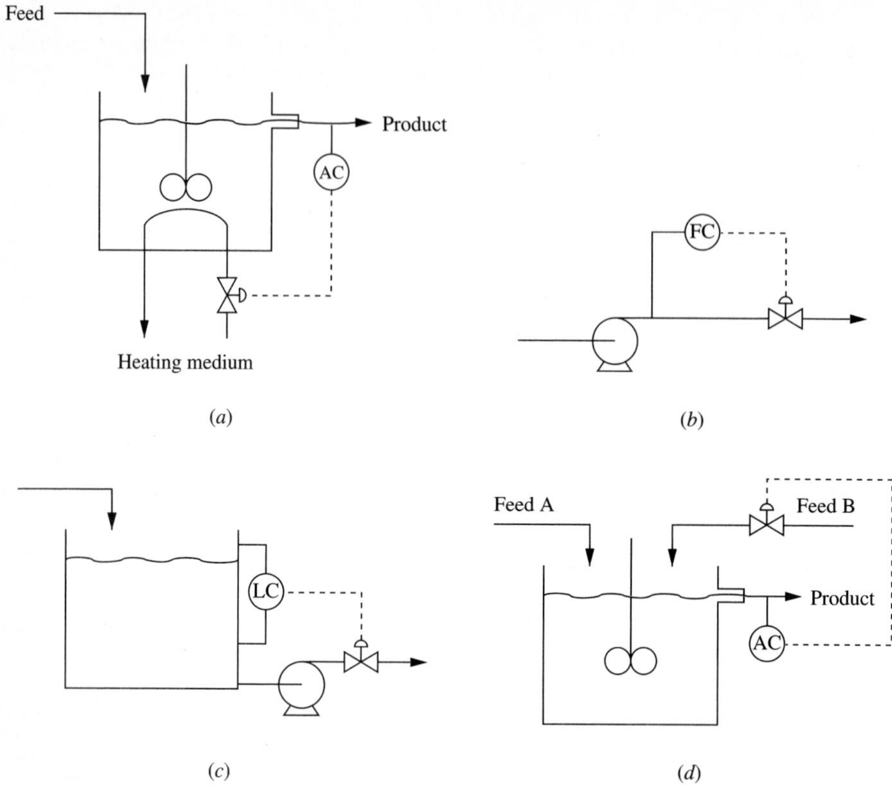

FIGURE 1.7
(*a*) Continuous stirred-tank reactor with composition control. (*b*) Flow controller. (*c*) Tank level with controller. (*d*) Mixing process with composition control.

Sample drawings are shown in Figure 1.7. All process equipment—piping, vessels, valves, and so forth—is drawn in solid lines. The symbols for equipment items such as pumps, tanks, drums, and valves are simple and easily recognized. Sensors are designated by a circle or "bubble" connected to the point in the process where they are located. The first letter in the instrumentation symbol indicates the type of variable measured; for example, "T" corresponds to temperature. Some of the more common designations are the following:

- A Analyzer (specific analysis is often indicated next to the symbol, for example, ρ (for density) or pH)
- F Flow rate
- L Level of liquid or solids in a vessel
- P Pressure
- T Temperature

Note that the symbol does not indicate the physical principle used by the sensor. Backup tabular documentation is required to determine such details.

The communication to the sensor is shown as a solid line. If the signal is used only for display to the operator, the second letter in the symbol is "I" for indicator. Often, the "I" is not used, so that a single letter refers to a measurement used for monitoring only, not for control.

If the signal is used in a calculation, it is also shown in a circle. The second letter in the symbol indicates the type of calculation. We consider only two possibilities in this book: "C" for feedback control and "Y" for any other calculation, such as addition or square root. The types of control calculations are covered later in the book. A noncontrol calculation might use the measured flow and temperatures around a heat exchanger to calculate the duty; that is, $Q = \rho C_p F(T_{in} - T_{out})$. For controllers, the communication to the final element is also shown as a dashed line when it is electrical, which is the mode communication considered in designs for most of this book.

The basic symbols with their meanings are documented in Appendix A. This simplified version of the Instrument Society of America standards are sufficient for this textbook and will provide an adequate background for more complex drawings. While using the standards may seem like additional work in the beginning, it should be considered a small investment leading to accurate communication, like learning grammar and vocabulary, used by all chemical engineers.

1.9 WHAT ARE SOME SAMPLE CONTROL STRATEGIES?

Some very simple example process control systems are given in Figures 1.7*a* through *d*. Each drawing contains a process schematic, a controller (in the instrumentation circle), and the connection between the measurement and the manipulated variable. As a thought exercise, you should analyze each process control system to verify the causal process relationship and to determine what action the controller would take in response to a disturbance or a change in desired value (set point). For example, in Figure 1.7*a*, with an increase in the inlet temperature, the control system would sense a decrease in the outlet composition of reactant. In response, the control system would adjust the heating coil valve, closing it slightly, until the outlet composition returned to its desired value.

Note that details about the equipment between the sensor and valve, such as the transmitters, filters, and converters from analog to digital and back, are not shown. This simplification of the instrumentation is common for the initial piping and instrumentation drawings. Naturally, a more detailed drawing is required for constructing the control strategy from the necessary components; a loop drawing defines every component and wiring connection (ISA, 1989). The additional equipment and its effect on the process dynamic response and control performance are explained in Parts II and III. A sample of a more complex process diagram, this one without the control design, is given in Figure 1.8. The process includes a chemical reactor, a flash separator, heat exchangers, and associated piping. Note that a control design engineer must select from a large number of possible measurements and valves to determine controller connections from an enormous number of possibilities! In Chapter 25 *you*

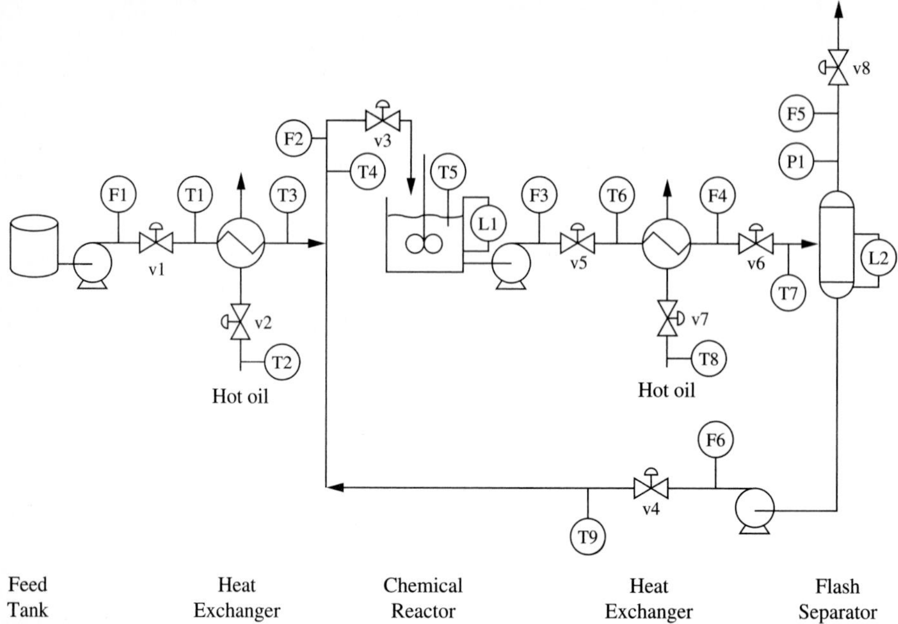

FIGURE 1.8
Integrated feed tank, reactor, and separator with recycle.

will design a control system for this process that controls the key variables, such as reactor level and separator temperature, based on specified control objectives.

1.10 CONCLUSIONS

The material in this chapter has presented a qualitative introduction to process control. You have learned the key features of feedback control along with the types of equipment (instruments and computers) required to apply process control. The importance of the process design on control was discussed several times in the chapter.

Based on this introduction, we are prepared to discuss more carefully the goals of process control in Chapter 2. Understanding the process control goals is essential to selecting the type of analysis used in control engineering design.

REFERENCES

ISA, *ISA-S5.3, Graphic Symbols for Distributed Control/Shared Display Instrumentation, Logic and Computer Systems,* Instrument Society of America, Research Triangle Park, NC, 1983.

ISA, *ISA-S5.1, Instrumentation Symbols and Identification,* Instrument Society of America, Research Triangle Park, NC, 1984.

ISA, *ISA-S5.5, Graphic Symbols for Process Displays,* Instrument Society of America, Research Triangle Park, NC, 1985.

ISA, *ISA-S5.4-1989, Instrument Loop Diagrams,* Instrument Society of America, Research Triangle Park, NC, July, 1989.

Mayer, Otto, *Origins of Feedback Control,* MIT Press, 1970.

ADDITIONAL RESOURCES

Process and control engineers need to refer to books for details on process control equipment. The following references provide an introduction to the resources on this specialized information.

Clevett, K., *Process Analyzer Technology,* Wiley-Interscience, New York, 1986.

Considine, R. and S. Ross, *Handbook of Applied Instrumentation,* McGraw-Hill, New York, 1964.

Liptak, B., *Instrument Engineers Handbook, Vol. 1: Process Measurements* and *Vol. 2: Process Control,* Chilton Book Company, Radnor, PA, 1985.

Driskell, L., *Control Valve Selection and Sizing,* ISA Publishing, Research Triangle Park, NC, 1983.

Hutchison, J. (ed.), *ISA Handbook of Control Valves* (2nd Ed.), Instrument Society of America, Research Triangle Park, NC, 1976.

ISA, *Standards and Practices for Instrumentation and Control* (11th Ed.), Instrument Society of America, Research Triangle Park, NC, 1992.

The following set of books gives a useful overview of process control, addressing both equipment and mathematical analysis.

Andrew, W. and H. Williams, *Applied Instrumentation in the Process Industries* (2nd Ed.), *Volume I: A Survey,* Gulf Publishing, Houston, 1979.

Andrew, W. and H. Williams, *Applied Instrumentation in the Process Industries* (2nd Ed.), *Volume II: Practical Guidelines,* Gulf Publishing, Houston, 1980.

Andrew, W. and H. Williams, *Applied Instrumentation in the Process Industries* (2nd Ed.), *Volume III: Engineering Data and Resource Manual,* Gulf Publishing, Houston, 1982.

Zoss, L., *Applied Instrumentation in the Process Industries, Volume IV: Control Systems Theory, Troubleshooting, and Design,* Gulf Publishing, Houston, 1979.

The following references provide clear introductions to general control methods and specific control strategies in many process industries, such as petrochemical, food, steel, paper, and several others.

Kane, L. (Ed.) *Handbook of Advanced Process Control Systems and Instrumentation,* Gulf Publishing, Houston, 1987.

Matley, J. (Ed.) *Practical Instrumentation and Control II,* McGraw-Hill, New York, 1986.

The following are useful references on drawing symbols for process and control equipment.

Austin, D., *Chemical Engineering Drawing Symbols,* Halsted Press, London, 1979.

Weaver, R., *Process Piping Drafting* (3rd Ed.), Gulf Publishing, Houston, 1986.

Finally, a good reference for terminology is

ISA, *Process Instrumentation Terminology,* ANSI/ISA S51.1-1979, Instrument Society of America, Research Triangle Park, NC, December 28, 1979.

QUESTIONS

1.1. Describe the four necessary components of a feedback control system.

1.2. Review the equipment sketches in Figure Q1.2*a* and *b* and explain whether each is or is not a level feedback control system. In particular, identify the four necessary components of feedback control, if they exist.

(*a*) The flow in is a function of the connecting rod position.

(*b*) The flow out is a function of the level (pressure at the bottom of the tank) and the resistance to flow.

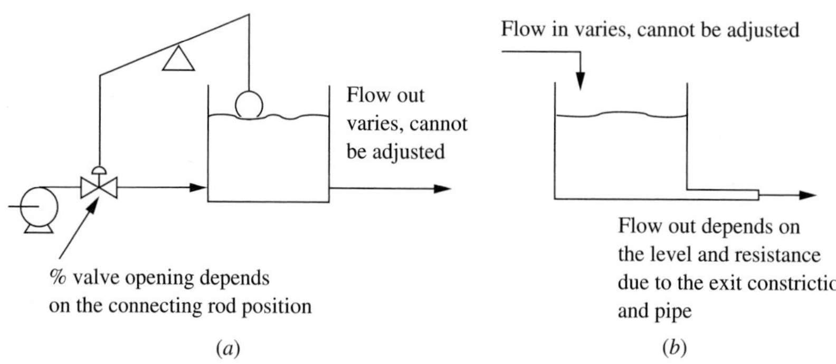

FIGURE Q1.2

1.3. Give some examples of feedback control systems in your everyday life, government, biology, and management. The control calculations may be automated or performed by people.

1.4. Discuss the advantages of having a centralized control facility. Can you think of any disadvantages?

1.5. Review the processes sketched in Figures 1.7*a* through *d* in which the controlled variable is to be maintained at its desired value.

(*a*) From your chemical engineering background, suggest the physical principle used by the sensor.

(*b*) Explain the causal relationship between the manipulated and controlled variables.

(*c*) Explain whether the control valve should be opened or closed to increase the value of the controlled variable.

(*d*) Identify possible disturbances that could influence the controlled variable. Also, describe how the process equipment would have to be sized to account for the disturbances.

1.6. The preliminary process designs have been prepared for the systems in Figure Q1.6. The key variables to be controlled for the systems are (*a*) flow rate, temperature, composition, and pressure for the flash system and (*b*) composition, temperature, and liquid level for the continuous-flow stirred-tank chemical reactor. For both processes, disturbances occur in the feed temperature and composition. Answer the following questions for both processes.

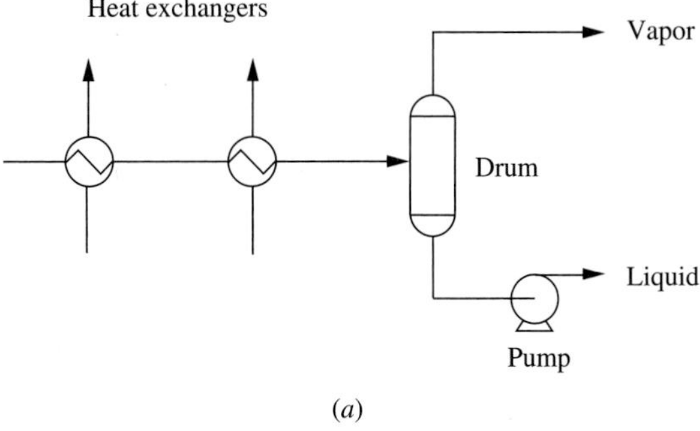

(*a*)

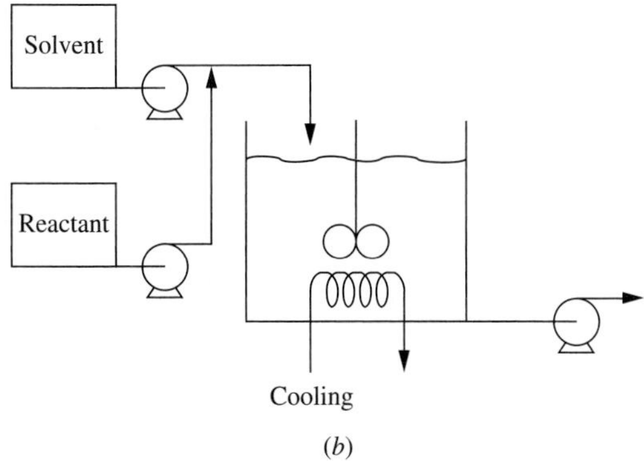

(*b*)

FIGURE Q1.6

(*a*) Determine which sensors and final elements are required so that the important variables can be controlled. Sketch them on the figure where they should be located.

(*b*) Describe how the equipment capacities should be determined.

(*c*) Select controller pairings; that is, select which measured variable should be controlled by adjusting which manipulated variable.

(These examples will be reconsidered after quantitative methods have been introduced.)

1.7. Consider any of the control systems shown in Figures 1.7*a* through *d*. Suggest a feedback control calculation that can be used to determine the proper value of the manipulated valve position. The only values available for the calculation are the desired value and the measured value of the controlled variable. (Do the best you can at this point. Control algorithms for feedback control are presented in Part III.)

1.8. Feedback control uses measurement of a system output variable to determine the value of a system input variable. Suggest an alternative control approach that uses a measured (disturbance) input variable to determine the value of a different (manipulated) input variable, with the goal of maintaining a system output variable at its desired value. Apply your approach to one of the systems in Figure 1.7. Can you suggest a name for your approach?

1.9 Evaluate the potential feedback control designs in Figure Q1.9. Determine whether each is a feedback control system. Explain why or why not, and explain whether the control system will function correctly as shown for disturbances and changes in desired value.

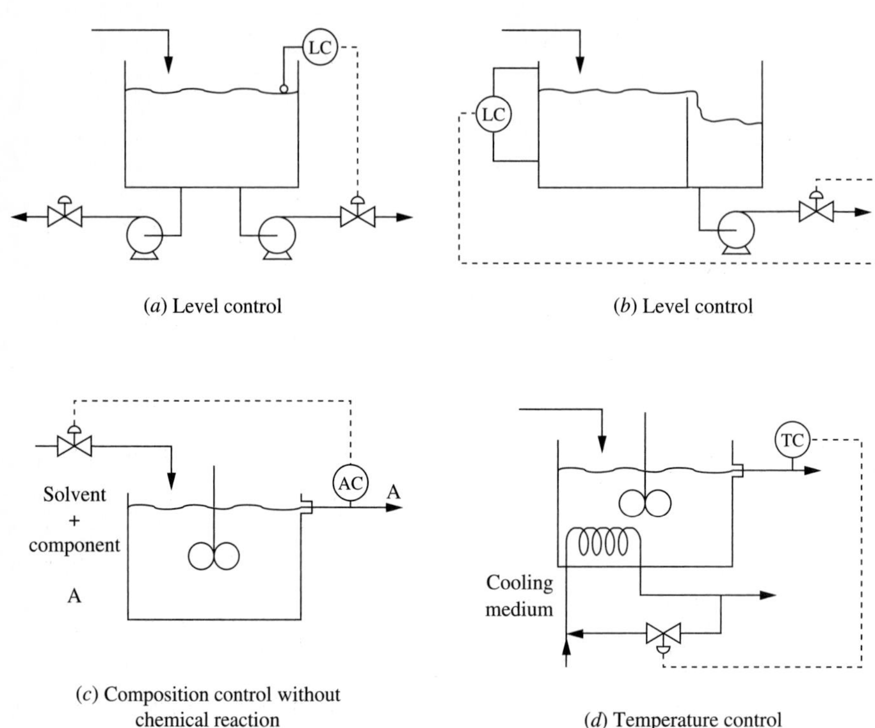

(*a*) Level control

(*b*) Level control

(*c*) Composition control without chemical reaction

(*d*) Temperature control

FIGURE Q1.9

CHAPTER 2

CONTROL OBJECTIVES AND BENEFITS

2.1 INTRODUCTION

The first chapter provided an overview of process control in which the close association between process control and plant operation was noted. As a consequence, control objectives are closely tied to process goals, and control benefits are closely tied to attaining these goals. In this chapter the control objectives and benefits are discussed thoroughly, and several process examples are presented. The control objectives provide the basis for all technology and design methods presented in subsequent chapters of the book.

While this book emphasizes the contribution made by automatic control, control is only one of many factors that must be considered in improving process performance. Three of the most important factors are shown in Figure 2.1, which indicates that proper equipment design, operating conditions, and process control should all be achieved simultaneously to attain safe and profitable plant operation. Clearly, equipment should be designed to provide good dynamic responses in addition to high steady-state profit and efficiency, as covered in process design courses and books. Also, the plant operating conditions, as well as achieving steady-state plant objectives, should provide flexibility for dynamic operation. Thus, achieving excellence in plant operation requires consideration of all factors. This book addresses all three factors; it gives guidance on how to design processes and select operating conditions favoring good dynamic performance, and it presents automation methods to adjust the manipulated variables.

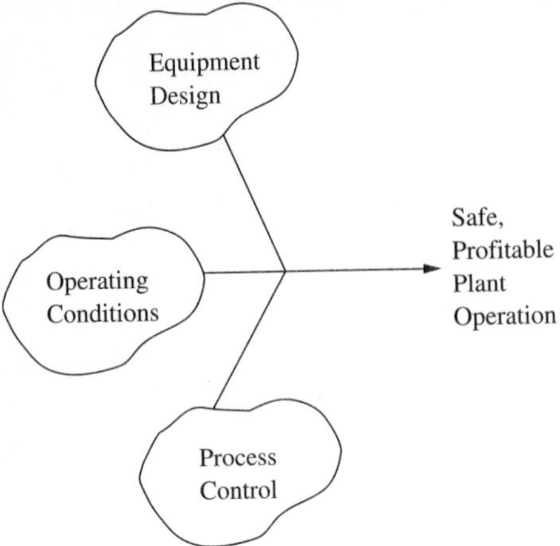

FIGURE 2.1
Schematic representation of three critical elements for achieving excellent plant performance.

2.2 CONTROL OBJECTIVES

The seven major categories of control objectives were introduced in Chapter 1. They are discussed in detail here, with an explanation of how each influences the control design for the example process shown in Figure 2.2. The process separates two components based on their different vapor pressures. The liquid feed stream, consisting of components A and B, is heated by two exchangers in series. Then the stream flows through a valve to a vessel at a lower pressure. As a result of the higher temperature and lower pressure, the material forms two phases, with most of the A in the vapor and most of the B in the liquid. The exact compositions can be determined from an equilibrium flash calculation, which simultaneously solves the material, energy, and equilibrium expressions. Both streams leave the vessel for further processing, the vapor stream through the overhead line and the liquid stream out from the bottom of the vessel. Although a simple process, the heat exchanger with flash drum provides examples of all control objectives, and this process is analyzed quantitatively with control in Chapter 24.

A control strategy is also shown in Figure 2.2. Since we have not yet studied the calculations used by feedback controllers, you should interpret the controller as a linkage between a measurement and a valve. Thus, you can think of the feedback pressure control (PC) system as a system that measures the pressure and maintains the pressure close to its desired value by adjusting the opening of the valve in the overhead vapor pipe. The type of control calculation, which will be covered in depth in later chapters, is not critical for the discussions in this chapter.

Safety

The safety of people in the plant and in the surrounding community is of paramount importance. While no human activity is without risk, the typical goal is that working at an industrial plant should involve much less risk than any other activity in a person's life. *No compromise with sound equipment and control safety practices is acceptable.*

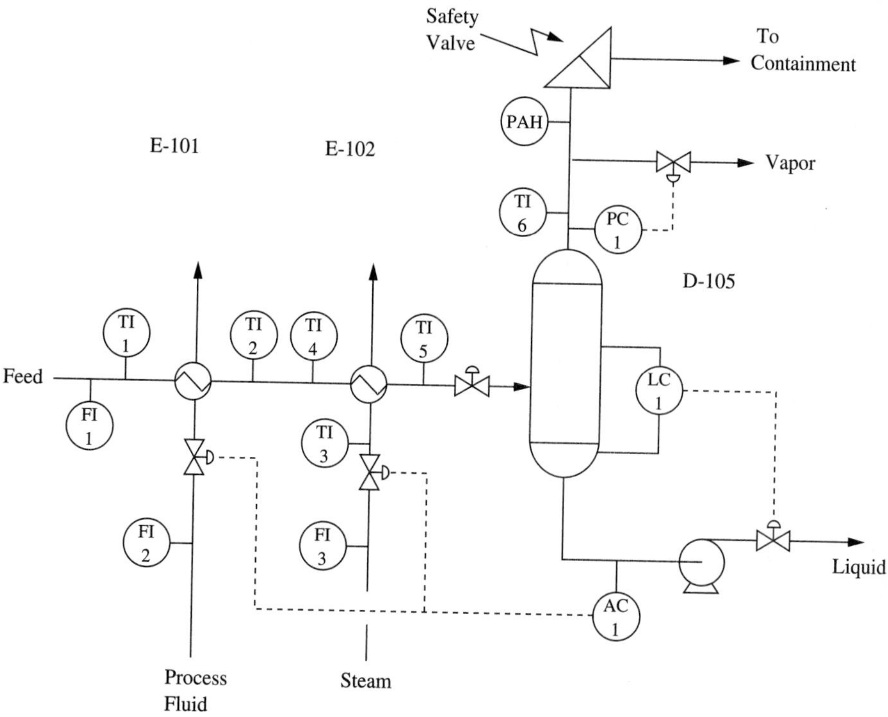

FIGURE 2.2
Flash separation process with control strategy.

Plants are designed to operate safely at expected temperatures and pressures; however, improper operation can lead to equipment failure and release of potentially hazardous materials. Therefore, the process control strategies contribute to the overall plant safety by maintaining key variables near their desired values. Since these control strategies are important, they are automated to ensure rapid and complete implementation. In Figure 2.2, the equipment could operate at high pressures under normal conditions. If the pressure were allowed to increase too far beyond the normal value, the vessel might burst, resulting in injuries or death. Therefore, the control strategy includes a controller labelled "PC-1" that controls the pressure by adjusting the valve position (i.e., percent opening) in the vapor line.

Another consideration in plant safety is the proper response to major incidents, such as equipment failures and excursions of variables outside of their acceptable bounds. Feedback strategies cannot *guarantee* safe operation; a very large disturbance could lead to an unsafe condition. Therefore, an additional layer of control, termed an *emergency system,* is applied to enforce bounds on key variables. Typically, this layer involves either safely diverting the flow of material or shutting down the process when unacceptable conditions occur. The control strategies are usually not complicated; for example, an emergency control might stop the feed to a vessel when the liquid level is nearly overflowing. Proper design of these emergency systems is based on a structured analysis of hazards (Battelle Laboratory, 1985;

Warren Centre ,1986) that relies heavily on experience about expected incidents and on the reliability of process and control equipment.

In Figure 2.2, the pressure is controlled by the element labelled "PC." Normally, it maintains the pressure at or near its desired value. However, the control strategy relies on the proper operation of equipment like the pressure sensor and the valve. Suppose that the sensor stopped providing a reliable measurement; the control strategy could improperly close the overhead valve, leading to an unsafe pressure. The correct control design would include an additional strategy using independent equipment to prevent a very high pressure. For example, the safety valve shown in Figure 2.2 is closed unless the pressure rises above a specified maximum; then, it opens to vent the excess vapor. It is important to recognize that this safety relief system is called on to act infrequently, perhaps once per year or less often; therefore, its design should include highly reliable components to ensure that it performs properly when needed.

Environmental Protection

Protection of the environment is critically important. This objective is mostly a process design issue; that is, the process must have the capacity to convert potentially toxic components to benign material. Again, control can contribute to the proper operation of these units, resulting in consistently low effluent concentrations. In addition, control systems can divert effluent to containment vessels should any extreme disturbance occur. The stored material could be processed at a later time when normal operation has been restored.

In Figure 2.2, the environment is protected by containing the material within the process equipment. Note that the safety release system directs the material for containment and subsequent "neutralization," which could involve recycling to the process or combusting to benign compounds. For example, a release system might divert a gaseous hydrocarbon to a flare for combustion, and it might divert a water-based stream to a holding pond for subsequent purification through biological treatment before release to a water system.

Equipment Protection

Much of the equipment in a plant is expensive and difficult to replace without costly delays. Therefore, operating conditions must be maintained within bounds to prevent damage. The types of control strategies for equipment protection are similar to those for personnel protection; that is, controls to maintain conditions near desired values and emergency controls to stop operation safely when the process reaches boundary values.

In Figure 2.2, the equipment is protected by maintaining the operating conditions within the expected temperatures and pressures. In addition, the pump could be damaged if no liquid were flowing through it. Therefore, the liquid level controller, by ensuring a reservoir of liquid in the bottom of the vessel, protects the pump from damage. Additional equipment protection could be provided by adding an emergency controller that would shut off the pump motor when the level decreased below a specified value.

Smooth Operation

A chemical plant includes a complex network of interacting processes; thus, the smooth operation of a process is desirable, because it results in few disturbances to all integrated units. Naturally, key variables in streams leaving the process should be maintained close to their desired values (i.e., with small variation) to prevent disturbances to downstream units. In Figure 2.2, the liquid from the vessel bottoms is processed by downstream equipment. The control strategy can be designed to make slow, smooth changes to the liquid flow. Naturally, the liquid level will not remain constant, but it is not required to be constant; the level must remain within specified limits. By the use of this control design, the downstream units would experience fewer disturbances, and the overall plant would perform better.

There are additional ways for upsets to be propagated in an integrated plant. For example, when the control strategy increases the steam flow to heat exchanger E-102, another unit in the plant must respond by generating more steam. Clearly, smooth manipulations of the steam flow require slow adjustments in the boiler operation and better overall plant operation. Therefore, we are interested in *both the controlled variables and the manipulated variables.* Ideally, we would like to have tight regulation of the controlled variables and slow, smooth adjustment of the manipulated variables. As we will see, this is not usually possible, and some compromise is required.

Product Quality

The final products from the plant must meet demanding quality specifications set by purchasers. The specifications may be expressed as compositions (e.g., percent of each component), physical properties (e.g., density), performance properties (e.g., octane number or tensile strength), or a combination of all three. Process control contributes to good plant operation by maintaining the operating conditions required for excellent product quality. Improving product quality control is a major economic factor in the application of digital computers and advanced control algorithms for automation in the process industries.

In Figure 2.2, the amount of component A, the material with the higher vapor pressure, is to be controlled in the liquid stream. Based on our knowledge of thermodynamics, we know that this value can be controlled by adjusting the flash temperature or, equivalently, the heat exchanged. Therefore, a control strategy would be designed to measure the composition in real time and adjust the heating medium flows that exchange heat with the feed.

Profit

Naturally, the typical goal of the plant is to return a profit. In the case of a utility such as water purification, in which no income from sales is involved, the equivalent goal is to provide the product at lowest cost. Before achieving the profit-oriented goal, selected independent variables are adjusted to satisfy the first five higher-priority control objectives. Often, some independent operating variables are not specified after the higher objectives (that is, including product quality but excepting profit) have been satisfied. When additional variables (degrees of freedom) exist, the control strategy can increase profit *while satisfying all other objectives.*

In Figure 2.2 all other control objectives can be satisfied by using exchanger E-101, exchanger E-102, or any combination of the two, to heat the inlet stream. Therefore, the control strategy can select the correct exchanger based on the cost of the two heating fluids. For example, if the process fluid used in E-101 were less costly, the control strategy would use the process stream for heating preferentially and use steam only when required for additional heating. How the control strategy would implement this policy, based on a selection hierarchy defined by the engineer, is covered in Chapter 22.

Monitoring and Diagnosis

Complex chemical plants require monitoring and diagnosis by people as well as excellent automation. Plant control and computing systems generally provide monitoring features for two sets of people who perform two different sets of functions: (1) the immediate safety and operation of the plant, usually monitored by plant operators, and (2) the long-term plant performance analysis, monitored by supervisors and engineers.

The plant operators require very rapid information so that they can ensure that the plant conditions remain within acceptable bounds. If undesirable situations occur—or, one hopes, before they occur—the operator is responsible for rapid recognition and intervention to restore acceptable performance. While much of this routine work is automated, the people are present to address complex issues that are difficult to automate, perhaps requiring special information not readily available to the computing system. Since the person may be responsible for a plant section with hundreds of measured variables, excellent displays are required. These are usually in the form of trend plots of several associated variables versus time and of indicators in bar-chart form for easy identification of normal and abnormal operation. Examples are shown in Figure 2.3.

Since the person cannot monitor all variables simultaneously, the control system includes an alarm feature, which draws the operator's attention to variables that

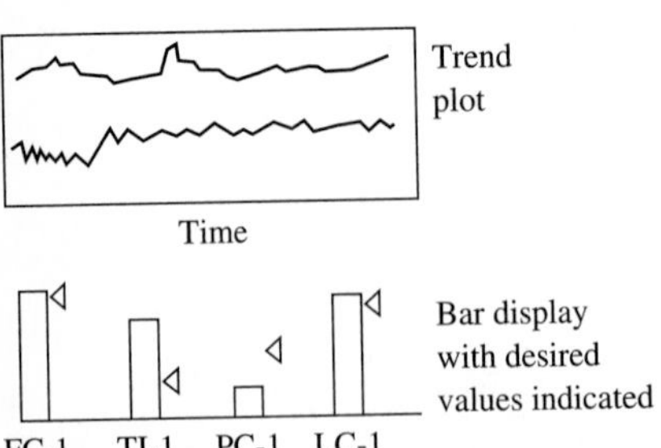

FIGURE 2.3
Examples of displays presented to a process operator.

are near limiting values selected to indicate serious maloperation. For example, a high pressure in the flash separator drum is undesirable and would at the least result in the safety valve opening, which is not desirable, because it diverts material and results in lost profit and because it may not always reclose tightly. Thus, the system in Figure 2.2 has a high-pressure alarm, PAH. If the alarm is activated, the operator might reduce the flows to the heat exchanger or of the feed to reduce pressure. This operator action might cause a violation of product specifications; however, maintaining the pressure within safe limits is more important than product quality. Every measured variable in a plant must be analyzed to determine whether an alarm should be associated with it and, if so, the proper value for the alarm limit.

Another group of people monitors the longer-range performance of the plant to identify opportunities for improvement and causes for poor operation. Usually, a substantial sample of data, involving a long time period, is used in this analysis, so that the effects of minor fluctuations are averaged out. Monitoring involves important measured and calculated variables, including equipment performances (e.g., heat transfer coefficients) and process performances (e.g., reactor yields and material balances). In the example flash process, the energy consumption would be monitored. An example trend of some key variables is given in Figure 2.4, which shows that the ratio of expensive to inexpensive heating source had an increasing trend. If the feed flow and composition did not vary significantly, one might suspect that the heat transfer coefficient in the first heat exchanger, E-101, was decreasing due to fouling. Careful monitoring would identify the problem and enable the engineer to decide when to remove the heat exchanger temporarily for mechanical cleaning to restore a high heat transfer coefficient.

Previously, this monitoring was performed by hand calculations, which was a tedious and inefficient method. Now, the data can be collected, processed if additional calculations are needed, and reported using digital computers. This combination of ease and reliability has greatly improved the monitoring of chemical process plants.

Note that both types of monitoring—the rapid display and the slower process analysis—require people to make and implement decisions. This is another form of feedback control involving personnel, sometimes referred to as having "a person in the loop," with the "loop" being the feedback control loop. While we will

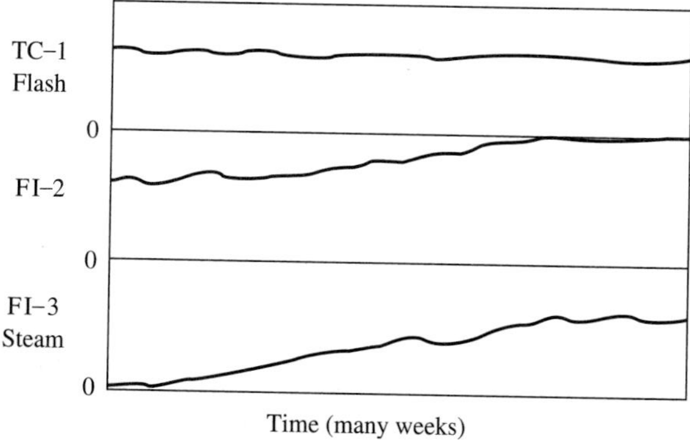

FIGURE 2.4
Example of long-term data, showing the increased use of expensive steam in the flash process.

concentrate on the automated feedback system in a plant, we must never forget that many of the important decisions in plant operation that contribute to longer-term safety and profitability are based on monitoring and diagnosis and implemented by people "manually."

Therefore,

> All seven categories of control objectives must be achieved simultaneously; failure to do so leads to unprofitable or, worse, dangerous plant operation.

The selection of controlled and manipulated variables is important for good control performance. Clearly, variables such as flow (production) rate and product quality are central to the success of the plant operation and should be maintained at desired values. In addition, variables such as temperature, pressure, and level are important, because they indicate the process environment, which should be maintained at specified conditions to ensure continually high-quality products. Principles of variable selection will be discussed throughout the book, along with many examples.

In this section, instances of all seven goals were identified in the simple heater and flash separator. The analysis of more complex process plants in terms of the goals is a challenging task, enabling engineers to apply all of their chemical engineering skills. Often a team of engineers and operators, each with special experiences and insights, performs this analysis. Again, we see that control engineering skills are needed by all chemical engineers in industrial practice.

2.3 DETERMINING PLANT OPERATING CONDITIONS

A key factor in good plant operation is the determination of the best operating conditions, which can be maintained within small variation by automatic control strategies. Therefore, setting the control objectives requires a clear understanding of how the plant operating conditions are determined. A complete study of plant objectives requires additional mathematical methods for simulating and optimizing the plant operation. For our purposes, we will restrict our discussion in this section to small systems that can be analyzed graphically.

Determining the best operating conditions can be performed in two steps. First, the region of possible operation is defined. The following are some of the factors that limit the possible operation:

- Physical principles; for example, all concentrations ≥ 0
- Safety, environmental, and equipment protection
- Equipment capacity; for example, maximum flow
- Product quality

The region that satisfies all bounds is termed the feasible operating region or, more commonly, the *operating window.* Any operation within the operating window is possible. Violation of some of the limits, called *soft constraints,* would lead to poor product quality or reduction of long-term equipment life; therefore, short-term violations of soft constraints are allowed but are to be avoided. Violation of critical bounds, called *hard constraints,* could lead to injury or major equipment damage; violations of hard constraints are not acceptable under any foreseeable circumstances. The control strategy must take aggressive actions, including shutting down the plant, to prevent hard constraint violations. For both hard and soft constraints, debits are incurred for violating constraints, so the control system is designed to maintain operation within the operating window. While any operation within the window is possible and satisfies minimum plant goals, a great difference in profit can exist depending on the conditions chosen. Thus, the plant economics must be analyzed to determine the best operation within the window. The control strategy should be designed to maintain the plant conditions near their most profitable values.

The example shown in Figure 2.5 demonstrates the operating window for a simple, one-dimensional case. The example involves a fired heater (furnace) with a chemical reaction occurring as the fluid flows through the pipe or, as it is often called, the *coil.* The temperature of the reactor must be held between minimum (no reaction) and maximum (metal damage or excessive side reactions) temperatures. When economic objectives favor increased conversion of feed, the profit function monotonically increases with increasing temperature; therefore, the best operation would be at the maximum allowable temperature. However, the dynamic data show that the temperature varies about the desired value because of disturbances such as those in fuel composition and pressure. Therefore, the effectiveness of the control strategy in maximizing profit depends on reducing the variation of the temperature. A small variation means that the temperature can be operated very close to, without exceeding, the maximum constraint.

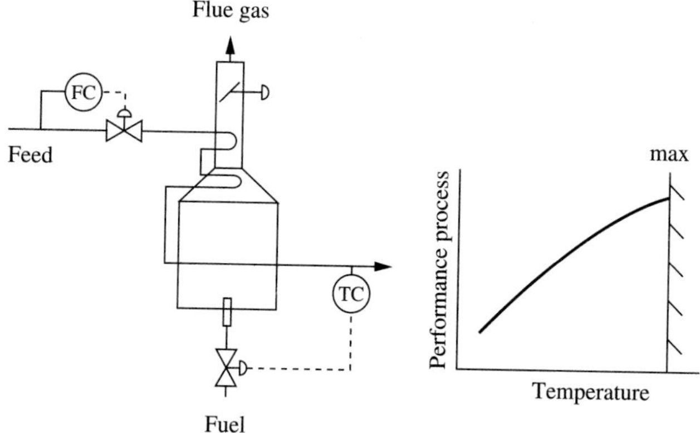

FIGURE 2.5
Example of operating window for fired-heater temperature.

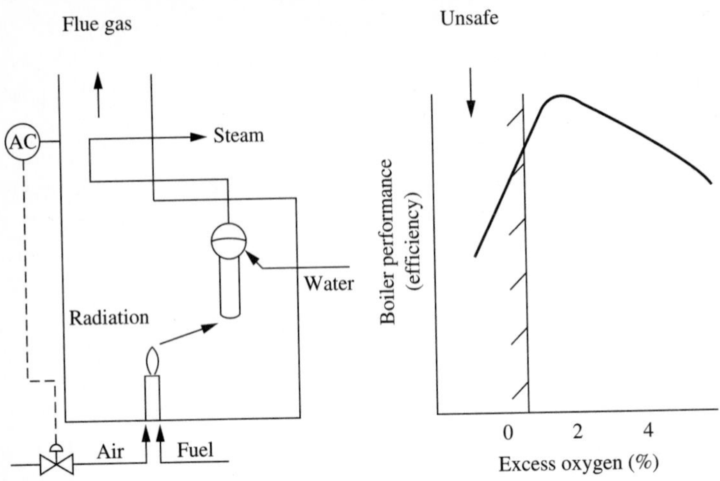

FIGURE 2.6
Example operating window for boiler combustion flue gas excess oxygen.

Another example is the system shown in Figure 2.6, where fuel and air are mixed and combusted to provide heat for a boiler. The ratio of fuel to air is important. Too little air (oxygen) means that some of the fuel is uncombusted and wasted, whereas excess air reduces the flame temperature and, thus, the heat transfer. Therefore, the highest efficiency and most profitable operation are near the stoichiometric ratio. (Actually, the best value is usually somewhat above the stoichiometric ratio because of imperfect mixing, leakage, and complex combustion chemistry.) The maximum air flow is determined by the air compressor and is usually not a limitation, but a large excess of air leads to extremely high fuel costs. Therefore, the best plant operation is at the peak of the efficiency curve. An effective control strategy results in a small variation in the excess oxygen in the flue gas, allowing operation near the peak.

However, a more important factor is safety, which provides another reason for controlling the excess air. A deficiency of oxygen could lead to a dangerous condition because of unreacted fuel in the boiler combustion chamber. Should this situation occur, the fuel could mix with other air (that leaks into the furnace chamber) and explode. Therefore, the air flow should never fall below the stoichiometric value. Note that the control sketch in Figure 2.6 is much simpler than actual control designs for combustion systems (for example, API, 1977).

Finally, a third example demonstrates that this analysis can be extended to more than one dimension. We now consider the chemical reactor in Figure 2.5 with two variables: temperature and product flow. The temperature bounds are the same, and the product flow has a maximum limitation because of erosion of the pipe at the exit of the fired heater. The profit function, which would be calculated based on an analysis of the entire plant, is given as contours in the operating window in Figure 2.7. In this example, the maximum profit occurs outside the operating window

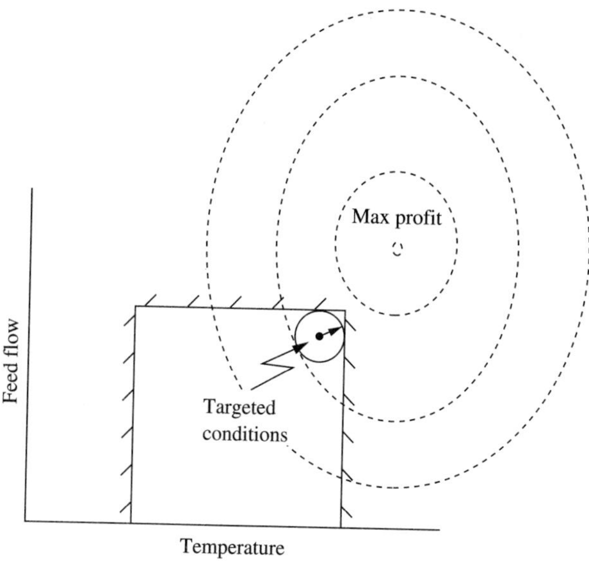

FIGURE 2.7
Example operating window for the feed and temperature of a fired-heater chemical reactor.

and therefore cannot be achieved. The best operation inside the window would be at the maximum temperature and flow, which are found at the upper right-hand corner of the operating window. As we know, the plant cannot be operated exactly at this point because of unavoidable disturbances in variables such as feed pressure and fuel composition (which affects heat of combustion). However, good control designs can reduce the variation of temperature and flow so that desired values can be selected that nearly maximize the achievable profit while not violating the constraints. This situation is shown in Figure 2.7, where a circle defines the variation expected about the desired values (Perkins, 1990; Narraway and Perkins, 1993). When control provides small variation, that is, a circle of small radius, the operation can be maintained closer to the best operation.

All of these examples demonstrate that

> Process control improves plant performance by reducing the variation of key variables. When the variation has been reduced, the desired value of the controlled variable can be adjusted to increase profit.

Quantitative analysis of operating windows begins in the next chapter.

Note that simply reducing the variation does not always improve plant operation. The profit contours within the operating window must be analyzed to determine the best operating conditions that take advantage of the reduced variation. Also, it is important to recognize that the theoretical maximum profit cannot usually be achieved because of inevitable variation due to disturbances. This situation should be included in the economic analysis of all process designs.

2.4 BENEFITS FOR CONTROL

The previous discussion of plant operating conditions provides the basis for calculating the benefits for excellent control performance. In all of the examples discussed qualitatively in the previous section, the economic benefit resulted from reduced variation of key variables. Thus, the calculation of benefits considers the effect of variation on plant profit. Before the method is presented, it is emphasized that the highest-priority control objectives—namely, safety, environmental protection, and equipment protection—are *not* analyzed by the method described in this section. Although the control designs for these objectives often reduce variation, they are not selected for increasing profit but rather for providing safe, reliable plant operation.

Once the profit function has been determined, the benefit method needs to characterize the variation of key plant variables. This can be done through the calculation shown schematically in Figure 2.8. The plant operating data, which is usually given as a plot or trend versus time, can be summarized by a frequency distribution. The frequency distribution can be determined by taking many sample measurements of the process variable, usually separated by a constant time period, and counting the number of measurements whose values fall in each of several intervals within the range of data values. The total time period covered must be long compared to the dynamics of the process, so that the effects of time correlation in the variable and varying disturbances will be averaged out.

The resulting distribution is plotted as frequency; that is, as fraction or percent of measurements falling within each interval versus the midpoint value of that interval. Such a plot is called a *frequency distribution* or *histogram.* If the variable were constant, perhaps due to perfect control or the presence of no disturbances, the distribution would have one bar, at the constant value, rising to 1.0 (or 100%).

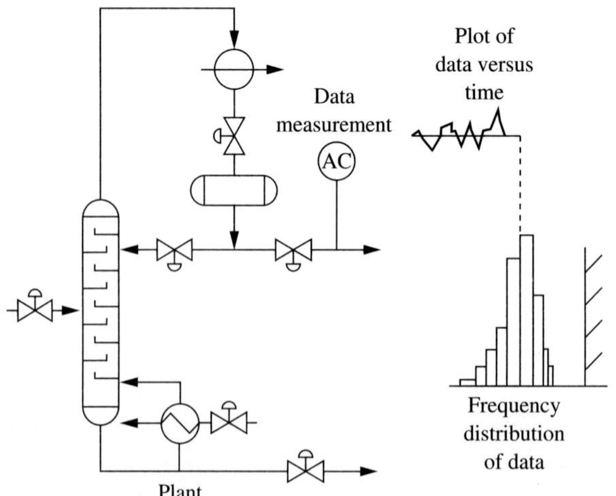

FIGURE 2.8
Schematic presentation of the method for representing the variability in plant data.

As the variation in the values increases, the distribution becomes broader; thus, the frequency distribution provides a valuable summary of the variable variation.

The distribution could be described by its *moments;* in particular, the mean and standard deviation are often used in describing the behavior of variables in feedback systems (Snedecor and Cochran, 1980; Bethea and Rhinehart, 1991). These values can be calculated from the plant data according to the following equations:

$$\text{Mean} = \bar{Y} = \frac{1}{n}\sum_{i=1}^{n} Y_i \tag{2.1}$$

$$\text{Standard deviation} = s_Y = \sqrt{\frac{\sum_{i=1}^{n}(Y_i - \bar{Y})^2}{n-1}} \tag{2.2}$$

where Y_i = measured value of variable
s_Y^2 = variance
n = number of data points

When the experimental distribution can be characterized by the standard normal distribution, the variation about the mean is characterized by the standard deviation as is shown in Figure 2.9. (Application of the central limit theorem to data whose underlying distribution is not normal often results in the valid use of the normal distribution.) When the number of data in the sample are large, the estimated (sample) standard deviation is approximately equal to the population standard deviation, and the following relationships are valid for the normally distributed variable:

About 68.2% of the variable values are within $\pm s$ of mean.

About 95.4% of the variable values are within $\pm 2s$ of mean.

About 99.7% of the variable values are within $\pm 3s$ of mean.

In all control performance and benefits analysis, the mean and standard deviation can be used in place of the frequency distribution when the distribution is

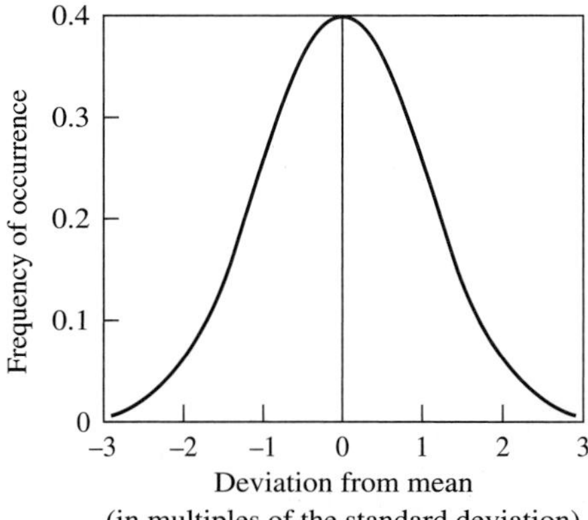

FIGURE 2.9
Normal distribution.

normal. As is apparent, a narrow distribution is equivalent to a small standard deviation. Although the process data can often be characterized by a normal distribution, the method for calculating benefits does *not depend on the normal distribution,* which was introduced here to relate the benefits method to statistical terms often used to describe the variability of data.

The empirical histogram provides how often—that is, what percentage of the time—a variable has a certain value, with the value for each histogram entry taken as the center of the variable interval. The performance of plant operation at each variable value can be determined from the *performance function.* Depending on the plant, the performance function could be reactor conversion, efficiency, production rate, profit, or other variable that characterizes the quality of operation. The average performance for a set of representative data (that is, frequency distribution) is calculated by combining the histogram and profit function according to the following equation (Bozenhardt and Dybeck, 1986; Marlin et al., 1991; and Stout and Cline, 1976).

$$P_{\text{ave}} = \sum_{j=1}^{M} F_j P_j \tag{2.3}$$

where P_{ave} = average process performance
F_j = fraction of data in interval $j = N_j/N_T$
N_j = number of data points in interval j
N_T = total number of data points
P_j = performance measured at the midpoint of interval j
M = number of intervals in the frequency distribution

This calculation is schematically shown in Figure 2.10. The calculation is tedious when done by hand but is performed easily with a spreadsheet or other computer program.

Note that methods for predicting how improved control affects the frequency distribution require technology covered in Part III of the book. These methods require a sound understanding of process dynamic responses and typical control calculations. For now, we will assume that the improved frequency distribution can be predicted.

Example 2.1. This example presents data for a reactor of the type shown in Figure 2.5. The reaction taking place is the pyrolysis of ethane to a wide range of products, one of which is the desired product, ethylene. The goal for this example is to maximize the conversion of feed ethane. This could be achieved by increasing the reactor temperature, but a hard constraint, the maximum temperature of 864°C, must not be exceeded, or damage will occur to the furnace. Control performance data is provided in Table 2.1.

In calculating benefits for control improvement, the calculation is performed twice. The first calculation uses the *base case* distribution, which represents the plant performance with poor control. The base case reactor temperature, shown as the top graph in Figure 2.11, might result from control via the plant operator occasionally adjusting the fuel flow. The second calculation uses the tighter distribution shown in the middle graph, which results from improved control using methods described in Parts

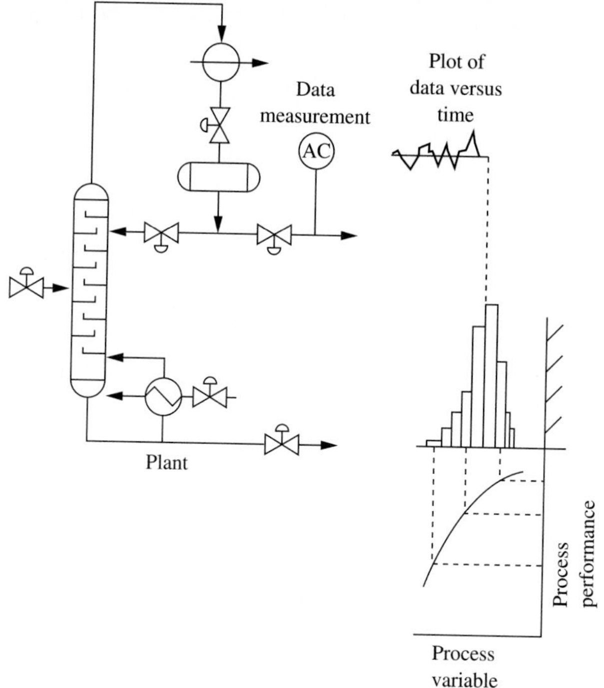

FIGURE 2.10
Schematic presentation of the method for calculating the average process performance from plant data.

TABLE 2.1
Frequency data for Example 2.1

Temperature midpoint (°C)	Conversion P_j (%)	Initial data		Data with improved control	
		F_j	$P_j * F_j$	F_j	$P_j * F_j$
842	50	0	0	0	0
844	51	0.0666	3.4	0	0
846	52	0.111	5.778	0	0
848	53	0.111	5.889	0	0
850	54	0.156	8.4	0	0
852	55	0.244	13.44	0	0
854	56	0.133	7.467	0	0
856	57	0.111	6.333	0	0
858	58	0.044	2.578	0.25	14.5
860	59	0.022	1.311	0.50	29.5
862	60	0	0	0.25	15
Average conversion (%) $= \sum P_j * F_j =$			54.6		59

III and IV. The process performance correlation, which is required to relate the temperature to conversion, is given in the bottom graph. The data for the graphs, along with the calculations for the averages, are given in Table 2.1.

The difference between the two average performances, a conversion increase of 4.4 percent, is the benefit for improved control. Note that the benefit is achieved by *reducing the variance* and *increasing the average temperature.* Both are required in this example; simply reducing variance with the same mean would not be a worthwhile achievement! Naturally, this benefit must be related to dollars and compared with the costs for equipment and personnel time when deciding whether this investment is justified. The economic benefit would be calculated as follows:

$$\Delta\text{profit} = (\text{Feed flow})\ (\Delta\ \text{conversion})\ (\$/\text{kg products}) \qquad (2.4)$$

In a typical ethylene plant, the benefits for even a small increase in conversion would be much greater than the costs. Additional benefits would result from fewer disturbances to downstream units and longer operating life of the fired heater due to reduced thermal stress.

Example 2.2. A second example is given for the boiler excess oxygen shown in Figure 2.6. The discussion in the previous section demonstrated that the profit is maximized when the excess oxygen is maintained slightly above the stoichiometric ratio, where the efficiency is at its maximum. Again, the process performance function, here efficiency, is used to evaluate each operating value, and frequency distributions are used to characterize the variation in performance.

The performance is calculated for the base case and an improved control case, and the benefit is calculated as shown in Figure 2.12 for an example with realistic data. The data for the graphs, along with the calculations for the averages are given in Table 2.2. The average efficiency increased by almost 1 percent with better control and would be related to profit as follows:

$$\Delta\text{profit} = (\Delta\ \text{efficiency}/100)\ (\text{Steam flow})\ (\Delta H_{\text{vap}})\ (\$/\text{Energy}) \qquad (2.5)$$

This improvement would result in fuel savings worth tens of thousands of dollars per year in a typical industrial boiler. In this case, the average of the process variable (excess oxygen) is the same for the initial and improved operations, because the improvement is *due entirely to the reduction in the variance* of the excess oxygen. The difference between the chemical reactor and the boiler results from the different process performance curves. Note that the improved control case has its desired value at an excess oxygen value slightly greater than where the maximum profit occurs, so that the chance of a dangerous condition is negligibly small.

A few important assumptions in this benefits calculation method may not be obvious, so they are discussed here. First, the frequency distributions can never be *guaranteed* to remain within the operating window. If a large enough data set were collected, some data would be outside of the operating window due to infrequent, large disturbances. Therefore, some small probability of exceeding the constraints always exists and must be accepted. For soft constraints, it is common to select an average value so that no more than a few percent of the data exceeds the constraint; often the target is two standard deviations from the limit. For important hard constraints, an average much farther from the constraint can be selected, since the emergency system will activate each time the system reaches a boundary.

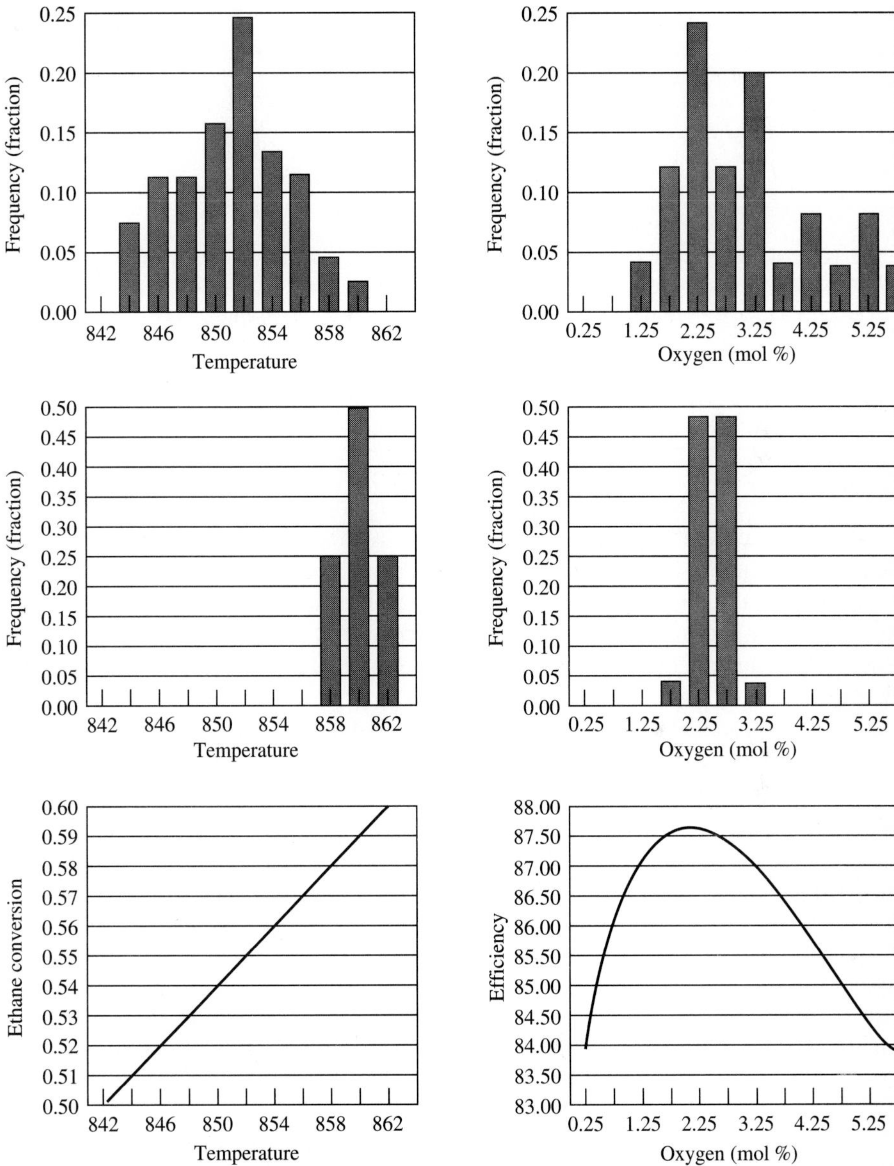

FIGURE 2.11
Data for Example 2.1 in which the benefits of reduced variation and closer approach to the maximum temperature limit in a chemical reactor are calculated.

FIGURE 2.12
Data for Example 2.2 in which the benefits of reducing the variation of excess oxygen in boiler flue gas are calculated.

TABLE 2.2
Frequency data for Example 2.2

Excess oxygen midpoint (mol fraction)	Boiler efficiency P_j (%)	Initial data		Data with improved control	
		F_j	$P_j * F_j$	F_j	$P_j * F_j$
0.25	83.88	0	0	0	0
0.75	85.70	0	0	0	0
1.25	86.85	0.04	3.47	0	0
1.75	87.50	0.12	10.50	0.250	2.19
2.25	87.70	0.24	21.05	0.475	41.66
2.75	87.54	0.12	10.50	0.475	41.58
3.25	87.10	0.20	17.42	0.025	2.18
3.75	86.48	0.04	3.46	0	0
4.25	85.76	0.08	6.86	0	0
4.75	85.02	0.04	3.40	0	0
5.25	84.36	0.08	6.75	0	0
5.75	83.86	0.04	3.35	0	0
Average efficiency (%) $= \sum P_j * F_j =$			86.77		87.70

A second assumption concerns the mixing of steady-state and dynamic relationships. Remember that the process performance function is developed from steady-state analysis. The frequency distribution is calculated from plant data, which is inherently dynamic. Therefore, the two correlations cannot strictly be used together as they are in equation (2.3). The difficulty is circumvented if the plant is assumed to have operated at quasi-steady state at each data point, then varied to the next quasi-steady state for the subsequent data point. When this assumption is valid, the plant data is essentially from a series of steady-state operations, and equation (2.3) is valid, because all data and correlations are consistently steady-state.

Third, the approach is valid for modifying the behavior of one process variable, with all other variables unchanged. If many control strategies are to be evaluated, the interaction among them must be considered. The alterations to the procedure depend on the specific plant considered but would normally require a model of the integrated plant.

The method explained in this section clearly demonstrates the importance of understanding the goals of the plant prior to evaluating and designing the control strategies. It also shows the importance of reducing the variation in achieving good plant operation and is a practical way to perform economic evaluations of potential investments.

2.5 IMPORTANCE OF CONTROL ENGINEERING

Good control performance yields substantial benefits for safe and profitable plant operation. By applying the process control principles in this book, the engineer will be able to design plants and control strategies that achieve the control objectives. Recapitulating the material in Chapter 1, control engineering facilitates good control by ensuring that the following criteria are satisfied.

Control Is Possible

The plant must be designed with control strategies in mind so that the appropriate measurements and manipulated variables exist. Control of the composition of the liquid product from the flash drum in Figure 2.2 requires the flexibility to adjust the valves in the heating streams. Even if the valve can be adjusted, the total heat exchanger areas and utility flows must be large enough to satisfy the demands of the flash process. Thus, the chemical engineer is responsible for ensuring that the process equipment and control equipment provide sufficient flexibility.

The Plant Is Easy to Control

Clearly, reduction in variation is desired. Typically, plants that are subject to few disturbances, due to inventory (buffer) between the disturbance and the controlled variable, are easier to control. Unfortunately, this is contradictory to many modern designs, which include energy-saving heat integration schemes and reduced plant inventories. Therefore, the dynamic analysis of such designs is important to determine how much (undesired) variance results from the (desired) lower capital costs and higher steady-state efficiency. Also, the plant should be "responsive"; that is, the dynamics between the manipulated and controlled variables should be fast—the faster the better. Plant design can influence this important factor substantially.

Proper Control Calculations Are Used

Properly designed control calculations can improve the control performance by reducing the variation of the controlled variable. Some of the desired characteristics for these calculations are simplicity, generality, reliability, and flexibility. The basic control algorithm is introduced in Chapter 8.

Control Equipment Is Properly Selected

Equipment for process control involves considerable cost and must be selected carefully to avoid wasteful excess equipment. Information on equipment cost can be obtained from the references in Chapter 1.

Example 2.3. Control performance depends on process and control equipment design. The plant section in Figures 2.13*a* and *b* includes different designs for a packed-bed chemical reactor and two distillation towers. The feed to the plant section experiences composition variation, which results in variation in the product composition, which should be maintained as constant as possible.

The lower-cost plant design in Figure 2.13*a* has no extra tankage and a low-cost analyzer that must be placed after the distillation towers. The more costly design has a feed tank, to reduce the effects of the feed compositions through mixing, and a more expensive analyzer located at the outlet of the reactor for faster sensing. Thus, the design in Figure 2.13*b* has smaller disturbances to the reactor and faster control. The dynamic responses show that the control performance of the more costly plant is much better. Whether the investment is justified requires an economic analysis of the

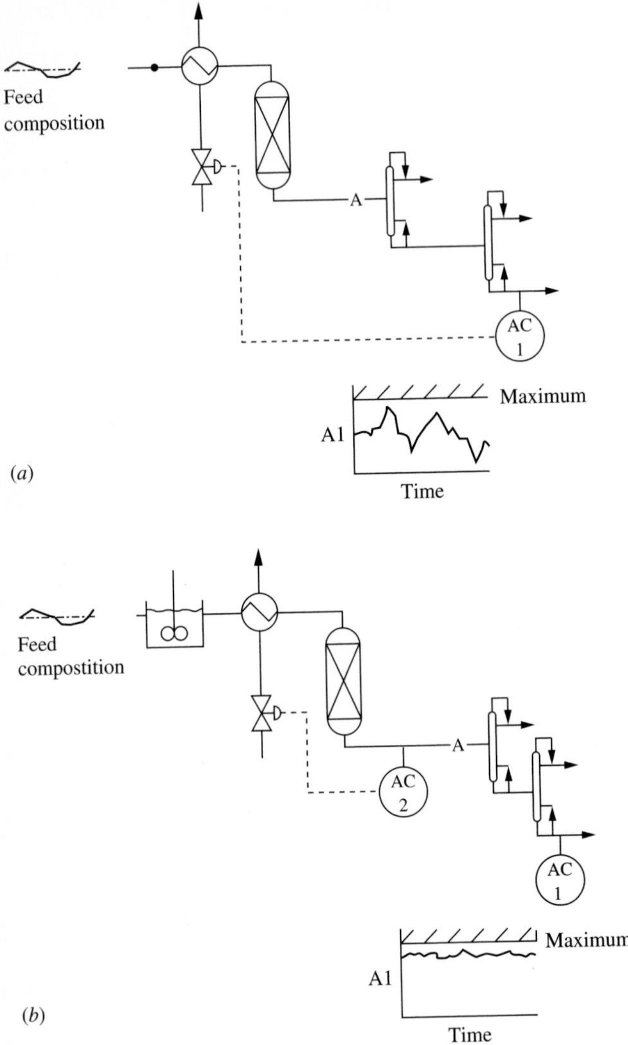

FIGURE 2.13
(*a*) Example of a process design that is difficult to control. (*b*) Example of a process that is easier to control.

entire plant. As this example demonstrates, good control engineering involves proper equipment design as well as control calculations.

Example 2.4. Control contributes to safety by maintaining process variables near their desired values. The chemical reactor with highly exothermic reaction in Figure 2.14 demonstrates two examples of safety through control. Many input variables, such as feed composition, feed temperature, and cooling temperature, can vary, which could lead to dangerous overflow of the liquid and large temperature excursions (runaway). The control design shown in Figure 2.14 maintains the level near its desired value by adjusting the outlet flow rate, and it maintains the temperature near its desired value

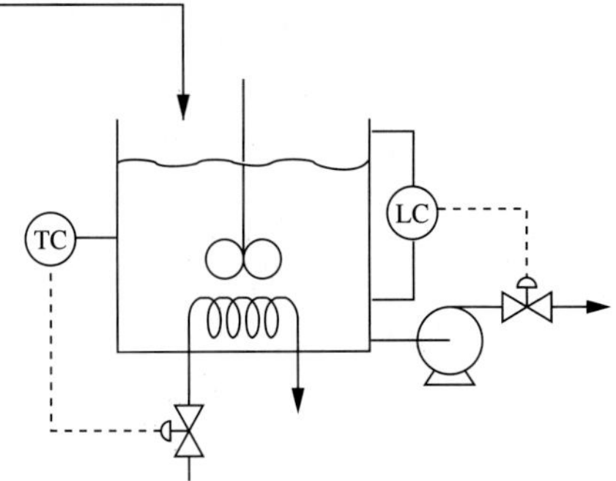

FIGURE 2.14
Control for stirred-tank reactor.

by adjusting the coolant flow rate. If required, these controls could be supplemented with emergency control systems.

Example 2.5. The type of control calculation can affect the dynamic performance of the process. Consider the system in Figures 2.15*a* through *c*, which has three different control designs, each giving a different control performance. The process involves mixing two streams to achieve a desired concentration in the exit stream by adjusting one of the inlet streams. The first design, in Figure 2.15*a*, gives the result of a very

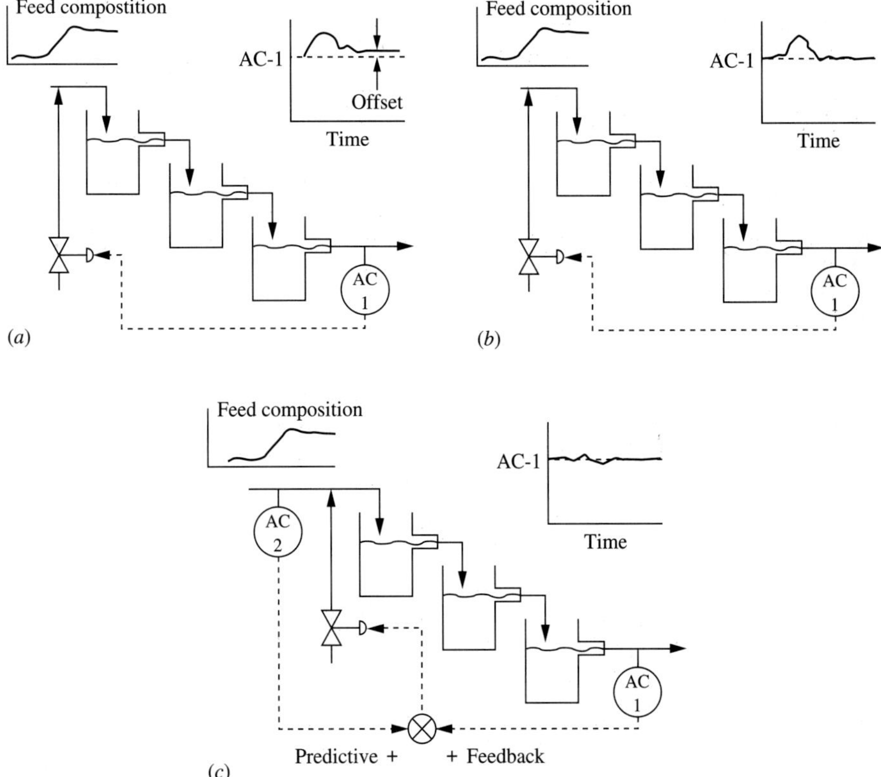

FIGURE 2.15

simple feedback control calculation, which keeps the controlled variable from varying too far from, but does not return the controlled variable to, the desired value; this deviation is termed *offset* and is generally undesirable. The second design, in Figure 2.15*b*, uses a more complex feedback control calculation, which provides response to disturbances that returns the controlled variable to its desired value. Since the second design relies on feedback principles, the controlled variable experiences a rather large initial deviation, which cannot be reduced by improved feedback calculations. The third design combines feedback with a predicted correction based on a measurement of the disturbance, which is called *feedforward.* The third design provides even better performance by reducing the magnitude of the initial response along with a return to the desired value. The calculations used for these designs, along with criteria for selecting among possible designs, are covered in later chapters. This example simply demonstrates that the type of calculation can substantially affect the dynamic response of a control system.

2.6 CONCLUSIONS

Good control design addresses a hierarchy of control objectives, ranging from safety to product quality and profit, which depend on the operating objectives for the plant. The objectives are determined by both steady-state and dynamic analysis of the plant performance. The steady-state feasible operating region is defined by the operating window; plant operation should remain within the window, because constraint violations involve severe penalties. Within the operating window, the condition that results in the highest profit is theoretically the best operation. However, because the plant cannot be maintained at an exact value of each variable due to disturbances, the variation must be considered in selecting an operating point that does not result in (unacceptably frequent) constraint violations yet still achieves a high profit. Process control reduces the variation and results in consistently high product quality and close approach to the theoretical maximum profit. Methods for quantitatively analyzing these factors are presented in this chapter.

As we have learned, good performance provides "tight" control of key variables; that is, the variables vary only slightly from their desired values. Clearly, understanding the dynamic behavior of processes is essential in designing control strategies. Therefore, the next part of the book addresses process dynamics and modelling. Only with a thorough knowledge of the process dynamics can we design control calculations that meet demanding objectives and yield large benefits.

REFERENCES

API, *American Petroleum Institute Recommended Practice 550* (2nd Ed.), *Manual on Installation of Refining Instruments and Control Systems: Fired Heaters and Inert Gas Generators,* API, Washington, DC, 1977.

Bethea, R., and R. Rhinehart, *Applied Engineering Statistics,* Marcel Dekker, New York, 1991.

Battelle Laboratory, *Guidelines for Hazard Evaluation Procedures,* American Institute for Chemical Engineering (AIChE), New York, 1985.

Bozenhardt, H., and M. Dybeck, "Estimating Savings From Upgrading Process Control," *Chem. Engr.,* 99–102 (Feb. 3, 1986).

Gorzinski, E., "Development of Alkylation Process Model," *European Conf. on Chem. Eng.,* 1983, pp. 1.89–1.96.

Marlin, T., J. Perkins, G. Barton, and M. Brisk, "Process Control Benefits, A Report on a Joint Industry–University Study," *Process Control, 1,* pp. 68–83 (1991).

Narraway, L., and J. Perkins, "Selection of Process Control Structure Based on Linear Dynamic Economics," *IEC Res., 32,* pp. 2681–2692 (1993).

Snedecor, G., and W. Cochran, *Statistical Methods,* Iowa State University Press, Ames, IA, 1980.

Stout, T., and R. Cline, "Control System Justification," *Instrument. Tech.,* Sept. 1976, 51–58.

Perkins, J., "Interactions Between Process Design and Process Control," in J. Rijnsdorp et al. (Ed.), *DYCORD+ 1990,* International Federation of Automatic Control, Pergamon Press, Maastricht, Netherlands, pp. 195–203 (1989).

Warren Centre, *Major Industrial Hazards,* Technical Papers, University of Sydney, Australia, 1986.

ADDITIONAL RESOURCES

In-depth benefits studies on seven industrial process plants are reported in the following reference, which also gives guidance on performing benefits studies.

Marlin, T., J. Perkins, G. Barton, and M. Brisk, *Advanced Process Control Applications—Opportunities and Benefits,* Instrument Society of America, Research Triangle Park, NC, 1987.

For further examples of operating windows and how they are used in setting process operating policies, see

Maarleveld, A., and J. Rijnsdorp, "Constraint Control in Distillation Columns," *Automatica, 6,* 51–58 (1970).

Roffel, B., and H. Fontien, "Constraint Control of Distillation Processes," *Chem. Eng. Sci., 34,* 1007–1018 (1979).

Morari, M., Y. Arkun, and G. Stephanopoulos, "Studies in the Synthesis of Control Structures for Chemical Processes, Part III," *AIChE J., 26,* 220 (1980).

Arkun, Y., and M. Morari, "Studies in the Synthesis of Control Structures for Chemical Processes, Part IV," *AIChE J., 26,* 975–991 (1980).

Fisher, W., M. Doherty, and J. Douglas, "The Interface Between Design and Control," *IEC Res., 27,* 597–615 (1988).

> These questions provide exercises in relating process variability to performance. Much of the remainder of the book addresses *how* process control can reduce the variability of key variables.

QUESTIONS

2.1. For each of the following processes, identify at least one control objective in each of the seven categories introduced in Section 2.2. Describe a feedback approach appropriate for achieving each objective.

(*a*) The reactor-separator system in Figure 1.8

(*b*) The boiler in Figure 14.17

(*c*) The distillation column in Figure 15.18

(*d*) The fired heater in Figure 17.17

2.2. The best distribution of variable values depends strongly on the performance function of the process. Three different performance functions are given in Figure Q2.2. In each case, the average value of the variable (x_{ave}) must remain at the specified value, although the distribution around the average is not specified. The performance function, P, can be assumed to be a quadratic function of the variable, x, in every segment of the distribution.

$$P_i = a + b\,(x_i - x_{\text{ave}}) + c\,(x_i - x_{\text{ave}})^2$$

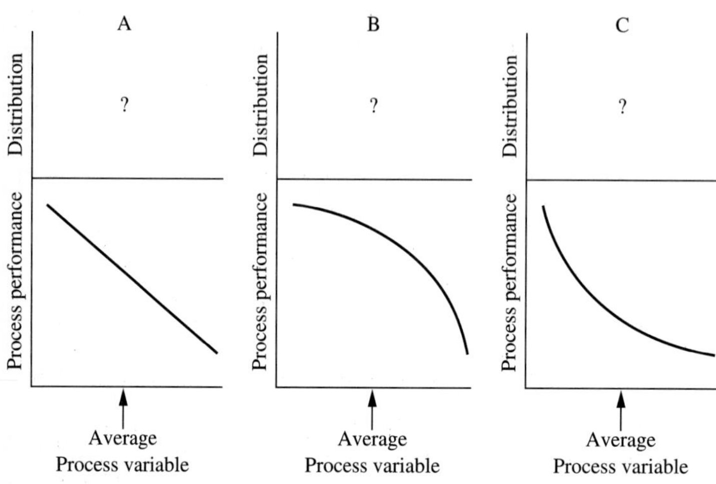

FIGURE Q2.2

For each of the cases in Figure Q2.2, discuss the relationship between the distribution and the average profit, and determine the distribution that will maximize the average performance function. Provide quantitative justification for your result.

2.3. The fired heater example in Figure 2.11 had a hard constraint.

(*a*) Sketch the performance function for this situation, including the performance when violations occur, on the figure.

(*b*) Assume that the distribution of the temperature would have 0.005 fraction of its operation exceeding the limit of 864°C and that each time the limit is exceeded, the plant incurs a cost of $1000 to restart the equipment. Can you calculate the total cost per year for exceeding the limit?

(*c*) Make any additional assumptions and complete the calculation.

2.4. Sometimes there is no active hard constraint. Assume that the fired heater in Figure 2.11 has no hard constraint, but that a side reaction forming undesired products begins to occur significantly at 850°C. This side reaction has an activation energy with larger magnitude than the product reaction. Sketch the shape of the performance function for this situation. How would you determine the best desired (average) value of the temperature and the best temperature distribution?

2.5. Sometimes engineers use a shortcut method for determining the average process performance. In this shortcut, the average variable value is used, rather than the full distribution, in calculating the performance. Discuss the assumptions implicit in this shortcut and when it is and is not appropriate.

2.6. A chemical plant produces vinyl chloride monomer for subsequent production of polyvinyl chloride. This plant can sell all monomer it can produce within quality specifications. Analysis indicates that the plant can produce 175 tons/day of monomer with perfect operation. A two-month production record is given in Figure Q2.6. Calculate the profit lost by not operating at the highest value possible. Discuss why the plant production might not always be at the highest possible value.

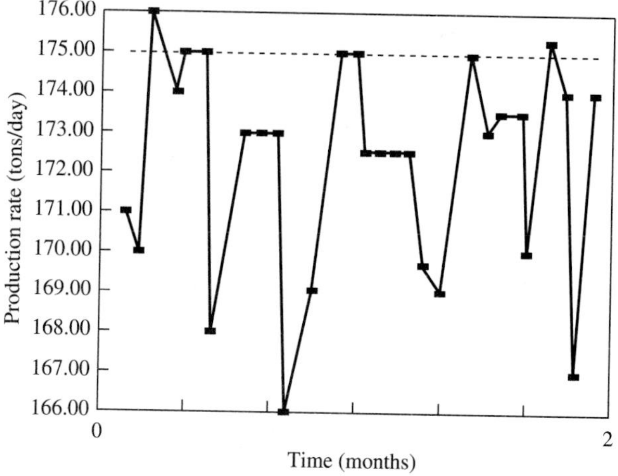

FIGURE Q2.6
(Reprinted by permission. Copyright ©1987, Instrument Society of America. From Marlin, T. et al., *Advanced Process Control Applications — Opportunities and Benefits*, ISA, 1987.)

2.7. A blending process, shown in Figure Q2.7, mixes component A into a stream. The objective is to maximize the amount of A in the stream without exceeding the upper limit of the concentration of A, which is 2.2 mole/m^3. The current operation is "open-loop," with the operator occasionally looking at the analyzer value and changing the flow of A. The flow during the period that the data was collected was essentially constant at 1053 m^3/hr. How much more A could have been blended into the stream with perfect control, that is, if the concentration of A had been maintained exactly at its maximum? What would be the improvement if the new distribution were normal with a standard deviation of 0.075 mole/m^3?

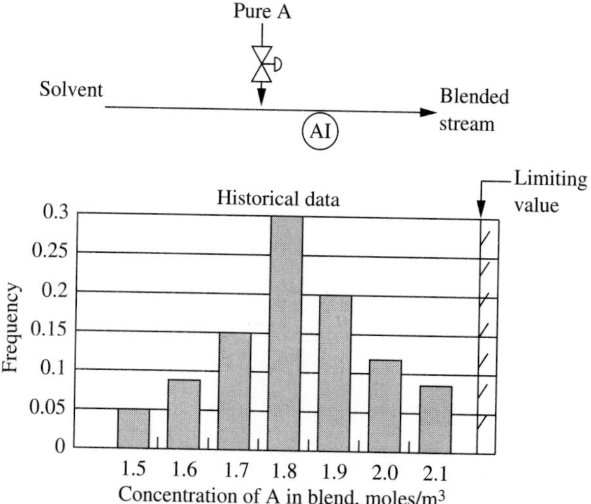

FIGURE Q2.7

2.8. The performance function for a distillation tower is given in Figure Q2.8 in terms of lost profit from the best operation as a function of the bottoms impurity, x_B (Stout and

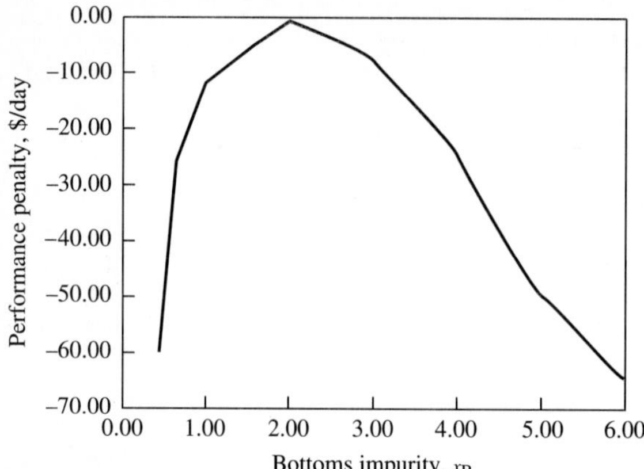

FIGURE Q2.8

(Reprinted by permission. Copyright ©1976, Instrument Society of America. From Stout, T., and R. Cline, "Control Systems Justification." *Instr. Techn.* September 1976, pp. 51–58.)

TABLE Q2.8

	Fraction of time at x_B			
x_B	A	B	C	D
0.25	0	0	0	0
0.5	0.25	0.05	0	0
0.75	0.50	0.05	0	0
1.0	0.25	0.10	0	0
1.5	0	0.20	0	0.333
2.0	0	0.30	0	0.333
3.0	0	0.20	0.25	0.333
4.0	0	0.10	0.50	0
5.0	0	0	0.25	0
6.0	0	0	0	0

Cline, 1978). Calculate the average performance for the four distributions (A through D) given in Table Q2.8 along with the average and standard deviation of the concentration, x_B. Discuss the relationship between the distributions and the average performance.

2.9. Profit contours similar to those in Figure Q2.9 have been reported by Gorzinski (1983) for a distillation tower separating normal butane and isobutane in an alkylation process for a petroleum refinery. Based on the shape of the profit contours, discuss the selection of desired values for the distillate and bottoms impurity variables to be used in an automation strategy. (Recall that some variation about the desired values is inevitable.) If only one product purity can be controlled tightly to its desired value, which would be the one you would select to control tightly?

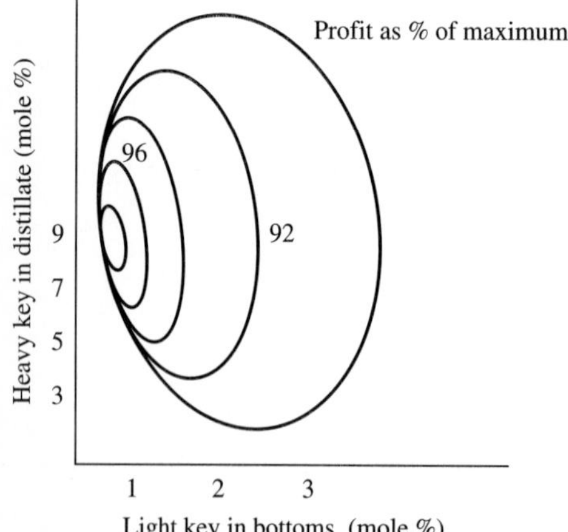

FIGURE Q2.9

PART II

PROCESS DYNAMICS

The engineer must understand the dynamic behavior of a physical system in order to design the equipment, select operating conditions, and implement an automation technique properly. The need for understanding dynamics is first illustrated through the discussion of two examples. The first involves the dynamic response of the bus and bicycle shown in Figure II.1. When the drivers wish to maneuver the vehicles, such as to make a 180° U turn, the bicycle can be easily turned in a small

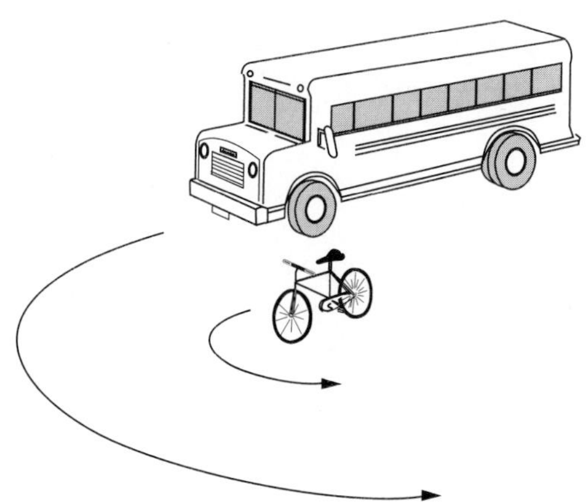

FIGURE II.1
Bus and bicycle maneuverability.

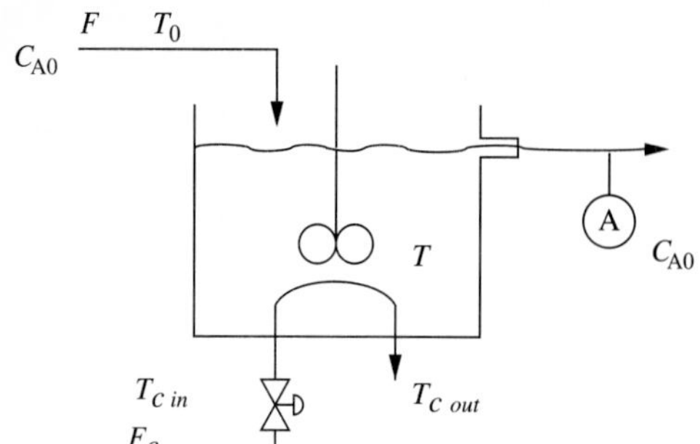

FIGURE II.2
Nonisothermal CSTR.

radius, while the bus requires an arc of considerably larger radius. Clearly, the design of the vehicle affects the possible maneuverability, even when the bus has an expert driver. Also, the driver of the bus and the rider of the bicycle must use different rules in steering. This simple example demonstrates that (1) a key aspect of automation is designing and building equipment that can be easily controlled, and (2) the design and implementation of an automation system requires knowledge of the dynamic behavior of the system.

These two important principles can be applied to the chemical reactor example shown in Figure II.2. The reactor operation can be influenced by adjusting the opening of the valve in the coolant pipe, and the outlet concentration is measured by an analyzer located downstream from the reactor outlet. Regarding the first principle (the effect of process design), it seems likely that the delay in measuring the outlet concentration would reduce the effectiveness of feedback control. Regarding the second principle (the effect of automation method), a very aggressive method for adjusting the coolant flow could cause a large overshoot or oscillations in returning the concentration to its desired value; thus, the feedback adjustments should be tailored to the specific process.

The knowledge of dynamic behavior required for process control is formalized in mathematical models. In fact, modelling plays such a central role in the theory and practice of process control that the statement is often made that modelling is the key element in the successful application of control. A complete explanation of the needs of process control cannot be presented until more detail is covered on feedback systems; however, the importance of the four basic questions to be addressed through modelling should be clear from the general discussion in the previous chapters, along with the examples in Figures II.1 and II.2.

1. *Which variables can be influenced?* Process control inherently involves some manipulated variables, which can be adjusted, and some controlled variables, which are affected by the adjustments. By turning the steering wheel, the driver can influence the direction of the bus, but not its speed. By changing the coolant valve opening in the reactor example, the reactor temperature and concentration can be influenced. The identification of variables will be addressed in this part through the analysis of degrees of freedom and cause-effect relationships, and the aspect of controllability will be introduced later in the book.

2. *Over what range can the variables be altered?* The acceptable range of process variables, such as temperature and pressure, and the limited range of the manipulated variables places bounds on the effects of adjustments. The bus wheels can only be turned a maximum amount to the right and left, and the coolant valve is limited between fully closed (no flow) and fully opened (maximum flow). The range of possible values is termed the *operating window,* and models can be used to determine the bounds or "frame" on this window quantitatively.
3. *How effectively can feedback maintain the process at desired conditions?* The following aspects of the process behavior are required to implement process control.
   a. *Sign and magnitude of response:* The bus driver must know how the bus will respond when the wheel is turned clockwise, and the operator needs to know whether temperature will increase or decrease when the valve is opened. It is essential that the sign does not change and is best if the magnitude does not vary greatly.
   b. *Speed of response:* The speed must be known to determine the manipulations that can be entered; if the manipulations are too aggressive, the system can oscillate and even become unstable. This can happen in driving a bus on a slippery road and in trying to control the concentration when there is a long delay between the adjusted variable and measurement.
   c. *Shape of response:* The shape of dynamic responses can vary greatly. For example, the two responses in Figure II.3 have the same "speed" as measured by the time to reach their final values, but the shapes are different. Response A, which gives an indication of the response without delay, is better for control than response B, which gives no output indication of the input change for a long time.
4. *How sensitive are the results?* Process control systems are usually applied in industrial-scale plants that change operations often and experience variation in operating conditions and equipment performance. This variation affects the dynamic behavior of the process, the items in the preceding question, which must be considered in process control. For example, the behavior of the chemical reactor could depend on an inhibitor in the feed and catalyst deactivation. The analysis

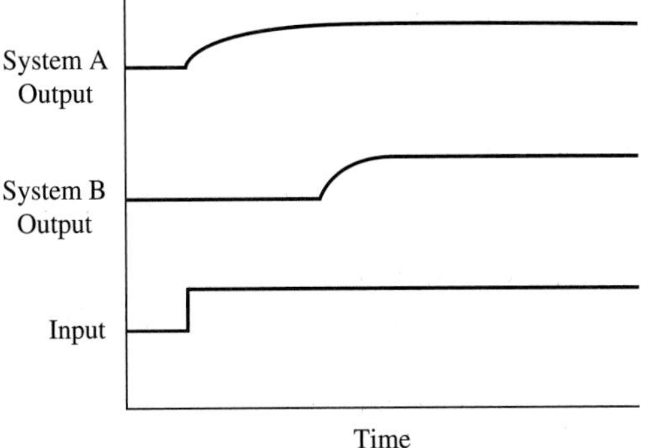

FIGURE II.3

of the possible variation in the system and sensitivity of the dynamic behavior to the variability begins in the modelling procedure.

In summary, the dynamic features most favorable to good control include (1) nearly constant sign and magnitude, (2) a fast response, (3) minimum delay, and (4) insensitivity to process changes. This good situation cannot always be achieved through process design, because processes are designed to meet additional requirements such as high pressures, volumes for reactor residence times, or area for mass transfer and heat transfer. However, the features that favor good control should be a consideration in the process design and must be known for the design of the process controls.

The modelling procedures in this part provide methods for determining these features and for relating them to process equipment design and operating variables. There are many types of models used by engineers, so important aspects of these models used in this book are briefly summarized and compared with alternatives.

1. *Mathematical models:* The following definition of a mathematical model was given by Denn (1986).

   > A mathematical model of a process is a system of equations whose solution, given specific input data, is representative of the response of the process to a corresponding set of inputs.

   We will deal exclusively with mathematical models for process analysis. In contrast, experimental or analog methods can use physical models, like a model airplane in a wind tunnel or an electrical circuit, to represent the behavior of a full-scale system empirically.
2. *Fundamental and empirical models:* Fundamental models are based on such principles as material and energy conservation and can provide great insight as well as predictive power. For many systems, fundamental models can be very complex, and simplified empirical models based on experimental dynamic data are sufficient for many process control tasks. Both types of models are introduced in Part II.
3. *Steady-state and dynamic models:* Both steady-state and dynamic models are used in process control analysis. Dynamic modelling is emphasized in this book because it is assumed that the reader has prior experience in steady-state modelling.
4. *Lumped and distributed models:* Lumped models are valid for systems in which the properties of a system do not depend on the position within the system. For lumped systems, steady-state models involve algebraic equations, and dynamic models involve ordinary differential equations. Distributed models are valid for systems in which the properties depend on position, and their dynamic models involve partial differential equations. To maintain a manageable level of mathematical complexity, essentially all models in this book will involve lumped systems, with the exception of a model for pure transportation delay in a pipe. Since many

chemical process designs involve inventories that are approximately well-mixed, lumped models are often sufficient, but each system should be evaluated for the proper modelling assumptions.

Finally, one must recognize that modelling is performed to answer specific questions; thus, no one model is appropriate for all situations. The methods in this part have been selected to provide the information required for the control analyses included in this book and provide only a limited introduction to the topic of process modelling. Many interesting modelling concepts, mathematical solution techniques, and results for important process structures are not included. Therefore, the reader is encouraged to refer to the references at the end of each chapter.

REFERENCES

Denn, M., *Process Modeling*, Pitman Publishing, Marshfield, MA, 1986.

CHAPTER

3

MATHEMATICAL MODELLING PRINCIPLES

3.1 INTRODUCTION

The models addressed in this chapter are based on fundamental theories or laws, such as the conservations of mass, energy, and momentum. Of many approaches to understanding physical systems, engineers tend to favor fundamental models for several reasons. One reason is the amazingly small number of principles that can be used to explain a wide range of physical systems; thus, fundamental principles *simplify* our view of nature. A second reason is the broad range of applicability of fundamental models, which allow extrapolation (with caution) beyond regions of immediate empirical experience; this enables engineers to evaluate potential changes in operating conditions and equipment and to design new plants. Perhaps the most important reason for using fundamental models in process control is the analytical expressions they provide relating key features of the physical system (flows, volumes, temperatures, and so forth) to its dynamic behavior. Since chemical engineers design the process, these relationships can be used to design processes that are as easy to control as possible, so that a problem created through poor process design need not be partially solved through sophisticated control calculations.

The presentation in this chapter assumes that the reader has previously studied the principles of modelling material and energy balances, with emphasis on steady-state systems. Those unsure of the principles should refer to one of the many introductory textbooks in the area (e.g., Felder and Rousseau, 1986; Himmelblau, 1982). In this chapter, a step-by-step procedure for developing fundamental models is presented that emphasizes dynamic models used to analyze the transient behavior of processes and control systems. The procedure begins with a definition of the goals and proceeds through formulation, solution, results analysis, and validation.

Analytical solutions will be restricted to the simple integrating factor for this chapter and will be extended to Laplace transforms in the next chapter.

Experience has shown that the beginning engineer is advised to follow this procedure closely, because it provides a road map for the sequence of steps and a checklist of issues to be addressed at each step. Based on this strong recommendation, the engineer who closely follows the procedure might expect a guarantee of reaching a satisfactory result. Unfortunately, no such guarantee can be given, because a good model depends on the insight of the engineer as well as the procedure followed. In particular, several types of models of the same process might be used for different purposes; thus, the model formulation and solution should be matched with the problem goals. In this chapter, the modelling procedure is applied to several process examples, with each example having a goal that would be important in its own right and leads to insights for the later discussions of control engineering. This approach will enable us to complete the modelling procedure, including the important step of results analysis, and learn a great deal of useful information about the relationships between design, operating conditions, and dynamic behavior.

3.2 A MODELLING PROCEDURE

Modelling is a task that requires creativity and problem-solving skills. A general method is presented in Table 3.1 as an aid to learning and applying modelling skills, but the engineer should feel free to adapt the procedure to the needs of particular problems. It is worth noting that the steps could be divided into two categories: steps 1 to 3 (model development) and steps 4 to 6 (model solution or simulation), because several solution methods could be applied to a particular model. All steps are grouped together here as an integrated modelling procedure, because this represents the

TABLE 3.1
Outline of fundamental modelling procedure

1. Define goals
   a. Specific design decisions
   b. Numerical values
   c. Functional relationships
   d. Required accuracy
2. Prepare information
   a. Sketch process and identify system
   b. Identify variables of interest
   c. State assumptions and data
3. Formulate model
   a. Conservation balances
   b. Constitutive equations
   c. Rationalize (combine equations and collect terms)
   d. Check degrees of freedom
   e. Dimensionless form
4. Determine solution
   a. Analytical
   b. Numerical
5. Analyze results
   a. Check results for correctness
      1. Limiting and approximate answers
      2. Accuracy of numerical method
   b. Interpret results
      1. Plot solution
      2. Characteristic behavior like oscillations or extrema
      3. Relate results to data and assumptions
      4. Evaluate sensitivity
      5. Answer "what if" questions
6. Validate model
   a. Select key values for validation
   b. Compare with experimental results
   c. Compare with results from more complex model

vernacular use of the term *modelling* and stresses the need for the model and solution technique to be selected in conjunction to satisfy the stated goal successfully. Also, while the procedure is presented in a linear manner from step 1 to step 6, the reality is that the engineer often has to iterate to solve the problem at hand. Only experience can teach us how to "look ahead" so that decisions at earlier steps are made in a manner that facilitate the execution of later steps. Each step in the procedure is discussed in this section and is demonstrated for a simple stirred-tank mixing process.

Define Goals

Perhaps the most demanding aspect of modelling is judging the type of model needed to solve the engineering problem at hand. This judgment, summarized in the goal statement, is a critical element of the modelling task. The goals should be specific concerning the type of information needed. A specific numerical value may be needed; for example, "At what time will the liquid in the tank overflow?" In addition to specific numerical values, the engineer would like to determine semi-quantitative information about the characteristics of the system's behavior; for example, "Will the level increase monotonically or will it oscillate?" Finally, the engineer would like to have further insight requiring functional relationships; for example, "How would the flow rate and tank volume influence the time that the overflow will occur?"

Another important factor in setting modelling goals is the accuracy of a model and the effects of estimated inaccuracy on the results. This factor is perhaps not emphasized sufficiently in engineering education—a situation that may lead to the false impression that all models have great accuracy over large ranges. The modelling and analysis methods in this book consider accuracy by recognizing likely errors in assumptions and data at the outset and tracing their effects through the modelling and later analysis steps. It is only through this careful analysis that we can be assured that designs will function properly in realistic situations.

Example 3.1.

Goal. The dynamic response of the mixing tank in Figure 3.1 to a step change in the inlet concentration is to be determined, along with the way the speed and shape of response depend on the volume and flow rate. In this example, the outlet stream cannot be used for further production until 90% of the change in outlet concentration has occurred; therefore, a specific goal of the example is to determine how long after the step change the outlet stream reaches this composition.

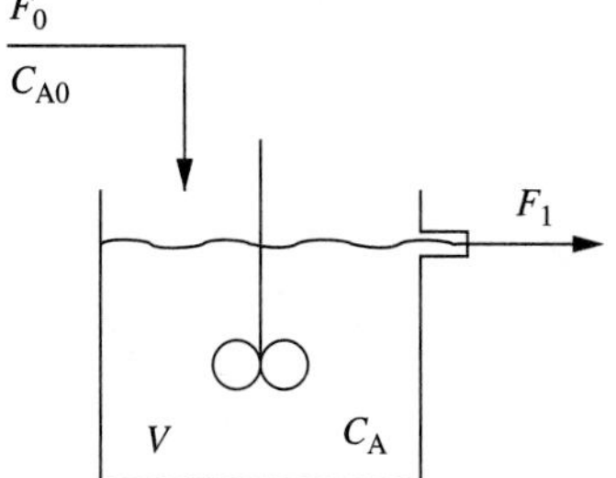

FIGURE 3.1
Continuous-flow stirred tank.

Prepare Information

The first step is to identify the system. This is usually facilitated by sketching the process, identifying the key variables, and defining the boundaries of the system for which the balances will be formulated. The system, or control volume, should be a volume within which the important properties do not vary with position. The assumption of a well-stirred vessel is often employed in this book because even though no such system exists in fact, many systems closely approximate this behavior. The reader should not infer from the use of stirred-tank models in this book that more complex models are never required. Modelling of systems via partial differential equations is required for many processes in which product quality varies with position; distributed models are required for many processes, such as paper and metals. Systems with no spatial variation in important variables are termed *lumped-parameter* systems, whereas systems with significant variation in one or more directions are termed *distributed-parameter* systems.

In addition to system selection, all models require information to predict a system's behavior. An important component of the information is the set of assumptions on which the model will be based; these are selected after consideration of the physical system and the accuracy required to satisfy the modelling goals. For example, the engineer usually is not concerned with the system behavior at the atomic level, and frequently not at the microscopic level. Often, but not always, the macroscopic behavior is sufficient to understand process dynamics and control. The assumptions used often involve a compromise between the goals of modelling, which may favor detailed and complex models, and the solution step, which favors simpler models.

A second component of the information is data regarding the physicochemical system (e.g., heat capacities, reaction rates, and densities). In addition, the external variables that are inputs to the system must be defined. These external variables, sometimes termed *forcing functions,* could be changes to operating variables introduced by a person (or control system) in an associated process (such as inlet temperature) or changes to the behavior of the system (such as fouling of a heat exchanger).

Example 3.1.

Information. The system is the liquid in the tank. The tank has been designed well, with baffling and impeller size, shape, and speed such that the concentration should be uniform in the liquid (Foust et al., 1980).

Assumptions.

1. Well-mixed vessel
2. Density the same for A and solvent
3. Constant flow in

Data.

1. $F_0 = 0.085\ \text{m}^3/\text{min}$; $V = 2.1\ \text{m}^3$; $C_{\text{Ainit}} = 0.925\ \text{mole/m}^3$; $\Delta C_{\text{A0}} = 0.925\ \text{mole/m}^3$; thus, $C_{\text{A0}} = 1.85\ \text{mole/m}^3$ after the step
2. The system is initially at steady state ($C_{\text{A0}} = C_{\text{A}} = C_{\text{Ainit}}$)

Note that the inlet concentration, C_{A0}, remains constant after the step change has been introduced to this two-component system.

Formulate the Model

First, the important variables, whose behavior is to be predicted, are selected. Then the equations are derived based on fundamental principles, which usually can be divided into two categories: conservation and constitutive. The *conservation balances* are relationships that are obeyed by all physical systems under common assumptions valid for chemical processes. The conservation equations most often used in process control are the conservations of material (overall and component), energy, and momentum.

These conservation balances are often written in the following general form for a system shown in Figure 3.2.

$$\text{Accumulation} = \text{In} - \text{Out} + \text{Generation} \tag{3.1}$$

For a well-mixed system, this balance will result in an ordinary differential equation when the accumulation term is nonzero and in an algebraic equation when the accumulation term is zero. General statements of this balance for the conservation of material and energy follow:

OVERALL MATERIAL BALANCE.

$$\{\text{Accumulation of mass}\} = \{\text{Mass in}\} - \{\text{Mass out}\} \tag{3.2}$$

COMPONENT MATERIAL BALANCE.

$$\begin{aligned}\{\text{Accumulation of component mass}\}\\ &= \{\text{Component mass in}\} - \{\text{Component mass out}\}\\ &\quad + \{\text{Generation of component mass}\}\end{aligned} \tag{3.3}$$

ENERGY BALANCE.

$$\begin{aligned}\{\text{Accumulation of } E + \text{PE} + \text{KE}\} &= \{E + \text{PE} + \text{KE in due to convection}\}\\ &\quad - \{E + \text{PE} + \text{KE out due to convection}\}\\ &\quad + Q - W\end{aligned} \tag{3.4}$$

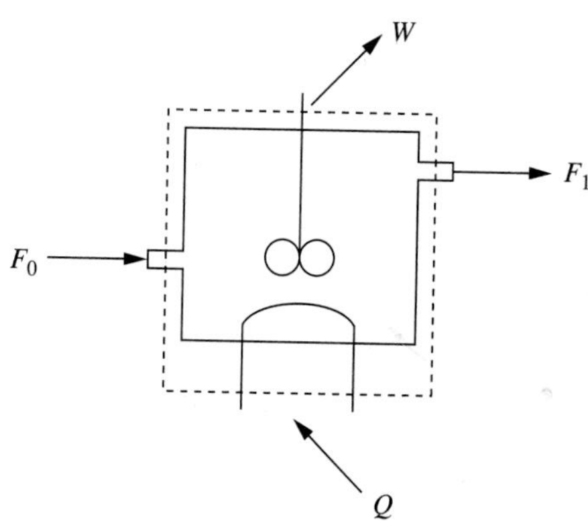

FIGURE 3.2
General lumped-parameter system.

which can be written for a system with constant volume as

$$\{\text{Accumulation of } E + \text{PE} + \text{KE}\} = \{H + \text{PE} + \text{KE in due to convection}\} \\ - \{H + \text{PE} + \text{KE out due to convection}\} \\ + Q - W_s \tag{3.5}$$

where E = internal energy
$H = E + pv$ = enthalpy
KE = kinetic energy
PE = potential energy
pv = pressure times specific volume (referred to as flow work)
Q = heat transferred to the system from the surroundings
W = work done by the system on the surroundings
W_s = shaft work done by the system on the surroundings

The selection of the balances relevant to a specific problem depends on the physical variables of interest. For example, pressures and levels depend on the total material inventory and require an overall material balance; compositions require a component material balance; and temperature requires an energy balance. Most realistic problems require several balances. Applications of these general statements to yield algebraic and differential equations are given in subsequent examples.

Often, the important unknown variables cannot be determined from the conservation balances alone. When this is the case, additional *constitutive equations* are included to provide sufficient equations for a completely specified model. Some examples of constitutive equations follow:

Heat transfer	$Q = hA(\Delta T)$
Chemical reaction rate	$r_A = k_0 e^{-E/RT} C_A$
Fluid flow	$F = C_v(\Delta P/\rho)^{1/2}$
Equation of state	$PV = nRT$
Phase equilibrium	$y_i = K_i x_i$

The constitutive equations provide relationships that are not universally applicable but are selected to be sufficiently accurate for the specific system being studied. The applicability of a constitutive equation is problem-specific and is the topic of a major segment of the chemical engineering curriculum.

An important issue in deriving the defining model equations is "How many equations are appropriate?" By that we mean the proper number of equations to predict the dependent variables. The proper number of equations can be determined from the recognition that the model is correctly formulated when the system's behavior can be predicted from the model; thus, a well-posed problem should have no degrees of freedom. The number of *degrees of freedom* for a system is defined as

$$DOF = NV - NE \tag{3.6}$$

with DOF equal to the number of degrees of freedom, NV equal to the number of variables, and NE equal to the number of independent equations. Not every symbol appearing in the equations represents a variable; some are parameters that have known constant values. Other symbols represent external variables (also called *exogenous* variables); these are variables whose values are not dependent on the behavior of the system being studied. External variables may be constant or vary with time in response to conditions external to the system, such as a valve that is opened according to a specified function (e.g., a step). The value of each external variable must be known. NV in equation (3.6) represents the number of remaining variables that depend on the behavior of the system and are to be evaluated through the model equations.

It is important to recognize that the equations used to evaluate NE must be *independent*; additional dependent equations, although valid in that they also describe the system, are not to be considered in the degrees-of-freedom analysis, because they are redundant and provide no independent information. This point is reinforced in several examples throughout the book. The three possible results in the degrees-of-freedom analysis are summarized in Table 3.2.

After the initial, valid model has been derived, a rationalization should be considered. First, equations can sometimes be combined to simplify the overall model. Also, some terms can be combined to form more meaningful groupings in the resulting equations. Combining terms can establish the key parameters that affect the behavior of the system; for example, control engineering often uses parameters like the time constant of a process, which can be affected by flows, volumes, temperatures, and compositions in a process. By grouping terms, many physical systems can be shown to have one of a small number of mathematical model structures, enabling engineers to understand the key aspects of these physical systems quickly. This is an important step in modelling and will be demonstrated through many examples.

TABLE 3.2
Summary of degrees-of-freedom analysis

	DOF = NV − NE
DOF = 0	The system is **exactly specified**, and the solution of the model can proceed.
DOF < 0	The system is **overspecified**, and in general, no solution to the model exists (unless all external variables and parameters take values that fortuitously satisfy the model equations). This is a symptom of an error in the formulation. The likely cause is either (1) improperly designating a variable(s) as a parameter or external variable or (2) including an extra, dependent equation(s) in the model. The model must be corrected to achieve zero degrees of freedom.
DOF > 0	The system is **underspecified**, and an infinite number of solutions to the model exists. The likely cause is either (1) improperly designating a parameter or external variable as a variable or (2) not including in the model all equations that determine the system's behavior. The model must be corrected to achieve zero degrees of freedom.

A potential final modification in this step would be to transform the equation into dimensionless form. A dimensionless formulation has the advantages of (1) developing a general solution in the dimensionless variables, (2) providing a rationale for identifying terms that might be negligible, and (3) simplifying the repeated solution of problems of the same form. A potential disadvantage is some decrease in the ease of understanding. Most of the modelling in this book retains problem symbols and dimensions for ease of interpretation; however, a few general results are developed in dimensionless form.

Example 3.1.

Formulation. Since this problem involves concentrations, overall and component material balances will be prepared. The overall material balance for a time increment Δt is

$$\{\text{Accumulation of mass}\} = \{\text{Mass in}\} - \{\text{Mass out}\} \tag{3.7}$$

$$(\rho V)_{(t+\Delta t)} - (\rho V)_{(t)} = F_0 \rho \Delta t - F_1 \rho \Delta t \tag{3.8}$$

with ρ = density. Dividing by Δt and taking the limit as $\Delta t \rightarrow 0$ gives

$$\frac{d(\rho V)}{dt} = V\overbrace{\frac{d\rho}{dt}}^{0} + \rho\frac{dV}{dt} = \rho F_0 - \rho F_1 \tag{3.9}$$

The flow in, F_0, is an external variable, because it does not depend on the behavior of the system. Because there is one equation and two variables (V and F_1) at this point, a constitutive expression is required for the flow out. Since the liquid exits by overflow, the flow out is related to the liquid level according to a weir equation, an example of which is given below (Foust et al., 1980).

$$F_1 = k_F \sqrt{L - L_W} \qquad \text{for} \qquad L > L_W \tag{3.10}$$

with k_F = constant, $L = V/A$, and L_W = level of the overflow weir. In this problem, the level is never below the overflow level, and the height above the overflow, $L - L_W$, is very small compared with the height of liquid in the tank, L. Therefore, we will assume that the liquid level in the tank is approximately constant, and the flows in and out are equal, $F_0 = F_1 = F$.

$$\frac{dV}{dt} = F_0 - F_1 = 0 \qquad \therefore V = \text{constant} \tag{3.11}$$

This result, stated as an assumption hereafter, will be used for all tanks with overflow, as shown in Figure 3.1.

The next step is to formulate a material balance on component A. Since the tank is well-mixed, the tank and outlet concentrations are the same:

$$\left\{\begin{matrix}\text{Accumulation of}\\ \text{component A}\end{matrix}\right\} = \left\{\begin{matrix}\text{Component}\\ \text{A in}\end{matrix}\right\} - \left\{\begin{matrix}\text{Component}\\ \text{A out}\end{matrix}\right\} + \left\{\begin{matrix}\text{Generation}\\ \text{of A}\end{matrix}\right\} \tag{3.12}$$

$$(\text{MW}_A V C_A)_{t+\Delta t} - (\text{MW}_A V C_A)_t = (\text{MW}_A F C_{A0} - \text{MW}_A F C_A)\Delta t \tag{3.13}$$

with C_A being moles/volume of component A and MW_A being its molecular weight, and the generation term being zero, because there is no chemical reaction. Dividing by Δt and taking the limit as $\Delta t \rightarrow 0$ gives

$$MW_A V\frac{dC_A}{dt} = MW_A F(C_{A0} - C_A) \tag{3.14}$$

One might initially believe that another balance on the only other component, solvent S, could be included in the model:

$$MW_S V\frac{dC_S}{dt} = MW_S F(C_{S0} - C_S) \tag{3.15}$$

with C_S the moles/volume and MW_S the molecular weight. However, equation (3.9) is the sum of equations (3.14) and (3.15); thus, only two of the three equations are independent. Therefore, only equations (3.11) and (3.14) are required for the model and should be considered in determining the degrees of freedom. The following analysis shows that the model using only independent equations is exactly specified:

Variables:	C_A and F_1	
External Variables:	F_0 and C_{A0}	DOF = NV − NE = 2 − 2 = 0
Equations:	(3.11) and (3.14)	

Note that the variable t representing time must be specified to use the model for predicting the concentration at a particular time.

The model is formulated assuming that parameters do not change with time, which is not exactly correct but can be essentially true when the parameters change slowly and with small magnitude during the time considered in the dynamic modelling problem. What constitutes a "small" change depends on the problem, and a brief sensitivity analysis is included in the results analysis of this example to determine how changes in the volume and flow would affect the answer to this example.

Mathematical Solution

Determining the solution is certainly of importance. However, the engineer should realize that the solution is implicitly contained in the results of the Information and Formulation steps; the solution simply "figures it out." The engineer would like to use the solution method that gives the most insight into the system. Therefore, analytical solutions are preferred in most cases, because they can be used to (1) calculate specific numerical values, (2) determine important functional relationships among design and operating variables and system behavior, and (3) give insight into the sensitivity of the result to changes in data. These results are so highly prized that we often make assumptions to enable us to obtain analytical solutions; the most frequently used approximation is linearizing nonlinear terms, as covered in Section 3.4.

In some cases, the approximations necessary to make analytical solutions possible introduce unacceptable errors into the results. In these cases, a numerical solution to the equations is employed, as described in Section 3.5. Although the numerical solutions are never exact, the error introduced can usually be made quite small, often much less than the errors associated with the assumptions and data in the model; thus, properly calculated numerical solutions can often be considered essentially exact. The major drawback to numerical solutions is loss of insight.

Example 3.1.

Solution. The model in equation (3.14) is a linear, first-order ordinary differential equation that is not separable. However, it can be transformed into a separable form by an integrating factor, which becomes more easily recognized when the differential equation is rearranged in the standard form as follows (see Appendix B).

$$\frac{dC_A}{dt} + \frac{1}{\tau}C_A = \frac{1}{\tau}C_{A0} \qquad \text{with } \frac{V}{F} = \tau \equiv \text{time constant} \tag{3.16}$$

The parameter τ is termed the *time constant* of the system and will appear in many models. The equation can be converted into separable form by multiplying both sides by the integrating factor, and the resulting equation can be solved directly:

$$\text{Integrating factor} = \text{IF} = \exp\left(\int \frac{1}{\tau}dt\right) = e^{t/\tau}$$

$$e^{t/\tau}\left(\frac{dC_A}{dt} + \frac{1}{\tau}C_A\right) = e^{t/\tau}\frac{dC_A}{dt} + C_A\frac{de^{t/\tau}}{dt} = \frac{d(e^{t/\tau}C_A)}{dt} = \frac{C_{A0}}{\tau}e^{t/\tau} \tag{3.17}$$

$$\int d(C_A e^{t/\tau}) = \int \frac{C_{A0}e^{t/\tau}}{\tau}dt = \frac{C_{A0}}{\tau}\int e^{t/\tau}dt$$

$$C_A e^{t/\tau} = \frac{C_{A0}\tau}{\tau}e^{t/\tau} + I$$

$$C_A = C_{A0} + Ie^{-t/\tau}$$

Note that the integration was simplified by the fact that C_{A0} is constant after the step change (i.e., for $t > 0$). The initial condition is $C_A(t) = C_{A\text{init}}$ at $t = 0$, which can be used to evaluate the constant of integration, I. This formulation implies that the time t is measured from the introduction of the step change.

$$I = C_{A\text{init}} - C_{A0} \qquad \therefore C_A = C_{A0} + (C_{A\text{init}} - C_{A0})e^{-t/\tau} \tag{3.18}$$

$$(C_A - C_{A\text{init}}) = [C_{A0} - (C_{A0})_{\text{init}}]\left(1 - e^{-t/\tau}\right)$$

The final equation has used the extra relationship that $(C_{A0})_{\text{init}} = C_{A\text{init}}$. Substituting the numerical values gives

$$C_A - 0.925 = (C_{A0} - 0.925)(1 - e^{-t/24.7})$$

Two important aspects of the dynamic behavior can be determined from equation (3.18). The first is the "speed" of the dynamic response, which is characterized by the time constant, τ. The second is the steady-state gain, which is defined as

$$\text{Steady-state gain} = K_p = \frac{\Delta\text{ output}}{\Delta\text{ input}} = \frac{\Delta C_A}{\Delta C_{A0}} = 1.0$$

Note that in this example the time constant depends on the equipment (V) and operation of the process (F), and the steady-state gain is independent of these design and operating variables.

Results Analysis

The first phase of the results analysis is to evaluate whether the solution is correct, at least to the extent that it satisfies the formulation. This can be partially verified by ensuring that the solution obeys some limiting criteria that are more easily derived than the solution itself. For example, the result

- Satisfies initial and final conditions
- Obeys bounds such as adiabatic reaction temperature
- Contains negligible errors associated with numerical calculations
- Obeys semi-quantitative expectations, such as the sign of the output change

Next, the engineer should "interrogate" the mathematical solution to elicit the information needed to achieve the original modelling goals. Determining specific numerical values is a major part of the results analysis, because engineers need to make quantitative decisions on equipment size, operating conditions, and so forth. However, results analysis should involve more extensive interpretation of the solution. When meaningful, results should be plotted, so that key features like oscillations or extrema (maximum or minimum) will become apparent. Important features should be related to specific parameters or groups of parameters to assist in understanding the behavior. Also, the sensitivity of the result to changes in assumptions or data should be evaluated. Sometimes this is referred to as *what-if* analysis, where the engineer determines what happens if a parameter changes by a specified amount. A thorough results analysis enables the engineer to understand the result of the formulation and solution steps.

Example 3.1.

Results analysis. The solution in equation (3.18) is an exponential curve as shown in Figure 3.3. The shape of the curve is monotonic, with the maximum rate of change occurring when the inlet step change is entered. The manner in which the variable changes from its initial to final values is influenced by the time constant (τ), which in this problem is the volume divided by the flow. Thus, the same dynamic response could be obtained for any stirred tank with values of flow and volume that give the same value of the time constant. It is helpful to learn a few values of this curve, which we will see so often in process control. The values for the change in concentration for several values of time after the step are noted in the following table.

Time from step	Percent of final steady-state change in output
0	0
τ	63.2
2τ	86.5
3τ	95.0
4τ	98.2

The specific quantitative question posed in the goal statement involves determining the time until 90 percent of the change in outlet concentration has occurred. This time can be calculated by setting $C_A = C_{Ainit} + 0.9(C_{A0} - C_{Ainit})$ in equation (3.18), which on rearrangement gives

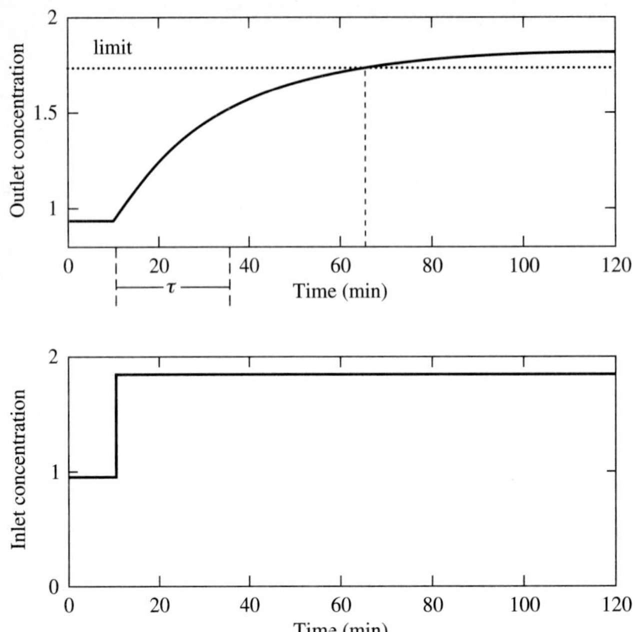

FIGURE 3.3
Dynamic result for Example 3.1.

$$t = -\tau \ln\left(\frac{0.1[(C_A)_{init} - C_{A0}]}{(C_A)_{init} - C_{A0}}\right) = -(24.7)(-2.30) = 56.8 \text{ min}$$

Note that this is time from the introduction of the step change, which, since the step is introduced at $t = 10$, becomes 66.8 in Figure 3.3. One should ask how important the specification is; if it is critical, a sensitivity analysis should be performed. For example, if the volume and flow are not known exactly but can change within ± 5 percent of their base values, the time calculated above is not exact. The range for this time can be estimated from the bounds on the parameters that influence the time constant:

$$\text{Maximum } t = -\frac{(2.1)(1.05)}{(0.085)(0.95)}(-2.30) = 62.8 \text{ min}$$

$$\text{Minimum } t = -\frac{(2.1)(0.95)}{(0.085)(1.05)}(-2.30) = 51.4 \text{ min}$$

Given the estimated inaccuracy in the data, one should wait at least 62.8 (not 56.8) minutes after the step to be sure that 90 percent of the concentration change has occurred.

Validation

Validation involves determining whether the results of steps 1 through 5 truly represent the physical process with the required fidelity for the specified range of conditions. The question to be evaluated is, "Does the model represent the data well enough that the engineering task can be performed using the model?" Since we know that all models are simplified representations of the true, complex physical world, this question must be evaluated with careful attention to the application of the model. We do not have enough background in control engineering at this point, so the sensitivity of process and control design to modelling errors must be deferred to a later point in the book; however, all methods will be based on models, so this question will be addressed frequently because of its central importance.

While the sensitivity analysis in step 5 could build confidence that the results are likely to be correct, a comparison with empirical data is needed to evaluate the validity of the model. One simple step is to compare the results of the model with the empirical data in a graph. If parameters are adjusted to improve the fit of the model to the data, consideration should be taken of the amount the parameters must be adjusted to fit the data; adjustments that are too large raise a warning that the model may be inadequate to describe the physical system.

It is important to recognize that no set of experiences can validate the model. Good comparisons only demonstrate that the model has not been invalidated by the data; another experiment could still find data that is not properly explained by the model. Thus, no model can be completely validated, because this would require an infinite number of experiments to cover the full range of conditions. However, data from a few experiments can characterize the system in a limited range of operating variables. Experimental design and modelling procedures for empirical models are the topic of Chapter 6.

Example 3.1.

Validation. The mixing tank was built, the experiment was performed, and samples of the outlet material were analyzed. The data points are plotted in Figure 3.4 along with the model prediction. By visual evaluation and considering the accuracy of each data

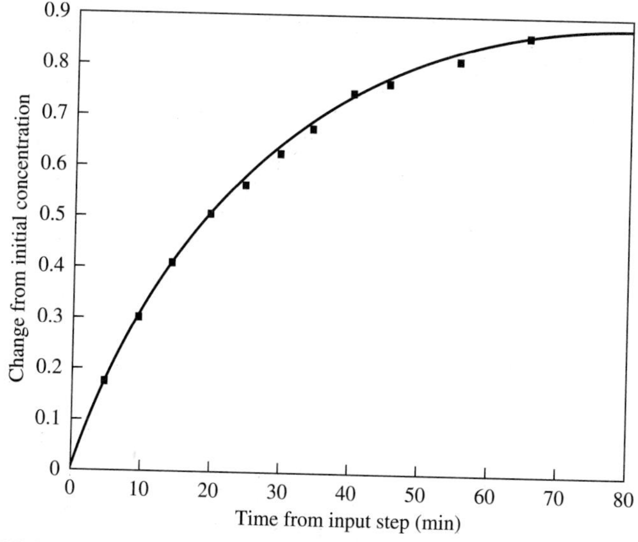

FIGURE 3.4
Comparison of empirical data (squares) and model (line) for Example 3.1.

point, one would accept the model as "valid" (or, more accurately, not invalid) for most engineering applications.

The modelling procedure presented in this section is designed to ensure that the most common issues are addressed in a logical order. While the procedure is important, the decisions made by the engineer have more impact on the quality of the result than the procedure has. Since no one is prescient, the effects of early assumptions and formulations may not be appropriate for the goals. Thus, a thorough analysis of the results should be performed so that the sensitivity of the conclusions to model assumptions and data is clearly understood. If the conclusion is unduly sensitive to assumptions or data, an iteration would be indicated, employing a more rigorous model or more accurate data. Thus, the procedure contains the essential opportunity for evaluation and improvement.

3.3 MODELLING EXAMPLES

Most people learn modelling by *doing modelling,* not observing results of others! The problems at the end of the chapter, along with many solved and unsolved problems in the references and resources, provide the reader with ample opportunity to develop modelling skills. To assist the reader in applying the procedure to a variety of problems, this section includes a few more solved example problems with solutions. In all examples, steps 1 to 5 are performed, but validation is not.

Example 3.2 Isothermal CSTR. The dynamic response of a continuous-flow, stirred-tank chemical reactor (CSTR) will be determined in this example and compared with the stirred-tank mixer in Example 3.1.

Goal. Determine the dynamic response of a CSTR to a step in the inlet concentration. Also, the reactant concentration should never go above 0.85 mole/m^3. If an alarm sounds when the concentration reaches 0.83 mole/m^3, would a person have enough time to respond? What would a correct response be?

Information. The process is the same as shown in Figure 3.1, and therefore, the system is the liquid in the tank. The important variable is the reactant concentration in the reactor.

Assumptions. The same as for the stirred-tank mixer.

Data. The flow, volume, and inlet concentrations (before and after the step) are the same as for the stirred-tank mixer in Example 3.1, and

1. The chemical reaction is first-order, $r_A = -kC_A$ with $k = 0.040\ \text{min}^{-1}$.
2. The heat of reaction is negligible, and no heat is transferred to the surroundings.

Formulation. Based on the model of the stirred-tank mixer, the overall material balance again yields $F_0 = F_1 = F$. To determine the concentration of reactant, a component material balance is required, which is different from that of the mixing tank because there is a (negative) generation of component A as a result of the chemical reaction.

$$\begin{Bmatrix}\text{Accumulation of}\\ \text{component A}\end{Bmatrix} = \begin{Bmatrix}\text{Component}\\ \text{A in}\end{Bmatrix} - \begin{Bmatrix}\text{Component}\\ \text{A out}\end{Bmatrix} + \begin{Bmatrix}\text{Generation}\\ \text{of A}\end{Bmatrix} \tag{3.19}$$

$$(\text{MW}_\text{A} V C_\text{A})_{t+\Delta t} - (\text{MW}_\text{A} V C_\text{A})_t = (\text{MW}_\text{A} F C_{\text{A}0} - \text{MW}_\text{A} F C_\text{A} - \text{MW}_\text{A} V k C_\text{A})\Delta t \tag{3.20}$$

Again, dividing by $\text{MW}_\text{A}(\Delta t)$ and taking the limit as $\Delta t \to 0$ gives

$$\frac{dC_\text{A}}{dt} + \frac{1}{\tau} C_\text{A} = \frac{F}{V} C_{\text{A}0} \qquad \text{with the time constant } \tau = \frac{V}{F + Vk} \tag{3.21}$$

The degrees-of-freedom analysis yields one equation, one variable (C_A), two external variables (F and $C_{\text{A}0}$), and two parameters (V and k). Since the number of variables is equal to the number of equations, the degrees of freedom are zero, and the model is exactly specified.

Solution. Equation (3.21) is a nonseparable linear ordinary differential equation, which can be solved by application of the integrating factor:

$$\begin{aligned}
\text{IF} &= \exp\left(\int \frac{1}{\tau} dt\right) = e^{t/\tau} \\
\frac{d(C_\text{A} e^{t/\tau})}{dt} &= \frac{F}{V} C_{\text{A}0} e^{t/\tau} \\
\int d(C_\text{A} e^{t/\tau}) &= \frac{F C_{\text{A}0}}{V} \int e^{t/\tau} dt \\
C_\text{A} e^{t/\tau} &= \frac{F C_{\text{A}0} \tau}{V} e^{t/\tau} + I \\
C_\text{A} &= \frac{F\tau}{V} C_{\text{A}0} + I e^{-t/\tau}
\end{aligned} \tag{3.22}$$

The initial condition gives the *inlet* concentration of 0.925 mole/m^3 at the time from the step, $t = 0$. The initial steady-state reactor concentration can be determined from the data and equation (3.21) with $dC_\text{A}/dt = 0$.

$$\begin{aligned}
(C_\text{A})_\text{init} &= \frac{F}{F + Vk} (C_{\text{A}0})_\text{init} \\
&= \frac{0.085}{0.085 + (2.1)(0.040)} 0.925 = 0.465 \frac{\text{mole}}{\text{m}^3}
\end{aligned}$$

The constant of integration can be evaluated to be

$$I = \frac{F[(C_{\text{A}0})_\text{init} - (C_{\text{A}0})]}{F + Vk} = \frac{-F(\Delta C_{\text{A}0})}{F + VK}$$

This can be substituted in equation (3.22) to give

$$\begin{aligned}
C_\text{A} &= \frac{F C_{\text{A}0}}{F + Vk} - \frac{F(\Delta C_{\text{A}0})}{F + Vk} e^{-t/\tau} \\
&= (C_\text{A})_\text{init} + \frac{F}{F + Vk} [C_{\text{A}0} - (C_{\text{A}0})_\text{init}] \left(1 - e^{-t/\tau}\right)
\end{aligned} \tag{3.23}$$

This can be rearranged with $K_p = F/(F + Vk)$ to give

$$\begin{aligned}
C_\text{A} - (C_\text{A})_\text{init} &= K_p \Delta C_{\text{A}0} \left(1 - e^{-t/\tau}\right) \\
\Delta C_\text{A} &= (0.503)(0.925)(1 - e^{-t/\tau})
\end{aligned}$$

Again, the time constant determines the "speed" of the response. Note that in this example, the time constant depends on the equipment (V), the operation (F), and the chemical reaction (k), and that by comparing equations (3.16) and (3.21) the time constant for the chemical reactor is always shorter than the time constant for the mixer, using the same values for F and V. The steady-state gain is

$$K_p = \frac{F}{F + VK} = 0.503 \frac{\text{mole/m}^3}{\text{mole/m}^3}$$

Thus, the steady-state gain in this example depends on equipment design and operating conditions.

Results analysis. First, the result from equation (3.23) is calculated and plotted. As shown in Figure 3.5*a*, the reactant concentration increases as an exponential function to its final value without overshoot or oscillation. In this case, the concentration exceeds its maximum limit; therefore, a corrective action will be evaluated. The concentration reaches the alarm limit in 19.6 minutes after the step (29.6 minutes in the figure) and exceeds the maximum limit after 22.5 minutes. The sensitivity of this result can be evaluated from the analytical solution; in particular, the dependence of the time constant on variables and parameters is given in equation (3.21). The time difference between the alarm and the dangerous condition is too short for a person to respond reliably, because other important events may be occurring simultaneously.

Since a response is required, the safety response should be automated; safety systems are discussed in Chapter 24. A proper response can be determined by considering equation (3.21). The goal is to ensure that the reactor concentration decreases immediately when the corrective manipulation has been introduced. One manner (for this, but not all processes) would be to decrease the inlet concentration to its initial value, so that the rate of change of C_A would be negative without delay. The transient response obtained by implementing this strategy when the alarm value is reached is shown in Figure 3.5*b*. The model for the response after the alarm value has been reached, 29.6

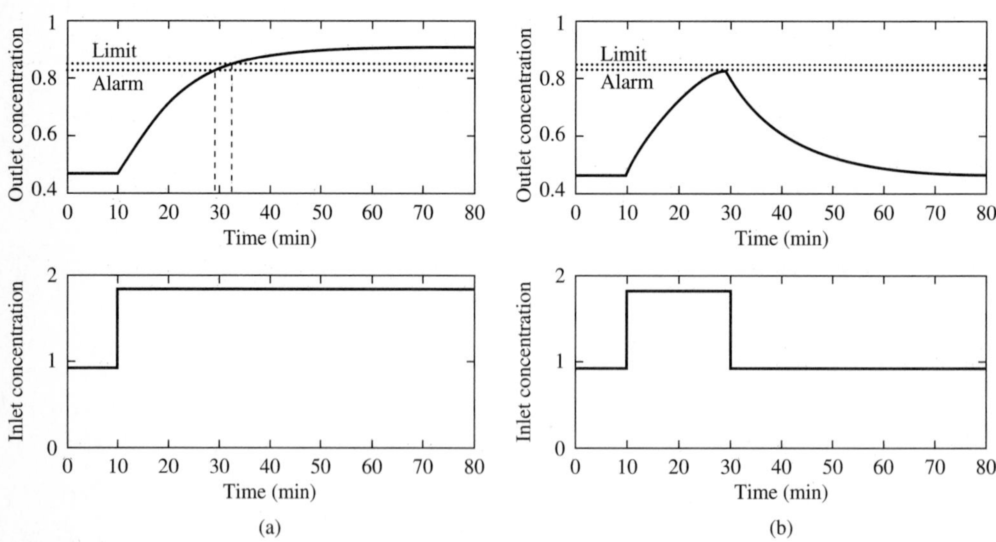

FIGURE 3.5
Results for Example 3.2: (*a*) without action at the alarm value; (*b*) with action at the alarm value.

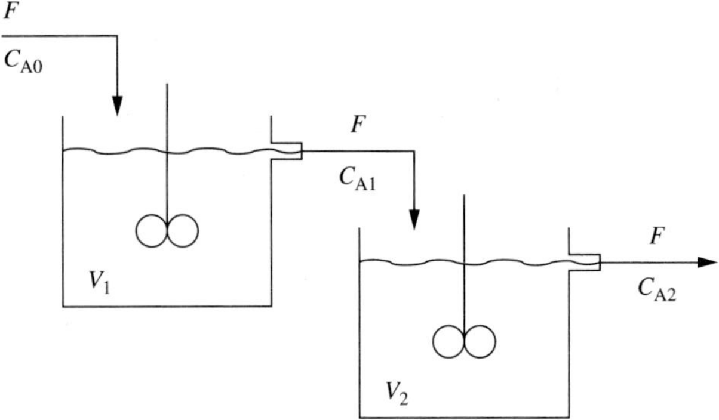

FIGURE 3.6
Two CSTRs in series.

minutes, is of the same form as equation (3.23), with the same time constant and gain.

Example 3.3 Two isothermal CSTR reactors. A problem similar to the single CSTR in Example 3.2 is presented, with the only difference that two series reactors are included as shown in Figure 3.6. Each tank is one-half the volume of the tank in Example 3.2.

Goal. The same as that of Example 3.2, with the important concentration being in the second reactor. Determine the time when this concentration exceeds 0.85 mole/m^3.

Information. The systems are the liquid in each tank. The data is the same as in Example 3.2, except that $V_1 = V_2 = 1.05$ m^3.

Formulation. Again, due to the assumptions for the overflow tanks, the volumes in the two tanks can be taken to be constant, and all flows are constant and equal. The value of the concentration in the second tank is desired, but it depends on the concentration in the first tank. Therefore, the component balances on both tanks are formulated.

$$V_1 \frac{dC_{A1}}{dt} = F(C_{A0} - C_{A1}) - V_1 k C_{A1} \tag{3.24}$$

$$V_2 \frac{dC_{A2}}{dt} = F(C_{A1} - C_{A2}) - V_2 k C_{A2} \tag{3.25}$$

The result is two linear ordinary differential equations, which in general must be solved simultaneously. Note that the two equations could be combined into a single second-order differential equation; thus, the system is second-order.

Solution. In this case, the balance on the first tank does not involve the concentration in the second tank and thus can be solved independently from the equation representing the second reactor. (More general methods for solving simultaneous linear differential equations, using Laplace transforms, are presented in the next chapter.) The solution for the first balance can be seen to be exactly the same form as the result for Example 3.2, equation (3.23). The analytical expression for the concentration at the outlet of the first tank can be substituted into equation (3.25) to give

$$\frac{dC_{A2}}{dt} + \frac{1}{\tau} C_{A2} = \frac{F}{V} C_{A1} = \frac{F}{V}\left(\frac{FC_{A0}}{F + Vk} - \frac{F\Delta C_{A0}}{F + Vk} e^{-t/\tau}\right) \tag{3.26}$$

Since the volumes and flows for the two reactors are identical,

$$\tau = \frac{V_1}{F + V_1 k} = \frac{V_2}{F + V_2 k} \tag{3.27}$$

Again, equation (3.26) can be solved by applying the integrating factor:

$$\text{IF} = \exp\left(\int \frac{1}{\tau} dt\right) = e^{t/\tau}$$

$$\frac{d(C_{A2} e^{t/\tau})}{dt} = \frac{F e^{t/\tau}}{V}\left(\frac{F C_{A0}}{F + Vk} - \frac{F \Delta C_{A0}}{F + Vk} e^{-t/\tau}\right) \tag{3.28}$$

$$C_{A2} = e^{-t/\tau}\left[\frac{F}{V}\left(\frac{F C_{A0}}{F + Vk}\int e^{t/\tau} dt - \frac{F \Delta C_{A0}}{F + Vk}\int e^{t/\tau} e^{-t/\tau} dt\right) + I\right]$$

$$= \frac{F}{V}\left(\tau \frac{F C_{A0}}{F + Vk} - \frac{F \Delta C_{A0}}{F + Vk} t e^{-t/\tau}\right) + I e^{-t/\tau}$$

The initial condition is $(C_{A2})_{\text{init}}$, which is determined by solving the steady-state model, equation (3.26), with $dC_{A2}/dt = 0$ using the value of C_{A0} before the step:

$$(C_{A2})_{\text{init}} = \frac{F}{V}\left(\tau \frac{F}{F + Vk}\right)(C_{A0})_{\text{init}} + I \tag{3.29}$$

$$= \left(\frac{F}{F + Vk}\right)^2 (C_{A0})_{\text{init}} + I = (C_{A2})_{\text{final}} + I$$

$$\therefore \quad I = (C_{A2})_{\text{init}} - (C_{A2})_{\text{final}}$$

Substituting this value gives the final expression for the second reactor concentration:

$$C_{A2} = (C_{A2})_{\text{final}} - \frac{F}{V}\frac{F \Delta C_{A0}}{F + Vk} t e^{-t/\tau} + [(C_{A2})_{\text{init}} - (C_{A2})_{\text{final}}]\, e^{-t/\tau} \tag{3.30}$$

The data can be substituted into this equation to give

$$C_{A2} = 0.828 - 0.050 t e^{-t/8.25} + (0.414 - 0.828) e^{-t/8.25}$$

Results analysis. The shape of the transient of the concentration in the second of two reactors in Figure 3.7 is very different from the transient for one reactor in Figure 3.3. The second-order response for this example has a sigmoidal or "S" shape, with a derivative that goes through a maximum at an inflection point and reduces to zero at the new steady state. Also, the total conversion of reactant is different from Example 3.2, although the total reactor volume is the same in both cases. The increased conversion in the two-reactor system is due to the higher concentration of the reactant in the first reactor. In fact, the concentration of the second reactor does not reach the alarm or limiting values after the step change for the parameters specified, although the close approach to the alarm value indicates that a slight change could lead to an alarm.

The action upon exceeding the alarm limit in the second reactor would not be as easily determined for this process, since equation (3.25) shows that decreasing the inlet concentration to the first reactor does not ensure that the derivative of the second reactor's concentration will be negative. The system has "momentum," which makes it more difficult to influence the output of the second reactor immediately.

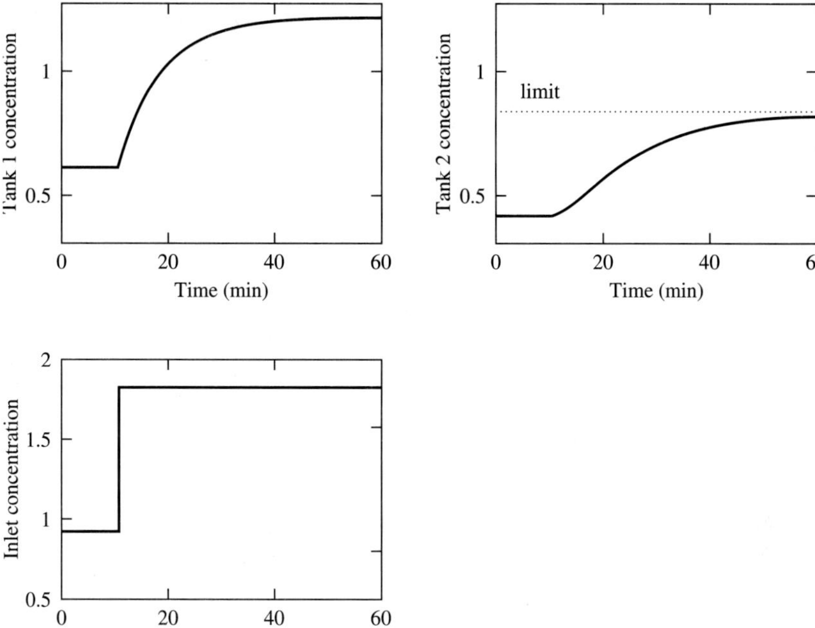

FIGURE 3.7
Dynamic responses for Example 3.3.

Example 3.4 On-off room heating. The heating of a dwelling with an on-off heater was discussed in Section 1.2. The temperature was controlled by a feedback system, and semi-quantitative arguments led to the conclusion that the temperature would oscillate. In this section, a very simple model of the system is formulated and solved.

Goal. Determine the dynamic response of the room temperature. Also, ensure that the furnace does not have to switch on or off more frequently than once per 3 minutes, to allow the combustion zone to be purged of gases before reignition.

Information. The system is taken to be the air inside the dwelling. A sketch of the system is given in Figure 1.2. The important variables are the room temperature and the furnace on-off status.

Assumptions.

1. The air in the room is well mixed.
2. No transfer of material to or from the dwelling occurs.
3. The heat transferred depends only on the temperature difference between the room and the outside environment.
4. No heat is transferred from the floor or ceiling.
5. Effects of kinetic and potential energies are negligible.

Data.

1. The heat capacity of the air C_V is 0.17 cal/(g°C), density is 1190 g/m^3.
2. The overall heat transfer coefficient, $UA = 45 \times 10^3$ cal/(°C hr).
3. The size of the dwelling is 5 m by 5 m by 3 m high.
4. The furnace heating capacity Q_h is either 0 (off) or 1.5×10^6 (on) cal/hr.
5. The furnace heating switches instantaneously at the values of 17°C (on) and 23°C (off).
6. The initial room temperature is 20°C and the initial furnace status is "off."
7. The outside temperature T_a is 10°C.

Formulation. The system is defined as the air inside the house. To determine the temperature, an energy balance should be formulated, and since no material is transferred, no material balance is required. The application of the energy balance in equation (3.5) to this system gives

$$\frac{dE}{dt} = (0) - (0) + Q - W_s \tag{3.31}$$

The shaft work is zero. From principles of thermodynamics and heat transfer, the following expressions can be used for a system with negligible accumulation of potential and kinetic energy:

$$\frac{dE}{dt} = \rho V C_v \frac{dT}{dt} \qquad Q = -UA(T - T_a) + Q_h \tag{3.32}$$

with

$$Q_h = \begin{cases} 0 & \text{when } T > 23°\text{C} \\ 1.5 \times 10^6 & \text{when } T < 17°\text{C} \\ \text{unchanged} & \text{when } 17 < T < 23°\text{C} \end{cases}$$

to give

$$\rho V C_V \frac{dT}{dt} = -UA(T - T_a) + Q_h \tag{3.33}$$

The degrees of freedom for this formulation is zero since the model has two equations, two variables (T and Q_h), four parameters (UA, C_v, V, and ρ), and one external variable (T_a). Thus, the system is exactly specified with equation (3.33), when the status of the heating has been defined by equation (3.32).

Solution. Rearranging equation (3.33) gives the following nonseparable linear ordinary differential equation, which can be solved by application of the integrating factor:

$$\frac{dT}{dt} + \frac{1}{\tau}T = \frac{UAT_a + Q_h}{V\rho C_v} \qquad \text{with } \tau = \frac{V\rho C_v}{UA} \tag{3.34}$$

$$T - T_{\text{init}} = (T_{\text{final}} - T_{\text{init}})(1 - e^{-t/\tau}) \tag{3.35}$$

where

t = time from step in Q_h

τ = time constant = 0.34 hr

T_{final} = final value of T as $t \to \infty = T_a + Q_h/UA$

= 10°C when $Q_h = 0$

= 43.3°C when $Q_h = 1.5 \times 10^6$

T_{init} = the value of T when a step in Q_h occurs

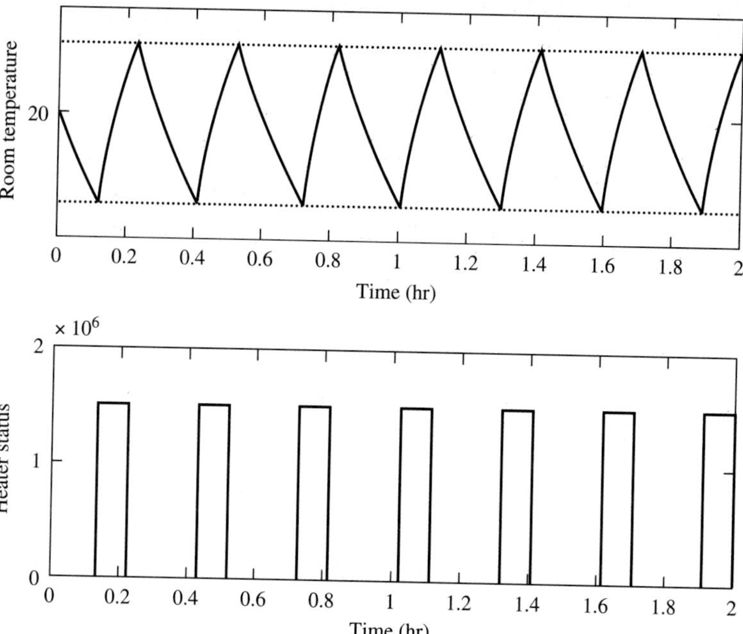

FIGURE 3.8
Dynamic response for Example 3.4.

Results analysis. First, the numerical result is determined and plotted in Figure 3.8. From the initial condition with the furnace off, the temperature decreases according to equation (3.35) until the switch value of 17°C is reached. Then, the furnace heating begins instantaneously (Q_h changes from 0 to 1.5×10^6), and since the system is first-order with no "momentum," the temperature immediately begins to increase. This procedure is repeated as the room temperature follows a periodic trajectory between 17 and 23°C.

The analytical solution provides insight into how to alter the behavior of the system. The time constant is proportional to the mass in the room, which seems reasonable. Also, it is inversely proportional to the heat transfer coefficient, since the faster the heat transfer, the more quickly the system reaches an equilibrium with its surroundings; therefore, insulating the house will decrease UA and increase the time constant. Finally, the time constant does not depend on the heating by the furnace, which is the forcing function of the system; therefore, increasing the capacity of the furnace will not affect the time constant, although it will affect the time between switches.

The goals of the modelling exercise have been satisfied. The temperature has been determined as a function of time, and the switching frequency of the furnace has been determined to be over 3 minutes; that is, longer than the minimum limit. However, a switch could occur much faster due to a sudden change in outside temperature or to a disturbance such as a door being opened, which would allow a rapid exchange of warm and cold air. Therefore, a special safety system would be included to ensure that the furnace would not be restarted until a safe time period after shutting off.

Building heating and air conditioning have been studied intensively, and more accurate data and models are available (McQuiston and Parker, 1988). Also, some extensions to this simple example are suggested in question 3.9 at the end of the chapter (adding capacitance, changing UA, and including ventilation).

This example is the first quantitative analysis of a continuous feedback control system. The simplicity of the model and the on-off control approach facilitated the solution while retaining the essential characteristics of the behavior. For most industrial processes, the oscillations associated with on-off control are unacceptable, and more complex feedback control approaches, introduced in Part III, are required to achieve acceptable dynamic performance.

3.4 LINEARIZATION

The models in the previous sections were easily solved because they involved linear equations, which were a natural result of the conservation balances and constitutive relationships for the specific physical systems. However, the conservation and constitutive equations are nonlinear for most systems, and general methods for developing analytical solutions for nonlinear models are not available. An alternative is numerical simulation, covered later in this chapter, which can provide accurate solutions for specific numerical values but usually offers much less understanding. Fortunately, methods exist for obtaining approximate linearized solutions to nonlinear systems, and experience over decades has demonstrated that linearized methods of control systems analysis provide very useful results for many (but not all) realistic processes. Therefore, this section introduces the important method for developing approximate linear models.

First, the concept of *linearity* needs to be formally defined. This will be done using the concept of an *operator*, which transforms an input variable into an output variable.

> An **operator** $\mathscr{F}$ is **linear** if it satisfies the properties of additivity and proportionality, which are included in the following **superposition**, where x_i are variables and a and b are constants:
>
> $$\mathscr{F}(ax_1 + bx_2) = a\mathscr{F}(x_1) + b\mathscr{F}(x_2) \tag{3.36}$$

We can test any term in a model using equation (3.36) to determine whether it is linear. A few examples are given in the following table.

Function	Check for linearity	Is check satisfied?
$\mathscr{F}(x) = kx$	$k(ax_1 + bx_2) \overset{?}{=} kax_1 + kbx_2$	Yes
$\mathscr{F}(x) = kx^{1/2}$	$k(ax_1 + bx_2)^{1/2} \overset{?}{=} k(ax_1)^{1/2} + k(bx_2)^{1/2}$	No

Next, it is worthwhile considering the dynamic behavior of a process, such as the stirred-tank heat exchanger shown in Figure 3.9, subject to changes in the feed temperature and cooling fluid flow rate. For a linear system, the result of the two changes is the sum of the results from each change individually. The responses to step changes in the feed temperature (at $t = 5$) and cooling medium flow rate (at $t = 20$)

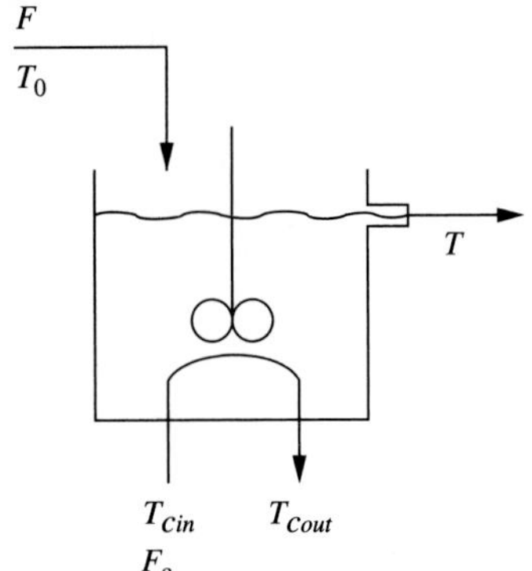

FIGURE 3.9
Stirred tank with heat exchanger.

are shown in Figure 3.10. The responses in parts *a* and *b* are the effects of each disturbance individually, and the response in part *c* is the total effect, which for this linear process is the sum of the two individual effects. Note that the true physical system experiences only the response in Figure 3.10*c*; the individual responses are the linear predictions for each input change. (The model for

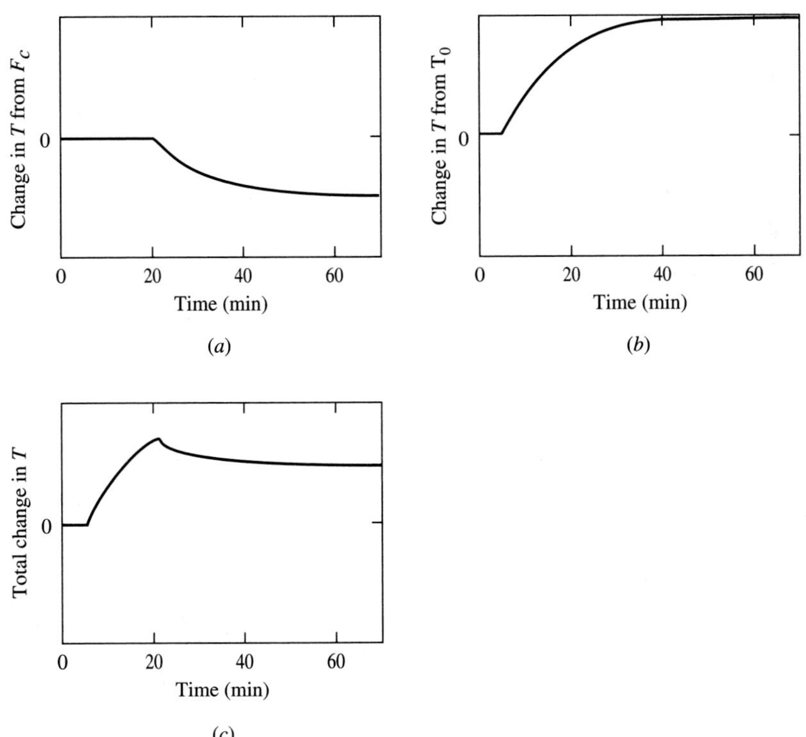

FIGURE 3.10
Response of the linear system in Figure 3.9 to two input changes.

this system will be derived in Example 3.7.) This concept, as an approximation to real nonlinear processes, is used often in analyzing process control systems.

A linearized model can be developed by approximating each nonlinear term with its linear approximation. A nonlinear term can be approximated by a Taylor series expansion to the nth order about a point if derivatives up to nth order exist at the point; the general expressions for functions of one and two variables are given in Table 3.3.

The term R is the remainder and depends on the order of the series. A few examples of nonlinear terms that commonly occur in process models, along with their linear approximations about x_s, are the following:

$$F(x) = x^{1/2} \qquad F(x) \approx x_s^{1/2} + \frac{1}{2} x_s^{-1/2}(x - x_s)$$

$$F(x) = \frac{x}{1 + ax} \qquad F(x) \approx \frac{x_s}{1 + ax_s} + \frac{1}{(1 + ax_s)^2}(x - x_s)$$

The accuracy of the linearization can be estimated by comparing the magnitude of the remainder, R', to the linear term. For a linear Taylor series approximation in one variable,

$$R' = \frac{1}{2}\frac{d^2F}{dx^2}\bigg|_{x=\xi}(x - x_s)^2 \qquad \text{with } \xi \text{ between } x \text{ and } x_s \tag{3.37}$$

The accuracy of a sample linearization is depicted in Figure 3.11. From this figure and equation (3.37), it can be seen that the accuracy of the linear approximation is relatively better when (1) the second-order derivative has a small magnitude (there is little curvature) and (2) the region about the base point is small. The successful application of linearization to process control systems is typically justified by the small region of operation of a process when under control. Although the uncontrolled system might operate over a large region because of disturbances in input variables, the controlled process variables should operate over a much smaller range, where the linear approximation often is adequate. Note that the accuracy of the linearization would in general depend on the normal operating point x_s.

Several modelling examples of linearized models are now given, with the linearized results compared with the nonlinear results. In all cases, the models will be

TABLE 3.3
Taylor series for functions of one and two variables

Function of one variable about x_s

$$F(x) = F(x_s) + \frac{dF}{dx}\bigg|_{x_s}(x - x_s) + \frac{1}{2!}\frac{d^2F}{dx^2}\bigg|_{x_s}(x - x_s)^2 + R \tag{3.38}$$

Function of two variables about x_{1s}, x_{2s}

$$\begin{aligned} F(x_1, x_2) = {} & F(x_{1s}, x_{2s}) + \frac{\partial F}{\partial x_1}\bigg|_{x_{1s},x_{2s}}(x_1 - x_{1s}) + \frac{\partial F}{\partial x_2}\bigg|_{x_{1s},x_{2s}}(x_2 - x_{2s}) \\ & + \frac{1}{2!}\frac{\partial^2 F}{\partial x_1^2}\bigg|_{x_{1s},x_{2s}}(x_1 - x_{1s})^2 + \frac{1}{2!}\frac{\partial^2 F}{\partial x_2^2}\bigg|_{x_{1s},x_{2s}}(x_2 - x_{2s})^2 \\ & + \frac{\partial^2 F}{\partial x_1 \partial x_2}(x_1 - x_{1s})(x_2 - x_{2s}) + R \end{aligned} \tag{3.39}$$

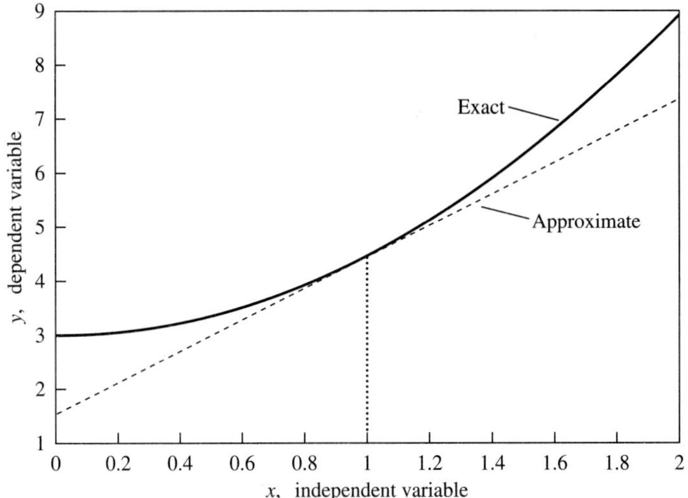

FIGURE 3.11
Comparison of a nonlinear function $y = (1.5x^2 + 3)$ with its linear approximation about $x_s = 1$.

expressed in deviation variables, such as $x - x_s$, where the subscript s represents the initial steady-state value of the variable. The deviation variable will always be designated with a prime (′).

Deviation variable: $(x - x_s) = x'$ with x_s = initial steady-state value

A deviation variable simply translates the total variable by a constant value, and the total value of the variable is easily recovered by adding the initial steady-state value x_s to its deviation value, x'. The use of deviation variables is not necessary and provides no advantage at this point in our analysis. However, expressing a model in deviation variables will be shown in Chapter 4 to provide a significant simplification in the analysis of dynamic systems; therefore, we will begin to use them here for all linear or linearized systems.

Example 3.5 Isothermal CSTR. The solution to the single-tank CSTR problem in Example 3.2 is now presented for a second-order chemical reaction.

Goal. Determine the transient response of the tank concentration in response to a step in the inlet concentration for the nonlinear and linearized models.

Information. The process equipment and flow are the same as shown in Figure 3.1. The important variable is the reactant concentration in the reactor.

Assumptions. The same as in Example 3.2.

Data. The same as in Example 3.2 except the chemical reaction rate is second-order, with $r_A = -kC_A^2$ and $k = 0.5$ [(mole/m^3)min]$^{-1}$.

Formulation. The formulation of the equations and analysis of degrees of freedom are the same as in Example 3.2 except that the rate term involves the reactant concentration to the second power.

$$V\frac{dC_A}{dt} = F(C_{A0} - C_A) - VkC_A^2 \tag{3.40}$$

The model equation is nonlinear because of the rate term. This term can be linearized by expressing it as a Taylor series and retaining only the linear terms:

$$C_A^2 \approx C_{As}^2 + 2C_{As}(C_A - C_{As}) \tag{3.41}$$

Recall that C_{As} is evaluated by setting the derivative to zero in equation (3.40) and solving for C_A, with C_{A0} having its initial value before the input perturbation, because the linearization is about the initial steady state. The approximation is now substituted in the process model.

$$V\frac{dC_A}{dt} = F(C_{A0} - C_A) - [VkC_{As}^2 + 2VkC_{As}(C_A - C_{As})] \tag{3.42}$$

The model can be expressed in deviation variables by first repeating the linearized model, equation (3.42), which is valid for any time, at the steady-state point, when the variable is equal to its steady-state value:

$$V\frac{dC_{As}}{dt} = F(C_{A0s} - C_{As}) - \left(VkC_{As}^2 + 2VkC_{As}(C_{As} - C_{As})\right) \tag{3.43}$$

Then equation (3.43) can be subtracted from equation (3.42) to give the equation in deviation variables:

$$V\frac{dC'_A}{dt} = F(C'_{A0} - C'_A) - 2VkC_{As}C'_A \tag{3.44}$$

The resulting model is a first-order, linear ordinary differential equation, which can be rearranged into the standard form.

$$\frac{dC'_A}{dt} + \frac{1}{\tau}C'_A = \frac{F}{V}C'_{A0} \qquad \text{with } \tau = \frac{V}{F + 2VkC_{As}} \tag{3.45}$$

Solution. Since the input forcing function is again a simple step, the analytical solution can be derived by a straightforward application of the integrating factor.

$$C'_A = C'_{A0}\left(\frac{F}{F + 2VkC_{As}}\right)(1 - e^{-t/\tau}) = \Delta C_{A0}K_p(1 - e^{-t/\tau})$$

$$\text{with } K_p = \frac{F}{F + 2VkC_{As}} \tag{3.46}$$

The data can be substituted into this expression to give

$$C'_A = (0.925)(0.146)\left(1 - e^{-t/3.62}\right)$$

Results Analysis. The linearized solution from equation (3.46) is plotted in Figure 3.12 in comparison with the solution to the original nonlinear differential equation, equation

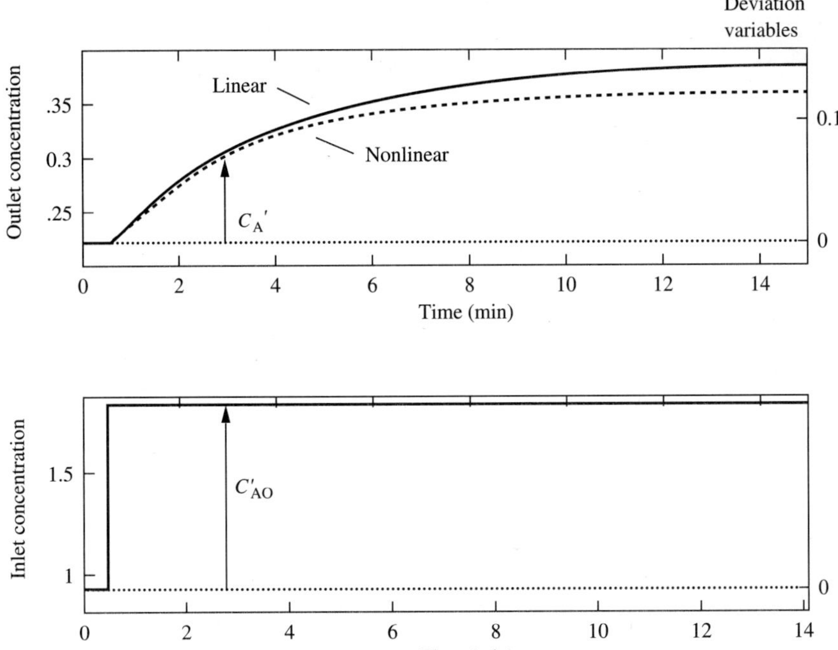

FIGURE 3.12
Dynamic responses for Example 3.5.

(3.40). The linear solution can be seen to give a good semi-quantitative description of the true process response.

An important advantage of the linearized solution is in the analytical relationships. For example, the time constants and gains of the three similar continuous-flow stirred-tank processes—mixer, linear reactor, and linearized model of nonlinear reactor—are summarized in Table 3.4. These results can be used to learn how process equipment design and process operating conditions affect the dynamic responses. Clearly, the analytical solutions provide a great deal of useful information on the relationship between design and operating conditions and dynamic behavior.

TABLE 3.4
Summary of linear or linearized models for single stirred-tank systems

Physical system	Is the system linear?	Time constant (T)	Steady-state gain, K_p
Example 3.1 (CST mixing)	Yes	V/F	1.0
Example 3.2 (CSTR with first-order reaction)	Yes	$V/(F + Vk)$	$F/(F + Vk)$
Example 3.5 (CSTR with second-order reaction)	No	$V/(F + 2VkC_{As})$ (linearized model)	$F/(F + 2VkC_{As})$ (linearized model)

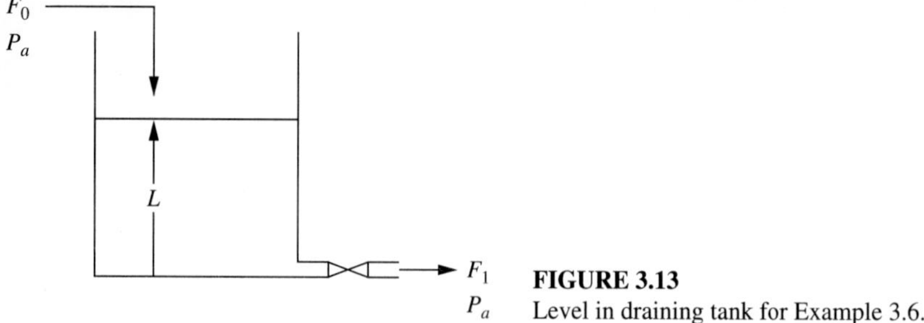

FIGURE 3.13
Level in draining tank for Example 3.6.

Example 3.6 Tank draining. The level and flow through a partially opened restriction out of the tank system in Figure 3.13 is considered in this example.

Goal. Determine a model for this system. Evaluate the accuracies of the linearized solutions for small (10 m^3/hr) and large (60 m^3/hr) step changes in the inlet flow rate.

Information. The system is the liquid in the tank, and the important variables are the level and flow out.

Assumptions.

1. The density is constant.
2. The cross-sectional area of the tank, A, does not change with height.

Data.

1. The initial steady-state conditions are (i) flows $= F_0 = F_1 = 100$ m^3/h and (ii) level $= L = 7.0$ m.
2. The cross-sectional area is 7 m^2.

Formulation. The level depends on the total amount of liquid in the tank; thus, the conservation equation selected is an overall material balance on the system.

$$\rho A \frac{dL}{dt} = \rho F_0 - \rho F_1 \tag{3.47}$$

This single balance does not provide enough information, because there are two unknowns, F_1 and L. Thus, the number of degrees of freedom (1) indicates that another equation is required. An additional equation can be provided to determine F_1 without adding new variables, through a momentum balance on the liquid in the exit pipe. In essence, another subproblem is defined to formulate this balance. The major assumptions for this subproblem are that

1. The system is at quasi-steady state, since the dynamics of the pipe flow will be fast with respect to the dynamics of the level.
2. The total pressure drop is due to the restriction.
3. Conventional macroscopic flow equations, using relationships for friction factors and restrictions, can relate the flow to the pressure driving force (Foust et al., 1980; Bird, Stewart, and Lightfoot, 1960).

With these assumptions, which relate the flow out to the liquid level in the tank, the balance becomes

$$F_1 = f(F_1)(P_a + \rho L - P_a)^{0.5} = k_{F1}L^{0.5} \tag{3.48}$$

with P_a constant. The system with equations (3.47) and (3.48) and with two variables, F_1 and L, is exactly specified. After the equations are combined, the system can be described by a single first-order differential equation:

$$A\frac{dL}{dt} = F_0 - k_{F1}L^{0.5} \tag{3.49}$$

The basic balance now has a nonlinear term, which can be linearized:

$$L^{0.5} \approx L_s^{0.5} + 0.5L_s^{-0.5}(L - L_s) \tag{3.50}$$

This expression can be used to replace the nonlinear term. The resulting equation, after subtracting the linearized balance at steady-state conditions and noting that the input is a constant step (i.e., $F_0' = \Delta F_0$), is

$$A\frac{dL'}{dt} = \Delta F_0 - [0.5k_{F1}L_s^{-0.5}]L' \tag{3.51}$$

Solution. The linearized differential equation can be rearranged and solved as before.

$$\frac{dL'}{dt} + \frac{1}{\tau}L' = \frac{1}{A}\Delta F_0 \qquad \text{with } \tau = \frac{A}{0.5k_{F1}L_s^{-0.5}} \tag{3.52}$$

giving the solution

$$L' = \frac{\tau \Delta F_0}{A} + Ie^{-t/\tau} \tag{3.53}$$

The initial condition is that $L' = 0$ at $t = 0$, with time measured from the input step; thus, $I = -\tau\Delta F_0/A$. Substitution gives

$$\begin{aligned} L' &= \frac{\tau\Delta F_0}{A}(1 - e^{-t/\tau}) \\ &= \Delta F_0 K_p(1 - e^{-t/\tau}) \qquad \text{with } K_p = \frac{\tau}{A} = \frac{1}{0.5k_{F1}L_s^{-0.5}} \end{aligned} \tag{3.54}$$

For this example,

$$k_{F1} = \frac{F_{1s}}{L_s^{0.5}} = 37.8\frac{\text{m}^3/\text{hr}}{\text{m}^{0.5}} \qquad \tau = 0.98 \text{ hr}$$

$$L' = 0.14\Delta F_0(1 - e^{-t/0.98})$$

Results Analysis. The solution of the linearized model indicates an exponential response to a step change. The results for the small and large step changes in flow in are plotted in Figures 3.14*a* and 14*b*, respectively. The solution to the approximate linearized model is quite accurate for the small step; however, it is inaccurate for a large step, even predicting an impossible *negative* level at the final steady state. The general trend that the linearized model should be more accurate for a small than for a large step conforms to the previous discussion of the Taylor series. Also, the large variation of the level, which for the larger input step is not maintained close to its initial condition as shown in Figure 3.14*b*, suggests that the linear solution might not be very accurate.

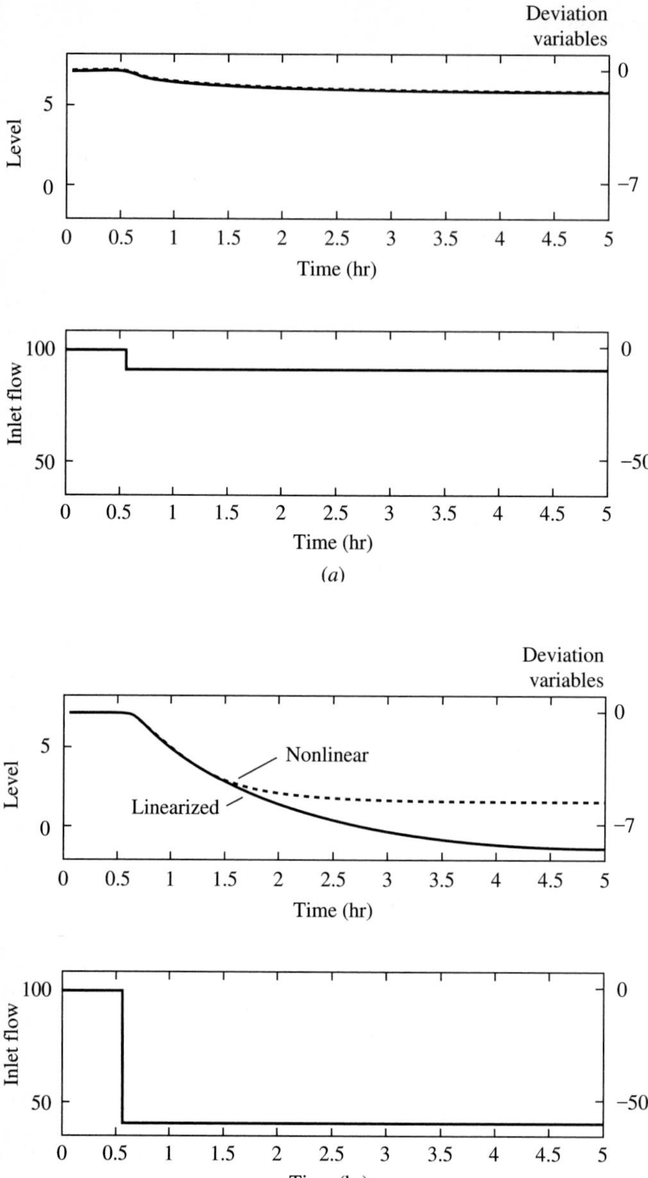

FIGURE 3.14
Dynamic responses for Example 3.6: (*a*) for a small input change (linearized and nonlinear essentially the same curve); (*b*) for a large input change.

Example 3.7 Stirred-tank heat exchanger. To provide another simple example of an energy balance, the stirred-tank heat exchanger in Figure 3.9 is considered.

Goal. The dynamic response of the tank temperature to a step change in the coolant flow is to be determined.

Information. The system is the liquid in the tank.

Assumptions.

1. The tank is well insulated, so that negligible heat is transferred to the surroundings.
2. The accumulation of energy in the tank walls and cooling coil is negligible compared with the accumulation in the liquid.
3. The tank is well mixed.
4. Physical properties are constant.
5. The system is initially at steady state.

Data. $F = 0.085$ m^3/min; $V = 2.1$ m^3; $T_s = 85.4$°C; $\rho = 10^6$ g/m^3; $C_p = 1$ cal/(g°C); $T_0 = 150$°C; $T_{cin} = 25$°C; $F_{cs} = 0.50$ m^3/min; $C_{pc} = 1$ cal/(g°C); $\rho_c = 10^6$ g/m^3.

Formulation. Overall material and energy balances on the system are required to determine the flow and temperature from the tank. The overall material balance is the same as for the mixing tank, with the result that the level is approximately constant and $F_0 = F_1 = F$. For this system, the kinetic and potential energy accumulation terms are zero, and their input and output terms cancel if they are not zero. The energy balance is as follows:

$$\frac{dE}{dt} = \{H_0\} - \{H_1\} + Q - W_s \tag{3.55}$$

Also, it is assumed (and could be verified by calculations) that the shaft work is negligible. Now, the goal is to express the internal energy and enthalpy in measurable variables. This can be done using the following thermodynamic relationships (Smith and Van Ness, 1987):

$$dE/dt = \rho V C_v dT/dt \approx \rho V C_p dT/dt \tag{3.56}$$

$$H_i = \rho C_p F_i [T_i - T_{\text{ref}}] \tag{3.57}$$

Note that the heat capacity at constant volume is approximated as the heat capacity at constant pressure, which is acceptable for this liquid system. Substituting the relationships in equations (3.56) and (3.57) into (3.55) gives

$$\rho V C_p \frac{dT}{dt} = \rho C_p F([T_0 - T_{\text{ref}}] - [T_1 - T_{\text{ref}}]) + Q \tag{3.58}$$

This is the basic energy balance on the tank, which is one equation with two variables, T and Q. To complete the model, the heat transferred must be related to the tank temperature and the external variables (coolant flow and temperature). Thus, a subproblem involving the energy balance on the liquid in the cooling coils is now defined and solved (Douglas, 1972). The assumptions are

1. The coil liquid is at quasi-steady state.
2. The coolant physical properties are constant.
3. The driving force for heat transfer can be approximated as the *average* between the inlet and outlet.

With these assumptions, the energy balance on the cooling coil is

$$T_{\text{cout}} = T_{\text{cin}} - \frac{Q}{\rho_c C_{pc} F_c} \tag{3.59}$$

The subscript c refers to the coolant fluid. Now, two constitutive relationships are employed to complete the model. The heat transferred can be expressed as

$$Q = -UA(\Delta T)_{\text{lm}} \approx -UA\left(\frac{(T - T_{\text{cin}}) + (T - T_{\text{cout}})}{2}\right) \tag{3.60}$$

The heat transfer coefficient would depend on both film coefficients and the wall resistance. For many designs the outer film resistance in the stirred tank and the wall resistance would be small compared with the inner film resistance; thus, $UA \approx h_{\text{in}}A$. The inner film coefficient can be related to the flow by an empirical relationship of the form (Foust et al., 1980)

$$UA = aF_c^b \tag{3.61}$$

Equations (3.59) to (3.61) can be combined to eliminate T_{cout} and UA to give the following expression for the heat transferred:

$$Q = -\frac{aF_c^{b+1}}{F_c + \dfrac{aF_c^b}{2\rho_c C_{pc}}}(T - T_{\text{cin}}) \tag{3.62}$$

This solution to the subproblem expresses the heat transferred in terms of the specified, external variables (F_c and T_{cin}) and the tank temperature, which is the dependent variable to be determined. Equation (3.62) can be substituted into equation (3.58) to give the final model for the stirred-tank exchanger.

$$V\rho C_p \frac{dT}{dt} = C_p \rho F(T_0 - T) - \frac{aF_c^{b+1}}{F_c + \dfrac{aF_c^b}{2\rho_c C_{pc}}}(T - T_{\text{cin}}) \tag{3.63}$$

The degrees-of-freedom analysis results in one variable (T), one equation (3.63), four external variables (T_{cin}, T_0, and F are assumed constant, and F_c can change with time), and seven parameters. Thus, the model is exactly specified. However, the model is nonlinear due to the variable F_c raised to the powers b and $b+1$ and to the product of variables F_c and T. Therefore, the heat transfer term must be linearized through a Taylor series in two variables.

$$Q = Q_s + K_T(T - T_s) + K_{Fc}(F_c - F_{cs}) \tag{3.64}$$

$$Q_s = \left(\frac{-aF_c^{b+1}(T - T_{\text{cin}})}{F_c + \dfrac{aF_c^b}{2\rho_c C_{pc}}} \right)_s \tag{3.65}$$

$$K_T = \left(\frac{-aF_c^{b+1}}{F_c + \dfrac{aF_c^b}{2\rho_c C_{pc}}} \right)_s \qquad K_{Fc} = \left(\frac{-abF_c^b \left(F_c + \dfrac{a}{b} \dfrac{F_c^b}{2\rho_c C_{pc}} \right)(T - T_{\text{cin}})}{\left(F_c + \dfrac{aF_c^b}{2\rho_c C_{pc}} \right)^2} \right)_s$$

The linear approximation can be used to replace the nonlinear term, and again the equation can be expressed in deviation variables:

$$VC_p\rho \frac{dT'}{dt} = F\rho C_p(-T') + K_T T' + K_{Fc}F_c' \tag{3.66}$$

Solution. The resulting approximate model is a linear first-order ordinary differential equation that can be solved by applying the integrating factor.

$$\frac{dT'}{dt} + \frac{1}{\tau}T' = \frac{K_{Fc}}{V\rho C_p}F_c' \qquad \text{with } \tau = \left(\frac{F}{V} - \frac{K_T}{V\rho C_p} \right)^{-1} \tag{3.67}$$

For a step change in the coolant flow rate at $t = 0$ and $T'(0) = 0$, the solution is given by

$$T' = \frac{K_{Fc}\Delta F_c \tau}{V\rho C_p}\left(1 - e^{-t/\tau}\right) = \Delta F_c K_p \left(1 - e^{-t/\tau}\right) \tag{3.68}$$

For this example, $b = 0.5$ and $a = 1.41 \times 10^5$ cal/(min °C), and the linearized coefficients can calculated to be $K_{Fc} = -5.97 \times 10^6$ ([cal/min]/[m^3/min]), $K_T = -9.09 \times 10^4$ ([cal/min]/°C). The steady-state gain and time constant can be determined to be

$$K_p = \frac{K_{Fc}\tau}{V\rho C_p} = -33.9 \frac{°\text{C}}{\text{m}^3/\text{min}} \qquad \tau = \left(\frac{F}{V} - \frac{K_T}{V\rho C_p} \right)^{-1} = 11.9 \text{ min}$$

Results analysis. The solution gives an exponential relationship between time and the variable of interest. The approximate linearized response is plotted in Figure 3.15 along with the solution to the nonlinear model. For the magnitude of the step change considered, the linearized approximation provides a good estimate of the true response.

The analytical linearized approximation provides relationships between the transient response and process design and operation. For example, since $K_T < 0$, equation (3.67) demonstrates that the time constant for the heat exchanger is always smaller than the time constant for the same stirred tank without heat exchanger, for which $\tau = V/F$.

Example 3.8 Flow manipulation. As explained briefly in Chapter 1, process control requires a manipulated variable that can be adjusted independently by a person or automation system. Possible manipulated variables include motor speed and electrical power, but the manipulated variable in the majority of process control systems is valve opening, which influences the flow of gas, liquid, or slurry. Therefore, it is worthwhile briefly considering a model for the effect of valve opening on flow. A simplified

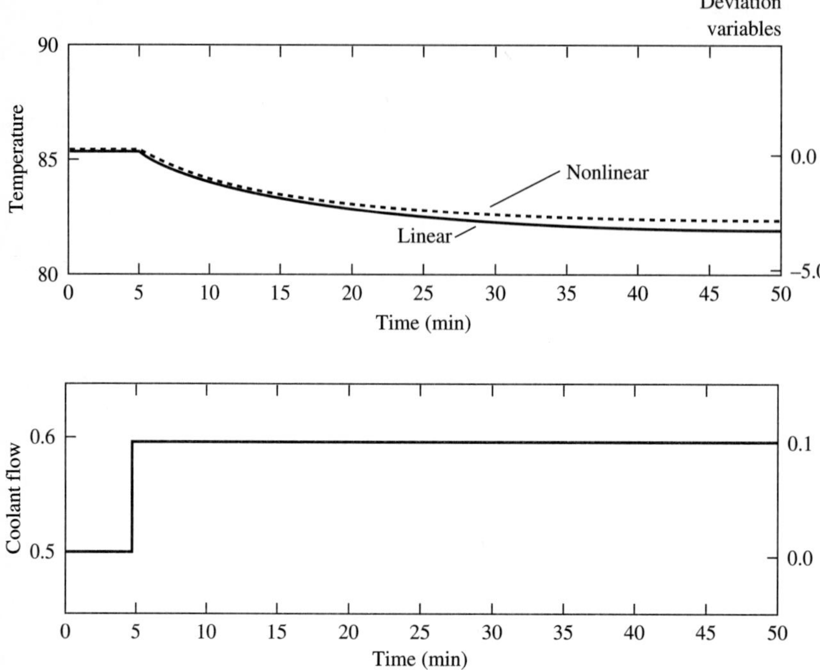

FIGURE 3.15
Dynamic response for Example 3.7.

system is shown in Figure 3.16, which is described by the following macroscopic energy balance (Foust et al., 1980; Hutchinson, 1976).

$$F = C_v v \sqrt{\frac{P_0 - P_1}{\rho}} \tag{3.69}$$

where C_v = inherent valve characteristic
v = valve stem position, related to percent open
F = volumetric flow rate

The valve stem position is changed by a person, as with a faucet, or by an automated system. The inherent valve characteristic depends in general on the stem position; also, the pressures in the pipe would depend on the flow and, thus, the stem position. For the present, the characteristic and pressures will be considered to be approximately constant. In that case, the flow is a linear function of the valve stem position:

$$F' = C_v \sqrt{\frac{P_0 - P_1}{\rho}} v' = K_v v' \qquad \text{with } K_v = C_v \sqrt{\frac{P_0 - P_1}{\rho}} \tag{3.70}$$

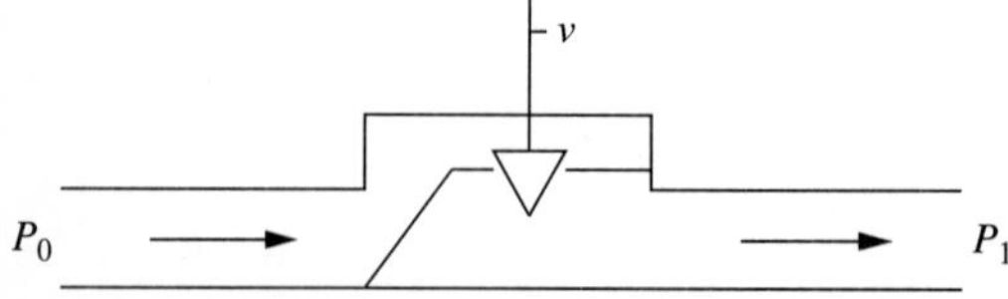

FIGURE 3.16
Simplified schematic of flow through valve.

Thus, linear or linearized models involving flow can be expressed as a function of valve position using equation (3.70). This is the expression used for many of the models in the next few chapters. More detail on the industrial flow systems will be presented in Chapters 7 (automated valve design) and 16 (variable characteristic and pressures).

The examples in this section have demonstrated the ease with which linearization can be applied to dynamic process models. As shown in equation (3.37), the second-order term in the Taylor series gives insight into the accuracy of the linear approximation. However, there is no simple manner for evaluating whether a linear approximation is appropriate, since the sensitivity of the modelling results depends on the formulation, input variables, parameters, and, perhaps most importantly, the goals of the modelling task. An analytical method for estimating the effects of the second-order terms in the Taylor series on the results of the dynamic model is available (Douglas, 1972); however, it requires more effort than the numerical solution of the original nonlinear equations. Therefore, the analytical method using higher-order terms in the Taylor series is not often used, although it might find application for a model solved frequently.

One quick check on the accuracy of the linearized model is to compare the final values, as time goes to infinity, of the nonlinear and linearized models. If they differ by too much, with this value specific to the problem, then the linearized model would be deemed to be of insufficient accuracy. If the final values are close enough, the dynamic responses could still differ and would have to be evaluated. Also, values of the time constants and gain at the initial and final conditions can be determined; if they are significantly different, the linearized model is not likely to provide adequate accuracy. The reader will be assisted in making these decisions by numerous examples in this book that evaluate linearized control methods applied to nonlinear processes.

The more complete approach for checking accuracy is to compare results from the linearized and full nonlinear models, with the nonlinear model solved using numerical methods, as discussed in the next section. Fundamental models can require considerable engineering effort to develop and solve for complex processes, so this approach is usually reserved for processes that are poorly understood or known to be highly nonlinear. In practice, engineers often learn by experience which processes in their plants can be analyzed using linearized models.

Again, this experience indicates that in the majority of cases, linear models are adequate for process control. An additional advantage of approximate linear models is the insight they provide into how process parameters and operating conditions affect the transient response.

3.5 NUMERICAL SOLUTIONS OF ORDINARY DIFFERENTIAL EQUATIONS

There are situations in which accurate solutions of the nonlinear equations are required. Since most systems of nonlinear algebraic and differential equations cannot be solved analytically, approximate solutions are determined using numerical methods. Many numerical solution methods are available, and a thorough coverage of the topic would require a complete book (for example, Carnahan et al., 1969, and Maron and Lopez, 1991). However, a few of the simplest numerical methods for solving ordinary differential equations will be introduced here, and they will be adequate, if not the most efficient, for most of the problems in this book.

Numerical methods do not find analytical solutions like the expressions in the previous sections; they provide a set of points that are "close" to the true solution of the differential equation. The general concept for numerical solutions is to use an initial value (or values) of a variable and an approximation of the derivative over a single step to determine the variable after the step. For example, the solution to the differential equation

$$\frac{dy}{dt} = f(y, t) \qquad \text{with } y\,|_{t=t_i} = y_i \tag{3.71}$$

can be approximated from $t = t_i$ to $t = t_{i+1}$, with $\Delta t = t_{i+1} - t_i$, by a linear Taylor series approximation to give

$$y_{i+1} \approx y_i + \left[\frac{dy}{dt}\right]_{t_i} (t_{i+1} - t_i) \tag{3.72}$$

$$y_{i+1} \approx y_i + f(y_i, t)\Delta t$$

The procedure in equation (3.72) is the *Euler* numerical integration method (Carnahan et al., 1969). This procedure can be repeated for any number of time steps to yield the approximate solution over a time interval.

Numerical methods can include higher-order terms in the Taylor series to improve the accuracy. The obvious method would be to determine higher-order terms in the Taylor series in equation (3.72); however, this would require algebraic manipulations that are generally avoided, although they could be practical with computer algebra. A manner has been developed to achieve the equivalent accuracy by evaluating the first derivative term at several points within the step. The result is presented here without derivation; the derivation is available in most textbooks on numerical analysis (Maron and Lopez, 1991). There are many forms of the solution, all of which are referred to as Runge–Kutta methods. The following equations are one common form of the Runge–Kutta fourth-order method:

$$y_{i+1} = y_i + \frac{\Delta t}{6}[m_1 + 2m_2 + 2m_3 + m_4] \tag{3.73}$$

with
$$\begin{aligned} m_1 &= f(y_i, t_i) \\ m_2 &= f(y_i + \frac{\Delta t}{2} m_1, t_i + \frac{\Delta t}{2}) \\ m_3 &= f(y_i + \frac{\Delta t}{2} m_2, t_i + \frac{\Delta t}{2}) \\ m_4 &= f(y_i + \Delta t m_3, t_i + \Delta t) \end{aligned}$$

All numerical methods introduce an error at each step, due to the loss of the higher-order terms in the Taylor series, and these errors accumulate as the integration proceeds. Since the accumulated error depends on how well the function is approximated, the Euler and Runge–Kutta methods have different accumulated errors. The Euler accumulated error is proportional to the step size; the Runge–Kutta error in equation (3.73) depends on the step size to the fourth power. Thus, the Euler method requires a smaller step size for the same accuracy as Runge–Kutta; this is partially offset by fewer calculations per step required for the Euler method. Since the errors from both methods increase with increasing step size, a very small step size might be selected for good accuracy, but a very small step size has two disadvantages. First, it requires a large number of steps and, therefore, long computing times to complete the entire simulation. Second, the use of too small a step size results in a very small change in y, perhaps so small as to be lost due to round-off. Therefore, an intermediate range of step sizes exists, in which the approximate numerical solution typically provides the best accuracy.

The engineer must choose the step size Δt to be the proper size to provide adequate accuracy. The proper step size is relative to the dynamics of the solution; thus, a key parameter is $\Delta t/\tau$, with τ being the smallest time constant appearing in a linear(ized) model. As a very rough initial estimate, this parameter could be taken to be approximately 0.01. Then, solutions can be determined at different step sizes; the region in which the solution does not change significantly, as compared with the accuracy needed to achieve the modelling goal, indicates the proper range of step size. There are numerical methods that monitor the error during the problem solution and adjust the step size during the solution to achieve a specified accuracy (Maron and Lopez, 1991).

Some higher-order systems have time constants that differ greatly (e.g., $\tau_1 = 1$ and $\tau_2 = 5000$); these systems are referred to as *stiff*. When explicit numerical methods such as Euler and Runge–Kutta are used for these systems, the step size must be small relative to the *smallest* time constant for good accuracy (and stability), but the total interval must be sufficient for the longest time constant to respond. Thus, the total number of time steps can be extremely large, and computer resources can be exorbitant. One solution method is to approximate part of the system as a quasi-steady state; this was done in several of the previous examples in this chapter, such as Example 3.7, where the coolant energy balance was modelled as a steady-state process. When this is not possible, the explicit numerical methods described above are not appropriate, and *implicit* numerical methods, which involve iterative calculations at each step, are recommended (Maron and Lopez, 1991).

Either the Euler or the Runge–Kutta method should be sufficient for the problems encountered in this book, but not for all realistic process control simulations. Recommendations on algorithm selection are available in the references already noted, and various techniques have been evaluated (Enwright and Hull, 1976). The numerical methods are demonstrated by application to examples.

Example 3.9 Isothermal CSTR. In Example 3.5 a model of an isothermal CSTR with a second-order chemical reaction was derived and an approximate linear model was solved. The nonlinear model cannot be solved analytically; therefore, a numerical solution is presented. The Euler method can be used, which involves the solution of the following equation at each step, i:

$$C_{Ai+1} = C_{Ai} + \Delta t\left(\frac{F}{V}(C_{A0i} - C_{Ai}) + kC_{Ai}^2\right) \tag{3.74}$$

An appropriate step size was found by trial and error to be 0.05. (Note that $\Delta t/\tau = 0.014$.) The numerical solution is shown in Figure 3.12 as the result from the nonlinear model.

Example 3.10 Nonisothermal CSTR. Many chemical reactors are not isothermal, because they involve significant heat transfer and heat of reaction. In this example, the model for a single, nonisothermal continuous stirred-tank chemical reactor (CSTR) is formulated and solved.

Goal. The temperature of a chemical reactor is to be raised to 395.3 K, without exceeding 395.3 K, by adjusting the coolant flow. How should the coolant flow be adjusted? A more fundamental question is the shape of the dynamic response; is it monotonic or oscillatory, and what design parameters and external variables influence this response?

Information. The process is shown in Figure 3.9, and the system is taken to be the liquid in the tank. The chemical reaction is first-order with Arrhenius temperature dependence.

Assumptions.

1. The tank is well mixed.
2. Physical properties are constant.
3. The shaft work is negligible.

Data.

1. $F = 1\ \text{m}^3/\text{min}$; $V = 1\ \text{m}^3$; $C_{A0} = 2.0\ \text{kmole/m}^3$; $T_0 = 323$ K; $C_p = 1$ cal/(g K); $\rho = 10^6\ \text{g/m}^3$; $k_0 = 1.0 \times 10^{10}\ \text{min}^{-1}$; $E/R = 8330.1$ K; $-\Delta H_{rxn} = 130 \times 10^6$ cal/(kmole); $T_{cin} = 365$ K; $(F_c)_s = 15\ \text{m}^3/\text{min}$; $C_{pc} = 1$ cal/(g K); $\rho_c = 10^6\ \text{g/m}^3$; $a = 1.678 \times 10^6$ (cal/min)/(K); $b = 0.5$.
2. For this data, the steady-state values of the dependent variables are $T_s = 394$ K and $C_{As} = 0.265\ \text{kmole/m}^3$.
3. The change in coolant flow is a step of $-1\ \text{m}^3/\text{min}$.

Formulation. The system is the liquid in the tank. The overall material balance, as in several previous examples, demonstrates that the mass in the tank is approximately constant; thus, $F_0 = F_1 = F$. The component material balance on the reactant gives

$$V\frac{dC_A}{dt} = F(C_{A0} - C_A) - Vk_0 e^{-E/RT} C_A \tag{3.75}$$

The energy balance for reacting system is not as straightforward to derive; thus, its derivation is outlined in Appendix C, with the following result:

$$V\rho C_p \frac{dT}{dt} = \rho C_p F(T_0 - T) - \frac{aF_c^{b+1}}{F_c + \dfrac{aF_c^b}{2\rho_c C_{pc}}}(T - T_{cin}) + (-\Delta H_{rxn})Vk_0 e^{-E/RT} C_A \tag{3.76}$$

These two nonlinear differential equations cannot be solved analytically. The linearized equations in deviation variables are as follows:

$$\frac{dC'_A}{dt} = a_{11}C'_A + a_{12}T' + a_{13}C'_{A0} + a_{14}F'_c + a_{15}T'_0 + a'_{16}F' \tag{3.77}$$

$$\frac{dT'}{dt} = a_{21}C'_A + a_{22}T' + a_{23}C'_{A0} + a_{24}F'_c + a_{25}T'_0 + a_{26}F' \tag{3.78}$$

with
$$\begin{aligned}
a_{11} &= -\frac{F}{V} - k_0 e^{-E/RT_s} \\
a_{12} &= -\frac{E}{RT_s^2} k_0 e^{-E/RT_s} C_{As} \\
a_{13} &= \frac{F}{V} \\
a_{14} &= 0 \\
a_{15} &= 0 \\
a_{16} &= \frac{(C_{A0} - C_A)_s}{V} \\
a_{21} &= \frac{-\Delta H_{rxn} k_0 e^{-E/RT_s}}{\rho C_p} \\
a_{22} &= -\frac{F}{V} - \frac{UA_s^*}{V\rho C_p} + (-\Delta H_{rxn}) \frac{\frac{E}{RT_s^2}}{\rho C_p} k_0 e^{-E/RT_s} C_{As} \\
a_{23} &= 0 \\
a_{24} &= \frac{-abF_{cs}^b \left(F_{cs} + \frac{a}{b}\frac{F_{cs}^b}{2\rho_c C_{pc}}\right)[T_s - (T_{cin})_s]}{(V\rho C_p)\left(F_{cs} + \frac{aF_{cs}^b}{2\rho_c C_{pc}}\right)^2} \\
a_{25} &= \frac{F}{V} \\
a_{26} &= \frac{(T_0 - T)_s}{V}
\end{aligned} \tag{3.79}$$

with $UA_s^* = a(F_c)_s^{b+1}/[(F_c)_s + a(F_c)_s^b/2\rho_c C_{pc}]$

The approximate model is derived about the steady-state operating point for the reactor. Note that the inlet concentration C'_{A0}, the coolant flow rate F'_c, the inlet temperature T'_0, and the feed flow rate F' are input variables in the foregoing linearized equations, although only the coolant flow varies in this example; this is done because changes in other input variables will be considered in later examples.

Solution. The analytical solution to the linearized model requires the simultaneous solution of equations (3.77) and (3.78), because C'_A and T' appear in both equations. Since the integrating factor method cannot be applied to this problem, the analytical solution to the linearized equations is deferred to the next chapter. To determine the behavior of the process and answer the specific question posed in this example, the solution of the nonlinear model will be determined via a numerical solution using an

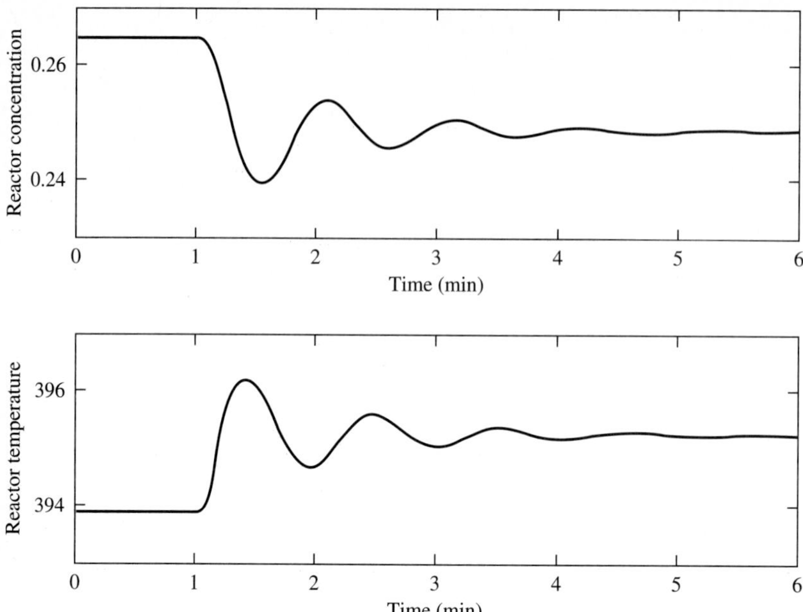

FIGURE 3.17
Dynamic response for Example 3.10 for step in cooling flow of $-1\ \text{m}^3/\text{min}$ at time $= 1$.

explicit method; the result for the Euler method with a step size of .005 minute is given in Figure 3.17. The solution is underdamped (i.e., oscillatory) for this model and set of design parameters and operating conditions. As a result, a single step in the coolant flow large enough to raise the temperature to its desired final value of 395.3 K leads to a response that exceeds this maximum value during the transient. Thus, it is not possible with one adjustment of the cooling flow to achieve the temperature specifications, although the temperature could be increased very close to, without exceeding, 395.3 K through a series of smaller adjustments to the coolant flow.

While the numerical solution provides the essential information to find a satisfactory set of coolant flow adjustments, it does not provide insight to the second goal, which addresses the aspects of the model structure and values of variables and parameters that lead to the oscillatory response. Perhaps one could learn the relationships through numerically solving many cases with different values for the external variables and parameters, but an approximate analytical solution would be more enlightening. The approximate solution to the linearized problem will be developed in the next chapter, after the Laplace transform method for analyzing and solving higher-order linear differential equations has been presented.

In summary, numerical methods provide the capability of solving complex, nonlinear ordinary differential equations. Thus, the engineer can formulate a model to satisfy the modelling goals without undue concern for determining an analytical solution. This power in developing specific solutions is achieved at a loss in engineering insight, so that the linearized solutions are often derived to establish relationships.

3.6 THE OPERATING WINDOW AND OPERATING POINT

The design of process control systems requires both steady-state and dynamic models. One use for the steady-state models is in determining the possible region of steady-state operation for a process that can be limited by constraints such as safety (e.g., pressure of a vessel), product quality (e.g., composition specifications), and equipment performance (e.g., maximum heat transfer or fluid flow). The region within which the process can be operated is called the operating window or feasible operating region, and any point within the region can be attained by adjusting external variables within a specified range. For continuous processes that operate at or near steady state most of the time, the economic optimum is usually based on the steady-state behavior of the process. Thus, a second use of steady-state models is for selecting the operating conditions in the window that yield the highest profit. Note that the task of process control is to maintain the process at or close to the best conditions during dynamic operation as the process is subjected to disturbances.

Steady-state models are derived using the procedure in Table 3.1. The only major difference from the previous examples in this chapter is in the accumulation term, which is zero for steady-state systems. For example, this approach can be applied to the nonisothermal chemical reactor in Example 3.10. The material and energy balance equations for the system are the dynamic balances, equations (3.75) and (3.76), with the left-hand-side derivatives set to zero. Note that

> It is good practice to use the nonlinear equations to determine the operating window, because the window normally covers a large range of values, over which the linearized model may not be accurate.

The operating window is a single point unless one or more of the external variables is allowed to change. Therefore, the window is defined for a specified range of these "input" variables. Typically, the output variables are plotted to describe the operation of the process in response to inputs. Determining the best conditions for a large plant can require complex analysis; two simple examples are presented here to illustrate the principles involved.

Example 3.11. Determine the steady-state operating window for the chemical reactor in Example 3.10. The initial analysis will assume that only the coolant flow rate can be changed; thus, other inlet flows, temperatures, and concentrations are constant. The solution to the nonlinear material and energy balance equations can be determined by many numerical methods, such as Newton–Raphson (Maron and Lopez, 1991); in this case, the equations can be rearranged to a form that can be solved by a simple trial-and-error method. The first equation is the steady-state component material balance based on equation (3.75), which can be solved analytically because it is linear:

$$C_A = \frac{F}{F + Vk} C_{A0} \tag{3.80}$$

The second equation, the steady-state energy balance based on equation (3.76), is nonlinear and can be separated into two terms, (Q_T) for "energy transfer" and (Q_R) for "release due to reaction," which sum to zero at steady state:

$$0 = Q_T + Q_R \tag{3.81}$$

with $Q_T = F\rho C_p(T_0 - T) - \dfrac{aF_c^{b+1}}{F_c + \dfrac{aF_c^b}{2\rho_c C_{pc}}}(T - T_{cin})$

$Q_R = (-\Delta H_{rxn})V k_0 e^{-E/RT} C_A$

To determine the steady-state solution for a value of coolant flow rate, the following procedure is followed:

1. Initialize a temperature, $T_1 <$ steady-state value of temperature.
2. Calculate $T_2 = T_1 + \Delta T$.
3. Calculate the concentration C_A at T_2.
4. Calculate the error $= Q_T + Q_R$, with $T = T_2$ in the energy balance.
5. If the error > 0, set $T_1 = T_2$; if the error < 0, set $\Delta T = \Delta T/2$.
6. Check convergence, $|\text{error}| <$ maximum. If not converged, return to (2); if converged, stop.

The values for the two important reactor variables C_A and T are plotted in Figure 3.18*a* and *b* as the coolant flow varies from 0.5 to its maximum value of 16 m^3/min, and the same results are plotted with the dependent variables on the axes in Figure 3.18*c*, which is the more conventional presentation of an operating window. Because the system has only one independent variable, the coolant flow, the window is a line. Therefore, *only one* of the two important dependent variables can be specified and attained independently by adjusting the coolant flow. Clearly, this result would have great importance when designing an automatic control system.

The operating window of the reactor could be "opened" by adding another adjustable external variable. For example, consider the case in which the coolant flow and the inlet concentration can be adjusted independently over 0.5 to 16 m^3/min and 1.0 to 2.0 moles/m^3, respectively. The steady-state solutions can be obtained as already described, with another iteration for various values of the inlet concentration. The operating window for this system is shown in Figure 3.19, where the region is larger due to the added input flexibility. Notice that the operating window in Figure 3.19 is not square; the process behavior imposes restrictions on the relationships between the reactant concentration and temperature. For example, a high reactor temperature causes a high reaction rate, so that simultaneously high values for concentration and temperature are not possible, given limits on the external variables.

The boundaries or frame of the window represent the extreme values of the *steady-state* operating conditions. The process cannot be maintained outside of those conditions at steady state by any allowable adjustments to the coolant flow and inlet concentration, for the assumed values of the external variables and parameters. It should be recognized that the temperature and concentration could temporarily exceed the boundaries of the operating window during dynamic operation, as shown in Example 24.4.

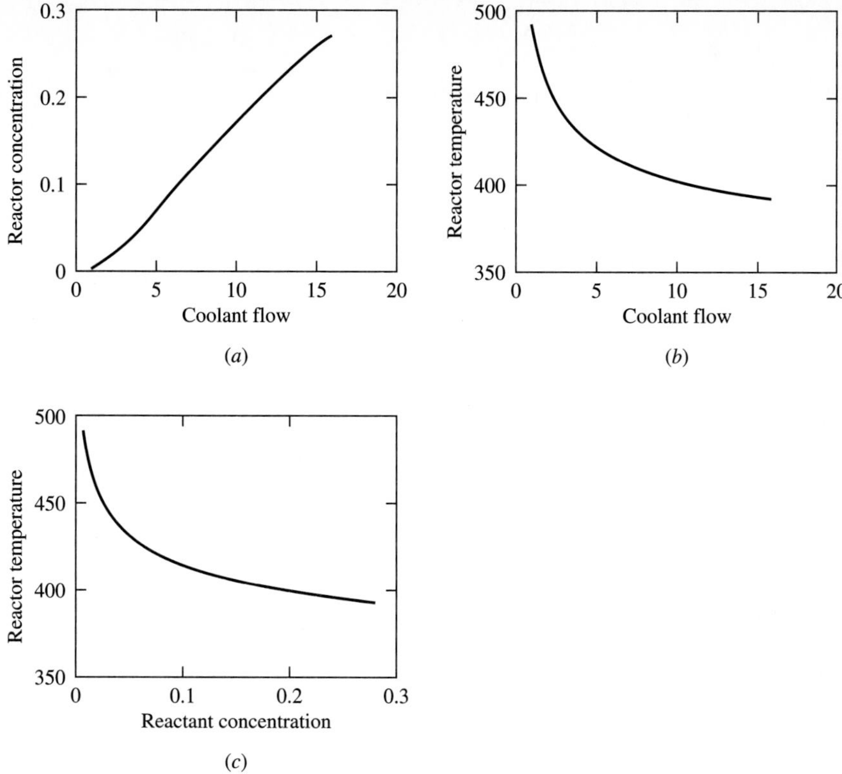

FIGURE 3.18
Operating window for reactor in Example 3.10 with cooling flow changing.

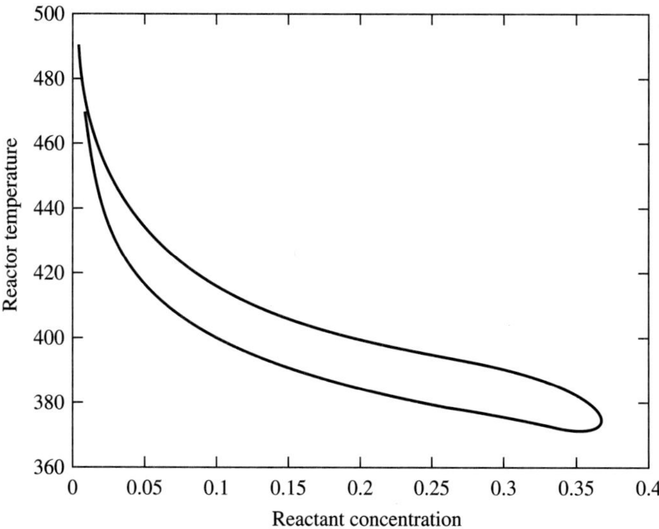

FIGURE 3.19
Operating window for reactor in Example 3.10 with cooling flow and inlet concentration changing.

The limitations to the operating window are often characterized as either *hard* or *soft.* Hard limits should never be violated; examples are concentrations, which must be nonnegative, and safety limits on pressure. Soft limits can be violated for short times, although economic penalties are incurred; an example is product quality or energy efficiency. Thus, a large region of plant operation is physically possible but excluded by demands for safety, high product quality, and efficient operation.

The analysis just presented, showing a single steady-state operating condition for each value of the input (external) variables, is typical for CSTRs and most other unit operations. However, a few important units, including some chemical reactors, can have multiple steady-state operating conditions for the same input values! An introduction to multiple steady states, with a chemical reactor example, is presented in Appendix C.

Since many (really, an infinite number of) possible operating points exist for most processes, an important question that often arises is, "At which point should the plant be operated?" Many issues should be considered in this decision, including safety, dynamic performance, product quality, and profit, and each is addressed in later chapters. To consider only profit briefly, the profit could be calculated for many points within the operating window, and the operating conditions with the highest profit could be selected.

Example 3.12 Optimum reactor operation. The yield of a desired product in a single stirred-tank reactor, component B, is to be maximized over a range of steady-state operating conditions.

Goal. The operating conditions that yield the maximum concentration of the desired product, component B, are to be determined for two cases: (1) the flow rate can be varied (V/F from 0 to 1.0) with the temperature constant at 330 K, and (2) both flow rate and temperature (300 to 360 K) can be varied.

Information.

1. The heat of reaction and work are negligible.
2. The volume is constant, and the contents are well mixed.
3. The temperature between 300 and 360 K can be achieved and maintained constant, by adjusting the heating medium flow, at negligible cost.

Data. The chemical reactions are

$$A \rightarrow B \rightarrow C$$

with
$$\begin{aligned} C_{A0} &= 3.0 \text{ mole/m}^3 \\ r_A &= -k_{10}e^{-E_1/RT}C_A \\ r_C &= k_{20}e^{-E_2/RT}C_B \\ k_{10} &= 17748.5 \text{ min}^{-1} \\ E_1/R &= 3000 \text{ K} \\ k_{20} &= 643048 \text{ min}^{-1} \\ E_2/R &= 4000 \text{ K} \end{aligned}$$

Formulation. The steady-state component material balances on the reactant A and the desired product B in the liquid in the reactor are

$$V\frac{dC_A}{dt} = 0 = F(C_{A0} - C_A) - Vk_{10}e^{-E_1/RT}C_A \tag{3.82}$$

$$V\frac{dC_B}{dt} = 0 = FC_B + Vk_{10}e^{-E_1/RT}C_A - Vk_{20}e^{-E_2/RT}C_B \tag{3.83}$$

These two algebraic equations can be solved for C_B to give

$$C_B = \frac{(V/F)k_{10}e^{-E_1/RT}C_{A0}}{\left(1 + (V/F)k_{10}e^{-E_1/RT}\right)\left(1 + (V/F)k_{20}e^{-E_2/RT}\right)} \tag{3.84}$$

Results analysis. For the first case, the temperature is constant and the maximum value of C_B can be determined analytically by taking the derivative of equation (3.84) with respect to (V/F) and setting it equal to zero. Alternatively, numerical values of C_B calculated from the equation can be plotted as in Figure 3.20. By either means, the optimum value of (V/F) is determined to be 0.378 minute, which leads to the maximum C_B of 0.556 mole/m^3.

For the second case, two independent variables can be adjusted to achieve the highest value for C_B. The results are plotted in Figure 3.21, with contours of C_B shown. The maximum value for this range occurs at the higher bound of V/F; for the range of variables in the figure, the maximum of $C_B = 0.581$ occurs at $V/F = 0.42$ and a temperature of 317.7 K. This result can be understood by recognizing that the activation energy for the conversion of desired B to byproduct C is higher than the activation energy for the conversion of A to B. Thus, the highest selectivity to B is at low temperatures, with compensating high residence times in the reactor to obtain a high conversion of component A.

For both cases, the optimum conditions would involve other issues, such as capital costs to build the equipment, heating costs, and downstream purification of B. Thus, this analysis is a simplification of a real optimization study, but it highlights the methods

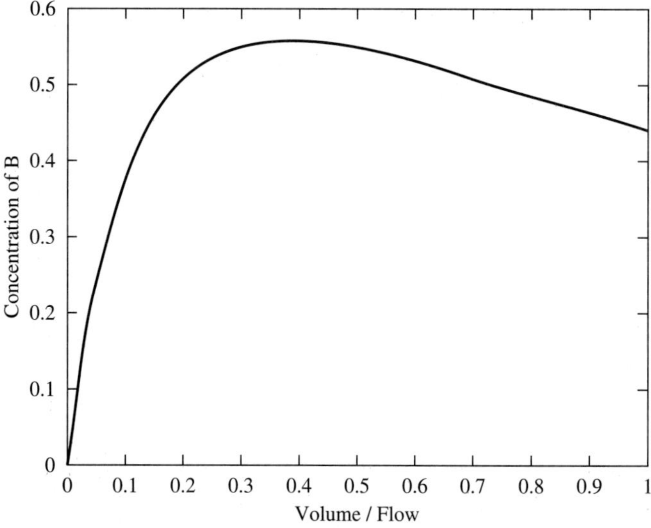

FIGURE 3.20
Results for Example 3.12 with temperature constant at 330 K.

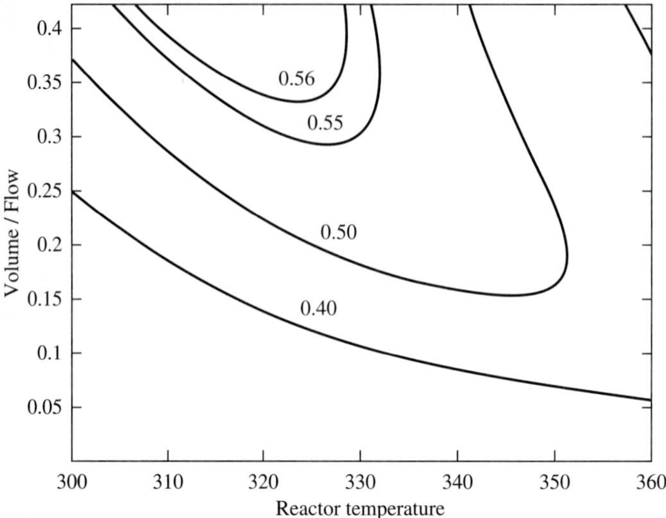

FIGURE 3.21
Results for Example 3.12 with two input variables changing. Contours are C_B.

and key tradeoffs. Further examples of the effects of operating conditions on reactor yields are given in Levenspiel (1972).

In conclusion, steady-state models can be used to establish the possible operating regions for continuous processes. The operating window can be evaluated to ensure that all desired steady-state operating conditions are possible and to locate the operating condition with the highest profit. Nonlinear models are typically used for this analysis, because of the wide range of conditions considered. However, steady-state models cannot provide essential information on the transient performance of a process, which can only be provided by using dynamic process models. Control systems that manipulate operating conditions to increase profit are described in Chapter 26.

3.7 CONCLUSIONS

The procedure in Table 3.1 provides a road map for developing, solving, and interpreting mathematical models based on fundamental principles. In addition to predicting specific behavior, these models provide considerable insight into the relationship between the process equipment and operating conditions and dynamic behavior. A thorough analysis of results is recommended in all cases so that the sensitivity of the solution to assumptions and data can be evaluated.

Perhaps the most important concept is

Modelling is a goal-oriented task, so the proper model depends on its application.

Thus, the same process may need several models. For example, a tray-to-tray distillation model with detailed thermodynamic correlations might be used for equipment design, while a much simpler linear model would be appropriate for control system design and implementation. The engineer learns model selection by paying attention to reported successes in the literature and personal experience.

Chemical engineers use fundamental mathematical models for many purposes, a few of which follow:

- *Process design:* The inherent dynamic response of the process without corrective action by a person or control system is important in the analysis of many process designs. The proper use and analysis of models will contribute to designing processes that are easily maintained near the desired operating conditions.
- *Process control design:* A natural application for modelling is in determining the effectiveness of process control. The engineer has many choices in designing a control system, as discussed briefly in Chapter 1 and presented in more detail later, and an understanding of the dynamic performance of systems under feedback control is required to make these choices properly. The use of modelling was discussed qualitatively in Section 1.2 and quantitatively for the on/off heating Example 3.4.
- *Operating conditions:* Steady-state models are used to define the operating window of a process and the optimum operating conditions within the window. These operating windows are useful in control design, because they define the range over which control can operate and equipment limitations that might change the number of adjustable variables. This analysis is demonstrated in Examples 2.1, 2.2, 3.11, and 3.12.
- *What-if analysis:* Engineers are able to evaluate many possible changes to the process equipment, feed materials, and operating conditions faster and at much lower costs through modelling than through experimentation. Fundamental models, which are based on solid physicochemical principles, are much more reliable than empirical models, discussed in Chapter 6, for extrapolating to new situations.
- *Training:* A dynamic model of a process can be used to train personnel on the proper operation of a plant. To make such training practical, the equations must be solved in *real time;* that is, plant variables in the model solution must respond at the same time (and in all other ways) as the true plant variables would respond. Also, an interface must be used for display of values and input of changes by the people being trained; a computer terminal with graphical display is often appropriate. This type of training is common in many industries, such as airlines, power generation and distribution plants, and chemical plants.

The examples in this chapter demonstrate the procedure through application to several realistic modelling goals. However, the examples were selected to limit the mathematical methods needed for solution to a small number of algebraic and ordinary differential equations. Models similar to the examples are often, but not always, sufficient for process dynamic analysis and control design. For example, a fundamental model of a distillation column with 40 trays and 10 components would involve at least 400 first-order ordinary differential equations to be solved simultaneously. Methods for determining approximate, low-order, linear models of processes from empirical data are presented in Chapter 6.

The observant reader may have noticed the similarities among the behaviors of many of the examples in this chapter. These similarities will lead to important generalizations, presented in Chapter 5, about the dynamics of processes that can be represented by simple sets of differential equations: one ordinary differential equation (first-order system), two equations (second-order system), and so forth. However, before exploring these generalities, some useful mathematical methods are introduced in Chapter 4. These mathematical methods are selected to facilitate the analysis of process control systems using models like the ones developed in this chapter and will be used extensively in the remainder of the book.

REFERENCES

Bird, R., W. Stewart, and E. Lightfoot, *Transport Phenomena,* Wiley, New York, 1960.

Carnahan, B., H. Luther, and J. Wilkes, *Applied Numerical Methods,* Wiley, New York, 1969.

Douglas, J., *Process Dynamics and Control, Volume I, Analysis of Dynamic Systems,* Prentice-Hall, Englewood Cliffs, NJ, 1972.

Felder, R. and R. Rousseau, *Elementary Principles of Chemical Processes* (2nd Ed.), Wiley, New York, 1986.

Himmelblau, D., *Basic Principles and Calculations in Chemical Engineering,* Prentice-Hall, Englewood Cliffs, NJ, 1982.

Hutchinson, J. (ed.), *ISA Handbook of Control Valves* (2nd Ed.), Instrument Society of America, Research Triangle Park, NC, 1976.

Enwright, W. and T. Hull, *SIAM J. Numer. Anal., 13,* 6, 944–961 (1976).

Levenspiel, O., *Chemical Reaction Engineering,* Wiley, New York, 1972.

Maron, M. and R. Lopez, *Numerical Analysis, A Practical Approach* (3rd Ed.), Wadsworth, Belmont, CA, 1991.

McQuiston, F. and J. Parker, *Heating, Ventilation, and Air Conditioning* (3rd Ed.), Wiley, New York, 1988.

Smith, J. and H. Van Ness, *Introduction to Chemical Engineering Thermodynamics* (4th Ed.), McGraw-Hill, New York, 1987.

Foust, A., L. Wenzel, C. Clump, L. Maus, and L. Andersen, *Principles of Unit Operations,* Wiley, New York, 1980.

ADDITIONAL RESOURCES

The following references, in addition to Douglas (1972), discuss goals and methods of fundamental modelling for steady-state and dynamic systems in chemical engineering.

Aris, R., *Mathematical Modelling Techniques,* Pitman, London, 1978.

Denn, M., *Process Modeling,* Pitman Publishing, Marshfield, MA, 1986.

Franks, R., *Modelling and Simulation in Chemical Engineering,* Wiley-Interscience, New York, 1972.

Friedly, J., *Dynamic Behavior of Processes,* Prentice-Hall, Englewood Cliffs, NJ, 1972.

Himmelblau, D. and K. Bishoff, *Process Analysis and Simulation, Deterministic Systems,* Wiley, New York, 1968.
Luyben, W., *Process Modelling, Simulation, and Control for Chemical Engineers* (2nd Ed.), McGraw-Hill, New York, 1989.

Guidance on the formulation, analysis, and efficient numerical computation of the sensitivity of the solution of differential equations to parameters is given in the following.

Leis, J. and M. Kramer, "The Simultaneous Solution and Sensitivity Analysis of Systems Described by Ordinary Differential Equations," *ACM Trans. on Math. Software, 14,* 1, 45–60 (1988).
Tomovic, R. and M. Vokobratovic, *General Sensitivity Theory,* Elsevier, New York, 1972.

The following reference presents methods for evaluating feasible operating conditions and economic optima in processes.

Edgar, T. and D. Himmelblau, *Optimization of Chemical Processes,* McGraw-Hill, New York, 1988.

The following reference discusses modelling as applied to many endeavors and gives examples in other disciplines, such as economics, biology, social sciences, and environmental sciences.

Murthy, D., N. Page, and E. Rodin, *Mathematical Modelling,* Pergamon Press, Oxford, 1990.

Stirred tanks are applied often in chemical engineering. Details on their design and performance can be found in the following reference.

Oldshue, J., *Fluid Mixing Technology,* McGraw-Hill, New York, 1983.

In answering the questions in this chapter (and future chapters), careful attention should be paid to the modelling methods and results. The following items are provided to assist in this analysis.

- Define the system and determine the balances and constitutive relations used.
- Analyze the degrees of freedom of the model.
- Determine how the design and operating values influence key results like gains and time constants.
- Determine the shape of the dynamic response. Is it monotonic, oscillatory, etc.?
- If nonlinear, estimate the accuracy of the linearized result.
- Analyze the sensitivity of the dynamic response to parameter values.
- Discuss how you would validate the model.

QUESTIONS

3.1. The chemical reactor in Example 3.2 is to be modelled, with the goal of determining the concentration of the product C_B as a function of time for the same input change. Extend the analytical solution to answer this question.

3.2. The series of two tanks in Example 3.3 are to be modelled with $V_1 + V_2 = 2.1$ and $V_1 = 2V_2$. Repeat the analysis and solution for this situation.

3.3. The step input is changed to an impulse for Example 3.3. An impulse is a "spike" with a (nearly) instantaneous duration and nonzero integral; physically, an impulse would be achieved by rapidly dumping extra component A into the first tank. Solve for the outlet concentration of the second tank after an impulse of M moles of A is put into the first tank.

3.4. A batch reactor with the parameters in Example 3.2 is initially empty and is filled at the inlet flow rate, with the outlet flow being zero. Determine the concentration of A in the tank during the filling process. After the tank is full, the outlet flow is set equal to the inlet flow; that is, the reactor is operated like a continuous-flow CSTR. Determine the concentration of A to the steady state.

3.5. The system in Example 3.1 has an input concentration that varies as a sine with amplitude A and frequency ω. Determine the outlet concentration for this input.

3.6. The level-flow system is Figure Q3.6 is to be analyzed. The flow F_0 is constant. The flow F_3 depends on the valve opening but not on the levels, whereas flows F_1 and F_2 depend on the varying pressures (i.e., levels). The system is initially at steady state, and a step increase in F_3 is made by adjusting the valve. Determine the dynamic response of the levels and flows using an approximate linear model. Without specific numerical values, sketch the approximate dynamic behavior of the variables.

3.7. A mathematical model of a CSTR with a first-order reaction and cooling is to be analyzed. The defining equations are the same as Example 3.10, but with no heat of reaction, $\Delta H_{rxn} = 0$. Determine the approximate dynamic response of the reactor concentration for a step change in the coolant flow rate. This question can be answered using a linearized model of the system and mathematical methods presented in this chapter.

3.8. The level-flow system in Figure Q3.8 is to be analyzed. The flow into the system, F_0, is independent of the system pressures. The feed is entirely liquid, and the first vessel is closed and has a nonsoluble gas in the space above the nonvolatile liquid. The flows F_1 and F_2 depend only on the pressure drops, because the restrictions in the pipes are fixed.

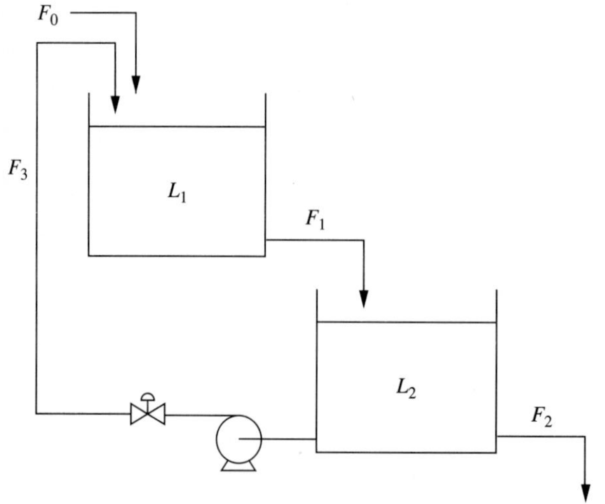

FIGURE Q3.6

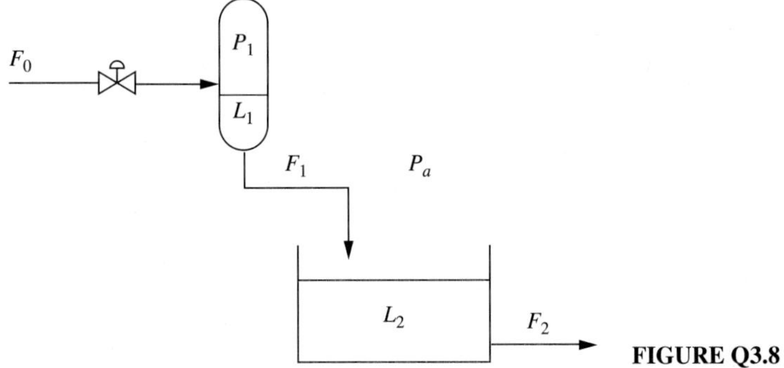

FIGURE Q3.8

Derive the linearized model for this system in response to a step change in F_0, solve the equations, and, without specific numerical values, sketch the dynamic responses.

3.9. The room heating Example 3.4 is reconsidered; for the following situations, each representing a single change from the base case, reformulate the model as needed and determine the dynamic behavior of the temperature and heating status.

(*a*) Due to leaks, a constant flow into and out of the room exists. Assume that the volume of air in the room is changed every hour with entering air at the outside temperature.

(*b*) A mass of material (e.g., furniture) is present in the room. Assume that this mass is always in equilibrium with the air; that is, the heat exchange is at quasi-steady state. The mass is equivalent to 200 kg of wood.

(*c*) The ambient temperature decreases to $-10°C$.

(*d*) The duty of the furnace is reduced to 0.50×10^6 when on.

(*e*) The heat transferred to the room does not change instantaneously when the furnace status changes. The relationship between the heat generated in the furnace (Q_f), which changes immediately when the switch is activated, and the heat to the room (Q_h) is

$$\tau_Q \frac{dQ_h}{dt} = Q_f - Q_h \qquad \text{with } \tau_Q = 0.10 \text{ hr}$$

3.10. Determine the dynamic responses for a +10 percent change in inlet flow rate in place of the original input change for one or more of Examples 3.2, 3.5, and 3.7. Determine whether the model must be linearized in each case. For cases that require linearization, estimate the errors introduced and compare a numerical solution with the approximate, linear dynamic response.

3.11. A stirred-tank heater could have an external jacket with saturated steam condensing in the jacket to heat the tank. Assume that this modification has been made to the system in Example 3.7 and derive an analytical expression for the response of the tank temperature to a step change in the steam pressure. Begin by sketching the system and listing assumptions.

3.12. The tank draining problem in Example 3.6 has been modified to remove the restriction (partially opened valve) in the outlet line. Now, the line is simply a pipe. Reformulate and solve the problem for the two following cases, each with a pipe long enough that end conditions are negligible.

(*a*) The flow in the outlet pipe is laminar.

(*b*) The flow in the outlet pipe is turbulent.

3.13. Answer the following questions.

(*a*) Explain what is meant by a stiff system of differential equations. Under what conditions (changing values of parameters) would the equations in Example 3.3 be stiff? If they were stiff, suggest several ways to solve them numerically. Would this stiffness affect the accuracy of the analytical solutions of the linearized model?

(*b*) The analysis of degrees of freedom suggests that terms that are constant in the current examples be separated into two categories: parameters and external variables. Why would this be useful for future analysis of feedback control systems? Suggest two subcategories for the external variables and why they might be useful for feedback control analysis.

(*c*) The degrees-of-freedom analysis should define the proper number of equations for a model. Suppose that the following model were proposed for Example 3.6.

$$A\frac{dL}{dt} = F_0 - F_1 \qquad (5)(2) = 10$$

When F_0 is constant, this model has two equations and two unknown variables, L and F_1. Explain why this model does not satisfy the degrees-of-freedom analysis and provide a mathematical test that can be applied to potential equation sets.

(*d*) Is it possible for a model to be linear for one external input perturbation and nonlinear for another? Explain and give examples.

(*e*) Give the equations to be solved at every time step for an Euler integration of the nonisothermal chemical reactor model in equations (3.75) and (3.76).

3.15. The chemical reactor in Example 3.3 is considered in this question. The only change to the problem is the input function; here, the inlet concentration is returned to its initial value in a step five minutes after the initial step increase.

(*a*) Determine the dynamic response of the concentration of both tanks.

(*b*) Compare your answer to the shape of the plot in Figure 3.5*b* and explain similarities and differences.

(*c*) Based on your results in (*a*) and (*b*), discuss how you would design an emergency system to prevent the concentration of A in the second tank from exceeding a specified maximum value. Discuss the variables F and C_{A0} as potential manipulated variables, and select the value to which the manipulated variable should be set when the action limit is reached. Also, discuss how you would determine the value of the action limit.

3.16. The dynamic response of the CSTR shown in Figure 3.1 is to be determined as follows.

Assumptions: (i) well mixed, (ii) isothermal, (iii) constant density, and (iv) constant volume.

Data: $V = 2\text{m}^3$; $F = 1\text{m}^3/\text{hr}$; $C_{A0}(0) = 0.5\text{mole/m}^3$.

Reaction: A → Products

with
$$r_A = -k_1 C_A/(1 + k_2 C_A) \text{ mole/(m}^3\text{hr)}$$
$$k_1 = 1.0 \text{ hr}^{-1}$$
$$k_2 = 1.0 \text{ m}^3\text{/mole}$$

(*a*) Formulate the model for the dynamic response of the concentration of A.

(*b*) Linearize the equation in (*a*).

(*c*) Analytically solve the linearized equation for a step change in the inlet concentration of A, C_{A0}.

(*d*) Give the equation(s) for the numerical solution of the "exact" nonlinear equation derived in (*a*). You may use any of the common numerical methods for solving ordinary differential equations.

(*e*) Calculate the transients for the (analytical) linearized and (numerical) nonlinear models. Graph the results for both the nonlinear and linearized predictions for two cases, both of which start from the initial conditions given above and have the magnitudes (1) $\Delta C_{A0} = 0.5$ and (2) $\Delta C_{A0} = 4.0$. Provide an annotated listing of your program or spreadsheet.

(*f*) Discuss the accuracy of the linearized solutions compared with solutions to the "exact" nonlinear equations for these two cases.

3.17. Discuss whether linearized dynamic models would provide accurate representations of the dynamic results for

(*a*) Example 3.2 with $\Delta C_{A0} = -0.925 \text{moles/m}^3$

(*b*) Example 3.7 for $\Delta F_c = -0.25 \text{m}^3\text{/min}$

(*c*) Example 3.10 as stated

(*d*) Example 3.11 about all steady states in the operating window.

3.18. A stirred-tank mixer has two input streams: F_A which is pure component A, and F_B, which has no A. The system is initially at steady state, and the flow F_A is constant. The flow of B changes according to the following description: From time $0 \rightarrow t_1$, $F'_B(t) = \alpha t$ (a ramp from the initial condition); and from time $t_1 \rightarrow \infty$, $F'_B(t) = \alpha t_1$ (constant at the value reached at t_1). The following assumptions may be used:

(1) The densities of the two streams are constant and equal, and there is no density change on mixing.

(2) The volume of the liquid in the tank is constant.

(3) The tank is well mixed.

(*a*) Sketch the process, define the system, and derive the basic balance for the weight fraction of A in the exit stream, X_A.

(*b*) Derive the linearized balance in deviation variables.

(*c*) Solve the equation for the forcing function, $F'_B(t)$, defined above. (*Hint*: You may want to develop two solutions, first from $0 \rightarrow t_1$ and then $t_1 \rightarrow \infty$.)

(*d*) Sketch the dynamic behavior of $F'_B(t)$ and $X'_A(t)$.

3.19. In the tank system in Figure 3.13, the outflow drains through the outlet pipe with a restriction as in Example 3.6, and in this question, a first-order chemical reaction occurs in the tank. Given the following data, plot the operating window with coordinates of level and concentration of A. Discuss the effect of changing reactor temperature on the operating window, if any.

Design parameters: Cross-sectional area $= 0.30 \text{ m}^2$, maximum level $= 4.0$ m. The chemical reaction is first-order with $k_0 = 2.28 \times 10^7 \text{ (hr}^{-1})$ and $E/R = 5000$ K. The base-case conditions can be used to back-calculate required parameters. The base case data are $T = 330$ K, $L = 3.33$ m, $F = 10 \text{ m}^3\text{/hr}$, and $C_A = 0.313 \text{ mole/m}^3$. The external variables can be adjusted over the following ranges; $0.20 \leq C_{A0} \leq 0.70$ and $3.0 \leq F \leq 12.5$.

3.20. A system of well-mixed tanks and blending is shown in Figure Q3.20. The delays in the pipes are negligible, the flow rates are constant, and the streams have the same density. Step changes are introduced in C_{A1} at t_1 and C_{A2} at t_2, with $t_2 > t_1$. Determine the transient responses of C_{A3}, C_{A4}, and C_{A5}.

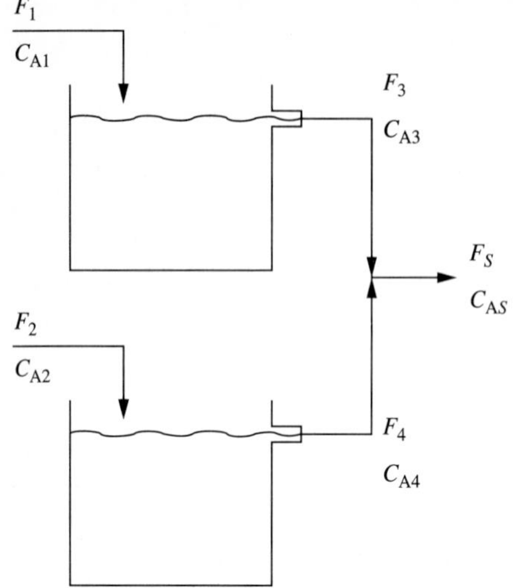

FIGURE Q3.20

TABLE Q3.22

Time	Temperature
0	103.5
.4	102
1.2	96
1.9	91
2.7	87
3.4	84
4.2	81
5.0	79
6.5	76
8.5	73

3.21. Determining the sensitivity of modelling results to parameters is a key aspect of results analysis. For the result from Example 3.2,

$$C_A = C_{Ainit} + \Delta C_{A0} K_p (1 - e^{-t/\tau})$$

(*a*) Determine analytical expressions for the sensitivity of the output variable C_A to small (differential) changes in the parameters, K_p, τ, forcing function magnitude ΔC_{A0}, and initial steady state, C_{Ainit}. These sensitivity expressions should be functions of time.

(*b*) For each result in (*a*), plot the sensitivities over their trajectories and discuss whether the answer makes sense physically.

3.22. Another experiment was performed to validate the fact that the vessel in Example 3.1 was well mixed. In this experiment, the vessel was well insulated and brought to steady state. Then a step change was introduced to the inlet temperature. The following data represents the operating conditions, and the dynamic data is given in Table Q3.22.

Data: $V = 2.7$ m^3, $F = 0.71$ m^3/min, $T_{0init} = 103.5$°C, $T_0 = 68$°C.

(*a*) Formulate the energy balance for this system, and solve for the expected dynamic response of the tank temperature.

(*b*) Compare your result prediction with the data.

(*c*) Given the two experimental results in Figure 3.4 and this question for the same equipment, discuss your conclusions on the assumption that the system is well mixed.

(*d*) Is there additional information that would help you in (*c*)?

3.23. The dynamic response of the reactant concentration in the reactor, C_A, to a change in the inlet concentration, C_{A0}, for an isothermal, constant-volume, constant-density CSTR with a single chemical reaction is to be evaluated. The reaction rate is modelled by

$$r_A = -\frac{k_1 C_A}{1 + k_2 C_A}$$

Determine how the approximate time constant of the linearized model of the process relating C_A to C_{A0} changes as k_1 and k_2 range from 0 to infinity. Explain how your answer make sense.

CHAPTER

4

MODELLING AND ANALYSIS FOR PROCESS CONTROL

4.1 INTRODUCTION

In the previous chapter, solutions to fundamental dynamic models were developed using analytical and numerical methods. The analytical integrating factor method was limited to sets of first-order linear differential equations that could be solved sequentially. In this chapter, an additional analytical method is introduced that expands the types of models that can be analyzed. The methods introduced in this chapter are tailored to the analysis of process control systems and provide the following capabilities:

1. The analytical solution of simultaneous linear differential equations with constant coefficients can be obtained using the Laplace transform method.
2. A control system can involve several processes and control calculations, which must be considered simultaneously. The overall behavior of a complex system can be modelled, considering only input and output variables, by the use of transfer functions and block diagrams.
3. The behavior of systems to sine inputs is important in understanding how the input frequency influences dynamic process performance. This behavior is most easily determined using frequency response methods.
4. A very important aspect of a system's behavior is whether it achieves a steady-state value after a step input. If it does, the system is deemed to be *stable*; if it does not, it is deemed *unstable*. Important control system analysis is based on this behavior, and the methods in this chapter are applied to determine the stability of feedback control systems in Chapter 10.

All of the methods in this chapter are limited to linear or linearized systems of ordinary differential equations. The source of the process models can be the fundamental modelling presented in Chapter 3 or the empirical modelling presented in Chapter 6.

The methods in this chapter provide alternative ways to achieve results that could, at least theoretically, be obtained for many systems using methods in Chapter 3. Therefore, the reader encountering this material for the first time might feel that the methods are redundant and unnecessarily complex. However, the methods in this chapter have been found to provide the best and *simplest* means for analyzing important characteristics of process control systems. The methods will be introduced in this chapter and applied to several important examples, but their power will become more apparent as they are used in later chapters. The reader is encouraged to master the basics here to ease the understanding of future chapters.

4.2 THE LAPLACE TRANSFORM

The Laplace transform provides the engineer with a powerful method for analyzing process control systems. It is introduced and applied for the analytical solution of differential equations in this section; in later sections (and chapters), other applications are introduced for characterizing important behavior of dynamic systems without solving the differential equations for the entire dynamic response.

The **Laplace transform** is defined as follows:

$$\mathcal{L}(f(t)) = f(s) = \int_0^\infty f(t)e^{-st}\,dt \tag{4.1}$$

Before examples are presented, a few important properties and conventions are stated.

1. Only the behavior of the time-domain function for times equal to or greater than zero are considered. The value of the time-domain function is taken to be zero for $t < 0$.
2. A Laplace transform does not exist for all functions. Sufficient conditions for the Laplace transform to exist are (i) the function $f(t)$ is piecewise continuous and (ii) the integral in equation (4.1) has a finite value; that is, the function $f(t)$ does not increase with time faster than e^{-st} decreases with time. Functions typically encountered in the study of process control are Laplace-transformable and are not checked. Further discussion of the existence of Laplace transforms is available (Boyce and Diprima, 1986).
3. The Laplace transform converts a function in the time domain to a function in the "s-domain," in which s can take complex values. Recall that a complex number x can be expressed in Cartesian form as $A + Bj$ or in polar form as $\text{R}e^{\phi j}$ with

$$A = \mathrm{Re}(x) \qquad B = \mathrm{Im}(x) \qquad R = \sqrt{A^2 + B^2} \qquad \phi = \tan^{-1}\left(\frac{B}{A}\right) \tag{4.2}$$

4. In this book, the Laplace transform of a function $T(t)$ will be designated by the argument s, as in $T(s)$. The function and its transform will be designated by the same symbol, which can be either a capital or a lowercase letter, and no overbar will be used for the transformed function. The function in the time domain will be designated as the variable (as T) or with the time shown explicitly [as $T(t)$], if needed for clarification.

5. The Laplace transform is a linear operator, because it satisfies the requirements specified in equation (3.36):

$$\mathcal{L}[aF_1(t) + bF_2(t)] = a\mathcal{L}[F_1(t)] + b\mathcal{L}[F_2(t)] \tag{4.3}$$

6. Tables of Laplace transforms are available, so the engineer does not have to apply equation (4.1) for many commonly occurring functions. Also, these tables provide the inverse Laplace transform,

$$\mathcal{L}^{-1}[f(s)] = f(t) \qquad \text{for } t \geq 0 \tag{4.4}$$

Since the Laplace transform is defined only for single-valued functions, the transform and its inverse are unique.

Before we proceed to the application of Laplace transforms to differential equations, equation (4.1) is applied to a few functions that will be used in later examples. A more extensive list of Laplace transforms is given in Table 4.1.

Constant

For $f(t) = C$,

$$\mathcal{L}(C) = \int_0^\infty Ce^{-st}\,dt = -\frac{C}{s}e^{-st}\Big|_0^\infty = \frac{C}{s} \tag{4.5}$$

Step of Magnitude C at $t = 0$

For $\qquad f(t) = CU(t) \qquad$ with $U(t) = \begin{cases} 0 & \text{at } t = 0^+ \\ 1 & \text{for } t > 0^+ \end{cases}$

$$\mathcal{L}[C(U(t))] = C\mathcal{L}[U(t)] = C\left(\int_0^\infty e^{-st}\,dt\right) = \frac{C}{s} \tag{4.6}$$

Exponential

For $f(t) = e^{-at}$,

$$\mathcal{L}\left(e^{at}\right) = \int_0^\infty e^{at}e^{-st}\,dt = \frac{1}{a-s}\,e^{-(s-a)t}\Big|_0^\infty = \frac{1}{s-a} \tag{4.7}$$

TABLE 4.1
Laplace transforms

No.	$f(t)$	$f(s)$
1	δ, unit impulse	1
2	$U(t)$, unit step or constant	$1/s$
3	$\dfrac{t^{n-1}}{(n-1)!}$	$1/s^n$
4	$\dfrac{1}{\tau}e^{-t/\tau}$	$\dfrac{1}{\tau s+1}$
5	$1+\dfrac{(a-\tau)}{\tau}e^{-t/\tau}$	$\dfrac{as+1}{s(\tau s+1)}$
6	$\dfrac{1}{\tau^n}\dfrac{t^{n-1}e^{-t/\tau}}{(n-1)!}$	$\dfrac{1}{(\tau s+1)^n}$
7	$\left(\dfrac{a}{\tau^2}+\dfrac{\tau-a}{\tau^3}t\right)e^{-t/\tau}$	$\dfrac{as+1}{(\tau s+1)^2}$
8	$1+\left(\dfrac{a-\tau}{\tau^2}t-1\right)e^{-t/\tau}$	$\dfrac{as+1}{s(\tau s+1)^2}$
9	$\dfrac{\tau_1-a}{\tau_1(\tau_1-\tau_2)}e^{-t/\tau_1}-\dfrac{\tau_2-a}{\tau_2(\tau_1-\tau_2)}e^{-t/\tau_2}$	$\dfrac{as+1}{(\tau_1+1)(\tau_2 s+1)}$
10	$1+\dfrac{\tau_1-a}{\tau_2-\tau_1}e^{-t/\tau_1}-\dfrac{\tau_2-a}{\tau_2-\tau_1}e^{-t/\tau_2}$	$\dfrac{as+1}{s(\tau_1 s+1)(\tau_2 s+1)}$
11	$\sin(\omega t)$	$\omega/(s^2+\omega^2)$
12	$\cos(\omega t)$	$s/(s^2+\omega^2)$
13	$e^{-at}\cos(\omega t)$	$\dfrac{s+a}{(s+a)^2+\omega^2}$
14	$e^{-at}\sin(\omega t)$	$\dfrac{\omega}{(s+a)^2+\omega^2}$
15	$\dfrac{C}{\tau}e^{-\xi t/\tau}\sin\left(\dfrac{\sqrt{1-\xi^2}}{\tau}t+\phi\right)$	$\dfrac{as+1}{\tau^2 s^2+2\xi\tau s+1}$
	$C=\sqrt{\dfrac{\frac{a^2}{\tau^2}-\frac{2\xi a}{\tau}+1}{1-\xi^2}}\qquad \phi=\tan^{-1}\left(\dfrac{\frac{a}{\tau}\sqrt{1-\xi^2}}{1-\frac{a\xi}{\tau}}\right)$	
16	$\dfrac{-1}{\tau^2\sqrt{1-\xi^2}}e^{-\xi t/\tau}\sin\left(\dfrac{\sqrt{1-\xi^2}}{\tau}t+\phi\right)$	$\dfrac{s}{\tau^2 s^2+2\xi\tau s+1}$
	$\phi=\tan^{-1}\left(\dfrac{\sqrt{1-\xi^2}}{\xi}\right)$	
17	$1-\dfrac{1}{\sqrt{1-\xi^2}}e^{-\xi t/\tau}\sin\left(\dfrac{\sqrt{1-\xi^2}}{\tau}t+\phi\right)$	$\dfrac{1}{s(\tau^2 s^2+2\xi\tau s+1)}$
	$\phi=\tan^{-1}\left(\dfrac{\sqrt{1-\xi^2}}{\xi}\right)$	
18	$f(t)=\begin{cases} f(t-a) & t\geq a \\ 0 & t<a\end{cases}$	$e^{-as}f(s)$

a, ω, and τ_i are real and distinct, $0<\xi\leq 1$, n = integer

Sine

For $f(t) = \sin(\omega t)$,

$$\mathcal{L}(\sin(\omega t)) = \int_0^\infty \sin(\omega t) e^{-st}\, dt = \int_0^\infty \left(\frac{e^{j\omega t} - e^{-j\omega t}}{2j} \right) e^{-st}\, dt \tag{4.8}$$

$$= \int_0^\infty \left(\frac{e^{-(s-j\omega)t} - e^{-(s+j\omega)t}}{2j} \right) dt$$

$$= \frac{1}{2j} \left[-\frac{e^{-(s-j\omega)t}}{(s-j\omega)} + \frac{e^{-(s+j\omega)t}}{(s+j\omega)} \right]_0^\infty = \frac{\omega}{s^2 + \omega^2}$$

Pulse

For $f(t) = C[U(0) - U(t_p)] = C/t_p$ for $t = 0$ to t_p, and $= 0$ for $t > t_p$, as graphed in Figure 4.1,

$$\mathcal{L}(f(t)) = \int_0^{t_p} \frac{C}{t_p} e^{-st}\, dt + \int_{t_p}^\infty 0 e^{-st}\, dt \tag{4.9}$$

$$= \frac{C}{t_p} \frac{(1 - e^{-st_p})}{s}$$

An *impulse* function, which has zero width and total integral equal to C, is a special case of the pulse. Its Laplace transform can be determined by taking the limit of equation (4.9) as $t_p \to 0$ (and applying L'Hospital's rule) to give

$$Y(s)\Big|_{t_p \to 0} = \lim_{t_p \to 0} \frac{C}{t_p} \frac{(1 - e^{-st_p})}{s} \tag{4.10}$$

$$= \lim_{t_p \to 0} \frac{-C(-se^{-st_p})}{s} = C$$

Derivative of a Function

To apply Laplace transforms to the solution of differential equations, the Laplace transform of derivatives must be evaluated.

$$\mathcal{L}\left(\frac{df(t)}{dt} \right) = \int_0^\infty \frac{df(t)}{dt} e^{-st}\, dt \tag{4.11}$$

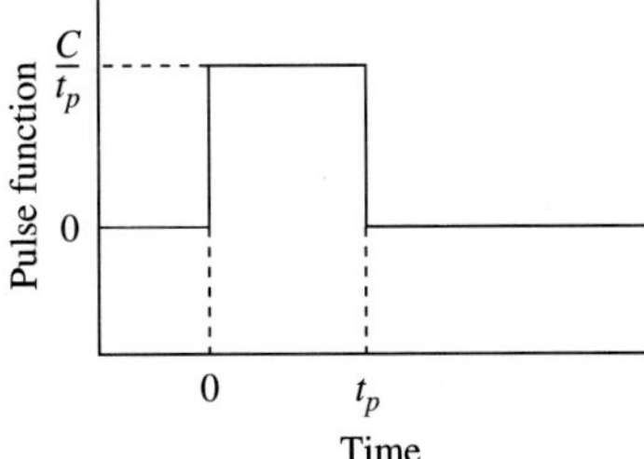

FIGURE 4.1
Pulse function.

This equation can be integrated by parts to give

$$\mathscr{L}\left(\frac{df(t)}{dt}\right) = -\int_0^\infty f(t)(-s)e^{-st}\,dt + f(t)e^{-st}\Big|_0^\infty \tag{4.12}$$
$$= sf(s) - f(t)\big|_{t=0}$$

The method can be extended to a derivative of any order by applying the integration by parts several times to give

$$\mathscr{L}\left(\frac{d^n f(t)}{dt^n}\right) = s^n f(s) - \left(s^{n-1} f(t)\big|_{t=0} + s^{n-2}\frac{df(t)}{dt}\bigg|_{t=0} + \cdots + \frac{d^{n-1}f(t)}{dt^{n-1}}\bigg|_{t=0}\right) \tag{4.13}$$

Integral

By similar application of integration by parts the Laplace transform of an integral of a function can be shown to be

$$\mathscr{L}\left(\int_0^t f(t')dt'\right) = \int_0^\infty \left(\int_0^t f(t')\,dt'\right)e^{-st}\,dt \tag{4.14}$$
$$= \int_0^\infty \frac{e^{-st}}{s} f(t)\,dt + \left[\left(\int_0^t f(t)\,dt\right)\frac{e^{-st}}{s}\right]_{t=0}^\infty = \frac{1}{s}f(s)$$

Differential Equations

One of the main applications for Laplace transforms is in the analytical solution of ordinary differential equations. The key aspect of Laplace transforms in this application is demonstrated in equation (4.13), which shows that the transform of a derivative is an algebraic term. Thus, a differential equation is transformed into an algebraic equation, which can be easily solved using rules of algebra. The challenge is to determine the inverse Laplace transform to achieve an analytical solution in the time domain. In some cases, determining the inverse transform can be complex or impossible; however, methods shown in this section provide a general approach for many systems of interest in process control. First, the solutions of a few simple models involving differential equations, some already formulated in Chapter 3, are presented.

Example 4.1. The continuous stirred-tank mixing model formulated in Example 3.1 (page 60) is solved here. The fundamental model in *deviation variables* is

$$V\frac{dC'_A}{dt} = F(C'_{A0} - C'_A) \tag{4.15}$$

The Laplace transform is taken of each term in the model:

$$V\left[sC'_A(s) - C'_A(t)\big|_{t=0}\right] = F\left[C'_{A0}(s) - C'_A(s)\right] \tag{4.16}$$

The initial value of the tank concentration, expressed as a deviation variable, is zero, and the deviation of the inlet concentration is constant at the step value for $t > 0$; that is, $C'_{A0}(s) = \Delta C_{A0}/s$. Substituting these values and rearranging equation (4.16) gives

$$C'_A(s) = \frac{\Delta C_{A0}}{s} \frac{1}{\tau s + 1} \qquad \text{with } \tau = \frac{V}{F} \tag{4.17}$$

The inverse transform of the expression in equation (4.17) can be determined from entry 5 of Table 4.1 to give the same expression as derived in Example 3.1.

$$C'_A(t) = \Delta C_{A0}(1 - e^{-t/\tau}) \tag{4.18}$$

Example 4.2. The model for the two chemical reactors in Example 3.3 (page 69) is considered here, and the time-domain response to a step change is to be determined. The Laplace transforms of the models in deviation variables, noting that the initial conditions are zero, are

$$sVC'_{A1}(s) = F\left(C'_{A0}(s) - C'_{A1}(s)\right) - VkC'_{A1}(s) \tag{4.19}$$

$$sVC'_{A2}(s) = F\left(C'_{A1}(s) - C'_{A2}(s)\right) - VkC'_{A2}(s) \tag{4.20}$$

These equations can be combined into one equation by eliminating $C'_{A1}(s)$ from the second equation. First, solve for $C'_{A1}(s)$ in equation (4.19):

$$C'_{A1}(s) = \frac{\left(\dfrac{F}{F + Vk}\right)}{\tau s + 1} C'_{A0} \tag{4.21}$$

This expression can be substituted into equation (4.20) along with the input step disturbance, $C'_{A0}(s) = \Delta C_{A0}/s$, to give

$$C'_{A2}(s) = \frac{K_P \Delta C_{A0}}{s(\tau s + 1)^2} \tag{4.22}$$

with $\tau = \dfrac{V}{F + VK}$

$K_P = \left(\dfrac{F}{F + Vk}\right)^2$

The inverse transform can be determined from entry 8 of Table 4.1 to give the resulting time-domain expression for the concentration in the second reactor.

$$C'_{A2} = \Delta C'_{A0} K_p \left[1 - \left(1 + \frac{t}{\tau}\right) e^{-t/\tau}\right] \tag{4.23}$$

This is the same result as obtained in Example 3.3, in slightly different form.

Example 4.3. The dynamic behavior of concentration transported by plug flow in a pipe is modelled in this example. The process is shown in Figure 4.2, where a step in concentration is entered at the inlet, and the outlet concentration is to be determined as a function of time. This system can be modelled using the procedure in Chapter 3.

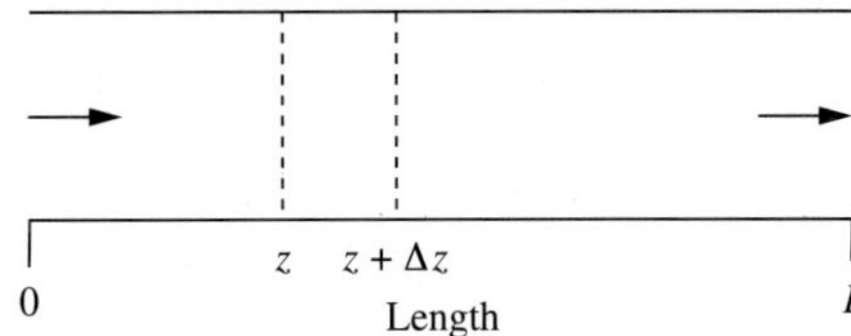

FIGURE 4.2
Schematic of plug flow process.

Information.

1. Plug flow applies; in other words, a flat velocity profile exists.
2. The flow rate is constant and unaffected by the concentration change.
3. The cross-sectional area is constant.

Formulation. Since the concentration changes in the z direction, the system is taken to be the differential volume shown in Figure 4.2. The material balance on C_A for a system of volume $A\Delta z$ is

$$\Delta zA\left(\frac{C_A|_{z,t+\Delta t} + C_A|_{z+\Delta z,t+\Delta t}}{2}\right) - \Delta zA\left(\frac{C_A|_{z,t} + C_A|_{z+\Delta z,t}}{2}\right) =$$
$$\Delta tF\left(\frac{C_A|_{z,t+\Delta t} + C_A|_{z,t}}{2}\right) - \Delta tF\left(\frac{C_A|_{z+\Delta z,t+\Delta t} + C_A|_{z+\Delta z,t}}{2}\right) \tag{4.24}$$

The limits are taken as $\Delta t \to 0$ and $\Delta z \to 0$ to give

$$\frac{\partial C_A(t,z)}{\partial t} + \frac{F}{A}\frac{\partial C_A(t,z)}{\partial z} = 0 \tag{4.25}$$

The model is a linear partial differential equation with constant coefficients, which can be expressed in deviation variables.

$$\frac{\partial C_A'(t,z)}{\partial t} + \frac{F}{A}\frac{\partial C_A'(t,z)}{\partial z} = 0 \tag{4.26}$$

Solution. An analytical solution can be determined by first taking the Laplace transform with respect to the time variable, which converts the partial differential equation to an ordinary differential equation (Jensen and Jeffreys, 1963). Since the system is initially at steady state, the initial condition is zero for all z; that is, $C_A'(t,z)\big|_{t=0} = 0$.

$$sC_A'(s,z) + \frac{F}{A}\frac{dC_A'(s,z)}{dz} = 0 \tag{4.27}$$

The resulting ordinary differential equation can be solved by separating the variables and integrating from $z = 0$ to $z = L$ to give

$$\ln\left(\frac{C_A'(s,L)}{C_A'(s,0)}\right) = -\frac{A}{F}Ls \tag{4.28}$$

This can be rearranged with $C_A(s,L) = C_A(s)$ and $C_A(s,0) = C_{A0}(s)$ to give

$$C_A'(s) = C_{A0}'(s)e^{-\theta s} \qquad \text{with } \theta = \frac{(A)(L)}{F} \tag{4.29}$$

The time-domain solution can be determined by taking the inverse Laplace transform of the foregoing expression, using entry 18 of Table 4.1, to give

$$C_A'(t) = C_{A0}'(t-\theta) \tag{4.30}$$

Results analysis. The result is that the outlet concentration at time t is equal to the inlet concentration at an earlier time, $(t-\theta)$. The difference is the time for a fluid element to flow the length of the pipe, which is (LA/F) and is termed the *transportation delay.* This conforms to our intuitive expectation for this process with plug flow.

Time Translation or Dead Time

The Laplace transform used in the solution of Example 4.3 involves the translation in time. It can be derived as follows:

$$\mathcal{L}[f(t-\theta)] = \int_0^\infty f(t-\theta)e^{-st}dt = e^{-\theta s}\int_0^\infty f(t-\theta)e^{-s(t-\theta)}d(t-\theta)$$

$$= e^{-\theta s}\int_0^\infty f(t')e^{-st'}dt' = e^{-\theta s}f(s) \tag{4.31}$$

When changing variables from $(t-\theta)$ to t', the lower bound of the integral remained at 0 (did not change to $t'-\theta$), because the function is defined $f(t) = 0$ for $t < 0$ for the Laplace transform. The expression in equation (4.31) is used often in process modelling to represent behavior in which the output variable does not respond immediately to a change in the input variable; this condition is often referred to as *dead time*. The effect of time translation is shown in Figure 4.3 for an arbitrary function.

Final Value Theorem

The final condition of the transient can be determined by applying the expression for the derivative of a function, equations (4.11) and (4.12), and taking the limit as $s \to 0$.

$$\lim_{s\to 0}\left[\int_0^\infty \frac{df(t)}{dt}e^{-st}dt\right] = \lim_{s\to 0}[sf(s) - f(t)]_{t=0} \tag{4.32}$$

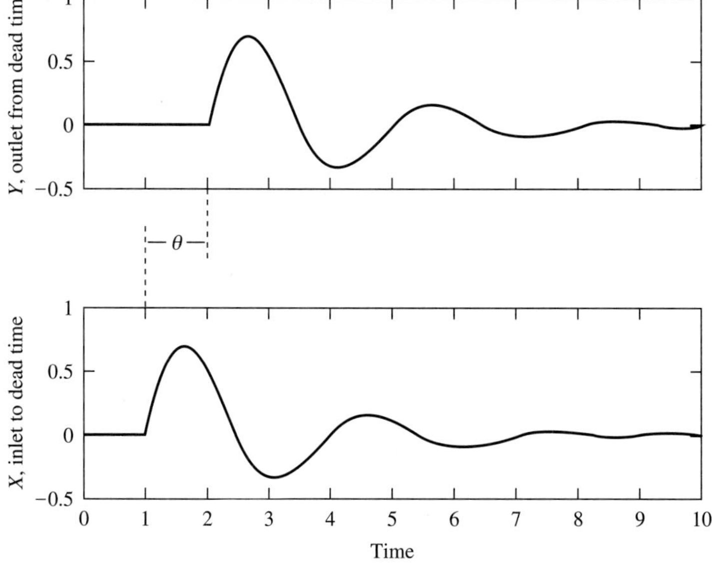

FIGURE 4.3
Input and output for dead time.

Changing the order of the limit and the integral gives

$$\int_0^\infty \frac{df(t)}{dt}dt = \lim_{s\to 0}\left[sf(s) - f(t)\right]\Big|_{t=0}$$

$$f(\infty) - f(t)\Big|_{t=0} = \lim_{s\to 0}\left[sf(s) - f(t)\right]\Big|_{t=0}$$

$$f(\infty) = \lim_{s\to 0} sf(s) \qquad (4.33)$$

Equation (4.33) provides an easy manner for finding the final value of a variable; however, one should recognize that a simpler method would be to formulate and solve the steady-state model directly. The final value theorem finds use because the dynamic models are required for process control, and the final value can be easily determined from the Laplace transform without further modelling effort. Also, it is important to recall that the final value is exact only for a truly linear process and is approximate when based on a linearized model of a nonlinear process.

Example 4.4. Find the final value of the reactor concentration, expressed as a deviation from the initial value, for the CSTR in Example 3.1 (page 60). The Laplace transform for the concentration in response to a step in the inlet concentration is given in equation (4.22). The final value theorem can be applied to give

$$\lim_{s\to 0}\left(s\frac{\Delta C_{A0}}{s}\frac{K}{(\tau s + 1)}\right) = K\Delta C_{A0} = \left(\frac{F}{F + Vk}\Delta C_{A0}\right)$$

Note that this is the final value, which gives no information about the trajectory to the final value.

The engineer must recognize a limitation when applying the final value theorem. The foregoing derivation is not valid for a Laplace transform $f(s)$ that is not continuous for all values of $s \geq 0$ (Churchill, 1972). If the transform has a discontinuity for $s \geq 0$, the time function $f(t)$ does not reach a final steady-state value, as will be demonstrated in the discussion of partial fractions. Therefore, *the final value theorem cannot be applied to unstable systems.*

Example 4.5. Find the final value for the following system.

$$Y(s) = (\Delta X)\frac{K}{s(\tau s - 1)} \qquad \text{with } \tau > 0$$

This transfer function has a discontinuity at $s = 1/\tau > 0$; therefore, the final value theorem does not apply. The analytical expression for $Y(t)$ is

$$Y(t) = (\Delta X)K(1 - e^{t/\tau})$$

The value of $Y(t)$ approaches negative infinity as time increases; this is not equal to the *incorrect* result from applying the final value theorem to the transfer function $-K(\Delta X)$.

Partial Fractions

The Laplace transform method for solving differential equations could be limited by the availability of entries in Table 4.1; with so few entries, it would seem that most models could not be solved. However, many complex Laplace transforms can be expressed as a linear combination of a few simple transforms through the use of partial fraction expansion. Once the Laplace transform can be expressed as a sum of simpler elements, each can be inverted using the entries in Table 4.1, thus greatly increasing the number of equations that can be solved. Also, the application of partial fractions provides very important generalizations about the forms of solutions to a wide range of differential equation models, and these generalizations enable us to establish important characteristics about a system's time-domain behavior *without* determining the complete transient solution.

The partial-fraction expansion can be applied to a Laplace transform that can be expressed as a ratio of polynomials in s. This does not pose a severe limitation, since many models have the following form for a specified input, $X(s)$.

$$D(s)Y(s) = F(s)X(s) = N(s) \tag{4.34}$$

$$Y(s) = \frac{N(s)}{D(s)}$$

with $Y(s)$ = Laplace transform of the output variable
$X(s)$ = Laplace transform of the input variable
$F(s)$ = Laplace transform of the function $F(t)$; $F(s)X(s)$ is the forcing function
$N(s)$ = numerator polynomial in s of order m
$D(s)$ = denominator polynomial in s of order n, termed the *characteristic polynomial*

The partial-fraction method requires that the order of the denominator be greater than the order of the numerator (i.e., $n > m$). As will be discussed later in this section, this requirement will be satisfied by models encountered in process control.

The Laplace transform in equation (4.34) can be expanded into an equivalent expression with simpler individual terms by the application of partial fractions.

Partial Fractions

$$Y(s) = \frac{N(s)}{D(s)} = \frac{C_1}{H_1(s)} + \frac{C_2}{H_2(s)} + \cdots \tag{4.35}$$

$$Y(t) = C_1\mathcal{L}^{-1}\left(\frac{1}{H_1(s)}\right) + C_2\mathcal{L}^{-1}\left(\frac{1}{H_2(s)}\right) + \cdots$$

The C_i are constants, and the $H_i(s)$ are low-order terms in s that represent the factors of the characteristic polynomial, $D(s) = 0$.

Initially, the C_i's are unknowns in equation (4.35) and must be determined so that the equation is satisfied. There are several ways to determine the constants, and the partial fraction expansions and the resulting Heaviside expansion formula are presented here for three types of factors of the characteristic polynomial: distinct, repeated, and complex, each with an example.

DISTINCT FACTORS. If the characteristic polynomial has a distinct root at α, the ratio of polynomials can be factored into

$$Y(s) = \frac{N(s)}{D(s)} = \frac{M(s)}{s - \alpha} = \frac{C}{s - \alpha} + H(s) \tag{4.36}$$

with $H(s)$ being the remainder. After multiplying equation (4.36) by $(s - \alpha)$ and setting $s = \alpha$ (resulting in the term $(s - \alpha)H(s)$ being zero), the constant can be determined to be $M(\alpha) = C$. This approach is performed individually for each distinct root, and the function of time, $Y(t)$, is the sum of the inverse Laplace transforms of all individual factors. The procedure is demonstrated in the following example.

Example 4.6. The model for the two chemical reactors in Example 3.3 (page 69) is considered in this example, but with the exception that the volumes of the two reactors are not the same volume, $V_1 = 1.4$ and $V_2 = 0.70$ m^3. (This alteration ensures that the roots of the characteristic polynomial are distinct.) Determine the time-domain response. The Laplace transforms of the models, in deviation variables, are

$$sV_1C'_{A1}(s) = F[C'_{A0}(s) - C'_{A1}(s)] - V_1kC'_{A1}(s) \tag{4.37}$$

$$sV_2C'_{A2}(s) = F[C'_{A1}(s) - C'_{A2}(s)] - V_2kC'_{A2}(s) \tag{4.38}$$

These equations can be combined into one equation by solving for $C'_{A1}(s)$ in equation (4.37) and substituting this into equation (4.38). Also, the input step disturbance can be substituted, $C'_{A0}(s) = \Delta C_{A0}/s$, to give

$$C'_{A2}(s) = \frac{K_pC'_{A0}(s)}{(\tau_1 s + 1)(\tau_2 s + 1)} = \frac{\frac{1}{\tau_1}\frac{1}{\tau_2}K_p\Delta C_{A0}}{s\left(s + \frac{1}{\tau_1}\right)\left(s + \frac{1}{\tau_2}\right)} \tag{4.39}$$

with
$$\tau_1 = \frac{V_1}{F + V_1k}$$
$$\tau_2 = \frac{V_2}{F + V_2k}$$
$$K_p = \left(\frac{F}{F + V_1k}\right)\left(\frac{F}{F + V_2k}\right)$$

The expression for C_{A2} can be expanded for the root at $s = -1/\tau_1$ as

$$C'_{A2}(s) = \frac{M_1(s)}{s + \frac{1}{\tau_1}} = \frac{C_1}{s + \frac{1}{\tau_1}} + H_1(s) \tag{4.40}$$

with

$$C_1 = M_1(s)\big|_{s=-1/\tau_1} = \frac{\left(\frac{1}{\tau_1}\right)\left(\frac{1}{\tau_2}\right)K_p\Delta C_{A0}}{-\frac{1}{\tau_1}\left(-\frac{1}{\tau_1} + \frac{1}{\tau_2}\right)} = K_p\Delta C_{A0}\left(\frac{\tau_1}{\tau_2 - \tau_1}\right) \tag{4.41}$$

Similar manipulations for the other two factors ($s = -1/\tau_2$) and $s = 0$) give the following results:

$$C_2 = K_p\Delta C_{A0}\left(\frac{\tau_2}{\tau_1 - \tau_2}\right) \qquad C_3 = K_p\,\Delta C_{A0} \tag{4.42}$$

Substituting the constants into equation (4.35) and taking the inverse transform for each term gives

$$
\begin{aligned}
C'_{A2}(t) &= K_p \Delta C_{A0}\left[\mathcal{L}^{-1}\left(\frac{1}{s}\right)+\left(\frac{\tau_1}{\tau_2-\tau_1}\right)\mathcal{L}^{-1}\left(\frac{1}{s+\dfrac{1}{\tau_1}}\right)-\left(\frac{\tau_2}{\tau_2-\tau_1}\right)\mathcal{L}^{-1}\left(\frac{1}{s+\dfrac{1}{\tau_2}}\right)\right]\\
&= K_p \Delta C_{A0}\left(1+\frac{\tau_1}{\tau_2-\tau_1}e^{-t/\tau_1}-\frac{\tau_2}{\tau_2-\tau_1}e^{-t/\tau_2}\right)
\end{aligned}
$$

This result is the same as could have been determined directly by the use of entry 10 in Table 4.1.

The expansion for distinct factors can be summarized in the following Heaviside expansion, which is a generalization of the technique just explained (Churchill, 1972).

$$
\mathcal{L}^{-1}\left(\frac{N(s)}{D(s)}\right)_{\left(\substack{n\ \text{distinct}\\ \text{factors}}\right)} = \sum_{i=1}^{n}\frac{N(s)\big|_{s=\alpha_i}}{\left.\dfrac{dD(s)}{ds}\right|_{s=\alpha_i}}e^{\alpha_i t} \tag{4.43}
$$

REPEATED FACTORS. A similar partial factor expansion can be applied for $n+1$ repeated factors (i.e., identical, real roots of the characteristic polynomial $D(s)$) as follows.

$$
Y(s) = \frac{N(s)}{D(s)} = \frac{M(s)}{(s-\alpha)^{n+1}} = \frac{C_1}{(s-\alpha)}+\frac{C_2}{(s-\alpha)^2}+\cdots+\frac{C_{n+1}}{(s-\alpha)^{n+1}}+H(s) \tag{4.44}
$$

The coefficients can be determined sequentially by

1. Multiplying equation (4.44) by $(s-\alpha)^{n+1}$ and setting $s = \alpha$ (determining C_{n+1})
2. Multiplying equation (4.44) by $(s-\alpha)^{n+1}$, taking the first derivative with respect to s, and setting $s = \alpha$ (determining C_n)
3. Continuing this procedure (with higher derivatives) until all coefficients have been evaluated

The time-domain function for repeated factors can be expressed as (Churchill, 1972)

$$
\mathcal{L}^{-1}\left(\frac{N(s)}{D(s)}\right)_{\substack{\text{repeated}\\ \text{factor}}} = \frac{1}{n!}\left[\frac{\partial^n}{\partial s^n}\left(M(s)e^{st}\right)\right]_{s=\alpha} \tag{4.45}
$$

where $M(s)$ is defined in equation (4.44). This technique is demonstrated in the following example.

Example 4.7. Determine the time-domain response of the two series CSTRs in Example 3.3 (page 69) to a step in inlet concentration. The Laplace transform of this system is similar to that in Example 4.6, but the time constants have identical values, as in the original Example 3.3.

$$C'_{A2}(s) = \frac{K_p \Delta C_{A0}}{s(\tau s + 1)^2} = \frac{1}{\tau^2} \frac{K_p \Delta C_{A0}}{s\left(s + \frac{1}{\tau}\right)^2} \tag{4.46}$$

This can be expanded into three factors, two of which are repeated:

$$C'_{A2} = \frac{C_1}{\left(s + \frac{1}{\tau}\right)^2} + \frac{C_2}{s} \tag{4.47}$$

Equation (4.45) can be applied to determine the inverse for the repeated roots.

$$M(s) = \frac{1}{\tau^2} \frac{K_p C_{A0}}{s} \tag{4.48}$$

$$\frac{1}{1!}\left[\frac{\partial M(s)}{\partial s} e^{st} + M(s) \frac{\partial \left(e^{st}\right)}{\partial s}\right]_{s=-1/\tau} = \left(-K_p C_{A0} - \frac{K_p C_{A0}}{\tau} t\right) e^{-t/\tau}$$

The final constant is evaluated as before for the distinct factor, $C_3 = K_p \Delta C_{A0}$. The time-domain solution is found to be the same result as could be obtained from entry 8 in Table 4.1.

$$C'_{A2}(t) = K_p \Delta C_{A0} \left[1 - \left(1 + \frac{t}{\tau}\right) e^{-t/\tau}\right] \tag{4.49}$$

It might be of interest to determine whether the step response could exhibit periodic behavior for some values of the time constant, τ. The values of time at which the concentration in the second reactor has a maximum or minimum can be determined by setting the derivative of equation (4.49) with respect to time equal to zero.

$$\frac{dC'_{A2}}{dt} = K_p \Delta C_{A0} \left(0 + \frac{1}{\tau} e^{-t/\tau} - \frac{1}{\tau} e^{-t/\tau} + \frac{t}{\tau^2} e^{-t/\tau}\right) = 0 \tag{4.50}$$

This equation is satisfied only at time = 0, and as time approaches infinity; therefore, the concentration cannot be periodic for any real value of the time constant. Thus, the structure of this process model precludes the possibility of oscillations in response to a step input for all physically possible values of F, V, k, and C_{A0}. This conclusion does not apply to all second-order systems, as is demonstrated in the next example.

COMPLEX FACTORS. The final possibility for the factors involves complex factors, and the analysis for a distinct, complex factor is given for the following system:

$$Y(s) = \frac{N(s)}{D(s)} = \frac{M(s)}{(s - \alpha)^2 + \omega^2} + H(s) \qquad \text{with } \alpha \text{ and } \omega \text{ real} \tag{4.51}$$

The complex roots can be expressed as two distinct roots, $\alpha \pm \omega j$, so that by applying equation (4.35) the Laplace transform and its inverse can be expressed as

$$\frac{N(s)}{D(s)} = \frac{M_1(s)}{s - \alpha + \omega j} + \frac{M_2(s)}{s - \alpha - \omega j} + H(s) \tag{4.52}$$

$$Y(t)_{\substack{\text{complex}\\ \text{factor}}} = [M_1(s)]_{s=\alpha-\omega j} e^{(\alpha - \omega j)t} + [M_2(s)]_{s=\alpha+\omega j} e^{(\alpha + \omega j)t} \tag{4.53}$$

The coefficients in equation (4.53) are complex conjugates and can be expressed as $M_1(+\alpha - \omega j) = (A + Bj)$ and $M_2(+\alpha + \omega j) = (A - Bj)$, respectively. These expressions can be substituted into equation (4.53) to give

$$
\begin{aligned}
Y(t)_{\substack{\text{complex}\\ \text{factor}}} &= (A + Bj)e^{(\alpha-\omega j)t} + (A - Bj)e^{(\alpha+\omega j)t} \\
&= e^{\alpha t}[A(e^{j\omega t} + e^{-j\omega t}) + jB(e^{-j\omega t} - e^{j\omega t})]
\end{aligned} \tag{4.54}
$$

Equation (4.54) can be modified to eliminate the complex terms by using the Euler relationships.

$$
\cos(\omega t) = \frac{e^{j\omega t} + e^{-j\omega t}}{2} \rightarrow 2\cos(\omega t) = e^{j\omega t} + e^{-j\omega t} \tag{4.55}
$$

$$
\sin(\omega t) = \frac{e^{j\omega t} - e^{-j\omega t}}{2j} \rightarrow 2\sin(\omega t) = j(e^{-j\omega t} - e^{j\omega t}) \tag{4.56}
$$

The resulting expression can be used to evaluate the inverse term for a complex conjugate pair of roots of the characteristic polynomial.

$$
Y(t)_{\substack{\text{complex}\\ \text{factor}}} = 2e^{\alpha t}\,[A\cos(\omega t) + B\sin(\omega t)] \tag{4.57}
$$

The proof of an alternative formulation, along with expressions for repeated complex factors, is available in Churchill (1972).

Example 4.8. A solution for the approximate linear model of the nonisothermal CSTR modelled in Example 3.10 (page 90) is developed in this exercise. The original model involved two nonlinear differential equations, which were linearized and expressed in deviation variables. In this example, the only input variable that changes is the coolant flow, which changes in a step; that is, $F_c'(s) = \Delta F_c/s$ and $C_{A0}'(s) = T_0'(s) = F'(s) = 0$. The Laplace transforms of equations (3.77) and (3.78) can be taken to give

$$
sC_A'(s) = a_{11}C_A'(s) + a_{12}T'(s) + a_{14}F_c'(s) \tag{4.58}
$$

$$
sT'(s) = a_{21}C_A'(s) + a_{22}T'(s) + a_{24}F_c'(s) \tag{4.59}
$$

Equations (4.58) and (4.59) can be combined algebraically. First, equation (4.58) is used to solve for $C_A'(s) = a_{12}T'(s)/(s - a_{11})$, since $a_{14} = 0$; this term is then substituted into equation (4.59) to give

$$
T'(s) = \frac{a_{24}s + (a_{21}a_{14} - a_{24}a_{11})}{s^2 - (a_{11} + a_{22})s + (a_{11}a_{22} - a_{12}a_{21})}F_c'(s) \tag{4.60}
$$

When the numerical values are substituted into equation (4.60), the result is

$$
F_c'(s) = \frac{-1}{s}
$$

$$
\begin{aligned}
a_{11} &= -7.55 & a_{12} &= -0.0931 & a_{14} &= 0.0 \\
a_{21} &= 852.02 & a_{22} &= 5.77 & a_{24} &= -6.07
\end{aligned}
$$

$$
T'(s) = \frac{(-1)(-6.07s - 45.83)}{s(s^2 + 1.79s + 35.80)} \tag{4.61}
$$

The method of partial fractions requires the roots of the characteristic polynomial, which are $-0.894 \pm 5.92j$ and 0.0; thus, two factors are complex. The inverse transform for the complex factors can be determined by using equation (4.57) with $\alpha = -0.894$ and $\omega = 5.94$.

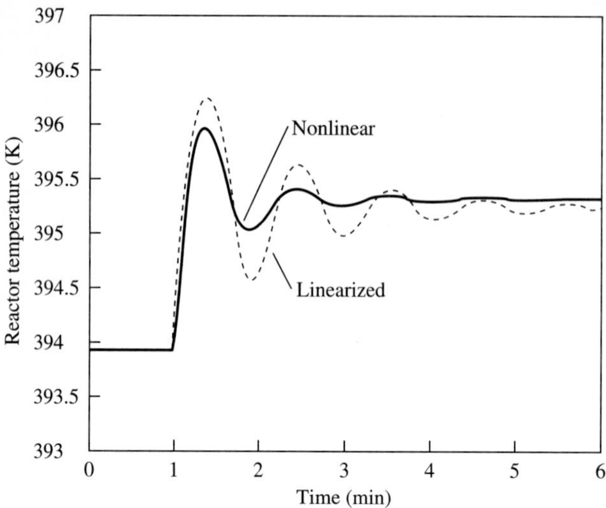

FIGURE 4.4
Linearized and nonlinear dynamic responses for Example 4.8.

$$M_1(s)\big|_{s=\alpha-\omega j} = \left(\frac{(-1)(-6.07s - 45.83)}{(s + 0.894 - 5.92j)(s)}\right)_{s=-0.894-5.92j} = -0.64 + 0.42j \quad (4.62)$$

$$T'(t)_{\substack{\text{complex}\\ \text{factor}}} = 2e^{-0.894t}[-0.64\cos(5.92t) + 0.42\sin(5.92t)] \quad (4.63)$$

The single distinct factor can be evaluated as described previously.

$$M(s)\big|_{s=0} = \left(\frac{(-1)(-6.07s - 45.83)}{s^2 + 1.789s + 35.80}\right)_{s=0} = 1.28 \quad (4.64)$$

$$T'(t)_{\text{distinct}} = 1.28e^{0t} = 1.28$$

The complete inverse transform is the sum of the two functions:

$$T'(t) = 1.28 + 2e^{-0.894t}[-0.64\cos(5.92t) + 0.42\sin(5.92t)] \quad (4.65)$$

The solution to the linear approximation is a damped oscillation. This behavior was observed in the numerical solution to the nonlinear equations in Example 3.10. A comparison of the solutions to the linearized and nonlinear equations in Figure 4.4 shows how the linearized model represents the essential characteristics of the true process response. Naturally, the accuracy depends on the size of the input change, which in this case is relatively small.

The analytical expression in equation (4.60) provides a manner for determining which aspects of the process design affect the process dynamic response, specifically the periodic nature of the response. For example, the oscillations are determined by the roots of the characteristic polynomial; thus, parameters that affect these roots influence the oscillations. It is easy to determine that the functional relationship between the coolant flow and the heat transfer coefficient (in a_{24}) does not appear in the denominator; thus, the periodicity does not depend on this function, assuming that the steady-state conditions (C_A and T) are unchanged. Although they give only approximations to the dynamic behavior, analytical expressions like equation (4.60) are invaluable in determining the effects of equipment design, operating conditions, and control system design on process behavior.

In summary, the Laplace transform and partial-fraction technique provide methods for determining the inverse transform of any Laplace transform that can be expressed as a ratio of polynomials with the order of the polynomial in the denominator greater than that in the numerator. In addition, an important characteristic of solutions to ordinary differential equations has been demonstrated. For any differential equation that can be arranged into the form of equation (4.34), the solution will be of the form

$$Y(t) = A_1 e^{a_1 t} + \cdots + \left(B_1 + B_2 t + \cdots\right) e^{a_p t} + \cdots + (C_1 \cos(\omega t) + C_2 \sin(\omega t))\, e^{a_q t} + \cdots \tag{4.66}$$

This equation includes distinct, repeated real, and complex roots, not all of which may appear in a specific solution. Note that the real parts of the roots of the characteristic polynomial $D(s)$ determine the exponents (α's) in the solution. These exponents determine whether the function approaches a constant value after a long time. For example, when all real parts of the roots (i.e., all $\text{Re}(\alpha_i)$) are negative, all terms on the right of equation (4.66) approach a constant value after an initial transient; this will be termed *stable* and be discussed in great detail later. If any $\text{Re}(\alpha_i)$ is greater than zero, the function $Y(t)$ will increase (or decrease) in an unbounded manner as time increases; this is termed *unstable*. Also, the nature of the roots of the characteristic polynomial determines whether the dynamic response will experience periodic behavior for nonperiodic inputs; complex roots of $D(s)$ lead to a periodic response.

Generally, a process will be easier to operate when all variables rapidly approach constant values and no variables tend to increase or decrease without limit (based on a linearized model). Also, while oscillations are not usually to be completely avoided, oscillations of large magnitude are generally undesirable. Thus, the nature of the roots of the characteristic polynomial, and how process design and control algorithms affect these roots, will be investigated thoroughly in Part III on feedback control. The material in this section provides the mathematical foundation for these important analyses.

4.3 INPUT-OUTPUT MODELS AND TRANSFER FUNCTIONS

In some cases the values for all dependent process variables need to be determined to meet modelling goals, and the fundamental models used to this point in the book, which provide expressions for all variables, can be used in these cases. For example, the model for two series CSTRs in Example 3.3 yields expressions for the concentrations in both reactors, and the model for the nonisothermal CSTR in Example 3.10 yields expressions for the temperature and concentration in the reactor. These models are not unduly complex; however, detailed models can involve a large number of equations. For example, a distillation tower with 40 trays and ten components would require over 400 differential equations.

A detailed fundamental model is often not required for process control, because the control system is principally involved with all input variables but only one or a few output variables. Thus, we need a method for "compressing" the model, which can be achieved by first grouping variables into three categories: input (causes), output (effects), and intermediate. Models that represent the input-output relationships must account for all facets of the process that affect this behavior. For *linear* dynamic models used in process control, it is possible to eliminate intermediate variables analytically to yield an input-output model, so that intermediate variables are considered in the model even though they are not explicitly calculated. Thus, no further assumptions or simplifications are involved in input-output modelling of linear systems.

Examples of this approach have already been encountered in this chapter. For example, the basic model for two series CSTRs in Example 4.6 included equations for the concentrations in both reactors, equations (4.37) and (4.38). After the Laplace transforms are taken and the equations are combined into one equation, the model in equation (4.39) involves only the input, $C'_{A0}(s)$, and the output, $C'_{A2}(s)$. The intermediate variable, $C'_{A1}(s)$, was eliminated, although all effects of the first reactor are represented in the model.

A very common manner for presenting input-output models, which finds considerable application in process control, is the transfer function. The *transfer function* is a model based on Laplace transforms with special assumptions, as follows.

The **transfer function** of a system is defined as the Laplace transform of the output variable, $Y(t)$, divided by the Laplace transform of the input variable, $X(t)$, with all initial conditions equal to zero.

$$\text{Transfer function} = G(s) = \frac{Y(s)}{X(s)} \tag{4.67}$$

The assumptions of $Y(0) = 0$ and $X(0) = 0$ are easily achieved by expressing the variables in the transfer function as deviations from the initial conditions. Thus, all transfer functions involve variables that are expressed as deviations from an initial steady state. All derivatives are zero if the initial conditions are at steady state. (Systems having all zero initial conditions are sometimes referred to as "relaxed"). These zero initial conditions are assumed for all systems represented by transfer functions used in this book; therefore, the prime symbol "$'$" for deviation variables is redundant and is not used here when dealing with transfer functions. Transfer functions will be represented by $G(s)$, with subscripts to denote the particular input-output relationship when more than one input-output relationship exists. Before proceeding with further discussion of transfer functions, a few examples are given.

Example 4.9. Derive the transfer functions for the systems in Examples 4.1 and 4.2. The Laplace transform of the model in Example 4.1 is in equation (4.16). This can be rearranged to give the transfer function for this system:

$$\frac{C_A(s)}{C_{A0}(s)} = \frac{1}{\tau s + 1} \quad \text{with} \quad \tau = \frac{V}{F} \tag{4.68}$$

The Laplace transform for the model in Example 4.2 is in equation (4.22). This can be rearranged to give the transfer function for this system:

$$\frac{C_{A2}(s)}{C_{A0}(s)} = \frac{K_p}{(\tau s + 1)^2} \tag{4.69}$$

with $\tau = \frac{V}{F+Vk}$

$K_p = \left(\frac{F}{F+Vk}\right)^2$

These equations could be used to form transfer functions, because they were in terms of deviation variables with zero initial conditions.

Note that the transfer function relates *one* output to *one* input variable. If more than one input or output exists, an individual transfer function is defined for each input-output relationship. Since the transfer function is a linear operator (as a result of the zero initial conditions), the effects of several inputs can be summed to determine the net effect on the output.

Example 4.10. Derive the transfer functions for the nonisothermal CSTR in Examples 3.10 (page 90) and 4.8 for changes in two input variables, the inlet concentration and the coolant flow, and one output variable, temperature in the tank.

The Laplace transforms of the two differential equations can be combined to express the temperature as a function of the two inputs. Again, equation (4.58) is used to eliminate $C_A(s)$ from equation (4.59) to give

$$T(s) = G_{TC}(s)C_{A0}(s) + G_{TF_c}F_c(s) \tag{4.70}$$

$$\text{with} \quad G_{TC}(s) = \frac{a_{23}s + (a_{21}a_{13} - a_{23}a_{11})}{s^2 - (a_{11} + a_{22})s + (a_{11}a_{22} - a_{12}a_{21})} \tag{4.71}$$

$$G_{TF_c}(s) = \frac{a_{24}s + (a_{21}a_{14} - a_{24}a_{11})}{s^2 - (a_{11} + a_{22})s + (a_{11}a_{22} - a_{12}a_{21})} \tag{4.72}$$

and the expressions for the a_{ij} given in Example 3.10. Note that equation (4.70) includes all effects in the two differential equations (i.e., the material and energy balances), although the reactor concentration $C_A(s)$ does not appear explicitly.

The transfer function clearly shows some important properties of the system briefly discussed below.

Order

The *order* of the system is the highest derivative of the output variable in the defining differential equation, when expressed as a combination of all individual equations. For transfer functions of physical systems, the order can be easily determined to be the highest power of s in the denominator.

Pole

A *pole* is defined as a root of the denominator of the transfer function; thus, it is the same as a root of the characteristic polynomial. Important information on the dynamic behavior of the system can be obtained by analyzing the poles, such as

1. The stability of the system
2. The potential for periodic transients, as shown clearly in equation (4.66)

The analysis of poles is an important topic in Part III on feedback systems, since feedback control affects the poles.

Zero

A *zero* is a root of the numerator of the transfer function. Zeros do not influence the exponents (Re(α)), but they influence the constants in equation (4.66). This can most easily be seen by considering a system with n distinct poles subject to an impulse input of unity. The expression for the output, since the Laplace transform of the unity input impulse is 1, is

$$Y(s) = G(s)X(s) = G(s) = \frac{N(s)}{D(s)} = \frac{M_i(s)}{s - \alpha_i} \qquad \text{for } i = 1, n \tag{4.73}$$

For a system with no zeros, the numerator would be equal to a constant, $N(s) = K$, and the constant associated with each root is

$$\text{with no zeros} \qquad C_i = \left(\frac{K}{D_i(s)}\right)_{s=-\alpha_i} \tag{4.74}$$

$D_i(s)$ is the denominator, with $(s - \alpha_i)$ factored out. For a system with one or more zeros, the constant associated with each root is

$$\text{with zeros} \qquad C_i' = \left(K\frac{N(s)}{D_i(s)}\right)_{s=-\alpha_i} \tag{4.75}$$

Thus, the numerator changes the weight placed on the various exponential terms. This demonstrates that the numerator of the transfer function cannot affect the stability of the system modelled by the transfer function, but it can have a strong influence on the trajectory followed by variables from their initial to final values. A simple, but less general, example to demonstrate the effect of numerator zeros is seen in the following transfer function.

$$G(s) = \frac{3s + 1}{(3s + 1)(2.5s + 1)} = \frac{1}{2.5s + 1} \tag{4.76}$$

The numerator zero cancelled one of the poles, with the result that the second-order system behaves like a first-order system. Important examples of how zeros occur in chemical processes and how they influence dynamic behavior are presented in the next chapter, Section 5.4.

Order of Numerator and Denominator

Physical systems conform to a specific limitation between the orders of the numerator and denominator; that is, the order of the denominator must be larger than the order of the numerator. This limitation results from the observation that real physical systems do not contain pure differentiation, as would be required for a system with a numerator order greater than the denominator order.

Causality

As discussed in Chapter 1 in the introduction of feedback control, the "direction" of the cause-effect relationship is essential to control system design. This direction is presented in the transfer function by identifying the variable in the denominator as an input (cause) and in the numerator as an output (effect). In designing feedback control strategies, the variable chosen to be adjusted must be an input, and the measured controlled variable used for determining the adjustment must be an output. When the physical system is causal, the order of the denominator is greater than that of the numerator, and the value of the transfer function as $s \to \infty$ is equal to 0. Such a transfer function is referred to as *strictly proper.*

Also, the current value of a system output variable can depend on past values of the output and inputs, but it cannot depend on future values of any variable. Therefore, the transfer function must not have prediction terms. By equation (4.31), the transfer function may not contain a term $e^{\theta s}$, which is a translation into the future (that cannot be eliminated by rearranging the transfer function). Such models are referred to as *noncausal or not physically realizable,* because they cannot represent a real physical system.

Steady-State Gain

The *steady-state gain* is the steady-state value of $\Delta Y/\Delta X$ for all systems whose outputs attain steady state after an input perturbation ΔX. The steady-state gain is normally represented by K, often with a subscript, and can be evaluated by setting $s = 0$ in the (stable) transfer function. This is exact for linear systems and gives the linearized approximation for nonlinear systems.

Example 4.11. Analyze the transfer function, which relates temperature to the coolant flow for the nonisothermal chemical reactor in Example 4.8, equation (4.60). Determine the order, the poles, the zeros, and steady-state gain.

The order is the highest power of s in the denominator, 2. This indicates that the system can be represented by two first-order ordinary differential equations. The poles are the roots of the denominator, which can be factored as follows.

$$s^2 + 1.788s + 35.80 = (s + 0.894 + 5.92j)(s + 0.894 - 5.92j) \tag{4.77}$$

Thus, the poles are $-0.894 \pm 5.92j$, and because they are complex, the system experiences periodic behavior. The zero, obtained by setting the numerator to zero, is -7.55.

The transfer function must be checked to ensure that it conforms to the limitations of a physical system. The order of the denominator, 2, is greater than the order of the numerator, 1. Also, the transfer function has no exponentials in s with a positive coefficient. Therefore, it is causal.

Since all poles have negative real parts, indicating that the system reaches a steady state after a step input, the system has a steady-state gain, which is the value of the transfer function with s set equal to zero, $-1.28°C/[m^3/hr]$.

In a specific situation the behavior of an output variable, from time 0 to completion of the response, depends on its initial conditions, input forcing, and transfer function (input-output) model. However, some very important properties of linear dynamic systems depend only on the transfer function, because the properties are independent of initial conditions and type of (bounded) forcing functions. For example, the stability of the system was shown in the previous section to be determined completely by the roots of the characteristic polynomial. The primary application of transfer functions is in the analysis of such properties of linear dynamic systems, and they are applied extensively throughout the remainder of the book.

4.4 BLOCK DIAGRAMS

The transfer function introduced in the previous section describes the behavior of the individual input-output system on which it is based. Often, several different individual systems are combined, and the behavior of the combined system is to be determined. For example, a control system could involve individual systems for a reactor, a distillation tower, a sensor, a valve, and a control algorithm. The overall model could be derived by writing all equations in a large set, taking the Laplace transforms, and combining into one transfer function. Another approach retains the distinct transfer functions of the individual systems and combines these transfer functions into an overall model. This second approach is usually preferred because

1. It retains individual systems, thereby simplifying model changes (e.g., a different sensor model).
2. It provides a helpful visual representation of the cause-effect relationships in the overall system.
3. It gives insight into how different components of the system influence the overall behavior (e.g., stability).

The block diagram provides the method for combining individual transfer functions into an overall transfer function. The three allowable manipulations in a block diagram are shown in Figure 4.5*a* through *c*. The first is the transform of an input variable to an output variable using the transfer function; this is just a schematic representation of the relationship introduced in equation (4.67) and discussed in the previous section. The second is the sum (or difference) of two variables; the third is splitting a variable for use in more than one relationship. These three manipulations can be used in any sequence for combining individual models. A more comprehensive set of rules based on these three can be developed (Distephano et al., 1976), but these three are usually adequate.

To clarify, a few illegal manipulations, which are sometimes mistakenly used, are shown in Figure 4.5*d* through *f*. The first two are not allowed because the transfer function is defined for a single input and output, and the third is not allowed because the block diagram is limited to linear operations.

The block diagram can be prepared based on linearized models (transfer functions) of individual units and the knowledge of their interconnections. Then an input-output model can be derived through the application of block diagram algebra, which uses the three operations in Figure 4.5*a* through *c*. The model reduction steps normally followed are

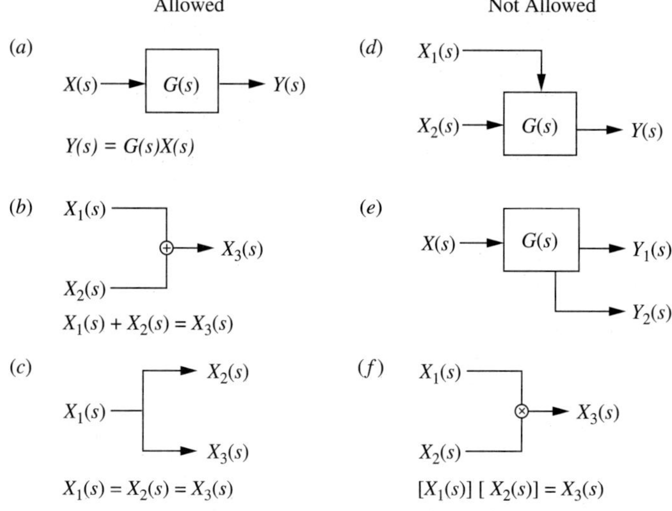

FIGURE 4.5
Summary of block diagram algebra: (*a–c*) allowed; (*d–f*) not allowed.

1. Define the input and output variables desired for the overall transfer function.
2. Express the output variable as a function of all variables directly affecting it in the block diagram. This amounts to working in the direction opposite to the cause-effect relationships (arrows) in the diagram.
3. Eliminate intermediate variables by this procedure until only the output and one or more inputs appear in the equation. This is the input-output equation for the system.
4. If a transfer function is desired, set all but one input to zero in the equation from step 3 and solve for the output divided by the single remaining input. This step may be repeated to form a transfer function for each input.

The following examples demonstrate the principles of block diagrams, and many additional applications will be presented in later chapters.

Example 4.12. Draw the block diagram for the two chemical reactors in Example 4.2 (the physical system in Example 3.3, page 69), and combine them into one overall block diagram and transfer function for the input C_{A0} and the output C_{A2}. The individual transfer functions are given below and shown in Figure 4.6*a*.

$$G_1(s) = \frac{C_{A1}(s)}{C_{A0}(s)} = \frac{K_1}{\tau s + 1} \quad \text{with } \tau = \frac{V}{F + Vk} \quad K_1 = \frac{F}{F + Vk} \tag{4.78}$$

$$G_2(s) = \frac{C_{A2}(s)}{C_{A1}(s)} = \frac{K_2}{\tau s + 1} \quad \text{with } \tau = \frac{V}{F + Vk} \quad K_2 = \frac{F}{F + Vk} \tag{4.79}$$

Block diagram manipulations can be performed to develop the overall input-output relationship for the system.

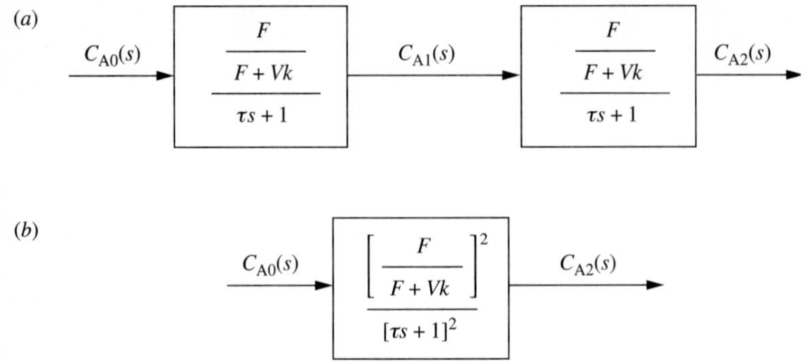

(a)

$C_{A0}(s)$ → $\dfrac{\frac{F}{F+Vk}}{\tau s+1}$ → $C_{A1}(s)$ → $\dfrac{\frac{F}{F+Vk}}{\tau s+1}$ → $C_{A2}(s)$

(b)

$C_{A0}(s)$ → $\dfrac{\left[\frac{F}{F+Vk}\right]^2}{[\tau s+1]^2}$ → $C_{A2}(s)$

FIGURE 4.6
Block diagrams for Example 4.12.

$$C_{A2} = G_2(s)\,C_{A1}(s) = G_2(s)\,[G_1(s)\,C_{A0}(s)] = G_2(s)\,G_1(s)C_{A0}(s) \tag{4.80}$$
$$= \frac{K_1K_2}{(\tau s+1)^2}C_{A0}(s)$$

This can be rearranged to give the transfer function and the block diagram in Figure 4.6*b*.

$$\frac{C_{A2}(s)}{C_{A0}(s)} = G(s) = \frac{K_1K_2}{(\tau s+1)^2} \tag{4.81}$$

Example 4.13. Derive the overall transfer functions for the systems in Figure 4.7. The system in part *a* is a series of transfer functions, for which the overall transfer function is the product of the individual transfer functions.

$$X_n(s) = G_n(s)\,X_{n-1}(s) = G_n(s)\,G_{n-1}(s)\,X_{n-2}(s) \tag{4.82}$$
$$= G_n(s)\,G_{n-1}(s)\,G_{n-2}(s)\cdots G_1(s)\,X_0(s)$$
$$\frac{X_n(s)}{X_0(s)} = \prod_{i=1}^{n} G_i(s)$$

The system in part *b* involves a parallel structure of transfer functions, and the overall transfer function can be derived as

$$X_3(s) = X_1(s) + X_2(s) = G_1(s)\,X_0(s) + G_2(s)\,X_0(s) \tag{4.83}$$
$$\frac{X_3(s)}{X_0(s)} = G_1(s) + G_2(s)$$

The system in part *c* involves a recycle structure of transfer functions, and the overall transfer function can be derived as

$$X_2(s) = G_1(s)\,X_1(s) = G_1(s)\,[X_0(s) + X_3(s)] = G_1(s)\,[X_0(s) + G_2(s)X_2(s)] \tag{4.84}$$
$$\frac{X_2(s)}{X_0(s)} = \frac{G_1(s)}{1 - G_1(s)G_2(s)}$$

Examples of processes that can be represented by these structures, along with the effects of the structures on dynamic behavior, will be presented in the next chapter.

It is perhaps worth noting that the block diagram is entirely equivalent to and provides no fundamental advantage over algebraic solution of the system's linear,

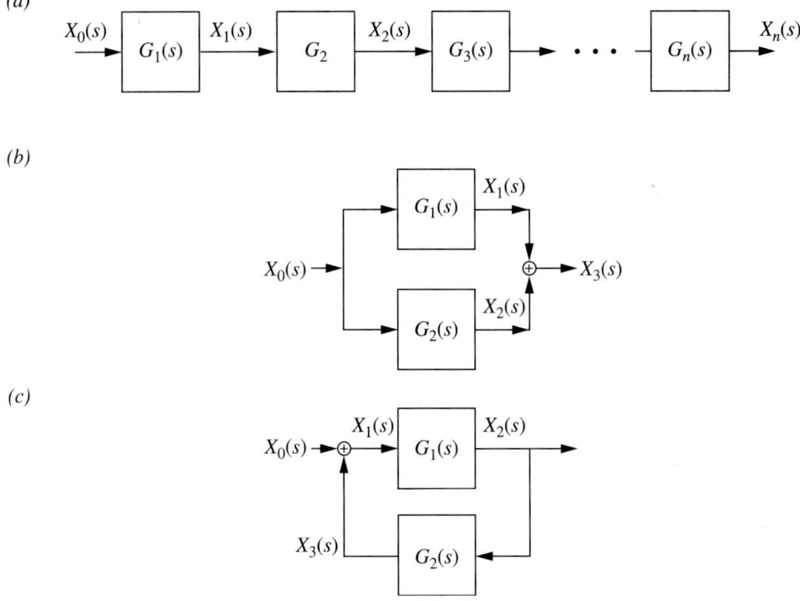

FIGURE 4.7
Three common block diagram structures considered in Example 4.13.

algebraic equations (in the s domain). Either algebraic or block diagram manipulations for eliminating intermediate variables to give the input-output relationship will result in the same overall transfer function. However, as demonstrated by the examples, the block diagram manipulations are easily performed.

Two further features of block diagrams militate for their extensive use. The first is the helpful visual representation of the integrated system provided by the block diagram. For example, the block diagram in Figure 4.7*c* clearly indicates a recycle in the system, a characteristic that might be overlooked when working with a set of equations. The second feature of the diagrams is the clear representation of the cause-effect relationship. The arrows present the direction of these relationships and enable the engineer to identify the input variables that influence the output variables. As a result, block diagrams are widely used and will be applied extensively in the remainder of this book.

4.5 FREQUENCY RESPONSE

An important aspect of process (and control system) dynamic behavior is the response to periodic input changes, most often disturbances. The range of possible dynamic behavior can be determined by considering cases (in thought experiments) at different input frequencies for an example system, such as the mixing tank in Figure 4.8. If an input variation is slow, with a period of once per year, the output response would be essentially at its steady-state value (the same as the input), with the transient response being insignificant. If the input changed very rapidly, say every nanosecond, the output would not be significantly influenced; that is, its output

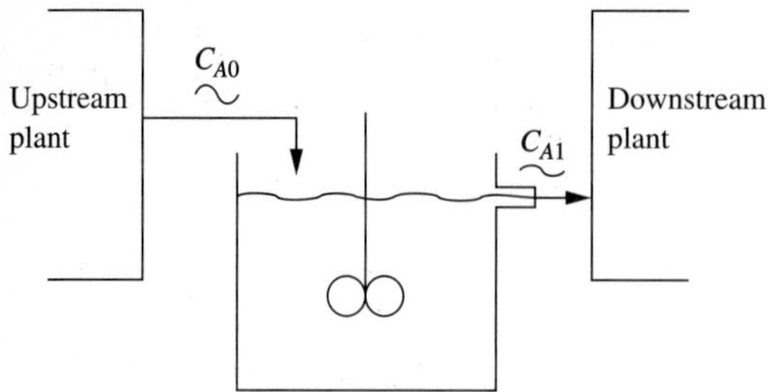

FIGURE 4.8
Intermediate inventory to attenuate variation.

amplitude would be insignificant. Finally, if the input varies at some *intermediate frequency* near the response time of the process, the output will fluctuate continuously at values significantly different from its mean value. The behavior at extreme frequencies is easily determined in this thought experiment, but the method for determining the system behavior at intermediate frequencies is not obvious and is useful for the design process equipment, selection of operating conditions, and formulation of control algorithms to give desired performance. Before presenting a simplified method for evaluating the effects of frequency, a process equipment design example is solved by determining the complete transient response to a periodic input.

Example 4.14. The feed composition to a reactor varies with an amplitude larger than acceptable for the reactor. It is not possible to alter the upstream process to reduce the oscillation in the feed; therefore, a drum is located before the reactor to reduce the feed composition variation, as shown in Figure 4.8. What is the minimum volume of the tank required to maintain the variation at the inlet to the reactor (outlet of the tank) less than or equal to ± 20 g/m^3?

Assumptions. The same assumptions apply as in Example 3.1 (page 60), with the addition that the input variation in concentration is well represented by a sine. Also, the system is initially at steady state.

Data.

1. $F = 1$ m^3/min.
2. C_{A0} is a sine with amplitude of 200 g/m^3 and period of 5 minutes about an average value of 200 g/m^3.

Solution. The model for this stirred-tank mixer was derived in Example 3.1 and applied in several subsequent examples. The difference in this example is that the input concentration is characterized as a sine rather than a step, $C'_{A0} = A \sin(\omega t)$. Thus, the model for the tank is

$$V \frac{dC'_{A1}}{dt} = F(A \sin(\omega t)) - FC'_{A1} \tag{4.85}$$

This equation could be solved by using either the integrating factor or Laplace transforms. Here, the Laplace transform of equation (4.85) is taken to give, after some rearrangement,

$$C'_{A1}(s) = \frac{A\omega}{\tau} \frac{1}{\left(s + \frac{1}{\tau}\right)(s^2 + \omega^2)} \qquad \text{with } \tau = \frac{V}{F} \tag{4.86}$$

This can be solved using a partial fraction expansion of equation (4.86) for the distinct factor $s = -1/\tau$ and complex factors $s = \pm\omega j$. The coefficient for the real, distinct root is

$$C_1 = \left(\frac{A\omega}{\tau}\frac{1}{s^2+\omega^2}\right)_{s=-1/\tau} = \frac{A\omega\tau}{1+\tau^2\omega^2} \tag{4.87}$$

The coefficients for the factors $s = \pm\omega j$ are

$$C_2 = \frac{A\omega}{\tau}\left(\frac{1}{s+\frac{1}{\tau}}\frac{1}{s+\omega j}\right)_{s=+\omega j} = \frac{A}{2j}\frac{1}{\sqrt{1+\omega^2\tau^2}}e^{\phi j} \tag{4.88}$$

$$C_3 = \frac{A\omega}{\tau}\left(\frac{1}{s+\frac{1}{\tau}}\frac{1}{s-\omega j}\right)_{s=-\omega j} = \frac{-A}{2j}\frac{1}{\sqrt{1+\omega^2\tau^2}}e^{-\phi j} \tag{4.89}$$

with $\phi = \tan^{-1}(-\omega\tau)$ and the final expressions derived by changing to polar form. The inverse Laplace transform for the three factors are summed to give the solution to the differential equation.

$$C'_{A1}(t) = \frac{A\omega\tau}{1+\tau^2\omega^2}e^{-t/\tau} + \frac{A}{\sqrt{1+\tau^2\omega^2}}\left(\frac{-e^{-(\omega t+\phi)j}+e^{+(\omega t+\phi)j}}{2j}\right) \tag{4.90}$$

The second term on the right-hand side of the equation can be combined using the Euler identity to give

$$C'_{A1}(t) = \frac{A\omega\tau}{1+\tau^2\omega^2}e^{-t/\tau} + \frac{A}{\sqrt{1+\tau^2\omega^2}}\sin(\omega t+\phi) \tag{4.91}$$

Results analysis. The first term in equation (4.91) tends to zero as time increases; thus, the response of the process after a long time of operation (about four time constants) is not affected by this term. The second term describes the "long-time" behavior of the concentration in response to a sine input. It is periodic, with the *same frequency* as the input forcing and an amplitude that depends on the input amplitude and frequency, as well as process design parameters. For this example, the output amplitude must be less than or equal to 20; by setting the amplitude equal to the limit, the time constant, and thus the volume, can be calculated.

$$|C_{A1}|_{max} = \frac{A}{\sqrt{1+\tau^2\omega^2}} = 20 \tag{4.92}$$

$$V = \tau F = F\frac{\sqrt{\left(\frac{A}{|C_{A1}|_{max}}\right)^2 - 1}}{\omega} = 1.0\frac{\sqrt{\left(\frac{200}{20}\right)^2 - 1}}{2\pi/5} = 7.9 \text{ m}^3 \tag{4.93}$$

Note that the analytical solution provides valuable sensitivity information, such as the amount the size of the vessel must be increased if the input frequency decreases.

For general frequency response analysis, periodic inputs will be limited to sine inputs, which will be a mathematically manageable problem. Also, only the "long-time" response (i.e., after the initial transient, when the output is periodic) is

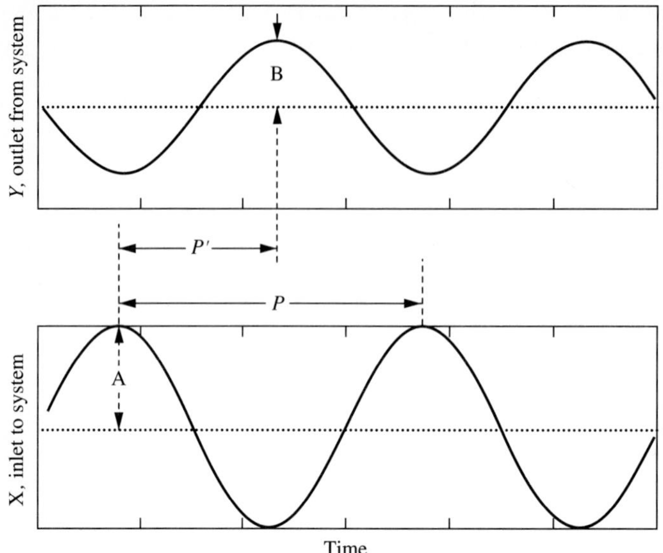

FIGURE 4.9
Frequency response for a linear system.

considered. The periodic behavior after a long time is sometimes referred to as "steady-state"; however, it seems best to restrict the term *steady-state* to describe systems with zero time derivatives.

The periodic behavior of the input and output after a long time—the frequency response—is shown in Figure 4.9, and frequency response is defined as follows:

> The **frequency response** defines the output behavior of a system to a sine input after a long enough time that the output is periodic. The output (Y') of a *linear* system will be a sine with the same frequency as the input (X'), and the relationship between input and output can be characterized by
>
> $$\textbf{Amplitude ratio} = \frac{\text{Output magnitude}}{\text{Input magnitude}} = \frac{|Y'(t)|_{\text{max}}}{|X'(t)|_{\text{max}}} \tag{4.94}$$
>
> **Phase angle** = Phase difference between the input and output

For the system in Figure 4.9 the amplitude ratio = B/A, and the phase angle = $-2\pi(P'/P)$ radians. Note that P' is the time difference between the input and its effect at the output and can be greater than P.

The usefulness of the amplitude ratio was demonstrated in Example 4.14, and the importance of the phase angle, while not apparent yet, will be shown to be very important in the analysis of feedback systems. Recalling that feedback systems adjust an input based on the behavior of an output, it is reasonable that the time (or phase) delay between these variables would affect the feedback system. The analysis of feedback systems using frequency response methods is introduced in Chapter 10 and used in many subsequent chapters.

Example 4.14 demonstrates that the frequency response of linear systems can be determined by the direct solution of the ordinary differential equations. However, this approach is time-consuming for complex systems. Also, the solution of the entire transient response provides information not needed, because only the behavior after the initial transient is desired. Now a simpler approach for determining frequency response is presented; it is based on the transfer function of the system.

The output of a general linear system with a transfer function $G(s)$ and sine input forcing can be expressed as

$$Y(s) = G(s)X(s) = G(s)A\frac{\omega}{s^2 + \omega^2} \tag{4.95}$$

with the sine magnitude A and frequency ω. The transfer function is assumed to be expressed as a ratio of polynomials, so that the solution can be analyzed using a partial-fraction expansion of the right-hand side of equation (4.95). In Section 4.2, the partial fraction method was explained and demonstrated on examples, all of which involved input forcing that attained a constant value after some time. In those cases the output was shown to approach a constant value if the roots of the characteristic polynomial (or the poles of the transfer function) had negative real parts. The difference in equation (4.95) is that the sine forcing introduces additional poles that have zero real parts.

The general form of the solution to equation (4.95) can be based on equation (4.66) with the effects of the additional poles from the sine input included.

$$\begin{aligned} Y(t) &= A_1 e^{\alpha_1 t} + \cdots + \left(B_1 + B_2 t + B_3 t^2 + \cdots\right) e^{\alpha_p t} \\ &\quad + \left(C_1 \cos(\omega t) + C_2 \sin(\omega t)\right) e^{\alpha_q t} + \cdots + D_1 e^{-j\omega t} + D_2 e^{j\omega t} \end{aligned} \tag{4.96}$$

As previously discussed, all but the last two terms tend toward zero as time increases, as long as $\text{Re}(\alpha_i) < 0$ for all i. Thus, only the last two terms in equation (4.96) affect the output behavior after a long time—in other words, the frequency response. The constants for the last two terms can be evaluated using the partial-fraction method for distinct roots, $\alpha = -j\omega$ and $\alpha = +j\omega$.

$$D_1 = \left(G(s)\frac{A\omega}{s - j\omega}\right)_{s=-j\omega} = G(s)\big|_{s=-j\omega}\left(\frac{A}{-2j}\right) = -A\frac{G(-j\omega)}{2j} \tag{4.97}$$

$$D_2 = \left(G(s)\frac{A\omega}{s + j\omega}\right)_{s=j\omega} = G(s)\big|_{s=j\omega}\left(\frac{A}{+2j}\right) = A\frac{G(j\omega)}{2j} \tag{4.98}$$

Since only these terms affect the long-time behavior, the output can be expressed (with the subscript FR for the frequency response) as

$$Y_{\text{FR}}(t) = -\frac{A}{2j}G(-j\omega)e^{-j\omega t} + \frac{A}{2j}G(j\omega)e^{j\omega t} \tag{4.99}$$

The transfer function, which involves complex numbers, can be expressed in polar form using

$$G(j\omega) = \left|G(j\omega)\right| e^{j\phi} \qquad \text{with } \phi = \tan^{-1}\left(\frac{\text{Im}[G(j\omega)]}{\text{Re}[G(j\omega)]}\right) \tag{4.100}$$

Substitution into equation (4.99) gives

$$Y_{FR}(t) = -\frac{A}{2j}\left|G(j\omega)\right| e^{-(\omega t+\phi)j} + \frac{A}{2j}\left|G(j\omega)\right| e^{(\omega t+\phi)j} \tag{4.101}$$

This result, along with Euler's identity to convert the exponential expressions to a sine, gives the final expression for the frequency response of a general linear system.

$$Y_{FR}(t) = A\left|G(j\omega)\right| \sin(\omega t + \phi) \tag{4.102}$$

The two key parameters of the frequency response can be determined from

$$\text{Amplitude ratio} = \text{AR} = \frac{\left|Y_{FR}\right|_{max}}{A} = \frac{A\left|G(j\omega)\right|}{A} = \left|G(j\omega)\right| \quad (4.103)$$

$$= \sqrt{\text{Re}[G(j\omega)]^2 + \text{Im}[G(j\omega)]^2} \quad (4.103)$$

$$\text{Phase angle} = \phi = \angle G(j\omega) = \tan^{-1}\left(\frac{\text{Im}[G(j\omega)]}{\text{Re}[G(j\omega)]}\right) \tag{4.104}$$

It is important to recognize that the frequency (ω) must be expressed as radians/time.

Thus, the frequency response can be determined by substituting $j\omega$ for s in the transfer function and evaluating the magnitude and angle of the resulting complex number! This is *significantly simpler* than solving the differential equation.

Note that the frequency response is entirely determined by the transfer function. This is logical because the initial conditions do not influence the long-time behavior of the system. Also, the derivation of the equations (4.103) and (4.104) clearly indicate that they are appropriate only for *stable systems.* If the system were unstable (i.e., if $\text{Re}(\alpha_i) > 0$ for any i), the output would increase without limit (for the linear approximation). Also, this analysis demonstrates that the output of a linear system forced with a sine approaches a sine after a sufficiently long time. How "long" this time is depends on all other terms in equation (4.96). For most of the transient to have died out (i.e., $e^{-\alpha t} < 0.02$), the time should satisfy $\alpha t = t/\tau > 4$. Thus, a long time can usually be taken to be about four times the longest time constant, or the smallest α.

Example 4.15. Repeat the frequency response calculations for Example 4.14, this time using the direct method based on the transfer function. The frequency response is determined by substituting $j\omega$ for s in the first-order transfer function with $\tau = 7.9$.

$$G(s) = \frac{1}{\tau s + 1}$$

$$G(j\omega) = \frac{1}{\tau\omega j + 1} = \frac{1}{\tau\omega j + 1}\frac{1 - \tau\omega j}{1 - \tau\omega j} = \frac{1 - \tau\omega j}{1 + \tau^2\omega^2}$$

$$\text{AR} = \left|G(j\omega)\right| = \frac{\sqrt{1 + \tau^2\omega^2}}{1 + \tau^2\omega^2} = \frac{1}{\sqrt{1 + \tau^2\omega^2}} = \frac{1}{\sqrt{1 + 62.4\omega^2}} \tag{4.105}$$

$$\phi = \angle G(j\omega) = \tan^{-1}(-\omega\tau) = \tan^{-1}(-7.9\omega) \tag{4.106}$$

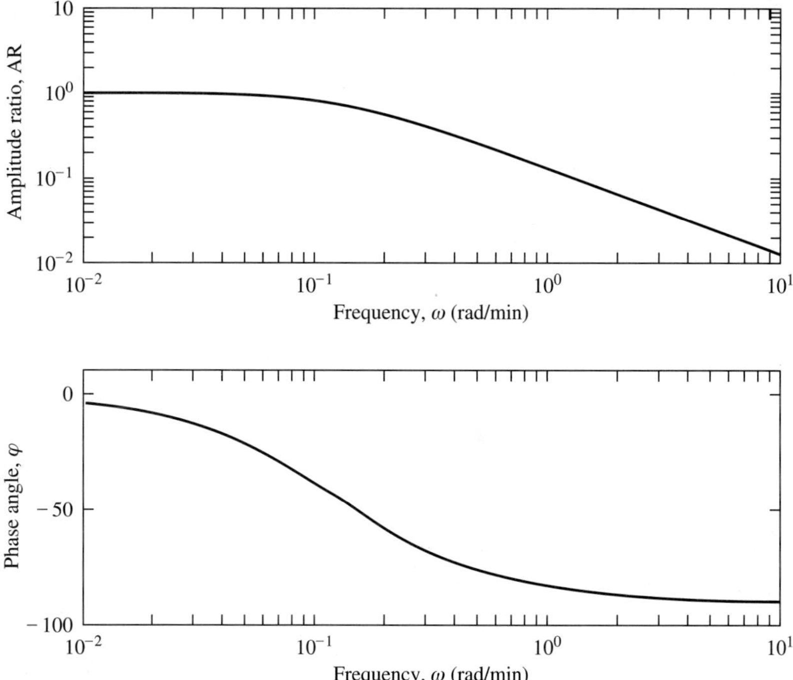

FIGURE 4.10
Frequency response for Example 4.15, $C_A(j\omega)/C_{A0}(j\omega)$, presented as a Bode plot.

A frequency response is often presented in the form of a *Bode plot,* in which the log of the amplitude and the phase angle are plotted against the log of the frequency. An example of the Bode plot for the system in Example 4.15 is given in Figure 4.10. From this result, it can be determined that the amplitude ratio is nearly 1.0 for all frequencies below about 0.10 rad/min for this example, and it decreases rapidly as frequency increases from this value. Also, the amplitude ratio at a frequency of $2\pi/5 = 1.26$ rad/min is the desired value of 0.10. Finally, this graph clearly indicates the sensitivity of the result to potential errors in time constant and frequency; for example, the output amplitude is insensitive to frequency at low frequencies and quite sensitive at high frequencies.

Example 4.16. The frequency response of the series CSTR process analyzed in Examples 4.2 and 4.9 is considered. In this example, the base inlet concentration is 0.925 mole/m^3, and the inlet to the first reactor varies as a sine with an amplitude of 0.925 mole/m^3 (the magnitude of the step in the previous example). Again, the key question is whether the concentration at the outlet of the second reactor will exceed the maximum limit of 0.85 mole/m^3 and if so, for which frequencies.

Solution. The frequency response can be calculated directly from the transfer function derived in Example 4.9.

$$\frac{C_{A2}(s)}{C_{A0}(s)} = G(s) = \frac{K_p}{(\tau s + 1)^2} \tag{4.107}$$

$$G(j\omega) = \frac{K_p}{(j\omega\tau + 1)^2} = \frac{K_p}{(1 - \tau^2\omega^2) + 2j\omega\tau}$$
$$= K_p \frac{(1 - \tau^2\omega^2) - 2\tau j\omega}{\left(1 + \tau^2\omega^2\right)^2}$$

$$\left|G(j\omega)\right| = \frac{K_p}{1 + \tau^2\omega^2} \qquad \phi = \angle G(j\omega) = \tan^{-1}\left(\frac{-2\tau\omega}{1 - \tau^2\omega^2}\right)$$

The output amplitude is the product of the input amplitude A and the amplitude ratio:

$$\left|C'_{A2}(j\omega)\right| = \frac{AK_p}{1 + \tau^2\omega^2} = \frac{(0.925)(0.448)}{1 + (8.25)^2\omega^2} = \frac{0.414}{1 + 68.1\omega^2} \tag{4.108}$$

The maximum value of the concentration in the second reactor is its steady-state value plus the amplitude of the deviation from steady state (i.e., the sine about the steady state).

$$C_{A2}(t) = (C_{A2})_s + C'_{A2}(t) = 0.414 + \left(\frac{0.414}{1 + \tau^2\omega^2}\right)\sin(\omega t + \phi) \tag{4.109}$$

From the form of the frequency response, the amplitude of the deviation variable is always less than or equal to 0.414, because $\sin(\omega t) \leq 1.0$. Therefore, the outlet concentration cannot exceed the maximum limit of 0.85 for any frequency given this data, although a small change in F, V, k, or input amplitude A could lead to a violation. Note that for this system the amplitude ratio is never greater than 1.0, so that the output amplitude is never more than the steady-state magnitude from a step input of the same size. This bound on the amplitude ratio does not hold for all systems, as the next example demonstrates.

Example 4.17. The frequency response of the temperature of the nonisothermal CSTR in Example 4.10 is to be determined for sine input forcing of the coolant flow rate. The frequency response can be determined directly from the transfer function:

$$G(s) = \frac{T(s)}{F_c(s)} = \frac{-6.07s - 45.83}{s^2 + 1.789s + 35.80} \tag{4.110}$$

$$G(j\omega) = \frac{(34.94\omega^2 - 1640.7) + (6.07\omega^3 - 135.3\omega)j}{(-\omega^2 + 35.8)^2 + (1.79\omega)^2}$$

The amplitude ratio and phase angle are plotted in a Bode plot in Figure 4.11. The amplitude ratio at zero frequency is equal to the steady-state gain and has essentially that value at very low frequencies. For the system in this example, the amplitude ratio increases above the steady-state gain for a range of frequencies, about 2 to 7 rad/min. Over this range of frequencies, the amplitude of the sine output (from its steady state) is greater than the output deviation resulting from a step input of the same input magnitude. This effect is attributed to resonance between the material and energy balances of the system.

The algebraic manipulations required to evaluate the amplitude ratio and phase angle can be tedious. However, relationships to ease hand calculations are provided in Chapter 10 for the commonly occurring series combinations of individual units. For more complex structures the frequency response can be easily evaluated using computer technology, because the amplitude ratio is the magnitude of the properly defined function of a complex variable; likewise, the phase angle is the argument of

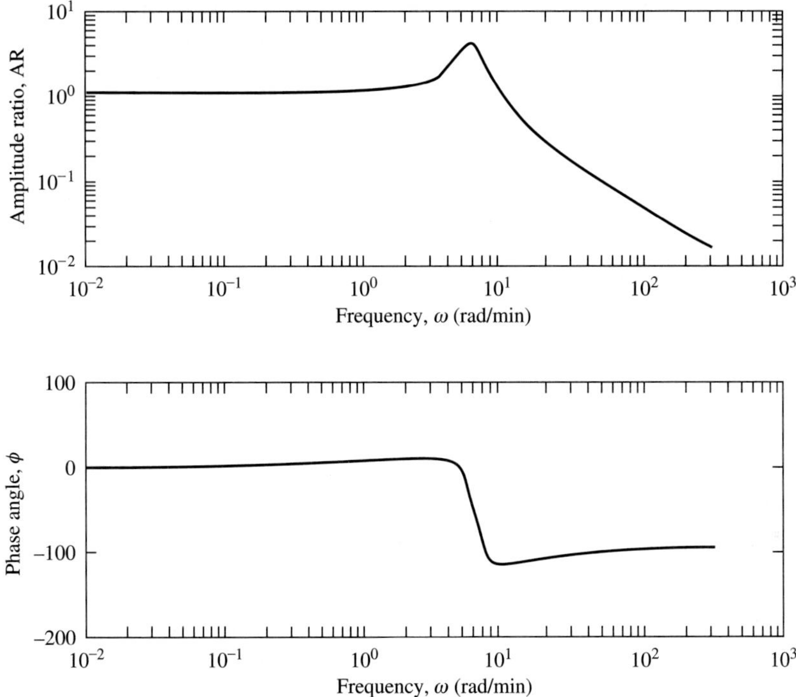

FIGURE 4.11
Frequency response for Exercise 4.17, $T(j\omega)/F_c(j\omega)$, presented as a Bode plot.

a complex variable. Many programming languages provide standard evaluations of these functions.

In conclusion, the frequency response of a linear system can be easily determined from the transfer function using equations (4.103) and (4.104). The frequency response gives useful information concerning how the process behaves for various input frequencies, and these results can be used for determining equipment design parameters, such as the size of a drum to attenuate fluctuations. The general frequency responses for some common systems are given in the next chapter for several common systems, such as first- and second-order, and important applications of frequency response to the analysis of feedback control systems are covered in Part III.

4.6 CONCLUSIONS

The methods in Chapters 3 and 4 can be combined in an approach, shown in Figure 4.12, designed to provide models in the format most useful for the analysis of process control systems. The initial steps involve the modelling procedure based on fundamental principles summarized in Table 3.1 (page 54). This procedure can be applied to each process in a complex plant. Then the transfer function of each system is determined by taking the Laplace transform of the linearized model. The block

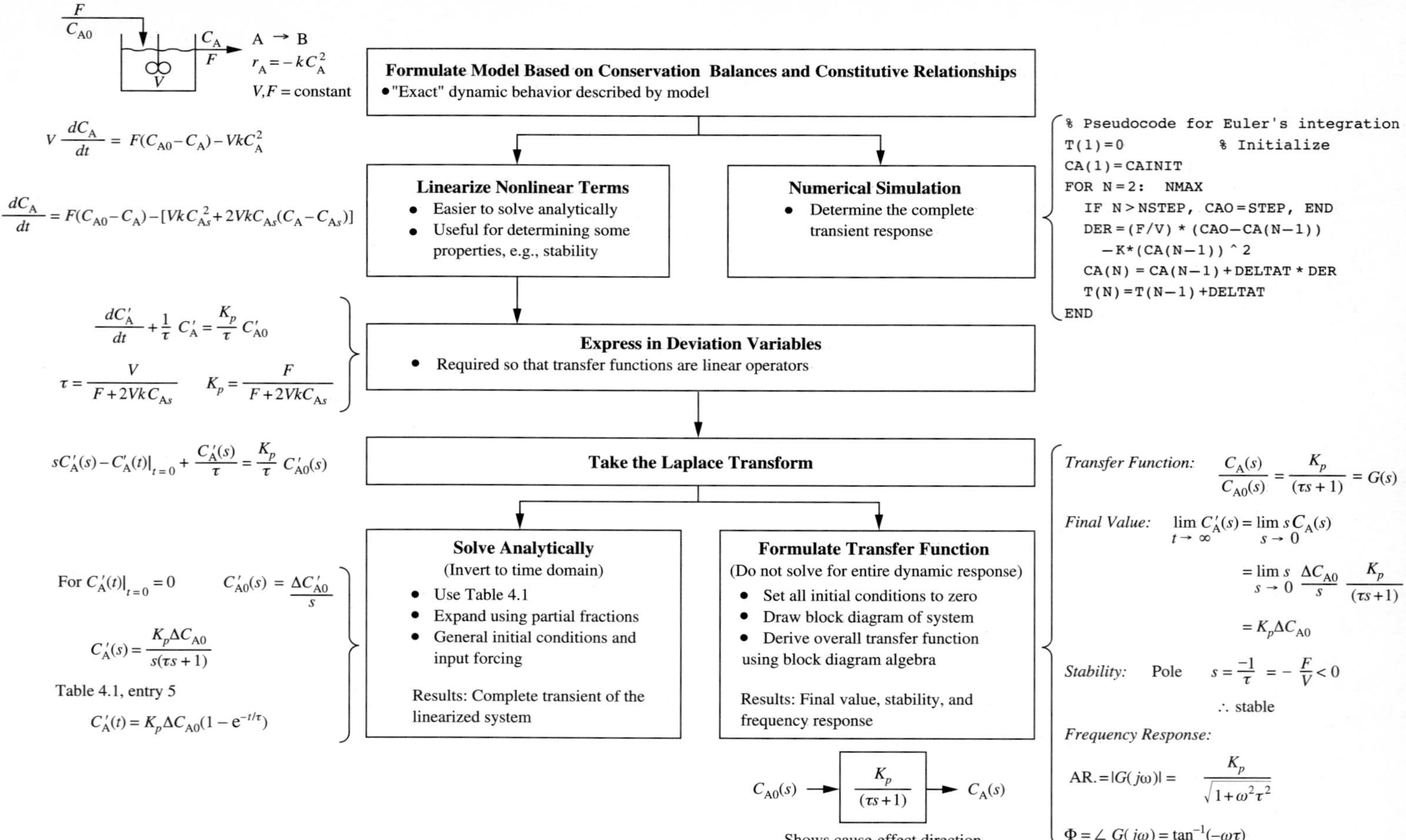

FIGURE 4.12
Steps in developing models for process control with results for Example 3.5 (page 77).

diagram can be constructed to present the interactions among the individual transfer functions, and the overall transfer function for the integrated system can be derived through block diagram manipulation.

The overall transfer functions can be used to determine some important properties of the system *without solving the defining differential equations*. These properties include

1. The final value of the output variable
2. The stability of the response
3. The response of the output to a sine input

Determining this information without the entire dynamic response has two advantages:

1. It reduces the effort to establish these system properties.
2. It assists in understanding the ways in which equipment design, operating conditions, and control systems affect these properties.

Naturally, information about the entire transient is not obtained by analyzing the poles of the transfer function or by the frequency response calculations. The complete transient response can be obtained if needed from analytical or numerical solution of the algebraic and differential equations.

As noted in the previous chapter, many different processes—heat exchangers, reactors, and so forth—behave in similar ways. The transfer function method presented in this chapter gives us a useful way to compare models for processes and recognize similarities and differences, which is the topic of the next chapter.

REFERENCES

Boyce, W. and R. Diprima, *Elementary Differential Equations,* Wiley, New York, 1986.

Caldwell, W., G. Coon, and L. Zoss, *Frequency Response for Process Control,* McGraw-Hill, New York, 1959.

Churchill, R., *Operational Mathematics,* McGraw-Hill, New York, 1972.

Distephano, S., A. Stubbard, and I. Williams, *Feedback Control Systems,* McGraw-Hill, New York, 1976.

Jensen, V., and G. Jeffreys, *Mathematical Methods in Chemical Engineering,* Academic Press, London, 1963.

Ogata, K., *Modern Control Engineering,* Prentice-Hall, Englewood Cliffs, 1990.

ADDITIONAL RESOURCES

In addition to Churchill (1972), the following references provide background on Laplace transforms and provide extensive tables.

Doetsch, G., *Introduction to the Theory and Application of Laplace Transforms*, Springer Verlag, New York, 1974.

Nixon, F., *Handbook of Laplace Transforms* (2nd Ed.), Prentice-Hall, Englewood Cliffs, NJ, 1965.

Spiegel, M., *Theory and Problems of Laplace Transforms,* McGraw-Hill, New York, 1965.

Frequency responses can be determined experimentally, although at the cost of considerable disturbance to the process. This was done to ensure the concepts applied to chemical processes, as discussed in the references below, but the practice has been discontinued.

Harriott, P., *Process Control,* McGraw-Hill, New York, 1964.
Oldenburger, R. (ed.), *Frequency Response,* Macmillan, New York, 1956.

For additional discussions on the solution of dynamic problems for other types of physical systems, see Ogata, 1990 (in the References) and

Ogata, K., *System Dynamics* (2nd Ed.), Wiley, New York, 1992.
Tyner, M. and May, F., *Process Control Engineering,* The Ronald Press, New York, 1968.

All of the questions in Chapter 3 relating to dynamics can be solved using methods in this chapter; thus, returning to those questions provides additional exercises. Also, when solving the questions in this chapter, it is recommended that the results be analyzed to determine

- The order of the system
- Whether the system can experience periodicity and/or instability
- The block diagram with arrows properly representing the causal relationships
- The final value

QUESTIONS

4.1. Several of the example systems considered in this chapter are analyzed concerning the violation of safety limits. A potential strategy for a safety system would be to monitor the value of the critical variable and when the variable approaches the safety limit (i.e., it exceeds a preset "action" value), a response is implemented to ensure safe operation. Three responses are proposed in this question to prevent the critical variable from exceeding a maximum-value safety limit, and it is proposed that each could be initiated when the measured variable reaches the action value. Critically evaluate each of the proposals, and if the proposal is appropriate, state the value of the action limit compared to the safety limit.

The proposed responses are

(i) Set the concentration in the feed (C_{A0}) to *zero.*

(ii) Set the inlet flow to zero.

(iii) Introduce an inhibitor that stops the chemical reaction.

The critical variables and reaction systems are

(*a*) C_A in the single chemical reactor in Example 4.1

(*b*) C_{A2} in the series of two chemical reactors in Example 4.2

(*c*) T in the nonisothermal chemical reactor in Example 4.8

4.2. Solve the following models for the time-domain values of the dependent variables using Laplace transforms.

(*a*) Example 3.2

(*b*) Example 3.2 with an impulse input and with a ramp input, $C'_{A0}(t) = at$ for $t > 0$ (with a an arbitrary constant)

(*c*) Example 3.3 with an impulse input

4.3 The room heating Example 3.4 is to be reconsidered. In this question, a mass of material is present in the room and exchanges heat with the air according to the equation $Q = UA_m(T - T_m)$, in which UA_m is an overall heat transfer coefficient between the mass and the room air, and T_m is the uniform temperature of the mass.

(*a*) Derive models for the temperatures of the air in the room and the mass. Combine them into one differential equation describing T.

(*b*) Explain how this system would behave with an on-off control and note differences, if any, with the result in Example 3.4.

4.4. Derive the dynamic response of a plug flow chemical reactor with a first-order reaction, $r_A = -kC_A$, from an initial steady state to a step change in inlet concentration.

4.5. A slightly modified version of the nonisothermal CSTR described and modelled in Example 3.10 is to be considered in this question. The system is the same except for the heat of reaction, ΔH_{rxn}, which is 0.0. You may use all of the results in the example, specifically equations (3.75) to (3.79), without deriving, and simply modify the results as appropriate. You do not have to substitute the numerical values to answer this question.

(*a*) The coolant flow experiences a single step change of magnitude ΔF_c. Derive a model that describes the response of the concentration of component A, $C'_A(t)$. The result should be in terms of the parameters of the process and can be expressed in terms of the a_{ij} coefficients in Example 3.10.

(*b*) Determine whether the response in part *a* of this question is stable or under what conditions it can be unstable.

(*c*) Describe the shape of the response to the step input for the case in which the system is stable. Under what conditions can it be periodic (or oscillatory) like the response in Figure 3.17?

4.6. For the following systems, (*a*) Apply the final value theorem and (*b*) calculate the frequency response.

(i) Example 3.2

(ii) Example 3.3

(iii) A level system with $L(s)/F_{in}(s) = -1/As$, with $F_{in}(s) = \Delta F_{in}/s$ and A = cross-sectional area [see equation (5.15)].

For each case, state whether the result is correct, and if not, why.

4.7. For the nonisothermal CSTR in Example 3.10,

(*a*) Determine the transfer functions relating $C_A(s)/F_c(s)$ and $C_A(s)/C_{A0}(s)$. These should be in terms of the a_{ij} coefficients in the linearized model. Compare the results with the numerators and denominators in equations (4.71) and (4.72) and comment.

(*b*) Determine $C'_A(t)$ for a step change in C_{A0} of $+0.05$.

(*c*) Determine the transfer function $T(s)/T_0(s)$.

4.8. Consider a modified version of the system in Example 4.14 with two tanks in series, each tank volume being one-half the original single-tank volume.

(*a*) Determine the transfer function relating the inlet and outlet concentrations.

(*b*) Calculate the amplitude ratio of the inlet and outlet concentration for the frequency response using equation (4.103).

(*c*) Determine whether either of the two designs is better (i.e., always provides the smaller amplitude ratio), for all frequencies. Explain your answer and discuss how this analysis would be used in equipment sizing.

4.9. The response of the two levels in Figure Q4.9 are to be determined. The system is initially at steady state, and a step change is made in F_0. Assume that F_0 is independent of the levels, that the flows F_1 and F_2 are proportional to the pressure differences between the ends of the pipes, and that P' is constant. Solve for the dynamic response of both levels.

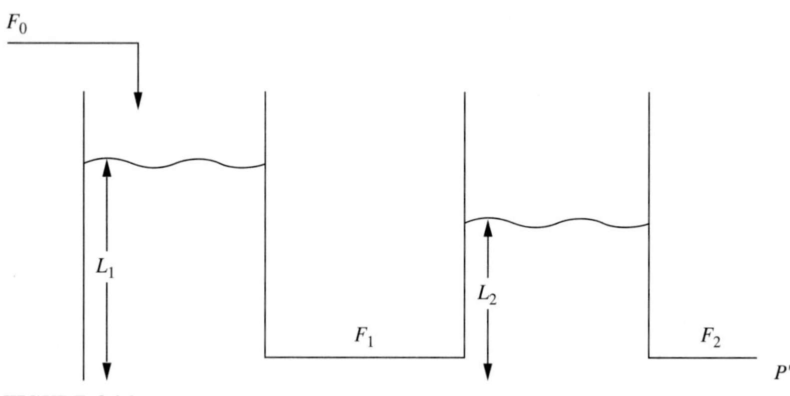

FIGURE Q4.9

4.10. For each of the block diagrams in Figure Q4.10, derive the overall input-output transfer function $X_1(s)/X_0(s)$. (Note that they are two of the most commonly occurring and important block diagrams used in feedback control.)

(*a*)

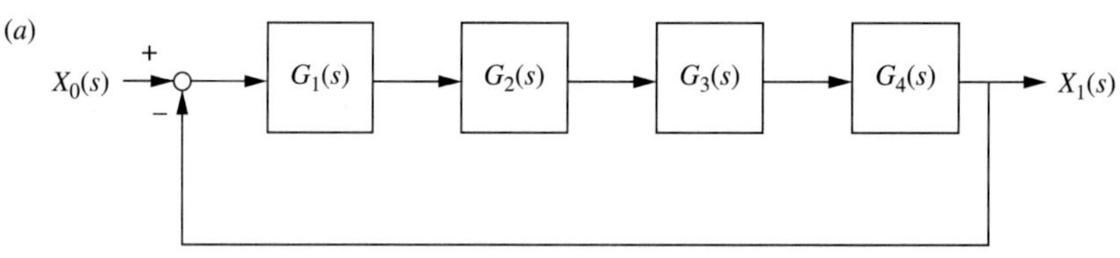

(*b*)

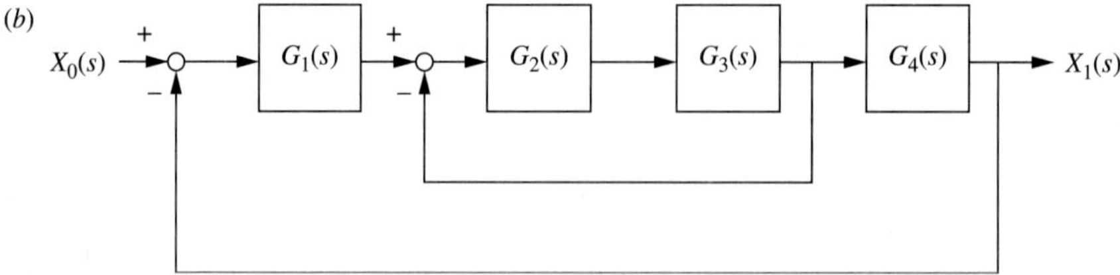

FIGURE Q4.10

4.11. The chemical reactor in Figure Q4.11 includes a liquid inventory in which the turbulent flow out depends on the liquid level. The chemical reaction is first-order with negligible heat of reaction, $A \rightarrow B$, and it occurs only in the tank, not in the pipe. The system is initially at steady state and experiences a step change in the inlet flow rate, with the inlet concentration constant.

(*a*) Derive the overall and component material balances.

(*b*) Linearize the equations and take the Laplace transforms.

(*c*) Determine the transfer function for $C_A(s)/F_0(s)$.

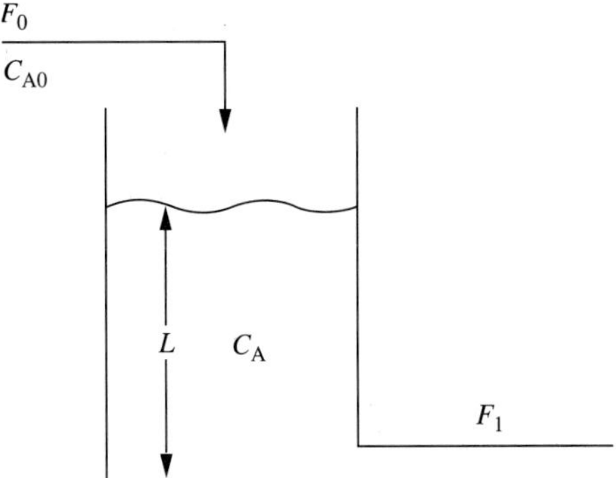

FIGURE Q4.11

4.12 The frequency response of a system can be determined empirically by introducing a sine to an input variable, waiting until the initial transient is negligible, and measuring the input and output amplitudes and the phase angle (see Figure 4.9). If this procedure were performed for several input frequencies, how could you determine whether the real physical system were first-order or second-order? After selecting the proper transfer function order, how could you determine the unknown parameters, gain, and time constant(s)? Also, discuss possible limitations to this empirical method.

4.13. A single, isothermal, well-mixed, constant-volume CSTR is considered in this question. The chemical reaction is

$$A \rightleftarrows B$$

which is first-order with the forward and reverse rate constants k_1 and k_2, respectively. Only component A appears in the feed. The system is initially at steady state and experiences a step in the concentration of A in the feed. Formulate a model to describe this system, and solve for the concentrations of A and B in the reactor.

4.14. Answer the following questions.

(*a*) The initial value of a variable can be determined in a manner similar to the final value. Derive the general expression for the initial value.

(*b*) The transfer function in equation (4.68) can be inverted to give

$$\frac{C_{A0}(s)}{C_A(s)} = \frac{\tau s + 1}{K_p}$$

Discuss whether this is also a transfer function describing the process.

(*c*) The transfer function is sometimes referred to as the *impulse response* of the (linear) system. Demonstrate why this statement is true.

(*d*) If only the input-output relationship is required, why are all equations for the system included in the model, rather than only those equations involving the input and output variables?

(*e*) Equations (4.71) and (4.72) differ in that (4.71) has a constant numerator (since $a_{23} = 0$), whereas the numerator in (4.72) includes s. Discuss why this occurs by referring to the type of interaction between the material and energy balances for the two inputs.

(*f*) Can a series of first-order systems like the chemical reactors in equation (4.69) experience resonance in its amplitude ratio?

4.15. A heat exchanger would be difficult to model, because of the complex fluid mechanics in the shell side. To develop a simple model, consider the two stirred tanks in Figure Q4.15, in which heat is transferred through the common wall, with $Q = UA(\Delta T)$ and UA being constant.

(*a*) Using typical assumptions for the stirred tanks and ignoring energy accumulation effects of the walls, derive an unsteady-state energy balance for the temperatures in both tanks.

(*b*) Solve for the analytical expression for both temperatures in response to a step in T_{h0}.

(*c*) Is it possible for this system to have periodic behavior?

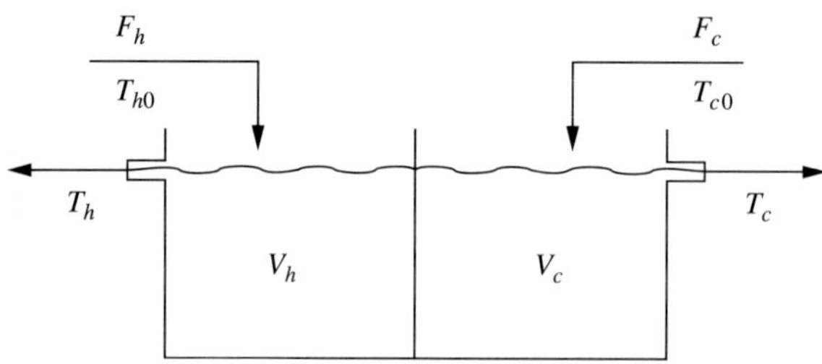

FIGURE Q4.15

4.16. For the series of isothermal CSTRs in Example 3.3:

(*a*) Derive the transfer function for $C_{A2}(s)/F(s)$.

(*b*) Use this result to determine the response of C_{A2} to an impulse in the feed rate F.

4.17. The system in Figure Q4.17 has a flow of pure A to and from a draining tank (without reaction) and a constant flow of B. Both of these flows go to an isothermal, well-mixed, constant-volume reactor with A + B → products and $r_A = r_B = -kC_AC_B$. Make any additional assumptions in determining analytical expressions for the dynamic responses from an initial steady state.

(*a*) Determine the flow of A to the chemical reactor in response to a flow step *into* the draining tank.

(*b*) Determine the concentration of A in the chemical reactor in response to (*a*).

4.18. The process in Figure Q4.18 involves a continuous-flow stirred tank with a mass of solid material. The assumptions for the system are:

1. The tank is well mixed.
2. The physical properties are constant, and $C_v \approx C_p$.
3. V = constant, F = constant [vol/time].
4. The solid material contributes a significant portion of the energy storage, and the temperature is uniform throughout the solid.
5. The heat transfer from the liquid to the metal is $UA(T - T_m)$.
6. Heat losses are negligible.
7. All variables are initially at steady state.

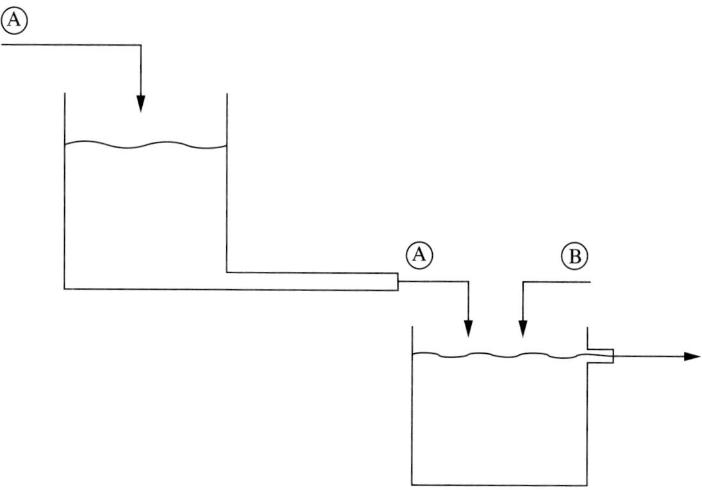

FIGURE Q4.17

(a) Determine the fundamental model equations that relate the behavior of $T(t)$ as $T_0(t)$ changes.

(b) Derive the Laplace transform of $T'(t)$ as a function of $T_0'(s)$. This involves the linear(ized) deviation variables. Identify the time constants and gains.

(c) Draw a block diagram of the system of equations and derive the transfer function $T(s)/T_0(s)$.

(d) State whether the system is stable or unstable and periodic or nonperiodic, and explain your answer.

(e) Solve the equations and sketch the dynamic response of $T'(t)$ for a step change in $T_0'(t)$.

(f) Describe briefly how the results in steps (c) through (e) would change as $UA \to \infty$.

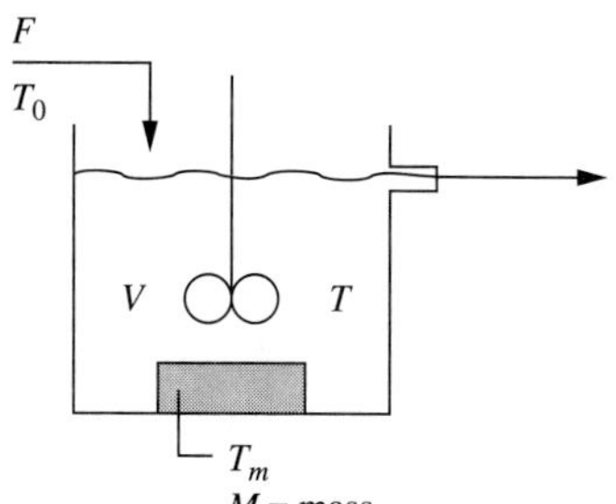

FIGURE Q4.18

CHAPTER 5

DYNAMIC BEHAVIOR OF TYPICAL PROCESS SYSTEMS

5.1 INTRODUCTION

Examples in the previous two chapters have demonstrated that physical systems, which involve very different physical principles, can have similar dynamic behavior. The concept that a single model type can apply to a wide range of entities, process plants, biological units, economic communities, and so forth provides the basis for "systems" analysis. Thus, it is possible to acquire understanding of a large number of systems from a thorough study of a much smaller number of basic models. In this chapter we study some fundamental model structures that occur frequently in process plants, along with their effects on dynamic behavior. This experience will enable us to recognize the effects of process designs on dynamic behavior.

First, the behavior of some simple, basic systems, such as first- and second-order and dead-time systems, is summarized using the results from previous chapters, with some extensions. Second, the behavior of these simple systems in series structures is determined. Third, the behavior of parallel structures of simple systems is derived. Fourth, the effects of recycle structures on dynamic responses are determined. The chapter concludes with an investigation of more complex physical systems of special importance in the process industries: staged systems and multiple input–multiple output systems.

In these sections, the manner in which the behavior of simple systems is altered by common process structures is derived for simple, idealized models but is demonstrated for important process examples involving levels, heat exchangers, chemical reactors, and distillation towers. This coverage demonstrates that the engineer must master both the physical principles of specific processes and systems analysis techniques to determine the dynamics of complex processes quantitatively.

5.2 BASIC SYSTEM ELEMENTS

The coverage of process dynamics begins with the simplest elements, which are often combined to model more complex systems. Since examples of most of these elements were included in previous chapters, the coverage here is concise. The basic model structure for each element is first defined, and several physical examples are given, with the system input designated by X and the output by Y. The chemical process principles should be apparent to the reader, while the electrical and mechanical models are based on Kirchhoff's and Newton's laws, and the reader is referred to Ogata (1992) and Weber (1973) for derivations. The graphical and analytical results of common inputs for several basic systems is summarized in Figure 5.1; the presentation of results in such a figure seems to have originated with Buckley (1964). Only the amplitude ratio is presented here, because more extensive frequency response analysis is presented in Chapter 10, where the importance of the phase behavior on stability is demonstrated and applied in control system analysis.

First-Order System

First-order systems occur as the result of a material or energy balance on a lumped (i.e., well-mixed) system, as demonstrated in Examples 3.1 and 3.6. Some further examples are given in Figure 5.2. The differential equation and transfer function for a first-order system are

$$\tau\frac{dY(t)}{dt} + Y(t) = K_pX(t) \qquad G(s) = \frac{Y(s)}{X(s)} = \frac{K_p}{\tau s + 1} \tag{5.1}$$

The step response is monotonic, with its maximum slope at the time of the step, and the time to reach 63.2% of its final change is one time constant. The final steady-state change is equal to $K_p(\Delta X)$.

$$Y'(t) = K_p(\Delta X)(1 - e^{-t/\tau}) \tag{5.2}$$

An impulse input occurs over a negligible time and transfers a finite amount into the system. For example, rapidly introducing a small amount of tracer into a stirred tank emulates a perfect impulse. The impulse response shows an immediate increase at the time of the impulse, which for the idealized stirred-tank example would mean that the concentration would change instantly by (Mass of tracer)/(Volume). After the impulse (C), the system follows an exponential path in return to its final condition.

$$Y'(t) = \frac{C}{\tau}e^{-t/\tau} \tag{5.3}$$

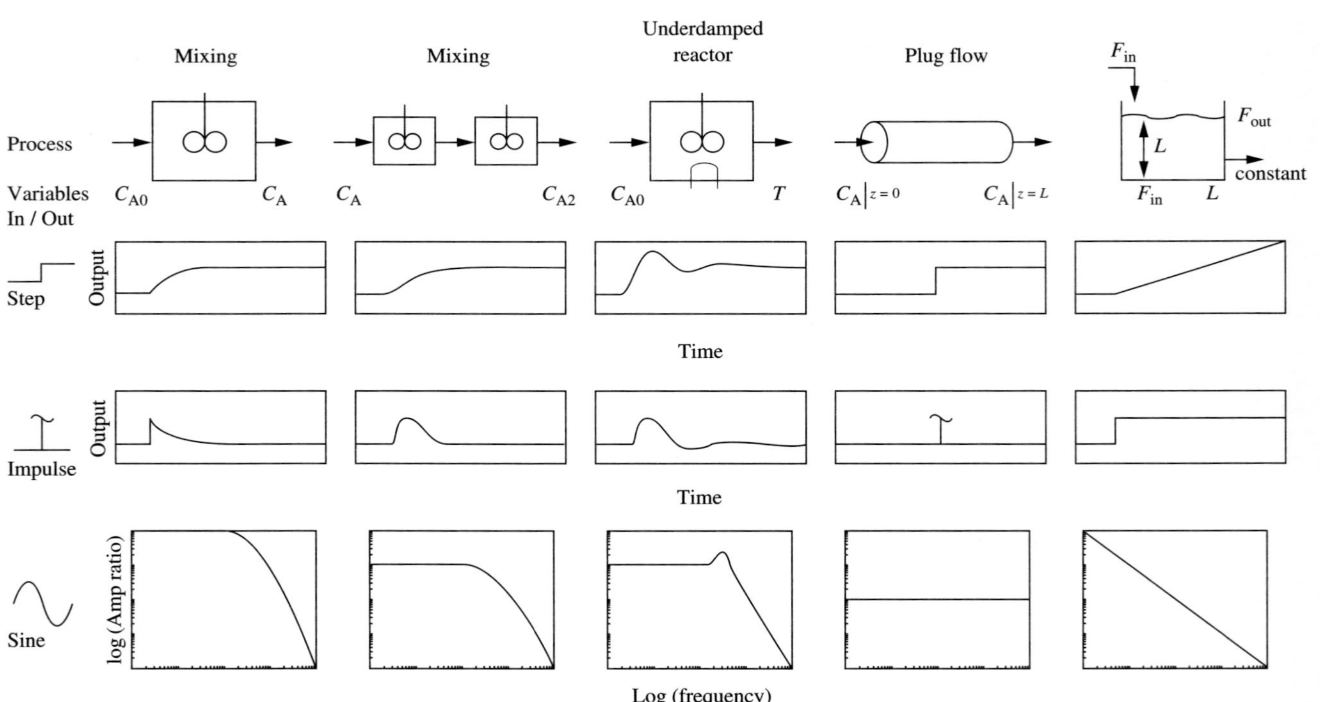

FIGURE 5.1
Dynamic responses for basic process-modelling elements.

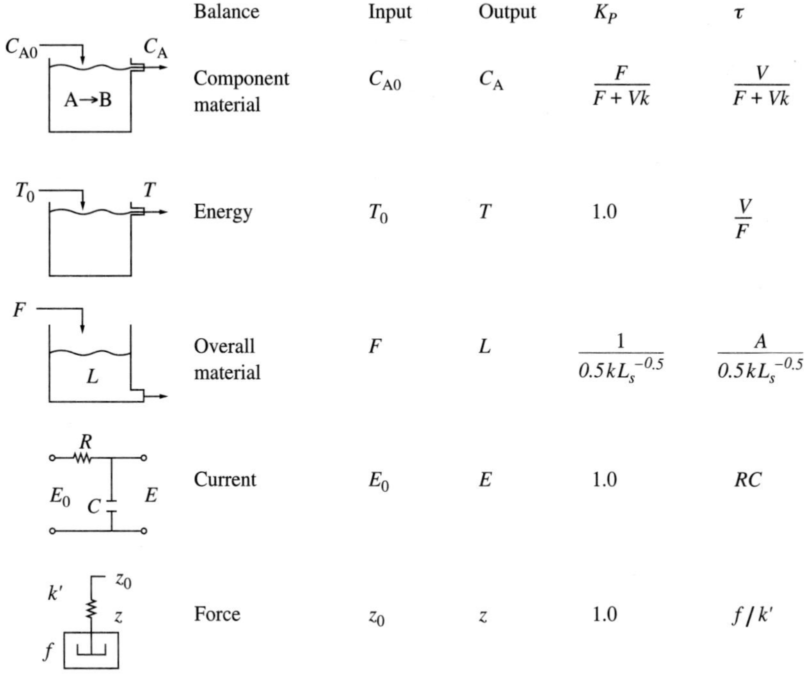

Balance	Input	Output	K_p	τ
Component material	C_{A0}	C_A	$\frac{F}{F+Vk}$	$\frac{V}{F+Vk}$
Energy	T_0	T	1.0	$\frac{V}{F}$
Overall material	F	L	$\frac{1}{0.5kL_s^{-0.5}}$	$\frac{A}{0.5kL_s^{-0.5}}$
Current	E_0	E	1.0	RC
Force	z_0	z	1.0	f/k'

FIGURE 5.2
First-order processes (E = voltage, z = position, k' = spring constant, and f = friction coefficient).

For the first-order system, the amplitude ratio is never greater than the process gain K_p, and it decreases monotonically as the frequency increases:

$$\text{AR} = |G(j\omega)| = \frac{|Y(j\omega)|}{|X(j\omega)|} = \frac{K_p}{\sqrt{1+\omega^2\tau^2}} \tag{5.4}$$

Second-Order System

The second-order system occurs when two first-order or one second-order ordinary differential equation is required to model the dynamic behavior. Some examples are given in Figure 5.3. The transfer function for the second-order system with a gain in the numerator (and no zeros) can be written as

$$\tau^2\frac{d^2Y(t)}{dt^2} + 2\xi\tau\frac{dY(t)}{dt} + Y(t) = K_pX(t) \tag{5.5}$$

$$G(s) = \frac{Y(s)}{X(s)} = \frac{K_p}{\tau^2s^2 + 2\xi\tau s + 1}$$

with $\alpha_{1,2} = \dfrac{-\xi \pm \sqrt{\xi^2-1}}{\tau}$

The parameter ξ is termed the damping coefficient, and $\alpha_{1,2}$ are the two roots of the characteristic polynomial, which determine the exponents of the time-domain

	Balance	Input	Output	K_P	τ^2	$2\xi\tau$
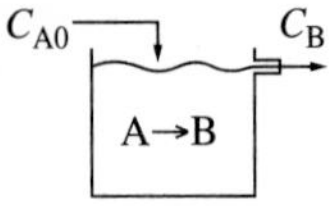 C_{A0}, C_B, A→B	Component material	C_{A0}	C_B	$\frac{Vk}{F+Vk}$	$\tau_A \tau_B$	$\tau_A + \tau_B$
T_0, T, T_c	Energy	T_0	T	[see question 5.2]		
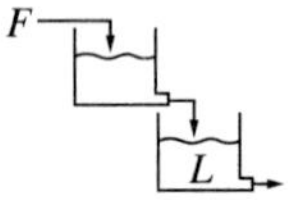 F, L	Overall material	F	L	$\frac{1}{0.5kL_s^{-0.5}}$	$\left[\frac{A}{0.5kL_s^{-0.5}}\right]^2$	2τ
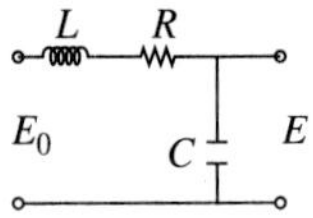 L, R, C, E_0, E	Current	E_0	E	1.0	LC	RC
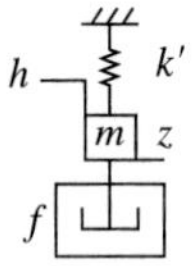 h, k', m, z, f	Force	h	z	$1/k'$	m/k'	f/k'

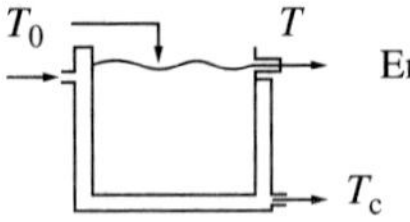

FIGURE 5.3
Second-order processes (E = voltage, z = position, k' = spring constant, f = friction coefficient, h = force, m = mass, $\tau_A = V/(F + Vk)$, and $\tau_B = V/F$).

output function. When the damping coefficient is less than 1.0, the system is termed *underdamped,* the roots of the characteristic polynomial are complex, and the system will have periodic behavior for a nonperiodic input. For example, the nonisothermal reactor system in Example 3.10, which exhibits oscillations for a step input, has a damping coefficient of 0.15. When the damping coefficient is greater than 1.0, the system is termed *overdamped,* the roots of the characteristic polynomial are real, and the system will have nonperiodic responses to nonperiodic inputs. Finally, the series reactor system in Example 3.3 has a damping coefficient of 1.0, which indicates real, repeated roots; this type of system is termed *critically damped.*

The possibility of periodic behavior is not affected by a numerator that could be a function of s; thus, the simplest numerator is considered here, K_p. Two entries are given in Figure 5.1 for second-order systems; one is for an overdamped system, and the other is for an underdamped system. The step response for the overdamped system initially at steady state is monotonic with an initial slope of zero and an inflection point. Note that the underdamped system experiences periodic behavior even for this simple input.

OVERDAMPED STEP RESPONSE ($\xi > 1$).

$$Y = K_p \Delta X \left(1 + \frac{\tau_1 e^{-t/\tau_1}}{\tau_2 - \tau_1} - \frac{\tau_2 e^{-t/\tau_2}}{\tau_2 - \tau_1}\right) \tag{5.6}$$

CRITICALLY DAMPED STEP RESPONSE ($\xi = 1$).

$$Y = K_p \Delta X \left(1 - \left(1 + \frac{t}{\tau}\right) e^{-t/\tau}\right) \tag{5.7}$$

UNDERDAMPED STEP RESPONSE ($\xi < 1$).

$$Y = K_p \frac{\Delta X}{\tau^2} - K_p \frac{\Delta X}{\tau^2 \sqrt{1-\xi^2}} e^{-\xi t/\tau} \sin\left(\frac{\sqrt{1-\xi^2}}{\tau} t + \phi\right) \qquad \phi = \tan^{-1}\left(\frac{\sqrt{1-\xi^2}}{\xi}\right) \tag{5.8}$$

OVERDAMPED IMPULSE RESPONSE ($\xi > 1$).

$$Y = C\left(\frac{e^{-t/\tau_1}}{\tau_1 - \tau_2} - \frac{e^{-t/\tau_2}}{\tau_1 - \tau_2}\right) \tag{5.9}$$

CRITICALLY DAMPED IMPULSE RESPONSE ($\xi = 1$).

$$Y = \frac{Ct}{\tau} e^{-t/\tau} \tag{5.10}$$

UNDERDAMPED IMPULSE RESPONSE ($\xi < 1$).

$$Y = \frac{C}{\tau \sqrt{1-\xi^2}} e^{-\xi t/\tau} \sin\left(\frac{\sqrt{1-\xi^2}}{\tau} t\right) \tag{5.11}$$

Both the step and impulse responses for a second-order system have initial responses that are more gradual than for a first-order system. The overdamped system approaches its final value smoothly, while the underdamped system experiences oscillations.

The amplitude ratio of the frequency response is monotonically decreasing for an overdamped system and begins to deviate substantially from K_p around the frequency equal to $1/\tau$. The amplitude ratio for second-order systems with a damping coefficient below 0.707 exceeds K_p over a limited frequency range around $1/\tau$. This resonance effect results from the inherent oscillatory tendency of the system reinforcing the input sine oscillations.

$$\text{AR} = |G(j\omega)| = \frac{|Y(j\omega)|}{|X(j\omega)|} = \frac{K_p}{\sqrt{(1-\omega^2\tau^2)^2 + (2\omega\tau\xi)^2}} \tag{5.12}$$

Dead Time

The dead time or transportation delay was introduced in Example 4.3 for plug flow of liquids and can also occur for transportation of solids along a conveyor belt. It was shown to have the following transfer function:

$$Y(t) = X(t-\theta) \qquad G(s) = \frac{Y(s)}{X(s)} = e^{-\theta s} \tag{5.13}$$

The step response, impulse response, and amplitude ratio can all be easily determined, because the output is the input translated in time by θ. For example, this leads to the conclusion that the amplitude ratio is equal to 1.0 for all frequencies, which can be demonstrated mathematically by

$$\mathrm{AR} = \left|e^{-j\omega\theta}\right| = \left|\cos(\omega\theta) - j\sin(\omega\theta)\right| = \sqrt{\cos^2(\omega\theta) + \sin^2(\omega\theta)} = 1 \qquad (5.14)$$

The dead time can be approximated by a transfer function composed of polynomials in the numerator and denominator; the results of this approach are referred to as Padé approximations (Coughanowr, 1991). This assisted design methods used in the past, which were limited to transfer functions that are ratios of polynomials in s. The Padé approximations are not widely used now in the analysis of process control systems.

The importance of dead time to feedback control can be understood by considering an example such as steering an automobile. With dead time, the automobile would not respond immediately after the change in steering wheel position. Clearly, such an automobile would be difficult to drive and would require a skilled and patient driver who could wait for the effect of a steering wheel change to occur.

Integrator

The integrator is a special type of first-order system; a process example of an integrator is a level system, which is modelled based on an overall material balance to give

$$\rho A \frac{dL}{dt} = \rho F_0 - \rho F_1 \qquad (5.15)$$

In many cases the inlet and outlet flows do not depend on the level (unlike the tank draining Example 3.6). When no causal relationship exists from the level to the flow, the model has the following general form:

$$\tau_H \frac{dY'}{dt} = X' \neq f(Y') \qquad \tau_H = \text{holdup time} \qquad (5.16)$$

$$G(s) = \frac{Y(s)}{X(s)} = \frac{1}{\tau_H s} \qquad (5.17)$$

The important difference between the integrator and the first-order system in equation (5.1) is the lack of dependence of the derivative on the output variable (Y'); that is, dY'/dt is independent of Y'. This results in a pole at $s = 0$ in the transfer function. The analytical expression for the output of the integrator is

$$Y'(t) = \int_0^t X'(t')dt' \qquad (5.18)$$

A system like this simply accumulates the net input: thus, the name *integrator.* If the deviation in the input remains nonzero and of the same sign, the magnitude of the idealized model output increases without limit as time increases toward infinity. For a step input,

$$Y' = \frac{\Delta X}{\tau_H} t \tag{5.19}$$

The impulse response also demonstrates that the system integrates the impulse (area under the impulse function), and then the output remains constant at its altered value when $X'(t)$ returns to zero. The value of the impulse response is $Y' = C/\tau_H$.

The amplitude ratio can be determined to be

$$\text{AR} = |G(j\omega)| = \left|\frac{1}{\tau_H j\omega}\right| = \left|\frac{-\omega j}{\tau_H \omega^2}\right| = \frac{1}{\tau_H \omega} \tag{5.20}$$

As the frequency decreases, the amount accumulated by the integrator each half period (which is related to the output amplitude) increases.

Self-Regulation

The unique behavior of the integrator demonstrates that not all (idealized) processes have the inherent characteristic of tending to a steady state after an input changes from its initial value to another constant value. To make the distinction, the term *self-regulation* is introduced here. An integrator system is often referred to as *non–self-regulating,* because the rate of change of the output is independent of the output. Recalling that the definition of feedback is the application of an output to influence an input, it is clear that there is no inherent feedback in this process. Any non–self-regulating process should be controlled, because even a small nonzero input over a long time will result in the output process variable deviating far from its proper value. The modelling and control of non–self-regulating liquid level systems is covered in Chapter 18.

By comparison, a flow in the level draining system in Example 3.6 depends on the level. Also, the accumulation terms of all first- and second-order systems considered to this point depend on the value of the system output. These systems are termed *self-regulatory* and tend to approach a steady state after a step input (if the roots of the characteristic polynomial have negative real parts). Self-regulating processes are generally easier to operate, because they tend toward a steady state, although the output variable can deviate from its acceptable range of values due to large input disturbances.

A stable self-regulatory process could be said to have inherent negative feedback. For example, the heat exchanger in Example 3.7 has inherent negative feedback, because an increase in the output (outlet temperature) causes a decrease in a model input term $-(F/V + UA/V\rho C_p)T$, which stabilizes the system by decreasing the derivative:

$$\frac{dT}{dt} = \underbrace{\left(\frac{F}{V}T_0 + \frac{UA}{V\rho C_p}T_{\text{cin}}\right)}_{\text{External inputs}} - \underbrace{\left(\frac{F}{V} + \frac{UA}{V\rho C_p}\right)T}_{\text{Inherent negative feedback}} \tag{5.21}$$

Some processes have inherent *positive* and negative feedback, for example, the nonisothermal chemical reactor with exothermic chemical reaction in Example 3.10.

$$\frac{dT}{dt} = \underbrace{\left(\frac{F}{V}T_0 + \frac{UA}{V\rho C_p}T_{\text{cin}}\right)}_{\text{External inputs}} - \underbrace{\left(\frac{F}{V} + \frac{UA}{V\rho C_p}\right)T}_{\text{Inherent negative feedback}} \tag{5.22}$$

$$+ \underbrace{\frac{(-\Delta H_{\text{rxn}})k_0 e^{-E/RT} C_{\text{A}}}{V\rho C_p}}_{\text{Inherent positive feedback}}$$

The reactor has a negative feedback term in its energy balance, the same as for the heat exchanger. However, the exothermic chemical reaction contributes positive feedback, because the input term $(-\Delta H_{\text{rxn}} k_0 e^{-E/RT} C_{\text{A}}/V\rho C_p)$ increases when the output temperature increases. For the parameter values in Example 3.10, the inherent negative feedback in the process dominates, and the process achieves a steady state after a step input. The positive feedback is substantial, however, which leads to the periodic behavior and complex poles. Additional comments on the behavior and stability of processes are given in Appendix C.

In summary, many different systems obeying the models of these basic elements behave in a similar manner. After the parameters have been determined, their behavior for specified inputs is well understood. Thus, the experience learned from a few examples can be extended, with care, to many other systems.

5.3 SERIES STRUCTURES OF SIMPLE SYSTEMS

A structure involving a series of systems occurs often in process control. As discussed in Chapter 2, this structure can occur because of a processing sequence — for example, feed heat exchange, chemical reactor, product cooling, and product separation. Also, a control loop involves a final element (valve), process, and sensor in a series, as will be more fully discussed in Part III. Therefore, the understanding of how series structures behave is important in the design of chemical plants and process control systems.

Noninteracting Series

There are two major categories of series systems, and the noninteracting system is covered first. It is worthwhile considering the mixing system, which conforms to the block diagram at the bottom of Figure 5.4*a,* in which each intermediate variable has physical meaning.

$$V\frac{dC'_{\text{A1}}}{dt} = FC'_{\text{A0}} - FC'_{\text{A1}} \tag{5.23}$$

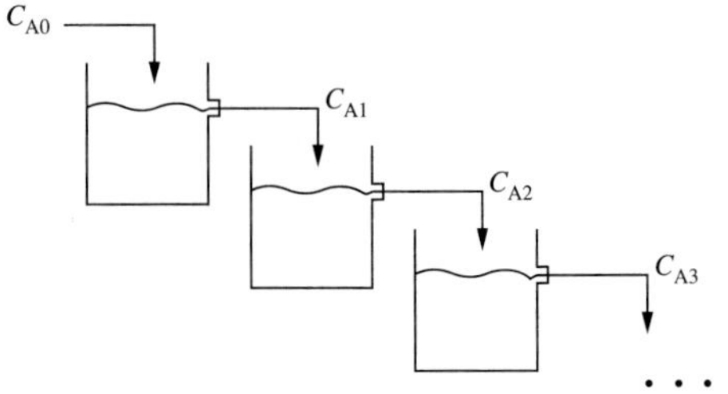

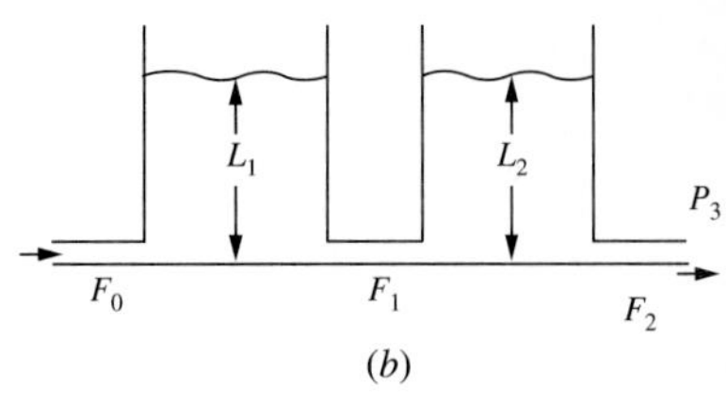

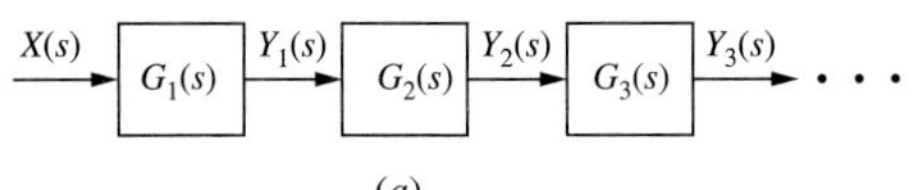

FIGURE 5.4
Series of processes: (*a*) noninteracting; (*b*) interacting.

$$V\frac{dC'_{A2}}{dt} = FC'_{A1} - FC'_{A2} \tag{5.24}$$

Note that the model equations have the general form

$$\tau_i \frac{dY'_i}{dt} = K_i Y'_{i-1} - Y'_i \qquad \text{for } i = 1, \ldots, n \qquad \text{with } Y'_0 = X' \tag{5.25}$$

Any system modelled with equations of this structure constitutes a noninteracting series system. Important features of the system follow from this model.

1. Only Y_{n-1} and Y_n (not Y_{n+1}) appear in the equation for dY_n/dt.
2. Following from (1), the downstream properties do not affect upstream properties; in the example, the concentration in tank 2 does not affect the concentration in tank 1 but does affect tank 3.
3. The model for the general noninteracting series of first-order systems can be developed by taking the Laplace transform of each equation (5.25) and combining them into one input-output expression. For a series of systems shown in Figure 5.4*a,* each represented by a transfer function $G_i(s)$, the overall transfer function

$$\frac{Y_n(s)}{X(s)} = G_n(s)G_{n-1}(s)\cdots G_1(s) = \prod_{i=0}^{n-1} G_{n-i}(s) \tag{5.26}$$

For n first-order systems in series, this gives

$$\frac{Y_n(s)}{X(s)} = \frac{\prod_{i=0}^{n-1} K_{n-i}}{\prod_{i=0}^{n-1}(\tau_{n-i}s + 1)} \qquad \text{with } K_{n-i} \text{ and } \tau_{n-i} \text{ for the individual systems} \tag{5.27}$$

The gains and time constants appearing in equation (5.27) are the same as the values for the individual systems, as in equation (5.25). Thus, the model of interacting systems can be determined directly from the individual models.

4. If each system is stable (i.e., $\tau_i > 0$ for all i), the series system is stable. This follows from the important observation that the poles (roots of the characteristic polynomial) of the series system are the poles of the individual systems.

Now the dynamic response of a series of noninteracting first-order systems can be considered. Since so many possibilities exist, the simplest case of *n identical systems,* all with unity gain, is considered. The response to a step in the input, $X'(s) = 1/s$, is plotted in Figure 5.5. Note that the time is divided by the order of the system (i.e., the number of systems in series), which time-scales the responses for easy comparison. We note that the shape of the response changes from the now-familiar exponential curve for $n = 1$. As n increases, the response begins to have an apparent dead time, which is the result of several first-order systems in series. For very large n, the output response has a very steep change at time equal to $n\tau$. Thus, we conclude that the series of identical noninteracting first-order systems approaches the behavior of a dead time with $\theta \approx n\tau$ for large n. Again looking ahead to feedback control, a system with several first-order systems in series would seem to be difficult to control, for the same reasons discussed for dead times.

A second observation is that the curves all reach 63% of their output change at approximately the same value of $t/n\tau$; this will be exploited later in the section. Finally, we note that the system is always overdamped, because the transfer function has n real poles, all at $-1/\tau$.

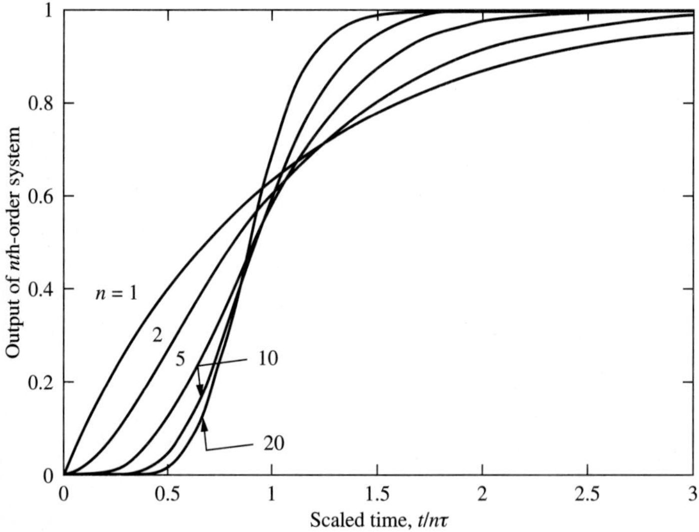

FIGURE 5.5
Responses of n identical noninteracting first-order systems with $K = 1$ in series to a unit step at $t = 0$.

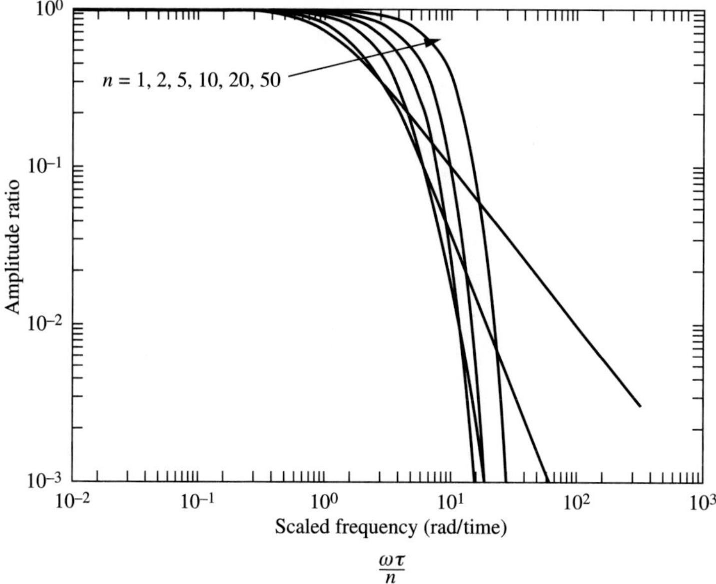

FIGURE 5.6
Frequency responses of n identical noninteracting first-order systems with $K = 1$ in series.

The amplitude ratio of the frequency response can be determined directly from the transfer function in equation (5.27) to be

$$\mathrm{AR} = \frac{|Y_n(j\omega)|}{|X(j\omega)|} = |G(j\omega)| = \left(\prod_{i=1}^{n} K_i\right)\left(\frac{1}{\sqrt{1+\omega^2\tau^2}}\right)^n \tag{5.28}$$

The amplitude ratio is always less than or equal to the overall gain, and it decreases rapidly as the frequency becomes large. Amplitude ratios for several series of identical first-order systems are shown in Figure 5.6; again, the frequency is scaled to the order of the system to provide time-scaling.

Interacting Series

The second major category of series systems is interacting systems. Again, it is worthwhile considering a physical example, this being the level-flow process in Figure 5.4*b*. Assuming that the flow through each pipe is a function of the pressure difference, the model can be derived based on overall material balance for each vessel to give

$$\begin{aligned} A_i \frac{dL_i}{dt} &= F_{i-1} - F_i \\ &= K_{i-1}(L_{i-1} - L_i) - K_i(L_i - L_{i+1}) \end{aligned} \tag{5.29}$$

because $F_i = K_i'(P_i - P_{i+1})$ for the linearized system, and the pressures are proportional to the liquid levels. These model equations have the following general form for a series of two interacting first-order systems:

$$H_1 \frac{dY_1'}{dt} = X' - K_1(Y_1' - Y_2') \tag{5.30}$$

$$H_2 \frac{dY_2'}{dt} = K_1(Y_1' - Y_2') - K_2(Y_2' - Y_3') \tag{5.31}$$

Many important physical systems, including that in Figure 5.4*b*, have structures described by equations (5.30) and (5.31); thus, these equations are considered representative of interacting systems for subsequent analysis. Some important features of these systems follow from their model structure:

1. The variables Y_{n-1}, Y_n, and Y_{n+1} appear in the equation for dY_n/dt.
2. Following from (1), the downstream properties affect upstream properties; for example, the exhaust pressure (P_3) influences *both* levels in Figure 5.4*b*.
3. The model for the general interacting series system of first-order systems can be developed by taking the Laplace transform of equations (5.30) and (5.31) and combining them into one input-output expression, which results in poles of the interacting system that are *different* from the poles of the individual systems.

The procedure for deriving the overall transfer function is shown in some detail, because the result is somewhat more complex than for a noninteracting system and because the procedure can be applied to systems of differing structures. First, the Laplace transform of equation (5.30) can be rearranged to give (with the primes deleted)

$$Y_1(s) = \frac{1/K_1}{\tau_{Y1}s + 1} X(s) + \frac{1}{\tau_{Y1}s + 1} Y_2(s) \qquad \text{with } \tau_{Y1} = \frac{H_1}{K_1} \tag{5.32}$$

The parameter τ_{Y1} is the time constant for the first system when considered *individually*. The Laplace transform of the second equation is

$$\tau_{Y2} s Y_2(s) = \frac{K_1}{K_2}[Y_1(s) - Y_2(s)] - [Y_2(s) - Y_3(s)] \qquad \text{with } \tau_{Y2} = \frac{H_2}{K_2} \tag{5.33}$$

Again, the parameter τ_{Y2} is the time constant for the second system when considered *individually*. The behavior of the combined system can be determined by substituting equation (5.32) into (5.33) to give, after some rearrangement,

$$Y_2(s) = \frac{(\tau_{Y1}s + 1)}{\tau_{Y1}\tau_{Y2}s^2 + \left(\tau_{Y1} + \tau_{Y2} + \tau_{Y1}\dfrac{K_1}{K_2}\right)s + 1} Y_3(s) + \frac{1/K_2}{\tau_{Y1}\tau_{Y2}s^2 + \left(\tau_{Y1} + \tau_{Y2} + \tau_{Y1}\dfrac{K_1}{K_2}\right)s + 1} X(s) \tag{5.34}$$

Several important conclusions on the effect of the series structure on the dynamic behavior can be determined from an analysis of the denominator of the transfer function. The time constants of the interacting system (τ_1 and τ_2), which are the inverses of the poles, can be determined by solving the quadratic equation for the roots of the characteristic polynomial to give

$$\alpha_{1,2} = \frac{1}{\tau_{1,2}} = \frac{-\left(\tau_{Y1} + \tau_{Y2} + \tau_{Y1}\dfrac{K_1}{K_2}\right) \pm \sqrt{\left(\tau_{Y1} + \tau_{Y2} + \tau_{Y1}\dfrac{K_1}{K_2}\right)^2 - 4\tau_{Y1}\tau_{Y2}}}{2\tau_{Y1}\tau_{Y2}} \tag{5.35}$$

Four characteristics of the dynamics of this type of series system are now established. First, the possibility of complex poles is determined to establish whether periodic behavior is possible. The expression within the square root in equation (5.35) can be rearranged to give

$$\left(\tau_{Y1} + \tau_{Y2} + \tau_{Y1}\frac{K_1}{K_2}\right)^2 - 4\tau_{Y1}\tau_{Y2} = (\tau_{Y1} - \tau_{Y2})^2 + \tau_{Y1}\frac{K_1}{K_2}\left(2\tau_{Y1} + 2\tau_{Y2} + \tau_{Y1}\frac{K_1}{K_2}\right) > 0 \tag{5.36}$$

Since both terms in the right-hand expression are greater than zero, the entire expression is greater than zero, and complex poles are not possible for this system. Therefore, periodic behavior cannot occur for nonperiodic inputs, such as a step.

Second, the stability of the process can be determined from equation (5.35). Note that the numerator has the form $-a \pm (a^2 - b)^{0.5}$, with a and b both positive. Therefore, the poles for both signs of the root are negative, and the system is stable.

Third, the "speed" of response of the interacting series system can be compared with the individual system responses. Since the poles are real, the characteristic polynomial in equation (5.34) can be written in an equivalent form as

$$(\tau_1 s + 1)(\tau_2 s + 1) = \tau_1 \tau_2 s^2 + (\tau_1 + \tau_2)s + 1 \tag{5.37}$$

Equating the coefficients of like powers of s in equations (5.34) and (5.37) gives

$$\tau_1 \tau_2 = \tau_{Y1}\tau_{Y2} \qquad \text{and} \qquad \tau_1 + \tau_2 = \tau_{Y1} + \tau_{Y2} + \tau_{Y1}\frac{K_1}{K_2} \tag{5.38}$$

Therefore, the sum of the time constants for the overall interacting system is greater than the sum of the individual systems. In other words, the interacting system is "slower," due to the interaction, than it would have been if the systems were noninteracting.

Fourth, equations (5.38) show that the product of the time constants is unchanged but the sum is greater. Therefore, the difference between the interacting system time constants $(\tau_1 - \tau_2)$ is greater than the difference between the individual time constants $(\tau_{Y1} - \tau_{Y2})$; that is, one time constant begins to dominate. This conclusion can be demonstrated by rearranging equations (5.38) to give

$$(\tau_1 - \tau_2)^2 = (\tau_{Y1} - \tau_{Y2})^2 + \tau_{Y1}\frac{K_1}{K_2}\left(2\tau_{Y1} + 2\tau_{Y2} + \tau_{Y1}\frac{K_1}{K_2}\right) \tag{5.39}$$

Since the noninteracting series system has been shown to have all real poles, the dynamic responses of an interacting system of first-order systems have many of the same characteristics as those of a noninteracting system; that is, they are stable and overdamped.

The previous results for interacting systems are applicable to (only) those systems that conform to the model; in addition to having variables Y_{n-1}, Y_n, and Y_{n+1} appear in the equation for dY_n/dt, the coefficients of each linearized term must conform to the structure and range of values in equations (5.30) and (5.31).

Many systems have the same model structures but different ranges for the values of the parameters. For example, the model of the nonisothermal CSTR in Example 3.10 has both output variables in both equations, but the coefficients, the a_{ij}, do not conform to the ranges of values of systems analyzed in this section. Therefore, the chemical reactor would not be considered an interacting series of first-order systems, and it can experience periodic behavior. If the type of system is not obvious from the structure of the equations and the values of the model parameters, the model can be analyzed using the procedure just applied to the equations (5.30) and (5.31) to determine important characteristics of its dynamic behavior.

Finally, some alternative terminology used in automatic control is introduced. What has been referred to here as a series system is sometimes termed a cascade system; however, *cascade* is used in process control for another topic covered in Chapter 14. Also, the two types of series systems are sometimes referred to as *nonloading* and *loading,* for noninteracting and interacting, respectively.

Noninteracting Series with Dead Time

As will become more apparent in the next chapter, we often use first-order-with-dead-time models to approximate more complex systems with monotonic step input responses. Therefore, noninteracting series of first-order-with-dead-time systems are considered to conclude this section. The direct application of equation (5.26) results in

$$\frac{Y(s)}{X(s)} = \prod_{i=0}^{n-1} G_{n-i}(s) = \frac{\left(\prod_{i=1}^{n} K_i\right)\exp\left(-\sum_{i=1}^{n}\theta_i s\right)}{\prod_{i=1}^{n}(\tau_i s + 1)} \qquad \text{with } G_i(s) = \frac{K_i e^{-\theta_i s}}{\tau_i s + 1} \tag{5.40}$$

This overall transfer function provides the basis for the following equations, which give values for key parameters of a noninteracting series of first-order-with-dead-time systems.

$$\text{Exact relationships} \qquad K = \prod_{i=1}^{n} K_i \qquad \theta = \sum_{i=1}^{n}\theta_i \tag{5.41a}$$

$$\text{Approximate relationship} \qquad t_{63\%} \approx \sum_{i=1}^{n}(\theta_i + \tau_i) \tag{5.41b}$$

The results for the overall gain and dead time follow directly from equation (5.40). The approximation for the time for the output response to a step input to reach 63% of its final value, $t_{63\%}$, is based on fitting an approximate model to the response of the series system, using the method of moments. The derivation of this expression is provided in Appendix D. The relationships in equations (5.41) are useful for quickly characterizing the approximate behavior of a noninteracting series system from the individual systems; comparison to solutions of noninteracting systems (e.g., Figure 5.5), shows that the expression for $t_{63\%}$ is a reasonable approximation but *not exact.*

Example 5.1. The four first-order-with-dead time systems, with parameters in the following table, are placed in a noninteracting series. Describe the output response of this system to a step change in the input to the series at time = 2.

System	**1**	**2**	**3**	**4**
Dead time, θ	0.40	0.90	1.2	1.70
Time constant, τ	1.5	3.3	5.2	0.95
Gain, K	1.0	0.25	3.0	1.33

The results in this section on noninteracting systems indicate that the output response will be an overdamped sigmoid. Equations (5.41) can be used to estimate key values of the response. Note that the input occurred at $t = 2$, so that the points indicated on Figure 5.7 are based on the following results as measured from $t = 2$.

$$K_p = 1.0 \qquad \theta - 2 = 4.2 \qquad \sum(\theta + \tau) = 15.1 \qquad \therefore (t_{63\%}) - 2 \approx 15.1$$

The overall response is compared with the approximation in Figure 5.7, which demonstrates the usefulness of the approximation for $t_{63\%}$, because it gives an approximate "time scale" for the response. However, many sigmoidal curves could be drawn through the two points in the figure. The entire curve can be determined through analytical or numerical solution of the defining equations.

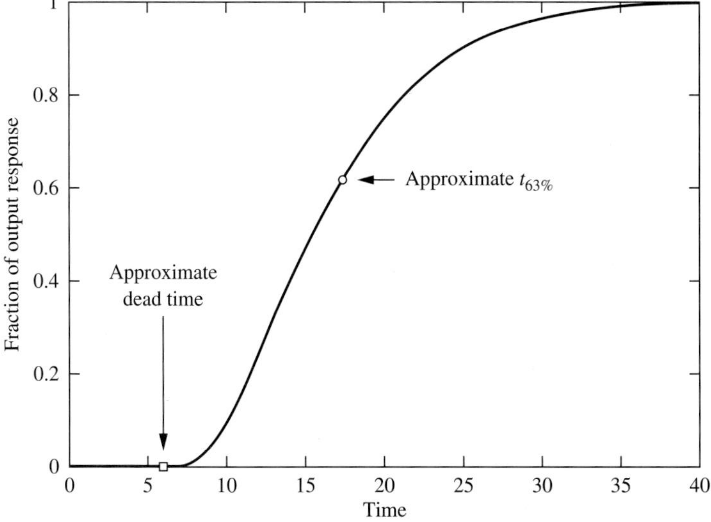

FIGURE 5.7
Dynamic response of series processes in Example 5.1 for a unit step at $t = 2$.

Example 5.2 Input-output response. Two series systems involve only transportation delays and mixing tanks. A step change is introduced into the input feed composition of each system with the flow rates constant. Determine and compare the dynamic responses of the output for each system. Since there is no chemical reaction, the systems have a gain of 1.0 and dynamic parameters given in the following table.

	θ_1	τ_1	θ_2	τ_2	θ_3	τ_3	θ_4	τ_4
Case 1	0	2	2	0	0	2	2	0
Case 2	0	2	2	2	1	0	1	0

The solution can be developed in several ways. The most general is to derive the overall input-output transfer functions for these systems.

$$Y_4(s) = G_4(s)Y_3(s) = \cdots = G_4(s)G_3(s)G_2(s)G_1(s)X(s)$$

$$\frac{Y_4(s)}{X(s)} = \frac{1.0e^{-(\theta_1+\theta_2+\theta_3+\theta_4)s}}{(\tau_1 s + 1)(\tau_2 s + 1)(\tau_3 s + 1)(\tau_4 s + 1)}$$

$$= \frac{1.0e^{-4s}}{(2s + 1)(2s + 1)}$$

Since the overall transfer functions are the same for the two systems, their dynamic input-output behaviors are *identical*. This is verified by the transient responses of the two cases for a step input at time= 2 in Figure 5.8, with each variable $Y_i(t)$ on a separate scale.

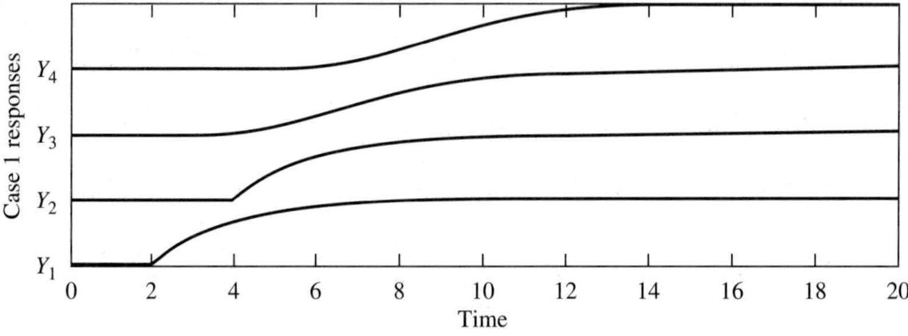

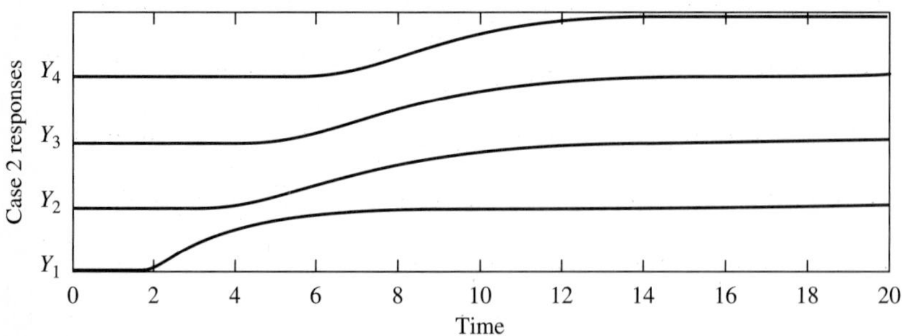

FIGURE 5.8
Dynamic responses for series system in Example 5.2 to a unit step at $t = 2$.

The responses in Figure 5.8 show that two systems can have the same input-output behavior with different values for intermediate variables.

In conclusion, the analysis in this section has demonstrated that both noninteracting and interacting series of n first-order systems can be modelled by a transfer function with a characteristic polynomial of order n. Much about the dynamic responses of the series systems can be determined from the models of the individual systems. The results are summarized in Table 5.1.

The series systems in this section provided additional reinforcement for the importance of transfer function poles. The strongest general conclusions were based on the manner in which the poles of the overall system were or were not affected by the series structure. These conclusions concerned stability and the related property of periodic behavior. Since these generalizations dealt with properties completely determined by the poles, they are independent of the numerators in the transfer functions. In fact, the generalizations on stability and periodicity can be extended to any series transfer functions with denominators expressed as a polynomial in s.

However, the values of the poles do not provide general conclusions for the time-domain responses to step and sine inputs. Since both the numerator and denominator of the transfer function influence the dynamic behavior, the more specific results on dynamic responses are valid only for systems consistent with the assumptions in the derivations — that is, with a constant for the numerator of each series transfer function element. In particular, Figures 5.4 and 5.5 and all conclusions on the step response and amplitude ratio are specific to systems whose component elements have constant numerators. Finally, such strong conclusions for an overall system, based on the individual elements, are not always possible, as demonstrated by the structures considered in the remainder of this chapter.

TABLE 5.1
Properties of series systems with first-order elements (responses between input, X, and output, Y_n)

Individual first-order systems	Noninteracting series system	Interacting series system, equations (5.30) and (5.31)
n first-order systems	nth-order system	nth-order system
Each is stable	Stable, not periodic	Stable, not periodic
Time constants, τ_i	Time constants are τ_i, $i = 1, \ldots, n$	Time constants are not τ_i's. They must be determined by solving the characteristic polynomial.
$t_{63\%}$	$t_{63\%} \approx \sum \tau_i$	$t_{63\%} > \sum \tau_i$
Step response	Overdamped, sigmoidal	Overdamped, sigmoidal
Frequency response	$\text{AR} \leq K_p$ for all ω	$\text{AR} \leq K_p$ for all ω

5.4 PARALLEL STRUCTURES OF SIMPLE SYSTEMS

Parallel structures involve systems with two or more paths between the input and output variables and thus are fundamentally different from series systems, which have only one path. Two process examples of parallel structures are considered in this section, from which general conclusions will be drawn about the unique features of these systems. The key difference between this and the previous sections is the concentration on the numerator, as well as the denominator, of the transfer function. As discussed in Chapter 4, the numerator contributes important characteristics to the dynamic behavior that cannot be determined from the transfer function denominator alone. The process examples in this section demonstrate the dramatic effects the numerator can have on the dynamic response.

Example 5.3 Heat exchanger with bypass. Often a process stream must be heated or cooled a variable amount using a heat exchanger. A common method for variable heating is a heat exchanger with bypass, as shown in Figure 5.9 for a cooler; the bypass provides the parallel structure in this example. The flows through the exchanger and through the bypass are adjusted while the total process flow is maintained constant. (How to achieve such flow behavior via the feedback control will be covered later.)

The behavior of an industrial shell-and-tube heat exchanger would be difficult to model, because it is a distributed-parameter system with complex flow patterns; therefore, the system is approximated as a stirred-tank heat exchanger, which retains the key properties of the system dynamics, in particular the response of the measured temperature signal to a step change in the flow to the exchanger.

Assumptions.

1. The same assumptions apply as in Example 3.7.
2. There is no transportation delay in short pipes.
3. The total flow (exchanger and bypass) is constant: $F_T = F_{\text{exch}} + F_{\text{by}} = \text{constant}$.

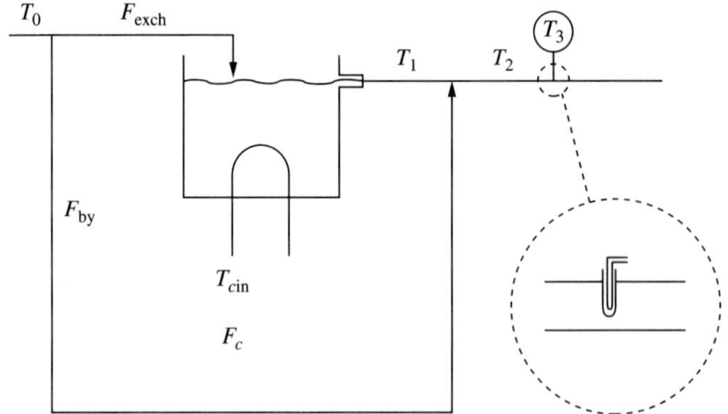

FIGURE 5.9
Heat exchanger with bypass and sensor.

Data. Note that these parameters are not realistic for a heat exchanger, although the dynamic response is reasonable because the increased fluid inventory takes the place of the substantial metal capacitance.

1. $T_0 = 100°\text{C}$; $\rho = 10^6\ \text{g/m}^3$; $C_p = 1\ \text{cal/(g°C)}$; $UA = 50 \times 10^6\ \text{cal/(min°C)}$; $T_{cin} = 60°\text{C}$; $\tau_3 = 0.5$ min; $V = 200\text{m}^3$.
2. Initial steady state: $F_{by} = 50\ \text{m}^3/\text{min}$; $F_{exch} = 50\ \text{m}^3/\text{min}$; $T_1 = 80°\text{C}$; $T_2 = T_3 = 90°\text{C}$.
3. Input change: $\Delta F_{exch} = -10\ \text{m}^3/\text{min}$ at $t = 10$ min; consequently, $\Delta F_{by} = +10\ \text{m}^3/\text{min}$.

Formulations. The fundamental model of the heat exchanger is the same as presented in Example 3.7, except that the feed flow rate, not the cooling medium flow, is changed in this example. Thus, model equations for the heat exchanger, bypass, and mixing are

$$\frac{dT_1}{dt} = \frac{F_{exch}}{V}(T_0 - T_1) - \frac{UA}{V\rho C_p}(T_1 - T_{cin}) \tag{5.42}$$

$$T_2 = \frac{F_{exch}T_1 + F_{by}T_0}{F_{exch} + F_{by}} = \frac{F_{exch}T_1 + (F_T - F_{exch})T_0}{F_T} \tag{5.43}$$

The temperature-measuring device is normally protected from contact with the process fluid by a metal sleeve called a *thermowell,* which introduces additional dynamic lag due to heat transfer dynamics associated with the thermowell. In this example, the thermowell dynamics are assumed to be well modelled by a first-order system with a time constant, τ_s, of 0.50 minutes (which is slower than most commercial sensor systems).

$$\tau_s \frac{dT_3}{dt} = T_2 - T_3 \tag{5.44}$$

with T_3 the signal from the sensor. These equations can be linearized, expressed in deviation variables, and transformed to the Laplace domain to give the individual transfer functions.

$$G_{ex} = \frac{T_1(s)}{F_{exch}(s)} = \frac{K_{exch}}{\tau_{exch}s + 1} \tag{5.45}$$

$$\text{with}\quad K_{exch} = \frac{(T_0 - T_1)_s \rho C_p}{(F_{exch})_s \rho C_p + UA} = 0.20\frac{°\text{C}}{\text{m}^3/\text{min}}$$

$$\tau_{exch} = \frac{V\rho C_p}{(F_{exch})_s \rho C_p + UA} = 2.0\ \text{min}$$

$$G_{FM}(s) = \frac{T_2(s)}{F_{exch}(s)} = \frac{(T_1 - T_0)_s}{(F_{exch} + F_{by})_s} = K_{FM} = -0.20\frac{°\text{C}}{\text{m}^3/\text{min}} \tag{5.46}$$

$$G_{TM}(s) = \frac{T_2(s)}{T_1(s)} = \frac{(F_{exch})_s}{(F_{exch} + F_{by})_s} = K_{TM} = 0.50\frac{\text{m}^3/\text{min}}{\text{m}^3/\text{min}} \tag{5.47}$$

$$G_s(s) = \frac{T_3(s)}{T_2(s)} = \frac{1.0}{\tau_s s + 1} = \frac{1.0}{0.5s + 1} \tag{5.48}$$

The block diagram for this model is shown in Figure 5.10, in which the parallel path is clearly evident, since the variable $F_{exch}(s)$ influences $T_2(s)$ through two paths. Note that for a parallel path to exist, a split must occur in the block diagram. The overall

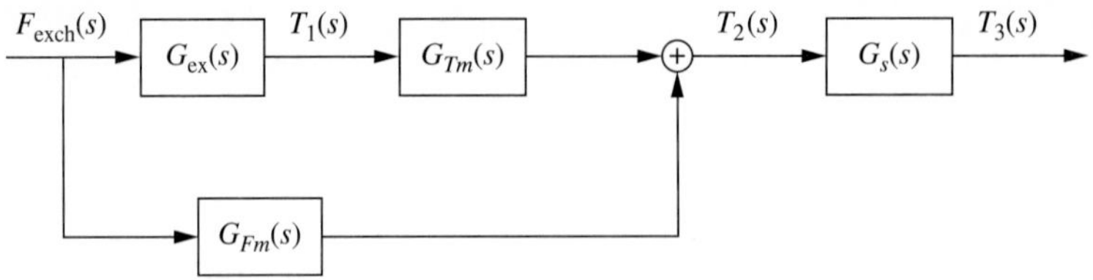

FIGURE 5.10
Block diagram of exchanger with bypass and sensor.

transfer function, relating the flow to the exchanger to the measured temperature, can be derived from block diagram algebra.

$$T_3(s) = G_s(s)[G_{FM}(s) + G_{TM}(s)G_{ex}(s)]F_{exch}(s) \tag{5.49}$$

$$\frac{T_3(s)}{F_{exch}(s)} = \frac{K_{FM}}{\tau_s s + 1} + \frac{K_{exch}K_{TM}}{(\tau_{exch}s + 1)(\tau_s s + 1)} \tag{5.50}$$

$$= \frac{(K_{FM} + K_{exch}K_{TM})\left[\left(\dfrac{K_{FM}\tau_{exch}}{K_{FM} + K_{exch}K_{TM}}\right)s + 1\right]}{(\tau_{exch}s + 1)(\tau_s s + 1)} = \frac{-0.1(4s + 1)}{(2s + 1)(0.5s + 1)}$$

From equation (5.50) we conclude that the poles of the overall system are the poles of the individual systems. In this example the system is second-order with real, distinct poles at $-1/\tau_{exch}$ and $-1/\tau_s$; thus, it is stable and is not periodic.

Due to the parallel structure, the transfer function has a zero in the numerator, which for this example is at $s = -(K_{FM} + K_{exch}K_{TM})/K_{FM}\tau_{exch}$ and is real and negative for this example. This zero can significantly affect the dynamic behavior of the system; therefore, the response of the system to a step input cannot be determined using Figure 5.5, which assumed a constant numerator. The dynamic response can be determined by inverting the Laplace transform of $T_4(s)$ for a step in $F_{exch}(s)$.

Solution. By substituting the data in the problem statement into equation (5.50), including the step input, $F_{exch}(s) = -10/s$, and determining the inverse using entry 10 in Table 4.1, the following analytical solution for the linear approximation can be found:

$$T_3'(t) = 1.0 - 2.333e^{-t/0.5} + 1.333e^{-t/2} \tag{5.51}$$

Results analysis. Dynamic responses are given in Figure 5.11 for the nonlinear and approximate linearized models. They both show that the system output, T_3, overshoots and then approaches its final value smoothly. Unlike the result in Example 4.6, which was a second-order system with distinct factors and a *constant numerator,* this response experiences a maximum. (The occurrence of the overshoot depends on the relative magnitudes of the numerator and denominator time constants.) The time at which the maximum occurs can be determined by setting the derivative of equation (5.51) to zero and solving for time, giving $t = 1.3$ minutes after the step. Thus, the parallel structure has fundamentally altered the dynamic behavior of this second-order system.

The reason for this behavior can be understood by considering the two parallel paths in the physical system. When the exchanger flow is decreased, temperature T_1 is initially unaffected, and the modified flow ratio to the mixing point results in an immediate increase in temperature T_2. However, the exchanger outlet temperature T_1 decreases with a first-order response because of the lower flow to the exchanger. As a

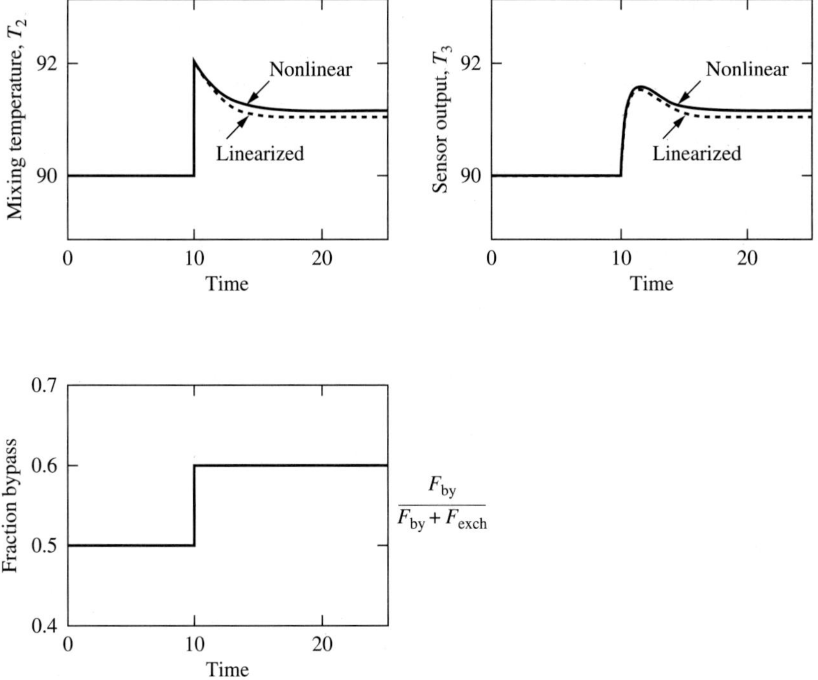

FIGURE 5.11
Nonlinear and linearized dynamic responses for Example 5.3.

result, the mixture temperature decreases from its initial peak to its final value with a first-order response. The measured temperature follows the mixture temperature after the sensor first-order lag. Note that the overshoot is not due to a complex pole and that the behavior is *not periodic*. Rather, the behavior is due to a parallel process path with significant differences in dynamics for the two paths.

Naturally, such behavior should be considered in designing and operating the process. Again, imagine driving an automobile that tends to overshoot the change in direction indicated by the steering wheel; a careful and skilled driver (or control algorithm) would be required.

Example 5.4 Series reactors. This example demonstrates that the parallel paths do not have to be external bypass streams but can be separate mechanisms within a single process. The process considered is a series of two CSTRs, shown in Figure 5.12, with the same vessel size, flow rate, and chemical reaction as in Example 3.3; thus, the reactor models are identical to those derived in Example 3.3, equations (3.24) and (3.25) (page 69). In this example the response of the reactant concentration at the outlet of the second reactor to a step change in the solvent flow is to be determined.

Formulation. Whereas in Example 3.3 the total flow was constant and the inlet concentration was changed, in this example the flows of the reactant and solvent can be changed independently. Also, the solvent flow is so much larger than the component A flow (F_A) that we assume that the total flow is the solvent flow; that is, $F \approx F_s$ and

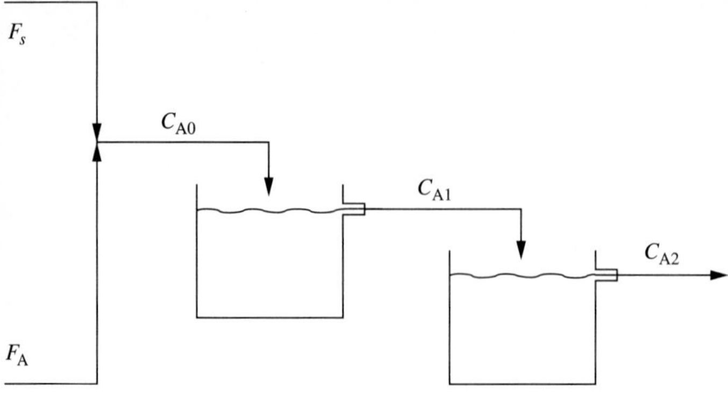

FIGURE 5.12
Series chemical reactors for Example 5.4.

$C_{A0} \approx (C_A F_A)/F_s$. (Note that $F_s = 0.085$ and $C_{A0} = 0.925$, so that $C_A F_A = 0.0786$ mole/min.) With these assumptions, the following transfer functions can be derived.

$$\frac{C_{A0}(s)}{F_s(s)} = G_{mix}(s) = -\frac{C_A F_A}{(F_s)^2} = K_{mix} = -10.9\frac{\text{mole/m}^3}{\text{m}^3/\text{min}} \tag{5.52}$$

$$\frac{C_{Ai}(s)}{F_s(s)} = G_{Fi}(s) = \frac{\dfrac{(C_{Ai-1}(s) - C_{Ai}(s))_s}{F_s + Vk}}{\left(\dfrac{V}{F_s + Vk}\right)s + 1} = \frac{K_{Fi}}{\tau s + 1}$$

$$G_{F1}(s) = \frac{2.41}{8.25s + 1} \qquad G_{F2}(s) = \frac{1.61}{8.25s + 1} \tag{5.53}$$

$$\frac{C_{Ai}(s)}{C_{Ai-1}(s)} = G_{Ai}(s) = \frac{\dfrac{F_s}{F_s + Vk}}{\left(\dfrac{V}{F_s + Vk}\right)s + 1} = \frac{K_{Ai}}{\tau s + 1} = \frac{0.669}{8.25s + 1} \qquad \text{for } i = 1, 2 \tag{5.54}$$

The linearized model is represented in the block diagram in Figure 5.13, which shows the parallel paths. In this example, the parallel paths result from the different effects of the solvent flow, through changes in the feed concentration and flow rate (residence time), on the outlet concentration of the second reactor. The overall

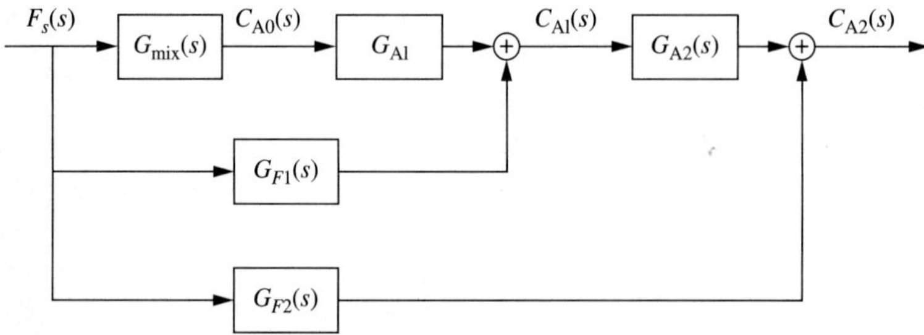

FIGURE 5.13
Block diagram of reactors in Example 5.4.

transfer function can be derived using the block diagram to give the overall input-output relationship.

$$\frac{C_{A2}(s)}{F_s(s)} = G_{F2}(s) + G_{A2}(s)G_{F1}(s) + G_{A2}(s)G_{A1}(s)G_{mix}(s) \tag{5.55}$$

This expression clearly shows that three separate effects of the input influence the output concentration. The first effect, $G_{F2}(s)$, is of the flow or residence time in the second reactor; this effect begins instantaneously and increases the concentration. The second effect, $G_{A2}(s)G_{F1}(s)$, is the residence time in the first reactor, which increases the feed concentration to the inlet to the second reactor. The third effect involves the decrease in the feed concentration, C_{A0}; this is slower but ultimately decreases the second reactor outlet concentration. The overall effect can be determined by substituting the individual transfer functions into equation (5.55) and rearranging to give

$$\frac{C_{A2}(s)}{F_s(s)} = \frac{K_{F2}(\tau s + 1) + K_{A2}K_{F1} + K_{A2}K_{A1}K_{mix}}{(\tau s + 1)^2} = \frac{-1.66(-8.0s + 1)}{(8.25s + 1)^2} \tag{5.56}$$

Again, the system is second-order and has the same poles as the individual elements in the system, but because of the numerator dynamics the response cannot be determined from a simple series system (i.e., Figure 5.5). Also, the result in this example is different from the previous example, because the transfer function in equation (5.56) has a *positive* numerator zero ($s = 1/8.0$). This is due to the last term in the numerator being large and negative, since K_{mix} is less than zero. This result indicates a mechanism for inverse response of the output variable, in which the initial response of the output can have the sign *opposite* to its final, steady-state change.

Solution. Again, the response can be determined by solving for the inverse Laplace transform using Table 4.1, entry 8. Substituting the data in the problem statement, including the input step of $F_S(s) = \Delta F_s/s = 0.0085/s$, gives

$$C_{A2}(t) = -0.0141 + (0.0141 + 0.00337t)e^{-t/8.25} \tag{5.57}$$

Results Analysis. The response to a step change in the solvent feed flow, with reactant flow unchanged, is shown in Figure 5.14 for the nonlinear and approximate linearized models. Note that the outlet reactant concentration initially *increases,* because of the decrease in residence time, which affects both reactors, including the last, immediately. However, the decreased feed concentration decreases the reactant concentration, initially in the first reactor and ultimately in the final reactor. Thus, the outlet concentration in the second reactor experiences an initial inverse response, because the fast effect of the residence time influences the output before the larger, slower feed concentration effect.

Behavior similar to this example is observed in other physical systems, especially tubular reactors. The series of CSTRs is selected in this example because the mathematical analysis is simpler, but lumped systems in series can serve as an approximation for the distributed system (Himmelblau and Bischoff, 1968). Modelling and experimental results for inverse responses in tubular reactors are presented by Silverstein and Shinnar (1982) and Ramaswamy et al. (1971).

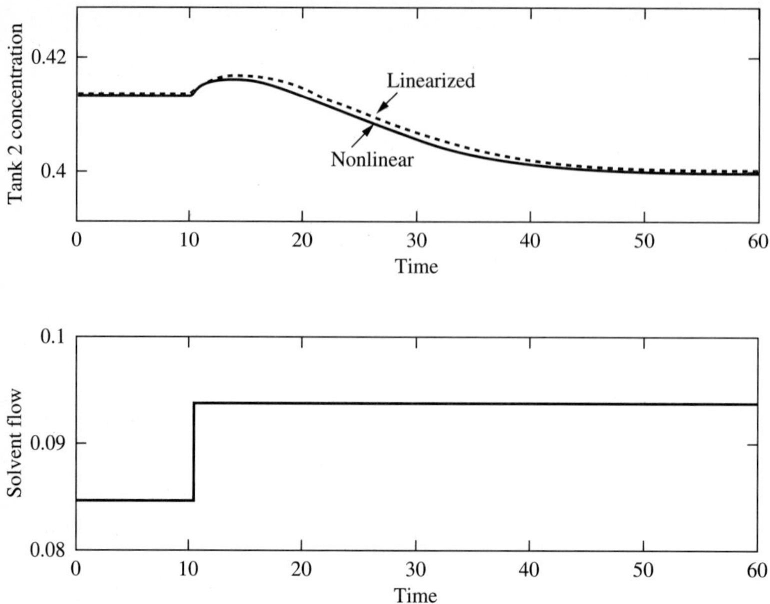

FIGURE 5.14
Response for series chemical reactor to step in solvent flow in Example 5.4.

The dynamic characteristics demonstrated in this example would be expected to have great influence on feedback control. Imagine driving an automobile that responded initially in the inverse direction to a change in steering! The most appropriate response would be to eliminate the inverse response by redesigning the process, if possible.

In summary, parallel paths exist in many processes either because of complex interconnecting flow structures of individual systems or because of parallel effects within a single process. Since the poles are unaffected by a parallel structure, stability and damping of the overall system are not affected. This can be seen from equations (5.50) or (5.56), in which the denominator of the overall transfer function has the poles of the individual transfer functions. However, the parallel paths can have a significant effect on the dynamic behavior of the system, and the most complex behavior—overshoot or inverse response—occurs when parallel paths have significantly *different speeds of response,* so that parallel responses from an input affect the output at different times. Also, the approximate time to reach 63 percent of the output change for a step input is affected by the numerator, and it is not simply the sum of the individual time constants.

The behavior of parallel systems of first-order individual systems is summarized in Table 5.2. Also, the model structures that yield numerator zeros and the ranges of parameter values giving various dynamic behaviors are addressed in questions 5.11 and 5.21.

The behavior presented in this section can cause some difficulty in terminology, since a stable overdamped system ($\xi \geq 1$) is usually thought to have a monotonic

TABLE 5.2
Properties of parallel systems with first-order elements

Individual first-order systems	Parallel system
Each is first order	Order of the highest order in a parallel path
Each is stable	Stable, not periodic
Poles are $1/\tau_i$	Poles are $1/\tau_i$, $i = 1, \ldots, n$
$t_{63\%}$	$t_{63\%} \neq \Sigma\tau_i$
Step response	Can be monotonic or experience overshoot or inverse response
Frequency response	Amplitude ratio can exceed steady-state process gain (for some frequency range)

response to a step input. This is true when the transfer function numerator is a constant, but it is not necessarily true when the numerator is a function of s. The potential dynamic behavior is summarized in the following table.

Poles	Response to nonperiodic input	Monotonic response to step
Complex	Periodic	Not possible
Real	Nonperiodic	Possible, depends on numerator

> The emphasis on complex dynamic responses in the examples in this section does not indicate that all systems with numerator zeros give unfavorable dynamics such as large overshoot or inverse response.

The engineer can analyze the physical process for possible parallel paths with different dynamics to identify potentially complex dynamics and then use quantitative methods to determine whether the behavior may cause difficulty for control. Each input must be considered separately, because the characteristics of the output dynamic response differ for different inputs.

5.5 RECYCLE STRUCTURES

Recycle structures are used often in process plants, to return valuable material for reprocessing and to recover energy from effluent streams through heat exchange. Such interconnections, termed *process integration,* are often cited as potential causes of difficulty in plant operations in spite of their advantages in the steady state; therefore, it is important to understand the effects of recycle on process dynamics. Again, this structure will be introduced through a process example and then will be generalized.

Example 5.5 Reactor with feed-effluent heat exchanger. The process design shown in Figure 5.15 has a feed-effluent heat exchanger and can be used for a chemical reactor with a high feed temperature and a need for cooling the product effluent stream.

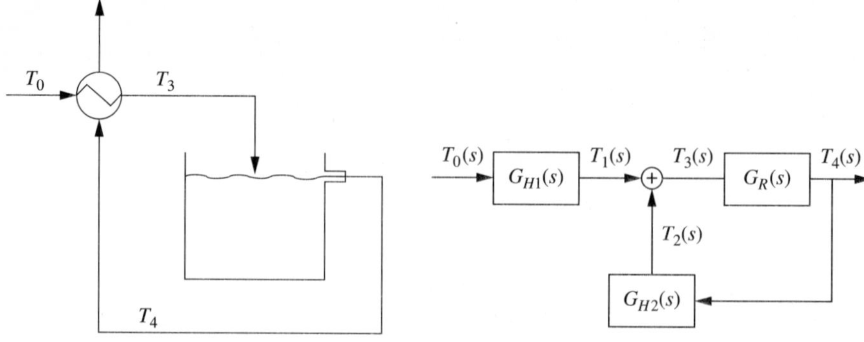

FIGURE 5.15
Reactor with feed-effluent heat exchanger in Example 5.5.

FIGURE 5.16
Block diagram of reactor-exchanger in Example 5.5.

Formulation. The analysis begins with the transfer functions of the following individual input-output relationships, represented in the block diagram in Figure 5.16.

$$\frac{T_1(s)}{T_0(s)} = G_{H1}(s) = \frac{K_{H1}}{\tau_{H1}s + 1} \qquad \frac{T_2(s)}{T_4(s)} = G_{H2}(s) = \frac{K_{H2}}{\tau_{H2}s + 1} \tag{5.58}$$

$$T_3(s) = T_1(s) + T_2(s) \qquad \frac{T_4(s)}{T_3(s)} = G_R(s) = \frac{K_R}{\tau_R s + 1}$$

The block diagram shows the output of the reactor returning to influence an input to the reactor. This is feedback that has been introduced into the process by a recycle of energy. To determine the behavior of the integrated system, the overall input-output transfer function must be determined using block diagram algebra.

$$\begin{aligned} T_4(s) &= G_R(s)T_3(s) = G_R(s)[T_1(s) + T_2(s)] \\ &= G_R(s)[G_{H2}(s)T_4(s) + G_{H1}(s)T_0(s)] \end{aligned} \tag{5.59}$$

$$\frac{T_4(s)}{T_0(s)} = \frac{G_R(s)G_{H1}(s)}{1 - G_R(s)G_{H2}(s)}$$

It is immediately apparent from the overall transfer function that recycle has fundamentally changed the behavior of the system, because the characteristic polynomial in equation (5.59) has been influenced and the poles of the overall system are not the poles of the individual units. Thus, the stability of the overall system cannot be guaranteed, even if each individual system is stable!

The chemical reaction could be either exothermic or endothermic. In either case, the result of an increase in reactor feed temperature T_3 is an increase in the reactor outlet temperature T_4. The difference is that the steady-state gain for $\Delta T_4/\Delta T_3$ is between 0.0 and 1.0 for an endothermic reaction and greater than 1.0 for an exothermic reactor. In this example, an exothermic reaction is considered. The parameters in the linearized transfer functions are as follows.

$$G_R(s) = \frac{3}{10s + 1} \qquad G_{H1}(s) = 0.40 \qquad G_{H2}(s) = 0.30$$

The transfer functions were selected to be representative but simple enough to yield a low-order overall model that can be understood easily. These transfer functions can be substituted into equation (5.59) to give the effect of the inlet temperature T_0 on the reactor temperature T_4 for the system with and without recycle.

With recycle.

$$\frac{T_4(s)}{T_0(s)} = \frac{\left(\dfrac{3}{10s+1}\right)(0.40)}{1 - \left(\dfrac{3}{10s+1}\right)(0.30)} = \frac{12}{100s+1} \tag{5.60}$$

Without recycle ($G_{H2}(s) = 0$).

$$\frac{T_4(s)}{T_0(s)} = G_{H1}(s)G_R(s) = \frac{1.2}{10s+1} \tag{5.61}$$

Results analysis. The foregoing expressions and the dynamic responses for a step input of 2°C in T_0 in Figure 5.17 show the dramatic effect of recycle on the steady-state gain and time constant; both increase by a factor of 10 due to recycle. This can be understood by analyzing the interaction between the exchanger and reactor during a transient; an increase in T_0 causes an increase in T_3 and then T_4, which causes an increase in T_2, which causes an increase in T_4, and so on; in short, the output change is reinforced through the recycle (feedback) exchanger. The system is still stable and self-regulatory, because of the dominant inherent negative feedback for the parameter values in this example, but the recycle has created an inherent *positive* feedback in the process, which has significantly affected the dynamic response. The potentially unfa-

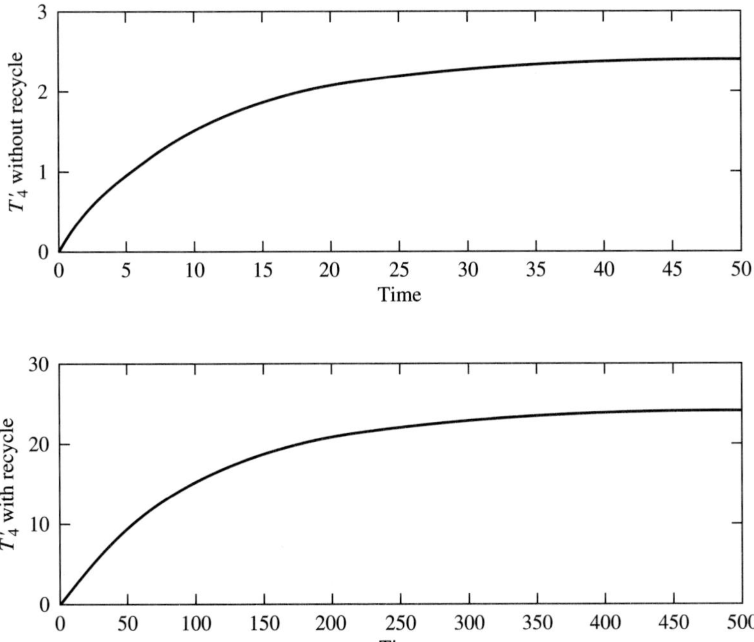

FIGURE 5.17
Dynamic responses for a 2°C step in T_0 at time = 0 Example 5.5 with and without recycle.

vorable dynamic effects of recycle can be reduced through automatic control strategies, which retain most of the process performance benefits, as demonstrated for this chemical reactor design in Figure 24.11.

The simple example in this section demonstrates the potential effects of recycle on dynamic behavior:

1. Recycle can alter the stability and possibility for periodic behavior of the overall system, because it affects the poles of the overall system.
2. The time constants and steady-state gain of the overall system with recycle can be changed substantially from their values without recycle.

Again, understanding the effect of recycle on dynamic responses is an important aspect of process dynamics, and the material in this section is enhanced by additional analysis in Appendix C and by reference to the studies of recycle in the Additional Resources at the end of this chapter.

Finally, the process recycle considered here has the characteristic of having a process output return to affect a process input, which is the definition of feedback introduced in Chapter 1. In later parts of the book, the negative-feedback principle is applied, to reduce the effects of disturbances, by means of control calculations that adjust a process input. Thus, recycle and feedback involve the same principle; the term *recycle* is usually used for process structures and *feedback* for external control calculations and the resulting input manipulations.

5.6 STAGED PROCESSES

Staged processes are used widely in the process industries for multiple contacting of streams and can be considered as a special interconnection of elements, in which an element exchanges material and energy with only the adjoining stages. Some common examples are vapor-liquid equilibrium (Treybal, 1955), multieffect evaporation (Nisenfeld, 1985), and flotation (Narraway et al., 1991). Staged systems can experience a wide variety of dynamic behavior depending on the physical processes (e.g., mass transfer, heat transfer, and chemical reaction) that occur at each stage.

The fundamental model for a staged system must include all significant balances on every stage. However, the variables at every stage are not always of great importance for the overall performance of the process, because only the properties of the streams leaving the process are usually of interest. In some cases, a few intermediate variables could be important; an example is the flows on stages of a stripping tower, which might approach or exceed the hydraulic limits for proper contacting efficiency. We will assume in this section that the only output properties of interest are in the product streams.

In this section the dynamics of a distillation tower, shown in Figure 5.18, are considered as an example of staged systems to introduce the modelling approach and describe typical dynamic behavior. An accurate model of a multicomponent

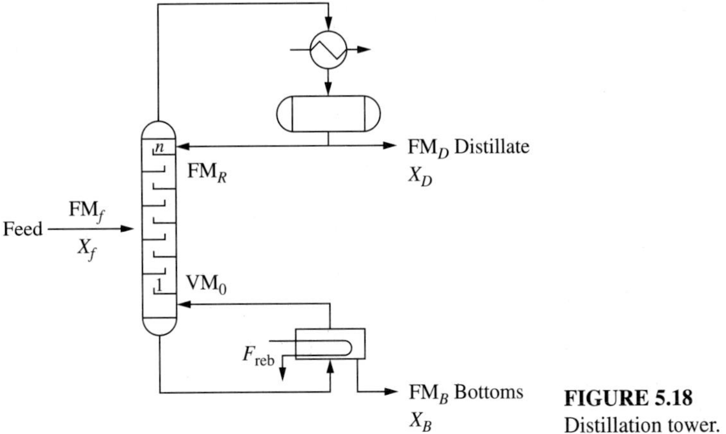

FIGURE 5.18
Distillation tower.

distillation tower must consider complex thermodynamic relationships and employ special numerical algorithms for the simultaneous solution of equilibrium expressions and material and energy balances. To simplify the presentation while maintaining a realistic model, the tower considered will separate only two components, and the phase equilibrium is assumed to be well represented by a constant relative volatility (Smith and Van Ness, 1987). Also, the energy balance at each stage can be simplified by the assumption of *equal molal overflow,* which implies that the heats of vaporization of both components are equal and mixing and sensible heat effects are negligible.

The assumptions are

1. The liquid level on every tray remains above the weir height.
2. Equal molal overflow applies.
3. Relative volatility α and heat of vaporization λ are constant.
4. Holdup in vapor phase is negligible.

The following nomenclature is used:

MM = molar holdup of liquid on tray
FM = molar flow rate of liquid
X = mole fraction of light component in liquid
λ = heat of vaporization
VM = molar flow rate of vapor
Y = mole fraction of light component in vapor

The schematic of a general tray in Figure 5.19 shows that every tray has the potential for feed and product flows and heat transfer. With the assumptions and the general tray structure, the basic overall and component balances for each stage or tray ($i = 1, \ldots, n$) can be formulated as

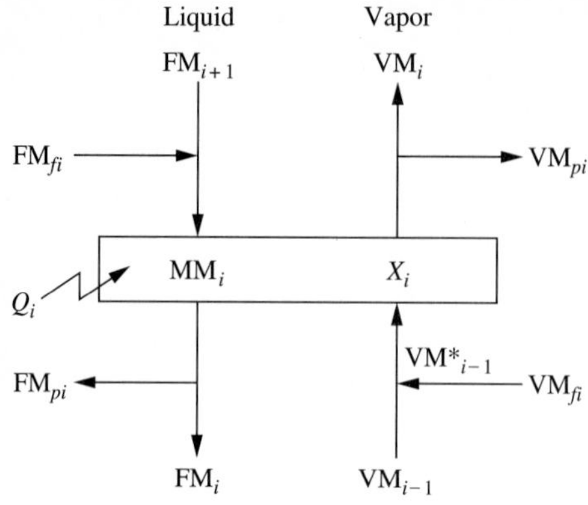

FIGURE 5.19
General tray used in modelling distillation.

Overall material (molar) balance on liquid phase:

$$\frac{d\text{MM}}{dt} = \text{FM}_{i+1} - \text{FM}_i + \text{FM}_{fi} - \text{FM}_{pi} - \frac{Q_i}{\lambda} \tag{5.62}$$

Quasi-steady-state overall material (molar) balance on vapor phase:

$$\text{VM}_i = \text{VM}^*_{i-1} - \text{VM}_{pi} + \frac{Q_i}{\lambda} \tag{5.63}$$

$$\text{VM}^*_{i-1} = \text{VM}_{i-1} + \text{VM}_{fi} \tag{5.64}$$

$$Y^*_{i-1} = \frac{\text{VM}_{i-1}Y_{i-1} + \text{VM}_{fi}Y_{fi}}{\text{VM}^*_{i-1}} \tag{5.65}$$

Light component balance on the tray:

$$\begin{aligned}\frac{d(\text{MM}_i X_i)}{dt} &= \text{FM}_{i+1}X_{i+1} + \text{FM}_{fi}X_{fi} - (\text{FM}_{pi} + \text{FM}_i)X_i \\ &\quad - (\text{VM}_i + \text{VM}_{pi})Y_i + \text{VM}^*_{i-1}Y^*_{i-1}\end{aligned} \tag{5.66}$$

This formulation is adequate for every equilibrium tray in the tower. For most trays, feed flows, product flows, and heat transferred are zero, while at least one tray has a nonzero feed. The top tray has a liquid feed, which is reflux, and its vapor stream goes to the total condenser. The bottom tray has its liquid go to the kettle reboiler, which is also an equilibrium stage. Note that although the equations can be formulated as shown, the computer implementation in this form would involve extensive multiplications for the zero streams; thus, an efficient implementation for a specific design would eliminate streams that are always zero.

Since there are many more variables than equations in the conservation balances, the model is not completely specified by these balances alone. The model requires constitutive expressions to relate liquid and vapor compositions. The phase equilibrium equation for a binary system with constant relative volatility α is

$$Y_i = \frac{\alpha X_i}{1 + (\alpha - 1)X_i} \tag{5.67}$$

The model also requires constitutive expressions to relate liquid flows and inventories on the trays. The liquid flow from a tray is related to the level ($L_i = \text{MM}_i/\rho_M A$) above the weir height, L_w, by (Foust et al., 1980)

$$\text{FM}_i = K_w \sqrt{\frac{\text{MM}_i}{\rho_M A} - L_w} \tag{5.68}$$

with A being the cross-sectional area and ρ_M moles/m^3. The modelling effort is not complete until models are developed for the associated equipment, which for this distillation tower includes the heat exchangers that vaporize part of the liquid accumulated in the bottom drum and condense the overhead vapor. The behavior of these is not particularly complex but requires feedback control to model properly. To maintain simple model structures without the need for control at this point, the reboiler duty is assumed to be proportional to the heating medium flow, and the vapor overhead is assumed to be completely condensed without subcooling, so that the pressure is maintained at a constant value by adjusting the condensing duty, thus

$$Q_{\text{cond}} = \text{VM}_n \lambda \tag{5.69}$$

$$Q_{\text{reb}} = K_{\text{reb}} F_{\text{reb}} \tag{5.70}$$

Also, the volumes in the overhead and bottom accumulators can be modelled by overall and component balances. In reality, the levels of these inventories would be controlled by adjusting the product flows; in this example, the levels are assumed exactly constant, so that the models become

$$\text{FM}_D = \text{VM}_n - \text{FM}_R \tag{5.71}$$

$$\text{FM}_B = \text{FM}_1 - \text{VM}_0 \tag{5.72}$$

The composition in the overhead accumulator ($X_{n+1} = X_D$) can be determined from a component material balance:

$$\text{MM}_\text{D} \frac{dX_D}{dt} = \text{VM}_n Y_n - X_D(\text{FM}_D + \text{FM}_R) = \text{VM}_n(Y_n - X_D) \tag{5.73}$$

Again, with the inventory constant, the kettle reboiler can be modelled with a component material balance ($X_0 = X_B$), equilibrium relationship, and a calculation of vapor flow based on heat transferred.

$$\text{MM}_B \frac{dX_B}{dt} = \text{FM}_1 X_1 - \text{FM}_B X_B - \text{VM}_0 Y_0 \tag{5.74}$$

$$Y_0 = \frac{\alpha X_B}{1 + (\alpha - 1)X_B} \tag{5.75}$$

$$\text{VM}_0 = \frac{Q_{\text{reb}}}{\lambda} \tag{5.76}$$

To specify the system completely, sufficient external input variables must be specified that the degrees of freedom are zero. The feed flow and composition must be specified along with two additional variables, here selected to be the distillate product flow F_D and the reboiler heating flow F_{reb}. With these external variables specified, the degrees-of-freedom analysis summarized in Table 5.3 shows that

TABLE 5.3
Distillation degrees of freedom for n trays

	Equations	Variables (dependent)	External specified variables (independent)
Trays	(5.62) to (5.68) for each tray (7n)	MM, FM, VM, X, Y, Y^*, V^* for each tray plus FM_{n+1}, X_{n+1}, V_0, Y_0 ($7n+4$)	FM_f, X_f, VM_f, Y_f, FM_p, VM_p, Q for each tray (7n)
Overhead	(5.69), (5.71), and (5.73) (3)	Q_{cond} (1)	FM_R or FM_D, MM_D (2)
Reboiler	(5.70), (5.72), (5.74), (5.75), and (5.76) (5)	X_B, FM_B, and Q_{reb} (3)	F_{reb}, MM_B (2)
Total	$7n+8$	$7n+8$	$7n+4$

the system is exactly specified. The number of equations is equal to the number of variables; thus, there are zero degrees of freedom. Note that the parameters (λ, α, K_w, MM_D, K_{reb}, MM_B, and L_w) were excluded from the analysis, because they are always constant. Also, the feed variables are determined by upstream process conditions. Typically, external variables like the reboiler heating flow rate and the distillate product flow rate are adjusted to achieve the desired product compositions; here, they are assumed known external variables. The model formulation included assumptions, like constant accumulator levels and pressure, that are not necessary but simplify the model and presentation.

Example 5.6. Determine the dynamic behavior of a binary distillation tower with the parameters in Table 5.4. The model equations can be integrated numerically to determine the response of the system from specified initial conditions for any values

TABLE 5.4
Base case design parameters for example binary distillation

Relative volatility	2.4
Number of trays	17
Feed tray	9
Analyzer dead times	2 min
Feed light key	$X_F = 0.50$
Distillate light key	$X_D = 0.98$ fraction
Bottoms light key	$X_B = 0.02$ fraction
Feed flow	$\mathrm{FM}_F = 10.0$ kmole/min
Reflux flow	$\mathrm{FM}_R = 8.53$ kmole/min
Distillate flow	$\mathrm{FM}_D = 5.0$ kmole/min
Vapor reboiled	$\mathrm{VM}_0 = 13.53$ kmole/min
Tray holdup	$\mathrm{MM}_i = 1.0$ kmole
Holdup in drums	$\mathrm{MM}_B = \mathrm{MM}_D = 10.0$ kmole

or functions of the external variables. The dynamic responses are obtained by establishing a steady-state operating condition and introducing a single step change to one of the external variables; each step is 1 percent of the base case input value. (This is exactly how the experiment would be performed on the physical tower, as explained in Chapter 6.) The results are shown in Figures 5.20*a* and *b*. The composition responses are smooth monotonic sigmoidal curves, in spite of the complexity of the process. Note that changing a single input affects both product compositions—an important factor in subsequent control design as discussed in Chapters 20 and 21.

Staged systems present a wide range of dynamic behavior and challenging control problems. Because of their industrial importance, they have been studied extensively. A few of the important results are summarized in the following list.

1. All model equations could be linearized, and the linearized model and transfer functions could be analyzed. An analytical solution has been developed for the response to a step input of a *linearized* binary tower section, without reflux or reboil with constant liquid holdup on all trays (Amundson, 1966).
2. The poles of the model described in this section have been demonstrated to be all real and distinct (Levy et al., 1969). More realistic models, which include energy balances, have complex conjugate poles; however, the constants associated with these poles are generally quite small (Levy et al., 1967; Heckle et al., 1975). Therefore, one can expect the real poles to dominate the behavior of the response, and expect *simple* distillation towers without control to experience nonperiodic behavior.
3. The numerator zeros of the system, when they exist, tend to cancel some of the poles approximately, so that the system for some inputs behaves like a lower-order process (Kim and Friedly, 1974). This explains the response of the bottom composition in Figure 5.20*a*, which appears close to a first-order response.
4. The dynamic response depends strongly on the operating conditions of the distillation tower (Fuentes and Luyben, 1983; Kapoor et al., 1986). Thus, an approximate linear model at one set of operating conditions can be a poor approximation for other operating conditions.
5. Although the liquid holdup on each tray is small, and thus the time constant of each tray when considered *individually* is on the order of tens of seconds, the dynamics of a distillation tower can be long, on the order of hours for some towers. Experience indicates that slow dynamics tend to occur for high-purity towers with low relative volatility. This large difference in dynamic response from the individual elements results from the tray structure with recycle in the design of distillation. This statement is supported by numerical results for towers with realistic numbers of trays in Fuentes and Luyben (1983) and Kapoor et al. (1986).

This summary presents a small sample of the results available on distillation dynamics. They have been presented as general guidelines for the behavior of two-product distillation with simple thermodynamics (e.g., no azeotropes) and no chemical reaction. The reader is encouraged to refer to the citations and Additional Resources for further details. This distillation example will be considered in later

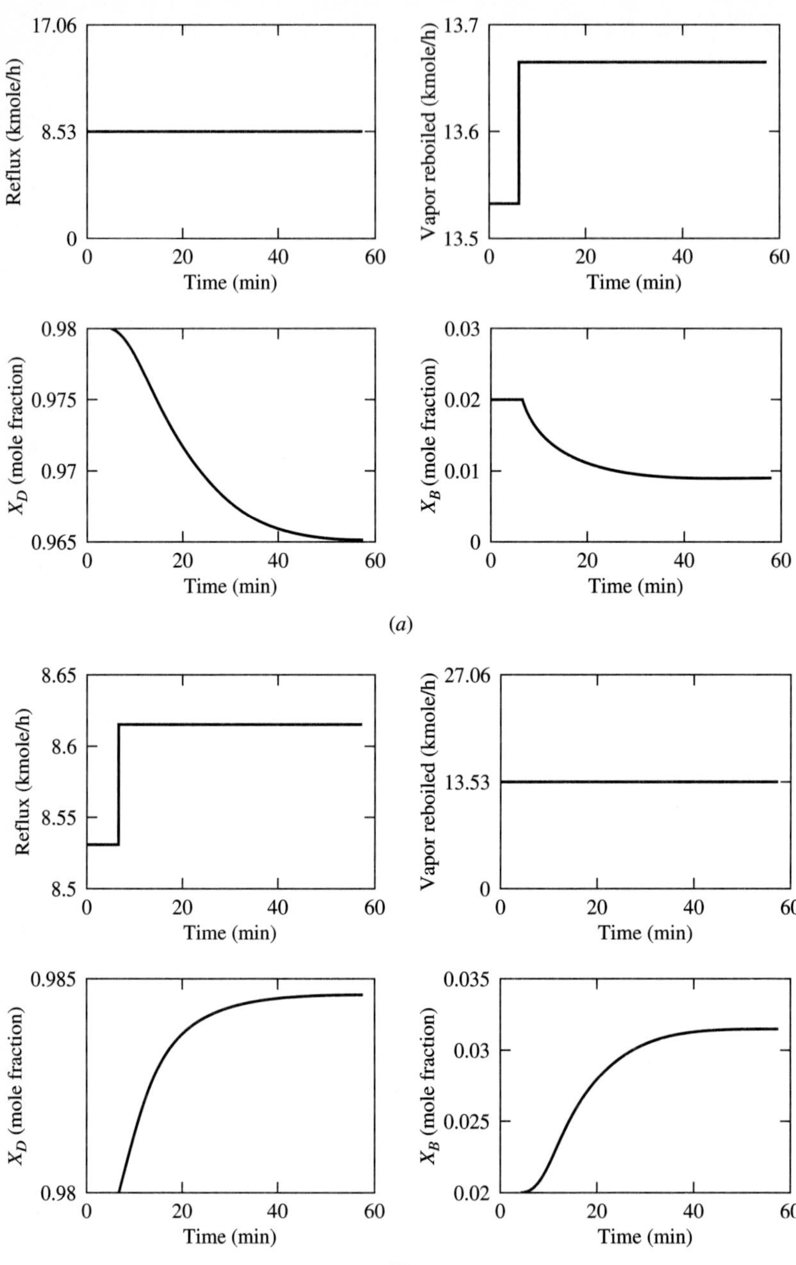

FIGURE 5.20
Response of distillate and bottoms products in Example 5.6 (without sensor dead times): (*a*) to reboiler step change; (*b*) to reflux step change.

chapters, where the control of the product compositions, through adjustments to such variables as the reboiler duty and reflux flow, will be investigated.

5.7 MULTIPLE INPUT–MULTIPLE OUTPUT SYSTEMS

Many, but not all, of the systems modelled in Chapters 3, 4, and 5 have involved a single input and output. If intermediate variables existed, they could be eliminated using transfer functions and block diagram algebra to develop a *single input–single output* (SISO) equation. This approach helped to simplify our task of learning how to model dynamic responses and is applicable to some realistic processes. However, the majority of processes have several inputs, and process operation is concerned with more than one output simultaneously. For example, the nonisothermal chemical reactor in Example 3.10 has coolant flow and inlet concentration as inputs and reactor concentration and temperature as outputs. Also, the distillation tower in the previous section has distillate product flow, reboiler flow, and all feed properties and flow rate as inputs and concentration of both product streams as outputs.

The methods described in the previous two chapters for developing fundamental models—linearization, transfer functions, block diagrams—are all applicable to these *multiple input–multiple output* (MIMO) systems. Again, we see that many intermediate variables can exist in a process; in the distillation tower, the tray compositions and holdups are intermediate variables. These intermediate variables are included in the fundamental model and eliminated algebraically from the linearized input-output relationship.

Example 5.7 MIMO block diagram. As an example, the block diagram for the nonisothermal chemical reactor in Example 3.10 is given in Figure 5.21.

Recall that the transfer function is defined for a single input-output relationship. Additional inputs, such as feed temperature and feed flow rate, could have been included but were omitted here to avoid cluttering the diagram. Also, since there is no interaction among the outputs (without feedback control), the dynamic responses and frequency responses of each output can be determined independently, using methods already derived and demonstrated.

Example 5.8 MISO response. Determine the response of the linearized nonisothermal chemical reactor in Example 3.10 to simultaneous step changes of -1 m^3/min in coolant flow rate and of $+0.05$ (mole/m^3) in feed composition.

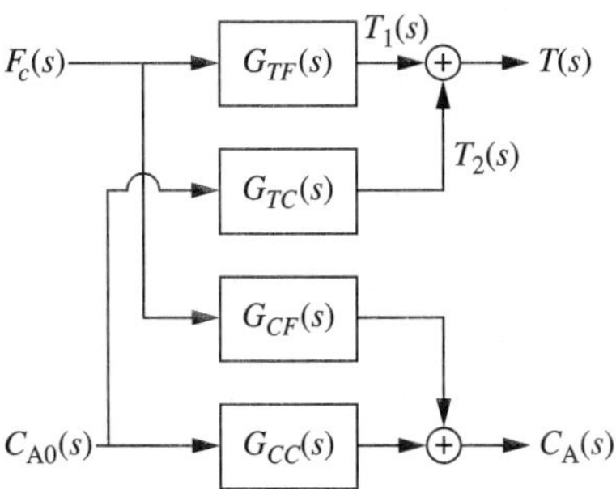

FIGURE 5.21
Block diagram of two-input–two-output system.

Solution. The solution requires the same approach for inverting Laplace transforms used in Example 4.8 except that an additional input variable must be considered. As shown in Figure 5.21, the response of the linearized system to each input can be determined separately, and then the two responses can be added to obtain the total change in reactor temperature. Again, this is an application of the superposition principle, which is valid only for linear systems. The transfer function for the effect of feed composition on reactor temperature was determined in Example 4.10 (page 125) from the linearized model (derived in Example 3.10) to be

$$G_{TC}(s) = \frac{T_2(s)}{C_{A0}(s)} = \frac{a_{23}s + (a_{21}a_{13} - a_{23}a_{11})}{s^2 - (a_{11} + a_{22})s + (a_{11}a_{22} - a_{12}a_{21})} \tag{5.77}$$

As expected, this transfer function has the same characteristic polynomial as $G_{TFc}(s)$ in Example 4.8. The data can be substituted, including $a_{23} = 0$, to give

$$T_2(s) = \frac{0.05}{s}\left(\frac{23.88}{s^2 + 1.788s + 35.8}\right) \tag{5.78}$$

The inverse of this Laplace transform can be determined using entry 17 in Table 4.1 to be

$$T_2'(t) = 1.19[1 - 1.01e^{-.894t}\sin(5.92t + 1.42)] \tag{4.79}$$

The response for the change in coolant flow rate, derived in Example 4.8 (pages 121–122), is repeated here.

$$T_1'(t) = 1.28 + 2[-0.64\cos(5.92t) + 0.42\sin(5.92t)]e^{-0.894t} \tag{5.80}$$

The dynamic responses from the linearized model ($T' = T_1' + T_2'$) and from the numerical solution of the nonlinear model are shown in Figure 5.22 for simultaneous

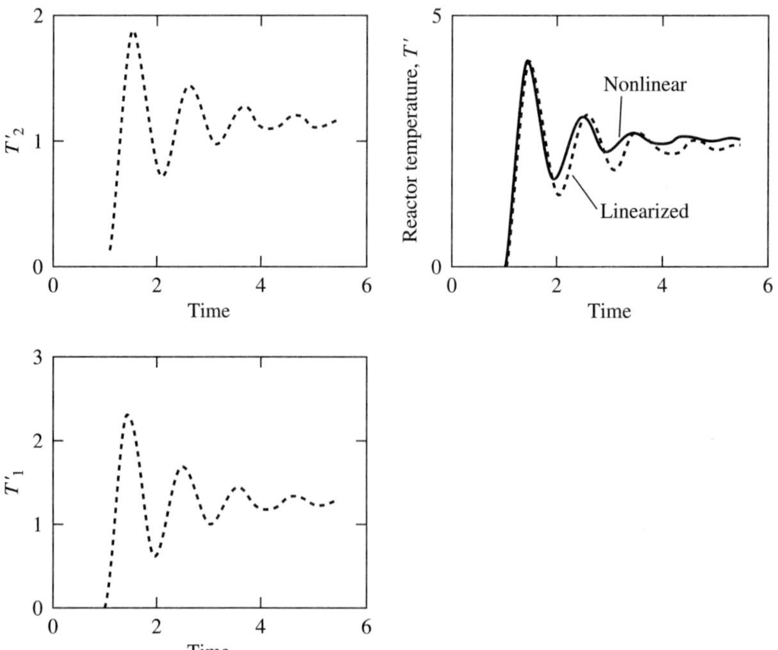

FIGURE 5.22
Dynamic responses for Example 5.8 in deviation variables.

step changes in the coolant flow and inlet concentration at $t = 1$; note that time in equations (5.79) and (5.80) is measured from the introduction of the step. Because of the complex poles, due to the inherent positive feedback effects of the exothermic reaction, both linearized outputs are periodic. It is important to recognize that the temperatures T_1 and T_2 do *not exist* physically; they are predictions of the separate effects of the two inputs on the single output in the linear model. This approach could be used to determine the response of the concentration to the same input changes.

5.8 CONCLUSIONS

The results of this chapter clearly demonstrate that process structures have strong effects on dynamic behavior and that these effects can be predicted using the methods presented in the previous chapters. Many of the strongest results relate to the "long-time" behavior of the systems, because they are determined by the poles of the transfer function and are independent of the numerator zeros. These properties involve stability and the related tendency for over- or underdamped behavior. However, the numerators also play an important role in the dynamic response, as shown by the examples in the section on parallel structures.

It is worth noting that each of these process structures is covered individually to clarify the analysis of their effects on dynamic behavior. Naturally, a process may contain several of these structures, all of which will influence its behavior. The study of complex processes is delayed until Parts V and VI, which address the control of multiple input–multiple output systems.

Finally, in the last three chapters, dynamic responses of many processes to a step input have been shown to have a sigmoidal shape. This means that these processes could be approximated by adjusting parameters in a model of simple structure. While this observation is not especially helpful for analytical modelling, it is very important for empirical modelling, which develops models based on experimental data. This is the topic of the next chapter.

REFERENCES

Amundson, N., *Mathematical Methods in Chemical Engineering, Matrices and Their Applications,* Prentice Hall, Englewood Cliffs, NJ, 1966.

Buckley, P., *Techniques in Process Control,* Wiley, New York, 1964.

Coughanowr, D., *Process Systems Analysis and Control,* McGraw-Hill, New York, 1991.

Foust, A., et al., *Principles of Unit Operations,* Wiley, New York, 1980.

Fuentes, C. and W. Luyben, "Control of High Purity Distillation Columns," IEC (Industrial and Engineering Chemistry) *Proc. Des. Devel., 22,* 361–366 (1983).

Heckle, M., B. Seid, and W. Gilles, "Conventional and Modern Control for Distillation Columns, Design and Operating Experience," *Chem. Ing. Tech.* (German), *47, 5,* 183–188 (1975).

Himmelblau, D. and K. Bischoff, *Process Analysis and Simulation; Deterministic Systems,* Wiley, New York, 1968.

Kapoor, N., T. McAvoy, and T. Marlin, "Effect of Recycle Structures on Distillation Time Constants," *A. I. Ch. E. J., 32, 3,* 411–418 (1986).

Kim, C. and Friedly, J., "Approximate Dynamics of Large Staged Systems," *IEC, Proc. Des. Devel., 13, 2,* 177–181 (1974).

Levy, R., A. Foss, and E. Grens, "Response Modes of Binary Distillation Columns," *IEC Fund., 8, 4,* 765–776 (1969).

Narraway, L., J. Perkins, and G. Barton, "Interaction Between Process Design and Process Control," *J. Proc. Cont., 1, 5,* 243–250 (1991).
Nisenfeld, E., *Industrial Evaporators, Principles of Operation and Control,* Instrument Society of America, Research Triangle Park, NC, 1985.
Ogata, K., *System Dynamics* (2nd Ed.), Prentice-Hall, Englewood Cliffs, NJ, 1992.
Ramaswamy, V., F. Stermole, and K. McKinstry, "Transient Response of a Tubular Reactor to Upsets in Flow Rate," *A. I. Ch. E. J., 17, 1,* 97–101 (1971).
Silverstein, J. and R. Shinnar, "Effect of Design on Stability and Control of Fixed Bed Catalytic Reactors with Heat Feedback. 1. Concepts," *IEC Proc. Des. Devel., 21,* 241–256 (1982).
Treybal, R., *Mass Transfer Operations,* McGraw-Hill, New York, 1955.
Smith, J. and H. Van Ness, *Chemical Engineering Thermodynamics* (4th Ed.), McGraw-Hill, New York, 1987.
Weber, T., *An Introduction to Process Dynamics and Control,* Wiley, New York, 1973.

ADDITIONAL RESOURCES

Recycle systems occur frequently and substantially affect process dynamics. Some studies on these effects are noted here.

Gilliland, E., L. Gould, and T. Boyle, "Dynamic Effects of Material Recycle," *Joint Auto. Cont. Conf.,* Stanford, CA, 140–146 (1964).
Rinard, I. and B. Benjamin, "Control of Recycle Systems, Part 1. Continuous Systems," *Auto. Cont. Conf.* 1982, WA5.
Luyben, W., "Dynamics and Control of Recycle Systems. 1. Simple Open-Loop and Closed-Loop Systems," *IEC Res., 32,* 466–475 (1993).
Douglas, J., J. Orcutt, and P. Berthiaume, "Design of Feed-Effluent Exchanger-Reactor Systems," *IEC Fund., 1, 4,* 253–257 (1962).

Inverse response can be a vexing problem for control. The engineer should understand the process causes of inverse response systems and modify the design to mitigate the effect.

Iionya, K. and R. Altpeter, "Inverse Response in Process Control," *IEC, 54, 7,* 39 (1962).

Modelling complex distillation columns is a challenging task that has received a great deal of study.

Holland, C., *Unsteady-State Processes with Applications in Multicomponent Distillation,* Prentice Hall, Englewood Cliffs, NJ, 1966.
Gilliland, E. and C. Reed, "Degrees of Freedom in Multicomponent Absorption and Rectification Columns," *IEC, 34, 5,* 551–557 (1942).
Howard, G., "Degrees of Freedom for Unsteady-State Distillation Processes," *IEC Fund., 6, 1,* 86–9 (1967).
Tyreus, B., W. Luyben, and W. Schiesser, "Stiffness in Distillation Models and the Use of Implicit Integration Method to Reduce Computation Time," *IEC Proc. Des. Devel., 14, 4,* 427–433 (1975).

The guidance before the questions in Chapters 3 and 4 is appropriate here as well. The key new issue introduced in this chapter and demonstrated in these questions is the effect of structure on the behavior of relatively simple individual elements.

QUESTIONS

5.1. A linearized model for a stirred-tank heat exchanger is derived in Example 3.7 for a change in the coolant flow rate. Extend these results by deriving the model for simultaneous changes in the coolant flow rate and inlet temperature. Also, determine an analytical expression for the outlet temperature $T'(t)$, for simultaneous step changes in the coolant flow and inlet temperature. (You may use all results from Example 3.7 without deriving.)

5.2. The jacketed heat exchanger in Figure Q5.2 is to be modelled. The input variable is T_0', and the output variable is T'. The inlet coolant temperature is constant. The following assumptions may be made:

1. Both vessels are well mixed.
2. Physical properties are constant.
3. Flows and volumes are constant.
4. $Q = UA(T - T_c)$
5. The dynamic balances on *both* volumes must be solved simultaneously.

(*a*) Write the basic balances for both volumes in deviation variables.

(*b*) Take the Laplace transforms.

(*c*) Combine into the transfer function $T'(s)/T_0'(s)$.

(*d*) Analyze this result to determine whether the dynamic behavior is (i) stable and (ii) periodic. Remember that these properties are defined by the denominator of the transfer function.

(*e*) The transfer function ignores initial conditions of the system. Briefly explain why the transfer function is useful — in other words, what properties can be determined easily using the transfer function?

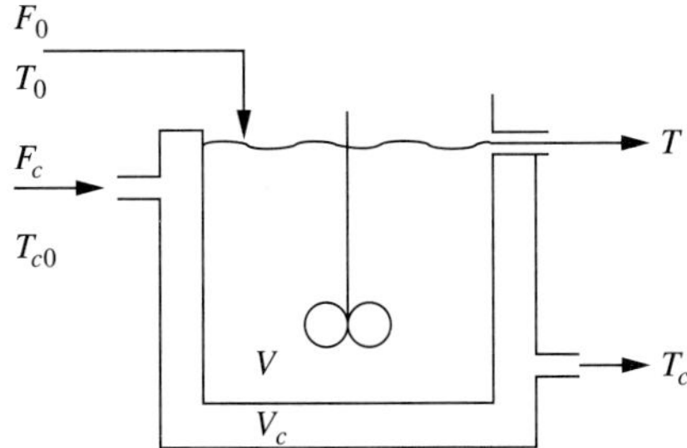

FIGURE Q5.2

5.3. The continuous-time systems of two stages shown in Figures Q5.3*a* and *b* are to be analyzed. Assumptions are the following:

1. Liquid holdups are constant $= M$.
2. Constant molal overflow; the liquid (L) and vapor (V) flows are constant.
3. The concentrations x_3 (and x_2^* in Figure Q5.3*b*) are constant.
4. The accumulation in the vapor phase is negligible.
5. Equilibrium can be modelled as $y_i = Kx_i$ for this binary system.

The nature of the dynamic behavior is to be determined for the input-output $x_2(s)/y_0(s)$.

(*a*) Derive the time-domain equations describing the dynamics of the concentrations on the two trays, $x_1'(t)$ and $x_2'(t)$, to the input variable $y_0'(t)$, in deviation variables.

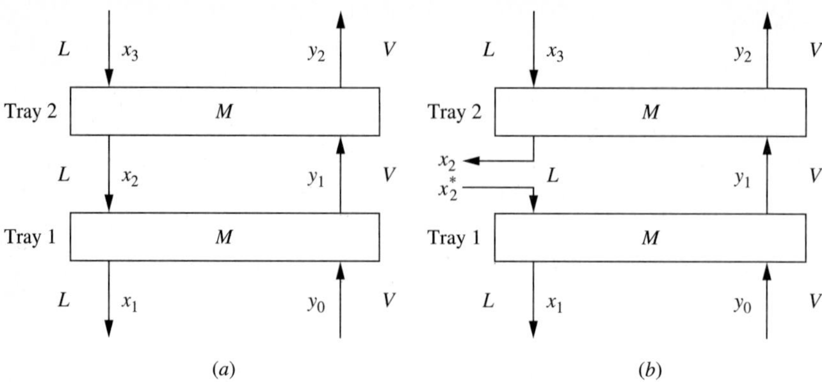

FIGURE Q5.3

(*b*) Combine the results of (*a*) into the single transfer function $x_2(s)/y_0(s)$.

(*c*) Determine the nature of the response. Is it (i) stable, (ii) over- or underdamped?

(*d*) Is the response of x_2 to a step change in y_0 in Figure Q5.3*a* faster or slower than in the system in Figure Q5.3*b* (with the same parameter values and x_2^* constant)?

5.4. A model and dynamic response are derived in Example 5.4 for a series of two chemical reactors. In the example, the solvent flow (F_s) is changed in a step, and the outlet concentration experiences an inverse response. Determine the dynamic response for a step change of ΔF_A, with all other inputs, including F_s, constant. All assumptions are the same as in Example 5.4, and you may use relevant results without deriving. The answer to this question should include an analytical expression for the response and a semi-quantitative description of the dynamic response of the concentration in the second reactor.

5.5. The recycle mixing system in Figure Q5.5 is to be considered. The feed flow is 1 unit, and the recycle flow is 9 units. The pipe has a dead time of 10 seconds, and the recycle has negligible dynamics. The system is initially at steady state with pure solvent entering as feed. At time $= 0$, the concentration of the feed increases to 10%A. Plot the concentration at the exit of the pipe from $t = 0$ to the new steady state.

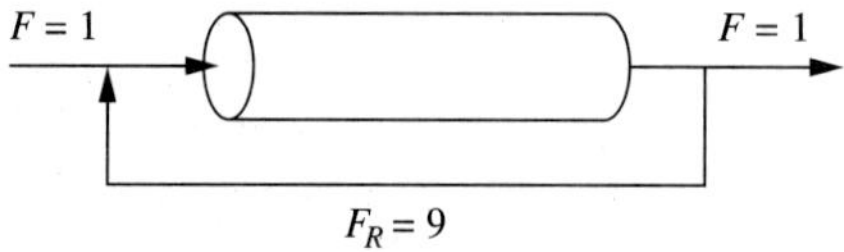

FIGURE Q5.5

5.6. The chemical reactor without control of temperature or concentration in Figure Q5.6 is to be modelled and analyzed. The assumptions are as follows:

1. $C_p(C_p = C_v)$, density, UA are constant.
2. $Q = UA(T - T_{cin})$
3. F, T_c, T_0, level are constant.
4. Disturbance is $C_{A0}(t)$.
5. Heat of reaction is significant.
6. Heat losses are insignificant.

7. System is initially at steady state.
8. Rate of reaction =

$$-r_A = k_0 e^{-E/RT} C_A \frac{\text{mole}}{(\text{m}^3)(\text{min})}$$

(*a*) Derive the material and energy balances for this reactor. Carefully define the system, state all assumptions, and show all steps, especially in the energy balance.

(*b*) Linearize the equations about their steady-state values and express them in deviation variables.

(*c*) Based on the linearized equations, state whether the system can experience overdamped behavior, and state mathematical criteria as a basis for your decision. (*Hint:* Solve for the terms that affect the exponents of the dynamic response, and establish criteria for the qualitative characteristics.)

(*d*) Repeat (*c*) for underdamped behavior.

(*e*) Repeat (*c*) for unstable behavior.

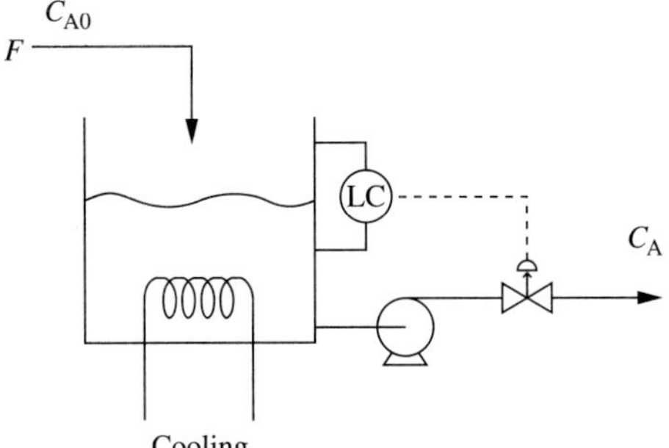

FIGURE Q5.6

5.7. A single isothermal CSTR has the following elementary reactions.

$$\text{Case I: } A \xrightarrow{k_A} B \qquad \text{Case II: } A \underset{k_{A'}}{\overset{k_A}{\rightleftarrows}} B$$

Only component A is in the feed stream, and its concentration, C_{A0}, can change as the input to the system. Answer the following questions for both Cases I and II.

(*a*) Derive the model describing the concentration of component B in the reactor.

(*b*) Which of the general system structures covered in this chapter describes this system?

(*c*) Determine whether the system can experience underdamped, overdamped, and unstable behavior for physically possible parameter values.

(*d*) Describe the response of this system to feed concentration step changes in C_{A0} and determine which system would have a faster response.

(*e*) Repeat all parts of this question, with the composition of A in the reactor being the output variable.

5.8. Figure 5.1 can be expanded to include more process systems and more inputs.

(*a*) Include the following systems, with a sketch of a physical process: (1) $1/(\tau s + 1)^3$ and (2) $e^{-\theta s}/(\tau s + 1)$.

(*b*) Include the following inputs for all systems: (1) ramp (Ct) and (2) pulse of finite duration.

5.9. The dynamic response of T_6 in the heat exchanger and stirred-tank system in Figure Q5.9 is to be determined for a step increase in the flow to the exchanger F_{ex}, with the total coolant flow F_c constant. (Assume that negligible transportation lag occurs in the pipes.)

(*a*) Derive the models for both stirred tanks.

(*b*) Determine the individual transfer functions.

(*c*) Derive the overall transfer function.

(*d*) Which of the general system structures covered in this chapter describes this system?

(*e*) Explain the numerator zeros (if any) and poles in the system.

(*f*) Describe the dynamic response of this system for the input step change in F_{ex}.

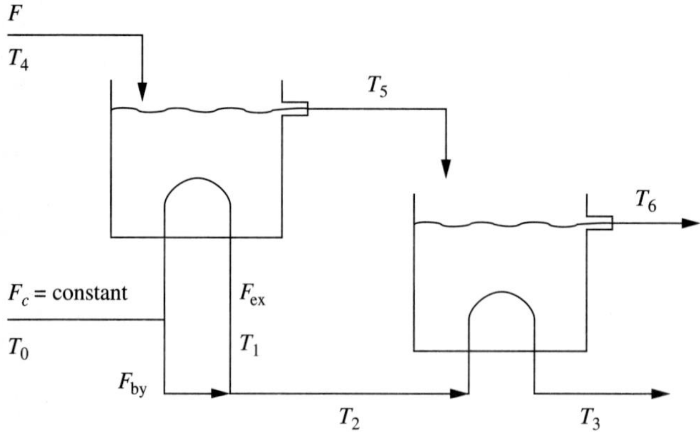

FIGURE Q5.9

5.10. The system of vessels in Figure Q5.10 has gas flowing through it, and F_0 is independent of P_1.

(*a*) Assume that the flow through the restrictions is subsonic.

1. Derive linearized models for the pressure in each system.
2. Determine the transfer function for $F_2(s)/F_0(s)$.
3. Describe the response of this system to a step in F_0.

(*b*) Repeat the analysis in part (*a*) for sonic flow through the restrictions.

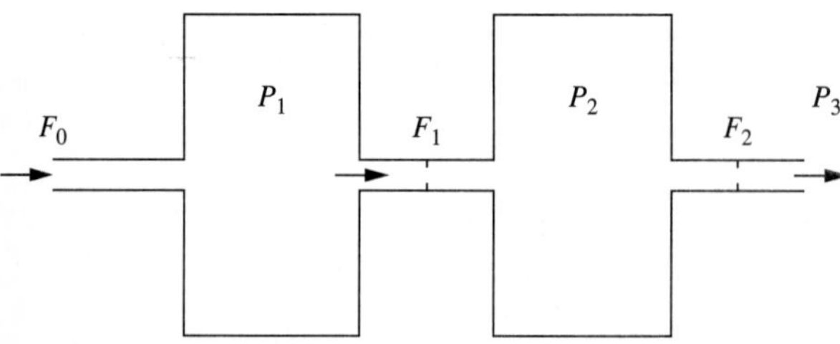

FIGURE Q5.10

5.11. Answer the following questions.

(*a*) Demonstrate that the dynamic behavior of a series of stable, first-order systems approaches the dynamic behavior of a dead time as the number of first-order systems becomes large, with $\tau_n = \tau_1/n$. Determine the value of the dead time.

(*b*) For the reactor with recycle in Example 5.5, determine the value of the heat exchanger gain, K_{H2}, that would cause the system to be unstable. Explain the expected dynamic response to an increase in the feed temperature.

(*c*) Discuss the manual control of a series of noninteracting time constants, a parallel system with overshoot, and a parallel system with inverse response. What would be your thought process for feedback control?

(*d*) For what range of sensor dynamics, τ_s, would the system in Example 5.3 experience an overshoot? Also, for what value of the numerator time constant in Example 5.4 would the system experience an inverse response?

(*e*) The interaction between material and energy balances can lead to a transfer function with a numerator zero. Compare transfer functions in equations (4.71) and (4.72) and explain the similarities and differences. Relate these to the physical system.

(*f*) What would be the order of the transfer function between the input FM_D and the output X_D for the distillation tower in Section 5.6?

(*g*) Sketch the general structure of a parallel system. What are the necessary conditions for a parallel system with two first-order components to experience an inverse response?

5.12. An autocatalytic system has a chemical reaction in which the product influences the rate; such kinetics occur in biological systems. Consider the following system occurring in a constant-volume, isothermal, well-stirred reactor.

$$\text{A} + \text{B} \rightarrow 2\text{B} + \text{other products} \qquad r_\text{A} = kC_\text{A}C_\text{B}$$

(*a*) Formulate a dynamic model of the reactor to predict the concentration of B in the reactor.

(*b*) Determine the possible steady-state values for C_B when only A is present in the feed. (*Hint:* Two possible steady states exist.)

(*c*) Under what conditions does the reactor go to each steady state?

(*d*) Reformulate the model and answer all questions for the case in which the product is separated and some pure B is returned to the reactor as a recycle. What would be the advantage of this recycle? How would the recycle affect the gain and time constant of C_B in response to a change in $C_{\text{A}0}$?

5.13. For each of the systems in Figure Q5.13, demonstrate through a fundamental model whether the system inventory is self-regulating or not for changes in flow in. In all cases, the flow in (F_in) can change independent of the inventory in the vessel.

(*a*) A heat exchanger in which the pure-component liquid entering at its boiling point in the vessel boils and the duty is proportional to the heat transfer area.

(*b*) A liquid-filled tank with a constant flow out.

(*c*) A gas-filled system with a moving roof and a constant mass on the roof. The gas exits through a partially open restriction.

(*d*) A gas-filled system with constant volume. The gas exits through a partially open restriction.

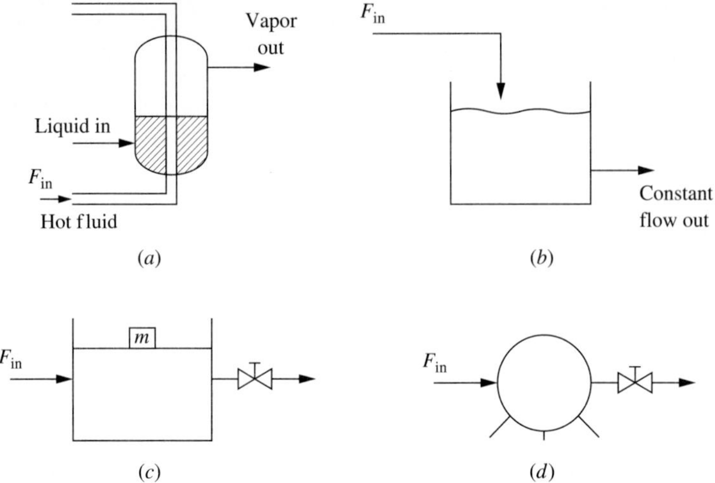

FIGURE Q5.13

5.14. The stirred-tank mixing process in Figure Q5.14 is to be analyzed. The system has a single feed, two tanks, and a single product. All flow rates, along with the levels, are constant. Answer the following questions completely. You may assume that (1) the tanks are well mixed, (2) the density is constant, and (3) transportation delays due to the pipes are negligible. For parts (*a*) through (*c*), $F_3 = F_0$.

(*a*) Derive the analytical model for the input-output system C_{A0} and C_{A2} with all flows constant.

(*b*) What is the general structure of the system in (*a*)?

(*c*) What conclusions can be determined for the system in (*a*) regarding the stability, periodicity, and either overshoot or inverse response for a step input?

(*d*) Determine the answers for (*a*) through (*c*) for (i) $F_3 = 0$ and (ii) $F_3 =$ very large.

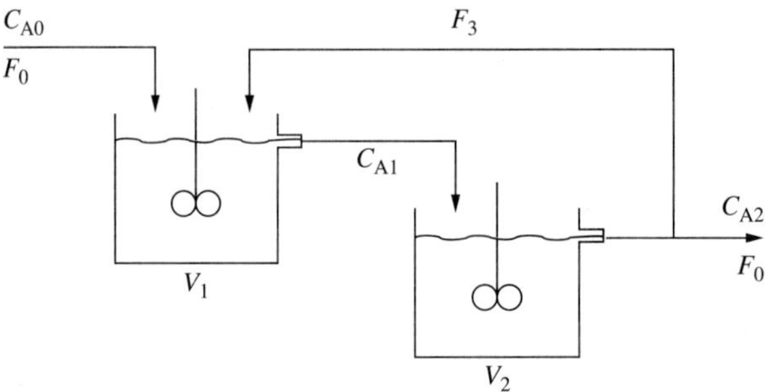

FIGURE Q5.14

5.15. The system in Figure Q5.15 has two stirred tanks; the first is a heat exchanger, and the second is a CSTR. The product of the reactor exchanges heat with the feed in the heat exchanger. A single, zeroth-order reaction of A→products occurs in the second reactor with a heat of reaction $(-\Delta H_{rxn})$.

(*a*) Formulate a model of the system to predict the temperature response in both tanks to a change in the feed temperature with all flows constant, and linearize the model. Determine to which process structure category this process belongs.

(*b*) Determine under what conditions the system would experience (i) periodic behavior and (ii) unstable behavior.

(*c*) Discuss your results and limitations in the model.

[*Hint:* This system is simpler than Example 3.10, in that the coolant flow is constant; thus, $UA = aF_c^b$ is constant. It is more complex in that the energy balances for the two tanks must be solved simultaneously.]

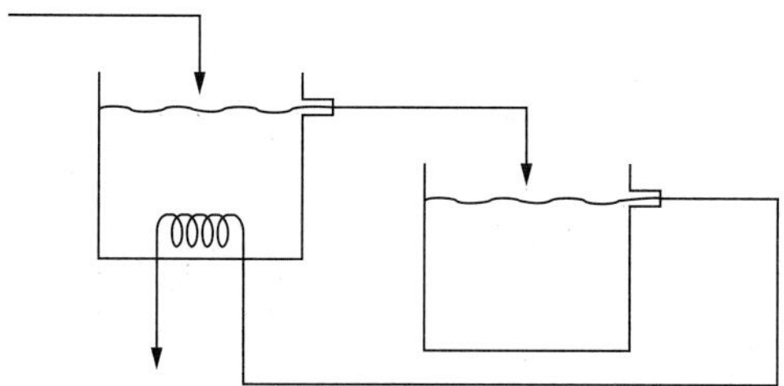

FIGURE Q5.15

5.16. The recycle system in Figure Q5.16 has a well-mixed, isothermal, constant-volume reactor and subsequent separation unit, in which the unreacted feed is separated from the product and returned to the reactor. A single step change occurs in the reactor temperature, which can be considered a step in the rate constant of the first-order reaction. Model the system and determine and compare the dynamics for two operating methods.

(*a*) The flow F_A is constant.

(*b*) The flow F_{AT} is constant.

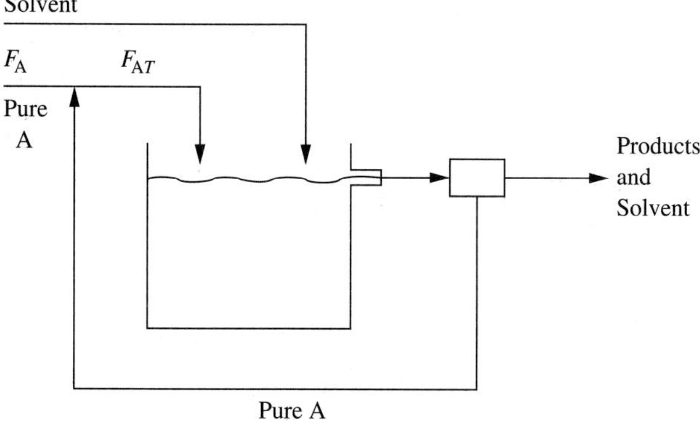

FIGURE Q5.16

5.17. A tubular heat exchanger with plug flow in the tube has steam at a constant temperature on the shell side. The system is initially at steady state with no temperature driving force, and the steam is introduced in a step to the shell.

(*a*) Determine the tube outlet temperature as a function of time. This will require analyzing a distributed-parameter model.

(*b*) Formulate a lumped-parameter model that would give an approximate result for the tube outlet temperature.

5.18. One way to account for imperfect mixing in a single stirred tank is to include commonly occurring *nonidealities* and fit parameters in a model to empirical data. For the nonideal model in Figure Q5.18, plot the shapes of the step and impulse responses for various values of the nonidealities. Could you fit an imperfect model using one of these sets of data?

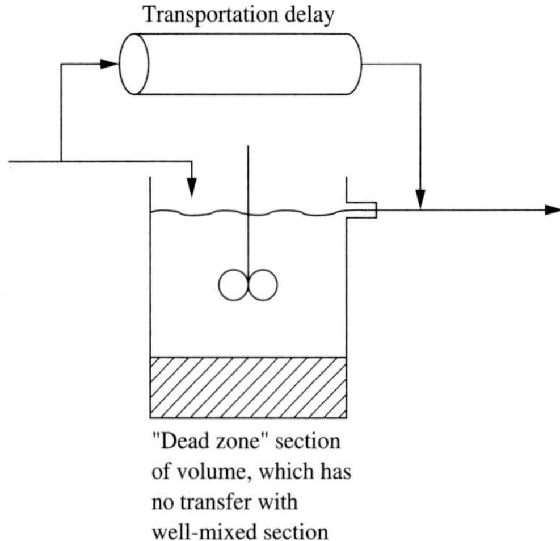

FIGURE Q5.18

5.19. Derive the models reported in Figures 5.2 and 5.3 for the electrical and mechanical systems.

5.20. From the principles in this chapter (and Appendix D), estimate the shape and $t_{63\%}$ of the step change for the following systems: (*a*) Example 3.3, (*b*) Example 3.10, (*c*) Question 4.15, and (*d*) Question 4.18.

5.21. Parallel structures were shown to lead to input-output transfer functions with numerator dynamics.

(*a*) For a system of two parallel first-order systems, draw a block diagram and derive the overall transfer function.

(*b*) Given the following transfer function, determine the relationship between the numerator time constant (τ_3) and the denominator time constants yielding each of the step responses in Figure Q5.21.

$$\frac{Y(s)}{X(s)} = \frac{K(\tau_3 s + 1)}{(s + 1)(2s + 1)}$$

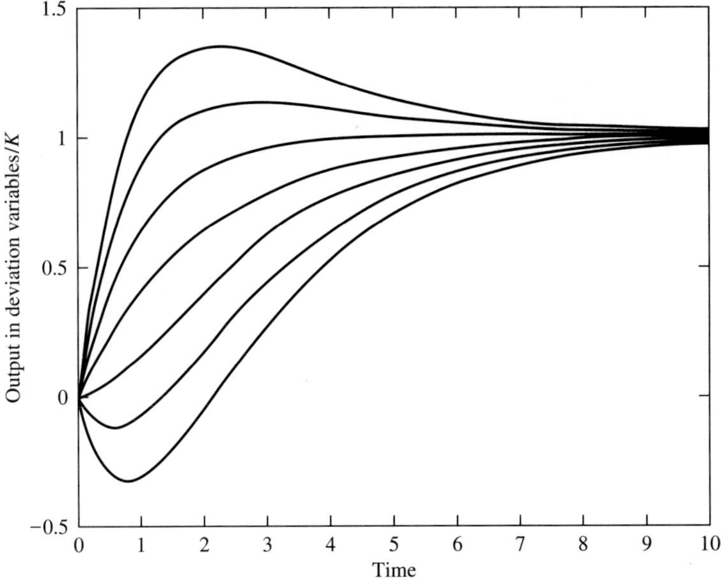

FIGURE Q5.21

5.22. A nonisothermal CSTR with heat transfer is modelled in Example 3.10. For each of the following situations, describe the possible shapes of the dynamic response of the concentration, C_A, to a step change in the coolant flow rate. There may be more than one per situation. Explain your answers by discussing, for example, the interaction between the material and energy balances.

(*a*) No chemical reaction, $k_0 = 0$

(*b*) Nonzero chemical reaction, but $\Delta H_{rxn} = 0$

(*c*) General case with nonzero reaction and heat of reaction

CHAPTER 6

EMPIRICAL MODEL IDENTIFICATION

6.1 INTRODUCTION

To this point, we have been modelling processes using fundamental principles, and this approach has been very valuable in establishing relationships between parameters in physical systems and the transient behavior of the systems. Unfortunately, this approach has limitations, which generally result from the complexity of fundamental models. For example, a fundamental model of a distillation column with 10 components and 50 trays would have on the order of 500 differential equations. In addition, the model would contain many parameters to characterize the thermodynamic relationships (equilibrium K values), rate processes (heat transfer coefficients), and model nonidealities (tray efficiencies). Therefore, modelling most realistic processes requires a large engineering effort to formulate the equations, determine all parameter values, and solve the equations, usually through numerical methods. This effort is justified when very accurate predictions of dynamic responses over a wide range of process operating conditions are needed.

This chapter presents a very efficient alternative modelling method specifically designed for process control, termed *empirical identification.* The models developed using this method provide the dynamic relationship between selected input and output variables. For example, the empirical model for the distillation column discussed previously could relate the reflux flow rate to the distillate composition. In comparison to this simple empirical model, the fundamental model provides information on how all of the tray and product compositions and temperatures depend on variables such as reflux. Thus, the empirical models described in this chapter, while tailored to the specific needs of process control, do not provide enough information to satisfy all process design and analysis requirements and cannot replace fundamental models for all applications.

In empirical model building, models are determined by making small changes in the input variable(s) about a nominal operating condition. The resulting dynamic response is used to determine the model. This general procedure is essentially an experimental linearization of the process that is valid for some region about the nominal conditions. As we shall see in later chapters, linear transfer function models developed using empirical methods are adequate for many process control designs and implementations. Because the analysis methods are not presented until later chapters, we cannot yet definitively evaluate the usefulness of the models, although we will see that they are quite useful. Thus, it is important to monitor the expected accuracy of the modelling methods in this chapter so that it can be considered in later chapters. As a rough guideline, the model parameters should be determined within ± 20 percent, although much greater accuracy is required for a few multivariable control calculations.

The empirical methods involve designed experiments, during which the process is perturbed to generate dynamic data. The success of the methods requires close adherence to principles of experimental design and model fitting, which are presented in the next section. In subsequent sections, two identification methods are presented. The first method is termed the *process reaction curve* and employs simple, graphical procedures for model fitting. The second and more general method employs statistical principles for determining the parameters. Several examples are presented with each method. The final section reviews some advanced issues and other methods not presented in this chapter so that the reader will be able to select the most appropriate technology for model building.

6.2 AN EMPIRICAL MODEL BUILDING PROCEDURE

Empirical model building should be undertaken using the six-step procedure shown in Figure 6.1. This procedure ensures that proper data is generated through careful experimental design and execution. Also, the procedure makes the best use of the data by thoroughly diagnosing and verifying results from the initial model parameter calculations. The schematic in Figure 6.1 highlights the fact that some a priori knowledge is required to plan the experiment and that the procedure can, and often does, require iteration, as shown by the dashed lines. At the completion of the procedure described in this section, an adequate model should be determined, or the engineer will at least know that a satisfactory model has not been identified and that further experimentation is required.

Throughout this chapter several examples are presented. The first example is shown in Figure 6.2, which has two stirred tanks. The process model to be identified relates the valve opening in the heating oil line to the outlet temperature of the second tank.

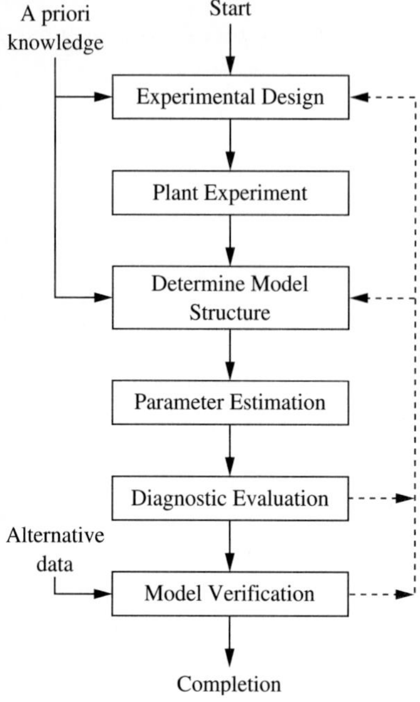

FIGURE 6.1
Procedure for empirical transfer function model identification.

Experimental Design

An important and often underestimated aspect of empirical modelling is the need for proper experimental design. Since every method requires some type of input perturbation, the design determines its shape and duration. It also determines the base operating conditions for the process, which essentially determine the conditions about which the process model is accurate. Finally, the magnitude of the input perturbation is determined. This magnitude must be small enough to ensure that the key safety and product quality limitations are observed. It is important to begin with a perturbation that is on the safe (small) side rather than cause a severe process disturbance.

Clearly, the design requires a priori information about the process and its dynamic responses. This information is normally available from previous operating

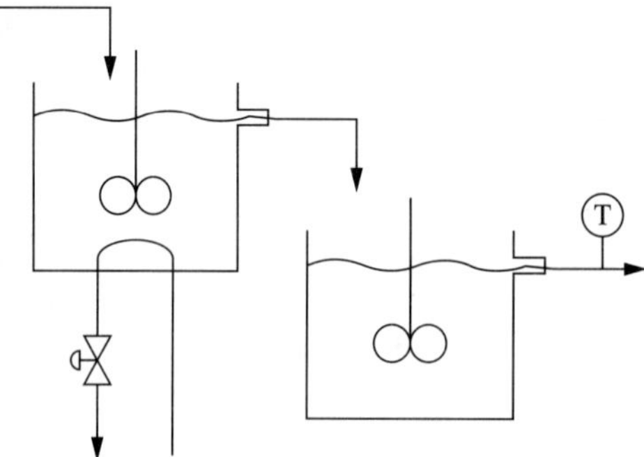

FIGURE 6.2
Example process for empirical model identification.

experience; if no prior information is available, some preliminary experiments must be performed. For the example in Figure 6.2, the time constants for each tank could be used to determine a first estimate for the response of the entire system.

The result of this step is a complete plan for the test which should include

1. A description of the base operating conditions
2. A definition of the perturbations
3. A definition of the variables to be measured, along with the measurement frequency
4. An estimate of the duration of the experiment

Naturally, the plan should be reviewed with all operating personnel to ensure that it does not interfere with other plant activities.

Plant Experiment

The experiment should be executed as close to the plan as possible. While variation in plant operation is inevitable, large disturbances during the experiment can invalidate the results; therefore, plant operation should be monitored during the experiment. Since the experiment is designed to establish the relationship between one input and output, changes in other inputs during the experiment could make the data unusable for identifying a dynamic model. This monitoring must be performed throughout the experiment, using measuring devices where available and using other sources of information, such as laboratory analysis, when process sensors are not available. For the example in Figure 6.2, variables such as the feed inlet temperature affect the outlet temperature of the second tank, and they should be monitored to ensure that they are approximately constant during the experiment.

Determining Model Structure

Currently, many methods are available to calculate the parameters in a model whose structure is set; however, few methods exist for determining the structure of a model (e.g., first- or second-order transfer function), based solely on the data. Typically, the engineer must assume a model structure and subsequently evaluate the assumption. The initial structure is selected based on prior knowledge of the unit operation, perhaps based on the structure of a fundamental model, and based on patterns in the experimental data just collected. The assumption is evaluated in the latter diagnostic step of this procedure.

The goal is *not* to develop a model that exactly matches the experimental data. Rather, the goal is to develop a model that describes the input-output behavior of the process adequately for use in process control.

> Empirical methods typically use low-order linear models with dead time. Often (but not always), first-order-with-dead-time models are adequate for process control analysis and design.

At times, higher-order models are required, and advanced empirical methods are available for determining the model structure (Box and Jenkins, 1976).

Parameter Estimation

At this point a model structure has been selected and data has been collected. Two methods are presented in this chapter to determine values for the model parameters so that the model provides a good fit to the experimental data. One method uses a graphical technique; the other uses statistical principles. Both methods provide estimates for parameters in transfer function models, such as gain, time constant(s), and dead time in a first-order-with-dead-time model. The methods differ in the generality allowed in the model structure and experimental design.

Diagnostic Evaluation

Some evaluation is required before the model is used for control. The diagnostic level of evaluation determines how well the model fits the data used for parameter estimation. Generally, the diagnostic evaluation can use two approaches: (1) a comparison of the model prediction with the measured data and (2) a comparison of the results with any assumptions used in the estimation method.

Verification

The final check on the model is to verify it by comparison with *additional* data not used in the parameter estimation. Although this step is not always performed, it is worth comparing the model to data collected at another time to be sure that typical variation in plant operation does not significantly degrade model accuracy. The methods used in this step are the same as in the diagnostic evaluation step.

It is appropriate to emphasize once again that the model developed by this procedure relates the input perturbation to the output response. The process modelled includes all equipment between the input and output; thus, the typical model includes the dynamics of valves and sensors as well as the process equipment. As we will see later, this is not a limitation; in fact, the empirical model provides the proper information for control analysis, because it includes the elements in the control loop.

Finally, two conflicting objectives must always be balanced in performing this experimental procedure. The first objective is the maintenance of safe, smooth, and profitable plant operation, for which a small experimental input perturbation is desired. However, the second objective is the development of an accurate model for process control design that will be improved by a relatively large input perturbation. The proper experimental procedure must balance these two objectives by allowing a short-term disturbance so that the future plant operation is improved through good process control.

6.3 THE PROCESS REACTION CURVE

The process reaction curve is probably the most widely-used method for identifying dynamic models. It is simple to perform, and although it is the least general method, it provides adequate models for many applications. First, the method is explained and demonstrated through an example. Then it is critically evaluated, with strong and weak points noted.

The process reaction curve method involves the following four actions:

1. Allow the process to reach steady state.
2. Introduce a single step change in the input variable.
3. Collect input and output response data until the process again reaches steady state.
4. Perform the graphical process reaction curve calculations.

The graphical calculations determine the parameters for a first-order-with-dead-time model: *the process reaction curve is restricted to this model.* The form of the model is as follows, with $X(s)$ denoting the input and $Y(s)$ denoting the output, both expressed in deviation variables.

$$\frac{Y(s)}{X(s)} = \frac{K_p e^{-\theta s}}{\tau s + 1} \tag{6.1}$$

There are two slightly different graphical techniques in common use, and both are explained in this section. The first technique, Method I, adapted from Ziegler and Nichols (1942), uses the graphical calculations shown in Figure 6.3 for the stirred-tank process in Figure 6.2. The intermediate values determined from the graph are

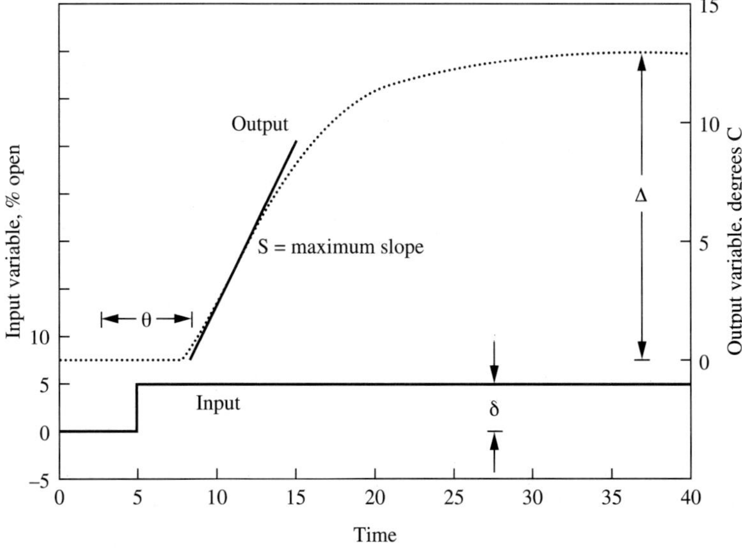

FIGURE 6.3
Process reaction curve, Method I.

the magnitude of the input change, δ; the magnitude of the steady-state change in the output, Δ; and the maximum slope of the output-versus-time plot, S. The values from the plot can be related to the model parameters according to the following relationships for a first-order-with-dead-time model. The general model for a step in the input with $t \geq \theta$ is

$$Y'(t) = K_p\delta(1 - e^{-(t-\theta)/\tau}) \tag{6.2}$$

The slope for this response at any time $t \geq \theta$ can be determined to be

$$\frac{dY'(t)}{dt} = \frac{d}{dt}\left\{K_p\delta(1 - e^{-(t-\theta)/\tau})\right\} = \frac{\Delta}{\tau}e^{-(t-\theta)/\tau} \tag{6.3}$$

The maximum slope occurs at $t = \theta$, so $S = \Delta/\tau$. Thus, the model parameters can be calculated as

$$\begin{aligned} K_P &= \Delta/\delta \\ \tau &= \Delta/S \\ \theta &= \text{intercept of maximum slope with initial value} \\ &\quad \text{(as shown in Figure 6.3)} \end{aligned} \tag{6.4}$$

A second technique, Method II, uses the graphical calculations shown in Figure 6.4. The intermediate values determined from the graph are the magnitude of the input change, δ; the magnitude of the steady-state change in the output, Δ; and the times at which the output reaches 28 and 63 percent of its final value. The values from the plot can be related to the model parameters using the general expression in equation (6.2). Any two values of time can be selected to determine the unknown

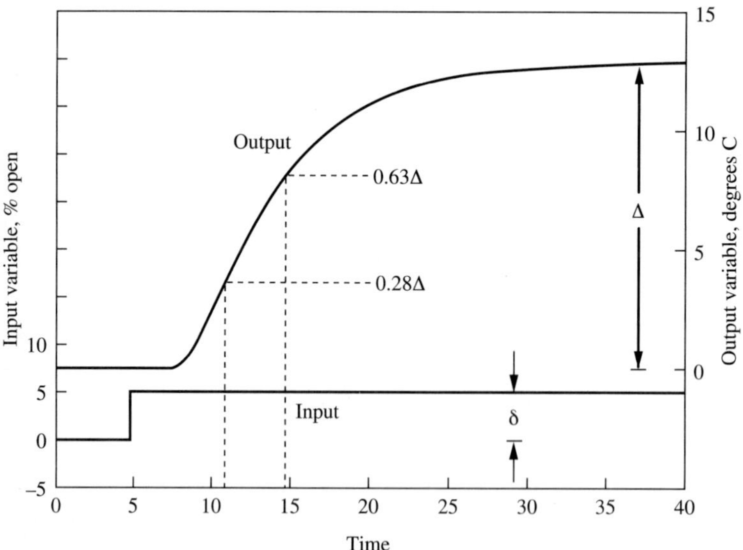

FIGURE 6.4
Process reaction curve, Method II.

parameters, θ and τ. The typical times are selected where the transient response is changing rapidly so that the model parameters can be accurately determined in spite of measurement noise (Smith, 1972). The expressions are

$$\begin{aligned} Y(\theta + \tau) &= \Delta(1 - e^{-1}) = 0.632\Delta \\ Y(\theta + \tau/3) &= \Delta(1 - e^{-1/3}) = 0.283\Delta \end{aligned} \tag{6.5}$$

Thus, the values of time at which the output reaches 28.3 and 63.2 percent of its final value are used to calculate the model parameters.

$$\begin{aligned} t_{28\%} &= \theta + \frac{\tau}{3} & \qquad t_{63\%} &= \theta + \tau \\ \tau &= 1.5(t_{63\%} - t_{28\%}) & \qquad \theta &= t_{63\%} - \tau \end{aligned} \tag{6.6}$$

Ideally, both techniques should give representative models; however, Method I requires the engineer to find a slope (i.e., a derivative) of a measured signal.

> Because of the difficulty in evaluating the slope, especially when the signal has high-frequency noise, Method I typically has larger errors in the parameter estimates; thus, Method II is preferred.

Example 6.1. The process reaction experiments have been performed on the stirred-tank system in Figure 6.2 and the data is given in Figures 6.3 and 6.4 for Methods I and II, respectively. Determine the parameters for the first-order-with-dead-time model.

Solution. The graphical calculations are shown in Figure 6.3 for Method I, and the calculations are summarized as

$$\begin{aligned} \delta &= 5.0\% \text{ open} \\ \Delta &= 13.1^\circ\text{C} \\ K_P &= \Delta/\delta = (13.1^\circ\text{C})/(5\% \text{ open}) = 2.6^\circ\text{C}/\% \text{ open} \\ S &= 1.40^\circ\text{C/min} \\ \tau &= \Delta/S = (13.1^\circ\text{C})/(1.40^\circ\text{C/min}) = 9.36 \text{ min} \\ \theta &= 3.3 \text{ min} \end{aligned}$$

The graphical results are shown in Figure 6.4 for Method II, and the calculations are summarized below. Note that the calculations for K_P, Δ, and δ are the same and thus not repeated. Also, time is measured from the input step change.

$$\begin{aligned} 0.63\Delta &= 8.3^\circ\text{C} & \qquad t_{63\%} &= 9.7 \text{ min} \\ 0.28\Delta &= 3.7^\circ\text{C} & \qquad t_{28\%} &= 5.7 \text{ min} \end{aligned}$$

$$\begin{aligned} \tau &= 1.5(t_{63\%} - t_{28\%}) = 1.5(9.7 - 5.7) \text{ min} = 6.0 \text{ min} \\ \theta &= t_{63\%} - \tau = (9.7 - 6.0) \text{ min} = 3.7 \text{ min} \end{aligned}$$

Further details for the process reaction curve method are summarized below with respect to the six-step empirical procedure.

Experimental Design

The calculation procedure is based on a perfect step change in the input as demonstrated in equation (6.2). The input can normally be changed in a step when it is a manipulated variable, such as valve percent open; however, some control designs will require models for inputs such as feed composition, which cannot be manipulated in a step, if at all. The sensitivity of the model results to deviations from a perfect input step are shown in Figure 6.5 for an example in which the true plant had a dead time of 0.5 and a process time constant (τ_{process}) of 1.0. The step change was introduced through a first-order system with a time constant (τ_{input}) that varied from 0.0 (i.e., a perfect step) to 1.0. This case study demonstrates that very small deviations from a perfect step input are acceptable but that large deviations lead to significant model parameter errors, especially in the dead time.

In addition to the input shape, the input magnitude is also important. As previously noted, the accuracy of the model depends on the magnitude of the input step change. The output change cannot be too small, because of noise in the measured output, which is caused by many small process disturbances and sensor nonidealities. The output signal is the magnitude of the change in the output variable. Naturally, the larger the input step, the more accurate the modelling results but the larger the disturbance to the process.

> A rough guideline for the process reaction curve is that the signal-to-noise ratio should be at least 5.

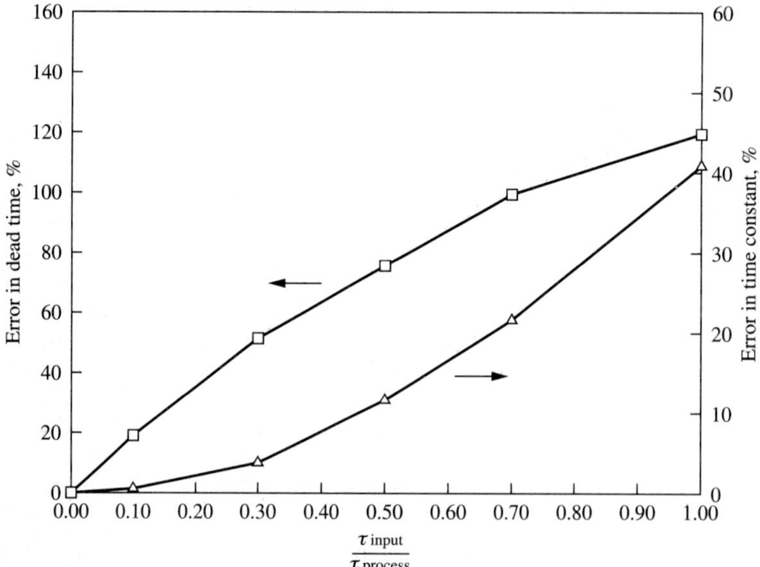

FIGURE 6.5
Sensitivity of process reaction curve to an imperfect step input, true process $\theta/(\theta + \tau) = 0.33$.

The noise level can be estimated as the variation experienced by the output variable when all measured inputs are constant. For example, if an output temperature varies $\pm 1°C$ due to noise, the input magnitude should be large enough to cause an output change Δ of at least 5°C.

Finally, the duration of the experiment is set by the requirement of achieving a final steady state after the input step. Thus, the experiment would be expected to last at least a time equal to the dead time plus four time constants, $\theta + 4\tau$. In the stirred-tank example, the duration of the experiment could be estimated from the time constants of the two tanks, plus some time for the heat exchanger and sensor dynamics. If the data is not recorded continuously, it should be collected frequently enough for the graphical analysis; 40 or more points would be preferable, depending on the amount of high-frequency noise.

Plant Experiment

Since model errors can be large if another, perhaps unmeasured, input variable changes, experiments should be designed to identify whether disturbances have occurred. One way to do this is to ensure that the final condition of the manipulated input variable is the same as the initial condition, which naturally requires more than one step change. Then, if the output variable also returns to its initial condition, one can reasonably assume that no long-term disturbance has occurred, although a transient disturbance could take place and not be identified by this checking method. If the final value of the output variable is significantly different from its initial value, the entire experiment is questionable and should be repeated. This situation is discussed further in Example 6.3.

Diagnostic Evaluation

The basic technique for evaluating results of the process reaction curve is to plot the data and the model predictions on the same graph. Visual comparison can be used to determine whether the model provides a good fit to the data used in calculating its parameters. This procedure has been applied to Example 6.1 using the results from Method II, and the comparison is shown in Figure 6.6. Since the data and model do not differ by more than about 0.5°C throughout the transient, the model would normally be accepted for most control analyses.

Most of the control analysis methods presented in later parts of the book require linear models, and information on strong nonlinearities would be a valuable result of empirical model identification. The linearity can be evaluated by comparing the model parameters determined from experiments of various magnitudes and directions, as shown in Figure 6.7. If the model parameters are similar, the process is nearly linear over the range investigated. If the parameters are very different, the process is highly nonlinear, and control methods described in Chapter 16 may have to be applied.

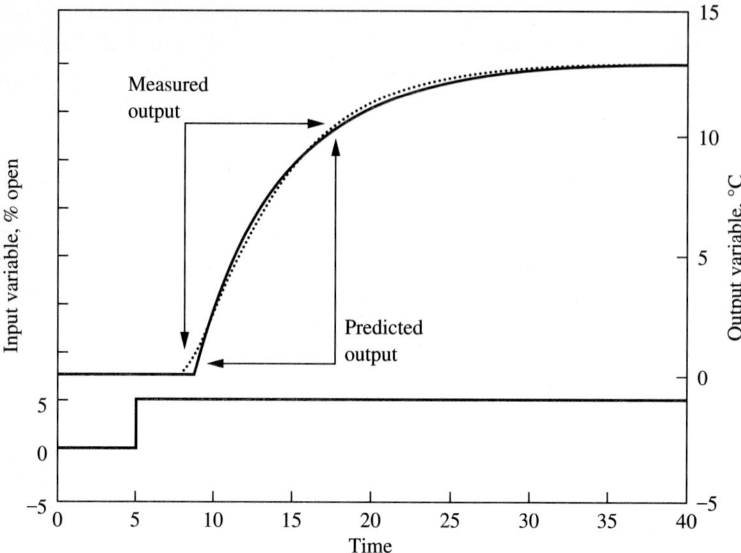

FIGURE 6.6
Comparison of measured and predicted outputs.

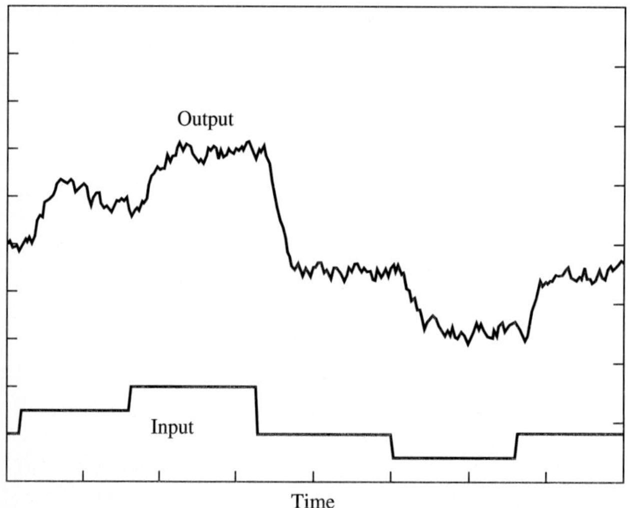

FIGURE 6.7
Example of experimental design to evaluate the linearity of a process.

Verification

If additional data is collected that is not used to calculate the model parameters, it can be compared with the model using the same techniques as in the diagnostic step.

Example 6.2. A more realistic set of data for the stirred-tank example is given in Figure 6.8. This data has noise, which could be due to imperfect mixing, sensor noise,

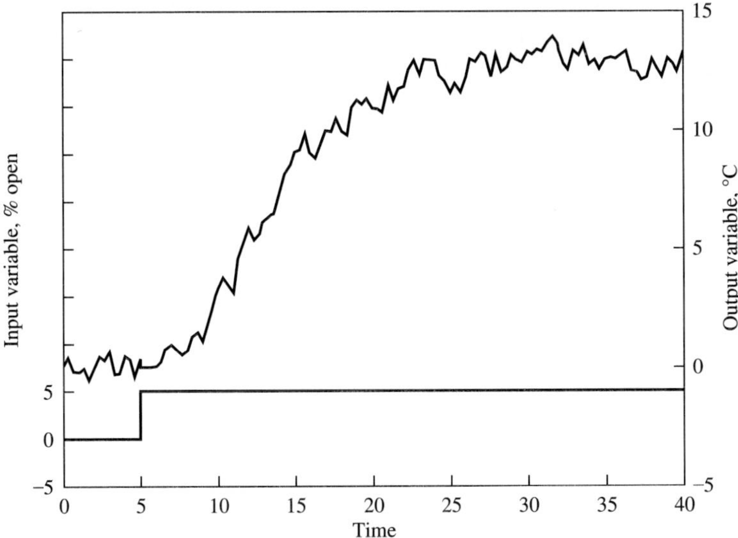

FIGURE 6.8
Process reaction curve for Example 6.2.

and variation in other input variables. The application of the process reaction curve requires some judgment. The reader should perform both methods on the data and note the difficulty in Method I. Typical results for the methods are given in the following table, but the reader can expect to obtain slightly different values due to the noise.

	Method I	Method II	
K_p	2.6	2.6	°C/%open
θ	2.4	3.7	min
τ	10.8	5.9	min

Example 6.3. Data for two step changes is given in Figure 6.9. Determine a dynamic model using the process reaction curve method.

Note that there is no difference between the initial and final values of the input valve opening. However, the output temperature does not return to its initial value. This is due to some nonideality in the experiment, such as an unmeasured disturbance or a sticky valve that did not move as expected. This data should not be used, and the experiment should be repeated.

Example 6.4. A fundamental model for a tank mixing process similar to Figure 6.10*a* will be developed in Chapter 7, where the time constant of each tank is shown to be volume/volumetric flow rate (V/F). Determine approximate models for this process at three flow rates of stream B given below (m^3/min) when each tank volume is 35 m^3.

This example demonstrates the usefulness of the insight provided from fundamental modelling, even though a simplified model is determined empirically. The process reaction curve experiment was performed for this process at the three flow

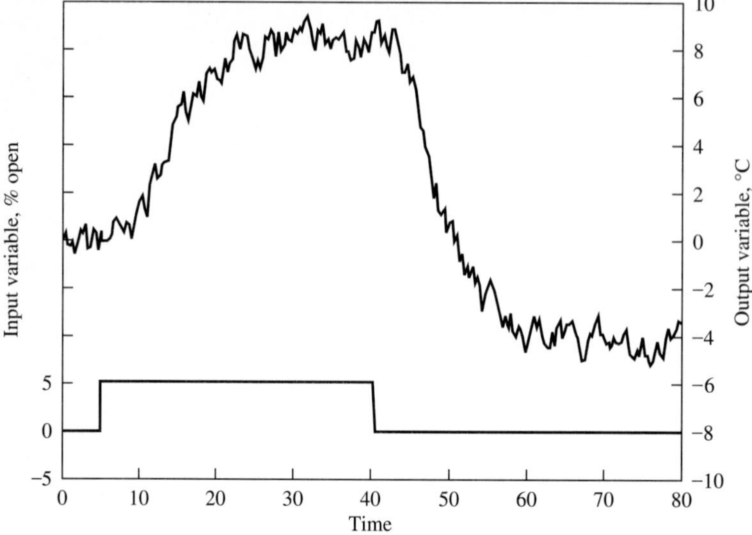

FIGURE 6.9
Experiment data for process reaction curve when input is returned to its initial condition.

rates, all at a base exit concentration of 3 percent A, and the results at the base case flow is shown in Figure 6.10*b*. The results are summarized in the following table.

	Simplified				Fundamental
Flow (m^3/min)	**K_p (%A/% open)**	**θ (min)**	**τ (min)**	**$\theta + \tau$ (min)**	**$\approx \sum \tau_i$ (min)**
5.1	.055	7.6	14.5	22.1	20.7
7.0	.04	5.5	10.5	16.0	15.0 ← base case
8.1	.036	4.7	9.1	13.8	12.9

The fundamental model demonstrates that the time constants ($\tau = V/F$) depend on the flow rate, decreasing as the flow increases. This trend is confirmed in the simplified model as well. Also, the approximate relationship for systems of noninteracting time constants in series, equation (5.41*b*), that the sum of the dead times plus time constants is unchanged by model simplification, is rather good for this process.

The most important characteristics of the process reaction curve method are summarized in Table 6.1. The major advantages of the process reaction curve method are its simplicity and short experimental duration, which result in its frequent application for simple control models.

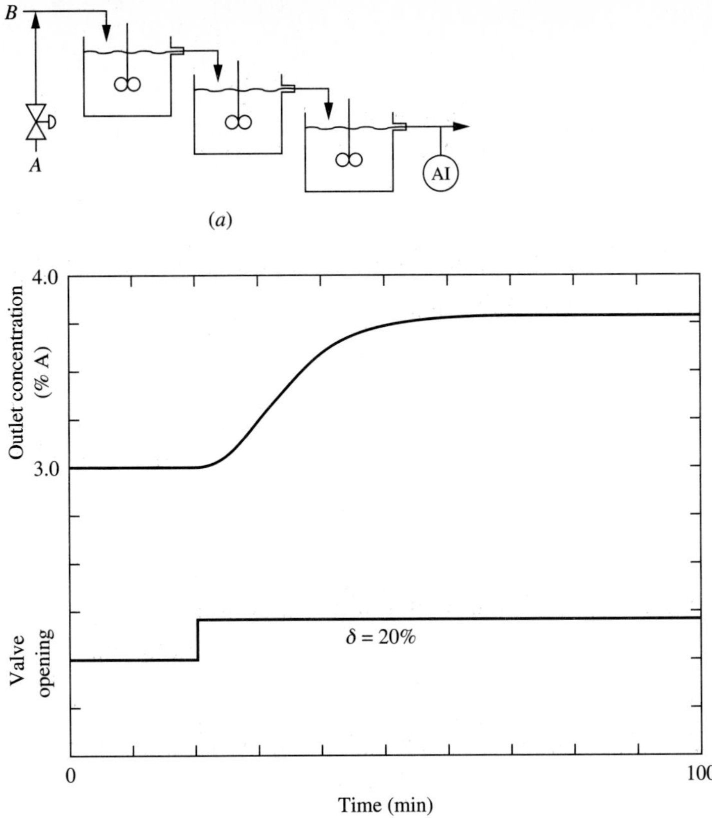

FIGURE 6.10
For Example 64.4: (*a*) Three-tank mixing process; (*b*) process reaction curve for base case.

TABLE 6.1
Summary of the process reaction curve

Characteristic	Process reaction curve
Input magnitude	Large enough to give an output signal-to-noise ratio greater than 5
Experiment duration	The process should reach steady state; thus the duration at least $\theta + 4\tau$
Input change	A nearly perfect step change is required
Model structure	First-order with dead time
Accuracy with unmeasured disturbances	Accuracy can be strongly affected (degraded) by significant disturbances
Diagnostics	Plot model versus data
Calculations	Simple hand and graphical calculations

6.4 STATISTICAL MODEL IDENTIFICATION

The previously described graphical method had two major limitations: a first-order-with-dead-time model and a perfect step input. Statistical model identification methods provide more flexible approaches to identification that relax these limits to model structure and experimental design and potentially lead to better model identification. A simple version of statistical model fitting is presented here to introduce the concept and provide another useful identification method. The same six-step procedure described in Section 6.2 is used with this method.

The statistical method introduced here involves the following four actions.

1. Allow the process to reach steady state.
2. Introduce a perturbation in the input variable. There is no restriction on the shape of the perturbation, but the effect on the output must be large enough to enable a model to be identified.
3. Collect input and output response data. It is not necessary that the process regain steady state at the end of the experiment.
4. Calculate the model parameters as described in the subsequent paragraphs.

The statistical method described in this section uses a regression method to fit the experimental data, and the closed-form solution method requires an algebraic equation with unknown parameters. Thus, the transfer function model must be converted into an algebraic model that relates the current value of the output to past values of the input and output. There are several methods for performing this transform; the most accurate and general for linear systems involves z-transforms, which serve a similar purpose for discrete systems as Laplace transforms serve for continuous systems (Franklin et al., 1990). The method used here is much simpler and is adequate for demonstrating the statistical identification method and fitting models of simple structure, such as first-order with dead time (see Appendix F).

The first-order-with-dead-time model can be written in the time domain according to the equation

$$\tau \frac{dY'(t)}{dt} + Y'(t) = K_p X'(t-\theta) \tag{6.7}$$

Again, the prime denotes deviation from the initial steady-state value. This differential equation can be integrated from time t_i to $t_i + \Delta t$ assuming that the input $X'(t)$ is constant over this period. Note that the dead time is represented by an integer number of sample delays (i.e., $\Gamma = \theta/\Delta t$). The resulting equation is

$$Y'_{i+1} = e^{-\Delta t/\tau} Y'_i + K_p \left(1 - e^{-\Delta t/\tau}\right) X'_{i-\Gamma} \tag{6.8}$$

In further equations the notation is simplified according to the equation

$$Y'_{i+1} = aY'_i + bX'_{i-\Gamma} \tag{6.9}$$

The challenge is to determine the parameters a, b, and Γ that provide the best model for the data. Then the model parameters K_p, τ, and θ can be calculated.

The procedure used involves linear regression, which is briefly explained here and is thoroughly presented in many references (e.g., Box et al., 1978). Assume for the moment that we know the value of Γ, the dead time (this assumption will be addressed later in the method). Typical data from the process experiment is given in Table 6.2; note that the measurements are provided at equispaced intervals. Since we want to fit an algebraic equation of the form in equation (6.9), the data must be arranged to conform to the equation. This is done in Table 6.2, where for every measured value of $(Y'_{i+1})_m$ the corresponding measured values of $(Y'_i)_m$ and $(X'_{i-\Gamma})_m$ are provided on the same line. Using the model it is also possible to predict the output variable at any time, with $(Y_{i+1})_p$ representing the predicted value, using the appropriate measured variables.

$$(Y'_{i+1})_p = a(Y'_i)_m + b(X'_{i-\Gamma})_m \tag{6.10}$$

Note that the subscript m indicates a measured value, and the subscript p indicates a predicted output value. The "best" model parameters a and b would provide an accurate prediction of the output at each time; thus, the goal is to calculate the values of the parameters a and b so that $(Y'_{i+1})_m$ and $(Y'_{i+1})_p$ are as nearly equal as possible. The common technique for determining the parameters is to apply the *least squares* method, which minimizes the sum of error squared between the measured and predicted values over all samples, $i = \Gamma + 1$ to n. The error can be expressed as follows.

$$\sum_{i=\Gamma+1}^{n} E_i^2 = \sum_{i=\Gamma+1}^{n} \left[(Y'_{i+1})_m - (Y'_{i+1})_p\right]^2 = \sum_{i=\Gamma+1}^{n} \left[(Y'_{i+1})_m - (a(Y'_i)_m + b(X'_{i-\Gamma})_m)\right]^2 \tag{6.11}$$

TABLE 6.2
Data for statistical model identification

Data in original format as collected in experiment			**Data in restructured format for regression model fitting, first-order-with-dead-time model with dead time of two sample periods**			
				z vector in equation (6.16)	**U matrix in equation (6.16)** $X' = X - X_s$ **with** $X_s = 50$; $Y' = Y - Y_s$ **with** $Y_s = 75$	
Time t	**Input,** X	**Output,** Y	**Sample no.** i	**Output** Y'_{i+1}	**Output,** Y'_i	**Delayed input,** X'_{i-2}
0	50	75	1			
0.2	50	75	2			
0.4	52	75	3	0	0	0
0.6	52	75	4	0.05	0	0
0.8	52	75.05	5	0.1	0.05	2
1.0	52	75.1	6	0.3	0.1	2
1.2	52	75.3	7	0.6	0.3	2
1.4	52	75.6	8	0.7	0.6	2
1.6	52	75.7	..	..	..	..

—— Table continued for duration of experiment ——

The minimization of this term requires that the derivatives of the sum of error squares with respect to the parameters are zero.

$$\frac{\partial}{\partial a}\left[\sum_{i=\Gamma+1}^{n} E_i^2\right] = -2\sum_{i=\Gamma+1}^{n} (Y_i')_m\left[(Y_{i+1}')_m - a(Y_i')_m - b(X_{i-\Gamma}')_m\right] = 0 \quad (6.12)$$

$$\frac{\partial}{\partial b}\left[\sum_{i=\Gamma+1}^{n} E_i^2\right] = -2\sum_{i=\Gamma+1}^{n} (X_{i-\Gamma}')_m\left[(Y_{i+1}')_m - a(Y_i')_m - b(X_{i-\Gamma}')_m\right] = 0 \quad (6.13)$$

Equations (6.12) and (6.13) are linear in the two unknowns a and b, as is perhaps more easily recognized when the equations are rearranged as follows.

$$a\sum_{i=\Gamma+1}^{n} (Y_i')_m^2 + b\sum_{i=\Gamma+1}^{n} (Y_i')_m(X_{i-\Gamma}')_m = \sum_{i=\Gamma+1}^{n} (Y_i')_m(Y_{i+1}')_m \quad (6.14)$$

$$a\sum_{i=\Gamma+1}^{n} (Y_i')_m(X_{i-\Gamma}')_m + b\sum_{i=\Gamma+1}^{n} (X_{i-\Gamma}')_m^2 = \sum_{i=\Gamma+1}^{n} (X_{i-\Gamma}')_m(Y_{i+1}')_m \quad (6.15)$$

The values of the unknowns can be determined using various methods for solving linear equations (Anton, 1987); however, a more convenient approach is to use a computer program that is designed to solve the least squares problem. With these programs, the engineer simply enters the data in the form of Table 6.2, and the program automatically sets up and solves equations (6.14) and (6.15) for a and b.

These programs are designed to solve the least squares method by matrix methods. The measured values for this problem can be entered into the following matrices.

$$\mathbf{U} = \begin{bmatrix} Y_3' & X_{3-\Gamma}' \\ Y_4' & X_{4-\Gamma}' \\ \vdots & \vdots \\ Y_{n-1}' & X_{n-\Gamma-1}' \end{bmatrix} \qquad \mathbf{z} = \begin{bmatrix} Y_4' \\ Y_5' \\ \vdots \\ Y_n' \end{bmatrix} \quad (6.16)$$

The least squares solution for the parameters can be shown to be (Graupe, 1972)

$$\begin{bmatrix} a \\ b \end{bmatrix} = (\mathbf{U}^{\mathrm{T}}\mathbf{U})^{-1}\mathbf{U}^{\mathrm{T}}\mathbf{z} \quad (6.17)$$

> Many computer programs exist for solving linear least squares, and simple problems can be solved easily using a spreadsheet program with a linear regression option.

Given this method for determining the coefficients a and b, it is necessary to return to the assumption that the dead time, $\Gamma = \theta/\Delta t$, is known. To determine the dead time accurately, it is necessary to solve the least squares problem in equations (6.14) and (6.15) for several values of Γ, with the value of Γ giving the lowest sum of error squared (more properly, the sum of error squared divided by the number of degrees of freedom, which is equal to the number of data points minus the number of parameters fitted) being the best estimate of the dead time. This approach, which is essentially a search in one direction, is required because the variable Γ is discrete (i.e., it takes only integer values), so that it is not possible to determine the analytical derivative of the sum of errors squared with respect to dead time. Caution should be used, because the relationship between the dead time and sum of errors squared may not be monotonic; if more than one minimum exists, the dead time resulting in the smallest sum of errors squared should be selected.

The statistical method presented in this section, minimizing the sum of errors squared, is an intuitively appealing approach to finding the best values of the parameters. However, it depends on assumptions that, if violated significantly, could lead to erroneous estimates of the parameters. These assumptions are completely described in statistics textbooks (Box, et al., 1978). The most important assumptions are the following:

1. The error E_i is an independent random variable with zero mean.
2. The model structure reasonably represents the true process dynamics.
3. The parameters a and b do not change significantly during the experiment.

The following assumptions are also made in the least squares method; however, the model accuracy is not as strongly affected when they are slightly violated:

4. The variance of the error is constant.
5. The input variable is known without error.

When all assumptions are valid, the least squares assumption will yield good estimates of the parameters. Note that the experimental and diagnostic methods are designed to ensure that the assumptions are satisfied.

Example 6.5. Determine the parameters for a first-order-with-dead-time model for the stirred-tank example data in Figure 6.3.

The data must be sampled at equispaced periods, which were chosen to be 0.333 minutes for this example. Since the data arrays are very long, they are not reported. The data was organized as shown in Table 6.2. Several different values of the dead time were assumed, and the regression was performed for each. The results are summarized in the following table.

Dead Time, Γ	a	b	$\sum E^2$
7	0.964	0.101	7.52
8	0.9605	0.108	6.33
9	0.9578	0.1143	5.86 ← (minimum)
10	0.9555	0.1196	6.21

The dead time is selected to be the value that gives the smallest sum of errors squared; thus, the estimated dead time is 3 minutes, $\theta = (\Gamma)(\Delta t) = 9(0.333)$. The other model parameters can be calculated from the regression results.

$$\tau = -\Delta t/(\ln a) = -0.333/(-0.0431) = 7.7 \text{ min}$$
$$K_p = b/(1 - a) = 0.1143/(1 - 0.9578) = 2.7°\text{C}/\%\text{open}$$

The comments in Section 6.3 regarding the process reaction curve and the six-step procedure are also relevant for this statistical method. Some additional comments specific to the statistical method are given here.

Experimental Design

The input change can have a general shape (i.e., a step is not required), although Example 6.5 demonstrates that the statistical method works for step inputs. This generality is very important, because it is sometimes necessary to build models for inputs that are not directly manipulated, such as measured disturbance variables.

Sufficient input changes are required to provide enough information to overcome random noise in the measurement. Also, the data selected from the transient for use in the least squares determines which aspects of the dynamic response are fitted best. For example, if the duration of the experiment is too short, the method will provide a good fit for the initial part of the transient, but not necessarily for the steady-state gain. For this method with one or a few input changes, the input changes should be large enough and of long enough duration that the output variable reaches *at least* 63 percent of its final value. Note that more sophisticated experimental design methods (beyond the treatment in this book) are available that require much smaller output variation at the expense of longer experiment duration (Box and Jenkins, 1976).

Finally, the dead time cannot be determined with accuracy greater than the data collection sample period Δt. Thus, this period must be small enough to satisfy control system design requirements explained in later parts of the book. For now, a rough guideline can be used that Δt should be less than 10 percent of the sum of the dead time plus time constant.

Plant Experimentation

The input variable must be measured without significant noise. If this is not the case, more sophisticated statistical methods must be used.

Model Structure

Equations have been derived for a first-order model in this chapter. Other models could be derived in the same manner. The simplest model structure that provides an adequate fit should be selected.

Diagnostic Procedure

One of the assumptions was that the error—the deviation between the model prediction and the measurement—is a random variable. The errors, sometimes referred to as the *residuals,* can be plotted against time to determine whether any unexpected, large correlation in time exists. This is done for the results of the following example.

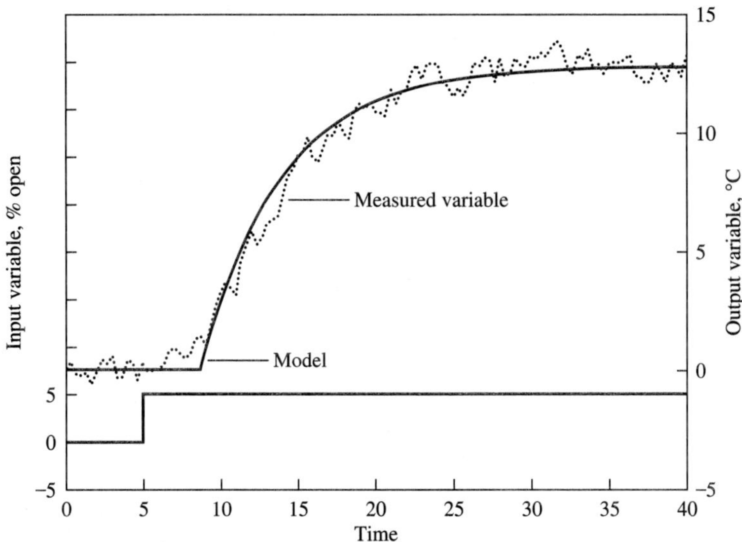

FIGURE 6.11
Comparison of measured and predicted output values from Example 6.6.

Example 6.6. Data has been collected for the same stirred-tank system analyzed in Example 6.2; however, the data in this example contains noise, as shown in Figure 6.8. Determine the model parameters using the statistical identification method.

The procedure for this data set is the same as used in Example 6.5. No judgment is required in fitting slopes or smoothing curves as was required with the process reaction curve method. The results are as follows, plotted in Figure 6.11:

$$\Delta t = 0.33 \text{ min} \qquad \Gamma = 11 \qquad a = 0.9384 \qquad b = 0.2578$$

$$\theta = 3.66 \text{ min} \qquad \tau = 5.2 \text{ min} \qquad K_p = 2.56°\text{C}/\% \text{ open}$$

Note that the model parameters are similar to the Method II results without noise, but that a slightly different value is determined for the dead time. The graphical comparison indicates a good fit to the experimental data.

Further diagnostic analysis is possible by plotting the residuals to determine whether they are nearly random. This is done on Figure 6.12. The plot shows little correlation; note that some correlation is expected, because the simple model structure selected will not often provide the best possible fit to a set of data. Since the errors are only slightly correlated and small, the model structure and dead time are judged to be valid.

Example 6.7. The dynamic data in Figure 6.13 was collected, showing the relationship between the inlet and outlet temperatures of the stirred tanks. Naturally, this data would require an additional sensor for the inlet temperature to the first tank. When this data was collected, the heating valve position and all other input variables were constant. Note that the input change was not even approximately a step, because the temperature depends on the operation of upstream units. Determine the parameters for a first-order-with-dead-time model.

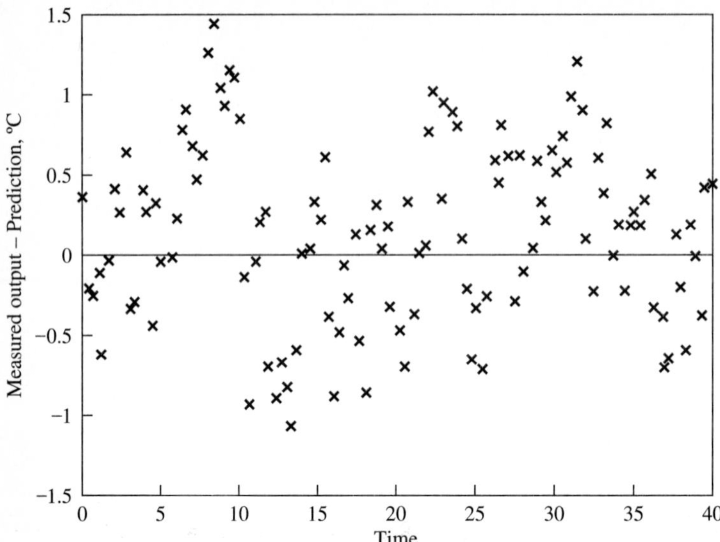

FIGURE 6.12
Plot of residuals between measured and predicted outputs from Example 6.6.

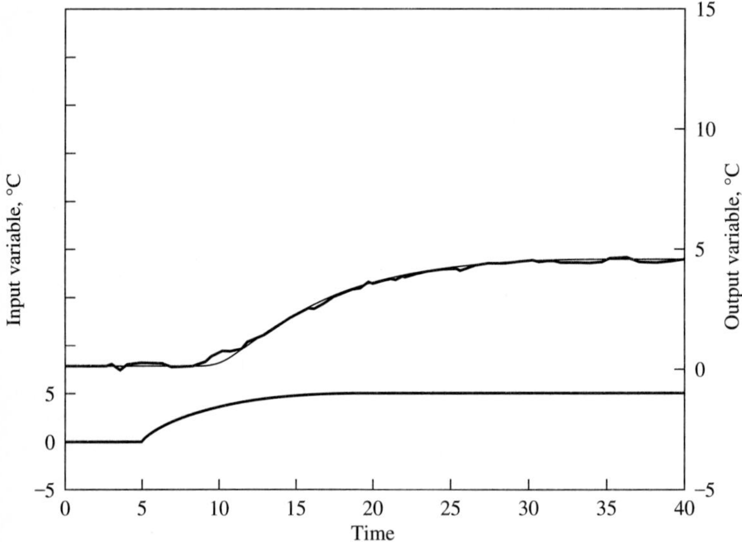

FIGURE 6.13
Experimental data for an imperfect input step for Example 6.7.

Again, the statistical procedure was used. The results are as follows:

$$\Gamma = 11 \qquad a = 0.9228 \qquad b = 0.0760$$
$$\theta = 3.66 \text{ min} \qquad \tau = 4.2 \text{ min} \qquad K_p = 0.98°\text{C}/°\text{C}$$

The model is compared with the data in Figure 6.13. The dynamic response is somewhat faster than the previous response, as might be expected because this model does not include the heat exchanger dynamics.

The linear regression identification method for a first-order-with-dead-time model is more general than the process reaction curve and can be used to fit important industrial processes. However, it also has limitations. Although it is easier to use and yields more accurate parameter values when the data has noise, it gives erroneous results when the noise is too large compared with the output change caused by the experiment—the same trend as with the process reaction curve.

Example 6.8. Figure 6.14 gives data recorded when a very small input change is introduced into the stirred tank system in Figure 6.2. The statistical method can be used, but the results (τ = 0.6 min, θ = 3.66 min, and K_p = 2.3° C/%open) deviate from the previously reported, more accurate results obtained with larger input disturbances. Clearly, a model from such a small input change is not reliable.

In addition, the simple statistical method used here is susceptible to unmeasured disturbances. The experimental design shown in Figure 6.9 is recommended to identify such disturbances. The statistical identification method described in this section is summarized in Table 6.3.

TABLE 6.3
Summary of the statistical identification method

Characteristic	Statistical identification
Input	If the input change approximates a step, the process output should deviate at least 63% of the potential steady-state change.
Experiment duration	The process does not have to reach steady state.
Input change	No requirement regarding the shape of the input.
Model structure	Model structures other than first-order-with-dead-time are possible, although the equations given here are restricted to first-order-with-dead-time.
Accuracy with unmeasured disturbances	Accuracy is strongly affected by significant disturbances.
Diagnostics	Plot model versus data, and plot residuals versus time.
Calculations	Calculations can be easily performed with a spreadsheet or special-purpose statistical computer program.

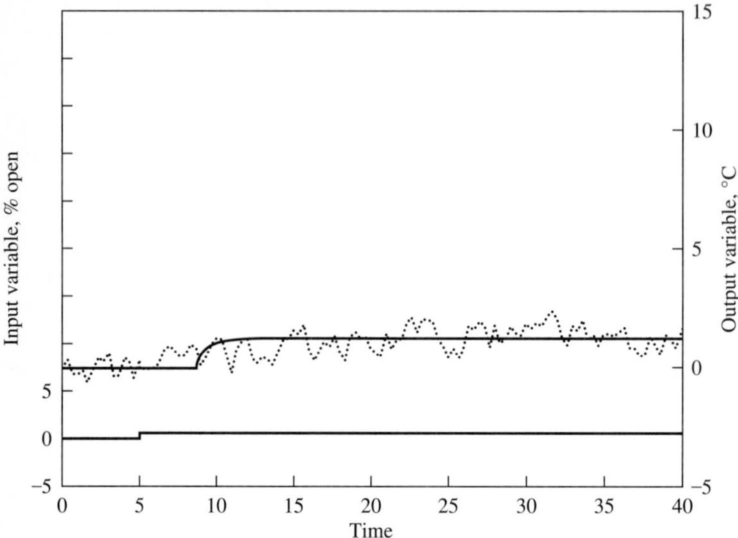

FIGURE 6.14
Example of empirical identification with input perturbation that is too small.

6.5 ADDITIONAL TOPICS IN IDENTIFICATION

Some additional topics in identification are addressed in this section. The topics relate to both the process reaction curve method and the statistical method, unless otherwise noted.

Other Model Structures

The methods presented here provide satisfactory models for processes that give smooth, sigmoidal-shaped responses to a step input. Most, but not all, processes are in this category. More complex model structures are required for the higher-order, underdamped, and inverse response systems. Graphical methods are available for second-order systems undergoing step changes (Graupe, 1972); however, the methods seem useful only when the output data has little noise, since they appear sensitive to noise.

Many advanced statistical methods are available for more complex model structures (Cryor, 1986; Box and Jenkins, 1976). The general concept is unchanged, but the major difference from the method demonstrated in this chapter is that the least squares equations, similar to equations (6.14) and (6.15), cannot be arranged into a set of linear equations in variables uniquely related to the model parameters; therefore, a nonlinear optimization method is required for calculating the parameters. Also, confidence intervals provide useful diagnostic information. Again, the engineer must assume a model structure and employ diagnostics to determine whether the assumed structure is adequate.

Multiple Variables

Sometimes models are desired between an input and several outputs. For example, we may need the transfer function models between the reflux and the distillate and bottoms product composition of a distillation column. These models could be determined from one set of experimental data in which the reflux flow is perturbed and both compositions are recorded, as shown in Figure 5.20*b* (page 182). Then each model would be evaluated individually using the appropriate method, such as the process reaction curve.

Operating Conditions

The operating conditions for the experiment should be as close as possible to the normal operation of the process when the control system, designed using the model, is in operation. This is only natural, because significant deviation could introduce error into the model and reduce the effectiveness of the control. For example, the dynamic response of the stirred-tank process in Figure 6.2 depends on the feed flow rate, as we would determine from a fundamental model. If the feed flow rate changes from the conditions under which the identification is performed, the linear transfer function model will be in error.

An associated issue relates to the status of the control system when the experiment is performed. A full discussion of this topic is premature here; however, the reader should appreciate that the process, including associated control strategies, must respond during the experiment as it would during normal operation. This topic is covered as appropriate in later chapters.

Frequency Response

As an alternative identification method, the frequency response of some physical systems, such as electrical circuits, can be determined experimentally by introducing input sine waves at several frequencies. Models can then be determined from the amplitude and phase angle relationships as a function of frequency. This method is not appropriate for complex chemical processes, because of the extreme disturbances caused over long durations, although it has been demonstrated on some unit operations (Harriott, 1964).

As a more practical manner for using the amplitude and phase relationships, the process frequency response can be constructed from a single input perturbation using Fourier analysis (Hougen, 1964). This method has some of the advantages of the statistical method (for example, it allows inputs of general shape), but the statistical methods are generally preferred.

Identification Under Control

The empirical methods presented in this chapter are for input-output relationships without control. After covering Part I on feedback control, you may wonder whether the process model can be identified when being controlled. The answer is yes, but only under specific conditions, as explained by Box and MacGregor (1976).

6.6 CONCLUSIONS

Transfer function models of most chemical processes can be identified empirically using the methods described in this chapter. The general, six-step experimental procedure should be employed, regardless of the calculation method used.

> It is again worth emphasizing that the vast majority of control strategies are based on empirical models; thus, the methods in this chapter are of great practical importance.

Model Error

Model errors result from measurement noise, unmeasured disturbances, imperfect input adjustments, and applying simple linear models to truly nonlinear processes. The examples in this chapter give realistic results, which indicate that model parameters are known only within ± 20 percent at best for many processes. However, these models appear to capture the dominant dynamic behavior. Engineers must always consider the sensitivity of their decisions and calculations to expected model errors to ensure good performance of their designs. We will investigate the effects of model errors in later chapters and will learn that moderate errors do not substantially degrade the performance of single-loop controllers. A summary of a few sensitivity studies, which are helpful when reviewing modelling and control design, are given in Table 6.4.

Experimental Design

The design of the experimental conditions, especially the input perturbation, has a great effect on the success of empirical model identification. The perturbation must be large enough, compared with other effects on the output, to allow accurate model parameter estimation. Naturally, this requirement is in conflict with the desire to minimize process disturbances, and some compromise is required. Model accuracy depends strongly on the experimental procedure, and no amount of analysis can compensate for a very poor experiment.

Six-Step Procedure

Empirical model identification is an iterative procedure that may involve several experiments and potential model structures before a satisfactory model has been determined. The procedure in Figure 6.1 clearly demonstrates the requirement for a priori information about the process to design the experiment. Since this information may

TABLE 6.4
Summary of sensitivity of control stability and performance to modelling errors

Case	Issue studied
Example 9.2	The effect on performance of using controller tuning parameters based on an empirical model that is lower-order than the true process
Example 9.5	The effect on performance of using controller tuning parameters based on an empirical model that is substantially different from the true process
Example 10.15	The effect of modelling error on the stability of feedback control, showing the change of model parameters likely to lead to significant differences in dynamic behavior
Example 10.18	The effect of modelling error on the stability of feedback control, showing the critical frequency range of importance
Figure 13.16 and discussion	The effect of modelling error on the performance of feedback control, showing the frequency range of importance

be inexact, the procedure may have to be repeated, perhaps using a larger perturbation, to obtain useful data. Also, the results of the analysis should be evaluated with diagnostic procedures to ensure that the model is accurate enough for control design. It is essential for engineers to recognize that the calculation procedure always yields parameter values and that they must judge the validity of the results based on diagnostics and knowledge of the process behavior based on fundamental models.

> No process is known exactly! Good results using models with (unavoidable) errors is not simply fortuitous; process control methods have been developed over the years to function well in realistic situations.

In conclusion, empirical models can be determined by a rather straightforward experimental procedure combined with either a graphical or a statistical parameter estimation method. Usually, the models take the form of low-order transfer functions with dead time, which, although not capable of perfect prediction of all aspects of the process performance, provide the essential input-output relationships required for process control. The important topic of model error is considered in many of the subsequent chapters, where it is shown that models of the accuracy achieved with these empirical methods are adequate for many control design calculations. However, the selection of algorithms and determination of adjustable parameters must be performed with due consideration for the likely model errors. Therefore, lessons learned in this chapter about accuracy are applied in many later chapters.

REFERENCES

Anton, H., *Elementary Linear Algebra,* Wiley, New York, 1987.
Box, G., W. Hunter, and J. Hunter, *Statistics for Experimenters,* Wiley, New York, 1978.
Box, G. and Jenkins, *Time Series Analysis: Forecasting and Control,* Holden Day, Oakland, CA, 1976.
Box, G. and J. MacGregor, "Parameter Estimation with Closed-Loop Operating Data," *Technometrics, 18, 4,* 371–380 (1976).
Cryor, *Time Series Analysis,* Duxbury Press, Boston, MA, 1986.
Despande, P. and R. Ash, *Elements of Computer Process Control,* Instrument Society of America, Research Triangle Park, NC, 1988.
Franklin, G., J. Powell, and M. Workman, *Digital Control of Dynamic Systems,* Addison-Wesley, Reading, MA, 1990.
Graupe, D. *Identification of Systems,* Van Nostrand Reinhold, New York, 1972.
Harriott, P., *Process Control,* McGraw-Hill, New York, 1964.
Hougen, J., *Experiences and Experiments with Process Dynamics,* Chem. Eng. Prog. Monograph Ser., 60, 4, 1964.
Smith, C., *Digital Computer Process Control,* Intext Education Publishers, Scranton, PA, 1972.
Ziegler J., and N. Nichols, "Optimum Settings for Automatic Controllers," *Trans. ASME, 64,* 759–768 (1942).

ADDITIONAL RESOURCES

Advanced statistical model identification methods are widely used in practice. The following reference provides further insight into some of the more popular approaches.

Vandaele, W., *Applied Time Series and Box-Jenkins Models,* Academic Press, New York, 1983.

The following proceedings give a selection of model identification applications.

Ekyhoff, P., *Trends and Progress in System Identification,* Pergamon Press, Oxford, 1981.

Computer programs are available to ease the application of statistical methods. The programs noted below can be applied to simple linear regression (Lotus and Quattro Pro), to general statistical model fitting (SAS), and to empirical dynamic modelling for process control (MATLAB).

Lotus® 1-2-3™, Lotus Development Corporation
MATLAB® and Identification Toolbox, The MathWorks
Quattro Pro®, Borland International
SAS®, SAS Institute

International standards have been established for testing and reporting dynamic models for process control equipment. A good summary is provided in

ISA-S26-1968 and ANSI MC4.1-1975, *Dynamic Response Testing of Process Control Instrumentation,* Instrument Society of America, Research Triangle Park, NC, 1968.

Good results from the empirical method depend on proper engineering practices in experimental design and results analysis. The engineer must always cross-check the empirical model against the possible models based on physical principles.

QUESTIONS

6.1. An experiment has been performed on a fired heater (furnace). The fuel valve was opened an additional increment of 2 percent in a step, giving the resulting temperature response in Figure Q6.1. Determine the model parameters using both process reaction curve methods and estimate the inaccuracies in the parameter values due to the data and calculation methods.

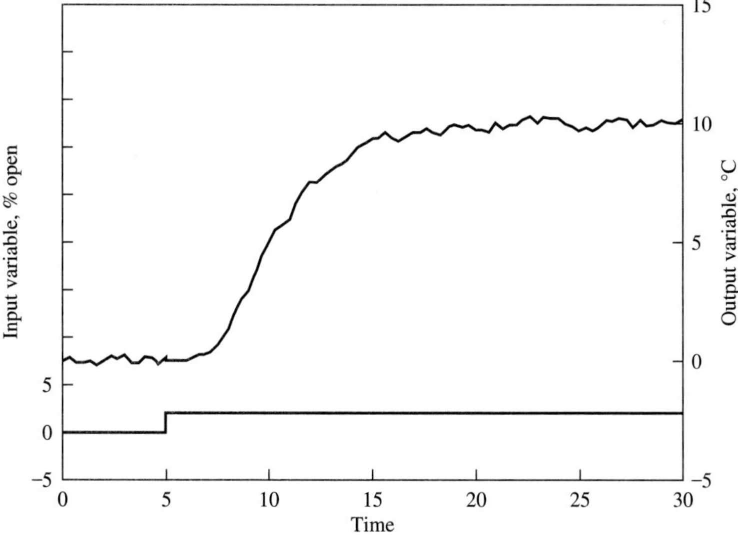

FIGURE Q6.1

6.2. Data has been collected from a chemical reactor. The inlet concentration was the only input variable that changed when the data was collected. The input and output data is given in Table Q6.2.

(*a*) Use the statistical identification method to estimate parameters in a first-order-with-dead-time model.

(*b*) Determine whether the model structure is adequate for this data.

(*c*) Estimate the inaccuracies in the parameter values due to the data and calculation method.

You may use a spreadsheet or statistical computer program. Note that the number of data points is smaller than desired for good estimation; this is solely to reduce the effort of typing the data into your program.

TABLE Q6.2

Time (min)	Input (% open)	Output (°C)	Time (min)	Input (% open)	Output (°C)	Time (min)	Input (% open)	Output (°C)
0	30	69.65	36	38	70.22	72	38	75.27
4	30	69.7	40	38	71.32	76	38	75.97
8	30	70.41	44	38	72.33	80	38	76.30
12	30	70.28	48	38	72.92	84	38	76.30
16	30	69.55	52	38	73.45	88	38	75.51
20	30	70.32	56	38	74.09	92	38	74.86
24	38	69.97	60	38	75.00	96	38	75.86
28	38	69.96	64	38	75.25	100	38	76.20
32	38	69.68	68	38	74.78	104	38	76.0

6.3. (*a*) The chemical reactor system in Figure Q6.3 is to be modelled. The relationship between the steam valve on the preheat exchanger and the outlet concentration is to be determined. Develop a complete experimental plan for a process reaction curve experiment. Include in your plan all actions, variables to be recorded or monitored, and any a priori information required from the plant operating personnel.

(*b*) Repeat the discussion for the experiment to model the effect of the flow of the reboiler heating medium on the distillate composition for the distillation tower in Figure 5.18.

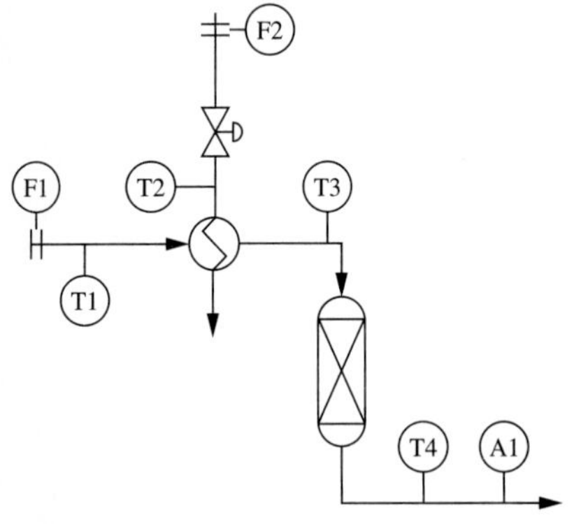

FIGURE Q6.3

6.4. Several experiments were performed on the chemical reactor shown in Figure Q6.3. In each experiment, the heat exchanger valve was changed and the reactor outlet temperature T4 was recorded. The dynamic data are given in Figure Q6.4*a* through *d*. Discuss the results of each experiment, noting any deficiencies and stating whether the data can be used for estimation and if so, which estimation method(s)—process reaction curve, statistical, or both—could be used.

6.5. Individual experiments have been performed on the process in Figure Q6.3. The following transfer function models were determined from these experiments.

$$\frac{T_3(s)}{T_2(s)} = \frac{0.55e^{-0.5s}}{2s+1} \qquad \frac{T_4(s)}{T_3(s)} = \frac{3.4e^{-2.1s}}{2.7s+1}$$

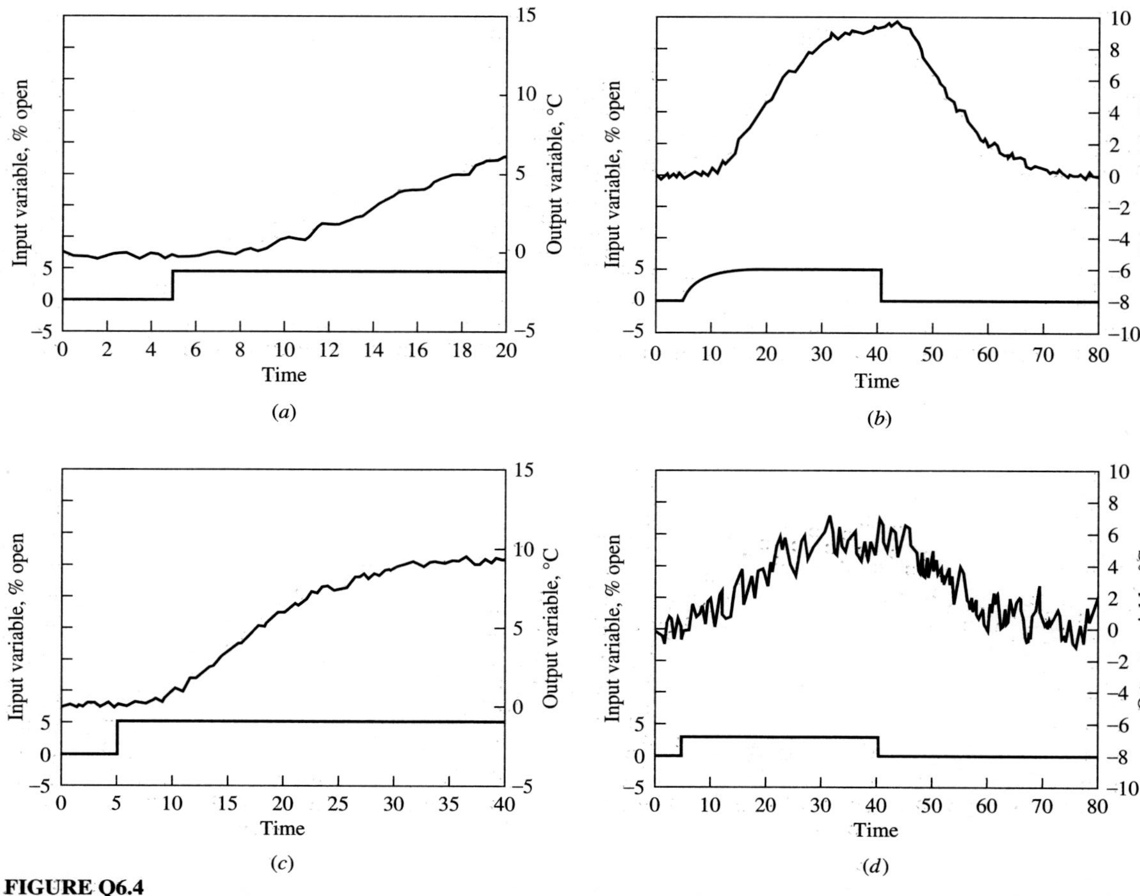

FIGURE Q6.4

(*a*) What are the units of the gains and do they make sense? Is the reaction exothermic or endothermic?

(*b*) Determine an approximate first-order-with-dead-time transfer function model for $T_4(s)/T_2(s)$.

(*c*) With better planning, could the model requested in (*b*) have been determined directly from the experimental data used to determine the models given in the problem statement?

6.6. Describe some physical processes for which the process reaction curve method is not appropriate. For each process, explain why the method will not provide an adequate model. Can you suggest an alternative method for empirically modelling each process? (*Hint:* Review the processes in Chapters 3 through 5.)

6.7. The difference equation for a first-order system was derived from the continuous differential equation in Section 6.4 by assuming that the input was constant over the sample period Δt. An alternative approach would be to approximate the derivative(s) by finite differences. Apply the finite difference approach to a first-order and a second-order model. Discuss how you would estimate the model parameters from a set of experimental data using least squares.

6.8. Although such experiments are not common for a process, frequency response modelling is specified for some instrumentation (ISA, 1968). Assume that the data in Table Q6.8 was determined by changing the fluid temperature about a thermocouple and thermowell in a sinusoidal manner. (Refer to Figure 4.9 for the meaning of frequency response.) Determine an approximate model by answering the following questions.

(*a*) Plot the amplitude ratio, and estimate the order of the model from this plot.

(*b*) Estimate the steady-state gain and time constant(s) from the results in (*a*).

(*c*) Plot the phase angle from the data and determine the value of the dead time, if any, from the plot.

TABLE Q6.8

Frequency	Amplitude ratio	Phase angle (°)
0.0001	1.0	−1
0.001	0.99	−7
0.005	0.85	−32
0.010	0.62	−51
0.015	0.44	−63
0.050	0.16	−80

6.9. It is important to use our knowledge of the process to design experiments and determine the range of applicability of the empirical models. Assume that the dynamic models for the following processes have been identified, for the input and output stated, using methods described in this chapter about some nominal operating conditions. After the experiments, the nominal operating conditions change as defined in the following table by a "substantial" amount, say 50%. You are to determine

(*a*) whether the input-output dynamic behavior would change as a result of the change in nominal conditions

(*b*) if so, which parameters would change and by how much

(*c*) whether the empirical procedure should be repeated to identify a model at the new nominal operating conditions

Process (all are worked examples)	**Input variable**	**Output variable**	**Process variable that changes for the new nominal operating condition**
Example 3.1: Mixing tank	C_{A0}	C_A	$(C_{A0})_s$
Example 3.1: Mixing tank	C_{A0}	C_A	F
Example 3.2: Isothermal CSTR	C_{A0}	C_A	T
Example 3.5: Isothermal CSTR	C_{A0}	C_A	$(C_{A0})_s$
Section 5.3: Noninteracting mixing tanks	C_{A0}	C_A	$(C_{A0})_s$
Section 5.3: Interacting levels	F_0	L_2	$(F_0)_s$

6.10. Use Method II of the process reaction curve to evaluate empirical models from the dynamic responses in Figure 5.20. Explain why you can obtain two models from one experiment.

6.11. The graphical methods could be extended to other models. Develop a method for estimating the parameters in a second-order transfer function with dead time and a constant numerator for a step input forcing function. The method should be able to fit both overdamped and underdamped systems. State all assumptions and explain all six steps.

6.12. The graphical methods could be extended to other forcing functions. For both first- and second-order systems with dead time, develop methods for fitting parameters from an impulse response.

6.13. We will be using first-order-with-dead-time models often. Sketch an ideal process that is exactly first-order with dead time. Derive the fundamental model and relate the equipment and operating conditions to the model parameters. Discuss how well this model approximates more complex processes.

6.14. Develop a method for testing whether the empirical data can be fitted using equation (6.2). The method should involve comparing calculated values to a straight-line model.

6.15. Both process reaction curve methods require that the process achieve a steady state after the step input. For both methods, suggest modifications that would relieve the requirement for a final steady state. Discuss the relative accuracy of these modified methods to those presented in the chapter. Could you apply your method to the first part of the transient response in Figure 3.10*c*?

6.16. Often, more than one input to a process changes during an experiment. For the process reaction curve and the statistical method:

(*a*) If possible, show how models for two inputs could be determined from such experiments. Clearly state the requirements of the experimental design and calculations.

(*b*) Assume that the model between one of the inputs and the output is known. Show how to fit the parameters for the remaining input.

6.17. For each of the processes and dynamic data, state whether the process reaction curve, the statistical model fitting method, or both can be used. Also, state the model form necessary to model the process adequately. The systems are Examples 3.3, 3.10, 5.1, 5.4, 5.5, and Figure 5.5 (with $n = 10$).

6.18. The residual plot provides a visual display of goodness of fit. How could you use the calculated residuals to test the hypothesis that the model has provided a good fit? What could you do if the result of this test indicates that the model is not adequate?

6.19. (*a*) Experiments were performed to obtain the process reaction curves in Figures 5.20*a* and *b*. How do you think that the results would change if

1. the step magnitudes were halved? doubled?
2. the step signs were inverted?
3. both steps were made simultaneously?

(*b*) Describe how the inventories (liquid levels) were controlled during the experiments.

(*c*) Would the results change if the inventories were controlled differently?

PART III

FEEDBACK CONTROL

To this point we have studied the dynamic responses of various systems and learned important relationships between process equipment and operating conditions and dynamic responses. In this part, we make a major change in perspective: we change from understanding the behavior of the system to *altering its behavior* to achieve safe and profitable process performance. This new perspective is shown schematically in Figure III.1 for a physical example given in Figure III.2. In discussing control, we will use the terms *input* and *output* in a specific manner, with input variables influencing the output variables as follows:

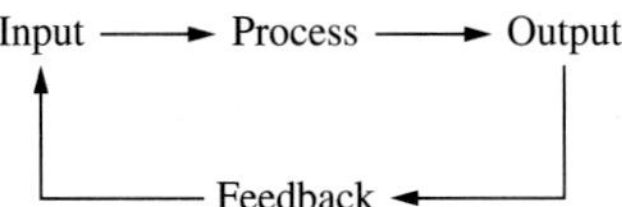

Here we see a difference in terminology between modelling and feedback control. In feedback control the input is the cause and the output is the effect, and there is no requirement that the input or output variables be associated with a stream passing through the boundary defining the system. For example, the input can be a flow and the output can be the liquid level in the system.

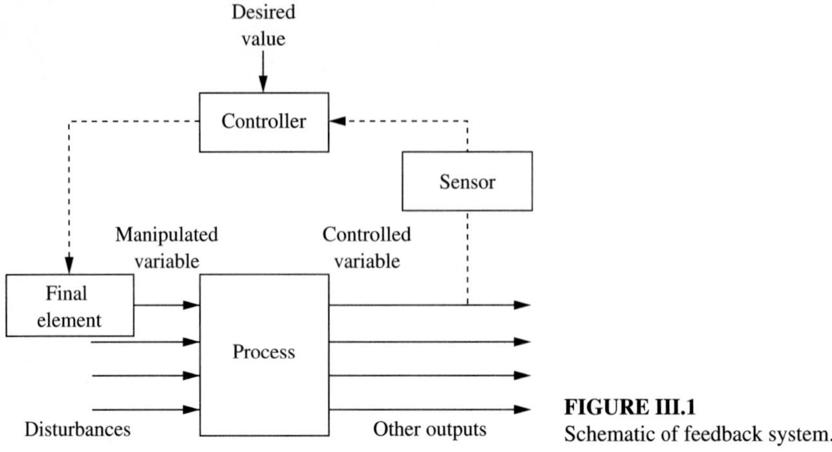

FIGURE III.1
Schematic of feedback system.

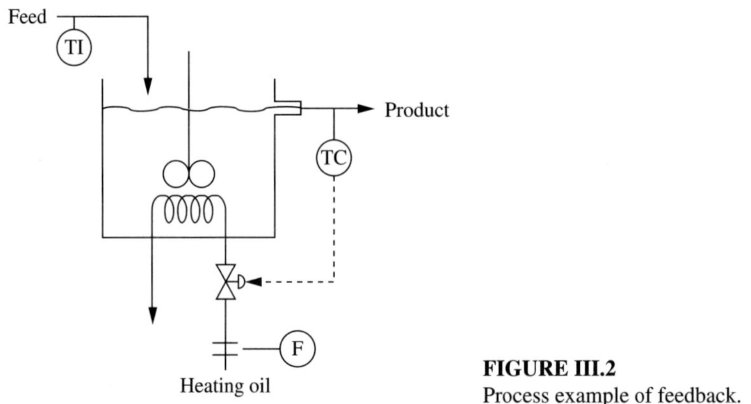

FIGURE III.2
Process example of feedback.

There is a cause/effect relationship in the process that cannot be directly inverted. In the process industries we usually desire to maintain selected output variables, such as pressure, temperature, or composition, at specified values. Therefore, feedback is applied to achieve the desired output by adjusting an input. This explains why the feedback control algorithm is sometimes described as the inverse of the process relationship.

First, the engineer selects the measured outlet variable whose behavior is specified; it is called the *controlled variable* and typically has a substantial effect on the process performance. In the example, the temperature of the stream leaving the stirred tank is the controlled variable. Many other output variables exist, such as the outlet flow rate and the exit heating oil temperature. Next, the variables that have been referred to as process inputs are divided into two categories: manipulated and disturbance variables. A *manipulated variable* is selected by the engineer for adjustment in a control strategy to achieve the desired performance in the controlled variable. In the physical example, the valve position in the heating oil is the manipulated variable, since opening the valve increases the flow of heating oil and results in greater heat transfer to the fluid in the tank. All other input variables that influence the controlled variable are termed *disturbances*. Examples of disturbances are the inlet flow rate and inlet temperature.

Since some specific dynamic performance has been specified, an additional component must be added to the system. Here we consider feedback control, which was introduced in Chapter 1 as a method for adjusting an input variable based on a measured output variable. In the simplest case, the feedback system could involve a person who observes a thermometer reading and adjusts the heating valve by hand. Alternatively, feedback control can be automated by providing a computing device with an algorithm for adjusting the valve based on measured temperature values. To automate the feedback, the sensor must be designed to communicate with the computing device, and the final element must respond to the command from the computing device.

Among the most important decisions made by the engineer are the selection of controlled and manipulated variables and the algorithm and parameters in the calculation. In this part, the greatest emphasis is placed on understanding the feedback principles through the analysis of particular feedback control algorithms. The selection of measured controlled variables and manipulated variables is provided in the examples in these chapters. Approaches for selecting these variables from among many candidates are introduced here and expanded in later chapters. While this part emphasizes the control algorithm, one must never lose sight of the fact that the process is part of the control *system*! Since chemical engineers are responsible for designing the process equipment and determining operating conditions to achieve good process performance, the material in this part provides qualitative and quantitative methods for evaluating the likely dynamic performance of process designs under feedback control.

CHAPTER 7

THE FEEDBACK LOOP

7.1 INTRODUCTION

Now that we are prepared with a good understanding of process dynamics, we can begin to address the technology for automatic process control. The goals of process control—safety, environmental protection, equipment protection, smooth operation, quality control, and profit—are achieved by maintaining the plant variables as close as possible to their best conditions. The variability of variables about their best values can be reduced by adjusting selected input variables using feedback control principles. As explained in Chapter 1, feedback makes use of an output of a system in deciding the way to influence an input to the system, and the technology presented in this part of the book explains how to employ feedback. This chapter builds on the chapters in Part I of the book, which were more qualitative and descriptive, by establishing the key quantitative aspects of a control system.

It is important to emphasize that we are dealing with the control *system,* which involves the process and instrumentation as well as the control calculations. Thus, this chapter has a section on the feedback loop in which all elements are discussed. Then, reasons for control are reviewed, and because engineers should always be prepared to define measures of the effectiveness of their efforts, quantitative measures of control performance are defined for key disturbances; these measures are used throughout the remainder of the book. Because the process usually has several input and output variables, initial criteria are given for selecting the variables for a control loop. These criteria provide minimum necessary features for the variables and will be augmented with additional guidelines as the performance of feedback control is more quantitatively analyzed in later chapters. Finally, several general approaches to feedback control, ranging from manual to automated methods, are discussed, along with guidelines for when to employ each approach.

7.2 PROCESS AND INSTRUMENT ELEMENTS OF THE FEEDBACK LOOP

All elements of the feedback loop can affect control performance. In this section, the process and instrument elements of a typical loop, excluding the control calculation, are introduced, and some quantitative information on their dynamics is given. This analysis provides a means for determining which elements of the loop introduce significant dynamics and when the dynamics of some fast elements can usually be considered negligible.

A typical feedback control loop is shown in Figure 7.1. This discussion will address each element of the loop, beginning with the signal, that is sent to the process equipment. This signal could be determined using feedback principles by a person or automatically by a computing device. Some key features of each element in the control loop are summarized in Table 7.1.

The feedback signal in Figure 7.1 has a range usually expressed as 0 to 100%, whether determined by a controller or set manually by a person. When the signal is transmitted electronically, it usually is converted to a range of 4 to 20 milliamps (mA) and can be transmitted long distances, certainly over one mile. When the signal is transmitted peumatically, it has a range of 3 to 15 psig and can only be transmitted over a shorter distance, usually limited to about 400 meters unless special signal reinforcement is provided. Pneumatic transmission would normally be used only when the controller is performing its calculations pneumatically, which is not common with modern equipment. Naturally, the electronic signal transmission is essentially

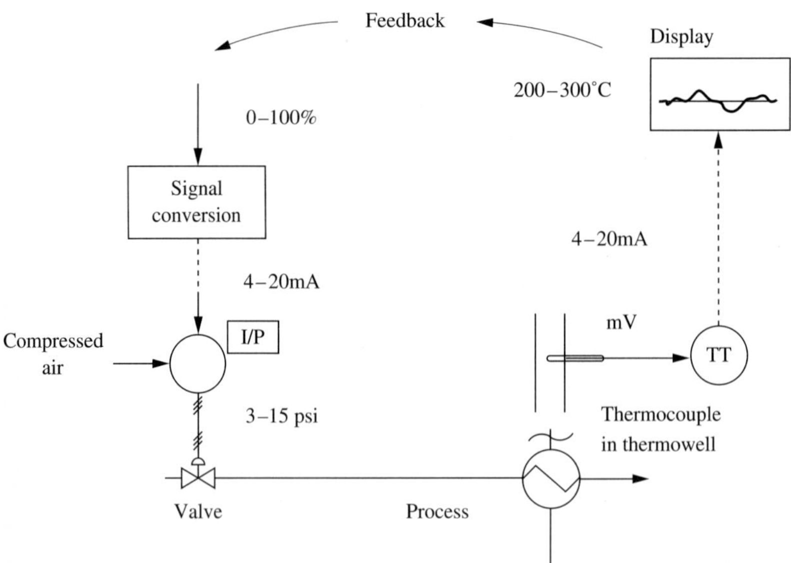

FIGURE 7.1
Process and instrument elements in a typical control loop.

TABLE 7.1
Key features of control loop elements, excluding the process

Loop element*	Function	Typical range	Typical dynamic response, $t_{63\%}$**
Controller output	Initiate signal at a remote location intended for the final element	Operator/ Controller use 0–100%	
Transmission	Carry signal from controller to final element and from the sensor to the controller	*Pneumatic:* 3–15 psig *Electronic:* 4–20 milliamp (mA)	*Pneumatic:* 1–5 s *Electronic:* Instantaneous
Signal conversion	Change transmission signal to one compatible with final element	*Electronic to pneumatic:* 4–20 mA to 3–15 psig *Sensor to electronic:* mV to 4–20 mA	0.5–1.0 s
Final control element	Implement desired change in process	*Valve*: 0–100% open	1–4 s
Sensor	Measure controlled variable	Scale selected to give good accuracy, e.g., 200–300°C	Typically from a few seconds to several minutes

* The terms *input* and *output* are with respect to a controller.

** Time for output to reach 63% after step input.

instantaneous; the pneumatic signal requires several seconds for transmission. Note that the standard signal ranges are very important so that equipment manufactured by different suppliers can be interchanged.

At the process unit, the output signal is used to adjust the final control element: the equipment that is manipulated by the control system. The final control element in the example, as in over 90 percent of process control applications, is a valve. The valve percent opening could be set by an electrical motor, but this is not usually done because of the danger of explosion with the high-amperage power supply a motor would require. The alternative power supply typically used is compressed air. The signal is converted from electrical to pneumatic; 3 to 15 psig is the standard range of the pneumatic signal. The conversion is relatively accurate and rapid, as indicated by the entry for this element in Table 7.1. The pneumatic signal is transmitted a short distance to the control valve, which is specially designed to adjust its percent opening based on the pneumatic signal. Control valves respond relatively quickly, with typical time constants ranging from 1 to 4 sec.

The general principles of a control valve are demonstrated in Figure 7.2. The process fluid flows through the opening in the valve, with the amount open (or resistance to flow) determined by the valve stem position. The valve stem is connected to the diaphragm, which is a flexible metal sheet that can bend in response to forces.

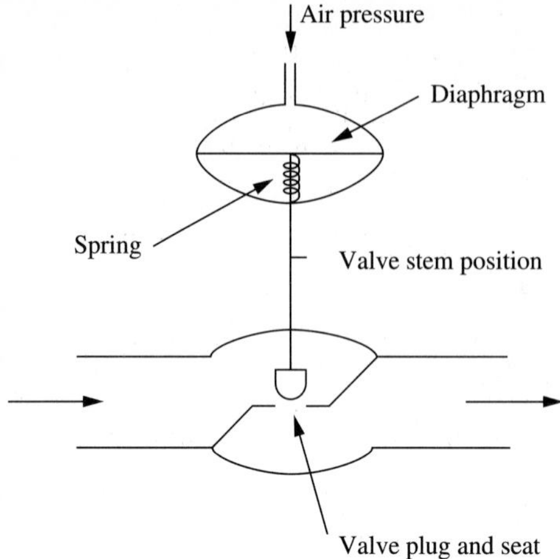

FIGURE 7.2
Schematic of control valve.

The two forces acting on the diaphragm are the spring and the variable pressure from the control signal. For a zero control signal (3 psig), the diaphragm in Figure 7.2 would be deformed upward because of the greater force from the spring, and the valve stem would be raised, resulting in the greatest opening for flow. For a maximum signal (15 psig), the diaphragm in Figure 7.2 would be deformed downward by the greater force from the air pressure, and the stem would be lowered, resulting in the minimum opening for flow. Other arrangements are possible, and selection criteria are presented in Chapter 12.

After the final control element has been adjusted, the process responds to the change. The process dynamics vary greatly for the wide range of equipment in the process industries, with typical dead times and time constants ranging from a few seconds (or faster) to many hours. When the process is by far the slowest element in the control loop, the dynamics of the other elements are negligible. This situation is common, but important exceptions occur, as demonstrated in Example 7.1.

The sensor responds to the change in plant conditions, preferably indicating the value of a single process variable, unaffected by all other variables. Usually, the sensor is not in direct contact with the potentially corrosive process materials; therefore, the protective equipment or sample system must be included in the dynamic response. For example, a thin thermocouple wire responds quickly to a change in temperature, but the metal sleeve around the thermocouple, the thermowell, can have a time constant of 5 to 20 sec. Most sensor systems for flow, pressure, and level have time constants of a few seconds. Analyzers that perform complex physicochemical analyses can have much slower responses, on the order of 5 to 30 minutes or longer; they may be *discrete,* meaning that a new analyzer result becomes available periodically, with no new information between results. Physical principles and performance of sensors are diverse, and the reader is encouraged to refer to information in the additional resources from Chapter 1 on sensors for further details.

The sensor signal is transmitted to the controller, which we are considering to be located in a remote control room. The transmission could be pneumatic (3 to 15 psig)

TABLE 7.2
Dynamic models for elements in Example 7.1

Element	Units*	Case A	Case B
Manual station	mA/% output	0.16	0.16
Transmission		1.0	1.0
Signal conversion	psi/mA	$0.75/(0.5s + 1)$	$0.75/(0.5s + 1)$
Final element	%open/psi	$8.33/(1.5s + 1)$	$8.33/(1.5s + 1)$
Process	°C/psi	$1.84e^{-1.0s}/(3s + 1)$	$1.84e^{-100s}/(300s + 1)$
Sensor	mV/°C	$0.11/(10s + 1)$	$0.11/(10s + 1)$
Signal conversion	mA/mV	$1.48/(0.51s + 1)$	$1.48/(0.51s + 1)$
Transmission		1.0	1.0
Display	°C/mA	$6.25/(1.0s + 1)$	$6.25/(1.0s + 1)$

* Time is in seconds.

or electrical (4 to 20 mA). The controller receives the signal and performs its control calculation. The controller can be an *analog* system; for example, an electronic analog controller consists of an electrical circuit that obeys the same equations as the desired control calculations (Hougen, 1972). For the next few chapters, we assume that the controller is a continuous electronic controller that performs its calculations essentially instantaneously, and we will see in Chapter 11 that essentially the same results can be obtained by a very fast digital computer, as is used in most modern control equipment.

Example 7.1. The dynamic responses of two process and instrumentation systems, *without the controller,* are evaluated in this exercise. The system involves electronic transmission, a pneumatic valve, a first-order-with-dead-time process, and a thermocouple in a thermowell. The dynamics of the individual elements are given in Table 7.2 with the time in seconds for two different systems, A and B. The dynamics of the entire loop are to be determined. The question could be stated, "How does a unit step change in the manual output affect the displayed variable, which is also the variable available for control, in the control house?" Note that the two systems are identical except for the process transfer functions.

The physical system in this problem and shown in Figure 7.1 is recognized as a series of noninteracting systems. Therefore, equation (5.40) can be applied to determine the transfer function of the overall noninteracting series system. The result for Case B is

$$\frac{Y(s)}{X(s)} = \frac{(0.16)(1.0)(0.75)(8.33)(1.84)(0.11)(1.48)(1.0)(6.25)e^{-100s}}{(0.5s + 1)(1.5s + 1)(300s + 1)(10s + 1)(0.51s + 1)(s + 1)}$$

Before the simulation results are presented for this example, it is worthwhile performing an approximate analysis, using the simple approximation introduced in Chapter 5 for series processes (page 162). The overall gains and approximate 63 percent times for both systems that relate the manual signal to the display are shown in the following table.

	Case A	Case B	
Process gain $K_p = \prod K_i$	1.84	1.84	K/(% controller output)
Time to 63% $\approx \Sigma(\tau_i + \theta_i)$	≈ 17.5	≈ 413.5	seconds

The two cases have been simulated, and the results are plotted in Figures 7.3*a* and *b*. The results of the approximate analysis compare favorably with the simulations. Note that for system A, which involves a fast process, the sensor and final element contribute significant dynamics, resulting in a substantial difference between the true process temperature and the displayed value of the temperature, which would be used for feedback control. In system B the process dynamics are much slower, and the dynamic effects of all other elements in the loop are negligible. This is a direct consequence of the time domain solution to the model of this process for a step ($1/s$) input, which has the form

$$Y'(t) = C_1 + C_2 e^{-t/\tau_2} + C_3 e^{-t/\tau_3} + \cdots$$

Clearly, a slow "mode" due to one especially long time constant will dominate the dynamic response, with the faster elements essentially at quasi-steady state. One

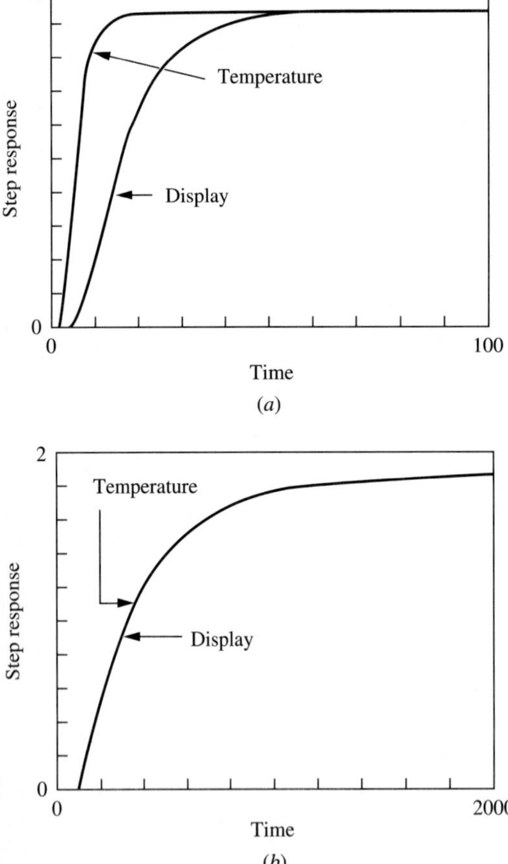

FIGURE 7.3
Transient response for Example 7.1 with a 1% step input change at time = 0. (*a*) Case A; (*b*) Case B.

would expect that a dynamic analysis that considered the process alone for control design would not be adequate for Case A but would be adequate for Case B.

It is worth recalling that the empirical methods for determining the "process" dynamics presented in Chapter 6 involve changes to the manipulated signal and monitoring the response of the sensor signal as reported to the control system. Thus, the resulting model includes all elements in the loop, including instrumentation and transmission. Since the experiments usually employ the same instrumentation used subsequently for implementing the control system, the dynamic model identified is between the controller output and input—in other words, the system "seen" by the controller. This seems like the appropriate model for use in design control systems, and that intuition will be supported by later analysis.

7.3 BLOCK DIAGRAM

The basic feedback control loop will be analyzed frequently in the subsequent chapters using block diagrams and transfer functions, which concisely summarize all information in the control loop. The overall system transfer function is derived in this section because it is often used to determine important properties of the feedback system, such as stability and performance. Note that the controller transfer function $G_c(s)$ has not yet been specified but will be in subsequent chapters; the results in this chapter are valid for any control calculation $G_c(s)$.

The block diagram is shown in Figure 7.4 with the terminology that will be used throughout the book. Notice that the elements in Table 7.1 are collected into three transfer functions: the valve or final element, $G_v(s)$; the process, $G_p(s)$; and the sensor, $G_s(s)$. The process output variable selected to be controlled is termed the *controlled variable*, CV(s), and the process input variable selected to be adjusted by the control system is termed the *manipulated variable*, MV(s). The desired value, which must be specified independently to the controller, is called the *set point*,

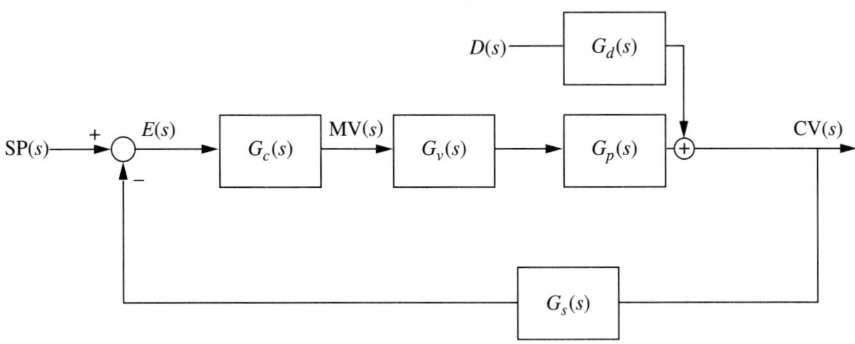

$G_c(s)$ = Controller
$G_v(s)$ = Transmission, transducer, and valve
$G_p(s)$ = Process
$G_s(s)$ = Sensor, transducer, and transmission
$G_d(s)$ = Disturbance

FIGURE 7.4
Block diagram of a feedback control system.

SP(s); it is also called the *reference value* in some books on automatic control. The difference between the set point and the measured controlled variable is termed the *error*, $E(s)$. An input that changes due to external conditions and affects the controlled variable is termed a *disturbance*, $D(s)$, and the relationship between the disturbance and the controlled variable is the *disturbance transfer function*, $G_d(s)$. First, the transfer function of the controlled variable to the disturbance variable, $\mathrm{CV}(s)/D(s)$, is derived, with the change in the set point, SP(s), taken to be zero.

The system involves a recycle, since the process output variable is used in determining the process input variable—our definition of feedback; therefore, special care must be taken in deriving the transfer function. The four-step procedure presented in Chapter 4 is used again here. The first step is to begin with the variable in the numerator of the transfer function, which in this case is CV(s). In the second step, the expression for this variable as a function of input variables is derived in reverse direction to the information flow in the block diagram. The result is

$$\begin{aligned} \mathrm{CV}(s) &= G_p(s)G_v(s)\mathrm{MV}(s) + G_d(s)D(s) \\ &= G_p(s)G_v(s)G_c(s)G_s(s)[\mathrm{CV}(s)] + G_d D(s) \end{aligned} \tag{7.1}$$

This procedure is followed until one of two situations is reached: the numerator variable can be expressed as a function of the denominator variable alone (which occurs for series systems), or the numerator variable can be expressed as a function of itself and the denominator variable (which occurs for a simple feedback system). The expression in equation (7.1) is clearly of the second type. The third step in the procedure is to rearrange the equation so that the variables are separated as follows:

$$[1 + G_p(s)G_v(s)G_c(s)G_s(s)]\mathrm{CV}(s) = G_d(s)D(s) \tag{7.2}$$

Equation (7.2) can be rearranged to yield the closed-loop disturbance transfer function, and the same procedure can be used to derive the set point transfer function.

Closed-loop transfer functions

$$\frac{\mathrm{CV}(s)}{D(s)} = \frac{G_d(s)}{1 + G_p(s)G_v(s)G_c(s)G_s(s)} \tag{7.3}$$

$$\frac{\mathrm{CV}(s)}{\mathrm{SP}(s)} = \frac{G_p(s)G_v(s)G_c(s)}{1 + G_p(s)G_v(s)G_c(s)G_s(s)} \tag{7.4}$$

In summary, the block diagram procedure for deriving a transfer function involves four steps:

1. Select the numerator of the transfer function.
2. Solve in reverse direction to the causal relationships (arrows) in the block diagram to eliminate all variables except the numerator and denominator in the transfer function.
3. Separate variables in the equation.
4. Divide by the denominator variable to complete the transfer function.

For simple systems like the one in Figure 7.3, the foregoing procedure will yield the transfer function. In more complex systems, it will not be possible to eliminate all intermediate variables immediately in step 2. Therefore, steps 2 and 3 must be performed several times, as will be demonstrated in later chapters.

Remember: Block diagrams and transfer functions are used to assist the engineer in determining the quantitative aspects of dynamic performance and in understanding the qualitative features of the system. The block diagram has no more information about the system than the linear equations in the time domain.

It is important again to recall that the Laplace transform model is expressed as a set of simultaneous linear algebraic equations. The block diagram is a useful way to represent, or picture, the equations. Also, block diagram manipulations are simply a structured manner to solve the linear equations algebraically for one variable in terms of another variable. Any other manner for performing this manipulation is acceptable; experience has shown that the procedure just described is generally easiest to learn and use.

The use of block diagrams entails one potential difficulty, especially for the person just learning process control. Since the block diagram represents the model of the system, there is no distinction in the symbols used for various physical components in the system. For example, the block diagram in Figure 7.3 represents a system composed of elements from the process, $G_p(s)$ and $G_d(s)$; instrumentation, $G_v(s)$ and $G_s(s)$; and a control calculation performed in some type of computing device, $G_c(s)$. The reader is encouraged to relate all block diagrams to process schematics such as Figure 7.1.

Two generalizations can be made about the closed-loop transfer functions to assist in checking the derived transfer functions using block diagram manipulations. First, the numerator is simply the product of all transfer functions between the input (denominator variable) and the output (numerator variable). Second, the denominator of the right-hand side is of the form $1 + G''(s)$. The term $G''(s)$ is the product of all elements in the feedback loop. These guidelines can be checked by applying them to equations (7.3) and (7.4).

Finally, the transfer function notation is often simplified by lumping all instrumentation and process dynamics into one term, $G_p(s)$. This is equivalent to the following expression.

$$G_p(s) = G'_p(s)\,G_v(s)G_s(s) \tag{7.5}$$

with $G'_p(s)$ being the process alone. This is a natural simplification, since the dynamics of all elements from the controller output to the controller input contribute to the control system performance. Also, when the dynamics are determined empirically, the only model determined is the overall product of all instrumentation and process elements, and the individual elements are not known. The resulting simplified transfer function is

$$\frac{\mathrm{CV}(s)}{D(s)} = \frac{G_d}{1 + G_p G_c} \quad \text{with} \quad G_p(s) = G'_p(s)\,G_v(s)\,G_s(s) \tag{7.6}$$

This simplification is not used when the effects of sensors and final elements are to be shown clearly; however, it is used often to simplify notation. If the process transfer function $G_p(s)$ is shown in a closed-loop block diagram or transfer function without the sensor and final element, the reader should assume that it includes the dynamics of the sensor and final element, since feedback control requires all elements in the loop.

7.4 CONTROL PERFORMANCE MEASURES FOR COMMON INPUT CHANGES

The purpose of the feedback control loop is to maintain a small deviation between the controlled variable and the set point by adjusting the manipulated variable. In this section, the two general types of external input changes are presented, and quantitative control performance measures are presented for each.

Set Point Input Changes

The first type of input change involves changes to the *set point:* the desired value for the operating variable, such as product composition. In many plants the set points remain constant for a long time. In other plants the values may be changed periodically; for example, in a batch operation the temperature may need to be changed during the batch.

Control performance depends on the goals of the process operation. Let us here discuss some general control performance measures for a change in the controller set point on the three-tank mixing process in Figure 7.5. In this process, two streams, *A* and *B*, are mixed in three series tanks, and the output concentration of component A is controlled by manipulating the flow of stream *A*. Here, we consider step changes to the set point; these changes represent the situation in which the plant operator occasionally changes the value and allows considerable time for the control system to respond. A typical dynamic response is given in Figure 7.6. This is somewhat idealized, because there is no measurement noise or effect of disturbances, but these effects will be considered later. Several facets of the dynamic response are considered in evaluating the control performance.

OFFSET. Offset is a difference between final, steady-state values of the set point and of the controlled variable. In most cases, a zero steady-state offset is highly desired, because the control system should achieve the desired value, at least after a very long time.

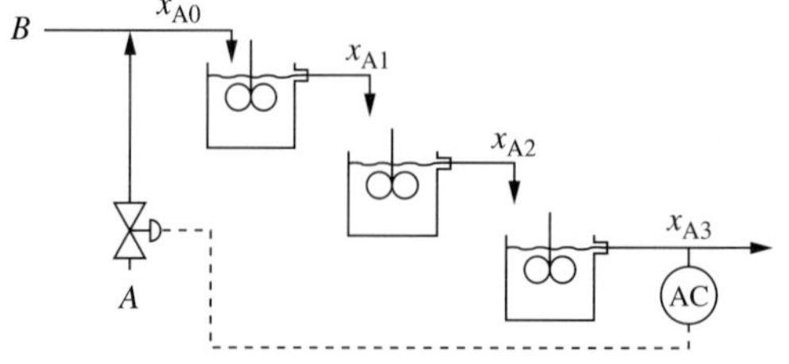

FIGURE 7.5
Example feedback control system, three-tank mixing process.

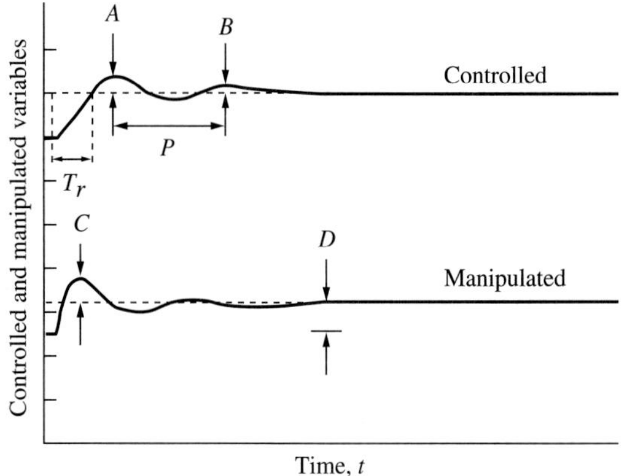

FIGURE 7.6
Typical transient response of a feedback control system to a step set point change.

RISE TIME. This (T_r) is the time from the step change in the set point until the controlled variable *first* reaches the new set point. A short rise time is usually desired.

INTEGRAL ERROR MEASURES. These indicate the cumulative deviation of the controlled variable from its set point during the transient response. Several such measures are used:

Integral of the Absolute value of the Error (IAE):

$$\text{IAE} = \int_0^\infty |\text{SP}(t) - \text{CV}(t)|\, dt \tag{7.7}$$

Integral of Square of the Error (ISE):

$$\text{ISE} = \int_0^\infty [\text{SP}(t) - \text{CV}(t)]^2 dt \tag{7.8}$$

Integral of product of Time and the Absolute value of Error (ITAE):

$$\text{ITAE} = \int_0^\infty t\, |\text{SP}(t) - \text{CV}(t)|\, dt \tag{7.9}$$

Integral of the Error (IE):

$$\text{IE} = \int_0^\infty [\text{SP}(t) - \text{CV}(t)]\, dt \tag{7.10}$$

The IAE is an easy value to analyze visually, because it is the sum of areas above and below the set point. It is an appropriate measure of control performance when the effect on control performance is linear with the deviation magnitude. The ISE is appropriate when large deviations cause greater performance degradation than small deviations. The ITAE penalizes deviations that endure for a long time. Note that IE is not normally used, because positive and negative errors cancel in the integral, resulting in the possibility for large positive and negative errors to give a small IE. A small integral error measure (e.g., IAE) is desired.

DECAY RATIO (*B*/*A*). The decay ratio is the ratio of neighboring peaks in an underdamped controlled-variable response. Usually, periodic behavior with large amplitudes is avoided in process variables; therefore, a small decay ratio is usually desired, and an overdamped response is sometimes desired.

THE PERIOD OF OSCILLATION (*P*). Period of oscillation depends on the process dynamics and is an important characteristic of the closed-loop response. It is not specified as a control performance goal.

SETTLING TIME. Settling time is the time the system takes to attain a "nearly constant" value, usually ±5 percent of its final value. This measure is related to the rise time and decay ratio. A short settling time is usually favored.

MANIPULATED-VARIABLE OVERSHOOT (*C*/*D*). This quantity is of concern because the manipulated variable is also a process variable that influences performance. There are often reasons to prevent large variations in the manipulated variable. Some large manipulations can cause long-term degradation in equipment performance; an example is the fuel flow to a furnace or boiler, where frequent, large manipulations can cause undue thermal stresses. In other cases manipulations can disturb an integrated process, as when the manipulated stream is supplied by another process. On the other hand, some manipulated variables can be adjusted without concern, such as cooling water flow. We will use the overshoot of the manipulated variable as an indication of how aggressively it has been adjusted. The overshoot is the maximum amount that the manipulated variable exceeds its final steady-state value and is usually expressed as a percent of the change in manipulated variable from its initial to its final value. Some overshoot is acceptable in many cases; little or no overshoot may the best policy in some cases.

Disturbance Input Changes

The second type of change to the closed-loop system involves variations in uncontrolled inputs to the process. These variables, usually termed *disturbances,* would cause a large, sustained deviation of the controlled variable from its set point if corrective action were not taken. The way the input disturbance variables vary with time has a great effect on the performance of the control system. Therefore, we must be able to characterize the disturbances by means that (1) represent realistic plant situations and (2) can be used in control design methods. Let us discuss three idealized disturbances and see how they affect the example mixing process in Figure 7.5. Several facets of the dynamic responses are considered in evaluating the control performance for each disturbance.

STEP DISTURBANCE. Often, an important disturbance occurs infrequently and in a sudden manner. The causes of such disturbances are usually changes to other parts of the plant that influence the process being considered. An example of a step upset in Figure 7.5 would be the inlet concentration of stream *B*. The responses of the outlet concentration, without and with control, to this disturbance are given

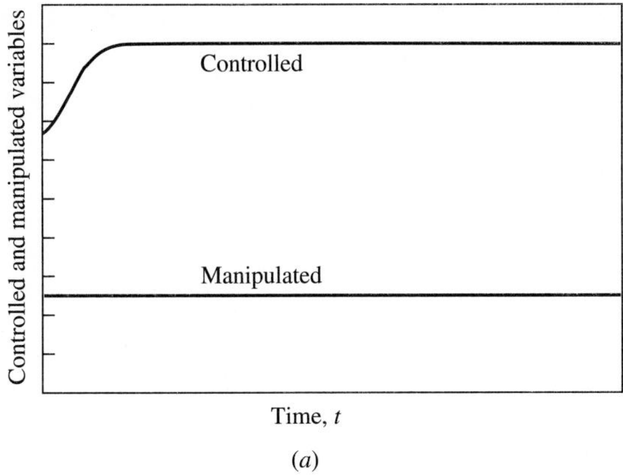

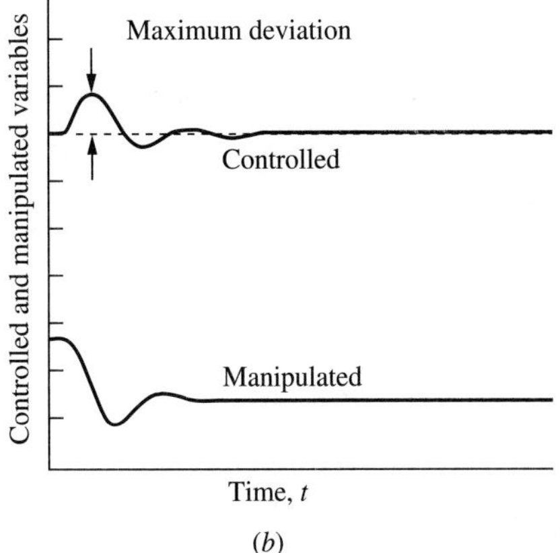

FIGURE 7.7
Transient response of the example process in Figure 7.5 in response to a step disturbance, (*a*) without feedback control; (*b*) with feedback control.

in Figure 7.7*a* and *b*. We will often consider dynamic responses similar to those in Figure 7.7 when evaluating ways to achieve good control that minimizes the effects of step disturbances. The explanations for the measures are the same as for set point changes except for rise time, which is not applicable, and for the following measure, which has meaning only for disturbance responses and is shown in Figure 7.7*b*:

The *maximum deviation* of the controlled variable from the set point is an important measure of the process degradation experienced due to the disturbance; for example, the deviation in pressure must remain below a specified value. Usually, a small value is desirable so that the process variable remains close to its set point.

STOCHASTIC INPUTS. As we recognize from our experiences in laboratories and plants, a process typically experiences a continual stream of small and large disturbances, so that the process is never at an exact steady state. A process that is subjected to such seemingly random upsets is termed a *stochastic system.* The response of the example process to stochastic upsets in all flows and concentrations is given in Figure 7.8*a* and *b* without and with control.

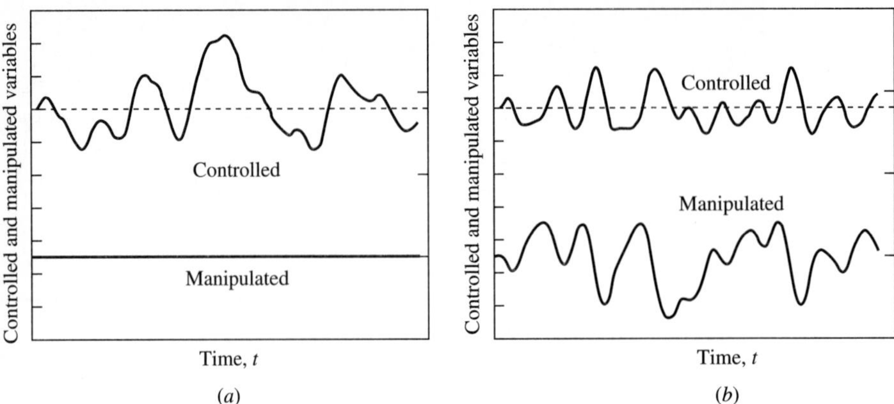

FIGURE 7.8
Transient response of the example process (*a*) without and (*b*) with feedback control to a stochastic disturbance.

The major control performance measure is the variance, σ_{CV}^2, or standard deviation, σ_{CV}, of the controlled variable, which is defined as follows for a sample of n data points:

$$\sigma_{\text{CV}} = \sqrt{\frac{1}{n-1}\sum_{i=1}^{n}(\overline{\text{CV}} - \text{CV}_i)^2} \tag{7.11}$$

$$\text{with the mean} = \overline{\text{CV}} = \frac{1}{n}\sum_{i=1}^{n}\text{CV}_i \tag{7.12}$$

This variable is closely related to the ISE performance measure for step disturbances. The relationship depends on the approximations that (1) the mean can be replaced with the set point, which is normally valid for closed-loop data, and (2) the number of points is large.

$$\frac{1}{n-1}\sum_{i=1}^{n}(\overline{\text{CV}} - \text{CV}_i)^2 \approx \frac{1}{T}\int_0^T (\text{SP} - \text{CV})^2\,dt \tag{7.13}$$

Since the goal is usually to maintain controlled variables close to their set points, a small value of the variance is desired. In addition, the variance of the manipulated variable is often of interest, because too large a variance could cause long-term damage to equipment (fuel to a furnace) or cause upsets in plant sections providing the manipulated stream (steam-generating boilers). We will not be analyzing stochastic systems in our design methods, but we will occasionally confirm that our designs perform well with example stochastic disturbances by simulation case studies. As you may expect, the mathematical analysis of these statistical disturbances is challenging and requires methods beyond the scope of this book. However, many practical and useful methods are available and should be considered by the advanced student (MacGregor, 1988; Cryor, 1986).

SINE INPUTS. An important aspect of stochastic systems in plants is that the disturbances can be thought of as the sum of many sine waves with different amplitudes

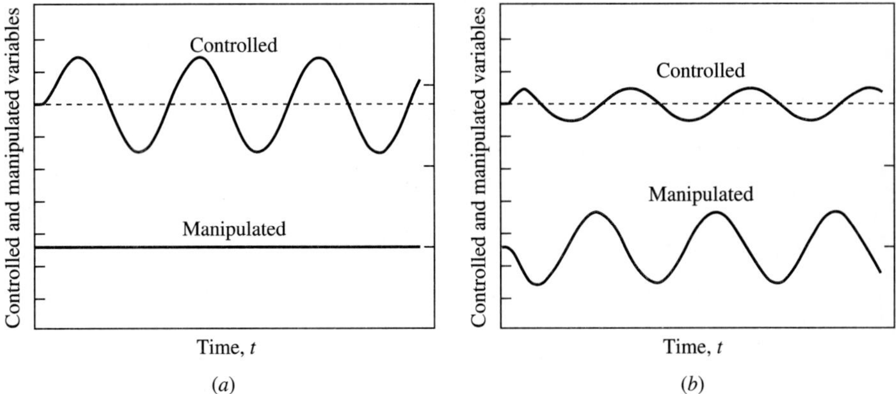

FIGURE 7.9
Transient response of the example system (*a*) without and (*b*) with control to a sine disturbance.

and frequencies. In many cases the disturbance is composed predominantly of one or a few sine waves. Therefore, the behavior of the control system in response to sine inputs is of great practical importance, because through this analysis we learn how the frequency of the disturbances influences the control performance. The responses of the example system to a sine disturbance in the inlet concentration of stream *B* with and without control are given in Figure 7.9*a* and *b*. Control performance is measured by the amplitude of the output sine, which is often expressed as the ratio of the output to input sine amplitudes. Again, a small output amplitude is desired. We shall use the response to sine disturbances often in analyzing control systems, using the frequency response calculation methods introduced in Chapter 4.

In summary, we will be considering two sources of external input change: set point changes and disturbances in input variables. Usually, we will consider the time functions of these disturbances as step and sine changes, because they are relatively easy to analyze and yield useful insights. The measures of control performance for each disturbance-function combination were discussed in this section.

It is important to emphasize two aspects of control performance. First, ideally good performance with respect to *all* measures is usually not possible. For example, it seems unreasonable to expect to achieve very fast response of the controlled variable through very slow adjustments in the manipulated variable. Therefore, control design almost always involves compromise. This raises the second aspect: that control performance must be defined with respect to the process operating objectives of a specific process or plant. It is not possible to define one set of universally applicable control performance goals for all chemical reactors or all distillation towers. Guidance on setting goals will be provided throughout the book via many examples, with emphasis on the most common goals.

> Feedback reduces the variability of the controlled variable at the expense of increased variability of the manipulated variable.

Finally, the responses to all changes have demonstrated by example an important point that will be proved in later chapters. The application of feedback control does not eliminate variability in the process plant; in fact, the "total variability" of the controlled and manipulated variables may not be changed. This conclusion follows from the observation that a manipulated variable must be adjusted to reduce the variability in the output controlled variable. If these variables are selected properly, the performance of the plant, as measured by safety, product quality, and so forth, improves. The availability of manipulated variables depends on a skillful process design that provides numerous utility systems, such as cooling water, steam, and fuel, which can be adjusted rapidly with little impact on the performance of the plant.

7.5 SELECTION OF VARIABLES FOR CONTROL

The chapters on modelling demonstrated that a complex process with many variables can be described by a model with one or several input variables and one or many output variables, as shown in Figure III.1. Selecting which output variables are to be controlled and which input variables are to be adjusted is an important control decision. From the definition of feedback, an important criterion for these variables is that a cause-effect relationship exists. This is often easily determined from a physical understanding of the process, and it can be verified through modelling to ensure that a path exists from the potential manipulated variable to the controlled variable in the direction of the arrows (causal direction) in the block diagram. Further criteria for variable selection will be developed as the mathematical analysis of feedback systems proceeds; however, a few guidelines are provided here.

Controlled Variable

The controlled variable should relate closely to the process objectives, such as operating at safe pressures and producing material with consistent composition. We shall assume for this introductory material that the appropriate variables are easily identified and can be measured accurately in real time. These assumptions will be relaxed in later parts of the book.

Manipulated Variable

The manipulated variable must be an external variable that can be adjusted independently. As stated previously, there must be a causal relationship between the manipulated variable and the controlled variable. The manipulated variable must have sufficient "range"; that is, the allowable adjustments must be large enough to achieve desired set points and compensate for expected disturbances. Also, there should be relatively little effect on plant performance (safety, product quality, profit, etc.) from increasing the variability of the manipulated variable.

Process Relationships

Feedback control bases the proper adjustment in the manipulated variable on the error between the set point and the current value of the controlled variable, as depicted in Figure 7.4. The direction of change is determined from a knowledge of the process; for example, a high concentration of A in the outlet of the three-tank mixer would be corrected by decreasing the feed flow of component A. Thus, the steady-state process relationship, or *gain,* must be known (approximately) and not change sign; otherwise the feedback controller would not know the proper direction of adjustment. Also, the dynamic response between the manipulated and controlled variables should be "favorable," which is generally interpreted as monotonic and fast, with little dead time or inverse response. The effects of process dynamics on the achievable response under feedback control response constitute a major part of the quantitative analysis in this part.

Example 7.2. One of the example processes analyzed several times in Part III is the three-tank mixing process in Figure 7.5. This process is selected for its simplicity, which enables us to determine many characteristics of the feedback system, although it is complex enough to exhibit realistic behavior. The process design and model are introduced here; the linearized model is derived; and the selection of variables is discussed.

Goal. The outlet concentration is to be maintained close to its set point. Derive the nonlinear and linearized models and select controlled and manipulated variables.

Assumptions.

1. All tanks are well mixed.
2. Dynamics of the valve and sensor are negligible.
3. No transportation delays (dead times) exist.
4. A linear relationship exists between the valve opening and the flow of component A.
5. Densities of components are equal.

Data.

V = Volume of each tank = 35 m^3

F_B = Flow rate of stream B = 6.9 m^3/min

x_{Ai} = Concentration of A in all tanks and outlet flow = 3% A (base case)

F_A = Flow rate of stream A = 0.14 m^3/min (base case)

$(x_A)_B$ = Concentration of stream B = 1% A (base case)

$(x_A)_A$ = Concentration of stream A = 100% A

v = Valve position = 50% open (base case)

Thus, the product flow rate is essentially the flow of stream B; that is, $F_B \gg F_A$.

Formulation. Since the variable to be controlled is the concentration leaving the last tank, component material balances on the mixing point and each mixing tank are given below.

$$x_{A0} = \frac{F_B(x_A)_B + F_A(x_A)_A}{F_B + F_A} \tag{7.14}$$

$$V_i \frac{dx_{Ai}}{dt} = (F_A + F_B)(x_{Ai-1} - x_{Ai}) \qquad \text{for } i = 1, 3 \tag{7.15}$$

Note that the differential equations are nonlinear, because the products of flow and concentrations appear. (If you need a refresher, see Section 3.4 for the definition of linearity.) We will linearize these equations and determine how the process gains and time constants depend on the equipment and operating variables. The linearized models are now summarized, with the subscript s representing the initial steady state and the prime representing deviation variables.

$$F'_A = K_v v' \qquad K_v = 0.0028 \frac{\text{m}^3/\text{min}}{\%\text{ open}} \tag{7.16}$$

$$x'_{A0} = \left[\frac{F_{Bs}\{(x_{AA})_s - (x_{AB})_s\}}{(F_{Bs} + F_{As})^2}\right] F'_A \tag{7.17}$$

$$\frac{dx'_A}{dt} + \frac{F_{As} + F_{Bs}}{V} x'_{Ai} = \frac{F_{As} + F_{Bs}}{V} x'_{Ai-1} \qquad \text{for } i = 1, 3 \tag{7.18}$$

assuming the total flow is approximately constant. By taking the Laplace transforms of these equations and performing standard algebraic manipulations, the feedback process transfer functions can be derived:

$$\frac{x_{A3}(s)}{v(s)} = G_p(s) = \frac{K_p}{(\tau s + 1)^3} \tag{7.19}$$

$$\text{with} \quad K_p = K_v \left[\frac{F_{Bs}(x_{AA} - x_{AB})_s}{(F_{As} + F_{Bs})^2}\right] = 0.039 \frac{\%\text{A}}{\%\text{ opening}} \tag{7.20}$$

$$\tau = \frac{V}{F_{Bs} + F_{As}} = 5.0 \text{ min} \tag{7.21}$$

It can be seen that the gain and all time constants are functions of the volumes and total flow. These expressions give an indication, which will be used in later chapters, of how the dynamic response changes as a result of changes in operating conditions.

The closed-loop block diagram also includes the disturbance transfer function $G_d(s)$: the effect of the disturbance if there were no control. This can be derived by assuming that the flows are all constant and that the important input variable that changes is $(x_A)_B$. The resulting model is

$$x'_{A0} = \left[\frac{F_B}{F_A + F_B}\right] x'_{AB} \approx x'_{AB} \tag{7.22}$$

This equation can be combined with equation (7.18) to give the disturbance transfer function,

$$\frac{x_{A3}(s)}{x_{AB}(s)} = G_d(s) = \frac{K_d}{(\tau s + 1)^3} \approx \frac{1.0}{(\tau s + 1)^3} \tag{7.23}$$

The system can be described by the block diagram in Figure 7.4, with the process transfer functions given in equations (7.19) and (7.23). Note that the sensor transfer function has been taken to be 1.0 and that the final element is a gain, as defined by the model in equation (7.16), and has been included in $G_p(s)$.

The choice for the controlled variable in this problem is essentially given in the goal statement; the concentration in the third tank. A preliminary selection for the manipulated variable is taken to be the valve in the component A line. The following checks were made to evaluate this manipulated-variable selection.

1. *External variable independent of tank concentration:* Yes.
2. *Causal relationship:* Yes; see block diagram in Figure 7.4 and equation (7.20); that is, $K_p \neq 0$.
3. *Degrees of freedom:* two parameters, (V and K_v); four external variables, v, F_B, $(x_A)_A$, and $(x_A)_B$; five variables, F_A and x_i for $i = 0$ to 3; and five equations. Therefore, the system is specified and the outlet concentration is uniquely determined when the external variables, including v, are specified. Also, the system is uniquely specified if one output variable (x_3) is specified and one external input variable is adjusted to satisfy the process model equations. Naturally, the output cannot be "set," because this would violate the cause-effect relationships, but feedback can achieve the desired value by adjusting an input.
4. *Range:* The ability of the manipulated variable to compensate for expected disturbances in the steady state is usually termed the *range* of the manipulated variable and is determined by the process equipment. In this case, the valve can be adjusted from 0 to 100% open, which means that F_A can vary from 0 to 0.28. From the steady-state material balance it can be demonstrated that the manipulated variable can successfully compensate for changes in the inlet concentration in stream B over the range of 0.0 to 3.0%, with all other conditions constant, and maintain the outlet at 3%. Alternatively, the manipulated variable can be adjusted to changes of 1.0 to 4.9% in the set point, with all other conditions constant.
5. *Variability:* The manipulated variable will be adjusted in response to changes. It is assumed that a storage tank of component A is available in this example so that such adjustments do not disturb other sections of the plant.
6. *Dynamic response:* The preliminary selection of variables is based on the transfer function $G_p(s)$. We do not have an analysis method for judging the adequacy yet, since that is one of the main objectives of this part. We will assume that it is a reasonable choice for now.

In conclusion, a model for the three-tank mixing process has been derived and controlled and manipulated variables have been selected.

Example 7.3. Assume that the feedback control has been implemented on the mixing tanks problem with the goal of maintaining the outlet concentration near 3.0%. As an example of the control performance measures, the previous example is controlled using feedback principles. The disturbance was a step change in the feed concentration, x_{AB}, of magnitude +1.0 at time = 20. A feedback control algorithm explained in the next chapter was applied to this process with two different sets of adjustable parameters in

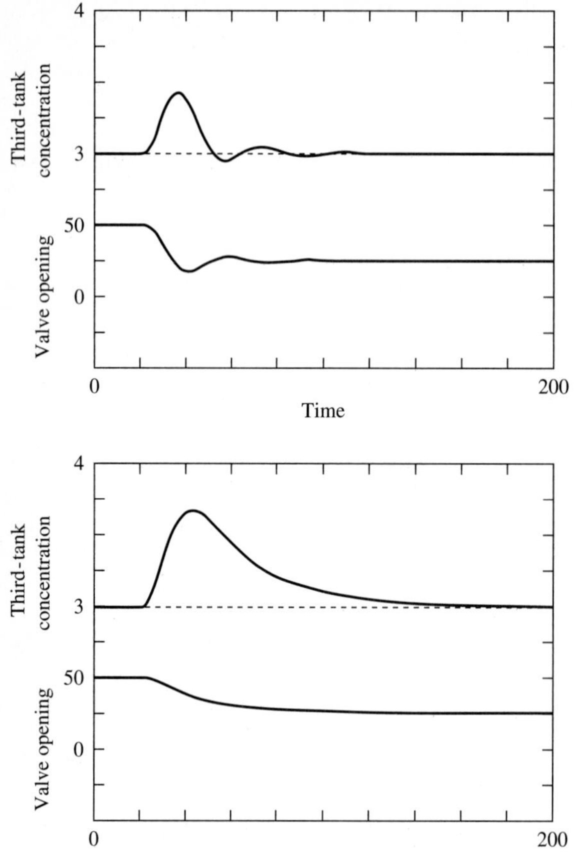

FIGURE 7.10
Feedback responses for Example 7.3. (*a*) Case A; (*b*) Case B.

Cases A and B, and the resulting control performance is shown in Figures 7.10*a* and *b* and summarized as follows.

Measure	Case A	Case B
Offset from SP	None	None
IAE	7.9	30.5
ISE	2.1	12.8
IE	−6.9	−30.5
CV maximum deviation	0.42	0.66
Decay ratio	$<.1$	(Overdamped)
Period (min)	37	(Overdamped)
MV maximum overshoot	6.9/25 = 28%	0% (expressed as % of steady-state change)

The controlled variable in Case A returns to its desired value relatively quickly, as indicated by the performance measures based on the error. This response requires a more "aggressive" (i.e., faster) adjustment of the manipulated variable. The general trend in feedback control is to require fast adjustments in the manipulated variable to achieve rapid return to the desired value of the controlled variable. One might be tempted to generally conclude that Case A provides better control performance, but there are instances in which Case B would be preferred. The final evaluation requires a more complete statement of control objectives.

Several useful measures of control performance were introduced in this section. Two important conclusions can be made that must be considered in future chapters.

1. The desired control performance must be matched to the process requirements.
2. Both the controlled and manipulated variables must be monitored in order to evaluate the performance of a control system.

7.6 APPROACHES TO PROCESS CONTROL

There could be many approaches to the control of industrial processes. In this section, five approaches are discussed so that the more common procedures are placed in perspective.

No Control

Naturally, the easiest approach is to do nothing other than to hold all input variables close to their design values. As we have seen, disturbances could result in large, sustained deviations in important process variables. This approach could have serious effects on safety, product quality, and profit and is not generally acceptable for important variables. However, a degrees-of-freedom analysis usually demonstrates that only a limited number of variables can be controlled simultaneously, because of the small number of available manipulated variables. Therefore, the engineer must select the most important variables to be controlled.

Manual Operation

When corrective action is taken periodically by operating personnel, the approach is usually termed *manual* (or *open-loop*) operation. In manual operation, the measured values of process variables are displayed to the operator, who has the ability to manipulate the final control element (valve) by making an adjustment in the control room to a signal that is transmitted to a valve, or, in a physically small plant, by adjusting the valve position by hand.

This approach is not always bad or "low-technology," so we should understand when and why to use it. A typical strategy used for manual operation can be related to the basic principles of statistical process control and can best be described with reference to the data shown in Figure 7.11. Along with the measured process variable, its desired value and upper and lower action values are plotted. The person observes the data and takes action only "when needed." Usually, the decision on when to take

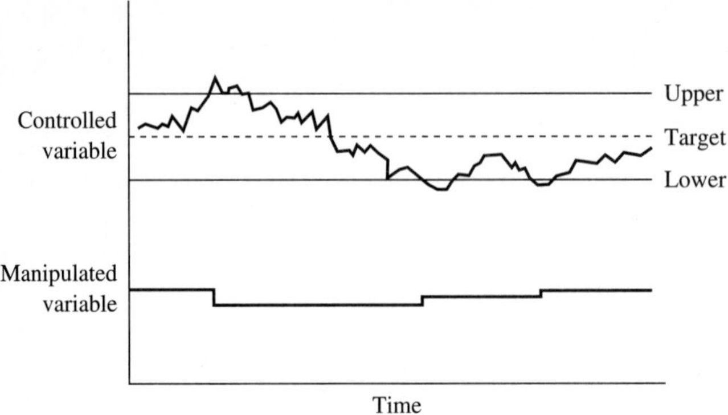

FIGURE 7.11
Transient response of a process under manual control to stochastic disturbances.

corrective action depends on the deviation from the desired value. If the process variable remains within an acceptable range of values defined by action limits, the person makes no adjustment, and if the process variable exceeds the action limits, the person takes corrective action. A slight alteration to this strategy could consider the consecutive time spent above (or below) the desired value but within the action limits. If the time continuously above is too long, a small corrective action can be taken to move the mean of the process variable nearer to the desired value.

This manual approach to process control depends on the person; therefore, the correct application of the approach is tied to the strengths and weaknesses of the human versus the computer. General criteria are presented in Table 7.3. They indicate that the manual approach is favored when the collection of key information is not automated and has a large amount of noise and when slow adjustments with "fuzzy," qualitative decisions are required. The automated approach is favored when rapid, frequent corrections using straightforward criteria are required. Also, the manual approach is favored when there is a substantial cost for the control effort; for example, if the process operation must be stopped or otherwise disrupted to effect the corrective action. In most control opportunities in the process industries, the corrective action, such as changing a valve opening or a motor speed, can be effected continuously and smoothly without disrupting the process.

Manual operation should be seen as complementary to the automatic approaches emphasized in this book. Statistical methods for monitoring, diagnosing, and continually improving process operation find wide application in the process industries (MacGregor, 1988; Oakland, 1986), and they are discussed further in Chapter 26.

On-Off Control

The simplest form of automated control involves logic for the control calculations. In this approach, trigger values are established, and the control manipulation changes state when the trigger value is reached. Usually the state change is between on and

TABLE 7.3
Features of manual and automatic control

Control approach	Advantages	Disadvantages
Manual operation	Reduces frequency of control corrections, which is important when control actions are costly or disruptive to plant operation	Control performance is usually far from the best possible
	Possible when control action requires information not available to the computer	Applicable only to slow processes
	Draws attention to causes of deviations, which can then be eliminated by changes in equipment or plant operation.	Personnel have difficulty maintaining concentration on many variables
	Keeps personnel's attention on plant operation	
Automated control	Good control perfomance for fast processes	Compensates for disturbances but does not prevent future occurrences
	Can be applied uniformly to many variables in a plant	Does not deal well with qualitative decisions
	Generally low cost	May not promote people's understanding of process operation

off, but it could be high or low values of the manipulated variable. This approach is demonstrated in Figure 7.12 and was modelled for the common example of on-off control in room temperature control via heating in Example 3.4. While appealing because of its simplicity, on-off control results in continuous cycling, and performance is generally unacceptable for the stringent requirements of many processes.

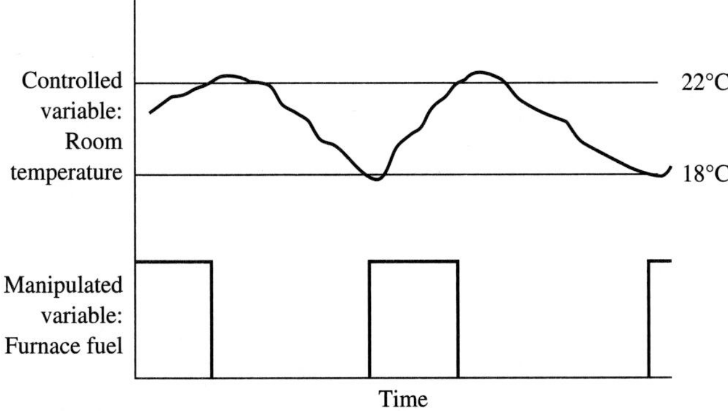

FIGURE 7.12
Example of a process under on/off control.

It is used in simple strategies such as maintaining the temperature of storage tanks within rather wide limits.

Continuous Automated Control

The emphasis of this book is on process control that involves the continuous sensing of process variables and adjustment of manipulated variables based on control calculations. This approach offers the best control performance for most process situations and can be easily automated using computing equipment. The types of control performance achieved by continuous control are shown in Figures 7.10*a* and *b*. The control calculation used to achieve this performance is the topic of the subsequent chapters in Part III. Since the control actions are performed continuously, the manipulated variable is adjusted essentially continuously. As long as the adjustments are not too extreme, constant adjustments pose no problems to valves and their associated process equipment that have been designed for this application.

Emergency Controls

Continuous control performs well in maintaining the process near its set point. However, continuous control does not ensure that the controlled variable remains within acceptable limits. A large upset can result in large deviations from the set point, leading to process conditions that are hazardous to personnel and can cause damage to expensive equipment. For example, a vessel may experience too high a pressure and rupture, or a chemical reactor may have too high a temperature and explode. To prevent safety violations, an additional level of control is applied in industrial and laboratory systems. Typically, the emergency controls measure a key variable(s) and take extreme action before a violation occurs; this action could include stopping all or critical flow rates or dramatically increasing cooling duty.

As an example of an emergency response, when the pressure in a vessel with flows in and out reaches an upper limit, the flow of material into the vessel is stopped, and a large outflow valve is opened. The control calculations for emergency control are usually not complex, but the detailed design of features such as sensor and valve locations is crucial to safe plant design and operation. The topic of emergency control is addressed in Chapter 24. You may assume that emergency controls are not required for the process examples in this part of the book unless otherwise stated.

In industrial plants all five control approaches are used concurrently. Plant personnel continuously monitor plant performance, make periodic changes to achieve control of some variables that are not automated, and intervene when equipment or controls do not function well. Their attention is directed to potential problems by audio and visual alarms, which are initiated when a process measurement exceeds a high or low limiting value. Continuous controls are applied to regulate the values of important variables that can be measured in real time. The use of continuous controls enables one person to supervise the operation of a large plant section with many variables. The emergency controls are always in reserve, ready to take the extreme but necessary actions required when a plant approaches conditions that endanger people, environment, or equipment.

7.7 CONCLUSION

A review of the elements of a control loop and of typical dynamic responses of each element, with an example of transient calculation, shows that all elements in the loop contribute to the behavior of the controlled variable. Depending on the dynamic response of the process, the contributions of the instrument elements can be negligible or significant. Material in future chapters will clarify and quantify the relationship between dynamics and performance of the feedback system.

The principles and methods for selecting variables and measuring control performance discussed here for a single-loop system can be extended to processes with several controlled and manipulated variables, as will be shown in later chapters.

A key observation is that feedback control does not reduce variability in a plant, but it moves the variability from the controlled variables to the manipulated variables. The engineer's challenge is to provide adequate manipulated variables that satisfy degrees of freedom and that can be adjusted without significantly affecting plant performance.

The techniques used for continuous automated rather than manual control are emphasized because:

1. As demonstrated by its wide application, it is essential for achieving good operation for most plants.

2. It provides a sound basis for evaluating the effects of process design on the dynamic performance. A thorough understanding of feedback control performance provides the basis for designing more easily controlled processes by avoiding unfavorable dynamic responses.

3. It introduces fundamental topics in dynamics, feedback control, and stability that every engineer should master. The study of automatic control theory principles as applied to process systems provides a link for communication with other disciplines.

In this chapter the controller has been left relatively loosely defined, merely conforming to Figure 7.4 and using variables consistent with feedback principles. This has allowed a general discussion of principles without undue regard for a specific approach. However, to build systems that function properly, the engineer will require greater attention to detail. Thus, the most widely used feedback control algorithm will be introduced in the next chapter.

REFERENCES

Cryor, J., *Time Series Analysis,* Duxbury Press, Boston, MA, 1986.

Hougen, J., *Measurements and Control—Applications for Practicing Engineers,* Cahners Books, Boston, MA, 1972.

MacGregor, J. M., "On-Line Statistical Process Control," *Chem. Engr. Prog. 84, 10,* 21–31 (1988).

Oakland, J., *Statistical Process Control,* Wiley, New York, 1986.

ADDITIONAL RESOURCES

Additional information on the dynamic responses of instrumentation can be found in

While, C., "Instrument Models for Process Simulation," *Trans. Inst. MC, 1, 4,* 187–194 (1979).

Additional references on the dynamic responses of pneumatic equipment can be found in

Harriott, P., *Process Control,* McGraw-Hill, New York, 1964, Chapter 10.

Instrumentation in the control loop performs many functions tailored to the specific process application. Therefore, it is difficult to discuss sensor systems in general terms. The reader is encouraged to refer to the instrumentation references provided at the end of Chapter 1.

The description of elements in the loop is currently accurate, but the situation is changing rapidly with the introduction of digital communication between the controller and the field instrumentation along with digital computation at the field equipment. For an introduction, see

Lindner, K., "Fieldbus—a Milestone in Field Instrumentation Technology," *Meas. and Cont., 23,* 272–277 (1990).

For a discussion of the interaction between the plant personnel and the automation equipment, see

Rijnsdorp, J., *Integrated Process Control and Automation,* Elsevier, Amsterdam, 1991.

Many important decisions can be made based on the understanding of feedback control, without consideration of the control calculation. These questions give some practice in thinking about the essential aspects of feedback.

QUESTIONS

7.1. Describe the features of a process for which on/off control would be appropriate. The description should include process dynamics, disturbance frequency and magnitude, and control objectives.

7.2. Elements in a control loop in Figure 1.7*d* are given in Table Q7.2 with their individual dynamics. The output signal is 0 to 100%, and the displayed controlled variable is 0 to 20 weight %. Determine the response of the indicator (or controller input) to a step change in the output signal from the manual station (or controller output).

(*a*) The time unit in the models is not specified. Using engineering judgment, what units would expect to be correct: seconds, minutes, or hours?

(*b*) First estimate the response, $t_{63\%}$, using an approximate method.

(*c*) Give an estimate for how much the sensor, transmission, and valve dynamics affect the overall response.

(*d*) Determine the response by solving the entire system numerically.

TABLE Q7.2
Dynamic models

Element	Units	Case A	Case B
Manual station	psi/% output	0.083	0.083
Transmission		$1.0/(1.3s + 1)$	1.0
Signal conversion	psi/mA	$0.75/(0.5s + 1)$	$0.75/(0.5s + 1)$
Final element	%open/psi	$8.33/(1.5s + 1)$	$8.33/(1.5s + 1)$
Process	m^3/psi	$0.50e^{-0.5s}/(30s + 1)$	$0.50e^{-20s}/(30s + 1)$
Sensor		$1.0/(1s + 1)$	$1.0/(10s + 1)$
Signal conversion	mA/mV	—	—
Transmission		1.0	1.0
Display	wt%/mA	$1.25/(1.0s + 1)$	$1.25/(1.0s + 1)$

7.3. The elements in several control systems are shown in Figure Q7.3. For each system, determine the transfer functions for CV(s)/SP(s) and CV(s)/D(s), where a disturbance is given.

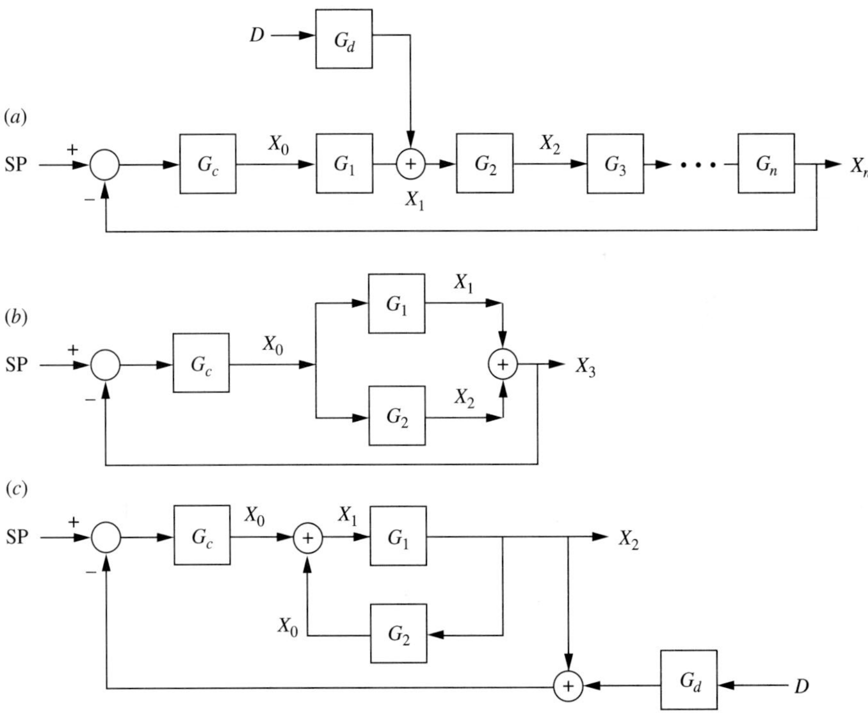

FIGURE Q7.3
Block diagrams for several control systems. All quantities are Laplace-transformed; the variable (s) is omitted for simplicity.

7.4. (a) Discuss the three types of disturbances described in this chapter and give a process example of how each could be generated by an upstream process.

(*b*) An alternative disturbance is a pulse function. Describe a pulse function, give control performance measures for a pulse disturbance, and give a process example of how it could be generated by an upstream process.

7.5. Dynamic responses for several different control systems in response to a change in the set point are given in Figure Q7.5. Discuss the control performance of each with respect to the measures explained in Section 7.4. (Note that the control performance cannot be evaluated exactly without a better definition of control objectives. Further exercises will be given in later chapters, when the objectives can be more precisely defined.)

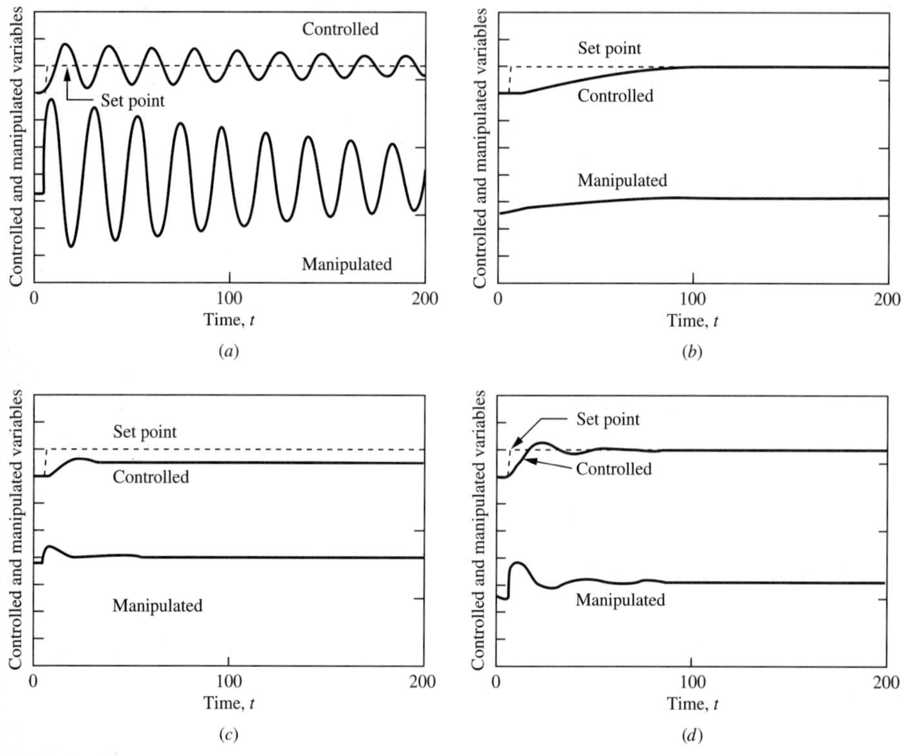

FIGURE Q7.5

7.6. A process with controls is shown in Figure Q7.6. The objective is to achieve a desired composition of B in the reactor effluent. The process consists of a feed tank of reactant A, which is maintained within a range of temperatures and is fed into the reactor, where the following reactions take place.

$$A \rightarrow B$$

$$A \rightarrow C$$

If the reactor level is too high, the pump motor should be shut off to prevent spilling the reactor contents. Identify at least one variable that is controlled by each of the five

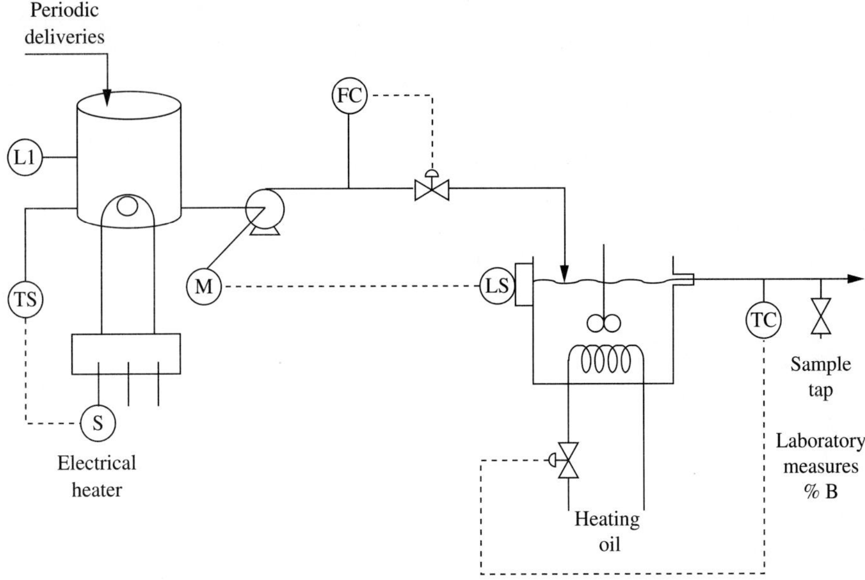

FIGURE Q7.6
Schematic drawing of process and control design.

approaches to control presented in this chapter. Discuss why the approach is (or is not) a good choice.

7.7. Note that the electrical and pneumatic transmission ranges have a nonzero value for the lowest value of the range. Why is this a good selection for the range; that is, what is the advantage of this range selection?

7.8. Confirm that the gains in the instrument models used in Example 7.1 are reasonable. The sensor is an iron-constantan thermocouple.

7.9. The proposal was made to make one of the control pairings for one single-loop controller for the nonisothermal CSTR in Example 3.10. Evaluate each using the criteria in Example 7.2.

(*a*) Control the reactor temperature by adjusting the coolant flow rate.
(*b*) Control the reactant concentration in the reactor by adjusting the coolant flow rate.
(*c*) Control the coolant outlet temperature by adjusting the coolant flow rate.

7.10. The proposal was made to make one of the control pairings for the binary distillation tower in Example 5.6. Evaluate each using the criteria in Example 7.2.

(*a*) Control the distillate composition by adjusting the reboiler heating flow.
(*b*) Control the distillate composition by adjusting the distillate flow.
(*c*) Control both the distillate and bottoms compositions simultaneously by adjusting the reboiler heating flow.

7.11. Answer the following questions, which address the range of a control system.

(*a*) The process in Example 5.3 is to control the process temperature after the mix by adjusting the flow ratio. Over what range of inlet temperatures T_0 can the outlet temperature T_3 be maintained at 90°C?

(*b*) The nonisothermal CSTR in Example 3.10 is to be operated at 420 K and 0.20 kmole/m^3. Can this condition be achieved for the range of inlet concentration (C_{A0}) and coolant flow rate (F_c) defined in Example 3.11? If not, which range has to be expanded and by how much?

(*c*) For the CSTR in Example 3.3, can the outlet concentration of reactant be controlled at 0.85 mole/m^3 by adjusting the inlet concentration? By adjusting the temperature of one reactor?

7.12. Answer the following questions on selecting control variables. Are there any limitations to the operating conditions for your answers?

(*a*) In Example 5.4, can the outlet concentration be controlled by adjusting the solvent flow rate?

(*b*) In Example 3.12, can the concentration of B be controlled by adjusting the reactor feed flow or by adjusting the temperature?

(*c*) In Figure 2.6, through adjustments of the air flow rate, can the (i) efficiency *and* (ii) the excess oxygen in the flue gas be controlled?

CHAPTER 8

THE PID ALGORITHM

8.1 INTRODUCTION

Continuous feedback control offers the potential for improved plant operation by maintaining selected variables close to their desired values. In this chapter we will emphasize the control algorithm, while remembering that all elements in the feedback loop affect control performance. Engineers should fully understand the algorithm for three reasons. First, the performance of the entire feedback system depends on the structure of the algorithm and the parameters used in the algorithm. Second, all other elements are process equipment and instrumentation, which are costly and time-consuming to alter, so a key area of flexibility in the loop is the control calculation. Third, while engineers use only a few algorithms, as will be explained, they are responsible for determining the values of adjustable parameters in the algorithms.

In this chapter, we will learn about the *proportional-integral-derivative (PID)* control algorithm. The PID algorithm has been successfully used in the process industries since the 1940s and remains the most often used algorithm today. It may seem surprising to the reader that one algorithm can be successful in many applications—petroleum processing, steam generation, polymer processing, and many more. This success is a result of the many good features of the algorithm, which are covered initially in this chapter and expanded on and evaluated in later chapters.

This algorithm is used for single-loop systems, also termed single input–single output (SISO), which have one controlled and one manipulated variable. Usually, many single-loop systems are implemented simultaneously on a process, and the performance of each control system can be affected by interaction with the other loops. However, the next few chapters will concentrate on ideal single-loop systems, in which interaction is negligible or nonexistent; extensions, including interaction, are covered in Parts V and VI.

As we cover the PID control algorithm here and in subsequent chapters, we will address important theoretical issues in feedback control including stability, frequency response, tuning, and control performance. Thus, by covering the PID controller in depth, we will acquire key analytical techniques applicable to all feedback control systems, including PID and alternative control algorithms, along with important knowledge about current practice.

8.2 DESIRED FEATURES OF A FEEDBACK CONTROL ALGORITHM

Many of the desired characteristics for feedback control were discussed in the previous chapter under quantitative measures of control performance. Here, a few of these characteristics are extended for use in this and upcoming chapters.

Key Performance Feature: Zero Offset

The performance measures discussed previously could be combined into two categories: dynamic (IAE, ISE, damping ratio, settling time, etc.) and steady-state. The steady-state goal—returning to set point—is further discussed here. This goal can be stated mathematically as follows by using the final value theorem,

$$\lim_{t \to \infty} E(t) = \lim_{s \to 0} sE(s) = 0 \tag{8.1}$$

with E denoting the *error:* the difference between the set point and (measured) controlled variable. The Laplace transform for $E(s)$ can be determined from block diagram algebra, resulting in an expression for the error that depends on the process $[G_v(s)G_p(s)G_s(s)]$, the control algorithm $[G_c(s)]$, and the *input* [SP(s) or $D(s)$]. It would seem unreasonable to demand that the control system return to set point for all fluctuations in inputs. Therefore, we select the most important, most often occurring input variation from among the following cases:

1. The input variable varies but ultimately returns to its initial value. For this case most (but not all) processes would require no feedback control to satisfy the condition in equation (8.1).
2. The input variable varies for some time and then attains a steady value different from its initial value; this type we shall term *steplike,* because the transition from initial to different final value does not have to be a perfect step. Feedback control is required to achieve zero steady-state offset.
3. The input variables never attain a steady state; for this discussion, a ramp input is often considered, $D(t) = at$, $D(s) = a/s^2$.

Case 2 is the most typical situation, while case 3 occurs occasionally, as in a batch system where the set point is changed as a ramp. For case 2, the expression in equation (8.1) becomes

$$\lim_{t \to \infty} E(t) = \lim_{s \to 0} sE(s) = \lim_{s \to 0} \left(\frac{\Delta X}{s} \right) G(s) = 0 \tag{8.2}$$

where $G(s) = E(s)/X(s)$, and $X(s)$ is the input disturbance $D(s)$ or set point change SP(s). By satisfying equation (8.2), the control algorithm is guaranteed to return the controlled variable to its set point for that particular process and input function. Note that systems satisfying equation (8.2) are not guaranteed to achieve zero steady-state offset for other inputs, such as a ramp. To evaluate the control performance in this chapter, a step input, $X(s) = 1/s$, will be used, because it represents the most commonly occurring situation; other inputs will be considered in later chapters. A more general coverage of this topic classifies systems with regard to their steady-state errors by defining system types (Distephano et al., 1976). Process systems that have zero offset for a step input are *Type 1*.

Insensitivity to Errors

As we learned in Part II, we can never model a process exactly. Because parameters in all control algorithms depend on process models, control algorithms will always be in error despite our best modelling efforts. Therefore, control algorithms should provide good performance when the adjustable parameters have "reasonable" errors. Naturally, all algorithms will give poor performance when the adjustable parameter errors are very large. The range of reasonable errors and their effects on control performance are studied in this and several subsequent chapters.

Wide Applicability

The PID control algorithm is a simple, single equation, but it can provide good control performance for many different processes. This flexibility is achieved through several adjustable parameters, whose values can be selected to modify the behavior of the feedback system. The procedure for selecting the values is termed *tuning,* and the adjustable parameters are termed *tuning constants*.

Simple Calculations

The control calculation is part of the feedback loop, and therefore it should be calculated rapidly and reliably. Excessive time for calculation would introduce an extra slow element in the control loop and, as we shall see, degrade the control performance. Iterative calculations, which might occasionally not converge, would result in a loss of control at unpredictable times. The PID algorithm is exceptionally simple—a feature that was crucial to its initial use prior to the availability of inexpensive digital computers for control. Because of its wide use, the PID controller is available in nearly all commercial digital control systems, so that efficiently programmed and well-tested implementations are available.

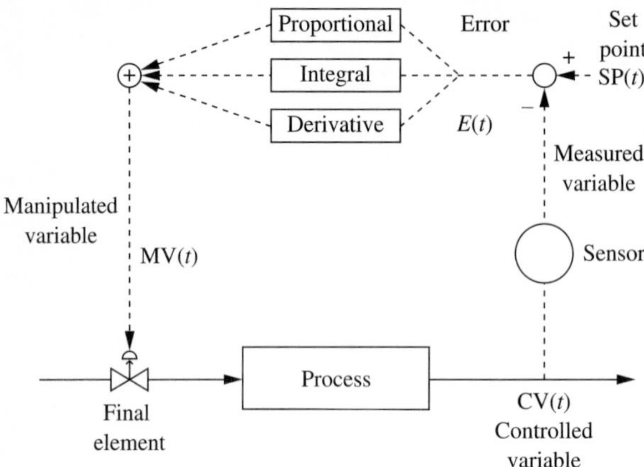

FIGURE 8.1
Overview schematic of a PID control loop.

Enhancements

No single algorithm can address all control requirements. A convenient feature of the PID algorithm is its compatibility with enhancements that provide capabilities not in the basic algorithm. Thus, we can enhance the basic PID without discarding it. Many of the common enhancements are presented in Part IV.

The main goal of this chapter is to explain the PID algorithm fully. Each element of the algorithm is termed a *mode* and uses the time-dependent behavior of the feedback information in a different manner, as indicated by the name *proportional-integral-derivative*. Each mode of the equation and the key capability it provides are discussed thoroughly. The complete PID equation, which is the sum of the three modes as shown in Figure 8.1, is then reviewed, and a few example control responses are presented. The reader is cautioned that there is no consistency in commercial control equipment regarding the sign of the subtraction when forming the error; the convention used in this book is $E(t) = \text{SP}(t) - \text{CV}(t)$. Some preprogrammed equipment uses the opposite sign, a factor that does not affect the principles of this book but certainly affects the performance of actual control systems! (Since the error is multiplied by one of the adjustable tuning constants, the sign of the constant can be adapted to the sign of the error to give the desired direction of the control manipulation.)

8.3 PROPORTIONAL MODE

It seems logical for the first mode to make the control action (i.e., the adjustment to the manipulated variable) proportional to the error signal, because as the error increases, the adjustment to the manipulated variable should increase. This concept is realized in the proportional mode of the PID controller:

$$\text{MV}_p(t) = K_c E(t) + I_p \tag{8.3}$$

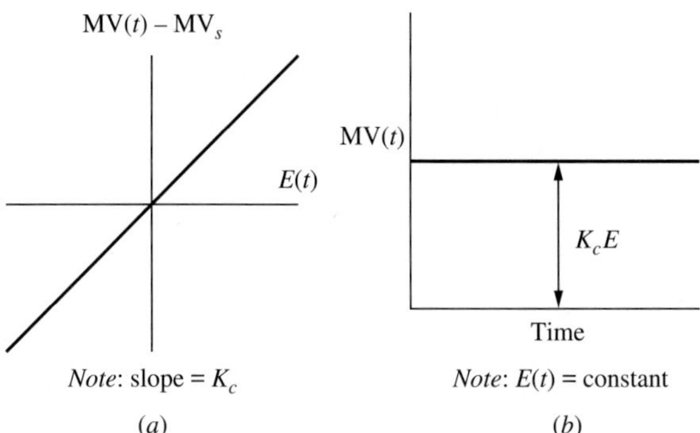

FIGURE 8.2
Summary of proportional mode.

The *controller gain* K_c is the first of three adjustable parameters that enable the engineer to tailor the PID controller to various applications. The controller gain has units of [manipulated]/[controlled] variables, which is the inverse of the process gain K_p. Note that the equation includes a constant term or bias, which is used during initialization of the algorithm I_p. During initialization the value of the manipulated variable should remain unchanged; therefore, the initialization constant can be calculated at the time of initialization as

$$I_p = [\mathrm{MV}(t) - K_c E(t)]\,|_{t=0} \tag{8.4}$$

The behavior of the proportional mode is summarized in Figures 8.2*a* and *b*. In deviation variables, a plot of manipulated variable versus error gives a straight line with slope equal to the controller gain and zero intercept. A plot of the manipulated variable versus time for constant error gives a constant value.

Although the concept seems logical, we do not yet know whether the control performance of the proportional controller satisfies the desired control performance goals presented in the previous chapter and section. To evaluate performance it is useful to have the closed-loop transfer function. The transfer function for the disturbance response of the system in Figure 8.1 is given in equation (7.3). Substituting the equation for a proportional controller, $G_c(s) = K_c$, gives the following transfer function.

$$\frac{\mathrm{CV}(s)}{D(s)} = \frac{G_d(s)}{1 + G_p(s)G_v(s)K_c G_s(s)} \tag{8.5}$$

One of the most important goals in control performance is zero offset at the final steady state. For a disturbance response, the zero steady-state offset requires $E'(t)\,|_{t\to\infty} = -\mathrm{CV}'(t)\,|_{t\to\infty} = 0$.

Example 8.1. The three-tank mixing process under control modeled in Example 7.2 (page 249) is now analyzed. Recall that the feedback and disturbance processes are third-order. The steady-state value for error under proportional control can be determined by rearranging equation (8.5), substituting the models for $G_p(s)$ and $G_d(s)$, and applying the final value theorem to the system with a steplike disturbance, $D(s) = \Delta D/s$. Recall that the valve transfer function is included in $G_p(s)$, and the sensor transfer function is assumed to be unity, implying instantaneous, error-free measurement.

$$\text{CV}'(t)\,|_{t\to\infty} = \lim_{s\to 0}\left[(s)(\Delta D/s)\frac{K_d\left(\dfrac{1}{\tau s+1}\right)\left(\dfrac{1}{\tau s+1}\right)\left(\dfrac{1}{\tau s+1}\right)}{1+K_cK_p\left(\dfrac{1}{\tau s+1}\right)\left(\dfrac{1}{\tau s+1}\right)\left(\dfrac{1}{\tau s+1}\right)}\right] \tag{8.6}$$

$$= \frac{K_d\Delta D}{1+K_cK_p}$$

Note that the feedback control system with proportional control does *not* achieve zero steady-state offset! This result can be understood by recognizing the proportional relationship between the error and the manipulated variable in the controller algorithm; the only way in which the control equation (8.3) can have the error return to zero is for the value of the manipulated variable to return to its initial condition. However, for the error to be zero in the process equation, the manipulated variable must be different from its initial value, because it must compensate for the disturbance. Thus, steady-state offset occurs with proportional-only control. This is a serious shortcoming, which must be corrected by one of the remaining two modes.

Example 8.2. Another important property of a control system is a fast response to a disturbance or set point change. The expression for a disturbance response is analyzed using equation (8.5) for a simple process with the disturbance and feedback processes being first-order with the same time constant. This system can be thought of as the heat exchanger in Example 3.7 and has been selected to simplify the analytical solution.

$$\frac{\text{CV}(s)}{D(s)} = \frac{\dfrac{K_d}{\tau s+1}}{1+\dfrac{K_cK_p}{\tau s+1}} = \frac{\dfrac{K_d}{1+K_cK_p}}{\left(\dfrac{\tau}{1+K_cK_p}\right)s+1} \tag{8.7}$$

with $K_cK_p > 0$ for negative feedback control. The analytical solutions for the step disturbance response, $D(s) = \Delta D/s$, for the process with and without proportional control are

$$\text{CV}'(t) = \Delta D K_d(1-e^{-t/\tau}) \qquad \text{(no control)} \tag{8.8}$$

$$\text{CV}'(t) = \frac{\Delta D K_d}{1+K_cK_p}(1-e^{-t/[\tau/(1+K_cK_p)]}) \qquad \text{(proportional control)} \tag{8.9}$$

Equation (8.9) demonstrates that the feedback controller alters both the time constant of the closed-loop system and the final deviation from set point by a factor of $1/(1 + K_cK_p)$ for a first-order process. This means that the feedback system responds faster than the open-loop system to a step disturbance and has a smaller deviation from set point. Both of these modifications to the system behavior are generally desired. The results in equation (8.9) indicate that as the controller gain is increased, the final value of the error decreases in magnitude and the system reaches steady state faster. We might be tempted to generalize this result (improperly) to all systems and apply high controller gains to all processes.

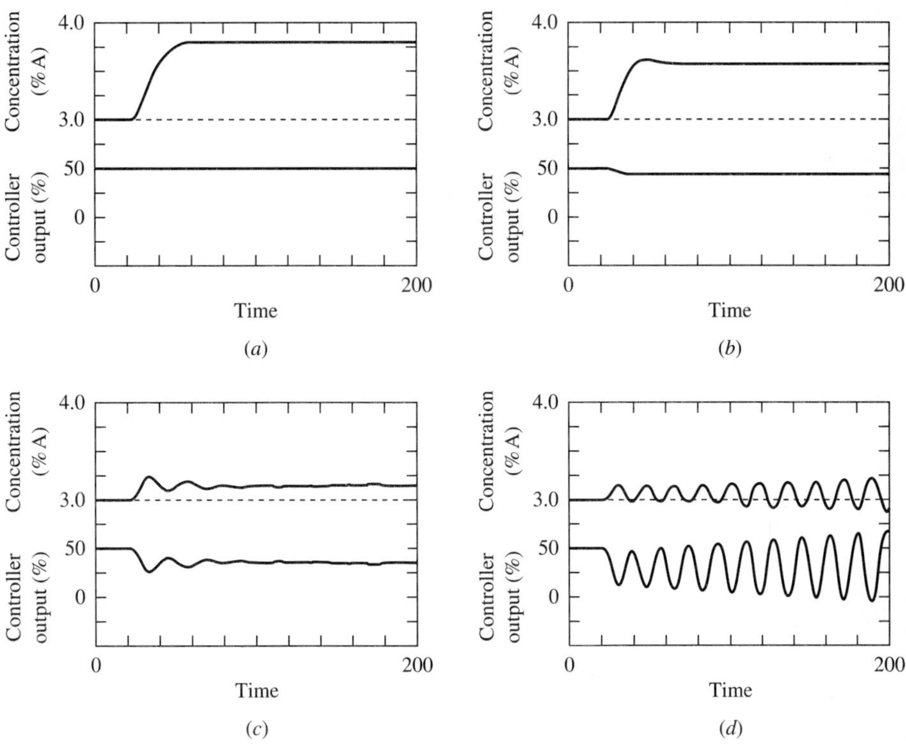

FIGURE 8.3
Disturbance responses for three-tank mixing process under proportional control subject to a disturbance in feed composition $(x_A)_B$ of 0.8%A and K_c = [%open/(%A)]: (*a*) without control; (*b*) with proportional control, $K_c = 10$; (*c*) with proportional control, $K_c = 100$; (*d*) with proportional control, $K_c = 220$.

To test this idea on a more complex process, several dynamic responses for the linearized model of the three-tank mixing process under proportional control are shown in Figures 8.3*a* through *d*. Again, the input is a step disturbance in the feed concentration. The case without control ($K_c = 0$) shows the response of a third-order system to a step input; it is overdamped and reaches a final value of the disturbance magnitude. As the gain is increased to 10, the final value of the error decreases, as predicted by equation (8.7). Also, the time to reach the steady state decreases; that is, the dynamic response becomes faster, as predicted. As the gain is increased to 100, the nature of the dynamic response changes from overdamped to underdamped. As the controller gain is increased further to 220, the system becomes *unstable!*

These results demonstrate an important feature of feedback control systems: the closed-loop response can become underdamped and ultimately unstable as the controller parameters are adjusted to make the controller very aggressive (here by increasing the controller gain, K_c). This example suggests, and later theoretical analysis will confirm, that it is generally not possible to maintain the controlled variable close to the set point by setting the controller gain to a very large value (although this approach would work for equation (8.9)). The reasons for the instability and methods for predicting the stability limits are presented in Chapter 10 after the control algorithm has been fully explained. In summary:

> The proportional mode is simple, provides a rapid adjustment of the manipulated variable, does not provide zero offset although it reduces the error, speeds the dynamic response, and can cause instability if tuned improperly.

8.4 INTEGRAL MODE

Since the proportional mode does not completely eliminate the effects of disturbances, the next mode should be "persistent" in adjusting the manipulated variable until the magnitude of the error is reduced to zero for a steplike input. This result is achieved by the integral mode:

$$\mathrm{MV}_I(t) = \frac{K_c}{T_I}\int_0^t E(t')dt' + I_I \tag{8.10}$$

The new adjustable parameter is termed the integral time, T_I, which has the units of time; it is combined with the controller gain in equation (8.10) because this is the conventional form of the PID controller in commercial equipment. This form is used throughout the book for consistency and so that later correlations for parameter values can be used. Again, the integral mode equation has a constant of initialization.

The behavior of the integral mode is summarized in Figure 8.4. For a constant error, the manipulated variable increases linearly with a slope of $E(t)K_c/T_I$. This behavior is different from the proportional mode, in which the value is constant over time for a constant error.

Example 8.3 The effect of the integral mode can be determined by evaluating the offset of the three-tank mixing process under integral control for a step disturbance, $D(s) = \Delta D/s$.

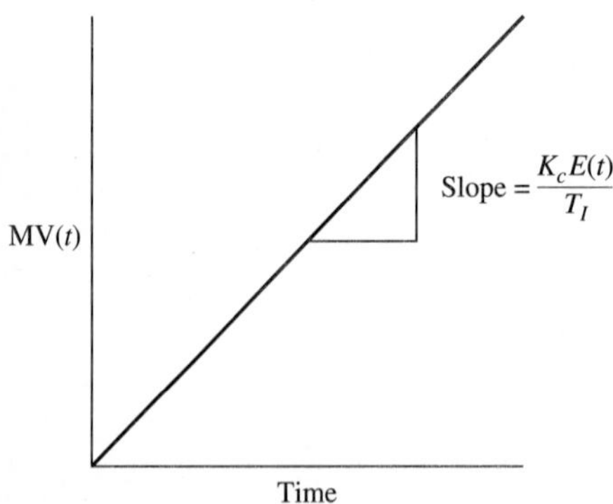

FIGURE 8.4
Summary of the behavior of the integral mode.

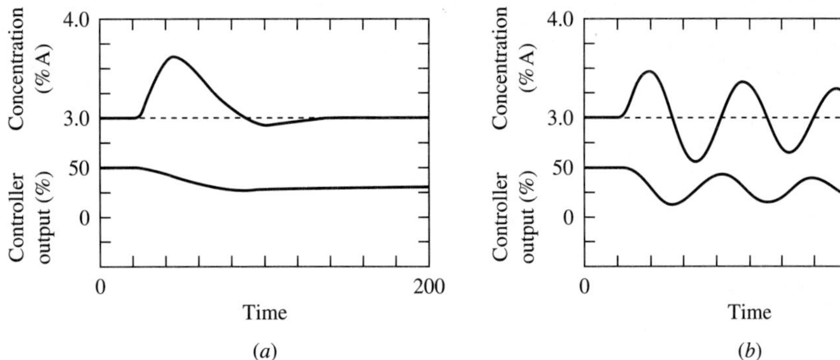

FIGURE 8.5
Three-tank mixing process under integral-only control subject to a disturbance in feed composition $(x_A)_B$ of 0.8%A and K_c = [%open/%A], T_I = [min] : (*a*) $K_c = 1,\ T_I = 1$; (*b*) $K_c = 1,\ T_I = 0.25$.

$$\mathrm{CV}'(t)\,|_{t=\infty} = \lim_{s\to 0}\left[\frac{(s)(\Delta D/s)K_d\left(\dfrac{1}{\tau s+1}\right)\left(\dfrac{1}{\tau s+1}\right)\left(\dfrac{1}{\tau s+1}\right)}{1+K_c\left(1+\dfrac{1}{T_I s}\right)K_p\left(\dfrac{1}{\tau s+1}\right)\left(\dfrac{1}{\tau s+1}\right)\left(\dfrac{1}{\tau s+1}\right)}\right] \tag{8.11}$$
$$= 0$$

The integral control mode achieves zero steady-state offset, which is the primary reason for including this mode.

Again, some dynamic responses of the three-tank mixing process are plotted, this time with an integral controller, in Figures 8.5*a* and *b*. As can be seen by comparing Figure 8.3*c* with Figure 8.5*a*, the manipulation of the controller output is slower for integral-only control than for proportional-only control. As a result, the controlled variable returns to the set point slowly and experiences a larger maximum deviation. If the integral time is reduced small enough, as in Figure 8.5*b*, the controller will be very aggressive, and the system will become highly oscillatory; further reduction in T_I can lead to an *unstable* system. Under integral-only control with properly selected tuning constants, the controlled variable returns to its set point, but the other aspects of control performance are usually not acceptable. In summary:

> The integral mode is simple; achieves zero offset; adjusts the manipulated variable in a slower manner than the proportional mode, thus giving poor dynamic performance; and can cause instability if tuned improperly.

8.5 DERIVATIVE MODE

If the error is zero, both the proportional and integral modes give zero adjustment to the manipulated variable. This is a proper result if the controlled variable is not changing; however, consider the situation in Figure 8.6 at time equal to t when the

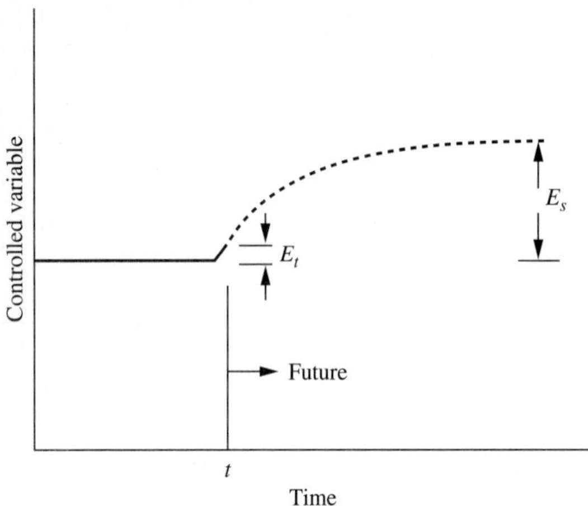

FIGURE 8.6
Assumed effect of disturbance on controlled variable.

disturbance just begins to affect the controlled variable. There, the error and integral error are nearly zero, but a substantial change in the manipulated variable would seem to be appropriate because the rate of change of the controlled variable is large. This situation is addressed by the derivative mode:

$$MV_d(t) = K_c T_d \frac{dE(t)}{dt} + I_d \tag{8.12}$$

The final adjustable parameter is the derivative time T_d, which has units of time, and the mode again has an initialization constant. Note that the proportional gain and derivative time are multiplied together to be consistent with the conventional PID algorithm.

Some further insight can be gained by examining the following development of a proportional-derivative controller (Rhinehart, 1991). Again consider the dynamic response in Figure 8.6, in which the data available at the current time t, which is at the beginning of the disturbance response, is shown by the solid line. The future response that would be obtained without feedback control is shown as the dotted line; note that this is simply the disturbance response. The value of the E_s, the total effect of the disturbance on the controlled variable as time approaches infinity, can be predicted using the assumption that the error is following a first-order response with a time constant equal to the disturbance process time constant:

$$\tau_d \frac{dE}{dt} + E = E_s \tag{8.13}$$

Since the error will increase to E_s ultimately, the manipulated variable will have to be adjusted by a value proportional to E_s, or $MV' = E_s/K_c$. Rather than wait until the error becomes large, when the proportional and integral modes would adjust the manipulated variable, the controller could anticipate the future error using the foregoing equation to give

$$MV = K_c \left(E + \tau_d \frac{dE}{dt} \right) + I_d \tag{8.14}$$

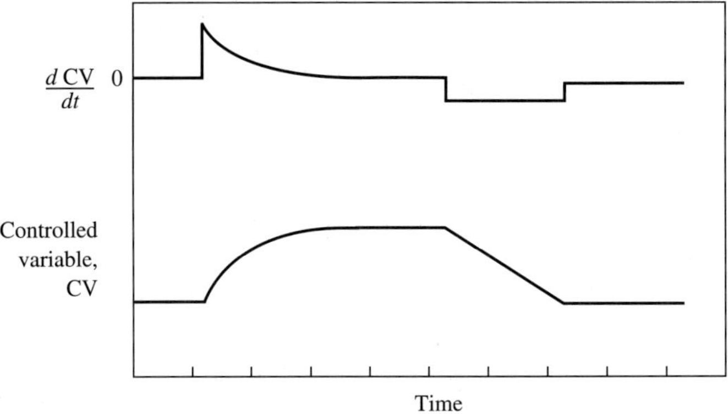

FIGURE 8.7
Example of the calculation of the derivative mode with constant set point.

Thus, the proportional-derivative modes are a natural result of the assumption that the error will respond as given in Figure 8.6. If the assumption is good, derivative mode may improve the control performance.

The behavior of the calculation for the derivative-only mode is shown in Figure 8.7. When the controlled variable is constant, the derivative mode makes no change to the manipulated variable. When the controlled variable changes, the derivative mode adjusts the manipulated variable in a manner proportional to the rate of change.

Example 8.4. The offset of a derivative controller can be determined by again applying the final value theorem to the three-tank mixing process for a step disturbance, $D(s) = \Delta D/s$.

$$\mathrm{CV}'(t)|_{t=\infty} = \lim_{s\to 0}\left[\frac{(s)(\Delta D/s)K_d\left(\frac{1}{\tau s+1}\right)\left(\frac{1}{\tau s+1}\right)\left(\frac{1}{\tau s+1}\right)}{1+K_cT_p s\left(\frac{1}{\tau s+1}\right)\left(\frac{1}{\tau s+1}\right)\left(\frac{1}{\tau s+1}\right)}\right] \tag{8.15}$$
$$= K_d\Delta D$$

As is apparent, the derivative mode does not give zero offset. In fact, it does not reduce the final deviation below that for a system without control for any disturbance whose derivative tends toward zero as time increases; thus, its only benefit can be in improving the transient response. Since the derivative is never used as the only controller mode, dynamic responses are not included in this section, but dynamic responses for the PID controller will be given.

The derivative mode amplifies sudden changes in the controller input signal, causing potentially large variation in the controller output that can be unwanted for two reasons. First, step changes to the set point lead to step changes in the error. The derivative of a step change goes to infinity or, in practical cases, to a completely open or closed control valve. This control action could lead to severe process upsets and even to unsafe conditions. One approach to prevent this situation is to alter the algorithm so that the derivative is taken on the controlled variable, not the error. The modified derivative mode, remembering that $E(t) = \mathrm{SP}(t) - \mathrm{CV}(t)$, is

$$\mathrm{MV}_d(t) = -K_c T_d \frac{d\mathrm{CV}(t)}{dt} + I_d \tag{8.16}$$

While equation (8.16) reduces the extreme variation in the manipulated variable resulting from set point changes, it does not solve the problem of high-frequency noise on the controlled-variable measurement, which will also cause excessive variation in the manipulated variable. An obvious step to reduce the effects of noise is to reduce the derivative time, perhaps to zero. Other steps to reduce the effects of noise are presented in Chapter 12. In summary:

> The derivative mode is simple; does not influence the final steady-state value of error; provides rapid correction based on the rate of change of the controlled variable; and can cause undesirable high-frequency variation in the manipulated variable.

8.6 THE PID CONTROLLER

Naturally, it is desired to retain the good features of each mode in the final control algorithm. This goal can be achieved by adding the three modes to give the final expression of the PID controller. Where the derivative mode appears, two forms are given: (*a*) the Instrument Society of America (ISA) standard and (*b*) the form recommended in this book because it prevents set point changes from causing excessive response, as described in the preceding section.

Time-Domain Controller Algorithms

PROPORTIONAL-INTEGRAL-DERIVATIVE.

$$\mathrm{MV}(t) = K_c \left(E(t) + \frac{1}{T_I} \int_0^t E(t')dt' + T_d \frac{dE(t)}{dt} \right) + I \qquad \text{(ISA standard)} \tag{8.17a}$$

$$\mathrm{MV}(t) = K_c \left(E(t) + \frac{1}{T_I} \int_0^t E(t')dt' - T_d \frac{d\mathrm{CV}(t)}{dt} \right) + I \qquad \text{(Recommended)} \tag{8.17b}$$

Again, the controller has an initialization constant. Depending on the desired performance, various forms of the controller are used. The proportional mode is normally retained for all forms, with the options being in the derivative and integral modes. The most common alternative forms are as follows:

PROPORTIONAL-ONLY CONTROLLER.

$$\mathrm{MV}(t) = K_c[E(t)] + I \tag{8.18}$$

PROPORTIONAL-INTEGRAL CONTROLLER.

$$\mathrm{MV}(t) = K_c\left(E(t) + \frac{1}{T_I}\int_0^t E(t')dt'\right) + I \tag{8.19}$$

PROPORTIONAL-DERIVATIVE CONTROLLER.

$$\mathrm{MV}(t) = K_c\left(E(t) + T_d\frac{dE(t)}{dt}\right) + I \qquad \text{(ISA standard)} \tag{8.20a}$$

$$\mathrm{MV}(t) = K_c\left(E(t) - T_d\frac{d\mathrm{CV}(t)}{dt}\right) + I \qquad \text{(Recommended)} \tag{8.20b}$$

Selection from among the four forms will be discussed after many features of the controllers have been introduced.

Laplace-Domain Transfer Functions

The control algorithms are used often in block diagrams and in closed-loop transfer functions. In these analyses the main purposes are to determine limiting behavior for control systems (stability and frequency response), usually for disturbance response; thus, the PID form with derivative on the error (ISA standard) is used for simplicity. The transfer functions for the common forms are as follows. Note that each transfer function is the output over the input, with the input and output taken with respect to the *controller,* which is the opposite of the process. Also, since transfer functions are always in deviation variables, the initialization constant does not appear.

PROPORTIONAL-INTEGRAL-DERIVATIVE.

$$\frac{\mathrm{MV}(s)}{E(s)} = K_c\left(1 + \frac{1}{T_I s} + T_d s\right) \tag{8.21}$$

PROPORTIONAL-ONLY.

$$\frac{\mathrm{MV}(s)}{E(s)} = K_c \tag{8.22}$$

PROPORTIONAL-INTEGRAL.

$$\frac{\mathrm{MV}(s)}{E(s)} = K_c\left(1 + \frac{1}{T_I s}\right) \tag{8.23}$$

PROPORTIONAL-DERIVATIVE.

$$\frac{\mathrm{MV}(s)}{E(s)} = K_c\,(1 + T_d s) \tag{8.24}$$

The reader is strongly encouraged to learn the various forms of the algorithms in the time and Laplace domains, because they will be used in all subsequent topics.

8.7 ANALYTICAL EXPRESSION FOR A CLOSED-LOOP RESPONSE

It is clear that the algorithm structure and adjustable parameters affect the closed-loop dynamic response. A straightforward method of determining how the parameters affect the response is to determine the analytical solution for the linear process with PID feedback. This is generally not done in practice, because of the complexity of the analytical solution for realistic processes, especially when the process has dead time. However, the analytical solution is derived here for a simple process, to aid in understanding the interplay between the process and the controller.

Example 8.5. To facilitate the solution, a simple process—the stirred-tank heater in Example 3.7 (page 83)—is selected, with the controlled variable being the tank temperature and the manipulated variable being the coolant flow valve, as shown in Figure 8.8. Since proportional control was considered in Example 8.2, a proportional-integral controller is selected, because this will ensure zero steady-state offset. The response to a step set point change will be determined.

Formulation. The model for this process was derived in Example 3.7. It is repeated here with the models for the other elements in the control loop: the valve and the controller (the sensor is assumed to be instantaneous).

$$V\rho C_p \frac{dT}{dt} = C_p\rho F(T_0 - T) - \frac{aF_c^{b+1}}{F_c + \dfrac{aF_c^b}{2\rho_c C_{pc}}}(T - T_{\text{cin}}) \tag{8.25}$$

$$F_c = \left(K_v\sqrt{\frac{\Delta P}{\rho_c}}\right)v \tag{8.26}$$

$$v = K_c\left((T_{\text{sp}} - T) + \frac{1}{T_I}\int_0^t (T_{\text{sp}} - T)dt'\right) + I \tag{8.27}$$

First, the degrees of freedom of the closed-loop control system will be evaluated.

Dependent variables: $T,\ F_c,\ v$

External variables: $T_0,\ F,\ T_{\text{cin}},\ T_{\text{sp}}$ $\qquad$ DOF $= 3 - 3 = 0$

Constants: $\rho,\ C_p,\ a,\ b,\ K_v,\ \Delta P,\ \rho_c,\ K_c,\ T_I,\ I,\ V$

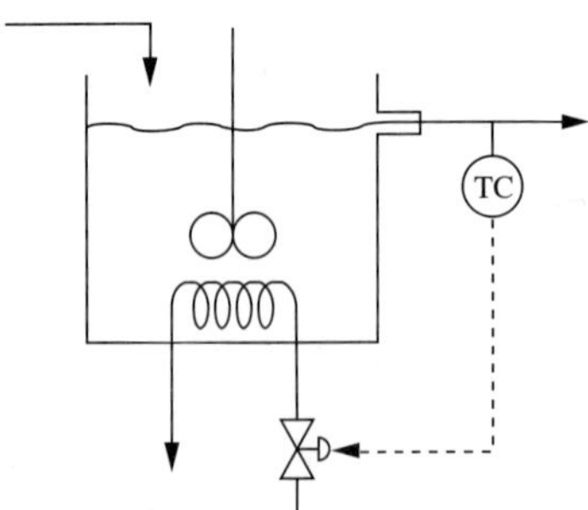

FIGURE 8.8
Heat exchanger control system in Example 8.5.

Thus, when the controller set point T_{sp} has been defined, the system is exactly specified. Note that the system without control requires the valve position to be defined, but that the controller now determines the valve opening based on its algorithm in equation (8.27). The three equations can be linearized and the Laplace transforms taken to obtain the following transfer functions:

$$G_p(s) = \frac{K_p}{\tau s + 1} \tag{8.28}$$

$$G_v(s) = K_v \sqrt{\frac{\Delta P}{\rho}} \approx 1.0 \tag{8.29}$$

$$G_c(s) = \frac{v(s)}{T_{sp}(s) - T(s)} = K_c\left(1 + \frac{1}{T_I s}\right) \tag{8.30}$$

The process gain and time constant are functions of the equipment design and operating conditions and are given in Example 3.7. We assume that the valve opening is expressed in fraction open and that $G_v(s) = 1$. The block diagram of the single-loop control system is given in Figure 7.4, and the closed-loop transfer function is

$$\text{CV}(s) = \frac{G_p(s)G_v(s)G_c(s)}{1 + G_p(s)G_v(s)G_c(s)G_s(s)}\text{SP}(s) \tag{8.31}$$

The general symbols are used for the controlled and set point variables, $\text{CV}(s) = T(s)$ and $\text{SP}(s) = T_{sp}(s)$. The transfer functions for the process, the PI controller, and the instrumentation ($G_s(s) = G_v(s) = 1$) can be substituted into equation (8.31) to give

$$\begin{aligned}
\text{CV}(s) &= \frac{G_p(s)G_c(s)}{1 + G_p(s)G_c(s)}\text{SP}(s) \\
&= \frac{\dfrac{K_p}{\tau s + 1} K_c\left(1 + \dfrac{1}{T_I s}\right)}{1 + \dfrac{K_p}{\tau s + 1} K_c\left(1 + \dfrac{1}{T_I s}\right)}\text{SP}(s) \\
&= \frac{T_I s + 1}{\dfrac{\tau T_I}{K_c K_p} s^2 + \dfrac{T_I(1 + K_c K_p)}{K_c K_p} s + 1}\text{SP}(s)
\end{aligned} \tag{8.32}$$

This can be rearranged to give the transfer function for the closed-loop system:

$$\frac{\text{CV}(s)}{\text{SP}(s)} = \frac{T_I s + 1}{(\tau')^2 s^2 + 2\xi\tau' s + 1} \tag{8.33}$$

This is presented in the standard form with the time constant (τ') and damping coefficient expressed as

$$\xi = \frac{1}{2}\sqrt{\frac{T_I}{K_c K_p}}\left(\frac{1 + K_c K_p}{\sqrt{\tau}}\right) \qquad \tau' = \sqrt{\frac{\tau T_I}{K_c K_p}} \tag{8.34}$$

Equation (8.33) can be rearranged to solve for CV(s) with SP(s) $= \Delta$SP/s (step change). This expression can be inverted using entries 15 and 17 in Table 4.1 to give, for $\xi < 1$,

$$T'(t) = \Delta \text{SP}\left[\frac{T_I}{\tau' \sqrt{1-\xi^2}} e^{-\xi t/\tau'} \sin\left(\frac{\sqrt{1-\xi^2}}{\tau'} t\right)\right]$$

$$+ \Delta \text{SP}\left[1 - \frac{1}{\sqrt{1-\xi^2}} e^{-\xi t/\tau'} \sin\left(\frac{\sqrt{1-\xi^2}}{\tau'} t + \phi\right)\right] \tag{8.35}$$

with $\phi = \tan^{-1}\left(\frac{\sqrt{1-\xi^2}}{\xi}\right)$

or using entry 10 in Table 4.1 to give, for $\xi > 1$,

$$T'(t) = \Delta \text{SP}\left[T_I \frac{\left(e^{-t/\tau_1'} - e^{-t/\tau_2'}\right)}{\tau_1' - \tau_2'} + 1 + \frac{\tau_1' e^{-t/\tau_1'} - \tau_2' e^{-t/\tau_2'}}{\tau_2' - \tau_1'}\right] \tag{8.36}$$

with τ_1' and τ_2' the real, distinct roots of the characteristic polynomial when $\xi > 1.0$.

Solution. Before an example response is evaluated, some important observations must be made:

1. The feedback system is second-order, although the process is first-order. Thus, we see that the integral controller increases the order of the system by 1.
2. The integral mode ensures zero steady-state offset, which can be verified by evaluating the foregoing expressions as time approaches infinity.
3. The response can be over- or underdamped, depending on the parameters in equation (8.34). Again, we see that feedback can change the qualitative characteristics of the dynamic response.
4. The response for this system is always stable (for negative feedback, $K_c K_p > 0$); in other words, the output cannot grow in an unbounded manner, because of the structure of the process and controller equations. This is not generally true for more complex and realistic process models (and essentially all control systems involving real processes), as will be explained in Chapter 10.

The final observation concerns the manipulated variable, which is also important in evaluating control performance. The transfer function for the manipulated variable can be derived from block diagram algebra to be

$$\frac{\text{MV}(s)}{\text{SP}(s)} = \frac{G_c(s)}{1 + G_p(s)G_c(s)G_v(s)G_s(s)} \tag{8.37}$$

The characteristic polynomials for the transfer functions in equations (8.31) and (8.37) are identical; thus, the periodic nature of the responses (over- or underdamped) of the controlled and manipulated variables are the same since they are affected by the same factors in the control loop. Thus, it would not be possible to obtain underdamped behavior for the controlled variable and overdamped behavior for the manipulated variable.

The close relationship between these variables is natural, because the manipulated variable is calculated by the PI controller based on the controlled variable.

Results analysis. A sample dynamic response is given in Figure 8.9 for this system with $K_p = -33.9°C/(m^3/min)$ and $\tau = 11.9$ min from Example 3.7 and tuning constant values of $K_c = -0.059\ (m^3/min)/°C$ and $T_I = 0.95$ min, giving $\tau' = 2.38$ min and $\xi = 0.30$, and $SP'(s) = 2/s$. The response is clearly underdamped, as indicated by the damping coefficient being less than 1.0. Also shown in the figure is the boundary defined by the exponential in the analytical solution, which determines the maximum amplitude of the oscillation at any time. Note that another set of controller tuning constants could yield overdamped behavior for the closed-loop system. The parameters used in this example were selected somewhat arbitrarily, and proper tuning methods are presented in the next two chapters.

Since both tuning constants, K_c and T_I, appear in τ' and ξ, it is not possible to attribute the damping or oscillations to a single tuning constant; they both affect the "speed" and damping of the response. It is apparent from the expression for ξ that the response becomes more oscillatory as K_c is increased and as T_I is decreased; the reason for the difference is that K_c is in the numerator of the controller, whereas T_I is in the denominator of the control algorithm. It is also apparent from equation (8.35) that the controlled-variable overshoot and decay ratio increase as the damping coefficient decreases.

This analysis could be extended to other simple systems, but it cannot be applied to most realistic systems, for which the inverse Laplace transform cannot be

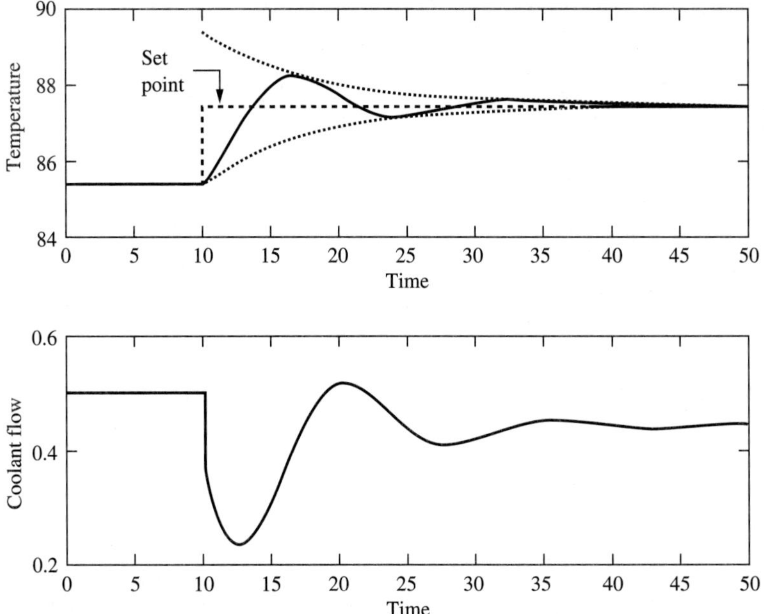

FIGURE 8.9
Dynamic response of feedback loop: set point (dotted), temperature (solid), and limits on magnitude (dashed).

evaluated. Therefore, the derivation of complete analytical solutions will not be extended here. However, the general principles learned in this example are applicable to the methods of analysis introduced in the next few chapters. Also, one important class of processes—inventories (levels)—is simple enough to allow process equipment and controller design based on analytical solution of the linearized models, as covered in Chapter 18.

8.8 IMPORTANCE OF THE PID CONTROLLER

The process industries, which operate equipment at high pressures and temperatures with potentially hazardous materials, needed reliable process control many decades before digital computers became available. As a result, the control methods developed many decades ago were tailored to the limited computing equipment available at that time. The main method of automated computing during this period, and one which continues to be used today, is analog computation. The principle behind analog computing is the design of a physical system that follows the same equations as the equations desired to be solved (Korn and Korn, 1972). Naturally, the computing system must be simple and should have easy ways to alter parameters. An example of an analog control system is shown schematically in Figure 8.10. Here the level in a tank is controlled by adjusting the flow into the tank. The sensor is a float in the tank, and the final control element is the valve stem position. The controller is a proportional-only algorithm, so that the controller output is proportional to the error signal. This algorithm is implemented in the figure by a bar that pivots on a fulcrum. As the level increases, the float rises and the valve closes, reducing flow into the tank. The control parameters can be changed by (1) increasing the height of the fulcrum to increase the set point (with an appropriate adjustment of the connecting bars) or (2) altering the fulcrum position along the bar to change the controller proportional gain.

Although a few systems like the one in Figure 8.10 are in use (indeed, a form of that system is found in domestic toilet tanks), most of the analog controllers in the process industries use more sophisticated pneumatic or electronic principles to automate the PID algorithm. The typical industrial implementation yields the following transfer function for an electronic analog controller calculation (Hougen, 1972):

$$\frac{\text{MV}(s)}{\text{CV}(s)} = K_c \left[\frac{1 + T_I s}{T_I s}\right]\left[\frac{1 + T_d s}{1 + \alpha T_d s}\right] \tag{8.38}$$

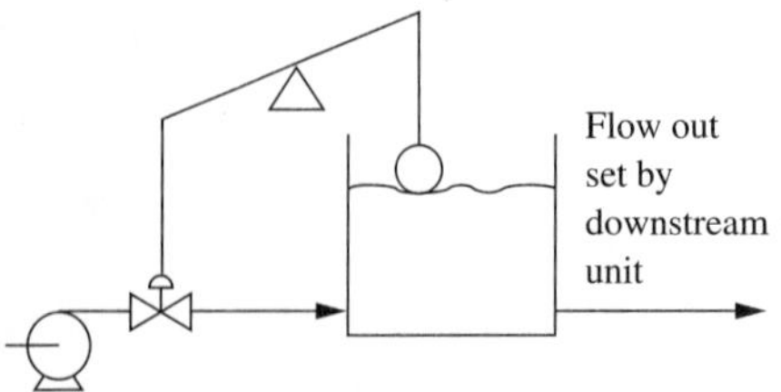

FIGURE 8.10
Example of an analog level controller.

Equation (8.38), often referred to as the *interactive PID algorithm,* is an approximation to the PID algorithm when α is small. The tuning constants are adjusted by changing values of resistors and capacitors used in the circuit. Note that since the equation structure is different from the forms already introduced, this equation would require different values of these tuning constants; the tuning rules in this book are for the forms in equations (8.17) and (8.21). Analog controllers were used for many decades prior to the introduction of digital controllers and continue to be used today. Pneumatic analog controllers use air pressure as the source of power for the calculation to approximate the PID calculation (Ogata, 1990).

The techniques in this book are based on the analysis of continuous systems, because we will be using Laplace transforms and similar mathematical methods. Most processes are continuous (e.g., stirred tanks and heat exchangers), and the controller is also continuous when implemented with analog computation. However, the controller is discrete when implemented by digital computation; discrete systems perform their function only at specific times. For most of this book, the assumption is made that the control calculations are continuous, and this assumption is generally very good for digital controllers as long as the time for calculation is short compared with the process dynamic response. Since this situation is satisfied in most process control systems, the approach taken here is usually valid. Special features of digital control systems are introduced in Chapter 11 and covered thereafter as appropriate for subsequent topics, and numerous books are dedicated entirely to the special aspects of digital control, for example, Franklin and Powell (1980) and Smith (1972).

8.9 CONCLUSION

In this chapter, the important proportional-integral-derivative control algorithm was introduced, and the key features of each mode were demonstrated. The proportional mode provides fast response but does not reduce the offset to zero. The integral mode reduces the offset to zero but provides relatively slow feedback compensation. The derivative mode takes action based on the derivative of the controlled variable but has no effect on the offset. The combination of the modes, or a subset of the modes, is required to provide good control in most cases.

It is clear that the PID controller can achieve good control performance with the proper choice of tuning constants. However, the control system can perform poorly, and even become unstable, if improper values of the controller tuning constants are used. An analytical method for determining good values for the tuning constants was introduced in this chapter for simple first-order processes with P-only and PI control. More general methods are presented for more complex systems in the next two chapters.

The dramatic influence of feedback on the dynamic behavior of a process was discussed in Chapter 7 and demonstrated mathematically in this chapter. Naturally, the ability to maintain the controlled variable near its set point is a desirable feature of feedback, but the potential change from an overdamped system to an underdamped or even unstable one is a facet of feedback that must be understood and monitored carefully to prevent unacceptable behavior. In Chapter 4, it was demonstrated that the key facets of periodicity and stability are determined by the roots of the characteristic equation; that is, by the poles of the transfer function. For the three-tank mixing process without control, the characteristic equation is

$$(\tau s + 1)^3 = 0 \tag{8.39}$$

giving the repeated poles $s = -1/\tau$. Since they are real and negative, the dynamic response is overdamped and stable. When proportional feedback is added, the transfer function is given in equation (8.6), and the characteristic equation is

$$(\tau s + 1)^3 + K_c K_p = 0 \tag{8.40}$$

Thus, the controller gain influences the poles and the exponents in the time-domain solution for the concentration. The influence of feedback control on stability is the major topic of Chapter 10.

Finally, it is important to note that the PID controller is emphasized in this book because of its widespread use and its generally good performance. The dominant position of this algorithm is not surprising, because it evolved over years of industrial practice. However, in nearly no case is it an "optimal" controller in any sense (i.e., minimizing IAE or maximum deviation). Thus, other algorithms can provide better performance in particular situations. Some alternative algorithms will be introduced in this book after the basic concepts of feedback control have been thoroughly covered.

REFERENCES

Distephano, S., A. Stubbard, and I. Williams, *Feedback Control Systems,* McGraw-Hill, New York, 1976.

Franklin, G. and J. Powell, *Digital Control of Dynamic Systems,* Addison-Wesley, Reading, MA, 1980.

Hougen, J., *Measurements and Control Applications for Practicing Engineers,* Cahners Books, Boston, MA, 1972.

Korn, G. and T. Korn, *Electronic Analog and Hybrid Computers,* McGraw-Hill, New York, 1972.

Rhinehart, R., personal communication, 1991.

Smith, C., *Digital Computer Process Control,* Intext, Scranton, PA, 1972.

Ogata, K., *Modern Control Engineering,* Prentice Hall, Englewood Cliffs, NJ, 1990.

ADDITIONAL RESOURCES

A brief history of operator interfaces for process control, showing the key graphical and pattern recognition features, is given in

Lieber, R., "Process Control Graphics for Petrochemical Plants," *Chem. Eng. Progr.* 45–52 (Dec. 1982).

Additional analytical solutions to low-order closed-loop systems can be found in

Weber, T., *An Introduction to Process Dynamics and Control,* Wiley, New York, 1973.

With models for the process and controller now available, the dynamic behavior of a closed-loop system can be analyzed quantitatively. These questions provide some learning examples while using the mathematical tools available; additional analytical methods are introduced in the next chapters. The key concept is the manner in which the process and controller both influence the feedback system.

QUESTIONS

8.1. Determine the analytical expression for a step set point change in the following processes under P-only and PI feedback control. You should select values for the tuning constant that give acceptable performance.

(*a*) Example 3.1 with C_A as the controlled variable, C_{A0} as the manipulated variable, and $\Delta SP = 0.1$ mole/m^3.

(*b*) Example 3.7 with T as the controlled variable, F as the manipulated variable, and $\Delta SP = 3°C$. (F_c is constant.)

(*c*) Example 3.3 with C_{A2} as the controlled variable, C_{A0} as the manipulated variable, and $\Delta SP = 0.05$ mole/m^3.

8.2. Program a dynamic simulation for the three-tank mixing system based on the equations derived in Example 7.2.

(*a*) Determine the open-loop responses in the third tank outlet concentration to a step change in

1. the inlet concentration of component A in stream B (1 to 1.5% A)
2. the valve position in the A stream (50 to 60% open)

(*b*) Determine the closed-loop (PID) responses of the third tank outlet concentration to

1. a step set point change (3 to 3.5% A)
2. a disturbance step change in the concentration of component A in stream B (1 to 1.5% A)

8.3. Using the appropriate transfer functions and applying the final value theorem, determine the final values of the error for a step set point change for the heater in Example 8.5 under P-only, PI, and PID control.

8.4. The control system given in Figure Q8.4 controls the level by adjusting the valve position of the flow out of the tank. Because of the pump, the flow out can be assumed to be a function of only the valve percent open and not of the level. Assume that the valve-flow relationship is linear (i.e., $F_{out} = K_v v$).

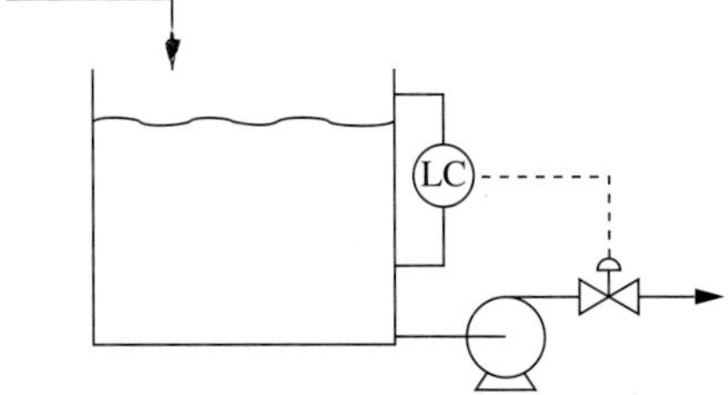

FIGURE Q8.4

(*a*) Derive the differential equation and transfer function relating the level to the flows in and out.

(*b*) For the process with feedback control, determine the final value of the error for a step change in the inlet flow for P-only and PI controllers. Are the criteria for zero steady-state offset the same as for the three-tank example? Explain why/why not.

(*c*) Discuss the differences between this and question 8.13.

8.5. The application to the final value theorem in equation (8.11) showed that the three-tank mixing system under I-only control has zero steady-state offset for a step disturbance. Is this a general conclusion for PID control for all (*a*) processes, (*b*) disturbance types, and (*c*) values of the tuning constants? Discuss the implications of your answers on the success of feedback control.

8.6. (*a*) The final value theorem seems to demonstrate that the offset tends to zero as the controller gain approaches infinity. Discuss this result, especially with regard to the definition of the Laplace transform and the dynamic responses shown in Figures 8.3*a* through *d*.

(*b*) The final value theorem provides one method for calculating the final value of a variable in a control system. Describe another way to determine the final value of variables without using the final value theorem. Use both methods to determine the final value of the manipulated variable in the three-tank mixing process for a step disturbance in the concentration of stream *B* (*a*) without control and (*b*) with P-only feedback control.

8.7. (*a*) Calculate the roots of the characteristic equations and relate them to the dynamic behaviors of the closed-loop systems in Figures 8.3*a* through *d*.

(*b*) Select different tuning constant values that yield substantially different dynamic behavior for the closed-loop system in Example 8.5. Describe the different time-domain behavior.

8.8. Answer the following questions.

(*a*) The transfer function of the PID controller in equation (8.21) has no initialization constant. Why?

(*b*) Describe how to calculate the initialization constant *I* in equation (8.17*a* and *b*) for a PID controller.

(*c*) The transfer functions $G_c(s) = \text{MV}(s)/\text{CV}(s)$ and $G_p(s) = \text{CV}(s)/\text{MV}(s)$. Why isn't $G_c(s) = G_p^{-1}(s)$? Why do they have units that are the inverse of one another?

(*d*) Verify the Laplace transform of the controller, equation (8.21), from equation (8.17*a*).

(*e*) Determine the final value for the three-tank mixing process under PI control for an impulse disturbance in the feed composition. Can you determine a conclusion generally applicable to all processes?

(*f*) Repeat part (*e*) for a ramp disturbance.

8.9. When designing the feedback control algorithm, why were the following modes not included, or when would they be applicable?

(*a*) $$\text{MV}(t) = K_c\left(E(t) + \frac{1}{T_I}\int_0^t\left[\int_0^{t'} E(t'')\,dt''\right]dt'\right) + I$$

(*b*) $$\text{MV}(t) = K_c(E(t))^2\left(E(t) + \frac{1}{T_I}\int_0^t E(t')\,dt'\right) + I$$

(c) $\text{MV}(t) = K_c\left((E(t))^2 + \frac{1}{T_I}\int_0^t (E(t'))^2\,dt'\right) + I$

8.10. The controller display for the plant personnel does not present all possible variables associated with the PID algorithm. For each variable, state whether or not it is displayed and why: (*a*) controlled variable, (*b*) error, (*c*) set point, (*d*) manipulated variable, (*e*) integral of the error, (*f*) derivative of the error, and (*g*) initialization constant.

8.11. Describe how you would calculate the PID algorithm in a digital computer. Prepare a flow chart of the calculations.

8.12. Consider the modified stirred-tank mixing system in Figure Q8.12. The original concentration of the third tank remains 3%.

(*a*) Derive the equations describing the system.

(*b*) Draw a block diagram of the system.

(*c*) Derive the transfer functions for each element in the block diagram.

(*d*) Derive the closed-loop transfer function, CV(*s*)/SP(*s*).

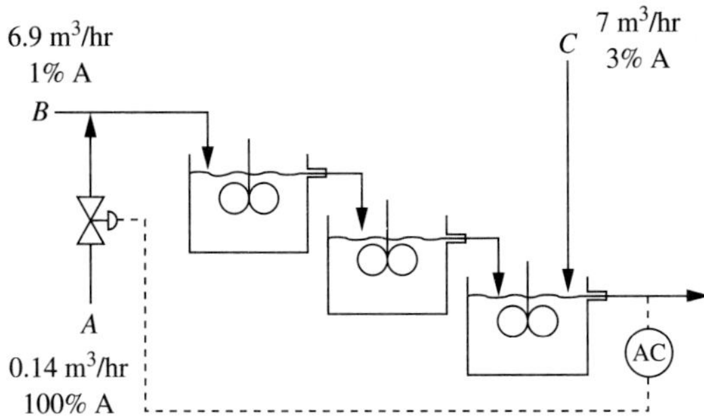

Disturbance is change in the concentration of stream *C* with the flow rate constant.

FIGURE Q8.12

8.13. The level control system with a proportional-only algorithm in Figure Q8.13 is to be analyzed; the inlet flow is a function of only the valve opening. The process is not typical; usually, the flow out would be pumped, but here it drains by gravity. However, this is a simple system to begin analyzing control systems; more realistic processes will be considered in subsequent chapters.

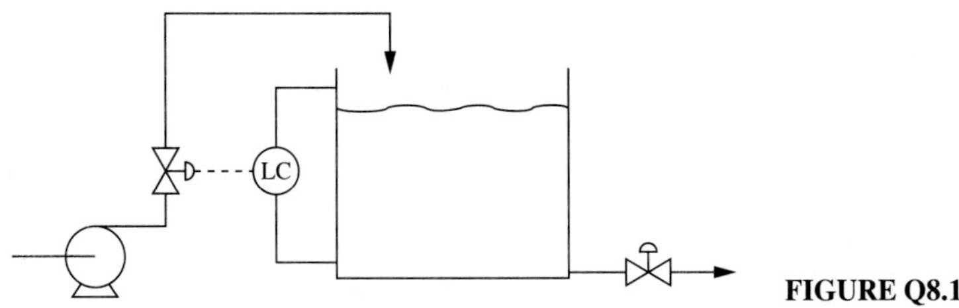

FIGURE Q8.13

(*a*) Derive a linearized model and transfer functions for the process and for the proportional-only controller.

(*b*) Draw a block diagram, and derive the closed-loop transfer function.

(*c*) Calculate the steady-state offset.

(*d*) Select an appropriate sign for the gain and calculate the time to reach 63% of the final steady-state error after a step disturbance in the outlet valve position.

(*e*) Discuss the differences between this and question 8.4.

8.14. Consider the PID algorithm in equation (8.17*a*). For each of the individual modes—proportional, integral and derivative—describe with a sketch the result of its calculation when the error is each of the following idealized functions: (*a*) a constant, (*b*) an impulse, and (*c*) a sine (consider one cycle). [This question provides a thought exercise to help understand the three PID modes; this type of analysis is not performed when monitoring a control system.]

8.15. A P-only controller is implemented on the nonisothermal CSTR in Examples 3.10, 4.10, and 5.8. The controlled variable is the temperature in the reactor, and the manipulated variable is the coolant flow. Select a good value for the controller gain (K_c), and determine the dynamic responses of T, F_c, and C_A for a step set point change of 2°C.

8.16. Analyze the following systems for the feasibility of feedback control.

(*a*) Example 5.3 with temperature T_3 as the controlled variable, F_{exch} as the manipulated variable, and $\Delta SP = 1°C$.

(*b*) Example 5.4 with C_{A2} as the controlled variable, F_s as the manipulated variable, and $\Delta SP = 0.01$ mole/m^3.

8.17. The continuous control system in Figure Q8.17 is to be tuned for an *underdamped* open-loop process, $\xi < 1.0$. As a physical example, you may think of the CSTR with underdamped temperature dynamics in response to a change in the coolant flow described in Examples 3.10 and 4.8. However, the question should be answered for the general system in Figure Q8.17.

(*a*) Determine the range of a P-only feedback controller gain that results in an *overdamped* closed-loop system. Discuss the implications of your results for the quality of feedback control performance.

(*b*) Repeat the analysis for a proportional-derivative controller and discuss the effect of the derivative mode on the closed-loop dynamic behavior, especially the periodicity.

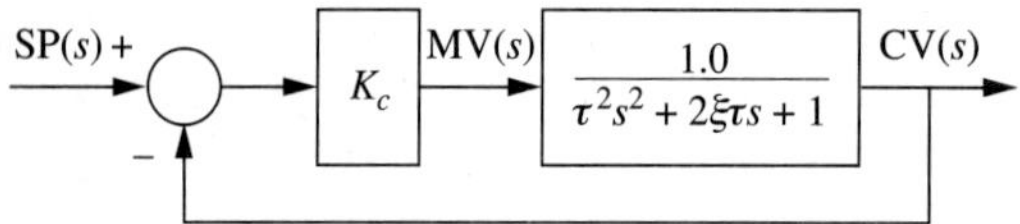

FIGURE Q8.17

8.18. (*a*) Determine the PID controller modes that are required for zero steady-state offset for an impulse disturbance for the following processes:

1. The three-tank mixing process in Examples 7.2 and 7.3 with x_{AB} an impulse
2. A non–self-regulating level system, like equation (5.15) with F_0 an impulse and F_1 adjusted by the controller

(*b*) Discuss the application of integral-only control to both processes.

CHAPTER 9

PID CONTROLLER TUNING FOR DYNAMIC PERFORMANCE

9.1 INTRODUCTION

As demonstrated in the previous chapter, the proportional-integral-derivative (PID) control algorithm has features that make it appropriate for use in feedback control. Its three adjustable tuning constants enable the engineer, through judicious selection of their values, to tailor the algorithm to a wide range of process applications. Previous examples showed that good control performance can be achieved with a proper choice of tuning constant values, but poor performance and even instability can result from a poor choice of values. Many methods can be used to determine the tuning constant values, such as the analytical method demonstrated for simple systems in the previous chapter. In this chapter a method is presented that is based on the time-domain performance of the control system. Controller tuning methods based on dynamic performance have been used for many decades (e.g., Lopez et al., 1969; Fertik, 1975; Zumwalt, 1981), and the method presented here builds on these previous studies and has the following features:

1. It addresses important performance issues that must be considered in controller tuning.
2. It provides easy-to-use correlations that are applicable to many controller tuning cases.
3. It provides a general calculation approach applicable to nearly any control tuning problem, which is important when the general correlations are not applicable.

4. It provides insight into important relationships between process dynamic model parameters and controller tuning constants.

The method in this chapter gives useful results and interpretations in the time domain, which are easy to interpret (since we live in the time domain!). However, it does not provide all important insights and theoretical foundations for control system analysis. The next chapter continues the same theme but applies a more mathematically fundamental approach. Thus, this and the next chapter present two complementary facets of the controller tuning issue.

9.2 FACTORS IN CONTROLLER TUNING

The entire control problem must be completely defined before the tuning constants can be determined and control performance evaluated. Naturally, the physical process is a key element of the system that must be defined. To consider the most typical class of processes, a first-order-with-dead-time plant model is selected here because this model can adequately approximate the dynamics of processes with monotonic responses to a step input, as shown in Chapter 6. Also, the controller algorithm must be defined; the form of the PID controller used here is

$$\mathrm{MV}(t) = K_c\left(E(t) + \frac{1}{T_I}\int_0^t E(t')\,dt' - T_d\frac{d\mathrm{CV}(t)}{dt}\right) + I \tag{9.1}$$

Note that the derivative term is calculated using the measured controlled variable, not the error. Many other variations of the PID algorithm are provided as preprogrammed calculations in commercial control equipment; as an example, the derivative may be calculated based on the error.

> The tuning constants must be derived using the *same* algorithm that is applied in the control system. The reader is cautioned to check the form of the PID controller algorithm used in developing tuning correlations and in the control system computation; these must be compatible.

The procedure in this chapter is based on a mathematical method for optimizing a quantitative measure of control performance, and the usefulness of the results depends on how well the control performance measure truly represents the goals of the plant operation. Therefore, we carefully define control performance by specifying several goals to be balanced concurrently. This definition provides a comprehensive specification of control performance that is flexible enough to represent most situations. The three goals are the following:

1. *Controlled-variable performance.* The well-tuned controller should provide satisfactory performance for one or more measures of the behavior of the controlled variable. As an example, we shall select to minimize the IAE of the controlled

variable. The meaning of the integral of the absolute value of the error, IAE, is repeated here.

$$\text{IAE} = \int_0^\infty \left|\text{SP}(t) - \text{CV}(t)\right| dt \tag{9.2}$$

Zero steady-state offset for a steplike system input is ensured by the integral mode appearing in the controller.

2. *Model error.* Linear dynamic models always have errors, because the plant is nonlinear and its operation changes. Since the tuning will be based on these models, the tuning procedure should account for the errors, so that acceptable control performance is provided as the process dynamics change. The changes are defined as ± percentage changes from the base-case or nominal model parameters. The ability of a control system to provide good performance when the plant dynamics change is often termed *robustness.*
3. *Manipulated-variable behavior.* The most important variable, other than the controlled variable, is the manipulated variable. We shall choose the common goal of preventing "excessive" variation in the manipulated variable by defining limits on its allowed variation, as explained shortly.

To evaluate the control performance, the goals and the scenario(s) under which the controller operates need to be defined. These definitions are summarized in Table 9.1; the general factors are in the second column, and the specific values used to develop correlations in this chapter are in the third column. This may seem like a rather lengthy list of factors to establish before tuning a controller, but they are essential

TABLE 9.1
Summary of factors that must be defined in tuning a controller

Major loop component	Key factor	Values used in this chapter for examples and correlations
Process	Model structure	Linear, first-order with dead time
	Model error	±25% in model parameters (structured so that all parameters increase and decrease the same %)
	Input forcing	step input disturbance with $G_d(s) = G_p(s)$ and step set point considered separately
	Measured variable	Unbiased controlled variable with high frequency noise
Controller	Structure	PID and PI
	Tuning constants	K_c, T_I, and T_d
Control Performance	Controlled-variable behavior	Minimize the total IAE for several cases spanning a range of plant model parameter errors
	Manipulated-variable behavior	Manipulated variable must not have variation outside defined limits, see Figure 9.4

to any proper tuning method. Fortunately, the rather standard set of specifications in the third column is appropriate for a wide range of applications, and therefore it is possible to develop correlations that can be used in many plants, *where this underlying specification of control performance is valid.* The entries in Table 9.1 will be further explained as they are encountered in the next section. All subsequent chapters in this book require a good understanding of the factors that affect control performance. The reader is encouraged to understand the factors in Table 9.1 thoroughly and to refer back to this section often when covering later chapters.

9.3 DETERMINING TUNING CONSTANTS THAT GIVE GOOD CONTROL PERFORMANCE

Given a complete definition of the process, controller, and control objectives, evaluating the tuning constants is a relatively straightforward task, at least conceptually. The "best" tuning constants are those values that satisfy the control performance goals. With our definitions of goals 1 through 3, the optimum is the best value of controlled-variable performance, here minimum IAE, for the selected plants (with variations in model parameters) that also satisfies the bounds imposed on the manipulated variable.

The control objectives have been defined so that they can be quantitatively evaluated from the dynamic response of a control system and the tuning constants determined through an optimization procedure. In general, the three tuning constants must be evaluated *simultaneously* in this optimization; however, it is worthwhile to begin by considering the goals and PID tuning constants sequentially to gain insight into how the goals (definition of good performance) influence the resulting tuning constant values and dynamic behavior. Therefore, we shall begin by determining the value of one tuning constant, K_c, which minimizes the simple performance goal 1 measure (IAE) with the other constants held at fixed values. Then the additional constants will be included in the optimization, so that we can determine the effect of optimizing the individual constants on the control performance as represented by goal 1. Finally, the complete definition of control performance, goals 1 through 3, is used to determine the best tuning constants. The sequential consideration of the additional goals enables us to learn the effects of modelling errors, limits on manipulated-variable variation, and noise on the tuning constant values. The main purpose of this section is to demonstrate the effects of tuning on control performance, not to cover optimization theory in detail. The reader is counseled to concentrate on the results of the calculations, with the objective of learning the *effects of tuning on control performance* in the time domain. More details of the calculations are provided in Appendix E.

The simple mixing process example shown in Figure 9.1 will be used throughout this chapter; it has a transportation delay and only one tank. The system is described by the following transfer function model.

$$\frac{\mathrm{CV}(s)}{D(s)} = \frac{G_d(s)}{1 + G_p'(s)G_v(s)G_c(s)G_s(s)} \tag{9.3}$$

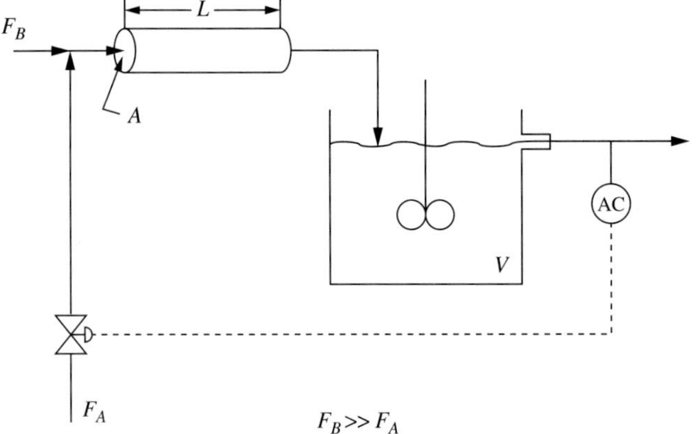

FIGURE 9.1
Process used for calculating example tuning constants for good control performance.

$$G_v(s)G'_p(s)G_s(s) = G_p(s) = \frac{K_p e^{-\theta s}}{1 + \tau s} \frac{\%\text{A}}{\%\text{ controller output}}$$

$$G_d(s) = \frac{1.0}{1 + \tau s} \frac{\%\text{A in outlet}}{\%\text{A in inlet}}$$

The dead time and time constant are functions of the feed flow rate and equipment size, so a range of dynamics can be represented by this example. The initial base-case values are given as follows, and other values for the dead time and time constant will be used in the chapter.

$$\text{Base case: } \theta = (A)(L)/F_B = 5.0 \text{ min} \qquad \tau = V/F_B = 5.0 \text{ min}$$

$$K_p = K_v[(x_A)_A - (x_A)_B]/F_B = 1.0\%\text{A}/\%\text{ open}$$

The disturbance considered in this section is a unit step in the inlet concentration, $D(s) = 1/s$ %A.

Goal 1: Controlled-Variable Performance (IAE)

Let us begin with a PID controller applied to the example process. We will start by optimizing only one controller constant. Recall that the integral mode is required so that the controlled variable returns to its set point. Therefore, the study will find the best value of the controller gain, K_c, with the integral time ($T_I = 10$ min) and derivative time ($T_d = 0$ min) temporarily maintained at fixed values. The value selected for the integral time (the sum of the dead time and time constant) is reasonable (although not optimum), as demonstrated by further results, and the derivative time of zero turns off the derivative mode. To simplify this first example, the goal in this analysis is temporarily limited to achieving the minimum value of the IAE for the base-case plant model.

Determining the best value of K_c would be made easier if it were possible to determine an analytical expression for the IAE as a function of K_c and apply the following criteria for a minimum:

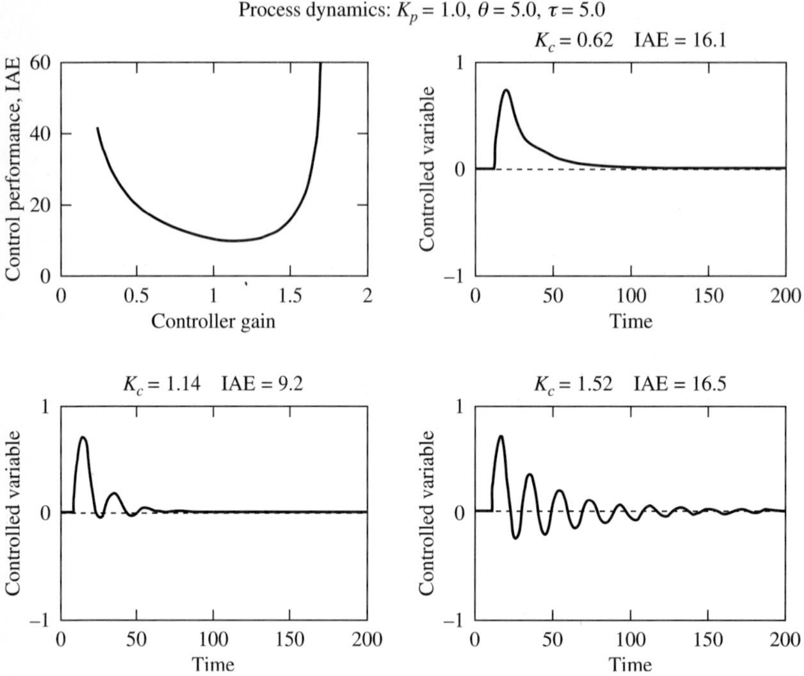

FIGURE 9.2
Relationship used to determine the best controller gain, K_c% open/%A, with T_I and T_D constant.

$$\text{IAE} = F(K_c) \qquad \frac{d(\text{IAE})}{dK_c} = 0 \qquad \frac{d^2(\text{IAE})}{dK_c^2} > 0 \tag{9.4}$$

Since an analytical expression cannot generally be developed for meaningful definitions of control performance, the procedure used here employs numerical methods to determine the value of K_c that satisfies equations (9.4) to a very close approximation. To determine the best controller gain, many values of the gain could be selected, and the value of IAE for each could be evaluated through *simulation;* that is, numerical solution of the differential equations that define the system. Although the IAE is defined for time from zero to infinity, the simulation is limited to a final time that is long enough for the controlled variable to return to its set point at steady state. The simulation employs a numerical method to solve the differential and algebraic equations with a small step size as described in Section 3.5; as a result, the numerical values are very close to the continuous system response.

The results of several transient responses are presented in Figure 9.2, with each case having a different value of the controller gain and constant values for the integral ($T_I = 10$) and derivative times ($T_d = 0$). The results show that the relationship between IAE and K_c is *unimodal;* that is, it has a single minimum that satisfies equations (9.4). The minimum IAE is at a controller gain value of about $K_c = 1.14\%/(\text{mole/m}^3)$ with an IAE of 9.1. For values of the controller gain smaller than the best value (e.g., $K_c = 0.62$), the controller is too "slow," leading to higher

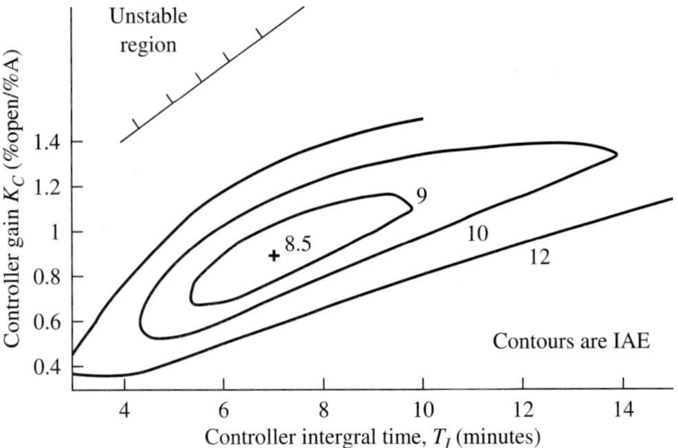

FIGURE 9.3
Contours of controller performance, IAE, for values of controller gain and integral time.

IAE. For values of the controller gain larger than the best value (e.g., $K_c = 1.52$), the controller is too "aggressive," leading to oscillations and higher IAE. Note that the optimum is somewhat "flat"; that is, the control performance does not change very much for a range (about $\pm 15\%$) about the optimum controller gain. However, if the controller gain is increased too much, the system will become unstable. (Determining the stability limit is addressed in the next chapter.)

The graphical presentation used for one constant can be extended to two constants by varying the controller gain and integral time simultaneously while holding the derivative time constant ($T_d = 0$). Again, many dynamic responses can be evaluated and the results plotted. In this case, the coordinates are the controller gain and integral time, with the IAE plotted as contours. The results are presented in Figure 9.3, where the optimum tuning is $K_c = 0.89$ and $T_I = 7.0$. Again, the same qualitative behavior is obtained, with very large or small values of either constant giving poor control performance. In addition, the contours show the interaction between the variables; for example, nearly the same control performance can be achieved by gain and integral time values of ($K_c = 0.6$ and $T_I = 4.5$) and ($K_c = 1.2$ and $T_I = 10$), respectively. Again, the control performance is not too sensitive to the tuning values, as shown by the large region (valley) in which the performance changes by only about 10 percent. Finally, the evaluations identified a region in which the control system is not stable; that is, where the IAE becomes infinite. It is interesting that the region of good control performance—the lower valley in the contour plot—runs nearly parallel to the stability bound. This result will be used in the next chapter, in which the stability of control systems is studied and tuning constant values are determined based on a margin from the stability bound.

When three or more values are optimized, as is the case for a three-mode controller, the results cannot be displayed graphically. One could take the same optimization procedure described for one- and two-variable problems, which is simply to evaluate the IAE over a grid of tuning constant values and estimate the best values from the results. This method is very wasteful, because it does not use

TABLE 9.2
Summary of tuning study

Case	Objective	Gain, K_c (%/%A)	Integral time, T_I (min)	Derivative time, T_d (min)	IAE[+]
Optimize K_c	Goal 1 (IAE)	1.14	10.0 (fixed)	0.0 (fixed)	9.2
Optimize K_c and T_I	Goal 1 (IAE)	0.89	7.0	0.0 (fixed)	8.5
Optimize K_c, T_I, and T_d	Goal 1 (IAE)	1.04	5.3	2.1	5.8
Optimize K_c, T_I, and T_d	Goal 1–3 simultaneously	0.88	6.4	0.82	7.4*

[+] Evaluated for nominal model (without error) without noise. Process parameters were the gain $K_p = 1.0\%\text{A}/\%$, the time constant $\tau = 5$ minutes, and the dead time $\theta = 5$ minutes.

* Greater than 5.8 because of additional goals 2 and 3.

intermediate results to concentrate on the most likely area of the best values; therefore, it would take an excessive number of evaluations of the IAE to determine the best tuning constant values accurately. A preferred method determines the gradient of the IAE with respect to the tuning constants and selects continually better values by moving in the direction of decreasing IAE. Details of the simulation approach and the optimization method are presented in Appendix E.

The application of an optimization to the example process yields values of all three parameters that minimize IAE, and the values are reported in Table 9.2. This table summarizes the results with one, two, and all three constants being optimized; clearly, as more constants are free for adjustment, the IAE controller performance measure improves (i.e., decreases). Also, the optimum values for the controller gain and integral time change when we include the derivative time as an adjustable variable in the optimization. This result again demonstrates the interaction among the tuning constants.

Minimizing the IAE is only the first of the three specified goals, which considers the behavior of only the controlled variable and assumes perfect knowledge (model) of the process. This preliminary result does *not* provide the best control performance according to our specified goals; therefore, we must continue to refine the procedure to determine the best tuning constant values.

Goal 2: Good Control Performance with Model Errors

To this point we have determined tuning constant values that minimize the IAE when the process dynamics are described *exactly* by the base-case dynamic model. However, the model is never perfect, because of errors in the model identification procedure, as demonstrated in Chapter 6. Also, plant operating conditions, such as production rate, feed composition, and purity level, change, and because processes are nonlinear, these changes affect the dynamic behavior of the feedback process. The effect of changing operating conditions can be estimated by evaluating the linearized models at different conditions and determining the changes in gain, time constant, and dead time from their base-case values. Since the true process dynamic behavior changes, a useful tuning procedure should determine tuning constants that give good performance for a range of process dynamics about the base-case or nominal model parameters, as required by the second control performance goal. When the tuning results in satisfactory performance for a reasonable range of process dynamics, the tuning is said to provide robustness.

In performing control and tuning analyses, the engineer must define the expected model error. The error estimate, usually expressed as ranges of parameters, can be based on the variation in plant operation and fundamental models from Chapters 3 through 5 or the results of several empirical model identifications using the methods in Chapter 6.

The size and type of model error is process-specific. For the purposes of developing correlations, the major source of variation in process dynamics is assumed to result from changes in the flow rate of the feed stream F_B in Figure 9.1 that cause $\pm 25\%$ changes in the parameters. While the range of parameters depends on the specific process, most processes experience parameter value changes of roughly this magnitude, and some have much larger variations. The resulting model parameters are given in Table 9.3; these values can be derived using the expressions already given relating the linearized model parameters to the process design and operation. Since in this example all parameters are proportional to the inverse of the feed flow, the parameters do not vary independently but in a *correlated* manner as a result of changes in input variables. Such correlation among parameter variation is typical, because the major cause of variation in process dynamics is nonlinearity, although the functional relationship depends on the process and is not always as shown in the table.

The goal is to provide good control performance for this range, and one way to consider the variability in dynamics is to modify the objective function to be the

TABLE 9.3
Model parameters for the three-tank process

Model parameters	Low flow, $i = 1$	Base case flow, $i = 2$	High flow, $i = 3$
K_p	1.25	1.0	0.75
θ	6.25	5	3.75
τ	6.25	5	3.75

sum of the IAE for the three cases, which include the base-case and the extremes of low and high flow rates in Table 9.3. The objective is stated as follows:

$$\text{Minimize} \qquad \sum_{i=1}^{3} \text{IAE}_i \tag{9.5}$$

$$\text{by adjusting} \qquad K_c, T_I, T_d$$

$$\text{IAE}_i = \int_0^\infty |\text{SP}(t) - \text{CV}_i(t)|dt$$

where $\text{CV}_i(t)$ is calculated using process parameters for $i = 1$ to 3.

This modification is very important, because tuning constants that yield good performance for the nominal model may give poor performance or even result in instability as the true process parameters vary. Next, the third goal is discussed; afterwards, the tuning constants satisfying all three goals are determined.

Goal 3: Manipulated-Variable Behavior

The third and final goal addresses the dynamic behavior of the manipulated variable by requiring it to observe a limitation. As previously discussed, its variation should not be too great, because of wear to control and process equipment and of disturbances to integrated units. There are many ways to define the variation of the manipulated variable. Here we will bound the allowed transient path of the manipulated variable to a specified region around the final steady-state value during the dynamic response as shown in Figure 9.4. This rather general limitation enables us to address two related issues in manipulated-variable variation:

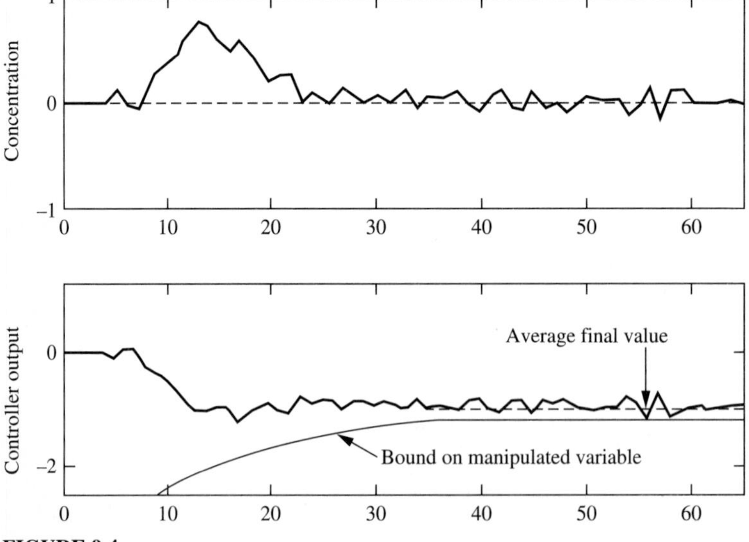

FIGURE 9.4
Dynamic response of a feedback control system showing the bound on allowable manipulated-variable adjustments.

1. The largest-magnitude variation in the manipulated variable in response to an upset
2. The high-frequency variation resulting from the small, continuous changes in the controlled variable often referred to as *noise*

The allowable manipulated-variable range is large during the initial part of the transient, where, in general, the manipulated variable should be able to overshoot its final value. The range is smaller after the effect of the step disturbance is corrected. Even after a long time, the manipulated variable cannot be required to be absolutely constant, because feedback control responds to the small, continuous changes in the controlled variable (i.e., the noise). The limitation on the manipulated variable is determined by parameters that define the bound shown in Figure 9.4. Simulations to evaluate a tuning for goals 1 through 3 include representative noise on the measured, controlled variable and a bound on the manipulated variable. A model for defining the bound on the path, along with parameters used in this book, is presented in Appendix E.

The proper values of the parameters used to define the allowed manipulated variable behavior should match the process application. The values in this study are good initial estimates for many process control designs. However, the specific parameter values are not the key concept in this goal statement; what is most important is this:

A properly defined statement of control performance includes a specification of acceptable manipulated-variable behavior.

Since both controlled- and manipulated-variable behaviors are important, most closed-loop transient responses in this book show both the controlled and manipulated variables; in general, it is not possible to evaluate control performance by observing only the controlled variable.

The controller constants in the example mixing process are optimized for the complete definition, and the results are $K_c = 0.88$, $T_I = 6.4$, and $T_d = 0.82$. The dynamic response is given in Figure 9.4 for the nominal plant response. (Recall that three dynamic responses, including model error, were considered concurrently in determining the optimum.) These tuning parameters satisfy goals 1 through 3 in our control performance definition. Note that compared to the results reported in Table 9.2, which satisfy only goal 1, the values satisfying all three goals have a lower gain, longer integral time, and shorter derivative time. Thus:

The controller is *detuned,* leading to less aggressive adjustments by the feedback controller, to account for modelling errors and to reduce the variation in the manipulated variable.

These tuning constants will not perform best when the model error is zero and no noise is present, but they will perform better over an expected range of conditions and are the values recommended for initial application.

Example 9.1. A modified process in Figure 9.1, with a shorter pipe and larger tank described by the nominal model in equation (9.6), is to be controlled by a PID controller. Determine the best initial tuning constant values for a PID controller based on (*a*) goal 1 alone and (*b*) goals 1 through 3.

$$G_v(s)G'_p(s)G_s(s) \approx G_p(s) = \frac{1.0e^{-2s}}{8s+1} \tag{9.6}$$

$$G_d(s) = \frac{1}{8s+1} \qquad \text{with } D(s) = \frac{1}{s}$$

$$G_c(s) = K_c\left(E(s) + \frac{E(s)}{T_I s} + T_d s\,\text{CV}(s)\right)$$

The mathematical optimization must be performed for the two cases. The results of the analysis are given in Table 9.4. The results are similar to the example discussed previously in that the controller gain is decreased, the integral time increased, and the derivative time is decreased—in this example to zero—as the additional goals are added. The net effect of adding goals 2 and 3 is that total deviation of the controlled variable from its set point (IAE) is larger than that achieved for the nominal process without modelling error. However, the performance indicated by the more comprehensive measure, considering all cases and behavior of both the controlled and manipulated variables, is the best possible with a PID control algorithm. Thus, the tuning from case (*b*) is more robust, as will be demonstrated in Example 9.5.

Again we see that there is interaction among the tuning constants. As demonstrated for a simple process in Example 8.5, each tuning constant affects many control performance measures, such as decay ratio and overshoot. Therefore, all tuning constants should be determined simultaneously to obtain the best possible performance within the capability of the PID algorithm.

In conclusion, a very general method has been presented in this section for evaluating controller tuning constants. The method can be applied to any process model and controller algorithm and was applied to the linear, first-order-with-dead-time process and PID controller in this section. The method addresses most control performance issues in a flexible manner, so that the engineer can adapt it to most circumstances by changing a few parameters in the control objective definition, such

TABLE 9.4
Results for Example 9.1

Case	Controller gain, K_c	Integral time, T_I	Derivative time, T_d	IAE+
(*a*) performance goal 1 alone	3.0	3.7	1.1	1.46
(*b*) performance goals 1–3	1.8	5.2	0.0	2.95

+ Evaluated for nominal model (without error) without noise.

as the magnitude of the model errors or the allowable variability of the manipulated variable. However, an optimization must be performed for each individual problem, which could be very time-consuming. Thus, the next section describes how controller tuning can be performed quickly in many situations using correlations developed with the optimization procedure.

9.4 CORRELATIONS FOR TUNING CONSTANTS

The purpose of tuning correlations is to enable the engineer to calculate tuning constants for many process applications that simultaneously achieve the three goals defined in Section 9.2 without performing the optimization. Correlations for tuning constants will reduce the engineering effort in controller tuning, and, perhaps more importantly, the correlations will show how the controller constants depend on feedback process dynamics. For the correlations developed in this section, the tuning goals will be those used in the previous example:

1. Minimize IAE
2. $\pm 25\%$ (correlated) change in the process model parameters
3. Limits on the variation of the manipulated variable

The correlation should provide values for K_c, T_I, and T_d based on values in a process dynamic model. The general approach is to select a model structure and determine the dimensionless parameters that define the closed-loop dynamic response. To provide simple, yet general correlations, the process model must have a small number of parameters. Modelling examples in Chapter 6 demonstrated that many processes can be represented by a first-order-with-dead-time transfer function; therefore, this model structure is used in developing the tuning correlations.

$$G_v(s)G'_p(s)G_s(s) \approx G_p(s) = \frac{K_p e^{-\theta s}}{1 + \tau s} \tag{9.7}$$

Since the control response is determined by the closed-loop transfer function, the form of the correlation is determined from this transfer function:

$$\frac{\mathrm{CV}(s)}{D(s)} = \frac{G_d(s)}{1 + G_c(s)G_p(s)} = \frac{G_d(s)}{1 + K_c\left(1 + \dfrac{1}{T_I s} + T_d s\right)\left(K_p \dfrac{e^{-\theta s}}{1 + \tau s}\right)} \tag{9.8}$$

Every process responds with a different "speed," which can be characterized by the time for a step response to achieve 63% of its final value. (This is the first moment of the impulse response; see Appendix D.) For a first-order-with-dead-time process, this time is $(\theta + \tau)$. Dividing the time by this value "scales" all processes to the same speed, so that one set of general correlations can be developed. The relationships are

$$t' = \frac{t}{\theta + \tau} \qquad s = \frac{s'}{\theta + \tau} \tag{9.9}$$

Substituting the modified Laplace variable for the time-scaled equation gives

$$\frac{CV(s')}{D(s')} = \frac{G_d(s')}{1 + K_c K_p \left(1 + \frac{1}{T_I s'/(\theta + \tau)} + \frac{T_d s'}{\theta + \tau}\right)\left(\frac{e^{-\theta s'/(\theta+\tau)}}{1 + \tau s'/(\theta + \tau)}\right)} \tag{9.10}$$

> The resulting equation has one parameter that characterizes the feedback process dynamics, $\theta/(\theta + \tau)$, which we shall term the *fraction dead time.*

This parameter indicates what fraction of the total time needed for the open-loop process step response to reach 63% of its final value is due to the dead time; it has values from 0.0 to 1.0. For example, the initial process data for Figure 9.1 had $\theta = 5$ and $\tau = 5$; thus, the fraction dead time was 0.5. Note that $\tau/(\theta + \tau)$ is not independent, because $\tau/(\theta + \tau) = 1 - \theta/(\theta + \tau)$.

Analysis of equation (9.10) also demonstrates that the controller tuning constants and process dynamic model parameters appear in the following *dimensionless* forms:

$$\begin{aligned} \text{Gain} &= K_c K_p \\ \text{Integral time} &= T_I/(\theta + \tau) \\ \text{Derivative time} &= T_d/(\theta + \tau) \end{aligned} \tag{9.11}$$

These relationships are consistent with a common-sense interpretation of the feedback controller relationships. The dimensionless gain involves the magnitude of the change in the manipulated variable to correct for an error and should be related to the process gain. Also, proportional mode has no time dependence. The dimensionless integral time and derivative times involve the time-dependent behavior of the controlled variable and should be related to the dynamics or "time scale" of the process.

The disturbance model is assumed to be the same as the feedback process model; that is, $G_d(s) = G_p(s)$. Noise is assumed to be present in the controlled variable, as discussed in Section 9.3 and defined in Appendix E. The resulting transfer function has only one parameter that is entirely a function of the process (i.e., the fraction dead time $\theta/(\theta + \tau)$); the tuning constants, expressed in the dimensionless forms in equation (9.11), also influence the dynamic performance. For the control objectives and process model (with error estimate) defined in Table 9.1, the tuning correlations are developed by (1) selecting various values of the fraction dead time in its possible range of 0 to 1 and (2) optimizing the control performance for each value by adjusting the tuning constants.

The results for the disturbance response are plotted in Figures 9.5*a* through *c*. The correlations indicate that a high controller gain is appropriate when the process has a small fraction dead time and that the controller gain generally decreases as the fraction dead time increases. This makes sense, because processes with longer dead times are more difficult to control; thus, the controller must be detuned. The dimensionless derivative time is zero for small fraction dead time and increases for

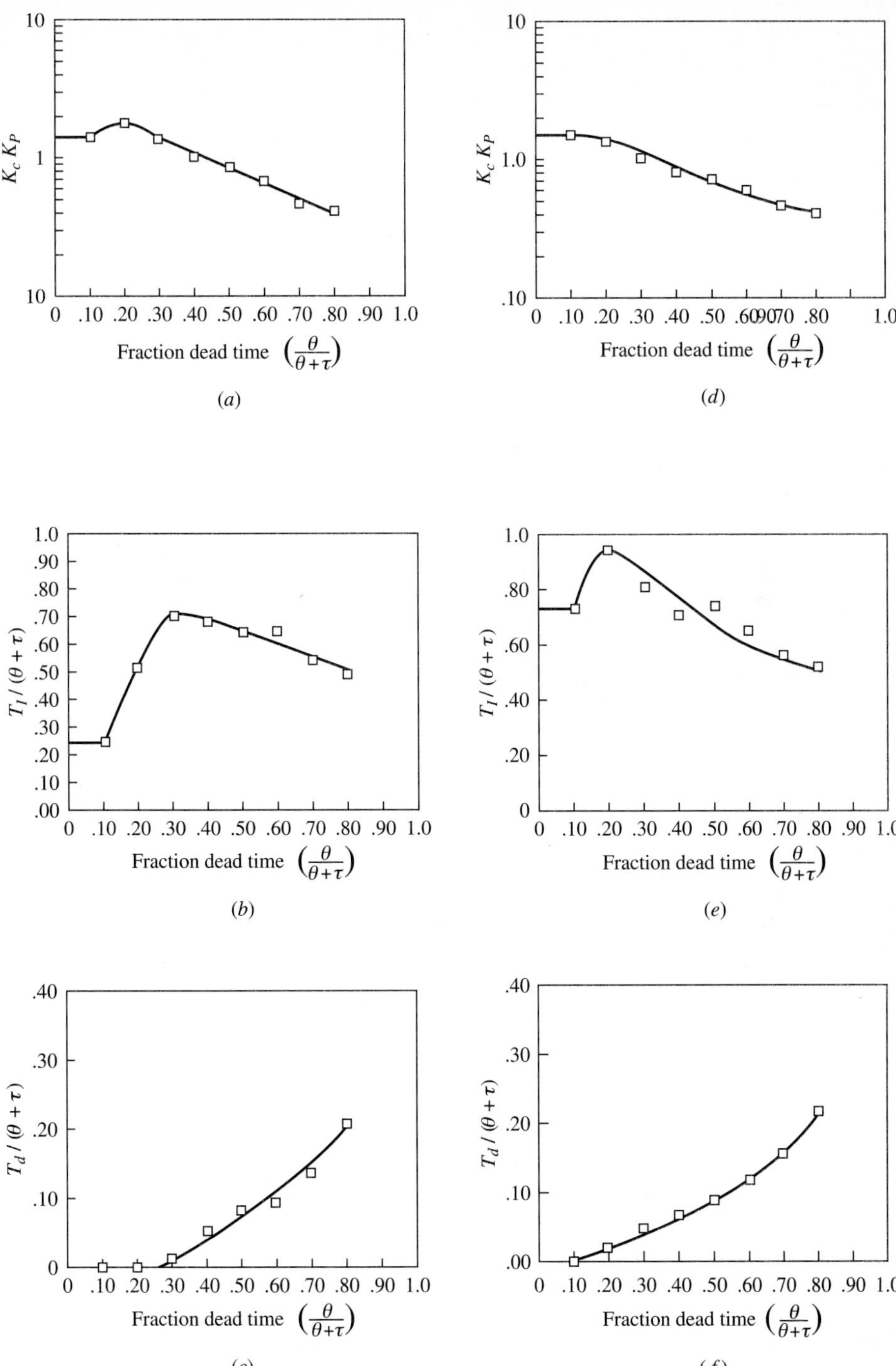

FIGURE 9.5
Ciancone correlations for dimensionless tuning constants, PID algorithm. For disturbance response: (*a*) control system gain, (*b*) integral time, (*c*) derivative time. For set point change: (*d*) gain, (*e*) integral time, (*f*) derivative time.

longer dead times to compensate for the lower controller gain. The dimensionless integral time remains in a small range as the fraction dead time increases.

The same procedure can be performed for the other major input forcing: set point changes. All of the assumptions and equation simplifications are the same, and the set point is assumed to change in a step. The resulting correlations are presented in Figures 9.5*d* through *f*. The tuning constants have the same general trends as the fraction dead time increases. The selection of whether to use the disturbance or set point correlations depends on the dominant input variation experienced by the control system.

The range of model errors, $\pm 25\%$, is reasonable when all parameters are significantly different from zero. However, when this percentage error is used, a very small dynamic parameter would also have a very small associated error, which may not be realistic. Because an underestimation of the error would generally lead to a controller that is too aggressive, and because the controller for $\theta/(\theta + \tau) = 0.10$ is already quite aggressive, the tuning correlations are not extended lower than 0.10, and the recommended tuning constant values are shown by the lines maintaining the constant values for $\theta/(\theta + \tau)$ from 0.10 to 0. These values can be improved through fine tuning, if required, as described later in this chapter.

The tuning correlations presented in this section were developed by Ciancone and Marlin (1992) and will be referred to subsequently as the *Ciancone correlations*. The use of the correlations is demonstrated in the following examples.

Example 9.2. Determine the tuning constants for a feedback PID controller applied to the three-tank mixing process for a disturbance response using the Ciancone tuning correlations.

The first step is to fit a first-order-with-dead-time model to the process, which was done using the process reaction curve method in Example 6.4 (page 207). The results were $K_p = 0.039$ %A/% valve opening; $\theta = 5.5$ min; and $\tau = 10.5$ min. Then, the independent parameter is calculated as $\theta/(\theta+\tau) = 0.34$. The dependent variables are determined from Figure 9.5*a* through *c*, and subsequent tuning constants are calculated as follows:

$$K_c K_p = 1.2 \qquad K_c = 1.2/.039 = 30\% \text{ open}/\%\text{A}$$
$$T_I/(\theta + \tau) = 0.69 \qquad T_I = 0.69(16) = 11 \text{ min}$$
$$T_d/(\theta + \tau) = 0.05 \qquad T_d = 0.05(16) = 0.8 \text{ min}$$

The dynamic response of the feedback system to a step feed composition disturbance of magnitude 0.80%A occurring at $t = 20$ is given in Figure 9.6, which results in an IAE of 7.4. The dynamic response is "well behaved"; that is, the controlled variable returns to its set point reasonably quickly without excessive oscillations, and the manipulated variable does not experience excessive variation.

The result in Example 9.2 shows that the correlations, which were developed for first-order-with-dead-time plants, provide reasonable tuning for plants with other structures as long as the feedback dynamics can be approximated well with a first-order-with-dead-time model.

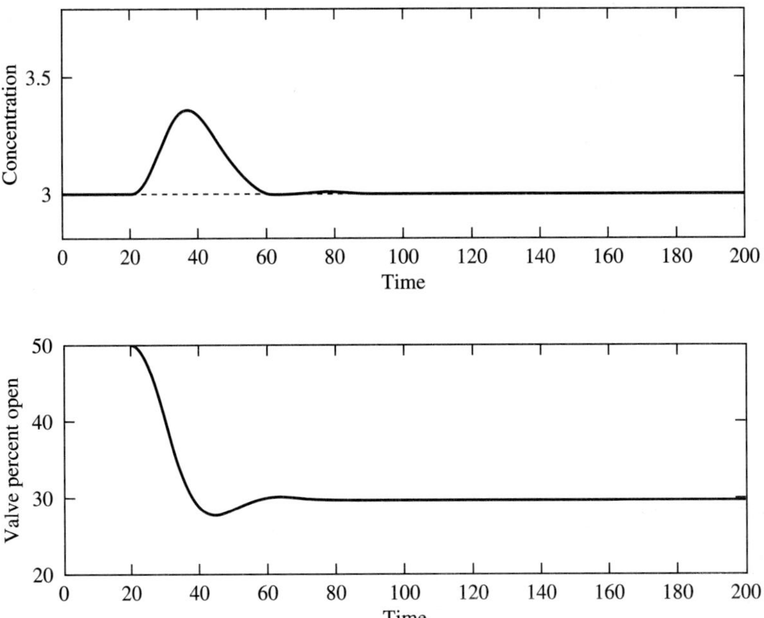

FIGURE 9.6
Dynamic response of three-tank process and PID controller with tuning from Example 9.2.

Example 9.3. When developing the correlations, the assumption was made that the disturbance transfer function was the same as the process feedback transfer function. Evaluate the tuning correlations for the same three-tank system considered in Example 9.2 with a different disturbance time constant.

Original disturbance transfer function:

$$G_d(s) = \frac{1}{(5s+1)^3}$$

Altered disturbance transfer function:

$$G_d(s) = \frac{1}{(5s+1)}$$

The altered transfer function would occur if the disturbance entered in the last tank of the three. The resulting transient of the system under closed-loop control is plotted in Figure 9.7. As would be expected, the response is different, with the faster disturbance resulting in poorer control with respect to the maximum deviation and IAE, which increased to 8.3. The slightly poorer control performance is the result of a more difficult process, due to the faster disturbance, being controlled. Note that the correlation tuning constants give reasonably good, although not "optimal," performance even when the disturbance transfer function differs significantly from the feedback transfer function.

Example 9.4. The correlations have been developed assuming that the process is linear, and it has accounted for changes in the process dynamics through the range of model error considered. In this example a process is considered in which the nonlinearities

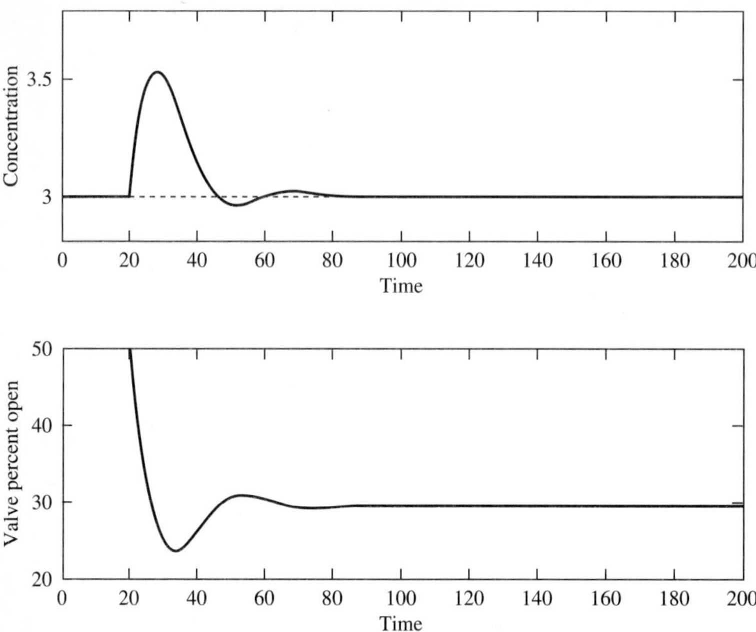

FIGURE 9.7
Dynamic response of three-tank mixing process with faster disturbance dynamics from Example 9.3.

influence the dynamics during the transient response. The three-tank mixer described in Example 7.2 (page 249) is nonlinear if the flow of stream *B* changes, as seen by the fact that the time constants and gain in the linearized model depend on F_B. Determine the tuning and dynamic response for the situation in which F_B changes from its base value of 6.9 m^3/min to 5.2 m^3/min and returns to its base value.

The tuning for the initial condition has been determined in Example 9.2. Before evaluating the dynamic response, it is worthwhile determining the change in the process dynamics, resulting from the change in F_B, which is summarized here for the models linearized about the base and disturbed steady states:

	Base case	Disturbed	
K_p eq.(7.20)	0.039	$0.0028(5.2)(100 - 1)/(5.2 + .14)^2 = 0.051$	%A/%open
τ eq.(7.21)	5.0	$35/(5.2 + 0.14) = 6.6$	min

The process model changes during the transient, and it would be proper to correct the tuning. However, it is not possible to change the tuning for all disturbances, many of which are not measured; thus, the base-case tuning is used during the entire transient in this example. The results are plotted in Figure 9.8. Note that the first transient in response to a decrease in flow experiences rather oscillatory behavior; this is because the process dynamics are slower because of the change in dynamics, and consequently the tuning is too aggressive. When returning to the base-case, the tuning is only slightly underdamped, because the conditions are close to the dynamics for which the tuning constants were determined. Even for this significant change in process dynamics, the PID algorithm with tuning from the Ciancone correlations provides acceptable

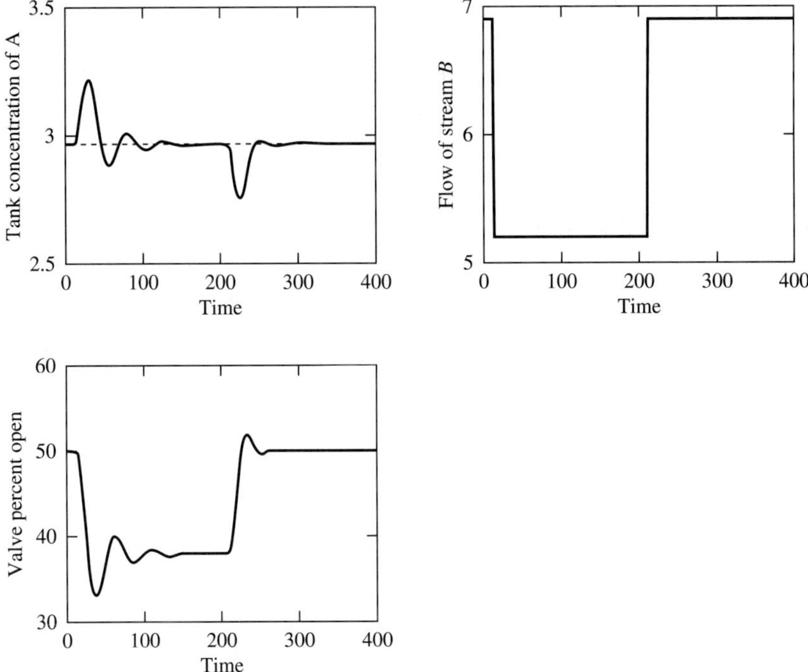

FIGURE 9.8
Dynamic response for Example 9.4.

performance. Thus, the system is robust to disturbances of the magnitude considered in this example. However, larger changes in process operation would result in larger model variation and could seriously degrade performance or even cause instability; therefore, a method for automatically modifying tuning constants based on measured process variables is presented in Section 16.3.

The results of the tuning studies lead to two important observations concerning the effects of process dynamics on tuning. First, the controller should be detuned; that is, the feedback adjustments should be reduced as the fraction dead time of the feedback process increases. Thus, we conclude that dead time in the feedback loop results in reduced or slower feedback adjustments and, presumably, poorer control. Theoretical justification for this result is presented in Chapter 10, and the effect on feedback performance is confirmed in Chapter 13. The second observation is that two models, the feedback process $G_p(s)$ and the disturbance process $G_d(s)$, both affect the tuning; this is determined by comparing the results for a process disturbance, which enters through a first-order time constant, with those for a set point change, which is a perfect step. However, the major influence on tuning is normally from the *feedback dynamics,* and again, theoretical justification for this result will be presented in the next chapter. Other studies by Hill et al. (1987) showed that the tuning is insensitive to the disturbance time constant when $\tau_d > \tau$; thus, the differences between Figures 9.5*a* through *c* and 9.5*d* through *f* typically represent the maximum change in tuning in response to different disturbance types.

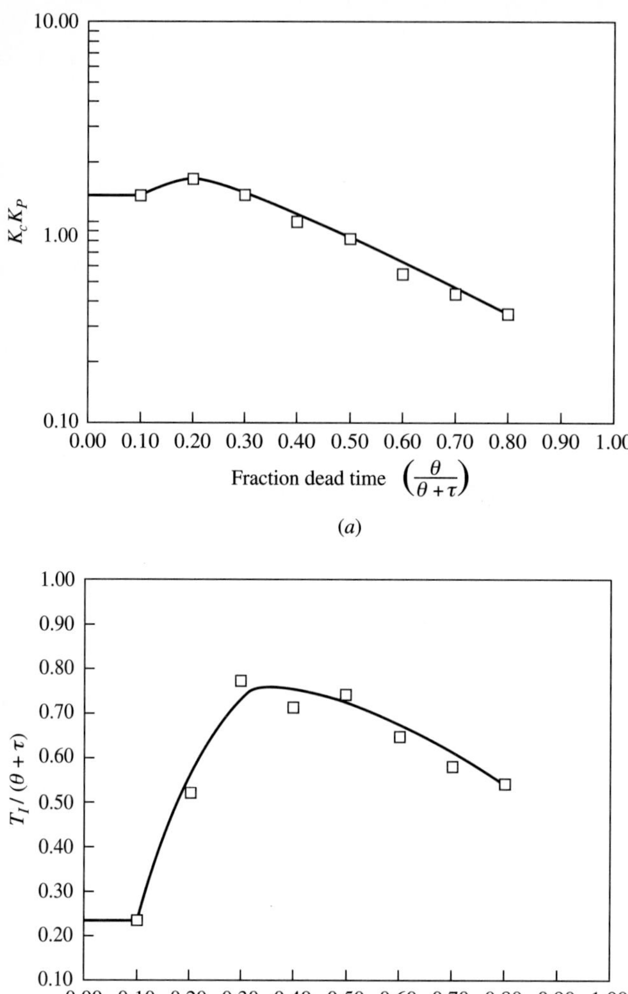

FIGURE 9.9
Ciancone correlations, PI disturbance response: (*a*) dimensionless gain, (*b*) dimensionless integral time.

In many control applications the derivative mode is not employed. This is the case if the measurement signal has considerable noise. Also, the tuning correlations demonstrate that the derivative time is very small when the fraction dead time is small. Thus, tuning correlations for a proportional-integral (PI) controller are provided in Figures 9.9*a* and *b* for a disturbance response. Note that it would not be correct to use the PID values and simply set the derivative time T_d to zero, because of the interaction between the tuning constant values, although the correlations in

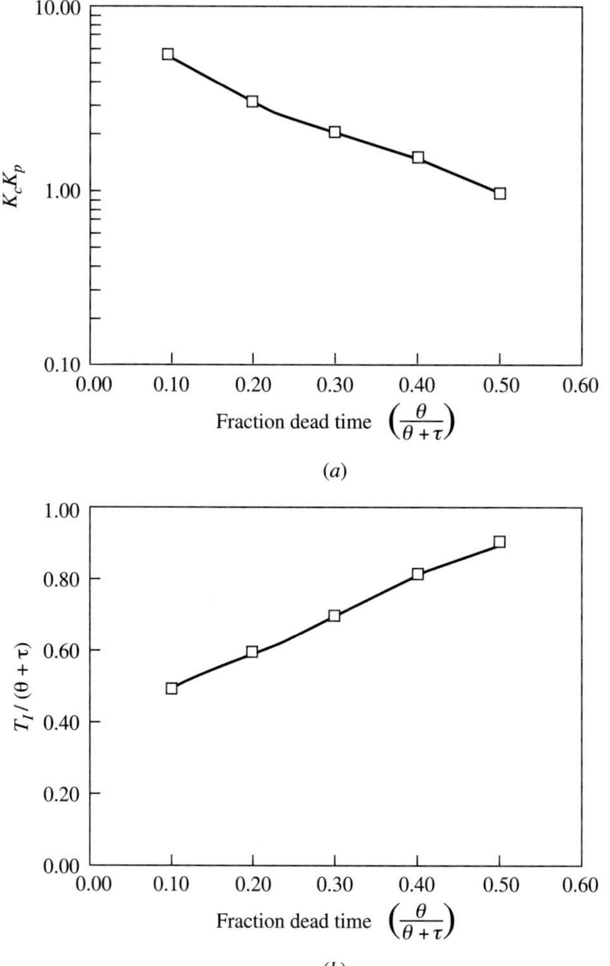

FIGURE 9.10
Lopez et al. (1969) tuning correlations for minimizing the IAE for a PI controller in response to a disturbance.

Figure 9.9 are close to those in Figure 9.5 because of the small values of the derivative time in Figure 9.5.

The tuning correlations presented in Figures 9.5 and 9.9 depend on the goals specified for the control performance. It is interesting to compare the results to a different set of goals. One of the earlier studies using an optimization procedure was performed by Lopez et al. (1969). In their study the goal was simply to minimize the IAE (our goal 1), without concern for potential variation in feedback dynamics or limitations on manipulated-variable transient behavior. Their results are presented in Figure 9.10*a* and *b* and are applied in the following example.

Example 9.5. The altered mixing process in Figure 9.1, with the transfer function given below, is to be controlled with a PI controller. Calculate the tuning constants according to correlations in Figure 9.9 and 9.10 using the nominal model given below. Calculate the transient responses to a step disturbance of 2%A in feed composition at $t = 7$ for (*a*) the nominal feedback process and (*b*) an altered plant as defined below. Note that the nominal and actual plants have the same steady-state gain and "speed of response," as measured by the time to reach 63% of their steady-state value to a step change input; they differ only in their fraction dead time.

Nominal plant:

$$G_p(s) = \frac{2.0e^{-2s}}{8s + 1}$$

$$G_d(s) = \frac{1.0}{8s + 1}$$

$$\theta + \tau = 10$$

$$\frac{\theta}{\theta + \tau} = 0.2$$

Altered plant:

$$G_p(s) = \frac{2.0e^{-3s}}{7s + 1}$$

$$G_d(s) = \frac{1.0}{7s + 1}$$

$$\theta + \tau = 10$$

$$\frac{\theta}{\theta + \tau} = 0.3$$

The tuning constant values can be calculated for each correlation from the charts using the nominal model as

	Ciancone	**Lopez**	
K_c	0.9	1.5	% open/%A
T_I	5.2	6.0	min

The closed-loop dynamic responses are given in Figures 9.11*a* through *d* (see pages 310 and 311), and the control performance measure of IAE (integrated from 0 to 100) is summarized as

	Ciancone	**Lopez**
IAE for nominal plant	5.9	4.0
IAE for altered plant	7.6	14.5

These results should be anticipated from the control objectives used to derive the correlations. The Lopez correlation minimized IAE without consideration for model error. Thus, it performs best when the plant model is known perfectly, but it is unacceptably oscillatory and tends toward instability for even the modest model error considered in this example. The Ciancone correlations determined the tuning to perform well over a range of process dynamics; thus, the performance does not degrade as rapidly with model error.

The results of this section show that simple PID tuning correlations can be developed for processes that can be approximated by a first-order-with-dead-time model. Selection of the proper correlation depends on the control performance goals. If the situation indicates that very accurate knowledge of the process is available and there is no concern for the manipulated-variable variation, the best performance (i.e., lowest IAE of the controlled variable with PI feedback) is obtained using the Lopez correlations; however, the control system with these tuning constants will not perform well if the process model has significant error or if the measurement has significant noise. As the control performance goals are defined more realistically for typical plant situations, the resulting tuning allows for more modelling error and for some limitation on the manipulated-variable variation, and the resulting correlations have a broader range of good performance. This is an important factor for control systems that function continuously for months or years as plant conditions change. Thus, the Ciancone correlations are recommended here as a starting point for most control systems.

> Tuning correlations have been developed as a function of fraction dead time for a PID controller, a first-order-with-dead-time process, and typical control objectives. These are recommended for obtaining initial tuning constant values when the plant situation matches the development assumptions in Table 9.1.

It is important to recognize that no claim is made for optimality in the real world, although an optimization method was used to determine the solution to the mathematical problem. The Ciancone correlations simply used a realistic definition of control performance to determine tuning. Also, while examples have shown that the correlations are valid for different disturbance model parameters and model errors, extrapolation beyond the defined conditions of the correlation (Table 9.1) must be done with care.

9.5 FINE-TUNING THE CONTROLLER TUNING CONSTANTS

The tuning constants calculated according to any method—optimization, correlations, or the stability analysis in the next chapter—should be considered to be *initial* values. These values can be applied to the process to obtain empirical information on closed-loop performance and modified until acceptable control performance is obtained. Determining modifications based on initial dynamic responses, often termed *fine tuning,* is necessary because of errors in the base-case process model and simplifications in the tuning method. A fine-tuning method is described here for a process being controlled by a PI control algorithm. This method is easy to perform and gives additional insight into the way the controller modes combine when controlling a process.

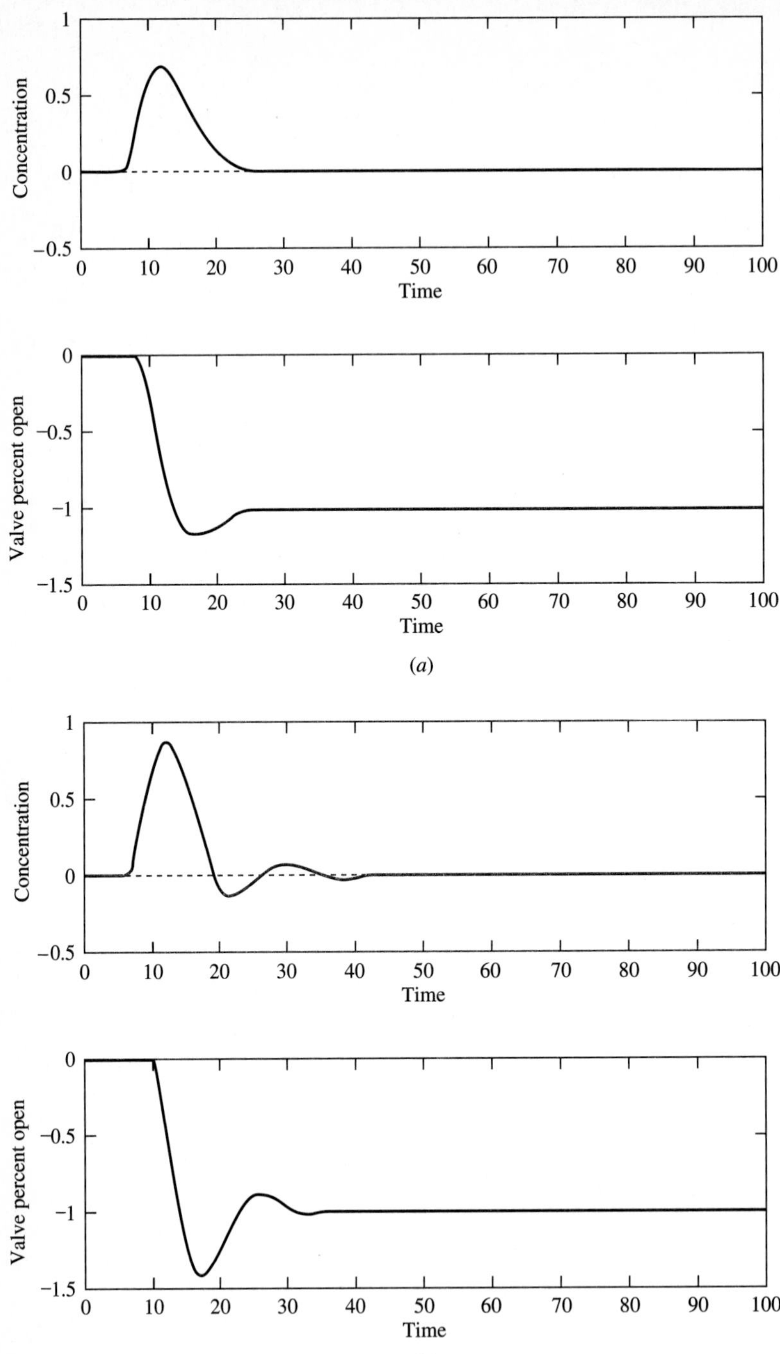

FIGURE 9.11
Dynamic responses. With Ciancone tuning: (*a*) nominal plant, (*b*) altered plant.

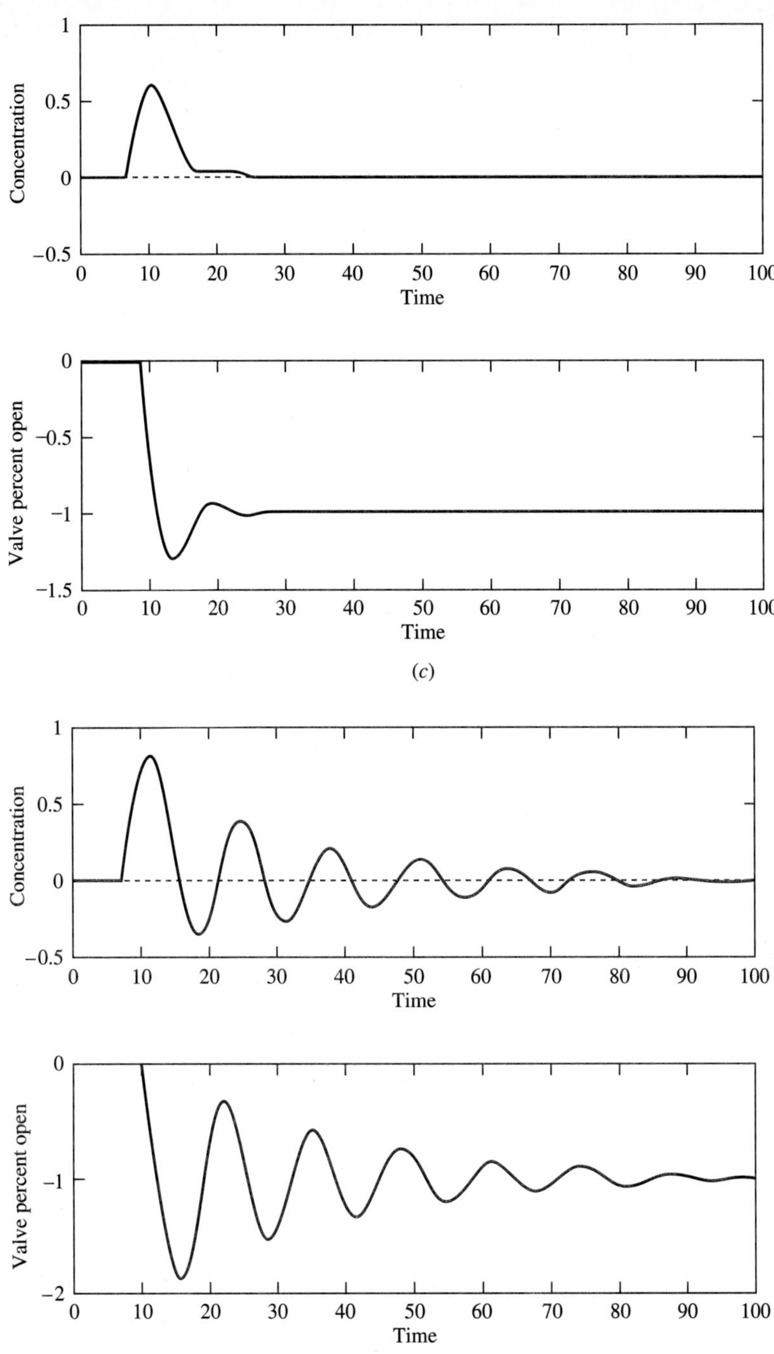

FIGURE 9.11 ***(continued)***
With Lopez tuning: (*c*) nominal plant, (*d*) altered plant.

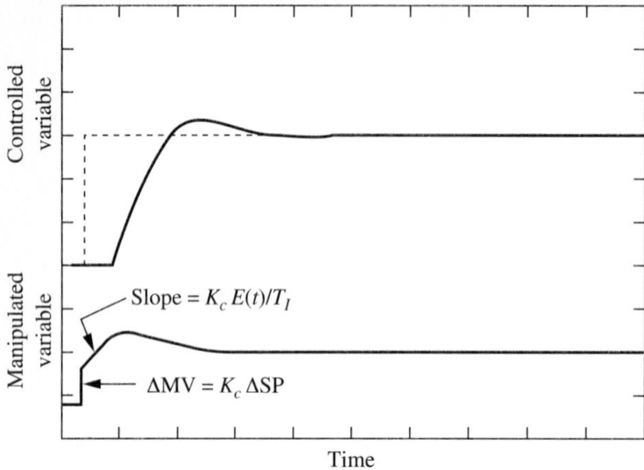

FIGURE 9.12
Typical set point response of a well-tuned PI control system.

After the initial tuning constants have been calculated and entered into the algorithm, the controller's status switch can be placed in the automatic position to allow the controller to perform its calculation and adjust the final element. Then, the response to a set point change is diagnosed to determine whether the tuning is satisfactory. A set point change is considered here because

1. It can be introduced when the diagnosis is performed.
2. A simple time-dependent input disturbance—a step—is easy to achieve.
3. The magnitude can be selected by the engineer.

Also, the effects of the proportional and integral modes on the initial part of the transient can be determined separately for a set point change; this greatly simplifies the fine-tuning analysis.

The step response of a control system with a well-tuned PI controller is given in Figure 9.12. The first important feature is the immediate change in the manipulated variable when the set point is changed. This is due to the proportional mode and is equal to $K_c \Delta E(t)$, which is equal to $K_c \Delta \mathrm{SP}(t)$. This initial change is typically 50 to 150% of the change at the final steady state. The second feature is the delay, due to the dead time, between when the set point is changed and when the controlled variable initially responds. No controller can reduce this delay to be less than the dead time. During the delay the error is constant, so that the proportional term does not change, and the magnitude of the integral term increases linearly in proportion to $K_c E(t)/T_I$. When the controlled variable begins to respond, the proportional term decreases, while the integral term continues to increase. At the end of the transient response the proportional term, being proportional to error, is zero, and the integral term has adjusted the manipulated variable to a value that reduces offset to zero.

The value of this interpretation can be seen when an improperly tuned controller, giving the response in Figure 9.13, is considered. The control response seems slow, resulting in a large IAE and long time to return to the set point. Analysis of

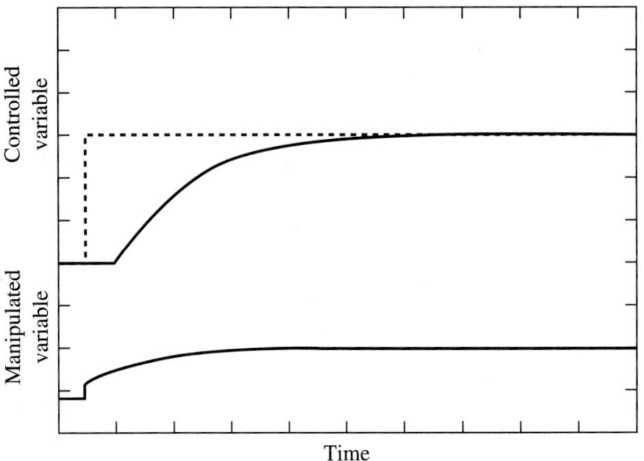

FIGURE 9.13
Example of a dynamic response of a PI control system with the controller gain too small.

the transient indicates that the initial change in the manipulated variable when the set point is changed, termed the proportional "kick," is only about 30% of the final value, which indicates too small a value for the controller gain. The conclusion for the diagnosis is that the control system performance can be improved by increasing the controller gain, most likely in several moderate steps, with a plant test at each step to monitor the results of the changes. The substantially improved performance of the control system with the controller gain increased by a factor of 2.5 is shown in Figure 9.12.

Example 9.6. A PI controller was not providing acceptable control performance. Preliminary analysis indicated that the sensor and control valve were functioning properly, so a step change was introduced to its set point. The response is given in Figure 9.14. Diagnose the performance, and suggest corrective action.

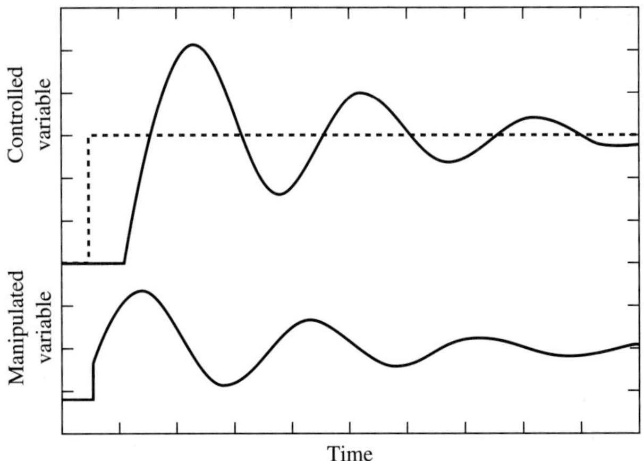

FIGURE 9.14
Dynamic response of the control system in Example 9.6.

Solution: The transient response is highly oscillatory, indicating a controller that is too aggressive. The cause could be too large a controller gain, too short an integral time, or both. The immediate proportional change is only about 70% of the final change in the manipulated variable; therefore, the controller gain is in a reasonable range, is certainly not too large, and should not cause oscillatory behavior. The conclusion is that the integral time is too short. The transient response with double the integral time is that shown in Figure 9.12, confirming that reasonably good control performance can be achieved by changing only the integral time.

Example 9.7. The three-tank mixing control system has been tuned initially, and the system's dynamic response to a set point change is given in Figure 9.15*a*. Note that

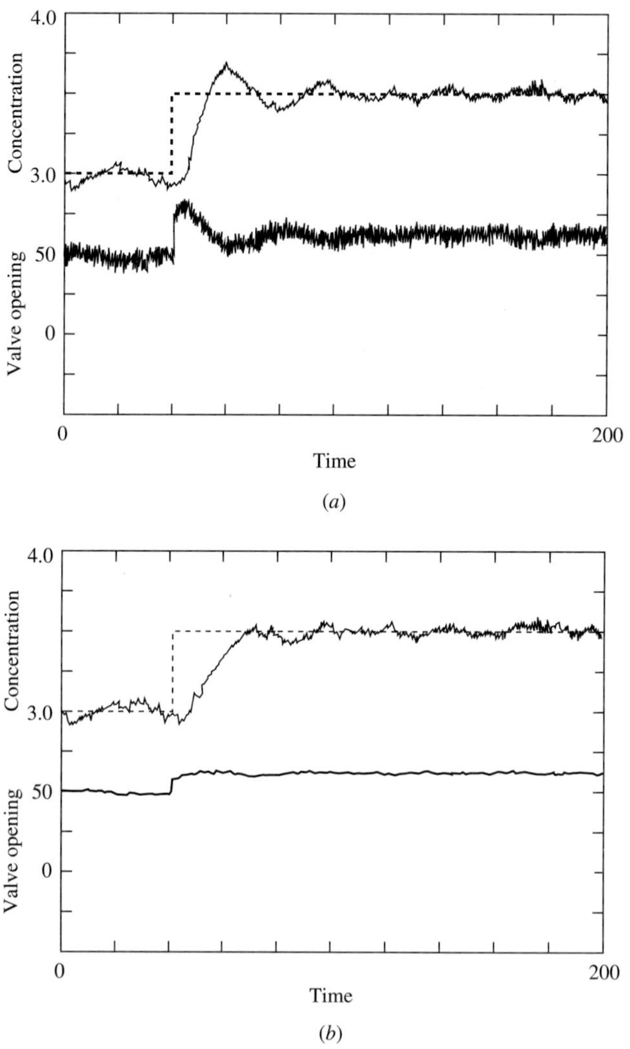

FIGURE 9.15
Dynamic responses of feedback control system in Example 9.7: (*a*) initial (IAE = 11.6); (*b*) after fine tuning (IAE = 12.9).

the measured concentration experiences many small disturbances because of changing inlet concentrations and flows in the process as well as measurement error. This noisy data more closely represents empirical data from process plants than do the ideal simulations in Figures 9.12 through 9.14. The control objectives have two unique aspects in this example, which are different from the general objectives considered so far but are not unusual in the process industries.

1. The downstream process is sensitive to oscillations in the concentration. Therefore, the controlled concentration should not experience overshoot.
2. The plant that supplies component A functions better with a smooth operation. Therefore, high-frequency variation in the manipulated variable is to be minimized.

The initial tuning constants are $K_c = 45\%$ opening/%A, $T_I = 11.0$ minutes, and $T_D = 0.8$ minutes. Suggest changes to the tuning constant values that will improve the performance.

Solution: The large, high-frequency variation in the manipulated variable is caused to a large extent by the noisy measurement and the derivative mode. Therefore, the first suggestion would be to reduce the derivative time to zero. Next, the controlled variable overshoots its set point, which can be prevented by making the controller feedback action less aggressive. Reducing the controller gain will slow the response and also slightly reduce the high-frequency variation of the manipulated variable, both desirable effects. The resulting tuning constants, which could be arrived at after several trials, are $K_c = 15, T_I = 11$, and $T_d = 0$. A much more satisfactory dynamic response—that is, one that more closely satisfies the stated objectives for this example—was obtained with these tuning constants, as shown in Figure 9.15*b*. Note that the much smoother performance was achieved with only a small increase in IAE, which changed from 11.6 to 12.9.

These fine-tuning examples demonstrate that

> Analysis of the responses of the controlled *and manipulated* variables to a step change in the set point provides valuable diagnostic information on the causes of good and poor control performance, allowing the performance to be tailored to unique control objectives.

Again, we see that both the controlled and manipulated variables must be observed when analyzing the performance of feedback control systems; complete diagnosis is not possible without information on both variables.

9.6 CONCLUSIONS

The starting point for feedback control consists of the control objectives, here specified as three goals. These goals encompass the major factors in process control performance; the specific parameters used (e.g., percent model error and limits on manipulated-variable variation) can be selected for a specific problem.

> Control performance must be defined with respect to all important plant operating goals. In particular, desired behavior of the controlled and manipulated variables must be defined for expected disturbances, model errors, and noisy measurements.

A simple variable reduction of the closed-loop transfer function, based on dimensional analysis, can be employed in extending the optimization to general tuning correlations. These correlations are applicable only to those systems for which the underlying assumptions are valid: The process should be well represented by a first-order-with-dead-time model, the model errors should be in the assumed range, and the desired controlled and manipulated behavior should be similar to the objectives stated in Table 9.1. Examples have demonstrated that the process does not have to be perfectly first-order with dead time to achieve acceptable dynamic responses using the tuning correlations.

A three-step tuning procedure would combine methods in previous chapters with methods in this chapter. The first step would be to determine the feedback process model $G'_p(s)G_v(s)G_s(s)$ by fundamental modelling or empirical modelling, using either the process reaction curve or a statistical identification method. Industrial controls are most often based on empirical models developed by one of the last two methods. In the second step, the initial tuning constant values would be determined; typically the values would be determined from the general correlations, but an optimization calculation could be performed for processes that are not adequately modelled by a first-order-with-dead-time model. The third step involves a test of the closed-loop control system and fine tuning, if necessary. The set point step change provides separate information on the proportional and integral modes to facilitate diagnosis and corrective action.

> The dynamic behavior of both the controlled and the manipulated variables is required for evaluating the performance of a feedback control system.

The reader should clearly recognize the meaning of the term *optimum.* It is used here to mean results (i.e., tuning constant values) that are determined so that certain mathematical criteria are satisfied. The criteria are goals 1 to 3. Naturally, the relationships in Table 9.1 were selected to represent the true control situation closely for the majority of cases. However, control performance has many facets, from safety through profit; therefore, it is sometimes difficult to condense all of the critical factors into one measure of control performance. Even if the mathematical objectives successfully represent the true desired performance, the results will be satisfactory only when the parameters in the mathematical formulation specify the desired behavior. These parameters, such as the proper controlled-variable measurement noise, the expected plant model error, and the allowable manipulated-variable variation, are never known exactly. Therefore, although the mathematical solution is "optimum," the usefulness of the results depends on the accuracy of the input data.

> Practically, the values from the optimization or correlations are used as *initial values* to be applied to the physical system and improved based on empirical performance during fine tuning.
>
> **Remember, when tuning a feedback controller, where you start is not as important as where you finish!**

Finally, the three tuning constants in the PID algorithm all influence the dynamic behavior of the closed-loop system. They must be determined simultaneously, because of this interaction.

It should be apparent that the tuning approach using optimization is not limited to PID controllers; if another algorithm were suggested, its parameters could be optimized by the same procedure. In fact, some results for other feedback controllers are presented in Chapter 19.

The techniques in this chapter provide practical methods for controller tuning that are applicable to many processes. However, they do not provide important explanations to key questions such as

1. Why do the tuning correlations have the shapes in Figure 9.5?
2. Why can a control system become unstable, and how can we predict when this will occur?
3. How does the controller change the dynamic behavior of an open-loop system to that of a closed-loop system?

Methods for answering these more fundamental questions are addressed in the next chapter.

REFERENCES

Ciancone, R. and T. Marlin, "Tune Controllers to Meet Plant Objectives," *Control, 5,* 50–57 (1992).

Edgar, T. and D. Himmelblau, *Optimization of Chemical Processes,* McGraw-Hill, 1988.

Hill, A., S. Kosinari, and B. Venkateshwa, "Effect of Disturbance Dynamics on Optimal Tuning," *Instrumentation in the Chemical and Petroleum Industries,* Vol. 19, Instrument Society of America, Research Triangle Park, NC, 89–97 (1987).

Fertik, H., "Tuning Controllers for Noisy Processes," *ISA Trans. 14, 4,* 292–304 (1975).

Lopez, A., P. Murrill, and C. Smith, "Tuning PI and PID Digital Controllers," *Instr. and Contr. Systems, 42,* 89–95 (Feb. 1969).

Zumwalt, R., *EXXON Process Control Professors' Workshop,* Florham Park, NJ, 1981.

ADDITIONAL RESOURCES

Other common forms of the PID control algorithm and conversions of tuning constants for these forms are given in

Witt, S. and R. Waggoner, "Tuning Parameters for Non-PID Three Mode Controllers," *Hydro. Proc., 69,* 74–78 (June 1990).

Analytical solutions for optimal tuning constant values for PID controllers can be obtained for some continuous control systems, specifically those involving processes without dead time. They can also be obtained for digital controllers for processes with dead time. References for analytical methods are given below; however, since such solutions are possible only with intensive analytical effort for limited control performance specifications, numerical methods are used in this chapter.

Jury, E. *Sample-Data Control Systems* (2nd Ed.), Krieger, 1979.
Newton, G., L. Gould, and J. Kaiser, *Analytical Design of Linear Feedback Controls,* Wiley, New York, 1957.
Stephanopoulos, G., "Optimization of Closed-Loop Responses," in Edgar, T. (ed.), *AIChE Modular Instruction Series, Vol 2, Module A2.5,* 26–38 (1981).

Background on mathematical principles and numerical methods of optimization can be obtained from many reference books, for example:

Reklaitis, G., A. Ravindran, and K. Ragsdell, *Engineering Optimization, Methods and Applications,* Wiley, New York, 1983.

Many other studies have been performed on optimizing time-domain control system performance. One further reference is

Bortolotto, G., A. Desages, and J. Romagnoli, "Automatic Tuning of PID Controllers through Response Optimization over Finite-Time Horizon," *Chem. Engr. Comm., 86,* 17–29 (1989).

The diagnostic fine-tuning method described in this chapter is limited to step changes in the controller set point. A powerful method for diagnosing feedback controller performance is based on statistical properties of the controlled and manipulated variables. The method, which establishes the approach to best possible control and identifies reasons for poor performance, is given in

Desborough, L. and T. Harris, "Performance Assessment for Univariate Feedback Control," *Can. J. Chem. Engr., 70,* 1186–1197 (1992).
Harris, T., "Assessment of Control Loop Performance," *Can. J. Chem. Engr., 67,* 856–861 (1989).
Stanfelj, N., T. Marlin, and J. MacGregor, "Monitoring and Diagnosing Control System Performance—SISO Case," *IEC Res., 32,* 301–314 (1993).

An alternative method of fine tuning is based on shapes or patterns of response to disturbances. Good and poor responses are identified, and tuning constants are altered accordingly. This method has been applied in an automatic tuning system. For an introduction, see

Kraus, T. and T. Myron, "Self-Tuning PID Controller Uses Pattern Recognition Approach," *Control Eng., 31,* 106–111 (June 1984).

The derivative mode can substantially improve the performance of control loops involving processes that are underdamped or unstable without control. For underdamped systems, see question 8.17. For open-loop unstable processes, see

Cheung, T. and W. Luyben, "PD Control Improves Reactor Stability," *Hydro. Proc., 58,* 215–218 (September, 1979).

These questions reinforce the key aspects of dynamic behavior that are considered in defining control performance and how the performance goals and process dynamics influence the controller tuning.

QUESTIONS

9.1. Given the results of the process reaction curve in Figure Q9.1, calculate the PI and PID tuning constants. The process was initially at steady state, and the manipulated variable was changed in a step at time = 0 by +7%.

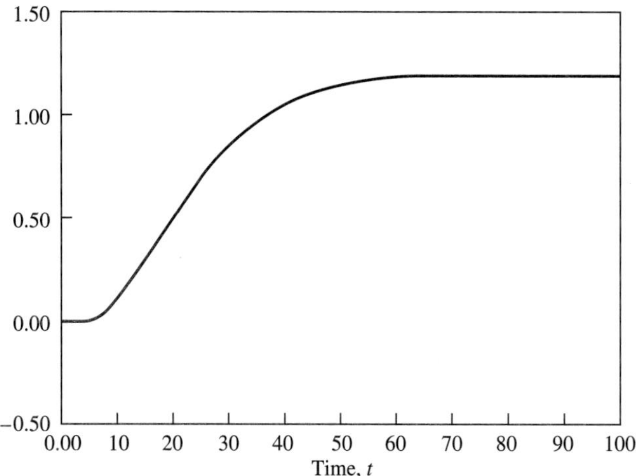

FIGURE Q9.1

9.2. Suppose that control goals different from those in Table 9.1 are specified for the tuning correlations. Predict the effect on the tuning constant values—that is, whether each would increase or decrease from the correlation values from Figure 9.5—for each set of goals.

(*a*) The only goal is to minimize the IAE for the base case model.

(*b*) The goals are to minimize IAE for ±25% change in model parameters, without concern for the manipulated-variable variation.

(*c*) The goals are to minimize IAE for ±50% change in model parameters, with concern for the manipulated-variable variation.

9.3. Confirm the correlation between the linearized model parameters and the process operating conditions in Table 9.3. Calculate the change in flow rate for the specified range of model parameters.

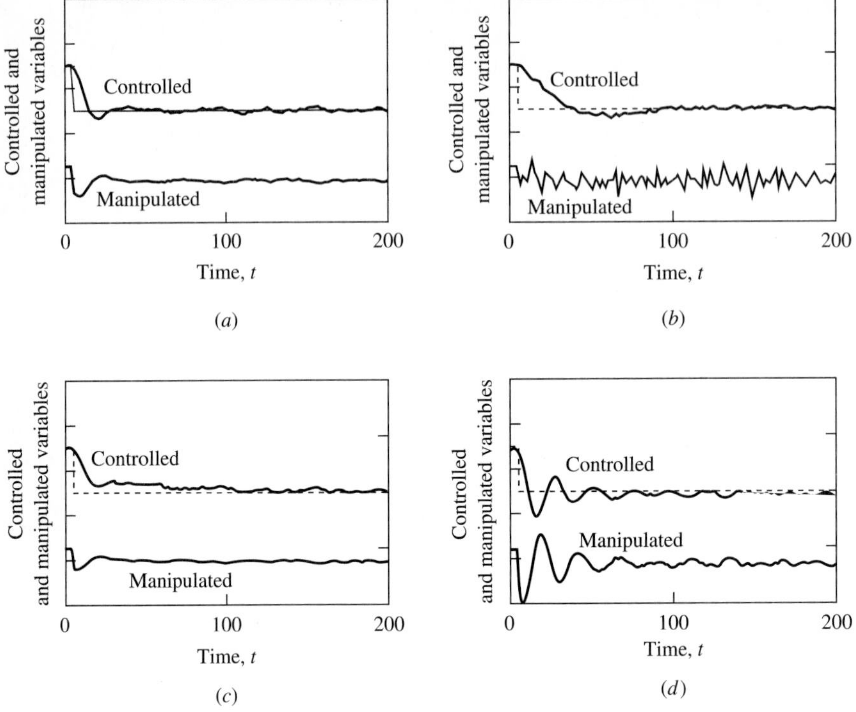

FIGURE Q9.4

9.4. The dynamic responses shown in Figure Q9.4 were obtained by introducing a step set point change to a PID controller. The dead time of the process is only a few minutes. For each case, determine whether the control is as good as possible and if not, what corrective steps should be taken. Note that the diagnosis of this data would require an exact specification of the control objectives. Use the general objectives considered in Table 9.1 and be as specific as possible regarding the change to the tuning constants.

9.5. The tuning constants for the three-tank control system are given in Example 9.2. Predict how the optimum tuning constants will change as the following changes are made to the control system. The analysis should be based on principles of process dynamics, tuning factors, and tuning correlations. Be as specific as possible without resolving the optimization problem for each case.

(*a*) A different control valve is installed whose maximum flow is 2.5 times greater than the original valve.

(*b*) The volume of each tank is reduced by a factor of 2.

(*c*) The temperature of stream *B* is increased by 20°C.

(*d*) The set point of the controller is increased to 3.5% of component A in the third-tank effluent.

(*e*) Substantial high-frequency noise is present in the measurement of the controlled variable.

9.6. Given the following process reaction curves, for which of the processes is it appropriate to use the general tuning charts in Figures 9.4*a* through *f*?

(*a*) Figure 3.7
(*b*) Figure 3.17
(*c*) Figure 5.5
(*d*) Figure 5.14
(*e*) Figure 7.3*a*
(*f*) Figure 8.3*a*
(*g*) Figure 5.20
Explain your answer for each case.

9.7. Explain in your own words why the dimensionless parameters are
(*a*) $K_c K_p$.
(*b*) $T_I/(\theta + \tau)$.
(*c*) $T_d/(\theta + \tau)$.

9.8. Derive the closed-loop transfer function for the three-tank mixing process using the analytical (third-order) linearized model in response to a change in the composition in the *A* stream from Example 7.2. Perform a dimensional analysis using the method demonstrated in Section 9.4, determine the key dimensionless parameters, and explain the form of tuning correlations for this model structure and how you would develop them.

9.9. For one or more of the following processes, calculate the PI controller tuning constants by two correlations: Ciancone and Lopez. Compare the expected control performance for both correlations in response to a step change in the controller set point. Under which circumstances would each correlation give the best constants?
(*a*) question 6.1
(*b*) question 6.2
(*c*) Example 3.10
(*d*) Example 5.1
(*e*) Example 5.4
(*f*) Example 6.4

9.10. The two series CSTRs in Example 3.3 with the reaction A → products

$$-r_A = 6.923 \times 10^5 e^{-5000/T} C_A$$

with T in K, has its outlet concentration of A, C_{A2}, controlled by adjusting the inlet concentration C_{A0}. The temperature varies slowly between 290 and 315 K. Would this temperature variation require a significant adjustment in controller tuning? Justify your answer with quantitative analysis.

9.11. The three cases used in the tuning optimization are selected to span the range of expected plant operation (i.e., the range of plant model parameters). Suppose that the control engineer knew what percentage of the time that the plant will operate at various operating conditions in the range. Suggest a modification to the optimization method, specifically the objective function, that would include the information on time at each operation in determining the optimum tuning constants.

9.12. The tuning optimization method integrates the equations over a finite time to evaluate the IAE.

(*a*) Write the equations that could be used to evaluate the IAE from the simulation results, and describe an optimization approach to find tuning constant values that satisfy equations (9.4).

(*b*) Write the equations for the ISE and ITAE that could be used with simulation results. For the ITAE, carefully define when the integration begins (i.e., where time equals zero).

(*c*) Examples in this chapter demonstrated that a poor choice of tuning constant values could lead to an unstable system, with the controlled variable diverging from the solution. What is the theoretical value of the IAE for an unstable control system? How would the optimization system described in this chapter respond if an intermediate set of tuning constants led to an unstable response?

(*d*) Determine the theoretical minimum IAE for controlling an ideal first-order process with dead time in response to a step disturbance.

(*e*) If an analytical expression were available for CV(t), it could be used in tuning. Determine the closed-loop transfer function for a PI controller and a first-order-with-dead-time process, $G_p(s) = K_p e^{-\theta s}/(\tau s + 1)$. For a step set point change, $\text{SP}(s) = \Delta\text{SP}/s$, solve for CV($s$) and invert the Laplace transform to obtain CV(t), if possible.

9.13. Control performance goals are defined in Table 9.1. Propose at least one alternative measure for every entry in the column labeled "Used in This Chapter." Each should involve a different performance measure and not be simply a different numerical value. Discuss the advantages of each entry, the original, and your proposed alternate.

9.14. Tuning constants for a PI controller for the following process are to be determined.

$$G'_p(s)G_v(s)G_s(s) = \frac{7.5e^{-2.3s}}{8.5s + 1} \qquad G_d(s) = \frac{100}{5s + 1}$$

The control objectives are essentially the same as used in this chapter. A colleague has calculated several sets of values for the controller gain and integral time. Determine which of these sets of constants, if any, is acceptable and explain why or why not.

Tuning	Case A	Case B	Case C	Case D
K_c	12	12	0.3	0.3
T_I	6	1	6	1

9.15. Rules for interpreting the control performance are presented in the section on fine tuning and summarized in Figure 9.12.

(*a*) Discuss the advantages of using a set point change response rather than the disturbance response.

(*b*) Prove the relationships given in Figure 9.12.

(*c*) Demonstrate why the initial change in the manipulated variable is about 50 to 150% of its final value. Does this tuning guideline depend on the tuning goals and correlations used?

9.16. Figure 9.2 gives the controlled variable behavior for various values of the controller gain. Sketch the behavior of the manipulated variable you would expect for each case and explain your answers. Also, sketch the variable given here as a function of the controller gain K_c, and explain your answer.

$$\int_0^\infty \left(\frac{d(\text{MV})}{dt}\right)^2 dt$$

CHAPTER

10

STABILITY ANALYSIS AND CONTROLLER TUNING

10.1 INTRODUCTION

To this point, we have developed a control algorithm (the proportional-integral-derivative controller) and a method for tuning its adjustable constants. One might ask, "Isn't this sufficient for designing feedback control systems?" The answer is a resounding "No!", because we do not have a general method for evaluating the effects of elements in the closed-loop system on dynamic stability and performance.

Through various examples and exercises, we have seen how feedback control can change the qualitative behavior of a process, introducing oscillations in an originally overdamped system and potentially causing instability. In fact, we shall see that the stability limit is what prevents the use of a very high controller gain to improve the control performance of the controlled variable. Therefore, a thorough understanding of the stability of dynamic systems is essential, because it provides important relationships among process dynamics, controller tuning, and achievable performance. These relationships are used in a variety of ways, such as selecting controller modes, tuning controllers, and designing processes that are easier to control.

As the reader will see, the concepts introduced in Chapter 4 on dynamics and modelling enable us to determine much about these issues without determining the analytical expression for the entire dynamic response, which would be very difficult for all but the simplest systems. This is where the effort invested previously in linearization, Laplace transforms, and frequency response really begins to pay dividends!

10.2 THE CONCEPT OF STABILITY

In vernacular English, the term "unstable" has a negative connotation. Certainly, no one would want to be described as unstable! This undesirable meaning extends to products of engineering design; we generally want our plants and control systems to be stable. To ensure consistency, we will use a clear and precise definition of stability, termed *bounded input–bounded output stability,* which can be employed in the design and analysis of process control systems.

> A system is **stable** if all output variables are bounded when all input variables are bounded. A system that is not stable is **unstable.**

A variable is *bounded* when it does not increase in magnitude to $\pm\infty$ as time increases. Typical bounded inputs are step changes and sine waves; an example of an unbounded input is a ramp function. Naturally, process output variables do not approach $\pm\infty$ in a chemical plant, but serious consequences occur when these variables *tend toward* $\pm\infty$ and reach large deviations from their normal values. For example, liquids overflow their vessels; vessels burst from high pressures; products degrade; and equipment is damaged by excessive temperatures. Thus, substantial incentives exist for maintaining plant variables, with and without control, at stable operating conditions.

As a further clarification, a chemical reactor would be stable according to our definition if a step increase of 1°C in its inlet temperature led to a new steady-state outlet temperature that was 100°C higher. Thus, systems that are very sensitive can be stable as long as they attain a steady state after a step change. The methods in this chapter determine stability strictly as defined here, which is required for good operation but clearly is not alone sufficient to ensure good control performance. Other aspects of achieving acceptable control performance will be addressed in Chapter 13.

10.3 STABILITY OF LINEAR SYSTEMS— A SIMPLE EXAMPLE

Since control system stability is the goal of this chapter, the definition will be reinforced through a process example that shows how the addition of feedback control changes the dynamic response of a linear process, to demonstrate clearly the factors affecting stability of a linear system. In the next section, the analysis is generalized to any linear system.

Example 10.1. The response of the non–self-regulating level process in Figure 10.1 to a step change in the inlet flow is to be determined for a case with proportional-only control.

The linear models for the process and the controller are

$$\begin{aligned} A\frac{dL}{dt} &= F_{\text{in}} - F_{\text{out}} \\ F_{\text{out}} &= K_c(\text{SP} - L) + (F_{\text{out}})_s \end{aligned} \tag{10.1}$$

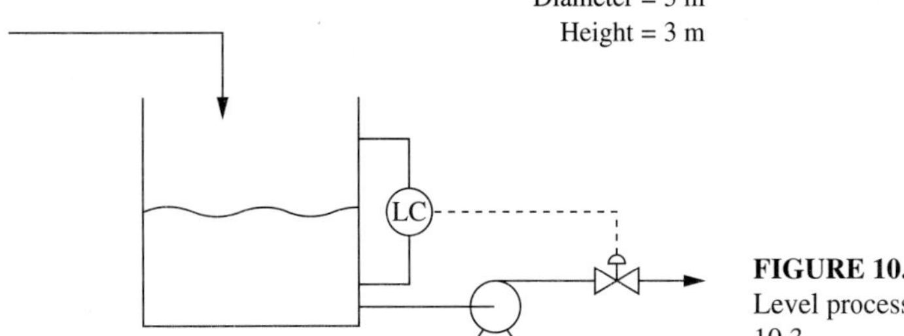

FIGURE 10.1
Level process for Examples 10.1 and 10.3.

Expressing variables in deviation form, equating the set point and initial steady state (i.e., $L' = L - L_s = L - \text{SP}$), and combining into one equation gives

$$A\frac{dL'}{dt} = F'_{\text{in}} + K_c L' \tag{10.2}$$

By taking the Laplace transform and rearranging, the transfer function for this system can be derived as

$$\frac{L(s)}{F'_{\text{in}}(s)} = \frac{-1/K_c}{\left(\frac{A}{-K_c}\right)s + 1} \tag{10.3}$$

Solution. Since the system is simple, the following analytical solution to the equations can be derived for a step change in the inlet flow, $F'_{\text{in}}(s) = \Delta F_{\text{in}}/s$.

$$L' = \frac{\Delta F_{\text{in}}}{-K_c}\left(1 - e^{-t/\tau}\right) \tag{10.4}$$

with $\tau = A/(-K_c)$. As can be seen, the controller gain affects the time constant of the feedback system. As observed in earlier examples, increasing the magnitude of the controller gain, which gives negative feedback control (which in this case is $K_c < 0$), decreases the time constant as well as reducing the steady-state offset.

Note that for this first-order system the controller gain can be set to a very large magnitude without causing instability. This conclusion can be demonstrated by analyzing the expression for the time constant, which would have to change sign to cause instability. Since the time constant is positive and the analytical solution has a negative exponent for all gains ($K_c < 0$), this idealized system is stable for any negative feedback controller gain. This result is not true for most processes, as will be demonstrated in later examples.

Recall that this analysis is valid only for the ideal, linear level control system described in equations (10.1), which has no sensor or final element dynamics and is perfectly linear. Also, this analysis ensures only that variables do not increase without bound; it does not ensure that the process variables in the real plant will remain within acceptable limits. Applying the final value theorem, the ultimate value of the level after a step change in the inlet flow is

$$\lim_{t\to\infty} L = \lim_{s\to 0} sL(s) = \lim_{s\to 0} s\frac{\frac{\Delta F_{\text{in}}}{(-K_c)s}}{\frac{A}{(-K_c)}s + 1} = \frac{\Delta F_{\text{in}}}{-K_c} \tag{10.5}$$

Substituting the process data into this expression for a 20 m^3/hr change in flow and a controller gain of -10 m^3/hr/m gives a final level deviation of 2 m, which, assuming that the level began in the middle of its range, is half a meter above the top of the tank wall! For this input the plant demonstrates nonlinear behavior by overflowing and is not modelled accurately by equations (10.1) when overflow occurs. Clearly, good control performance requires more than stability; however, stability is one essential component of a well-performing control system.

This example demonstrates that the stability of the level system depends on the sign of the exponential term in the solution and that the feedback controller affects the exponential term. In the next section, the relationship of the exponential term to stability is generalized to address a set of ordinary differential equations of arbitrary order.

10.4 STABILITY ANALYSIS OF LINEAR AND LINEARIZED SYSTEMS

Essentially all chemical processes are nonlinear. Since no general stability analysis of nonlinear systems is available, the *local* stability of the linearized approximation about a steady state is evaluated. The local linear analysis is valid only in a very small region (theoretically, a differential region) about the linearization conditions. We will assume that a differential region exists about the steady-state operating conditions within which stability can be investigated, and Perlmutter (1972) gives a thorough justification of the linearized analysis, sometimes referred to as *Liapunov's first method.*

Since the control system reduces variability in the controlled variables, the linear stability analysis is often adequate for making the control design and tuning decisions. However, we must recognize that the analysis is valid only at a point and that no rigorous conclusions can be drawn for a finite distance from this point. This is not a desirable situation, especially for the beginning engineer, but it is the standard approach used by practicing engineers using the best current technology (and the 80/20 rule for obtaining necessary results from limited effort). The successes of the vast majority of process control strategies designed using linear methods attest to the validity of the approach, when applied judiciously.

To develop a general stability analysis for linearized systems, the following nth-order linear dynamic model with a forcing function $f(t)$ is considered.

$$\frac{d^nY}{dt} + a_1\frac{d^{n-1}Y}{dt^{n-1}} + \cdots + a_nY = f(t) \tag{10.6}$$

Note that we often formulate the model as a set of first-order differential equations, which can be combined in the form of equation (10.6) by any of several procedures, such as taking the Laplace transform of the original models and combining algebraically, as in Example 4.8.

The solution to equation (10.6) is composed of two terms: the *particular* solution, which depends on the forcing function, and the *homogeneous* solution, which is independent of the forcing function (Boyce and Diprima, 1986). The forcing functions for process control systems are set point changes and disturbances in process variables such as feed composition, which, since they are bounded, cannot cause instability in an otherwise stable system. Thus, we conclude that the particular solution of a stable system with bounded inputs must be stable. Therefore, the stability analysis concentrates on the homogeneous solution, which determines whether the system is stable, with or without forcing, as long as the inputs are bounded (Willems, 1970).

The Laplace transform of the homogeneous part of equation (10.6), with all initial conditions equal to zero, is

$$(s^n + a_{n-1}s^{n-1} + \cdots + a_1 s + a_0)Y(s) = 0 \tag{10.7}$$

As demonstrated in Chapter 4, the solution to equation (10.7) is of the form

$$Y(t) = A_1 e^{\alpha_i t} + \cdots + (B_1 + B_2 t + \cdots)e^{\alpha_p t} \cdots + (C_1 \cos(\omega t) + C_2 \sin(\omega t))e^{\alpha_q t} + \cdots \tag{10.8}$$

where
α_i = the ith real distinct root of the characteristic polynomial
α_p = repeated real root of the characteristic polynomial
α_q = real part of complex root of the characteristic polynomial
A, B, C = constants depending on the initial conditions

The stability of the linearized system is entirely determined by the values of the exponents (the α's). When all of the exponents have negative real parts, the solution cannot increase in an unlimited fashion as time increases. However, if one or more exponents have positive real parts, variables in the system will be unbounded as time increases, and the system will be unstable by our definition. The special case of a zero real part is considered in Example 10.3, where it is shown that a system with one or more zero real parts is bounded input–bounded output unstable. Thus, a test for stability involves determining all exponential terms and can be summarized in the following principle.

> The *local* stability of a system about a steady-state condition can be determined from a linearized model.
>
> The linear approximation of the system is bounded input–bounded output stable if all exponents have negative real parts and is unstable if any exponential real part is zero or positive.

The linear approximation is valid only at the point of linearization. If the process operation changes significantly, the stability can be determined for several points with different operating conditions. However, the fact that a system may be stable for many points does not ensure that it is stable for conditions between these stable points. This is sometimes referred to as *pointwise* or *local* stability determination.

Example 10.2. Determine the stability of the variable $T'(t)$ from the following model.

$$\frac{d^2T'}{dt^2} - 1.23\frac{dT'}{dt} - 1.38T' = 0 \tag{10.9}$$

The exponential terms can be evaluated according to the following procedure.

$$
\begin{aligned}
(s^2 - 1.23s - 1.38)T'(s) &= 0 \\
s^2 - 1.23s - 1.38 &= 0 \\
s = -0.71 \qquad s &= 1.94 \\
T'(t) &= A_1 e^{-0.71t} + A_2 e^{1.94t}
\end{aligned}
\tag{10.10}
$$

It is clear that $T'(t)$ is locally unstable about the steady state, because one of the exponential terms has a real part greater than zero. Insight into the cause of instability in a process without feedback control is given in Appendix C, where a chemical reactor is analyzed. For several reactor designs, the linear (local) stability analysis shows that the nonlinear reactor cannot return to the unstable steady state, but we know from physical arguments that the reactor temperature does not approach infinity. In the example the temperature either oscillates continuously (in a limit cycle) or approaches a different steady state that is stable. (The numerical values for this example are from Case II in Appendix C, Table C.1.)

Example 10.3. The stability of the level process *without* control ($K_c = 0$) shown in Figure 10.1 is to be determined. The vessel size and steady-state flow are the same as in Example 10.1. A material balance on the vessel results in the following model:

$$
A\frac{dL(t)}{dt} = F_{\text{in}}(t) - F_{\text{out}}(t) \tag{10.11}
$$

The model can be written in deviation variables and in transfer function form for the case with the outlet flow constant:

$$
A\frac{dL'(t)}{dt} = F'_{\text{in}}(t) \tag{10.12}
$$

$$
\frac{L(s)}{F_{\text{in}}(s)} = \frac{1}{As} \tag{10.13}
$$

The solution to this equation has a real part of the exponential equal to zero. We will assume that the process is initially at steady state and investigate the behavior of the level for two different input flows. First, assume that the flow in varies around its steady-state value according to a sine, $M \sin(\omega t)$, and the system is initially at steady state. The analytical solution for the level is as follows, and the dynamic behavior is shown in Figure 10.2 with $M = 2$ and $\omega = 1$.

$$
L'(t) = \frac{M}{A\omega}[1 - \cos(\omega t)] \tag{10.14}
$$

For this bounded input function, the output of the linearized system is bounded; therefore, the system is stable in this case. The second case involves a step function in the inlet flow, which increases by 2 m^3/h at time $= 0$. The analytical solution for the level subject to a step change of magnitude M from an initial steady state is as follows, and the dynamic behavior is shown in Figure 10.3.

$$
L'(t) = \frac{M}{A}t \tag{10.15}
$$

For this bounded input, the output of the linearized model is unbounded (although the true nonlinear level is bounded because the maximum level is reached and the liquid overflows). Thus, the result of the stability analysis indicates a serious deficiency in the process behavior without control, which should be modified through feedback.

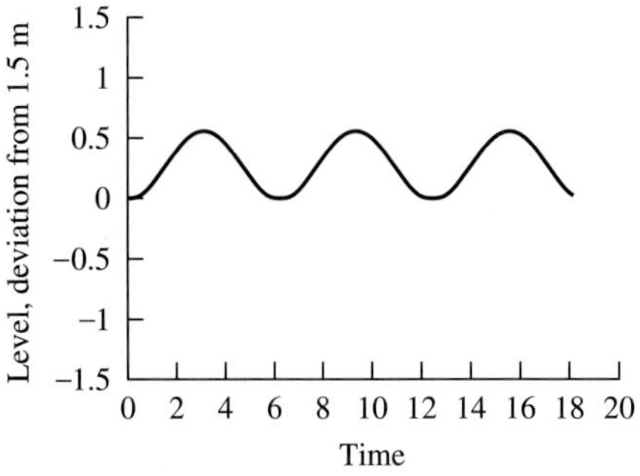

FIGURE 10.2
Response of the level in Example 10.3 to a sine flow disturbance.

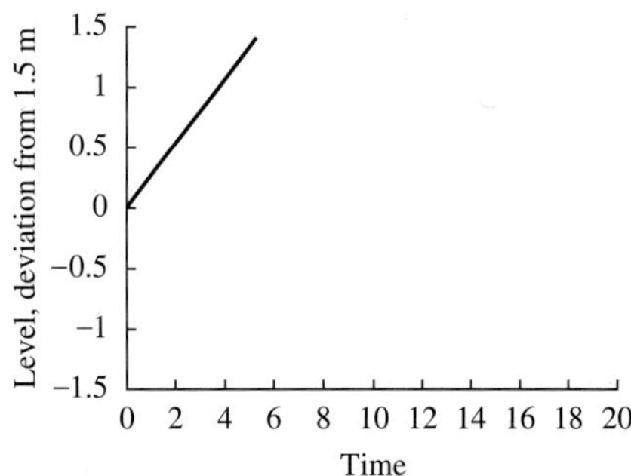

FIGURE 10.3
Response of the level in Example 10.3 to a step flow disturbance.

The difference between the behavior of the levels in these two cases is due to the nature of the forcing functions. The sine variation in deviation variables has a zero integral over any multiple of its period; thus, the level increases and decreases but does not accumulate. The step forcing function has a nonzero integral that increases with time, and the level, which integrates the difference between input and output, increases monotonically toward infinity. Since we are interested in general statements on stability that are valid for *all bounded inputs,* we shall consider a system with a zero real part in its exponential to be unstable, because it is unstable for some bounded input functions.

> Local stability analysis using linearized models determines stability at the steady state; no rigorous information about behavior a finite deviation from the steady state is obtained.

10.5 STABILITY ANALYSIS OF CONTROL SYSTEMS: PRINCIPLES

Again, the local stability of a system will be evaluated by analyzing the linearized model. The analysis method for linear systems can be tailored to feedback control systems by considering the models in transfer function form. The resulting methods will be useful in (1) determining the stability of control designs, (2) selecting tuning constant values, and (3) gaining insight into how process characteristics influence tuning constants and control performance. We begin by considering a general transfer function for a linear control system in Figure 10.4.

$$\frac{\mathrm{CV}(s)}{\mathrm{SP}(s)} = \frac{G_p(s)G_v(s)G_c(s)}{1 + G_p(s)G_v(s)G_c(s)G_s(s)}$$

$$\frac{\mathrm{CV}(s)}{\mathrm{D}(s)} = \frac{G_d(s)}{1 + G_p(s)G_v(s)G_c(s)G_s(s)} \tag{10.16}$$

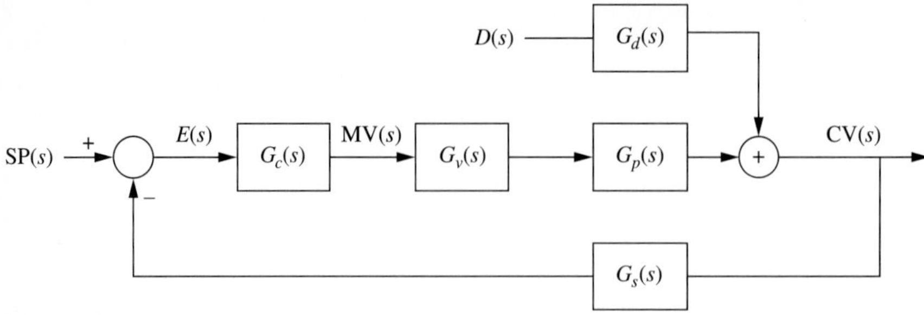

$G_c(s)$ = Controller
$G_v(s)$ = Transmission, transducer, and valve
$G_p(s)$ = Process
$G_s(s)$ = Sensor, transducer, and transmission
$G_d(s)$ = Disturbance

FIGURE 10.4
Block diagram of a feedback control system.

For the present, we will consider only the disturbance transfer function and will assume that the transfer function can be expressed as a polynomial in s as follows:

$$\left(1 + G_p(s)G_v(s)G_c(s)G_s(s)\right)\mathrm{CV}(s) = G_d(s)D(s) \tag{10.17}$$

$$\left(s^n + a_1 s^{n-1} + a_2 s^{n-2} + \cdots\right)\mathrm{CV}(s) = (s - \beta_1)(s - \beta_2)\cdots(s - \beta_m)D(s)$$

The right-hand side (the numerator of the original transfer function) represents the forcing function, which is always bounded (by our choice), because physical variables cannot take unbounded values.

The essential information on stability is in the left-hand side of equation (10.17), called the *characteristic polynomial,* which is the denominator of the closed-loop transfer function. In the system being considered, Figure 10.4, the characteristic polynomial is $1 + G_p(s)G_v(s)G_c(s)G_s(s)$. Setting the characteristic polynomial to zero produces the *characteristic equation.*

Before continuing, it is important to note that either transfer function in equation (10.16) could be considered, because the characteristic equations of both are identical. Thus, the stability analyses for set point changes and for disturbances yield the same results. Examination of the characteristic equation demonstrates that the equation contains all elements in the feedback control loop: process, sensors, transmission, final elements, and controller. As we would expect, all of these terms affect stability. The disturbances and set point changes are not in the characteristic equation, because they affect the input forcing; therefore, they do not affect stability. Naturally, the numerator terms affect the dynamic responses and control performance and must be considered in the control performance analysis, although not in this part, which establishes stability.

Continuing the stability analysis, the solution to the homogeneous solution is evaluated to determine stability. For the transfer function, the exponents can be determined by the solution of the following equation resulting from equation (10.17):

$$(s^n + a_1 s^{n-1} + a_2 s^{n-2} + \cdots) = 0 \tag{10.18}$$

As before, if any solution of equation (10.18) has a real part greater than or equal to zero, the linearized system is unstable, because the controlled variable increases without limit as time increases. The stability test is summarized as follows:

> A linearized closed-loop control system is **locally stable** at the steady-state point if all roots of the characteristic equation have negative real parts. If one or more roots with positive or zero real parts exist, the system is **locally unstable.**

Recall that the roots of the characteristic equation are also referred to as the *poles* of the closed-loop transfer function, such as $G_d(s)/[1+G_p(s)G_v(s)G_c(s)G_s(s)]$. This approach to determining stability is applied to two examples to demonstrate typical results.

Example 10.4. The stability of the series chemical reactors shown in Figure 10.5 is to be determined. The reactors are well mixed and isothermal, and the reaction is first-order in component A. The outlet concentration of reactant from the second reactor is controlled with a PI feedback algorithm that manipulates the flow of the reactant, which is very much smaller than the flow of the solvent. The sensor and final element are assumed fast, and process data is as follows.

Process.

$$\begin{aligned} V &= 5\,\text{m}^3 \qquad F_s = 5\,\text{m}^3/\text{min} >> F_A \\ v_s &= 50\%\ \text{open} \\ C_{A0} &= 20\ \text{mole/m}^3 \qquad k = 1\ \text{min}^{-1} \\ C_{A0}(s)/v(s) &= K_v = 0.40\ (\text{mole/m}^3)/(\%\ \text{open}) \end{aligned}$$

Controller.

$$\begin{aligned} K_c &= 15(\%\ \text{open})/(\text{mole/m}^3) \\ T_I &= 1.0\ \text{min} \end{aligned}$$

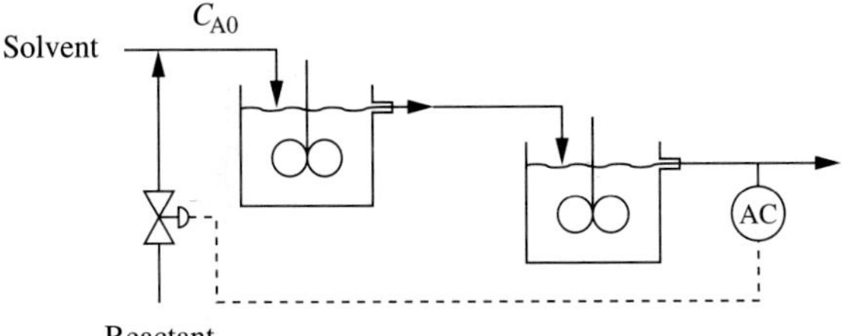

FIGURE 10.5
Series chemical reactors analyzed in Example 10.4.

Formulation. The model structure for this system is the same as for Example 3.3, but the data is different and the valve gain is included. The transfer functions for the process and controller are

$$G_p(s) = \frac{K_p}{(\tau s + 1)(\tau s + 1)} \tag{10.19}$$

$$G_c(s) = K_c\left(1 + \frac{1}{T_I s}\right)$$

with

$$K_p = K_v\left(\frac{F}{F + VK}\right)^2 = 0.10\frac{\text{mole/m}^3}{\%}$$

$$\tau = \left(\frac{V}{F + VK}\right) = 0.50 \text{ min}$$

The individual transfer functions can be combined to give the closed-loop transfer function for a set point change, which includes the characteristic equation.

$$\frac{\text{CV}(s)}{\text{SP}(s)} = \frac{G_p(s)G_v(s)G_c(s)}{1 + G_p(s)G_v(s)G_c(s)G_s(s)} = \frac{15\left(1 + \dfrac{1}{1.0s}\right)\dfrac{0.10}{(0.5s + 1)^2}}{1 + 15\left(1 + \dfrac{1}{s}\right)\left(\dfrac{0.1}{(0.5s + 1)^2}\right)} \tag{10.20}$$

Characteristic equation.

$$0 = 1 + 15\left(1 + \frac{1}{s}\right)\left(\frac{0.1}{(0.5s + 1)^2}\right)$$

$$0 = 0.25s^3 + 1.0s^2 + 2.5s + 1.5$$

The solution to this cubic equation gives the exponents in the time-domain solution. These values are

$$\alpha_{1,2} = -1.60 \pm 2.21j \qquad \alpha_3 = -0.81$$

Since all roots have negative real parts, this system is stable. Remember, we still do not know how well the closed-loop control system performs, although the complex poles indicate that the system is underdamped and the integral mode indicates that the controlled variable will return to its set point for a steplike disturbance.

Example 10.5. The stability of the three-tank mixing process in Example 7.2 (page 249) is to be evaluated under feedback control with a proportional-only controller.

Assuming that the sensor is fast, $G_s(s) = 1$, the closed-loop transfer function is

$$\frac{\text{CV}(s)}{D(s)} = \frac{G_d(s)}{1 + G_p(s)G_v(s)G_c(s)G_s(s)} = \frac{\dfrac{1}{(5s + 1)^3}}{1 + K_c\dfrac{0.039}{(5s + 1)^3}} \tag{10.21}$$

Characteristic equation.

$$125s^3 + 75s^2 + 15s + (1 + 0.039K_c) = 0$$

The solutions to the characteristic equation determine whether the system is stable or unstable. Solutions have been determined for several values of the controller gain (with the proper sign for negative feedback control), and the results are plotted in

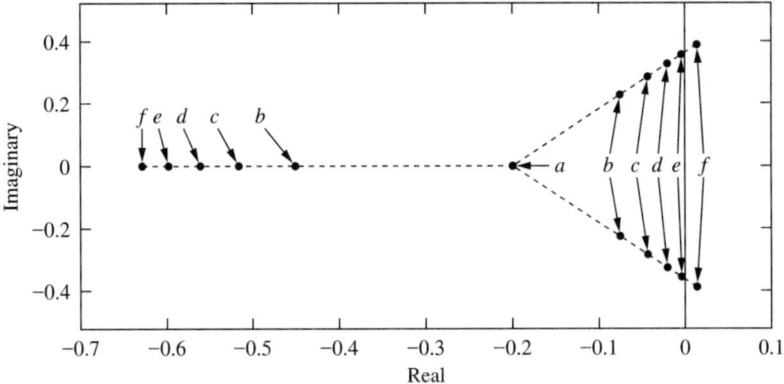

FIGURE 10.6
Root locus plot for Example 10.5 for controller gain values of (*a*) 0, (*b*) 50, (*c*) 100, (*d*) 150, (*e*) 200, and (*f*) 250.

Figure 10.6. Since the characteristic equation is cubic, three solutions exist. The system without control, $K_c = 0$, is stable, because all roots (i.e., exponential terms) have the same negative real value (-0.2). As the controller gain is increased, two poles begin to move toward the imaginary axis and begin to have imaginary parts. This result is typical of many systems and has two important interpretations in the time domain. First, the imaginary parts indicate that the response has an oscillatory nature. This can be seen from the Euler identity,

$$\cos \omega t = \frac{e^{j\omega t} + e^{-j\omega t}}{2} \tag{10.22}$$

Thus, the solution of a system with complex conjugate exponents has the form

$$\begin{aligned} Y(t) &= A_1 e^{(\alpha + j\omega)t} + A_2 e^{(\alpha - j\omega)t} + A_3 e^{\alpha_3 t} \\ &= (A_1' \sin \omega t + A_2' \cos \omega t) e^{\alpha t} + A_3 e^{\alpha_3 t} \end{aligned} \tag{10.23}$$

As the controller gain is increased from 0 to 250 in increments of 50, the poles approach, and then cross, the imaginary axis. This path can be interpreted as the solution becoming more oscillatory, due to the increasing size of the imaginary parts, and finally becoming unstable, since the exponents have zero and then positive real parts. Based on this analysis, the three-tank mixing process is found to be (barely) stable (and periodic) for $K_c < 200$ and unstable for $K_c > 250$; further study shows that the stability limit is about $K_c = 208$. The control performance would be clearly unacceptable when the system is unstable, but again, we do not yet know for what range of controller gain the control performance is acceptable.

The results of Example 10.5 can be generalized to establish relationships between locations of roots of the characteristic equation (poles of the closed-loop transfer function). In addition, features of dynamic responses can be inferred from the poles if a constant transfer function numerator is assumed. These generalizations are sketched in Figure 10.7, which shows the nature of the dynamic responses for various pole locations. Clearly, the numerical values of the poles (or equivalently, their location in the complex plane) are very important for the dynamic response of a closed-loop system.

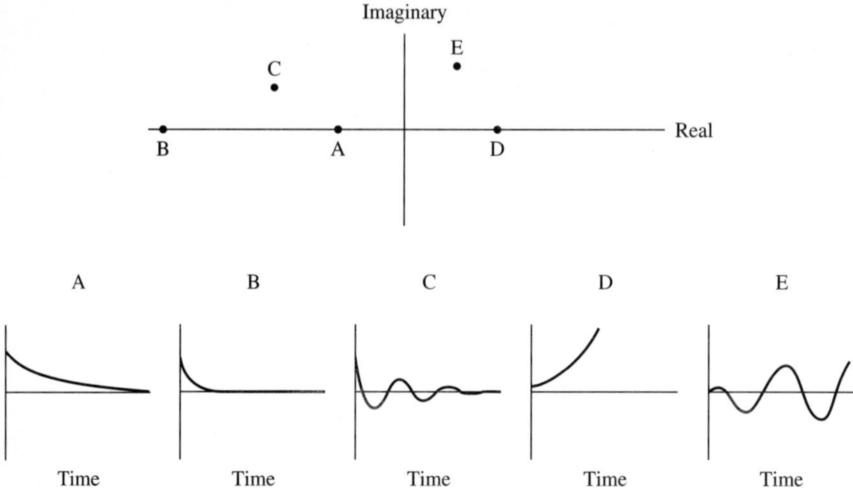

FIGURE 10.7
Examples of the relationship between the locations of the exponential terms and the dynamic behavior.

The method of plotting the roots of the characteristic equation as a function of the controller tuning constant(s) is termed *root locus* analysis and has been used for decades. Note that a root-solving computer program is required to facilitate the construction of the plots. We will use another stability analysis method in further studies, but we directly calculated the poles of the closed-loop transfer function here because of the excellent visual display of the effect of the tuning constants on the exponential terms and therefore, on stability. In summary, for a linearized model (which determines local properties):

> Application of the general stability analysis method to feedback control systems demonstrates that the roots of the characteristic equation determine the stability of the system.
>
> When the characteristic equation is a polynomial, a straightforward manner of determining the stability is to calculate the roots of the characteristic equation.
>
> If all roots have negative real parts, the system is bounded input–bounded output stable; if any root has a positive or zero real part, the system is unstable.

10.6 STABILITY ANALYSIS OF CONTROL SYSTEMS: THE BODE METHOD

The method presented in the previous section presents the principles of stability analysis of transfer functions and provides a vivid picture of the effects of controller tuning on the stability of control systems. However, we would like to have a method for analyzing control systems that

1. Involves simple calculations
2. Addresses most processes of interest
3. Gives information on the *relative stability* of the system (i.e., how much a parameter must change to change the stability of the system)
4. Yields insight into how various process and controller characteristics affect tuning and control performance

The most commonly used stability analysis methods are summarized in Table 10.1. Since many plants in the process industries have dead time, the methods that require polynomial transfer functions (root locus and Routh) will not be considered further. Of the two remaining, the Nyquist method is the most general. However, in spite of a few limitations, the Bode method of stability analysis is selected for emphasis in this book, because it involves simple calculations and, more importantly in the age of computers, gives more easily understood insights into the effect of process and controller elements on the stability of closed-loop systems.

The basis of the Bode method is first explained with reference to the system in Figures 10.8*a* and *b*; then, a simple calculation procedure is presented with several worked examples. Suppose that a sine wave is introduced into the set point with the loop maintained open as in Figure 10.8*a,* with $G_p(s)$ including the process without sensor and final element. Because the system is linear, all variables oscillate in a sinusoidal manner. After some time, the system attains a steady state, a standing wave in which the amplitudes do not change. The sine frequency can be selected so that the output signal, CV(t), lags the input signal, SP(t), by 180°. Note that the relative

TABLE 10.1
Summary of stability analysis methods

Method	Plant model	Stability results	Results display
Root locus (Franklin et al., 1991)	Polynomial in s	Relative	Graphical
Routh (Willems, 1970)	Polynomial in s	Yes or no	Tabular
Bode	(1) Open loop-stable (2) Monotonic decreasing amplitude ratio (AR) and phase angle (ϕ) as frequency increases	Relative	Graphical
Nyquist (Dorf, 1986)	Linear	Relative	Graphical

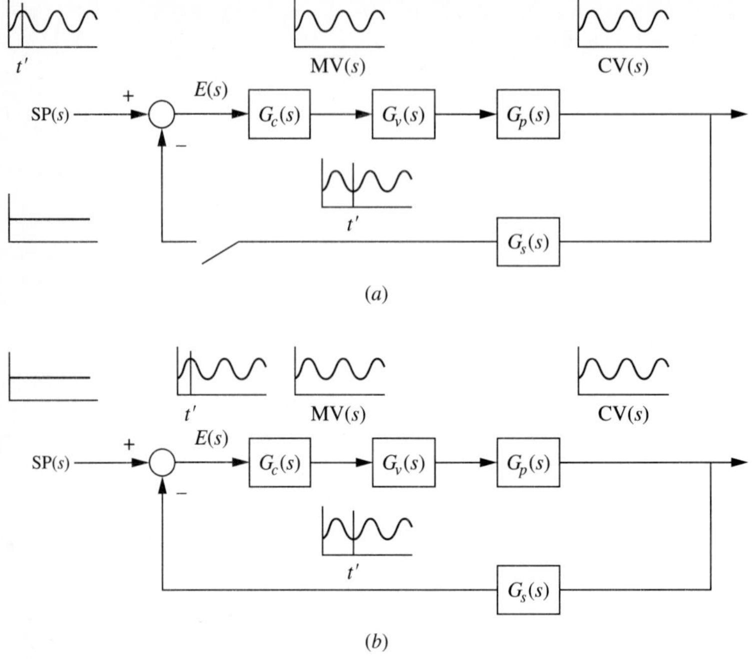

FIGURE 10.8
Bode stability analysis. (*a*) Behavior of open-loop system with sine forcing; (*b*) Behavior of system after the forcing is stopped and the loop is closed.

amplitudes of the various signals in Figure 10.8*a* would normally be different but are shown to be equal here because the process and controller transfer functions have not yet been specified.

After steady state has been attained, the set point is changed to a constant value and the loop is closed, as shown in Figure 10.8*b*. Since this is a closed-loop system, the sine affects the process output, which is fed back via the error signal to the process input. For the frequency selected with a phase difference of 180°, the returning signal *reinforces* the previous error signal because of the negative sign of the comparator.

A key factor that determines the behavior of this closed-loop system is the amplification as the sine wave travels around the control loop once. If the signal decreases in magnitude every pass, it will ultimately reduce to zero, and the system is stable. If the signal increases in amplitude every pass, the wave will grow without limit and the system is unstable. This analysis leads to the Bode stability criterion. Two important factors need to be emphasized. First, the analysis is performed at the frequency at which the feedback signal lags the input signal by 180°; this is termed the *critical* or *crossover* frequency. Naturally, the critical frequency depends on all of the dynamic elements in the closed-loop system. Second, for the amplitude of the wave to increase, the *gain* of the elements in the loop must be greater than 1. This gain depends on the amplitude ratios of the process, instrument, and controller elements in the loop *at the critical frequency.* The result is the Bode stability criterion for linear systems, which gives local results for a nonlinear system.

> The Bode stability criterion states that a closed-loop linear system is stable when its amplitude ratio is less than 1 at its critical frequency. The system is unstable if its amplitude ratio is greater than 1 at its critical frequency.

From this analysis, it is clear that a system with an amplitude ratio of exactly 1.0 would be at the stability limit, with a slight increase or decrease resulting in instability or stability, respectively. Because of small inaccuracies in modelling and nonlinearities in processes, no real process can be maintained at its stability limit.

Note that the Bode method considers all elements in the feedback loop: process, sensors, transmission, controller, and final element. Naturally, some of these may contribute negligible dynamics and can be lumped into a smaller number of transfer functions. By convention, the transfer function used in the Bode analysis is termed the *open-loop* transfer function and is represented by the symbol $G_{\text{OL}}(s)$.

$$G_{\text{OL}}(s) = G_p(s)G_v(s)G_c(s)G_s(s) \tag{10.24}$$

Before the Bode method is discussed further, limitations are pointed out. The Bode method *cannot* be applied to a few systems in which $G_{\text{OL}}(s)$ has particular features:

1. Unstable without control
2. Nonmonotonic phase angles or amplitude ratios at frequencies higher than the first crossing of $-180°$

The Bode method is not appropriate for these systems because

1. The experiment in Figure 10.8 cannot be performed for an unstable process.
2. Nonmonotonic behavior in the Bode diagram of $G_{\text{OL}}(s)$ could lead to a higher harmonic of the critical frequency for which the magnitude is greater than 1.0.

For processes with these features, the Nyquist stability analysis is recommended (Dorf, 1986).

The amplitude ratio could be determined by simulating the system with a sine input or through analytical relationships introduced in Chapter 4. Analytical relationships are much preferred, because we can not only solve specific problems with simple calculations but also gain insight into the effects of terms in the transfer function on the stability of control systems. The important relationships are summarized below for a general transfer function; these were applied to process transfer functions in Chapter 4 and will be extended here to $G_{\text{OL}}(s)$. As a brief summary of results in Chapter 4,

1. The frequency response relates the long-time output response to input sine forcing of the system.
2. The frequency response of a linear system can be easily calculated from any stable transfer function, $G(s)$, as $G(j\omega)$.

3. The amplitude ratio is the ratio of the output over the input sine magnitudes and can be calculated as

$$\text{AR} = |G(j\omega)| = \sqrt{(\text{Re}[G(j\omega)])^2 + (\text{Im}[G(j\omega)])^2} \tag{10.25}$$

4. The phase angle gives the amount that the output sine lags the input sine and can be calculated as

$$\phi = \angle G(j\omega) = \tan^{-1}\left(\frac{\text{Im}[G(j\omega)]}{\text{Re}[G(j\omega)]}\right) \tag{10.26}$$

Another important simplification provides a way for the frequency response of a series of transfer functions to be calculated from the individual frequency responses. First, each individual transfer function can be represented in polar form by

$$G_i(j\omega) = |G_i(j\omega)|\, e^{-\phi_i j} \tag{10.27}$$

The series transfer function can then be expressed as

$$G(j\omega) = \prod_{i=1}^{n} G_i(j\omega) = \left(\prod_{j=1}^{n} |G_i(j\omega)|\right)\exp\left(-\sum_{i=1}^{n}\phi_i j\right) = \text{AR}e^{-\Phi j} \tag{10.28}$$

with

$$\text{AR} = \prod_{i=1}^{n} |G_i(j\omega)| \qquad \Phi = \sum_{i=1}^{n}\phi_i$$

These are especially useful relationships, because the individual transfer functions used in the Bode method, $G_{\text{OL}}(s)$, are often in series as shown in Figure 10.4 and equation (10.24).

In addition to the simplifications in the calculation, the frequency response of a transfer function can be presented in a clear graphical manner using Bode plots. These plots, introduced in Chapter 4, present the amplitude ratio and the phase angle as a function of the frequency; an example Bode plot is given in Figure 4.10, and many additional Bode plots are given in examples in this chapter. The log scales are used to cover larger ranges of variables with reasonable accuracy. The reason for the inclusion of the phase angle plot was not obvious in Chapter 4 but becomes apparent when stability of feedback systems is evaluated, as the next few examples demonstrate.

Example 10.6. The single-tank mixing process with proportional control shown in Figure 10.9 is considered. This is the same as the three-tank mixer in Example 7.2 with the last two tanks removed. The process transfer function, which includes an ideal sensor and fast final element dynamics, is given as

$$G_p(s)G_v(s)G_s(s) = \frac{0.039}{5s+1} \tag{10.29}$$

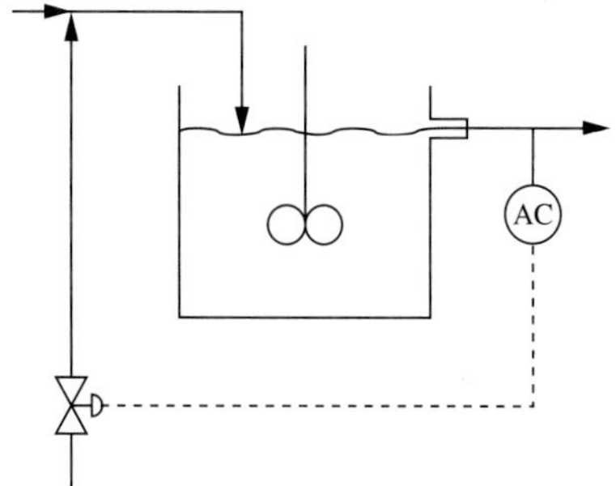

FIGURE 10.9
Mixing process analyzed in Example 10.6.

with time in minutes. Note that the process is stable without control, since it has one pole at $(-0.2, 0)$ in the real-imaginary plane, so that it satisfies the criteria in Table 10.1 for the Bode method. The stability is to be determined by the Bode method.

First, $G_{OL}(s)$ must be determined. This is the product of the valve, process, sensor, and controller transfer functions; $G_{OL}(s)$ with proportional-only control can be written as

$$G_{OL}(s) = \frac{0.039K_c}{5s + 1} \tag{10.30}$$

The magnitude and phase angle of $G_{OL}(s)$ can be calculated from $G_{OL}(j\omega)$:

$$\begin{aligned} \text{AR} &= |G_{OL}(j\omega)| && (10.31) \\ &= \left|(0.039K_c)\left(\frac{1}{1+5j\omega}\right)\left(\frac{1-5j\omega}{1-5j\omega}\right)\right| \\ &= \frac{(0.039K_c)}{\sqrt{1+25\omega^2}} \\ \Phi &= \angle G_{OL}(j\omega) \\ &= \tan^{-1}(-5\omega) \end{aligned}$$

These expressions are presented in Bode plots in Figure 10.10 for $K_c = 1$. Since the phase angle for this first-order system does not decrease below $-90°$ for any controller gain, the phase angle never reaches $-180°$, and the feedback signal cannot reinforce oscillations in the control loop. As a result, this *idealized* control system is stable for all negative feedback proportional-only controller gains ($K_c > 0$ in this case). Note that this result is consistent with the analytical dynamic response derived in Section 8.3 (page 270). As the next example illustrates, nearly every realistic system can be made unstable with improper feedback control.

Example 10.7. The mixing process and proportional controller in Example 10.6 are considered here, with the modification that the valve and sensor dynamics are modelled according to the following first-order transfer functions with short time constants:

$$G_v(s) = \frac{1}{0.033s + 1} \qquad G_s(s) = \frac{1}{0.25s + 1} \tag{10.32}$$

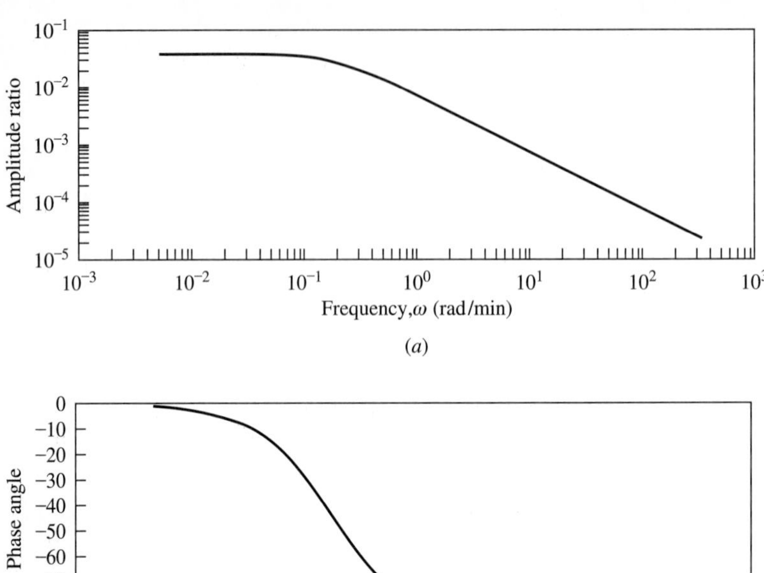

FIGURE 10.10
Bode plot for the $G_{OL}(j\omega)$ in Example 10.6, with $K_c = 1.0$.

Equations (10.28) can be used to determine the amplitude ratio and phase angle for this series system, and the results are

$$G_{OL}(s) = (0.039K_c)\frac{1}{1+5s}\,\frac{1}{1+0.25s}\,\frac{1}{1+0.033s} \tag{10.33}$$

$$\left|G_{OL}(j\omega)\right| = \frac{0.039K_c}{\sqrt{1+25\omega^2}}\,\frac{1}{\sqrt{1+0.0625\omega^2}}\,\frac{1}{\sqrt{1+0.0011\omega^2}}$$

$$\angle G_{OL}(j\omega) = \tan^{-1}(-5\omega) + \tan^{-1}(-0.25\omega) + \tan^{-1}(-0.033\omega)$$

The amplitude ratio and phase angle are plotted in Figure 10.11 for a controller gain of 1.0. Because of the added dynamic elements in $G_{OL}(s)$, the phase angle exceeds $-180°$. At the critical frequency (11.6 rad/min), the following values for the amplitude ratio are determined:

$K_c = 1.0$	Amplitude ratio $= 0.0002 < 1.0$	Stable
$K_c = 500$	Amplitude ratio $= 0.10 \quad < 1.0$	Stable
$K_c = 6000$	Amplitude ratio $= 1.2 \quad > 1.0$	Unstable

As can be seen by applying the Bode stability criterion, the system is stable for controller gain values of 1.0 and 500 because the amplitude ratios at the critical frequencies are less than 1.0, and the system is unstable for a controller gain of 6000, which has an amplitude ratio greater than 1.0 at the critical frequency.

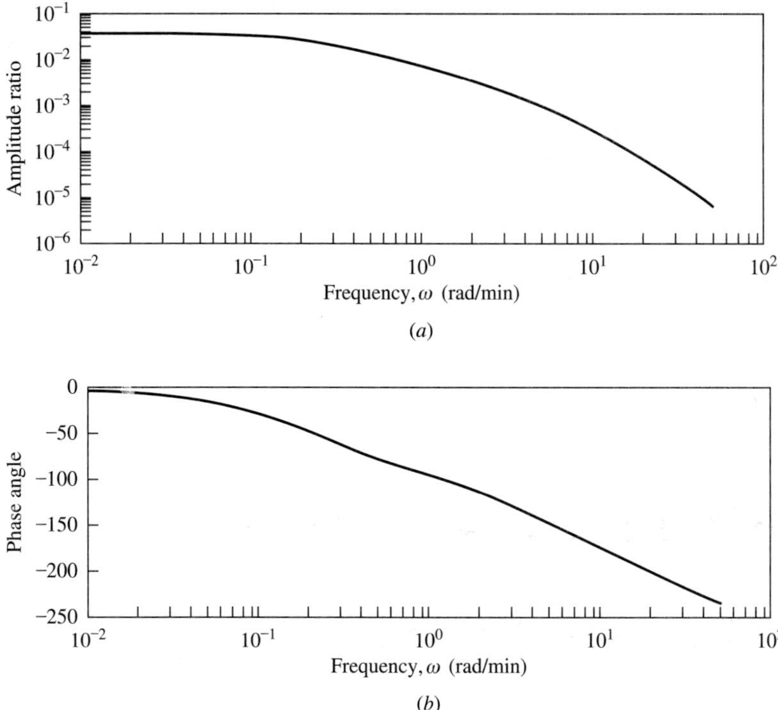

FIGURE 10.11
Bode plot of $G_{OL}(j\omega)$ for the system in Example 10.7 with $K_c = 1.0$.

Two important lessons have been learned from the last examples. The first lesson is that in theory, a stable transfer function $G_{OL}(s)$ that is first- or second-order cannot be made unstable with proportional-only feedback control, because its phase angle is never less than $-180°$. The second lesson demonstrates that all real systems have additional dynamic elements in the control loop (e.g., valve, sensor, transmission) that contribute additional phase lag and result in a phase angle less than $-180°$, albeit at a very high frequency. Thus, essentially all real process control systems can be made locally unstable (locally, since the analysis considers the linearized representation about the steady-state point) simply by increasing the magnitude of the negative feedback controller gain.

Example 10.8. The chemical reactor process and control system in Example 10.4 are changed slightly. In this case, a transportation delay of 1 min exists between the mixing point and the first stirred-tank reactor, with no reaction in the transport delay. Therefore, the process transfer function is modified to include the dead time. A proportional-integral controller is proposed to control this process with the same tuning as Example 10.4; $K_c = 15$ and $T_I = 1$. Determine whether this system is stable.

The Bode method can be applied to this example with the new aspect that dead time exists in the process. The first task is to determine $G_{OL}(s)$. As explained above, this transfer function contains all elements in the feedback loop; therefore, $G_{\mathrm{OL}}(s)$ is

$$G_{\mathrm{OL}}(s) = 15\left(1 + \frac{1}{s}\right)\frac{0.10e^{-s}}{(0.50s + 1)^s} \tag{10.34}$$

The amplitude ratio and phase angle for each element can be combined to give the amplitude ratio and phase angle of $G_{\mathrm{OL}}(j\omega)$.

$$\begin{aligned} \mathrm{AR} &= \left|15\left(1 + \frac{1}{j\omega}\right)\right| \left|\frac{0.10}{(0.50j\omega + 1)^2}\right| \left|e^{-j\omega}\right| \\ &= 15\sqrt{1 + \frac{1}{\omega^2}}\,\frac{0.10}{(1 + 0.25\omega^2)}\,(1.0) \\ \Phi &= \angle 15\left(1 + \frac{1}{j\omega}\right) + \angle\left(\frac{0.10}{(0.50j\omega + 1)^2}\right) + \angle e^{-j\omega} \\ &= \tan^{-1}(-1/\omega) + 2\tan^{-1}(-0.5\omega) - 1.0\omega\frac{360}{2\pi} \end{aligned} \tag{10.35}$$

These terms are plotted in Figure 10.12. Since the amplitude ratio is greater than 1 (1.32) at the critical frequency of 1.31 rad/min, the system is unstable. Note that the dead time introduced additional phase lag in the feedback system and caused the system

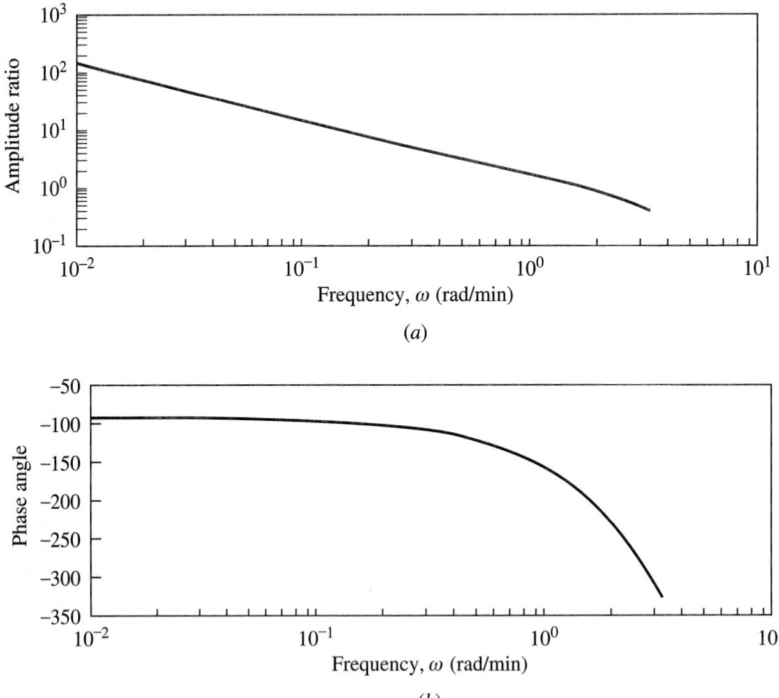

FIGURE 10.12
Bode plot of $G_{\mathrm{OL}}(j\omega)$ for Example 10.8.

to become unstable. This result agrees with our qualitative understanding that processes with dead time are more difficult to control via feedback. Stable control could be obtained by adjusting the tuning constant values.

The preceding examples have demonstrated interesting results. To expand on these experiences, it would be valuable to understand the contributions of commonly occurring process models and controller modes to the stability of a feedback control system. Also, it would be useful, when performing calculations, to have analytical and sample graphical frequency responses for these common elements. Both of these goals are satisfied by the analytical expressions and Bode plots presented to complete this section. The plots for the key process components—gain, first-order, second-order, pure integrator, and dead time—are presented in Figures 10.13*a* through *e*; these were developed from the transfer functions and expressions for amplitude ratio and phase angle in Table 10.2. The plots for the PI, PD, and PID controllers are presented in Figures 10.13*f* through *h* and were also developed from the analytical expressions in Table 10.2. Note that these plots are presented in dimensionless parameters, so that they can be used to determine the frequency responses quickly for

TABLE 10.2
Summary of amplitude ratios and phase angles for common transfer functions (ω is in rad/time, n is a positive integer)

Transfer function	Amplitude ratio	Phase angle (°)
K	K	0
$\dfrac{K}{\tau s + 1}$	$\dfrac{K}{\sqrt{\tau^2\omega^2 + 1}}$	$\tan^{-1}(-\omega\tau)$
$\dfrac{K}{\tau^s s^2 + 2\tau\xi s + 1}$	$\dfrac{K}{\sqrt{(1 - \tau^2\omega^2)^2 + (2\tau\omega\xi)^2}}$	$\tan^{-1}\left(\dfrac{-2\tau\omega\xi}{1 - \tau^2\omega^2}\right)$
$\dfrac{K}{(\tau s + 1)^n}$	$K\left(\dfrac{1}{\sqrt{\tau^2\omega^2 + 1}}\right)^n$	$-n\tan^{-1}(-\omega\tau)$
$e^{-\theta s}$	1	$-\theta\omega\left(\dfrac{360}{2\pi}\right)$
$\dfrac{1}{As}$	$\dfrac{1}{A\omega}$	−90
$K_c\left(1 + \frac{1}{T_I s}\right)$	$K_c\sqrt{1 + \dfrac{1}{\omega^2 T_I^2}}$	$\tan^{-1}\left(\dfrac{-1}{\omega T_I}\right)$
$K_c(1 + \tau_d s)$	$K_c\sqrt{1 + (T_d\omega)^2}$	$\tan^{-1}(T_d\omega)$
$K_c\left(1 + \frac{1}{T_I s} + T_d s\right)$	$K_c\sqrt{1 + \left(T_d\omega - \dfrac{1}{T_I\omega}\right)^2}$	$\tan^{-1}\left(T_d\omega - \dfrac{1}{T_I\omega}\right)$

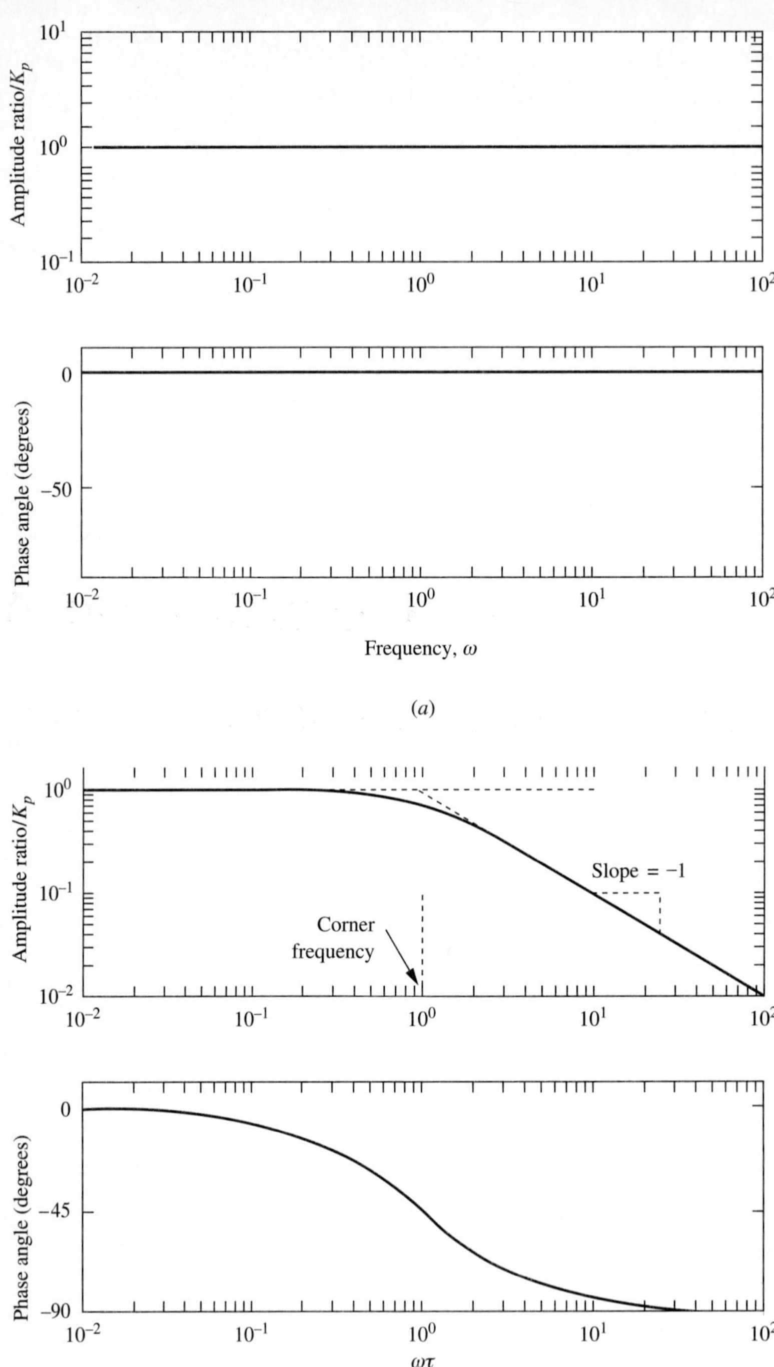

FIGURE 10.13
Generalized Bode plots: (*a*) gain; (*b*) first-order system.

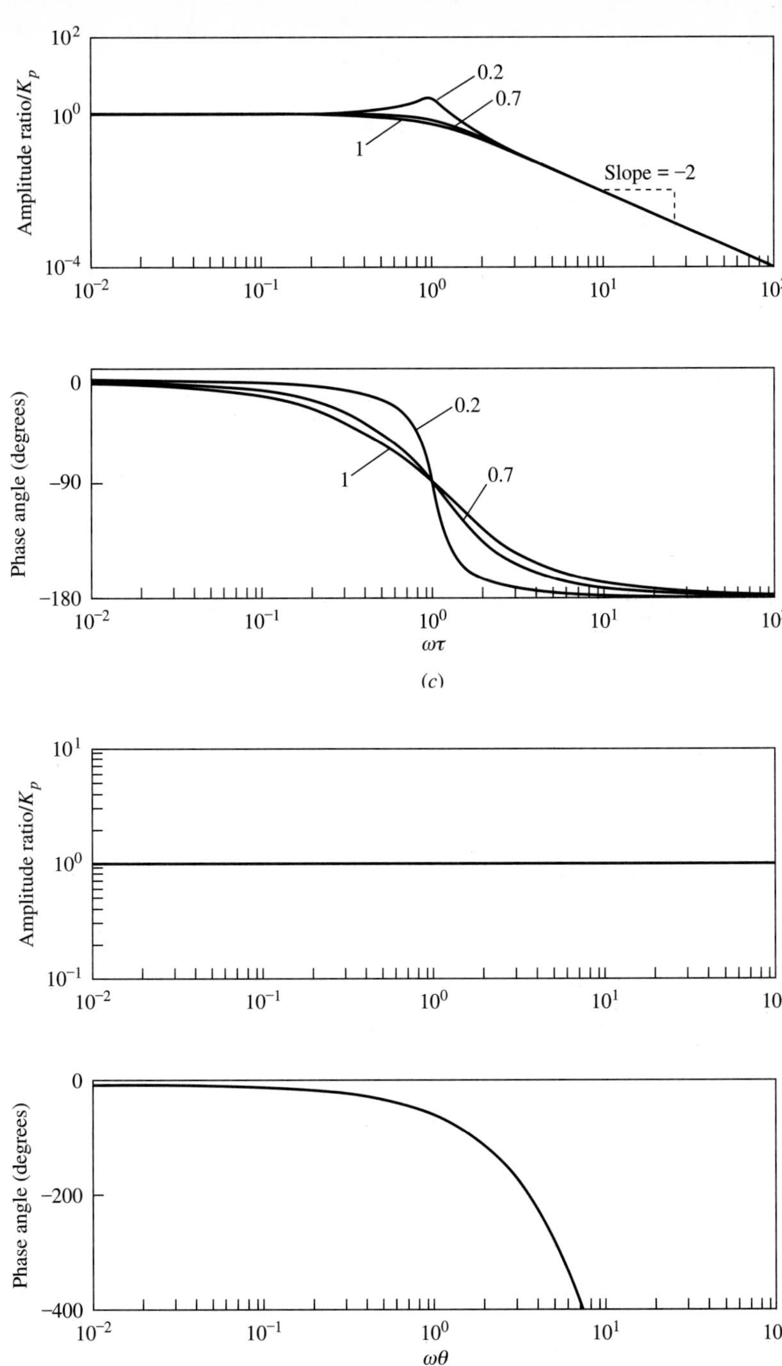

FIGURE 10.13 ***(continued)***
Generalized Bode plots: (*c*) second-order system (the parameter is the damping coefficient ξ); (*d*) dead time.

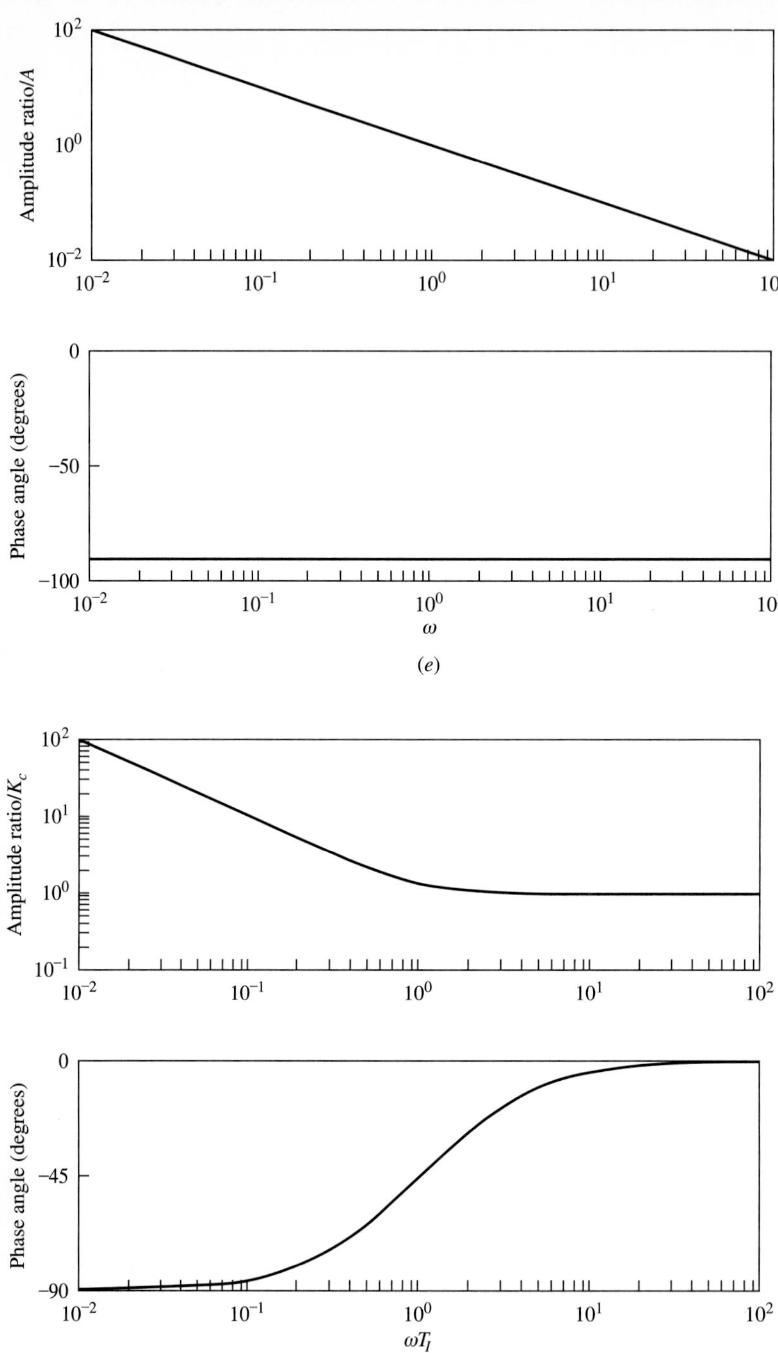

FIGURE 10.13 ***(continued)***
Generalized Bode plots: (*e*) integrator; (*f*) proportional-integral controller.

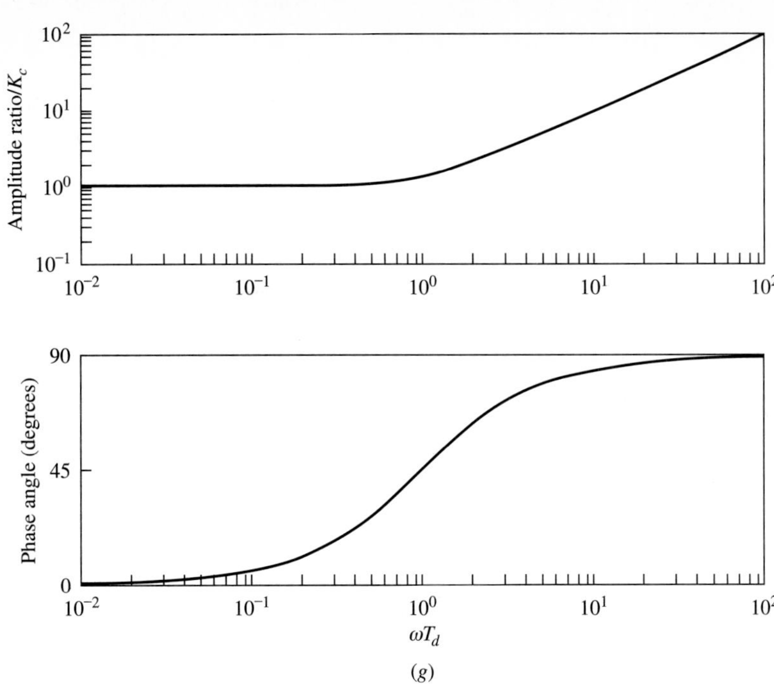

(g)

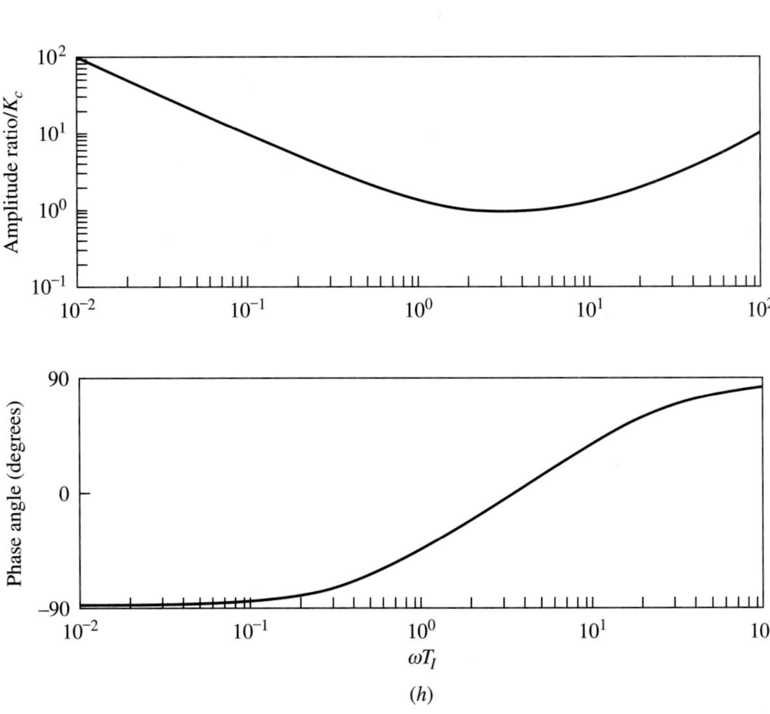

(h)

FIGURE 10.13 ***(continued)***
Generalized Bode plots: (*g*) proportional-derivative controller; (*h*) proportional-integral-derivative controller for which the derivative time is one-tenth of the integral time.

a system conforming to one of the general models. As an example of the preparation of the dimensionless plots, the expressions for the amplitude ratio and phase angle for a first-order system are given in Table 10.2 and repeated here:

$$\text{AR} = \frac{K_p}{\sqrt{\omega^2\tau^2 + 1}} \qquad \frac{\text{AR}}{K_p} = \frac{1}{\sqrt{\omega^2\tau^2 + 1}} \tag{10.36}$$

$$\phi = \tan^{-1}(-\omega\tau)$$

Noting that the two variables ω and τ always appear as a product, they can be combined into one variable, $\omega\tau$, and the Bode plots expressed as a function of this single variable. Also, the amplitude ratio can be normalized by dividing by the process gain K_p. Similar manipulations are possible for the transfer functions of the other building blocks.

Example 10.9. Determine the amplitude ratio and phase angle of the following transfer function at a frequency of 0.40 rad/min.

$$G(s) = \frac{0.039}{(1 + 5s)^2} \tag{10.37}$$

The first step is to calculate the parameters in the generalized Figure 10.13*c*. The results can be calculated as follows:

$$\tau^2 = 25 \qquad \tau = 5.0 \qquad \xi = \frac{10}{2\tau} = 1.0 \tag{10.38}$$

From the generalized charts, $\text{AR}/K_p = 0.2$; $\text{AR} = 0.2(0.039) = 0.0078$; and $\phi = -125$. The same answers can be determined directly using the equations in Table 10.2.

The Bode plot of any $G_{\text{OL}}(j\omega)$ for a system consisting of a series of common elements can be easily prepared by using the expressions for these individual elements and equation (10.28). The usefulness of the general plots is not primarily in simplifying the calculations, because the calculations are not difficult by hand and computer programs are available to automate the calculations and plot the results. The real importance is in highlighting the contributions of various components to the stability of a feedback system. For example, note that an element in the feedback path that has a large phase angle contributes to lowering the critical frequency. Since most process models have amplitudes that decrease with increasing frequency, a lower critical frequency yields a higher amplitude ratio for $G_{\text{OL}}(j\omega)$. Since a lower amplitude ratio is desired to maintain the amplitude ratio below 1.0 for stability, elements with the larger phase angle tend to destabilize a feedback control system. Some of the key features of the most important transfer functions are summarized in Table 10.3. The readers are encouraged to compare the entries in the table with the Bode figures so that they understand the major contributions of each transfer function.

Before we move on to controller tuning, a word of caution regarding terminology is provided. The common term for the expression in equation (10.24), $G_{\text{OL}}(s)$, is the open-loop transfer function; hence, the subscript OL. The term refers to Figure 10.8, where the feedback loop was temporarily opened. Unfortunately, the term

TABLE 10.3
Summary of key features of process transfer function frequency responses

Transfer function	Amplitude ratio, AR	Phase angle, ϕ	Key feature
Gain	Constant	0	
First-order	Monotonically decreases with increasing frequency, limiting slope $= -1$	0 to $-90°$	At corner frequency ($\omega = 1/\tau$), AR = 0.707, and $\phi = -45°$
Second-order	(1) Shape depends on the damping ratio, can be nonmonotonic (2) Limiting slope $= -2$	0 to $-180°$	(1) AR is not monotonic for small damping coefficients (2) Key frequency is $\omega = 1/\tau$
nth order from n first order in series	Monotonically decreases with increasing frequency, limiting slope $= -n$	0 to $(-90)n°$	
Dead time	1.0	0 to $-\infty$	At $\omega = 1/\theta$, $\phi = -57.3$ and decreases rapidly as ω increases
Integrator	Straight line with a slope of -1 from $-\infty$ to $+\infty$ through ($\omega = 1$, AR $= 1$)	$-90°$	At $\omega = 1/A$, AR $= 1$

Notes:

1. All slopes refer to the Bode diagram ($\Delta \log(\text{AR})/\Delta \log(\omega)$).
2. The phase angles for all transfer functions in this table decrease monotonically as frequency increases.

open-loop is also used for the response of a process to an input change without control. In this second case, the transfer function being considered is either the process transfer function $G_p(s)$ or the disturbance transfer function $G_d(s)$, depending on which input-output relationship is being considered. To avoid misinterpretation, it is best to relate the subscript OL to Figure 10.8 and to recognize that $G_{\text{OL}}(s)$ contains all elements in the feedback loop, *including the controller*. The conventional terminology, although not as clear as desired, is used in this book to prevent confusion when consulting other references.

In summary, Bode stability analysis provides a method for determining the stability of most feedback control systems that include dead time. The calculations are relatively simple by hand when $G_{\text{OL}}(s)$ involves a series of individual transfer functions, and a computer can be programmed to perform the calculations automatically. In addition to providing a quantitative test, the Bode analysis yields insight into the effects on stability of various elements in the feedback loop.

10.7 CONTROLLER TUNING BASED ON STABILITY: ZIEGLER-NICHOLS CLOSED-LOOP

The Bode stability analysis provides a way to determine whether a process and feedback controller, with all elements completely specified, is stable. It is possible to alter the procedure slightly to determine, for a given process, the value of the gain for a *proportional-only* controller that results in a desired amplitude ratio for $G_{OL}(j\omega)$ at its critical frequency. In particular, it is straightforward to determine the controller gain that would result in the system being on the margin just between stable and unstable behavior. Note that the proportional-only controller affects the amplitude ratio but not the phase angle, thus making the calculation easier.

The importance of this approach is that the results of the calculation (the controller ultimate gain and critical frequency) can be used with tuning rules presented in this section to determine initial tuning for P, PI, and PID controllers. This tuning method is an alternative to the method presented in the previous chapter. While the tuning rules do not generally give as good performance as the Ciancone correlations for simple first-order-with-dead-time processes, the method in this section has two advantages:

1. It can be applied to processes that are not well modelled by first-order-with-dead-time models.
2. It provides considerable insight into the effects of all loop elements (process, instrumentation, and control algorithm) on stability and proper tuning constant values.

As with most tuning methods, the starting point is a process model that can be determined by fundamental modelling or by empirical model identification. The method then follows four steps.

1. Plot the amplitude ratio and the phase angle in the form of a Bode plot for $G_{OL}(s)$. At this step, the controller is a proportional-only algorithm with the gain K_c set to 1.0.
2. Determine the critical frequency ω_c and the amplitude ratio at the critical frequency, $|G_{OL}(j\omega_c)|$.
3. Calculate the value of the controller gain for a proportional-only controller that would result in the feedback system being at the stability margin. Since the stability margin is characterized by an amplitude ratio of 1.0 for $G_{OL}(j\omega_c)$, and K_c does not influence the critical frequency, the controller gain at the stability limit can be determined by first calculating the critical frequency and then calculating the controller gain.

$$\angle G_{OL}(j\omega_c) = \angle G_p(j\omega_c)G_v(j\omega_c)G_s(j\omega_c) = -180° \quad (10.39)$$

$$\left|G_{OL}(j\omega_c)\right| = K_u \left|G_p(j\omega_c)G_v(j\omega_c)G_s(j\omega_c)\right| = 1.0$$

$$\therefore K_u = \frac{1}{\left|G_p(j\omega_c)G_v(j\omega_c)G_s(j\omega_c)\right|}$$

$$P_u = \frac{2\pi}{\omega_c} \quad (10.40)$$

K_u, termed the *ultimate gain,* is the controller gain that brings the system to the margin of stability at the critical frequency. P_u, termed the *ultimate period,* is the period of oscillation of the system at the margin of stability. Note that K_u has the units of the inverse of the process gain $(K_p K_v K_s)^{-1}$ and that P_u has the units of time.

4. Calculate the controller tuning constant values according to the Ziegler-Nichols closed-loop tuning correlations given in Table 10.4 (Ziegler and Nichols, 1942). The description "closed-loop" indicates that the analysis is based on the stability of the closed-loop feedback system, $G_{OL}(s)$. These correlations have been developed to provide acceptable control performance (they selected a 1 : 4 decay ratio) with reasonably aggressive feedback action; they believed that this also maintains the system a *safe margin from instability.*

Example 10.10. Calculate controller tuning constants for the three-tank mixing process (page 249) by using the Ziegler-Nichols closed-loop method.

The transfer function for this process has already been developed, and the Bode plot of the transfer function with ($K_c = 1$) is presented in Figure 10.14 based on

$$\angle G_{OL}(j\omega) = 3\tan^{-1}(-5\omega)$$

$$\left|G_{OL}(j\omega)\right| = 0.039\left(\frac{1}{\sqrt{1+5^2\omega^2}}\right)^3$$

TABLE 10.4
Ziegler-Nichols closed-loop tuning correlations

Controller	K_c	T_I	T_d
P-only	$K_u/2$	—	—
PI	$K_u/2.2$	$P_u/1.2$	—
PID	$K_u/1.7$	$P_u/2.0$	$P_u/8$

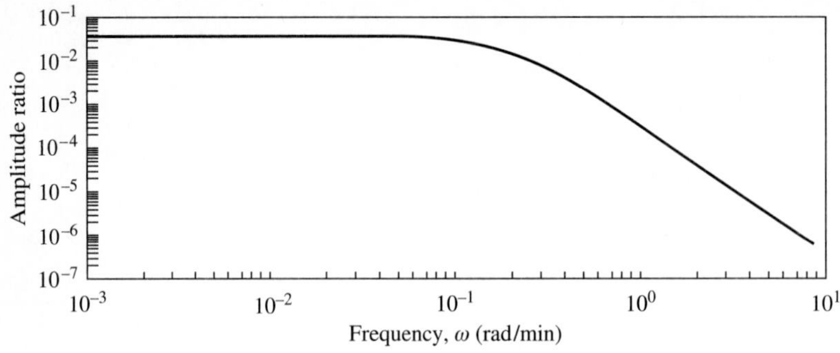

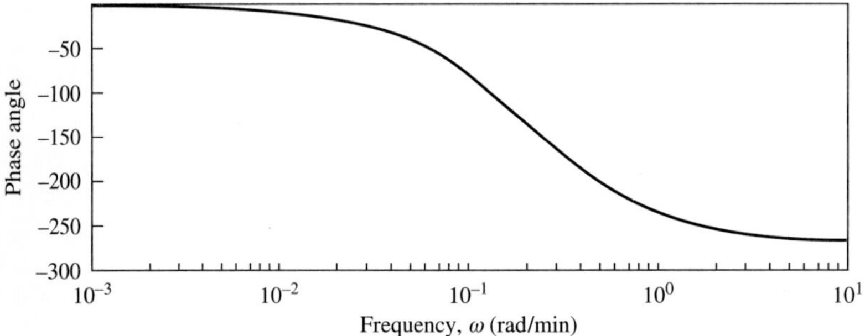

FIGURE 10.14
Bode plot of $G_{OL}(j\omega)$ for Example 10.10 with $K_c = 1$.

If the plot were not available, the calculations would have to be performed by hand. They involve a trial-and-error procedure to determine the critical frequency and are often arranged in a table similar to the following results.

Frequency ω (rad/min)	Phase angle $\phi(°)$	Amplitude ratio AR
0.10	−79.7	0.0279
0.20	−135	0.0138
0.35	−180.8 (critical frequency)	0.0048
0.40	−190.3	0.0035

From the results in the table, the ultimate gain and period can be determined to be $P_u = 2\pi/\omega_c = 17.9$ min and $K_u = 1/\text{AR}_c = 208$. The tuning constants for P, PI, and PID controllers according to the Ziegler-Nichols correlations are

Controller	K_c	T_I	T_d
P-only	104	—	—
PI	94.5	14.9	—
PID	122.4	8.95	2.2

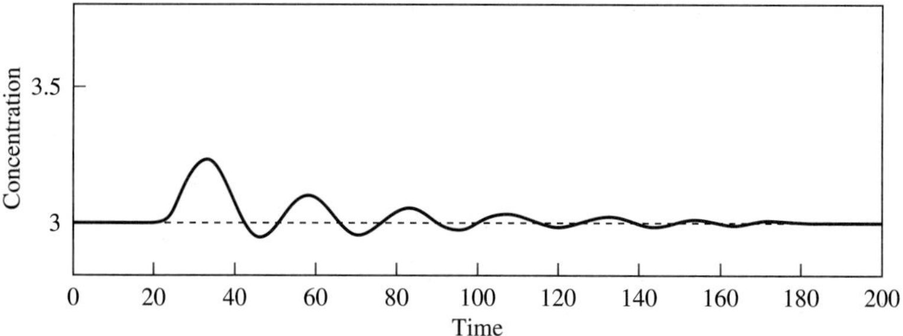

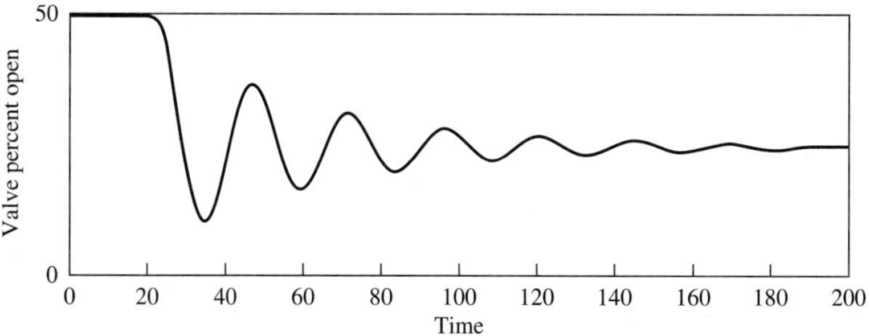

FIGURE 10.15
Dynamic response of three-tank mixing control system in Example 10.10 with Ziegler-Nichols tuning.

A sample of the transient response for a step change of +0.8%A in the feed concentration under PI control is given in Figure 10.15. As can be seen, the control performance is quite oscillatory, resulting in large variation in the manipulated variable and in a long settling time. For most plant situations, this is too oscillatory, and control performance for this system similar to Figure 9.6 would be preferred. The engineer could fine-tune the controller constants using the concepts presented in Section 9.6.

Example 10.11. Calculate tuning for a PI controller applied to the series chemical reactors in Example 10.8. Recall that this is a second-order-with-dead-time process.

The Bode plot for $G_{OL}(j\omega)$ with $K_c = 1$ is given in Figure 10.16. Note that the contribution of the individual elements in $G_{OL}(s)$ can be determined using the following relationships for transfer functions in series, equation (10.28):

$$\angle G_{OL}(j\omega) = \angle \frac{1}{1+0.5j\omega} + \angle \frac{1}{1+0.5j\omega} + \angle e^{-1j\omega} + \angle(0.10) + \angle K_c|_{=1}$$

$$= \tan^{-1}(-0.5\omega) + \tan^{-1}(-0.5\omega) - \omega\frac{360}{2\pi} + 0 + 0$$

$$|G_{OL}(j\omega)| = \left|\frac{1}{1+0.5j\omega}\right|\left|\frac{1}{1+0.5j\omega}\right|\left|e^{-j\omega}\right||0.10|\left|K_c|_{=1}\right|$$

$$= \sqrt{\frac{1}{1+0.25\omega^2}}\sqrt{\frac{1}{1+0.25\omega^2}}(1.0)(0.10)(1.0)$$

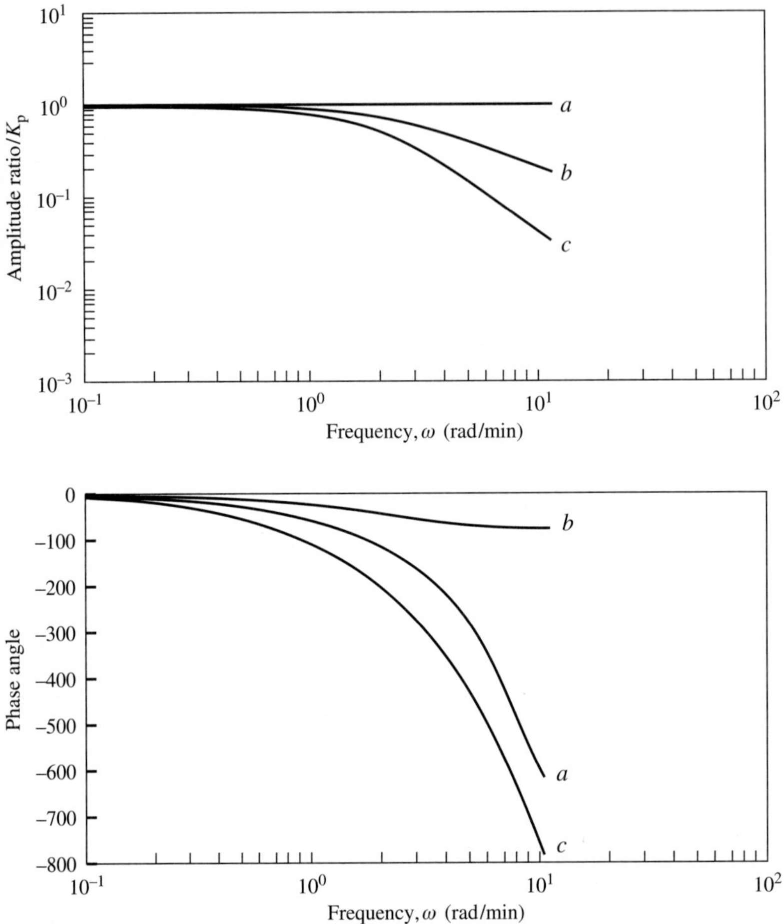

FIGURE 10.16
Bode plot of Example 10.11 with $K_c = 1$; for (*a*) the dead time, (*b*) one first-order system, and (*c*) the entire transfer function $G_{OL}(j\omega)/K_p$.

The results in Figure 10.16 are presented so that the effects of the individual process elements are clearly displayed. The dead time and one first-order system are designated as *a* and *b*, respectively. The overall amplitude ratio and phase angle for $G_{OL}(j\omega)$ can be determined from the foregoing equations. When the frequency responses of the individual elements are presented in the Bode plot, the overall amplitude ratio is the *sum* of the distances on the plot of the individual deviations from 1.0, since the amplitude ratio is plotted on a log scale. Also, the total phase angle is the *sum* of the distances on the plot of the individual deviations from zero degrees, since the phase angle is plotted on a linear scale. These rules are not particularly important as far as simplifying the calculations, which are easily programmed; however, they help the engineer visualize the effects on stability of individual elements in the feedback loop. For example, any element that contributes a large phase lag itself will cause a large phase lag for $G_{OL}(j\omega)$. From this figure, the critical frequency is 1.73 rad/min and the magnitude at this frequency is 0.057; thus, the controller tuning would be, according to the Ziegler-Nichols tuning correlations, $K_c = 8.0\%$ open/(mole/m^3) and $T_I =$ 3.0 min.

Before this section is concluded, two common questions are addressed. First, the novice often has difficulty in selecting an initial frequency for the trial-and-error calculation for the critical frequency. Since an exact guess is not required, a good initial estimate can usually be determined from the relationships in Tables 10.2 and 10.3, along with the plots in Figure 10.13. Basically, the initial frequency should be taken in the region where the Bode diagrams of the individual elements change greatly with frequency. Rough initial estimates for the frequency are given by the following expressions:

$$\omega\tau = 1 \qquad \text{(first-order system)}$$

$$\omega\tau = 1 \qquad \text{(second-order system)}$$

$$\omega\theta = 1 \qquad \text{(dead time)}$$

When these calculations give very different results, use the lowest of the estimated frequencies to begin the trial-and-error calculations, which usually converge quickly.

The second common question regards the required accuracy of the converged answer. The engineer must always consider the accuracy of the information used in a calculation when interpreting the results. In Chapter 6, the results of empirical model fitting were found to have significant errors, usually 10 to 20% in all parameters. Therefore, it is not necessary to determine the critical frequency so that the phase angle deviation from $-180°$ is less than $0.001°$! A few degrees error is usually acceptable. In addition, our application of the results in determining tuning constants must consider the likely error in the model, as discussed in the next section.

10.8 CONTROLLER TUNING AND STABILITY—SOME IMPORTANT INTERPRETATIONS

Analysis using the Bode plots provides a quantitative method for evaluating how elements in the control loop influence stability and tuning. The principles and examples presented so far have demonstrated important results, which are reinforced in the following six interpretations, discussed with further examples. (To shorten the presentations, some examples are based on transfer function models, but the gains, time constants, and dead times are always related to physical properties.) The reader is advised that these interpretations are very important, not only in tuning single-loop controllers but also in designing more complex control strategies and process modifications to achieve desired control performance.

Interpretation I: Effect of Process Dynamics on Tuning

Clearly, the types of process and instrument equipment in the control loop affect the system stability and feedback tuning constants. It is worthwhile determining how process dynamics affect feedback control, specifically the gain and integral time of a PI controller. Since the ultimate gain of the proportional-only controller is the inverse of the amplitude ratio at the critical frequency, a higher controller gain for a stable system is achieved by decreasing the amplitude ratio at the critical frequency. Also, the amplitude ratio generally decreases for process elements as the frequency increases. Therefore, smaller time constants and dead times lead to a larger allowable controller gain. By the same logic, smaller values of the time constants and dead times lead to a smaller integral time, which, since integral time appears in the denominator, has the effect of giving stronger control action. The general conclusion is that more and longer time constants and dead times lead to detuning of the PID controller and that fewer and shorter time constants and dead times lead to larger controller gain, smaller integral time, and stronger feedback action. We expect that stronger feedback action will give better control performance, as is discussed in depth in Chapter 13.

Example 10.12. Consider a set of processes with one to seven first-order systems in series, each with a gain of 1.0 and a time constant of 5.0. Determine the PI tuning for each of these systems.

The expressions for the amplitude ratio and phase angle for a series of n first-order systems can be developed using equations (5.40) and (10.36) and are given as

$$\text{AR} = \left(\frac{K_p}{\sqrt{1 + \omega^2 \tau^2}} \right)^n \qquad \phi = n \tan^{-1}(-\omega\tau) \qquad \text{with } K_p = 1.0 \quad \text{and} \quad \tau = 5.0$$

The Ziegler-Nichols closed-loop tuning for these systems is as follows

n	ω_c	$\text{AR}\vert_{\omega c}$	K_c	T_I
1	∞	—	∞	—
3	0.35	0.122	3.72	15.0
5	0.145	0.348	1.31	36.1
7	0.096	0.484	0.94	54.5

Clearly, the controller must be detuned as the feedback dynamics become slower.

The previous example clearly demonstrates that time constants affect feedback tuning and stability. Next, we would like to learn the relative importance of dead times and time constants. Since many processes can be represented by a first-order-with-dead-time model, the key relationships between tuning and fraction dead time $\theta/(\theta + \tau)$ is investigated for this model and Ziegler-Nichols PID tuning. In fact, correlations similar to those developed in Chapter 9 can be calculated using the Bode stability and Ziegler-Nichols methods. The PID controller gain correlations for Ciancone and Ziegler-Nichols are compared in Figure 10.17. The correlations have the same general shape, which points to the importance of the stability limit in determining the most aggressive control action. Recall that stability was not explicitly considered in the Ciancone method, although tuning that gave unstable or oscillatory systems would have a large IAE and thus would not have been selected as optimum. Note that the Ciancone gain values are lower, partly because of the objectives

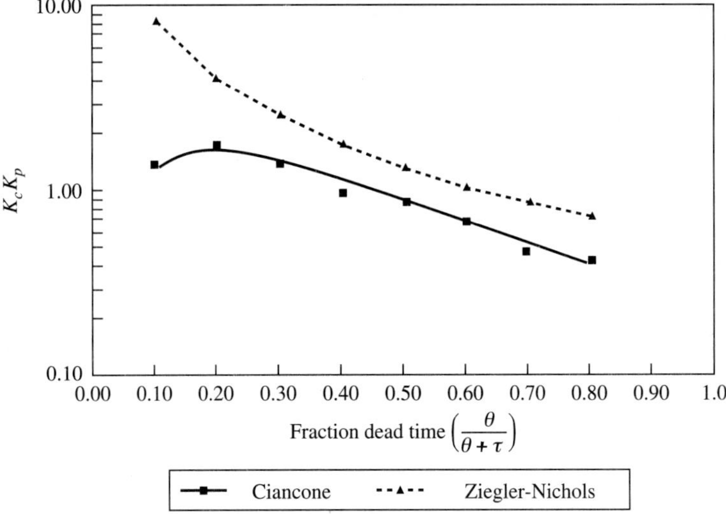

FIGURE 10.17
The effect of fraction dead time on PID controller gain with $\theta + \tau$ constant.

of robust performance with model errors and partly because of the limitation on manipulated-variable variation with a noisy measured controlled variable. We would expect the Ciancone correlations to yield controllers that are more robust than those developed with Ziegler-Nichols tuning and thus perform better when realistic model errors occur.

The Bode analysis demonstrates the fundamental relationship between fraction dead time and tuning; the controller gain must be decreased to maintain stability as the fraction dead time increases (at constant $\theta + \tau$). Finally, it is important to reiterate that only the terms in the characteristic equation influence stability. Therefore, the disturbance transfer function $G_d(s)$ and the manner in which the set point is changed do not influence the stability of the feedback control system.

Increasing time constants and dead times requires detuning of the PID controller. The dead time has a greater effect on the phase lag and tuning. Therefore, increasing the fraction dead time, $\theta/(\theta + \tau)$, at constant $\theta + \tau$ requires detuning of the PID controller.

Example 10.13. The two following different first-order-with-dead-time processes are to be controlled by PI controllers. Calculate the tuning constants for each and compare the results.

	Plant A	Plant B
K_p	1.0	1.0
τ	8.0	2.0
θ	2.0	8.0

For each plant the Bode stability and Ziegler-Nichols tuning calculations are summarized as

	Plant A	Plant B
ω_c	0.86	0.32
AR_c	0.144	0.84 (P-only with $K_c = 1$)
K_u	4.08	1.19
P_u	7.3	19.60
K_c	4.1	0.70
T_I	3.65	9.8
T_d	0.91	2.45

Note that the two plants have the same time to reach 63% of their open-loop response after a step change: $\theta + \tau$. Even though they have the same "speed" of response, Plant B, with the higher fraction dead time, $\theta/(\theta + \tau)$, has a much smaller controller gain and larger integral time. The difference in controller tuning constants, resulting from the different stability bound, certainly will result in poorer control performance for Plant B. (Naturally, the longer dead time for plant B also degrades the control performance.)

Interpretation II: Effect of Controller Modes on Stability

Each mode of the PID controller affects the stability of the feedback system. As shown in Figure 10.13*a*, a gain in $G_{OL}(s)$ does not affect the phase angle, although it affects the amplitude ratio. Therefore, increasing the magnitude of the controller gain tends to *destabilize* the system; that is, move it toward an amplitude ratio greater than 1. The proportional-integral controller shown in Figure 10.13*f* affects both the amplitude ratio and the phase angle; it increases the amplitude ratio beyond the proportional-only controller and increases the phase lag. Thus, increasing the gain and decreasing the integral time tend to *destabilize* the feedback system. The proportional-derivative controller shown in Figure 10.13*g* increases the amplitude ratio but contributes negative phase lag, referred to as *phase lead.* Therefore, the derivative mode tends to *stabilize* the feedback system. These qualitative results are reflected in the Ziegler-Nichols tuning rules, which show the controller gain decreasing from P-only to PI control and increasing from PI to PID control.

Example 10.14. The stability of the three-tank mixing process is to be determined for two cases: (*a*) under proportional-only feedback control ($K_c = 122$) and (*b*) under proportional-integral feedback control ($K_c = 122$ and $T_I = 8$). Note that the controller gain is the Ziegler-Nichols value for the PID controller from Example 10.10, but the integral time is slightly different and the derivative time is 0.

The Bode plots are presented in Figures 10.18*a* and *b*. From Figure 10.18*a*, it is determined that case (*a*) is stable, since the amplitude ratio (0.60) is less than 1.0 at the critical frequency (0.35 rad/min). From Figure 10.18*b*, it is determined that case (*b*) is unstable, because it has an amplitude ratio greater than 1.0 (1.3) at its critical

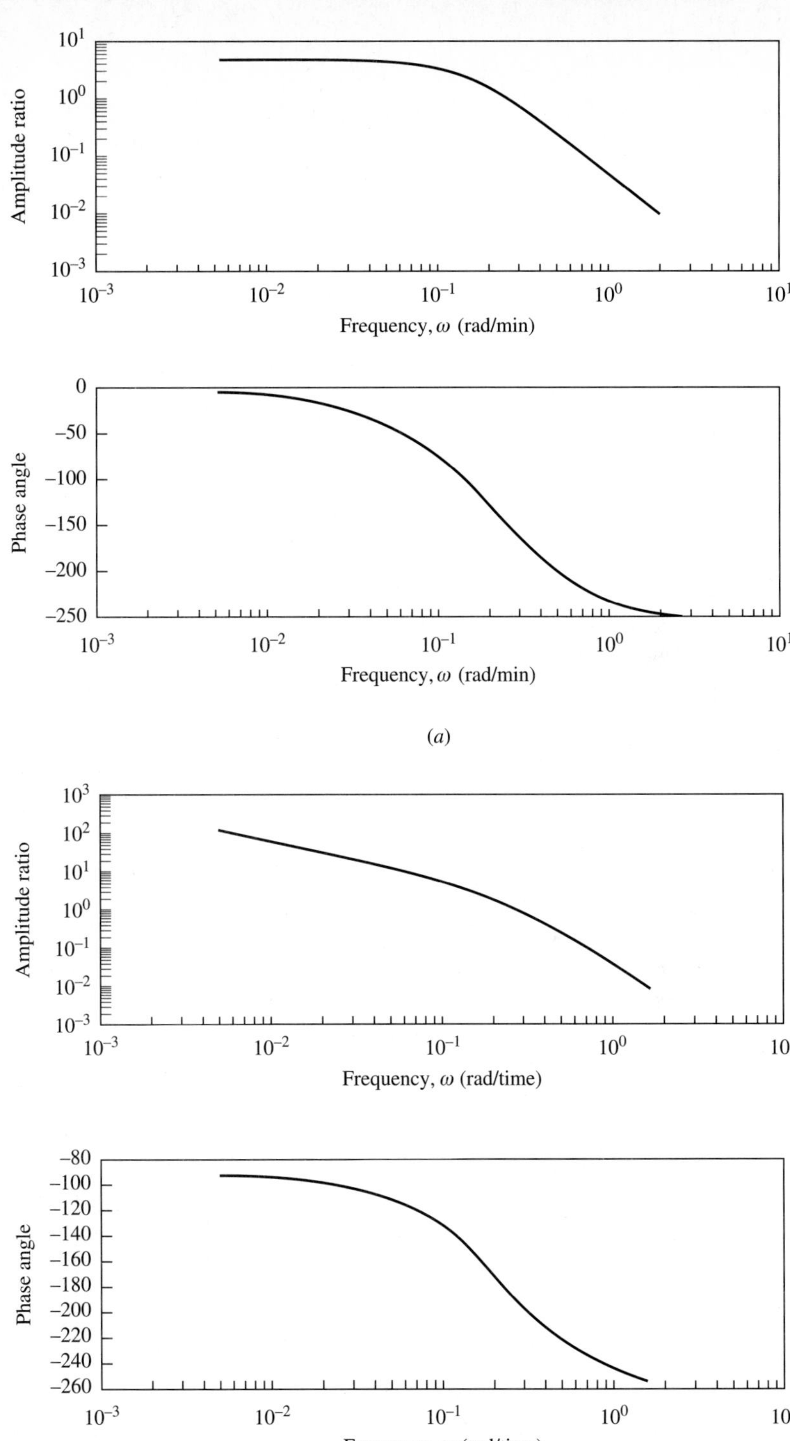

FIGURE 10.18
Bode plots of $G_{OL}(j\omega)$ for Example 10.14.

frequency (0.25 rad/min). This result clearly demonstrates the effect of the integral mode, which tends to destabilize the control system, since it contributes phase lag. Remember that the integral mode is nearly always retained, in spite of its tendency to destabilize the control system, because it ensures *zero steady-state offset.*

Interpretation III: Effect of Modelling Errors on Stability

The preceding examples in this chapter have assumed that the models of the process were known exactly. Since the true dynamic response is never known exactly, it is important to determine how model errors affect stability. The best estimate of the dynamics will be called the *nominal model.* The general trends are relatively easy to ascertain based on the Bode stability analysis; plants with amplitude ratios and phase lags greater than their nominal models will be closer to the stability margin than the nominal model. As a example, consider a first-order-with-dead-time process. Assuming that a nominal model is used to calculate the tuning constants, the system will tend to be closer to the stability margin than predicted if (1) the actual process dead time is greater than the nominal model, (2) the process gain is greater than the nominal model, or (3) the process time constant is greater than the nominal model.

A consideration of modelling errors should be an integral part of any controller tuning method. The time-domain Ciancone method in Chapter 9 specified modelling errors and optimized the dynamic responses for several cases simultaneously, and the Ziegler-Nichols correlations included a factor for model error by reducing the amplitude ratio at the critical frequency to about 0.5. As a result, a combination of model errors would have to cause the actual amplitude ratio at the critical frequency to be about twice the nominal model value for the system to be unstable. An alternative to the Ziegler-Nichols guideline for tuning based on the stability limit explicitly considers a measure of potential error. This method adjusts the controller tuning constant values so that the system is on the stable side of the limit by a specified amount. Either of the following specifications is used.

GAIN MARGIN. The amplitude ratio of $G_{OL}(j\omega)$ at the critical frequency is equal to 1/GM, where GM is called the *gain margin* and should be greater than 1. This ensures that the system is stable for any process modelling error that increases the actual AR of the process by less than a factor of GM. A typical value for GM is 2.0, but a larger value would be appropriate if large modelling errors that primarily influenced the amplitude ratio were anticipated.

PHASE MARGIN. The phase angle of $G_{OL}(j\omega)$ where the amplitude ratio is 1.0 is equal to $(-180° + \text{PM})$, with PM a positive number referred to as the *phase margin.* A positive phase margin ensures that the system is stable for model errors that decrease the phase angle. A typical value for the phase margin is 30°, but a larger value would be appropriate if larger modelling errors were anticipated.

Even if the models were perfect, the values of the gain margin and phase margin should not be reduced much below 2.0 and 30°, respectively. If they were reduced further, the performance of the feedback control system would be poor (i.e., highly oscillatory), because the roots of the characteristic equation would be too near the imaginary axis. Thus, these margins can be used as a way to include additional conservatism in the Ziegler-Nichols tuning methods if large model errors are expected.

Example 10.15. A nominal model for a process is as follows along with parameters defining Processes I and II, which represent the range of the true process dynamics experienced as operating conditions vary. Naturally, we never know the true process, but we can usually estimate the potential deviations between the nominal model and true process from an analysis of repeated model identification experiments and from fundamental models, which indicate how the process dynamics change with, for example, the flow rate.

(*a*) Determine values for the PI tuning constants based on the Ziegler-Nichols method for the nominal model and determine the resulting gain and phase margins.

(*b*) Determine the stability of the true process at the extremes of its parameter ranges using the tuning based on the nominal model.

		True Process	
	Nominal Model	**I**	**II**
K_p	1.0	1.0	1.0
τ	9.0	9.5	8.0
θ	1.0	0.5	2.0

Tuning can be determined for the nominal model using the Bode and Ziegler-Nichols closed-loop methods, giving the following results:

$$G_{OL}(s) = K_c \left(\frac{1.0e^{-s}}{9s + 1} \right) \qquad \text{with } K_c = 1.0$$

$$\omega_c = 1.65 \qquad |G_{OL}(j\omega)| = 0.067 \qquad K_u = 14.9$$

$$K_c = 6.8 \qquad T_I = 3.2 \qquad \text{Gain margin} = 2.0 \qquad \text{Phase margin} = 30^\circ$$

The gain and phase margins for the nominal model using the Ziegler-Nichols tuning, which satisfies the general guidelines for gain and phase margins, are now applied to the extremes of the dynamics of the true process.

$$G_{OL}(s) = 6.8 \left(1 + \frac{1}{3.2s} \right) \frac{1.0e^{-\theta s}}{\tau s + 1}$$

	True process with PI control	
	I	**II**
ω_c	3.1	0.66
AR_c	$0.23 < 1$	$1.39 > 1$

Note that Process I is stable with the nominal tuning, whereas Process II is unstable. The general trend should be expected, since Process II has a longer dead time, which contributes substantial phase lag and is more difficult to control. Process I has a shorter dead time, which contributes less phase lag and is easier to control. The key point is that the control system would become unstable for the moderate amount of variation of Process II from the nominal model.

Thus:

> The control engineer should not rely exclusively on general tuning guidelines but should include information on the expected variation in process dynamics when tuning controllers.

The goal is normally for the *worst-case* model error to be stable and to give an acceptable (usually stable and not too oscillatory for the worst case) closed-loop dynamic response. Further calculations for Example 10.15 indicate that gain and phase margins for the *nominal model* of 4 and 60°, respectively, were required to give satisfactory performance for Process II. (This tuning gave gain and phase margins of 2 and 40°, respectively, for Process II.)

The need for a larger stability margin can be understood when the Bode plot is prepared using the entire range of models possible, not just the nominal model. The range of possible models depends on the reasons for model errors; here the simplest approach is taken, with the process models I and II defining the extremes of the amplitude and phase angles possible. The Bode plot of $G_{OL}(s) = G_c(s)G_p(s)$, with the PI controller tuning for the nominal plant from Example 10.15, gives the range of values in Figure 10.19. Any amplitude ratio and phase angle within the two lines are

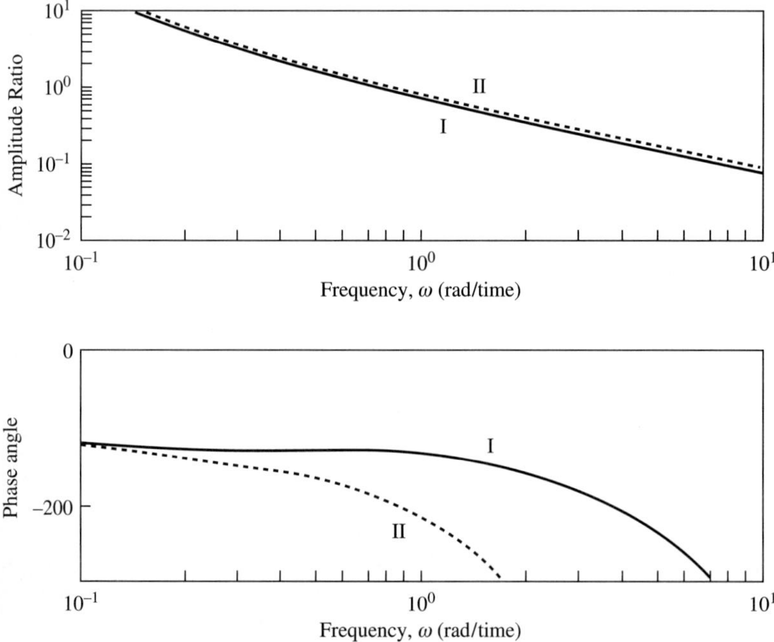

FIGURE 10.19
Uncertainty in Example 10.15 defined by models I and II with tuning for the nominal model.

possible for the assumed uncertainty. This plot clearly shows the effects of model errors, the possibility for instability in this case, and the need for a (larger) safety margin to account for the error. (Other ways to characterize the model error link the variation in process operation to the change in dynamics; for example, see Example 9.4 and Chapter 16.)

Interpretation IV: Experimental Tuning Approach

The Bode tuning method enables the engineer to calculate the proportional controller gain that brings the system to the stability limit. The same principle could be used to determine the ultimate gain experimentally through a simple trial-and-error procedure called *continuous cycling*. The real physical system would be controlled by a proportional-only controller, the set point perturbed slightly, and the transient response of the controlled variable observed. If the system is stable, either overdamped or oscillatory, the gain is increased; if unstable, the gain is decreased. The iterative procedure is continued, changing K_c until after a set point perturbation, the system oscillates with a constant amplitude. This behavior occurs when the system has exponential terms with (very nearly) zero values for their real parts, indicating that the system is at the stability margin. The gain at this condition is the ultimate gain, and the frequency of the oscillation is the critical frequency. These values, which in the continuous cycling procedure have been determined empirically, can be used with the Ziegler-Nichols closed-loop tuning correlations in Table 10.4 for calculating the PID constants. From this explanation, it should be clear why the correlations used in this section are called the "closed-loop" continuous cycling correlations. This experimental method is *not recommended,* because of the significant, prolonged disturbances introduced to the process. It is presented here to give a physical, time-domain meaning to the Bode stability calculations.

Example 10.16. Perform the empirical continuous cycling tuning method on the three-tank mixing process.

The resulting dynamic response at the stability limit is given in Figure 10.20. The controller gain was found by trial and error to be 206 and the period to be about 18 minutes. These are essentially the same answers as found in Example 10.10, where the three-tank mixing process was analyzed using the Bode method.

Interpretation V: Relationship between Stability and Performance

The analysis of roots of the characteristic equation $1+G_{\text{OL}}(s) = 0$ and, equivalently, Bode plots of $G_{\text{OL}}(s)$ provide methods for determining the stability of linear systems. Naturally, any feedback control system must be stable if it is to provide good control performance. However, stability is not sufficient to guarantee good performance. To see why, consider the closed-loop transfer function for a disturbance response:

$$\frac{\text{CV}(s)}{D(s)} = \frac{G_d(s)}{1+G_c(s)G_p(s)} \qquad \text{or} \qquad \text{CV}(s) = \frac{G_d(s)}{1+G_c(s)G_p(s)}D(s) \quad (10.41)$$

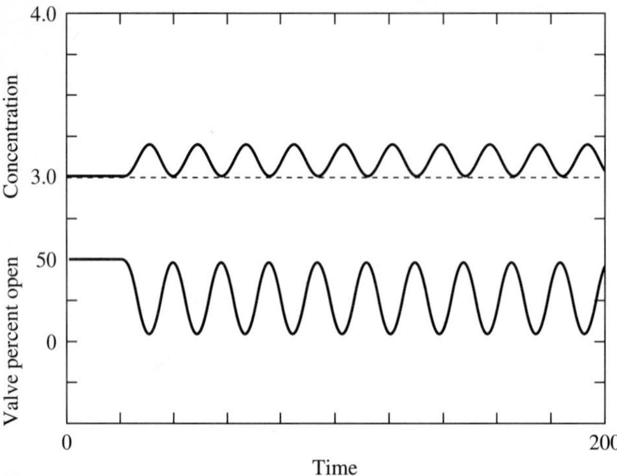

FIGURE 10.20
Dynamic response of three-tank mixing process with proportional-only controller and $K_c = 206$, the ultimate gain.

The stability analysis considers the denominator in the characteristic equation, $1 + G_c(s)G_p(s)$. Naturally, control performance also depends on the disturbance size and dynamics that appear in the numerator of the transfer function. For example, the three-tank mixing process would certainly remain closer to the set point for an inlet concentration disturbance of 0.01% in stream B compared to a 1% disturbance. Also, the system is stable when the feedback controller gain has a value of 0.1, which would give very poor control performance compared with the tuning determined in Example 10.10 for this process ($K_c = 94.5$). Clearly, the methods in this chapter, while providing essential stability information, do not provide all the information required for process control design. Control system performance is covered in more detail in Chapter 13.

> Stability is required for good control system performance. However, a control system can be stable and perform poorly.

Example 10.17. Determine how the control performance changes for the following process with different disturbance dynamics.

$$G_p(s) = \frac{0.039}{(1 + 5s)^3} \qquad G_d(s) = \frac{1}{(1 + \tau_d s)^n} \tag{10.42}$$

with $\tau_d = 5$ and n equal to (a) 3 and (b) 1.

The system was simulated with a PI controller using the tuning from Example 10.10. The two different disturbance transfer functions given here were considered. The first case (a) is the standard three-tank mixing system, and the dynamic response is given in Figure 10.15. The results for the faster disturbance, case (b), are given in Figure 10.21. As expected, the faster the disturbance enters the process, the poorer the feedback control system performs. Remember, the two cases considered in this

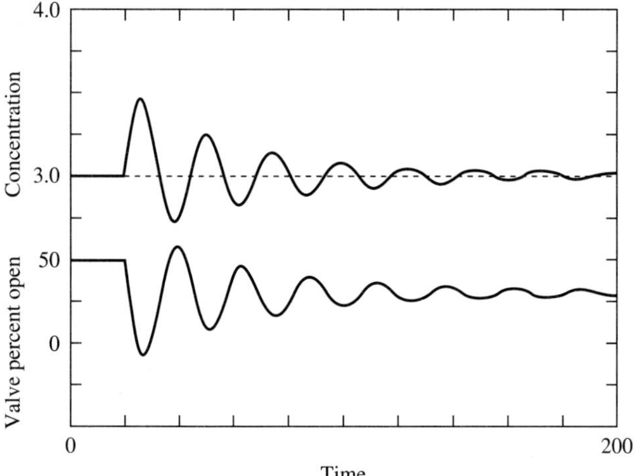

FIGURE 10.21
Dynamic response for the system in Example 10.17, case (*b*) (faster disturbance).

example have the same relative stability because the feedback dynamics $G_p(s)$ and the controller G_c are identical; only the disturbances are different. (Also, note that the valve goes below 0% open in the simulation of the linearized model, which is not physically possible; a nonlinear simulation should be performed.)

Interpretation VI: Modelling Requirement for Stability Analysis

We use approximate models for control system analysis and design, and we should select the model that provides an adequate representation of the dynamic behavior required by the analysis method. The Bode stability analysis has pointed out the extreme importance of model accuracy *near the critical frequency*. Thus, we do not require a model that represents the process accurately at high frequencies—that is, those frequencies much higher than the critical frequency.

Example 10.18. Compare the frequency responses for the three-tank mixing process derived from (*a*) fundamentals and (*b*) empirical model fitting.

The linearized fundamental model derived in Example 7.2 and repeated in equation (10.42) is third-order, and the empirical model is a first-order-with-dead-time (approximate) model in Example 6.4. Their frequency responses, which equal $G_{OL}(s)$ with $G_c(s) = K_c = 1$, are given in Figure 10.22. Note that the two frequency responses are quite close at low frequencies, since they have the same steady-state gains. At very high frequencies, they differ greatly, but we are not interested in that frequency range. Near the critical frequency ($\omega_c \approx 0.35$), the models do not differ greatly, which indicates that the two models give similar, but not exactly the same, tuning constants. Since essentially no model is perfect, we conclude that the error introduced by using a first-order-with-dead-time model approximation is often acceptably small for the purposes of calculating initial tuning constant values. Recall that further tuning improvements are made through fine-tuning.

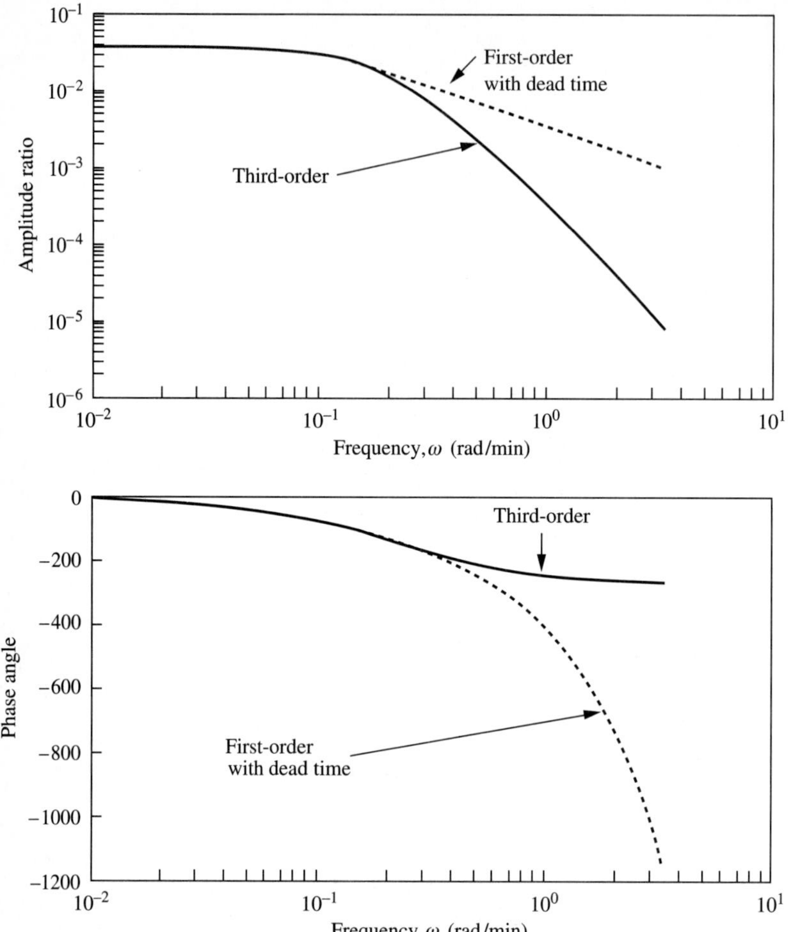

FIGURE 10.22
Comparison of Bode plots for exact and approximate process models.

In summary, tuning methods were presented in this section that are based on margins from the stability limit. The method can be applied to any stable process with a monotonic relationship between the phase angle and frequency. The methods in this section are especially helpful in determining the effects of various process and controller elements on the tuning constants.

10.9 ADDITIONAL TUNING METHODS IN COMMON USE, WITH A RECOMMENDATION

To this point, two controller tuning methods have been presented. The Ciancone correlations were based on a comprehensive definition of control performance in the time domain, whereas the Ziegler-Nichols closed-loop method was based on stability margin. Many other tuning methods have been developed and reported in the

TABLE 10.5
Ziegler-Nichols open-loop tuning based on process reaction curve

	K_c	T_I	T_d
P-only	$(1/K_p)/(\tau/\theta)$	—	—
PI	$(0.9/K_p)(\tau/\theta)$	$3.3\,\theta$	—
PID	$(1.2/K_p)(\tau/\theta)$	$2.0\,\theta$	$0.5\,\theta$

literature and textbooks. A few of the better known are summarized in this section, along with a recommendation on the methods to use.

One well-known method, known as the Ziegler-Nichols *open-loop* method (Ziegler and Nichols, 1942), provides correlations that can be used with simplified process models developed from such sources as an open-loop process reaction curve. The objective of these correlations is a 1 : 4 decay ratio for the controlled variable. The tuning constants are calculated from the experimental model parameters according to the expressions in Table 10.5. Notice that the dead time is in the denominator of the calculation for the controller gain. This indicates that the controller gain should decrease as the dead time increases, a result consistent with other tuning methods already considered. However, the open-loop Ziegler-Nichols correlation predicts a very large controller gain for processes with small dead times and an infinite gain for processes with no dead time. These results will lead to excessive variation in the manipulated variable and to a controller with too small a stability margin. Therefore, these correlations should not be used for processes with small fraction dead times.

Many other tuning methods have been developed, generally based on either stability margins or time-domain performance. A summary of the methods is presented in Table 10.6, which gives the main objectives of each method, along with a reference, either in this book or in the literature. Note that the IMC method is covered in Chapter 19.

With such a large selection available, some recommendations are needed to assist in the proper choice of tuning method. Before presenting recommendations, a few key factors should be reiterated. First, most tuning methods rely on a simplified dynamic model of the open-loop process. As a result, good control performance from the tuning depends on reasonably accurate model identification. Tuning calculations cannot correct for modelling errors; they can only reduce the detrimental effects of such errors. Second, the tuning constants should be determined so that the control system achieves desired performance objectives relevant to the process. Because each method has different objectives, each provides somewhat different dynamic performance, which should be matched to the process requirements. Third, all methods provide initial values, which should be fine-tuned based on plant experience; the tuning procedure shown in Figure 10.23 should be used. The tuning methods being discussed appear as the "initial tuning" that relies on the identification and is modified by fine-tuning, which corrects for modelling errors and adapts the performance to that desired for the process.

TABLE 10.6
Summary of PID tuning methods

Tuning method	Stability objective	Objective for CV(t)	Objective for MV(t)	Model error	Noise on CV(t)	Input SP = set point D = disturbance
Ciancone (chapter 9)	None explicit	Min IAE	Overshoot and variation with noise	±25%	Yes	SP and D individually
Fertik (1974)	None explicit	Min ITAE with limit on overshoot	None	None explicit	No	SP and D individually
Gain/phase margin (Section 10.8)	Gain margin or phase margin	None	None	Depends on margins	No	n/a
IMC tuning (Section 19.7)	For specified model error	ISE (robust performance)	None	Tune λ, see Morari and Zafiriou (1989)	No	SP and D (step and ramp) individually
Lopez et al. (1969)	None explicit	IAE, ISE, or ITAE	None	None	No	SP and D individually
Ziegler-Nichols closed-loop (Section 10.7)	Implicit margin for stability (GM ≈ 2)	4 : 1 decay ratio	None	None explicit	No	n/a
Ziegler-Nichols open-loop (Section 10.9)	Implicit margin for stability (GM ≈ 2)	4 : 1 decay ratio	None	None explicit	No	n/a

The proper selection for a particular application should follow from the information in the table. In other words:

> The best choice for the initial tuning correlation is the method that was developed for objectives conforming most closely to those of the actual situation for which the controller is being tuned.

The following ranking, with the first entry being the preferred method, represents the author's personal preference for calculating initial tuning.

1. Ciancone tuning correlations from Chapter 9
2. Bode/(closed-loop) Ziegler-Nichols when process cannot be satisfactorily fitted by a first-order-with-dead-time model
3. Nyquist/gain margin when the process does not satisfy the Bode criteria

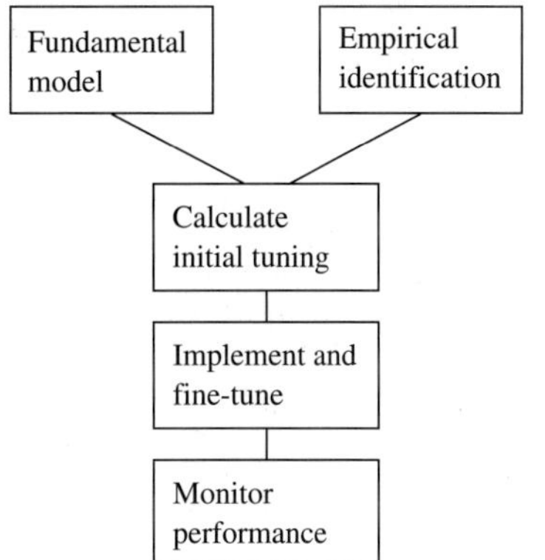

FIGURE 10.23
Major steps in the tuning procedure.

4. Any of the other correlations as appropriate for the application scenario
5. Detailed analysis of the robustness of the system, through either the optimization method in Chapter 9 or the robust performance applied to the IMC as described in Morari and Zafiriou (1989)

Approach 5 would always be the best, but it requires more effort than is usually justified for initial tuning. However, it may be required for systems involving complex dynamics and large model errors.

10.10 CONCLUSION

Several important topics have been covered in this chapter that are essential for a complete understanding of dynamic systems. We have learned

1. A useful definition of stability related to poles of the transfer function; i.e., the exponents in the solution of a set of linear differential equations
2. The effects of process and control elements in the feedback path that affect stability, such as dead times and time constants
3. Tuning methods based on a margin from the stability limit
4. That model errors must always be considered in tuning and that this results in detuned (i.e., less aggressive) feedback control action

All of these results are consistent with the experience gathered in Chapter 9, which was restricted to first-order-with-dead-time processes and PID control. The methods in this chapter provide a valuable theoretical basis that helps us understand time-domain behavior and that can be applied for quantitatively analyzing stability and determining tuning for a wide range of systems. Numerical examples in this chapter, as well as Chapter 9, have demonstrated that simple linear models are often adequate for calculating initial tuning constants. These results confirm that the first-order-with-dead-time models from empirical model fitting provide satisfactory accuracy for this control analysis.

The stability analysis methods presented in this chapter are summarized in Figure 10.24, which gives a simple flowchart for the selection of the appropriate method for a particular problem. Note that the direct analysis of the roots of the characteristic equation is applicable to either open- or closed-loop systems that have polynomial characteristic equations. The Bode method can be applied to most closed-loop systems, and the Nyquist method is the most general.

Many controller tuning methods have been presented in these two chapters. The correct method for a particular application depends on the objectives of the control system. The information in Table 10.6 will enable you to match the tuning with the control objectives. If no specific information is available, the Ciancone tuning correlations in Chapter 9 are recommended for initial tuning constant values.

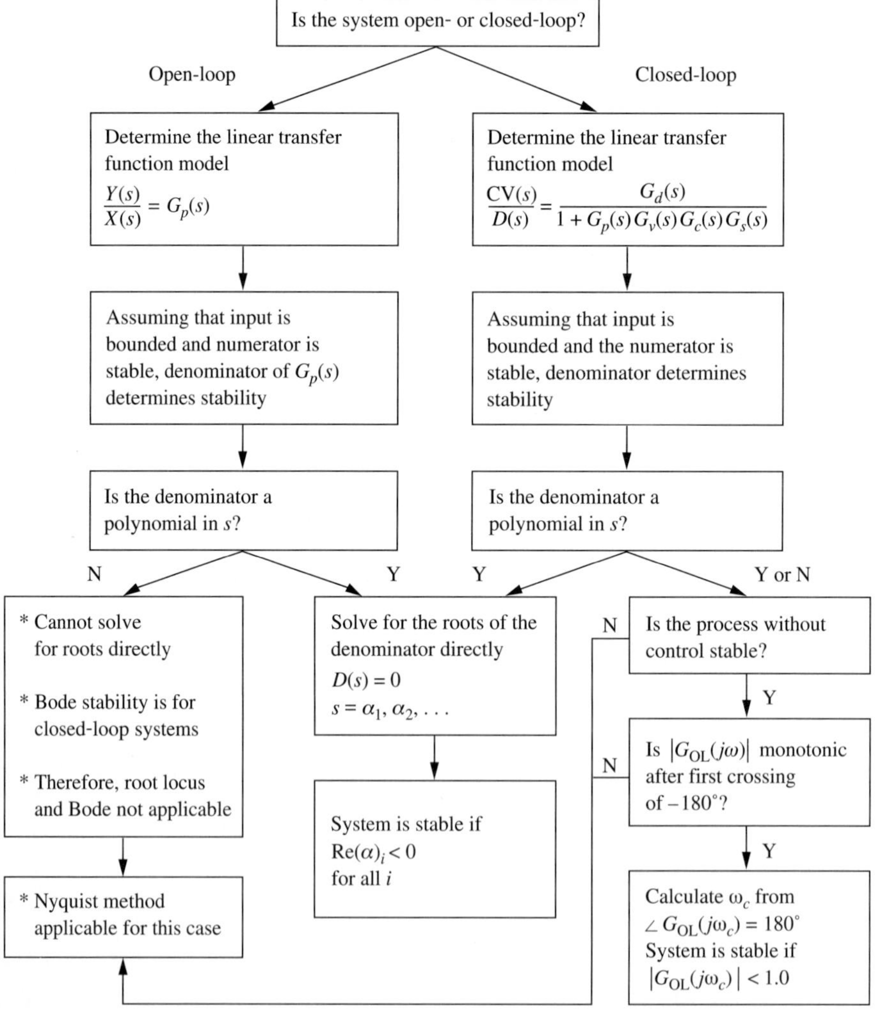

FIGURE 10.24
Flowchart for selecting the stability analysis method for local analysis using linearized models.

REFERENCES

Boyce, W. and R. Diprima, *Elementary Differential Equations,* Wiley, New York, 1986.

Dorf, R., *Feedback Control Systems Analysis and Synthesis,* McGraw-Hill, New York, 1986.

Fertik, H., "Tuning Controllers for Noisy Processers," *ISA Trans., 14, 4,* 292–304 (1974).

Franklin, G., J. Powell, and A. Emami-Naeini, *Feedback Control of Dynamic Systems,* (2nd Ed.), Addison-Wesley, Reading, MA, 1991.

Lopez, A., P. Murrill, and C. Smith, "Tuning PI and PID Digital Controllers," *Intr. and Contr. Systems,* 89–95 (Feb 1969).

Morari, M. and E. Zafiriou, *Robust Process Control,* Prentice-Hall, Englewood Cliffs, NJ, 1987.

Perlmutter, D., *Stability of Chemical Reactors,* Prentice-Hall, Englewood Cliffs, NJ, 1972.

Willems, J., *Stability Theory of Dynamical Systems,* Thomas Nelson and Sons, London, 1970.

Ziegler, J. and N. Nichols, *Trans ASME, 64,* 759–768 (1942).

OTHER RESOURCES

For a more detailed analysis of the root locus method, see the reference below, which gives design rules and applications.

Douglas, J., *Process Dynamics and Control, Vols. I and II,* Prentice-Hall, Englewood Cliffs, NJ, 1972.

The next two references give further details on frequency response. The first is introductory, and the second uses more challenging mathematical methods.

Caldwell, W., G. Coon, and L. Zoss, *Frequency Response for Process Control,* McGraw-Hill, New York, 1959.

MacFarlane, A. (Ed.), *Frequency-Response Methods in Control Systems,* IEEE Press, New York, 1979.

The following references present some historical background and some of the key milestones in control systems engineering.

MacFarlane, A., "The Development of Frequency Response Techniques in Automatic Control," *IEEE Trans. Auto Contr., AC-24,* 250 (1979).

Oldenburger, R. (ed.), *Frequency Response,* Macmillan, New York, 1956.

The stability of a nonlinear system in a defined region can be determined for some systems and regions using the (second) method of Liapunov, which is presented in Perlmutter (1972) and

LaSalle, J. and S. Lefschetz, *Stability of Liapunov's Direct Method with Applications,* Academic Press, New York, 1961.

> The methods introduced in this chapter provide a theoretical basis for determining the effects of all elements in the feedback loop, process, instrumentation, and control algorithm on stability and tuning. These questions ask you to apply these methods.

QUESTIONS

10.1. Consider the three-tank mixing process with a proportional-only controller in Example 10.5. Recalculate the root locus for the case with the three tank volumes reduced from 35 to 17.5 m^3. Determine the controller gain for a proportional-only algorithm at which the system is at the stability limit. Compare your result with Example 10.5 and discuss.

10.2. Example 10.4 established the stability of a system when operated at a temperature $T = 320\,\text{K}$. Given the expression for the reaction rate constant of $k = 6.63 \times 10^8 e^{-6500/T}\,\text{min}^{-1}$, determine if the system is stable at 300 K and 340 K. Explain the trend in your results and determine which of the three cases is the worst case from a stability point of view.

10.3. Answer the following questions, which revisit the interpretations (I–VI) in Section 10.8.

(*a*) (I) For the process in Example 5.2, determine the PI controller tuning constants using the Ziegler-Nichols closed-loop method. The manipulated variable is the inlet feed concentration, and the controlled variable is (i) Y_1, (ii) Y_2, (iii) Y_3, and (iv) Y_4. Answer for both cases 1 and 2.

(*b*) (II) Discuss the effect of the derivative mode on the stability of a closed-loop control system. Explain the results with respect to a Bode stability analysis.

(*c*) (III) A linearized model is derived for the process in Figure 9.1. The model is to be used for controller tuning. Model errors are estimated to be 30% in A, V, and F_B, and they can vary independently. Estimate the worst-case dynamic model that is possible within the estimated errors.

(*d*) (IV) Assume that experimental data indicates that a closed-loop PI system experienced sustained oscillations with constant amplitude at specified values of their tuning constants K_c' and T_I'. Estimate proper, new values for the tuning constants.

(*e*) (V) Determine the range of tuning constant values that result in stability for the following systems and plot the region with K_c and T_I as axes. Locate good tuning constant values within this region: (1) the level system in Example 10.1 for P-only and PI controllers; (2) the three-tank mixing system with a PI controller; (3) Figure 9.1 with $\theta = 5$ and $\tau = 5$.

(*f*) (VI) Using arguments relating to stability and Bode plots, determine model simplifications in the system in Figure 7.1 and Table 7.2, to give the lowest-order system needed to analyze stability and tuning with acceptable accuracy.

10.4. Given the process reaction curve in Figure Q10.4, calculate initial PID controller tuning constants using the Ziegler-Nichols tuning rules. Compare the results to values

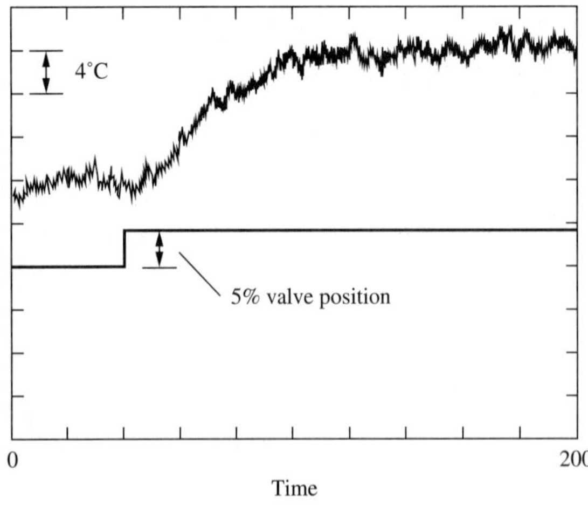

FIGURE Q10.4

from the Ciancone correlations and predict which set of values would provide more aggressive control.

10.5. Given a feedback control system such as the three-tank mixing process, determine the effect of the following equipment changes on the tuning constants.

(*a*) Installing a faster-responding control valve.

(*b*) Installing a control valve with a larger maximum flow.

(*c*) Installing a faster-responding sensor.

10.6. Without calculating the exact values, sketch the Bode plots for the following transfer functions using approximations.

$$(a)\ G_{OL}(s) = \frac{5.3}{4.5s + 1} \qquad (b)\ G_{OL}(s) = \frac{1}{5s} \qquad (c)\ G_{OL}(s) = \frac{2.0e^{-2s}}{(3s + 1)^2}$$

10.7. Determine the root locus plot in the complex plane from a zero controller gain to instability for the following processes: (*a*) example heater in Section 8.7; (*b*) Example 10.1; (*c*) Example 10.8. For the systems with PI controllers, assume that the integral time is fixed at the value in the original solution.

10.8. (*a*) Is the Bode stability criterion necessary, sufficient, or necessary and sufficient?

(*b*) Is it possible to determine the stability of a feedback control system with non–self-regulating process using the Bode stability criterion?

(*c*) Explain the limitations on the process transfer function imposed for the use of the Bode method.

(*d*) Determine the stability of the system in Example 10.4 using the Bode method.

10.9. Confirm the expressions for the amplitude ratios and phase angles given in Table 10.2.

10.10. Prove the following statements and give an explanation for each in your own words by referring to a sample physical system.

(*a*) The phase lag for a gain is zero.

(*b*) The amplitude ratio for a first-order system goes to zero as the frequency goes to infinity.

(*c*) The amplitude ratio for a second-order system with small damping coefficient is not monotonic with frequency.

(*d*) The phase angle decreases without limit as the frequency increases for a dead time.

(*e*) For an integrator, the amplitude ratio becomes very large for low frequencies and becomes very small for large frequencies.

(*f*) The amplitude ratio for a PI controller becomes very large at low frequencies.

10.11. For each of the physical systems in Table Q10.11, explain whether it can experience the dynamic responses shown in Figure Q10.11 for a step input (not necessarily at $t = 0$). The systems are to be considered idealized; in other words, the mixing is perfect, final element and sensor dynamics are negligible unless otherwise stated, and so forth. Give two answers for each situation: (*a*) for the system with the original parameters and (*b*) for the system with any physically possible parameters (flows, ΔH_{rxn}, etc.). Provide quantitative support for each answer based on the model structure of the system.

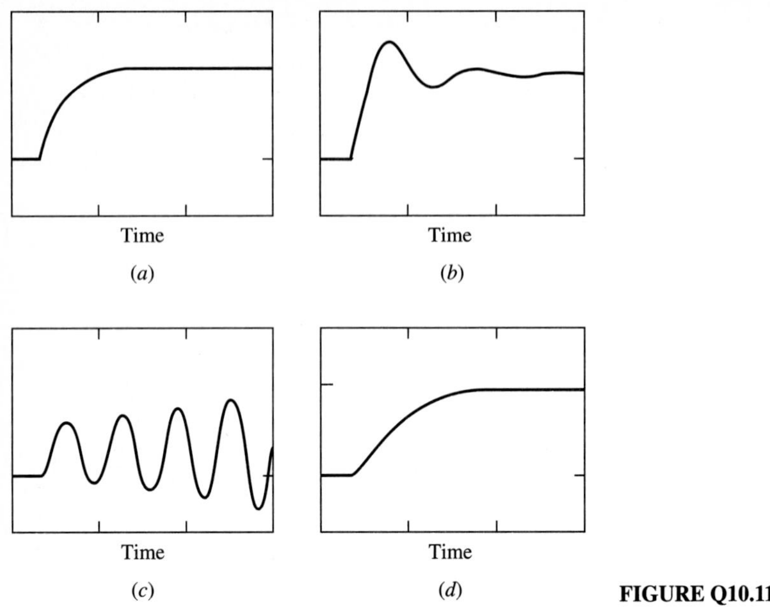

FIGURE Q10.11

TABLE Q10.11

System	Input variable (separate answer for each)	Output variable (separate answer for each)	Control (separate answer for each)
Figure 7.1 and Example 7.1	Signal to value	Measured temperature	None
Example 8.5	Set point	Tank temperature	(i) P-only (ii) PI
Examples 3.10, 4.8 and 5.8	F_c	Reactor temperature	None
Example 5.4	(i) F_s (ii) F_A	(i) C_{A1}, (ii) C_{A2}	None
Example 3.3	(i) C_{A0} (ii) F	(i) C_{A1}, (ii) C_{A2}	None
Example 9.1	Set point	C_A	PID
Example 7.2	(i) Signal to valve (ii) Set point of controller in Example 9.2	(i) C_{A1}, (ii) C_{A2}, and (iii) C_{A3}	(i) none (ii) P-only (iii) PI

10.12. Consider the three-tank mixing process in Example 7.2 with feedback control, but not using the PID algorithm.

(*a*) Determine the form of the controller transfer function that would give a single, stable root of the characteristic equation with a value α that could be selected by the engineer to give good control.

(*b*) The ability to specify the characteristic equation by deriving a control algorithm seems like a powerful control design method. Can you think of any difficulties if the controller in (*a*) were used?

10.13. The stability analysis methods introduced in this chapter are for linear systems, which give local results for nonlinear systems. What conclusions can be drawn from the linear analysis at the extremes of the ranges given about the intermediate stability of the following systems? (*a*) Example 9.4 with F_B varying from 6.9 to 5.2 m^3/min and (*b*) Example 9.1 with the volume of the tank and pipe varying by $\pm 30\%$.

10.14. Given the systems with roots of the characteristic equation shown in Figure Q10.14, sketch the transient responses to a step input for each, assuming the numerator of the transfer function is 1.0.

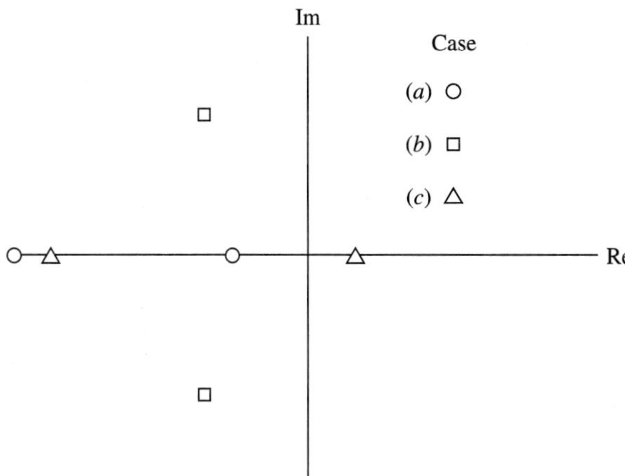

FIGURE Q10.14

10.15. Prove all of the statements in Table 10.3.

10.16. Answer the following questions regarding the derivative mode.

(*a*) Based on the Bode plot of a PID controller, what is the effect of high-frequency measurement noise on the manipulated variable?

(*b*) Redraw Figure 10.13*h* for $T_d = T_I/4$.

(*c*) We have considered a PID controller that uses error in the derivative for the stability analysis. However, the controller algorithm used commonly in practice uses the controlled variable in the derivative mode. How should the stability analysis be altered to account for the use of the controlled variable in the derivative?

10.17. Consider the three-tank mixing process in Example 7.2 with the same three 5-minute time constants and with a transportation delay of 4.3 min between the mixing point and the entrance to the first tank.

(*a*) Calculate initial tuning parameters using the Bode stability method and the Ziegler-Nichols correlations.

(*b*) Explain the changes in the tuning constant values from those in Example 10.10.

(*c*) Would you expect the control performance for the system with transportation delay to be better or worse than the system without transportation delay?

10.18. (*a*) The dynamic performance of the system in Example 10.10 was deemed too oscillatory with the initial tuning calculated via Ziegler-Nichols correlations. How would you change the tuning constants, which constants, and by how much to achieve reasonably good performance with little oscillation?

(*b*) Given the results in Example 10.13 which showed that the Ziegler-Nichols tuning correlations do not seem to yield robust control performance for low fraction dead time, how would you modify the Ziegler-Nichols correlations for $\theta/(\theta+\tau) < 0.2$?

CHAPTER 11

DIGITAL IMPLEMENTATION OF PROCESS CONTROL

11.1 INTRODUCTION

As we have seen in the previous chapters, PID feedback control can be successfully implemented using continuous (analog) calculating equipment. This conclusion should not be surprising, given the 50 years of good industrial experience with process control and given the fact that digital computers were not available for much of this time. However, digital computers have been applied to process control since the 1960s, as soon as they provided sufficient computing power and reliability. Most, but not all, new control-calculating equipment uses digital computation; however, the days of analog controllers are not over, for at least two reasons. First, control equipment has a long lifetime, so that equipment installed 10 or 20 years ago can still be in use; second, analog equipment has cost and reliability advantages in selected applications. Therefore, most plants have a mixture of analog and digital equipment, and the engineer should have an understanding of both approaches for control implementation. The basic concepts of digital control implementation are presented in this chapter.

The major motivation for using digital equipment is the greater computing power and flexibility it can provide for controlling and monitoring process plants. To perform feedback control calculations via analog computation, an electrical circuit must be fabricated that obeys the PID algebraic and differential equations. Since each circuit is constructed separately, the calculations are performed rapidly in parallel, with no interaction between what are essentially independent analog computers. Analog equipment can be designed and built for a simple, standard calculation such as a PID controller, but it would be costly to develop analog systems for a wide range of controller equations, and each system would be inflexible: the algorithm could not be changed; only the parameters could be adjusted.

In comparison, digital computation uses an entirely different concept. By representing numbers in digital (binary) format and solving equations numerically to represent behavior of the control calculation of interest, the digital computer can easily execute a wide range of calculations on the same equipment, hardware, and basic software. Two differences between analog and digital systems are immediately apparent. First, the digital system performs its function periodically, which, as we shall see, affects the stability and performance of the closed-loop system. Often, we refer to this type of control as *discrete* control, because control adjustments occur periodically or discretely. Second, the digital computer performs calculations in series; thus, if time-consuming steps are involved in the control calculations, digital control might be too slow. Fortunately, modern digital computers and associated equipment are fast enough that they do not normally impose limitations related to execution speed.

Digital computers also provide very important advantages in areas not emphasized in this book but crucial to the successful operation of process plants. One area is minute-to-minute monitoring of plant conditions, which requires plant operators to have rapid access to plant data, displayed in an easily analyzed manner. Digital systems provide excellent graphical displays, which can be tailored to the needs of each process and person. Another area is the longer-term monitoring of process performance. This often involves calculations based on process data to report key variables such as reactor yields, boiler efficiencies, and exchanger heat transfer coefficients. These calculations are easily programmed and are performed routinely by the digital computer.

The purpose of this chapter is to provide an overview of the unique aspects of digital control. The approach taken here is to present the most important differences between analog and digital control that would affect the application of the control methods and designs covered in this book. This coverage will enable the reader to implement digital PID controllers as well as enhancements, such as feedforward and decoupling, and new algorithms, such as Internal Model Control, covered later in the book.

11.2 STRUCTURE OF THE DIGITAL CONTROL SYSTEM

Before investigating the key unique aspects of digital control, we shall quickly review the structure of the control equipment when digital computing is used for control and display. The components of a typical control loop, without the control calculation, were presented in Figure 7.2. Note that the sensor and transmission components are analog devices and can remain unchanged with digital control calculations. The loop with digital control is shown in Figure 11.1, where the unique features are highlighted. First, the signal of the controlled variable is converted from analog (e.g., 4–20 mA) to a digital representation. Then the control calculation is performed, and finally, the digital result is converted to an analog signal for transmission to the final control element.

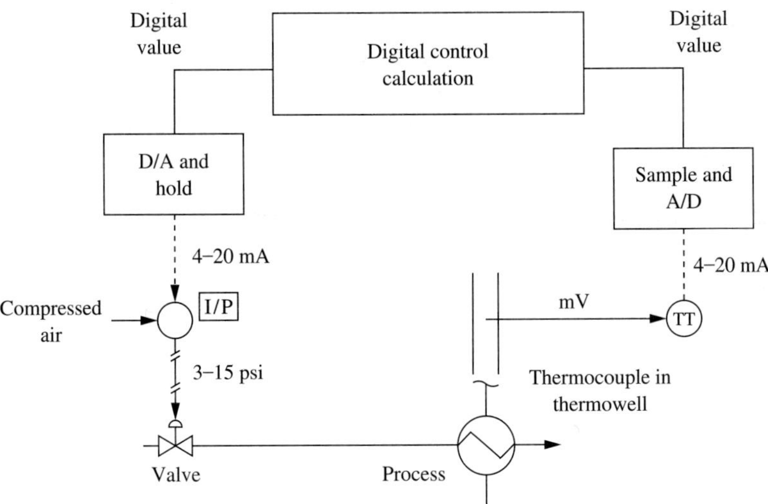

FIGURE 11.1
Schematic of single feedback control loop using digital calculation.

Process plants usually involve many variables, which are controlled and monitored from a centralized location. A digital control system to achieve these requirements is shown in Figure 11.2. Each measurement signal for control and monitoring is sent through an analog-to-digital (A/D) converter to a digital computer (or microprocessor). The results of the digital control calculations are converted for transmission in a digital-to-analog (D/A) converter. The system may have one processor

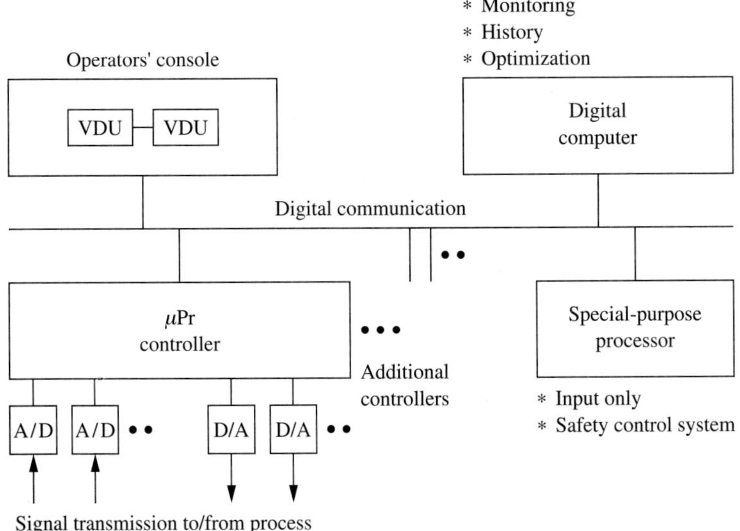

FIGURE 11.2
Schematic of a distributed digital control system.

per control loop; however, most industrial systems have several measurements and controller calculations per processor. Systems with 32 input measurements and 16 controller outputs per processor are not uncommon. This design is less costly, although it is somewhat less reliable, because several control loops would be affected should a processor fail.

Some data from each individual processor is shared with other processors to enable proper display and human interaction. The information exchange is performed via a digital communication network (local area network, LAN), which enables data sharing among processors and between each processor and the unit that provides operator interface, usually called the operator console. An operator console is required so that a person can monitor the process and intervene to make changes in variables such as a valve opening, controller set point, or controller status (automatic or manual). Thus, the controller set point and tuning constants must be communicated from the console, where they are entered by a person, to the processor, where the control calculation is being performed. Also, the values of the controlled and manipulated variables should be communicated from the controller to the console for display to the person. Some data that is *not* typically communicated from the individual control processors would be intermediate values, such as the integral error used in the controller calculation. The operator console has its own processor and data storage and has visual displays (video display units, VDUs), audio annunciators, and a means, such as a keyboard, for the operator to interact with the control variables. Graphical display of variables, which is easier to interpret, is used along with digital display, which is more precise. Also, variables can be superimposed on a schematic of the process to aid operators in placing data in context.

To add flexibility, more powerful processors can be connected to the local area network so that they can have access to the process data. These processors can perform tasks that are not time-critical. Examples are process-monitoring calculations and process optimization, which may adjust variables infrequently (e.g., once every few hours or shift).

Since each digital processor performs its functions serially, it must have a means for deciding which task from among many to perform first. Thus, each processor has a *real-time operating system,* which organizes tasks according to a defined priority and schedule. For example, the control processor would consider its control calculations to be of high priority, and the operators' console would consider a set point change to be of high priority. Lower-priority items, such as monitoring calculations, are performed when free time is available. An important aspect of real-time calculating is the ability to stop a lower-priority task when a high-priority task appears. This is known as a *priority interrupt* and is an integral software feature of each processor in a digital control system.

The goal, which is nearly completely achieved, is that the integrated digital system responds so fast that it is indistinguishable from an instantaneous system. Since each function is performed in series, each step in the control loop must be fast. For most modern equipment, the analog-to-digital (A/D) and digital-to-analog (D/A) conversions are very fast with respect to other dynamics in the digital equipment or the process. Each processor is designed to guarantee the execution of high-priority control tasks within a specified period, typically within 0.1 to 1 second.

When estimating the integrated system response time, it is important to consider all equipment in the loop. For example, response to a set point change, after it is entered by a person, includes the execution periods of the console processor, digital communication, control processor, and D/A converter with hold circuit and the dynamic responses of the transmission to the valve and of the valve. This total system might involve several seconds, which is not significant for most process control loops but may be for very fast processes, such as machinery control.

Another important factor in the control equipment is the accuracy of many signal conversions and calculations, which should not introduce errors that significantly influence the accuracy of the control loop. The values in the digital system are communicated with sufficient resolution (16 or more bits) that errors are very small. Typically, the A/D converter has an error on the order of $\pm 0.05\%$ of the sensor range, and the D/A converter has an error on the order of $\pm 0.1\%$ of the final element range. In older digital control computers, calculations were performed in fixed-point arithmetic; however, current equipment uses floating-point arithmetic, so that roundoff errors are no longer a significant problem. As a result, the errors occurring in the digital system are not significant when compared to the inaccuracies associated with the sensors, valves, and process models in common use.

The system in Figure 11.2 and described in this section is a network of computers with its various functions distributed to individual processors. The type of control system is commonly called a *distributed control system* (DCS). Today's digital computers are powerful enough that one central computer could perform all of these functions. However, the distributed control structure has many advantages, some of the most important of which are presented in Table 11.1. These advantages militate for the continued use of the distributed structure for control equipment design, regardless of future increases in computer processing speed.

TABLE 11.1
Features of a distributed control system (DCS)

Feature	Effect on process control
Calculations performed in parallel by numerous processors	Control calculations are performed faster than if by one processor.
Limited number of controller calculations performed by a single processor	Control system is more reliable, because a processor failure affects only few control loops.
Control calculations and interfacing to process independent of other devices connected to the LAN	Control is more reliable, because failure of other devices does not immediately affect a control processor.
Small amount of equipment required for the minimum system	Only the equipment required must be purchased, and the system is easily expanded at low cost.
Each type of processor can have different hardware and software	Hardware and software can be tailored to specific applications like control, monitoring, operator console, and general data processing.

The major disadvantage of modern digital systems, which is not generally true for analog systems, is that few standards for design or interfacing are being observed. As a result, it is difficult to mix the equipment of two or more digital equipment suppliers in one control system.

In conclusion, the control system in Figure 11.2 is designed to provide fast and reliable performance of process control calculations and interactions with plant personnel. Clearly, the computer network is complex and requires careful design. However, the plant operations personnel interact with the control equipment as though it were one entity and do not have to know in which computer a particular task is performed. Also, considerable effort is made to reduce the computer programming required by process control engineers. For the most part, the preparation of control strategies in digital equipment involves the selection and integration of preprogrammed algorithms. This approach not only reduces engineering time; it also improves the reliability of the strategies. While distributed digital systems are the predominant structure for digital control equipment, the principles presented in the remainder of the chapter are applicable to any digital control equipment.

11.3 EFFECTS OF SAMPLING A CONTINUOUS SIGNAL

The digital computer operates on discrete numerical values of the measured controlled variables, which are obtained by sampling from the continuous signal and converting this signal to digital form via A/D conversion. In this section, the way that the sampling is performed and the effects of sampling on process control are reviewed. As one might expect, some information is lost when a continuous signal is represented by periodic samples, as shown in Figures 11.3*a* through *c*. These figures show the results of sampling a continuous sine function in Figure 11.3*a* at a constant period, which is the common practice in process control and the only situation considered in this book. The sampled values for a small period (high frequency) in Figure 11.3*b* appear to represent the true, continuous signal closely, and the continuous signal could be reconstructed rather accurately from the sampled values. However, the sampled values for a long period in Figure 11.3*c* appear to lose important characteristics of the continuous measurement, so that a reconstruction from the sampled values would not accurately represent the continuous signal. The effects of sampling shown in Figure 11.3 are termed *aliasing,* which refers to the loss of high-frequency information due to sampling.

An indication of the information lost by the sampling process can be determined through *Shannon's sampling theorem,* which is stated as follows and is proved in many textbooks (e.g., Astrom and Wittenmark, 1990).

> A continuous function with all frequency components at or below ω' can be represented uniquely by values sampled at a frequency equal to or greater than $2\omega'$.

The importance of this statement is that it gives a quantitative relationship for how the sampling period affects the signal reconstruction. The relationship stated is not

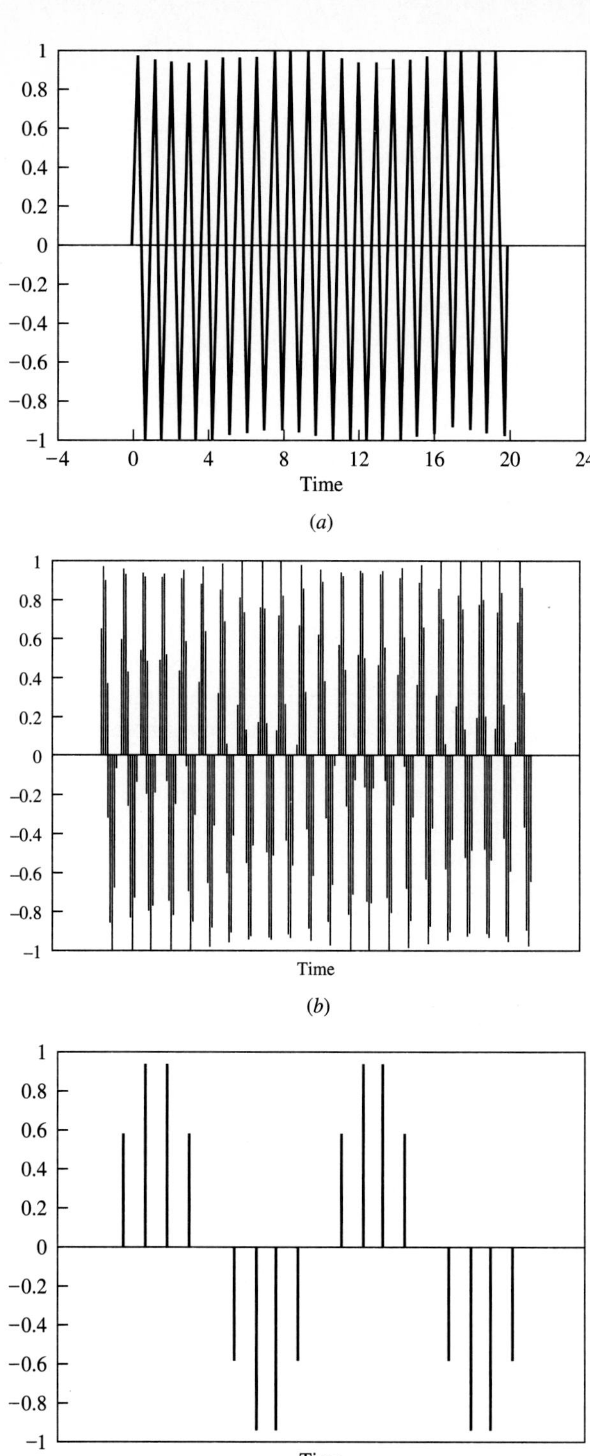

FIGURE 11.3
Digital sampling. (*a*) Example continuous measurement signal; (*b*) results of sampling of the signal with a period of 2; (*c*) results of alternative sampling of the signal with a period of 12.8.

exactly applicable to process control, because the reconstruction of the signal for any time t' requires data after t', which would introduce an undesirable delay in the reconstructed signal being available for feedback control. However, the value given by the statement provides a useful bound that enables us to estimate the frequency range of the measurement signal that is lost when sampling at a specific frequency.

Example 11.1. The composition of a distillation tower product is measured by a continuous sensor, and the variable fluctuates due to many disturbances. The dominant variations are of frequencies up to 0.1 cycle/min (0.628 rad/min). At what frequency should the signal be sampled for complete reconstruction using the sampled values?

If the signal has no frequency components above 0.1 cycles/sec, the sampling frequency should be 1.256 rad/min for complete reconstruction. However, most signals have a broad range of frequency components, including some at very high frequencies. Thus, a very high sampling frequency would be required for complete reconstruction of essentially all process measurement signals.

Fortunately, our goal is *not* to reconstruct the signal perfectly but to provide sufficient information to the controller to achieve good dynamic performance. Thus, it is often possible to sample *much less frequently* than specified by Shannon's theorem and still achieve good control performance (Gardenhire, 1964). If the signal has substantial high-frequency components with significant amplitudes, the continuous signal may have to be filtered, as discussed in Chapter 12.

There are many options for using the sampled values to reconstruct the signal approximately. Two of the most common, zero- and first-order holds, are considered here. The simplest is the zero-order hold, which assumes that the variable is constant between samples. The first-order hold assumes that the variable changes in a linear fashion as predicted from the most recent two samples. These two methods are compared in Figures 11.4 and 11.5, where the main difference is the amplification in the magnitude caused by the first-order hold. Also, the first-order hold has a larger

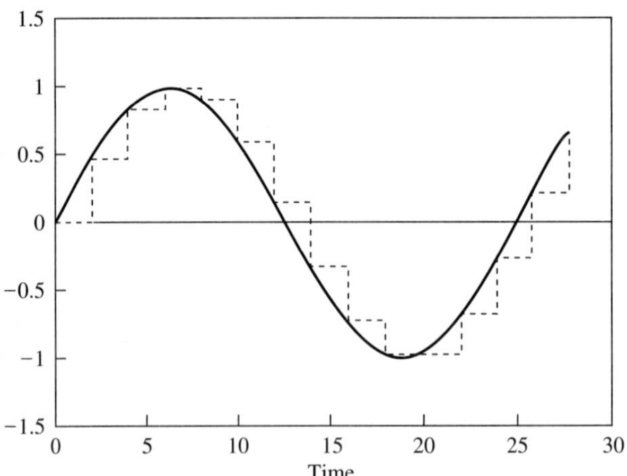

FIGURE 11.4
Zero-order hold.

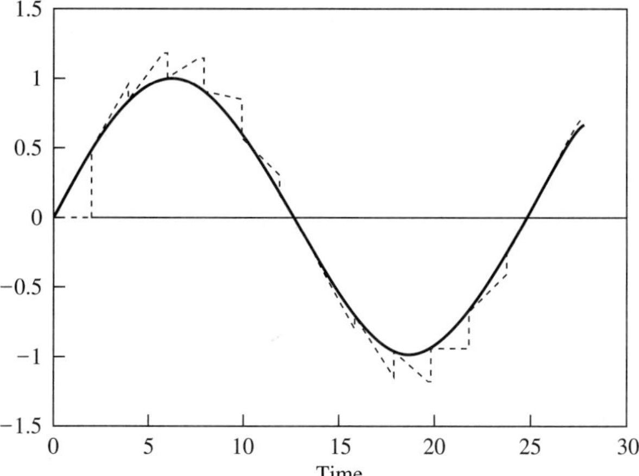

FIGURE 11.5
First-order hold.

phase lag, which is undesirable for closed-loop control. For both of these reasons, the simpler zero-order hold is used almost exclusively for process control.

The effect of the zero-order hold on the dynamics can be seen clearly if we reconstruct the original signal as shown in Figure 11.6. In the figure, the reconstructed signal is a smooth curve through the midpoint of the zero-order hold. It is apparent that the reconstructed signal after the zero-order hold is identical to the original signal after being passed through a dead time of $\Delta t/2$, where the sample period is Δt. This explains the rule of thumb that the major effect on the stability and control performance of sampling can be estimated by adding $\Delta t/2$ to the dead time of the system. Since any additional delay due to sampling is undesirable for feedback control and process monitoring, feedback control performance degrades as the process

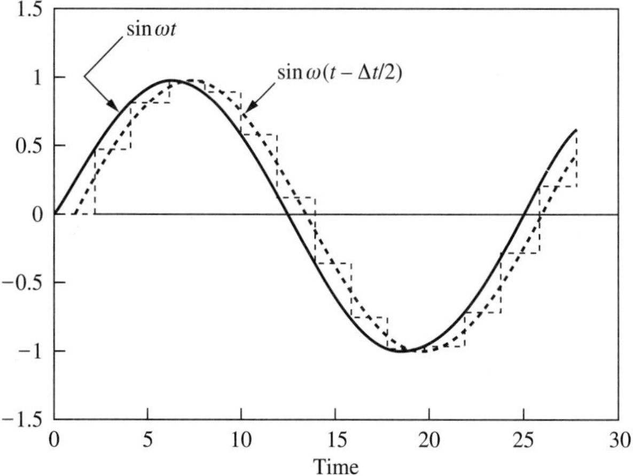

FIGURE 11.6
Reconstruction of signal after zero-order hold.

dynamics, including sampling, become slower. Therefore, the controller execution period should generally be made short.

In some cases, monitoring process operations requires high data resolution, because short-term changes in key variables can significantly influence process safety and profit. However, process monitoring also involves variables that change slowly with time, such as a heat transfer coefficient, and the data collected for this purpose does not have to be sampled rapidly.

In conclusion, sampling is the main difference between continuous and digital control. Since process measurements have components at a wide range of frequencies, some high-frequency information is lost by sampling. The effect of sampling on control performance, with a zero-order hold used for sampling, is addressed after the digital controller algorithm is introduced in the next section.

11.4 THE DISCRETE PID CONTROL ALGORITHM

The proportional-integral-derivative control algorithm presented in Chapter 8 is continuous and cannot be used directly in digital computations. The algorithm appropriate for digital computation is a modified form of the continuous algorithm that can be executed periodically using sampled values of the controlled variable to determine the value for the controller output. The controller output passes through a digital-to-analog converter and a zero-order hold; therefore, the signal to the final control element is changed to the result of the last calculation and held at this value until the next controller execution.

The digital calculation should approximate the continuous PID algorithm:

$$\mathrm{MV}(t) = K_c\left(E(t) + \frac{1}{T_I}\int_o^t E(t')dt' - T_d\frac{d\mathrm{CV}(t)}{dt}\right) + I \tag{11.1}$$

The method for approximating each mode is presented in equations (11.2) to (11.4). In these equations the value at the current sample is designated by the subscript N and the ith previous sample by $N - i$. Thus the current values of the controlled variable, set point, and controller output are CV_N, SP_N, and MV_N, respectively. The error is defined consistently with continuous systems as $E_N = \mathrm{SP}_N - \mathrm{CV}_N$.

Proportional mode:

$$(\mathrm{MV}_N)_{\mathrm{prop}} = K_c E_N \tag{11.2}$$

Integral Mode:

$$(\mathrm{MV}_N)_{\mathrm{int}} = \frac{K_c(\Delta t)}{T_I}\sum_{i=1}^{N} E_i \tag{11.3}$$

Derivative mode:

$$(\mathrm{MV}_N)_{\mathrm{der}} = -K_c\frac{T_d}{\Delta t}(\mathrm{CV}_N - \mathrm{CV}_{N-1}) \tag{11.4}$$

The proportional term is self-explanatory. The integral term is derived by approximating the continuous integration with a simple rectangular approximation. Those familiar with numerical methods recognize that this is not as accurate an approximation as possible with other integration methods used in numerical analysis (Gerald and Wheatley, 1989). However, small numerical errors in this calculation are not too important, because the integral mode continues to make changes in the output until the error is zero. Thus, zero steady-state offset for steplike inputs is not compromised by small numerical errors. Note that all past values of the error do not have to be stored, because the summation can be calculated recursively according to the equation

$$S_N = \sum_{i=1}^{N} E_i = E_N + S_{N-1} \tag{11.5}$$

where $S_{N-1} = \sum_{i=1}^{N-1} E_i$ and is stored from the previous controller execution.

The derivative is approximated by a backward difference. This approximation provides some smoothing; for example, the derivative of a perfect step is not infinite using equation (11.4), since Δt is never zero.

The three modes are combined into the full-position PID control algorithm:

$$\text{MV}_N = K_c\left(E_N + \frac{\Delta t}{T_I}\sum_{i=1}^{N} E_i - \frac{T_d}{\Delta t}(\text{CV}_N - \text{CV}_{N-1})\right) + I \tag{11.6}$$

Note that the constant of initialization is retained so that the manipulated variable does not change when the controller initiates its calculations.

Equation (11.6) is referred to as the *full-position algorithm* because it calculates the value to be output to the manipulated variable at each execution. An alternative approach would be to calculate only the change in the controller output at each execution, which is achieved with the *velocity* form of the digital PID:

$$\Delta\text{MV}_N = K_c\left(E_N - E_{N-1} + \frac{\Delta t}{T_I}E_N - \frac{T_d}{\Delta t}(\text{CV}_N - 2\text{CV}_{N-1} + \text{CV}_{N-2})\right) \tag{11.7}$$

$$\text{MV}_N = \text{MV}_{N-1} + \Delta\text{MV}_N \tag{11.8}$$

This equation is derived by subtracting the full-position equation (11.6) at sample $N-1$ from the equation at sample N. Either the full-position or the velocity form can be used, and many commercial systems are in operation with each basic algorithm.

The digital PID controller, either equation (11.6) or (11.8), can be rapidly executed in a process control computer. Only a few multiplications of current or recent past values times parameters and a summation are required. Also, little data storage is required for the parameters and few past values.

In conclusion, simple numerical methods are adequate for approximating the integral and derivative terms in the PID controller. As a result, the controller modes, set point, and tuning constants are the same in the digital PID algorithm as they are in the continuous algorithm. This is very helpful, because we can apply what we have learned in previous chapters about how the modes affect stability and performance to the digital algorithm. For example, it can be shown for the digital controller that the integral mode is required for zero steady-state offset and that the derivative mode amplifies high-frequency noise.

11.5 EFFECTS OF DIGITAL CONTROL ON STABILITY, TUNING, AND PERFORMANCE

The tuning of continuous control systems is presented in Chapter 9, and stability analysis is presented in Chapter 10. A similar, mathematically rigorous analysis of the stability of digital control systems can be performed, but it would require developing principles of the z transform, which serves a similar function for discrete systems as the Laplace transform does for continuous systems (Franklin, Powell, and Workman, 1990). Since an exposition on z transforms would require considerable time and space, this section provides the essential results without detailed mathematical proofs. The major differences in digital systems are highlighted, modifications to existing tuning guidelines are provided, and examples are presented to demonstrate the results. The measures of control performance and the definition of stability are the same as introduced in previous chapters of this part for continuous systems.

As described in Section 11.3, sampling introduces an additional delay in the feedback system, and this delay is similar to, but not the same as, a dead time. Thus, we expect that longer sampling will tend to destabilize a feedback system and degrade its performance.

Example 11.2. As an example, we consider a feedback control system for which the transfer functions for the process and disturbance are as follows and the disturbance is a step of magnitude 3.6.

$$G_p(s) = \frac{-1.0e^{-2s}}{(10s+1)(0.2s+1)} \approx \frac{-1.0e^{-2.2s}}{(10s+1)} \tag{11.9}$$

$$G_d(s) = \frac{1}{(5s+1)(10s+1)} \qquad D(s) = \frac{3.6}{s}$$

The performance of the system under continuous PI feedback control is given in Figure 11.7a using the Ciancone tuning from Figure 9.9. The performance is given in Figure 11.7b under discrete PI control with an execution period of 9, using the *same tuning* as in Figure 11.7a. We notice that the discrete response is more oscillatory and gives generally poorer performance. Several other responses were simulated, and their results are summarized in Table 11.2. When the execution period was made long, in this case 10 or greater, the control system became unstable!

This example shows that the control performance generally degrades for increasing sample periods and that the system can become unstable at long periods.

Since sampling introduces a delay in the feedback loop, we would expect that the tuning should be altered for digital systems to account for the sampling. The result in Section 11.3 indicated that the sample introduced an additional delay, which can be approximated as a dead time of $\Delta t/2$. Thus, one common approach for tuning PID feedback controllers and estimating their performance is to add $\Delta t/2$ to the feedback process dead time and use methods and guidelines for continuous systems (Franklin, Powell, and Workman, 1990). The tuning rules developed in Chapters 9 and 10 can be applied to digital systems with the dead time used in the calculations equal to the process dead time plus one-half of the sample period (i.e., $\theta' = \theta + \Delta t/2$). Thus, the

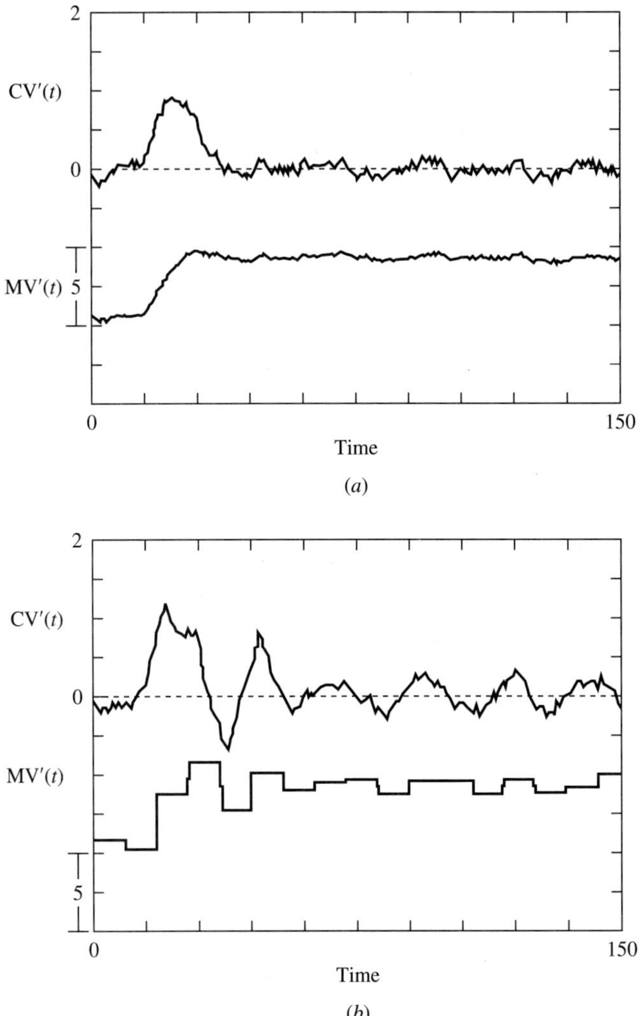

FIGURE 11.7
Example process: (*a*) under continuous control; (*b*) under digital PI control with $\Delta t = 9$ using continuous tuning.

tuning guidelines must be modified from those for continuous controllers as follows to achieve good control performance.

1. Obtain an empirical model.
2. Determine the sample period (e.g., $\Delta t \approx 0.05(\theta + \tau)$).
3. Determine the tuning constants using appropriate methods (e.g., Ciancone correlation or Ziegler-Nichols method) with $\theta' = \theta + \Delta t/2$.
4. Implement the initial tuning constants and fine-tune.

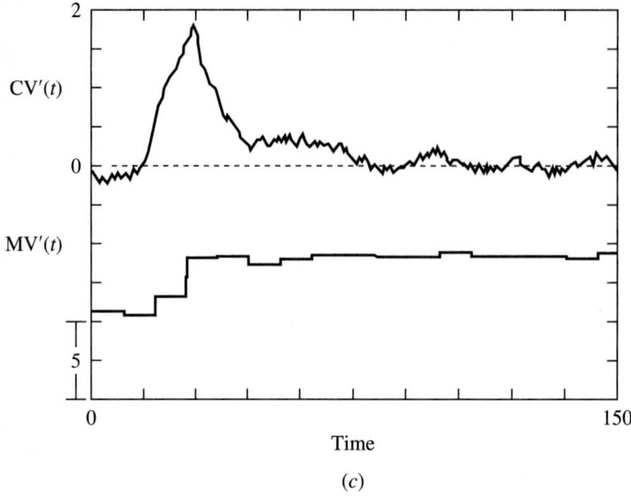

FIGURE 11.7 ***(continued)***
Example process: (*c*) under digital control with $\Delta t = 9$ using altered tuning from Table 11.3.

As stated previously, the values obtained from these guidelines should be considered initial estimates of the tuning constants, which are to be evaluated and improved based on empirical performance through fine tuning.

Example 11.3. Apply this method to tune a digital controller for the process defined in equation (11.9). Results for several execution periods are given in Table 11.3, with the tuning again from the Ciancone correlations in Figure 9.9, and the control performance is shown in Figure 11.7*c* for an execution period of 9. Note that in all cases, including

TABLE 11.2
Example of the performance of PI controllers for various execution periods with $K_c = -1.7$ and $T_I = 5.5$

Execution Period	IAE
Continuous	18.9
0.5	18.9
1.0	19.1
3.0	19.9
7.0	25.8
9.0	32.0
10.0	Unstable

TABLE 11.3
Example of the performance of PI controllers for various execution periods with tuning adjusted accordingly

Execution period Δt	Dead time $\theta' = \theta + \Delta t/2$	Fraction dead time $\theta'/(\theta' + \tau)$	K_c	T_I	IAE
Continuous	2.2	0.18	1.7	5.5	18.9
1	2.7	0.21	1.76	6.1	19.2
3	3.7	0.27	1.50	8.9	26.8
5	4.7	0.32	1.23	10.3	36.0
7	5.2	0.34	1.2	10.6	37.2
9	6.7	0.40	1.05	11.0	42.7
10	7.2	0.42	1.0	11.1	44.8

that with an execution period of 10, the dynamic performance is stable and well behaved (not too oscillatory). Recall that the performance could be improved (IAE reduced) with some fine tuning, but at the expense of robustness.

It is apparent that the dynamic response is well behaved, with a reasonable damping ratio and moderate adjustments in the manipulated variable, when the digital controller is properly tuned. Also, it is clear that the performance of the digital controller is not as good as that of the continuous controller. In fact,

> The performance of a continuous process under digital PID control is nearly always worse than under continuous control. The difference depends on the length of the execution period relative to the feedback dynamics.

Note that the execution period is related to the dynamics of the feedback process, since "fast" and "slow" must be relative to the process. A guideline drawn from Table 11.3 is that the effect of sampling and digital control is not usually significant when the sample period is less than about $0.05(\theta + \tau)$ to $0.10(\theta + \tau)$. For a summary of many other guidelines, see Seborg, Edgar, and Mellichamp (1989).

Occasionally, controllers give poor performance that is a direct result of the digital implementation. This type of performance is shown in Figure 11.8*a*, in which a digital PI controller with a relatively slow execution period is controlling a process with very fast dynamics. The models for the process and controller are as follows:

$$\text{Process:} \qquad \frac{-1.0}{0.5s + 1} \tag{11.10}$$

$$\text{Controller:} \qquad \text{MV}_N = K_c\left(E_N + \frac{\Delta t}{T_I}\sum_{i=1}^{N} E_i\right) \qquad \text{with } \Delta t = 0.5 \tag{11.11}$$

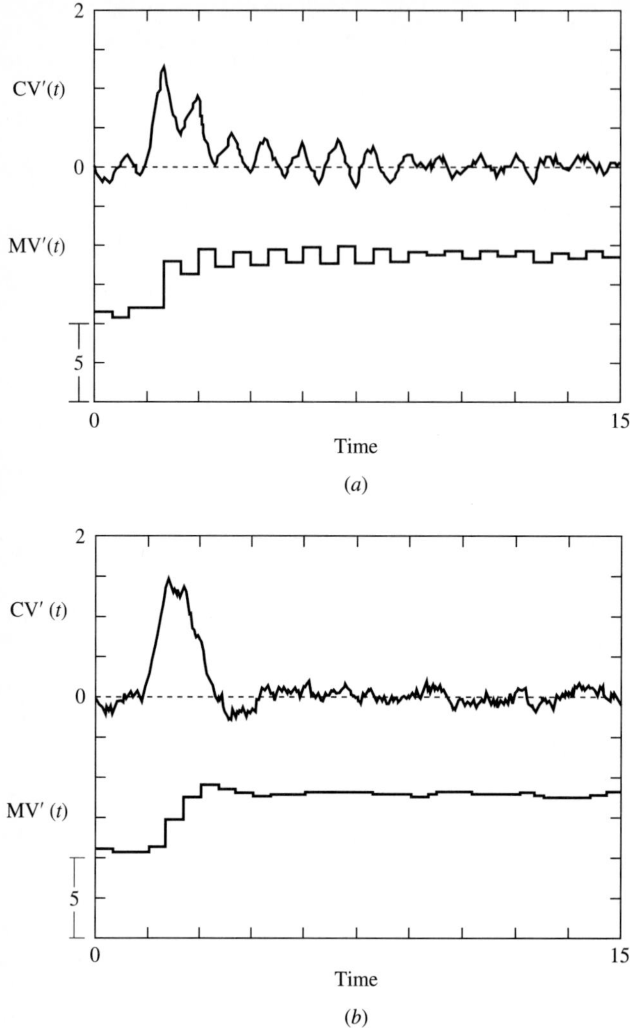

FIGURE 11.8
Digital control of a fast process (*a*) with $K_c = -1.4$ and $T_I = 7.0$; (*b*) with $K_c = -0.14$ and $T_I = 0.64$.

The oscillations in the manipulated variable are known as *ringing*. Diagnosing the causes of ringing requires mathematics (z-transforms) not covered in this book. However, the cause of this poor performance can be understood by considering the digital controller equation (11.11). The controller adjusts the manipulated variable to correct an error (e.g., a large positive adjustment). If a large percentage of the effect appears in the measured control variable at the next execution, the current error E_N can be small while the past error, E_{N-1}, will be large with a negative sign, causing a large negative adjustment in the manipulated variable. The net effect is an oscillation in the manipulated variable every execution period, which is very undesirable. In this case, the oscillations can be reduced by decreasing the controller gain and decreasing the integral time, so that the controller behaves more like an integral controller. The improved performance for the altered tuning is given in Figure 11.8*b*. This type of correction is usually sufficient to reduce ringing for PID control.

We have seen that the digital controller generally gives poorer performance than the equivalent continuous controller, although the difference is not significant if the controller execution is fast with respect to the feedback process.

11.6 EXAMPLE OF DIGITAL CONTROL STRATEGY

To demonstrate the analysis of a control system for digital control, the execution periods for the flash system in Figure 11.9 are estimated. The process associated with the flow controller is very fast; thus, the execution period should be fast, and perhaps, the controller gain may have to be decreased due to ringing. The level inventory would normally have a holdup time (volume/flow) of about five minutes, so that very frequent level controller execution is not necessary.

Let us assume that the analyzer periodically takes a sample from the liquid product and determines the composition by chromatography. In this case, the analyzer provides new information to the controller at the completion of each batch analysis, which can be automated at a period depending on the difficulty of separation. For example, a simple chromatograph might be able to send an updated measured value of the controlled variable every 2 min. Since the analyzer controller should be executed only when a new measured value is available, the controller execution period should be 2 min.

The execution periods can be approximated using the guideline of $\Delta t = 0.05(\theta + \tau)$ for PID controllers, with the process parameters determined by one of the empirical methods described in Chapter 6. Modern digital controllers typically execute most loops very frequently, usually with a period under one second, unless the engineer specifies a longer period. The results for the example are summarized in Table 11.4.

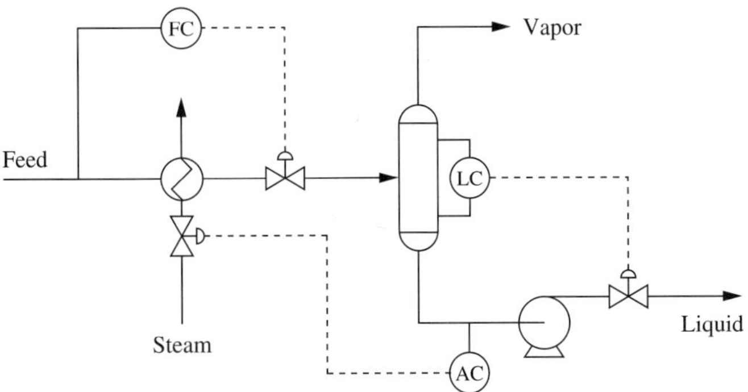

FIGURE 11.9
Example process for selecting controller execution periods.

TABLE 11.4
Digital controller execution periods for the example in Figure 11.9

Controlled variable	Maximum execution period	Typical execution period in commercial equipment
Flow	0.2 sec	0.1 to 0.5 sec
Level	15 sec	0.1 to 0.5 sec
Analyzer	2 min	2 min

Notice that the conventional digital systems might not satisfy the guideline for very fast processes, but the resulting small degradation of the control performance is not usually significant for most flows and pressures. When the variable is extremely important, as is the case in compressor surge control, which prevents damage to expensive mechanical equipment, digital equipment with faster execution (and sensors and valves with faster responses) should be used (Staroselsky and Ladin, 1979).

The analyzer has a very long execution period; therefore, it would be best to select the execution immediately when the new measured value becomes available, rather than initiate execution every 2 min whether the updated measurement is available or not. Thus, it is common practice for the controller execution to be synchronized with the update of a sensor with a long sample period; this is achieved through a special signal that indicates that the new measured value has just arrived.

11.7 TRENDS IN DIGITAL CONTROL

The basic principles presented in this chapter should not change as digital control equipment evolves. However, many of the descriptions of the equipment will undoubtedly change; in fact, the simple descriptions here do not attempt to cover all of the newer features being used. A few of the more important trends in digital control are presented in this section.

Signal Transmission

The equipment described in this chapter involves analog signal transmission between the central control room and the sensor and valve. It is possible to collect a large number of signals at the process equipment and transmit the information via a digital communication line. This digital communication would eliminate many—up to thousands—of the cables and terminations and result in great cost savings. The reliability of this digital system might not be as good, because the failure of the single transmission line would cause a large number of control loops to fail simultaneously. However, the potential economic benefit provides a driving force for improved, high-reliability designs. This is a rapidly changing area for which important standards are being developed that should facilitate the integration of equipment from various suppliers (Lidner, 1990; ISA, 1993).

It is possible to communicate without physical connections, via *telemetry*. This method is now used to collect data from remote process equipment such as crude petroleum production equipment over hundreds of kilometers. When telemetry is sufficiently reliable, some control could be implemented using this communication method.

Smart Sensors

Microprocessor technology can be applied directly at the sensor and transmitter to provide better performance. An important feature of these sensors is the ability for *self-calibration*—that is, automatic corrections for environmental changes, such as temperature, electrical noise, and process conditions.

Operator Displays

Excellent displays are essential so that operating personnel can quickly analyze and respond to ever-changing plant conditions. Current displays consist of multiple cathode ray tubes (CRTs). Future display technology is expected to provide flat screens of much larger area. These larger screens will allow more information to be displayed concurrently, thus improving process monitoring.

Controller Algorithms

The flexibility of digital calculations eliminates a restriction previously imposed by analog computation that prevented engineers from employing complex algorithms for special purpose applications. Some of the most successful new algorithms use explicit dynamic models in the controller. These algorithms are presented in this book in Chapter 19 on predictive control and in Chapter 23 on centralized multivariable control.

Monitoring and Optimization

The large amount of data collected and stored by digital control systems provides an excellent resource for engineering analysis of process performance. The results of this analysis can be used to adjust the operating conditions to improve product quality and profit. This topic is addressed in Chapter 26.

11.8 CONCLUSIONS

Digital computers have become the standard equipment for implementing process control calculations. However, the trend toward digital control is not based on better performance of PID control loops. In fact, the material in this chapter demonstrates that most PID control loops with digital controllers do not perform as well as those with continuous controllers, although the difference is usually small.

The sampling of a continuous measured signal for use in feedback control introduces a limit to control performance, because some high-frequency information is lost through sampling. Shannon's theorem provides a quantitative estimate of the frequency range over which information is lost.

Sampling and discrete execution introduce an additional dynamic effect in the feedback loop, which influences stability and performance. Guidelines are provided that indicate how the PID controller tuning should be modified to retain the proper margin from the stability limit while providing reasonable control performance. As we recall, the stability margin is desired so that the control system performs well when the process dynamic response changes from its estimated value—in other words, so that the system performance is robust.

A major conclusion from this chapter is that

> The characteristics of the modes and tuning constants for the continuous PID controller can be interpreted in the same manner for the digital PID controller. The digital PID controller must use modified tuning guidelines to achieve good performance and robustness.

This valuable result enables us to apply the same basic concepts to both continuous analog and digital controllers.

The power of digital computers is in their flexibility to execute other control algorithms easily, even if the computations are complex.

> All control methods described in subsequent chapters can be implemented in either analog or digital calculating equipment, unless otherwise stated. Where the digital implementation is not obvious, the digital form of the controller algorithm is given.

This power will be capitalized on when applying advanced methods such as nonlinear control (Chapters 16 and 18), inferential control (Chapter 17), predictive control (Chapters 19 and 23), and optimization and statistical monitoring (Chapter 26).

REFERENCES

Astrom, K. and B. Wittenmark, *Computer Controlled Systems,* Prentice-Hall, Englewood Cliffs, NJ, 1990.

Franklin, G., J. Powell, and M. Workman, *Digital Control of Dynamic Systems* (2nd Ed.), Addison-Wesley, Reading, MA, 1990.

Gardenhire, L., "Selecting Sampling Rates," *ISA J.* 59–64 (April 1964).

Lidner, K-P., "Fieldbus—A Milestone in Field Instrumentation Technology," *Meas. Control, 23,* 272–277 (1990).

ISA, *SP50 Field Bus Standard,* Instrument Society of America, Research Triangle Park, NC, 1993 (in development).

Seborg, D., T. Edgar, and D. Mellichamp, *Process Dynamics and Control,* Wiley, New York, 1989.

Staroselsky, N. and L. Ladin, "Improved Surge Control for Centrifugal Compressors," *Chem. Engr., 86,* 175–184 (May 21, 1979).

Gerald, C. and P. Wheatley, *Applied Numerical Analysis* (4th Ed.), Addison-Wesley, Reading, MA, 1989.

ADDITIONAL RESOURCES

Each commercial digital control system has an enormous array of features, making comparisons difficult. A summary of the equipment for some major suppliers is provided in the manual

Wade, H. (ed.), *Distributed Control Systems Manual,* Instrument Society of America, Research Triangle Park, NC, 1992 (with periodic updates).

In addition to the references by Astrom and Wittenmark (1990) and by Franklin et al. (1990), the following book gives detailed information on z-transforms and digital control theory.

Ogata, K., *Discrete-Time Control Systems,* Prentice-Hall, Englewood Cliffs, NJ, 1987.

For an analysis of digital controller execution periods that considers the disturbance dynamics for statistical, rather than deterministic, disturbances, see

MacGregor, J., "Optimal Choice of the Sampling Interval for Discrete Process Control," *Technometrics, 18,* 2, 151–160 (May 1976).

QUESTIONS

11.1. Answer these questions about the digital PID algorithm.

(*a*) Give the equations for the full-position and velocity PID controllers if a trapezoidal numerical integration were used for the integral mode.

(*b*) The digital controller can be simplified to the following form to reduce real-time computations. Determine the values for the constants (the A_is) in terms of the tuning constants and execution period for (i) PID, (ii) PI, and (iii) PD controllers.

$$\Delta \text{MV}_N = A_1 E_N + A_2 E_{N-1} + A_3 \text{CV}_N + A_4 \text{CV}_{N-1} + A_5 \text{CV}_{N-2}$$

11.2. Many tuning rules were designed for continuous control systems, such as Ziegler-Nichols, Ciancone, and Lopez.

(*a*) Describe the conditions, including quantitative measures, for which these tuning rules could be applied to digital controllers without modification.

(*b*) How could you adjust the rules to systems that had longer execution periods than determined by the approximate guidelines given in part (*a*) of this question?

11.3. Develop a simulation of a simple process under digital PID control. Equations for the process are given below. The calculations can be performed using a spreadsheet or a programming language. The input change is a step set point change from 1 to 2.0 at time $= 1.0$. The process parameters can be taken from the system in Section 9.3; $K_p = 1.0$, $\tau = 5.0$, and $\theta = 5.0$, and the controller and simulation time steps can be taken to be equal; that is, $\delta t = \Delta t = 1.0$.

$$\text{CV}_N = (e^{-\delta t/\tau})\text{CV}_{N-1} + K_p(1 - e^{-\delta t/\tau})\text{MV}_{N-\Gamma-1} \qquad \Gamma = \frac{\theta}{\delta t}$$

$$\text{MV}_N = \text{MV}_{N-1} + K_c\left(E_N - E_{N-1} + \frac{\Delta t}{T_I}E_N\right)$$

with δt = step size for the numerical solution of the process equation
Δt = execution period of the digital controller

(*a*) Verify the equations for the process and controller and determine the initial conditions for MV and CV.

(*b*) Repeat the study summarized in Figure 9.2 for a set point change.

(*c*) Use the tuning in Table 9.2 to obtain the IAE for set point changes.

(*d*) Select tuning from points on the response surface in Figure 9.3. Obtain the dynamic responses and explain the behavior: oscillatory, overdamped, and so forth.

11.4. Repeat question 11.3 for the system in Example 8.5 and obtain the dynamic response given in Figure 8.9. You must determine all parameters in the equations, including appropriate values for the process simulation step size and the execution period of the digital controller. Solve this problem by simulating (1) the linearized process model and (2) by simulating the nonlinear process model.

11.5. State for each of the controller variables in the following list

(*a*) its source (e.g., from an operator, from process, or from a calculation)

(*b*) whether the variable would be transferred to the operator console for display

1. SP_N, the controller set point
2. CV_N, the current value of the controlled variable
3. K_c, the controller gain
4. S_N, the sum of all past errors used in approximating the integral error
5. MV_N, the current controller output
6. M/A, the status of the controller (M=manual or off, A=automatic or on)
7. ΔMV_N, the current change to the manipulated variable
8. E_N, the current value of the error

11.6. A process control design is given in Figure Q11.6. The process transfer functions $G_p(s)$ follow, with time in minutes.

$$(G_p(s))_T = \frac{T_1(s)}{v_1(s)} = \frac{3e^{-1.2s}}{1+2s} \quad \left(\frac{°\text{C}}{\%\text{ open}}\right)$$

$$(G_p(s))_A = \frac{A_1(s)}{v_2(s)} = \frac{1.3e^{-0.5s}}{1+14s} \quad \left(\frac{\text{wt\%}}{\%\text{ open}}\right)$$

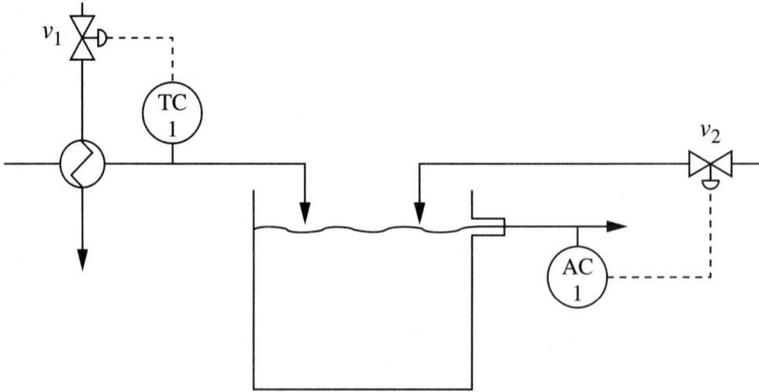

FIGURE Q11.6

(*a*) For each controller, determine the maximum execution period so that digital execution does not significantly affect the control performance.

(*b*) Determine the PID controller tuning for each controller for two values of the execution period:

(1) The result in (*a*) and (2) a value of 3 minutes

11.7. In the chapter it was stated that the digital controller should not be executed faster than the measured controlled variable is updated. In your own words, explain the effect of executing the controller faster than the measurement update and why this effect is undesirable.

11.8. An example of ringing occurs when a digital proportional-only controller is applied to a process that is so fast that it reaches steady state within one execution period, Δt. The following calculations, which are simple enough to be carried out by hand, will help explain ringing.

(*a*) Calculate several steps of the response of a control system with a steady-state process with $K_p = 1.0\,(\theta = \tau = 0)$ and a proportional-only controller, $K_c = 0.8$. Assume that the system is initially at steady state and a set point change of 5 units is made.

(*b*) Repeat the calculation for an integral-only controller, equation (8.101). Find a value of the parameter $(K_c\Delta t/T_I)$ by trial and error that gives good dynamic performance for the controlled and manipulated variables.

(*c*) Generalize the results in (*a*) and (*b*) and give a tuning rule for integral-only, digital control of a fast process.

11.9. Some example process dynamics and associated digital feedback execution periods are given in the following table. For each, calculate the PI controller tuning constants, assuming standard control performance objectives.

	Process transfer function $G_p(s)$	**Execution period** Δt
(*a*)	Three-tank mixer, Example 7.2	Selected by reader
(*b*)	Recycle system in Equation (5.60)	Selected by reader
(*c*)	Reactor in Example 5.8	Selected by reader
(*d*)	$\dfrac{1.2e^{-0.1s}}{1+0.5s}$	0.25
(*e*)	$\dfrac{1.2e^{-0.1s}}{1+0.5s}$	5.0
(*f*)	$\dfrac{2.1e^{-20s}}{1+100s}$	30
(*g*)	$\dfrac{2.1e^{-100s}}{1+20s}$	30

11.10. Considering the description of a distributed digital control system, determine which processors, signal converters, and transmission equipment must act and in what order for (*a*) the result of an operator-entered set point change to reach the valve; (*b*) a process change to be detected and acted upon by the controller so that the valve is adjusted.

11.11. Consider a signal that is a perfect sine with period T_{signal}, and is sampled at period T_{sample}, with $T_{\text{sample}} < T_{\text{signal}}$. Determine the primary aliasing frequency (the sample frequency at which the sampled values are periodic with a period a multiple of the true signal sine frequency) as a function of the two periods.

11.12. (*a*) Determine bounds on the error between the continuous signal and the output of the sample/hold for a zero-order and a first-order hold. (*Hint:* Consider the rate of change of the continuous signal.)

(*b*) Apply the results in (*a*) to a continuous sine signal and determine the errors for various values of the sample period to sine period.

(*c*) Which hold gives a smaller error in (*b*)?

11.13. Answer the following questions regarding the computer implementation of the digital PID controller.

(*a*) Can the controller tuning constants be changed while the controller is functioning without disturbing the manipulated variable? (Consider the velocity and full-positional forms separately.)

(*b*) For the velocity form of the PID, what is the value for MV_{N-1} for the first execution of the controller?

(*c*) For the full-position form of the PID, the sum of the error term might become very large and overflow the word length. Is this a problem likely to occur?

(*d*) Discuss how the calculations could be programmed to introduce limits on the change of the manipulated variable (ΔMV_N), the set point (SP_N), and the manipulated variable (MV_N).

(*e*) Can you anticipate any performance difficulties when the limitations in (*d*) are implemented? If yes, suggest modifications to the algorithm.

CHAPTER 12

PRACTICAL APPLICATION OF FEEDBACK CONTROL

12.1 INTRODUCTION

The major components of the feedback control calculations have been presented in previous chapters in this part. However, much more needs to be done to ensure the successful application of the principles already covered. Practical application of feedback control requires that equipment and calculations provide accuracy and reliability and also overcome a few shortcomings of the basic PID control algorithm. Some of these requirements are satisfied through careful specification and maintenance of equipment used in the control loop. Other requirements are satisfied through modifications to the control calculations. Some of the more important topics, involving both physical equipment and calculations, are addressed in this chapter.

A typical digital PID control loop may execute tens to hundreds of lines of computer code; this is much more than the single PID control calculation in equations (8.17) and (11.6). Fortunately, the modifications described in this chapter, and many more, are available in most commercial control equipment as preprogrammed algorithms that can be linked together. Therefore, application of the modifications described here is generally straightforward once the engineer fully understands their proper use. The application issues will be discussed with reference to the control loop diagram in Figure 12.1, which shows that many of the calculations can be grouped into three categories: input processing, control algorithm, and output processing. As shown in Table 12.1, most of the calculation modifications are available in both analog and digital equipment; however, a few are not available on standard analog equipment, because of excessive cost. The application requirements are discussed

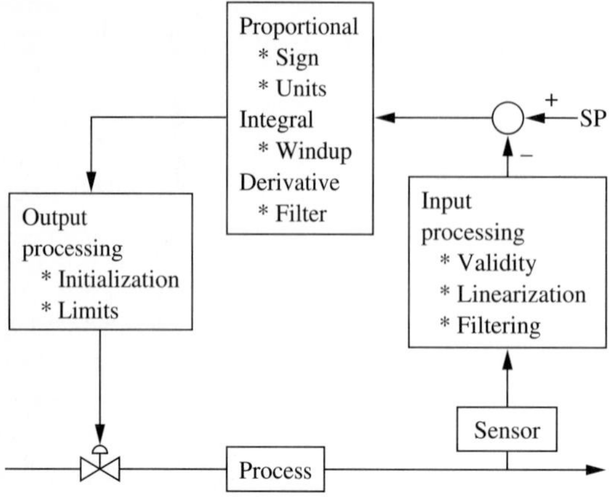

FIGURE 12.1
Simplified control loop drawing, showing application topics.

in the order of the four major topics given in Table 12.1. A few key equipment specifications are presented first, followed by input processing calculations, performed before the control calculation. Then, modifications to the PID control calculation are explained. Finally, a few issues related to output processing are presented. The topics in this chapter are by no means a complete presentation of practical issues for successful application of control; they are limited to the most important issues for single-loop control. Further topics, addressing design, reliability, and safety, are covered in Part VI after multiple-loop processes and controls have been introduced.

TABLE 12.1
Summary of application issues

Application topic	Available in either analog or digital equipment	Typically available only in digital equipment
Equipment specification		
Measurement range	†	
Final element capacity	†	
Failure mode	†	
Input processing		
Input validity		X
Engineering units		X
Linearization	X	
Filtering	X	
Control algorithm		
Sign	X	
Dimensionless gain	X	
Anti–reset windup	X	
Derivative filter	X	
Output processing		
Initialization	X	
Bounds on output variable	X	

† Involves field control equipment that is independent of analog or digital controllers.

12.2 EQUIPMENT SPECIFICATION

Proper specification of process and control equipment is essential for good control performance. In this section, the specification of sensors and final control elements is discussed. Sensors are selected to provide an indication of the true controlled variable and are selected based on accuracy, reproducibility, and cost. The first two terms are defined here as paraphrased from ISA (1979).

Accuracy is the degree of conformity to a standard (or true) value when the device is operated under specified conditions. This is usually expressed as a bound that errors will not exceed when a measuring device is used under these specified conditions, and it is often reported as inaccuracy as a percent on the instrument range.

Reproducibility is the closeness of agreement among repeated sensor outputs for the same process variable value. Thus, a sensor that has very good reproducibility can have a large deviation from the true process variable; however, the sensor is consistent in providing (nearly) the same indication for the same true process variable.

Often, deviations between the true variable and the sensor indication occur as a "drift" or slow change over a period of time, and this drift contributes a bias error. In these situations, the accuracy of the sensor may be poor, although it may provide a good indication of the change in the process variable, since the sensitivity relationship (Δ sensor signal)/(Δ true variable) may be nearly constant. Although a sensor with high accuracy is always preferred because it gives a close indication of the true process variable, cases will be encountered in later chapters in which reproducibility is acceptable as long as the sensitivity is unaffected by the drift. For example, reproducibility is often acceptable when the measurement is applied in enhancing the performance of a control design in which the key output controlled variable is measured with an accurate sensor. The importance of accuracy and reproducibility will become clearer after advanced control designs such as cascade and feedforward control are covered; therefore, this topic is returned to in Chapter 24.

Often, inaccuracies can be corrected by periodic calibration of the sensor. If the period of time between calibrations is relatively long, a drift from high accuracy over days or weeks could result in poor control performance. Thus, critical instruments deserve more frequent maintenance. If the period between calibrations is long, some other means for compensating the sensor value for a drift from the accurate signal may be used; often, laboratory analyses can be used to determine the bias between the sensor and true (laboratory) value. If this bias is expected to change very slowly, compared with laboratory updates, the corrected sensor value, equalling measurement plus bias, can be used for real-time control. Further discussion on using measurements that are not exact, but give approximate indications of the process variable over limited conditions, is given in Chapter 17 on inferential control.

Sensor Range

An important factor that must be decided for every sensor is its range. For essentially all sensors, accuracy and reproducibility improve as the range is reduced, which means that a small range would be preferred. However, the range must be large enough to span the expected variation of the process variable during typical conditions, including disturbances and set point changes. Also, the measurement ranges are selected for easy interpretation of graphical displays; thus, ranges are selected that are evenly divisible, such as 10, 20, 50, 100, or 200. Naturally, each measurement must be analyzed separately to determine the most appropriate ranges, but some typical examples are given in the following table.

Variable	Typical set point	Sensor range
Furnace outlet temperature	600°C	550–650°C
Pressure	50 bar	40–60 bar
Composition	0.50 mole %	0–2.0 mole %

Levels of liquids (or solids) in vessels are typically expressed as a percent of the span of the sensor rather than in length (meters). Flows are often measured by pressure drop across an orifice meter. Since orifice plates are supplied in a limited number of sizes, the equipment is selected to be the smallest size that is (just) large enough to measure the largest expected flow. The expected flow is always greater than the design flow; as a result of the limited equipment and expected flow range, the flow sensor can usually measure at least 120 percent of the design value, and its range is essentially never an even number such as 0 to 100 m^3/day.

These simple guidelines do not satisfy all situations, and two important exceptions are mentioned here. The first special situation involves nonnormal operations, such as startup and major disturbances, when the variable covers a much greater range. Clearly, the suppressed ranges about normal operation will not be satisfactory in these cases. The usual practice is to provide an *additional* sensor with a much larger range to provide a measurement, with lower accuracy and reproducibility, for these special cases. For example, the furnace outlet temperature shown in Figure 12.2, which is normally about 600°C, will vary from about 20 to 600°C during

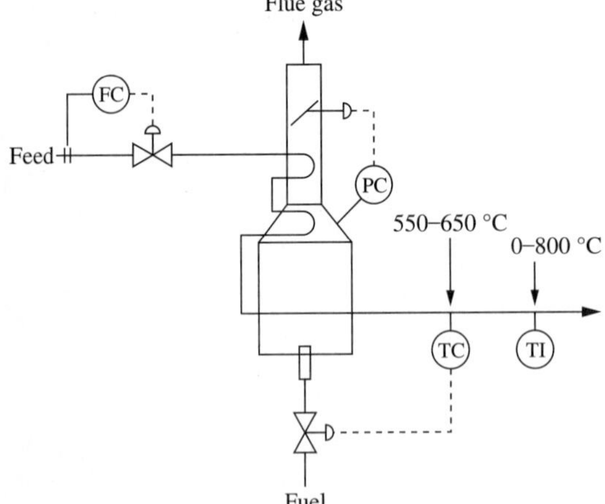

FIGURE 12.2
Fired heater with simple control strategy.

startup and must be monitored to ensure that the proper warm-up rate is attained. An additional sensor with a range of 0 to 800°C could be used for this purpose. The additional sensor could be used for control by providing a switch, which selects either of the sensors for control. Naturally, the controller tuning constants would have to be adapted for the two types of operation.

A second special situation occurs when the accuracy of a sensor varies over its range. For example, a flow might be normally about 30 m^3/hr in one operating situation and about 100 m^3/hr in the other. Since a pressure drop across an orifice meter does not measure the flow accurately for the lower one-third of its range, two pressure drop measurements are required with different ranges. For this example, the meter ranges might be 0 to 40 and 0 to 120 m^3/hr, with the smaller range providing good accuracy for smaller flows.

Control Valve

The other critical control equipment item is the final element, which is normally a control valve. The valve should be sized just large enough to handle the maximum expected flow at the expected pressure drop and fluid properties. Oversized control valves (i.e., valves with maximum possible flows many times larger than needed) would be costly and might not provide precise maintenance of low flows. The acceptable range for many valves is about 25 : 1; in other words, the valve can regulate the flow smoothly from 4 to nearly 100 percent of its range, with flows below 4 percent having unacceptable variation. (Note that the range of stable flow depends on many factors in valve design and installation; the engineer should consult specific technical literature for the equipment and process design.) Valves are manufactured in specific sizes, and the engineer selects the smallest valve size that satisfies the maximum flow demand. If very tight regulation of small changes is required for a large total flow, a typical approach is to provide two valves, as shown in Figure 12.3. This example shows a pH control system in which acid is adjusted to achieve the desired pH. In this design, the position of the large valve is changed infrequently by the operator, and the position of the small valve is changed automatically by the controller. Strategies for the controller to adjust both valves are presented in Chapter 22 on variable-structure control.

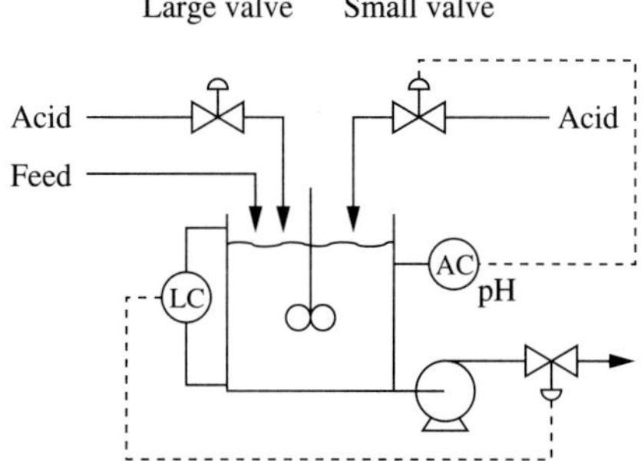

FIGURE 12.3
Stirred-tank pH control system with two manipulated valves, of which only one is adjusted automatically.

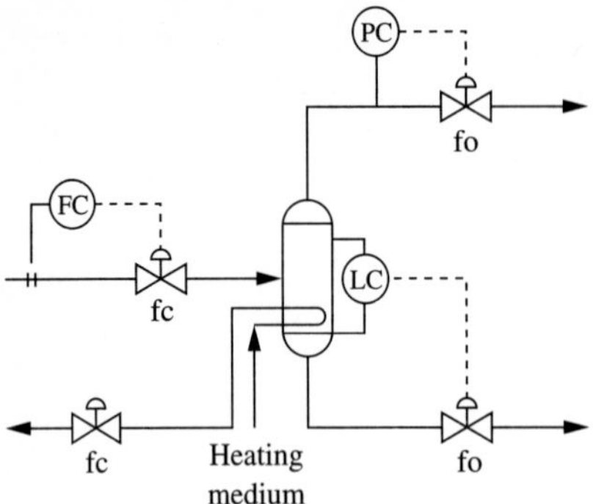

FIGURE 12.4
A flash separation unit with the valve failure modes.

> Sensors and final elements are sized to (just) accommodate the typical operating range of the variable. Extreme oversizing of a single element is to be avoided; a separate element with larger range should be provided if necessary.

Another important issue is the behavior of control equipment when power is interrupted. Naturally, a power interruption is an infrequent occurrence, but proper equipment specification is critical so that the system responds safely in this situation. Power is supplied to most final control elements (i.e., valves) as air pressure, and loss of power results from the stoppage of air compressors or from the failure of pneumatic lines. The response of the valve when the air pressure, which is normally 3 to 15 psig, decreases below 3 psig is called its *failure mode*. Most valves fail open or fail closed, with the selection determined by the engineer to give the safest process conditions after the failure. Normally, the safest conditions involve the lowest pressures and temperatures. As an example, the flash drum in Figure 12.4 would have the valve failure modes shown in the figure, with "fo" used to designate a fail-open valve and "fc" a fail-closed valve. (An alternative designation is an arrow on the valve stem pointing in the direction that the valve takes upon air loss.) The valve failure modes in the example set the feed to zero, the output liquid flow to maximum, the heating medium flow to zero, and the vapor flow to its maximum. All of these actions tend to minimize the possibility of an unsafe condition by reducing the pressure. However, the proper failure actions must consider the integrated plant; for example, if a gas flow to the process normally receiving the liquid could result in a hazardous situation, the valve being adjusted by the level controller would be changed to fail-closed.

The proper failure mode can be ensured through simple mechanical changes to the valve, which can be made after installation in the process. Basically, the failure mode is determined by the spring that directs the valve position when no external air pressure provides a counteracting force. This spring can be arranged to ensure either a fully opened or fully closed position. As the air pressure is increased, the force on the restraining diaphragm increases, and the valve stem (position) moves against the spring.

> The failure mode of the final control element is selected to reduce the possibility of injury to personnel and of damage to plant equipment.

The selection of a failure mode also affects the normal control system, because the failure mode is the position of the valve at 0 percent controller output. As the controller output increases, a fail-open valve closes and a fail-closed valve opens. As a result, the failure mode affects the sign of the process transfer function expressed as CV(s)/MV(s), which is the response "seen" by the controller. As a consequence, the controller gain used for negative feedback control is influenced by the failure mode. If the gain for the process CV(s)/$F(s)$, with $F(s)$ representing the flow through the manipulated valve, is K_p^*, the correct sign for the controller gain is given by

Failure mode	Sign of the controller gain considering the failure mode
Fail closed	$\text{sign}(K_p^*)$
Fail open	$-\text{sign}(K_p^*)$

This brief introduction to determining sensor ranges, valve sizes, and failure modes has covered only a few of the many important issues. These topics and many more are covered in depth in many references and instrumentation handbooks, which should be used when designing control systems (see references in Chapter 1).

12.3 INPUT PROCESSING

The general control system, involving the sensor, signal transmission, control calculation, and transmission to the final element, was introduced in Chapter 7. In this section, we will look more closely at the processing of the signal from the completion of transmission to just before the control algorithm. The general objectives of this signal processing are to (1) improve reliability by checking signal validity, (2) perform calculations that improve the relationship between the signal and the actual process variable, and (3) reduce the effects of high-frequency noise.

Validity Check

The first step is to make a check of the validity of the signal received from the field instrument via transmission. As we recall, the electrical signal is typically 4 to 20 mA, and if the measured signal is substantially outside the expected range, the logical conclusion is that the signal is faulty and should not be used for control. A faulty signal could be caused by a sensor malfunction, power failure, or transmission cable failure. A component in the control system must identify when the signal is outside of its allowable range and place the controller in the manual mode *before the value is used for control.* An example is the furnace outlet temperature controller in Figure 12.2. A typical cause of a sensor malfunction is for the thermocouple measuring the

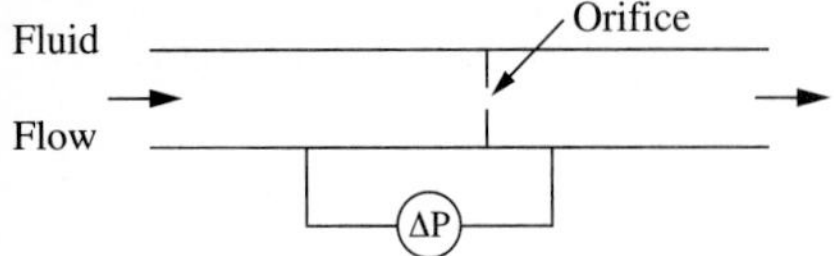

FIGURE 12.5
Flow measurement by sensing the pressure difference about an orifice in a pipe.

temperature to break physically, opening the circuit and resulting in a signal, after conversion from voltage to current, below 4 mA. If this situation were not recognized, the temperature controller would receive a measurement equal to the lowest value in the sensor range and, as a result, increase the fuel flow to its maximum. This action could result in serious damage to the process equipment and possible injury to people. The input check could quickly identify the failure and interrupt feedback control. An indication should be given to the operators, because the controller mode would be changed without their intervention. Because of the logic required for this function, it is easily provided as a preprogrammed feature in many digital control systems, but it is not a standard feature in analog control because of its increased cost.

Conversion for Nonlinearity

The next step in input processing is to convert the signal to a better measure of the actual process variable. Naturally, the physical principles for sensors are chosen so that the signal gives a "good" measure of the process variable; however, factors such as reliability and cost often lead to sensors that need some compensation. An example is a flow meter that measures the pressure drop across an orifice, as shown in Figure 12.5. The flow and pressure drop are ideally related according to the equation

$$F = K\sqrt{\frac{\Delta P}{\rho}} \tag{12.1}$$

with F = volumetric flow rate
ρ = density
ΔP = pressure difference across the orifice

Typically, the sensor measures the pressure drop, so that

$$F = \frac{K}{\sqrt{\rho}}\sqrt{(S_1 - S_{10})(R_1) + Z_1} \tag{12.2}$$

with S_1 = signal from the sensor
S_{10} = lowest value of the sensor signal
R_1 = range of the true process variable measured by the sensor
Z_1 = value of the true process variable when the sensor records its lowest signal (S_{10})
ρ = constant

Thus, using the sensor signal directly (i.e., without taking the square root) introduces an error in the control loop. The accuracy would be improved by using the square root of the signal, as shown in equation (12.2), for control and also for process monitoring. In addition, the accuracy could be improved further for important flow measurements by automatically correcting for fluid density variations as follows:

$$F = K\sqrt{\frac{(S_1 - S_{10})(R_1) + Z_1}{(S_2 - S_{20})(R_2) + Z_2}} \tag{12.3}$$

with the subscript 1 for the pressure difference sensor signal and 2 for the density sensor signal. By far the most common flow measurement approach used commercially is equation (12.2), with equation (12.3) used only when the accurate flow measurement is important enough to justify the added cost of the density analyzer.

Another common example of sensor nonlinearity is the thermocouple temperature sensor. A thermocouple generates a millivolt signal that depends on the temperature difference between the two junctions of the bimetallic circuit. The signal transmitted for control is either in millivolts or linearly converted to milliamps. However, the relationship between millivolts and temperature is not linear. Usually, the relationship can be represented by a polynomial or a piecewise linear approximation to achieve a more accurate temperature value; the additional calculations are easily programmed as a function in the input processing to achieve a more accurate temperature value.

These orifice flow and thermocouple temperature examples are only a few of the important relationships that must be considered in a plantwide control system. Naturally, each relationship should be evaluated based on the physics of the sensor and the needs of the control system. Standard handbooks and equipment supplier manuals provide invaluable information for this analysis. The importance of the analysis extends beyond control to monitoring plant performance, which depends on accurate measurements to determine material balances, reactor yields, energy consumption, and so forth. Thus, many values are corrected for nonlinearities even when they are not used for closed-loop control.

Engineering Units

Another potential input calculation expresses the input in engineering units, which greatly simplifies the analysis of data by operations personnel. This calculation is possible only in digital systems, as analog systems perform calculations using voltage or pressure. Recall that the result of the transmission and any correction for nonlinearity in digital systems is a signal in terms of the instrument range expressed as a percent (0 to 100) or a fraction (0 to 1). The variable is expressed in engineering units according to the following equation:

$$\text{CV} = Z + R(S_3 - S_{30}) \tag{12.4}$$

with S_3 the signal from the sensor after correction for nonlinearity.

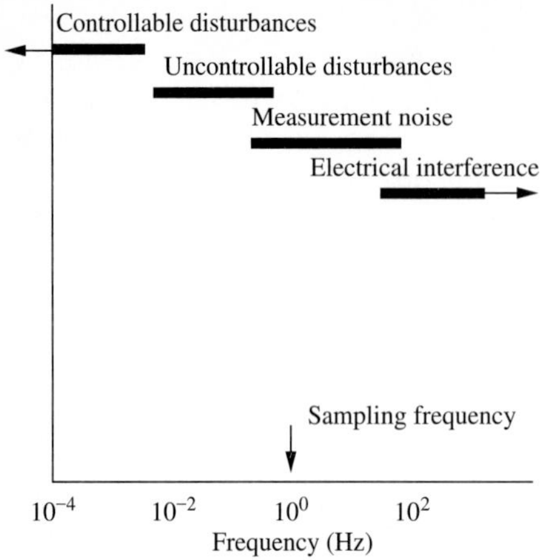

FIGURE 12.6
Example frequency ranges for components in the measurement. (Reprinted by permission. Copyright©1966, Instrument Society of America. From Goff, K., "Dynamics of Direct Digital Control, Part I," *ISA J., 13, 11,* 45–49.)

Filtering

An important feature in input processing is filtering. The transmitted signal represents the result of many effects; some of these effects are due to the process, some are due to the sensor, and some are due to the transmission. These contributions to the signal received by the controller vary over a wide range of frequencies, as presented in Figure 12.6. The control calculation should be based on only the responses that can be affected by the manipulated variable, because very high-frequency components will result in high-frequency variation of the manipulated variable, which will not improve and may degrade the performance of the controlled system.

Some noise components are due to such factors as electrical interference and mechanical vibration, which have a much higher frequency than the process response. (This distinction may not be so easy to make in controlling machinery or other very fast systems.) Other noise components are due to changes such as imperfect mixing and variations in process input variables such as flows, temperatures, and compositions; some of these variations may be closer to the critical frequency of the control loop. Finally, some measurement variations are due to changes in flows and compositions that occur at frequencies much below the critical frequency; the effects of these disturbances can be attenuated effectively by feedback control.

The very high-frequency component of the signal cannot be influenced by a process control system, and thus is considered "noise"; the goal, therefore, is to remove the unwanted components from the signal, as shown in Figures 12.7 and 12.8. The filter is located in the feedback loop, and dynamics involved with the filter, like process dynamics, will influence the stability and control performance of the closed-loop system. This statement can be demonstrated by deriving the following transfer function, which shows that the filter appears in the characteristic equation.

$$\frac{\mathrm{CV}(s)}{\mathrm{SP}(s)} = \frac{G_p(s)G_v(s)G_c(s)}{1 + G_p(s)G_v(s)G_c(s)G_f(s)G_s(s)} \tag{12.5}$$

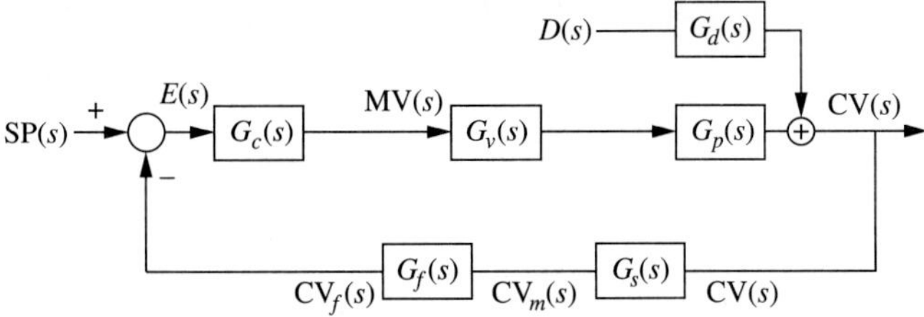

FIGURE 12.7
Block diagram of a feedback loop with a filter on the measurement.

If it were possible to separate the signal ("true" process variable) from the noise, the perfect filter in Figure 12.8 would transmit the unaltered "true" process variable value to the controller and reduce the noise amplitude to zero. In addition, the perfect filter would do this without introducing phase lag! Unfortunately, there is no clear distinction between the "actual" process performance, which can be influenced by adjusting the manipulated variable, and the "noise," which cannot be influenced and should be filtered. Also, no filter calculation exists that has the features of a perfect filter in Figure 12.8.

The filter calculation usually employed in the chemical process industries is a first-order transfer lag:

$$\mathrm{CV}_f(s) = \frac{1}{\tau_f s + 1}\mathrm{CV}_m(s) \tag{12.6}$$

with $\mathrm{CV}_f(s)$ = value after the filter
$\mathrm{CV}_m(s)$ = value before the filter
τ_f = filter time constant

The gain is unity because the filter should not alter the actual signal at low frequency, including the steady state. The frequency response of the continuous filter, since it is a first-order system, has been derived several times in this book. The amplitude

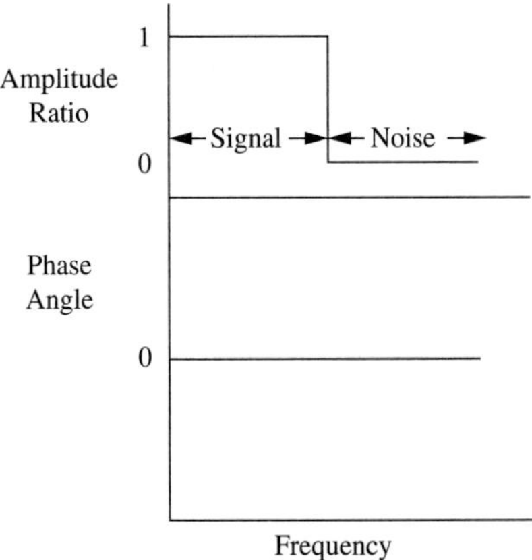

FIGURE 12.8
The amplitude ratio and phase angle of a perfect filter, which *cannot* be achieved exactly.

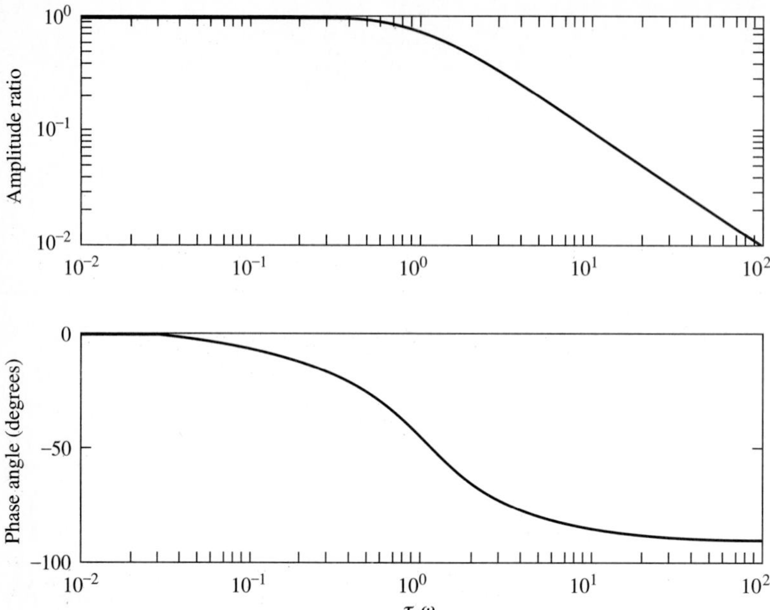

FIGURE 12.9
Bode plot of first-order filter.

ratio and phase angle of the filter are given by the following equations and shown in Figure 12.9.

$$\text{AR} = \frac{1}{\sqrt{1 + \omega^2 \tau_f^2}} \qquad (12.7)$$

$$\phi = \tan^{-1}(-\omega \tau_f)$$

The filter time constant, τ_f, is a tuning parameter that is selected to approximate the perfect filter shown in Figure 12.8; this goal requires that it be small with respect to the dominant process dynamics so that feedback control performance is not significantly degraded. Also, it should be large with respect to the noise period (inverse of frequency) so that noise is attenuated. These two requirements cannot usually be satisfied perfectly, because the signal has components of all frequencies and the cutoff between process and noise is not known. As seen in Figure 12.9, the amplitude of high-frequency components decreases as the filter time constant is increased. In the example, signal components with a frequency smaller than $0.5/\tau_f$ are essentially unaffected by the filter, while components with a much higher frequency have their magnitudes reduced substantially. This performance leads to the name *low-pass filter,* which is sometimes used to describe the filter that does not affect low frequencies—lets them pass through—while attenuating the high-frequency components of a signal. A simple case study has been performed to demonstrate the trade-off between filtering and performance. The effect of filtering on a first-order-

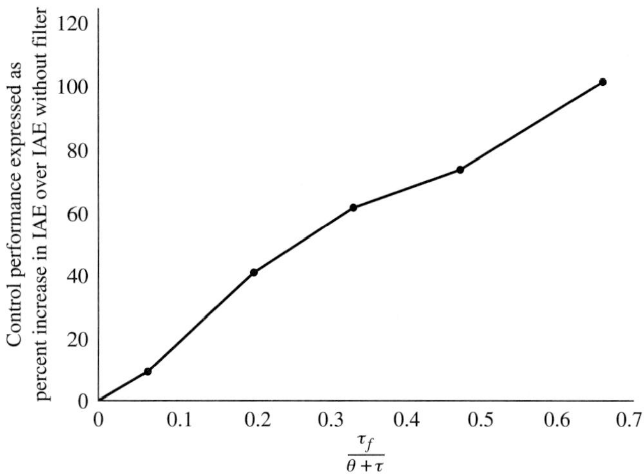

FIGURE 12.10
The effect of measurement filtering on feedback control performance ($\theta/(\theta + \tau) = 0.33$).

with-dead-time plant is given in Figure 12.10. The controlled-variable performance, measured simply as IAE in this example, degrades as the filter time constant is increased. The results are given in Figure 12.10, which shows the percent increase in IAE over control without the filter as a function of the filter time constant. This case study was calculated for a plant with fraction dead time of 0.33 under a PI controller with tuning according to the Ciancone correlations. Thus, the results are *typical but not general;* similar trends can be expected for other systems.

Based on the goals of filtering, the guidelines in Table 12.2 are recommended for reducing the effects of high-frequency noise for a typical situation. These steps should be implemented in the order shown until the desired control performance is achieved. Normally, step 2 will take priority over step 3, because the controlled-variable performance is of greater importance. If reducing the effects of high-frequency noise is an overriding concern, the guidelines can be altered accordingly,

TABLE 12.2
Guidelines for reducing the effects of noise

Step	Action	Justification
1. Reduce the amplification of noise by the control algorithm	Set derivative time to zero, $T_d = 0$	Prevent amplification of high-frequency component by controller
2. Allow only a slight increase in the IAE of the controlled variable	Select a small filter τ_f, e.g., $\tau_f < 0.05(\theta + \tau)$	Do not allow the filter to degrade control performance
3. Reduce the noise effects on the manipulated variable	Select filter time constant to eliminate noise, e.g., $\tau_f > 5\omega_n$, where ω_n is the noise frequency	Achieve a small amplitude ratio for the high-frequency components

such as achieving step 3 while allowing some degradation of the controlled-variable control performance.

The final issue in filtering relates to digital implementation. A digital filter can be developed by first expressing the continuous filter in the time domain as a differential equation:

$$\tau_f \frac{d\text{CV}_f(t)}{dt} + \text{CV}_f(t) = \text{CV}_m(t) \tag{12.8}$$

leading to the digital form of the first-order filter,

$$(\text{CV}_f)_n = A(\text{CV}_f)_{n-1} + (1 - A)(\text{CV}_m)_n \qquad \text{with } A = e^{-\Delta t/\tau_f} \tag{12.9}$$

This equation can be derived by solving the differential equation defined by equation (12.8) and assuming that the measured value $(\text{CV}_m)_n$ is constant over the filter execution period Δt. The digital filter also has to be initialized when the calculations are first performed or when the computer is restarted. The typical filter initialization sets the initial filtered value to the value of the initial measurement.

$$(\text{CV}_f)_1 = (\text{CV}_m)_1 \tag{12.10}$$

As is apparent, the first-order filter can be easily implemented in a digital computer. However, the digital filter does not give exactly the same results as the continuous version, because of the effects of sampling. As discussed in Chapter 11 on digital control, sampling a continuous signal results in some loss of information. Shannon's theorem shows us that information in the continuous signal at frequencies above about one-half the sample frequency cannot be reconstructed from the sampled data. For example, sampled data taken at a period of one minute could not be used to determine a sinusoidal variation in the continuous signal with a period of one second. As a result, the *digital filter cannot attenuate higher-frequency noise.*

This is potentially a serious problem, because very high-frequency noise is possible due to mechanical vibrations of the sensor and electrical interference in signal transmission, as shown in Figure 12.6. Since a digital filter alone at a relatively long period cannot provide adequate filtering, most commercial digital control equipment has *two filters in series:* an analog filter before the analog-to-digital (A/D) conversion and an (optional) digital filter after the conversion, as shown in Figure 12.11. The purpose of the analog filter is to reduce high-frequency components of the signal substantially, and typically, it has a time constant on the order of the sample period. The

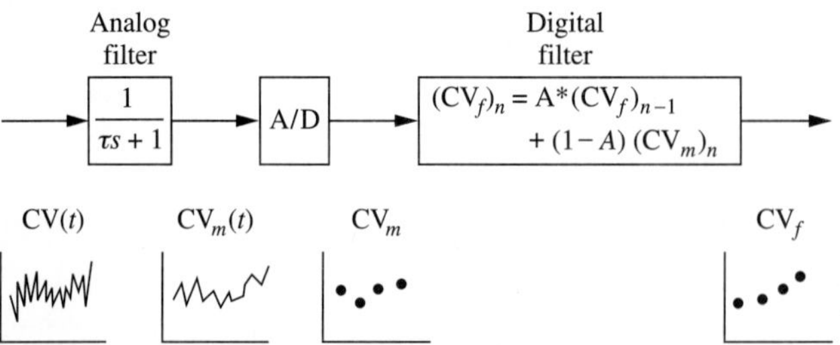

FIGURE 12.11
Schematic of the effects of analog and digital filters in series.

analog filter in this configuration is sometimes referred to as an *antialiasing* filter, since it reduces potential errors resulting from slowly sampling a signal with high-frequency components. The digital filter in the design, if needed, would be tuned according to the guidelines in Table 12.2 to further attenuate variations of higher frequencies.

There is a tendency to overfilter signals used for control. Thus, the following recommendation should be considered:

> Since the filter is a dynamic element in the feedback loop, signals used for control should be filtered no more than the minimum required to achieve good control performance.

Not all measurements are used for control; in fact, a rough estimate is that less than one-third of the signals transmitted to a central control room are used for control. The other signals serve the important purpose of enabling plant personnel to monitor the process. For displaying the current status of the process, these signals should not be filtered, except for the analog filter before the A/D converter, because any filter would delay the information display, which could confuse the plant operator.

Much of this information is also stored for later process analysis. Since high-frequency data is usually not required, a typical approach is to store data consisting of averages of several samples of the measured variable within meaningful time periods such as hour, shift (8 hours), day, and week. This data concentration approach represents a *filter* that reduces the effects of high-frequency noise and short-term plant variations. Assuming that the values used to calculate the average are taken infrequently enough to be independent, the effect of the number of values used in the average on the standard deviation is given as

$$\sigma_{\text{aver}} = \frac{\sigma_m}{\sqrt{n}} \tag{12.11}$$

with σ_{aver} = the standard deviation of the average

σ_m = the standard deviation of the individual measurements used in calculating the average

n = number of measurements used to calculate the average

This filtering is desired for the purpose of long-term process analysis, such as detecting slow changes in heat transfer coefficients or catalyst activity, which in many cases change slowly over weeks or months.

Example 12.1. The measurement of the controlled variable in the three-tank mixer feedback control system in Examples 7.2 (page 249) and 9.2 (page 302) is modified to have higher-frequency sensor noise. Determine how a filter affects (*a*) the open-loop response of the controlled variable after the filter and (*b*) the control performance of the feedback system.

Typical dynamic data of the controlled variable without control is shown in Figure 12.12, along with the responses of the signal after filters with two different time

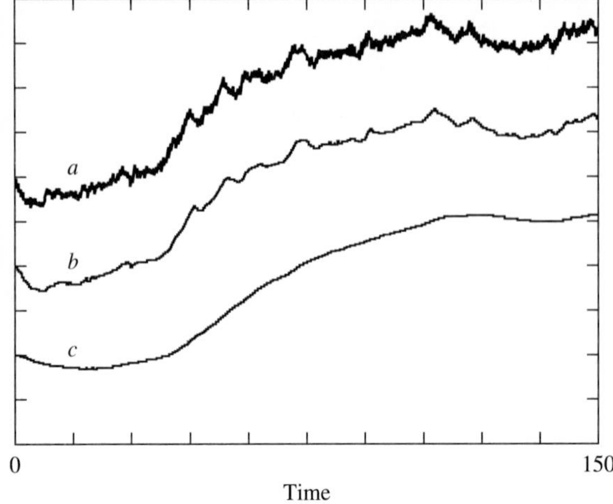

FIGURE 12.12
Open-loop dynamic data for Example 12.1 with τ_f equal to (*a*) 0.0, (*b*) 3.0, and (*c*) 10.0.

constants; the mean values are the same, but the plots are displaced for clearer comparison. As expected, the filters reduce the high-frequency variation in the unfiltered signal. The other key issue is the effect of the filter on the control performance. The dynamic responses of the control system with and without the derivative mode for various filter time constants are shown in Figures 12.13*a* through *c;* in all of these figures, the value of the controlled variable plotted is *before* the filter; thus, this signal is modulated before being used in the controller. The amplification of the measurement noise by the derivative mode is apparent by comparing Figure 12.13*a* and Figure 12.13*b*. In fact, simply eliminating the derivative might be sufficient in this case. The addition of the filter further smooths the manipulated variable but worsens the performance of the

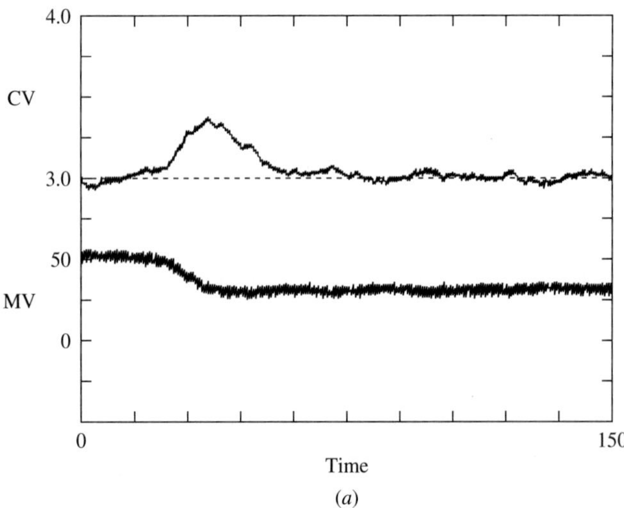

FIGURE 12.13
Closed-loop dynamic data for the system in Example 12.1: (*a*) PID without filtering.

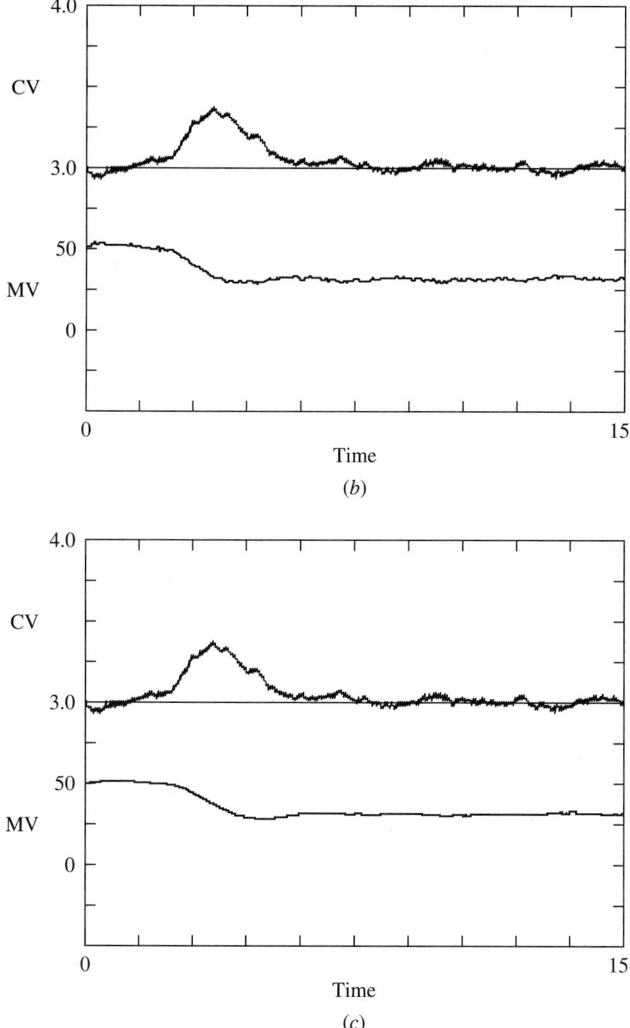

FIGURE 12.13 (*continued*)
Closed-loop dynamic data for the system in Example 12.1: (*b*) PI without filtering; (*c*) PI with filtering ($\tau_f = 3.0$).

controlled variable. A measure of the controlled-variable performance is summarized in Table 12.3, which includes the need to change the controller tuning because of the addition of the filter in the control loop. The results are in general agreement with the guidelines shown in Figure 12.10.

Set-Point Limits

Often, limits are placed on the set point. Without a limit, the set point can take any value in the controlled-variable sensor range. Since the controlled-variable sensor range is selected to provide information during upsets and other atypical operations,

TABLE 12.3
Results from Example 12.1

K_c	T_I	T_d	τ_f	IAE
30	11	0.88	0	9.4
30	11	0	0	9.5
29	12	0	1	10.3
26	14	0	3	12.5
22	23	0	10	21.2

it may include values that are clearly undesirable but not entirely preventable. Limits on the set point prevent an incorrect value being introduced (1) inadvertently by the operator or (2) by poor control of a primary in a cascade control strategy (see Chapter 14).

12.4 FEEDBACK CONTROL ALGORITHM

Many features and options are included in commercial PID control algorithms. In this section, some selected features are introduced, because they are either required in many systems or are optional features used widely. The features are presented according to the mode of the PID controller that each affects.

Controller Proportional Gain

Throughout the previous section, we have allowed the controller gain to be either positive or negative as required to achieve negative feedback. In many control systems that use preprogrammed algorithms, the controller gain is required to be *positive.* Naturally, another option must be added; this is a "sense switch" that defines the sign of the controller output. The effect of the sense switch is

$$\mathrm{MV}(t) = (K_{\text{sense}})K_c\left(E + \frac{1}{T_I}\int_0^t E\,dt' - T_d\frac{d\mathrm{CV}}{dt}\right) + I \qquad (12.12)$$

The sense switch has two possible positions, which are defined in the following table using two common terminologies.

Value of K_{sense}	Position	Position
+1	Direct-acting	Increase/increase
−1	Reverse-acting	Increase/decrease

This approach is not necessary, but it is used so widely that control engineers should be aware of the practice. We will continue to use controller gains of either sign in subsequent chapters unless otherwise specified.

Example 12.2. What is the correct sense switch position for the temperature feedback controller in Figure 12.2?

Note that the process gain and failure mode of the control valve must be known to determine the proper sense of the controller. In this example, the failure mode is fail-closed. Therefore, an increase in the controller output signal results in (1) the valve opening, (2) the fuel flow increasing, (3) the heat transferred increasing, and (4) the temperature increasing. Therefore, the process gain ($K_p = \Delta T/\Delta \text{MV}$) is positive, and the controller gain, being proportional to the inverse of the process gain, should be positive. As a result, K_{sense} is +1, and the sense is direct-acting.

Another convention in commercial control systems is the use of dimensionless controller gains. This is required for analog systems, which perform calculations in scaled voltages or pressures, and it is retained in most digital systems. The scaling in the calculation is performed according to the following equation:

$$\frac{\text{MV}}{\text{MV}_r} = (K_c)_s \left(\frac{E}{\text{CV}_r} + \frac{1}{T_I} \int_0^t \frac{E}{\text{CV}_r} dt' - T_d \frac{d\left(\dfrac{\text{CV}}{\text{CV}_r}\right)}{dt} \right) + I'' \tag{12.13}$$

with $(K_c)_s$ = dimensionless (scaled) controller gain = $K_c(\text{CV}_r/\text{MV}_r)$

MV_r = range of the manipulated variable [0 to 100% for a control valve]

CV_r = range of the sensor measuring the controlled variable in engineering units

The range of values for the unscaled controller gain K_c is essentially unlimited, because the value can be altered by changing the units of the measurement. For example, a controller gain of 1.0 (weight%)/(% open) is the same as 1.0×10^6(ppm)/(% open). However, the scaled controller gain has a limited range of values, because properly designed sensors and final elements have ranges that give good accuracy. For example, a very small dimensionless controller gain indicates that the final control element would have to be moved very accurately for small changes to control the process. In this case, the final element should be changed to one with a smaller capacity. A general guideline is that the scaled controller gain should have a value near 1.0. Scaled controller gains outside the range of 0.01 to 10 suggest that the range of the sensor or final element may have been improperly selected.

Example 12.3. For the three-tank mixing process, the concentration sensor has a range of 0 to 5%, and the control valve is fail-closed. Determine the dimensionless controller gain. In this example, the ranges are MV_r = 0 to 100% valve opening and CV_r = 0 to 5% concentration. The dimensionless controller gain is

$$(K_c)_s = 30(\%\ \text{opening}/\%\text{A})(5\%\text{A}/100\%\ \text{opening}) = 1.5$$

By analysis similar to the previous example, the controller sense is determined to be direct-acting.

Reset (Integral) Windup

The integral mode is included in the PID controller to eliminate steady-state offset for steplike disturbances, which it does satisfactorily as long as it has the freedom to adjust the final element. If the final element cannot be adjusted because it is fully open or fully closed, the control system cannot achieve zero offset. This situation is not a deficiency of the control algorithm; it represents a shortcoming of the process and control equipment. The condition arises because the equipment capacity is not sufficient to compensate for the disturbance, which is presumably larger than the disturbances anticipated during the plant design. The fundamental solution is to increase the equipment capacity.

However, when the final element (valve) reaches a limit, an additional difficulty is encountered that is related to the controller algorithm and must be addressed with a modification to the algorithm. When the valve cannot be adjusted, the error remains nonzero for long periods of time, and the standard PID control algorithm (e.g., equation (12.12) or (11.6)) continues to calculate values for the controller output. Since the error cannot be reduced to zero, the integral mode integrates the error, which is essentially constant, over a long period of time; the result is a controller output value with a very large magnitude. Since the final element can change only within a restricted range (e.g., 0 to 100% for a valve), these large magnitudes for the controller output are meaningless, because they do not affect the process, and should be prevented.

The situation just described is known as *reset* (integral) *windup.* Reset windup causes very poor control performance when, because of changes in plant operation, the controller is again able to adjust the final element and achieve zero offset. Suppose that reset windup has caused a very large positive value of the calculated controller output because a nonzero value of the error occurred for a long time. To reduce the integral term, the error must be negative for a very long time; thus, the controller maintains the final element at the limit for a long time simply to reduce the (improperly "wound-up") value of the integral mode.

The improper calculation can be prevented by many modifications to the standard PID algorithm that do not affect its good performance during normal circumstances. These modifications achieve *anti–reset windup.* The first modification explained here is termed *external feedback* and is offered in many commercial analog and digital algorithms. The external feedback PI controller is shown in Figure 12.14. The system behaves exactly like the standard algorithm when the limitation is not active, as is demonstrated by the following transfer function, which can be derived by block diagram manipulation based on Figure 12.14.

$$\frac{\mathrm{MV}^*(s)}{E(s)} = K_c\left(1 + \frac{1}{T_I s}\right)$$

$$\mathrm{MV}^*(s) = \mathrm{MV}(s) \tag{12.14}$$

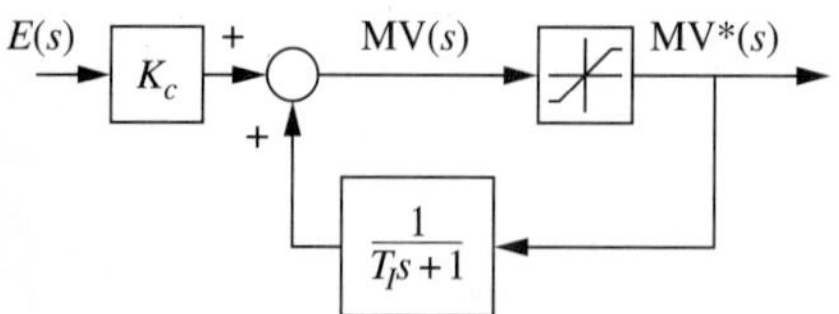

FIGURE 12.14
Block diagram of a PI control algorithm with external feedback.

However, the system with external feedback behaves differently from the standard PI controller when a limitation is encountered. When a limitation is active in Figure 12.14, the following transfer function defines the behavior.

$$\begin{aligned} \mathrm{MV}^*(s) &= \text{Constant} \\ \mathrm{MV}(s) &= K_c E(s) + \frac{\mathrm{MV}^*(s)}{T_I s + 1} \end{aligned} \tag{12.15}$$

with $\mathrm{MV}^*(s)$ at its upper or lower limit. In this case, the controller output approaches a finite, reasonable limiting value of $K_c E(s) + \mathrm{MV}^*(s)$. Thus, external feedback is successful in providing anti–reset windup. These calculations can be implemented in either analog or digital systems.

The second, alternative anti–reset windup modification can be implemented in digital systems. Reset windup can be prevented by using the velocity form of the digital PID algorithm, which is repeated here.

$$\begin{aligned} \Delta \mathrm{MV}_n &= K_c \left(E_n - E_{n-1} + \frac{\Delta t E_n}{T_I} - \frac{T_d}{\Delta t}(\mathrm{CV}_n - 2\mathrm{CV}_{n-1} + \mathrm{CV}_{n-2}) \right) \\ \mathrm{MV}_n &= \mathrm{MV}_{n-1} + \Delta \mathrm{MV}_n \end{aligned} \tag{12.16}$$

This algorithm does not accumulate the integral as long as the past value of the manipulated variable, MV_{n-1}, is evaluated *after* the potential limitation. When this convention is observed, any difference between the previously calculated MV and the MV actually implemented (final element) is not accumulated.

Many other methods are employed to prevent reset windup. The two methods described here are widely used and representative of the other methods. The key point of this discussion is that

> Anti–reset windup should be included in every control algorithm that has integral mode, because limitations are encountered, perhaps infrequently, by essentially all control strategies due to large changes in operating conditions.

Reset windup is relatively simple to recognize and correct for a single-loop controller outputting to a valve, but it takes on increasing importance in more complex control strategies such as cascade and variable-structure systems, which are covered later in this book. Also, the general issue of reset windup exists for any controller that provides zero offset when no limitations exist. For example, reset windup is addressed again when the predictive control algorithms are covered in Chapter 19.

Example 12.4. The three-tank mixing process in Examples 7.2 and 9.2 initially is operating in the normal range. At a time of about 20 minutes, it experiences a large increase in the inlet concentration that causes the control valve to close and thus reach a limit. After a period of about 140 minutes, the inlet concentration returns to its original value. Determine the dynamic responses of the feedback control system with and without anti–reset windup.

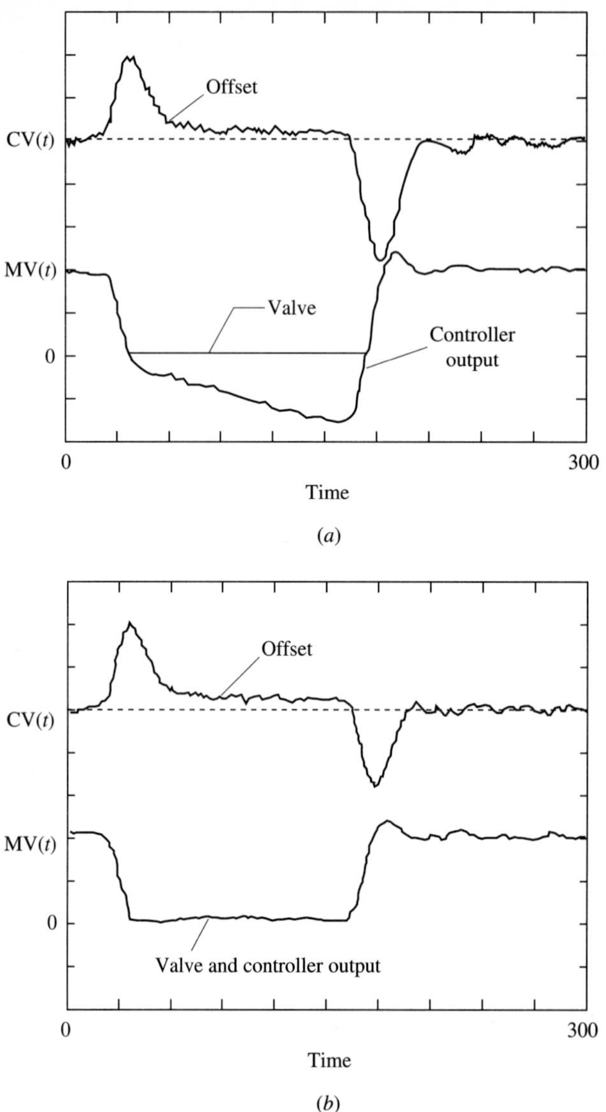

FIGURE 12.15
Dynamic response of the three-tank mixing system (*a*) without anti–reset windup; (*b*) with anti–reset windup.

The results of simulations are presented in Figures 12.15*a* and *b*. In Figure 12.15*a* the dynamic response of the system without anti–reset windup is shown. As usual, the set point, controlled variable, and manipulated variable are plotted. In addition, the calculated controller output is plotted for assistance in analysis, although this variable is not normally retained for display in a control system. After the initial disturbance, the valve position is quickly reduced to 0 percent open. Note that the calculated controller output continues to decrease, although it has no additional effect on the valve. During the time from 20 to 160 minutes, the controlled variable does not return to its set point because of the limitation in the range of the manipulated variable. When the inlet concentration returns to its normal value, the outlet concentration initially falls below its set point. The controller detects this situation immediately, but it cannot adjust the valve until the calculated controller output increases to the value of zero. This delay, which would be longer had the initial disturbance been longer, is the cause of a rather large disturbance. Finally, the PI controller returns the controlled variable to its set point, since the manipulated variable is no longer limited.

The case with anti–reset windup is shown in Figure 12.15*b*. The initial part of the process response is the same. However, the calculated controller output does not fall below the value of 0 percent; in fact, it remains essentially equal to the true valve position. When the inlet concentration returns to its normal value, the controller output is at zero percent and can rapidly respond to the new operating conditions. The second disturbance is much smaller than in Figure 12.15*a*, showing the advantage of anti–reset windup.

Derivative Filter

An additional modification of the PID algorithm addresses the effect of noise on the derivative mode. It is clear that the derivative mode will amplify high-frequency noise present in the measured controlled variable. This effect can be reduced by decreasing the derivative time, perhaps to zero. Unfortunately, this step also reduces or eliminates the advantage of the derivative mode. A compromise is to filter the derivative mode by using following the equation:

$$\frac{T_d s}{\alpha_d T_d s + 1} \tag{12.17}$$

The result of this modification is to reduce the amplification of noise while retaining some of the good control performance possible with the derivative mode. As the factor α_d is increased from 0 to 1, the noise amplification is decreased, but the improvement in control performance due to the derivative mode decreases. This parameter has typical values of 0.1 to 0.2 and is not normally tuned by the engineer for each individual control loop. Since the PID control algorithm has been changed when equation (12.17) is used for the derivative mode, the controller tuning values must be changed, with the Ciancone correlations no longer being strictly applicable. Tuning correlations for the PID controller with $\alpha = 0.1$ are given by Fertik (1974).

Initialization

The PID controller requires special calculations for initialization. The specific initialization required depends upon the particular form of the PID control algorithm; typical initialization for the standard digital PID algorithm is as follows.

$$\begin{aligned}
\Delta \text{MV}_n &= K_c\left(E_n - E_{n-1} + \frac{\Delta t E_n}{T_I} - \frac{T_d}{\Delta t}(\text{CV}_n - 2\text{CV}_{n-1} + \text{CV}_{n-2})\right)\\
\text{MV}_n &= \text{MV}_{n-1} + \Delta \text{MV}_n\\
\text{MV}_1 &= \text{MV}_0 \qquad \text{that is, } \Delta \text{MV}_1 = 0 \quad \text{for } n = 1 \quad \text{for initialization}
\end{aligned}$$

$$\begin{aligned} E_{n-1} &= E_n && \text{for } n = 1 \\ \text{CV}_{n-2} &= \text{CV}_{n-1} = \text{CV}_n && \text{for } n = 1 \end{aligned} \tag{12.18}$$

This initialization strategy ensures that no large initial change in the manipulated variable will result from outdated past values of the error or controlled variables.

12.5 OUTPUT PROCESSING

The standard PID controller has no limits on output values, nor does it have special considerations when the algorithm is first used, as when the controller is switched from manual to automatic. As already described, the calculated controller output is initialized so that the actual valve position does not immediately change on account of the change in controller mode.

In addition to initialization, the PID algorithm can be modified to limit selected variables. The most common limitation is on the manipulated variable, as is done when certain ranges of the manipulated variable are not acceptable. Thus, the manipulated variable is maintained within a restricted range less than 0 to 100%.

$$\text{MV}_{\text{min}} < \text{MV}(t) < \text{MV}_{\text{max}} \tag{12.19}$$

An example of limiting the manipulated variable is the damper (i.e., valve), position in the stack of a fired heater as shown in Figure 12.2. The stack damper is adjusted to control the pressure of the combustion chamber. Since the stack is the only means for the combustion product gases to leave the combustion chamber, it should not be entirely blocked by a closed valve. However, the control system could attempt to close the damper completely due to a faulty pressure measurement or poor controller tuning. In this case, it is common to limit the controller output to prevent a blockage in the range of 0 to 80% (not 20 to 100%, because the damper is fail-open, so that a signal of 100% would close the valve).

Sometimes the rate of change of the manipulated variable is limited using the following expression.

$$\Delta\text{MV}_n = \min(|\Delta\text{MV}|, \Delta\text{MV}_{\text{max}}) \left(\frac{\Delta\text{MV}}{|\Delta\text{MV}|}\right) \tag{12.20}$$

This modification is appropriate when a rapid adjustment of the manipulated variable can disturb the operation of a process.

12.6 CONCLUSIONS

Clearly, the simple, single PID equation, while performing well under limited conditions, is not sufficient to provide feedback control under the various conditions experienced in realistic plant operation. Some of the most important modifications have been presented in this chapter, and many more modifications are described in publications noted in the references and additional resources.

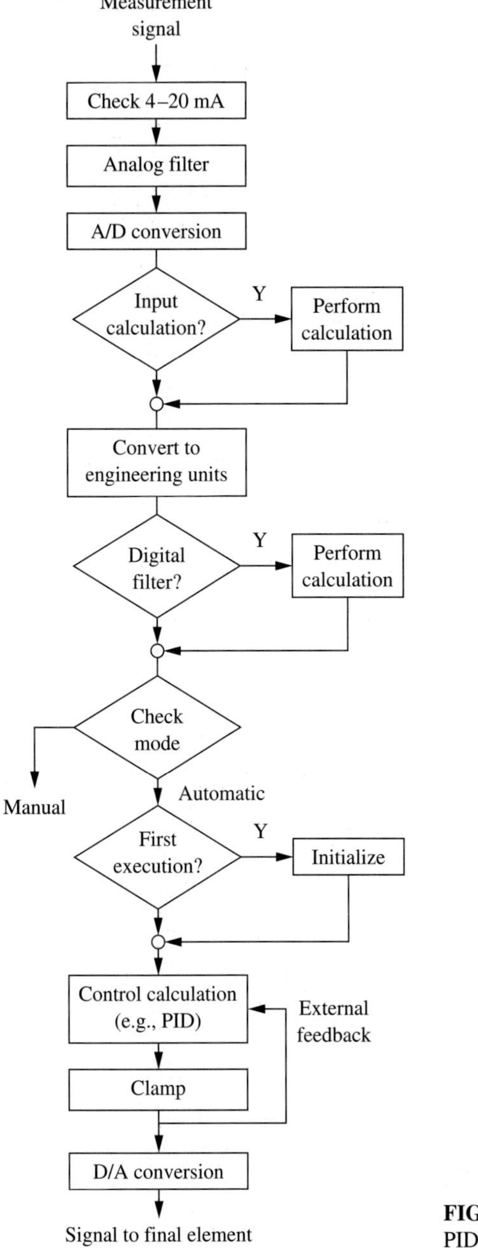

FIGURE 12.16
PID calculation flowchart.

To complete this chapter, the flowchart for a PID controller that includes the modifications described in this chapter is given in Figure 12.16. The added complexity is apparent. However, the computations are readily packaged in preprogrammed algorithms and performed rapidly by powerful microprocessor-based instrumentation. A wise and productive engineer uses these programs and does not attempt to develop all real-time calculations from scratch, although doing limited algorithm programming is a useful learning exercise for the student.

REFERENCES

Fertik, H., "Tuning Controllers for Noisy Processes," *ISA Trans., 14, 4,* 292–304 (1974).

ISA, *Process Instrumentation Terminology,* ISA-S51.1-1979, Instrument Society of America, Research Triangle Park, NC, 1979.

Goff, K.W., "Dynamics in Direct Digital Control, Part I," *ISA J., 13, 11,* 45–49 (November, 1966); "Part II," *ISA J. 13, 12,* 44–54 (December, 1966).

Mellichamp, D., *Real-Time Computing,* Van Nostrand, New York, 1983.

ADDITIONAL RESOURCES

There are many technical references available for determining the performance of sensors and final control elements, including the references in Chapter 1. Also, equipment manufacturers provide information on the performance of their equipment. The following references, along with the references for Chapter 18 on level control, provide additional information on key sensors.

DeCarlo, J., *Fundamentals of Flow Measurement,* Instrument Society of America, Research Triangle Park, NC, 1984.

Miller, R., *Flow Measurement Engineering Handbook,* McGraw-Hill, New York, 1983.

Pollock, D., *Thermocouples, Theory and Practice,* CRC Press, Ann Arbor, 1991.

The following references discuss many options for anti–reset windup.

Gallun, S., C. Matthews, C. Senyard, and B. Slater, "Windup Protection and Initialization for Advanced Digital Control," *Hydrocarbon Proc., 64,* 63–68 (1985).

Khandheria, J. and W. Luyben, "Experimental Evaluation of Digital Control Algorithms for Anti-reset-windup," *IEC Proc. Des. Devel., 15, 2,* 278–285 (1976).

Many calculations in commercial instrumentation use scaled variables, because they are performed in analog systems using voltages or pressures. For an introduction to scaling, see

Gordon, L., "Scaling Converts Process Signals to Instrument Ones," *Chem. Engr., 91,* 141–146 (June 25, 1984).

For an introduction to some of the causes of high-frequency noise and means of its prevention, see

Hazlewood, L., "Getting the Noise Out," *Chem. Engr., 95,* 105–108 (November 21, 1988).

For a very clear discussion of filtering, in addition to Goff (1966) noted above, see

Corripio, A., C. Smith, and P. Murrill, "Filter Design for Digital Control Loops," *Instr. Techn.,* 33–38 (January 1973).

For a concise description of many PID controller enhancements in analog and digital form, see

Clark, D., "PID Algorithms and Their Computer Implementation," *Trans. Inst. Meas. and Cont., 6, 6,* 305–316 (1984).

QUESTIONS

12.1. Many filtering algorithms are possible. For each of the algorithms suggested below, describe its open-loop frequency response and sketch its Bode plot. Also, discuss its advantages and disadvantages as a filter in a closed-loop feedback control system.

(*a*) $\dfrac{1}{\tau^2 s^2 + 0.4\tau s + 1}$

(*b*) $\dfrac{1}{(\tau s + 1)^n}$ with n = positive integer

(*c*) Averaging filter with m values in average

$$(\text{CV}_f)_n = \frac{\text{CV}_n + \text{CV}_{n-1} + \cdots + \text{CV}_{n-m+1}}{m}$$

(*d*) $\dfrac{0.707}{\tau_f s + 1}$

12.2. Answer the questions in Table Q12.2 for each PID controller mode or tuning constant associated with each mode. Explain every entry completely, giving theoretical justification as well as the brief answer indicated. Answer this question on the basis of a commercial control system in which all control calculations are performed in scaled variables.

12.3. You have been given three control systems to analyze. Each has the *dimensionless* controller gain given below. From this information alone, what can you determine about each control system? $(K_c)_s$ = (*a*) 0.02, (*b*) 0.75, and (*c*) 123.00.

12.4. The control systems with the processes given below are to be tuned (1) without a filter and with a first-order filter with (2) τ_f = 0.5 min and (3) τ_f = 3.0 min. Determine the PI tuning constants for all three cases using the Bode stability analysis and Ziegler-Nichols correlations. Also, state whether you expect the control performance, as measured by IAE, to be better or worse with the filter (after retuning). Why?

(*a*) The empirical model derived in question 6.1 for the fired heater.

(*b*) The empirical model for the packed-bed reactor in Figure 6.3 from the data in Figure Q6.4*c*.

(*c*) The linearized, analytical model for the stirred-tank heater in Example 8.5.

12.5. For the process in Figure 2.2, answer the following questions.

(*a*) Determine the proper failure modes for all valves. Also, give the proper controller sense for each controller, assuming that commercial controllers are being used ($K_c > 0$).

(*b*) What type of input processing would be appropriate for each measurement? Why?

(*c*) The following alterations are made after the process has been operating successfully. Determine any other changes that must be made as a consequence of each alteration. Your answers should be as specific and quantitative as possible.

TABLE Q12.2

	P	I	D
(*a*) Which modes eliminate offset?			
(*b*) Describe the speed of response for an upset (fastest, middle, slowest).			
(*c*) Compare the propagation of high-frequency noise from controlled to manipulated variable (most, middle, least).			
(*d*) As process dead time increases (with $\theta + \tau$ constant), the tuning constant (increases, decreases, unchanged)?			
(*e*) Does the mode cause windup (Y, N)?			
(*f*) Should tuning constant be changed when filter is added to loop (Y, N)?			
(*g*) Is tuning constant affected by limits on the manipulated variable (Y, N)?			
(*h*) Should tuning constant be altered if the sensor range is changed (Y, N)?			
(*i*) Does tuning constant depend on the failure mode of the final element (Y, N)?			
(*j*) Does tuning constant depend on the linearization performed in the input processing (Y, N)?			
(*k*) Should tuning constant be altered if the final element capacity is changed (Y, N)?			
(*l*) Should tuning constant be changed if the digital controller execution period is changed (Y, N)?			

(1) The control valve for a steam heat exchanger is increased to accommodate a flow 50% greater than the original. (2) The failure mode of the control valve in the liquid product stream changed from fail-open to fail-closed. (3) The range of the temperature sensor is changed from 50–100°C to 75–125°C.

12.6. Answer Questions 12.5 (*a*) and (*b*) for the CSTR in Figure 2.14.

12.7. Answer Questions 12.5 (*a*) and (*b*) for the boiler oxygen control in Figure 2.6.

12.8. In the discussion on external feedback, equations (12.14) and (12.15) were given to prove that reset windup would not occur.

(*a*) Derive these equations based on the block diagram and explain why reset windup does not occur.

(*b*) Prepare the equations in their proper sequence for the digital implementation of external feedback.

12.9. An alternative anti–reset windup method is to use logic to prevent "inappropriate" integral action. This logic is based on the status of the manipulated variable. Develop a flow chart or logic table for this type of anti–reset windup and explain how it would work.

12.10. The goal of initializing the PID controller is to prevent a "bump" when the mode is changed and to prepare the controller for future calculations. Determine the proper initialization for the full-position digital PID controller algorithm in equation (11.6) and explain each step.

12.11. A process uses infrequent laboratory analyses for control. The period of the analyses is much longer than the dynamics of the process. Due to the lack of accuracy in the laboratory method, the result has a relatively large standard deviation, resulting in noise in the feedback loop. Describe steps you would take to reduce this noise by a factor of 2. (For the purposes of this problem, you may not change the frequency for collecting one or a group of samples from the process.)

12.12. A signal to a digital controller has considerable high-frequency noise in spite of the analog filter before the A/D converter. The controller is being executed according to the rule that $\Delta t/(\theta + \tau) = 0.05$, and the manipulated variable has too large a standard deviation. Explain what steps you would take in the digital PID control system to reduce the effects of noise on the manipulated variable and yet to have minimal effect on the control performance as measured by IAE of the controlled variable.

12.13. Answer the following questions regarding filtering.

(*a*) Confirm the transfer function in equation (12.5).

(*b*) The equation for the digital first-order filter is presented in equation (12.9). Confirm this equation by deriving it from equation (12.8).

(*c*) Discuss the behavior of a low-pass filter and give examples of its use in process control.

(*d*) A *high-pass* filter attenuates the low-frequency components. Describe an algorithm for a high-pass filter and give examples of its use.

12.14. Consider an idealized case in which process data consists of the sum of a constant true signal plus purely random (white) noise with a mean of 0 and a standard deviation of 0.30.

(*a*) Determine the value of the parameter A in the digital filter equation (12.9) that reduces the standard deviation of the filtered value to 0.1. You might have to build a simulation in a spreadsheet with several hundred executions and try several values of A.

(*b*) Determine the number of duplicate samples of the variable to be taken every execution so that the average of these values will have a standard deviation of 0.10.

12.15. Consider the situation in which the measured controlled variable consisted of nearly all noise, with very infrequent changes in the true process variable due to slowly varying disturbances. Suggest a feedback control approach, not a PID algorithm, that would reduce unnecessary adjustments of the manipulated variable.

12.16. Many changes have been proposed to the standard digital PID controller, and we have considered several, such as the derivative on measured variable rather than error. For each of the following proposed modifications in the PID algorithm, suggest a reason for the modification (that is, what possible benefit it would offer and under what circumstances) and any disadvantages.

(*a*) The proportional mode is calculated using the measured variables rather than the error.

$$MV_n = K_c\left(CV_n + \frac{\Delta t}{T_I}\sum_{i=0}^{n}(SP_i - CV_i) - \frac{T_d}{\Delta t}(CV_n - CV_{n-1})\right) + I$$

(*b*) The controller gain is nonlinear; for example,

$$\text{For} \quad (SP_n - CV_n) > 0 \qquad K_c = K'$$

$$\text{For} \quad (SP_n - CV_n) < 0 \qquad K_c = K' + K''\left|SP_n - CV_n\right|$$

(*c*) The rate of change of the manipulated variable is limited, $|\Delta MV| < \max$.

(*d*) The allowable set point is limited, $SP_{min} < SP < SP_{max}$.

CHAPTER 13

PERFORMANCE OF FEEDBACK CONTROL SYSTEMS

13.1 INTRODUCTION

As we have learned, feedback control has some very good features and can be applied to many processes using simple control algorithms like the PID controller. We certainly anticipate that a process with feedback control will perform better than one without feedback control, but how well do feedback systems perform? There are both theoretical and practical reasons for investigating control performance at this point in the book. First, engineers should be able to predict the performance of control systems to ensure that all essential objectives, especially safety but also product quality and profitability, are satisfied. Second, good performance estimates can be used to evaluate potential investments associated with control. Only those control strategies or process changes that provide sufficient benefits beyond their costs, as predicted by quantitative calculations, should be implemented. Third, an engineer should have a clear understanding of how key aspects of process design and control algorithms contribute to good (or poor) performance. This understanding will be helpful in designing process equipment, selecting operating conditions, and choosing control algorithms. Finally, after understanding the strengths and weaknesses of feedback control, it will be possible to enhance the control approaches introduced to this point in the book to achieve even better performance. In fact, Part IV of this book presents enhancements that overcome some of the limitations covered in this chapter.

Two quantitative methods for evaluating closed-loop control performance are presented in this chapter. The first is frequency response, which determines the response of important variables in the control system to sine forcing of either the disturbance or the set point. Frequency response is particularly effective in determining and displaying the influence of the frequency of an input variable on control performance. The second quantitative method is simulation, involving numerical solution of the equations defining all elements in the system. This method is effective in giving the entire transient response for important changes in the forcing functions, which can be any general function. Both of these methods require computations that are easily defined but very time-consuming to perform by hand. Fortunately, the calculations can be programmed using simple concepts and executed in a short time using digital computers.

After the two methods have been explained and demonstrated, they are employed to develop further understanding of the factors influencing control performance. First, a useful performance bound is provided that defines the best performance possible through feedback control. Then, important effects of elements in the feedback system are analyzed. In one section the effects of feedback and disturbance dynamics on performance are clarified. In another section the effects of control elements, both physical equipment and algorithms, on control performance are evaluated. The chapter concludes with a table that summarizes the salient effects of control loop elements on control performance.

13.2 CONTROL PERFORMANCE

Many measures of control performance are possible, and each is appropriate in particular circumstances. The important measures are listed here, and the reader is referred to Chapter 7 to review their meanings, if necessary.

- Integral error (IAE, ISE, etc.)
- Maximum deviation of controlled variable
- Maximum overshoot of manipulated variable
- Decay ratio
- Rise time
- Settling time
- Standard deviation of controlled and manipulated variables
- Magnitude of the controlled variable in response to a sine disturbance

Two additional factors should be achieved for control performance to be acceptable; generally, they are not difficult to achieve but are included here for completeness of presentation. The first is zero steady-state offset of the controlled variable from the set point for steplike input changes. For nearly all control systems, zero offset is a desirable feature, and control systems must use a controller with an integral mode to achieve this objective. An important exception where zero offset is not required occurs with some level controllers. Level control is addressed in Chapter 18, where different control performance criteria from those used in this chapter are introduced.

The second factor is stability. Clearly, we want every control system to be stable; therefore, control algorithms and tuning constants are selected to give stable performance over a range of operating conditions. It is very important to recognize that stability places a limit on the maximum controller gain and, in a sense, the control system performance. Without this limit, proportional-only controllers with very high gains might provide tight control of the controlled variable in many applications. Therefore, in this chapter we will confine our discussion to control systems that require zero offset and to controller tuning constant values that provide good performance over a reasonable range of operating conditions.

Also, we recognize that no general boundary exists between good and poor process performance. A maximum controlled-variable deviation of 5°C may be totally unacceptable in one case and result in essentially no detriment to operation in another case. In this chapter we identify the key factors influencing control performance and develop quantitative methods for predicting performance measures that can be applied to a wide range of processes; the desired value or limit for each measure will depend on the particular process being considered. In evaluating control performance, we will use the following definition.

> **Control performance** is the ability of a control system to achieve the desired dynamic responses, as indicated by the control performance measures, over an expected range of operating conditions.

This definition of performance includes both set point changes and disturbances. The phrase "over an expected range of operating conditions" refers to the fact that we never have perfect information on the process dynamics or disturbances. Differences between model and plant are inevitable, whether the models were derived analytically from first principles or were developed from empirical data such as the process reaction curve. In addition, differences occur because the plant dynamics change with process operating conditions (e.g., feed flow rate and catalyst activity). Since any model we use has some error, the control system must function "well" over an expected range of errors between the real plant and our expectation, or model, of the plant. The expected range of conditions can be estimated from our knowledge of the manner in which the plant is being operated (values of feed flow, reactor conversion, and so forth).

The ability of a control system to function as the plant dynamics change is sometimes referred to as *robust* control. However, throughout this book we will consider performance to include this factor implicitly without expressly including the word *robust* every time. To reiterate, *we must always consider our lack of perfect models and changing process dynamics when analyzing control performance.*

It is important to emphasize that the performance of control systems depends on all elements of the system: the process, the sensor, the final element, and the controller. Thus, all elements are included in the quantitative methods described in the next two sections, and important effects of these elements on performance are explored further in subsequent sections.

13.3 CONTROL PERFORMANCE VIA CLOSED-LOOP FREQUENCY RESPONSE

Continuously operating plants experience frequent, essentially continuous, disturbances, so predicting the control system performance for this situation is very important. The approach introduced here is very general and can be applied to any linear plant, not just first-order-with-dead-time, and any linear control algorithm. Also, it provides great insight into the influence of the frequency of the input (set point and disturbance) changes on the effectiveness of feedback control.

The approach is based on the frequency response methods introduced in previous chapters. Frequency response calculates the system output in response to a sine input; we will use this approach in evaluating control system performance by assuming that the input variable—set point change or disturbance—is a sine function. While this is never exactly true, often the disturbance is periodic and behaves approximately like a sine. Also, a more complex disturbance can often be well represented by a combination of sines (e.g., Kraniauskas, 1992); thus, frequency response gives insight into how various frequency components in a more complex input affect performance.

The control performance measure in this section is the amplitude ratio of the controlled variable, which can be considered the deviation from set point because the transfer function equations are in deviation variables. As demonstrated in equations (4.103) and (4.104), the frequency response of a stable, linear control system can be calculated by replacing the Laplace variable s with $j\omega$ in its transfer function. The resulting expressions describe the amplitude ratio and phase angle of the controlled variable after a long enough time that the nonperiodic contribution to the solution is negligible. The control system in Figure 13.1 is the basis for the analysis, and this system has the following transfer function in response to a disturbance:

$$\frac{\text{CV}(s)}{D(s)} = \frac{G_d(s)}{1 + G_p(s)G_v(s)G_c(s)G_s(s)} \tag{13.1}$$

It is helpful to consider the amplitude ratio of the controlled variable to the disturbance in equation (13.1), which can be expressed as the product of two factors:

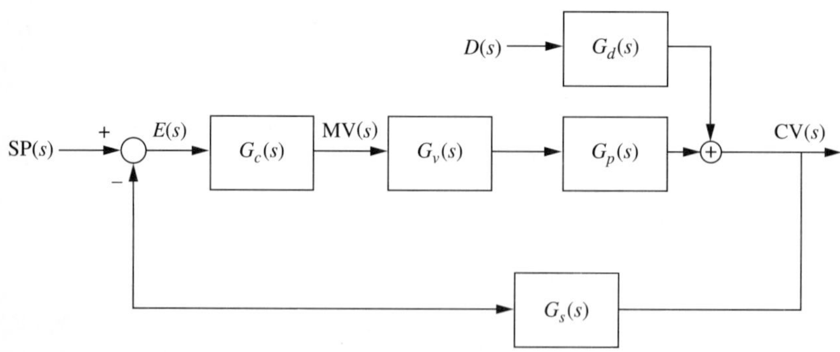

FIGURE 13.1
Block diagram of feedback control system.

$$\frac{|\mathrm{CV}(j\omega)|}{|D(j\omega)|} = \left[|G_d(j\omega)| \left| \frac{1}{1 + G_p(j\omega)G_v(j\omega)G_c(j\omega)G_s(j\omega)} \right| \right] \tag{13.2}$$

The two factors in the brackets represent the amplitude ratio of the transfer function. The first factor of the amplitude ratio is the numerator, which contains the open-loop process disturbance model. The second factor is the contribution from the feedback control system. The frequency responses of the factors are given in Figure 13.2*a* and *b* and are referred to in analyzing the frequency response of the closed-loop system. The results in Figure 13.2 are for the (arbitrary) system

$$G_p(s)G_v(s)G_s(s) = \frac{1.0e^{-15s}}{20s + 1} \qquad G_c = 0.60\left(1 + \frac{1}{30s}\right) \qquad G_d = \frac{0.48}{20s + 1}$$

When interpreting these plots, it is helpful to remember that (unachievable) perfect control would result in no controlled-variable deviation for all frequencies; in other words, the magnitude would be zero for all frequencies. The closed-loop system is first considered at limits of very low and very high frequency. This analysis makes use of equation (13.2) and Figure 13.2*a* and *b*. For disturbances with a very low frequency, the first factor (i.e., the process through which the disturbance travels) does not attenuate the disturbance; thus, its magnitude is large. (The disturbance dynamics are assumed similar to the feedback dynamics for this example.) However, the relatively fast feedback control loop will effectively attenuate a disturbance in this frequency range; thus, the magnitude of the feedback factor is small. The control system response is the product of the two magnitudes; therefore, the control system provides good performance at input frequencies much lower than the critical frequency, because of feedback control. Note that the integral mode of the PI controller is especially effective in rejecting slow disturbances and that in general, feedback control systems provide good control performance at very low disturbance frequencies.

For disturbances at the other extreme of very high frequency, the feedback controller is not effective, because the disturbance is faster than the control loop can respond. In this case the magnitude of the second factor is nearly 1. However, the disturbance process, as long as it consists of first- or higher-order time constants (and not simply gains and dead times), filters the high-frequency disturbance. This filter results in a small magnitude of $|G_d(j\omega)|$, reducing the magnitude of the controlled variable substantially. Therefore, the feedback control system provides good control performance for very high frequencies as well. Note that the good performance is not due to feedback control but rather to the disturbance time constant(s), which in this range is much larger than the disturbance period (i.e., $1/\tau_d \ll \omega$).

For intermediate frequencies, a harmonic or resonant peak occurs. This peak represents the most difficult frequencies for the feedback control system. In fact, for some systems the control system can perform *worse* than the same plant without control, indicating that disturbances can be slightly amplified by the feedback control loop around the harmonic frequency.

The general shape of the closed-loop frequency response to a disturbance for most feedback controller systems is similar to the curve in Figure 13.3. It is important

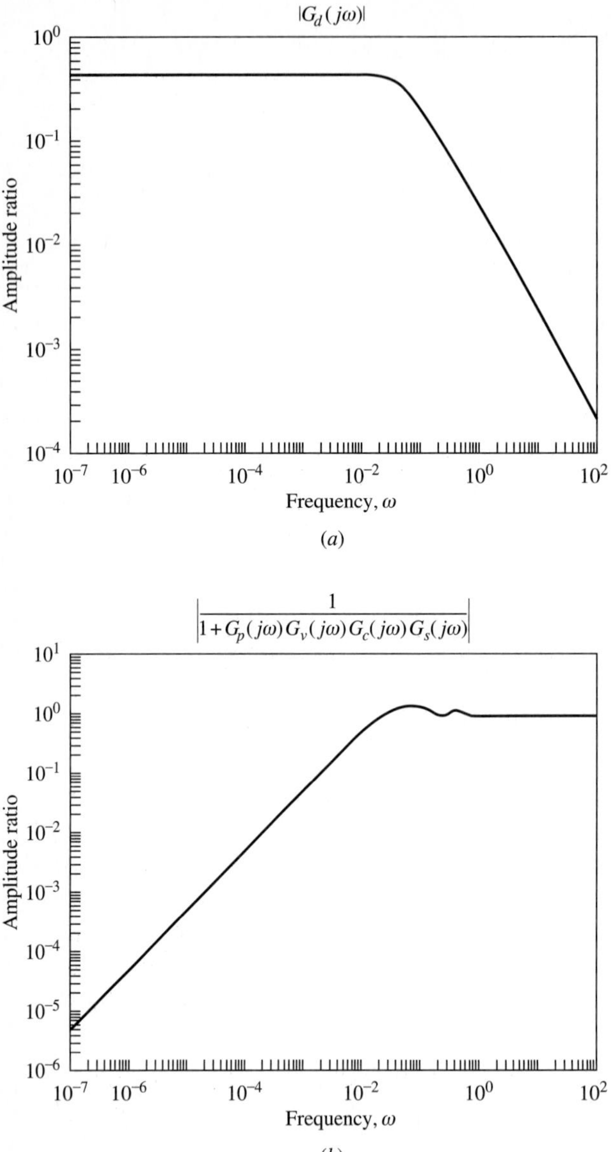

FIGURE 13.2
Amplitude ratios in equation (13.2): (*a*) numerator; (*b*) denominator.

that the engineer understand the reasons for the behavior in the low-, intermediate-, and high-frequency regions. Many disturbances in process plants have low frequencies, because they result from the changing operation of slowly responding systems such as the composition of flows from large upstream feed tanks. Many very fast disturbances occur due to imperfect mixing and high-frequency pressure disturbances.

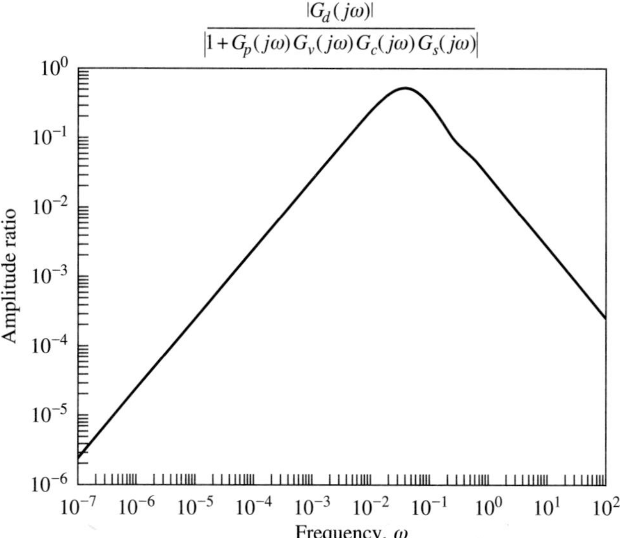

FIGURE 13.3
Frequency response of feedback-controlled variable to disturbance.

For both disturbances, feedback control performance tends to be good. However, many disturbances typically occur around the critical frequency of a feedback loop, because oscillations caused by an integrated process under feedback control tend to be in the same frequency range.

> Disturbances around the closed-loop resonant frequency are essentially uncontrollable with *any single-loop feedback controller,* and therefore such disturbances should be prevented by changes to the process design or attenuated using enhancements discussed in Part IV.

Example 13.1. The plants presented in Figure 13.4 are subject to periodic disturbances. All plants have the same equipment structure, but they have different equipment sizes. They can all be modelled as first-order-with-dead-time processes, and the dynamics of the sensor and valve are negligible. Determine the control performance in response to a disturbance (D) possible with the four designs and rank them according to the amplitude ratios achieved by PI controllers.

The solution to the example involves calculating the closed-loop frequency response for each case. The calculations are based on equation (13.2), with the appropriate transfer functions for the individual elements—in this case, a first-order-with-dead-time process, a first-order disturbance, and a PI controller. The calculation of the amplitude ratio follows the same procedure used in Chapter 10, where s is replaced by $j\omega$ in the transfer function; then the magnitude of the complex expression is determined. The results of the algebraic manipulations for this example are given in equation (13.3).

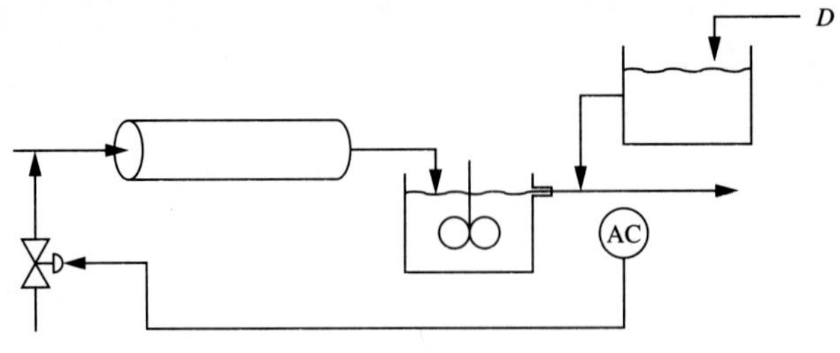

Case	K_P	θ	τ	τ_d
A	1.0	1.0	1.0	1.0
B	1.0	4.0	4.0	1.0
C	1.0	0.5	1.5	1.0
D	0.1	0.5	1.5	1.0

FIGURE 13.4
Schematic of process with model parameters for Example 13.1.

$$\text{Amplitude Ratio} = |G_d(j\omega)| \left| \frac{1}{1 + G_c(j\omega)G_p(j\omega)} \right| \tag{13.3}$$

where $|G_d(j\omega)| = \frac{K_d}{\sqrt{1+\omega^2\tau_d^2}}$ with $K_d = 1$

$$\left| \frac{1}{1 + G_c(j\omega)G_p(j\omega)} \right| = \frac{\sqrt{(AC + BD)^2 + (BC + AD)^2}}{C^2 + D^2}$$

$$A = -T_I\tau_p\omega^2 \qquad B = T_I\omega$$

$$C = K_pK_c[\cos(-\theta\omega) - T_I\omega\sin(-\theta\omega)] - T_I\tau_p\omega^2$$

$$D = K_pK_c[\sin(-\theta\omega) + T_I\omega\cos(-\theta\omega)] + T_I\omega$$

In each case, the PI controller has to be tuned; the tuning for this example is given below based on the Ciancone correlations in Figure 9.9*a* and *b* (page 306).

Case	$\boldsymbol{\theta/(\theta + \tau)}$	$\boldsymbol{K_cK_p}$	$\boldsymbol{T_I/(\theta + \tau)}$	$\boldsymbol{K_c}$	$\boldsymbol{T_I}$
A	0.5	0.85	0.75	0.85	1.5
B	0.5	0.85	0.75	0.85	6.0
C	0.25	1.70	0.65	1.70	1.3
D	0.25	1.70	0.65	17.0	1.3

The best control performance has the smallest amplitude ratio (i.e., the smallest deviation from set point). These calculations have been performed, and the results are given in Figure 13.5, which shows that the best performance is possible with designs C and D. The next best is Case A, and the worst is Case B.

Since the disturbance transfer function is the same for all cases, the processes with the longest dead time and the longest dead time plus time constant in the feedback path are more difficult to control; this explains why Case B has the poorest performance and why Case A is not as good as B and C. Note that processes C and D have the same dynamics and differ only in their gains. Thus, the controller gain can be

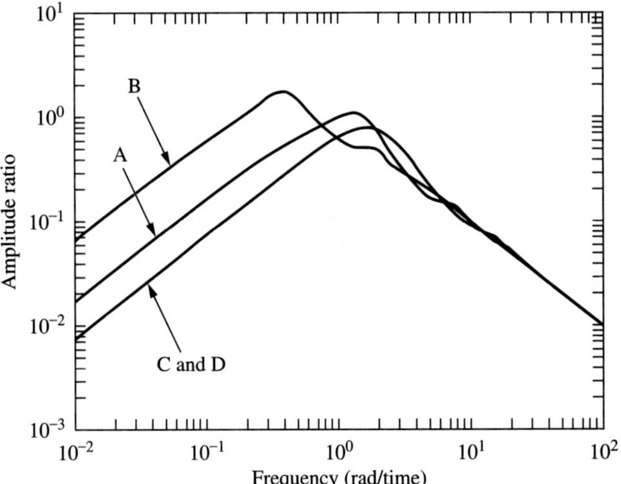

FIGURE 13.5
Closed-loop frequency responses for the cases in Example 13.1.

selected to achieve the same $K_p K_c$ and the same control performance. (This result assumes that the manipulated variable can be adjusted over a larger range for the process with the smaller process gain.) In addition to finding the best process, we have identified a region of disturbance frequency for which feedback control will not function well. Process changes or control enhancements would be in order if disturbances with large magnitudes were expected to occur in this frequency range.

Example 13.2. Normal plant disturbances have many causes with different frequencies. This example presents a simple case of two disturbances. As depicted in Figure 13.6, the input disturbance is the sum of two sine waves that have the same phase and have the amplitudes and frequencies given in the following table. The input disturbances are not measured, but sample open-loop dynamic data of the output variable (i.e., $G_d(s)D(s)$) are given in Figure 13.7*a*. What is the magnitude of the sine wave of the controlled variable when PI feedback control is implemented for the same disturbance?

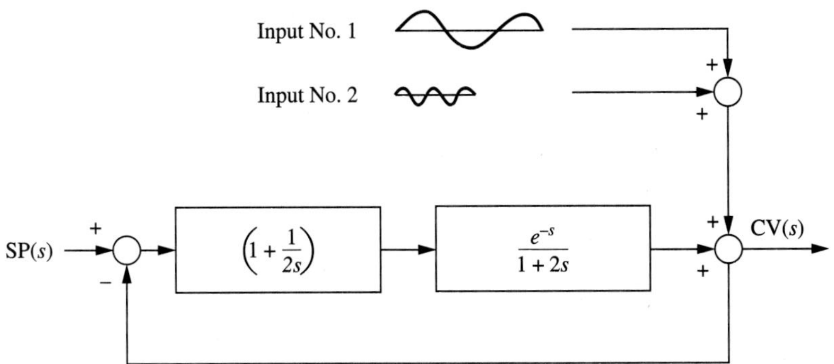

FIGURE 13.6
Schematic showing the system and disturbances considered in Example 13.2.

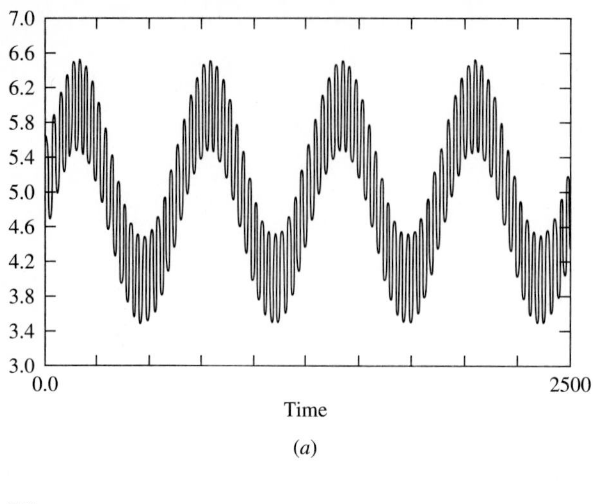

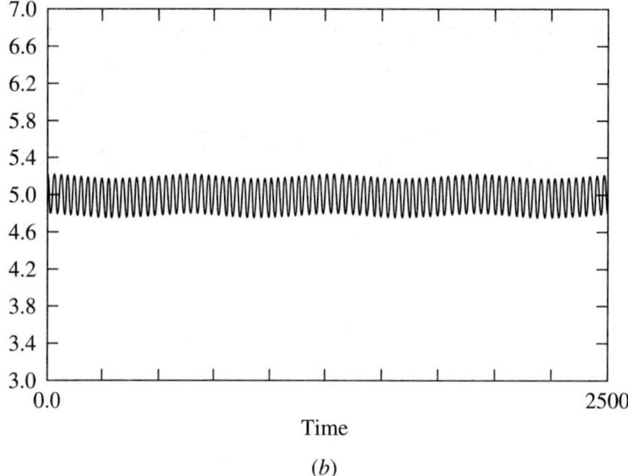

FIGURE 13.7
Results for Example 13.2: (*a*) disturbance without control; (*b*) closed-loop dynamic response with PI control.

	Input No. 1	**Input No. 2**
Frequency (rad/min)	0.010	0.20
Amplitude	1.0	0.50

The first step in the solution is to calculate the closed-loop frequency response for this process with PI control. The process is first-order with dead time, and the calculations employ equation (13.3) with the following parameters:

$$K_p = 1.0 \qquad \tau = 2.0 \qquad \theta = 1.0 \qquad G_d(s) = 1$$
$$K_c = 1.0 \qquad T_I = 2.0$$

The amplitude ratio of each input considered individually can be determined as shown in Figure 13.8. The lower-frequency disturbance (Input No. 1) has a very small amplitude ratio. Thus, the control performance for this part of the disturbance is good.

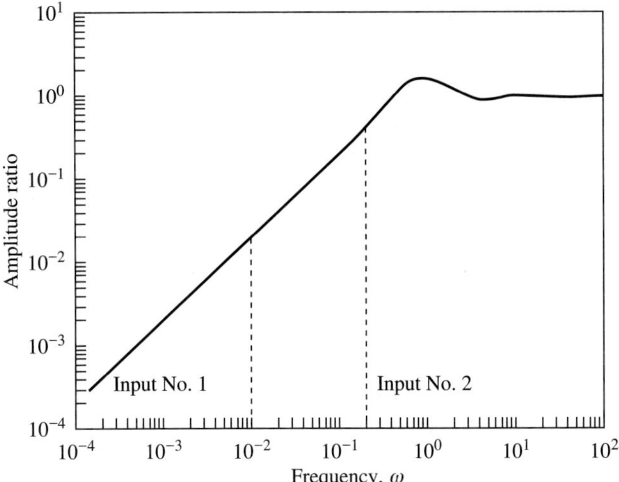

FIGURE 13.8
Closed-loop amplitude ratio for Example 13.2.

The amplitude ratio for the higher-frequency input (Input No. 2) is not small and is about 0.50, because it is in the region of the resonant frequency. Therefore, Input No. 2 contributes most of the deviation for the closed-loop feedback control system.

This analysis can be compared with the dynamic response of the closed-loop control system with the two sine disturbances given in Figure 13.7*b*. The response shows almost no effect of the slow sine disturbance and a significant effect from the faster sine disturbance. The magnitude of the closed-loop simulation, about 0.25, is the same as the prediction from the frequency response analysis, 0.5×0.5. We can conclude from this example that the frequency response method provides valuable insight into which disturbance frequencies will and will not be attenuated significantly by feedback control.

Most process control systems are primarily for disturbance response, but some have frequent changes to their set points. The frequency response approach developed for disturbance performance analysis can be extended to set point response to determine how well the control system can follow, or *track,* its set point. The following transfer function relates the controlled variable to the set point for the system in Figure 13.1.

$$\frac{\mathrm{CV}(s)}{\mathrm{SP}(s)} = \frac{G_p(s)G_v(s)G_c(s)}{1 + G_p(s)G_v(s)G_c(s)G_s(s)} \tag{13.4}$$

The amplitude ratio of this transfer function can be calculated using standard procedures (setting $s = j\omega$) and plotted versus frequency of the set point variation. Perfect control would maintain the controlled variable exactly equal to the set point; in other words, the amplitude ratio would be equal to one (1.0) for all frequencies. Very good control performance is achieved for very low frequencies, when the feedback control system has time to respond to slow set point change. As the frequency increases, the control performance becomes poorer, because the set point variations become too fast for the feedback control system to track closely. Again, a resonant peak can occur at intermediate frequencies.

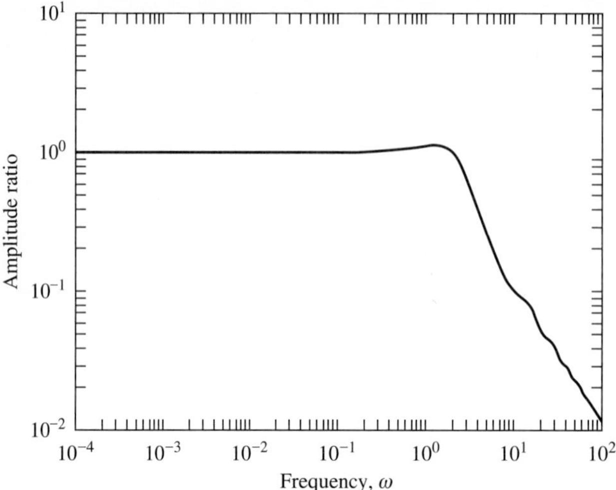

FIGURE 13.9
Closed-loop frequency response for the set point response in Example 13.3.

Example 13.3. Calculate the set point frequency response for the plant in Example 13.1, case C.

The transfer functions of the process and controller are given in Example 13.1. The result of calculating the amplitude ratio of equation (13.4) is given in Figure 13.9. As shown in the figure, the control system would provide good set point tracking (i.e., an amplitude ratio close to 1.0) for a large range of frequencies. The frequency range for which the amplitude ratio responds satisfactorily is often referred to as the *system bandwidth;* taking a typical criterion that the amplitude ratio of 1.0 to 0.707 is acceptable, the bandwidth of this system is frequencies from 0.0 to about 3 rad/time.

The calculation of the frequency response for the closed-loop system is performed by applying the same principles as for open-loop systems. However, the calculations are much more complex. The frequency response for closed-loop systems requires that the transfer function be solved for the magnitude, and the results must be derived for each system individually, as was done analytically in equation (13.3). Clearly, this amount of analytical manipulation could inhibit the application of the frequency response technique.

In the past, graphical correlations have been used to facilitate the calculations for a limited number of process and controller structures. The Nichols charts (Edgar and Hougen, 1981) are an example of a graphical correlation approach to calculate the closed-loop from the open-loop frequency response. These charts are not included in this book because closed-loop calculations are not now performed by hand.

Since the advent of inexpensive digital computers, the calculations have been performed with the assistance of digital computer programs. Most higher-level languages (e.g., FORTRAN) provide the option for defining variables as complex and solving for the real and imaginary parts; thus, the computer programming is straightforward, basically programming equation (13.2) with complex variables. An extension to the programming approach is to use one of many software packages that

TABLE 13.1
Example MATLAB program to calculate a closed-loop frequency response

```
%  ** EXAMPLE 13.3 FREQUENCY RESPONSE ***
%  this MATLAB M-file calculates and plots for Example 13.3
%  ********************************************
%  parameters in the linear model
%  ********************************************
kp = 1.0 ; taup = 1.5 ; thetap = 0.5;
kc = 1.7  ;  ti = 1.3;
%  ********************************************
%  simulation parameters
%  ********************************************
wstart = .0001  ;   % the smallest frequency
wend   = 100   ;   %  the highest frequency
wtimes = 800  ;   %  number of points in frequency range
omega = logspace ( log10(wstart), log10(wend), wtimes);
jj = sqrt(-1) ;   %  define the complex variable
%  ********************************************
%  put calculations here
%  ********************************************
for kk = 1:wtimes
     s = jj*omega(kk)          ;
     Gp(kk) = kp * exp (- thetap * s) /( ( taup*s + 1)) ;
     Gc(kk) = kc*(1 + 1/ (ti * s));
     G (kk) = Gc(kk)*Gp(kk)/(1 + Gc(kk)*Gp(kk));
     AR(kk) = abs (G(kk));
end  % for cnt
%  **************************************************
%  plot the results in Bode plot
%  **************************************************
loglog( omega, AR)
axis ([ -4 2 -2 1])
xlabel ('frequency, rad/time ')
ylabel ('amplitude ratio')
```

are designed for control system analysis, such as MATLAB™. An example of a simple MATLAB program to calculate the frequency response in Figure 13.9 is given in Table 13.1. For simple models, the approach in Example 13.1 can be used, but computer methods are recommended over algebraic manipulation for closed-loop frequency response calculations.

The frequency response approach presented in this section is a powerful, general method for predicting control system performance. The method can be applied to any stable, linear system for which the input can be characterized by a dominant sine. The calculations of the amplitude ratio for a closed-loop system are usually too complex to be performed by hand but are easily performed via digital computation.

> The great strength of frequency response is that it provides a clear indication of the control performance for an input (disturbance or set point change) at various frequencies.

13.4 CONTROL PERFORMANCE VIA CLOSED-LOOP SIMULATION

Solution of the time-domain equations defining the dynamic behavior of the system is another valuable method for evaluating the expected control performance of a design. Unfortunately, the differential and algebraic equations for a realistic control system are usually too complex to solve analytically, although that would be preferred so that analytical performance relationships could be determined. However, numerical solution of the algebraic and differential equations is possible and usually provides an excellent approximation to the behavior of the exact equations.

One reason for using simulation is that control performance specifications are defined in the time domain. The comparison of the predicted performance to the specifications often requires the entire dynamic response—the variables over the entire transient response—to ensure proper dynamic behavior. Thus, the solution to the complete model is required. Also, the engineer likes to see the entire transient response to evaluate all factors, such as maximum deviation, decay ratio, and settling time. The simulation approach is particularly useful in determining the response of a system to a worst-case disturbance. This largest expected disturbance can be introduced, and the resulting response will indicate whether or not all process variables can be maintained within their specified limits.

Numerical methods used to solve ordinary differential equations were described briefly in Chapter 3. Note that equations for all elements in the system—process, instrumentation, and controller—must be solved *simultaneously.* Also, since the solution is numerical, there is no requirement to linearize the equations, although insight from the analysis of linear models is always helpful. It should be clear that simulation methods have been used to prepare most of the closed-loop dynamic responses in figures for this book.

Example 13.4. Determine the dynamic response of the three-tank mixing process defined in Example 7.2 (page 249) under PID control to a disturbance in the concentration in stream B of $+0.8\%$.

This is the case considered in Example 9.2 (page 302), in which the PID tuning was first determined from a process reaction curve. The dynamic response of the closed-loop control system was then determined by solving the algebraic and differential equations describing the system, equations (7.14) and (7.15), along with the algorithm for the feedback controller. The following equations summarize the model.

$$
\begin{aligned}
E &= \text{SP} - x_{\text{A3}} \qquad (13.5)\\
v &= K_c\left(E + \frac{1}{T_I}\int_0^t E(t')dt' - T_d\frac{dx_{\text{A3}}}{dt}\right) + 50\\
F_A &= 0.0028v\\
x_{\text{A0}} &= \frac{F_B(x_{\text{A}})_B + F_A(x_{\text{A}})_A}{F_B + F_A}\\
V_i\frac{dx_{\text{A}i}}{dt} &= (F_A + F_B)(x_{\text{A}i-1} - x_{\text{A}i}) \qquad \text{for } i = 1, 3
\end{aligned}
$$

The PID controller can be formulated for digital implementation as described in Chapter 11. Also, the differential equations can be solved by many methods; here they are formulated in the discrete manner using the Euler integration method. Both the process and the controller are executed at the period Δt.

$$\begin{aligned}
E_n &= \mathrm{SP}_n - (x_{A3})_n \\
(v)_n &= (v)_{n-1} + K_c\left(E_n - E_{n-1} + \frac{\Delta t E_n}{T_I} + \frac{T_d}{\Delta t}(-(x_{A3})_n + 2(x_{A3})_{n-1} - (x_{As})_{n-2})\right) \\
(F_A)_n &= 0.0028(v)_n \\
(x_{A0})_n &= \left(\frac{F_B(x_A)_B + F_A(x_A)_A}{F_B + F_A}\right)_n \\
(x_{Ai})_{n+1} &= (x_{Ai})_n + \frac{\Delta t(F_B + F_A)_n}{V_i}[(x_{Ai-1})_n - (x_{Ai})_n] \qquad \text{for } i = 1, 3
\end{aligned} \tag{13.6}$$

The initial conditions are $(x_{Ai})_0 = 3.0\%$ A for $i = 0, 3$ and $(v)_0 = 50\%$ open. For the example considered, the system is nearly linear with time constants for all tanks of about 5 minutes, and the controller tuning constants are $K_c = 30$, $T_I = 11$, and $T_d = 0.8$. The disturbance was a step in $(x_A)_B$ from its initial value of 1.0 to 1.8 at time 20. The execution period was selected to be small relative to the time constants of the process, 0.1 minute. The result of executing the equations (13.6) recursively is the entire transient response. The manipulated and controlled variables are plotted in Figure 9.6 (page 303). Note that the numerical solution is not limited to linear cases; for example, the same numerical model was used to solve Example 9.4, which involved a change in the flow rate of stream B, F_B.

The simulation method is not restricted to simple input forcing functions, and this flexibility is very useful in estimating likely improvements in control performance. As demonstrated in the previous example, the control performance can be determined based on a model of the feedback process and a model of the disturbance. If the disturbance is a complicated function, a representative sample of the effect of the disturbance on the variable to be controlled can be used as a "model" of the disturbance. The effect of the disturbance(s) can be obtained by collecting *open-loop* data of the variable to be controlled as typical variabilities in plant operation occur.

Example 13.5. PI control is to be applied to the plant with feedback dynamics characterized by a dead time and single time constant, similar to Figure Q6.3 (page 224). In the plant an undesirable feed component is reacted to a benign effluent component. The outlet concentration is to be controlled by adjusting the feed preheat. The control objective is to maintain the outlet concentration just below its maximum value. Too low a concentration leads to costly side reactions and byproducts; thus, the goal is to reduce the variance. The model, determined by empirical identification, and the controller tuning are as follows:

$$G_p(s)G_v(s)G(s) = \frac{\mathrm{AC}(s)}{v(s)} = \frac{1.0e^{-2s}}{1+2s} \qquad G_c = 1.0\left(1 + \frac{1}{2.5s}\right) \tag{13.7}$$

A sample of representative dynamic data of the reactor effluent without control is presented in Figure 13.10a. Note that some of the variation is of low frequency;

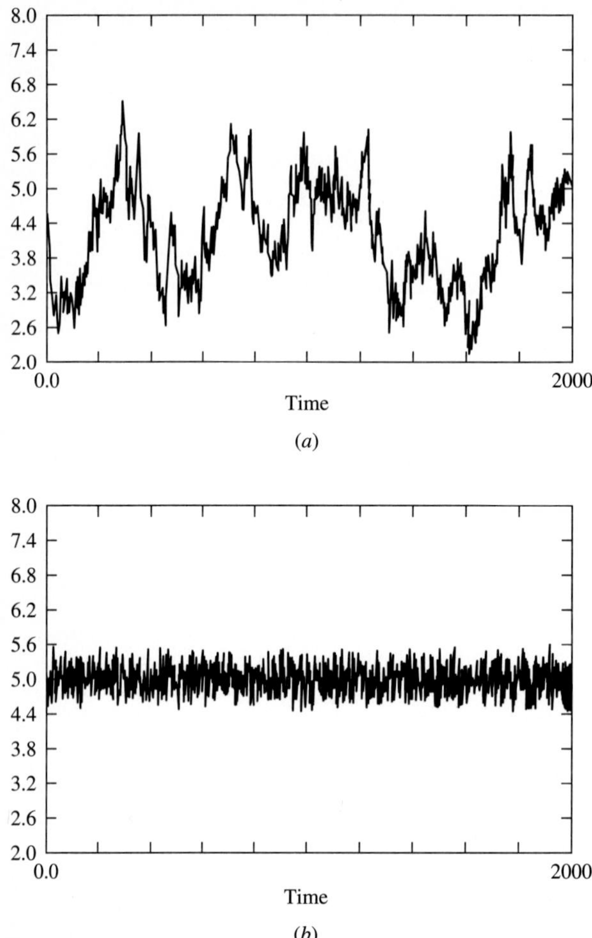

FIGURE 13.10
Reactor outlet concentration, Example 13.5: (*a*) effect of disturbance without control; (*b*) dynamic response with feedback control.

feedback control would be expected to be successful in attenuating these low-frequency components. Also, some of the variation is relatively high-frequency, which, we expect, would be difficult to reduce with feedback control.

To predict the performance of the control system, a simulation can be performed using the plant model with the sample disturbance data. This approach is shown schematically in Figure 13.11, where the digital simulation would introduce the disturbance data collected from the process, Figure 13.10*a*, as the forcing function. Naturally, the controller calculation, here a proportional-integral algorithm, receives the controlled process output, which is the sum of the effects from the manipulated variable and the disturbance. The results of the simulation are given in Figure 13.10*b*. The variability of the controlled variable, measured by standard deviation (see Chapter 2), has been reduced substantially by feedback control. Analysis of a larger set of data

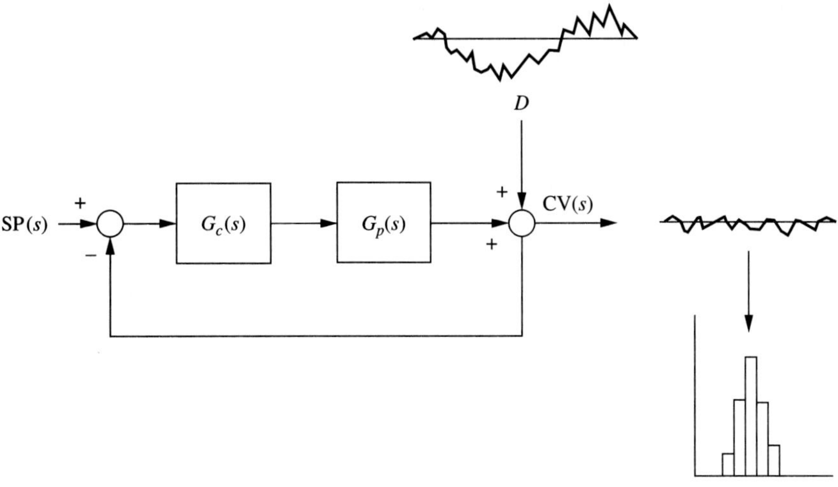

FIGURE 13.11
Schematic of the calculation method for predicting control performance with a complex disturbance model by a simulation method.

than shown in the figure, which gives a more reliable indication of performance, shows that the standard deviation is reduced by a factor of 5. As expected, the high-frequency components are not substantially reduced by the feedback control system. Because of the smaller variation, the average value of the concentration (i.e., the controller set point) could be increased to realize the benefits from improved control performance.

This example clearly demonstrates the improvement possible with feedback control and provides a simple, simulation-based method for estimating control performance. The method requires a process model, a controller equation, and a sample of the output variable without control; it provides a prediction of the standard deviation of the manipulated and controlled variables. It can be used in conjunction with the benefits calculations from Chapter 2 to estimate control benefits quantitatively, as shown in Figure 13.11.

The material in this section has demonstrated that:

> Dynamic simulation via numerical solution of the system equations provides a manner for determining the dynamic performance of a closed-loop process control system. The approach can (1) provide a solution for nonlinear as well as linear systems; (2) consider any input forcing functions; and (3) provide detailed information on all variables throughout the transient response.

Frequency response and dynamic simulation, along with stability analysis, provide methods required to analyze control systems quantitatively. These methods are applied in the next sections to develop understanding of general concepts and trends.

13.5 PROCESS FACTORS INFLUENCING SINGLE-LOOP CONTROL PERFORMANCE

Because the process ($G_p(s)$ and $G_d(s)$), instrumentation ($G_v(s)$ and $G_s(s)$), and the controller ($G_c(s)$) appear in the closed-loop transfer function in equation (13.1), all elements in the feedback system influence its dynamic response and control performance. It is tempting to believe that a cleverly designed controller algorithm can compensate for a difficult process; however, the process imposes limitations on the achievable feedback control performance, *regardless of the feedback algorithm used.* An understanding of the effects of process dynamics on control performance enables us to design plants that are easier to control, recognize limits to the performance of single-loop feedback control, and design enhancements. The next topic establishes a bound on the best achievable feedback control performance that gives valuable insight into the effects of process dynamics.

A Bound on Achievable Performance

The first topic introduced in this section is the performance bound (i.e., the best achievable performance) for a feedback system. The best performance is explained with reference to the process shown in Figure 13.4, where the control system is subjected to a step change disturbance. (Note that this concept is applicable to more general processes than Figure 13.4.) The dynamic responses of the controlled and manipulated variables are graphed versus time in Figure 13.12, and several important features of the response are highlighted. First, note that the effect of the feedback adjustment has no influence on the controlled variable for a period of time equal to the dead time in the feedback loop. Therefore, the integral error and maximum deviation shown in Figure 13.12 cannot be reduced lower than the open-loop response for time from zero (when the disturbance first affects the controlled variable) to the dead time. For the special case of a step disturbance with magnitude ΔD and a first-order disturbance transfer function with gain K_d and time constant τ_d, the limiting integral error and maximum deviation can be simply derived by the equations

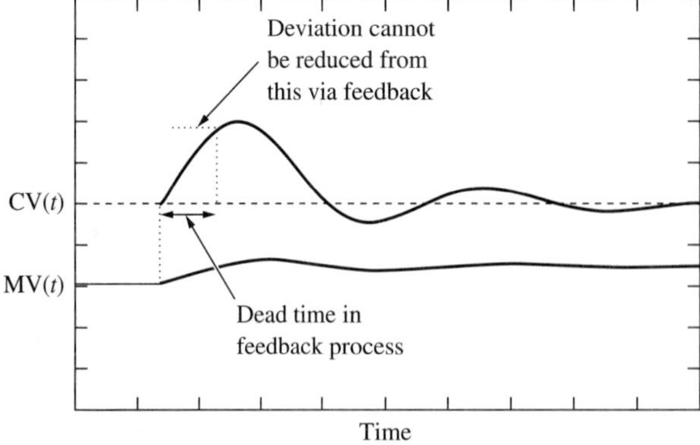

FIGURE 13.12
Typical dynamic response for a feedback control system.

$$E = K_d(1 - e^{(-t/\tau_d)})\Delta D \qquad \text{for } 0 \le t \le \theta \tag{13.8}$$

$$\begin{aligned} \text{IAE}_{\min} &= \int_0^\theta \left|E\right| dt \qquad (13.9) \\ &= \left|K_d \Delta D\right| \int_0^\theta \left|\left(1 - e^{-(t/\tau_d)}\right)\right| dt \\ &= \left|K_d \Delta D\right| \left(\theta + \tau_d \left(e^{-\theta/\tau_d} - 1\right)\right) \end{aligned}$$

$$(E_{\max})_{\min} = \left|K_d \Delta D\right| \left(1 - e^{-(\theta/\tau_d)}\right) \tag{13.10}$$

$\text{IAE}_{\min}$ represents the minimum IAE possible, and $(E_{\max})_{\min}$ represents the minimum value possible for the maximum deviation for a feedback system with dead time θ, a step disturbance, and a disturbance time constant of τ_d. *No single-loop feedback controller* can reduce the values further. As shown in the figure, these values provide a useful bound with which to evaluate control performance. The important conclusion from this discussion is that

> The dead time in the feedback path is the facet of the process that usually limits the control performance.

The theoretical best achievable control performance cannot usually be realized with a PID control algorithm, although the PID often provides entirely satisfactory performance. Methods exist for deriving the control algorithms giving the theoretical best or "optimal" control, with *optimal* defined several ways, such as minimum integral of error squared (Newton, Gould, and Kaiser, 1958; Astrom and Wittenmark, 1984). It is important to recognize that these controllers can result in excessive variation in the manipulated variable, and their performance can be very sensitive to model errors. Therefore, the "optimal" algorithms are not often applied in the process industries, although their concepts are useful in determining the achievable performance bounds in equations (13.9) and (13.10).

A Control Performance Correlation

The second topic in this section builds on the result that the dead time is a key factor in control performance to develop approximate correlations that indicate the quantitative effects of key parameters on control performance. Many investigators (e.g., Astrom, 1992; Jeffreson, 1976) have noted the usefulness of such correlations in understanding how process characteristics affect control system performance. Since our goal is to obtain a simple relationship between a few model parameters and control performance, we will restrict the analysis to a simple, general process model. The natural choice is the first-order-with-dead-time model. As we have seen, this is a widely applicable model; therefore, the results will be useful for a great many, but not all, plants. Our goal is to determine how the control performance is affected by the plant model parameters: feedback process dead time, time constant, and gain, along with the disturbance time constant. To develop the correlation, we will extend the dimensional analysis used in Chapter 9 on PID tuning (page 299).

Here, the closed-loop response of a control system to a disturbance shown in Figure 13.1 is analyzed.

$$\frac{CV(s)}{D(s)} = \frac{G_d(s)}{1 + G_p(s)G_c(s)} \quad \text{with } G_p(s) \approx G'_p(s)G_v(s)G_s(s) \tag{13.11}$$

It is assumed that the process can be modelled as first-order with dead time and that the control algorithm is proportional-integral. The resulting transfer function is

$$\frac{CV(s)}{D(s)} = \frac{K_d/(1 + \tau_d s)}{1 + K_c\left(1 + \dfrac{1}{T_I s}\right)\left(\dfrac{K_p e^{-\theta s}}{1 + \tau s}\right)} \tag{13.12}$$

This transfer function defines the dynamic response of the control system to a disturbance from an initial steady state. It has a large number of parameters, so any generalization would be difficult. The challenge is to reduce the number of independent parameters; this is accomplished by eliminating several parameters that are dependent on a smaller number of truly independent parameters in the equation.

Since our goal is to develop a general correlation, we must first address the differences in process "speed" of response, which can be addressed by time scaling. A natural choice for the scaling variable is the process dead time plus time constant, which gives the time for the open-loop process to attain 63 percent of its response to an input step change. In fact, this variable is the first moment of the process to an impulse input, as discussed in Appendix D. The first moment, as defined by the following equation, gives a good characterization of the "speed" of the system.

$$t' = \frac{t}{\theta + \tau} \qquad s' = s(\theta + \tau) \tag{13.13}$$

By dividing the actual time by the plant first moment, all plants are placed on a time scale that allows their responses to be compared. The resulting transfer function is

$$\frac{CV(s')}{D(s')} = \frac{\dfrac{K_d}{\left(1 + \dfrac{\tau_d s'}{(\theta + \tau)}\right)}}{1 + K_c K_p\left(1 + \dfrac{1}{T_I s'/(\theta + \tau)}\right)\left(\dfrac{e^{\theta s'/(\theta+\tau)}}{1 + \tau s'/(\theta + \tau)}\right)} \tag{13.14}$$

The variables in equation (13.14) are listed in three categories: independent, dependent, and determined by tuning.

Independent		**Dependent**	
$\theta/(\theta + \tau)$	(fraction dead time)	$\tau/(\theta + \tau)$	(1 − fraction dead time)
ΔD	(disturbance magnitude)		
K_d	(disturbance gain)		
$\tau_d/(\theta + \tau)$	(disturbance "speed")		

Determined by tuning	
$K_p K_c$	(closed-loop gain)
$T_I/(\theta + \tau)$	(dimensionless integral time)

For this performance analysis, the tuning correlations introduced in Chapter 9 are used. As demonstrated in Chapter 9, the tuning parameters, $K_c K_p$ and $T_I/(\theta + \tau)$, depend on the plant fraction dead time, $\theta/(\theta + \tau)$. Therefore, the controller tuning parameters depend on process parameters in the transfer function. Note that the parameter $\tau/(\theta + \tau) = 1 - \theta/(\theta + \tau)$ is dependent when the independent parameters are specified. Also, the process and controller gains appear only as a product, sometimes referred to as the steady-state closed-loop gain; the control system response depends on this product, not on either gain alone. Thus, the feedback fraction dead time, $\theta/(\theta + \tau)$, is the remaining independent feedback process parameter for the time-scaled process.

The disturbance appears in three terms: the forcing function $D(s)$, the gain K_d, and the dimensionless time constant $\tau_d/(\theta + \tau)$. To develop correlations, the "shape" (the time dependence) of the input function is required. For simple performance comparisons, we will assume a unit step disturbance. Also, the effect of any disturbance can be normalized by dividing by the magnitude of the disturbance gain; after this division, disturbances of any magnitude can be compared. For the purposes of this correlation, the disturbance time constant remains a variable.

As a result of the dimensional analysis and simplifying assumptions, only two parameters remain that affect the control system dynamic response; these are the feedback fraction dead time, $\theta/(\theta + \tau)$, and the disturbance speed, $\tau_d/(\theta + \tau)$. Therefore, the performance of a control system under PI control, with the simplifications discussed, can be plotted as a function of only two parameters!

Two measures of control performance are selected for graphing: maximum deviation and integral of the absolute value of the error (IAE). These measures are scaled so that the correlations can be applied to many processes. The results of simulations of different systems are given in Figures 13.13*a* and *b*. Similar correlations have been developed by Lopez et al. (1969) and by Jeffreson (1976); they used more aggressive tuning rules, which led to larger manipulated-variable changes and less robust systems, but obtained results with the same general trends.

> For the performance correlations in Figures 13.13*a* and *b*, the system was a first-order-with-dead-time feedback process with first-order disturbance dynamics, and the controller is a proportional-integral algorithm. It was simulated without noise or model error in response to a single step disturbance. Tuning was from the Ciancone correlations in Figures 9.9*a* and *b*, which provide reasonable robustness, and the controller execution period was 10% of $(\theta + \tau)$.

As expected, the control performance is very good for small fraction dead time and becomes worse as the fraction dead time increases. The poorer performance is the result of two factors related to the dead time. First, the controller must be detuned as

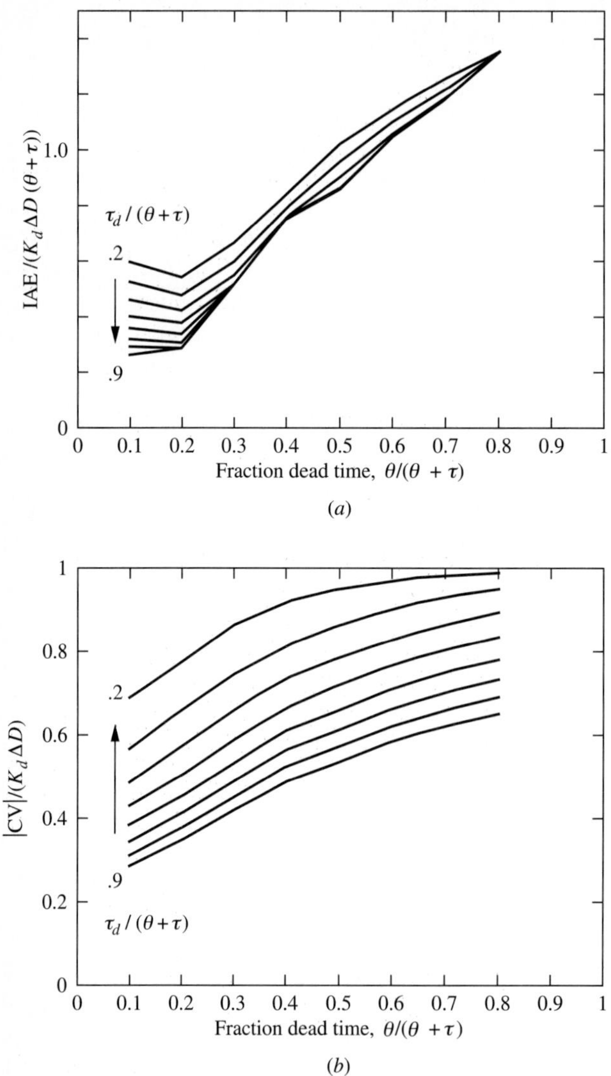

FIGURE 13.13
Correlations for PI controllers as a function of process fraction dead time: (*a*) of IAE; (*b*) maximum deviation from set point.

the fraction dead time increases, as dictated by stability considerations. Second, the manipulated variable is unable to influence the controlled variable for the dead time, as shown in Figure 13.12. Thus, the fraction dead time, $\theta/(\theta + \tau)$, is a useful parameter in determining the ease with which a process can be controlled. Also apparent in the figures is the effect of the disturbance time constant. As the disturbance time constant increases (i.e., as the disturbance becomes slower), the control performance improves, with the largest effect on the maximum deviation. The correlations in this section are valuable in evaluating processes and estimating approximate control system performance. Some applications are demonstrated in the following examples.

Example 13.6. The four potential process designs shown in Figure 13.4 have been proposed for a plant. It is expected that all designs have about the same capital and operating costs. For the process dynamics given in the figure, which plant with PI feedback control would give the best performance for a unit step disturbance?

The control performances are first evaluated qualitatively using the trends in Figures 13.13*a* and *b*. All potential designs have the same disturbance time constant, so this factor will not affect their relative control performances. Therefore, we turn our attention to the feedback processes. A preliminary analysis notes that Case B has the highest $\theta/(\theta + \tau)$ and the highest $(\theta + \tau)$. It is both the slowest and the most difficult to control as a result of its large fraction dead time; therefore, it must provide the poorest control performance. Cases A, C, and D have the same $(\theta + \tau)$, which means that they would reach 63% of their open-loop responses at the same time. However, Cases C and D have a lower $\theta/(\theta + \tau)$. Therefore, the latter two cases should provide better control. The only difference between Cases C and D is their values of process gain, K_p. As we know, the value of process gain can be compensated by the controller gain to give the desired K_pK_c, as long as the manipulated variable has sufficient range (which can be determined from a calculation of the steady-state operating window). Therefore, cases C and D would provide equally good control performance.

The foregoing discussion is confirmed by determining the control performance from Figure 13.13. The results are consistent with the conclusions based on frequency response in Example 13.1.

Case	$IAE/K_d\Delta D$	$(\text{Max dev})/(K_d\,\lvert\Delta D\rvert)$	Minimum $IAE/K_d\Delta D$	Minimum $(\text{Max dev})/(K_d\,\lvert\Delta D\rvert)$
A	1.6	0.68	0.37	0.63
B	> 8.0	> 0.95	3.02	0.98
C and D	0.8	0.50	0.11	0.39

The best achievable feedback control performance for this process and control design are also given based on equations (13.9 and 13.10). These results show the same ranking of control performances.

Example 13.7. The maximum deviation for the outlet concentration in Example 13.6, Case C, should never exceed the limit of 0.38 after a step disturbance that would cause a deviation of 1.0 if it were not controlled—that is, for $K_d(\lvert\Delta D\rvert) = 1$. How can the control or process be altered to observe the limit?

It is important to recognize that PI feedback control, or feedback using any other algorithm, cannot achieve the specified performance, because of the *process dynamics*. The results based on equation (13.10) demonstrate this fact, because the minimum value of the maximum deviation exceeds the specified limit: $0.39 > 0.38$. There are two ways to improve the control performance by changing the process. One way to improve the control performance is to reduce the feedback dead time. The relationship between the maximum deviation and the dead time is given in Figure 13.13*b*. For the shortest fraction dead time covered in the figure, 0.10, the maximum deviation remains too large to satisfy the performance requirement. Thus, the dead time would have to be reduced below 0.17 in order to approach the requirement. The dead time might be decreased by reducing the transportation delay; however, the required reduction may not be possible.

Another process alteration that would reduce the maximum deviation would be to increase the disturbance time constant (i.e., slow the disturbance effect on the controlled variable), leaving the feedback dynamics unchanged. From Figure 13.13*b*, the required disturbance time constant to satisfy the performance requirement at the same fraction feedback dead time would be about 1.8, calculated as the product of $0.9(\theta + \tau)$. One example of how this change might be accomplished is by introducing tankage upstream of the process, which would attenuate concentration and temperature variations in the inlet streams.

A better way to achieve the desired performance would be to improve the process control in the unit creating the disturbance, if possible; this would reduce ΔD and improve the operation of the other unit as well. It is also possible to use a sensor measurement indicating the disturbance to improve control performance, as explained in Part IV.

The Effect of Inverse Response

Inverse response is an important characteristic of the feedback process dynamics that, when it exists, has a major effect on control performance. A physical process that experiences an inverse response is described in Example 5.4 (page 169). In that example, the parallel process structure resulted in the concentration first increasing, then decreasing in response to a step in the solvent flow rate. (The reader may want to review this example before proceeding.) Clearly, such a process is difficult to control, because the initial response of the controlled variable is in the "wrong" direction. The initial inverse response imposes a limit to the achievable control performance in a way similar to dead time.

Example 13.8. The inverse response process for the reactor without control in Example 5.4 is shown in Figure 5.14. The feedback control system shown in Figure 13.14 is

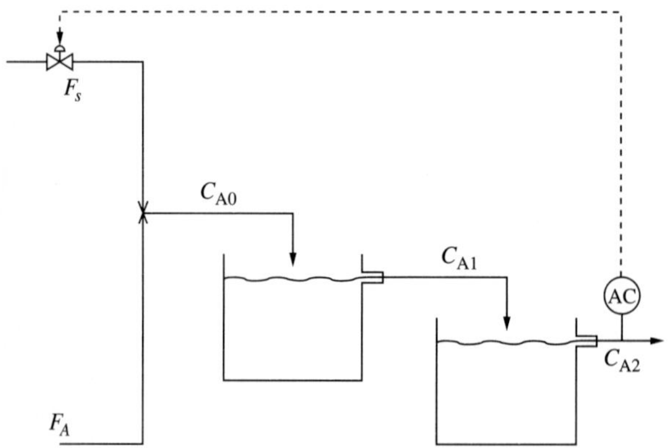

FIGURE 13.14
Feedback control design for Example 13.8.

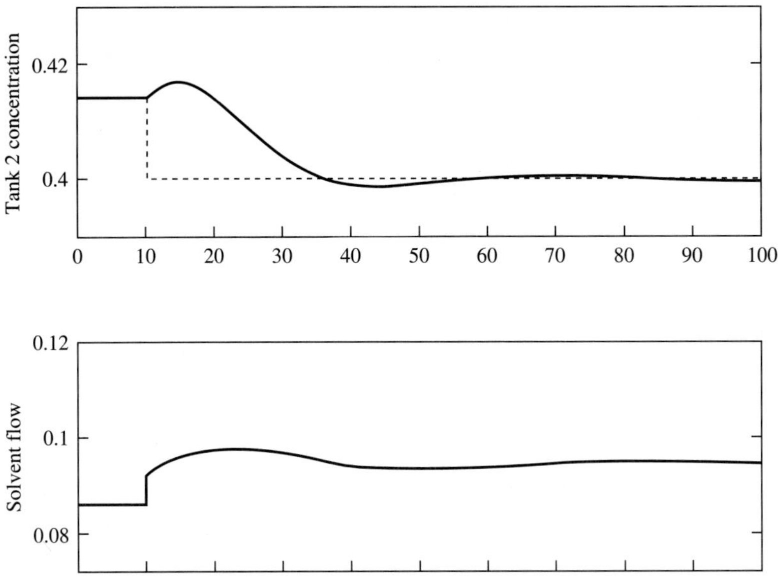

FIGURE 13.15
Closed-loop response of the inverse response process in Example 13.8.

proposed for this reactor. Determine the control performance for this system in response to a step change in the set point of a PI controller.

The model for this process, linearized about the initial steady state, is repeated here; however, this model is not exact for the transient considered, because the gain and time constants depend on the flow of solvent, which changes through the transient:

$$G_p(s) = \frac{-1.66(-8.0s + 1)}{(8.25s + 1)^2} \tag{13.15}$$

The tuning for the PI controller was determined by trial and error to be $K_c = -0.45\ \text{m}^3/\text{min(mole/m}^3)$ and $T_I = 13.0$ min, which resulted in the transient response in Figure 13.15. This transient was evaluated by a numerical solution of the nonlinear differential equations. The control performance is less than ideal, because the initial response of the controlled variable is inverse to the change in the set point. However, the response is stable, returns to the set point, and is "well behaved" (i.e., not unduly oscillatory or slow to return to the set point).

It is important to recognize that this second-order process *without dead time* cannot be controlled tightly, because of the inverse response, regardless of the feedback control algorithm.

Again, we see the influence of feedback dynamics on control performance.

Model Requirements for Predicting Control Performance

Throughout this book, we have monitored the effects of modelling errors on design decisions such as tuning and on the resulting control performance. Here the effects of modelling errors on the accuracy of control performance predictions are considered. Two linear models for the three-tank mixing process have been developed; one involves a third-order system, and the other involves a first-order-with-dead-time approximation. How well does the performance predicted using the approximate model compare with the performance using the "exact" third-order model? To answer the question for this example, the closed-loop frequency responses have been calculated for both cases. The controller is a PI algorithm with the tuning constants from Example 9.2 (page 302, with the small derivative time set to zero). The closed-loop transfer functions for the two cases are as follows:

Exact third-order model.

$$\frac{\mathrm{CV}(s)}{D(s)} = \frac{\dfrac{1}{(5s+1)^3}}{1 + \dfrac{0.039}{(5s+1)^3} 30\left(1 + \dfrac{1}{11s}\right)} \tag{13.16}$$

Approximate first-order-with-dead-time model.

$$\frac{\mathrm{CV}(s)}{D(s)} = \frac{\dfrac{1e^{-5.5s}}{(10.5s+1)}}{1 + 0.039\dfrac{e^{-5.5s}}{(10.5s+1)} 30\left(1 + \dfrac{1}{11s}\right)} \tag{13.17}$$

The results of the analysis are plotted in Figure 13.16. The approximate first-order-with-dead-time model represents the system with sufficient accuracy to predict the control performance, especially for the low-frequency disturbances, which is the range for which feedback control is designed and effective. The predictions differ in the high-frequency range, but they both predict very good disturbance attenuation. The approximate model leads to some error in the region of the resonance peak; however, both models identify the proper resonance frequency and properly predict that feedback is not effective in this frequency region.

The results of this example on control performance, along with Examples 9.2 and 9.3 on tuning and Example 10.17 on stability analysis, lead to a very important conclusion:

> An approximate first-order-with-dead-time model typically provides sufficient accuracy for single-loop control tuning and performance analysis when the open-loop process has an overdamped, sigmoidally shaped response between the manipulated and controlled variables.

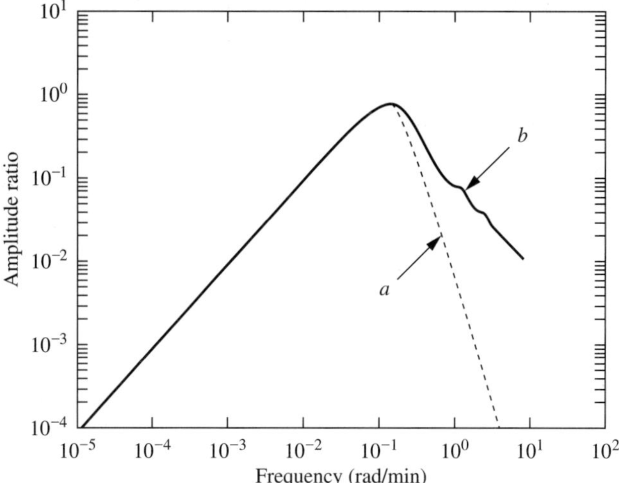

FIGURE 13.16
Comparison of closed-loop frequency response for (*a*) exact third-order, equation (13.16), and (*b*) approximate process models, equation (13.17).

Since many processes have such well-behaved dynamic responses, the first-order-with-dead-time models are used frequently in the process industries.

The topics in this section demonstrate some key limitations imposed on control performance by process dynamics and provide some quantitative estimates of how various process parameters affect performance. From these results, it becomes clear that many deficiencies in control performance cannot be corrected by improving the single-loop control algorithm or tuning. Finally, the sensitivity of control design methods to modelling errors has been analyzed, and the results in this section, in conjunction with previous chapters, confirm the usefulness of approximate models.

13.6 CONTROL SYSTEM FACTORS INFLUENCING CONTROL PERFORMANCE

The goal of the control instrumentation and algorithm is to achieve, as closely as is practically possible, the best control performance (for the controlled and manipulated variables) for the existing process dynamics. The effect of controller algorithm and tuning constants on the system's stability has been covered extensively in Chapters 9 and 10 and will not be repeated here. Suffice it to say that the controller tuning is selected to provide a compromise that gives acceptable behavior over a range of process dynamics. Several other important control system factors are discussed in this section.

Manipulated-Variable Behavior

As emphasized in Chapter 9, the behavior of the manipulated variable is also considered when evaluating control system performance. The effect of feedback control can be determined from the block diagram in Figure 13.1.

$$\frac{\mathrm{MV}(s)}{D(s)} = \frac{-G_d(s)G_s(s)G_c(s)}{1 + G_p(s)G_v(s)G_c(s)G_s(s)} \tag{13.18}$$

The numerator includes the product of the disturbance and controller transfer functions. As the controller tuning is selected for more aggressive control (i.e., the gain is increased or integral time decreased), the magnitude of the manipulated-variable variation is increased. In contrast, maintaining the controlled variable close to its set point requires aggressive control, as limited by feedback dynamics. Thus, the tuning is often selected as a compromise of these two concerns, manipulated- and controlled-variable performance.

Example 13.9. Evaluate the frequency response of the controlled and manipulated variables for the system in Example 13.1, Case C. Evaluate three values of the controller gain relative to the base case: (*a*) 75%, (*b*) 100%, and (*c*) 125%.

The magnitude of the controlled variable is determined from equation (13.2), and the magnitude of the manipulated variable is determined from the following equation:

$$\frac{|\mathrm{MV}(j\omega)|}{|D(j\omega)|} = \left| \frac{G_d(j\omega)G_s(j\omega)G_c(j\omega)}{1 + G_p(j\omega)G_v(j\omega)G_c(j\omega)G_s(j\omega)} \right| \tag{13.19}$$

The results are given in Figures 13.17*a* and *b*. Note that the manipulated-variable variation at low frequencies is nearly independent of the controller gain, since the manipulated variable is adjusted slowly, in quasi-steady state, in response to the disturbance. However, at higher frequencies a smaller controller gain results in a smaller manipulated-variable magnitude (variation). As expected, the smaller controller gain also results in an increased controlled-variable magnitude (variation).

Sensor and Final Element Dynamics

The dynamics of the final control element, usually but not always a valve, and the sensor appear in the feedback path. Therefore, they influence the stability and control performance. The closed-loop transfer function, including the instrument elements, for the system was derived in Chapter 7 and is repeated here:

$$\frac{\mathrm{CV}(s)}{D(s)} = \frac{G_d(s)}{1 + G_p(s)G_v(s)G_c(s)G_s(s)} \tag{13.20}$$

Example 13.10. Calculate the frequency response of the controlled variable to a disturbance input for the system in Example 7.1, Case A (page 237), (*a*) when the sensor and final element dynamics are as given in the Example, and (*b*) when these dynamics are negligible (i.e., all instrument dead times and time constants are reduced to zero, so that the only significant dynamics in the feedback path are from the process). For both cases, the disturbance time constant is 3 minutes.

The controller tuning has to be determined individually for (*a*) and (*b*). The dynamics can be approximated from the process reaction curves in Figure 7.3*a* using the graphical Method II explained in Chapter 6, and the tuning can be calculated from the Ciancone correlations.

	K_p	θ	τ	K_c	T_I
Example 13.10(*a*)	1.84	5.5	13.5	0.65	13.3
Example 13.10(*b*)	1.84	1.0	3	0.65	2.8

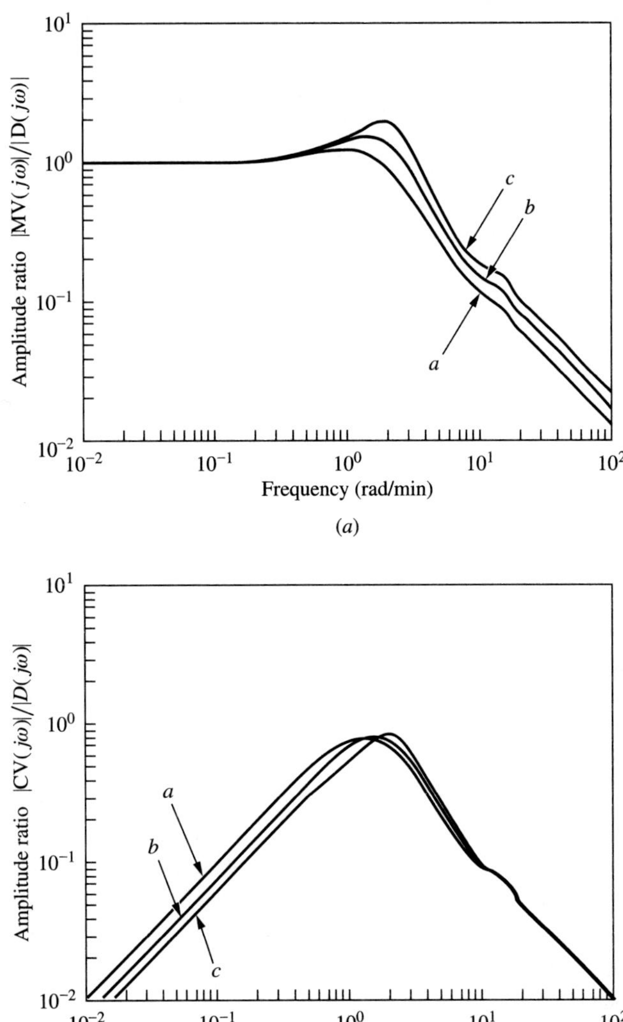

FIGURE 13.17
Amplitude ratios for disturbance input for Example 13.9: (*a*) of manipulated variable; (*b*) of controlled variable.

The results of the frequency response calculations are given in Figure 13.18. Clearly, the control performance is better for (*b*), where the instrumentation dynamics are negligible, because the instrumentation dynamics in (*a*) are substantial compared with the process.

Recall that the dynamic model determined through empirical identification includes all elements in the feedback path, $G_p(s)G_v(s)G_c(s)G_s(s)$. When the control system uses the same instrumentation, the identified model provides the information needed for tuning and control performance assessment.

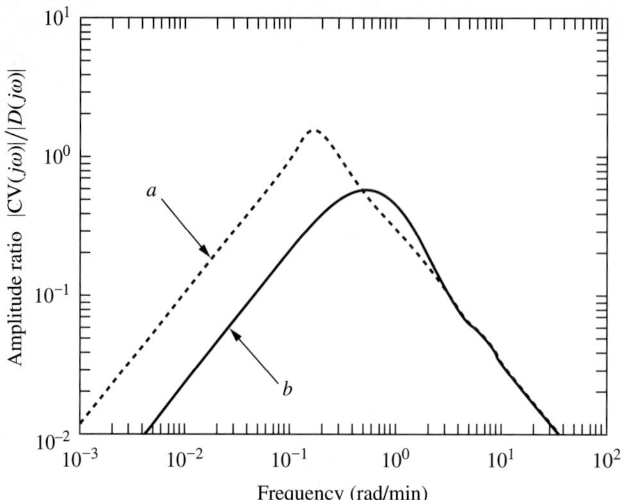

FIGURE 13.18
Amplitude ratio of controlled variable to disturbance for Example 13.10.

Digital PID Controllers

The PID algorithm can be implemented in a digital, or discrete manner, where the calculation is performed periodically. The effects of the execution period on tuning and control performance were covered in Chapter 11, where $\Delta t/(\theta + \tau)$ was identified as the parameter indicating the change from a continuous system. When this parameter is small, approximately 0.05 to 0.10, the system behavior is similar to that with a continuous controller; as the parameter increases, the control performance degrades from that achieved with a continuous controller. The digital control system can be easily simulated by executing the appropriate number of process simulation time steps between successive controller executions to provide an accurate representation of the process dynamics. The magnitude of the controlled variable in response to a sine input (i.e., the amplitude ratio of the frequency response) can be obtained, but the calculations require mathematical methods for discrete systems (z-transforms) not covered in this book (Ogata, 1987).

PID Mode Selection

With detailed analysis of controller tuning and control system performance, it is possible to discuss the selection of controller modes—proportional, integral, and derivative—for various applications. Naturally, the appropriate selection depends on the control objectives. For the vast majority of applications, zero offset is desired for steplike inputs, and an integral mode is required, as was demonstrated in Chapter 8. A few control strategies do not require zero offset, and proportional-only control is possible for these. The most common instances are some, but not all, level controllers, which are described in Chapter 18. Also, the proportional mode is nearly always used with the integral mode, because control systems with integral-only controllers tend to have slow, oscillatory dynamic responses.

Therefore, the proportional and integral modes are used for nearly all controllers, and the only choice regards the use of the derivative mode. The tuning correlations in Chapter 9 show that the derivative time (i.e., the contribution from the derivative mode) should be small for small fraction dead times and increase as the fraction dead time increases. A rationale for this trend is that the derivative is a "predictive" mode and that prediction is needed because of the dead time in the closed-loop system. A quantitative explanation is that the phase lead provided by the derivative mode allows a higher controller gain and shorter integral time, resulting in better control performance.

As previously discussed, the derivative mode amplifies high-frequency noise in the measured variable. If the difference between the noise and process response frequencies is large, the noise can be attenuated by filtering (see Chapter 12). If this is not the case, the controller derivative time must be reduced, perhaps to zero, to observe the limitation on the high-frequency variation of the manipulated variable.

Example 13.11. Select appropriate modes for the PID controller applied to the process shown in Figure 13.19.

LI-1 and LI-5. The feed tanks have periodic, rather than continuous, supply flows. As a result, their levels must vary with time, and their total volumes must be large enough to contain the change in inventory accumulated between supply or delivery flows. Therefore, their levels are not controlled. Level indication allows plant operating personnel to monitor the levels.

FC-1 and FC-2. Flow controllers should maintain the flows at their set points. The flow process has little dead time and a relatively noisy measurement signal. Therefore, a PI controller is used. Since the flow process is so fast, the PI is sometimes tuned with a small gain and small integral time so that it performs closer to an integral-only controller. This tuning further reduces the effects of noise.

LC-2. The reactor level influences the residence time and, therefore, the reaction conversion. The level should be maintained at its set point, but extremely rapid changes to the manipulated flow are not desirable. A PI controller is used.

TC-1. The reactor temperature is also a key variable in determining the reaction conversion. The controller would be PID or PI, depending on the fraction dead time.

TC-2. The flash drum temperature is an important variable in controlling the separation. The controller would be PID or PI, depending on the fraction dead time.

LC-3. There is no incentive to maintain the flash drum level at a specific value as long as the level remains within its allowed range. Also, flow variation to downstream units should be small. Therefore, a P-only controller could be used. A PI controller is also allowable in this case.

PC-1. The pressure of the flash drum is important for safety. It is also important for product quality, because the pressure affects the components in the flash vapor and

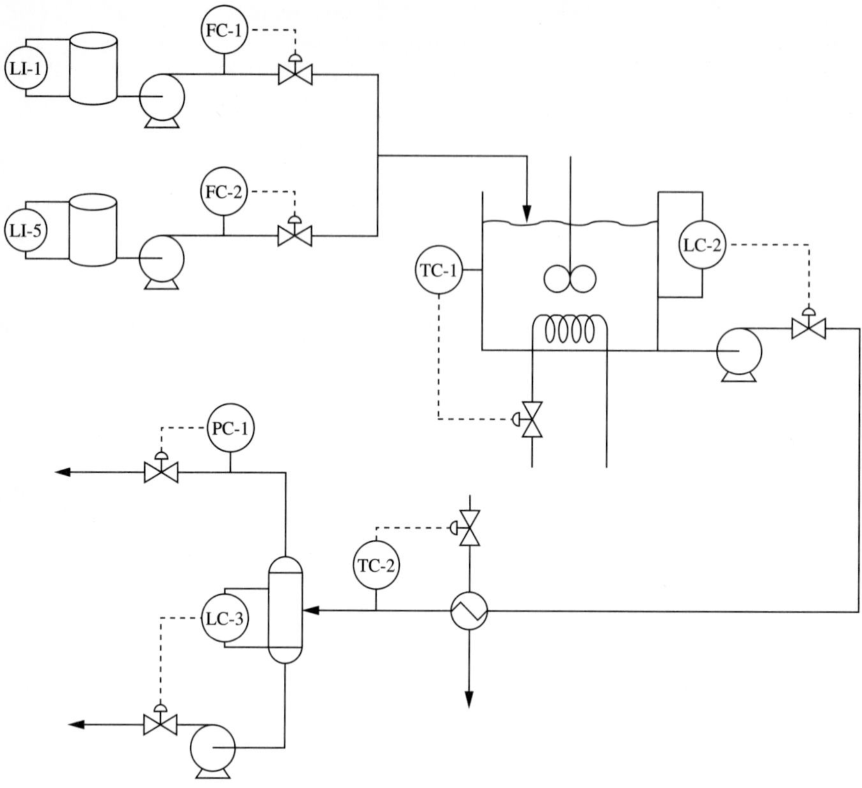

FIGURE 13.19
Schematic of process and controllers considered in Example 13.11.

liquid phases. The pressure dynamics should have essentially no dead time. Therefore, a PI controller is selected.

13.7 CONCLUSIONS

Two general, quantitative methods—frequency response and dynamic simulation—have been introduced for analyzing the control performance of feedback control systems. Each has specific strengths. Frequency response clearly shows the effects of the input frequency on the closed-loop performance, as indicated by the magnitude of important variables; it is applicable to stable, linear systems. Dynamic simulation provides detailed information on the performance of variables throughout a transient for any time-varying input function and can be applied to any system, linear or nonlinear. Both of these methods require extensive computation and are implemented using computer calculations.

The two quantitative analysis methods have been used to develop insights and generalizations about control performance. Many general conclusions have been developed about the effects of process and controller parameters on control performance, and they are summarized in Table 13.2.

TABLE 13.2
Summary of factors affecting single-loop PID controller performance

Key factor	Typical parameter	Effect on control performance
Feedback process gain	K_p	The key factor is the product of the process and controller gains. For example, a small process gain can be compensated by a large controller gain. Note that the manipulated variable must have sufficient range.
Feedback process "speed"	$\theta + \tau$	Control performance is always better when this term is small.
Feedback fraction process dead time	$\dfrac{\theta}{\theta + \tau}$	Control performance is always better when this term is small.
Inverse response	Numerator term in transfer function, $(\tau s + 1)$ with $\tau < 0$	Control performance degrades for large inverse response.
Magnitude of disturbance effect	$K_d \lvert \Delta D \rvert$	Control performance is always better when this term is small.
Disturbance dynamics	τ_d	Control performance is best when the disturbance is slow (the time constant is large).
	ω_d	Feedback control is effective for low-frequency disturbances and is least effective at the resonant frequency.
	θ_d	Disturbance dead time does not influence performance.
Sensor		Measurement should be accurate. Dynamics should be fast with little noise.
Filter	$\tau_f/(\theta + \tau)$	Attenuates higher-frequency components of measurement. Reduces the variability of the manipulated variable, but degrades controlled-variable performance as filter time constant is increased.
Final element		Dynamics should be fast without sticking or hysteresis. Range should be large enough for response to demands.
Controller execution period	$\dfrac{\Delta t}{\theta + \tau}$	Control performance is best when this parameter is small. Continuous PID tuning correlations can be used by modifying the dead time, $\theta' = \theta + \Delta t/2$.
Controller tuning	$K_c K_p$ $\dfrac{T_I}{(\theta + \tau)}$ $\dfrac{T_D}{(\theta + \tau)}$	These terms are determined from tuning correlations based on control objectives (see Chapters 7, 9, and 10).
Modelling errors		Errors in identifying the process model parameters lead to poorer control performance and, potentially, instability. Tuning should consider the estimate of model errors.
Limitations on manipulated variables	$\min < MV(t) < \max$	Limitations on manipulated variables reduce the operating window (the range of achievable conditions). An active limit would cause steady-state offset from the set point.

The analysis of controller modes, tuning, and stability in Chapters 8 through 10 emphasized the feedback process dynamics. In fact, it was demonstrated in Chapter 10 that the stability of linear systems is independent of the type of input, so long as it is bounded. In contrast:

> Control system performance depends on the dynamics of both the feedback and the disturbance and depends critically on the frequency and magnitude of the disturbance.

Although generally giving good performance, the PID controller does not provide the best performance in all cases. The performance of a single-loop PID control system can be improved in some cases by using additional measurements, modified PID algorithms, or entirely new feedback algorithms. Some of the most successful enhancements for single-loop control are described in Part IV of this book.

REFERENCES

Astrom, K., "Towards Intelligent PID Control," *Automatica, 28, 1,* 1–9 (1992).
Astrom, K., *Introduction to Stochastic Control Theory,* Academic Press, New York, 1970.
Astrom, K. and B. Wittenmark, *Computer Controlled Systems,* Prentice-Hall, Englewood Cliffs, NJ, 1984.
Edgar, T. and J. Hougen, "Other Synthesis Methods: Feedback Controller Design," in Edgar, T. (ed.), *AIChE Modular Instruction, Series A, Vol 2,* AIChE, New York, 1981, 53–61.
Jeffreson, C., "Controllability of Process Systems," *IEC Fund., 15, 3,* 171–179 (1976).
Kraniauskas, P., *Transforms in Signals and Systems,* Addison-Wesley, Reading, MA. 1992.
Lopez, A., P. Murrill, and C. Smith, "Tuning PI and PID Digital Controllers," *Instr. and Contr. Sys.,* 89-95 (February, 1969).
MATLAB™, The MathWorks, Inc., Cochituate Place, 24 Prime Park Way, South Natick, MA, USA, 01760.
Ogata, K. *Digital Control Systems,* Prentice-Hall, Englewood Cliffs, NJ, 1987.
Newton, G., L. Gould, and J. Kaiser, *Analytical Design of Feedback Controls,* Wiley, New York, 1957.

ADDITIONAL RESOURCES

A nice introduction to a rigorous analysis of stochastic control systems, like those in Figure 13.11, is given by

Koenig, D., *Control and Analysis of Noisy Processes,* Prentice-Hall, Englewood Cliffs, NJ, 1991.

The performance of control systems can be monitored and diagnosed with the goal of improving performance.

Smith, K., *A Methodology for Applying Time Series Analysis Techniques,* ISA paper no. 92-0087, 1992.
Weinstein, B., "A Sequential Approach to the Evaluation and Optimization of Control System Performance," *Proc. Amer. Contr. Conf.,* 2354–2358 (1992).

Many examples of control performance have been published, such as the following:

Marlin, T., J. Perkins, G. Barton, and M. Brisk, *Advanced Process Control Applications–Opportunities and Benefits,* Instrument Society of America, Research Triangle Park, NC, 1987.
Bellingham, B., *Energy Conservation via Distributed Process Control,* ISA paper no. 85-0716, 1985.
Leonard, D. and T. Kehoe, "Automatic Control Ups Heater Combustion Efficiency," *Oil Gas J.,* 134–138 (September 1981).
Smith, D., W. Stewart, and D. Griffin, "Distill with Composition Control," *Hydrocarbon Proc.,* 57, 2, 99–107 (1978).

Control performance depends on all elements in the feedback loop and the disturbance path. The following questions require you to (1) apply general principles to evaluate designs and (2) apply quantitative analysis to answer analytical or numerical questions.

QUESTIONS

13.1. The mixing process in Figure Q13.1 is to be analyzed in this question. The concentration at the outlet is controlled by adjusting a mixing stream at the inlet of three tanks. The main disturbance is the concentration of a stream flowing through a long pipe and a single stirred tank. Assume that in the base case the feedback PI controller is well tuned. For each of the following changes (*a*) through (*f*) from the base case answer the following questions and explain your answer.

(i) How should the two tuning constants be changed (increased, decreased, or unchanged) to maintain good control performance?

(ii) After the tuning has been adjusted, when necessary, how would the control performance, as measured by maximum deviation of the controlled variable in response to a step disturbance, differ from the base case (larger, smaller, same value)? *Hint:* It would help to identify the feedback and disturbance paths, which elements are in each, and how each is affected by the changes considered.

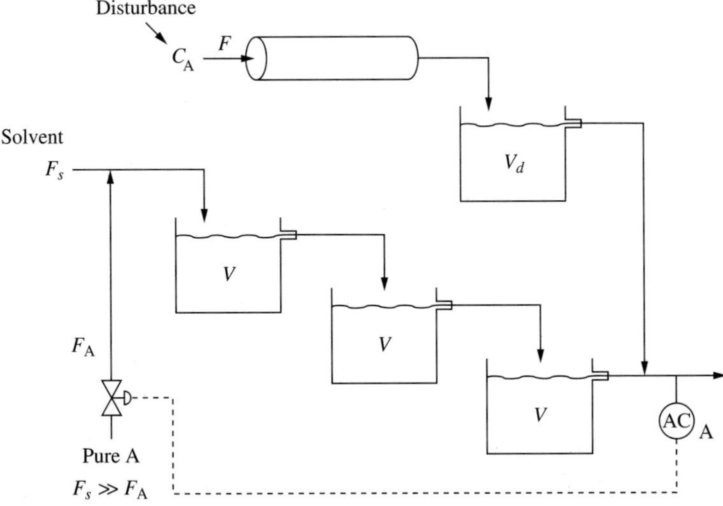

FIGURE Q13.1

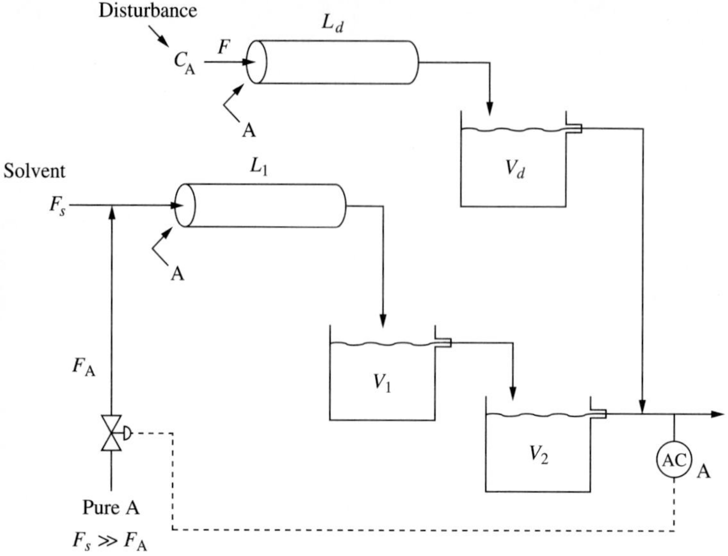

FIGURE Q13.2

Process changes (considered individually)

(*a*) The volume of each of the three tanks, V, is increased by 50%.

(*b*) The volume of the single tank, V_d, is increased by 50%.

(*c*) The initial operating condition (controller set point) is increased from 1% to 2% of A in the product.

(*d*) The length of pipe is doubled.

(*e*) The maximum capacity of the control valve is doubled.

(*f*) The solvent flow, F_s, is reduced by 50%.

13.2. Five mixing process designs, all having the process structure shown in Figure Q13.2, are to be analyzed in this question. The concentration at the outlet is controlled by adjusting a mixing stream at the inlet. The main disturbance is the concentration of a stream flowing through a pipe and a single stirred tank. The key parameters for each design are given in the following table, with all values in minutes.

Design	$L_1A/(F_A + F_S)$	$V_1/(F_A + F_S)$	$V_2/(F_A + F_S)$	$L_d/(F/A)$	$V_d/(F)$
I	1.0	1.0	1.0	1.0	1.0
II	0.5	1.0	1.0	1.0	1.0
III	1.0	0.5	1.0	1.0	1.0
IV	1.0	1.0	1.0	0.5	1.0
V	1.0	1.0	1.0	1.0	0.5

Rank the five designs from best to worst control performance in response to a step disturbance in C_A shown in the figure. Maximum deviation of the controlled variable from its set point is the measure of control performance. Assume that the feedback control system in the figure is used without change, but properly retuned for each plant.

Hint: It would help to identify the feedback and disturbance paths, which elements are in each, and how each is affected by the process designs considered.

13.3. Assume that the process transfer function $G_p(s)$ used in deriving equations (13.9) and (13.10) is unchanged but that the disturbance transfer function was modified to second-order of the form that follows. Derive expressions for the minimum values for the IAE and the maximum deviation of the controlled variable equivalent to equations (13.9) and (13.10).

$$\frac{\mathrm{CV}(s)}{D(s)} = \frac{K_d}{(\tau^2 s^2 + 2\tau\xi s + 1)}$$

13.4. In this chapter the statement is made that the integral mode is particularly effective in reducing the effect of sine disturbances with low frequencies. Evaluate this statement by comparing the closed-loop frequency responses for PI and P-only controllers in the very low-frequency region. Is the P-only controller as effective? Explain your answer.

13.5. Concerning the frequency response equation (13.3):

(*a*) Verify that the equations are correct.

(*b*) Determine the modifications for the second-order disturbance model in question 13.3 being used in place of the first-order model. How would this change affect the general shape of the closed-loop frequency response?

13.6. For the following process control designs, select the proper feedback controller modes and discuss the proper execution periods for digital implementation. (*a*) Figure Q7.6, (*b*) Figure 2.2, and (*c*) Figure Q1.9 (those designs for which feedback control is possible).

13.7. (*a*) A plant with the process configuration of Figure 13.4 is analyzed in this question. Calculate closed-loop frequency responses of the controlled-variable response to a disturbance. The plant transfer functions follow with all time units in minutes, and the controller algorithm is a PI, with tuning to be determined by you. You may use equation (13.3) or use a computer program to perform the complex manipulations.

$$G_p = \frac{2.2e^{-3s}}{1 + 3.4s} \qquad G_d = \frac{1.0}{1 + 2s}$$

(*b*) The process requires a deviation from set point of less than 1.0 for the dominant disturbance, which has a magnitude of 1.5 at a frequency of 0.40 rad/min. Determine whether the PI controller can achieve the performance. If not, how should the disturbance and process feedback transfer functions be changed to satisfy the control objective?

(*c*) How would the answers to parts (*a*) and (*b*) of this question change if the disturbance transfer function $Gd(s)$ had an additional dead time of 3 min?

13.8. (*a*) In your own words, describe why processes with large dead times are difficult to control.

(*b*) Sketch a typical closed-loop frequency response and explain the three major sections of the curve at low, intermediate, and high frequencies. Perform this exercise for both set point and disturbance inputs.

(*c*) As discussed in previous chapters and reiterated here, controller tuning is selected to be somewhat conservative to ensure stability as the process dynamics change. Discuss how this tuning practice influences controller performance.

(*d*) Place each of the following factors in one of two categories, labelled "favorable" and "unfavorable" for control performance: disturbance frequency near critical

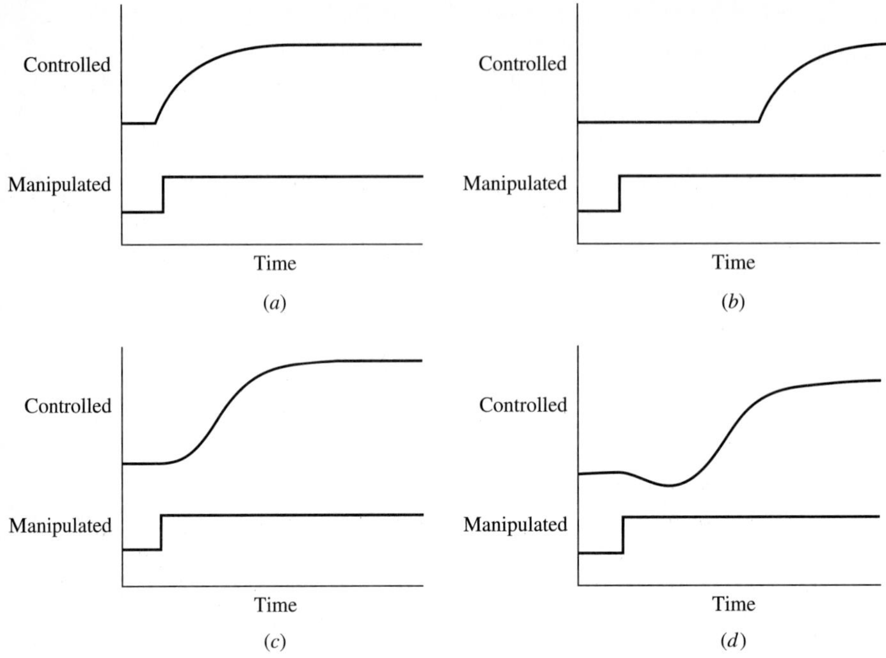

FIGURE Q13.9

frequency; small fraction dead time; large disturbance dead time; large process steady-state gain; ratio of digital execution period to feedback dynamics greater than 0.20; detuning controller gain for robustness; large value of $(\theta + \tau)$.

13.9. Open-loop responses between the manipulated and controlled variables for four potential process designs are given in Figure Q13.9, all having the same scales.

(*a*) Rank the processes for the expected control performance for set point changes.

(*b*) Rank the processes for the expected control performance for disturbance response.

13.10. Based on the model of the feedback process, how would the control performance change for the system in Example 8.5 for each of the following changes, made individually, to the initial steady-state operating conditions? Calculate the modification of tuning in response to the operating condition change and assume that this tuning change has been made.

(*a*) Determine the PI tuning that would give "good" control performance for the initial plant operating conditions in Example 8.5.

(*b*) The flow through the heat exchanger is reduced from 0.085 to 0.0425 m^3/min.

(*c*) The volume in the tank is increased from 2.1 to 3.0 m^3.

(*d*) The temperature set point is changed from 85.4 to 90°C.

13.11. The closed-loop frequency response calculated using equation (13.2) for a process with the structure in Figure 13.4 and with the following process parameters is given in Figure Q13.11. Results are shown for several values of the controller gain, all with the integral time at a value of 6.0. Critically discuss these calculations and select from the three alternatives the value of the controller gain that would give the best control performance.

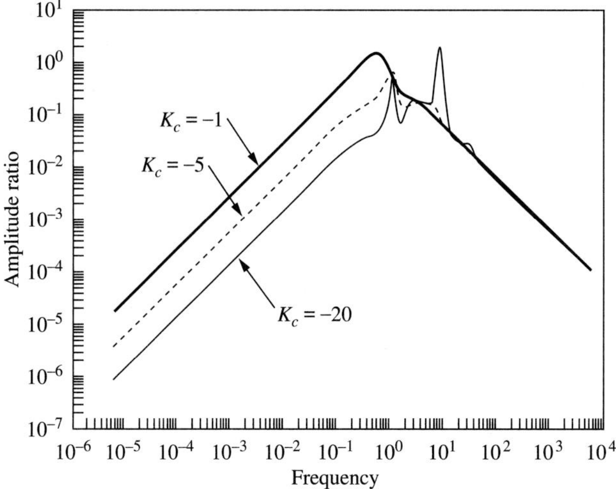

FIGURE Q13.11

$$G_p = \frac{-2.0e^{-2.5s}}{1+4s} \qquad G_d = \frac{1.0}{1+1.8s} \qquad G_c = K_c\left(1 + \frac{1.0}{6s}\right)$$

13.12. (*a*) One rule of thumb for quickly estimating the standard deviation of a sample of process data is that it is equal to $\frac{1}{8}$ of the difference between the maximum and minimum values in the sample. Discuss the basis and validity of this rule of thumb.

(*b*) Apply this rule of thumb to the data in Figures 13.10*a* and *b*.

(*c*) Assume that the goal is to increase the average concentration without exceeding the value of 6.2. Evaluate the performance of the system in Figure 13.10*b* and suggest any changes to the set point that are appropriate.

(*d*) Discuss some of the factors you would consider in selecting "representative" open-loop dynamic data that could be used in estimating feedback control performance for a potential PI controller by the method in Example 13.5.

13.13. Based on the model of the feedback process, answer the following questions for the three-tank system in Examples 7.2 and 9.2 for the situation in which each tank volume is increased from 35 to 105 m^3.

(*a*) Describe the control performance you would expect with the original tuning constants from Example 9.2 applied to the modified process.

(*b*) If necessary, modify the PID controller tuning.

(*c*) Compare the control performance for the original system and the modified system after tuning changes in (*b*). Consider the IAE and maximum deviation for a step inlet concentration disturbance.

13.14. Discuss how the process structure in the following systems would affect the feedback control performance: (*a*) Example 5.3 (overshoot); (*b*) Example 5.5 (recycle); and (*c*) Example 5.8 (underdamped).

13.15. The tradeoff between manipulated- and controlled-variable behaviors has been discussed frequently.

(*a*) Describe the behavior of the manipulated variable for the system in Figures 9.2 and 9.3. On each figure, sketch an approximate plot of the variability of the manipulated variable, showing where the variability is high and low as a function of the variable tuning constant(s). Either of the following measures of the variability can be used.

$$\int_0^\infty \left(\frac{d\text{MV}(t)}{dt}\right)^2 dt \qquad \text{or} \qquad \int_0^\infty \left|\frac{d\text{MV}(t)}{dt}\right| dt$$

(*b*) Recalculate Figure 13.7 with a PID controller and discuss the difference.

13.16. The system in Example 13.10(*a*) evaluated the closed-loop amplitude ratio of the controlled to disturbance variables. For the same system, calculate the amplitude ratio of the measured variable to the disturbance. Recalling that only the measured variable is known to the plant personnel, discuss the differences in the results and their importance in analyzing plant performance.

13.17. The transfer function between the set point and the controlled variable is given in equation (13.4). Apply the following controller design method to arrive at an algorithm other than PID. Assume the input-output response is defined at some good performance (i.e., $\text{CV}(s)/\text{SP}(s) = T(s)$ is specified). Solve for the controller transfer function that would give this performance. Discuss whether this controller can be implemented in analog or digital form.

13.18. Confirm that the dimensionless variables in Figures 13.13*a* and *b* are proper for correlating the performance of various first-order-with-dead-time systems.

13.19. The process design in Example 5.4 with a parallel structure is considered in this question. The concentration at the outlet of the second reactor is to be controlled as in Example 13.8, except that the flow rate of stream A (not the solvent) is to be manipulated.

(*a*) Based on the different dynamics between the manipulated and controlled variables, predict the control performance and whether it would be better than the system in Example 13.8. (*Hint*: The results from end-of-chapter question 5.4 will help in answering this question.)

(*b*) Develop a dynamic simulation for this design, tune the feedback PI controller, and compare the control performance with Example 13.8.

13.20. The process with recycle was analyzed in question 5.14. Determine the value of the recycle for which a feedback PI control system, controlling the outlet composition by adjusting C_{A0}, would give the best performance.

13.21. Chemical reactors were analyzed in question 5.7 for two different reaction kinetics. For both kinetics (answered separately), determine which feedback control system, controlling C_A or C_B by adjusting C_{A0}, would provide the best performance. Base your answer entirely on the feedback dynamics, not the process gain.

PART IV

ENHANCEMENTS TO SINGLE-LOOP PID FEEDBACK CONTROL

As we have seen, single-loop PID feedback control often provides good control performance and always yields zero steady-state offset for steplike inputs. The controller is easy to use, because the PID control algorithm can be applied on nearly all processes without alteration. The performance is very good, considering the little information required for design and tuning. As might be expected, the simplicity of the PID controller, while reducing engineering effort and computer calculations, results in control performance that is not always the best possible. The key advantages and disadvantages of single-loop feedback control are summarized in Table IV.1. The methods in Part IV are designed to partially overcome these disadvantages.

The best approach for improving control performance is to eliminate or reduce the disturbances by improving the operation of upstream processes. The next best solution is to eliminate the difficult factors from the feedback dynamic responses by changing the process design. For example, the dead time could be reduced by relocating sensors or eliminating sample systems by placing the sensor *in situ*. If the

TABLE IV.1
Summary of single-loop feedback control

Advantages	Disadvantages
Achieves zero steady-state offset for all step-like inputs	Process output must be upset before feedback action begins.
Uses only one measurement	Feedback control performance can be poor for some combinations of disturbance frequencies and feedback dynamics.
Algorithm and tuning rules available	Poor feedback can cause instability.
	PID does not provide the best possible control for all processes.

dead time could be reduced sufficiently, much improved control performance could be achieved. However, changing the process design is not always possible or the best economic decision.

We assume here that all reasonable process modifications have been made and that further enhancements are to be achieved through control modifications. To improve the feedback performance, the controller design must be changed in a manner that takes advantage of additional knowledge about the process dynamics or control objectives through one or more of the following steps:

- Use additional measures of process outputs.
- Use additional measurements of the process inputs.
- Use explicit modelling in the control calculation.
- Modify the PID algorithm and tuning to match the control objective.

To achieve enhancements, the engineer requires additional process insight, which is developed though increased engineering analysis and effort. The additional effort can be richly rewarded, because the enhancements can substantially improve control performance, reducing the integral errors and maximum deviations by more than a factor of ten, in some situations. The success of the enhancement depends on the quality of the engineering analysis—that is, the accuracy of the process insight and the application of the design principles. It is important to remember that regardless of the complexity of the enhancements, feedback control from the controlled variable should always be retained (when the sensor exists) so that zero steady-state offset is achieved.

Several control enhancements are presented in this part of the book. They were chosen based on the following criteria:

- *Reinforce principles.* Each of these enhancements partially overcomes one or more of the causes for poor control performance. Thus, we have the opportunity to reconsider these process-related limitations as we learn how an enhancement improves performance. This perspective is important because

TABLE IV.2
Summary of single-loop control enhancements

Enhancement	Key issue	New input measurement	New output measurement	Process modelling	Standard or modified PID	New control algorithm
Cascade	Feedback dynamics		X		X	
Feedforward	Feedback dynamics	X				X
Nonlinear processes	Changing process dynamics	X		X	X	
Inferential	Lack of online sensor		X	X	X	
Level	Different control objective				X	
Predictive control	Complex feedback dynamics			X		X

the enhancements are designed based on sound control theory and are not a collection of "ad hoc tricks."

- *Demonstrate practice.* The enhancements are some of the most frequently applied designs in the process industries and can be implemented in commercial control equipment at low cost.
- *Apply process insight.* The proper use of the enhancements requires sound understanding of dynamics and operating goals of the process. Thus, this part strengthens our understanding of how the process equipment and operating conditions influence designs to achieve specified control performance.

Each enhancement is briefly summarized in Table IV.2; the reader may find it helpful to refer to this table after covering each chapter in this part.

The single-loop enhancements covered in this part are used widely in process control. They become especially important when designing control strategies for complex units with many controlled variables, sensors, and final control elements. Therefore, it is essential that the student master these enhancements before progressing to the more advanced, multivariable design topics.

CHAPTER 14

CASCADE CONTROL

14.1 INTRODUCTION

Cascade control is one of the most successful methods for enhancing single-loop control performance. It can dramatically improve the performance of control strategies, reducing both the maximum deviation and the integral error for disturbance responses. Since the calculations required are simple, cascade control can be implemented with a wide variety of analog and digital equipment. This combination of ease of implementation and potentially large control performance improvement has led to the widespread application of cascade control for many decades. In this chapter, cascade control is fully explained with special emphasis placed on clear guidelines that, when followed, ensure that the cascade method is properly designed and is employed only where appropriate.

As explained in the introduction to this part, single-loop enhancements take advantage of extra information to improve on the performance of the PID feedback control system. Cascade uses an additional measurement of a process variable to assist in the control system. The selection of this extra measurement, which is based on information about the most common disturbances and about the process dynamic responses, is critical to the success of the cascade controller. Therefore, insight into the process operation and dynamics is essential for proper cascade control design.

The basic concepts of cascade control are presented in the next section. Subsequent sections provide concise explanations of the design criteria, performance expectations, tuning methods, and implementation issues. All of the methods and guidelines are presented for continuous systems but are applicable to digital control systems when the execution period is small with respect to the feedback dynamics. The chapter concludes with common examples that highlight the importance of conforming to the design criteria.

14.2 AN EXAMPLE OF CASCADE CONTROL

The best way to introduce cascade control is with reference to a simple process example, which will be the stirred-tank heat exchanger shown in Figure 14.1. The goal is to provide tight control of the exit temperature. The conventional feedback controller, with integral mode, attempts to maintain the exit temperature near its set point in response to all disturbances and ensures zero steady-state offset for steplike disturbances. Suppose that one particularly frequent and large disturbance is the heating oil pressure. When this pressure increases, the initial response of the oil flow and the heat transferred is to increase. Ultimately, the tank exit temperature increases, and the feedback controller reduces the control valve opening to compensate for the increased pressure. While the effect of the disturbance is ultimately compensated by the single-loop strategy, the response is slow, because the exit temperature must be disturbed before the feedback controller can respond.

Cascade control design considers the likely disturbances and tailors the control system to the disturbance(s) that strongly degrades performance. Cascade control uses an additional, "secondary" measured process input variable that has the important characteristic that it indicates the occurrence of the key disturbance. For the stirred-tank heat exchanger, all measured variables are shown in Figure 14.1. The secondary variable is selected to be the heating oil flow, because it responds in a predictable way to the disturbances in the oil pressure, which is not measured in this case. The control objective (tight control of the outlet temperature) and the final element are unchanged.

The manner in which the additional measurement is used is shown in Figure 14.2. The control system employs two feedback controllers, both of which can use the standard PID controller algorithm. The important feature in the cascade structure is the way in which the controllers are connected. The output of the exit temperature controller adjusts the set point of the flow controller in the cascade structure; that is, the secondary controller set point is equal to the primary controller output. Thus, the secondary flow control loop is essentially the manipulated variable for the primary temperature controller. The net feedback effect is the same for single-loop or cascade control; in either case, the heating oil valve is adjusted ultimately by the feed-

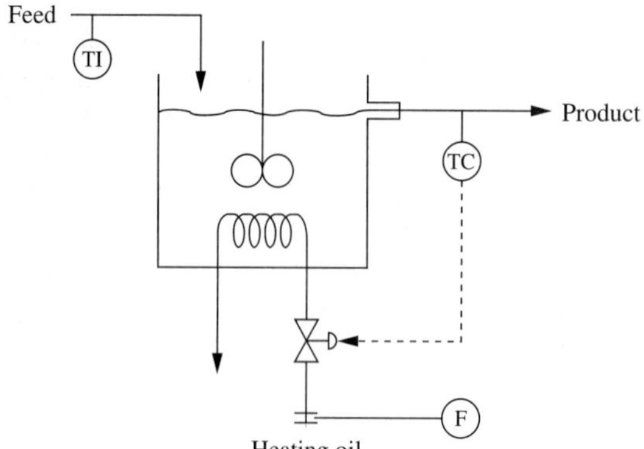

FIGURE 14.1
Stirred-tank heat exchanger with single-loop temperature control.

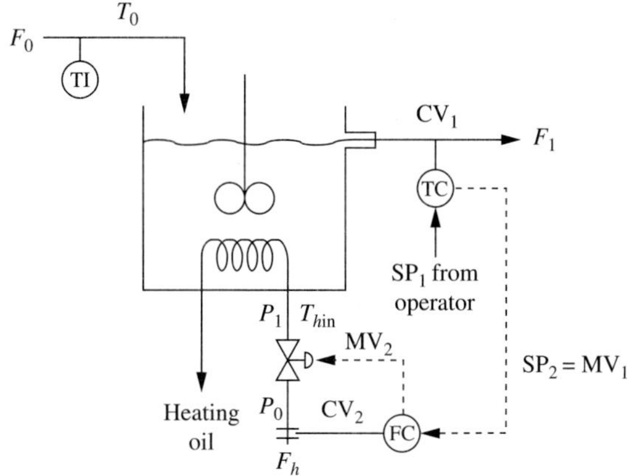

FIGURE 14.2
Stirred-tank heat exchanger with cascade control.

back. Therefore, the ability to control the exit temperature has not been changed with cascade.

As described previously, the single-loop structure makes no correction for the oil pressure disturbance until the tank exit temperature is upset. The cascade structure makes a much faster correction, which provides better control performance. The reason for the better performance can be seen by analyzing the initial response of the cascade system to an oil pressure increase. The valve position is initially constant; therefore, the oil flow increases. The oil flow sensor quickly detects the increased flow. Since the flow controller set point would be unchanged, the controller would respond by closing the valve to return the flow to its desired value. Because the sensor and valve constitute a very fast process, the flow controller can rapidly achieve its desired flow of oil. By responding quickly to the pressure increase and compensating by closing the control valve, the secondary controller corrects for the disturbance before the tank exit temperature is significantly affected by the disturbance. Typical dynamic responses of the single-loop and cascade control systems are given in Figure 14.3*a* and *b* for a decrease in oil pressure.

A few important features of the cascade structure should be emphasized. First, the flow controller is much faster than the temperature controller. The improvement results from the much shorter dead time in the secondary loop than in the original single-loop system; as discussed in Chapter 13, shorter dead times improve single-loop control. If the flow controller were not faster, the cascade design would have no advantage. Second, the temperature controller with an integral mode remains in the design to ensure zero offset for all disturbance sources. The primary controller is essential, because (1) the secondary variable may not totally eliminate the effect of the disturbance, (2) other disturbances that are not affected by the cascade will also occur, and (3) the ability to change the primary set point must be retained. Remember that the secondary variable is selected for one (or a few) common disturbances; in the example, a heat exchanger feed temperature disturbance would affect the tank outlet temperature but does not influence the heating oil measurement. Finally, the judicious selection of the secondary variable has made the improvement possible *without using a model* of the effect of pressure on exit temperature in the control calculation; the only models used were

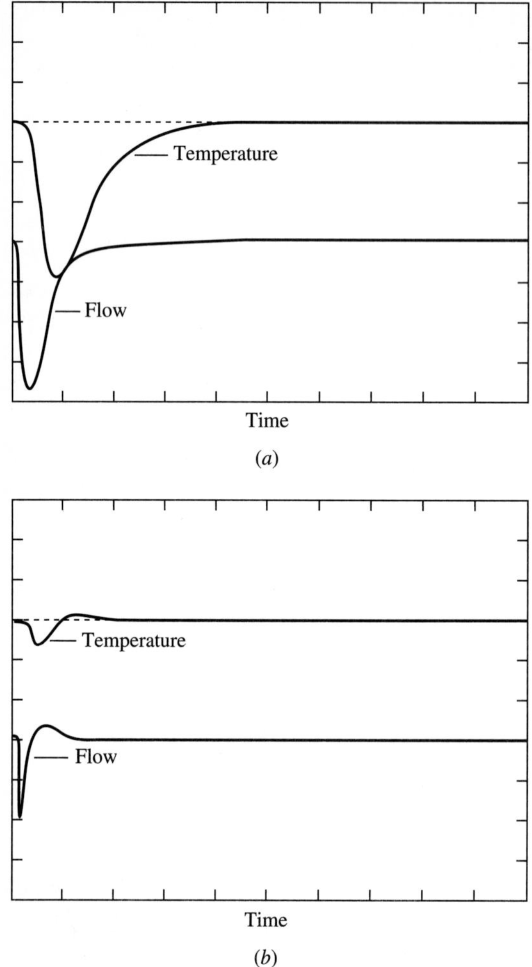

FIGURE 14.3
Dynamic response of stirred-tank heat exchanger to a disturbance in oil pressure: (*a*) with single-loop control; (*b*) with cascade control.

the process models used to tune the two feedback controllers. As a result, cascade control is not strongly sensitive to modelling errors, although large errors could lead to oscillations or instability in one of the feedback controllers.

The two controllers in the cascade are referred to by various names. The three pairs of names in the most commonly used terminology are presented as they would be applied to the stirred-tank heat exchanger:

Temperature	Flow
Primary	Secondary
Outer	Inner
Master	Slave

At first encounter, it may seem improper to use two feedback controllers to achieve one objective; however, the propriety of cascade control can be established by analyzing the degrees of freedom of the system. For the heat exchanger in Figure 14.2, the material and energy balances were derived in Example 3.7 (for cooling) and are repeated here for heating.

$$F_0 = F_1 \tag{14.1}$$

$$V\rho C_p \frac{dT}{dt} = F_1 \rho C_p (T_0 - T) - \frac{aF_h^{b+1}}{F_h + \dfrac{aF_h^b}{2\rho_h C_{ph}}}(T - T_{hin}) \tag{14.2}$$

The heating flow is related to the valve position (v) according to the following general equation:

$$F_h = C_v v \sqrt{\frac{P_0 - P_1}{\rho_h}} \tag{14.3}$$

where we assume that the pressures and the coefficient C_v are constant, although they can be variables (see Chapter 16). The final equations are the two cascade controllers:

$$F_{hsp} = K_{c1}\left((T_{sp} - T) + \frac{1}{T_{I1}}\int_0^t (T_{sp} - T)\,dt'\right) + I_{Fh} \tag{14.4}$$

$$v = K_{c2}\left((F_{hsp} - F_h) + \frac{1}{T_{I2}}\int_0^t (F_{hsp} - F_h)\,dt'\right) + I_v \tag{14.5}$$

Variables: $F_1, F_h, T, (F_h)_{sp}, v$ DOF $= 5 - 5 = 0$

External variables: $F_0, T_0, T_{hin}, T_{sp}$

Parameters: $V, \rho, C_p, \rho_h, C_{ph}, a, b, C_v, P_0, P_1, K_{c1}, T_{I1}, K_{c2}, T_{I2}, I_{Fh}, I_v$

The number of degrees of freedom is equal to the number of variables minus the number of equations; thus, the system is exactly specified when the temperature controller set point has been defined. Note that the cascade secondary controller was placed between the primary controller output and the valve, which added one variable (F_{hsp}) and one equation (14.5).

14.3 CASCADE DESIGN CRITERIA

The principles of cascade control have been introduced with respect to the example stirred-tank heater. In Table 14.1, the design criteria are summarized in a concise form so that they can be applied in general. Adherence to these criteria ensures that cascade control is designed properly and used only where appropriate. The first two items address the selection of cascade control. Naturally, only when single-loop control does not provide acceptable control performance is an enhancement such as cascade control necessary. As described in Chapter 13, single-loop control provides good performance when the dynamics are fast, the fraction dead time is small, and disturbances are small and slow. Also, the second criterion requires an acceptable measured secondary variable to be available or added at reasonable cost.

TABLE 14.1
Cascade control design criteria

Cascade control is desired when
1. Single-loop control does not provide satisfactory control performance.
2. A measured secondary variable is available.
A secondary variable must satisfy the following criteria:
1. The secondary variable must indicate the occurrence of an important disturbance.
2. There must be a causal relationship between the manipulated and secondary variables.
3. The secondary variable dynamics must be faster than the primary variable dynamics.

A potential secondary variable must satisfy three criteria. First, it must indicate the occurrence of an important disturbance; that is, the secondary variable must respond in a predictable manner every time the disturbance occurs. Naturally, the disturbance must be important (i.e., have a significant effect on the controlled variable and occur frequently), or there would be no reason to attenuate its effect. Second, the secondary variable must be influenced by the manipulated variable. This causal relationship is required so that a secondary feedback control loop functions properly. Finally, the dynamics between the final element and the secondary must be much faster than the dynamics between the secondary variable and the primary controlled variable. The secondary must be relatively quick so that it can attenuate a disturbance before the disturbance affects the primary controlled variable. A general guideline is that the secondary should be three times as fast as the primary. This could be roughly interpreted as the secondary reaching its steady state in one-third the time of the primary after an open-loop step change in the manipulated variable. A more proper comparison is the critical frequency of each loop; cascade is recommended when the critical frequency of the secondary is at least three times that of the primary. Using critical frequencies accounts for differences in the fraction dead time as well as the speed of response.

A cascade control strategy combines two feedback controllers, with the primary controller's output serving as the secondary controller's set point. The design should conform to the design criteria in Table 14.1, which provide a simple, step-by-step procedure for selection.

14.4 CASCADE PERFORMANCE

In the introduction to this chapter, cascade was described as simple and effective. The foregoing material has demonstrated how simple a cascade strategy is to design. In this section, its effectiveness is shown by calculating its performance using simulation and frequency response for a few cascade systems and comparing with single-loop control performance on the same systems. Because the number of

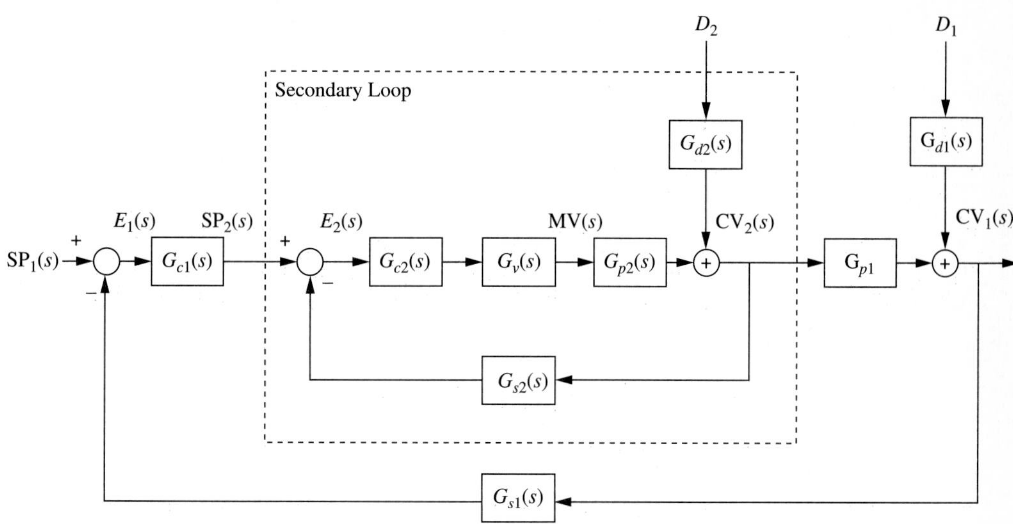

FIGURE 14.4
Block diagram of cascade control.

parameters in a cascade system—primary dynamics, secondary dynamics, disturbance dynamics—make general performance correlations intractable, this section presents sample results for typical process dynamics. The general trends showed by these results should be expected for most realistic processes.

The block diagram in Figure 14.4 presents the structure of a cascade control system, which summarizes the flow of information and can be used to evaluate important properties such as stability and frequency response. Transfer functions can be derived from this block diagram for the relationships between the primary controlled variable $CV_1(s)$ and the secondary disturbance $D_2(s)$, the primary disturbance $D_1(s)$, and the primary set point $SP_1(s)$, as follows:

$$\frac{CV_1(s)}{D_2(s)} = \frac{G_{d2}G_{p1}(s)}{1 + G_{c2}(s)G_v(s)G_{p2}(s)G_{s2}(s) + G_{c1}(s)G_{c2}(s)G_v(s)G_{p1}(s)G_{p2}(s)G_{s1}(s)} \tag{14.6}$$

$$\frac{CV_1(s)}{D_1(s)} = \frac{G_{d1}(s)[1 + G_{c2}(s)G_v(s)G_{p2}(s)G_{s2}(s)]}{1 + G_{c2}(s)G_v(s)G_{p2}(s)G_{s2}(s) + G_{c1}(s)G_{c2}(s)G_v(s)G_{p1}(s)G_{p2}(s)G_{s1}(s)} \tag{14.7}$$

$$\frac{CV_1(s)}{SP_1(s)} = \frac{G_{c1}(s)G_{c2}(s)G_v(s)G_{p2}(s)G_{p1}(s)}{1 + G_{c2}(s)G_v(s)G_{p2}(s)G_{s2}(s) + G_{c1}(s)G_{c2}(s)G_v(s)G_{p1}(s)G_{p2}(s)G_{s1}(s)} \tag{14.8}$$

As apparent from the introductory example, a key factor in cascade control is the relative dynamic responses of the secondary and primary processes. Since the main reason for cascade is secondary disturbances, the studies in this section evaluate the responses to secondary disturbances: step, sine, and stochastic. For these simulation studies, the models for the sensors $G_{si}(s)$ and valve $G_v(s)$ were taken to be unity, and the dynamics of the plant models and disturbance model are given below, with all times scaled so that the process models have a common value of the fraction dead time. The relative dynamics between the secondary and primary are defined by a variable η, which will be allowed to vary in the following models:

Cascade System

	Process	Control
Secondary:	$G_{p2}(s) = \dfrac{1.0e^{(-0.3/\eta)s}}{1 + (0.7/\eta)s}$	PI controller tuned accordingly
Primary:	$G_{p1} = \dfrac{1.0e^{(-0.3)s}}{1 + 0.7s}$	PI controller tuned accordingly

Single-Loop System

Process	Control
$G_p = \dfrac{1.0e^{-(0.3+0.3/\eta)s}}{(1 + (0.7/\eta)s)(1 + 0.7s)}$	PI controller tuned accordingly

For All Cases

$$\text{Disturbance:} \quad G_{d2}(s) = \frac{1.0}{1 + (0.7/\eta)s}$$

$$\text{Instrumentation:} \quad G_{s1}(s) = G_{s2}(s) = G_v(s) = 1.0$$

Response to Step Disturbance in D_2

In the first cascade system studied, a step disturbance was introduced in the secondary loop, and no noise was added to the measurements, so that only the effect of the cascade could be determined. Both primary and secondary controllers used PI algorithms with conventional tuning. The control performance measure is the integral of the absolute value of the error, IAE, of the primary controlled variable; it is reported as a ratio of cascade to single-loop IAE to characterize the improvement achieved through cascade. The resulting control performance is shown in Figure 14.5 as a function of the relative secondary/primary process dynamics, η. As expected, the performance is very good when the secondary is fast. For example, the integral error is reduced by 95% or more for cascade versus single-loop control when the secondary is more than 20 times faster. This large ratio in primary to secondary dynamics is typical when the secondary is a fast loop such as a flow or pressure controller, which is often the case. However, many cascade control systems cannot achieve such a remarkable improvement, because the secondary loop is not so fast, and some potential secondary loop dynamics are so slow as to prohibit cascade control.

Sample dynamic responses from cascade control are shown in Figures 14.6*a* and *b* for a step disturbance in the secondary loop, $D_2(s) = -1/s$ at time=10. The case with a very fast secondary demonstrates how quickly the secondary controller

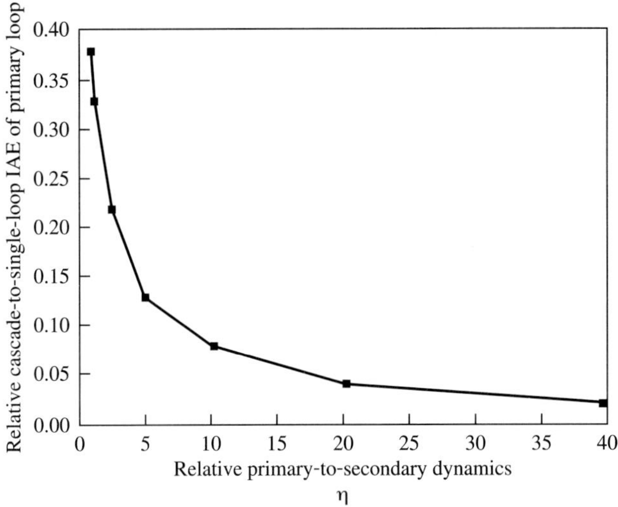

FIGURE 14.5
Relative performance (IAE_{casc}/IAE_{sl}) of cascade and single-loop control for a step disturbance in the secondary loop.

attenuates the effect of the disturbance. The case of a much slower secondary shows much poorer performance, especially the highly oscillatory response. These oscillations, which are more troublesome with the continuous disturbances experienced in industrial plants, usually prohibit the use of cascades with PID controllers when η is less than about 3, although Figure 14.5 shows that some improvement in performance may be possible. (See Chapter 19 for the use of predictive controllers in cascade control, which can increase the region of acceptable cascade performance.)

Response to Stochastic Disturbance in D_2

In a second study, the same process was considered with a stochastic disturbance, which is more representative of the responses encountered in a continuously operating plant. The block diagram and models were the same as for the step disturbance, and the disturbance enters in the secondary loop. Again, the system was simulated with single-loop PI control and cascade control tuning. The control performance in Figure 14.7 is expressed as standard deviation from the set point

$$\sigma_{sp} = \sqrt{\sum_{i=0}^{n} \frac{(SP_i - CV_i)^2}{n}}$$

The standard deviation of the primary variable is plotted as a function of the relative secondary/primary dynamics, η. Again, the faster the secondary, the better the performance of the cascade. Dynamic responses for this system are given for $\eta = 10$, in Figure 14.8*a* through *c* for open-loop, single-loop, and cascade

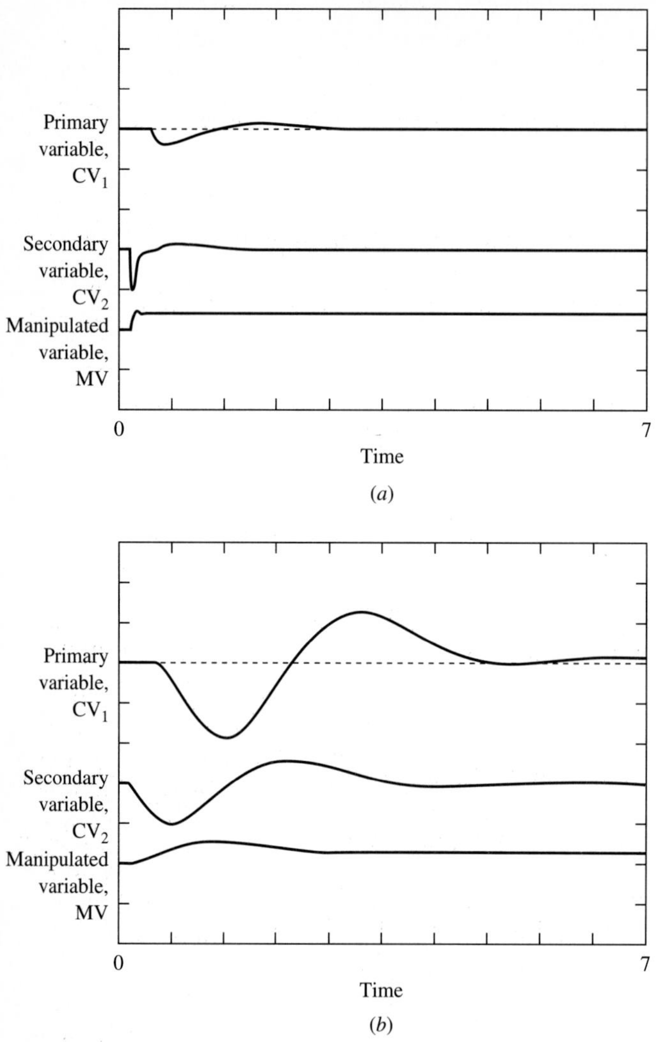

FIGURE 14.6
Performance of cascade control for a disturbance in the secondary loop (*a*) with $\eta = 10$; (*b*) with $\eta = 1.0$. (Scales for the plots: One tick (10%) is 0.15 for primary, 0.50 for secondary, and 2.5 for manipulated variable.)

control, respectively. It is important to recognize that the results in Figure 14.7 are limited to the *specific process and disturbance studied;* other disturbances, with different frequency components, would give different results, although the general trend would be unchanged.

Response to Sine Disturbance in D_2

The third cascade study investigates the frequency response, which evaluates the control performance of a cascade control system for a range of disturbance frequencies.

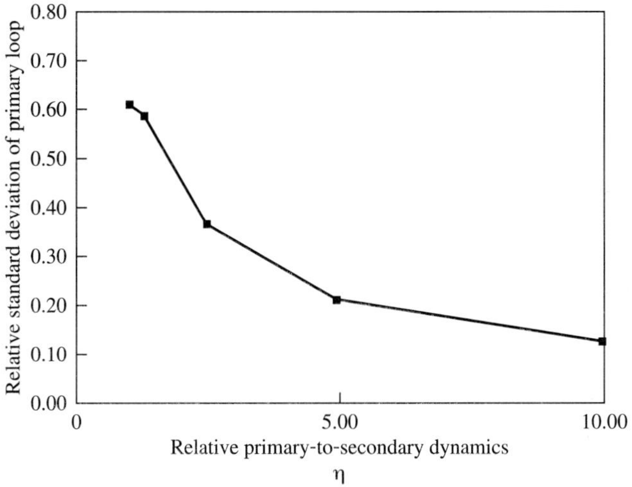

FIGURE 14.7
Relative control performance of single-loop and cascade ($\sigma_{casc}/\sigma_{sl}$) for a stochastic disturbance in the secondary loop.

As described in Chapter 13, the amplitude ratio gives the magnitude of the variation in the controlled variable for a unit sine input; thus, the smaller the amplitude ratio for a disturbance response, the better the control performance. The amplitude ratios for the cascade control system were calculated for a range of frequencies using equation (14.6). Because of the complexity of the algebra, the amplitude ratios were evaluated using a computer program similar to the one in Table 13.1, and the results are plotted in Figure 14.9.

The smaller amplitude ratio for cascade clearly demonstrates the advantage of cascade control, especially when the secondary process is much faster than the

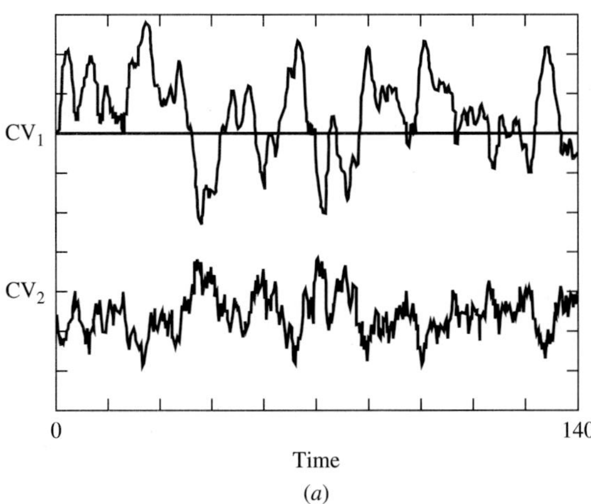

FIGURE 14.8
Dynamic responses for stochastic secondary disturbance with η = 10: (*a*) open-loop process.

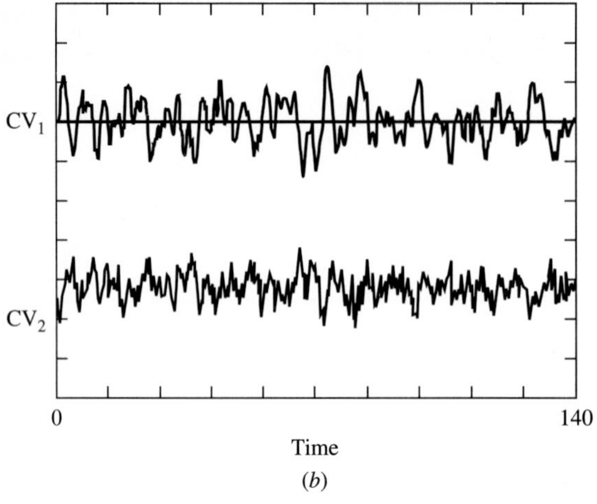

(b)

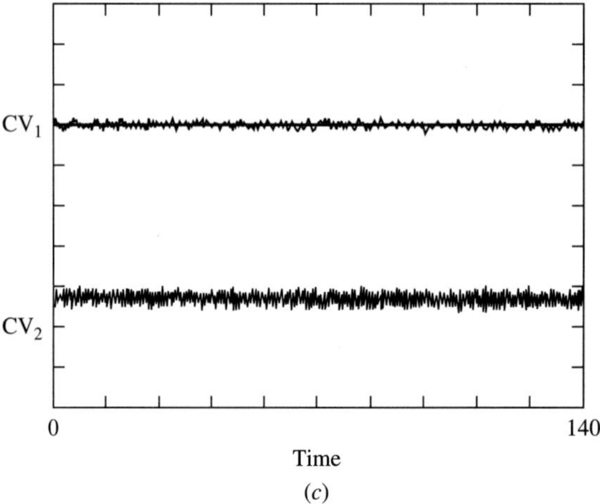

(c)

FIGURE 14.8 ***(continued)***
Dynamic responses for stochastic secondary disturbance with $\eta = 10$: (*b*) single-loop control; (*c*) cascade control.

primary (here, $\eta = 10$). The cascade system is very effective for slower disturbance frequencies. Both systems have little deviation for very fast disturbances, because the process attenuates these disturbances. Also, the effect of the resonant frequency, which was discussed in Chapter 13, is attenuated but not eliminated by the cascade system.

Finally, the performance of a cascade control strategy must be evaluated for circumstances for which this enhancement was not specifically designed—that is, primary disturbances that do not directly affect the secondary variable and changes to the primary set point. By analyzing the cascade block diagram, it is apparent that the primary controller can respond to other types of disturbance in the cascade design; the only difference is that it manipulates the secondary set point rather than the valve directly. One would expect that the responses to unmeasured disturbances and set

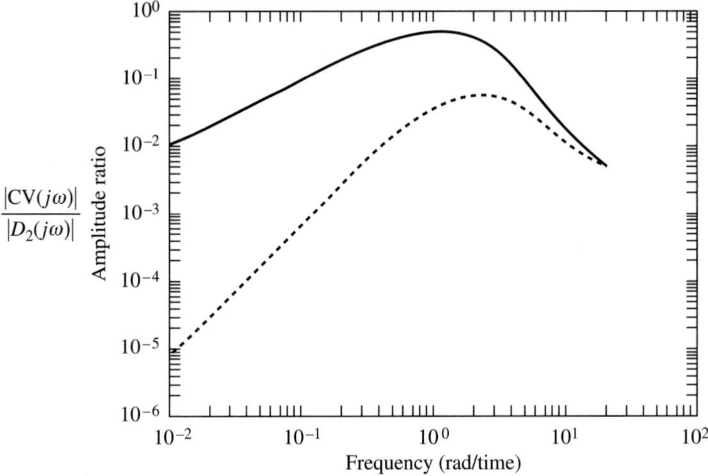

FIGURE 14.9
Closed-loop frequency responses for single-loop (solid curve) and cascade (dotted curve) control with $\eta = 10$.

points are not substantially changed. This is the case, with cascade providing slightly better performance because it increases the critical frequency of the secondary loop (Krishnaswamy et al., 1990). In conclusion,

> Cascade control can substantially improve control performance for disturbances entering the secondary loop and is recommended for use when the secondary loop is much faster than the primary loop.

14.5 CONTROLLER ALGORITHM AND TUNING

Cascade control can use the standard feedback control PID algorithm; naturally, the correct modes must be selected for each controller. The secondary must have the proportional mode, but it does not require the integral mode, because the overall control objective is to maintain the *primary* variable at its set point. However, integral mode is often used in the secondary, for two reasons. First, since a proportional-only controller results in offset, the secondary must have an integral mode if it is to attenuate the effect of a disturbance completely, preventing the disturbance from propagating to the primary. Second, the cascade is often operated in a partial manner with the primary controller not in operation, for example, when the primary sensor is not functioning or is being calibrated. A negative side of including integral mode in the secondary controller is that it tends to induce oscillatory behavior in the cascade system, but the result is not significant when the secondary is much faster than the primary. Studies have demonstrated the effectiveness of the integral mode in the secondary loop (Krishnaswamy et al., 1992). The secondary may have derivative mode if required, but the fast secondary loop almost never has a large enough fraction dead time to justify a derivative mode.

The modes of the primary controller are selected as for any feedback PID controller. It is again emphasized that the integral mode is essential for zero offset of the primary variable.

The cascade strategy is tuned in a sequential manner. The secondary controller is tuned first, because the secondary affects the open-loop dynamics of the primary, $CV_1(s)/SP_2(s)$. During the first identification experiment (e.g., process reaction curve), the primary controller is not in operation (i.e., the primary controller is in manual or the cascade is "open"), which breaks the connection between the primary and secondary controllers. The secondary is tuned in the conventional manner as described in Chapter 9. This involves a plant experiment, initial tuning calculation, and fine tuning based on a closed-loop dynamic response.

When the secondary has been satisfactorily tuned, the primary can be tuned. The initial plant experiment perturbs the variable that the primary controller adjusts; in this case, the *secondary set point* is perturbed, such as in a step for the process reaction curve. The calculation of the initial tuning constants and the fine tuning follow the conventional procedures. Naturally, the secondary must be tuned satisfactorily before the primary can be tuned.

> Tuning a cascade control system involves two steps; first the secondary controller is tuned; then, the primary controller is tuned. Conventional initial tuning guidelines and fine-tuning heuristics apply.

14.6 IMPLEMENTATION ISSUES

When properly displayed for the operator, cascade control is very easy to understand and to monitor. Since it uses standard PID control algorithms, the operator displays do not have to be altered substantially. The secondary controller requires one additional feature: a new status termed "cascade" in addition to automatic and manual. When the status switch is in the cascade position (cascade closed), the secondary set point is connected to the primary controller output; in this situation the operator cannot adjust the secondary set point. When the status switch is in the automatic or manual positions (cascade open), the secondary set point is provided by the operator; in this situation the cascade is not functional.

Cascade control is shown in a very straightforward way in engineering drawings. Basically, each controller is drawn using the same symbols as a single-loop controller, with the difference that the primary controller output is directed to the secondary controller as shown in Figure 14.2. Often, the signal from the primary controller output is annotated with "reset" or "SP" to indicate that it is adjusting or resetting the secondary set point.

The calculations required for cascade control, basically a PID control algorithm, are very simple and can be executed by any commercial analog or digital control system. Two special features contribute to the success of cascade. The first is anti–reset windup. The potential exists for any controller in a cascade to experience integral windup due to a limitation in the control loop. Analysis for the secondary is the same as for a single-loop design; however, reset windup can occur for one of several reasons in the primary controller. The primary controller output can fail to move the valve because of limits on (1) the secondary set point, (2) the secondary controller output, or (3) the valve (fully opened or closed). Thus, the potential for reaching limits and encountering reset windup, along with the need for anti–reset windup, is *much greater in cascade designs.* Standard anti–reset windup methods described in Chapter 12 provide satisfactory anti–reset windup protection.

The second feature is "bumpless" initialization. Note that changing the secondary status switch to and from the cascade position could immediately change the value of the secondary set point, which is not desired. The desired approach is to recalculate the primary controller output to be equal to the secondary set point on initialization. Many commercial controllers include calculations to ensure that the secondary set point is not immediately changed (bumplessly transferred) when the secondary mode switch is changed.

Digital control equipment can use the standard forms of the PID algorithm presented in Chapter 11 for cascade control. In addition to the execution period of each controller, the scheduling of the primary and secondary influences cascade control performance. To reduce delays due to control processing, the secondary should be scheduled to execute immediately after the primary. Naturally, it makes no sense to execute the primary controller at a higher frequency (i.e., with a shorter period) than the secondary, because the primary can affect the process (move the valve) only when the secondary is executed.

The cascade control system uses more control equipment—two sensors and controllers—than the equivalent single-loop system. Since the cascade requires all of this equipment to function properly, its reliability can be expected to be lower than the equivalent single-loop system, although the slightly lower reliability is not

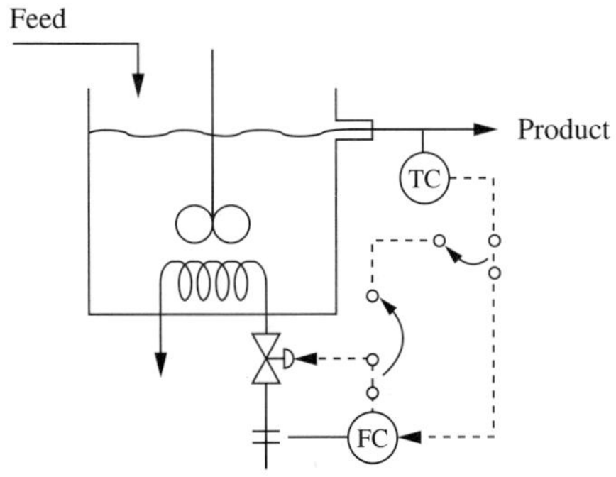

FIGURE 14.10
Control design with a switch to bypass the cascade and convert to single-loop control.

usually a deterrent to the use of cascade. If feedback control must be maintained when the secondary sensor or controller is not functioning, the flexibility to bypass the secondary and have the primary output directly to the valve can be included in the design. This option is shown in Figure 14.10, where the positions of both switches are coordinated.

Since the cascade involves more equipment, it costs slightly more than the single-loop system. The increased costs include a field sensor and transmission to the control house (if the variable were not already available for monitoring purposes), a controller (whose cost may be essentially zero if a digital system with spare capacity is used), and costs for installation and documentation. These costs are not usually significant compared to the benefits achieved through a properly designed cascade control strategy.

> Cascade control, where applicable, provides a simple method for substantial improvements in control performance. The additional costs and slightly lower reliability are not normally deterrents to implementing cascade control.

14.7 FURTHER CASCADE EXAMPLES

The concept of cascade control is consolidated and a few new features are presented through further examples in this section.

Example 14.1 Packed-bed reactor. The first example is the packed-bed reactor shown in Figure 14.11. The goal is to tightly control the exit concentration measured by AC-1. Suppose that the single-loop controller does not provide adequate control performance and that the most significant disturbance is the heating medium temperature, T2. The goal is to design a cascade control strategy for this process using the sensors and manipulated variables given. The reader is encouraged to design a cascade control system before reading further.

Since we are dealing with a cascade control strategy, the key decision is the selection of the secondary variable. Therefore, the first step is to evaluate the potential measured variables using the design criteria in Table 14.1; the results of this evaluation are summarized in Table 14.2, with Y(N) indicating that the item is (is not) satisfied. Since *all of the criteria must be satisfied* for a variable to be used as a secondary, only the reactor inlet temperature, T3, is a satisfactory secondary variable.

The resulting cascade control strategy is shown in Figure 14.12. Given the cascade design, an interesting and important question is, "How well does it respond to other disturbances for which it was not specifically designed?" Several disturbances are discussed qualitatively in the following paragraphs.

Feed temperature, T1. A change in the feed temperature affects the outlet concentration through its influence on the reactor inlet temperature, T3. Therefore, the cascade controller is effective in attenuating the feed temperature disturbance.

Heating oil pressure (not measured). A change in the oil pressure influences the oil flow and, therefore, the heat transferred. As a result the reactor inlet temperature, T3,

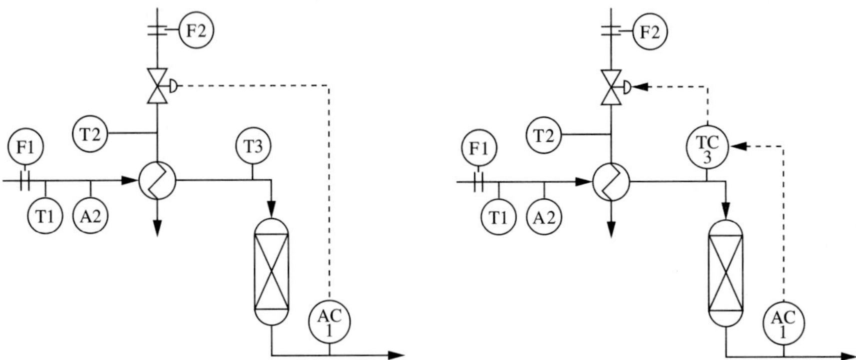

FIGURE 14.11
Single-loop packed-bed reactor control.

FIGURE 14.12
Cascade packed-bed reactor control.

is affected. Again, the cascade controller is effective in attenuating the oil pressure disturbance.

Feed flow rate, F1. A change in the feed flow rate influences the reactor outlet concentration in two ways: it changes the inlet temperature T3, and it changes the residence time in the reactor. The cascade controller is effective in attenuating the effect of the disturbance on T3 but is not effective in compensating for the residence time change. The residence time effect must be compensated by the primary controller, AC-1.

Feed composition, A2. A change in the feed composition clearly changes the reactor outlet concentration. The cascade has no effect on the feed composition disturbance, because the composition does not influence T3. Therefore, this disturbance must be totally compensated by the primary feedback controller, AC-1.

Conclusions and Tuning.

A single cascade control system can be effective in compensating for the effects of several disturbances, and given several possible secondary variables, the one that attenuates the most important disturbances is the best choice.

TABLE 14.2
Evaluation of potential secondary variables

Criterion	A2	F1	F2	T1	T2	T3
1. Single-loop control is not satisfactory	Y	Y	Y	Y	Y	Y
2. Variable is measured	Y	Y	Y	Y	Y	Y
3. Indicates a key disturbance	N	N	N	N	Y	Y
4. Influenced by MV	N	N	Y	N	N	Y
5. Secondary dynamics faster	N/A	N/A	Y	N/A	N/A	Y

In some cases, the attenuation is complete (at least in the steady-state sense); in other cases, the attenuation is partial. Thus, a well-designed cascade strategy can produce a major improvement in the control system performance. This example is completed by describing the tuning procedure.

1. Both controllers are placed in the manual mode, and a process reaction curve experiment is performed to obtain a model for tuning the secondary controller. The model parameters and tuning based on Figure 9.9*a* and *b* are

Model relating valve to T_3	**T_3 controller**
$K_{p2} = 0.57°C/\%$	$K_{c2} = 2.4\%/°C$
$\theta_2 = 8$ s	$T_{I2} = 23$ s
$\tau_2 = 20$ s	

2. The tuning constants are entered into the secondary controller, which is fine-tuned by placing it in automatic and entering small set point changes.
3. A process reaction curve experiment is performed to obtain the model for tuning the primary controller. The model and parameters are

Model relating T_3 set point to A_1	**A_1 controller**
$K_{p1} = -0.19$ mole/m^3/°C	$K_{c1} = -7.9$ °C/mole/m^3
$\theta_1 = 20$ s	(changed to -3.7 in fine-tuning)
$\tau_1 = 50$ s	$T_{I1} = 54$ s (changed to 70 in fine-tuning)

4. The tuning constants are entered into the primary controller, and the controller is fine-tuned by placing it in automatic and entering small set point changes. Note that the primary loop was somewhat oscillatory, so that the primary controller gain and time constant were modified as noted in the foregoing list.
5. The response to a disturbance is observed and further fine-tuning is applied to improve the response, if necessary. A dynamic response to a disturbance in the secondary loop is shown in Figure 14.13: $D_2(s) = -1.8/s$; $G_{d2}(s) = 1/(1 + 20s)$.

Finally, we note that the secondary variable is sensed at the outlet of the heat exchanger. This contributes to its effectiveness, because it can sense the influence of many inlet temperature and flow disturbances. However, the heat exchanger dynamics make the secondary dynamic response somewhat slow. One improvement in the design is to add another level of cascade to compensate for the oil pressure disturbance, which can be sensed by the flow sensor F2. The three-level cascade is shown in Figure 14.14. Industrial designs with three to four cascade levels (and sometimes more) are not unusual and function well.

There is no theoretical or practical limit to the number of cascade levels used as long as each level conforms to the design criteria in Table 14.1.

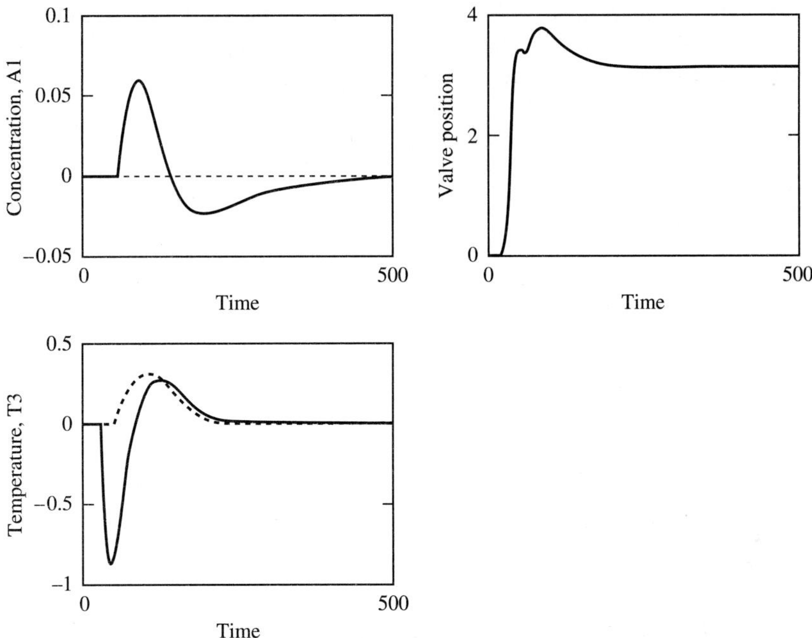

FIGURE 14.13
Dynamic response of the reactor cascade controller to a disturbance in the heating medium temperature in Example 14.1.

Cascade control can be applied to a variety of processes. A few more examples are presented briefly to demonstrate the diversity of the cascade approach. In each case, an analysis similar to the method shown in Table 14.2 was performed to design the cascade strategy.

Example 14.2 Fired heater. Another typical cascade design is given in Figure 14.15 for furnace control. A single-loop temperature controller would adjust the fuel valve directly, making the fuel flow subject to pressure disturbances. A cascade control strategy is possible that satisfies all of the design criteria. In the cascade, the outlet temperature

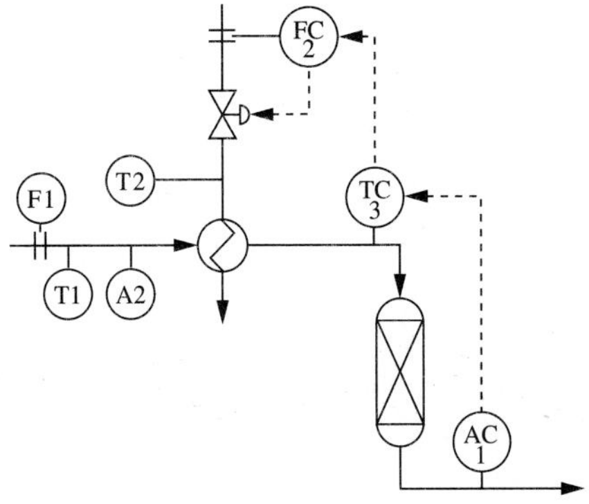

FIGURE 14.14
Three-level cascade control for packed-bed reactor.

of the fluid in the coil is controlled tightly by adjusting the fuel flow controller set point, which adjusts the valve position. An additional advantage of the cascade becomes apparent when considering the performance of many real control valves; the valve does not always move exactly the amount directed by the controller, because friction occasionally causes sticking, which degrades control performance. The cascade design with a flow controller as its fast secondary corrects quickly for both fuel pressure disturbances and the effects of a sticking valve and substantially improves control performance over the single-loop strategy.

Example 14.3 Valve positioner. Cascade control principles are used to enhance the performance of control valves when fast secondary variables cannot be included in the design. Because of a difference between static and dynamic friction, valves often stick and do not exactly achieve the percent stem position demanded by the controller output. The result is that the valve may remain stationary and then "jump" to a value beyond necessary to bring the controlled variable to its set point. A standard control valve can have a dead zone of up to 3% (Buckley, 1970), which can lead to poor control and cycling in many control systems. If the valve is being adjusted by a fast control loop (and the process is not sensitive to high-frequency cycling), no corrective action may be necessary; however, valve sticking can lead to severe control performance degradation.

The effects of valve sticking can be reduced by including a cascade controller called a *valve positioner,* which is included as part of the valve equipment as shown in Figure 14.16. The primary controller, which can be controlling any measured process variable, sends a signal, which is interpreted as the desired valve stem position (0–100%), to the valve positioner. The positioner senses the *actual* valve stem position and adjusts the air pressure until the desired stem position is (nearly) achieved. Since a valve positioner uses a proportional-only algorithm, it does not give perfect compensation for sticking, but the fast dynamics allow a very high controller gain, which reduces the dead zone to about one-tenth of that experienced without a positioner. It is worth noting that this improvement is achieved with minimal investment.

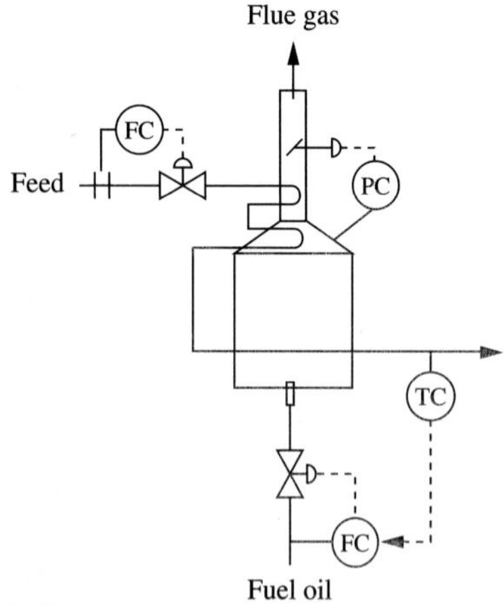

FIGURE 14.15
Cascade control design for outlet temperature.

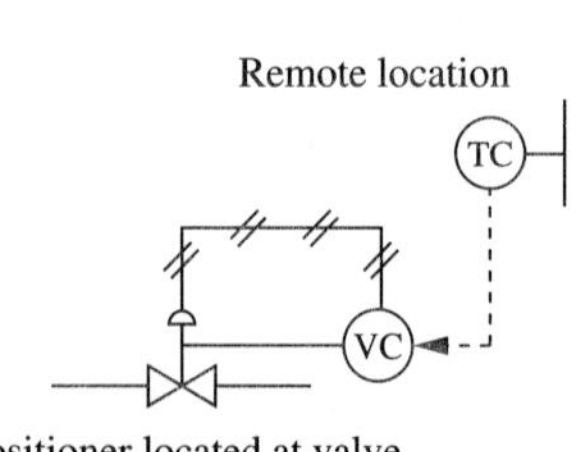

FIGURE 14.16
Schematic of a valve positioner.

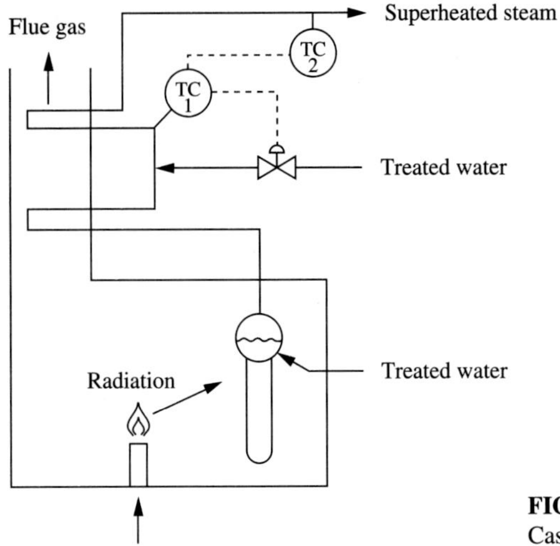

FIGURE 14.17
Cascade control design for boiler superheated steam temperature control.

Other advantages are provided by valve positioners, such as faster valve dynamics and overcoming large pressure drops. They are also used when split range control (see Chapter 22) or changes in the valve characteristic (see Chapter 16) are required. There is no consensus concerning the application of positioners on fast loops; some practitioners recommend them on essentially all control valves, whereas others recommend them only on slow loops.

Example 14.4 Steam superheater. Industrial processes consume large quantities of steam for heating, and machinery consumers require superheated steam for power. As shown in Figure 14.17, steam is generated by vaporizing water in a boiler where the heat transfer is by radiation. The saturated steam temperature is then raised further by convective heat exchange with the hot combustion flue gases. The final steam temperature is controlled by injecting water in the steam. The primary temperature controller could adjust the spray water valve directly, but the control performance would not be good, because of the long dynamic response to disturbances in water flow and heat transfer. The control performance is good for the cascade control with a secondary that responds quickly to both types of disturbances.

Example 14.5 Heater exchanger. A process stream can be cooled by exchanging heat with a refrigerant, which is vaporized in the exchanger. An example is shown schematically in Figure 14.18, which shows that the rate of heat transfer can be controlled by adjusting the heat transfer area (i.e., the liquid refrigerant level in the exchanger). This operating policy is implemented by the cascade control strategy in the figure, where the secondary controller responds quickly to disturbances in liquid flow resulting from pressure variations. An additional advantage is that the level controller maintains the liquid level within acceptable limits; in contrast, a single-loop temperature controller directly adjusting the valve might cause liquid refrigerant to carry over and damage downstream equipment.

Example 14.6 Pressure control. Normally, pressure control involves fast process responses and does not require cascade control. However, processes sometimes have

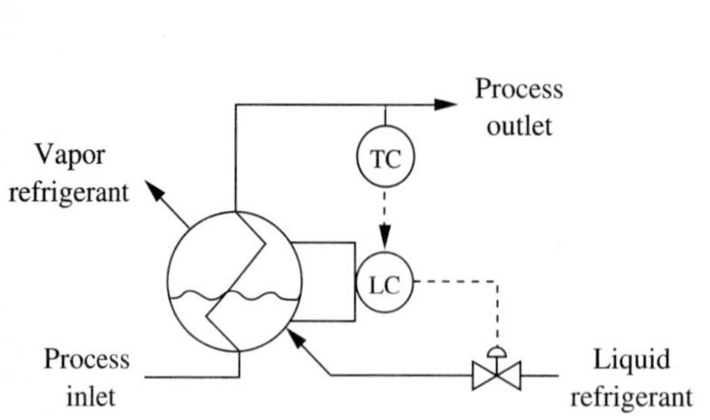

FIGURE 14.18
Cascade control design for temperature control.

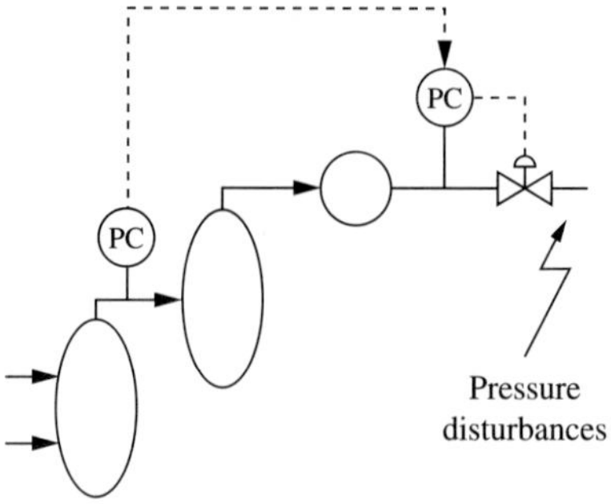

FIGURE 14.19
Cascade control design for pressure control.

large, integrated systems with a single valve regulating the pressure. An example is shown in Figure 14.19, where the most important pressure is at the initial unit, and the pressure control valve is located far downstream. In this case, pressure disturbances near the control valve can cause a relatively large, prolonged disturbance in the initial unit's pressure. The cascade strategy shown in the figure rapidly senses and corrects for downstream disturbances before they upset the integrated upstream unit.

Example 14.7 Jacketed CSTR. An often-noted example of cascade control is the jacketed continuous-flow stirred-tank reactor shown in Figure 14.20. The dynamics related to the thermal capacitance of the jacket fluid and metal could lead to poor control performance with single-loop control from the reactor temperature directly to the valve that controls the inlet temperature of the jacket fluid. A cascade controller uses a secondary variable to sense and quickly correct a disturbance in the jacket fluid inlet temperature.

14.8 CASCADE CONTROL INTERPRETED AS DISTRIBUTED DECISION MAKING

Cascade control is essentially a way to delegate decision making to the lowest level possible. In the packed bed reactor (Figure 14.12), the outlet analyzer AC-1 deter-

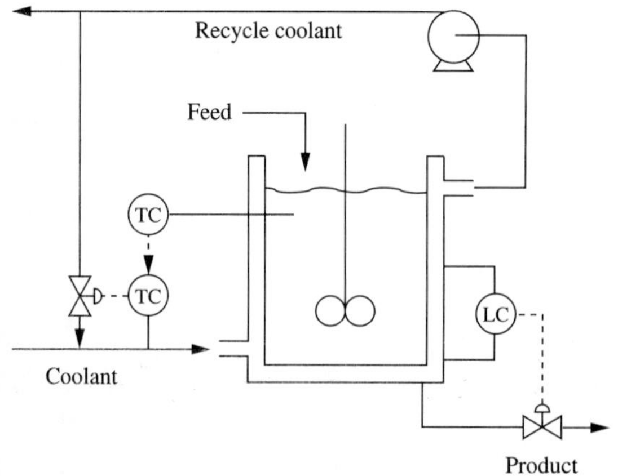

FIGURE 14.20
Cascade control for a stirred-tank reactor with cooling jacket.

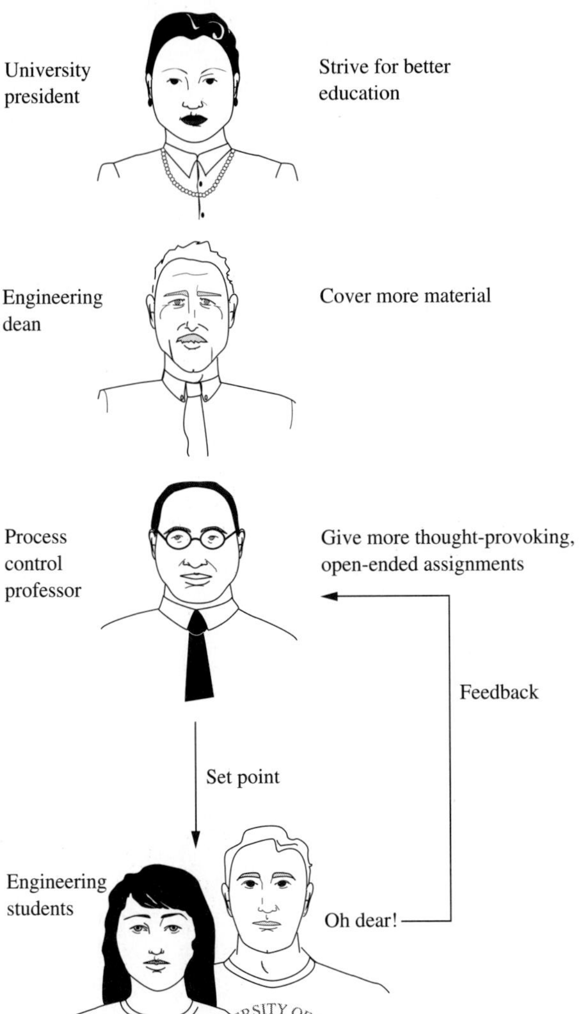

FIGURE 14.21
A cascade process at a university.

mines the desired value for the inlet temperature TC-3. The inlet temperature controller is free to determine the valve position that is required to achieve the desired value of TC-3. Since the T3 measurement provides rapid indication of every effect on the reactor inlet temperature, it can achieve the required inlet temperature control better than AC-1.

Cascade control concepts are not limited to engineering control systems. Social and business organizations also benefit from distributed decision making. A hypothetical example of university decision making is given in Figure 14.21. The president of a university decides to improve the education of the engineering students. Rather than tell each student how and what to learn, the president informs the Dean of Engineering. The Dean of Engineering gives directions to the Process Control professor, who finally gives directions to the students. The students then implement the decision by adjusting their studying to satisfy the requirements set by the professor. This distribution allows quick response to disturbances, such as competing demands of other courses, and provides frequent feedback from class discussion and course quizzes. The distributed system surely functions better than the single-loop feedback approach in which the president would obtain feedback every few years and then give directions to every student in the university. It also might clarify why the secondary controller is sometimes called the "slave"!

14.9 CONCLUSIONS

In this chapter, the principles of cascade control have been presented, and the excellent performance of cascade control for disturbance rejection has been established. Cascade control employs the principle of *feedback* control, since the secondary variable is a process output that depends on the manipulated variable in a causal manner. Cascade control can improve performance when the dynamics, mainly the dead time, of the secondary loop is much shorter than in the primary. In this situation, some disturbances can be measured and compensated quickly. As shown in Figures 14.3 *a* and *b*, this improved performance of the controlled variable is achieved without significantly increasing the variability of the manipulated variable. Based on this performance improvement and simplicity of implementation, the engineer is well advised to evaluate cascade control as the first potential single-loop enhancement.

The first few times new control engineers evaluate cascades, they should perform a careful study like the one in Table 14.2, but after some experience they will be able to design cascade controls quickly by applying the design principles without explicitly writing the criteria and table.

However, cascade control is not universally applicable; the design criteria in Table 14.1 can be used to determine whether cascade is appropriate and if so, select the best secondary variable. If it is not immediately possible and a significant improvement in control performance is desired, the engineer should investigate the possibility of adding the necessary secondary sensor. Even with improved sensors, cascade is not always possible; for example, a causal relationship between the manipulated variable and a measurement indicating the disturbance may not exist. Thus, while cascade is usually the preferred choice for enhancing control performance, further enhancement approaches are often required, and some of these are introduced in subsequent chapters.

REFERENCES

Buckley, P. "A Control Engineer Looks at Control Valves," in Lovett, O. (Ed.), Paper 2.1, *Final Control Elements, First ISA Final Control Elements Symposium, May 14–16, 1970.*

Krishnaswamy, P., G. Rangaiah, R. Jha, and P. Despande, "When to Use Cascade Control," *IEC Res., 29*, 2163–2166 (1990).

Krishnaswamy, P., G. Rangaiah, "Role of Secondary Integral Action in Cascade Control", *Trans. IChemE.*, 70, 149–152 (1992).

Verhaegen, S., "When to Use Cascade Control," *In. Tech.*, 38–40 (October 1991).

ADDITIONAL RESOURCES

Cascade control has been used for many decades and seems to have been developed by industrial practitioners, who often do not publish their results. Therefore, the inventor(s) of cascade are not known. A few of the earlier papers are listed here.

Franks, R. and C. Worley, "Quantitative Analysis of Cascade Control," *IEC, 48*, 1074 (1956).
Webb, P., "Reducing Process Disturbances with Cascade Control," *Control Eng., 8, 8*, 63 (1961).

QUESTIONS

14.1. (*a*) In your own words, discuss each of the cascade design criteria. Give a process example in which cascade control is appropriate.

(*b*) Identify the elements of the cascade block diagram in Figure 14.4 that are process, instrumentation, and control calculations.

(*c*) Discuss the topics in Table 13.2 that are influenced by cascade control, explaining how cascade improves performance for each.

(*d*) For the mixing system in Figure 13.4 and a disturbance in the feed concentration, discuss how you would add one sensor to improve control performance through cascade control.

14.2. Derive the transfer function in equations (14.6 to 14.8) based on the block diagram in Figure 14.4.

14.3. Figure 14.9 presents the frequency responses for single-loop and cascade control with a disturbance in the secondary loop.

(*a*) Sketch the general shapes and discuss the frequency responses for cascade and single-loop control for (1) a set point change and (2) a disturbance in the primary loop.

(*b*) Calculate the frequency responses for (*a*), with $G_{d1}(s) = G_{p1}(s)$ and $\eta = 10$.

14.4. Based on the transfer function in equation (14.6), mathematically demonstrate the assertion made in this chapter that cascade control performance for a secondary disturbance improves as the secondary becomes faster. For this question, assume that the secondary controller is a PI. (*Hint:* Evaluate the amplitude ratio as a function of frequency.)

14.5. Answer the following questions based on the transfer function in equation (14.6).

(*a*) Mathematically demonstrate the assertion made in this chapter that integral mode in the secondary controller is not required for zero offset in the primary.

(*b*) Demonstrate the assertion that the secondary controller must be tuned before the primary is tuned.

(*c*) This question addresses why the integral mode is often included for the secondary controller. Consider the secondary controller and its initial response to a disturbance before any feedback from the primary. Calculate the amount that a P-only and a PI controller attenuate the same disturbance at the limit of low frequency (i.e., at steady state). Based on this analysis, which controller is more effective in attenuating a disturbance?

(*d*) When the secondary has an integral mode, the integral error of the primary is zero for a step disturbance. The oscillatory effect of the integral error being zero is apparent in Figure 14.6*b*. For the transfer function in equation (14.6), prove this statement. You may use the following relationship (see Appendix D):

$$\int_0^{\infty} E_1(t')dt' = E_1(s)|_{s=0}$$

14.6. Discuss the proposed cascade control designs. In particular, apply the cascade control criteria to each proposed design, and estimate whether the cascade design would provide better performance than single-loop for disturbance response. Consider each of the disturbances separately. To assist in the analysis, prepare a block diagram for each process showing the appropriate cascade control systems.

(*a*) Jacketed stirred-tank reactor in Figure 14.20; disturbances are (1) coolant pressure, (2) coolant temperature, (3) recycle pump outlet pressure, (4) reaction rate (e.g., feed concentration), and (5) feed flow rate.

(*b*) Furnace coil outlet temperature control in Figure 14.15; disturbances are (1) fuel pressure, (2) fuel density (composition), (3) valve sticking, and (4) feed temperature.

(*c*) Repeat (*b*) with the temperature controller cascaded to a valve positioner, without the flow controller.

(*d*) Figure Q1.9*a* modified for cascade control with a level to flow to valve control structure; disturbances are (1) pump outlet pressure and (2) second outflow valve percent open.

(*e*) Analyzer to reboiler flow cascade for distillation in Figure Q14.6; disturbances are (1) heating medium temperature, (2) feed temperature, (3) tower pressure, and (4) heating medium downstream pressure.

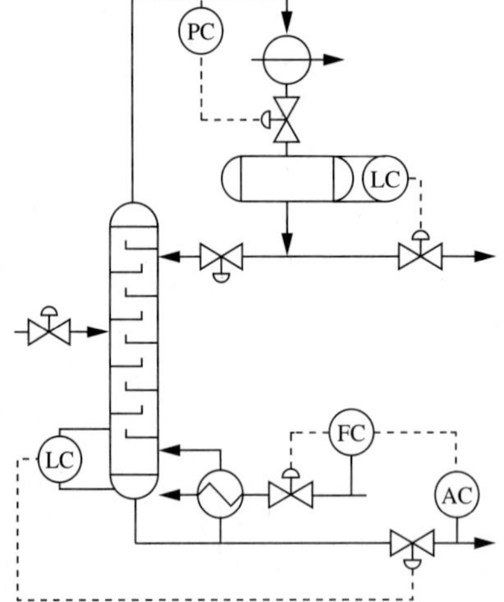

FIGURE Q14.6

14.7. Assume that the dynamic behavior in Figures 14.8*a* through *c* are for a fired heater in Example 2.1 that has the goal of maximizing temperature without exceeding a maximum constraint of 864°C; a performance correlation is provided in the example. The primary variable is temperature, which is plotted as the top variable in Figures 14.8*a* through *c*. Assume that 10% of the scale represents 5°C and that the top of the scale is 864°C.

(*a*) Estimate the benefit for (1) single-loop and (2) cascade control due to the reduction in the variability in the temperature using the data in Figures 14.8*a* through *c*.

(*b*) Suggest changes to the operating conditions (set points) for both control designs and repeat the estimation done in (*a*).

14.8. The cascade control design shown in Figure 14.12 should have anti–reset windup protection.

(*a*) Discuss the potential causes for integral windup in this strategy.

(*b*) Assume that the feedback algorithms are of the form that use external feedback, as described in Chapter 12. Which signal should be used for external feedback for each controller? Explain your answer.

(*c*) In Chapter 11, the use of digital PI algorithms was explained. Discuss the performance of the incremental (velocity) algorithm as the primary temperature controller when the valve reaches its maximum or minimum position. Does the velocity form of the PI controller satisfactorily prevent reset windup?

14.9. The initial design for the packed-bed reactor, Figure 14.11, included a controller for the outlet concentration and a single control valve. The cascade control design *added a controller* but did not change the number of control valves. This might seem to violate the degrees of freedom of the process.

(*a*) In your own words, discuss why the cascade is possible.

(*b*) Perform a degrees-of-freedom analysis to demonstrate that the cascade control is possible. You may use the transfer function model of the reactor for this analysis.

14.10. For the following control designs determine which valves should have positioners and explain why: (*a*) Figure 13.19, (*b*) Figure 2.2, (*c*) Figure 14.1, and (*d*) Figure 14.2.

14.11. A stirred-tank chemical reactor is shown in Figure Q14.11 with the following reaction.

$$A \rightarrow B \qquad \Delta H_{rxn} = 0$$

$$r_A = -k_0 e^{-E/RT} \frac{C_A}{1 + kC_x}$$

The available sensors and control valves are shown in the figure, and no changes to these are allowed. The goal is to control the reactor concentration tightly by single-loop or cascade control, whichever is better. For each of the following disturbances, design the best control system and explain your design: (*a*) the heating medium pressure (P_1), (*b*) the solvent feed temperature (T_s), (*c*) the reactant feed pressure (P_2), and (*d*) the inhibitor concentration (C_x) which enters in the solvent. (To assist in the analysis, prepare a block diagram for each case showing the appropriate single-loop or cascade control system.)

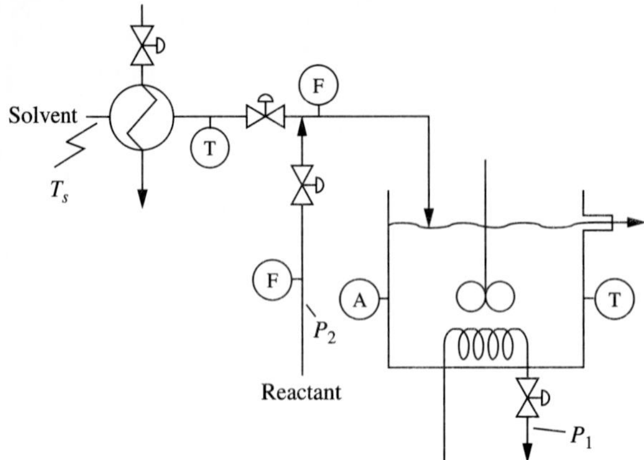

FIGURE Q14.11

14.12. A set of cascade design criteria are presented in an article by Verhaegen (1991). Discuss the similarities and differences between the criteria in this chapter and in the article.

14.13. The chemical reactor with separate solvent and reactant feed flows similar to Example 5.4, but with *one vessel,* has the following properties: well mixed, isothermal, constant volume, constant density, and $F_s \approx F_A$ (the solvent flow is *not* much larger). The chemical reaction occurring is A → B, with the reaction rate $r_A = -kC_A$. The concentration of the product can be measured in the reactor without delay.

(*a*) The total feed flow (F) and the feed concentration (C_{A0}) are the potential *manipulated* variables for the reactor effluent composition control. Design a regulatory control scheme that will control these two variables (F, C_{A0}) simultaneously to independent set point values. Sketch the controls on a process diagram and place the sensors and final elements for these variables where you think appropriate.

(*b*) Derive the dynamic model for $C_B(s)/C_{A0}(s)$ under the control in (*a*). Analyze the model regarding (1) order, (2) stability, (3) periodicity, and (4) step response characteristics.

(*c*) Derive the dynamic model for $C_B(s)/F(s)$ under the control in (*a*). Analyze the model regarding (1) order, (2) stability, (3) periodicity, and (4) step response characteristics.

(*d*) Based on the results in (*b*) and (*c*), recommend which of these manipulated variables would provide the best feedback control for C_B for a set point change using a PI controller. Sketch the feedback (cascade) structure on the figure prepared in (*a*). Compare performance with Example 13.8.

14.14. In the packed-bed reactor example (14.1) an alternative approach would be possible. In the alternative approach, the oil temperature (T2) would be measured and the effect on the outlet analyzer due to changes in oil temperature could be calculated and used to adjust the valve. Which design—cascade control or the alternative described here—would you prefer? Discuss why.

14.15. Prepare a digital computer program to perform the control calculations for the cascade control system in Figure 14.12. Include initialization, reset windup, and other factors required for good implementation.

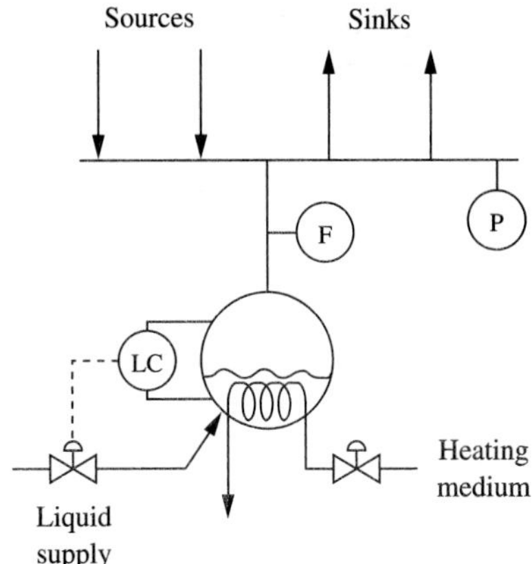

FIGURE Q14.16

14.16. A vaporizer process is shown in Figure Q14.16. The gas pipe (header) has several sources and sinks of gas, and the pressure in the pipe is to be controlled by adjusting the amount vaporized.

(*a*) A cascade control design has been suggested from the pressure to the flow of vapor to the heating medium valve. Evaluate this design using the cascade design criteria. Correct it if necessary.

(*b*) Discuss the response of this design to a disturbance in the heating medium pressure, upstream of the control valve.

(*c*) Discuss the response of this system to a disturbance in one of the source or sink flows.

(*d*) Discuss the response of this system to a disturbance in the liquid temperature.

(*e*) Generalize the results from (*b*) through (*d*) and develop a further cascade design criterion to be added to those in Table 14.1.

CHAPTER
15

FEEDFORWARD CONTROL

15.1 INTRODUCTION

Feedforward uses the measurement of an input disturbance to the plant as additional information for enhancing single-loop PID control performance. This measurement provides an "early warning" that the controlled variable will be upset some time in the future. With this warning the feedforward controller has the opportunity to adjust the manipulated variable before the controlled variable deviates from its set point. Note that the feedforward controller does not use an output of the process! This is the first example of a controller that does not use feedback control; hence the new name *feedforward.* As we will see, feedforward is usually combined with feedback so that the important features of feedback are retained in the overall strategy.

Feedforward control is effective in reducing the influences of disturbances, although not usually as effective as cascade control with a fast secondary loop. Since feedforward control also uses an additional measurement and has design criteria similar to cascade control, engineers often confuse the two approaches. Therefore, the reader should pay careful attention to the design criteria presented in this chapter.

15.2 AN EXAMPLE AND CONTROLLER DERIVATION

The process example used in this introduction is the same stirred-tank heat exchanger considered in Chapter 14 for cascade control. The control objective is still the maintenance of the outlet temperature very close to its set point, and the manipulated variable is still the heating medium valve position. The only difference is that the heating oil pressure is not varying significantly; thus, the cascade controller is not

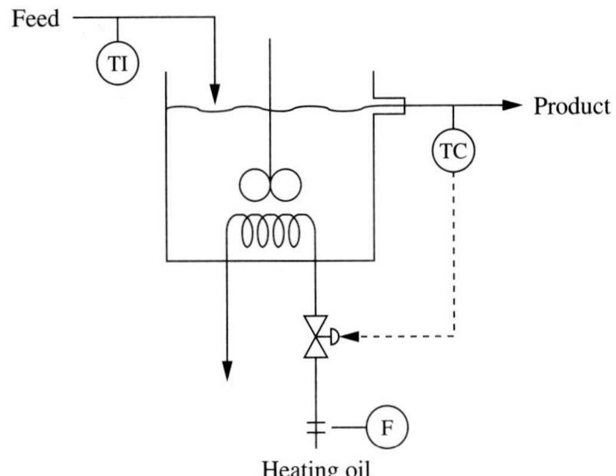

FIGURE 15.1
Stirred-tank heat exchanger with single-loop feedback temperature control.

required, as shown in Figure 15.1. In this case, the inlet temperature varies with sufficient amplitude to disturb the outlet temperature significantly. The challenge is to design a feedforward controller that reduces or, in the ideal case, eliminates the effect of the inlet temperature on the outlet temperature, by adjusting the heating oil valve.

The approach to designing a feedforward controller is based on completely cancelling the effect of the disturbance. This concept is sketched in Figure 15.2. The disturbance is shown as a step change to simplify the discussion, but the analysis and resulting feedforward controller are applicable to any disturbance of arbitrary time dependence. The change in the outlet temperature in response to the inlet temperature change, shown as curve A, is the response that would occur without control. For perfect control, the outlet temperature would not change; this is shown as curve B. To achieve perfect control the manipulated variable must be adjusted to compensate for the disturbance—that is, to cause the mirror image of the disturbance so that the sum of the two effects is zero. Thus, curve C shows the effect of the manipulated variable on the outlet temperature; the sum of curves A and C is a zero disturbance

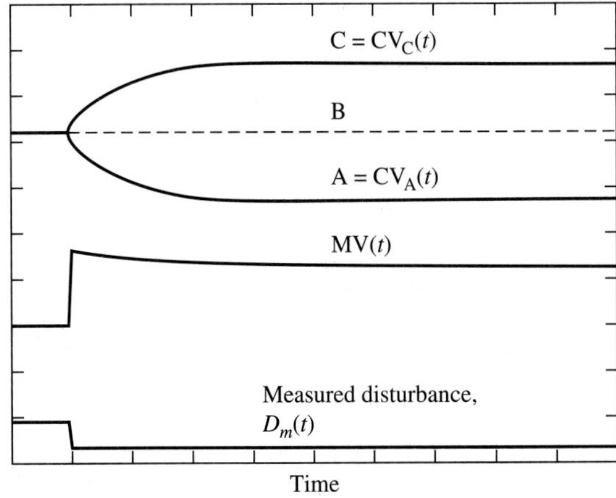

FIGURE 15.2
Time domain plot showing perfect feedforward compensation for a measured disturbance.

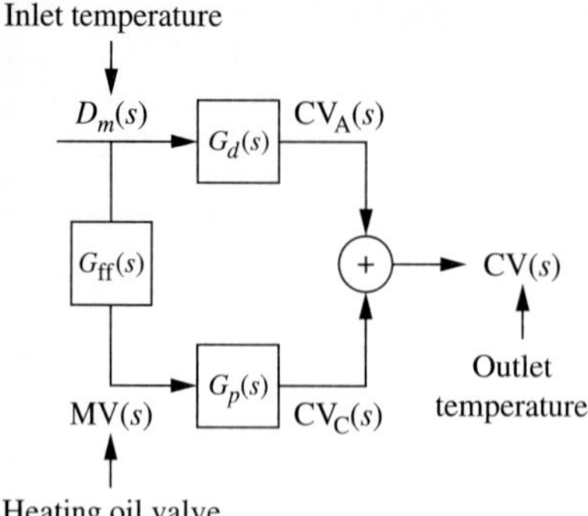

FIGURE 15.3
Simplified block diagram of feedforward compensation.

to the controlled variable, which gives perfect feedforward control. The feedforward control algorithm uses the measurement of the disturbance to calculate the manipulated variable with the goal of perfect feedforward compensation as shown in Figure 15.2.

The control calculation that achieves this goal can be derived by analyzing the block diagram of the feedforward control system in Figure 15.3. The individual blocks account for the process $G_p(s)$, the disturbance $G_d(s)$, and the feedforward controller $G_{ff}(s)$. The equation that relates the measured disturbance to the outlet variable is

$$\text{CV}(s) = [G_d(s) + G_{ff}(s)G_p(s)]D_m(s) \tag{15.1}$$

Since equation (15.1) involves deviation variables, the goal is to maintain the outlet temperature at zero, $\text{CV}(s) = 0$ ($T'_{\text{out}}(s) = 0$). The only unknown, which is the controller, $G_{ff}(s)$, can be determined by rearranging equation (15.1). The result is

$$G_{ff}(s) = -\frac{G_d(s)}{G_p(s)} \tag{15.2}$$

It is important to note that the feedforward controller depends on the models for the disturbance and the process. The feedforward controller is never a PID algorithm—a result that should not be surprising, because we have a new control design goal that does not apply feedback principles.

Equation (15.2) provides the general feedforward control equation. Typical transfer functions for the disturbance and the process are now substituted to derive the most common form of the controller. Assume that the transfer functions have the following first-order-with-dead-time forms

$$\frac{\text{CV}(s)}{\text{MV}(s)} = G_p(s) = \frac{K_p e^{-\theta s}}{\tau s + 1} \qquad \frac{\text{CV}(s)}{D_m(s)} = G_d(s) = \frac{K_d e^{-\theta_d s}}{\tau_d s + 1} \tag{15.3}$$

By substituting equations (15.3) into equation (15.2), the feedforward controller would have the form

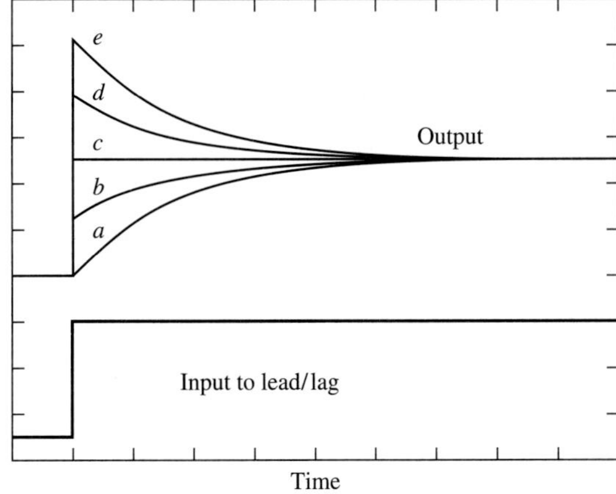

FIGURE 15.4
Example of dynamic responses for a lead/lag algorithm. Ratios of lead to lag times are (*a*) 0, (*b*) 0.5, (*c*) 1.0, (*d*) 1.5, (*e*) 2.0.

$$G_{ff} = \frac{MV(s)}{D_m(s)} = -\frac{G_d(s)}{G_p(s)} = K_{ff}\left(\frac{T_{ld}s + 1}{T_{lg}s + 1}\right)e^{-\theta_{ff}s} \tag{15.4}$$

with

$$\begin{aligned}
\text{lead/lag algorithm} &= (T_{ld}s + 1)/(T_{lg}s + 1)\\
\text{feedforward controller gain} &= K_{ff} = -K_d/K_p\\
\text{feedforward controller dead time} &= \theta_{ff} = \theta_d - \theta\\
\text{feedforward controller lead time} &= T_{ld} = \tau\\
\text{feedforward controller lag time} &= T_{lg} = \tau_d
\end{aligned}$$

In most (but not all) cases, this form of the feedforward controller provides sufficient accuracy; usually, second- or higher-order terms in the controller do not improve the control performance, especially because the models are not known exactly.

The special form of the feedforward controller in equation (15.4) consists of a gain, dead time, and a factor called a *lead/lag*. The dynamic behaviors of gains and dead times are well known by this point in the book, but lead/lag is new, so a few typical dynamic responses are presented in Figure 15.4. Each result uses the same lead/lag algorithm with different parameters as indicated. Again, for simplicity, the input is a step change, but the feedforward controller with a lead/lag performs well for any input function. The analytical expression for the output of a lead/lag, here represented as $y(t)$, for a *unit step input* can be determined from entry 5 in Table 4.1 to be

$$y(t) = 1 + \frac{T_{ld} - T_{lg}}{T_{lg}} e^{-t/T_{lg}} \tag{15.5}$$

As seen in the figure of dynamic responses, when the lead time is less than the lag time, the manipulated variable rises to the steady-state value as a first-order response from an initial step that does not reach the final steady state. This is consistent with a process whose disturbance time constant is greater than its process time constant, requiring the control action to be slowed so that the effects of the disturbance and the feedforward controller cancel. When the lead and lag times are equal, the manipulated variable immediately attains its steady-state value. This is consistent

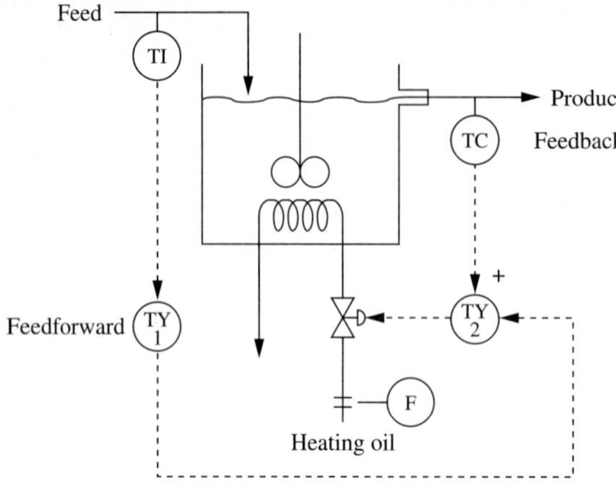

FIGURE 15.5
Stirred-tank heat exchanger with feedforward-feedback control strategy.

with a process that has equal disturbance and process time constants. Finally, when the lead time is greater than the lag time, the manipulated variable initially exceeds, then slowly returns to, its steady-state value. This is consistent with a process whose disturbance time constant is smaller than its process time constant.

The derivation of the feedforward controller ensures perfect control if (1) the models used are perfect, (2) the measured disturbance is the only disturbance experienced by the process, and (3) the control calculation is realizable; that is, capable of being implemented as discussed in Section 15.7. Neither of conditions 1 or 2 is generally satisfied. Therefore, feedforward control is always combined with feedback control, when possible, to ensure zero steady-state offset! Since the process and control calculations are considered to be linear, the adjustments to the manipulated variable from the feedforward and feedback controllers can be added. A typical feedforward-feedback control system is given in Figure 15.5 for the stirred-tank heat exchanger.

15.3 FEEDFORWARD CONTROL DESIGN CRITERIA

The principles of feedforward control have been introduced with respect to the stirred-tank heater. In Table 15.1 the design criteria are summarized in a concise form so that they can be applied in general. Adherence to these criteria ensures that feedforward control is used when appropriate.

The first two items in the table address the application of feedforward control. Naturally, only when feedback control does not provide acceptable control performance is an enhancement like feedforward control employed. The second criterion requires that an acceptable measured feedforward variable be available or that it can be added at reasonable cost.

A potential feedforward variable must satisfy three criteria. First, it must indicate the occurrence of an important disturbance; that is, there must be a direct, reproducible correlation between the process disturbance and the measured feedforward variable, and the measured variable should be relatively insensitive to other

TABLE 15.1
Feedforward control design criteria

Feedforward control is desired when
1. Feedback control does not provide satisfactory control performance.
2. A measured feedforward variable is available.
A feedforward variable must satisfy the following criteria:
3. The variable must indicate the occurrence of an important disturbance.
4. There must *not* be a causal relationship between the manipulated and feedforward variables.
5. The disturbance dynamics must not be significantly faster than the manipulated–output variable dynamics (when feedback control is also present).

changes in operation. Naturally, the disturbance must be important (i.e., change frequently and have a significant effect on the controlled variable), or there would be no reason to attenuate its effect. Second, the feedforward variable must *not* be influenced by the manipulated variable, because the feedback principle is not used. Note that this requirement provides a clear distinction between variables used for cascade and feedforward. Finally, the disturbance dynamics should not be faster than the dynamics from the manipulated to the controlled variable.

This final requirement is related to combined feedforward-feedback control systems. Should the effect of the disturbance on the controlled variable be very fast, feedforward could not affect the output variable in time to prevent a significant deviation from the set point. As a result, the feedback controller would sense the deviation and adjust the manipulated variable. Unfortunately, the feedback adjustment would be in addition to the feedforward adjustment; thus, a double correction would be made to the manipulated variable; remember, the feedforward and feedback controllers are independent algorithms. The double correction would cause an overshoot in the controlled variable and poor control performance. In conclusion, feedforward control should not be used when the disturbance dynamics are very fast and PID feedback control is present. Naturally, if feedback is not present (perhaps due to the lack of a real-time sensor), feedforward can be applied regardless of the disturbance dynamics.

Feedforward and Feedback Are Complementary

Each has important advantages that compensate for deficiencies of the other, as summarized in Table 15.2. The major advantage of feedback control is that it reduces steady-state offset to zero for all disturbances. As we have seen, it can provide good control performance in many cases but requires a deviation from the set point before it takes corrective action. However, feedback does not provide good control performance when the feedback dynamics are unfavorable; a summary of the factors influencing the performance of feedback control is given in Table 13.2 (page 463). In addition, feedback control can cause instability if not correctly tuned.

Feedforward control acts before the output is disturbed and is capable of very good control performance with an accurate model. Another advantage is that a stable

TABLE 15.2
Comparison of feedforward and feedback principles

	Feedforward	Feedback
Advantages	Compensates for a disturbance before the process output is affected	Provides zero steady-state offset
	Does not affect the stability of the control system	Effective for all disturbances
Disadvantages	Cannot eliminate steady-state offset	Does not take control action until the process output variable has deviated from its set point
	Requires a sensor and model for each disturbance	Affects the stability of the closed-loop control system

feedforward controller cannot induce instability in a system that is stable without feedforward control. This fact can be demonstrated by analyzing the transfer function of a feedforward-feedback system shown in Figure 15.6, which accounts for sensors and the final element explicitly:

$$\frac{\mathrm{CV}(s)}{D_m(s)} = \frac{G_v(s)G_p(s)G_{\mathrm{ffs}}(s)G_{\mathrm{ff}}(s) + G_d(s)}{1 + G_v(s)G_p(s)G_{\mathrm{fbs}}(s)G_c(s)} \tag{15.6}$$

As long as the numerator is stable, which is normally the case, stability is influenced by the terms in the characteristic equation, which contain terms for the feedback process, instrumentation, and controller. The disturbance process, feedforward instrumentation, and feedforward controller appear only in the numerator. Therefore, a (stable) feedforward controller cannot cause instability, although it can lead to poor performance if improperly designed and tuned. The major limitation to feedforward control is its inability to reduce steady-state offset to zero. As explained, this limitation is easily overcome by combining feedforward with feedback.

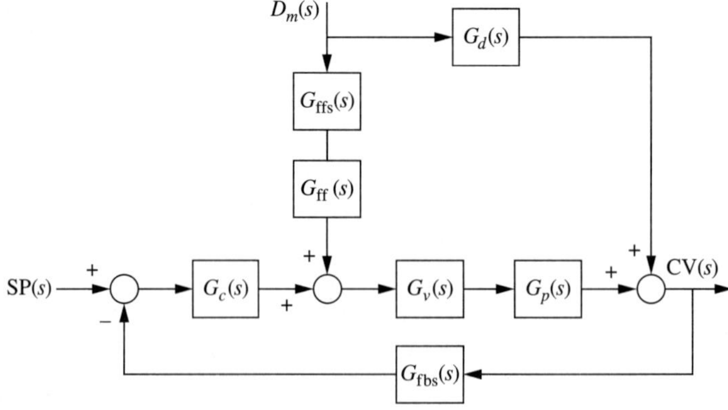

FIGURE 15.6
Block diagram of feedforward-feedback control system with sensors and final element.

Feedforward control uses a measured input disturbance to determine an adjustment to an input manipulated variable. All feedforward control strategies should conform to the design criteria in Table 15.1.

15.4 FEEDFORWARD PERFORMANCE

In the introduction to this chapter, feedforward control was described as simple and effective. The foregoing material has demonstrated how simple a feedforward strategy is to evaluate and design. In this section, the performance is calculated for a few sample systems and compared to the single-loop performance to demonstrate the effectiveness of feedforward. Due to the large number of process and controller parameters, no general correlations concerning feedforward control performance are available. The general trends in these examples should be applicable to most realistic processes.

Based on the feedforward design method, perfect control performance is theoretically possible; however, it is never achieved, because of model errors. Therefore, a key factor in feedforward control performance is model accuracy. A typical feedforward-feedback system consistent with the block diagram in Figure 15.6 was simulated for various cases; this system can be thought of as the heat exchanger system in Figure 15.5 with the following process and controller models:

$$
\begin{aligned}
G_p(s) &= \frac{e^{-15s}}{20s+1} \qquad G_d(s) = \frac{e^{-30s}}{20s+1} \\
G_{ff}(s) &= -1.0e^{-15s}\frac{20s+1}{20s+1} = -1.0e^{-15s} \qquad (15.7)\\
G_{ffs}(s) &= 1 \qquad G_{fbs}(s) = 1 \qquad G_v(s) = 1 \qquad G_c(s) = 0.9\left(1+\frac{1}{20s}\right)
\end{aligned}
$$

The upset was a single step change and no noise was added to the measurements, so that the effect of the control alone could be determined. The actual process and disturbance responses remain unchanged for all cases; the feedforward controller tuning parameters are changed to determine the effect of controller model errors on performance. The feedback PI controller was tuned by conventional means for good regulation of the controlled variable without excessive variation in the manipulated variable. The control performance measure was the integral of the absolute value of the error (IAE).

The resulting control performances are shown in Figure 15.7 as a function of the feedforward model error. Note that the results are reported relative to the feedback-only performance, so any value less than 100% represents an improvement through feedforward control; recall that the system returns to the set point, even with feedforward errors, because of the feedback controller's integral mode. This provides a simple comparison of performances in a manner useful for answering the key question of whether or not to use feedforward to enhance feedback. Separate plots provide the control performance with errors in the gain, dead time, and ratio

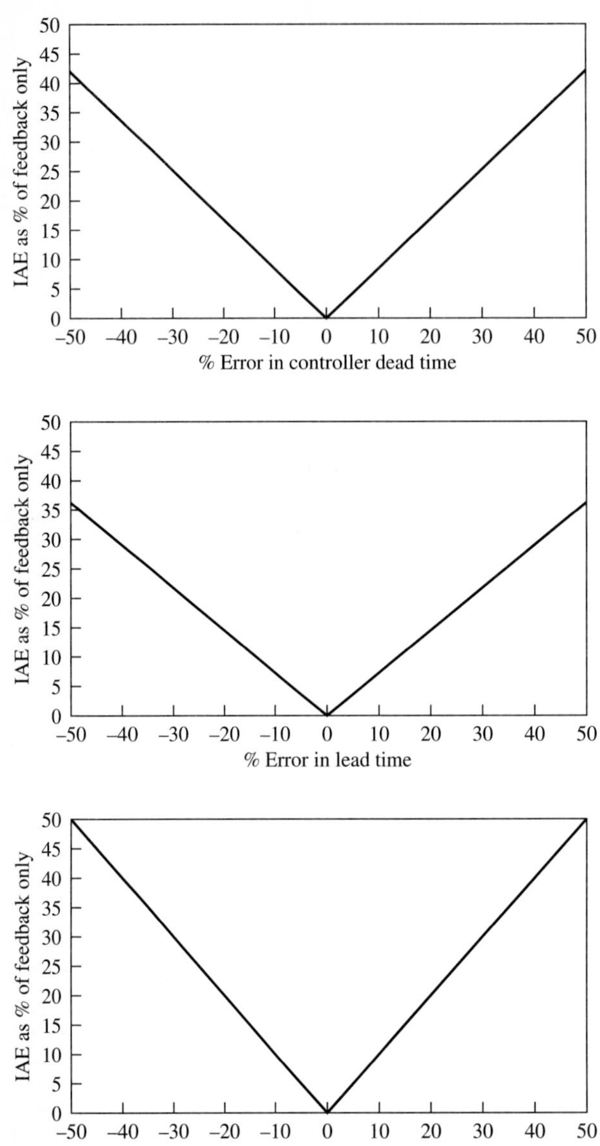

FIGURE 15.7
Example of the effect of errors in the feedforward controller on the performance, reported as a percent of feedback-only IAE.

of lead to lag times. For each of these plots the other model parameters matched the process exactly.

The results demonstrate that feedforward control can substantially improve control performance, even with significant errors in the model used. For this process studied, feedforward would provide substantial improvement, maintaining the IAE much lower than that achieved by feedback only for the large range of model errors considered. This insensitivity of performance to model error leads to robust control over a large range of process dynamics without updating feedforward controller parameters.

Typical transient responses with feedback/ feedforward control are given in Figure 15.8*a* through *e* for the example system subject to a unit step disturbance. Figure 15.8*a* shows the performance of feedback-only, and the next three parts show the performance of the feedforward-feedback control system with model errors as indicated in the caption. These sample results, along with Figure 15.7, demonstrate the general insensitivity of feedforward control to model errors, which is an important property contributing to its successful application. The final sample result, Figure 15.8*e*, shows the performance with feedforward-only control, which gives steady-state offset unless the feedforward gain is perfect—a highly unlikely situation. The steady-state offset could be determined by applying the final value theorem to equation (15.1).

The results in Figures 15.7 and 15.8 support a frequently used simplification to feedforward control. Often the lead/lag and dead time elements are eliminated from the feedforward controller; the resulting controller is usually called *steady-state feedforward.* This simplification does not substantially degrade control performance when the feedforward controller dead time is small and the lead and lag times are nearly equal. In conclusion:

> Feedforward control can substantially improve control performance of processes for which feedback alone does not provide acceptable control, and its performance does not degrade rapidly with model errors.

15.5 CONTROLLER ALGORITHM AND TUNING

The approach to deriving the feedforward controller algorithm was described along with the first example in Section 15.2. The controller is expressed as a transfer function in that section. Analog implementation would require an electrical circuit that closely approximates the transfer function. Such a circuit would be costly and is seldom made for a range of model structures, but it is available for the lead/lag with gain. To clarify the application of feedforward, the *digital* implementation of a typical feedforward controller is developed here. The programming of the controller is shown schematically in Figure 15.9. The gain is simply a multiplication. The dead time can be implemented by using a table of data whose length times the sample period equals the dead time. The data location (or pointer) is shifted one step every time that the controller is executed. The lead/lag element must be transformed into a digital algorithm. One way to do this is to convert the lead/lag into a differential equation by remembering that multiplication of the Laplace transform by the variable s is equivalent to differentiation. The resulting equation is

$$T_{\text{lg}} \frac{dY(t)}{dt} + Y(t) = T_{\text{ld}} \frac{dX(t)}{dt} + X(t) \tag{15.8}$$

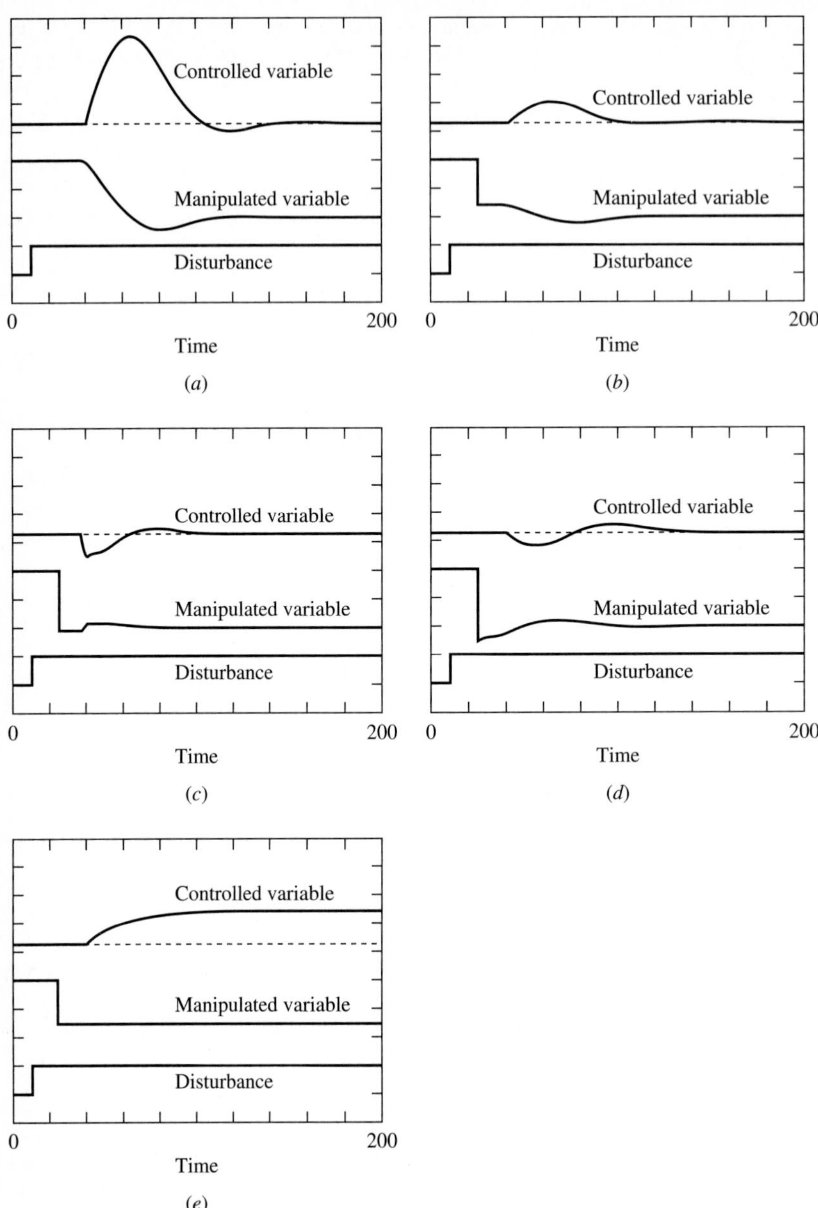

FIGURE 15.8

Transient responses: (*a*) feedback-only, (*b*) feedforward-feedback with −25% error in K_{ff}, (*c*) feedforward-feedback with −20% error in θ_{ff}, (*d*) feedforward-feedback control with +25% error in the τ_{ld}, (*e*) feedforward-only control with −25% error in K_{ff}. One tick (10% of scale) is 0.2 for the controlled variable, 0.50 for the manipulated variable, and 1.0 for the disturbance.

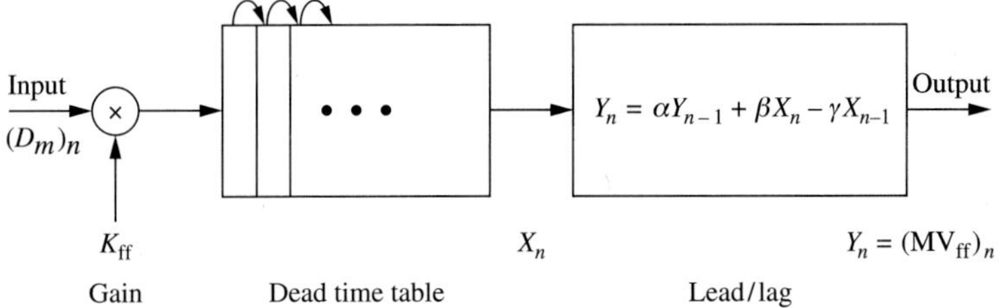

FIGURE 15.9
Schematic diagram of a digital feedforward controller.

with X the input to the lead/lag and Y the output from the lead/lag algorithm. The differential equation can be transformed into a difference equation by approximating the derivative by a backward difference, as follows:

$$\frac{dY(t)}{dt} \approx \frac{Y_n - Y_{n-1}}{\Delta t} \qquad \frac{dX(t)}{dt} \approx \frac{X_n - X_{n-1}}{\Delta t} \tag{15.9}$$

The resulting equation can be rearranged to yield the following equation, which can be used in a digital computer to implement the digital lead/lag.

$$Y_n = \left(\frac{\frac{T_{\text{lg}}}{\Delta t}}{\frac{T_{\text{lg}}}{\Delta t} + 1} \right) Y_{n-1} + \left(\frac{\frac{T_{\text{ld}}}{\Delta t} + 1}{\frac{T_{\text{lg}}}{\Delta t} + 1} \right) X_n - \left(\frac{\frac{T_{\text{ld}}}{\Delta t}}{\frac{T_{\text{lg}}}{\Delta t} + 1} \right) X_{n-1} \tag{15.10a}$$

with Y_n = the output signal from the lead/lag

X_n = the input signal to the lead/lag

which can be combined with the gain and dead time for the digital form of a feedforward controller with lead/lag:

$$\text{MV}_n = \left(\frac{\frac{T_{\text{lg}}}{\Delta t}}{\frac{T_{\text{lg}}}{\Delta t} + 1} \right) \text{MV}_{n-1} + K_{\text{ff}} \left(\frac{\frac{T_{\text{ld}}}{\Delta t} + 1}{\frac{T_{\text{lg}}}{\Delta t} + 1} \right) (D_m)_{n-\Gamma} - K_{\text{ff}} \left(\frac{\frac{T_{\text{ld}}}{\Delta t}}{\frac{T_{\text{lg}}}{\Delta t} + 1} \right) (D_m)_{n-\Gamma-1} \tag{15.10b}$$

with $\Gamma = \theta/\Delta t$. The reader should note that the method in equation (15.10) is not the best, most general method for converting the algorithm to digital form. Limitations are presented by the delay table, which requires the dead time divided by the execution period to be an integer. In addition, the difference approximation is accurate only for execution times that are small compared to the lead and lag times. More general methods (which require the use of z-transforms) for deriving digital algorithms are available (Smith, 1972).

Tuning the feedforward-feedback control system follows a simple, stepwise procedure. Either controller may be tuned first; assume that the feedback is tuned first, which requires the identification of the feedback process model $G_p(s)$. Because the tuning parameters for the feedforward controller are derived from both the disturbance and process models, the disturbance model must also be identified through plant experiments, as described in Chapter 6. The disturbance variable cannot normally be changed in a perfect step; thus, the statistically based methods are usually required for identifying $G_d(s)$. The feedforward control performance can be tested through application of feedforward-only control (i.e., with the feedback controller temporarily in manual mode). A typical transient result is given in Figure 15.8*e*. The steady-state offset gives an indication of the error in the feedforward gain K_{ff}, which can be further adjusted until the desired accuracy is achieved. Some information on the dynamic tuning parameters can be deduced from feedforward-only control. Should the controlled variable initially respond in the direction indicating too rapid a change in the manipulated variable, either the feedforward controller dead time is too short or the lead/lag time constant ratio is too high. Trial and error are required to establish the improved values. A method for adjusting the lead and lag times is available (Shinskey, 1988), but it requires a perfect step change in the disturbance variable. The disturbance is not usually controlled independently (if it were controlled, it would not be a disturbance), so the method is of limited applicability.

Finally, some common sense is required when tuning the lead/lag times. First, the effect of high-frequency noise in the feedforward measurement should be considered. The lead/lag calculation can amplify noise when the lead time is much greater than the lag time. This effect can be understood by noting that the lead/lag calculation approaches a proportional-derivative calculation as the lead time increases (i.e., $T_{\mathrm{lg}} \approx 0$):

$$\frac{T_{\mathrm{ld}}s + 1}{T_{\mathrm{lg}}s + 1} \approx T_{\mathrm{ld}}s + 1 \tag{15.11}$$

Even without high-frequency noise, the lead/lag could make large changes in the manipulated variable when the lead time is much larger than the lag time, as shown in Figure 15.4 and in equation (15.5). To reduce the effect of noise and limit the overshoot in the manipulated variable, the ratio of lead to lag times should not exceed about 2:1, unless plant experience indicates otherwise.

> Tuning a feedforward-feedback control system requires that each controller be tuned independently, following individual initial and fine-tuning methods.

15.6 IMPLEMENTATION ISSUES

Feedforward control involves a new algorithm for which there is no accepted standard display used in commercial equipment. Since the feedforward controller responds to disturbances, it has no set point—a factor that changes the display significantly. One feature that should be provided in the display is the ability for the operator to turn the feedforward and the feedback on and off separately. Also, the operator should have a display of the result after the feedforward and feedback signal have been combined, because the operator always wants to know the signal sent to the final control element.

As shown, the calculations for feedforward, equations (15.4) and (15.10*b*), are simple and can be performed with standard algorithms available in most commercial control equipment. The engineer normally connects or "configures" the preprogrammed algorithms and enters the tuning constants. An important feature that must be included is smooth (i.e., "bumpless") transfer when feedforward or feedback controllers are turned on and off. One approach to bumpless transfer is to use incremental or velocity forms of the feedforward and feedback control equations. Whenever one or both of the controllers is turned off (i.e., put in manual), the change in its output becomes zero. When it is turned on, or put in automatic, its output calculation resumes. This is an example of an approach to bumpless transfer, other approaches are possible; for example, see (Gallun et al., 1985).

The feedforward-feedback control system uses more control equipment—two sensors and controllers—than the equivalent single-loop system. Since the system performance requires all of this equipment to function properly, its reliability can be expected to be lower than that of the equivalent single-loop system. However, it is important to note that feedback control is not dependent on the feedforward; should any component in the feedforward controller fail, the feedforward part can be turned off, and the feedback controller will function properly. Usually, the lower reliability does not prevent the use of feedforward.

Since the feedforward-feedback involves more equipment, it costs slightly more than the single-loop system. The increased costs include a field sensor and transmission to the control house (if the variable is not already available for monitoring purposes), a controller (whose cost may be essentially zero if a digital system with spare capacity is used), and costs for installation and documentation. These costs are not usually significant compared to the benefits achieved through a properly designed feedforward control strategy, except that expensive analyzers for feedforward are often not economically justified.

> Feedforward control, where applicable, provides a simple method for substantial improvement in control performance. The additional costs and slightly lower reliability are not normally deterrents to implementing feedforward control.

15.7 FURTHER FEEDFORWARD EXAMPLES

In this section the concept of feedforward control is consolidated, and a few new features are presented through further examples.

Example 15.1. Packed-bed chemical reactor. For the first example the packed-bed chemical reactor analyzed in Chapter 14 is considered again (page 490). The process with its feedback control strategy is shown in Figure 15.10. The control objective is still the maintenance of the outlet concentration close to its set point by adjusting the preheat. The cascade control strategy is retained for its good responses to disturbances in heating oil pressure, heating oil temperature, and inlet temperature. Suppose that the feed composition is another significant disturbance. The goal is to design a feedforward control strategy for this process using the sensors and manipulated variables given. (The reader is encouraged to design a control system before reading further.)

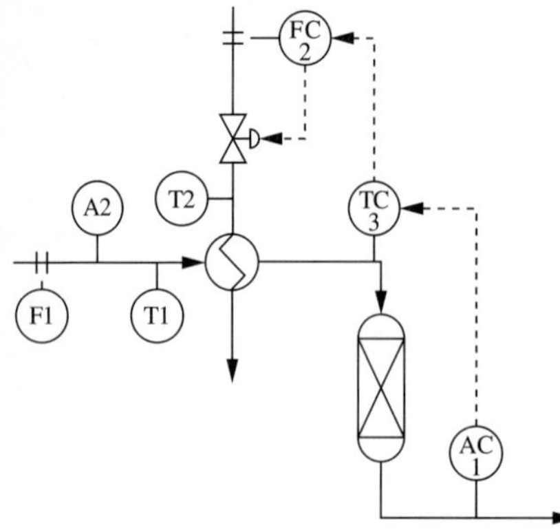

FIGURE 15.10
Packed-bed chemical reactor with feedback control.

Since we are dealing with a feedforward control strategy, the key decision is the selection of the feedforward variable. Therefore, the first step is to evaluate the potential measured variables using the design criteria in Table 15.1. The results of this evaluation are summarized in Table 15.3. Since all of the criteria must be satisfied for a variable to be used for feedforward, only the reactor inlet concentration, A2, is a satisfactory variable. The resulting control strategy is shown in Figure 15.11.

Signal combination. In this example the feedback controller involves a cascade strategy; therefore, an important additional issue is how the feedforward and feedback signals should be combined. First, we assume that the process behaves in (approximately) a linear manner, so that the feedback and feedforward corrections can be calculated independently and added. Second, the correct location for combining the signals can be determined by referring to the feedforward and feedback equations:

$$\mathrm{MV}_{\mathrm{ff}}(s) = G_{\mathrm{ff}}(s)D_{A2}(s) \qquad \mathrm{MV}_{\mathrm{fb}}(s) = G_c(\mathrm{SP}_{\mathrm{A1}}(s) - \mathrm{CV}_{\mathrm{A1}}(s)) \tag{15.12}$$

The outputs of the two controllers can be combined when they both manipulate the same variable, that is, if $\mathrm{MV}_{\mathrm{ff}}(s)$ and $\mathrm{MV}_{\mathrm{fb}}(s)$ represent the same manipulated variable. This demonstrates that the feedforward controller output can be added to the output of the feedback controller which is regulating the same controlled variable. In this example, the feedforward controller output is added to the output of the outlet analyzer controller, AC-1, as shown in Figure 15.11. The combined signal is sent to the set point of the inlet temperature controller. For the feedforward controller to function properly in this design, it must be derived using the temperature set point as its manipulated variable:

TABLE 15.3
Evaluation of potential feedforward variables

Criterion	A2	F1	F2	T1	T2	T3
1. Single-loop control not satisfactory	Y	Y	Y	Y	Y	Y
2. Variable measured	Y	Y	Y	Y	Y	Y
3. Indicates key disturbance	Y	N	N	N	N	N
4. Not influenced by MV	Y	Y	N	Y	Y	N
5. Suitable disturbance dynamics	Y	N/A	N/A	N/A	N/A	N/A

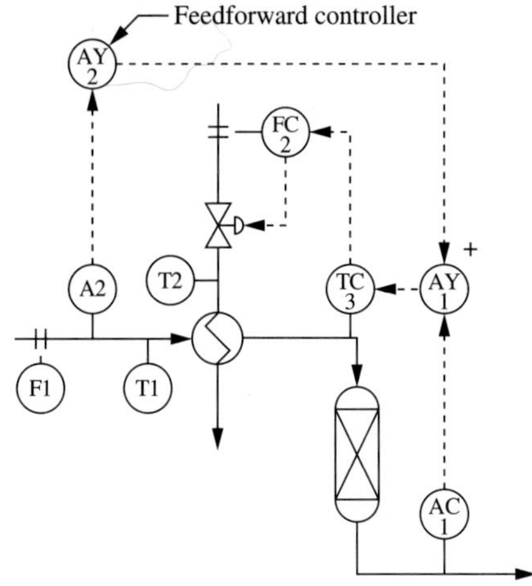

FIGURE 15.11
Packed-bed chemical reactor with feedforward-feedback control.

$$G_{\text{ff}}(s) = \frac{\text{A1}(s)/\text{A2}(s)}{\text{A1}(s)/\text{T3}_{\text{sp}}(s)} = \frac{T3_{\text{sp}}(s)}{\text{A2}(s)} \tag{15.13}$$

Two other locations for combining the feedforward and feedback are possible, but incorrect, in this example; they are at the manipulated variable of the inlet temperature controller, TC-3, and at the manipulated variable of the flow controller, FC-2. (As a thought exercise, the reader should describe the effect of an inlet composition disturbance for these two *incorrect* feedforward-feedback connections and why the performance of the feedforward control would not be good.) In conclusion, care must be taken to combine the feedforward and cascade feedback controllers properly, and the feedforward controller must be derived to be compatible with the cascade control manipulated variable.

Solution. To complete this example, the feedforward controller tuning constants are calculated from the following empirically determined disturbance and process models:

$$G_d(s) = \frac{\text{A1}(s)}{\text{A2}(s)} = \frac{0.3e^{-18s}}{35s+1}\left[\frac{\text{outlet mole/m}^3}{\text{inlet mole/m}^3}\right] \tag{15.14}$$

$$G_{p_1}(s) = \frac{\text{A1}(s)}{\text{T3}_{\text{sp}}(s)} = \frac{-0.19e^{-20s}}{50s+1}\left[\frac{\text{outlet mole/m}^3}{\text{inlet °C}}\right]$$

The resulting controller parameters are determined by applying equations (15.4):

$$\begin{aligned}
\text{Feedforward gain} &= -(0.3/(-0.19)) = 1.58 \;\; [\text{°C/input mole/m}^3]\\
\text{Feedforward lead time} &= \tau_p \qquad\qquad\; = 50 \;\; \text{min}\\
\text{Feedforward lag time} &= \tau_d \qquad\qquad\; = 35 \;\; \text{min}\\
\text{Feedforward dead time} &= \theta_d - \theta_p \qquad = 18\text{–}20 = -2 \;\; \text{min} < 0 \text{ (not possible)}
\end{aligned}$$

Note that the disturbance dead time is smaller than the process dead time. As a result, the feedforward controller requires a negative dead time for perfect compensation. A negative dead time is not possible, because it requires a prediction of the future disturbances; this situation is termed *not physically realizable.* However, since the negative number is small compared to the process dynamics, we can set it equal to the smallest feasible number, which is zero. Based on the example sensitivity of feedforward performance to errors in this chapter (Figure 15.7), an error in the feedforward dead time, as long as it is small, should not significantly degrade the performance. Note that a better way to resolve this problem would be to relocate the inlet analyzer farther upstream; this preferred solution, if possible, would provide an earlier warning and give a longer disturbance dead time.

Example 15.2 Multiple feedforward measurements. In Chapter 14 we learned that a single cascade controller could attenuate the effects of several disturbances. Since feedforward must sense the disturbance to be effective, a separate feedforward controller is required for *each* disturbance. Assuming linearity, the resulting calculations from all feedforward controllers can be added. An example of two feedforward controllers is shown in Figure 15.12 for the stirred-tank heat exchanger. In this case, both the inlet temperature and the inlet flow change significantly. Two separate feedforward controllers calculate individual adjustments for the heating oil flow. They are both added to the feedback signal in the completed strategy.

Sometimes the effects of several measured disturbance variables can be combined into a single feedforward controller. The combination relies on insight into the underlying process models. In the case of the stirred-tank heat exchanger, the following linearized model can be written (see Example 3.7):

$$\rho C_v \frac{dT}{dt} = \rho C_p F(T_{in} - T) + K F_{oil} \tag{15.15}$$

It is clear that the steady-state effects of the disturbances appear in the first term on the right-hand side of the equation. This can be rearranged to yield

$$\Delta F_{oil} = \frac{\rho C_p}{K} \Delta[F(T^* - T_{in})] \tag{15.16}$$

with T^* = the outlet temperature filtered or averaged so that this calculation does not give an improper response to an unmeasured disturbance

Therefore, the steady-state flow change required to compensate for a disturbance can be calculated directly and, assuming it is proportional to the heating valve position, output as the feedforward signal. The controller is shown schematically in Figure 15.13.

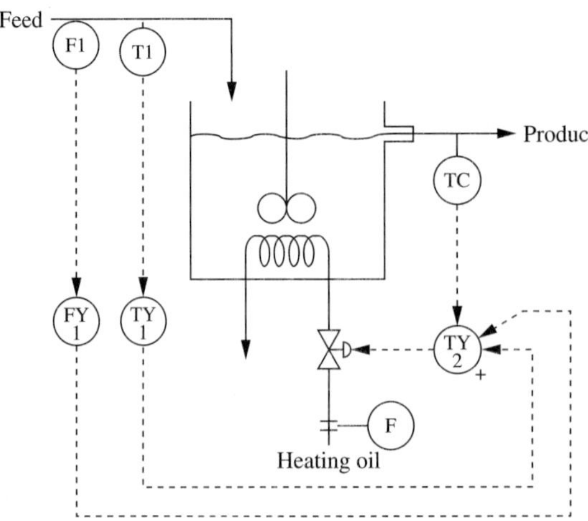

FIGURE 15.12
Stirred-tank heat exchanger with two feedforward controllers.

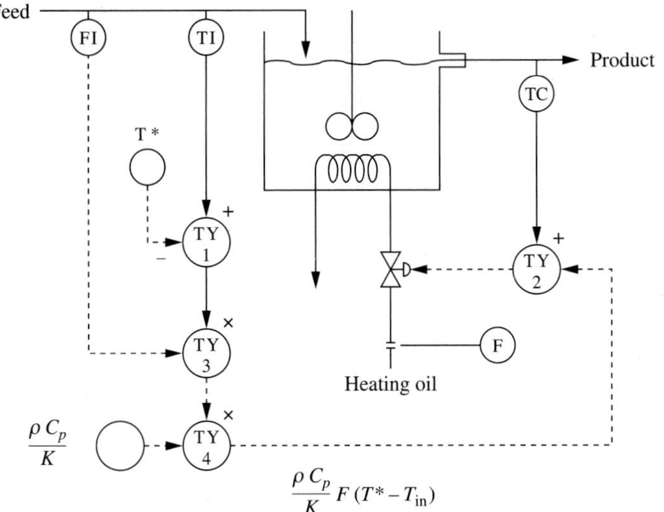

FIGURE 15.13
Stirred-tank heat exchanger with two disturbance variables and one feed-forward controller.

Example 15.3 Feedforward-only control. The derivation for the stirred-tank heat exchanger might lead one to propose a feedforward-only controller derived by setting equation (15.15) to zero and solving for F_{oil}. As mentioned several times already and demonstrated in Figure 15.8*e*, feedforward-only control cannot eliminate steady-state offset. Thus, it should be used only when feedback is not possible.

Example 15.4 Ratio control. One particularly simple form of feedforward control is widely used to maintain flows at desired proportions. The process situation is shown in Figure 15.14*a*, where one of the flows is controlled by another strategy; as far as this process is concerned, it is uncontrolled or *wild.* The other stream can be manipulated

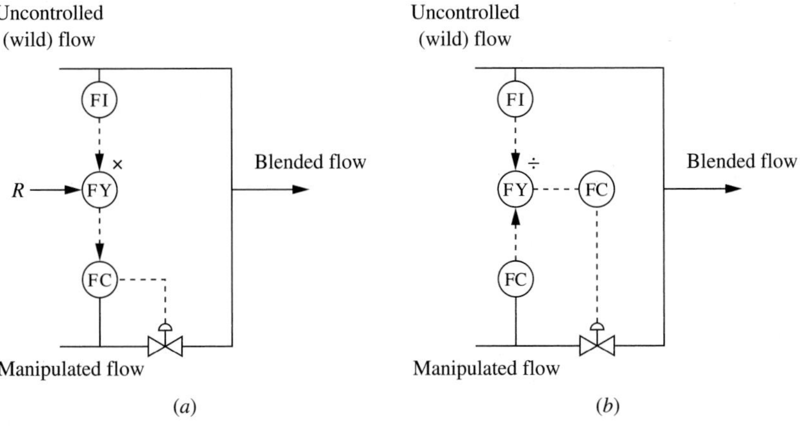

FIGURE 15.14
Flow ratio control: (*a*) steady-state feedforward, (*b*) feedback.

with a valve to achieve the desired composition of the blended stream. The feedforward/feedback strategy measures the flow rate of the uncontrolled stream and adjusts the flow of the manipulated stream to maintain the desired ratio. The feedforward controller uses the measurement of the uncontrolled flow, multiplied by a gain, and outputs to the set point of the feedback flow controller. Because of the fast dynamics, no dead time or lead/lag is required. Note that the ratio control provides feedforward-only compensation; if strict composition control is required, a composition sensor can be placed in the mixed stream and used with a PID controller to achieve zero steady-state offset by adjusting the ratio R.

An alternative approach is also used in practice. This approach achieves the same goal, but it does not satisfy the criteria for a feedforward controller. The ratio controller shown in Figure 15.14*b* uses the two flow measurements to calculate the actual ratio and adjusts the valve to achieve the desired value. The control calculation in this design could be a feedback PI controller with a calculated controlled variable rather than a single measured variable. Again, this ratio design does not guarantee zero steady-state offset of the composition.

Example 15.5 Flow disturbances. As the material passes through the plant, the flow rate is varied to control inventories. As a result, the flow may not be as constant throughout the plant as it is at the inlet. This situation is further explained in Chapter 18 on level control. Feedforward control is very effective in attenuating disturbances resulting from flow rate disturbances. An example of fired-heater control is given in Figure 15.15. The temperature of the fluid in the coil at the outlet of the heater is to be controlled. The flow rate sensor is a reliable, inexpensive feedforward measurement, and the combined feedforward-feedback strategy is very effective. Similar flow rate feedforward can be applied to other processes such as distillation and chemical reactors.

Example 15.6 Fired heaters. Several types of feedforward control to improve the control performance of a fired heater are possible. One approach, shown in Figure

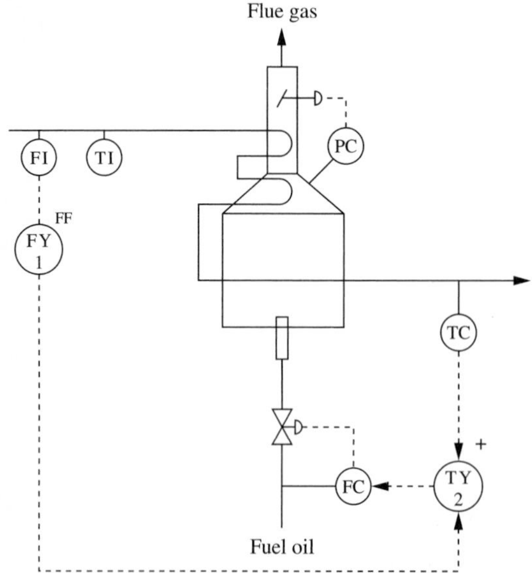

FIGURE 15.15
Example of feed rate feedforward control applied to a fired heater.

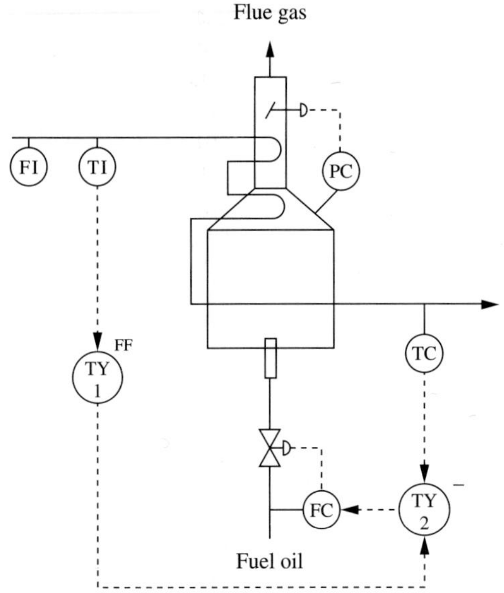

FIGURE 15.16
Example of inlet temperature feedforward control applied to a fired heater.

15.16, measures the inlet temperature. If this temperature varies significantly and tight outlet temperature control is important, the feedforward strategy shown can be used to compensate for the disturbance.

Another example of feedforward control is given in Figure 15.17 for a heater with two fuels. In this case, one fuel is not controlled by the heater system; this can occur when the fuel is a byproduct in another section of the plant and large economic incentives exist for consuming the fuel, here designated as A. To prevent the variations in the byproduct fuel from upsetting the outlet temperature, a feedforward controller adjusts the manipulated fuel flow (B) to maintain the *heat fired* (i.e., the sum of the fuel

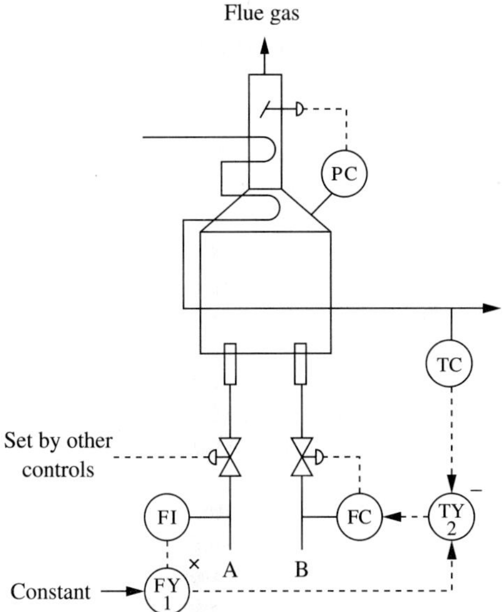

FIGURE 15.17
Example of feedforward compensation for a wild fuel being consumed in a fired heater.

rates times their heats of combustion) at the desired value. The feedforward controller usually needs no dynamic elements but must consider the differences in heats of combustion in its calculation. Control designs like the one in Figure 15.17 are widely used in petrochemical plants, which have large fuel byproduct streams.

Also, the general principle demonstrated in the two-fuel furnace can be applied to any process that has two potential manipulated variables of which one is adjusted by another control strategy (i.e., a wild stream). Other examples include (1) the use of two reboilers in distillation, with one (wild) reboiler duty varied to maximize heat integration and the other manipulated to control product purity, and (2) balancing electrical demand with varying (wild) in-plant generation and manipulated purchases.

Example 15.7 Distillation. Distillation columns can have slow dynamics with long dead times and analyzer delays. Therefore, distillation is a good candidate for feedforward control when product composition control is important. In addition, a distillation column has two products, so a disturbance can affect two different controlled variables. The feedforward controller in Figure 15.18*a* provides compensation for changes in the feed flow rate by adjusting the reflux and reboiler flows. The feedforward controller shown in Figure 15.18*b* provides compensation for feed composition. (Note that the feedforward controllers for multivariable systems cannot be designed using equation (15.2) for each controller; interaction must be considered. See question 21.17.) The disturbance models for this controller must be identified empirically, as explained in Section 15.5. If both the feed flow rate and composition vary significantly, the two feedforward controllers can be implemented and their outputs summed.

Feedforward Is Not Used Everywhere

Sometimes engineers have the impression that because feedforward is generally a good idea, it should be applied in all process control strategies. This is not the case. As strongly emphasized in the first design criterion, feedforward is applied *when feedback control does not provide satisfactory control performance.* Thus, feedforward is not used if tight control is not needed or if feedback control provides good control. An example of the first situation is given in Figure 15.19, where the level can vary within limits without influencing the plant economics or safety; thus, feedforward is not applied. An example of the second situation is given in Figure 15.20, where tight control of the mixing process is possible with feedback-only control because the process has almost no dead time. Refer to Table 13.2 (page 463) for the factors that affect feedback control performance.

15.8 FEEDFORWARD CONTROL IS GENERAL

Feedforward control is a way to take corrective action as soon as information on a disturbance is available. In the packed-bed reactor (Figure 15.12), the inlet analyzer AC-2 provides an early warning of a disturbance. The feedforward controller adjusts the inlet temperature without interfering with the feedback controller.

Feedforward control concepts are not limited to engineering control systems. Social organizations also benefit from early response to events. In business, feedforward may be termed "positive preactions"; whatever the name, the improved performance can be dramatic. A hypothetical example of university decision making is

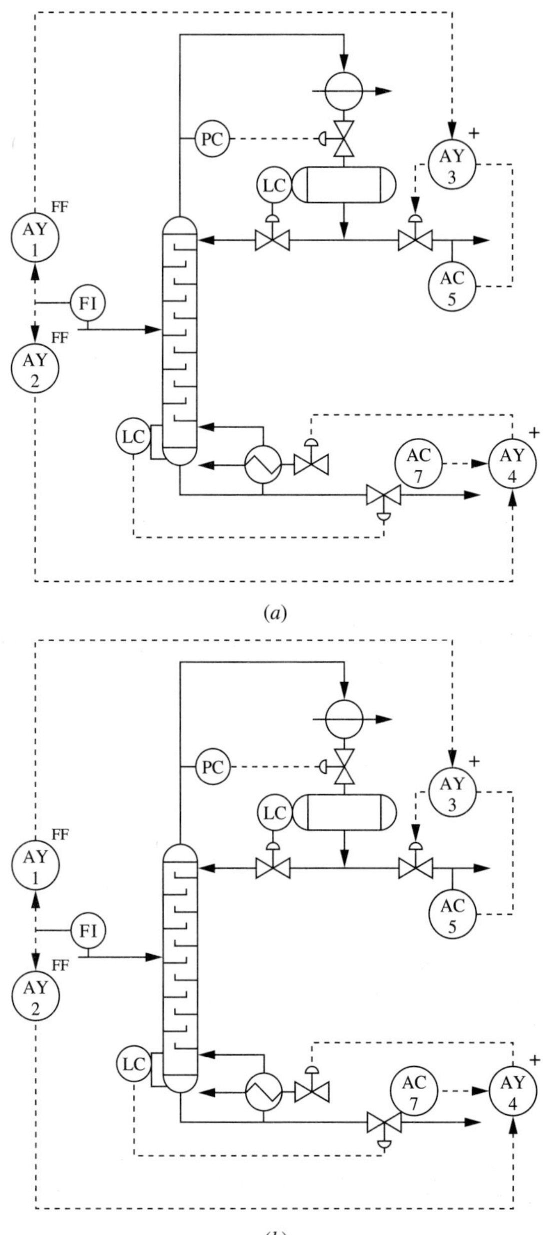

FIGURE 15.18
Feedforward control in a two-product distillation tower: (*a*) from feed rate; (*b*) from feed composition.

given in Figure 15.21. The goal is to have needed faculty, staff, and buildings available for all of the students attending the university. A major variable is the number of students. Therefore, the total number of young people (e.g., 14 years old) in the population can be measured or estimated. Should this number increase significantly, the facilities can be increased over several years so that the university is able to accommodate the demand when it occurs.

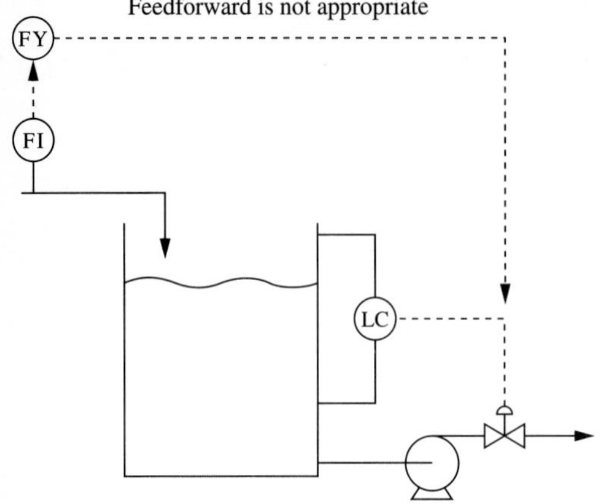

FIGURE 15.19
Feedforward control is not generally required for level control, where the outlet flow manipulations should be smooth.

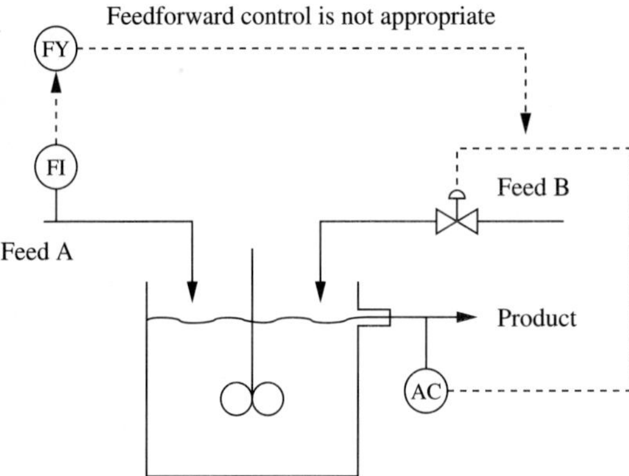

FIGURE 15.20
Feedforward control is not generally required for nearly linear processes with little dead time.

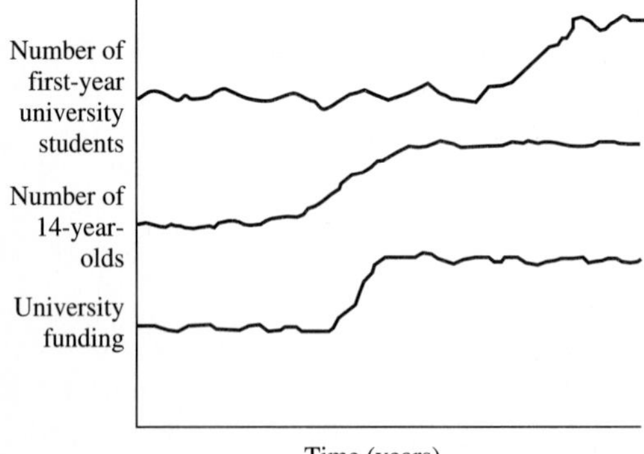

FIGURE 15.21
Example of feedforward control applied to a planning decision in a university.

15.9 CONCLUSIONS

Feedforward control does not employ the feedback principle; it manipulates a system input based on the measured value of a different system input. This approach to control requires new algorithms, with the proper algorithm depending on models of the disturbance and feedback dynamics. As shown in Figures 15.8*a* through *e*, improved performance is achieved without increased variation in the manipulated variable and without the requirement of highly accurate models. Based on this performance improvement and simplicity of implementation, the engineer is well advised to evaluate potential feedforward controls for important controlled variables.

The first few times engineers evaluate feedforward, they must perform careful studies like the one in Table 15.3, but after gaining some experience they will be able to design feedforward control strategies quickly without explicitly writing the criteria and table.

Feedforward control is not universally applicable; the design criteria in Table 15.1 can be used to determine whether feedforward is appropriate and, if so, to select the best feedforward variable. If it is not immediately possible and improved performance is required, the engineer should investigate the possibility of adding the necessary sensor. However, feedforward control is effective only for the measured disturbance(s); thus, additional enhancements, such as cascade and feedback from the final controlled variable, should be used in conjunction with feedforward.

REFERENCES

Gallun, S., C. Matthews, C. Senyard, and B. Slater, "Windup Protection and Initialization for Advanced Digital Control," *Hydro. Proc.,* 63–68, June 1985.

Shinskey, F.G., *Process Control Systems,* McGraw-Hill, 1988.

Smith, C., *Digital Computer Process Control,* Intext, Scranton, NJ, 1972.

QUESTIONS

15.1. (*a*) In your own words, discuss the feedforward control design criteria. Give process examples in which feedforward control is appropriate and not appropriate.

(*b*) One of the design criteria for feedforward control requires that the feedforward variable not be affected by the manipulated variable. Why is this required? If the variable were influenced by the manipulated variable, what control design would be appropriate?

(*c*) In a feedforward-feedback control strategy, which controller should be tuned first? What would be the effect of reversing the order of tuning? Clearly state any assumptions you have used.

(*d*) Describe how the addition of feedforward control to an original feedback-only system affects the resonant frequency and the amplitude ratio (controlled to measured disturbance) at the resonant frequency.

(*e*) Discuss why the last design rule in Table 15.1 is valid when feedforward is applied in conjunction with feedback. Is it also valid for feedforward-only?

(*f*) Review the factors in Table 13.2 and determine which factors are influenced by feedforward control.

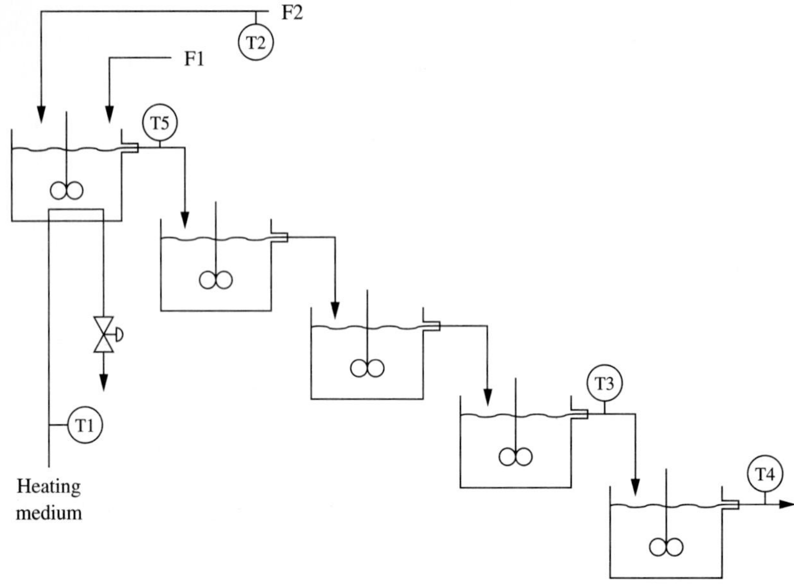

FIGURE Q15.2

15.2. In this question, you will design control strategies for the system of stirred tanks in Figure Q15.2. The measurements and manipulated variable are shown in the figure; you may not alter them and need not use them all. The following information will help you design the strategy.

1. The goal is to control the temperature T_4 at the outlet of the last tank tightly. The major disturbance is the temperature of one of the feed streams, T_2.
2. The flow rates to the stirred tanks cannot be changed by your control strategy and are essentially constant; $F_1 = 5\ \text{m}^3/\text{min}$ and $F_2 = 5\ \text{m}^3/\text{min}$.
3. The volume of each tank is $10\ \text{m}^3$.
4. At the flow rates given, the steady-state gain for the heating coil is 1°C/% (change in first tank temperature per change in % valve opening). The sensor, valve, and heating coil dynamics are negligible, and the heat losses are small but not negligible.

(*a*) Decide whether a single-loop feedback control strategy is possible. Explain your answer. If yes, draw the single-loop control system on the process figure and define each control algorithm.

(*b*) Decide whether a cascade control strategy is possible, yes or no. Explain your answer. If yes, draw the best cascade control system on the process figure and define each controller algorithm.

(*c*) Decide whether a feedforward/feedback control strategy is possible, yes or no. Explain your answer. If yes, draw the best feedforward/feedback control system on the process figure and define each controller algorithm.

(*d*) Rank the strategies in (*a*) through (*c*) that are possible according to their control performance; that is, the ability to control the outlet temperature T_4. Explain the ranking.

(*e*) For the *best* strategy, calculate all parameters for the control algorithms: gains, integral times, leads, lags, dead times, and so forth. (*Hint:* You must develop analytical models and transfer functions for the relevant input-output relationships.)

15.3. The feedforward-feedback strategy has an additional sensor and controller. How is it possible to add these and not violate the degrees of freedom of the system? For the heat exchanger example in Section 15.2,

(*a*) Derive all equations describing the process and the feedforward-feedback controllers.

(*b*) Analyze the degrees of freedom to verify that the system is exactly specified.

(*c*) Discuss how you would solve the equations in (*a*) numerically for a dynamic response (simulate the process with a digital control system).

15.4. Propose feedforward/feedback control designs for the following systems, where possible. Draw the design on a sketch of the process and verify the design using the feedforward design criteria. The processes, with [controlled/disturbance] variables, are

(*a*) Example 14.1 [A1/T2]

(*b*) Figure 14.17 (T2/fuel flow)

(*c*) Figure 14.20 [tank temperature/fresh coolant temperature]

(*d*) Figure Q13.2 [outlet concentration/C_A]

(*e*) Figure 7.5 and Example 9.2 [$x_{A3}/(x_A)_B$]

(*f*) Figure Q8.12 [outlet concentration (AC)/ flow of stream C]

15.5. Derive the transfer function in equation (15.6) based on the block diagram in Figure 15.6.

15.6. Verify that a feedforward-feedback control system has zero steady-state offset for a measured disturbance. What restrictions must you place on the disturbance, feedback process, and control algorithms in your derivation?

15.7. The following transfer functions have been evaluated for the process in Figure 15.15, with time in min:

$$\frac{T_{\text{out}}(s)}{T_{\text{in}}(s)} = \frac{0.40e^{-1.8s}}{3.5s + 1} \qquad \frac{T_{\text{out}}(s)}{T_{\text{fuel}}(s)} = \frac{0.1e^{-1.1s}}{4.2s + 1}$$

(*a*) Determine the continuous feedforward and feedback algorithms and the values of all adjustable parameters.

(*b*) Determine the digital feedforward and feedback algorithms and the values of all adjustable parameters, including the execution period.

15.8. (*a*) Describe how to program a digital feedforward-feedback controller so that the automatic/manual status of each controller can be changed independently.

(*b*) Describe how to initialize the feedforward controller.

(*c*) Derive the algorithm for an incremental (or velocity) form of the feedforward algorithm that calculates the change in the manipulated variable at each execution.

(*d*) Discuss the possibility of integral windup caused by feedforward control.

15.9. In Example 15.2, the tank temperature was replaced with a "filtered" value, T^*. Explain why this was done. Can this analysis be generalized to an additional criterion for feedforward control with calculated variables?

15.10. In the description of the control design for a packed-bed reactor in Example 15.1, the correct location for combining the feedforward and feedback controllers is explained. Discuss the behavior of the control system for the two improper locations, adding the feedforward to (*a*) the outlet of the T_3 controller and (*b*) the outlet of the F_2 controller. How would the control system respond to a disturbance for each of the improper connections?

15.11. Assume that the feedback process dynamics has an inverse response and the disturbance has second-order dynamics. This is the situation that occurs for the series reactors in Examples 5.4 and 13.8 when the solvent flow is the manipulated variable and the flow of A is the measured disturbance. The structures for the models can be simplified to the forms

$$G_d(s) = \frac{C_{A2}(s)}{F_A(s)} = \frac{K_d}{(\tau_{d1}s + 1)(\tau_{d2}s + 1)}$$

$$G_p(s) = \frac{C_{A2}(s)}{F_s(s)} = \frac{K_p(-\tau_{1d}s + 1)}{(\tau_1 s + 1)(\tau_2 s + 1)}$$

with all $\tau > 0$. The model structure and parameters for $C_{A2}(s)/F_s(s)$ can be found in Example 5.4 and the model structure for $C_{A2}(s)/F_A(s)$ in the answer to question 5.4. Answer the following questions about this system.

(*a*) Is perfect feedforward control (no deviation in the controlled variable) possible for this system?

(*b*) Derive the feedforward controller transfer function for this system. Sketch the shape of the response of the manipulated and controlled variables to a step change in the measured disturbance with feedforward-only control.

(*c*) Based on the answers in (*a*) and (*b*), propose a modified feedforward controller that provides acceptable performance. Substitute numerical values for the reactor process.

(*d*) Answer parts (*a*) through (*c*) for the system in which the process and disturbance transfer functions given previously are switched. This situation would occur in the series reactors if the flow of reactant A were the manipulated variable and the solvent flow were the disturbance.

(*e*) Can the results in (*a*) through (*d*) be used to develop a general conclusion on the effects of right-half-plane zeros on feedforward-feedback control performance?

15.12. Given the processes in Figure Q15.12, place them in order of how much each would benefit from feedforward control for a disturbance measured by analyzer A. Explain your ranking.

15.13. The feedforward control from set point given in Figure Q15.13 has been suggested.

(*a*) Derive the transfer function for the set point feedforward controller, $G_{sp}(s)$.

(*b*) Discuss this controller. Is it possible to implement, and how would it affect the dynamic response of the controlled and manipulated variables?

(*c*) Discuss the need for a set point feedforward if the feedback controller uses a PID algorithm.

15.14. The feedforward controller was derived to provide perfect control. Using the block diagram in Figure 7.4, derive the *feedback* controller that gives perfect control. Are there any reasons why this controller is not practical?

15.15. (*a*) Verify that all designs in Section 15.7 satisfy the feedforward design criteria.

(*b*) In the description of flow ratio control, it was not specified whether the orifice ΔP measurements were used or their square roots were used. Which is correct and why?

(*c*) Derive the analytical relationship in equation (15.5) for the output of a lead/lag element when the input experiences a step change.

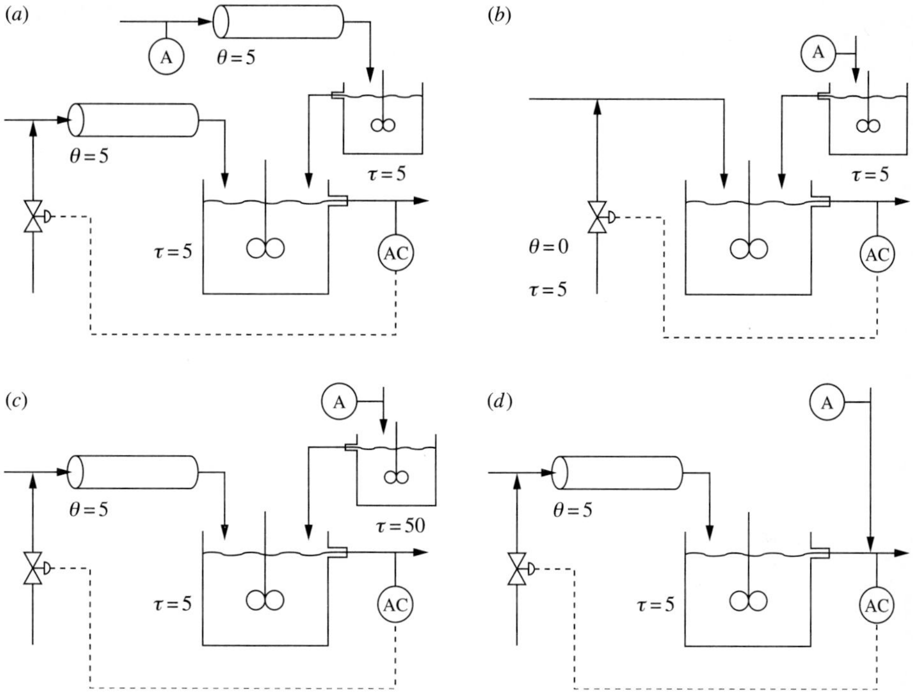

FIGURE Q15.12
The solid lines contribute negligible dead time.

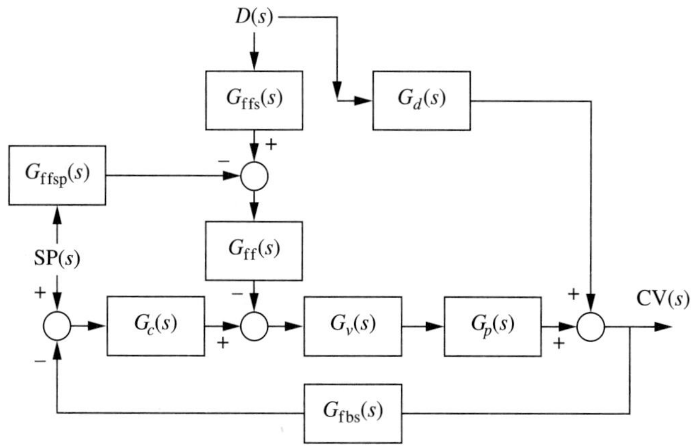

FIGURE Q15.13

(*d*) Explain the feedforward calculation for Figure 15.17. Give the equations and the physical property data required.

15.16. Discuss one example of feedforward control in each of the following categories: university, government, and business organizations.

CHAPTER
16

ADAPTING SINGLE-LOOP CONTROL SYSTEMS FOR NONLINEAR PROCESSES

16.1 INTRODUCTION

Linear control theory provides methods for the analysis and design of many successful control strategies. Control systems based on these linear methods are generally successful in the process industries because (1) the control system maintains the process in a small range of operating variables, (2) many processes are not highly nonlinear, and (3) most control algorithms and designs are not sensitive to reasonable ($\pm 20\%$) model errors due to nonlinearities. These three conditions are satisfied for many processes, but they are not satisfied by all; therefore, the control of nonlinear processes must be addressed.

It is possible that the response of a nonlinear system could give better performance than a linear system and, therefore, a nonlinear control calculation might be better than any linear algorithm. However, there is no recognized, *general* nonlinear control theory that has been successfully applied in the process industries. (An example of a nonlinear algorithm applied to level control is given in Chapter 18.) Therefore, the goal of the approaches in this chapter is to attain the performance achieved with a well-tuned linear controller. To reach this goal, the control methods in this chapter attempt to achieve a system that has a linear closed-loop relationship. If an element in the control loop is nonlinear, the approach applied here is to introduce a *compensating nonlinearity,* so that the overall closed-loop system behaves approximately linearly. This compensating nonlinearity may be introduced in the control algorithm or in physical equipment, such as a sensor or final element.

The next section begins the analysis by introducing a method for determining when nonlinearities significantly affect a control system. This analysis is extended to evaluate the proper fixed set of tuning constants for a linear PID controller applied to a nonlinear process. If a fixed set of tuning constants and linear instrumentation are not satisfactory, improvements can be achieved by adapting either the control calculation or the equipment responses. First, a common method for adapting the controller tuning in real time to compensate for nonlinearities is presented. Then the same concept is applied to introduce compensating nonlinearities in selected instrumentation, such as the control valve, to improve performance.

16.2 ANALYZING A NONLINEAR PROCESS WITH LINEAR FEEDBACK CONTROL

A relatively simple process is analyzed in this section so that analytical models can be derived; the general approach is applicable to more complex processes. The process is shown in Figure 16.1*a*, which is the three-tank mixer considered in Examples 7.2 and 9.4 (pages 249 and 303). The outlet concentration of the last tank is controlled

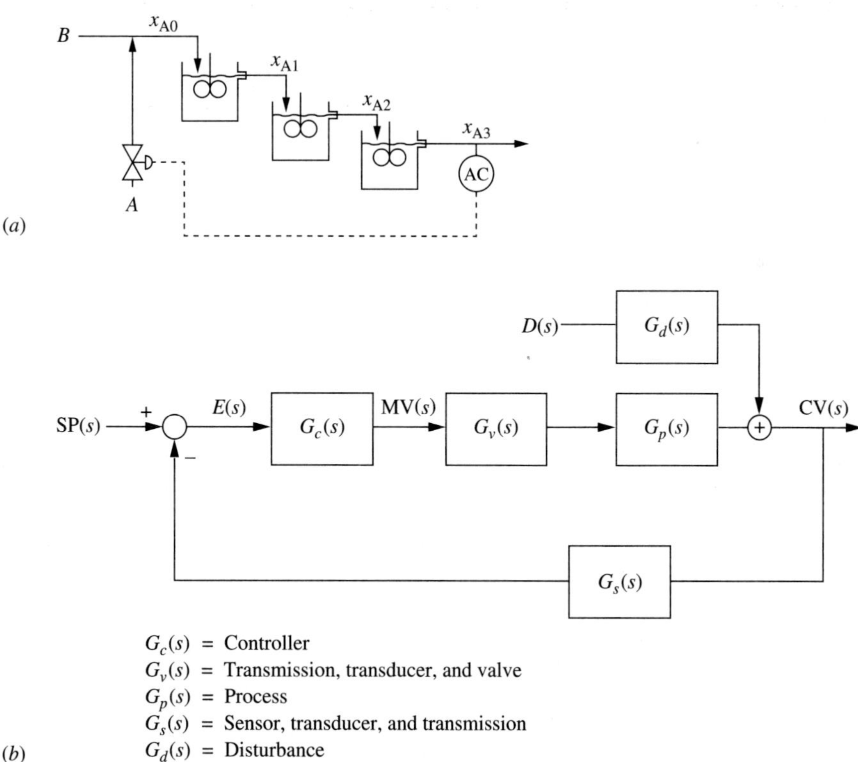

FIGURE 16.1
Mixing process. (*a*) Schematic; (*b*) control system block diagram.

by adjusting the addition of component A to the feed to the first reactor. The relevant modelling assumptions are the same as given in Example 7.2, except that we now allow the flow of the main stream, F_B, to vary. The equations describing the system are derived in Example 7.2 and summarized as follows:

$$x_{A0} = \frac{F_B(x_A)_B + F_A(x_A)_A}{F_B + F_A} \tag{16.1}$$

$$V_i \frac{dx_{Ai}}{dt} = (F_A + F_B)(x_{A(i-1)} - x_{Ai}) \qquad \text{for } i = 1, 3 \tag{16.2}$$

Note that the differential equations are nonlinear. (See Section 3.4 for a refresher on the definition of linearity.) We can linearize these equations, express the variables as deviations from the initial steady state, and take the Laplace transforms to yield the transfer function model:

$$\frac{x_{A3}}{\text{SP}(s)} = \frac{G_p(s)G_v(s)G_c(s)}{1 + G_p(s)G_v(s)G_c(s)G_s(s)} \approx \frac{G_p(s)G_c(s)}{1 + G_{OL}(s)} \tag{16.3}$$

with the valve transfer function a constant lumped into $G_p(s)$ and the sensor $G_s(s) = 1.0$.

$$G_{OL}(s) = G_p(s)G_v(s)G_c(s)G_s(s) \approx G_p(s)G_c(s) \tag{16.4}$$

$$= \frac{K_p}{(\tau s + 1)^3} G_c(s)$$

with

$$K_p = K_v \left[\frac{F_{Bs}(x_{AA} - x_{AB})_s}{(F_{As} + F_{Bs})^2} \right] \qquad K_v = 0.0028 \tag{16.5}$$

$$\tau = \frac{V}{F_{Bs} + F_{As}} \tag{16.6}$$

This linearized model clearly demonstrates how the gain and time constants depend on the volumes, total flow, and compositions. We will consider the response of the system for various values of one operating variable, the total flow rate $(F_A + F_B)$, which has the greatest variability for the situation considered here. In the scenario, the production rate changes periodically and remains nearly constant for a long time (relative to the feedback dynamics) at each production rate. The process dynamics are summarized in Table 16.1 for the range of flow variability (i.e., production rates)

TABLE 16.1
Summary of process dynamics and tuning for the three-tank mixing process*

Case	Process parameters			Controller parameters		
	F_B	K_p	τ	K_c	T_I	T_d
A	3.0	0.087	11.4	13.8	25.1	1.82
B	4.0	0.064	8.6	18.6	19.0	1.4
C	5.0	0.052	6.9	23.1	15.2	1.10
D	6.0	0.043	5.7	27.9	12.7	0.92
E	6.9	0.039	5.0	30.0	11.0	0.80

*For the combinations of process dynamics and tuning in this table, the gain margin for each case is 1.7.

expected. The variation in the process dynamics due to the nonlinearity is not randomly distributed, because in this example the effect of an increase in flow rate is to decrease the process gain and time constants concurrently. This type of correlation is typical for nonlinear processes and demonstrates the need for careful analysis of the dynamic responses at different operating conditions. Naturally, other factors, such as incorrect data in fundamental models or noise in empirical models, contribute to the modelling errors, but change in operating conditions is often the dominant factor causing the difference between models and true process behavior.

> The values in Table 16.1 demonstrate that the changes in dynamic model parameters due to nonlinearity can be *highly correlated.*

The effects of nonlinearity on two important system characteristics, stability and performance, are now investigated. As demonstrated in Chapter 10, the process dynamics influence the stability of a closed-loop system, and to achieve a stable control system the tuning parameters are adjusted to be compatible with the process dynamics. The PID feedback controller tuning for this process has been determined for five different flow rates that span the expected range of operation. (Note that Case E in Table 16.1 is the same as Example 9.2.) The tuning was determined by evaluating the process reaction curve, fitting an approximate first-order-with-dead-time model, and using the Ciancone correlations (see Chapter 9). Similar trends would be obtained for other tuning methods such as Ziegler-Nichols. It is important to recognize that the tuning reported in Table 16.1 has a reasonable margin from the stability limit for every case. In fact, the gain margin for all cases is about 1.7. The results in the table clearly indicate that the values of "good" tuning constants change significantly, over 50%, for the range of process operating conditions considered. This analysis indicates that the nonlinearity is significant for the changes in flow considered in this scenario.

The simplest control design approach would be to use a *single set* of tuning constants for all the operating conditions. The results in Table 16.1 provide the basic information needed to decide whether to use this tuning approach. If the tuning constants were not very different, it would be concluded that either the nonlinearities are mild or the operating conditions do not change much from the base case. For either situation, a constant set of tuning constants, which could be taken as the average values, would yield good PID feedback control performance.

If the proper values of the tuning constants differ significantly, as they do in Table 16.1, further analysis is necessary. Recall that the tuning for each case was determined to give good dynamic response and a proper gain margin for the nominal process model in that case. The single set of tuning constants to be used for all process models in the table must provide acceptable (if not good) feedback control performance for all cases. Since the process dynamics change, the stability margin of the closed-loop system can change, and the closed-loop system can become highly oscillatory or unstable for an improper choice of fixed tuning. Since instability and severe oscillations are to be avoided, the overriding concern is maintaining a reasonable stability margin for all expected process dynamics.

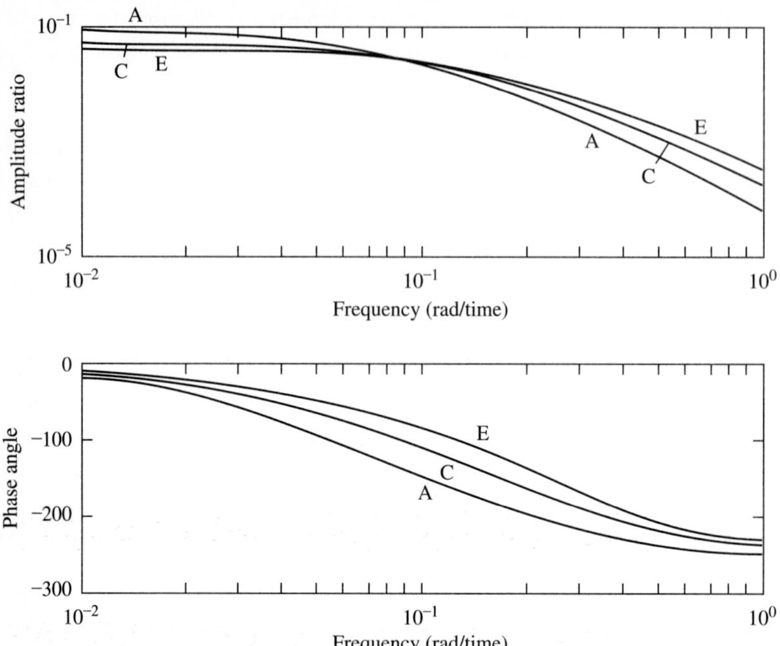

FIGURE 16.2
Bode plot for three-tank mixing system (cases defined in Table 16.1).

To ensure that the control system with varying process dynamics performs acceptably over the expected range of operation, the *worst-case dynamics* must be identified. This worst case gives the poorest control performance under the feedback controller and is usually the closed-loop system closest to the stability limit. The Bode plots of $G_p(s)$ for three of the cases in Table 16.1 are given in Figure 16.2. The results show that Case A has the lowest critical frequency and the highest amplitude ratio at its critical frequency. This result conforms to our experience that processes with longer time constants are more difficult to control. Thus, Case A would be selected as the most difficult process operation, or the worst case, within the scenarios.

The Bode analysis of $G_p(s)$ is substantiated by the results in Table 16.1, which indicate that Case A has the least aggressive feedback controller, because the controller gain is smallest and integral time is largest. Applying the controller tuning from Case A would result in a stable system for all cases, albeit with large controlled-variable deviations from set point and large IAE for some cases. Using a more aggressive set of tuning constants, Case E for example, would lead to good performance in some cases, but the closed-loop performance would be very poor, and perhaps unstable, for other cases.

Dynamic simulations of closed-loop systems with various tunings are shown in Figures 16.3 and 16.4. The results in Figure 16.3*a* and *b* give the dynamic responses of the closed-loop system, with controller tuning from Case A, for two different

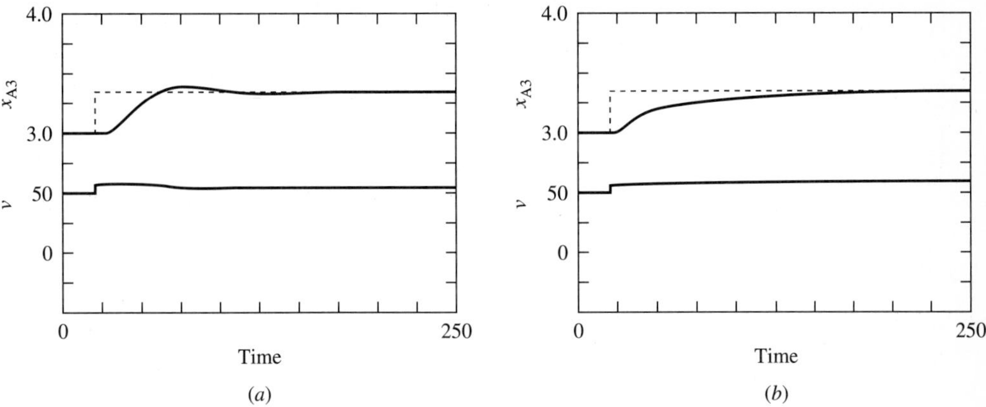

FIGURE 16.3
Dynamic responses for the mixing system with tuning from Case A ($K_c = 13.8$, $T_I = 25.1$, and $T_d = 1.82$). (*a*) Case A process dynamics, $F_B = 3$, gain margin = 1.7; (*b*) Case E process dynamics, $F_B = 6.9$, gain margin = 4.5.

process dynamics. Note that the response, when controlling the plant with dynamics for Case A (the most difficult plant to control), is well behaved. The performance when controlling process E is rather poor, with a long time required to return to set point, but at least the response is stable.

The results in Figure 16.4*a* and *b* give the closed-loop dynamic responses for the controller tuning from Case E and the same two plant dynamics. Although the performance for the process dynamics from Case E is good, the performance for the process dynamics from Case A is unacceptable because the system is *unstable.* Since excessive oscillations and instability are to be avoided at all cost, the controller tuning used in Figure 16.4, based on the dynamics in Case E, is deemed unacceptable.

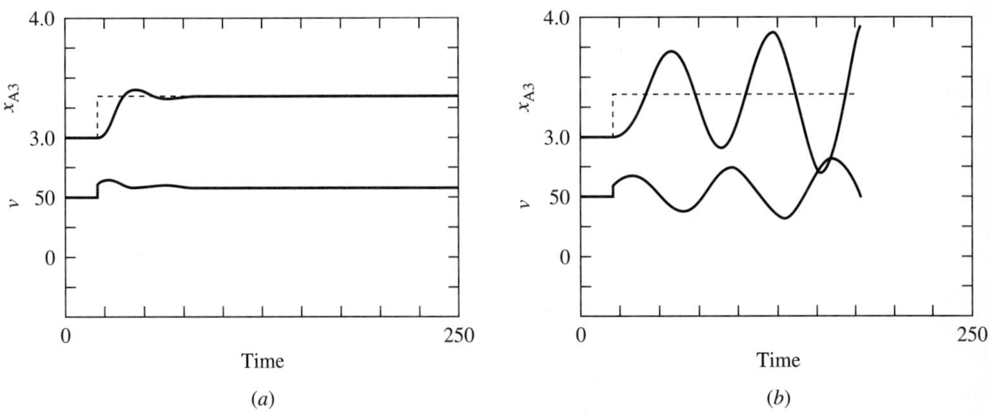

FIGURE 16.4
Dynamic responses for the mixing system with tuning from Case E ($K_c = 30.0$, $T_I = 11.0$, and $T_d = 0.80$). (*a*) Case E process dynamics, $F_B = 6.9$, gain margin = 1.7; (*b*) Case A process dynamics, $F_B = 4.0$, gain margin < 1.0, indicating instability.

> When the feedback controller tuning constants are fixed and the process dynamics change, the fixed set of tuning constants selected should have the proper gain margin for the most difficult process dynamics in the range considered. This approach will ensure stability, but it may not provide satisfactory performance.

This section has presented a manner for determining whether nonlinearities significantly affect stability and control performance. The method is based on the stability analysis of the system linearized about various operating conditions. Also, a tuning selection criterion is given that is applicable when a fixed set of tuning constants is used as the process dynamics vary. The goal of this criterion is to provide the best possible control performance, with constant tuning, while preventing instability or excessive oscillations. The resulting control performance may be unacceptably poor, providing sluggish compensation for some cases; therefore, the next sections present common methods for improving performance, while preventing instability, by compensating for the nonlinearity.

16.3 IMPROVING NONLINEAR PROCESS PERFORMANCE THROUGH DETERMINISTIC CONTROL LOOP CALCULATIONS

The approach described in the previous section can lead to poor control performance for two reasons. First, some process operating conditions lead to poor performance because of increased feedback dynamics (e.g., longer dead time and time constants). Second, the fixed values for the feedback controller tuning constants are too "conservative" for some process operations. Clearly, modifying the tuning cannot prevent degradation in feedback control performance arising from the changes in plant dynamics. However, modifying the tuning to be compatible with the current process dynamics can maintain the feedback control performance close to the best possible with the PID algorithm for whatever plant dynamics exist.

The approach for modifying the controller tuning constants through deterministic calculations can be applied to improve the control of some nonlinear processes. The term "deterministic" is used to designate an unchanging relationship between the operating condition and the tuning constant values. The operating condition is determined by measuring a process variable that is directly related to the feedback dynamics. Then the control constants can be expressed as a function of this measured variable, PV, as shown in the following equation.

$$\mathrm{MV} = K_c(\mathrm{PV})\left(E + \frac{1}{T_I(\mathrm{PV})}\int_0^t E(t')dt' + T_d(\mathrm{PV})\frac{d\mathrm{CV}}{dt}\right) + I \qquad (16.7)$$

The resulting controller is *nonlinear.* The stability analysis presented in Chapter 10 can be applied to this system assuming that the value of PV in equation (16.7) changes slowly; that is, it has a much lower frequency than the closed-loop critical frequency. When this condition is satisfied, the tuning can be considered constant for the stability calculation.

This approach is demonstrated by applying it to the three-tank mixing system introduced in the previous section. The correlations between the tuning constants and the measured variable that indicates the change in process dynamics—in this example the flow—can be fitted by an equation or arranged in a lookup table. Equations for the example are given below for the range of operation in Table 16.1 $(3.1 < (F_A + F_B) < 7.0)$ with the parameters determined by a least squares fit.

$$\begin{aligned} K_c &= -5.64 + 7.368(F_A + F_B) - 0.3135(F_A + F_B)^2 \\ T_I &= 50.37 - 10.626(F_A + F_B) + 0.7164(F_A + F_B)^2 \\ T_d &= 3.66 - 0.776(F_A + F_B) + 0.0525(F_A + F_B)^2 \end{aligned} \tag{16.8}$$

The controller using tuning calculated by equations (16.8) can be applied to the nonlinear mixing process. The resulting dynamic responses are essentially the same as in Figure 16.3*a* and Figure 16.4*a*. The control performance is good for the different flow rates because the tuning is modified to be compatible with the process dynamics. Note that the performance in Figure 16.4*a* is better than in Figure 16.3*a*, even though the controllers in both systems are well tuned, because the feedback dynamics in Case E are faster. Comparison with Figures 16.3*b* and 16.4*b* demonstrates the potential performance advantage of this approach over maintaining the tuning constants at fixed values. The procedure introduced in this example is summarized in Table 16.2.

The use of controller tuning modifications described in this section is often referred to as *gain scheduling,* because early applications adjusted only the controller gain. With digital computers, all tuning constants can be easily adjusted when required to achieve the desired control performance.

If adequate control performance is achieved through adapting only the controller gain and the controller gain should be proportional to feed flow, gain scheduling can be implemented as part of modified feedforward/feedback control design. An example is given in Figure 16.5*a* for the feedforward and feedback control of the simple mixing system. The model for the system is

$$x_m = \frac{F_A x_A + F_B x_B}{F_A + F_B} \tag{16.9}$$

TABLE 16.2
Criteria for the deterministic modification of controller tuning

Deterministic modification of tuning is appropriate when
1. Constant controller tuning values do not provide satisfactory control performance because of significant changes in operating conditions.
2. The nonlinearity can be predicted based on a process variable measured in real time.
3. The relationship between the measurement and the process dynamics can be determined either from a fundamental model or from empirically developed models.
4. The changes in the process dynamics are at a frequency much lower than the critical frequency of the control system.

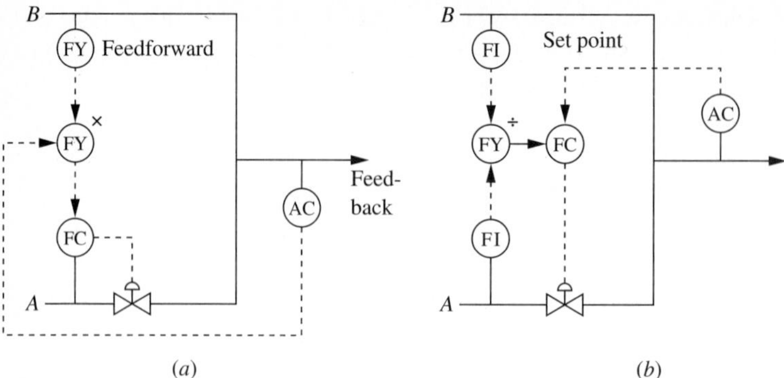

FIGURE 16.5
Examples of gain scheduling through feedforward control.

The flows and compositions for this mixing process are assumed to be the same as for the three-tank mixing process. The steady-state equations describing the system are

$$G_p(s) = \frac{F_B[(x_A)_A - (x_A)_B]}{(F_A + F_B)^2} \approx \frac{K_p}{F_B} \tag{16.10}$$

$$G_{cff}(s) = F_B \qquad G_{cfb}(s) = K_c\left(1 + \frac{1}{T_I s}\right)$$

with K_p and K_c constant. The stability margin is determined from the Bode analysis by referring to $G_{OL}(s)$, which follows for the example:

$$\begin{aligned} G_{OL}(s) &= G_p(s) K_{cff} K_c\left(1 + \frac{1}{T_I s}\right) \\ &= \frac{K_p}{F_B} F_B K_c\left(1 + \frac{1}{T_I s}\right) \end{aligned} \tag{16.11}$$

To include the modification of the loop gain as indicated in equation (16.11), the outputs of the feedforward and feedback controllers are *multiplied*, rather than added as described in Chapter 15. Thus, as the feed flow increases, the effective gain of the feedback controller ($K_c F_B$) increases to compensate for the decrease in the process gain (K_p/F_B). Note that this design is an extension of the feedforward design shown in Figure 15.14*a* to include feedback and therefore retains the good disturbance response through feedforward control. Also, this approach to controller gain modification is a simplification of the general approach described in Table 16.2.

16.4 IMPROVING NONLINEAR PROCESS PERFORMANCE THROUGH CALCULATIONS OF THE MEASURED VARIABLE

In addition to the controller calculation, other elements in the control loop can also be modified in response to nonlinearities. Relationships between the sensor signal and the true process variable sometimes involve particularly simple nonlinearities that can be addressed by programmed calculations during the input processing phase of the control loop. Some examples are temperature (polynomial fit of thermocouple) and flow orifice (square root, density correction on ΔP). In addition to the linearization in the control loop, the availability of more accurate measured values for use in control and process monitoring is another important benefit of these calculations.

16.5 IMPROVING NONLINEAR PROCESS PERFORMANCE THROUGH FINAL ELEMENT SELECTION

The same results in the previous sections (introducing a compensating nonlinearity in the control loop) can be achieved by selecting appropriate control equipment to compensate for nonlinearities. The final control element, usually a valve, is the control loop element that is often modified in the process industries, because the modifications involve little cost. Again, the explanation in this section assumes that the desired closed-loop relationship is linear; if another relationship is required, the approach can be altered in a straightforward manner.

Since the valve is normally very fast relative to other elements in the control loop, only the gains of the elements are considered to vary. The linearized control system is shown in Figure 16.1*b*, and the loop gain for this system depends on the product of the individual gains.

$$
\begin{aligned}
G_{OL}(s) &= G_p(s)G_v(s)G_c(s)G_s(s) \qquad (16.12)\\
&= K_p K_v K_c K_s G_p^*(s)G_v^*(s)G_c^*(s)G_s^*(s)
\end{aligned}
$$

where the gains (K_i) may be a function of operating conditions, and the dynamic elements of the transfer functions($G_i^*(s)$ with $G_i^*(0) = 1.0$) do not change significantly with operating conditions. The manipulated variable in the majority of control loops is a valve stem position (v), also referred to as the *valve lift,* which influences a flow rate. The feedback system behaves as though it is linear if $G_{OL}(s)$ does not change as plant operating conditions change. Thus, linearity can be achieved, even if the process gain (K_p) changes, as long as the changes due to nonlinearities in the individual gains cancel. In this section, a method is described in which the valve nonlinearity is designed to cancel an undesirable process nonlinearity, with the controller (K_c) and sensor (K_s) gains assumed to be constant.

The final element selection is introduced through an example of flow control. The relationship between the controller output and the flow is often desired to be linear, so that the control system is linear. The relationship between the valve stem position and the flow is given below (Foust et al., 1960; Hutchinson, 1976).

$$
F = F_{\max}\left(\frac{C_v(v)}{100}\right)\sqrt{\frac{\Delta P_v}{\rho}} \qquad (16.13)
$$

with F = flow

$F_{\max}$ = maximum flow through system with valve fully open

$C_v(v)$ = inherent valve characteristic, which is a function of v

v = valve stem position (% open or closed)

ΔP_v = pressure difference across the valve
ρ = fluid density

This is simply the expression for the flow through a restriction, with the variable v representing the valve stem position expressed in percent. The driving force for the flow is the difference between the pressures immediately before and after the valve, ΔP_v. The factor C_v is called the *inherent valve characteristic* and represents the percentage of maximum flow at any given valve stem position at a *constant pressure drop,* usually the design value. The C_v is a function of the valve design, basically the size and shape of the opening and plug, which can be linear or any of a selection of standard nonlinear relationships at the choice of the engineer. Three common inherent valve characteristics are shown in Figure 16.6.

In the typical process design, the pressures just before and after the valve change as the flow rate changes, as shown in Figure 16.7 (Quance, 1976). Typically, the pressure at the pump outlet is not constant; it decreases as the flow through the pump increases (Labanoff and Ross, 1985; Karassik, 1981). Also, the pressure drop from the valve to the pipe outlet increases as the flow increases. For the example process in Figure 16.7, the pressure drop from the valve outlet to the end of the pipe could be calculated from the energy balance on the fluid, with losses determined from friction factor correlations (Foust et al., 1960):

$$P_2 = P_{out} + \Delta P_e + \sum_{i=1}^{2} \Delta P_{Hxi} + \Delta P_{pipe} + \Delta P_{fit} \tag{16.14}$$

where P_{out} = outlet pressure (constant in this example)
ΔP_e = pressure drop due to change in elevation
$\Delta P_{Hxi}(F)$ = pressure drop due to heat exchangers (i = 1, 2)
$\Delta P_{pipe}(F)$ = pressure drop in the pipe due to skin friction
$\Delta P_{fit}(F)$ = pressure drop in elbows and expansions due to form friction

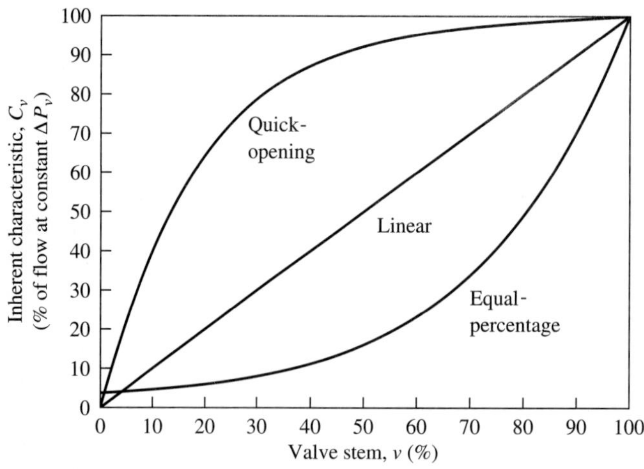

FIGURE 16.6
Three standard control valve inherent characteristics (Reprinted by permission. Copyright ©1976, Instrument Society of America. From Hutchinson, J., ed. *ISA Handbook of Control Valves,* 2nd Ed.)

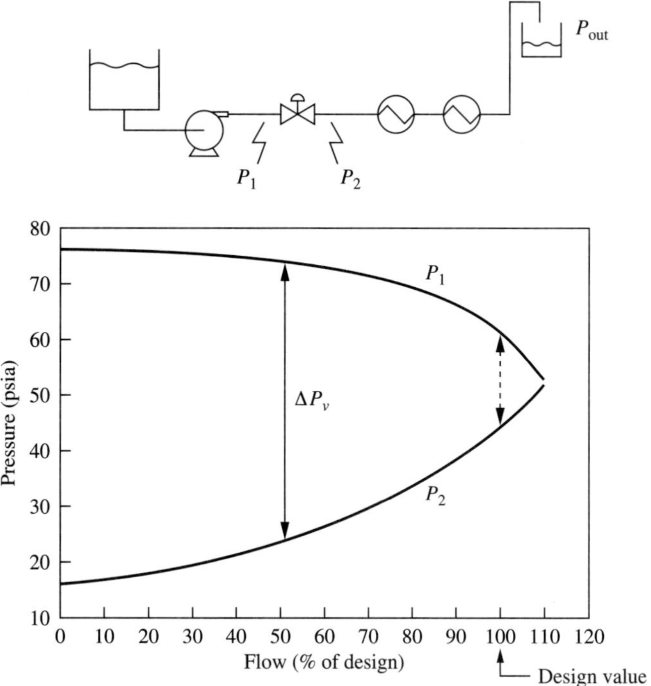

FIGURE 16.7
Relationship between pressures and flow for a typical system. (Reprinted by permission. Copyright ©1979, Instrument Society of America. From Quance, R., "Collecting Data for Control Valve Sizing," *In. Tech., 55.*)

Note that the last three pressure drop terms are functions of the flow rate (F). Due to the functional relationships for P_1 and P_2, the pressure drop across the valve ($\Delta P_v = P_1 - P_2$), decreases as the flow increases. This demonstrates that only part of the total pressure drop from the pump to the outlet is due to the valve; a considerable amount of the pressure drop is due to other frictional losses.

The goal of a linear system—a constant closed-loop relationship, $G_{OL}(s)$—is achieved when the relationship between the controller output and controlled variable is linear. In the case of flow control, the controller output can be taken to be the valve position, and the controlled variable is the measured flow rate. Since the pressure drop across the valve shown in Figure 16.7 is not constant, the relationship between the controller output and the valve opening must introduce a compensating nonlinearity for the overall gain to be constant. The nonlinearity can be introduced at low cost by selecting the appropriate valve characteristic C_v. The typical nonlinearity applied for situations in Figure 16.7 is the equal-percentage characteristic curve shown in Figure 16.6. The use of an equal-percentage valve in a process similar to that shown in Figure 16.7, in which the pressure drop decreases with flow, usually results in an approximately linear relationship between the valve stem position and flow. An experimental investigation of the application of an equal-percentage valve for the process described in Figure 16.7 resulted in the desired linear relationship shown in Figure 16.8. Note that the valve is about 85% open at the design flow; this

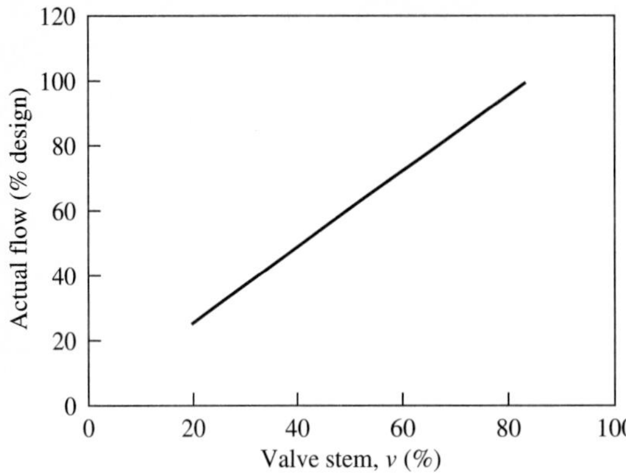

FIGURE 16.8
Linear overall relationship between v and flow for the example flow process with an equal-percentage valve. (Reprinted by permission. Copyright ©1979, Instrument Society of America. From Quance, R., "Collecting Data for Control Valve Sizing", *In. Tech.*, *55*.)

is a bit high, since most designers specify that the valve be about 70 to 80% open at the design flow. In all cases, the valve opening at design must be such that the required maximum flow can be achieved when the valve is fully open.

The general method employed for linearizing the flow control loop is now extended to a more complex process: a stirred-tank heat exchanger, shown in Figure 16.9. The energy balance for this system was derived in Example 3.7 (page 83) and used in Example 8.5 (page 276), and this example uses the same design parameters. The model is repeated here:

$$V\rho C_p \frac{dT}{dt} = F\rho C_p(T_0 - T) - \frac{aF_c^{b+1}}{F_c + \dfrac{aF_c^b}{2\rho_c C_{pc}}}(T - T_{cin}) \tag{16.15}$$

$$F_c = F_{max}\left(\frac{C_v}{100}\sqrt{\frac{\Delta P}{\rho_c}}\right)v \tag{16.16}$$

In this example the pressure drop across the valve is assumed constant so that the analysis will highlight the effects of other process nonlinearities; however, if this were not the case, the same approach could be used, with an appropriate model for the coolant pressures included. This model can be used to evaluate the linearity of

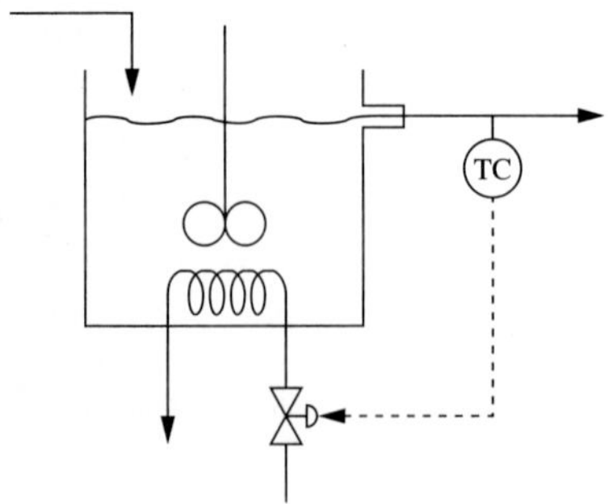

FIGURE 16.9
Heat exchanger control system.

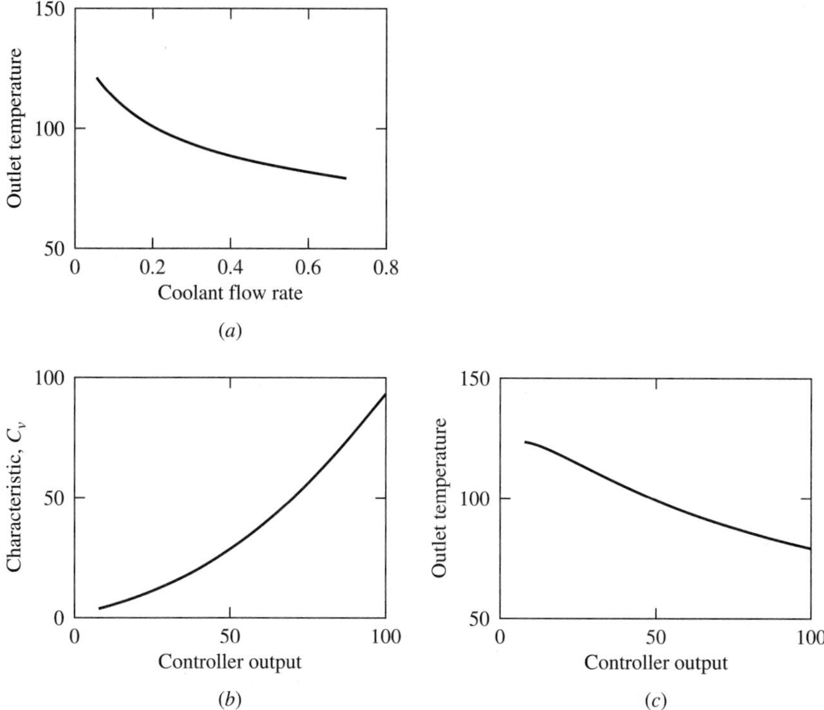

FIGURE 16.10
Summary of nonlinear process behavior and compensating characteristic.

the steady-state process by calculating the steady-state value of the temperature at various coolant flow rates by setting $dT/dt = 0$. The results of this calculation are given in Figure 16.10*a*. This plot clearly shows the nonlinearity in the process gain, which changes by a factor of more than 5 over the range of operation considered.

The goal of compensation would be to achieve a linear relationship between the controller output and the temperature. The proper linear relationship would be

$$(K_p)_{\text{ave}} = \frac{\Delta T}{\Delta v} \approx \frac{79.8 - 122.5°\text{C}}{100 - 5\%} = -0.45\frac{°\text{C}}{\%} \tag{16.17}$$

which is the total change in temperature over the total change in valve position over the controllable range. To maintain the loop gain at this value while the process gain changes, the valve characteristic must change. From the value of the desired average gain in equation (16.17) and the process gain, $\Delta T/\Delta F_c$ in Figure 16.10*a*, the value of the characteristic can be evaluated as $C_v = (K_p)_{\text{ave}}/K_p$. The results of this calculation are given in Figure 16.10*b*. As expected, the plot of C_v versus valve stem has a slope with small magnitude where the process has a gain with large magnitude (i.e., at low coolant flows). Also, the figures show that the process gain, and therefore the slope of C_v versus valve stem, is nearly constant over the higher range of flows. The steady-state behavior of the system with the linearizing C_v installed gives Figure

TABLE 16.3
Method for achieving a linear control system by selecting the proper valve characteristic

Goal: A linear relationship (i.e., constant $G_{OL}(s)$) between the controller output and the controlled variable. The valve stem position is assumed to be equal to the controller output.

1. Determine the relationship between the pressure drop and the flow for the specific process system considered, K_{p1}.
2. Determine the relationship between the flow and controlled variable (if not flow rate), K_{p2}.
3. Calculate the C_v based on the results in (1) and (2) so that $C_v K_{p1} K_{p2}$ = constant. This will ensure that the steady-state gain of the process, as "seen" by the controller, is constant.
4. Select the commercial valve with the inherent characteristic, C_v, closest to the function determined in (3).

16.10*c*; the linear relationship between the controller output and temperature indicates that a PID feedback controller with constant tuning parameters would be adequate.

It is important to understand the approach just demonstrated through these flow and heat exchanger examples: therefore, it is summarized in Table 16.3. The correct application of the procedure in this table frequently, but not always, results in an equal-percentage characteristic. An example of an exception occurs when the pressure drop across the valve is constant and the objective is flow control; then a linear valve characteristic is required to achieve a linear relationship between the controller output and the flow. Another exception occurs when a nonlinear relationship between the controller output and controlled variable is required in selected situations. For example, a cooling medium flow may normally be small or zero but need to be increased to a large value quickly upon demand. This situation would benefit from a nonlinear relationship between the controller output and the flow, which is provided by the "quick-opening" characteristic. Both of these characteristics are shown in Figure 16.6, and many other characteristics are commercially available (Hutchinson, 1976).

There are many physical designs of the flow patterns, orifice shape, and valve plug shape that are used to achieve the desired relationship. The specific design selected depends on many factors (Hutchinson, 1976), such as the desire to

1. Have tight closing (i.e., no flow) when the controller output is 0% (or 100% for a fail-open valve)
2. Prevent sticking or clogging when the fluid is viscous or is a slurry
3. Accurately control the flow over a specified range
4. Reduce the pressure loss due to the valve, to conserve energy

The reader is cautioned that the selection of the proper control valve requires more information than is provided in this brief introduction. Details of typical valves, along with pictures of the internal details, are available and should be consulted (Hutchinson, 1976; Andrew and Williams, 1979, 1980; Driskell, 1983). Also, engineering standards for sizing calculations and selection are available for many common situations (ISA, 1992).

Finally, it should be noted that a nonlinearity can be added to the controller calculation in place of the nonlinear valve characteristic. Many commercial digital controllers have the facility to introduce a nonlinearity after the control algorithm, in the output processing phase, via a general polynomial. However, the use of the valve characteristic is still the most common means in practice for compensating for simple process gain nonlinearities.

16.6 IMPROVING NONLINEAR PROCESS PERFORMANCE THROUGH CASCADE DESIGN

Other particularly simple nonlinearities can be addressed through cascade design that compensates for nonlinearities in the secondary, resulting in a (nearly) linear primary control system. One example encountered often in the process industries is maintaining two quantities in a desired proportion. An example of blending is shown in Figure 16.5*b*, although the concept applies to other proportions, such as reboiler to feed in a distillation tower or reactant ratio in a chemical reactor. The feedback controller can adjust the set point of the ratio controller as shown in Figure 16.5*b*. This is really another example of feedforward and feedback being combined as a product rather than a sum; thus, Figures 16.5*a* and *b* are alternative solutions to the same control design problem.

A cascade can also provide compensation for nonlinearities in other control designs. An example is shown in Figure 16.11, in which a linear relationship between the level and the level controller output is desired to be linear. Following the arguments in this section, the valve in Figure 16.11*a* would normally have an equal-percentage characteristic so that the relationship between the controller output and

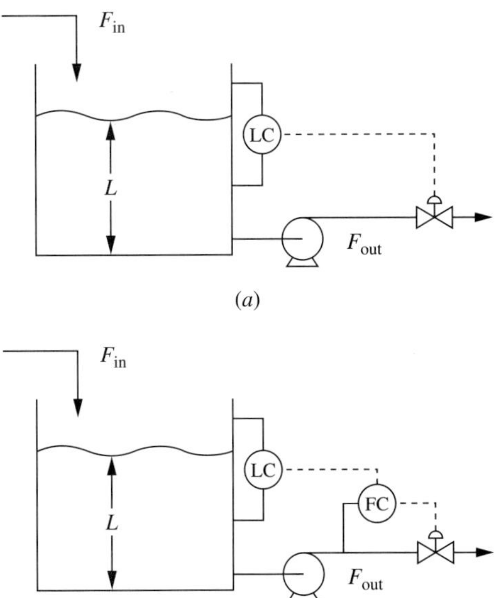

FIGURE 16.11
Example of cascade control applied to linearize the loop.

flow would be approximately linear. However, since the flow controller in Figure 16.11*b* is a fast loop, the relationship between the primary controller output and the flow would be linear, regardless of how well the characteristic compensated for other nonlinearities. Thus, the level control system is linearized as a result of the cascade design. Notice that the cascade strategy retains the advantages of improved disturbance response explained in Chapter 14.

16.7 REAL-TIME IMPLEMENTATION ISSUES

Adaptive methods involving real-time calculations are relatively straightforward to implement; however, a few special considerations should be included. Some of the methods for adapting tuning are based on one or more measurements, and should a measurement not represent the true process conditions because of a sensor failure, the resulting tuning constants could be far from the proper values, leading to poor or even unstable performance. Thus, the measurement(s) used in the updating calculations should be checked for validity before being used to calculate the tuning. An example is checking the consistency of a flow measurement with associated flow rates in the process to ensure that a realistic flow is being used to update tuning. In addition, the value of the measured variable used in the correlations, as in equations (16.8), should be limited to the range over which the correlation is valid. This practice serves two purposes:

1. Error due to an unrecognized sensor failure is limited.
2. Extrapolation of a correlation beyond its region of applicability is prevented.

An issue that may not have been apparent in the previous sections is the ever-present need for defining the desired control performance. The tuning correlations must reflect the performance desired; thus, the tuning correlations selected must be based on control objectives consistent with the performance desired in the plant. As will be explained in Parts V and VI, tight control of one variable may degrade the control performance of another, more important variable because of process interaction. Thus, the performance goals of all control loops must be determined considering the overall process performance, which may lead to loose tuning for selected loops.

Also, it would be wise to provide the facility to fine-tune the controller tuning constants while retaining the correlations. One simple method would be to provide an adjustable parameter in equations (16.8). The engineer could adjust the parameter to achieve improved performance at one operating condition, and the parameter would be unchanged for other operating conditions.

16.8 ADDITIONAL TOPICS IN CONTROL LOOP ADAPTATION

All of the methods described in detail in this chapter are predicated on the assumption that the change in process dynamics can be predicted. This assumption, which leads to the compensating calculations and equipment designs, is not always valid. For example, the effect of acid flow on pH (i.e., the shape of a pH curve) can change substantially due to changes in the buffering agents present; the effect of temperature on reactor conversion can depend on the activity of the catalyst. Therefore, there are situations in which deterministic methods are not appropriate. One response to this situation would be to detune the controller substantially and accept the performance degradation. Better performance would be possible with an adaptive method that could "learn" the process dynamics from the real-time system behavior and retune the controller based on the updated knowledge of process dynamics. Two general retuning approaches are used in this situation:

1. Periodic adaptive tuning at the request of a person, which is applicable when the dynamics change infrequently
2. Continuous adaptive retuning, which is applicable when the dynamics change frequently

The analysis of these approaches require more advanced mathematics than is consistent with the level of this book; however, a few of the methods are introduced in the following paragraphs.

Periodic Retuning Based on Model Identification

In this approach an empirical model identification method is implemented to determine the dynamic model of the process, $G_p(s)G_v(s)G_s(s)$. The model fitting could use one of the methods described in Chapter 6 or other statistically based methods. Based on this model, the method can automatically introduce updated tuning using an appropriate controller tuning method. Note that this method introduces perturbations in the manipulated variable, which will disturb the process, but only when a person requests a retuning.

Periodic Retuning Based on Empirical Identification of the Critical Conditions

The Bode stability criterion highlighted the importance of the feedback system at the critical frequency. The feedback system's stability and controller tuning can be based on the amplitude ratio at the critical frequency, $|G_{OL}(\omega_c)|$. Thus, some methods of adaptive tuning determine the critical conditions empirically. One possible approach would be to automate the Ziegler-Nichols closed-loop tuning experiment described in Section 10.8, Interpretation IV; however, this approach would introduce large, prolonged disturbances. A more successful approach uses this principle with a relay in place of the controller to determine the same information, with smaller disturbances to the plant (Astrom and Hagglund, 1984).

Continuous Retuning Based on Statistics

It is possible to identify the process dynamics and determine how to modify the tuning without introducing external perturbations, as long as some disturbances occur in the process. Approaches to formulating and solving this problem are given in Astrom and Wittenmark (1989).

Continuous Retuning Based on Rules (Bristol, 1977)

Fine-tuning of closed-loop systems based on the response to set point changes was discussed in Chapter 9. This concept can be applied to disturbance responses so that external perturbations are not necessary; then the method can be automated to achieve continuous retuning. In one commercial system the control performance is defined by the engineer in terms of (1) controlled-variable damping and overshoot, (2) expected noise levels, and (3) bounds on the controller tuning constants (Kraus and Myron, 1984). The retuning method uses rules to adjust the tuning constants to achieve the desired performance.

16.9 CONCLUSIONS

Modification to an element in the closed-loop system may be needed to attain high-performance feedback control when the feedback dynamics change. The three major steps are given in Table 16.2 for evaluating deterministic approaches for compensating for nonlinearities. The first step is to determine whether the process dynamics change significantly over the range of operation. If a fundamental analytical model is available, the linearized expression can be evaluated through the range of operating variables to determine whether the gain, dead time, and time constants change significantly. If no analytical model is available, several linear models can be determined through empirical identification at various operating conditions. The variability in the tuning and degradation in performance due to the nonlinearities can be determined as explained in Section 16.2. Since control objectives are different from plant to plant, it is not possible to give a generally applicable guideline for when the nonlinearity is "significant." However, since modelling errors of $\pm 20\%$ are expected in identification, nonlinearities causing model parameter variations of this magnitude or less would normally not be significant.

The approaches presented in this chapter are summarized in Table 16.4. The order of presentation is from simplest and most reliable to most complex and challenging to implement. Generally, the engineer will apply the methods in the order presented in the table, proceeding only to the method needed to achieve acceptable performance.

If the variability is significant and it can be predicted based on real-time measurements, an element can be introduced to linearize the control loop by compensating for the nonlinearity. The compensating element can be in any of the three categories of the control calculation: input processing, control algorithm, or output processing. It can also be included in the control equipment, specifically in the final control element.

If the variability in dynamics is significant but cannot be predicted using correlations, one approach is to detune the feedback controller so that it is stable for all dynamics encountered. Naturally, this approach will result in a degradation in

TABLE 16.4
Summary of methods to compensate control systems for nonlinearities

Description	Compensation for nonlinearity	Example	Additional effects on control performance
Measurement	Calculation to compensate for nonlinear sensor	Square root of orifice flow meter (Figure 12.7)	Improved accuracy for process monitoring
Final element	Final element selected to compensate for process nonlinearity	Valve characteristic to account for changes in pressure (Figures 16.6 and 16.7)	
Cascade control	Select secondary set point that has linear relationship with primary	Level-flow cascade (Figure 16.11)	Improved response to secondary disturbances
Detune	Determine single set of tuning constants for the range of operating conditions	Three-tank mixing process (Table 16.1, Case A tuning)	Poor performance can result
Gain schedule	Calculate the controller gain based on real-time measurement		
	Multiply feedforward and feedback (where this leads to proper gain scheduling)	Figure 16.5	Feedforward compensation for measured disturbance
Controller tuning	Calculate tuning based on a process model and real-time measurement	Three-tank mixing process [equation (16.8)]	
Occasionally retune	Empirically determine key model characteristics and tune controller according to preselected performance criteria	Relay method of finding critical conditions (Astrom and Hagglund, 1984)	Undesired variation during (infrequent) retuning
Continuously retune	Empirically determine key model characteristics and tune controller according to preselected performance criteria	On-line identification	

performance. Another approach is to modify the tuning of the controller based on some information of the real-time dynamic behavior of the system. Various methods are available, and references are provided.

Finally, the limits of the adapting approach should be recognized. First, a great strength of feedback control is that it does not require a highly accurate model. Thus, reasonable model errors can be tolerated with little degradation in control performance. Second, the adaptations require some time for the method to recognize the change in process behavior and introduce the compensation to the tuning (or other element of the loop). Thus, if the process dynamics are changing with a frequency near the critical frequency of the feedback control loop, an adaptive approach will not be able to introduce the compensation quickly enough. This limitation also holds when an infrequent change in process dynamics is large and abrupt: adaptation may not be able to detect the situation rapidly enough. Finally, the reader is advised to establish the potential improvement using the first entries in Table 16.4 before attempting the substantially more complex approaches in later table entries.

REFERENCES

Astrom, K. and T. Hagglund, "Automatic Tuning of Simple Regulators with Specifications on Phase and Amplitude Margins," *Automatica, 20,* 645–651 (1984).

Astrom, K. and B. Wittermark, *Adaptive Control,* Addison-Wesley, Reading, MA, 1989.

Andrews, W. and H. Williams, *Applied Instrumentation in the Process Industries* (2nd Ed.), 2 vols., Gulf Publishing, Houston, TX, 1979, 1980.

Bristol, E., "Pattern Recognition: An Alternative to Parameter Identification in Adaptive Control," *Automatica, 13,* 197–202 (1977).

Chalfin, S., "Specifying Control Valves," *Chem. Eng., 81,* 105–114 (October, 1974).

Driskell, L., *Control Valve Selection and Sizing,* Instrument Society of America, Research Triangle Park, NC, 1983.

Foust, A., in chapter 3, *Principles of Unit Operations,* Wiley, New York, 1960.

Hutchinson, J. (ed.), *ISA Handbook of Control Valves* (2nd Ed.), Instrument Society of America, Research Triangle Park, NC, 1976.

ISA, *ISA Standards and Practices* (11th Ed.), Instrument Society of America, Research Triangle Park, NC, 1992.

Karassik, I., *Centrifugal Pump Clinic,* Marcel Dekker, New York, 1981.

Kraus, T. and T. Myron, "Self-Tuning PID Controller Uses Pattern Recognition Approach," *Cont. Eng., 31,* 106–111 (June 1984).

Labanoff, V. and R. Ross, *Centrifugal Pumps, Design and Applications,* Gulf Publishing, Houston, TX, 1985.

Quance, R., "Collecting Data for Control Valve Sizing," *In. Tech.* 55 (November, 1979).

ADDITIONAL RESOURCES

Computer programs and exercises for several adaptive tuning methods have been prepared by

Roffel, B., P. Vermeer, and P. Chin, *Simulation and Implementation of Self-Tuning Controllers,* Prentice-Hall, Englewood Cliffs, NJ, 1989.

Adaptive tuning of PID and other feedback control algorithms are presented at an advanced level in

Goodwin, G. and K. Sin, *Adaptive Filtering, Prediction, and Control,* Prentice-Hall, Englewood Cliffs, NJ, 1984.

Astrom, K. and T. Haggland, *Automatic Tuning of PID Controllers,* ISA, Research Triangle Park, NC, 1988.

A particularly challenging situation occurs when the process gain changes *sign* as operating conditions change. An industrial process where this occurs is discussed in

Dumont, G. and Astrom, K., "Wood Chip Refiner Control," *IEEE Cont. Sys., 8,* 38–43 (1988).

Some industrial examples of adaptive control are given in the following references.

Piovoso, M. and J. Williams, *Self-Tuning Control of pH,* ISA Paper no. 0-87664-826-x, 1984.
Vermeer, P., B. Roffel, and P. Chin, "An Industrial Application of an Adaptive Algorithm for Overhead Composition Control," *Proc. Auto. Cont. Conf.,* St. Paul, MN, 1987.
Whately, M. and D. Pott, "Adaptive Gain Improves Reactor Control," *Hydro. Proc.,* 75–78 (May, 1988).

QUESTIONS

16.1. Consider the same three-tank mixing process, but with the outlet concentration of component A changed to 50% in all cases. Recalculate the values in Table 16.1 for the same changes in the flow rate of stream *B*. Compare and comment on the similarities and differences.

16.2. Answer each of the following questions, with a full explanation of your answer.

(*a*) Could closed-loop frequency response, as explained in Section 13.3, be used to determine when feedback controller tuning should be adapted for changes in operating conditions?

(*b*) Review all cascade examples in Section 14.7 and determine whether each results in a (nearly) linear relationship between the secondary and primary. Would the single-loop control (primary to valve) be significantly nonlinear?

(*c*) Review all of the feedforward-feedback control designs in Section 15.7 and for each, recommend how to combine the feedforward and feedback signals (add, multiple, divide, other) to provide the best tuning compensation for the measured disturbances.

(*d*) The discussion and examples in this chapter involved feedback control. Discuss whether there is any advantage to adapting the adjustable parameters in a feedforward controller. If yes, discuss how this could be evaluated and the proper values determined.

16.3. Recalculate the tuning in Table 16.1 using the Ziegler-Nichols closed-loop tuning method. Compare the similarities and differences of the effect of F_B on the tuning for the two tuning methods. Which tuning would you recommend using?

16.4. Based on the information in question 9.10, would you recommend automatic deterministic retuning of the feedback control system? If yes, determine the measured variable and the tuning constants as a function of the measured variable.

16.5. Based on the information in question 10.2, would you recommend automatic deterministic retuning of the feedback control system? If yes, determine the measured variable and the tuning constants as a function of the measured variable.

16.6. You have been given the task of developing a rule-based adaptive tuning method for use with a PID controller. Also, the introduction of any perturbations by the method has been prohibited. Develop a set of rules that can be applied to normal operating data (with disturbances) to improve the performance gradually by adjusting the controller tuning constants. Remember to consider both the controlled and manipulated variables when evaluating performance.

16.7. The chemical reactor in Example 3.12 experiences $\pm 30\%$ changes in the flow rate and inlet concentration of component A, and its volume is constant. Discuss a controller tuning method for a feedback controller that would adjust the temperature to maintain the constant value of the concentration of reactant B.

16.8. The stirred-tank heat exchanger in Section 16.5 experiences changes in the feed inlet temperature of 120 to 170°C and in the coolant inlet temperature of 20 to 30°C. These temperature changes occur independently, and the feed flow and temperature set point remain constant at their base-case values. Discuss the need for adapting the feedback PI controller tuning constants and, if necessary, provide correlations for the valve characteristic and tuning as appropriate.

16.9. Design feedforward-feedback control for the chemical reactor in Example 15.1 for input disturbances in both feed composition, A2, and feed flow, F1. Pay particular attention in how the feedforward and feedback signals should be combined. Is there a need for adapting the feedback tuning for disturbances? Can this be achieved in combination with the feedforward control? Is there a need to adapt the feedforward controller parameters?

16.10. The behavior of the heat exchanger in the recycle system in Example 5.5 varies due to fouling. Experience has shown that G_{H2} changes within the range of 0.20 to 0.32 about its nominal value of 0.30. Determine whether this change is significant. If so, how could deterministic controller adaptation be implemented?

16.11. Some process equipment has to be removed from service occasionally for maintenance. Consider a multiple-tank mixing process that is basically the mixing tank process in Example 7.2, but modified to have between two and four tanks, depending on the equipment availability. Determine how the feedback controller tuning has to be modified for the situations of two, three, and four tanks in the feedback process. Also, compare the control performance for these three situations.

16.12. Level control is to be added to the draining tank process in Example 3.6. The controller adjusts the opening of a valve in the exit pipe at the base of the tank, and essentially all of the pressure drop in the pipe and valve occurs across the valve. Determine the valve characteristic that will yield a linear relationship between the controller output and the level. The inlet flow varies from 50 to 150 m^3/h.

16.13. In some feedback control systems the manipulated variable can be changed, usually by selecting the position of a switch at the controller output that directs the signal to one of the possible manipulated variables. For the following cases, determine appropriate PI tuning for each of the two possible manipulated variables and whether the difference is significant enough to require changing the tuning depending on the manipulated variable selected for the controller output.

(*a*) Consider the reactor control system in Example 13.8 and question 13.19 in which two possible manipulated variables exist: the reactant flow (F_A) and the solvent flow (F_s).

(*b*) The nonisothermal CSTR in Examples 3.10 and 4.8 in which the two possible manipulated variables for temperature control are F_c and T_{cin}.

16.14. Question 13.1 describes a process with feedback control and changes in operating conditions, (*a*) through (*f*). For each change in operating conditions, determine whether it is necessary to adapt the feedback controller tuning, and if so, how the adaptation could be implemented automatically.

16.15. Consider a series of *three* isothermal CSTRs, each with the physical design parameters of the process in Example 3.5. The base case operating conditions are the same as the example: $F = 0.085$, $C_{A0} = 0.925$, and $k = 0.50$. The composition of reactant A in the third reactor is controlled by adjusting the feed composition, C_{A0}. Determine (*a*) the steady-state operating conditions for this base case, (*b*) the linearized model for the system, and (*c*) PID feedback tuning for this base case system. Then determine whether the controller tuning must be adjusted if the feed flow rate changes from 0.085 to 0.20.

CHAPTER 17

INFERENTIAL CONTROL

17.1 INTRODUCTION

In all of the control methods considered to this point, the important variables have been measured, a situation which is desirable and most often possible. However, not all important variables can be measured in real time, that is, fast enough that timely control actions can be based on their measurements. There are various reasons for the lack of key measurements. First, some sensitive analyses have not been sufficiently automated to provide accurate, reliable measurements without human management of the procedure; thus, these measurements can be obtained only infrequently from a laboratory. There are even some properties that cannot be determined from intermediate material properties in a plant. Usually, these properties relate to the final use of the material; for example, some qualities of products such as soap, food products, or polymers depend on their application as final products and cannot be measured until the products are formulated and used. Second, even if the real-time measurement is possible, the cost of installing a sensor in the plant may not be justified by the potential benefits derived from the additional sensor, especially considering the alternative methods in this chapter. The cost is not typically high for conventional sensors for measuring temperature, pressure, flow, and level, but it may be prohibitive for an expensive analyzer with sample system and ongoing maintenance. Third, the sensor may not provide information in a timely manner. There are several reasons for slow feedback; for example, the analyzer may have a very long dead time because it must be located far downstream. Also, an analyzer may have a long processing time—one hour or longer—which would delay the feedback information. Finally, there may be no directly measurable quantity; for example, the controlled variable may be the heat transferred in an exchanger.

The lack of measurements of key variables in a timely manner certainly makes automated control difficult, but not always impossible. As with many feedback enhancements in Part IV, *inferential control* uses extra information to improve this situation. In this case, the extra information is additional measured variables that, while not giving a perfect indication of the key unmeasured variable, provide a valuable inference. The selection and use of these additional inferential variables requires process insight and adherence to the methods described in this chapter. Since inferential control is widely applied with great success, the analysis and design of inferential variables is important for engineers who design and operate plants, as well as for control specialists.

Since the characterization of variables as inferential may initially seem somewhat arbitrary, the general concept is explained here. All sensors depend on physical principles that relate the process variable to the sensor output, and thus no sensor "directly" measures the process variable. For example, a thermocouple temperature sensor provides a millivolt signal that is related to temperature (and the reference junction temperature), and an orifice flow sensor provides a pressure difference signal that is related to the flow (and fluid density). We normally consider the standard sensors for temperature, pressure, flow, and level as direct measurements, not inferential variables, because (1) they provide reasonably good accuracy and reproducibility, (2) they do not usually require corrections (e.g., for reference junction temperature), and, most importantly, (3) the relationship between the sensor signal and the process variable is not specific to a particular process. For example, essentially the same relationship between the pressure difference across an orifice and the flow through the orifice is used in thousands of plants. In contrast, a relationship between a reactor temperature and conversion is clearly specific to a particular process (flow patterns, reaction rate, heat of reaction, catalyst activity, feed conditions, and so forth) and is considered an inferential variable.

Since there is no generally accepted naming convention, we will refer to the variable we would like to control as the "true" controlled variable, $CV_t(t)$. The inferential variable, $CV_i(t)$, can be used because of a *process-dependent* relationship, which must be determined by the engineer. For example, a good inferential variable in Figure 17.1 is closely related to the true variable so that controlling $CV_i(t)$ will maintain $CV_t(t)$ close to its desired value. In most cases, the inferential variable is not as accurate as an on-stream sensor of the true variable. Also, the approximate relationship used for the inferential variable has a limited range, beyond which the inferential variable might not be satisfactory. It is important to remember that zero steady-state offset for the true variable is possible only when it is measured, perhaps infrequently, and used in the control system to adjust the set point of the inferential controller, $SP_i(s)$.

Figure 17.1 can be used to determine the relationship necessary for good inferential control. First, the response of the true controlled variable to a disturbance can be evaluated.

$$\begin{aligned} CV_t(s) &= G_{dt}(s)D(s) + G_{pt}(s)MV(s) \\ &= G_{dt}(s)D(s) - \frac{G_{pt}(s)G_c(s)G_{di}(s)}{1 + G_{pi}(s)G_c(s)}D(s) \end{aligned} \tag{17.1}$$

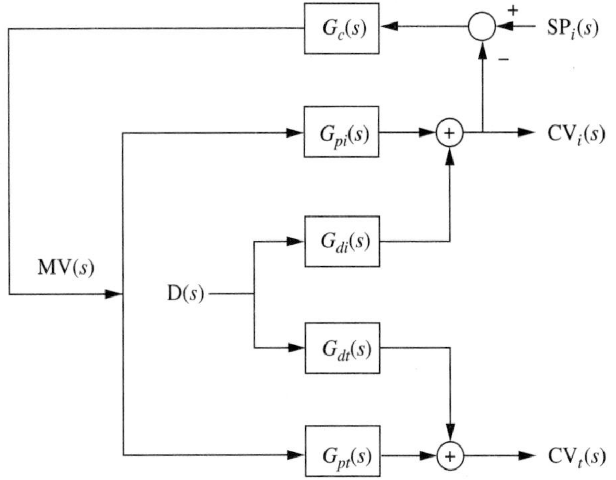

FIGURE 17.1
Block diagram of a feedback control system with a true controlled variable, $CV_t(s)$, and an inferential controlled variable, $CV_i(s)$.

A key goal of the control system is to maintain zero steady-state deviation in the controlled variable. This can be evaluated by applying the final value theorem to equation (17.1) with a step disturbance and PI feedback controller to give

$$\lim_{t\to\infty} \mathrm{CV}(t) = \lim_{s\to 0} \mathrm{CV}(s) = K_{dt}\Delta D - \frac{(K_{pt}K_{di}K_c)/T_I}{(K_{pi}K_c)/T_I}\Delta D = 0 \qquad (17.2)$$

> Thus, the criterion for perfect steady-state inferential control in response to a disturbance is that $K_{dt}/K_{pt} = K_{di}/K_{pi}$.

As the process relationships deviate from this criterion, the performance of the inferential controller degrades. Thus, an important engineering decision is the selection of a proper inferential measured or calculated variable.

17.2 AN EXAMPLE OF INFERENTIAL CONTROL

Application to a flash separator demonstrates the typical analysis steps for inferential control, along with a very common inferential variable. The process is shown in Figure 17.2 where a stream of light hydrocarbons is heated, the pressure of the stream is lowered, and the liquid and vapor phases are separated in a drum. The base-case compositions of all three streams are given in Table 17.1. The true controlled variable is the ethane concentration in the drum liquid; however, an analyzer is not available, perhaps because of cost. (Accurate on-stream analyzers are commercially available for such a measurement.) The goal is to infer the concentration of ethane in the liquid stream leaving the drum, using the sensors shown in the figure. This goal may or may not be possible within the accuracy required; therefore, an analysis of the system is performed.

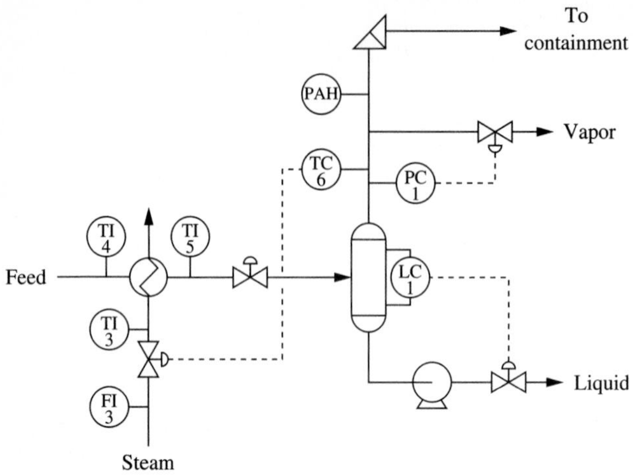

FIGURE 17.2
Flash separator considered for inferential control of ethane composition in the liquid from the drum.

From a knowledge of vapor-liquid equilibrium, we expect that the temperature of the drum and the compositions will be related. In fact, the following model of the flash shows the relationship.

$$\begin{aligned} \text{FM}_{\text{feed}} &= \text{FM}_L + \text{FM}_V \\ \text{FM}_{\text{feed}} Z_i &= \text{FM}_L X_i + \text{FM}_V Y_i \\ Y_i &= K_i X_i \end{aligned} \qquad (17.3)$$

where FM = molar flow
X, Y, Z = mole fractions for liquid, vapor, and feed
K_i = vaporization equilibrium constant depending on T, P
P = pressure
T = temperature

From equations (17.3) it can be seen that the liquid ethane composition is a function of the feed composition and the temperature and pressure in the flash vessel. (Further details on the flash calculation and the data used in this example can be found

TABLE 17.1
Base-case data for flash process in mole percent

Component	Feed	Liquid	Vapor
Methane	10	1.3	19.8
Ethane	20	10.0	31.2
Propane	30	30.2	29.8
i-Butane	15	20.1	8.8
n-Butane	20	30.0	9.3
n-Pentane	5	8.4	1.1

in Smith and Van Ness, 1987.) Let us assume that the drum pressure is controlled at essentially a constant value by adjusting a valve in the vapor line and that the temperature can be maintained at its desired value by manipulating the steam flow. If the feed had only two components, the temperature and pressure would uniquely determine the liquid and vapor composition; however, the feed has six components. Therefore, the pressure and temperature do not exactly define the compositions in the two phases. The essential question to answer is how closely the temperature is related to the liquid ethane composition, that is, how accurate an inference of liquid ethane concentration is supplied by the temperature when changes in the process operation occur.

The proposed inferential system is summarized by the following variables:

True variable	$= x_e =$	liquid composition of ethane to be controlled at 0.10 ± 0.02 mole fraction
Inferential variable	$= T =$	temperature
Manipulated variable	$=$	heating medium flow
Disturbance	$=$	feed composition (as subsequently defined)

Inferential relationship: $$x_e = \alpha T + \beta \tag{17.4}$$

An analysis is performed to establish whether the relationship between the temperature and the liquid ethane concentration is satisfactory for inferential control. It is not possible to develop a closed-form analytical model of this process; therefore, the inferential model will be developed based on data representing the process. This data could be developed from mathematical simulation or plant experimentation, so that the general method explained here is applicable to either. In this case, where excellent data exists for the vapor-liquid equilibrium, a simulation was performed to generate the relationship shown as the "base-case" line in Figure 17.3. The first step in evaluating the potential inferential relationship involves determining whether the sensitivities are appropriate. Figure 17.3 shows that the slope is about −0.0027 mole fraction per °C, which means that the expected errors in the temperature measurement and control, here estimated to be ±0.5°C, will not introduce a significant error in the calculated estimate of the ethane concentration.

To expand on this sensitivity analysis, three hypothetical curves are shown in Figure 17.4. Curve A represents an acceptable sensitivity. Curve B has a very small slope; this type of relationship would not be satisfactory, because the temperature measured over its normal range would not indicate a significant change in the ethane concentration. Curve C has a slope with a large magnitude; this type of relationship would not be satisfactory, because the temperature would have to be measured extremely accurately to provide an acceptable inference of ethane concentration, and any errors (bias or variance) in the temperature measurement would lead to large errors in the inferred composition. Thus, an "appropriate" sensitivity involves a tradeoff of instrument range and accuracy to yield an inference that is acceptable for the plant needs.

Since the temperature has passed the first step, the analysis is extended to the second step by including *disturbances:* unmeasured input operating variables that are

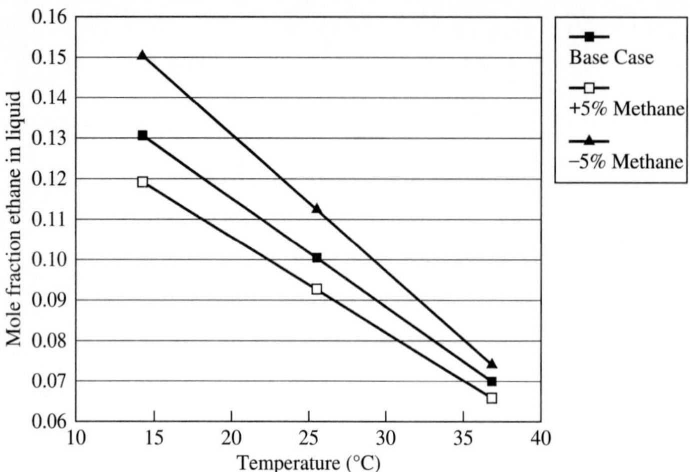

FIGURE 17.3
The relationship between the flash temperature and the concentration of ethane in the liquid at the base-case pressure (1000 kPa). Changes in methane are compensated by changes in butane of equal magnitude and opposite sign.

expected to change significantly. In this example, the feed composition is the major disturbance. The question is whether the temperature remains a satisfactory inferential variable when the feed composition changes; to answer this question, additional cases that characterize typical plant variability have to be included in the analysis. In this example, the expected feed composition change from the upstream units involves offsetting differences in the methane and butane, which can be up to 5 mole %. The relationships between temperature and composition for the extremes of feed composition variation are shown in Figure 17.3. Clearly, holding the temperature constant is not equivalent to maintaining the ethane concentration constant. For the expected changes in feed composition and the expected accuracy in controlling the

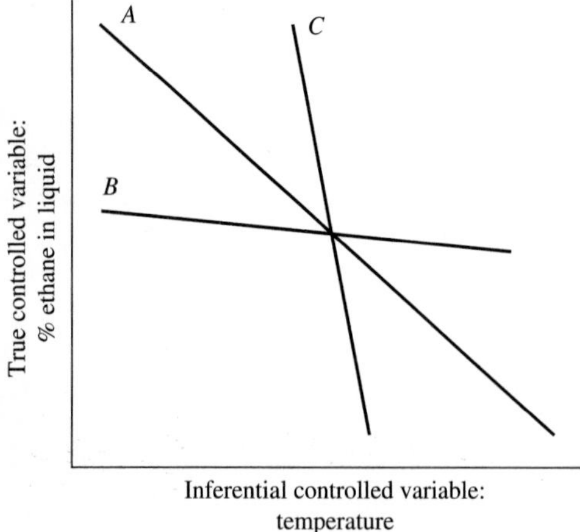

FIGURE 17.4
Hypothetical inferential variables with acceptable (A) and unacceptable (B and C) sensitivities.

temperature, ±0.5°C, the range of ethane liquid composition is from 0.091 to 0.117 mole fraction when the measured temperature is maintained at the proper value for no model error and nominal feed composition (25.5°C). Whether this accuracy is acceptable depends on the plant requirements; for this example it satisfies the stated objectives of inferential control (±0.02 maximum error). Since the accuracy with the inferential variable is acceptable, the temperature provides an acceptable steady-state inferential measure of ethane concentration in the liquid stream, and the control strategy in Figure 17.2 could be appropriate. If it were not, perhaps due to a narrowing of the acceptable ethane concentration variation, an on-stream analyzer would be required.

If the steady-state accuracy is satisfactory, the dynamics of the potential inferential control system must be evaluated. Good dynamic responses, as discussed in Chapter 13, would have such characteristics as a fast response with a short dead time. For this example, the temperature could be controlled by adjusting the heating medium flow. Therefore, the dynamics seem favorable because the response would be fast. This judgement is supported by the dynamic response of this system presented in Chapter 24.

Recall that the temperature controller set point must be corrected based on a measure of ethane concentration to achieve zero offset. The composition feedback could involve the temperature set point being occasionally corrected by the operator based on infrequent measurements in the laboratory performed on samples taken from the drum liquid. An alternative design, shown in Figure 17.5, involves the temperature set point being adjusted by a downstream analyzer controller in a *cascade* strategy. In this design, the analyzer is located downstream because of cost; the single downstream analyzer can measure components for distillation control as well as the flash drum control. Note that the cascade strategy adjusting the inferential temperature would be advisable in Figure 17.5, because of the slow dynamic response between the heating medium and the analyzer. With the inferential temperature

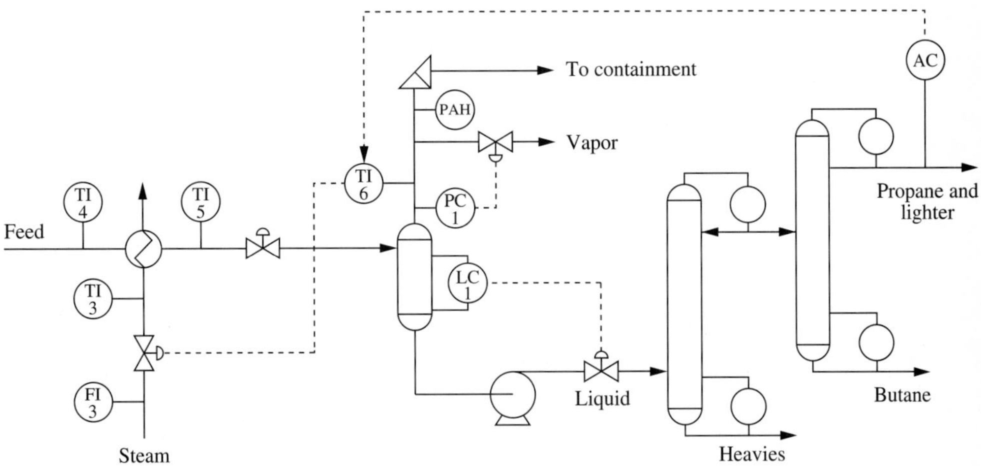

FIGURE 17.5
Flash inferential temperature controller reset in a cascade design by a downstream analyzer.

controller, reasonably tight control of the ethane concentration can be achieved without installing an additional on-stream analyzer at the outlet of the flash drum.

Since the inferential set point is ultimately reset based on a measure of the true controlled variable, the inferential measure need not be extremely accurate. However, it should be *reproducible;* that is, the inferential sensor should provide essentially the same signal for the same process conditions. Then the slower feedback based on the true variable would correct for inaccuracies in the inferential relationship and ultimately return the true controlled variable to its desired value.

The control example in this section, using temperature of a flash equilibrium to infer composition, is a standard practice in many industries; in fact, it is so common that the term *inferential* may seem exaggerated. However, it provides an excellent initial example. The next section provides a summary of the general design method for inferential control, which can be applied to more challenging cases.

17.3 INFERENTIAL CONTROL DESIGN CRITERIA

The preceding example addressed all of the major design criteria, which are summarized in Table 17.2. First, an analysis of the process economics and expected disturbances is performed to determine whether an inferential variable is appropriate. If yes, the process must be analyzed to identify a measurable variable with an acceptable relationship to the unmeasured, true controlled variable. It is especially important to ensure that the inferential variable is adequate for the expected range

TABLE 17.2
Design criteria for inferential control

Inferential control is appropriate when

1. Measurement of the true controlled variable is not available in a timely manner because
   - An on-stream sensor is not possible.
   - An on-stream sensor is too costly.
   - Sensor has unfavorable dynamics (e.g., long dead time or analysis time) or is located far downstream.
2. A measured inferential variable is available.

An inferential variable must satisfy the following criteria:

1. The inferential variable must have a good relationship to the true controlled variable for changes in the potential manipulated variable.
2. The relationship in criterion 1 must be insensitive to changes in operating conditions (i.e., unmeasured disturbances) over their expected ranges.
3. Dynamics must be favorable for use in feedback control.

Correction of inferential variable

1. By primary controller in automated cascade design
2. By plant operator manually, based on periodic information
3. When inferential variable is corrected frequently, the sensor for the inferential variable must provide good reproducibility, not necessarily accuracy

of plant operating conditions. Usually, the initial selection is based on a steady-state analysis, and the dynamic response is subsequently evaluated.

Two similar approaches are used for designing inferential controllers. Both approaches are described in this chapter, along with industrial examples. One approach determines the best inferential variables based on data (experimental or simulation) from the process; this type will be referred to as the *empirical* approach. The inferential temperature in the previous flash control is an example. The other approach uses closed-form analytical models as a basis for inferential relationships. An example of this approach, which will be referred to as the *analytical* approach, is applied to a chemical reactor in Section 17.6. The application of either approach involves nearly the same steps to yield an inferential model for control. The analysis steps for each method are summarized in Table 17.3.

Application of the design criteria in Table 17.2 and the steps in Table 17.3 ensures that a proper inferential variable is selected, if one exists. These approaches are usually adequate, because inferential variables are employed to reduce, although not eliminate, large offsets due to disturbances. To reiterate, an inferential strategy can achieve zero offset only when the true controlled variable is ultimately measured and used to adjust the set point of the inferential controller.

TABLE 17.3
Steps required to design an inferential controller

Step	Empirical approach	Analytical approach
1	Select one or a few measured variables for evaluation based on process insight.	Select one or a few measured variables for evaluation.
2	Develop a *representative* set of data that contains typical changes in the manipulated and disturbance variables.	Derive the analytical model from fundamentals.
3	Develop a correlation between the measured inferential and true controlled variables by fitting the model to the data to determine the unknown parameters.	The analytical model provides the necessary correlations.
4	Evaluate the accuracy and reproducibility of the correlation against process needs. This evaluation should consider realistic levels of noise on the inferential variable.	Same, although sensitivity information can be obtained directly from the model.
5	Select the best of the inferential variables and evaluate the dynamic response for use in feedback control.	
6	If the best inferential variable is acceptable, design the control system including ultimate feedback from the true variable.	
7	If no measured variable has both acceptable accuracy and acceptable dynamics, then inferential control is not possible. An on-stream sensor should be purchased and installed, if available. If no sensor is available, then the control objectives cannot be achieved unless other steps, such as reducing upstream disturbances, can be taken to reduce the variation in the true controlled variable.	

17.4 IMPLEMENTATION ISSUES

An inferential controller using a single measured variable is basically the same as any other single-loop or cascade controller, and no special implementation considerations are necessary. If the correction from the true controlled variable is made manually by the operator, a simple correlation is helpful in deciding the necessary change in inferential controller set point. In the flash separator example, the slope of the correlation in Figure 17.3 indicates that the temperature should be changed +1°C for a change of −0.0027 mole fraction ethane. As an example of how the person would use the correlation, if the laboratory analysis were 0.0040 mole fraction below the desired ethane concentration, the operator would implement a −1.5°C change in the temperature set point based on the correlation.

The situation changes when additional variables are used in the inferential relationship. In the flash separator, the strategy in Figure 17.2 might not be adequate if the drum pressure varied significantly, which can occur when the pressure is not controlled at the drum but varies with downstream units. A simple manner for considering this change would be to add an additional term, which would account for changing pressure, to the inferential correlation used to calculate the ethane concentration. The enhanced inferential relationship would be

$$x_e = \alpha T + \gamma P + \beta' \tag{17.5}$$

Again, this expanded relationship would be developed based on representative data for the system over the expected range of pressures. The effect of pressure for the base case feed concentration is shown in Figure 17.6, which would provide information for an additional linear term that would be valid over a limited range. A correlation using two measured variables in the inferential control strategy is shown in Figure 17.7. This is often referred to as a *pressure-corrected temperature,* which refers to the correction of the relationship between temperature and composition to account for pressure changes.

The reliability of inferential controllers is the same as other similar systems. Controllers using additional variables would be expected to have lower reliability.

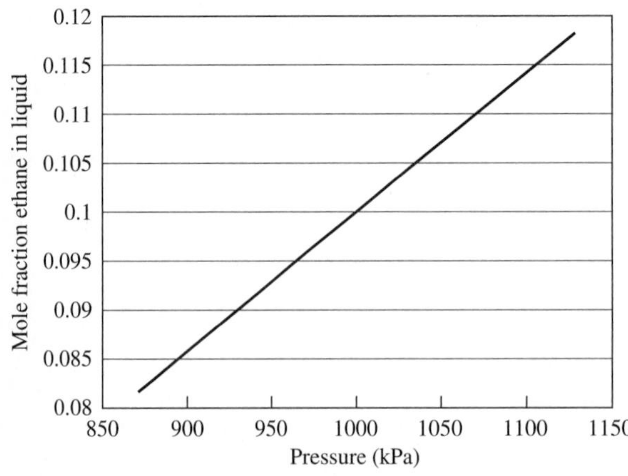

FIGURE 17.6
The effect of pressure on ethane concentration in the liquid from the flash process at the base case temperature and feed composition.

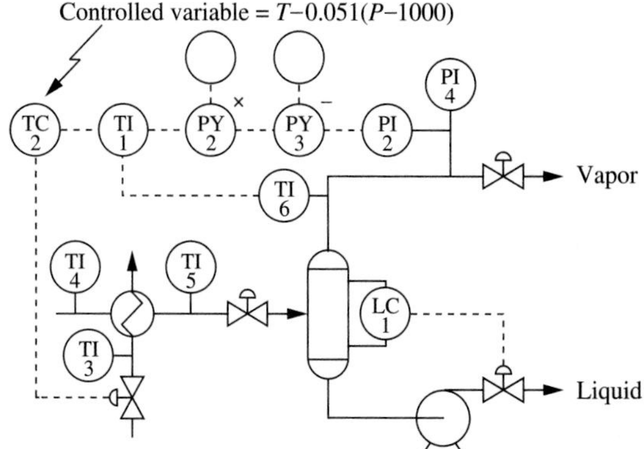

FIGURE 17.7
Enhanced inferential controller with compensation for changes in pressure.

For example, the pressure-corrected temperature controller in Figure 17.7 uses two measurements, and its reliability would be lower than the temperature-only design in Figure 17.2. Since sensors used in inferential control tend to have high reliability (their purpose is to replace the expensive and less reliable sensors), the slight loss in reliability is not usually a significant concern.

17.5 INFERENTIAL CONTROL EXAMPLE: DISTILLATION

This example extends the concept of the flash separator to a distillation tower. In operating a distillation tower, the product purities are achieved by adjusting manipulated variables such as the reboiler heating medium and distillate product flows. On-stream analyzers can be used successfully to control distillation; however, each analyzer is expensive, and not all towers require such accurate control of both product qualities. Therefore, an important question arises concerning which tray temperature, if any, can be used to infer the product composition. An analysis will be described here that, by following the general inferential design criteria, provides an answer to this question. This example considers the distillation tower in Figure 17.8, where the top product composition is to be controlled, but no analyzer is available. The tower separates a feed that contains benzene, toluene, and xylene. The top product contains benzene and toluene and 1 mole % xylene, and the bottom product contains xylene and 2.4 mole % toluene. The temperature profile is given in Figure 17.9 for the base-case operation. The goal is to control the inferred top composition by adjusting the distillate flow. The potential inferential control strategy is summarized as follows:

True variable	$= x_D =$	heavy key in distillate = 1 mole %
Inferential variable	$= T =$	tray temperature
Manipulated variable	=	distillate flow rate
Disturbances	=	reboiler duty, feed composition
Parameters	=	tray efficiency, thermodynamics

Inferential relationship: $x_D = \alpha T + \beta$ (17.6)

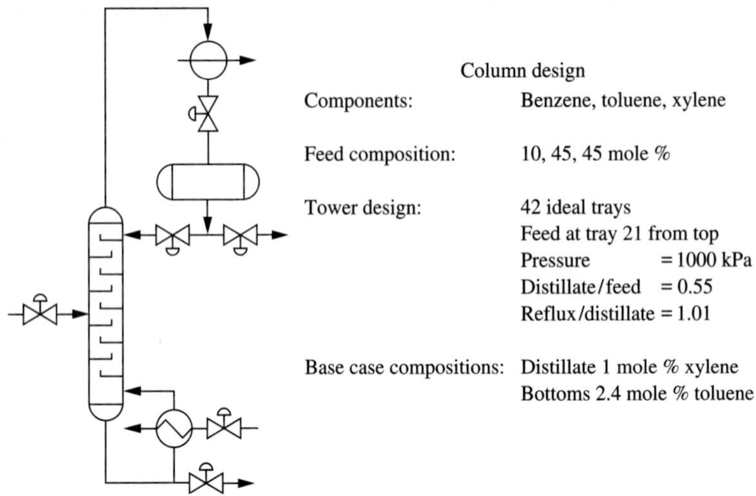

FIGURE 17.8
Parameters for the distillation tower investigated for inferential tray temperature control.

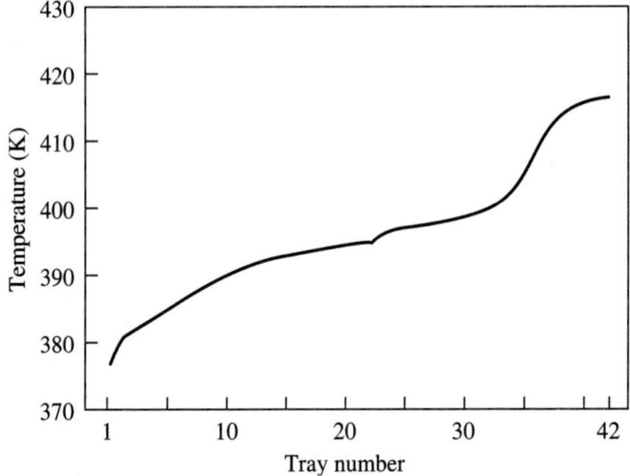

FIGURE 17.9
Tray temperature profile for the base case distillation tower.

A procedure similar to the flash example is followed, except that several tray temperatures are initially considered, with the goal of selecting the best single temperature. The trays considered are numbered 1, 5, 10, and 30 from the top; all trays could be included in this analysis, but that would expand the number of graphs. As we learned in Chapter 16, transformations of highly nonlinear relationships can often improve the performance of linear control systems. In this example, the log of the composition is controlled to linearize the feedback loop; this feature is not required for inferential control but is a good practice in distillation control (e.g., Koung and Harris, 1987) and is included in the control design. Potential relationships between tray temperatures and overhead composition for changes in operating conditions are evaluated in Figures 17.10*a* through *c* for changes in the manipulated variable (distillate flow) and in the disturbances (reboiler duty and feed composition). The distillation tower is too complex to use an analytical model to determine the relationships.

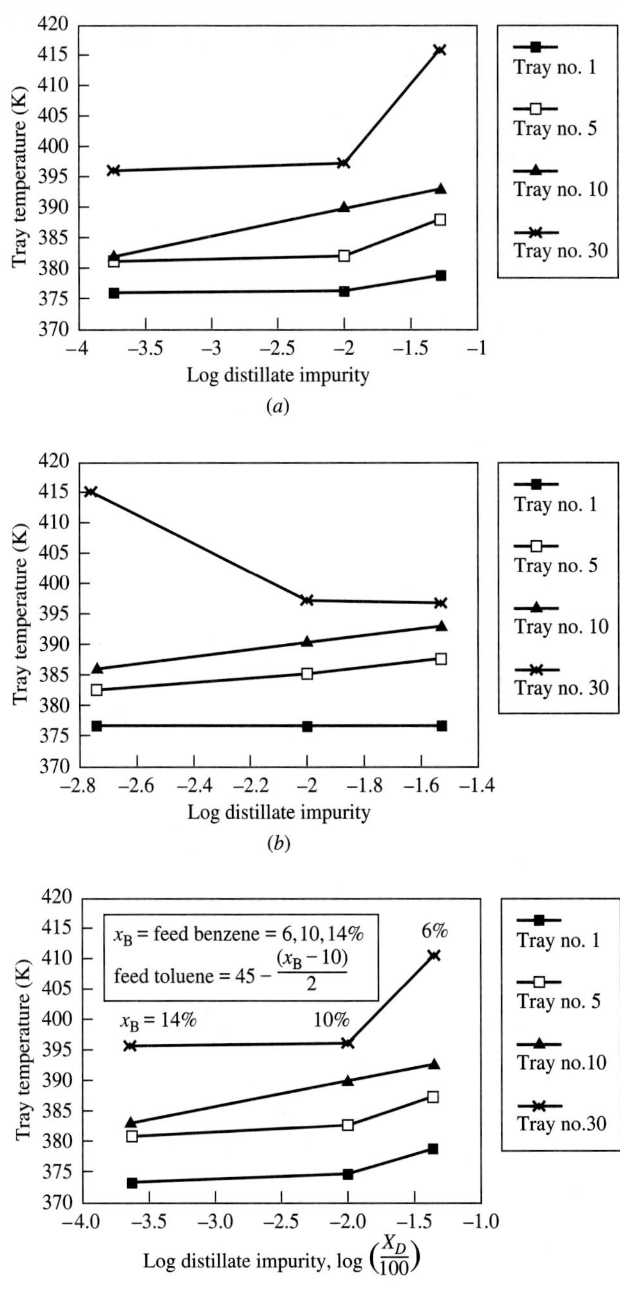

FIGURE 17.10
Relationship between tray temperature and distillate composition (*a*) for ±5% changes in the distillate flow with the feed composition and reboiler duty constant; (*b*) for ±5% changes in the reboiler duty with distillate flow and feed composition constant; (*c*) for changes in feed composition with distillate flow and reboiler flow constant.

Therefore, the values in these figures were obtained by detailed steady-state simulations of the tray-by-tray model with accurate thermodynamic data (Kresta, 1992).

For good inferential control, the selected tray temperature would have nearly the same, constant slope for all figures. Each of the candidate tray temperatures is evaluated individually to determine whether it satisfies the design criteria. The results in Figure 17.10*a* show the relationship as the manipulated variable changes, and the results in Figure 17.10*b* and *c* show the relationship as disturbances occur. All figures show clearly that the tray 1 temperature does not change significantly even though the tower operation and top product purity change. Thus, the top tray temperature would be a very poor inferential variable, because the sensor errors and low-magnitude noise would invalidate any correlation drawn from these simulations; therefore, tray 1 will not be considered further. Additional analysis of the figures reveals that tray 30 is not acceptable, because the slope changes sign between Figure 17.10*a* and *b*. For this tray temperature, a temperature increase would indicate an increase in top purity for some situations and a decrease in top purity for other situations. This would not be a good inferential variable—a result that might be expected, because the feed tray is between tray 30 and the top product, which is usually not advisable in distillation tray temperature control. Of the remaining trays, both trays 5 and 10 have reasonably linear responses, with sensitivities much greater than the noise in the temperature sensors and not changing greatly for the three figures. Thus, the temperatures for trays 5 and 10 satisfy the steady-state criteria based on this open-loop data. The preliminary conclusion is that either tray 5 or 10 would be an acceptable inferential temperature.

To evaluate this preliminary conclusion, tray 7 was chosen as representative of either tray 5 or 10 and was controlled by adjusting the distillate flow as shown in Figure 17.11 for feed composition disturbances. The steady-state errors in top product composition are plotted in Figure 17.12 for the case without an analyzer resetting the inferential controller. This measure of performance is used to evaluate the reduction in steady-state offset from perfect control that could be achieved with

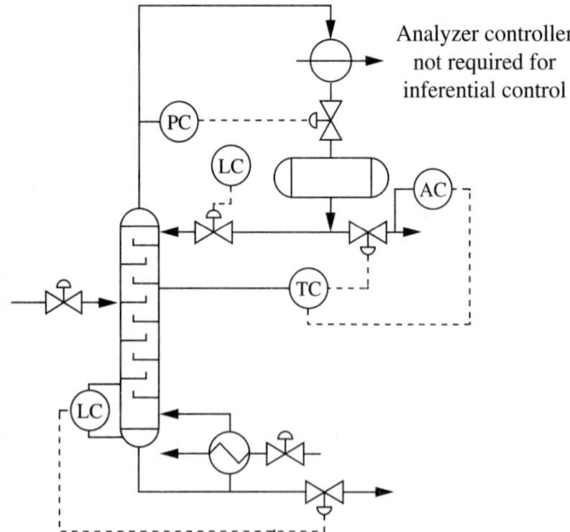

FIGURE 17.11
Control strategy for tray temperature inferential control.

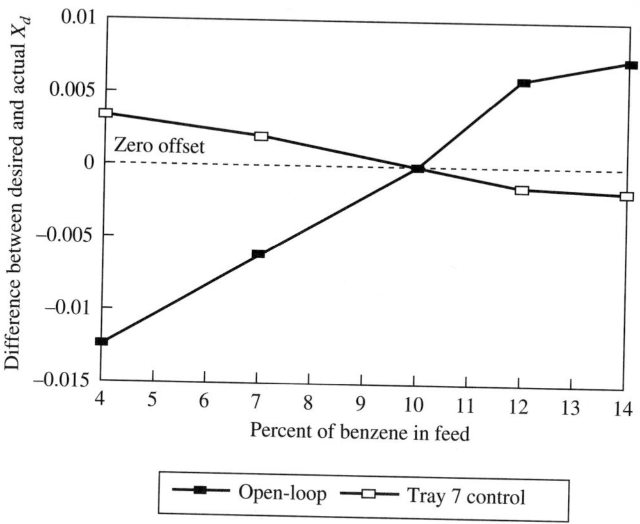

FIGURE 17.12
Steady-state offset for the distillation tower without control (open-loop) and with tray 7 inferential control (without analyzer feedback).

inferential tray temperature control. As can be seen, the top composition remains much closer to its desired value compared with the results without inferential control (open-loop), indicating that, in this case, the tray 7 temperature is a good single-tray inferential variable. Thus, inferential control offers the potential for much improved control performance.

The dynamic response of the inferential controller should also be evaluated. In this case, the tray temperature, being in the top section of the tower, introduces only a few trays between the controlled and manipulated variables. The dynamic response between the manipulated distillate flow and the controlled temperature is expected to be fast. Thus, the selection would appear to be appropriate from a dynamic viewpoint.

To achieve the zero offset performance at steady state shown in Figure 17.12, the tray temperature must be adjusted to correct small errors in the inferential relationship. This can be done by an operator, who would make manual changes to the set point based on periodic laboratory analyses. Alternatively, the tray temperature controller can be a secondary that is reset by an analyzer feedback controller. Such an approach is shown in Figure 17.11.

The procedure just described does not always identify a good tray temperature, because in some distillation towers no single tray temperature is a good inference of product composition. An example of this situation occurs when the key components have nearly the same volatility. The tray temperatures are not very different, so that the temperature variation due to composition changes is within the measurement accuracy of the sensor; in this situation the tray temperatures would not be expected to correlate with product composition. This situation occurs in the separation of propylene and propane by distillation, which demands a high-purity top product with a relative volatility of about 1.1 (Finco et al., 1989). To provide good product composition in these distillation towers, on-stream analyzers are usually provided.

The development of an empirical inferential model in this section followed the same steps used for the flash separator. Inferential tray temperature controllers designed using methods similar to the analysis in this section are widely applied in the process industries; in fact, far more distillation tower product composition controllers use tray temperature inference than use on-stream analyzers.

17.6 INFERENTIAL CONTROL EXAMPLE: CHEMICAL REACTOR

The inferential control examples for the flash and distillation processes demonstrated the empirical inferential method, in which the model is based on fitting representative data. In this section the analytical method is demonstrated for a process that can be represented by a simple closed-form model. The example in this section is the packed-bed reactor with an exothermic reaction shown in Figure 17.13. The goal is to control the moles reacted without an on-stream analyzer. Simplified steady-state material and energy balances, assuming no heat loss, for the packed-bed reactor with a single reaction occurring are

$$\mathrm{A} \rightarrow \mathrm{B}$$

$$\Delta T = -\Delta H_{\mathrm{rxn}} \frac{C_{\mathrm{Ain}}}{\rho C_p} X_{\mathrm{A}} = \frac{-\Delta H_{\mathrm{rxn}}}{\rho C_p} \Delta C_{\mathrm{A}} \tag{17.7}$$

$$\Delta T = \mathrm{T4} - \mathrm{T3}$$

where
C_{A} = concentration of A, moles/volume
ρ = density, mass/volume
C_p = heat capacity, energy/(°C · mass)
ΔH_{rxn} = heat of reaction, energy/mole
X_{A} = fraction of feed reacted = $(C_{\mathrm{Ain}} - C_{\mathrm{Aout}})/C_{\mathrm{Ain}} = \Delta C_{\mathrm{A}}/C_{\mathrm{Ain}}$

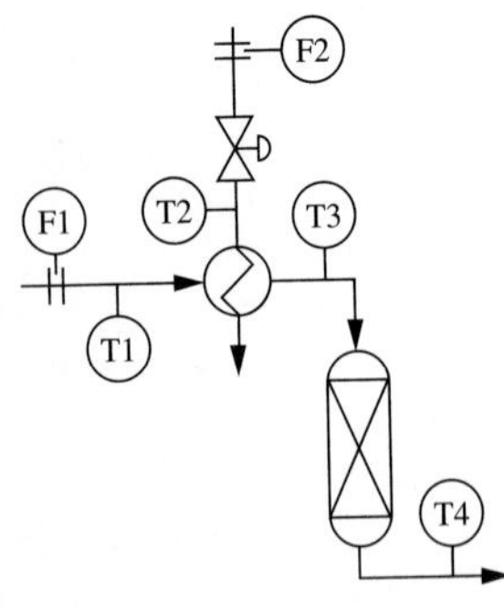

FIGURE 17.13
Packed-bed chemical reactor considered for inferential control.

A brief summary of the inferential system being evaluated is

True variable = ΔC_A = moles of A reacted

Inferential variable = ΔT = temperature difference

Manipulated variable = heating medium flow

Disturbances = inlet concentration, feed flow rate

Parameters = ρ, C_p, ΔH_{rxn}

Inferential relationship: $$\Delta C_A = \alpha \Delta T + \beta \quad (17.8)$$

To evaluate the inferential measurement, the design criteria are applied; they require a good relationship between the true variable and the inferential variable when the manipulated variable is changed and little modification to the relationship when disturbances occur. On the first issue, there is clearly a strong relationship between temperature difference and amount reacted, which could provide a reliable inference as the inlet temperature changes. The success of this approach depends on the temperature difference being much larger than the sensor error and noise in the temperature sensors, as is often, but not always, the case. On the second issue, the relationship is insensitive to changes in operating variables such as feed rate and inlet composition as seen in equation (17.7). However, the relationship is dependent on parameters such as heat capacities and heat of reaction; if these parameters are relatively constant, they will not influence the accuracy of the inferential measurement. Therefore, controlling temperature difference across the reactor could provide good inferential control of amount reacted.

Notice that the analysis to this point is for steady-state conditions. As previously mentioned, the control system dynamics must also be investigated. A typical dynamic response of the inlet and outlet temperatures and the instantaneous temperature difference to a step increase in the heating medium valve position are given in Figure 17.14. The inlet temperature responds quickly, while the outlet temperature responds slowly, because of the time required to heat the catalyst. Therefore,

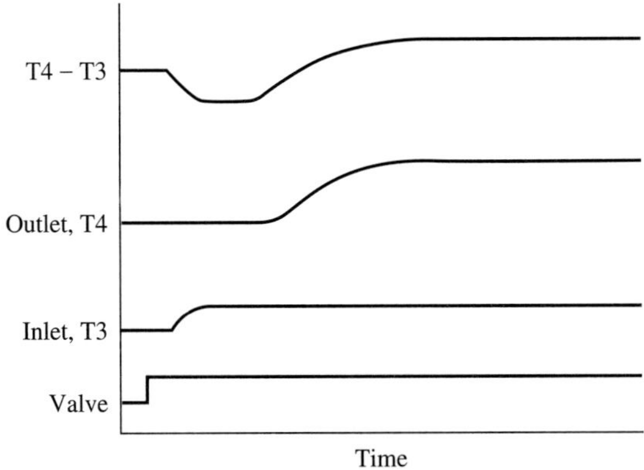

FIGURE 17.14
Plot of key variables for packed-bed reactor inferential control. Note the significant inverse response of the instantaneous temperature difference.

the *instantaneous* temperature difference is not a good inference of reactor performance, even though the steady-state temperature difference is an acceptable inferential variable. The figure demonstrates the complex inverse response between the manipulated (valve position) and potential controlled (ΔT) variables that results from this seemingly simple inferential control design. The initial inverse response can be a large multiple of the final change, and a PID controller might not perform well for this response, depending on the extent of the inverse response, as demonstrated in Example 13.8.

One method for using the available measurements is to wait for the process to achieve steady state before calculating a correction in the heating medium flow. This approach would result in very slow feedback and poor performance if frequent disturbances occur. A better control design for this example would compensate the temperatures used in the difference to account for the dynamics. One approach for this is shown in Figure 17.15, in which the inlet temperature is passed through a dynamic element that matches the outlet temperature response. The element TY-2 in the control strategy has the dynamics of the transfer function T4(s)/T3(s). Then, the two temperatures can be compared and used for control with a PID control algorithm, which would not "see" the inverse response. Another approach would be to use a predictive control algorithm in place of the PID; predictive control, which employs a simple dynamic model in the control calculation and is able to control processes with complex responses like the one in Figure 17.14, is presented in Chapter 19.

If the goal in this example were to control the outlet concentration C_{Aout} rather than the conversion, the analysis would have to be repeated for this different true controlled variable. The relationship between outlet concentration and temperature difference is unchanged as equations (17.7); however, a key operating variable that might change significantly—inlet concentration—appears in the relationship, as follows.

$$\Delta T = \frac{-\Delta H_{\text{rxn}}}{\rho C_p}(C_{\text{Ain}} - C_{\text{Aout}}) \tag{17.9}$$

Therefore, maintaining the temperature difference constant does not ensure constant outlet concentration when the inlet concentration changes. Further study would have to be performed to determine the typical variation in inlet concentration and whether this variation would introduce unacceptable errors in the inferential calculation of C_{Aout}.

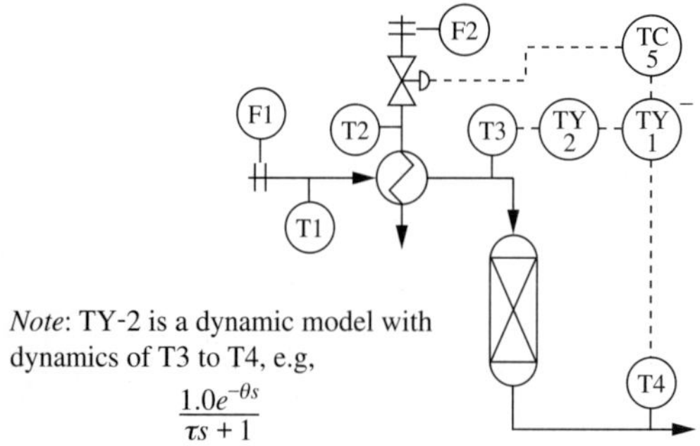

FIGURE 17.15
Design for packed-bed reactor inferential temperature difference control that does not have an inverse response.

One more possibility can be explored in the reactor example. In chemical reacting systems with multiple reactions, it is often important to control the selectivity of feed to the more valuable product as well as controlling the total conversion. Thus, we investigate here whether the temperature difference can be used to infer selectivity. The steady-state energy balance follows for a reactor with two parallel reactions in which the feed can react to either product B or product C.

$$A \rightarrow B \text{ with moles reacted } = \xi_B \tag{17.10}$$

$$A \rightarrow C \text{ with moles reacted } = \xi_c$$

$$\Delta T = \frac{(-\Delta H_B)_{rxn}\xi_B + (-\Delta H_C)_{rxn}\xi_C}{\rho C_p}$$

It can be seen from the equations that the selectivity is *not* uniquely determined when the temperature difference is specified. A measured temperature difference could be the result of many ratios of the products B and C. Therefore, the temperature difference is not a satisfactory inferential variable for selectivity in this case. In fact, if the ratio of ξ_B/ξ_C changes significantly during plant operation and the heats of reaction are different, the temperature difference is not even a good inference of the total conversion of reactant A.

The development of an inferential model based on fundamental modelling principles was demonstrated in this section.

> When possible, the inferential control model should be based on fundamental modelling principles.

This method provides excellent insight into the variables included in the model as well as the model structure. The model also provides insight into the accuracy of the inferential estimate for changes in the operating variables and physical properties.

17.7 INFERENTIAL CONTROL EXAMPLE: FIRED HEATER

As another example, inferential control can be combined with cascade control to improve the performance of the fired heater shown in Figure 17.16. The outlet temperature of the fluid in the coil is to be controlled tightly, and a primary sensor is available for this purpose. As discussed in Chapter 14, this strategy benefits from a cascade design with a secondary flow controller that corrects for some disturbances. However, the cascade does not correct completely for the effect of changing fuel gas density. The upset occurs because the heat of combustion changes as a result of changing fuel gas composition (density); thus, the heat transferred to the coil is disturbed. An improvement to the cascade control design in Figure 17.16 involves an inferential variable as the secondary of the cascade that indicates the heat released through combustion of the fuel gas. The best inferential variable of the heating value depends on the gas composition; a good inferential measure for a mixture of light hydrocarbons without hydrogen, which is a common industrial fuel gas, is the mass

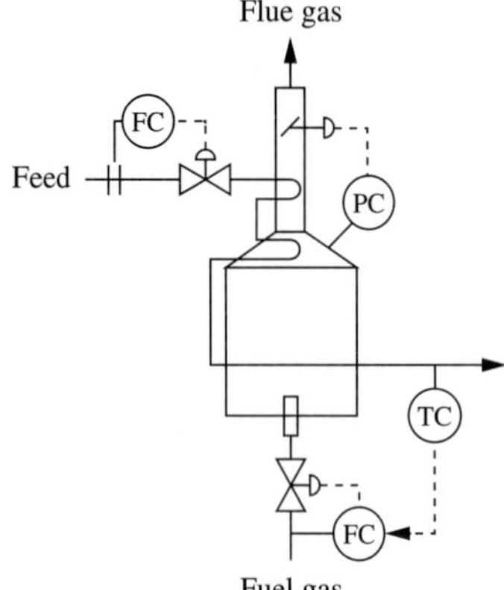

FIGURE 17.16
Fired heater process with basic controls considered for enhancement by inferential control.

flow rate of fuel (see question 17.7). To improve the response to a composition disturbance, the secondary controller in the cascade design could be altered to ensure that the mass flow, rather than the pressure difference across the orifice, is maintained constant. The potential inferential system is summarized as follows:

True variable	$= Q$	$=$ actual rate of heat released via combustion
Inferential variable	$= F_m$	$=$ mass flow rate
Manipulated variable $=$		fuel valve position
Disturbance $=$		fuel composition

Inferential relationship: $$Q = \alpha F_m + \beta \tag{17.11}$$

The mass flow can be calculated as the product of the volumetric flow rate and the density according to the following equation:

$$F_m = K\sqrt{\frac{\Delta P}{\rho}}\rho = K\sqrt{\rho}\sqrt{\Delta P} \tag{17.12}$$

where $\Delta P =$ the pressure difference across the orifice
$\rho =$ the density of the fuel gas

The inferential calculation requires an additional measurement: the density of the fuel gas at the stream conditions. This measure can be used so that the secondary controller maintains the heat fired, rather than the ΔP, at the desired value. The improved control strategy shown in Figure 17.17 provides superior performance for disturbances in fuel composition, because it rapidly adjusts the fuel flow so that the total heat fired is maintained at its desired value. Control designs based on this principle have proved to be extremely successful in industrial applications (API, 1977). The reader should be aware that the other combustion inferential measurements and control calculations should be used for different fuel compositions, such as when hydrogen or inert gases are present (Duckelow, 1981); thus, the control design that is satisfactory for this example is not generally applicable for all combustion systems.

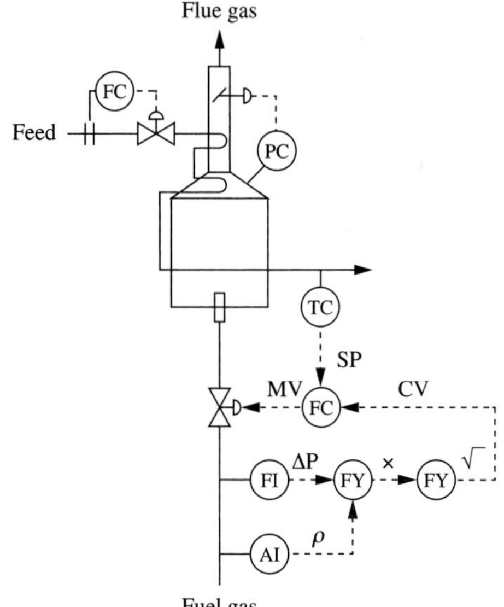

FIGURE 17.17
Fired heater with inferential control for better performance as fuel gas composition changes.

17.8 ADDITIONAL TOPICS IN INFERENTIAL CONTROL

Application of the method described in the previous sections often leads to an adequate inferential model if one can be found. Several alternative approaches to inferential control require advanced mathematics to cover completely; thus, only the basic concepts are introduced here, along with references for further study.

Multiple Measurements

Often there are many measurements available for use in an empirical inferential model. If the measurements have independent effects on the true controlled variable, the method explained in this chapter can be used. However, the measurements may have correlated effects on the true controlled variable. In the correlated case, caution must be used when fitting the model to the empirical model. For example, an inferential model for the distillation example could be formulated using many (even all) tray temperatures and flows as follows:

$$(x_D)_i = \alpha_1 T_1 + \alpha_2 T_2 + \cdots + \alpha_{n+1} F_R + \alpha_{n+2} F_D + \cdots + \alpha_m P \tag{17.13}$$

where $(x_D)_i$ = calculated estimate of the true controlled variable
T_i = temperature of ith tray
F_R = reboiler heating medium flow rate
F_D = distillate product flow rate
P = pressure

The coefficients α_i could be determined from plant data using linear regression (e.g., Draper and Smith, 1981); however, the strong correlation among the input variables can lead to a model with poor predictive ability. Note that the tray temperatures will be strongly correlated among themselves, since adjacent tray temperatures tend to increase or decrease as the product purity changes. The difficulty arises because the large number of parameters enables the model to fit much of the "noise" in the data. More advanced statistical model building and diagnostic methods based on multivariate statistics are recommended when correlated inputs are used (Kresta et al., 1994; Mejdell and Skogestad, 1991).

Plant Conditions

It is important to recognize that the empirical model represents correlation between the inputs (inferential variables) and output (true variable) in a base-case set of data used in model building. This empirical model should be used only within the range of plant operating conditions used for building the model. Operating conditions could be feed rates, feed compositions, product quality specifications, or control strategies. The empirical inferential model could give poor predictions when used outside the base-case conditions and should be reestimated when plant operations, including control structure, change (Kresta et al., 1994).

Kalman Filter

A powerful method exists when a fundamental dynamic model is available. The *Kalman filter* provides a method for using measured variables to update the fundamental model and provide a dynamic estimate of the unmeasured true controlled variable (Grewal and Andrews, 1993). This method requires mathematics beyond the general level in this book; considerable engineering effort; and, when applied, more intensive real-time computing. It should be considered when a dynamic inferential variable is required.

17.9 CONCLUSIONS

The importance of inferential control cannot be exaggerated. Many variables are difficult or impossible to measure on-stream for use in automatic, real-time control. To counter this shortcoming, inferential control is widely applied in the process industries. It may seem surprising that most of the analysis in this chapter involved steady-state relationships. This situation results from two causes. First, the major benefits for inferential control often result from a substantial reduction of the steady-state offset of the true controlled variable from its desired value. To achieve this goal, the inferential variable with the most accurate steady-state relationship is desired, even if the dynamics of the inferential controller are not the best. This situation is demonstrated in the chemical reactor example, where the inverse response dynamics are not desirable.

Another reason for the emphasis on steady-state analysis is the lack of a generally accepted design method for dynamic inferential control based on empirical models. Some initial developments in this area are noted by Kresta (1992). Note that the Kalman filter also addresses dynamic control of unmeasured variables when fundamental models exist.

When engineers first encounter inferential strategies, they often believe that the designs were based on trial-and-error methods or perhaps developed through years of observing process behavior. To the contrary; the evaluation of inferential variables follows the procedure presented in this chapter. However, the insight required for selecting the proper measurements and process relationships cannot be condensed into a simple procedure. This is a critical step, because inferential relationships can be developed over time using plant data only if the design engineer has provided the appropriate sensors. An additional challenge is to determine the proper candidates from among the numerous existing sensors—a decision requiring process knowledge, tied to the understanding of the final application with noisy sensors and process disturbances. Engineers should view this situation as an opportunity to apply their technical and problem-solving skills to this important aspect of process monitoring and control, recalling that "engineering insight" usually comes from application of fundamental principles, quantitative analysis, and hard work.

REFERENCES

American Petroleum Institute, *API Recommended Practice 550, Part III* (3rd Ed.), *Fired Heaters and Inert Gas Generators,* 1977.

Draper, N. and H. Smith, *Applied Regression Analysis* (2nd Ed.) Wiley, New York, 1981.

Duckelow, S., "Trying to Save Energy? Don't Forget Your Process Heaters," *In. Tech.* 35–39 (1981).

Finco, M., W. Luyben, and R. Pollock, "Control of Distillation Columns with Low Relative Volatilities," *IEC Res., 28,* 75–83 (1989).

Grewal, M. and A. Andrews, *Kalman Filtering: Theory and Practice,* Prentice-Hall, Englewood Cliffs, NJ, 1993.

Koung, C-W. and T. Harris, "Analysis and Control of High Purity Distillation Columns Using Nonlinearly Transformed Composition Measurements," paper presented at Can. Chem. Eng. Conf., Montreal, October 1987.

Kresta, J., "Applications of Partial Least Squares," Ph.D. thesis, McMaster University, Hamilton, Ontario, 1992.

Kresta, J., T. Marlin, and J. MacGregor, "Development of Inferential Process Models Using Partial Least Squares," *Comp. Chem. Eng., 8,* 597–612 (1994)

Mejdell, T. and S. Skogestad, "Estimation of Distillation Compositions from Multiple Temperature Measurements Using PLS Regression," *IEC Res., 30,* 2543–2555 (1991).

Smith, J. and H. Van Ness, *Introduction to Chemical Engineering Thermodynamics* (4th Ed.), McGraw-Hill, New York, 1987.

ADDITIONAL RESOURCES

For additional examples of selecting a single tray temperature for distillation control, see the following:

Tolliver, T. and L. McCune, "Finding the Optimum Temperature Control Trays for Distillation Columns," *In. Tech.,* 75–80 (September 1980).

Luyben, W., "Profile Position Control of Distillation Columns with Sharp Temperature Profiles," *AIChE J., 18, 1,* 238–240 (1972).

Experimental design is a crucial step in collecting data for model structure selection and parameter estimation. Only the most rudimentary data was used in the examples in this chapter; experimental design is covered in

Box, G., S. Hunter, and J. Hunter, *Statistics for Experimenters,* Wiley, New York, 1987.

QUESTIONS

17.1. (*a*) Discuss the inferential design criteria in your own words.

(*b*) Why are cases with changes in disturbance and manipulated variables included when selecting an inferential variable?

(*c*) Suppose that a recommendation were made to select an inferential tray temperature that had a large slope ($\Delta T/\Delta$ (tray number)); would you use this method?

(*d*) Discuss how specifications on product quality and economic values for energy and product quality would be used when evaluating an inferential variable.

(*e*) Complete the block diagram in Figure 17.1 for closed-loop, feedback control of the true controlled variable, $CV_t(s)$, using cascade control. Give the required modes for both controllers in the cascade to ensure that there would be zero steady-state offset in the true variable for a steplike disturbance.

17.2. When the inferential controller is a secondary in an automated cascade design, the primary controller can be thought of as correcting the inferential model.

(*a*) Given the following model for the control strategy in Figure 17.5, explain how the primary controller corrects the inference; that is, which parameter(s) are essentially modified through the feedback.

$$x_e = \alpha T + \beta$$

(*b*) How does this feedback affect the stability of the secondary loop?

17.3. The measured variables used directly or in calculations for inferential variables described in this chapter have been outputs (causes) from the process. It would be possible to measure inputs, both manipulated variables and disturbances, and build an inferential model using process input variables.

(*a*) Describe the similarities between inferential control using input variables and other enhancements covered in Part IV.

(*b*) Discuss the differences between using process output and input variables for inferential control and when each would be preferred.

17.4. An analyzer feedback control system that adjusts the reboiler heating medium flow, as in Figure Q14.6, is subject to disturbances in heating medium temperature. Design an *inferential* controller, implemented as a cascade secondary, that would improve control performance for the disturbances noted. State the assumptions you have made in the design.

17.5. Consider the following questions for the flash process in Section 17.2.

(a) How well would the temperature inferential controller perform if the feed had only two components: ethane and propane?

(b) For the original feed composition and operating conditions in Table 17.1, how well would the temperature perform as an inference of the ratio of *n*-butane to *n*-propane in the liquid phase?

17.6. The series of two chemical reactors in Example 3.3 are considered in this question. The reaction is A → B, and because of the cost of sensors, the measurements available are the feed flow, tank temperature, and second-tank composition of component A. Evaluate the use of these measurements for inferential control of the composition of component B in the second tank, which should be maintained within ± 0.05 mole/m^3.

17.7. Collect data on the heats of combustion for light hydrocarbons (C1 to C4), hydrogen, and carbon monoxide.

(*a*) In Section 17.7, the proposal was made that the mass flow is an acceptable inferential variable for the rate of heat release upon combustion for a stream of light hydrocarbons only. Evaluate this statement for significant changes in the stream composition.

(b) Reconsider (*a*) when significant hydrogen has been added to the stream. Is mass flow an acceptable inference? If not, what measured or calculated flow quantity is an acceptable inferential variable?

(*c*) Reconsider (*b*) with significant carbon monoxide.

17.8. Implementing an inferential controller using several measured variables should involve special care.

(*a*) Provide a detailed description of the calculations required to implement the digital inferential controller using temperature and pressure shown in Figure 17.7. You should consider initialization, calculation of the controlled variable, the feedback controller, and reset windup protection.

(*b*) Assume that it is possible to check the validity of all measured signals used in (*a*); this might be achieved by ensuring that the signal is within the allowable range. Add the logic used to respond to an invalid measurement for pressure and temperature. (*Hint:* The logic should be different for the two measurements.)

(*c*) Discuss the use of filtering the measured variables in inferential control.

17.9. A criterion for perfect steady-state inferential control in response to a disturbance is given in Section 17.1. Extend this approach to determine the criterion for perfect steady-state inferential control in response to a step change in the inferential controller set point. How would you determine which of these criteria is important for a potential application?

17.10. The concentration of component B (C_B) in the reactor system in question 5.12 is to be controlled. It cannot be measured, but the feed concentration of component A (C_{A0}), volume, and inlet flow can be measured. Propose an inferential variable for this system and discuss its strengths and weaknesses.

17.11. The concentration of component C (C_C) in the reactor system in Example 3.12 is to be controlled.

(*a*) It cannot be measured, but the feed concentration of A (C_{A0}), the volume, temperature, and flow rate can be measured. Propose an inferential variable for this system and discuss its strengths and weaknesses.

(*b*) Repeat part (*a*) with the addition of a measurement of the concentration of B (C_B) in the reactor.

(*c*) Repeat part (*b*) with the addition of a measurement of the concentration of A (C_A) in the reactor.

17.12. Derive the model used for the inferential control of the fixed-bed reactor, equation (17.7).

(*a*) Discuss how you would evaluate (1) the required accuracy (or reproducibility) for the temperature sensors, (2) the effects of heat transfer to the surroundings, and (3) the sensitivity of the inferential variable to changes in the feed flow rate. How would the results differ if a new, more active catalyst were used in the reactor?

(b) Suggest a modification to the control system design in Figure 17.15, employing an enhancement presented in Part IV, that would provide better performance for disturbances in T2.

17.13. For the inferential control system with closed-loop analyzer feedback in Figure 17.5:

(*a*) Can the concentration of the ethane and lighter components in the propane product be controlled by adjusting variables in the final distillation column?

(*b*) The same analyzer could measure the amount of butane in the propane product. Could this variable be controlled by adjusting a manipulated variable in the distillation column? If yes, which variables(s)?

17.14. The concentration of component A (C_A) is to be controlled in the nonisothermal CSTR in Example 3.10. It is not measured, but the following measurements are available: F, C_{A0}, T_0, T, V, F_c, T_{cin}, and T_{cout}. Propose an inferential variable for this system and discuss its strengths and weaknesses.

CHAPTER

18

LEVEL AND INVENTORY CONTROL

18.1 INTRODUCTION

Level control is extremely important for the successful operation of most chemical plants, because it is through the proper control of flows and levels that the desired production rates and inventories are achieved. Since some level processes are non–self-regulatory, automatic control is required to prevent the levels from overflowing or emptying completely when flow disturbances occur. Furthermore, the performance of some processes, such as chemical reactors, depends critically on the residence time in the vessel, which in turn depends on the level. In addition, the study of level control is helpful at this point because it emphasizes the importance of *control objectives* in controller design and tuning. Contrary to the situation with most control loops, the behavior of the manipulated variable—a flow in or out of the vessel—often is of as much importance as is the controlled variable itself! Thus, we have to modify some of the approaches developed in previous chapters to achieve the desired dynamic performance. As should be expected, these modifications are based on the principles of dynamic modelling and control system stability and performance already covered.

In this chapter we will first review the types of inventory processes and their process dynamics. Liquid levels are used throughout this chapter, but the results are also applicable to the control of inventories of solids and gases, although the process equipment and sensors must be modified. As we will see, level is one of the few industrially important processes for which the closed-loop dynamic response can be determined analytically. Based on this analysis, the dynamic performances of standard feedback controllers are evaluated, and the tuning rules and feedback controller algorithms to meet new objectives are developed. Finally, some additional application issues, such as selecting manipulated variables for levels in series, are discussed.

18.2 REASONS FOR INVENTORIES IN PLANTS

There are many good reasons to include inventories in plants. First, inventories are provided to enable plant operation to continue when some flows temporarily decrease, perhaps to zero. Some examples of periodic fluctuations in selected flows are feed material delivery, product shipping, and individual unit shutdown for maintenance. Inventories to account for these discontinuous flows can be quite large—on the order of hours or days of processing—so that plant operation can be maintained for periods when one or a few flows are zero. For example, a petroleum refinery which processes 700 m^3/hr of crude oil and receives deliveries every three days requires over 50,000 m^3 of inventory and usually has much more, to store different crude oils separately and to account for delays in feed delivery.

Another important use of inventories is to ensure liquid flow to a pump. If the vessel were to empty, liquid flow would be interrupted to the pump. Many pumps cannot automatically resume flow after the flow has stopped; even worse, many pumps can be damaged if they remain in operation without flow. Therefore, a liquid inventory is required at all times. For most units, an inventory with a *holdup time* (τ_H= maximum volume divided by normal flow rate) of 5 to 10 min can attenuate normal flow variations.

Finally, inventories can be placed between a disturbance source and a sensitive unit to attenuate variation in stream properties and flow rate in input flows, so that the disturbance magnitude to the sensitive unit is significantly decreased. Vessel sizing to reduce disturbances, using frequency response principles covered in Parts II and III, is demonstrated in the following example.

Example 18.1. The concentration of a feed stream to a chemical reactor, C_{A0}, experiences significant variation due to upstream process operation. The liquid flow rate is 2 m^3/min, and the variation can be closely approximated as a sine wave with an amplitude of 20 g/m^3 and a period of 6 min/cycle. Analysis has determined that the disturbance cannot be reduced further in the upstream unit. The chemical reactor can tolerate inlet concentration variation C_A of no more than 2.0 g/m^3. Determine the size of a well-mixed vessel to be placed before the reactor. Assume that the vessel volume is controlled at a constant value.

The transfer function for the concentration in a well-mixed vessel was derived in Examples 3.1 and 4.9 (page 124) to be

$$\frac{C'_A(s)}{C'_{A0}(s)} = \frac{1}{\tau s + 1}$$

For this system, the time constant τ is equal to V/F. The amplitude ratio can be determined to be

$$\text{AR} = \frac{|C'_A(j\omega)|}{|C'_{A0}(j\omega)|} = \frac{1}{\sqrt{\omega^2\tau^2 + 1}}$$

The value for the time constant and the volume can be calculated from these relationships:

$$\omega = 2\pi/\text{period} = (6.28 \text{ rad/cycle})/(6 \text{ min/cycle}) = 1.047 \text{ rad/min}$$

$$\text{AR} = 2/20 = 0.1$$

$$\tau = \frac{1}{\omega}\sqrt{\frac{1}{\text{AR}^2} - 1} = 9.50 \text{ min}$$

$$V = \tau F = (9.5 \text{ min})(2 \text{ m}^3/\text{min}) = 19 \text{ m}^3$$

In spite of the many helpful aspects of inventories, there are several reasons to minimize or eliminate them. First is the cost of the vessels themselves, along with the land or building space and maintenance. Second is the cost of material inventory, which is money invested in feedstock rather than distributed as profit. Third is the potential quality degradation from storing material. Finally, and often most important, is safety; the net effect of any accident can be much worse when a large inventory of flammable or hazardous material is involved.

> Thus, only the minimum inventory is provided in a plant to achieve the desired dynamic operation.

As is apparent by now, control objectives play a major role in the design and tuning of feedback strategies. Levels are normally controlled by adjusting a flow in or out of the vessel. (The selection is discussed later in the chapter.) Assume that the level in Figure 18.1 is to be controlled by adjusting the flow out and that the flow in experiences flow rate disturbances. Analysis of the entire process is required to determine the control objectives, and two distinct situations commonly occur. The first, referred to as *tight* level control, is where the level is very important and variation in the manipulated flow is not of great importance; for example, this situation occurs when the vessel is a chemical reactor, with the manipulated flow going to a storage tank. The second situation, referred to as *averaging* level control, occurs when variation in the level is not important, as long as the value remains within specified limits, but the manipulated flow should not experience rapid variations with a significant magnitude. This situation occurs in controlling the level of a storage drum upstream of

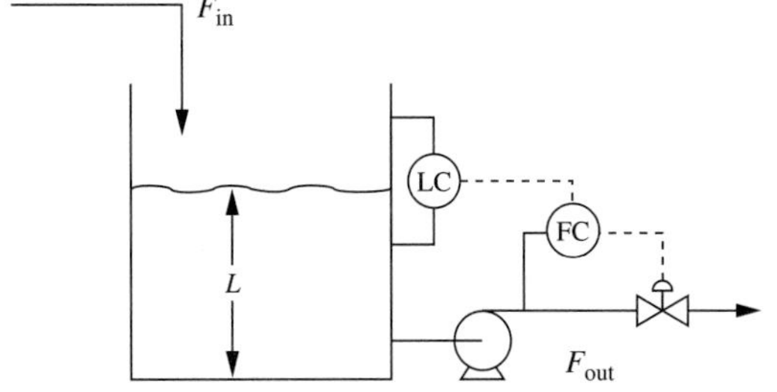

FIGURE 18.1
Typical level control system.

TABLE 18.1
Comparison of tight and averaging level control

Variable	Tight level control	Averaging level control
Controlled variable: level	Fluctuations should be reduced to a small magnitude	Fluctuations within specified limits, e.g., 20 to 80%, are allowed
Manipulated variable: flow	Fluctuations required to achieve desired level performance are accepted	Fluctuations are to be minimized, consistent with maintaining the level within limits

a critical unit. These two different control objectives are summarized in Table 18.1 with their common designations, tight and averaging level control, as just given.

18.3 LEVEL PROCESSES AND CONTROLLERS

The level processes must be understood before controller algorithms can be selected. Plant vessels are built in many different shapes, such as vertical and horizontal drums and spherical and cylindrical tanks. To simplify the mathematical analysis, only cylindrical tanks with straight sides are considered in this chapter, but all results can be extended to more complex designs, although many vessels do not significantly deviate from these assumptions in their normal range of operation. Most of the level processes can be characterized by one of the four process designs shown in Figure 18.2. Each of these processes is briefly described here, and models are derived for the industrially important designs.

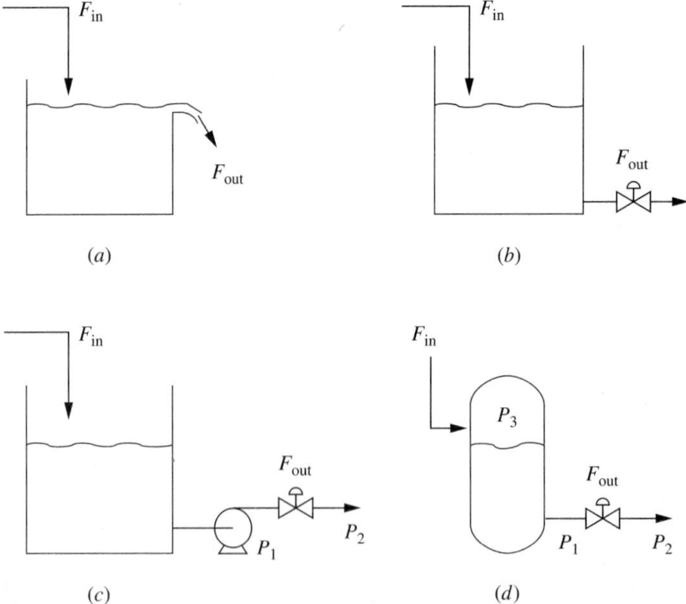

FIGURE 18.2
Various common level processes.

The overflow process in Figure 18.2*a* is seldom used in chemical plants because of its inflexibility in changing the level; however, it is used for large flows where gravity can be used as the driving force (e.g., in wastewater treatment plants). The gravity flow process in Figure 18.2*b* is not used frequently in process plants either, because it also requires a plant to flow downhill. Therefore, the process designs in Figures 18.2*a* and *b* will not be considered further in this chapter.

The level with flow out via a pump shown in Figure 18.2*c* is a very common design. The flow out depends on the valve position v and the pressure drop; here, the valve characteristic is assumed linear, so that $C_v = K$. When a pump supplies the driving force for flow, the pump outlet pressure is relatively constant; thus, the flow is independent of the level.

$$A\frac{dL}{dt} = F_{\text{in}} - F_{\text{out}} \tag{18.1}$$

$$F_{\text{out}} = K(v)\sqrt{\frac{P_1 - P_2}{\rho}} \qquad \text{with } P_1 \approx \text{constant} \tag{18.2}$$

The flow from a high-pressure to a much lower-pressure system in Figure 18.2*d* also involves a nearly constant pressure drop, since the effect of the head of liquid is very small. Thus, it is independent of the liquid level.

$$A\frac{dL}{dt} = F_{\text{in}} - F_{\text{out}} \tag{18.3}$$

$$F_{\text{out}} = K(v)\sqrt{\frac{P_1 - P_2}{\rho}} \qquad \text{with } P_1 = P_3 + \rho L\frac{g}{g_c} \approx P_3 \tag{18.4}$$

The models derived in equations (18.1) to (18.4) demonstrate that these two processes are non–self-regulating, because the derivative of the level (the flows in and out) is not significantly influenced by the liquid level. The responses of such levels without control to two common input flow disturbances are given in Figure 18.3*a* and *b*. As is apparent, the level without control can exceed its limits for all disturbances, depending on magnitude, and will definitely exceed limits for a step change (see Example 10.3, page 328).

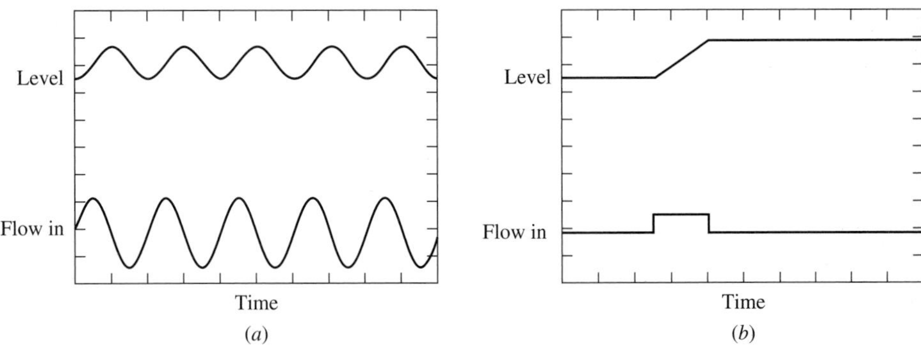

FIGURE 18.3
Response of a non–self-regulating level without control: (*a*) to sine flow variation; (*b*) to a pulse flow variation.

Based on the open-loop responses, one would conclude that feedback control is essential. The process has no dead time and a phase lag of only 90°, indicating that feedback control would be straightforward for tight level control. This is actually the case in many systems, since the sensor and valve dynamics are usually negligible. The characteristics of several common level feedback control systems are now considered. The derivations involve the flow as the manipulated variable in a cascade structure as shown in Figure 18.1, which is essentially the same as manipulating the valve for the levels under consideration.

We begin by considering proportional-only feedback control. For the non–self-regulating process, the following derivation provides the transfer function for the closed-loop system.

$$A\frac{dL'}{dt} = F'_{\text{in}} - F'_{\text{out}} \tag{18.5}$$

with $L' = L - L_s$ and $F' = F - F_s$. Substituting the control equation ($F'_{\text{out}} = K_c(L_{\text{SP}} - L) = -K_c L'$), with $K_c < 0$ for negative feedback, A the constant cross-sectional area, and $L_s = L_{\text{SP}}$, and taking the Laplace transform yields the following transfer function:

$$\frac{L(s)}{F_{\text{in}}(s)} = \frac{1/(-K_c)}{\dfrac{A}{(-K_c)}s + 1} \tag{18.6}$$

Note that the closed-loop system is first-order, clearly self-regulating. As a result, the response of the level and the outlet flow to a step change in the inlet flow would be overdamped. As expected, the level is not necessarily controlled to its set point; the steady-state offset for a step flow disturbance (ΔF_{in}) can be determined from the final value theorem to be $\Delta F_{\text{in}}/(-K_c)$.

Next, proportional-integral control is considered. The process model in equation (18.5) is unchanged, and the controller equation becomes

$$F'_{\text{out}} = -K_c\left(L' + \frac{1}{T_I}\int_0^t L'dt'\right) \tag{18.7}$$

Substituting this expression into equation (18.5) and taking the Laplace transform yields the transfer function for the closed-loop system.

$$\frac{L(s)}{F_{\text{in}}(s)} = \frac{\left(\dfrac{T_I}{(-K_c)}\right)s}{\tau^2 s^2 + 2\tau\xi s + 1} \tag{18.8}$$

with $$\tau = \sqrt{\frac{AT_I}{(-K_c)}} \quad \text{and} \quad \xi = \frac{1}{2}\sqrt{\frac{T_I(-K_c)}{A}} \tag{18.9}$$

By applying the final value theorem, it can be shown that the system is self-regulating with zero offset for a step disturbance. The response is now second-order and can be either overdamped or underdamped, depending on the value of the damping coefficient ξ. As shown in equation (18.9), the damping coefficient depends on controller parameters K_c and T_I and the vessel area.

Important qualitative features of the dynamic response and the steady-state offset for the level control system depend on the process design and controller algorithm and its tuning.

Before we determine how to match these factors to the control objectives, a modification to the linear PI controller is considered.

18.4 A NONLINEAR PROPORTIONAL-INTEGRAL CONTROLLER

Looking ahead to the application of averaging level control, we anticipate the need for an algorithm that makes small flow adjustments for small level deviations from set point and large adjusts for large deviations. Thus, a nonlinear algorithm seems appropriate. Many nonlinear modifications have been proposed; only one of the more common is discussed in this section (Shunta and Feherari, 1976). The algorithm is given as follows, and the relationship of the proportional mode between the level and manipulated flow is shown in Figure 18.4.

$$F'_{\text{out}} = -K_c\left(L' + \frac{1}{T_I}\int_0^t L'\,dt'\right) \tag{18.10}$$

with

$$K_c = \begin{Bmatrix} K_{cS} & \text{when } |L'| < L'_B \\ K_{cL} & \text{when } |L'| > L'_B \end{Bmatrix} \qquad r_K = \frac{K_{cL}}{K_{cS}}$$

Along with the integral time and gain, K_{cL}, the algorithm has two additional tuning parameters: the "break" point between the large- and small-controller-gain

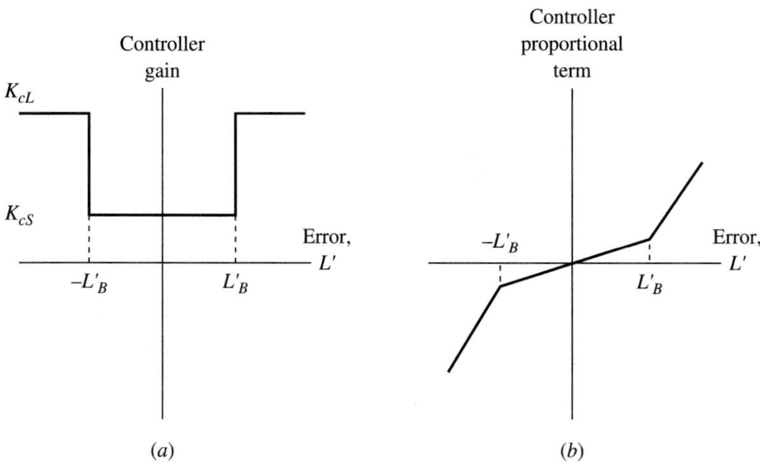

FIGURE 18.4
Graphical display of the nonlinear PI control algorithm for level control.

regions, L_B', and the ratio of the large and small gains, r_K. Note that if the ratio is 1, the controller in equation (18.10) simplifies to a linear algorithm. If the ratio is infinity, the nonlinear controller takes no action for small deviations; that is, it has a "dead band" for an error $\pm L_B'$. The integral mode ensures that the level ultimately reaches its set point, whereas an infinite value for T_I would result in a proportional-only controller with steady-state offset.

18.5 MATCHING CONTROLLER TUNING TO PERFORMANCE OBJECTIVES

The two sets of control objectives in Table 18.1 require different approaches, and each is presented separately in this section. The approach for determining the tuning constants for this simple process is to specify some key characteristics of the closed-loop transient response to a step flow disturbance and then to calculate tuning constants that achieve the specified characteristics. As with all tuning calculations, the resulting constants should be considered initial estimates, which can be fine-tuned based on plant performance.

Tight Level Control

We will begin by considering the case of tight level control, where the performance of the level is of greatest importance. As mentioned, the control problem is not difficult, because of the lack of dead time (or inverse response) in the process. As a result, a linear controller is adequate. The key variables used to characterize the system are the level process design and the maximum step disturbance in the uncontrolled flow. The desired transient response can be characterized by the maximum allowable level deviation in response to the disturbance and the damping coefficient ξ. A good starting value for the damping coefficient is 1.0, but the method presented here can be used for any other damping coefficient. The following expression gives the dynamic response of a level under PI control to a step flow disturbance when the damping coefficient is 1.0. With the step inlet flow, $\Delta F_{in}/s$, the expression for the level in equation (18.8) can be determined by inverting the Laplace transform using entry 6 in Table 4.1.

$$L' = \frac{\Delta F t}{A} e^{-t(-K_c)/2A} \tag{18.11}$$

The time when the maximum occurs can be determined by differentiating equation (18.11) and setting the result equal to zero, which gives a unique value of $t_{max} = 2A/(-K_c)$ because the system is not underdamped. This time can be substituted into equation (18.11) to determine the maximum level deviation for a step input.

$$\Delta L_{max} = 0.736 \frac{\Delta F_{max}}{(-K_c)} \tag{18.12}$$

The tuning constants K_c and T_I can be calculated from equations (18.9) and (18.12) using specified values for the control performance: the magnitude of the disturbance, ΔF_{max}, and desired values for $\xi(= 1.0)$ and ΔL_{max}.

An alternative tuning approach, using specifications for the maximum level deviation and maximum rate of change for the manipulated flow, is given by Cheung and Luyben (1979). Their approach requires a trial-and-error solution, for which they have prepared graphical correlations.

Example 18.2. The level in a vessel with a volume of 20 m^3, a cross-sectional area of 10 m^2, and a normal flow of 2 m^3/min is to be controlled tightly with a PI controller. The expected maximum step change in the uncontrolled flow rate, based on plant experience, is 0.2 m^3/min (i.e., 10% of normal). Tight level control requires a small level deviation, so that the maximum allowable change in the level is selected to be 0.05 m (i.e., ±2.5% of the range). Estimate the tuning constants for PI and P-only controllers.

Solution. The damping coefficient is selected to be 1.0. Using equations (18.9) and (18.12), the tuning constants for PI control are

$$K_c = \frac{-0.736\Delta F_{\max}}{\Delta L_{\max}} = \frac{-0.736(0.2\ \text{m}^3/\text{min})}{0.05\ \text{m}} = -2.94\frac{\text{m}^3/\text{min}}{\text{m}}$$

$$T_I = \frac{4\xi^2 A}{(-K_c)} = \frac{(4)(1^2)\ 10\text{m}^2}{2.94\ \dfrac{\text{m}^3/\text{min}}{\text{m}}} = 13.6\ \text{min}$$

and, for P-only control,

$$K_c = \frac{-\Delta F_{\max}}{\Delta L_{\max}} = -\frac{0.20}{0.05} = -4.0\frac{\text{m}^3/\text{min}}{\text{m}}$$

The dynamic response for the level under tight PI control subject to the step disturbance is given in Figure 18.5*a*.

Linear Averaging Level Control

Averaging level control can be achieved with either a linear or a nonlinear controller. Both are discussed here, with the linear given first. Before presenting tuning methods, it is worth noting that averaging level control is improved by providing a large inventory (i.e., vessel volume). Thus, the performance of the averaging level system depends on the *process,* algorithm, and tuning—which is naturally true for all control systems.

The approach for the linear controller tuning is the same as for the tight control, except that the value for the allowable deviation would be much larger, to provide as much attenuation in the manipulated variable as possible.

Example 18.3. Calculate the tuning constants for Example 18.2 for a linear averaging level controller. All physical parameters are the same; however, the maximum level change is selected to be 0.8 m, which is ±40% of the level range, to allow inlet flow variations to be attenuated.

Solution. The same equations as in Example 18.2 are used. For PI control,

$$K_C = \frac{-0.736\Delta F_{\max}}{\Delta L_{\max}} = \frac{-0.736(0.2\ \text{m}^3/\text{min})}{0.8\ \text{m}} = -0.184\frac{\text{m}^3\ \text{min}}{\text{m}}$$

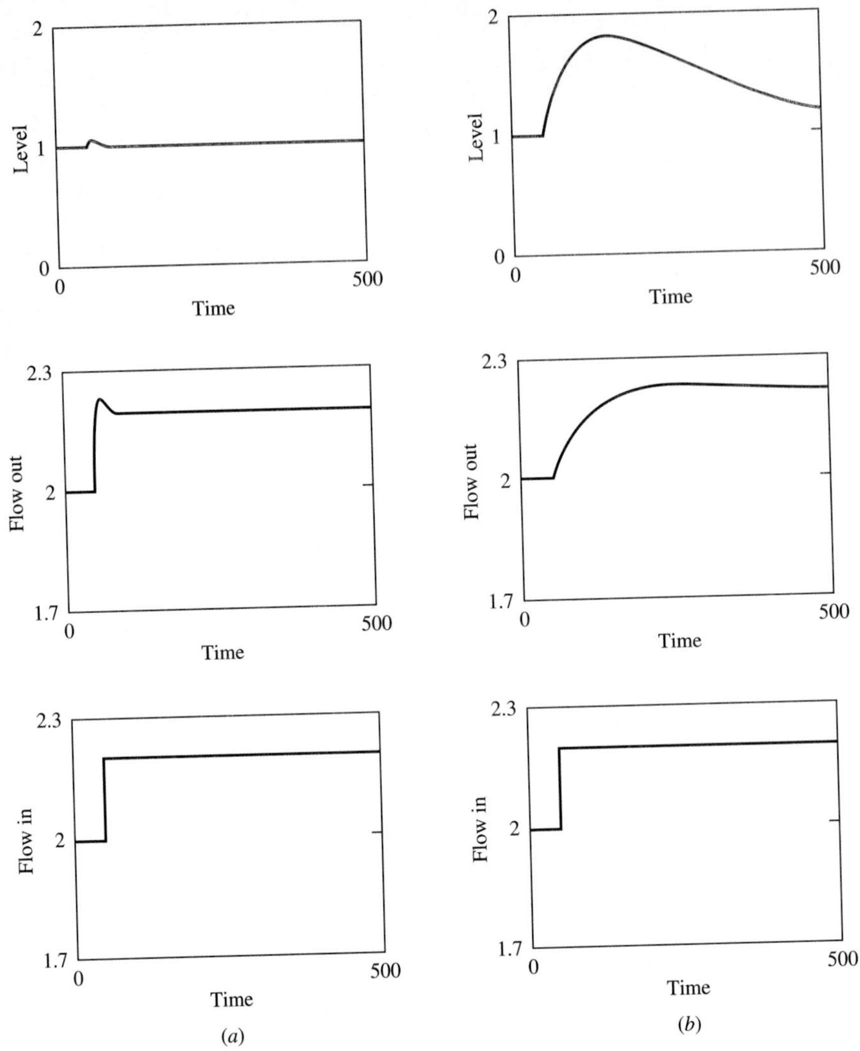

FIGURE 18.5
PI level control for Examples 18.2, and 18.3: (*a*) tight, (*b*) linear averaging.

$$T_I = \frac{4\xi^2 A}{(-K_c)} = \frac{(4)(1^2)10\ \text{m}^2}{0.184\dfrac{\text{m}^3/\text{min}}{\text{m}}} = 217\ \text{min}$$

and, for P-only control,

$$K_c = \frac{-\Delta F_{\max}}{\Delta L_{\max}} = -0.25\frac{\text{m}^3/\text{min}}{\text{m}}$$

A dynamic response for the level under averaging PI control subject to the step disturbance is given in Figure 18.5*b*. The slower response of the flow out is obvious, and the maximum rate of change of the manipulated flow is about 1/15 the value for the tight level control response, which was achieved with the same vessel and control algorithm through modified tuning.

Nonlinear Averaging Level Control

The nonlinear controller has two additional parameters to specify. With proper values for these parameters, the nonlinear controller can provide better performance (i.e., make smaller manipulations) when the system experiences frequent, small flow disturbances. The value of L'_B is selected to be smaller than the maximum level deviation but to be larger than most level variations experienced in normal operation. The value for the gain ratio is selected to provide small corrections for the small deviations; a value of 20 is usually a good starting point. To simplify the calculations for the initial estimates, the proportional gain is calculated so that the proportional term *alone* can correct for the largest expected flow disturbance. The proportional term can be calculated as follows by conforming to Figure 18.4*b*.

$$\Delta F_{\max} = -K_{cS}L'_B - K_{cL}(\Delta L_{\max} - L'_B) = \left(\frac{L'_B}{r_K} + \Delta L_{\max} - L'_B\right)(-K_{cL}) \tag{18.13}$$

Then the integral time is calculated so that the damping coefficient is 1.0 for the small-gain region, which ensures that the damping coefficient is greater than one in the large-gain region.

Example 18.4. Calculate the tuning constants for Example 18.3 with a nonlinear averaging controller.

Solution. The nonlinear controller requires two additional parameters. The guidelines suggest that $r_K = 20$, and we select L'_B to be relatively large, to provide small outlet flow variations for most inlet flow oscillations. Thus, $L'_B = 0.7$ m, which is ±35% of the level range. For the PI controller,

$$K_{cL} = \frac{-\Delta F_{\max}}{\dfrac{L'_B}{r_K} + \Delta L_{\max} - L'_B} = \frac{-0.2 \text{ m}^3/\text{min}}{\dfrac{0.7\text{m}}{20} + 0.1 \text{ m}} = -1.48\frac{\text{m}^3/\text{min}}{\text{m}}$$

$$K_{cS} = \frac{K_{cL}}{20} = -0.074\frac{\text{m}^3/\text{min}}{\text{m}}$$

$$T_I = \frac{4\xi^2 A}{(-K_{cL}/r_K)} = \frac{(4)(1^2)10 \text{ m}^2}{0.074 \text{ m}^3/\text{min}} = 540 \text{ min}$$

Now that we have tuned the linear and nonlinear controllers, it is worthwhile comparing their performance for a periodic input disturbance, because plants often experience such variation. The responses to sine disturbances are given in Figures 18.6*a* through *c* for the tunings determined in Examples 18.2 through 18.4, with the input flow disturbance a sine with magnitude 0.2 m^3/min and period of 80 min. The results in Figure 18.6*a* demonstrate the performance of the tight level controller, which maintains the level close to its set point but has a large maximum rate of change in the output flow, 1.8×10^{-2} (m^3/min)/min. Recall that it is not possible to achieve tight level control with small flow manipulations simultaneously.

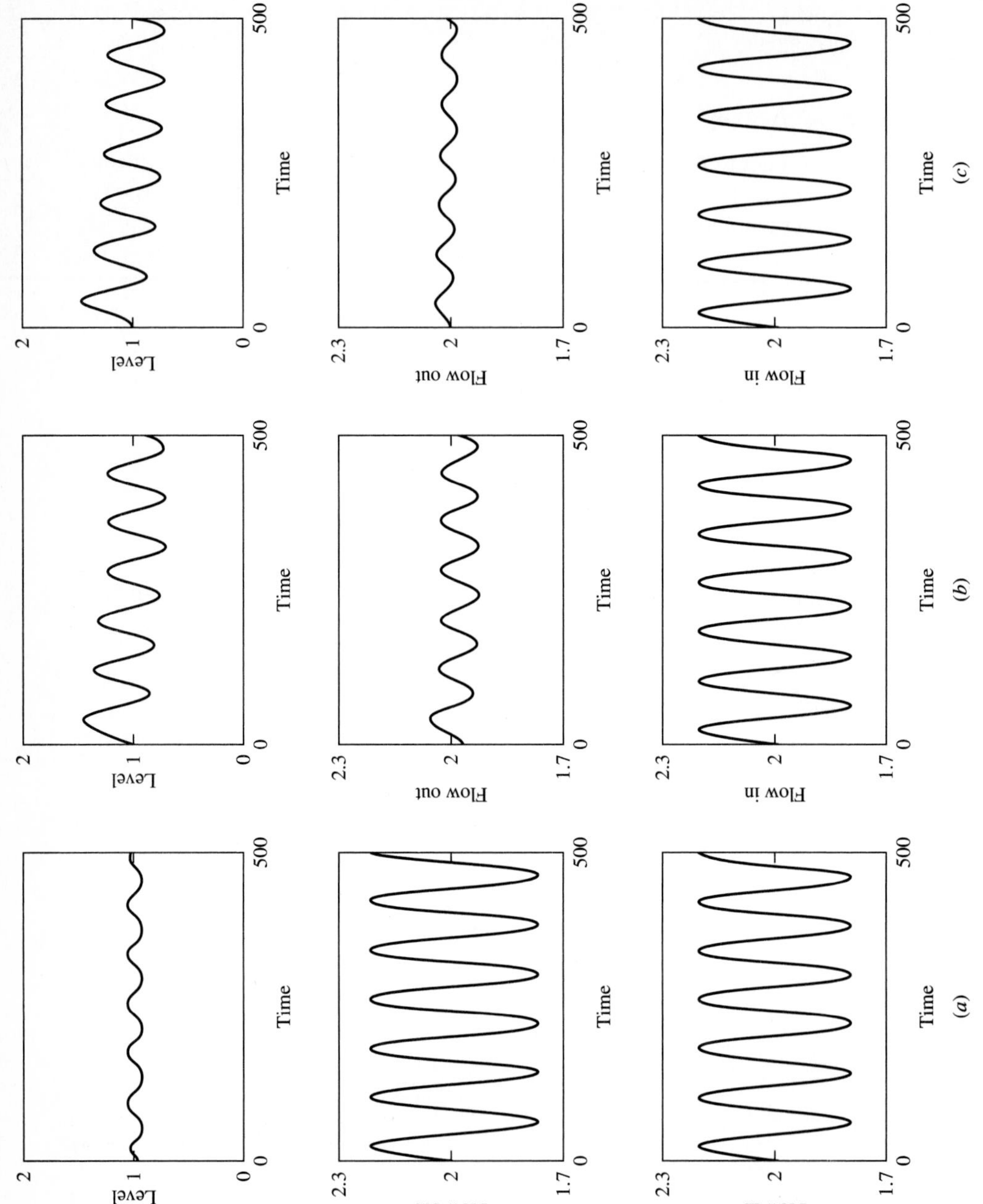

FIGURE 18.6
Level control for an input sine flow disturbance: (*a*) tight PI control with tuning from Example 18.2; (*b*) linear averaging control with tuning from Example 18.3; (*c*) nonlinear averaging control with tuning from Example 18.4.

A linear PI controller provides excellent performance when tight level control is required. The alternative design, using a proportional-only controller with a high controller gain, is also acceptable.

The performance for averaging level control demonstrates that both linear and nonlinear approaches provide flow attenuation; in other words, the manipulated flow varies substantially less than the inlet flow. The response for the linear averaging PI controller is given in Figure 18.6*b*, which demonstrates the smaller variability in the manipulated flow (the maximum rate of change is 0.40×10^{-2} (m^3/min)/min), and a larger variability in level. The response for the nonlinear averaging PI controller is given in Figure 18.6*c*, which demonstrates the even smaller variability in the manipulated flow (the maximum rate of change is 0.16×10^{-2} (m^3/min)/min) and a yet larger variability in level. Note that the nonlinear averaging level controller reduced the maximum rate of change of the manipulated flow by an order of magnitude when compared with the tight controller for the *same inventory volume.*

The nonlinear level controller is preferred for averaging control when the flow variations and vessel volume are such that the level remains within $\pm L'_B$ for most of the time.

The level algorithms and tuning in this section have provided the flexibility to use the existing inventory to the greatest advantage. However, acceptable performance for averaging level control requires sufficient inventory; therefore, determining the proper inventory is addressed in the next section.

18.6 DETERMINING INVENTORY SIZE

Naturally, the control performance is influenced by the vessel holdup time, so that an important task of the engineer is to determine inventory sizes when designing or modifying the plant. Given the flow rate disturbance, the performance specification, and the controller tuning method, the holdup time can be determined using the results from previous sections. For a step disturbance, the calculations would involve the relationships already derived and used in tuning calculations to determine the volume required to maintain the level within $\pm\Delta L_{max}$ and the maximum rate of change of the manipulated variable at or below a specified value. It is assumed that the damping coefficient should be 1.0, although the approach can be adapted for other values.

The calculation of the inventory size can be performed in a noniterative manner by using the analytical expression of the manipulated flow to a step change in the in flow. First, the transfer function relating the flows in and out is derived using equation (18.8) and the PI controller transfer function:

$$\frac{F_{out}(s)}{F_{in}(s)} = \frac{L(s)}{F_{in}(s)}\frac{F_{out}(s)}{L(s)} \tag{18.14}$$

$$= \frac{\dfrac{T_I}{-K_c}s}{\tau^2 s^2 + 2\xi\tau s + 1}\left[-K_c\left(1 + \frac{1}{T_I s}\right)\right] = \frac{T_I s + 1}{\tau^2 s^2 + 2\xi\tau s + 1}$$

Then the step input is substituted ($F_{in}(s) = \Delta F_{in}/s$) and the inverse Laplace transform is determined from entry 8 in Table 4.1 to give

$$F'_{out}(t) = \Delta F_{in}\left[1 + \left(\frac{T_I - \tau}{\tau^2}t - 1\right)e^{-t/\tau}\right] \tag{18.15}$$

The derivative of the flow rate can then be taken to give

$$\frac{dF_{out}}{dt} = \Delta F_{in}\left[\left(\frac{T_I - \tau}{\tau^2} - \frac{T_I - \tau}{\tau^3}t + \frac{1}{\tau}\right)e^{-t/\tau}\right] \tag{18.16}$$

It is clear from this result (noting that $T_I > \tau$ for the tuning selected) that the maximum rate of change occurs at $t = 0$. Setting $t = 0$ and substituting the value of τ from equation (18.9) gives

$$\left.\frac{dF_{out}}{dt}\right|_{max} = \left(\frac{\Delta F_{in}}{A}\right)(-K_c) \tag{18.17}$$

The value of the controller gain from equation (18.12) can be substituted to give

$$\left.\frac{dF_{out}}{dt}\right|_{max} = \frac{0.736(\Delta F_{in})^2}{A(\Delta L_{max})} \tag{18.18}$$

The product $A(\Delta L_{max})$ represents the allowable variability in the inventory above (or below) the set point. If the level is allowed to vary ±40%, $A(\Delta L_{max}) = 0.40\ V$. Thus, the final expression for the inventory volume for linear averaging level control with conventional tuning is

$$V = \frac{1.84(\Delta F_{max})^2}{\left.\dfrac{dF_{out}}{dt}\right|_{max}} \tag{18.19}$$

Example 18.5. A flow into a vessel has a base value of 2.0 and a maximum step disturbance of 0.20 m³/min. The flow out should have a rate of change that does not exceed 1.0×10^{-3} (m³/min)/min, and the level can vary within ±40% of its middle value. Determine the inventory size to satisfy this requirement when the flow out is manipulated by a PI controller.

Solution. Equation (18.19) can be used directly to calculate the volume to be

$$V = \frac{1.84(0.20\ \text{m}^3/\text{min})^2}{1 \times 10^{-3}\ \text{m}^3/\text{min}^2} = 73.6\ \text{m}^3$$

The area and height can be selected to satisfy this volume (e.g., $A = 36.8 \text{ m}^2$ and $L = 2$ m). The tuning for this controller can then be calculated for $\Delta L_{max} = 0.8$ m to be

$$K_c = 0.736\frac{\Delta F_{max}}{\Delta L_{max}} = \frac{(0.736)(0.20 \text{ m}^3/\text{min})}{0.8 \text{ m}} = -0.184\frac{\text{m}^3/\text{min}}{\text{m}}$$

$$T_I = \frac{4\xi A}{-K_c} = \frac{4(1)(36.8 \text{ m}^2)}{0.184(\text{m}^3/\text{min})/\text{m}} = 800 \text{ min}$$

The result of this example is a level process and tuning that (just) satisfy the objective on the outlet flow behavior for the specified input step disturbance.

18.7 IMPLEMENTATION ISSUES

Level control is generally quite straightforward to implement. Many different sensors can be used to determine the inventory in a vessel. The most common is the pressure difference measurement, which is shown in Figure 18.1. Assuming a constant liquid density, the difference in pressure is proportional to the level in the vessel between the two measuring points, called *taps.* Note that the lower tap is usually placed somewhat above the bottom of the vessel, to prevent plugging from a small accumulation of solid contaminants. The level displayed to the operating personnel could be expressed in units of length; however, this would require the people to remember the maximum level in each individual vessel. Therefore, the level is normally displayed as a percentage of the maximum value between the taps.

Many other types of level sensors are possible (e.g., Blickley, 1990; Cho, 1982; and Cheremisinoff, 1981). An example is a float that remains at the interface and indicates the level by its physical position as transmitted by a connecting rod. Levels of materials that do not rest evenly in the vessel, such as granular solids, or of very corrosive materials can be measured by sound waves directed at the material from above a vessel. For some accurate measurements, the entire vessel and contents can be weighed.

Level control often uses cascade principles by resetting a flow controller, as shown in Figure 18.1. Usually, this is not to improve the dynamic response to disturbances but to make the operation easier for the operator when the cascade is opened. Level control can be implemented with either linear or nonlinear proportional-only or proportional-integral control algorithms. Both are available as preprogrammed options in most digital equipment.

18.8 VESSELS IN SERIES

In many chemical plants, units are arranged in series as shown in Figure 18.7. Plants do not usually have many simple tanks in series, but units such as reactors, flash drums, and distillation towers are generally in series and have liquid inventories. The behavior of these systems is investigated here by considering the simpler, but representative, system of tanks. We will consider two important questions:

1. How can the throughput and levels be controlled?
2. How does a series of levels respond dynamically?

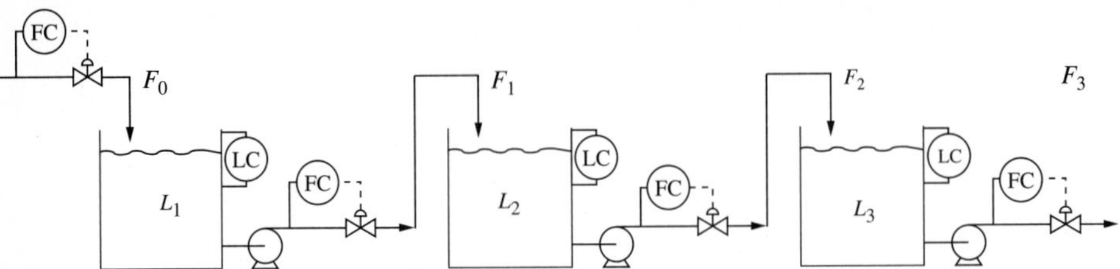

FIGURE 18.7
Design for three levels in series.

We can answer the first question by analyzing the degrees of freedom in the system. For simplicity, proportional-only controllers are considered, but the results are equally valid for other controller algorithms. The system in Figure 18.8 can be modelled according to the following equations:

For each level ($n = 1$ to 3):

$$A\frac{dL'_n}{dt} = F'_{n-1} - F'_n \tag{18.20}$$

and *one* of either of these controller equations for each level:

$$F'_n = -K_c L'_n \qquad \text{or} \qquad F'_{n-1} = -K_c L'_n \tag{18.21}$$

Note that there are six equations and seven variables (three levels and four flows). Thus, one flow rate can be set independently. This result should be not be surprising, since level control requires the inlet and outlet flows to be equal at steady state.

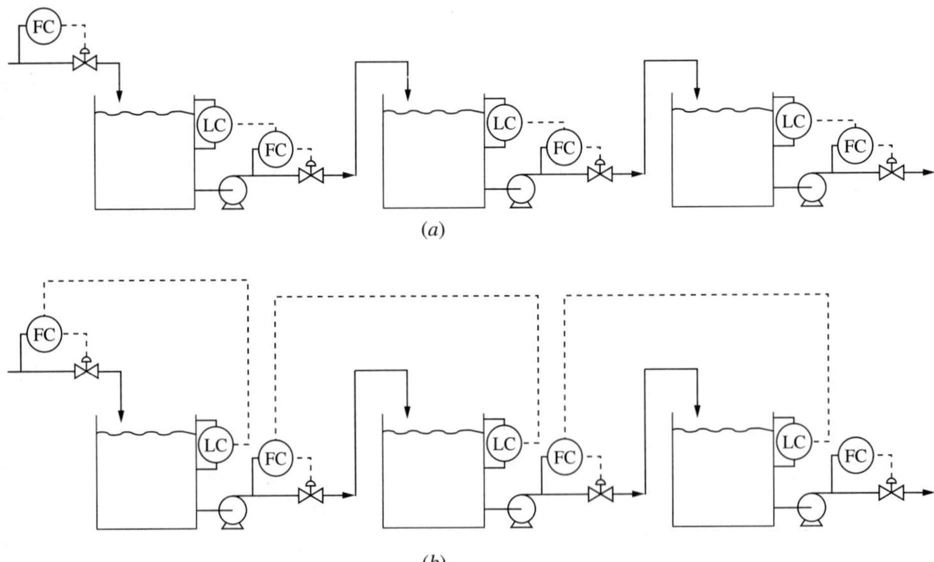

FIGURE 18.8
Two possible control designs for levels in series.

Another question to answer is which flow should be set to determine the flow rate. The degrees-of-freedom analysis cannot provide further insight, because any flow is acceptable; thus, this more detailed design decision requires more information on the control objectives and process equipment. If no constraints are encountered in the plant, the inlet or feed rate is often set independently, as shown in Figure 18.8*a*. If the production rate should be held constant, the outlet flow is set independently, as shown in Figure 18.8*b*. If an intermediate flow should be constant, as is the case if a constraint like pump capacity or heat exchanger duty is encountered in an intermediate unit, the intermediate flow can be set independently. An interesting control strategy that controls all levels and maximizes the flow rate is given by Shinskey (1981).

Now that the control structure has been determined, the second question about dynamic response can be addressed (Cheung and Luyben, 1979). Based on equation (18.14), the series of three identical level systems shown in Figure 18.8*a* can be combined in the following overall transfer function:

$$\frac{F_3(s)}{F_0(s)} = \left(\frac{T_I s + 1}{\tau^2 s^2 + 2\tau\xi s + 1} \right)^3 \tag{18.22}$$

Since the poles of the individual level control systems are the poles of the series system, if each individual system is overdamped, the overall system is overdamped. However, if the systems are underdamped, the overall system will be underdamped. Dynamic responses of the manipulated flows are given in Figure 18.9*a* through *c* for

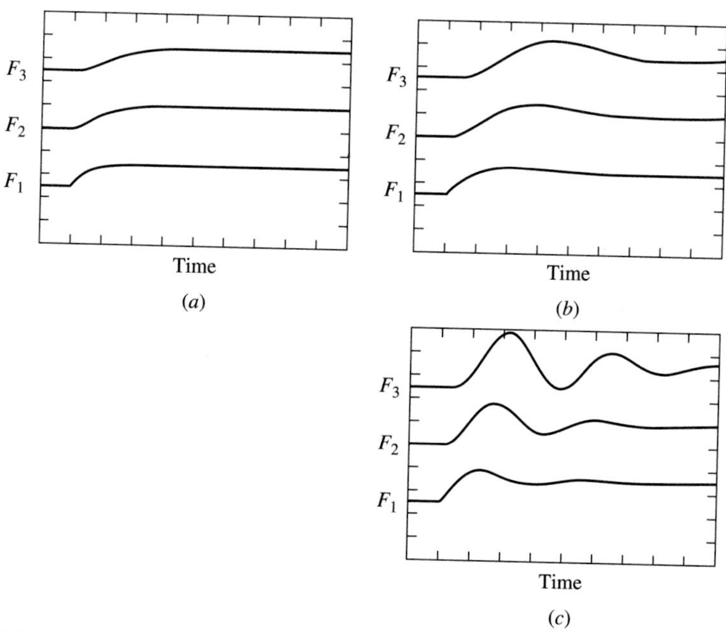

FIGURE 18.9
Response to step of input flow of three series level controllers: (*a*) P-only; (*b*) PI (individual ξ's = 1.0); (*c*) PI level (individual ξ's = 0.5).

the system with different damping coefficients in response to a step change in the inlet flow F_0.

The flow adjustments are monotonic for the proportional-only controllers, but the adjustments result in overshoot for all proportional-integral controllers, even those that are critically (or over) damped.

> It is important to note that for a step response (1) the manipulated flow for PI control always overshoots its final value and (2) the magnitude of the oscillations *increases* in series systems when each element in the series is underdamped!

A relatively small oscillation at the first level can be magnified, leading to very poor performance, by other downstream levels in the series. Thus, a series process structure of inventories heightens the importance of careful algorithm selection and tuning for each level controller.

18.9 CONCLUSIONS

The key features of inventory control are the range of control objectives and the need to match the control algorithm with the relevant objective. Feedback control provides excellent tight level control performance, because the system has little or no dead time. Proportional-only or proportional-integral control with simple tuning guidelines is adequate for tight level control.

Analysis of plant requirements indicates that averaging control is appropriate for many level systems. The linear P-only and PI algorithms can achieve averaging control with proper tuning. Improved averaging control can be achieved using a nonlinear PI algorithm when most flow disturbances are of the magnitude and frequency to allow moderate flow manipulations and have the level remain within an acceptable range. This modification is especially advantageous when the system experiences high-frequency disturbances. One should never lose sight of the fact that the performance of averaging level control improves with a large vessel inventory, which must be provided when the process is being designed.

We can derive analytical expressions for the time-domain behavior of level processes and can determine proper tuning rules to achieve specified behavior based on these expressions. The approach used for levels would have been valuable for all feedback systems because of its excellent specification of closed-loop performance. Unfortunately, the approach would not be successful for more complex processes, for which analytical models for closed-loop response cannot be developed. Thus, this excellent approach is limited to a few simple processes.

Smooth overall operation often requires that all flows in the series system have little oscillation. We have seen how levels in series can potentially increase oscillations and have derived models for predicting the responses. These results demonstrate the importance of ensuring that level systems not have small damping coefficients.

Since controlling flows and inventories is an essential aspect of designing controls for multiple units, the material covered in this chapter provides an essential foundation for the control design topics in Part VI.

REFERENCES

Blickley, G., "Level Measured Many Ways," *Cont. Eng., 37,* 35–44 (August, 1990).

Cheung, T.F., and W. Luyben, "Liquid-Level Control in Single Tanks and Cascades of Tanks with Proportional-Only and Proportional-Integral Feedback Controllers," *IEC Fund., 18, 1,* 15–21 (1979).

Shinskey, F., *Controlling Multivariable Processes,* Instrument Society of America, Research Triangle Park, NC, 1981.

Shunta, J., and W. Feherari, "Non-Linear Control of Liquid Level," *Instr. Techn.,* 43–48 (January, 1976).

ADDITIONAL RESOURCES

In addition to the general references cited in Chapter 1, the following books provide specialized information on level sensors and control.

Cho, C., *Measurement and Control of Liquid Level,* Instrument Society of America, Research Triangle Park, NC, 1982.

Cheremisinoff, N., *Process Level Instrumentation and Control,* Marcel Dekker, New York, 1981.

Many other linear and nonlinear controllers similar in purpose to the algorithm presented in Section 18.4 are in use. For a review of the performance of several, see

Cheung, T.F., and W. Luyben, "Nonlinear and Non-Conventional Liquid Level Controllers," *IEC Fund., 19,* 93–98 (1980).

Many other approaches to level control have been proposed. An interesting method that derives an averaging level control algorithm to minimize the maximum rate of change of the manipulated flow is given in

MacDonald, K., T. McAvoy, and A. Tits, "Optimal Averaging Level Control," *AIChE J., 32,* 75–86 (1986).

An alternative to the nonlinear PI algorithm using signal selects (see Chapter 22) has been suggested in

Buckley, P., "Recent Advances In Averaging Level Control," in *Productivity through Control Technology,* April 18–21, 1983, Houston, ISA Paper no. 0-87664-783-2/83/075-11.

The control structure for flows and levels in a system with recycle is shown in Figure Q18.13 and discussed in

Buckley, P., "Material Balance Control of Recycle Systems," *Instr. Techn.,* 29–34 (May, 1974).

The sizing of many inventories in a complex plant is discussed in

Hiester, A., S. Melsheimer, and E. Vogel, "Optimum Size and Location of Surge Capacity in Continuous Chemical Processes," *AIChE Annual Meet,* Nov. 15–20, 1987, paper 86c.

Model predictive control methods are introduced in the next chapter. Additional approaches to level control using model predictive control (see Chapter 19) are given in

Campo, P., and M. Morari, "Model Predictive Optimal Averaging Level Control," *AIChE J., 35, 4,* 579–591 (1989).

Cutler, C., "Dynamic Matrix Control of Imbalanced Systems," *ISA Trans., 21,* 1–6 (1982).

> Level control gives the engineer opportunity to match key closed-loop performance measures to the analytical solution to the transient response. This approach enables the engineer to tailor the performance to a wide range of control objectives.

QUESTIONS

18.1. Two tanks in series are placed upstream of a chemical reactor that is sensitive to feed concentration disturbances. Each tank has a holdup of 19 m^3, which is controlled approximately constant, and the design feed rate is 2 m^3/min. If the concentration of the inlet to the first tank has a concentration variation that can be approximated as $20 \sin(1.05t)$, what is the variation in the feed concentration to the reactor?

18.2. Two tanks are placed in series to attenuate flow rate disturbances. Each has a holdup time of τ_H minutes and is controlled by a linear PI controller. If the inlet flow variation is $A \sin(\omega t)$, what is the minimum variation in the flow rate leaving the second tank?

18.3. It was stated that the controller algorithm introduced in Section 18.4 is nonlinear. Using the definition of linearity (see Section 3.4), prove that the algorithm is nonlinear.

18.4 (*a*) Demonstrate that a proportional-only controller for a single level with a holdup time of 5 min and no instrumentation dynamics can have an arbitrarily large controller gain and remain stable.

(*b*) If the system in (*a*) has sensor dynamics of a first-order system with a time constant of 10 sec and valve dynamics of a first-order system with a time constant of 3 sec, what is the ultimate gain of the proportional-only controller? What would be a good choice for the controller gain?

18.5. Averaging level control implements relatively detuned feedback control. Since the integral mode is the "slow" mode, it might seem as though it should be used for control. To investigate why level controllers are predominantly proportional controllers, carry out the following development: Derive the transfer function for a level process under integral-only feedback control. Determine the dynamic response of the level for a step change in the uncontrolled flow. Is this good control performance?

18.6. The derivative mode does not seem to be used in level control. State whether you agree with this decision and why.

18.7. For each of the systems in Figure Q18.7, the flow in (F_{in}) can change independently of the inventory in the vessel. Each is described briefly:

(*a*) A heat exchanger in which the liquid in the vessel boils and the duty is proportional to the heat transfer area

(*b*) A tank with a constant flow out

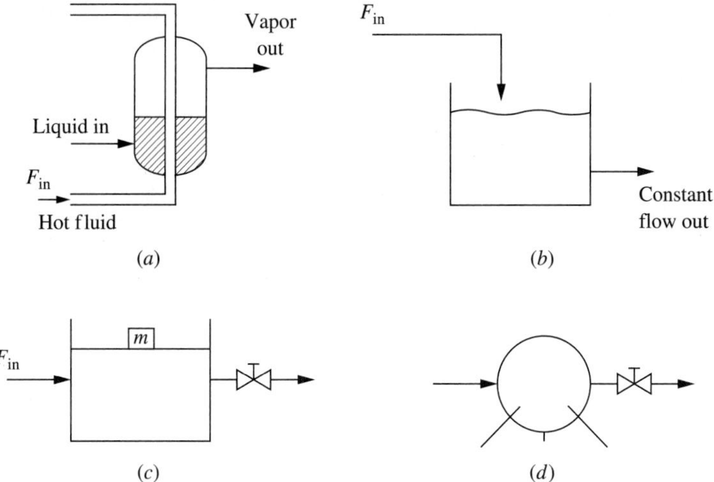

FIGURE Q18.7

(*c*) A gas-filled system with a moving roof and a constant mass on the roof; the gas exits through a partially open restriction

(*d*) A gas-filled system with constant volume; the gas exits through a partially open restriction

(i) For all systems without feedback control ($K_c = 0$), assume that the material balance was initially at steady state, and derive the response to a step change in the inlet flow rate. Is each system self-regulatory or not?

(ii) Determine the proper variable to measure to determine the inventory in each system, and describe how it should be controlled.

18.8. The closed-loop dynamic responses for the manipulated flow of a level process under PI control experience overshoot of their final steady-state values in response to a step in flow disturbance.

(*a*) Describe why this occurs and determine steps to prevent this overshoot.

(*b*) In Chapter 5, criteria were derived for transfer function's numerator zero that would lead to an overshoot of the output in response to an input step change. Verify that the criteria are met for $\xi = 1$.

18.9 The value of the small controller gain in the nonlinear level control was recommended to be about 1/20 of the large gain. Describe the performance of the nonlinear level control system with $K_{cS} = 0$ to

(*a*) A large step change in the uncontrolled flow

(*b*) A sine of *small* amplitude in the uncontrolled flow

(*c*) Based on these results, would you support the general recommendation of a zero value for the small controller gain? Under what special circumstances would this be advisable?

18.10. In Section 18.7, control of levels in series was discussed. Sketch on Figure 18.7 the control design when the flow leaving the second vessel is set (constant) by flow control.

18.11. Feedforward control was not considered in this chapter. Discuss whether feedforward control would improve (1) tight level control and (2) averaging level control.

18.12. The system of vessels in series (e.g., Figure 18.7) experiences periodic changes to the operating conditions of upstream units, during which the feed composition from upstream units changes substantially. The amount of mixed material produced during these infrequent and planned changes is to be minimized. What steps would you suggest to minimize the mixing without changing the equipment given in the figure?

18.13. The system of units with a recycle solvent stream is shown in Figure Q18.13. Solvent is added to the main process stream before the stirred-tank reactor and is separated in the flash drum. The solvent is collected, purified in the fixed-bed chemical reactor, and stored. The solvent is heated prior to being mixed with the feed. The feed flow rate is determined elsewhere and can be considered uncontrollable for this question. Also, the maximum purge and makeup flows are 1/10 of the normal solvent flow rate, and the material sent to purge cannot be recycled to the process.

(*a*) Design a control system that (1) ensures solvent addition at the desired ratio in the feed flow and (2) maintains all inventories in acceptable ranges. You may add sensors but make no other changes to the process equipment.

(*b*) Discuss the data and computations required to determine the size of the tanks, especially the middle solvent storage tank.

(*c*) Discuss how to determine the proper flow rates for the purge and makeup flows. Could they both properly be nonzero concurrently?

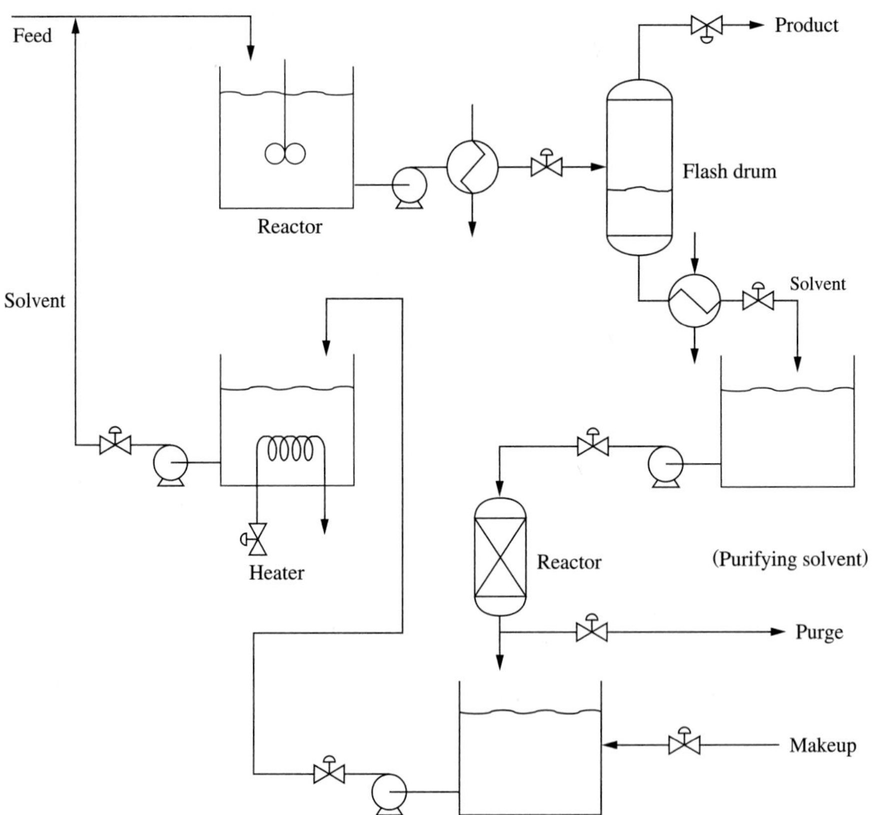

FIGURE Q18.13

18.14. Level controller tuning was not based on the methods and guidelines developed in Chapters 9 and 10. Why?

18.15. Verify the derivation of equations (18.8), (18.9), and (18.11) for the closed-loop response to a step disturbance for a level under PI control.

18.16. For both averaging and tight level control, sketch three examples of processes that should have this type of control and explain why.

18.17. Proposed steps for digital implementation of the nonlinear proportional-integral controller are given below. Discuss whether this implementation satisfies the algorithm described in the chapter, and if not, prescribe modifications.

1. Read measurement L_n and operator entry L_{SPn}.
2. Retrieve parameters K_{cS}, K_{cL}, L'_B, and T_I.
3. Retrieve stored value; S^*.
4. Set $K_c = K_{cL}$.
5. If $|L_{SP} - L_n| < L'_B$, then set $K_c = K_{cS}$.
6. Set $\text{MV}_n = K_c\{(L_{SPn} - L_n) + \frac{1}{T_I}(S^* + \Delta t(L_{SPn} - L_n))\}$.
7. Store L_n and $\sum_{i=0}^{n}(\Delta t)(L_{SPi} - L_n) = S^*$.
8. Wait Δt, then go to step 1.

18.18. Develop a method for determining the size of an inventory for averaging control based on the response of the system to a sine flow rate disturbance using frequency response principles.

CHAPTER

19

SINGLE-VARIABLE MODEL PREDICTIVE CONTROL

19.1 INTRODUCTION

Most modifications to single-loop feedback control presented in this part of the book have used additional measurements to improve control performance. In contrast, the emphasis in this chapter will be on an alternative to the proportional-integral-derivative (PID) feedback algorithm. The PID controller was introduced in Chapter 8 by explaining the features associated with each mode and by demonstrating that the combined modes could provide reasonable control performance. In subsequent chapters the applications of PID in feedback, cascade, and combined feedforward/feedback have indicated that the adoption of PID as the standard algorithm in the 1940s was an appropriate choice. Perhaps the most remarkable feature of the PID is the success of this single algorithm in so many different applications.

However, the development of the PID lacked a fundamental structure from which the algorithm could be derived, limitations could be identified, and enhancements could be developed. In this chapter a general development is presented that gives great insight into the roles of both the control algorithm and the process in the behavior of feedback systems. This development also provides a method for tailoring the feedback control algorithm to each specific application. Because a model of the process is an integral part of the control algorithm, the controller equation structure depends on the process model, in contrast to the PID controller, which has only one equation structure.

Although the control algorithm is different, the feedback concept is unchanged, and the selection criteria for manipulated and controlled variables are the same as explained in Chapters 1 and 7. In fact, the algorithms presented in this chapter could be used as replacements for the PID controller in nearly all applications so far discussed. Generally, the PID controller is considered the standard algorithm; an alternative algorithm is selected only when the alternative provides better performance.

The derivation of control algorithms is based on the predictive control structure introduced in the next section. Many methods are possible for deriving practical control algorithms to be implemented within the predictive structure, and two of these—Internal Model Controller (IMC) and Smith predictor—are explained in detail, along with guidance on implementation issues. Finally, some applications are presented in which predictive controllers offer potential improvements over PID. In addition to introducing some very useful single-loop control methods, this chapter offers an opportunity for another perspective on the fundamentals of feedback control and an introduction to the predictive control structure shown to be well suited to multivariable control in Chapter 23.

19.2 THE MODEL PREDICTIVE CONTROL STRUCTURE

The predictive control structure is based on a very natural manner of interpreting feedback control. Before the general predictive structure is developed, it is worthwhile to consider the typical thought process used by a human operator implementing feedback control manually. Assume that the three-tank mixing process in Figure 7.5 (page 242) is initially at steady state, and the goal is to reduce the outlet concentration by adjusting the flow of component A. First, the operator estimates the amount of change in the valve position (controller output) required to achieve the desired steady-state change in the controlled variable. This estimate requires an estimate of the steady-state model of the process (i.e., K_p). The operator can then estimate the proper adjustment in the valve position to be $\Delta v = (\Delta x_A)_1/K_p$.

Next, the operator would decide whether to implement this entire adjustment in one step or to introduce the change in several smaller steps. If the decision were to introduce the entire adjustment in one step, the dynamic response might look like the initial transient in Figure 19.1. The person waits until steady state is achieved to observe the response and determine whether the estimate was correct. In this example, the concentration change was too large in magnitude, as is shown by a difference between the actual and predicted changes in the concentration. As a result of this error, the operator would have to make another change in the valve position. A clever

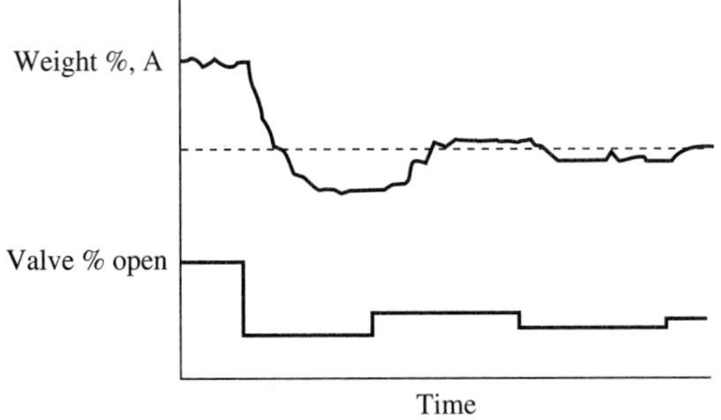

FIGURE 19.1
Example of manual control.

operator might conclude that the assumed gain is incorrect and modify the estimate of K_p; however, the operator in this example applies a more straightforward approach, in which the next correction is based on the same value of the process gain; that is, $\Delta v = (\Delta x_A)_2/K_p$, where $(\Delta x_A)_2$ is the difference between the predicted and actual Δx_A. Several iterations of the procedure result in the transient response achieving steady state, as given in Figure 19.1.

The approach used by the operator has three important characteristics:

1. It uses a model of the process to determine the proper adjustment to the manipulated variable, because the future behavior of the controlled variable can be *predicted* from the values of the manipulated variable.
2. The important feedback information is the difference between the predicted model response and the actual process response. If this difference were zero, the control would be perfect, and no further correction would be needed.
3. This feedback approach can result in the controlled variable approaching its set point after several iterations, even with modest model errors.

These characteristics provide the basis for the predictive control structure.

A continuous version of the approach just described can be automated with the general predictive control structure given in Figure 19.2. Three transfer functions represent the true process with the final element and sensor, $G_p(s)$; the controller, $G_{cp}(s)$; and a dynamic model of the process, $G_m(s)$. To avoid confusion, the term *predictive control algorithm* will be used to denote the calculation represented by $G_{cp}(s)$, which is used for the controller in the block diagram in Figure 19.2. The term *predictive control system* will be used to denote all calculations in the control system, which includes the predictive control algorithm, the predictive model, and two differences. All calculations in the predictive control system must be executed every time a value of the final element is determined.

The feedback signal E_m is the difference between the measured and predicted controlled variable values. The variable E_m is equal to the effect of the disturbance,

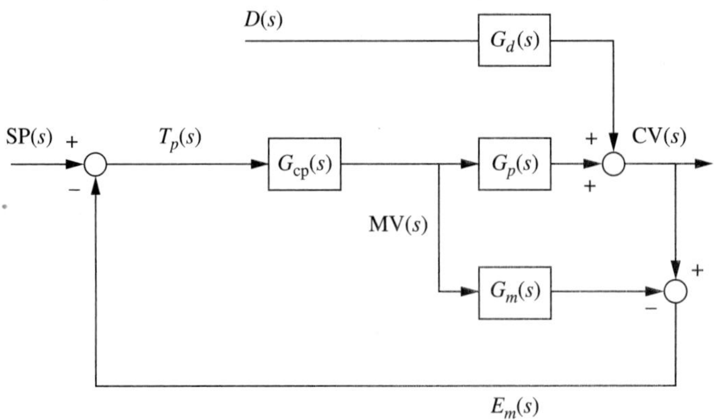

FIGURE 19.2
Predictive control structure.

$G_d(s)D(s)$, if the model is perfect (if $G_m(s) = G_p(s)$); thus, the structure highlights the disturbance for feedback correction. However, the model is essentially never exact, so that the feedback signal includes the effect of the disturbance and the model error, or *mismatch*. The feedback signal can be considered as a model correction; it is used to correct the set point so as to provide a better target value, $T_p(s)$, to the predictive control algorithm. The controller calculates the value of the manipulated variable based on the corrected target.

The following closed-loop transfer functions for responses to set point and disturbances can be derived through standard block diagram algebra.

$$\frac{\mathrm{CV}(s)}{\mathrm{SP}(s)} = \frac{G_{\mathrm{cp}}(s)G_v(s)G_p'(s)}{1 + G_{\mathrm{cp}}(s)[G_v(s)G_p'(s)G_s(s) - G_m(s)]} \approx \frac{G_{\mathrm{cp}}(s)G_p(s)}{1 + G_{\mathrm{cp}}(s)[G_p(s) - G_m(s)]} \tag{19.1}$$

$$\frac{\mathrm{CV}(s)}{D(s)} = \frac{[1 - G_{\mathrm{cp}}(s)G_m(s)]G_d(s)}{1 + G_{\mathrm{cp}}(s)[G_v(s)G_p'(s)G_s(s) - G_m(s)]} \approx \frac{[1 - G_{\mathrm{cp}}(s)G_m(s)]G_d(s)}{1 + G_{\mathrm{cp}}(s)[G_p(s) - G_m(s)]} \tag{19.2}$$

In all further transfer functions in this chapter, the dynamics of the sensor are considered negligible, and the overall model of the final element and process is taken to be $G_p(s)$. A linear dynamic process model, $G_m(s)$, can be determined using fundamental (Chapters 3 through 5) or empirical (Chapter 6) modelling methods. The controller algorithm, $G_{\mathrm{cp}}(s)$, for the predictive structure is as yet unknown and will be determined to give good dynamic performance.

A few properties of the predictive structure are now determined that establish important general features of its performance and give guidance for designing the controller, $G_{\mathrm{cp}}(s)$. Normally, a very important control performance objective is to ensure that the controlled variable returns to its set point in steady state. This objective can be evaluated from the closed-loop transfer functions by applying the final value theorem and determining whether the final value of the controlled variable, expressed as a deviation variable from the initial set point, reaches the set point. The application of the final value theorem for this purpose is performed for the following conditions:

1. The input is steplike, in that it reaches a steady state after a transient, $\mathrm{SP}(s) = \Delta\mathrm{SP}/s$ and $D(s) = \Delta D/s$.
2. The process without control reaches a steady state after a steplike input, $G_p(0) = K_p$ and $G_m(0) = K_m$.
3. The closed-loop system is stable, which can be achieved via tuning.

Note that the use of the steady-state gain of the process, $G_p(0)$, limits the results to stable processes without control.

In fact, the results in this chapter are limited to these stable processes. Under these conditions, application of the final value theorem yields

$$\lim_{t\to\infty} \mathrm{CV}(t) = \lim_{s\to 0} s\mathrm{CV}(s) = s\frac{\Delta \mathrm{SP}}{s}\frac{G_{\mathrm{cp}}(0)G_p(0)}{1+G_{\mathrm{cp}}(0)[G_p(0)-G_m(0)]}$$
$$= \Delta \mathrm{SP} \quad \text{if and only if } G_{\mathrm{cp}}(0) = G_m^{-1}(0) \tag{19.3}$$

$$\lim_{t\to\infty} \mathrm{CV}(t) = \lim_{s\to 0} s\mathrm{CV}(s) = s\frac{\Delta D}{s}\frac{[1-G_{\mathrm{cp}}(0)G_m(0)]G_d(0)}{1+G_{\mathrm{cp}}(0)[G_p(0)-G_m(0)]}$$
$$= 0 \qquad \text{if and only if } G_{\mathrm{cp}}(0) = G_m^{-1}(0) \tag{19.4}$$

Therefore, the predictive control system will satisfy both of the foregoing equations, thus providing zero steady-state offset for a steplike input, if

$$G_{\mathrm{cp}}(0) = G_m^{-1}(0) \qquad \text{or} \qquad K_{\mathrm{cp}} = 1/K_m \tag{19.5}$$

Equation (19.5) requires that the steady-state gain of the controller algorithm must be the inverse of the steady-state gain of the *dynamic model* used in the predictive system. This important requirement can be easily achieved, because the engineer has perfect knowledge of the model, although certainly not of the process $G_p(s)$ itself.

> A stable predictive system does not require a perfect model; it must only satisfy equation (19.5) to return the controlled variable to the set point at steady state.

To gain further insight into the predictive structure, the next control performance objective considered is perfect control. Here, the term *perfect control* is taken to mean that the controlled variable never deviates from the set point. As we have seen, this performance is not possible with feedback control and might not generally be desired because of other control performance considerations. However, it is considered here to provide insight into the predictive system and to give further guidance on control algorithm design. The closed-loop transfer functions in equations (19.1) with $\mathrm{CV}(s)/D(s) = 0$ and (19.2) with $\mathrm{CV}(s)/\mathrm{SP}(s) = 1$ provide the basis for the following condition, required for the controlled variable to be equal to the set point at all times during the transient response:

$$G_{\mathrm{cp}}(s) = G_m^{-1}(s) \tag{19.6}$$

Thus, perfect control performance would be achieved if the controller could be set equal to the inverse of the dynamic model in the predictive system. This might seem to be a simple requirement, since any model, even a constant, could be used for the model, and the controller would be easily evaluated as the inverse. However, block diagram algebra can be applied to derive the following condition for the behavior of the manipulated variable under perfect control:

$$\frac{\mathrm{MV}(s)}{D(s)} = \frac{-G_d(s)G_{\mathrm{cp}}(s)}{1+G_{\mathrm{cp}}(s)[G_p(s)-G_m(s)]}$$
$$= \frac{-G_d(s)G_{\mathrm{cp}}(s)}{1+G_{\mathrm{cp}}(s)G_p(s)-1} = -\frac{G_d(s)}{G_p(s)} \tag{19.7}$$

This shows that the perfect control system must invert the *true process* in some manner. The following are four reasons why an exact inverse of the process is not possible:

1. *Dead time.* In most physical processes, the feedback transfer function includes dead time in the numerator. The application of equations (19.6) and (19.7) to a typical process model with dead time gives, when the model is factored into two terms with $g_m(s)$ all polynomial terms in s,
$$G_m(s) = g_m(s)e^{-\theta s} \qquad G_{cp}(s) = [G_m(s)]^{-1} = [g_m(s)]^{-1}e^{\theta s} \qquad (19.8)$$
The perfect controller in this situation would have to include the ability to use *future* information in determining the current manipulated-variable value, as indicated by the predictive element $e^{\theta s}$. As discussed in Section 4.3, such noncausal models are not physically realizable—such behavior cannot occur (except in science fiction).
2. *Numerator dynamics.* As demonstrated in Section 5.4 on parallel process structures, some process models have dynamic elements in the numerators of the feedback transfer functions. Application of equation (19.6) to an example gives
$$G_m(s) = K\frac{\tau_2 s + 1}{(\tau_1 s + 1)^2} \qquad G_{cp}(s) = [G_m(s)]^{-1} = \frac{1}{K}\frac{(\tau_1 s + 1)^2}{\tau_2 s + 1} = \frac{\text{MV}(s)}{T_p(s)} \quad (19.9)$$
For all values of τ_2 the controlled-variable behavior would be stable, because the product $G_{cp}(s)G_m(s) = 1$. However, the controller algorithm alone would be stable only for $\tau_2 \geq 0$ and would be unstable for $\tau_2 < 0$. (This is termed a right-half-plane zero in $G_m(s)$, leading to a right-half-plane (unstable) pole in $G_{cp}(s)$.) An unstable controller would be expected to cause the manipulated variable to behave in an unstable manner, as is demonstrated in Example 19.2. Thus, the controller in equation (19.9) would not be able to achieve the "perfect" performance when $\tau_2 < 0$.
3. *Constraints.* The manipulated variable must observe constraints. These could be physical constraints, such as a valve, which is limited to 0 to 100% open, or more limiting constraints, such as the fuel to a furnace, which must be above a minimum limit greater than zero to maintain a stable flame. There is no guarantee that the controller defined in equation (19.6), which was derived using linear equations that did not consider constraints, would observe the constraints. Thus, in some cases values of the manipulated variable that are required to achieve perfect control performance would not be possible. In such cases, the resulting control performance would not be perfect, and the controlled variable would deviate from its set point.
4. *Model mismatch.* The model used in the predictive system will almost certainly be different from the true process. If this difference is large, the closed-loop system could be unstable, a situation that precludes acceptable control performance. (Recall that the final value theorem assumes stability of the system.)

Thus, the predictive control system clearly shows that dead times, certain numerator process dynamics (right-half-plane zeros), constraints, and model mismatch all prevent perfect feedback control performance.

These results are not new; they were discussed in Part III and summarized in Table 13.2 (page 463). However, this development reinforces the importance of the process in determining the achievable feedback control performance. It also provides a unified approach to developing these conclusions.

Example 19.1. Feedback control was introduced using the classical structure (e.g., Figure 7.4 on page 239). Determine the relationship between the controllers in the classical structure $G_c(s)$ and the predictive system $G_{cp}(s)$.

Solution. Block diagram algebra can be applied to reduce the predictive controller and model into one transfer function, which gives

$$\frac{\text{MV}(s)}{\text{SP}(s) - \text{CV}(s)} = G_c(s) = \frac{G_{cp}(s)}{1 - G_m(s)G_{cp}(s)} \tag{19.10}$$

Therefore, there is an equivalence between the classical and predictive structures, and a control system can be represented by either block diagram, as long as the proper controller transfer function is used. It is important to note that the conversion of a predictive system into a classical system does not necessarily result in a PID controller in the classical system; thus, the behavior of the two closed-loop systems could, and in general would, differ. In this chapter, the predictive controllers will be represented by the block diagram in Figure 19.2 to show the use of an explicit model in the control system clearly; also, there are advantages in performing the calculations in this manner, as will become clear in later sections.

Example 19.2. When the predictive model is perfect (i.e., $G_m(s) = G_p(s)$), what else is required for the closed-loop system to be stable?

Solution. We would like both the controlled *and* manipulated variables to be stable (referred to as *internal stability* by Morari and Zafiriou, 1989), which requires that the following transfer functions be stable:

$$\frac{\text{CV}(s)}{\text{SP}(s)} = G_{cp}(s)G_p(s) \tag{19.11}$$

$$\frac{\text{CV}(s)}{D(s)} = G_d(s) \tag{19.12}$$

$$\frac{\text{MV}(s)}{\text{SP}(s)} = G_{cp}(s) \tag{19.13}$$

Thus, the product of the controller and the process, the disturbance, and the controller itself must be stable for the entire control system to behave in a stable manner. Clearly, the manipulated variable would be stable only if the controller is stable. (Recall that the controller could be unstable, while the product $G_{cp}(s)G_p(s)$ is stable.) Also, the final value theorem in equations (19.3) and (19.4) involves the terms $G_p(0)$ and $G_{cp}(0)$, the transfer functions evaluated at $s = 0$, which were taken to be constant values. This result is valid only when the transfer functions are stable; if they are unstable, the final value theorem is not applicable.

So far, the predictive concept has been introduced, the block diagram structure presented, and the closed-loop transfer function derived. The starting points for the predictive control algorithm design are the requirement for zero steady-state offset in equation (19.5) and the definition of the perfect controller in equation (19.6). Since the perfect controller is not possible even if it were desirable, a manner for deriving an *approximate inverse* of the model is required, with an approximate inverse being a $G_{cp}(s)$ that does not exactly satisfy equation (19.6) but contains the important features for control performance. Many methods exist for developing an approximate inverse, and each would result in a different controller algorithm giving different control performance. In the next sections, two methods for designing single-loop predictive control algorithms are presented. They have been selected because they involve straightforward mathematics, are simple to implement in a digital computer, yield good control performance in many cases, and have been applied industrially.

19.3 THE IMC CONTROLLER

The system in Figure 19.2 has been described by several investigators, who have used different terminology for what is now generally referred to as the predictive structure. The publications by Brosilow (1979) and Garcia and Morari (1983), in which they introduced the terms *inferential control* and *internal model control,* respectively, sparked considerable interest in the chemical engineering community. The controller design approach presented in this section follows the developments of these publications, which is generally referred to as the IMC method.

Since an exact inverse is not possible, the IMC approach segregates and eliminates the aspects of the model transfer function that make calculation of a realizable inverse impossible. The first step is to factor the model into the product of the two factors

$$G_m(s) = G_m^+(s)G_m^-(s) \tag{19.14}$$

with

$G_m^+(s)$ The *noninvertible* part has an inverse that is not causal or is unstable. The inverse of this term includes predictions ($e^{\theta s}$) and unstable poles ($1/(1+\tau s)$, with $\tau < 0$) appearing in $G_{cp}(s)$. The steady-state gain of this term must be 1.0.

$G_m^-(s)$ The *invertible* part has an inverse that is causal and stable, leading to a realizable, stable controller. The steady-state gain of this term is the gain of the process model K_m.

> The IMC controller eliminates all elements in the process model $G_m(s)$ that lead to an unrealizable controller by taking the inverse of only the invertible factor to give
>
> $$G_{cp}(s) = \left[G_m^-(s)\right]^{-1} \tag{19.15}$$

This design equation ensures that the controller is realizable and that the system is internally stable (at least with a perfect model), but it does not explicitly guarantee that the behavior of the control system is acceptable. However, the performance of such controllers, as modified shortly, will be seen to be acceptable in many cases. Before we proceed, this procedure is applied to two examples.

Example 19.3. Apply the IMC procedure to design a controller for the three-tank mixing process in Example 7.2 (page 249).

Solution. The IMC controller design requires a transfer function model of the process. The linearized third-order model derived in Example 7.2 will be used. In this case,

$$G_m(s) = \frac{K_m}{(\tau s + 1)^3} = \frac{0.039}{(5s + 1)^3} = G_m^-(s) \qquad G_m^+(s) = 1.0$$

Thus, the model can be inverted directly to give

$$G_{cp}(s) = \left[G_m^-(s)\right]^{-1} = \frac{(\tau_m s + 1)^3}{K_m} = \frac{(5s + 1)^3}{0.039}$$

This controller in the predictive structure in Figure 9.2 could theoretically provide good control of the controlled variable. However, there are several drawbacks with this design. First, the controller involves first, second, and third derivatives of the feedback signal. These derivatives cannot be calculated exactly, although they can be estimated numerically. Second, the appearance of high-order derivatives of a noisy signal could lead to unacceptably high variation and large overshoot in the manipulated variable. Finally, these high derivatives could lead to extreme sensitivity to model errors. Therefore, this controller would not be used without modification.

Example 19.4. Design an IMC controller for the process in Example 19.3, using the alternative first-order-with-dead-time approximate model for the process that was determined using the process reaction curve in Example 9.2 (page 302):

$$G_m(s) = \frac{K_m e^{-\theta_m s}}{\tau_m s + 1} = \frac{0.039 e^{-5.5s}}{10.5s + 1}$$

This model must be factored into invertible and noninvertible parts:

$$G_m^-(s) = \frac{K_m}{\tau_m s + 1} = \frac{0.039}{10.5s + 1}$$

$$G_m^+(s) = e^{-\theta_m s} = e^{-5.5s}$$

The invertible part is then employed in deriving the controller:

$$G_{cp}(s) = \left[G_m^-(s)\right]^{-1} = \frac{\tau_m s + 1}{K_m} = \frac{10.5s + 1}{0.039}$$

This controller is a proportional-derivative algorithm, which still might be too aggressive but will be modified to give acceptable performance in Example 19.6.

As discussed in Section 4.3, all realistic processes are modelled by transfer functions having a denominator order greater than the numerator order. Thus, the controller according to equation (19.15), which is the inverse of the process model, will have a numerator order greater than the denominator order. This results in first-

or higher-order derivatives in the controller, which generally lead to unacceptable manipulated-variable behavior and, thus, poor performance and poor robustness when model errors occur.

Achieving good control performance requires modifications that modulate the manipulated-variable behavior and increase the robustness of the system. The IMC design method provides one feature to account for both of these concerns: a filter of the feedback signal. The filter can be placed before the controller, as shown in Figure 19.3, so that the closed-loop transfer functions for the controlled and manipulated variables become

$$\frac{\text{CV}(s)}{\text{SP}(s)} = \frac{G_f(s)G_{\text{cp}}(s)G_p(s)}{1 + G_f(s)G_{\text{cp}}(s)\left[G_p(s) - G_m(s)\right]} \tag{19.16}$$

$$\frac{\text{MV}(s)}{\text{SP}(s)} = \frac{G_f(s)G_{\text{cp}}(s)}{1 + G_f(s)G_{\text{cp}}(s)\left[G_p(s) - G_m(s)\right]} \tag{19.17}$$

$$\frac{\text{CV}(s)}{D(s)} = \frac{G_d(s)\left[1 - G_f(s)G_{\text{cp}}(s)G_m(s)\right]}{1 + G_f(s)G_{\text{cp}}(s)\left[G_p(s) - G_m(s)\right]} \tag{19.18}$$

$$\frac{\text{MV}(s)}{D(s)} = \frac{-G_d(s)G_{\text{cp}}(s)G_f(s)}{1 + G_f(s)G_{\text{cp}}(s)\left[G_p(s) - G_m(s)\right]} \tag{19.19}$$

Now, four desirable properties of the filter are determined as a basis for selecting the filter algorithm. First, the steady-state value of the filter needs to be determined. Application of the final value theorem to the closed-loop transfer function in equation (19.16) with the requirement of zero steady-state offset yields

$$\begin{aligned} \lim_{t\to\infty} \text{CV}(t) &= \lim_{s\to 0} s\frac{\Delta\text{SP}}{s}\left[\frac{G_f(0)G_{\text{cp}}(0)G_p(0)}{1 + G_f(0)G_{\text{cp}}(0)[G_p(0) - G_m(0)]}\right] \\ &= \Delta\text{SP} \qquad \text{only if } G_{\text{cp}}(0) = \left[G_f(0)G_m(0)\right]^{-1} \end{aligned} \tag{19.20}$$

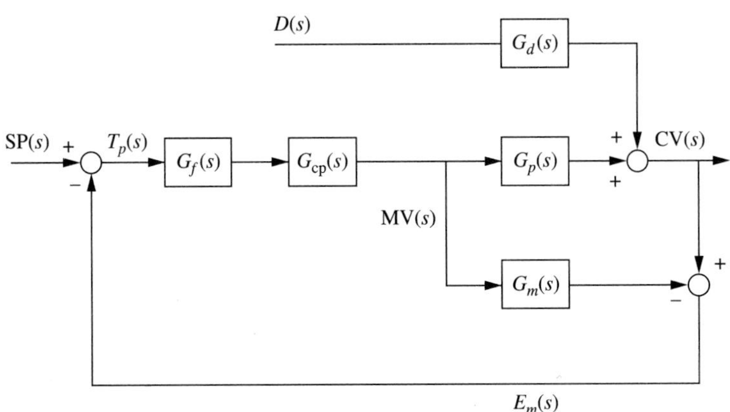

FIGURE 19.3
Predictive structure with single filter.

By convention, the controller gain is required to be the inverse of the process model; therefore, the steady-state gain of the filter must be unity; that is, $G_f(0) = K_f = 1.0$.

Second, a desired effect of the filter on the manipulated-variable behavior must be decided. Generally, the filter should reduce unnecessary high-frequency fluctuations due to noise. Since $G_f(s)$ appears in the numerator of equations (19.17) and (19.19), the magnitude of the filter magnitude should decrease with increasing frequency. The filter with the proper amplitude ratio attenuates the effects of high-frequency variation in the controlled variable (and set point) on the variation in the manipulated variable while it transmits the lower-frequency variation essentially unchanged. The term introduced in Chapter 12 for this behavior was *low-pass filter.*

Third, the filter influences the controlled-variable performance. Its appearance in the numerators of equations (19.16) and (19.18) indicates that filters with monotonically decreasing amplitude with increasing frequency degrade the performance of the controlled variable: filters lead to larger deviations from set point during transients. Thus, too much damping through the filter is not desirable.

Fourth, the effect of the filter on stability can be interpreted by analyzing the closed-loop transfer function, which has $G_{OL}(s) = G_f(s)G_{cp}(s)\left[G_p(s) - G_m(s)\right]$ for the predictive system. Clearly, the system is always stable if the model is perfect (and the controller is stable). However, the model is essentially never perfect, and the filter is required to ensure stability for a reasonable range of model error. Recalling that stability is improved as the magnitude of $G_{OL}(j\omega_c)$ is decreased, a filter that has decreasing magnitude as frequency increases will reduce the effects of model mismatch on $|G_{OL}(j\omega_c)|$ and stabilize the closed-loop system.

In summary, filters with a steady-state gain of 1.0 and decreasing magnitudes as frequencies increase satisfy the general requirements of increased robustness and noise attenuation. Many potential filter transfer functions satisfy the requirements just developed.

In the single-loop IMC design, it is conventional to use the following filter equation to improve robustness and manipulated-variable behavior.

$$G_f(s) = \left[\frac{1}{\tau_f s + 1}\right]^N \tag{19.21}$$

In this equation, the exponent N is selected to be large enough that the product $G_f(s)G_{cp}(s)$ has a denominator polynomial in s of order at least as high as its numerator polynomial. For further examples in this chapter, the model $G_m(s)$ will be first-order with dead time and the filter will be a first-order system ($N = 1$), but this is not always the case for other process models. The filter time constant can be adjusted to satisfy the performance specifications. Increasing the filter time constant modulates the manipulated-variable fluctuations and increases robustness at the expense of larger deviations of the controlled variable from its set point during the transient response.

Example 19.5. The filter location in Figure 19.3 influences the behavior of the control system for both disturbance and set point responses. Develop an alternative structure to separate these effects, so that the disturbance and set point responses can be influenced independently.

To achieve robustness, one filter must be located within the feedback loop. A design is shown in Figure 19.4, which has one filter, $G_{fF}(s)$, in the feedback path and a second filter, $G_{fS}(s)$, for set point. The advantage of this design is the ability to modify the set point and disturbance responses independently. This design is sometimes referred to as a *two-degree-of-freedom controller.*

The predictive control system is difficult to implement in analog computing equipment because of the dead time in the model $G_m(s)$, but it is straightforward with digital computers, regardless of the model structure. The simple models considered in this chapter can be expressed in discrete form by methods already introduced in Chapters 6 and 15 and in Appendix F. The IMC system in Figure 19.3 with a single filter will be considered, and the dynamic model will be assumed to be first-order with dead time. Thus, the predictive control system equations in continuous form are

$$\frac{\text{CV}_m(s)}{\text{MV}(s)} = G_m(s) = \frac{K_m e^{-\theta_m s}}{\tau_m s + 1} \tag{19.22}$$

$$G_m^-(s) = \frac{K_m}{\tau_m + 1} \tag{19.23}$$

$$\frac{\text{MV}(s)}{T_p(s)} = G_f(s)G_{\text{cp}}(s) = \frac{1}{K_m}\frac{\tau_m s + 1}{\tau_f s + 1} \tag{19.24}$$

with $\text{CV}_m(s)$ the predicted value of the controlled variable, that is, the output from the model $G_m(s)$. The dynamic model can be simulated in discrete form, as explained in Appendix F.

$$(\text{CV}_m)_n = \left[e^{-\Delta t/\tau_m}\right](\text{CV}_m)_{n-1} + K_m\left[1 - e^{-\Delta t/\tau_m}\right]\text{MV}_{n-\Gamma-1} \tag{19.25}$$

with Δt the digital controller execution period and the dead time modelled as $\Gamma = \theta_m/\Delta t$, an integer value.

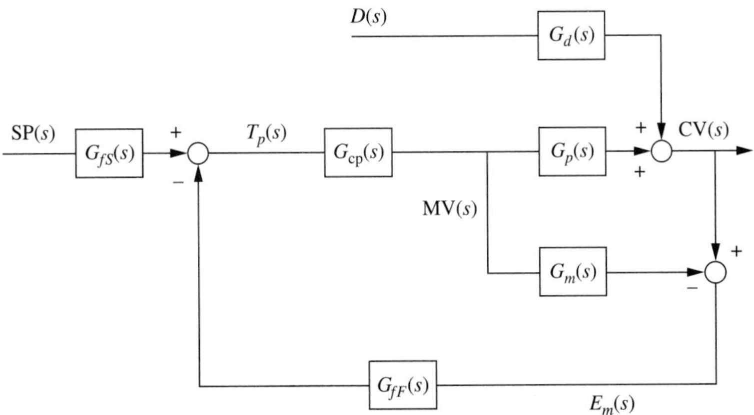

FIGURE 19.4
Two-degree-of-freedom predictive controller.

Note that the product of $G_f(s)G_{cp}(s)$ can be implemented as one algorithm in this case: a lead-lag transfer function, which was expressed in discrete form in Section 15.5.

$$\mathrm{MV}_n = \left[\frac{\frac{\tau_f}{\Delta t}}{\frac{\tau_f}{\Delta t}+1}\right]\mathrm{MV}_{n-1} + \frac{1}{K_m}\left[\frac{\frac{\tau_m}{\Delta t}+1}{\frac{\tau_f}{\Delta t}+1}\right](T_p)_n - \frac{1}{K_m}\left[\frac{\frac{\tau_m}{\Delta t}}{\frac{\tau_f}{\Delta t}+1}\right](T_p)_{n-1} \tag{19.26}$$

with T_p the *target*—that is, the set point as corrected by the feedback signal; the difference between the measured and predicted values of the controlled variable.

In summary, the predictive control system execution at step n involves the following:

1. Calculate the predicted controlled variable, equation (19.25).
2. Calculate the difference between the measured and model-predicted controlled variables, $(E_m)_n = \mathrm{CV}_n - (\mathrm{CV}_m)_n$.
3. Correct the set point with the feedback signal, $(T_p)_n = \mathrm{SP}_n - (E_m)_n$.
4. Calculate the manipulated-variable value, equation (19.26).

Example 19.6. Simulate the dynamic response of the linearized three-tank mixing process in Example 19.4, operating at the base-case inlet flow rate, under IMC feedback control.

The true process $G_p(s)$ is taken as the linear, third-order system, and the controller and dynamic model $G_m(s)$ will be based on the approximate first-order-with-dead-time model. This structural mismatch, which is typical of realistic applications, precludes perfect control; thus, the results of this exercise give a realistic evaluation of the performance of IMC controllers.

The controller with filter and model transfer functions are

$$G_p(s) = \frac{0.039}{(5s+1)^3} \qquad G_m(s) = \frac{0.039e^{-5.5s}}{10.5s+1} \qquad G_f(s)G_{cp}(s) = \frac{1}{0.039}\frac{10.5s+1}{\tau_f s+1}$$

The controller calculations can be converted to discrete form with $\Delta t = 0.10$ to give

$$\begin{aligned}(\mathrm{CV}_m)_n &= \left[e^{-0.1/10.5}\right](\mathrm{CV}_m)_{n-1} + 0.039\left[1 - e^{-0.1/10.5}\right]\mathrm{MV}_{n-55-1}\\ &= 0.9905(\mathrm{CV}_m)_{n-1} + 0.000388\,\mathrm{MV}_{n-56}\end{aligned}$$

$$\mathrm{MV}_n = \left[\frac{\frac{\tau_f}{0.1}}{\frac{\tau_f}{0.1}+1}\right]\mathrm{MV}_{n-1} + \frac{1}{0.039}\left[\frac{\frac{10.5}{0.1}+1}{\frac{\tau_f}{0.1}+1}\right](T_p)_n - \frac{1}{0.039}\left[\frac{\frac{10.5}{0.1}}{\frac{\tau_f}{0.1}+1}\right](T_p)_{n-1}$$

In this example, the closed-loop simulation is performed using the foregoing equations for the controller (based on an approximate model) and the linearized third-order model for the plant; thus, significant model mismatch exists between the process and the model. The results are given in Figure 19.5*a* for a feed composition disturbance of magnitude of 0.80 and in Figure 19.5*b* for a set point change, each for three values of the filter time constant. As the filter time constant increases, the aggressiveness of the controller decreases, as indicated by the slower response of the manipulated variable and slower return to the set point. It is noteworthy that the disturbance response appears acceptable, albeit slow, for all values of the filter tuning, while the set point response experiences extreme manipulated-variable variability for the lowest filter value.

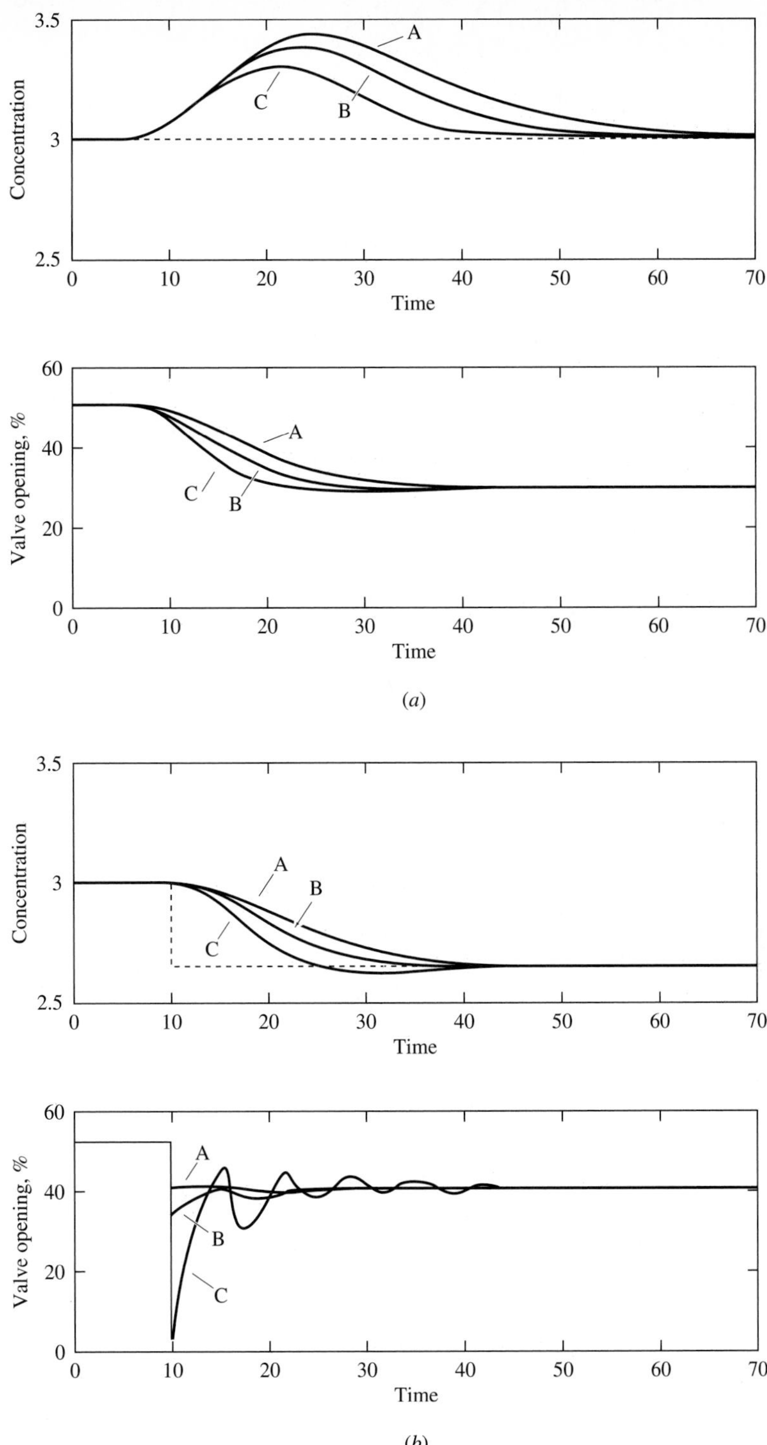

FIGURE 19.5
(a) Disturbance responses and (b) set point responses for Example 19.6. The values of the filter time constant τ_f are 10 min (case A), 6.1 min (case B), and 2.0 min (case C).

This comparison demonstrates the disadvantage for a single filter and the potential for improvement by using separate filters, as shown in Figure 19.4, to influence the disturbance and set point responses separately.

Based on the results in Example 19.6 and our previous experience with the PID controller, we would expect that the performance of predictive control depends on a proper choice of all parameters in the system. In general, all parameters appearing in the IMC model and the control algorithm could be tuned, but it is common practice to use the best estimates for the dynamic model. Thus, only the filter time constant, τ_f, is considered available for tuning.

The considerations for controller tuning were thoroughly discussed in Chapter 9. Clearly, no one value or correlation will suit all situations, but a few studies have been performed to provide initial tuning values, which are applicable to many situations and can be fine-tuned based on empirical experience. One tuning guideline, due to Brosilow (1979), suggests that the filter time constant be related to the likely model error, $\tau_f = 0.25(\delta\theta)$, with $\delta\theta$ the maximum likely error in the estimated dead time; Morari and Zafiriou (1989) recommend that a thorough robust tuning analysis be performed.

The method for tuning discussed in Chapter 9 and summarized in Appendix E for PID controllers, which minimized the IAE of the noisy controlled variable subject to limitations on variations in the manipulated variable over a range of model mismatch, has been applied to the IMC design as well (Ciancone et al., 1993). The results for a first-order-with-dead-time process model are given in Figure 19.6 for good performance for a step disturbance. The filter tuning constant has a large value for small fraction dead times, although one might initially expect the opposite correlation because systems that are easier to control require more filtering. The reason for these results is the need to moderate the high-frequency variation in the manip-

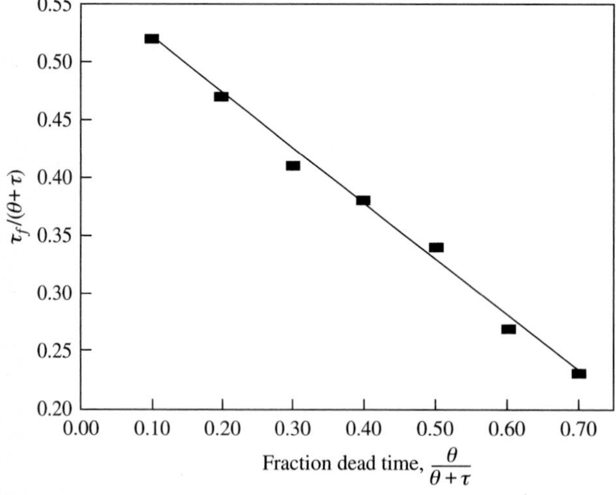

FIGURE 19.6
Tuning correlation for single-loop IMC disturbance response on a first-order-with-dead-time process.

ulated variable. Thus, the ratio of process time constant to filter time constant in the lead-lag element in the controller should not be too large; these results indicate that a reasonable ratio is around 2. A smaller filter time constant would be allowable for stability and give good controlled-variable performance, but the variability in the manipulated variable would be unacceptably large for many applications.

Example 19.7. Calculate the filter tuning constant for the IMC controller applied to the three-tank mixing process.

Applying the correlations noted for Brosilow's approach, and assuming that the likely dead time error is 35%, gives

$$\tau_f = (0.35)(5.5) \approx 2 \text{ min}$$

The Ciancone correlation assumes that all three model parameters can be in error by ±25% in the structured manner described in Chapter 9:

$$\theta/(\theta + \tau) = 5.5/(5.5 + 10.5) = 0.34$$

$$\tau_f/(\theta + \tau) = 0.38 \qquad \text{from Figure 19.6}$$

$$\tau_f = (0.38)(16) = 6.1 \text{ min}$$

The dynamic responses using these tuning values are cases B and C in Figures 19.5*a* and *b*. The controlled-variable IAE for the IMC disturbance response is 9.1, which is slightly larger than the value obtained under PID control for the same disturbance in Example 9.2.

Example 19.8. Evaluate the robustness for the IMC controllers implemented in Example 19.7.

Assuming that the system closely approximates a continuous system, the analysis could be performed using methods introduced in Chapter 10. However, root locus is not applicable, because the characteristic equation involves exponentials in s. Also, the Bode method is *not generally applicable* because the Bode plots for predictive systems do not always conform to the requirements noted in Table 10.1 (i.e., monotonically decreasing amplitude and phase behavior after the critical frequency). The stability could be determined using the Nyquist method applied to $G_{OL}(s) = G_f(s)G_{cp}(s)\left[(G_p(s) - G_m(s)\right]$; however, this method has not been stressed in this book.

Therefore, the robustness of this example will be evaluated by simulating cases with a fixed value of the filter time constant and different operating conditions. The range of process operations for the three-tank process is the same as considered in Section 16.2, where the flow rate has its base-case value in case E; is decreased by about 30% in case C; and is decreased by about 55% in case A. The parameters for the process model of the true process are given in Table 16.1 (page 534).

Case	K_p (%A /% open)	τ_i (i = 1, 3 for third-order system) (min)
A	0.087	11.4
C	0.052	6.9
E	0.039	5.0

In this example, the controller parameters are fixed at values appropriate for the base-case approximate first-order-with-dead-time empirical model, as determined in Example 19.6; thus, the model gain was 0.039, the model time constant was 10.5, and the model dead time was 5.5. The filter time constant was determined to be 6.1 min in Example 19.7.

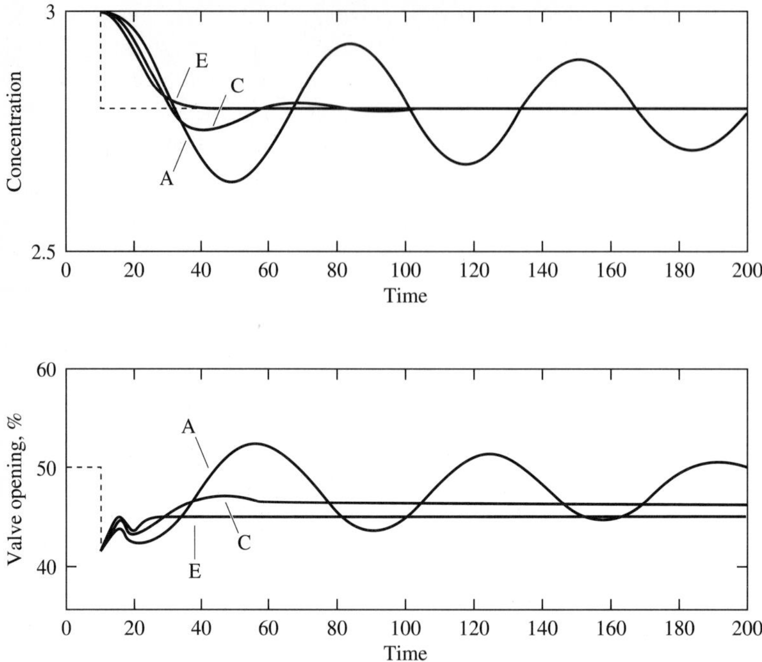

FIGURE 19.7
Dynamic response of three-tank system with IMC control for Example 19.8 (cases A, C, E are from Table 16.2).

The results are presented in Figure 19.7. The performance is acceptable for the base case and +30% change in process behavior. For the largest model error, the system has very poor performance and appears on the limit of stability. These results can be compared with performance achieved with a PI algorithm with fixed tuning for the same range of process operating conditions, shown in Figures 16.4*a* and *b* (page 537). The control performance degrades substantially for both control systems when the process dynamics change significantly from the nominal conditions.

In conclusion, the IMC controller is based on the general predictive control structure. The controller design method adheres to criteria that ensure zero offset for steplike disturbances, and it employs a factorization approach to obtain a realizable approximate inverse that gives good feedback control performance. An adjustable filter (tuning parameter) was introduced to enable the engineer to moderate the feedback action to maintain good performance of the controller and manipulated variables in the presence of measurement noise and model error.

19.4 THE SMITH PREDICTOR

The control design by O. Smith (1957) preceded much of the general analysis of predictive systems; in fact, it predated the application of digital computers to process control, so that widespread implementation of Smith's results was delayed until real-time digital control computers became commercially available. Smith's approach,

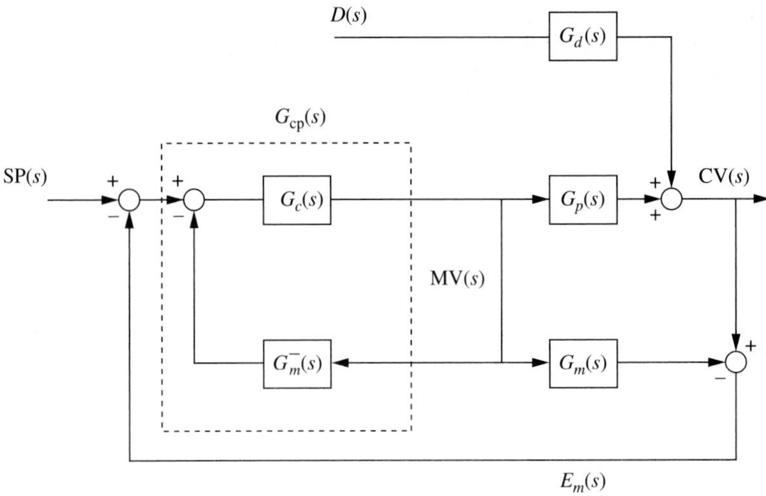

FIGURE 19.8
Block diagram of Smith predictor.

shown in Figure 19.8, relies on the general predictive structure in which the controller is calculated by the elements in the dashed box; these elements perform the function of the predictive control algorithm, $G_{cp}(s)$, in Figure 19.2.

Smith reasoned that "eliminating the dead time" from the control loop would be beneficial, which is certainly true but not possible via a feedback controller; only physical changes in the process can affect the feedback dead time. Therefore, Smith suggested that controlling a model of the process, without the dead time (or other noninvertible element), would provide a better calculation of the manipulated variable to be implemented in the true process. He retained the conventional PI control algorithm; thus, the system in Figure 19.8 consists of a feedback PI algorithm $G_c(s)$ that controls a simulated process, $G_m^-(s)$, which is easier to control than the real process. $G_m^-(s)$ has the same meaning here as for IMC control in equation (19.14), and the absence of dead time or inverse response (right-half-plane zero) in the model $G_m^-(s)$ allows much more aggressive control of the model than of the true plant.

The calculated manipulated variable resulting from controlling the model is implemented in the true process, which could yield good control as long as the model were perfect. Naturally, the model will not be perfect, and some form of feedback is required to achieve zero steady-state offset. Smith recognized the value of the predictive structure and, as shown in Figure 19.8, proposed correcting the model with the difference between the measured and the predicted controlled variables. Note that the prediction is determined using the complete linear dynamic model $G_m(s)$, including any noninvertible dynamics. The feedback signal $E_m(s)$ can be interpreted as a correction to the model $G_m^-(s)$.

The closed-loop transfer function of the system in Figure 19.8 is

$$\frac{\text{CV}(s)}{\text{SP}(s)} = \frac{G_c(s)G_p(s)}{1 + G_cG_m^-(s) + G_c(s)\left[G_p(s) - G_m(s)\right]} \tag{19.27}$$

If the model were perfect, the characteristic equation would not contain a dead time, because $G_m(s)$ and $G_p(s)$ would cancel. Thus, for the case with a perfect model, the characteristic equation involves only the expression $1 + G_c(s)G_m^-(s)$, which is easier to control and allows a more aggressive adjustment of the manipulated variable. Naturally, the true process is never known exactly, and the actual behavior and stability depend on all terms without cancellation. Application of the final value theorem to equation (19.27), for a step change in the set point and a PI algorithm for the controller, gives

$$\lim_{t\to\infty} \text{CV}(t) = \lim_{s\to 0} s\frac{\Delta\text{SP}}{s}\frac{G_c(s)G_p(s)}{1 + G_cG_m^-(s) + G_c(s)\left[G_p(s) - G_m(s)\right]} \tag{19.28}$$

For a stable process, $G_p(0) = K_p$ and $G_m(0) = K_m = G_m^-(0) = K_m^-$,

$$\lim_{t\to\infty} \text{CV}(t) = \lim_{s\to 0} \Delta\text{SP}\frac{K_pK_c\left(1 + \dfrac{1}{T_Is}\right)}{1 + K_c\left(1 + \dfrac{1}{T_Is}\right)K_m^- + K_c\left(1 + \dfrac{1}{T_Is}\right)(K_p - K_m)} = \Delta\text{SP} \tag{19.29}$$

> Thus, zero steady-state offset for a step input with Smith predictor control does not require a perfect model; it requires only that the steady-state gains for the two models be identical ($K_m = K_m^-$) and that the controller algorithm $G_c(s)$ have an integral mode.

Again, the performance and robustness of the Smith predictor control system depend on the controller tuning. The reader is cautioned that the PI controller in the Smith predictor should not be tuned using correlations from Part III, which were developed for the conventional control structure, using $G_m^-(s)$ for the feedback dynamics. The purpose of the PI controller is to calculate an approximate inverse rapidly, as demonstrated by the following:

$$G_{\text{cp}} = \frac{\text{MV}(s)}{\text{SP}(s) - E_m(s)} = \frac{G_c(s)}{1 + G_c(s)G_m(s)} \approx \frac{1}{G_m}(s) \qquad \text{for “large” } G_c(s) \tag{19.30}$$

Thus, the inverse would be approximated by a tightly tuned controller. A proper tuning procedure should consider the behavior of the controlled and manipulated variables as well as robustness for the model mismatch expected to be encountered. The proper tuning can be related to the IMC tuning by recognizing the equivalence of the IMC and Smith predictor for application to a process with first-order-with-dead-time feedback dynamics:

$$\text{Smith predictor: } \frac{\text{MV}(s)}{T_p(s)} = \frac{G_c(s)}{1 + G_p(s)G_c(s)} = \frac{K_c\left(1 + \dfrac{1}{T_Is}\right)}{1 + K_c\left(1 + \dfrac{1}{T_Is}\right)\dfrac{K_m}{1 + \tau_m s}} \tag{19.31}$$

$$\text{IMC controller:} \quad \frac{\text{MV}(s)}{T_p(s)} = \frac{1}{K_m}\frac{\tau_m s + 1}{\tau_f s + 1} \tag{19.32}$$

These two expressions can be shown to be equal when

$$K_c = \frac{\tau_m}{\tau_f K_m} \qquad T_I = \tau_m \tag{19.33}$$

Thus, the tuning correlations in Figure 19.6 along with equations (19.33) can be used to estimate initial tuning for the Smith predictor with a first-order-with-dead-time process model. Alternative guidelines are provided by Laughlin and Morari (1987).

The Smith predictor is easily programmed in a digital system. The digital form of the PI controller was presented in Chapter 11, and for a first-order-with-dead-time model, the digital models are programmed using equation (19.25) for $G_m(s)$ and the same equation with no dead time, $\Gamma = 0$, for $G_m^-(s)$.

Example 19.9. Apply the Smith predictor to the same process as considered in Example 19.7, the three-tank mixing process.

Again, the approximate first-order-with-dead-time model will be used as given in Example 19.7. The PI tuning can be estimated using Figure 19.6 and equations (19.33) to give

$$\theta/(\theta + \tau) = 0.34 \qquad \tau_f/(\theta + \tau) = 0.38 \qquad \tau_f = 6.1 \text{ min}$$

$$K_c = \tau/(\tau_f K_m) = 10.5/[(6.1)(.039)] = 44.1(\%\text{ open})/(\%\text{A}) \qquad T_I = \tau = 10.5 \text{ min}$$

The models $G_m(s)$ and $G_m^-(s)$ can be converted to digital approximations, as demonstrated in Example 19.8, and the PI controller can be programmed digitally, as shown in Chapter 11. The dynamic response of the control system, with the controller implemented as a digital algorithm, is essentially identical to the response for the IMC controller shown in Figure 19.5, case B, so that plot is not repeated. The controller can be fine-tuned using the same approach as described in Section 9.5. For example, the controlled-variable performance for the base-case model can be improved by increasing the controller gain to 88% open/%A, as shown in Figure 19.9, which gives an IAE of 6.8. Since this tuning is more aggressive than the correlations in Figure 19.6, it is less robust and would not normally be used initially, but it could be reached through fine tuning if empirical experience indicated that the actual model errors and noise were smaller than anticipated in deriving the initial tuning correlation.

In conclusion, the Smith predictor conforms to the general principles of the predictive control structure. It employs a unique method for calculating an approximate model inverse: by controlling a model consisting of the invertible part of the model. This structure can achieve zero steady-state offset for steplike disturbances by conforming to easily achieved criteria. Again, the Smith predictor system is simple to implement in digital control and generally yields good control performance. The tuning of the PI controller must be appropriate for the predictive structure and can be adjusted to make the Smith predictor control more or less aggressive to provide the desired controlled- and manipulated-variable performance for the expected range of model mismatch.

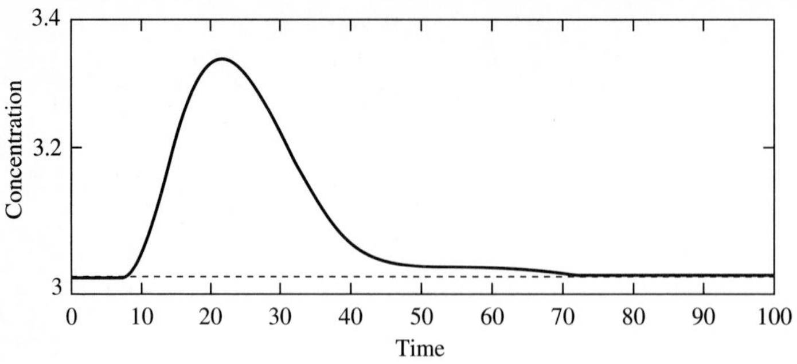

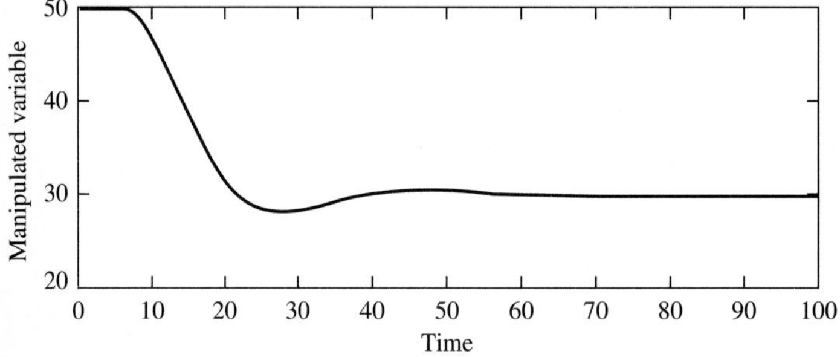

FIGURE 19.9
Dynamic response for three-tank system under Smith predictor control for Example 19.9 with $K_c = 88\%$ open/%A.

19.5 IMPLEMENTATION GUIDELINES

It is important to remember that predictive controllers employ the same feedback principles as classical structures and involve basically the same tasks to design, implement, and operate. The engineering tasks include selecting the feedback measurement, selecting the manipulated variable, determining an appropriate model structure with parameters, selecting an algorithm, and establishing the tuning constants. In operating the system, process personnel must decide on the status of the controller—automatic or manual—and enter the set point value. Thus, predictive controllers can be presented to plant operating personnel in exactly the same manner as classical PID controllers, so that displays and faceplates need not be altered.

One programming detail is important for proper implementation: the variable used as the input to the model $G_m(s)$. This variable should have the value of the actual process input variable and must observe any limitations that exist in the plant, such as the valve being limited to 0 to 100% open. If this guideline is not observed, the control system will be subject to the undesirable integral (reset) windup, described in Chapter 12. When a limitation is reached in the manipulated variable, the controlled variable cannot be returned to its set point, regardless of the control algorithm used. In this situation, the magnitude of the controller output must not increase without limit (the symptom of integral windup). The behavior of the controller output can be determined by applying the final value theorem to the Laplace trans-

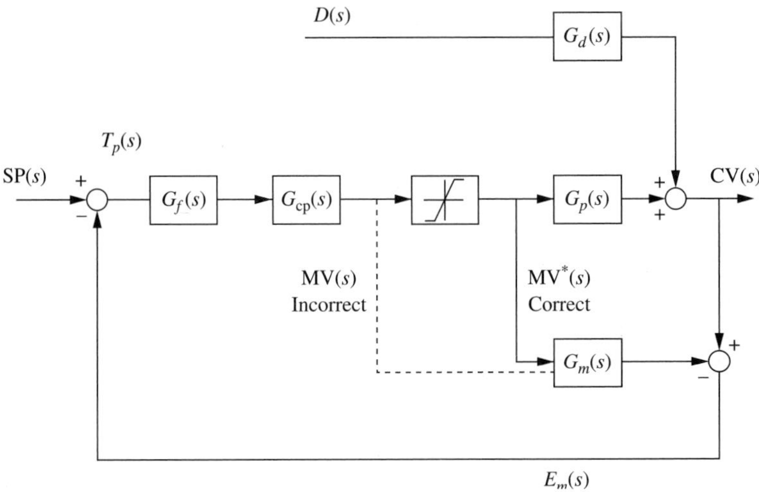

FIGURE 19.10
Predictive control structure with correct (solid) and incorrect (dashed) inputs to the model.

form of the controller output, MV(s), for the IMC control system in Figure 19.10. This is done in the following paragraphs for incorrect and correct implementations for a step disturbance; in both, MV(s) is the output of the controller.

Incorrect (windup occurs): The IMC model input is the signal before any limitation, MV(s), which can differ from the true value of the input variable to the process, MV*(s). The closed-loop transfer function for this system, when the manipulated variable is at an upper or lower limit, is

$$\lim_{t\to\infty} \mathrm{MV}(t) = \lim_{s\to 0} s\mathrm{MV}(s) = \lim_{s\to 0} s \frac{-G_d(s)G_f(s)G_{cp}(s)\frac{\Delta D}{s}}{1 - G_f(s)G_{cp}(s)G_m(s)}$$
$$= \frac{-K_d K_f \dfrac{1}{K_m}\Delta D}{1 - K_f\left(\dfrac{1}{K_m}\right)K_m} \to \infty \tag{19.34}$$

Clearly, this implementation suffers from integral windup. This undesirable behavior results because of the requirements that $K_f = 1.0$ and $K_{cp} = K_m^{-1}$ for zero steady-state offset in the normal, unconstrained situation. The cause of reset windup could also be interpreted as the model $G_m(s)$ ceasing to represent the causal relationship between the calculated controller output and the measured controlled variable, which is properly zero when the limitation occurs.

Correct (windup prevented): The IMC model input is the true value of the variable in the process, MV*(s), which is affected by the process limitations. When the manipulated variable reaches a limit, it is constant, and the system behaves like an open-loop system.

$$\begin{aligned}\lim_{t\to\infty} \mathrm{MV}(t) &= \lim_{s\to 0} s\mathrm{MV}(s) \\ &= \lim_{s\to 0} s\left(-G_f(s)G_{\mathrm{cp}}(s)G_d(s)\frac{\Delta D}{s}\right) \\ &= -K_f\left(\frac{1}{K_m}\right)K_d\Delta D \end{aligned} \tag{19.35}$$

This approach achieves the proper behavior because the model $G_m(s)$ represents a causal relationship that is valid for all situations. When the values of the predicted and the measured controlled variables reach essentially constant values, the feedback signal is constant except for disturbances, which vary over a limited range of values. The feedback signal in the correct implementation results in a value of the controller output that is limited to proportional responses to disturbances, as is proper to prevent integral windup.

Predictive controllers can be applied as primaries in cascades. PID can be used as the secondary, because the simple secondary loops do not usually contain major noninvertible elements of dead time and inverse response, which seriously degrade control performance of PID control loops. The implementation of the predictive IMC system as a primary in a cascade is shown schematically in Figure 19.11. The important detail is the use of the measured value of the secondary variable, rather than its set point (which is the primary controller output) as the input to the model. The selection is based on the purpose of the model, which is to provide a prediction of the primary controlled variable, and this prediction should respond to the important input actually occurring in the process, not to the set point, which does not directly affect the process. Use of the set point as input to the model will result in integral windup, for the reasons just explained; use of the secondary controlled variable prevents windup.

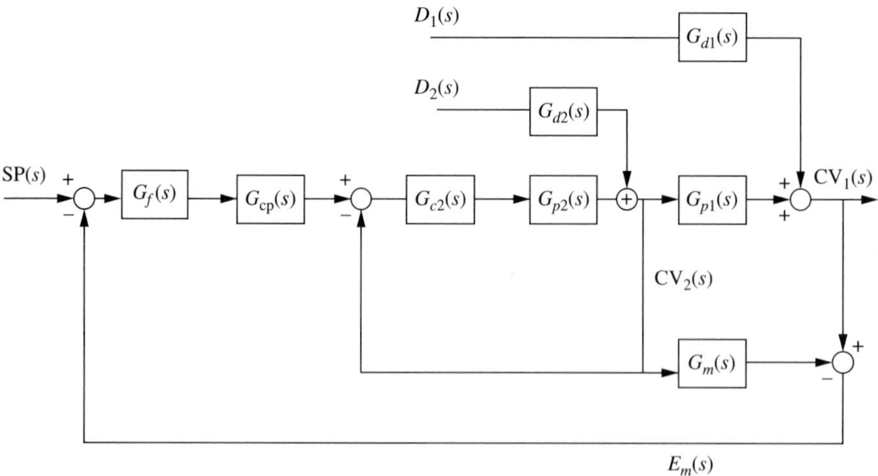

FIGURE 19.11
Cascade control with a primary predictive controller.

With proper design and care for implementation details, digital implementation of predictive controllers is straightforward; in fact, the algorithms can be preprogrammed so that engineers need only select from a set of possible model structures and enter values for the model parameters and tuning constants. Thus, a predictive controller should not require more effort to implement than a standard PID algorithm.

19.6 ALGORITHM SELECTION GUIDELINES

To this point, single-loop predictive control has been introduced, the IMC and Smith predictor control algorithms have been presented, and tuning and programming guidelines have been provided. These controllers can be used in place of any PID controller; however, since the PID is the standard algorithm selected, a predictive control system is normally selected only when it performs better than a PID algorithm. In this section four applications are discussed in which predictive control offers potential advantages; the IMC controller will be used in all examples, but similar results can be obtained with a Smith predictor.

Long Dead Times

The Smith predictor is often referred to as a *dead time compensator* because Smith's original goal was to improve the performance of feedback control systems with long dead times in the feedback processes. When the model is perfect in Figure 19.3, the predictive system behaves like a feedforward controller; thus, the control action in this ideal situation can be as aggressive as required for the desired performance, without concern for stability. Even with modest model errors, the predictive system has the potential for improved performance when applied to processes with large fraction dead times. Selection of the proper algorithm (PID or predictive) depends on the particular situation; the advantages of predictive control are greater as the model is more accurate, the noise is small, and the feedback fraction dead time is large, usually greater than about 0.70.

Inverse Response

As explained in Section 13.5, an inverse response in the feedback process degrades the performance of a feedback controller in a manner similar to dead time. The PID algorithm has particular difficulty, because its error signal—the difference between the set point and the measured variable—initially increases in magnitude in spite of a proper initial feedback adjustment to the manipulated variable. The predictive controller has been reported to perform well, because its feedback signal—the difference between the predicted and the measured values—does not experience an inverse response (Iionya and Altpeter, 1962; Shunta, 1984).

Cascade Control

One of the design criteria presented in Chapter 14 for cascade control is that the secondary control loop must be much faster than the primary loop. There are situations in which a cascade is desirable for disturbance response, but the dynamic response of the secondary is not substantially faster than the primary. An example is a distillation composition controller that acts as a primary by resetting a tray temperature controller set point (e.g., Fuentes and Luyben, 1983). If the appropriate cascade criterion is not satisfied, significant fluctuation of the relatively slow secondary controlled variable causes a transient disturbance in the primary. If a PID control algorithm is used as the primary controller, it unnecessarily adjusts the set point of the secondary, although the secondary would ultimately compensate for the disturbance, as shown in Figure 14.6*b*.

A predictive control system for the primary in a cascade offers a distinct advantage, because the feedback signal is the sum of the model error in the primary, along with primary disturbances (Bartman, 1981). Secondary disturbances, which cause deviations in the secondary measurement, appear in both the measured and predicted primary variable at about the same time and magnitude (if the model is reasonably accurate). As a result, the secondary disturbances have little or no effect on the feedback signal, $E_m(s)$. Naturally, this behavior is valid only when the model input is the measured secondary variable, as already discussed. Therefore, the dynamic behavior of some cascade systems can potentially be better (i.e., much less oscillatory) when the primary is a predictive control system.

Feedforward

As described in Chapter 15, a feedforward-feedback control system can perform quite well when the process dynamics allow a complete compensation of the disturbance by the feedforward controller. This criterion is satisfied when the disturbance dead time is longer than the feedback process dead time (i.e., $\theta_d \geq \theta_p$). When this criterion is not satisfied, the feedforward controller changes the manipulated variable enough to compensate fully for the disturbance when the steady state is achieved; however, the initial effect of the disturbance appears in the measured controlled variable prior to the effect of the manipulated variable. A PID feedback controller cannot recognize that the feedforward compensation has been introduced and makes an unnecessary *additional* change to the manipulated variable.

The essential deficiency in the feedforward-feedback design is the feedback PID controller, which cannot determine that the proper manipulation has been entered by the feedforward controller. This deficiency can be overcome through the use of a predictive control system for feedback, as shown in Figure 19.12. In this design, the predictive model includes relationships for the manipulated and measured disturbance variables, and the value of the manipulated variable used in the model includes the changes from both feedforward and feedback controllers. Again, the feedback signal is the difference between the measured and predicted controlled variables. As long as the models $G_{\text{dm}}(s)$ and $G_m(s)$ are reasonably accurate, the feedback signal, $E_m(s)$, does not change during the transient resulting from imperfect feedforward dynamics. In the situation of "slow" feedforward control ($\theta_d < \theta_p$), the model predictive feedback will not introduce additional adjustments to the manipulated variable.

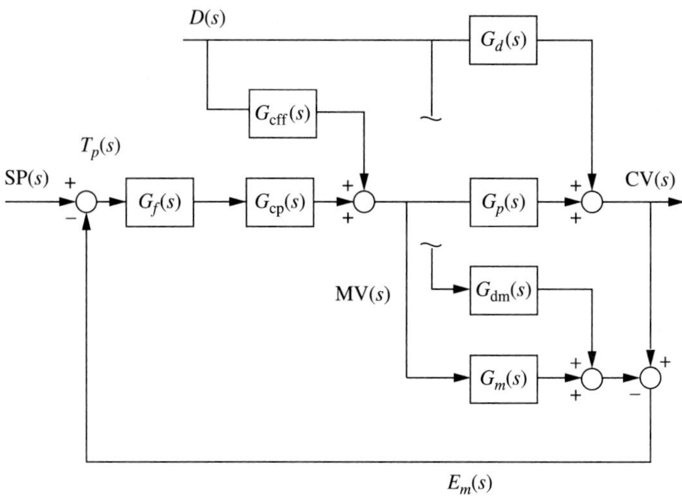

FIGURE 19.12
Feedforward with predictive feedback control.

The guidelines in this section indicate applications in which predictive control is likely to perform better than single-loop PID controllers. In many other applications, PID and predictive control systems give equivalent performance, and the usual selection is PID.

19.7 ADDITIONAL TOPICS IN SINGLE-LOOP MODEL PREDICTIVE CONTROL

The presentation in this chapter provides only an overview of predictive control, which is still a developing topic. A few important additional issues are introduced briefly in this section with reference to the IMC control system.

Digital Implementation

The procedure used here was to perform the design assuming that all variables were continuous, allowing Laplace transform methods, and subsequently, to convert the resulting model and controller to discrete form. A more general approach is to convert the models to the discrete form before performing the controller design, which enables more general model structures to be used. This approach requires the use of z-transforms (Ogata, 1987) and is presented in Morari and Zafiriou (1989).

Controller Design

The method for calculating the controller was based on the goal of perfect control (zero deviation of the controlled variable from its set point). Since this goal is not possible, the controller design method factored the model and used only the invertible part, $G_{cp}(s) = [G_m^-(s)]^{-1}$. This approach leads to a realizable controller, but there is no guarantee that it is the "best" in any sense. Alternative controller algorithm design methods are based on other goals. For example, the controller could be designed to minimize the integral of the error squared, ISE, of the controlled variable during the disturbance response. This approach is presented by Newton et al. (1957) and applied to IMC control by Morari and Zafiriou (1989). Naturally, a tunable filter remains in the design to achieve the desired robustness and manipulated-variable behavior.

Filter Design

The filter described in this chapter improves the robustness of the predictive system at the expense of increased deviation of the controlled variable from its set point. Alternative filter designs can be selected to improve the response of the system to a specific disturbance. For example, the response of predictive systems to disturbances, as described in this chapter, can be somewhat slow, but the disturbance response can be improved by designing a filter that infers the disturbance variable $D(s)$ from the feedback signal $E_m(s)$. When there is no model error, this calculation requires that the filter be related to the inverse of the disturbance transfer function. While this concept is theoretically sound, it can lead to aggressive controllers that are tailored to a specific disturbance and may not respond well to other disturbance types. Again, some of these ideas are in Morari and Zafiriou (1989).

Robustness and Tuning

A vast literature is developing in methods for designing controllers using knowledge of the likely model errors, or mismatch. The key aspect of the design methods is to provide not only stability, but also the best performance possible, over the likely model errors. If the mismatch characterization is simple, like the gain margin used in Ziegler-Nichols, the methods are easily applied, but they can yield conservative feedback performance. This approach was promoted by the discussions and methods in Doyle and Stein (1981).

PID Tuning

The IMC controller can be expressed as an equivalent classical controller design, and this equivalence can be used to express PID tuning as a function of only one parameter: the IMC filter τ_f. Results have been developed by Rivera et al. (1986) and are summarized here for the "improved PI" tuning:

$$K_p K_c = \frac{2\tau + \theta}{2\tau_f} \qquad T_I = \tau + \frac{\theta}{2} \tag{19.36}$$

The recommendation is that $\tau_f \geq \max\,(1.7\theta, 0.2\tau)$ (Morari and Zafiriou, 1989).

19.8 CONCLUSIONS

In this chapter, an alternative feedback control structure and algorithm were introduced for processes that are open-loop stable. This predictive structure employs an explicit model of the process in the control calculations. In addition, the controller

$G_{cp}(s)$ is designed to be an approximation of the process model inverse. Since the feedback signal is the disturbance (when $G_m(s)$ is a perfect model), the predictive controller functions somewhat like a feedforward controller and potentially can implement more aggressive adjustments to the manipulated variables.

Analysis of the predictive control structure provides a unified viewpoint for evaluating the effects of dynamic elements in the feedback process on control performance. In particular, the importance of dead time, inverse response (right-half-plane zeros), constraints, and model error are clearly identified. Recall that the controller cannot eliminate dead time or inverse response from the process, and the overall improvement achieved by employing predictive control is limited. Substantial improvement in control performance requires process changes to reduce these dynamic characteristics.

The predictive structure has the feature of removing process elements that are difficult to control from the calculation of the feedback adjustment, when the model is perfect. In this situation, the feedback signal E_m does not depend on the controller output MV, because it is affected only by disturbances. In single-loop control, predictive control offers potential for improved performance in feedback processes with large fraction dead times, inverse responses, or both. Also, predictive control can be employed in cascade control with similar secondary and primary dynamics and in feedforward-feedback control with a disturbance dead time less than the feedback dead time.

In addition to providing useful single-loop control algorithms, the material in this chapter provides a general manner for analyzing feedback systems. Specifically, the predictive structure is the basis for the powerful multivariable control algorithm presented in Chapter 23.

REFERENCES

Bartman, R., "Dual Composition Control of C_3/C_4 Splitter," *CEP, 77, 9,* 58–62 (1981).

Brosilow, C., "The Structure and Design of Smith Predictors from the Viewpoint of Inferential Control," *Proc. Joint Auto. Cont. Conf., Denver, Colorado*, 1979.

Ciancone, R., S. Sampath, and T. Marlin, "An Unbiased Method for Tuning Process Controllers," *Can. Soc. Chem. Engr. Ann. Conf.*, Ottawa, 1993.

Doyle, J. and G. Stein, "Multivariable Feedback Design: Concepts for a Classical/ Modern Synthesis," *IEEE Trans. Auto. Cont., AC-26,* 4 (1981).

Fuentes, C. and W. Luyben, "Control of High Purity Distillation Columns," *IEC Proc. Des. Devel., 22,* 361–366 (1983).

Iionya, K. and R. Altpeter, "Inverse Response in Process Control," *IEC, 54, 7*, 39 (1962).

Laughlin, D. and M. Morari, "Smith Predictor Design for Robust Performance," *Proc. ACC, paper WP8-2:30*, 637–642 (1987).

Morari, M. and E. Zafiriou, *Robust Process Control*, Prentice-Hall, Englewood Cliffs, NJ, 1989.

Newton, G., L. Gould, and J. Kaiser, *Analytical Design of Feedback Controls,* Wiley, New York, 1957.

Ogata, K., *Discrete-Time Control Systems,* Prentice-Hall, Englewood Cliffs, NJ, 1987.

Parrish, J. and C. Brosilow, "Inferential Control Applications," *Automatica, 21, 5*, 527–538 (1985).

Rivera, D., S. Skogestad, and M. Morari, "Internal Model Control: 4. PID Controller Design," *IEC Proc. Des. Devel., 25,* 252–265 (1986).

Shunta, J., "A Compensator for Inverse Response," *Proc. ISA Meet. October 22, 1984,* pp. 771–793.

Smith, O., "Closer Control of Loops with Dead Time," *Chem. Eng. Prog., 53, 5,* 217–219 (1957).

ADDITIONAL RESOURCES

The results in this chapter are limited to stable processes, which eliminates unstable processes such as the chemical reactors in Appendix C and non–self-regulating levels. The approach can be extended, as described in Morari and Zafiriou (1989).

QUESTIONS

19.1. For the following processes, design IMC and Smith predictor model predictive controllers. Specify all parameters and give all equations for digital implementation. Simulate each for a set point change.

(*a*) The process in question 6.1, controlling temperature by adjusting the valve.

(*b*) The process in question 6.2, controlling temperature by adjusting the valve.

(*c*) The chemical reactor in Examples 3.10 and 4.8, controlling the temperature by adjusting the coolant flow.

(*d*) The chemical reactors in Example 3.3, controlling outlet composition C_{A2} by adjusting the inlet composition C_{A0}. Use an approximate first-order-with-dead-time model for $G_m(s)$ and in designing $G_{cp}(s)$.

19.2. The three-tank mixing process with IMC control was investigated in Example 19.8 for various flow rates. Using the deterministic-calculation approach introduced in Section 16.3, determine a method for maintaining good IMC control performance as the measured flow rate changes over the range in Table 16.1.

19.3. The following process models have been identified for processes that conform to the block diagram in Figure 19.11. For each process, determine whether cascade control or single-loop control is appropriate, assuming that $D_2(s)$ is a significant disturbance. For cascades, decide whether the performance might be improved by using an IMC controller as the primary. Assume that $G_{d1}(s) = G_{p1}(s)$ and $G_{d2}(s) = G_{p2}(s)$.

(*a*) $G_{p2}(s) = \dfrac{0.5}{\tau s + 1} \qquad G_{p1}(s) = \dfrac{2.3}{(\tau s + 1)^5}$

(*b*) $G_{p2}(s) = \dfrac{0.5}{(\tau s + 1)^3} \qquad G_{p1}(s) = \dfrac{2.3}{(\tau s + 1)^3}$

(*c*) $G_{p2}(s) = \dfrac{0.5}{(\tau s + 1)^5} \qquad G_{p1}(s) = \dfrac{2.3}{(\tau s + 1)}$

19.4. (*a*) Verify the equalities given in equations (19.31) to (19.33), relating the IMC and Smith predictor approximate inverses for $G_m^-(s) = K_p/(1 + \tau s)$.

(*b*) Verify the relationship in equation (19.10) relating the model predictive and classical controllers.

(*c*) In Figures 19.10 to 19.12, which of the transfer functions represent control calculations and which represent process behavior?

(*d*) Determine the criteria for zero steady-state offset with model predictive control of a stable process with an impulse input change.

19.5. Describe a proper method for providing anti–reset windup for the Smith predictor. Include a block diagram and apply the final value theorem to prove that your design is adequate.

19.6. Perform a similar analysis as in Section 8.7 for the stirred-tank heat exchanger under IMC feedback control. To simplify the analysis, assume a perfect model when determining the analytical solution and that $G_d(s) = G_p(s)$.

(*a*) Analyze the degrees of freedom.

(*b*) Derive the linearized model for the process and controller.

(*c*) Determine the analytical solution for the controlled variable for a step set point change. Assume that $\tau_f = \tau$.

(*d*) Determine the analytical solution for the manipulated variable for a step set point change. Assume that $\tau_f = \tau$.

(*e*) Recalculate the results in (*c*) and (*d*) for $\tau_f = \beta\tau$, with $\beta = 0.5$ and 0.1. Sketch the shape of the dynamic responses of the controlled and manipulated variables for these three values of τ_f.

(*f*) Select the best value of τ_f for the heat exchanger, not necessarily one of the values considered in previous parts of this question.

19.7. The selection of the manipulated and controlled variables are discussed in Section 7.5. Discuss how these criteria should be modified for feedback control using model predictive control.

19.8. A mixing process with the structure in Figure 13.4 and with the following feedback and disturbance transfer functions is to be controlled with an IMC controller. The controlled variable is to be maintained within ±0.37 of the set point for a unit step disturbance. What value of the IMC controller filter is required to achieve this performance?

$$G_p(s) = \frac{1.0e^{-0.5s}}{1 + 1.5s} \qquad G_d(s) = \frac{1.0}{1 + 1.0s}$$

19.9. The chemical reactor process described in Examples 5.4 and 13.8 has a feedback system with the outlet concentration controlled by adjusting the solvent flow rate. Design IMC and Smith predictor model predictive controllers for feedback control. Program one of them and compare the control performance with that achieved with PI feedback in Example 13.8. Discuss the relative performances and steps required to substantially improve the control performance.

19.10. Draw a block diagram for the Smith predictor control system. In each block that involves controller calculations, show the equations solved in the digital implementation. Assume that an adequate process model is first-order with dead time.

19.11. The results of the tuning study given in Figure 19.6 show that the IMC filter factor decreases as the fraction dead time increases. Since the filter was introduced to make the feedback adjustments less aggressive and more robust, one might expect larger filter values to be necessary at large fraction dead times. Discuss the effects of the filter and reconcile the tuning correlations with the robustness expectations just stated. (*Hint:* Consider all control performance criteria and process conditions involved in determining the tuning correlations. They are described in Sections 9.2 and 9.3.)

19.12. Analyze the control performance for IMC (or Smith predictor) feedback control of the three-tank mixing process using closed-loop frequency response.

(*a*) Derive the expression used for this calculation assuming that the controller uses the first-order-with-dead-time approximation in Example 19.5 and $\tau_f = 6.1$. (Do not solve for the real and imaginary parts analytically.)

(*b*) Use a computer program to evaluate the magnitude of the transfer function over a range of frequencies; that is, determine $|CV(j\omega)|/|D(j\omega)|$.

(*c*) Compare the results in (*b*) with the equivalent results in Figure 13.16 (curve *a*) for PI control, discussing similarities and differences.

19.13. A method for analyzing the stability of the model predictive control systems should be available. Perform the following analysis for the three-tank mixing process under IMC control for two cases: $\tau_f = 0$ and $\tau_f = 6.1$. Use the continuous transfer functions from Example 19.6.

(*a*) Determine the expression for $G_{\text{OL}}(s)$ that could be used to analyze stability.

(*b*) Determine the magnitude and phase angle of $G_{\text{OL}}(j\omega)$ for several decades around the critical frequency. Present the results in Bode plots.

(*c*) Evaluate the Bode plot for the assumptions required for use in stability analysis.

(*d*) Based on the results of (*b*) and (*c*), discuss the use of the Bode method for stability analysis of model predictive control systems. Suggest a more general method that is applicable.

19.14. Consider a process with first-order-with-dead-time feedback dynamics and first-order disturbance dynamics (e.g., the process in Figure 13.4). The system is to be controlled with an IMC system and is subject to an unmeasured step disturbance. Assume the model predictive control calculations can be designed with perfect models. Answer the following questions for two cases: (1) zero feedback dead time and (2) nonzero feedback dead time.

(*a*) Define the best, physically possible feedback control performance. For this question alone, control performance is determined completely by the ISE of the controlled variable.

(*b*) Determine the transient response of the manipulated variable that would result in the behavior of the controlled variable determined in (*a*).

(*c*) Derive the IMC controller and filter $G_f(s)G_{\text{cp}}(s)$ that would give the performance defined in (*a*) and (*b*) using the feedback measurement only.

19.15. Using the IMC tuning rules for a PI controller in equation (19.36) and Figure 19.6, develop graphs of K_pK_c and $T_I/(\theta + \tau)$ versus the fraction feedback dead time, $\theta/(\theta + \tau)$. Compare these graphs with Figure 19.9.

PART V

MULTIVARIABLE CONTROL

In this part we continue the trend of addressing increasingly complex process control systems. Although some of the control systems in Part IV (e.g., cascade and feedforward control) involved more than one measured variable, we considered these to be single-variable control because they had the ultimate objective of maintaining only one variable near its set point. By contrast, *multivariable control* involves the objective of maintaining several controlled variables at independent set points.

The simple chemical reactor process shown in Figure V.1 is considered first to introduce the concept of a multivariable process. The control objectives depend on the goals of the entire plant and of the design of associated equipment, but typical objectives would be to control the level, temperature, and outlet concentration at independent set points, which would be achieved by adjusting selected manipulated variables in the process. Again, the variability of the controlled variables is reduced through actions that increase the variability of the manipulated variables. In Part V the complexity of multivariable systems is reduced by assuming (for the most part) that the process design, measurements, and final elements cannot be changed; thus, the process dynamics and control calculations are addressed. These restrictions will be relieved in Part VI, when process control design is addressed.

Control of multivariable systems requires more complex analysis than that of single-variable systems, as summarized in Table V.1. Fortunately, essentially all methods and results learned for single-variable systems are applicable to multivariable

TABLE V.1
Characteristics of multivariable control systems

Single-loop characteristics that generally lead to good control performance in multivariable systems	Characteristics unique to multivariable systems
1. Fast feedback processes (small $\theta + \tau$) 2. Feedback processes with a small fraction dead time ($\theta/(\theta + \tau)$) and no inverse response 3. Disturbances with small magnitudes far from the critical frequency 4. No limitations encountered in the manipulated variable 5. Digital controllers with relatively fast execution periods 6. Controllers based on accurate models 7. Controllers using appropriate enhancements from Part IV	1. Interaction between variables influences control stability and performance. 2. Feasibility of control depends on overall process, not just individual cause-effect relationships. 3. The source of the disturbance, not just the magnitude, must be considered in designing the control strategy. 4. The pairing of measured variables and final elements via control is a design decision. 5. Some processes have an unequal number of controlled and manipulated variables. 6. Some multivariable control designs are very sensitive to modelling errors.

systems. Thus, aspects of a single-variable system that make it easy or difficult to control have generally the same effect for multivariable systems. However, in multivariable systems new characteristics due to *interaction* must be considered. Interaction results from process relationships that cause a manipulated variable to affect more than one controlled variable. In Figure V.1 the heating oil valve position influences both the temperature and, through the reaction rate constant, the concentration. This is the major difference from single-loop systems and has a profound effect on the steady-state and dynamic behavior of a multivariable system.

Thus, it is not possible to analyze each manipulated–controlled variable connection individually to determine its performance; the integrated control system must be considered simultaneously. A closely related new issue is the disturbance source,

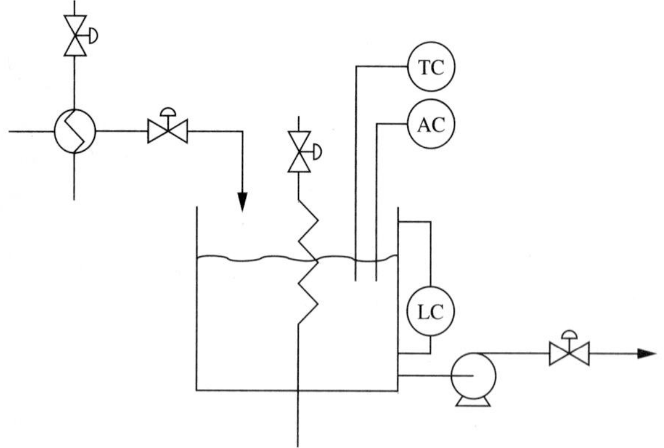

FIGURE V.1
Multivariable process.

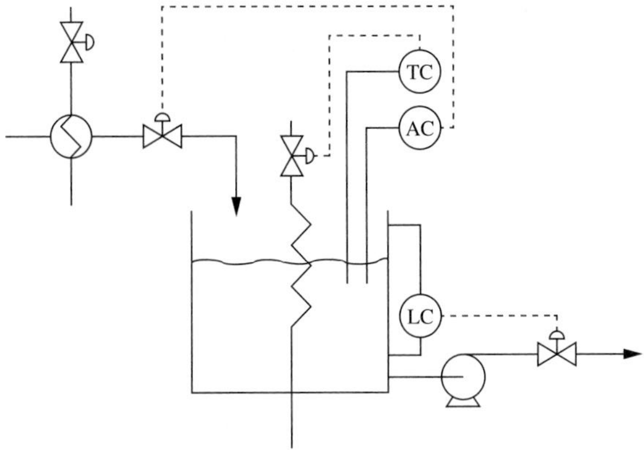

FIGURE V.2
Example of multiloop control design.

because multivariable systems respond differently to different disturbances. For example, the chemical reactor responds differently to disturbances in feed composition and feed temperature, and, as we shall see, these differences must be considered in designing a multivariable control system.

Another realistic issue is the number of controlled and manipulated variables, which may not be equal. Note that the system in Figure V.1 has four manipulated variables, which can be adjusted to control three measured variables. Multivariable control methods presented in this part are able to utilize all flexibility available in the process.

There are two basic multivariable control approaches. The first is a straightforward extension of single-loop control to many controlled variables in a single process, as shown in Figure V.2. This is termed *multiloop* control and has been applied with success for many decades. The second main category is *coordinated* or *centralized* control, in which a single control algorithm uses all measurements to calculate all manipulated variables simultaneously, as shown in Figure V.3. Algorithms for this approach have been available for several decades and have been widely applied for a considerable time in the process industries.

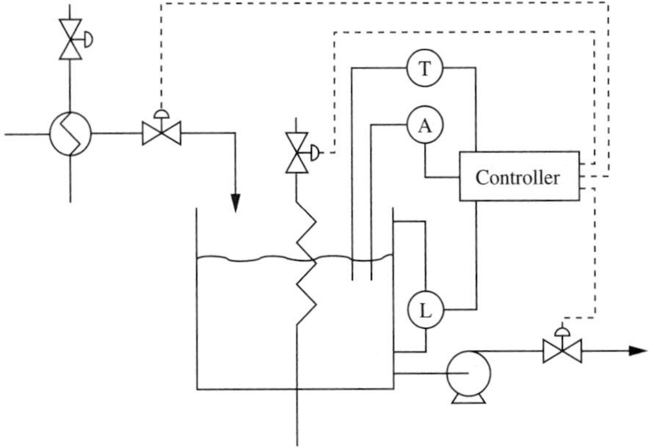

FIGURE V.3
Example of multivariable control design.

At the conclusion of this part, the unique characteristics of multivariable process systems and how these chracteristics affect process control will have been presented. The reader is cautioned that this is a complex topic, worthy of an entire book, and that the presentation here is elementary. However, it presents the major issues, along with some of the more common analysis methods and control approaches.

CHAPTER 20

MULTILOOP CONTROL: EFFECTS OF INTERACTION

20.1 INTRODUCTION

Multivariable control occurs in nearly all processes, because production rate (flow), inventory (level and pressure), process environment (temperature), and product quality are normally controlled simultaneously. The *multiloop* approach, using multiple single-loop controllers, was the first approach used for multivariable control in the process industries. Through decades of research and experience, many successful multiloop strategies have been developed and continue to be used.

One advantage of multiloop control is the use of simple algorithms, which is especially important when the control calculations are implemented with analog computing equipment. A second advantage is the ease of understanding by plant operating personnel, which results from the simplicity of the control structure. Since each controller uses only one measured controlled variable and adjusts only one manipulated variable, the actions of the controllers are relatively easy to monitor. A third advantage is that standard control designs have been developed for the common unit operations, such as furnaces, boilers, compressors, and simple distillation towers. This does *not* mean that a single control design functions well for all unit operations of the same type. However, several general structures are in common use, and selection among alternatives can be based on analysis and experience. Considering these advantages, one could conclude that multiloop designs will continue to be used extensively, although not exclusively.

An example of multiloop control of a flash process is given in Figure 20.1. Let us consider the behavior of the system when the feed flow rate increases. An initial effect is an increase in the amount of vapor entering the drum, although the

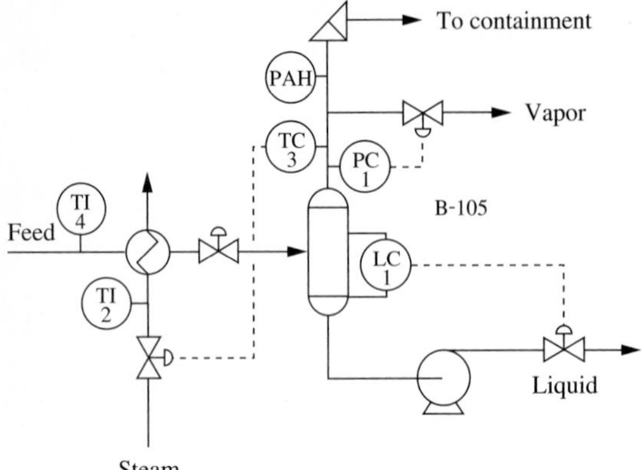

FIGURE 20.1
Example of a multiloop control system.

percentage feed vaporized decreases because of a slight decrease in inlet temperature. The pressure in the drum increases because of the additional vapor; therefore, the pressure controller PC-1 takes action by increasing the percent opening of the valve in the vapor line. Another effect is a decrease in the temperature after the heat exchanger, which is sensed by TC-3. This feedback controller increases the steam flow to the exchanger, which returns the temperature to its set point and causes even more feed to be vaporized. This additional vapor causes the pressure to increase, and the pressure controller has to respond to this change as well. The increase in feed rate and changes in percent vaporized introduce changes in the liquid rate into the liquid inventory in the drum. The level controller increases the opening of the valve in the liquid product line to maintain the level near its set point.

Two important features of this system become clear when observing its behavior:

1. The single-loop controllers are completely independent algorithms that do not communicate directly among themselves.
2. The manipulations made by one controller can influence other controlled variables; that is, there can be *interaction through the process* among the individual control loops.

The interaction is the key effect addressed in this chapter, where we will demonstrate that several single-loop controllers on a process should not generally be analyzed as though each were a single-loop system.

We shall use the following definition of interaction.

> A multivariable process is said to have **interaction** when process input (manipulated) variables affect more than one process output (controlled) variable.

This definition is consistent with the use of the word in the vernacular and will serve us in the study of multivariable systems. However, the definition does not distinguish between various important properties that will be introduced in this chapter. Thus, careful attention must be paid to the effects of various types of interaction on control stability and performance.

In this chapter the basic principles of multiloop control are presented, with the goal of understanding multivariable systems. As with single-loop control, we start with the process by reviewing modelling approaches for multivariable processes and developing models for two sample systems, which will be used in later examples. Then the concept of interaction is discussed to highlight its effects on system behavior, and a quantitative measure of interaction is introduced. Finally, some approaches for tuning multiloop controllers are presented. All of the concepts developed in this chapter are employed in the next chapter, which addresses the performance of multiloop control systems.

20.2 MODELLING AND TRANSFER FUNCTIONS

As explained in Part II, process models can be derived from fundamental principles or can be estimated based on empirical data. Regardless of the modelling method used, the analysis, design, and tuning of multiloop controllers will be based on linear input-output models employing block diagram manipulation, stability analysis, and frequency response. The following two examples demonstrate the modelling approaches applied to blending and distillation, and the resulting models will be employed in several subsequent examples.

Example 20.1 Blending is an important unit operation and is employed in a wide variety of industries, as in production of gasoline (Stadnicki and Lawler, 1985) and cement (Sakr, et al., 1988). Typically, the controlled variables in a blending process are production rate and blended product composition. The blending process in Figure 20.2 is modelled with the following assumptions:

1. The inlet concentrations are constant.
2. Mixing where the flows merge is perfect.
3. The densities of the solvent and component A are equal.

The overall and component A material balances at the point of mixing are

$$F_m = F_A + F_S \tag{20.1}$$

$$F_m X_m = F_A X_A + F_S X_S \tag{20.2}$$

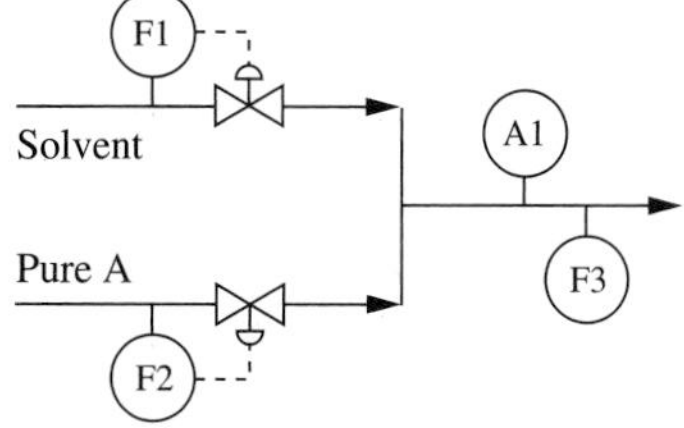

FIGURE 20.2
Example blending process.

with F_m = flow rate of mixed liquid (mass/time)
X_A = mass fraction of component A in pure A = 1.0
X_S = mass fraction of component A in solvent = 0.0
X_m = mass fraction of component A in the mixed liquid

Equation (20.2) can be linearized about the steady state to give

$$X'_m(t) = \left(\frac{F_S}{(F_S + F_A)^2}\right)_s F'_A(t) + \left(\frac{-F_A}{(F_S + F_A)^2}\right)_s F'_S(t) \tag{20.3}$$

with the prime indicating deviation variables. The system is liquid-filled; thus, there is essentially no delay between a change in a component flow rate and a change in the mixed-product flow rate. It is also assumed that the concentration at the location of the analyzer is essentially the same as at the mixing point; that is, there is no transportation dead time. Also, the inlet flow measurements are assumed to be exactly and instantaneously equal to the actual flows, $F_1 = F_S$ and $F_2 = F_A$. The dynamics of the mixed stream flow and concentration sensors are not instantaneous and are characterized by a first-order-with-dead-time model with gains of 1.0 and the following dynamic parameters:

	Dead time	**Time constant**
Flow	θ_F	τ_F
Concentration	θ_A	τ_A

Thus, the measured controlled variables are related to the instantaneous process variables in equations (20.1) and (20.3) by

$$\tau_A \frac{dA'_1(t)}{dt} = X'_m(t - \theta_A) - A'_1(t) \tag{20.4}$$

$$\tau_F \frac{dF'_3(t)}{dt} = F'_m(t - \theta_F) - F'_3(t) \tag{20.5}$$

Equations (20.1) and (20.3) to (20.5) can be combined to give the following linearized dynamic model:

$$A_1(s) = \frac{\left(\dfrac{-F_A}{(F_S + F_A)^2}\right)_s e^{-\theta_A s}}{1 + \tau_A s} F_1(s) + \frac{\left(\dfrac{F_S}{(F_S + F_A)^2}\right)_s e^{-\theta_A s}}{1 + \tau_A s} F_2(s) \tag{20.6}$$

$$F_3(s) = \frac{1.0e^{-\theta_F s}}{1 + \tau_F s} F_1(s) + \frac{1.0e^{-\theta_F s}}{1 + \tau_F s} F_2(s) \tag{20.7}$$

Clearly, interaction is present in this process, because each output is affected by both inputs. Numerical values will be determined for different operating conditions later in this chapter.

Example 20.2 The empirical identification procedures described in Chapter 6 can be applied to the distillation process shown in Figure 20.3. Design and operating data are given in Example 5.6 (page 180). (This design was originally suggested by McAvoy and Weischedel (1981) and was approximated for constant relative volatility by Sampath (1991).) The manipulated variables are reflux and reboiler flow rates, and the controlled variables are distillate and bottoms composition. Other important variables, such as pressure and levels, are controlled tightly as shown.

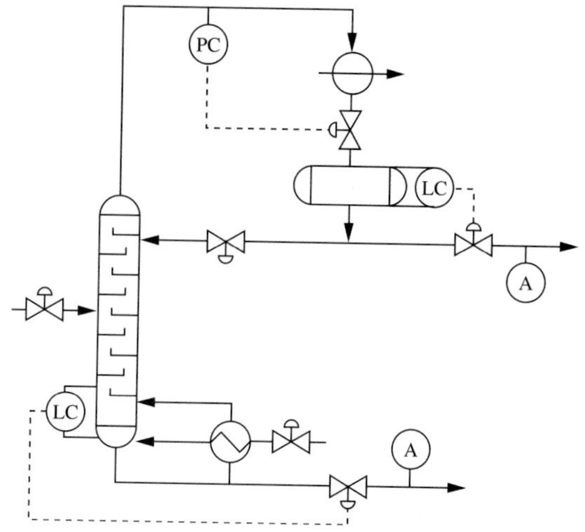

FIGURE 20.3
Example distillation tower.

One experiment must be performed for each input variable, and the responses of all output variables (after 2 min analyzer dead time) are recorded. Either the process reaction curve or statistical methods can be used to fit parameters in the transfer functions. The models derived by this empirical procedure are as follows:

$$X_D(s) = \frac{0.0747e^{-3s}}{12s+1}F_R(s) - \frac{0.0667e^{-2s}}{15s+1}F_V(s) + \frac{0.70e^{-5s}}{14.4s+1}X_F(s) \qquad (20.8)$$

$$X_B(s) = \frac{0.1173e^{-3.3s}}{11.7s+1}F_R(s) - \frac{0.1253e^{-2s}}{10.2s+1}F_V(s) + \frac{1.3e^{-3s}}{12s+1}X_F(s) \qquad (20.9)$$

As reported in Table 5.4, the units are mole fractions of the light key components for the compositions, kmole/min for the flows, and min for time. Note that the reflux flow (F_R) and amount vaporized in the reboiler (F_V) are potential manipulated variables, and the feed composition (X_F) is a disturbance, because it depends on upstream operations and is assumed not free to be adjusted.

Finally, the linearized models in Examples 20.1 and 20.2 will be used in subsequent system analysis examples. When the dynamic responses are determined via simulation, the linearized distillation model will be used, but the nonlinear blending model will be used because of the large range of operating conditions considered in the blending examples.

Linearized models, whether derived from fundamental balances or from experiments, can be used to analyze the system with and without control. To understand the entire system, it is helpful to present the process in a block diagram. The block diagram of a general 2×2 system, recalling that each process transfer function relates one input to one output, is shown in Figure 20.4. Each term $G_{ij}(s)$ relates manipulated input j to output i, and the terms $G_{di}(s)$ relate the effects of a disturbance on each process output. If more than one important disturbance is to be considered, additional disturbance transfer functions can be included. Note that if both $G_{12}(s)$ and $G_{21}(s)$ (or alternatively $G_{11}(s)$ and $G_{22}(s)$) are zero, the process has no interaction, because one input affects only one output. In such a case, the system behaves like two independent processes, and the behavior of each control loop is independent.

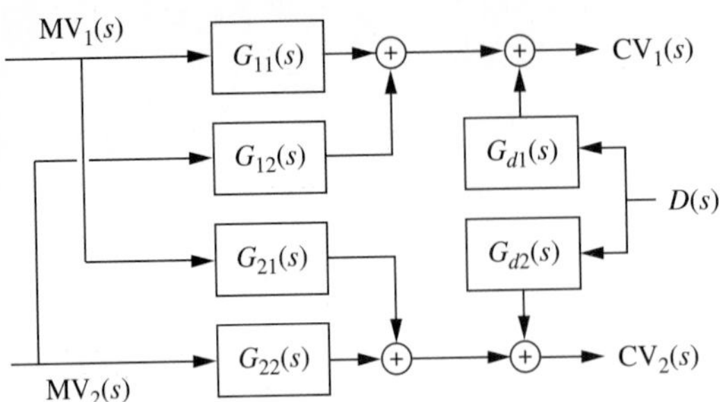

FIGURE 20.4
Block diagram of 2 × 2 open-loop system.

The set of simultaneous equations relating inputs to outputs in Figure 20.4 are often presented in matrix form as follows:

$$\begin{bmatrix} \mathrm{CV}_1(s) \\ \mathrm{CV}_2(s) \end{bmatrix} = \begin{bmatrix} G_{11}(s) & G_{12}(s) \\ G_{21}(s) & G_{22}(s) \end{bmatrix} \begin{bmatrix} \mathrm{MV}_1(s) \\ \mathrm{MV}_2(s) \end{bmatrix} + \begin{bmatrix} G_{d1}(s) \\ G_{d2}(s) \end{bmatrix} D(s) \qquad (20.10)$$

Each element of the matrix is a transfer function relating one input to one output. Thus:

> Linear models for multivariable systems can be developed using the same analytical and empirical procedures as for single-variable systems.

20.3 INFLUENCE OF INTERACTION ON THE POSSIBILITY OF FEEDBACK CONTROL

In Section 7.5, some basic requirements are stated for the variables involved in a single-loop feedback control system. Briefly, the controlled variable should be closely related to process performance; the manipulated variable should be independently adjustable; there should be a causal relationship between the manipulated and controlled variables; and the dynamics should be favorable. These guidelines are still useful, but a somewhat more complex analysis is required for multivariable systems, because range and controllability are influenced by process interactions.

Operating Window

The first issue is the control system's range of attainable variable values. This topic was introduced in Section 3.6, where the term *operating window* was used for the range of possible (or feasible) steady-state process values. The operating window can be sketched using different variables as coordinates; in one approach, the controlled variables are used to characterize the range of possible set points, with all inputs other than the manipulated variables (i.e., disturbances) constant. Another common approach is to use the disturbance variables as coordinates to characterize the range of disturbance values that can be compensated by the control system (i.e.,

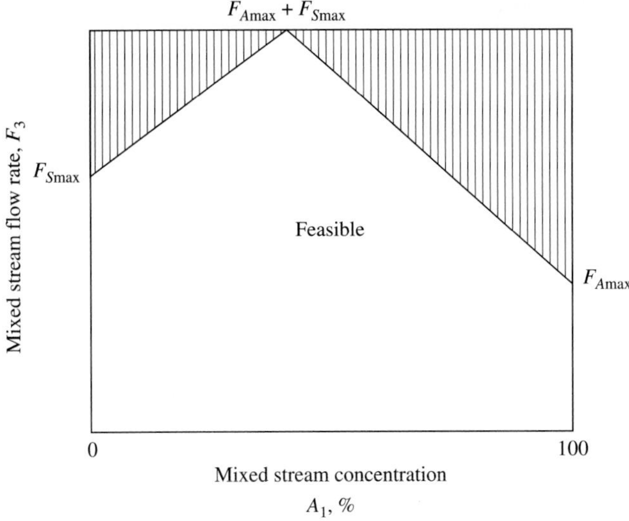

FIGURE 20.5
Operating window for blending with controlled variables as coordinates.

for which the controlled variables can be maintained at constant set points). The two approaches are demonstrated in the following examples.

Example 20.3. The component flow rates in the blending example can be adjusted continuously from zero to maximum rates, $F_{A\max}$ and $F_{S\max}$. Draw the operating window of attainable total flow rate and composition, assuming that the component compositions remain unchanged.

The attainable total flow F_3 and composition A_1 are shown in Figure 20.5. The limiting values are easily determined by solving equations (20.1 and 20.2) for various values of one flow, with the other flow at its maximum value. The interaction between variables is clear, because the value of one variable influences the range of the other variable. If the variables were independent and no interaction occurred, the operating window would be rectangular, which it clearly is not.

Example 20.4. The feed flow rate and composition to the distillation tower in Example 20.2 change over ranges of 8 to 12 kmole/min and mole fraction 0.4 to 0.6, respectively. Also, the vapor condensed in the condenser cannot be greater than 15.0 kmole/min. Determine the range of disturbances for which the product qualities can be maintained at 0.98 and 0.02 mole fraction.

The method for calculating the operating window for this example depends on the equation-solving methods available. A trial-and-error method could be used to specify the disturbances and simulate the tower with X_D and X_B at their set points. This trial-and-error procedure, involving many simulations, would be executed until the disturbance value that resulted in the maximum overhead vapor flow was found. A direct method of solving this problem would be to specify X_D and X_B and calculate the feed composition X_F that resulted in the overhead vapor flow meeting its maximum limit; this approach is possible with a steady-state model solved using an equation-based approach (Perkins, 1984). The results of the analysis, performed by either method, are

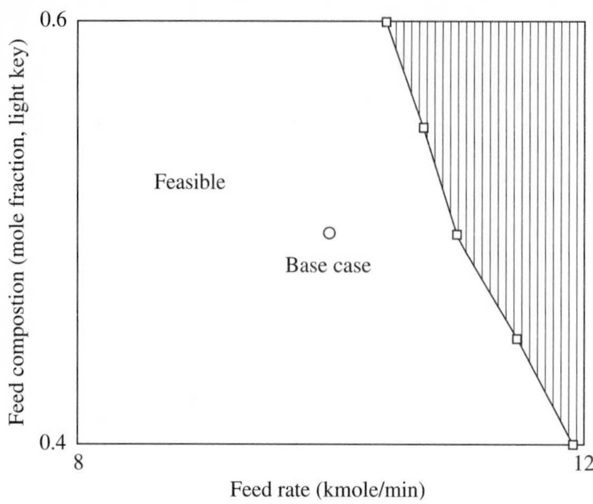

FIGURE 20.6
Operating window for distillation with disturbance variables as coordinates.

the feasible values of feed rate and composition, with X_D and X_B maintained at their desired values; the operating window is given in Figure 20.6. Again, the interaction is apparent by the shape of the operating window. The maximum feed rate is attainable with a feed containing the least light key, because the least amount of distillate product is generated by this feed and the least distillate requires the minimum overhead vapor.

Controllability

Another important issue in multivariable control is the independence of the input-output process relationships between selected manipulated variables (MV_j's) and controlled variables (CV_i's); a process in which the relationships are independent is termed *controllable*. Many definitions for the term *controllability* are used in automatic control (e.g., Franklin et al., 1990); for the purposes of this book we will use the following definition (a somewhat less restrictive version of Rosenbrock's (1974) "functionally controllable (f)"):

> A system is **controllable** if the controlled variables can be maintained at their set points, in the steady state, in spite of disturbances entering the system.

Controllability is defined for a selected set of manipulated and controlled variables, and a system may be controllable for one selection and uncontrollable for another selection. A system's controllability is not always easy to determine by observation; thus, a quantitative method for determining controllability is presented in this section. There is no general method for nonlinear systems; therefore, the controllability of the locally linearized system will be analyzed to evaluate the system. As a result, the results of the controllability test are strictly valid only at the operating point at which the linear model is evaluated.

The multivariable dynamic system can be described by a model of the form given in equation (20.10); only a 2×2 system is given, but the extension to higher orders is straightforward. We will assume that the system begins at steady state. The definition of controllability will be met if the controlled variables can be maintained at their set points, so that their deviation variables are zero, by adjusting the specified manipulated variables in the presence of steplike disturbances, which achieve a constant value, at least asymptotically. The behavior of the system at steady state can be determined through the final value theorem. As noted in Chapter 4, the final value theorem can be applied if the output is bounded, which excludes bounded input–bounded output unstable systems. Applying the final value theorem to equation (20.10), with $\text{CV}_i(s) = 0$ for all i, gives

$$\begin{bmatrix} 0 \\ 0 \end{bmatrix} = \begin{bmatrix} K_{11} & K_{12} \\ K_{21} & K_{22} \end{bmatrix} \begin{bmatrix} \text{MV}_1' \\ \text{MV}_2' \end{bmatrix} + \begin{bmatrix} K_{d1} \\ K_{d2} \end{bmatrix} D' \tag{20.11}$$

with $K_{ij} = \lim_{s \to 0} G_{ij}(s)$

The system is controllable if there is a solution for this set of linear algebraic equations for arbitrary nonzero values of K_{d1}, K_{d2}, and D' (i.e., all possible input variables as disturbances).

> A solution exists for a square system of linear equations (20.11) when an inverse to the matrix of feedback process gains (**K**) exists; thus, the system is controllable if the determinant of the gain matrix is nonzero.

A square physical system (numbers of manipulated and controlled variables are constant) is not controllable if any of the following conditions occurs:

1. Any two process inputs are linearly dependent (giving dependent columns).
2. Any two process outputs are linearly dependent (giving dependent rows).
3. A process output is not influenced by any input (giving a column of zeros).
4. A process input does not influence any output (giving a row of zeros).

The controllability test is applied to the two processes in the following example to ensure that they are controllable, and it is extended and applied in Part VI on control design.

Example 20.5. Evaluate the controllability of the blending and distillation processes. The gain matrices and their determinants are

Blending $\begin{bmatrix} \dfrac{-F_A}{(F_S + F_A)^2} & \dfrac{F_S}{(F_S + F_A)^2} \\ 1.0 & 1.0 \end{bmatrix}$ Determinant: $\dfrac{-F_S + F_A}{(F_S + F_A)^2} \neq 0.0$

Distillation $\begin{bmatrix} 0.0747 & -0.0667 \\ 0.1173 & -0.1253 \end{bmatrix}$ Determinant: $-0.001536 \neq 0.0$

Since each determinant is nonzero, each process is controllable for the selected manipulated and controlled variables.

Note that a controllable system indicates that the manipulated variables can compensate for effects of disturbances on selected controlled variables for some small region over which the linearization is valid and constraints are not encountered in the manipulated variables. In contrast, the operating window, which is evaluated using the nonlinear steady-state models including constraints, defines the entire possible region of operation. Both analyses should be made to ensure the possibility of multivariable control.

Finally, the controllability and range of the system are affected by the process design and operating conditions, along with the selected controlled variables. Therefore, deficiencies in controllability and range must be compensated through changes to the equipment or process operating point, not control algorithms.

20.4 PROCESS INTERACTION: IMPORTANT EFFECTS ON MULTIVARIABLE SYSTEM BEHAVIOR

We now continue investigating the effects of interaction on multivariable system behavior, assuming that the process has a controllable input-output selection. The goal of this section is to demonstrate how the responses of a control system are influenced by interaction. To simplify the analysis, only relationships for two-input, two-output systems are considered, but the results obtained can be extended to control systems of higher order. Insights will be provided in this section through analyzing several examples and are formalized in the next section.

The first step is to derive the transfer function for the multiloop feedback control system and determine the main differences from single-loop control. We begin this procedure by considering the same system (1) without control, (2) with one controller, and finally (3) with two controllers. First, suppose that a single controller were to be implemented on the system in Figure 20.4, with the goal of controlling $\text{CV}_1(s)$ by adjusting $\text{MV}_1(s)$. The transfer function $G_{11}(s)$ would have to be considered when tuning the controller, as demonstrated by the transfer function:

$$\frac{\text{CV}_1(s)}{\text{MV}_1(s)} = G_{11}(s) \qquad \text{no control} \tag{20.12}$$

In this case, the control loop could be considered a single-loop system; however, changes in $\text{MV}_1(s)$ caused by the controller would affect $\text{CV}_2(s)$ because of interaction.

Next, we consider a more complex structure to determine whether it affects the first loop. The block diagram for a multivariable process with one single-loop controller is given in Figure 20.7. This example is considered to demonstrate the effects of interaction on closed-loop systems. The transfer function relating $\text{MV}_1(s)$ with $\text{CV}_1(s)$ would have to be considered when tuning the controller using these measured and manipulated variables. This transfer function follows for the case with $G_{c2}(s)$ implemented:

$$\frac{\text{CV}_1(s)}{\text{MV}_1(s)} = G_{11}(s) - \frac{G_{12}(s)G_{21}(s)G_{c2}(s)}{1 + G_{c2}(s)G_{22}(s)} \qquad G_{c2}(s) \text{ implemented} \tag{20.13}$$

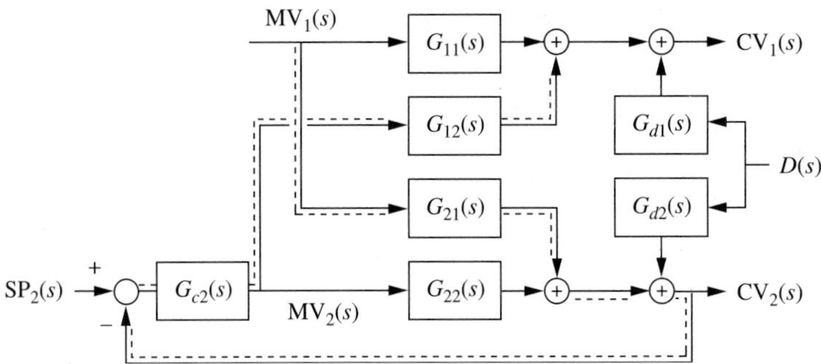

FIGURE 20.7
Block diagram of 2 × 2 system with one single-loop controller.

This equation differs from the transfer function with no control of $CV_2(s)$, equation (20.12), by the second term, and the path represented by the second term is shown as a dashed line in Figure 20.7. Clearly, this path results from the process interaction and the second controller. The second term on the right-hand side in equation (20.13) would be zero if either or both $G_{12}(s)$ and $G_{21}(s)$ were zero, in which case the controller $G_{c2}(s)$ would have no effect on the transfer function for $CV_1(s)/MV_1(s)$. The path shown with the dashed line will be referred to as *transmission* interaction and will be seen to have an important influence on stability.

> **Transmission interaction** exists when a change in the set point of a controller affects its controlled variable through a path that includes another controlled variable and controller.

Note that it is possible to have process interaction (i.e., only $G_{12}(s)$ *or* $G_{21}(s)$ nonzero) without having transmission interaction, which requires *both* to be nonzero.

The control design can be completed by applying two single-loop controllers to the process, as shown in Figure 20.8. The following closed-loop transfer functions can be determined from block diagram manipulation. (The results for the other controlled variable, $CV_2(s)$, can be obtained by transposing the subscripts.)

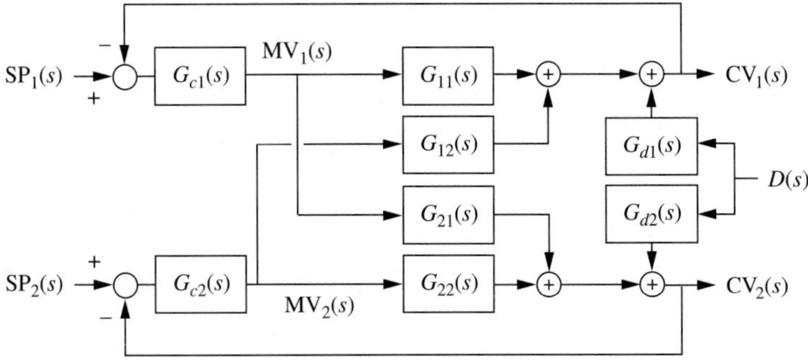

FIGURE 20.8
Block diagram of 2 × 2 system with two single-loop controllers.

$$\frac{\mathrm{CV}_1(s)}{\mathrm{SP}_1(s)} = \frac{G_{c1}(s)G_{11}(s) + G_{c1}(s)G_{c2}\left[G_{11}(s)G_{22}(s) - G_{12}(s)G_{12}(s)\right]}{\mathrm{CE}(s)} \tag{20.14}$$

$$\frac{\mathrm{CV}_1(s)}{\mathrm{SP}_2(s)} = \frac{G_{c2}(s)G_{12}(s)}{\mathrm{CE}(s)} \tag{20.15}$$

$$\frac{\mathrm{CV}_1(s)}{D(s)} = \frac{\left(G_{d1}(s) - \dfrac{G_{d2}G_{21}(s)G_{c2}(s)}{[1 + G_{c2}(s)G_{22}(s)]}\right)[1 + G_{c2}(s)G_{22}(s)]}{\mathrm{CE}(s)} \tag{20.16}$$

with the characteristic expression CE(s), which is the same for equations (20.14) through (20.16),

$$\begin{aligned}\mathrm{CE}(s) &= 1 + G_{c1}(s)G_{11}(s) + G_{c2}(s)G_{22}(s)\\ &\quad + G_{c1}(s)G_{c2}(s)[G_{11}(s)G_{22}(s) - G_{12}(s)G_{21}(s)]\end{aligned} \tag{20.17}$$

When both interaction terms $G_{12}(s)$ and $G_{21}(s)$ are nonzero, the dynamic response of a single-loop controller between $\mathrm{CV}_1(s)$ and $\mathrm{MV}_1(s)$ depends on all terms in the closed-loop transfer function. As a result, the stability and performance of loop 1 depend on the tuning of loop 2. By a similar argument, the stability and performance of loop 2 depend on the tuning of loop 1. Therefore, the two controllers must be tuned *simultaneously* to achieved desired stability and performance.

Further insight can be obtained by considering the steady-state behavior of the multivariable system. In particular, the necessary adjustments in the manipulated variables can be used as an indication of how interaction changes the system's behavior. The general steady-state relationship for a 2×2 system is expressed here in deviation variables:

$$\mathrm{CV}_1' = K_{11}\mathrm{MV}_1' + K_{12}\mathrm{MV}_2' \tag{20.18}$$

$$\mathrm{CV}_2' = K_{21}\mathrm{MV}_1' + K_{22}\mathrm{MV}_2' \tag{20.19}$$

These equations are often written in matrix form as

$$\begin{bmatrix}\mathrm{CV}_1'\\ \mathrm{CV}_2'\end{bmatrix} = \mathbf{K}\begin{bmatrix}\mathrm{MV}_1'\\ \mathrm{MV}_2'\end{bmatrix} \quad \text{with} \quad \mathbf{K} = \begin{bmatrix}K_{11}K_{12}\\ K_{21}K_{22}\end{bmatrix} \tag{20.20}$$

Equation (20.20) can be rearranged to give

$$\begin{bmatrix}\mathrm{MV}_1'\\ \mathrm{MV}_2'\end{bmatrix} = \mathbf{K}^{-1}\begin{bmatrix}\mathrm{CV}_1'\\ \mathrm{CV}_2'\end{bmatrix} \tag{20.21}$$

where $\mathbf{K}^{-1}$ is the inverse of the steady-state gain matrix and exists for a controllable system. Note that equation (20.21) represents the calculation performed by the controller with zero steady-state offset. For example, equation (20.21) could be used to determine the steady-state changes in MV_1' and MV_2' for any specified changes in CV_1' and CV_2' (i.e., set point changes). Several hypothetical systems are considered first so that the extent of interaction can be changed incrementally from the base model; then some realistic processes are considered.

The process gain matrices in Table 20.1 represent hypothetical systems with various extents of interaction: A has no interaction ($K_{12} = K_{21} = 0$); B has moderate transmission interaction; C has strong transmission interaction; D is not

TABLE 20.1
Summary of manipulated-variable changes for example systems with differing amounts of interaction

System	Process gain matrix K	Inverse gain matrix K^{-1}	Changes for case I $CV_1' = 1.0$ $CV_2' = 1.0$ No model error	case II $CV_1' = 1.0$ $CV_2' = 0.0$ No model error	case III $CV_1' = 1.0$ $CV_2' = 0.0$ Slight model difference $K_{11} = K_{22} = 0.98$ with K_{12}, K_{21} unchanged
A No interaction	$\begin{bmatrix} 1.0 & 0.0 \\ 0.0 & 1.0 \end{bmatrix}$	$\begin{bmatrix} 1.0 & 0.0 \\ 0.0 & 1.0 \end{bmatrix}$	$MV_1' = 1.0$ $MV_2' = 1.0$ Same as single-loop	$MV_1' = 1.0$ $MV_2' = 0.0$ Same as single-loop	$MV_1' = 1.02$ $MV_2' = 0.0$ Same sensitivity as single-loop
B Moderate transmission interaction	$\begin{bmatrix} 1.0 & 0.75 \\ 0.75 & 1.0 \end{bmatrix}$	$\begin{bmatrix} 2.29 & -1.71 \\ -1.71 & 2.29 \end{bmatrix}$	$MV_1' = 0.56$ $MV_2' = 0.56$ Less than single-loop	$MV_1' = 2.29$ $MV_2' = -1.71$ Larger than single-loop	$MV_1' = 2.46$ $MV_2' = -1.88$ Some increase in sensitivity
C Strong transmission interaction	$\begin{bmatrix} 1.0 & 0.90 \\ 0.90 & 1.0 \end{bmatrix}$	$\begin{bmatrix} 5.26 & -4.74 \\ -4.74 & 5.26 \end{bmatrix}$	$MV_1' = 0.52$ $MV_2' = 0.52$ Less than single-loop	$MV_1' = 5.26$ $MV_2' = -4.74$ Much larger than single-loop	$MV_1' = 6.52$ $MV_2' = -5.98$ Large increase in sensitivity
D Not controllable	$\begin{bmatrix} 1.0 & 1.0 \\ 1.0 & 1.0 \end{bmatrix}$	Singular; inverse does not exist			
E One-way interaction	$\begin{bmatrix} 1.0 & 1.0 \\ 0.0 & 1.0 \end{bmatrix}$	$\begin{bmatrix} 1.0 & -1.0 \\ 0.0 & 1.0 \end{bmatrix}$	$MV_1' = 0.0$ $MV_2' = 1.0$ Different from single-loop	$MV_1' = 1.0$ $MV_2' = 0.0$ Same as single-loop	$MV_1' = 1.02$ $MV_2' = 0.0$ Same sensitivity as single-loop

controllable (the determinant of the gain matrix is zero), and E has one-way interaction ($K_{21} = 0$). Thus, their behaviors are expected to vary. In particular, multivariable control is not possible with system D, because it is not possible to control CV_1' and CV_2' independently. This system is not considered further, because a process design change would be required to control the selected controlled variables. (Chapter 24 presents an example of an initial design selection that is noncontrollable.)

The responses of the manipulated variables are evaluated using the base model for two cases that differ only in the values of the set point changes. For comparison, note that introducing a unity change in a set point of a single-variable control system with a process gain of 1.0 would cause a steady-state manipulated-variable change of magnitude 1.0. For case I the changes in the manipulated variables are smaller for the interacting systems B and C than for system A, which represents independent single-loop processes. For case II the changes in the manipulated variables are greater for the interacting systems B and C than for the single-loop system A. Three trends are apparent from these cases:

1. The difference from single-loop behavior increases as the interaction, as indicated by the magnitude of the off-diagonal gains, increases.
2. Whether the manipulated variables move a greater or lesser amount in multivariable control compared with single-loop depends on the *direction* of the set point changes. Note that this behavior is different from a single-loop case, in which only the input *magnitude* is important.
3. System E, with one-way interaction, deviates less from single-loop behavior than the systems with moderate to strong transmission interaction.

The sensitivity of the example results to model changes is determined by comparing cases II and III, which differ only by 2% in the gains of the off-diagonals. For system A with no interaction and system E with one-way interaction, the difference in the manipulated-variable changes is 2%; thus, the model change is reflected by a change of the manipulated variables of the same relative magnitude. However, the manipulated-variable changes are much greater for systems B and C, with the difference about 20% in system C. We conclude that the proper manipulated-variable adjustments can be very sensitive to the model for systems with strong transmission interaction. This result has great significance for controllers that use a model to calculate the manipulated-variable adjustments; here is a case where the controller may be extremely sensitive to modelling errors—an undesirable situation.

The major differences between single-loop and interactive steady-state multiloop systems are summarized as follows:

1. The values of the manipulated variables that satisfy the desired controlled variables must be determined simultaneously.
2. Differences between single-variable and multivariable behavior increase as the transmission (two-way) interaction increases.
3. The sensitivity of the adjustments in manipulated variables to model changes can be much greater in multiloop systems than in single-loop systems.

Before we conclude this section, two examples of process gain matrices are considered. These examples demonstrate that the behavior shown in Table 20.1 occurs in realistic chemical processes.

Example 20.6. The first is the blending system shown in Figure 20.2, where the product flow and composition are controlled by adjusting the flows of the two component streams. The gains are determined from the linearized model in equations (20.6) and (20.7). The base conditions are taken to be

$$\begin{aligned} F_1 &= 95.0 \text{ kg/min} \qquad F_2 = 5.0 \text{ kg/min} \\ A_1 &= 0.05 \text{ wt fraction A}, \qquad F_3 = 100 \text{ kg/min} \end{aligned} \tag{20.22}$$

The gain matrix and its inverse for these conditions are

$$\begin{bmatrix} A_1 \\ F_3 \end{bmatrix} = \begin{bmatrix} -0.0005 & 0.0095 \\ 1.0 & 1.0 \end{bmatrix} \begin{bmatrix} F_1 \\ F_2 \end{bmatrix} \qquad \mathbf{K}^{-1} = \begin{bmatrix} -100 & 0.95 \\ 100 & 0.05 \end{bmatrix} \tag{20.23}$$

The gain and inverse matrices have one element that is nearly zero. Thus, the system is likely to behave similar to system E in Table 20.1. As a result, this system is not expected to experience very strong departures from the single-variable behavior in manipulated-variable adjustment magnitudes or to have extreme sensitivity to model errors.

Example 20.7. The second example is the binary distillation tower in Figure 20.3, where the product compositions are controlled by adjusting the reflux and reboiler flows. The steady-state gains can be taken from the transfer function matrix in equations (20.8) and (20.9).

$$\mathbf{K} = \begin{bmatrix} 0.0747 & -0.0667 \\ 0.1173 & -0.1253 \end{bmatrix} \qquad \mathbf{K}^{-1} = \begin{bmatrix} 81.58 & -43.42 \\ 76.36 & -48.63 \end{bmatrix} \tag{20.24}$$

The distillation tower appears highly interactive in the two-way manner similar to systems B and C. To complete this distillation example, steady-state changes in manipulated variables are calculated for single-loop and multivariable control. In both cases, the bottoms mole fraction of light key is to be decreased by 0.01. In the first case, only the bottoms mole fraction is specified and the distillate mole fraction is not controlled. This is single-loop control, and the necessary change in vaporization is

$$\text{Single-loop} \qquad \Delta F_V = \frac{\Delta X_B}{K_{XB,v}} = \frac{-0.01}{-0.1253} = 0.0798 \text{ kmole/min}$$

Since the bottoms composition is not controlled, $\Delta F_R = 0$ and $\Delta X_D \neq 0$. In the alternative multivariable case, the distillate mole fraction is maintained unchanged ($\Delta X_D = 0$), while the bottoms composition is changed by -0.01.

$$\text{Multivariable} \qquad \begin{bmatrix} \Delta F_R \\ \Delta F_V \end{bmatrix} = \begin{bmatrix} 81.53 & -43.42 \\ 76.36 & -48.63 \end{bmatrix} \begin{bmatrix} 0 \\ -0.01 \end{bmatrix} \begin{bmatrix} 0.4343 \\ 0.4863 \end{bmatrix}$$

The results demonstrate that the change in the vaporization in the reboiler was much larger in magnitude for the multivariable system (0.4863 compared with 0.0798 kmole/min), and in addition, a large change in reflux was required. Clearly, the interaction has strongly affected the steady-state behavior of the system.

In conclusion, interaction can strongly influence the steady-state and dynamic behavior of multivariable systems. There exists a range of interaction from completely independent through nearly dependent (i.e., nearly singular), with this interaction dependent on the *process characteristics*, not on control. In general, the closer the system approaches singularity (system D in Table 20.1), the more its behavior differs from that of independent loops. The final two process examples demonstrated that real processes can have interaction similar to the range of examples in Table 20.1. In the next section, a quantitative measure of interaction is introduced.

20.5 PROCESS INTERACTION: THE RELATIVE GAIN ARRAY (RGA)

As shown in the previous section, process interaction is an important factor influencing the behavior of multivariable systems. A quantitative measure of interaction is needed to proceed with a multiloop analysis method, and the relative gain array, which has proved useful in control system analysis, is introduced to meet this need. The relative gain array was developed by Bristol (1966) and extended by many engineers, most notably Shinskey (1988) and McAvoy (1983*b*). In this section, the relative gain is defined, special properties and methods for calculation are given, and interpretations for control analysis are presented. The relative gain array (RGA) is a matrix composed of elements defined as ratios of open-loop to closed-loop gains as expressed by the following equation, which relates the *j*th input and the *i*th output.

$$\lambda_{ij} = \frac{\left(\dfrac{\partial \mathrm{CV}_i}{\partial \mathrm{MV}_j}\right)_{\mathrm{MV}_k=\mathrm{const},\, k\neq j}}{\left(\dfrac{\partial \mathrm{CV}_i}{\partial \mathrm{MV}_j}\right)_{\mathrm{CV}_k=\mathrm{const},\, k\neq i}} = \frac{\left(\dfrac{\partial \mathrm{CV}_i}{\partial \mathrm{MV}_j}\right)_{\text{other loops open}}}{\left(\dfrac{\partial \mathrm{CV}_i}{\partial \mathrm{MV}_j}\right)_{\text{other loops closed}}} \tag{20.25}$$

Consistent with prior terminology, the open-loop gain (K_{ij}) is the change in output i for a change in input j with all other inputs constant (for stable processes). By *closed-loop gain* we mean the steady-state relationship between MV'_j and CV'_i with all other control loops closed (i.e., in automatic). In this definition, it is assumed that the controllers have an integral mode so that the steady-state values of the controlled variables are maintained constant (i.e., $\mathrm{CV}'_k = 0$ for those under feedback control). If the relative gain is 1.0, the process gain is unaffected by the other control loops and no (transmission) interaction exists. Thus, the amount that the relative gain deviates from 1.0 indicates, in some sense, the "extent" of transmission interaction in a quantitative manner.

Before control-relevant interpretations of the relative gain are developed, some important properties must be noted:

1. The relative gain is scale-independent. This is important because rules for interpretation do not change when the units of a variable change (e.g., from percent to parts per million).
2. The expression in equation (20.25) suggests that both open- and closed-loop data is required to determine the relative gain. However, the relative gain can be calculated from the open-loop data alone, which can be demonstrated by rearranging equation (20.25) to give

$$\lambda_{ij} = \left(\frac{\partial \mathrm{CV}_i}{\partial \mathrm{MV}_j}\right)_{\mathrm{MV}_k=\mathrm{const},\, k\neq j} \left(\frac{\partial \mathrm{MV}_j}{\partial \mathrm{CV}_i}\right)_{\mathrm{CV}_k=\mathrm{const},\, k\neq i} \tag{20.26}$$

The procedure for calculating the relative gain array is to evaluate the open-loop gain matrix $\mathbf{K}$; calculate its inverse transposed $(\mathbf{K}^{-1})^{\mathrm{T}}$; and multiply them in an *element-by-element manner*. This type of matrix multiplication is referred to as the *Hadamard product* (McAvoy, 1983*b*). The following expression gives the result for each element in the relative gain array with K_{ij} being the elements in the gain matrix and KI_{ij} being the elements of the inverse of the gain matrix,

$$\lambda_{ij} = K_{ij}\mathrm{KI}_{ji} \tag{20.27}$$

For a 2×2 system, the (1,1) element of the relative gain array can be shown to be

$$\lambda_{11} = \frac{1}{1.0 - \dfrac{K_{12}K_{21}}{K_{11}K_{22}}} \tag{20.28}$$

3. The rows and columns of the relative gain array sum to 1.0. This property enables 2×2 systems to be characterized by the λ_{11} element, as follows.

$$\begin{array}{c|cc} & \mathrm{MV}_1 & \mathrm{MV}_2 \\ \hline \mathrm{CV}_1 & \lambda_{11} & 1-\lambda_{11} \\ \mathrm{CV}_2 & 1-\lambda_{11} & \lambda_{11} \end{array} \tag{20.29}$$

4. The relative gain calculation can be very sensitive to errors in the gain calculation. As an example, consider the following relative gain for a 2×2 process, and assume that each process gain can be in error by a factor ϵ_{ij}, which is 1.0 for no error.

$$\lambda_{11} = \frac{1}{1.0 - \dfrac{(K_{12}\epsilon_{12})(K_{21}\epsilon_{21})}{(K_{11}\epsilon_{11})(K_{22}\epsilon_{22})}} \tag{20.30}$$

When the relative gain element has a large magnitude, the relative gain can take widely varying values and can even change sign for small errors in individual process gains, as shown by the following example cases. In this example, the actual values for the gains are $K_{11} = K_{22} = 1.0$ and $K_{12} = K_{21} = 0.949$, and the erroneous relative gain is shown for a few example sets of gain errors.

True λ_{11}	ϵ_{11}	ϵ_{12}	ϵ_{21}	ϵ_{22}	**λ_{11} calculated with model errors**	
10	1.0	1.0	1.0	1.0	10.0	no error
10	1.0	1.1	1.0	1.0	100.0	
10	1.0	1.2	1.0	1.0	−16.6	
10	0.97	1.03	1.03	0.97	−7.8	only 3% errors

Since the sign of the relative gain is of great importance in control design decisions, the sensitivity to model errors demonstrated as the foregoing property 4 must be considered, to prevent incorrect results. Thus, great accuracy is required in the process gains used for calculating the relative gain. Probably the best method is to derive an analytical model and evaluate the process gains from analytical derivatives. This can be done for the blending example using the linearized model and the foregoing property 2:

$$[\lambda_{ij}] = \begin{bmatrix} \dfrac{F_1}{F_1+F_2} & \dfrac{F_2}{F_1+F_2} \\ \dfrac{F_2}{F_1+F_2} & \dfrac{F_1}{F_1+F_2} \end{bmatrix} \tag{20.31}$$

However, few complex industrial processes can be accurately modelled by sets of equations small enough to be conveniently manipulated analytically by hand, although advances in algebraic processing by computers could change this situation in the future. Thus, numerical differentiation using steady-state process simulators is a common approach to evaluating process gains. In this procedure, a separate simulation is performed at the base case and at a case with each input MV_j changed a small amount from the base case. The process gains are calculated using the equation below, and the relative gain array is determined from equation (20.27):

$$K_{ij} \approx \frac{CV_i(MV_1, MV_2, \ldots, MV_j + \Delta MV_j, \ldots) - CV_i(MV_1, MV_2, \ldots)}{\Delta MV_j} \qquad (20.32)$$

Special care is required when using this method because of the accuracy required for the relative gain. When numerical differentiation is used, two potential causes of errors are introduced: the convergence tolerances in solving the equations and the use of approximate rather than exact derivatives. As demonstrated by McAvoy (1983*b*), the convergence tolerances and ΔMVs used in equation (20.33) in calculating the approximate gain must be reduced until the estimated process gains are not significantly influenced by further reductions in their values. The conclusion of this analysis can be stated as follows:

> The gains K_{ij} used for calculating the relative gains must be accurate; the use of gains from linearized fundamental models is recommended. Given the typical errors in empirical model identification, the use of empirically determined process gains using methods in Chapter 6 is *not recommended* for calculating relative gains.

Some very useful control-related interpretations based on the RGA are summarized as follows and will be used in a hierarchical analysis procedure in the next chapter.

$\lambda_{ij} < 0$ In this case, the open- and closed-loop process gains are of different signs. In a 2 × 2 process, if the single-loop ($CV_i - MV_j$) controller gain were positive for stable feedback control, the same controller gain would have to be negative for stable multiloop feedback control. Thus, the sign of the controller gain to retain stability would depend on the mode of other controllers in the multiloop system—not a desirable situation.

$\lambda_{ij} = 0$ One situation in which the relative gain is zero occurs when the open-loop process gain ($\Delta CV_i/\Delta MV_j$ with the other loops open) is zero, which indicates no steady-state relationship between the input and output variables. Thus, the controller with this pairing can function, if at all, only when other controllers are in automatic. Again, this is not generally a desirable situation but is acceptable in special circumstances, as explained in the next chapter.

$0 < \lambda_{ij} < 1$ From equation (20.25), the steady-state loop process gain with the other loops closed (e.g., equation (20.13)) is larger than the same process gain with the other loops open.

$\lambda_{ij} = 1$ In this situation, there is no transmission interaction, in the sense that the *product* of $K_{12}K_{21}$ is zero, but either one of the terms may be nonzero. Thus, a change in $MV_j(s)$ is transmitted to $CV_i(s)$ only through $G_{ij}(s)$. Note that this does not preclude the possibility that the manipulated variable might affect another controlled variable (i.e., one-way interaction).

$\lambda_{ij} > 1$ From equation (20.25), the steady-state loop process gain with the other loops closed (e.g., equation (20.13)) is smaller than the same process gain with the other loops open.

$\lambda_{ij} = \infty$ When the process gain is zero with the other loops closed, it is not possible to control the variable in a multiloop system.

As examples, the relative gains for all cases in Table 20.1 and the two process examples are reported here. (Note that the model for the distillation tower was developed from very small perturbations in the nonlinear model without noise; the probability of obtaining an accurate relative gain value from empirical model fitting is quite small.)

System	Relative gain, λ_{11}	
A	1.0	
B	2.29	
C	5.26	
D	∞	
E	1.0	
Blending	$\lambda_{A1-F2} = 0.95$	operating conditions in equation (20.22)
Distillation	$\lambda_{XD-FR} = 6.09$	operating conditions in Table 5.4

These values are consistent with the previous, qualitative evaluations of interaction in that systems with relative gain deviating most from 1.0 deviate most from single-loop behavior. Note that system E with only one-way interaction has $\lambda_{11} = 1.0$; in general, the relative gain array is the identity matrix for systems with a steady-state gain matrix that is lower (or upper) diagonal (i.e., with nonzero entries only on and below (or above) the diagonal).

Finally, the relative gain can be related directly to the closed-loop transfer function of a 2×2 system. To do this, the definition of relative gain has been extended by Witcher and McAvoy (1977) to include frequency-dependent terms by replacing the steady-state gains with the corresponding transfer functions. Thus, the frequency-dependent relative gain is

$$\lambda_{11}(s) = \frac{1}{1 - \dfrac{G_{12}(s)G_{21}(s)}{G_{11}(s)G_{22}(s)}} \tag{20.33}$$

This expression can be used, by setting $s = j\omega$, to evaluate the magnitude of the relative gain elements at various frequencies.

Using the foregoing expression, the characteristic expression (20.17) can be rewritten as

$$\text{CE}(s) = 1 + G_{c1}(s)G_{11}(s) + G_{c2}(s)G_{22}(s) + \frac{G_{c1}(s)G_{c2}(s)G_{11}(s)G_{22}(s)}{\lambda_{11}(s)} \quad (20.34)$$

This analysis demonstrates the fundamental nature of the relative gain and the close relationship between the relative gain and system stability for 2 × 2 systems.

A summary of the key results for the relative gain array follows:

1. The deviation from single-loop behavior, specifically the transmission interaction, is related to the difference of the relative gain element from the value of 1.0.
2. The condition of $\lambda_{ij} \leq 0$ results in multiloop systems that, to maintain acceptable performance, must alter the $(\text{CV}_i - \text{MV}_j)$ controller gain or automatic status, depending on the status of other controllers.
3. A direct relationship between frequency-dependent relative gain and control system stability has been demonstrated for 2 × 2 systems.

20.6 EFFECT OF INTERACTION ON STABILITY AND TUNING OF MULTILOOP CONTROL SYSTEMS

The final topic that needs to be covered before analyzing multiloop performance is controller tuning. Analysis of the closed-loop transfer function demonstrates that interaction influences the characteristic equation and, therefore, stability; thus, controller tuning must consider interaction as well as the single-loop feedback process dynamics. The following example provides further insight into the effect of interaction on stability and tuning.

Example 20.8. A dynamic system with the following model is to be controlled by two PI controllers. The input-output pairings are 1–1 and 2–2 as shown in Figure 20.8. Determine the allowable range of tuning constants that yield a stable system.

$$\begin{bmatrix} \text{CV}_1(s) \\ \text{CV}_2(s) \end{bmatrix} = \begin{bmatrix} \dfrac{1.0e^{1.0s}}{1+2s} & \dfrac{0.75e^{-1.0s}}{1+2s} \\ \dfrac{0.75e^{-1.0s}}{1+2s} & \dfrac{1.0e^{-1.0s}}{1+2s} \end{bmatrix} \begin{bmatrix} \text{MV}_1(s) \\ \text{MV}_2(s) \end{bmatrix} \quad (20.35)$$

The example system has transmission interaction, because both off-diagonal elements are nonzero; thus, it would not be correct to tune each controller independently. The stability limit is determined by the characteristic expression, given in equation (20.17). Finding the limiting values of the tuning constants would be an arduous task because all four controller tuning constants (K_{c1}, T_{I1}, K_{c2}, and T_{I2}) appear in the characteristic equation and, therefore, all affect stability *simultaneously*. To simplify the calculations and allow graphical presentation of the results, the integral times of the controllers will be held constant at 3.0 min, which are reasonable values, being the sum of the dead time and time constant of each transfer function. Note that this selection will not necessarily yield the best performance, but it is a reasonable choice for this example calculation.

With the integral times fixed, the characteristic equation has two remaining tuning parameters, the controller gains.

$$\text{CE}(s) = 1 + G_{\text{OL}}(s) \tag{20.36}$$

with

$$G_{\text{OL}}(s) = K_{c1}\left(1 + \frac{1}{3s}\right)\left(\frac{1.0e^{-1.0s}}{1+2s}\right) + K_{c2}\left(1 + \frac{1}{3s}\right)\left(\frac{1.0e^{-1.0s}}{1+2s}\right)$$
$$+ K_{c1}\left(1 + \frac{1}{3s}\right)K_{c2}\left(1 + \frac{1}{3s}\right)\left(\frac{1.0e^{-1.0s}}{1+2s}\frac{1.0e^{-1.0s}}{1+2s} - \frac{0.75e^{-1.0s}}{1+2s}\frac{0.75e^{-1.0s}}{1+2s}\right)$$

To calculate the stability region, one gain (e.g., K_{c2}) was given a value, and the Bode stability analysis was performed to determine the ultimate value of K_{c1} that defines the stability limit. These calculations involve extensive manipulations of complex numbers and were therefore performed using a computer program. The results of the calculations are displayed in Figure 20.9. If there had been no interaction, the stability region would have encompassed the entire box defined by values of the controller gains of (0,0) and (3.76,3.76) shown in the figure, because the tuning of one controller would not have influenced the tuning of the other. As can be seen, the interactions in this example *reduced* the allowable values for the controller gains.

Example 20.9. Although all tuning within the defined region yields a stable system, the control performance is different for various tunings chosen from within the stable region. To investigate by example the effect of tuning on performance, three sets of tuning constants were chosen for the system in Example 20.8 from within the stable area shown in Figure 20.9. The tuning was selected to have a reasonable gain margin (i.e., margin from the stability boundary). The simulation results for multiloop PI

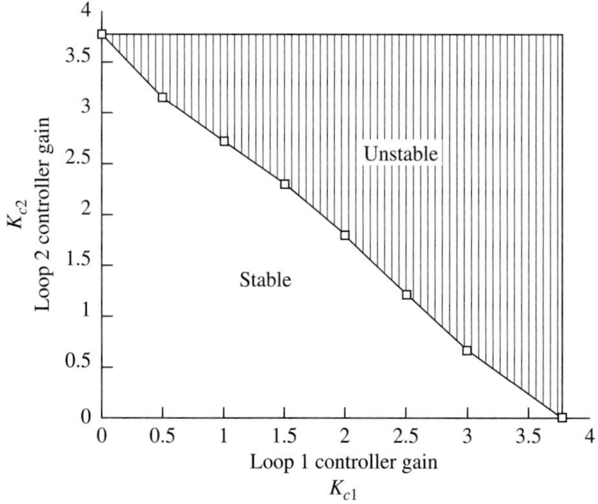

FIGURE 20.9
Map of stable and unstable controller gain regions for Example 20.8 with $T_{I1} = T_{I2} = 3.0$.

TABLE 20.2
Summary of example tuning for 2 × 2 system

Case	K_{c1}	T_{I1}	K_{c2}	T_{I2}	IAE_1	IAE_2	$IAE_1 + IAE_2$
Figure 20.10*a*	0.95	3.0	0.95	3.0	7.22	5.41	12.63
Figure 20.10*b*	1.40	3.0	0.50	3.0	4.90	10.3	15.2
Figure 20.10*c*	0.50	3.0	1.40	3.0	13.7	3.67	17.37
Figure 20.10*d*	1.23	1.76	0.89	1.06	3.46	2.46	5.92

controllers responding to a CV_1 set point change of 1.0 for three different tuning constants are given in Figures 20.10*a* through *c* and tabulated in Table 20.2. Figure 20.10*a* gives equal weight to both controlled variables. Figure 20.10*b* gives more importance to controlled variable 1, whereas Figure 20.10*c* gives more importance to controlled variable 2. These results demonstrate that controller tuning influences multiloop system performance, so tuning can be used as a method for adapting system performance to conform to specific priorities in the importance of controlled variables. This result will be exploited in the next chapter.

Since tuning influences performance, the engineer should be able to use this flexibility to obtain good control system performance. Three approaches are typically used for tuning multiloop systems, and each is described here.

Trial and Error

Although potentially tedious, a trial-and-error method is often used in practice. Initial tuning constant values are typically the single-loop values altered for stability, perhaps with the gains reduced by a factor of 2 or more. These initial values are adjusted through fine-tuning, as described in Chapter 9, with trials performed on a simulation or directly on the process. The final tuning must be conservative (i.e., not too close to the stability margin) to account for changes in process operating conditions that would occur after the trial-and-error procedure has been completed. Naturally, the success of this approach depends on the expertise of the engineer, but the approach can reach reasonable results quickly when transmission interaction is not too strong.

Optimization

An optimization approach, similar to the approach described in Chapter 9 that optimized a simulated transient response, can be implemented to automate the trial-and-error procedure. This approach would require a computer optimization of the simulated transient response to obtain good initial values for each control system (Edgar and Himmelblau, 1988). Optimization is justified when process interaction is strong and the trial-and-error method would be time-consuming or result in severe process disturbances.

As an example, the tuning for the control system considered in Example 20.9 was optimized for a unit step change in controlled variable 1, assuming equal importance of the two controlled variables and no other objectives; therefore, the objective was to minimize the total integral of absolute value of errors ($IAE_1 + IAE_2$). (As we know, this is not an adequate definition for general use, because it does not consider potential model errors or manipulated-variable behavior, but it is used here for simplicity.) The tuning and transient response are given in Figure 20.11 and included in Table 20.2. The optimization method yielded initial estimates with little engineering effort and modest computing resources.

Approximate, Noniterative Approach

A few methods have been proposed for estimating the tuning for multiloop systems without the time-consuming iterations associated with trial and error or the computer computations associated with the optimization approach. The goal of these methods is to provide initial tuning constants that are much closer than single-loop tuning constants to the "best" multiloop values. Naturally, fine-tuning based on plant experience is still required. Unfortunately, there is no generally accepted method for quickly estimating multiloop tuning. The method explained here is selected because it provides insight and introduces some key process-related issues. It also provides a useful correlation for many 2×2 systems; however, it is not easily extended to higher-order systems.

The method takes advantage of simplifications to determine the tuning for three cases of limiting process dynamics for 2×2 systems with PI multiloop controllers (McAvoy, 1983*a* and Marino-Galarraga et al., 1987). In all of these cases, the relative importances of the controlled variables are considered equal; this is the most demanding case for tuning, but other situations are considered in the next chapter as we tailor the performance to control objectives. The general approach is to establish how much the PI controller tuning must be changed from single-loop values when applied in a multiloop system.

The basis of the analysis is the closed-loop characteristic expression (20.34) divided by $1 + G_{c2}(s)G_{22}(s)$, which does not change the stability limit:

$$\text{CE}(s) = 1 + G_{c1}(s)G_{11}(s)\left(\frac{1 + G_{c2}(s)G_{22}(s)/\lambda_{11}(s)}{1 + G_{c2}(s)G_{22}(s)}\right) \tag{20.37}$$

As demonstrated in Chapter 10, the closed-loop characteristic expression given by equation (20.37) determines the stability of the control system. To evaluate potential simplifications, the relative importance of each term must be determined at the critical frequency of the loop. Since the approach is based on stability analysis, which considers only the denominator of the closed-loop transfer function, the same tuning is obtained for all disturbances and set point changes. The method considers three limiting cases for tuning loop 1: loop 1 much faster than loop 2; loop 1 much slower; and both loops having the same dynamics. (The following analysis considers loop 1, but the same results can be obtained for loop 2 by simply transposing the subscripts.)

LOOP 1 MUCH FASTER THAN LOOP 2. When the loop 1 process is much faster than loop 2, the term $G_{c2}(j\omega)G_{22}(j\omega)$ is very small at the loop 1 critical frequency because of the tendency of processes to have amplitude ratios that decrease rapidly

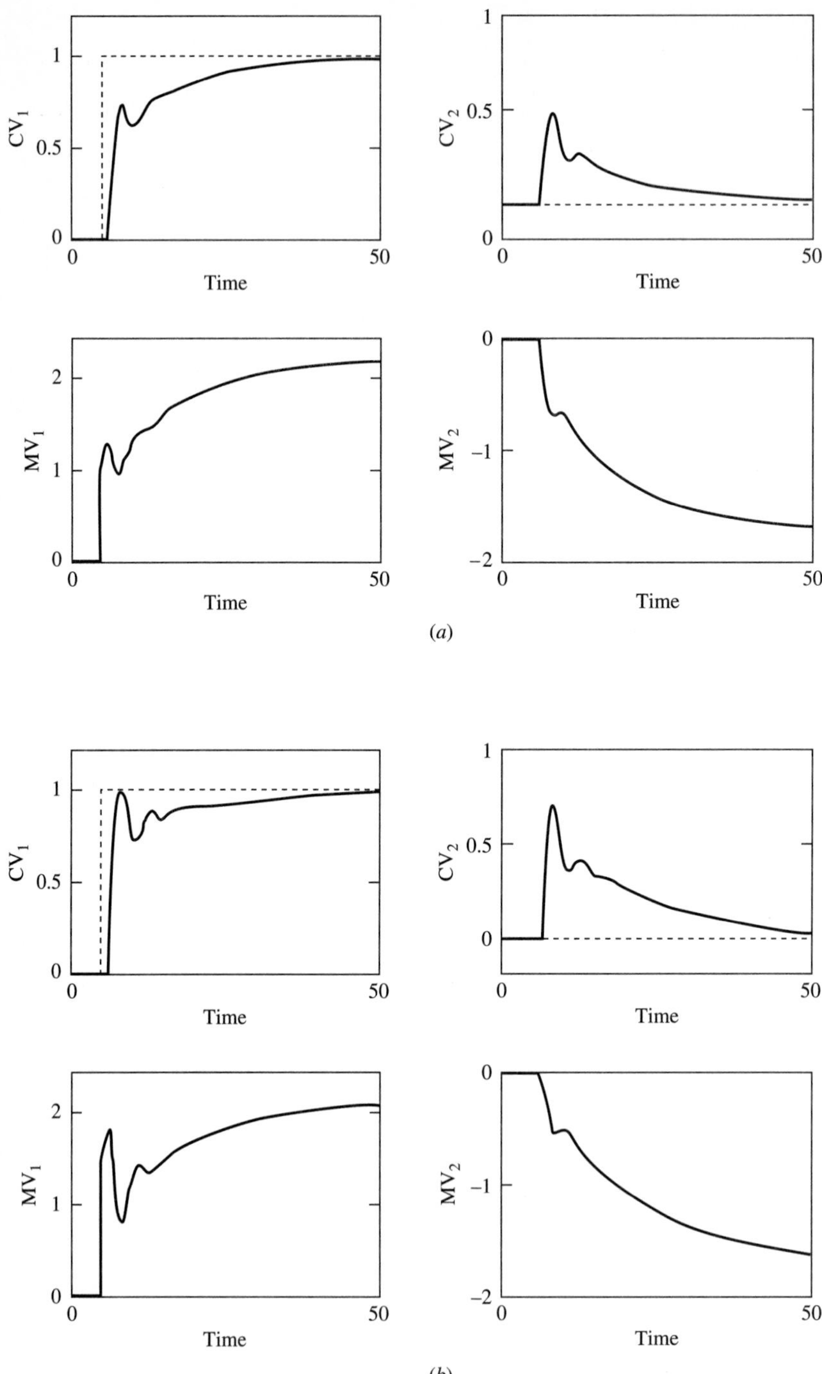

FIGURE 20.10
Multiloop control (*a*) with the same gains for both controllers; (*b*) with loop 1 gain higher.

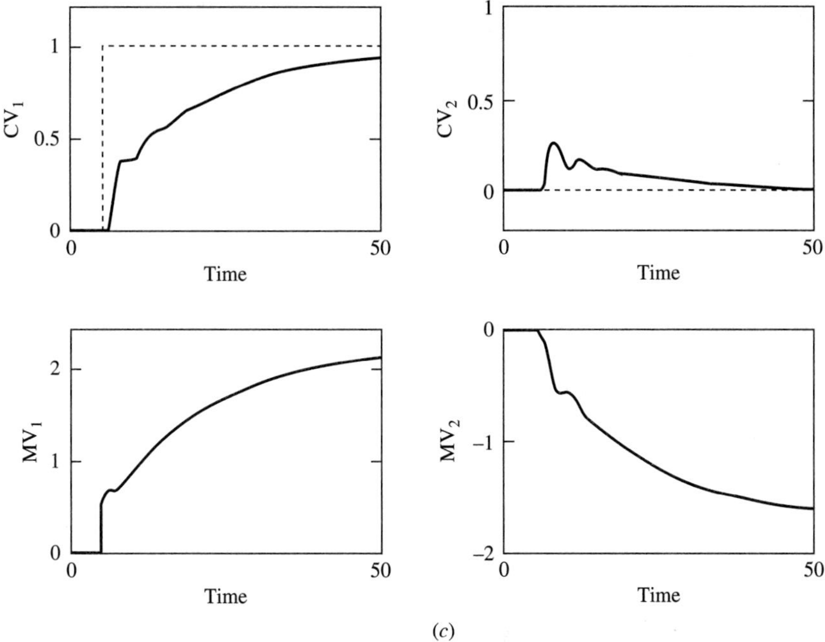

FIGURE 20.10 (*continued*)
Multiloop control (*c*) with loop 2 gain higher.

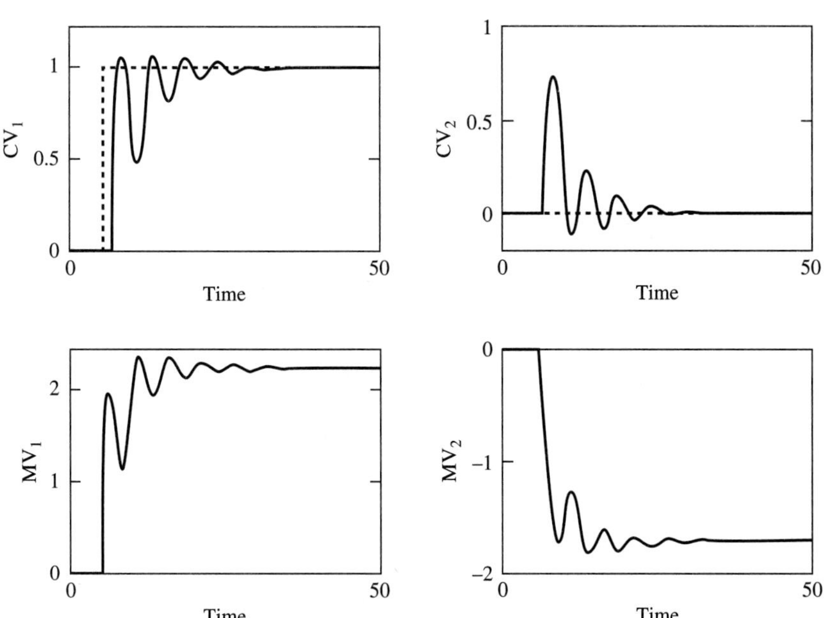

FIGURE 20.11
Multiloop control with PI tuning that minimizes $\sum$ IAE.

after the corner frequency (see Figure 10.13*b*). Assuming that λ_{11} is not a strong function of frequency, as is most often true,

$$\left| \frac{1 + G_{c2}(j\omega)G_{22}(j\omega)/\lambda_{11}}{1 + G_{c2}(j\omega)G_{22}(j\omega)} \right| = 1.0 \tag{20.38}$$

which gives

$$\text{CE}(j\omega) \approx 1 + G_{c1}(j\omega)G_{11}(j\omega) \tag{20.39}$$

Therefore, the very fast loop 1 in this case can be tuned like a single-loop controller without interaction. This result confirms a qualitative argument in which we would consider the interaction from the slow loop to be a slow disturbance to the very fast loop 1, which could be tuned using single-loop methods.

LOOP 1 MUCH SLOWER THAN LOOP 2. When loop 1 is much slower, the term for the fast controller, $G_{c2}(j\omega)$, would have a very large magnitude at the critical frequency of loop 1, because the amplitude ratio of the integral mode in $G_{c2}(j\omega)$ will have a very high value at a frequency much less than the loop 2 critical frequency (see Figure 10.13*f*). Therefore, $|G_{c2}(j\omega)| \gg 1.0$, which leads to the following simplification in the characteristic equation.

$$\text{CE}(j\omega) \approx 1 + G_{c1}(j\omega)G_{11}(j\omega)\left(\frac{G_{c2}(j\omega)G_{22}(j\omega)/\lambda_{11}}{G_{c2}(j\omega)G_{22}(j\omega)}\right) \tag{20.40}$$

$$\text{CE}(j\omega) \approx G_{c1}(j\omega)\frac{G_{11}(j\omega)}{\lambda_{11}(j\omega)} \approx G_{c1}(j\omega)\frac{G_{11}(j\omega)}{\lambda_{11}} \tag{20.41}$$

As noted, the steady-state relative gain has been used as an approximation for the frequency-dependent relative gain. In this case, the gain of the process "seen" by the controller 1 in the multiloop system is changed by $1/\lambda_{11}$ from the single-loop gain (K_{11}). Therefore, the controller gain has to be adjusted by approximately the inverse, λ_{11}, to maintain the proper stability margin. Since the phase lag is not affected, the integral time is unchanged. Again, this result seems consistent with a qualitative argument that a very fast associated loop would "become part of the process" and affect only the closed-loop process gain.

Since the relative gain can take a wide range of values for different processes, the implication of equation (20.41) is that multiloop controller systems with fast and slow processes could have controller gains that change by a large amount from single-loop to multiloop. This situation might be necessary but would generally be undesirable; therefore, this property influences the selection criteria for control structures, as will be developed further in the next chapter.

LOOPS 1 AND 2 HAVE THE SAME DYNAMICS. The entire closed-loop characteristic equation (20.34), as follows, must be considered.

$$\text{CE}(s) \approx 1 + 2\Lambda(s) + \frac{\Lambda^2(s)}{\lambda_{11}} \tag{20.42}$$

with $\Lambda(s) = G_{c1}(s)G_{11}(s) = G_{c2}(s)G_{22}(s)$, because the loop dynamics are equal.

The roots of the characteristic equation, which influence stability, can be solved to be

$$\Lambda = \lambda_{11} + \sqrt{\lambda_{11}^2 - \lambda_{11}} \tag{20.43}$$

with the selection of the proper root in the quadratic discussed by McAvoy (1981). In this case, the interaction can influence both the magnitude and phase lag, so the controller gain and integral times have to be altered to retain the proper stability margin, as shown by Marino-Galarraga et al. (1987). The solution to this problem provides the proper stability margin, but it requires calculations that are not significantly easier than the optimization-based tuning method already described.

With the simplification that all transfer functions in the process, $G_{ij}(s)$, have similar dynamics, the effects of interaction on tuning are completely represented by the relative gain, and the results of equation (20.43) can be condensed into detuning correlations in Figures 20.12*a* and *b* (Marino-Galarraga et al., 1987). These figures show how single-loop tuning must be altered for 2 × 2 multiloop control when all input-output dynamics are similar. The controller gain is reduced by about a factor of 2.0 as the relative gain changes from 1.0. Also, the integral time increases by a factor of about 2.0 as the relative gain decreases to 0.5.

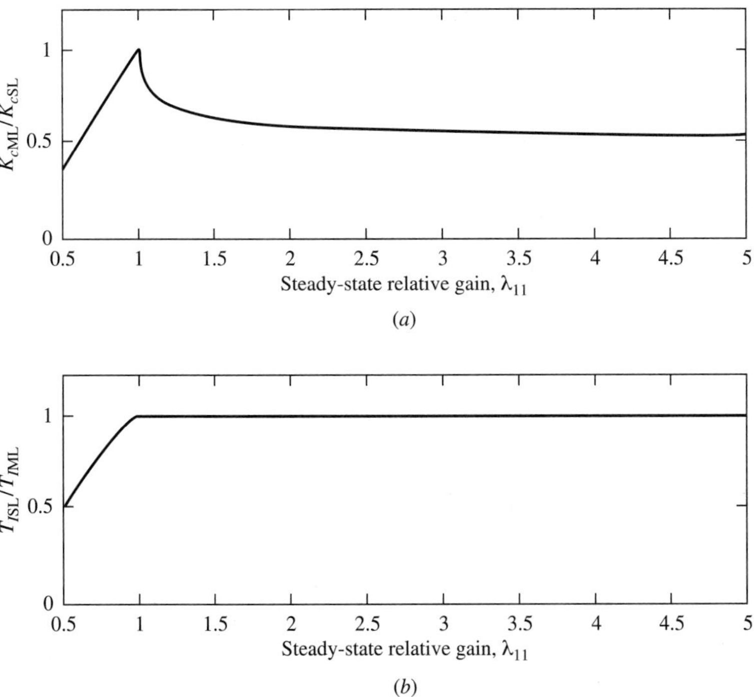

FIGURE 20.12
Relationship between single-loop (SL) and multiloop (ML) PI controller tuning when both loops have similar dynamics.

TABLE 20.3
Summary of example tuning for 2 × 2 system

Situation	Characteristic expression	Interaction effect
General	$1 + G_{c1}(s)G_{11}(s) + G_{c2}(s)G_{22}(s)$ $+ G_{c1}(s)G_{11}(s)G_{c2}(s)G_{22}(s)/\lambda_{11}(s)$	Transmission interaction affects stabililty
Loop 1 much faster	$1 + G_{c1}(s)G_{11}(s)$	Loop 1 stability is not strongly affected by interaction; use single-loop tuning
Loop 1 much slower*	$1 + G_{c1}(s)G_{11}(s)/\lambda_{11}$	Loop 1 stablility is affected by the change in close-loop process gain; multiply single-loop controller gain by λ_{11}
Both loops with equal dynamics	$1 + 2\Lambda(s) + \Lambda(s)^2/\lambda_{11}$ with $\Lambda(s) = G_{c1}(s)G_{11}(s) = G_{c2}(s)G_{22}(s)$	Loop 1 stability is affected by changes in gain and phase; use Figure 20.12

*This approach will lead to a very large controller gain for large λ. If the interacting controller is switched to manual, loop 1 could become unstable. Thus, the additional limit $(K_c)_{ML} \leq (K_c)_{SL}$ is often applied to ensure stability for both single- and multiloop systems.

Two important conclusions for systems with similar dynamics become apparent from this plot:

1. The multiloop controllers must be detuned from their single-loop tuning over the entire range of relative gain.
2. The change in tuning constants is not very large.

Thus, interaction results in controller detuning, which slows feedback action for most 2×2 multiloop systems. In addition to providing useful initial tuning guidelines, the relationship in Figure 20.12 is used in analyzing control performance. The tuning results for 2×2 PI control presented in this section are summarized in Table 20.3.

Note that these results are appropriate for systems that satisfy the assumptions employed. At the current time, there is no approximate method for the general case with very different dynamics of all paths, $G_{ij}(s)$. The trial-and-error or optimization-based methods must be used in these cases. Also, the importances of the controlled variables have been assumed to be relatively equal; the case for unequal importances is covered in the next chapter. The next two examples apply the tuning approach to realistic processes.

Example 20.10. Determine initial tuning constants for multiloop PI controllers applied to the blending system operating at the conditions given in equations (20.22), 5% A in the product, and the following sensor dynamics:

	Dead time	Time constant
Flow	1	2 sec
Concentration	15	30 sec

TABLE 20.4
Tuning for the blending system with dilute product ($x_m = 0.05$, $\lambda = 0.95$)

	A_1–F_2 controller (slow loop)		F_3–F_1 controller (fast loop)	
Tuning term	**Single-Loop**	**Multiloop**	**Single-Loop**	**Multiloop**
Process Gain	$K_{11} = 0.0095$	$K_{11}/\lambda_{11} = 0.01$	$K_{22} = 1.0$	$K_{22} = 1.0$
$\theta/(\theta + \tau)$	0.333	0.333	0.333	0.333
$K_c K_p$	1.0	1.0	1.0	1.0
$T_I/(\theta + \tau)$	0.85	0.85	0.85	0.85
K_c (kg/min/wt fraction)	105.0	100.0	1.0	1.0
T_I (sec)	38.0	38.0	2.6	2.6

Consider first the A_1–F_2 and F_3–F_1 controlled-manipulated variable pairing. The basis for the tuning values is the linear transfer function models in equations (20.6) and (20.7) with gain values from equation (20.23), and any single-loop tuning method could be used. The dynamics above indicate that this case fits the situation having one fast and one slow loop. Referring to Table 20.3, and noting that $\lambda_{11} = \lambda_{22} = 0.95 \approx 1.0$, both the fast and slow loops can be tuned very close to their single-loop values. The tuning results using the Ciancone single-loop correlations are summarized in Table 20.4.

A transient response of this system, simulated using the linearized equations, for a set point change of 0.01 in the mixed concentration, is given in Figure 20.13. This is a reasonably well-behaved response, which could be fine-tuned as needed. An important result of this analysis is that the tuning for this loop pairing does not change

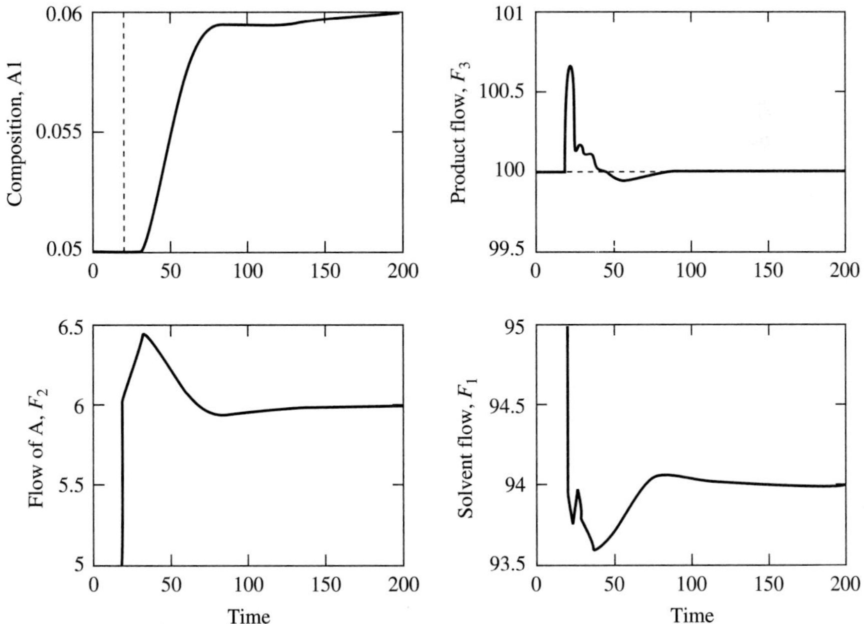

FIGURE 20.13
Set point response for multiloop blending system in Example 20.10.

TABLE 20.5
Tuning for the blending system with dilute product ($x_m = 0.05$, $\lambda = 0.05$)

Tuning term	A_1–F_1 pairing (slow loop) Single-loop	Multiloop	F_3–F_2 pairing (fast loop) Single-loop	Multiloop
Process gain	$K_{11} = -0.0005$	$K_{11}/\lambda_{11} = -0.01$	$K_{22} = 1.0$	$K_{22} = 1.0$
$\theta/(\theta+\tau)$	0.333	0.333	0.333	0.333
$K_c K_p$	1.0	1.0	1.0	1.0
$T_I/(\theta+\tau)$	0.85	0.85	0.85	0.85
K_c (kg/min/wt fraction)	−2000.	−100.	1.0	1.0
T_I (sec)	38.	38.	2.6	2.6

significantly from single-loop to multiloop; in other words, the tuning of the controllers does not depend on the control status (automatic/manual) of the other controller. This is a good situation.

Now consider the alternative loop pairing, A_1–F_1 and F_3–F_2. Again, the system consists of a fast and slow loop, so that the same approach can be used. However, in this system, the relative gain has a value far from unity, $\lambda_{11} = \lambda_{22} = 0.05$. Therefore, the response of the slow loop (A_1–F_1), which has an effective process gain of K_{11}/λ_{11}, is significantly altered by interaction. The results, using the recommendations in Table 20.3 and the Ciancone single-loop tuning correlations, are summarized in Table 20.5.

The transient response of the multiloop system with the multiloop tuning given in Table 20.5 is essentially the same as that for the previous pairing and is not shown. However, the single-loop and multiloop tunings are very different in Table 20.5, because the relative gain is much different from 1.0. If both loops are in automatic, the A_1 controller gain must be the (small) multiloop value given in the table. When the F_3 controller is in manual, the effective process gain for the A_1 controller changes to its single-loop value (which is lower by a factor of about 20).

A summary of the implications of the multiloop system in Table 20.5 follows:

Tuning of A_1	Single-loop (A_1) system	Multiloop system
Single-loop ($K_c = -2000$)	Good performance	Unstable system
Multiloop ($K_c = -100$)	Poor performance (very slow)	Good performance

Thus, the controller tuning in Table 20.5 must be matched to the status of the controllers—a *situation to be avoided if possible.* This complexity in updating tuning online suggests that the pairing in Table 20.4, which can have the same tuning for any combination of loop statuses (since $\lambda \approx 1.0$), is a *much better choice.*

Example 20.11. Determine initial tuning constants for the distillation tower with the pressure and level controller pairings given in Figure 20.3, resulting in the model in equations (20.8) and (20.9). Evaluate the dynamic behavior for a step change in the feed light key of −0.04 mole fraction light key.

This process has similar dynamics for both loops, so that the results in Table 20.3 recommend the tuning correlations in Figure 20.12. The large value of the relative gain (6.09) indicates that the controller gains must be reduced by a factor of 2.0 from their single-loop values, and the integral times can remain unchanged. The

TABLE 20.6
Tuning analysis for distillation control system

Tuning term (λ = 6.09)	X_D–F_B controller		X_B–F_V controller	
	Single-Loop	Multiloop	Single-Loop	Multiloop
Process Gain	0.0747		$K_{22} = -0.1253$	
$\theta/(\theta + \tau)$	0.20		0.16	
$K_c K_p$	1.55		1.7	
$T_I/(\theta + \tau)$	0.60		0.50	
K_c	20.75	$K_{cSL}/2 = 10.4$	−13.6	$K_{cSL}/2 = -6.8$
T_I	9.0	9.0	6.1	6.1

results from applying this approach are given in Table 20.6, and a dynamic response of the multiloop system, using the multiloop tuning from the table, is shown in Figure 20.14. The response is well behaved, because the controlled variables return to their set points reasonably quickly and the manipulated variables experience moderate adjustments. Thus, the correlations provide acceptable initial tuning, which can be tailored to specific objectives through fine-tuning.

20.7 ADDITIONAL TOPICS IN INTERACTION ANALYSIS

The material on interaction in this chapter is only introductory, and a coverage of much more material would be required for a mastery of the topic. Some of the key additional topics are reviewed briefly in this section.

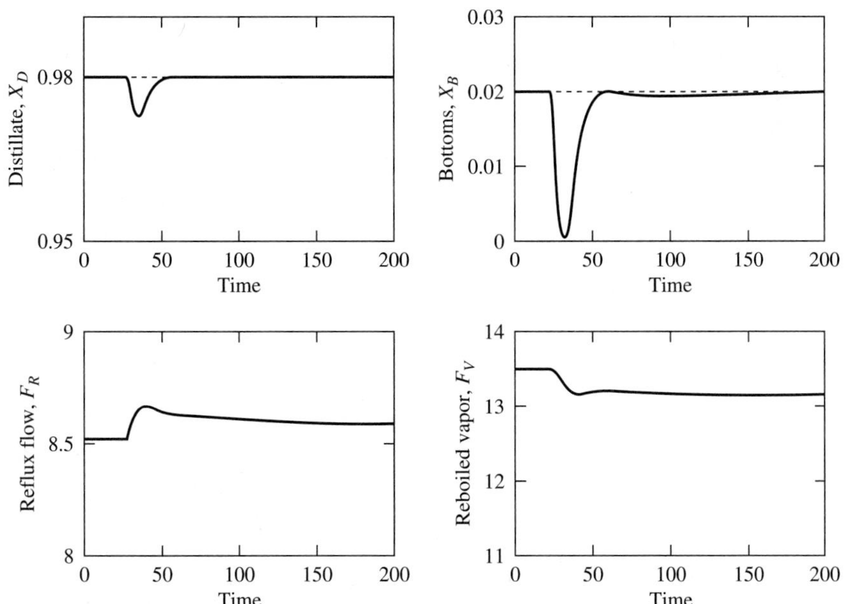

FIGURE 20.14
Example disturbance response for multiloop distillation in Example 20.11.

Feasible Pairings

An important property of a multiloop pairing is proper performance without excessive monitoring programs. Consider the situation in which the controllers have integral modes and the controller gains have the proper sign for negative feedback in the single-loop case. The multiloop system will be *integral-controllable* if all the controllers' gains can be decreased (in the same proportion) from some initial values to zero and the system remains stable. Certainly this is a desirable feature. Since we have established that the sign of the steady-state input-output gain depends on the multiloop system and can change sign (because relative gains can be negative), it is not entirely unexpected that not all multiloop systems are integral-controllable. The following conditions provide quantitative tests for integral-controllability, assuming the system is controllable.

A 2 × 2 SYSTEM.

If loops are paired on (i, j) elements with $\lambda_{ij} < 0$	the system is not integral-controllable
If loops are paired on (i, j) elements with $\lambda_{ij} \geq 0$	the system is integral-controllable

This condition is necessary and sufficient.

***n* × *n* SYSTEM (NIEDERLINSKI CRITERION).** The variable numbers can be rearranged for any system so that the control loop pairing involves 1–1, 2–2, …, *i*–*i*. With this convention, and **K** being the gain matrix with elements $K_{ij} = G_{ij}(0)$,

$$\text{If } \left(\frac{\det \mathbf{K}}{\prod_{i-1}^{n} K_{ii}}\right) < 0 \qquad \text{the system is not integral-controllable}$$

This condition is sufficient but is not necessary (except for 2 × 2 systems). The proofs for these conditions, along with limitations on the process dynamics $G_{ij}(s)$, are given by Grosdidier et al. (1987). These criteria can be used to eliminate some possible control designs from further consideration, which will simplify the analysis. Note that this useful test has been performed using only steady-state information!

Modelling

Models for multivariable control should be developed with their ultimate use in mind. Recent results on model consistency (Skogestad, 1991; Haggblom and Waller, 1988) give useful relationships that can be used to verify that linearized models observe fundamental properties of the nonlinear system. Also, new experimental designs (Kwong and MacGregor, 1994) could be of use in obtaining better empirical estimates of process gains, but even with these careful experimental steps the use of empirical models for calculating relative gains with large magnitudes is problematical.

Interaction Measures

The important features of systems with transmission interaction discussed in Section 20.3 can be developed through singular-value analysis for systems of arbitrary size and dynamics. The relevant matrix, here the process gain matrix **K**, can be decomposed into three matrices, which can be used to determine the directions in the CVs that cause the manipulated variables to change the "most" and the "least" (as measured by the root sum of squares of the changes in the MVs). Also, the ratio of the largest to the smallest changes in these two directions can be determined and is called the *condition number.* Clearly, the larger the condition number, the more interaction affects the multiloop system. Also, the condition number indicates the sensitivity of the calculation to model errors. The basic mathematics of this analysis is presented in Ortega (1987), and control applications are given in Barton et al. (1991) and Arkun (1984). The relationship between the relative gain and condition number is given by McAvoy (1983*b*) and Grosdidier et al. (1985). An alternative measure of interaction has been proposed by Grosdidier and Morari (1987). Finally, the controllability and relative gain calculations can be extended to systems with pure integration, such as liquid levels, by replacing the derivative of the variable (dL/dt) with a surrogate variable ξ and proceeding with the standard method thereafter (McAvoy, 1983*b*).

Frequency-Dependent Measures

The material in this chapter on controllability and interaction relied principally on steady-state measures. The definition of controllability used here involves steady-state behavior. An alternative frequency-dependent definition involves the ability to influence the dynamic trajectory of the output variables and requires that det $\mathbf{G}(s) \neq 0$ (Rosenbrock, 1974). Calculation of the magnitude of the elements of the individual transfer functions for the chemical reactor in question 20.18 gives the results in Figure 20.15. These results indicate that the system is not controllable at steady state, because both gains for the flow are zero. However, the system is controllable at higher frequencies. Since this book deals mainly with continuous processes operated at specified steady-state conditions, the definition of controllability used here involves steady-state ($\omega = 0$) controllability.

In addition, the effects of interaction should be evaluated near the critical frequencies of the control loops. The definition of the frequency-dependent relative gain in equation (20.33) can be applied to the distillation model in equations (20.8) and (20.9), giving the result in Figure 20.16. The value of the relative gain around a frequency of 0.5 rad/min shows that the interaction is only somewhat less than at steady state; thus, the steady-state values indicate the interaction for this example. Frequency-dependent interaction is discussed by McAvoy (1981).

Tuning

Another approach to tuning multiloop PID controllers that seems to have met with success is presented by Monica et al. (1988). This method can be extended to higher-order systems with frequency response calculations. The definition of modelling

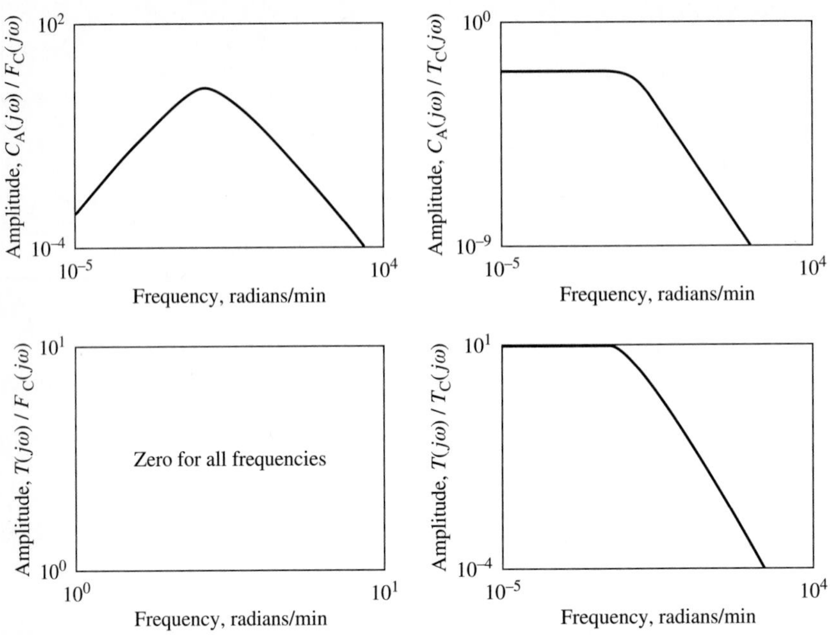

FIGURE 20.15
The magnitudes of the individual transfer functions for the chemical reactor in question 20.18.

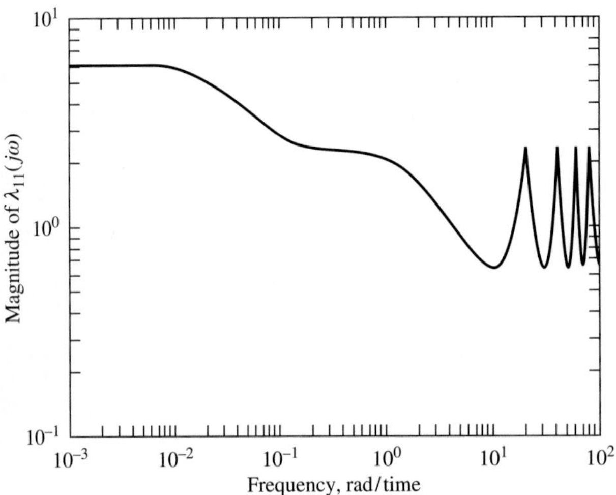

FIGURE 20.16
Frequency-dependent relative gain for distillation.

errors to be considered in tuning multivariable systems is much more difficult, because errors in the individual transfer functions and parameters within an individual transfer function are not independent.

20.8 CONCLUSIONS

Multiloop process control systems have been introduced, and the important concept of process interaction defined. Standard modelling methods can be used to represent the input-output behavior of the process without control. Interaction—one input affecting more than one output—is seen to influence the behavior of multivariable systems. Using the convention that the single-loop controllers are paired on the 1–1 and 2–2 elements in a two-variable process, interaction occurs when at least one of the interacting terms, $G_{12}(s)$ or $G_{21}(s)$, is nonzero. The process model can be employed to determine a useful measure of interaction: the relative gain array.

Requirements of controllability and values for relative gain are really extensions of conditions that are required for good single-loop feedback control, as summarized in the following table.

Required condition	**Single-loop system**	**Multiple-loop system**
Controllability	A causal relationship exists below the manipulated and controlled variables, $K_p \neq 0$.	n independent, causal relationships exist between the manipulated and controlled variables, det $\mathbf{K} \neq 0$.
Monotonic process gain	K_p does not change sign	The sign of the feedback gain "seen" by each controller does not depend on the status of other controllers; i.e., the system is integral-controllable

Since the requirements are less obvious in multiloop systems, the rigorous mathematical tests are provided.

Transmission interaction—the additional connection path between an input and output through an interacting controller—occurs when both interacting terms in a 2×2 system are nonzero. Transmission interaction can strongly affect the behavior of a multivariable system. First, depending on the directions of the desired changes in controlled variables, it can substantially increase or decrease the adjustments required in the manipulated variables. Second, it can influence the sensitivity of the system to modelling errors. Third, it can influence the system's stability and proper controller tuning.

Some of the results introduced in this chapter are general for all multiloop systems of any order ($n \times n$), while some are restricted to two-variable (2×2) systems. The following summary is provided to help the reader.

$n \times n$ systems	2×2 systems only
Modelling Controllability Relative gain array definition, equation (20.25) Relative gain calculation, equation (20.27)	Closed-loop transfer function, equations (20.14) to (20.17) Relationship between RGA and stability, equation (20.34) PI tuning, Section 20.5
Interpretations of relative gain in Section 20.5	Condition for 2×2 integral controllability
Condition for $n \times n$ integral controllability	

Finally, an important interpretation concerning control performance can be reached from these tuning results by considering a system having similar dynamics and a relative gain much larger than 1.0. (Many important processes have large relative gains.) In this system, the multiloop process gain is *smaller* than the single-loop process gain by a factor of about $1/\lambda$, as shown in equation (20.25). However, stability and tuning analysis indicated that the controller gain in the multiloop system must be *reduced* from its single-loop value, as shown in Figure 20.12! As a result, the reduction in effective process gain caused by interaction in the multiloop system *cannot* be compensated by an increase in the controller gain; if an attempt is made to increase the controller gain to improve control performance, the system will become unstable! Thus, the product $K_p K_c$ can be small (i.e., very much less than 1.0) in the interactive system, and feedback adjustments in response to some disturbances can be very slow because of this "detuning" effect of interaction. This stability limit for multiloop systems accounts for the very slow return to set point experienced by some processes with large relative gains.

To this point, general interpretations of multiloop system behavior have been developed. The many useful insights and quantitative expressions for the effects of interaction on multivariable behavior and stability will be exploited in the next chapter on multiloop control performance, in which specific methods for tailoring control design to performance goals are presented.

REFERENCES

Arkun, Y., B. Monousiouthakis, and A. Palazuglu, "Robustness Analysis of Process Control Systems, A Case Study of Decoupling in Distillation," *IEC Proc. Des. Devel., 23*, 93–101 (1984).

Barton, G., W. Chan, and J. Perkins, "Interaction Between Process Design and Process Control: The Role of Open-loop Indicators," *J. Proc. Cont., 1*, 161–170 (1991).

Bristol, E., "On a New Measure of Interaction for Multivariable Process Control," *IEEE Trans. Auto. Control, AC-11*, 133–134 (1966).

Edgar, T. and D. Himmelblau, "Optimization of Chemical Processes," McGraw-Hill, New York, 1988.

Franklin, G., J. Powell, and M. Workman, *Digital Control of Dynamic Systems* (2nd Ed.), Addison-Wesley, Reading, MA, 1990.

Grosdidier, P., M. Morari, and B. Holt, "Closed-Loop Properties from Steady-State Gain Information," *IEC Fund., 24,* 221–235 (1985).

Grosdidier, P. and M. Morari, "The μ Interaction Measure," *IEC Res. 26,* 1193–1202 (1987).

Haggblom, K. and K. Waller, "Transformations and Consistency Relations of Distillation Control Structures," *AIChE J., 34, 10,* 1634–1648 (1988).

Kwong, C.W., and J. MacGregor, "Identification for Robust Multivariable Control: The Design of Experiments," *Automatica, 30,* 1541–1554 (1994).

Marino-Galarraga, M., T. McAvoy, and T. Marlin, "Short-Cut Operability Analysis. 2. Estimation of f_i Detuning Parameter for Classical Control Systems," *IEC Res. 26,* 511–521 (1987).

Marlin, T., McAvoy, T., M. Marino-Galarraga, and N. Kapoor, "A Short-Cut Method For Process Control and Operability Analysis," in Morari, M. and T. McAvoy, *Chemical Process Control III,* Elsevier, Amsterdam, 1986, pp. 369–419.

McAvoy, T., "Connection Between Relative Gain and Control Loop Stability and Design," *AIChE J., 27, 4,* 613–619 (1981).

McAvoy, T., "Some Results on Dynamic Interaction Analysis of Complex Systems," *IEC Proc. Des. Devel., 22,* 42–49 (1983*a*).

McAvoy, T., *Interaction Analysis,* Instrument Society of America, Research Triangle Park, NC, 1983*b*.

McAvoy, T. and K. Weischedel, "A Dynamic Comparison of Material Balance Control of Distillation Columns," *Proc. Seventh IFAC Congress, Kyoto, Japan,* paper 107.2, 1981.

Monica, T., C. Yu, and W. Luyben, "Improved Multiloop Single-Input /Single-Output Controllers for Multivariable Processes," *IEC Res., 27,* 969–973 (1988).

Ortega, *Matrix Theory, A Second Course,* Plenum Press, New York, 1987.

Perkins, J., "Equation Oriented Flowsheeting," in Westerberg, A. and H. Chien (Ed.), *Proc. Second Int. Conf. Found. Computer Aided Process Design, CACHE,* University of Michigan, Ann Arbor, MI, 1984.

Rosenbrock, H., *Computer-Aided Control System Design,* Academic Press, New York, 1974.

Sakr, M., A. Bahgat, and A. Sakr, "Computer-Based Raw Material Blending Optimization in a Cement Manufacturing Plant," *Control and Computers, 16, 3,* 75–78 (1988).

Sampath, S., course project, McMaster University, 1991.

Shinskey, F.G., *Process Control Systems* (3rd Ed.), McGraw-Hill, New York, 1988.

Skogestad, S., "Consistency of Steady-State Models Using Insight about Extensive Variables," *IEC Res., 30,* 654–661 (1991).

Stadnicki, S. and M. Lawler, "An Integrated Planning and Control Package for Refining Product Blending," *Contr. Eng. Conf.,* pp. 315–322 (1985).

Waller, K., K. Haggblom, P. Sandelin, and D. Finnerman, "Disturbance Sensitivity of Distillation Control Structures," *AIChE J., 34, 5,* 853–858 (1988).

Witcher, M. and T. McAvoy, "Interacting Control Systems: Steady-State and Dynamic Measurement of Interaction," *ISA Trans., 16, 3,* 35–41 (1977).

ADDITIONAL RESOURCES

The effects of process interaction were investigated earlier by several researchers, including

Rijnsdorp, J., "Interaction for Two-Variable Control Systems in Distillation Columns," I, *Automatica, 1,* 15–29 (1965) and II, *Automatica, 1,* 29–51 (1965).

Further information on the effects of interaction on multiloop systems is available in

Despande, P. (ed.), *Multivariable Process Control,* Instrument Society of America, Research Triangle Park, NC, 1989.

Shinskey, F. G., *Controlling Multivariable Processes,* Instrument Society of America, Research Triangle Park, NC, 1981.

A rigorous criterion for the stability of linear multivariable systems is available in Despande (1989, just cited) and

Luyben, W., *Process Modelling, Simulation, and Control for Chemical Engineers* (2nd Ed.), McGraw-Hill, New York, 1990.

The material in the chapter enables the engineer to evaluate the suitability of candidate processes and variables for multiloop control quantitatively. Specifically, controllability and operating window (or range of operation) can be used to establish the feasibility (or infeasibility) of feedback control for potential process designs. Interpretations of the relative gain suggest that only variable pairings with $\lambda_{ij} > 0$ for 2×2 systems should normally be considered further (but see Chapter 21 for important exceptions). Also, the effects of interaction on tuning are demonstrated by some preliminary tuning rules for 2×2 systems. The methods in this chapter enable the engineer to eliminate some candidate designs as infeasible for multiloop control, so that future effort can be directed toward evaluating the remaining feasible candidates.

QUESTIONS

20.1. For the blending process in Figure 20.2, design a control system to control the following three product variables at independent values: (*a*) the total flow (F_3), (*b*) the mass fraction of component A, and (*c*) the mass fraction of component *S*. You may assume that both mass fractions can be measured by the analyzer A_1.

20.2. Answer the following questions for two physical processes: (1) the chemical reactor described in Example 3.10 and (2) the same chemical reactor with no heat of reaction. Both processes have two feedback PI controllers: $T \rightarrow F_c$ and $C_A \rightarrow C_{A0}$ (with F unchanged).

(*a*) Does process interaction influence the stability of the closed-loop system? Provide quantitative analysis to support your conclusion.

(*b*) Does process interaction influence the dynamic behavior of the closed-loop system? Explain your answer briefly.

20.3. Prove the statements made in this chapter about the relative gain array: (*a*) The elements are scale-independent. (*b*) The sum of values in a row or column is 1.0. (*c*) the λ_{ij} in equation (20.28).

20.4. Verify the closed-loop transfer functions in equations (20.12) through (20.17).

20.5. Answer the following question about controllability.

(*a*) How must the controllability test be modified when a constraint is encountered in one or more manipulated variables?

(*b*) Develop an alternative definition of controllability and develop a mathematical test for the situation in which the controlled variables must only achieve specified values at a single point in time. This might be valid for batch control or for intercepting a missile.

(*c*) Relate the definition of controllability used in this chapter to the relative gain array.

(*d*) How would the test for controllability in Section 20.3 be modified if the control algorithms were implemented via digital calculation?

(*e*) How far can one extrapolate the conclusions of the controllability test to other operating conditions? In your answer, consider the process in Example 3.12.

20.6. Determine the controllability and possible loop pairings ($\lambda > 0$) for the process in Figure Q20.6 for the following two situations. The feed consists of only solvent and component A. The manipulated variables are the valves, and the controlled variables are the level and the composition of A, C_A.

(*a*) The situation without chemical reaction (i.e., a mixing tank).

(*b*) The situation with a single chemical reaction $A \rightarrow B$, $r_A = -kC_A$.

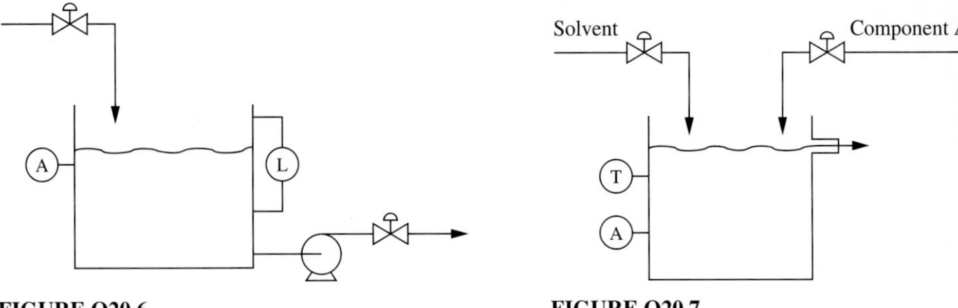

FIGURE Q20.6

FIGURE Q20.7

20.7 Consider the CSTR in Figure Q20.7 in which solvent and component A in solvent (C_{A0}) are mixed. The two streams can be at different temperatures. A single reaction A → B occurs in the reactor. The rate expression is $r_A = -kC_A$, and the heat of reaction can be nonzero. The manipulated variables are the flow rates of the two inlet streams, and the controlled variables are the temperature and concentration of A in the reactor.

(*a*) Determine under what conditions the system is controllable.

(*b*) For the conditions which are controllable, if any, determine allowable loop pairings ($\lambda_{ij} > 0$).

20.8. Answer the following questions for a 2 × 2 control system with PI controllers.

(*a*) Is it possible for tuning values to exist that would yield a stable multiloop system and an unstable single-loop system for the same process?

(*b*) Is it possible for tuning values to exist that would yield an unstable multiloop system and a stable single-loop system for the same process?

(*c*) State the criteria for the single-loop system in Figure 20.7 to be stable.

(*d*) Suggest a manner for using the results in Example 20.8 in tailoring the dynamic performance to control system goals.

20.9. The following transfer function was provided by Waller et al. (1987) for a distillation column with the levels and pressure controlled with single-loop controllers as in Figure 20.3. The product qualities were not measured directly; they were inferred from tray temperatures (°C) near the top, T_4, and near the bottom, T_{14}, trays. The manipulated variables are the reflux, F_R, and reboiler steam, F_S, both in kg/hr. Time is in minutes.

$$\begin{bmatrix} T_4(s) \\ T_{14}(s) \end{bmatrix} = \begin{bmatrix} \dfrac{-0.045e^{-0.5s}}{8.1s+1} & \dfrac{0.048e^{-0.5s}}{11s+1} \\ \dfrac{-0.23e^{-1.5s}}{8.1s+1} & \dfrac{0.55e^{-0.5s}}{10s+1} \end{bmatrix} \begin{bmatrix} F_R(s) \\ F_s(s) \end{bmatrix}$$

Answer the following questions for this system.

(*a*) Determine whether the input-output combination is controllable.

(*b*) Determine if either loop pairing can be eliminated based on the relative gains ($\lambda_{ij} > 0$).

(*c*) Determine the initial tunings for PI controllers for all allowable loop pairings.

(*d*) Estimate whether the interaction affects the magnitude of the manipulated variable changes for a set point change between single-loop and multiloop control.

20.10. The outlet temperature of the process fluid and the oxygen in the flue gas can be controlled in the fired heater in Figure Q20.10 by adjusting the fuel pressure (flow)

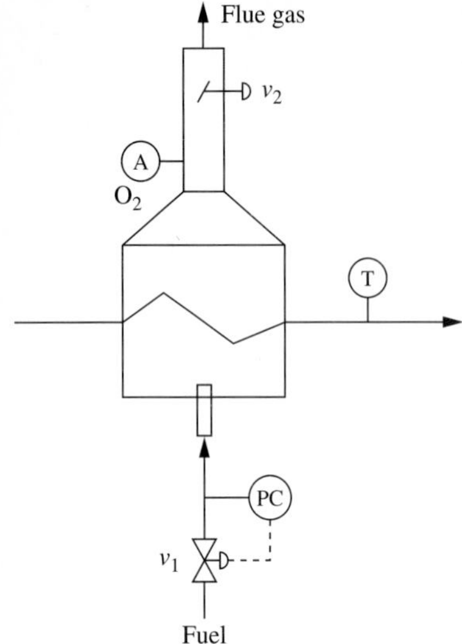

FIGURE Q20.10

and the stack damper % open. A dynamic model for the fired heater in Figure Q20.10 was reported by Zhuang et al. (1987) and repeated here:

$$\begin{bmatrix} T(s) \\ A(s) \end{bmatrix} = \begin{bmatrix} \dfrac{0.6}{2400s^2 + 85s + 1} & \dfrac{-0.04}{3000s^2 + 90s + 1} \\ \dfrac{-1.1}{70s + 1} & \dfrac{0.30}{70s + 1} \end{bmatrix} \begin{bmatrix} P_{sp}(s) \\ v_2(s) \end{bmatrix}$$

The inputs and outputs are in percent of the range of each instrument, and the time is in seconds.

(*a*) Determine whether the input-output combination is controllable.

(*b*) Estimate whether the interaction changes the magnitude of the manipulated variable changes for a set point change between single-loop and multiloop control.

(*c*) Determine if either loop pairing can be eliminated based on the relative gains ($\lambda_{ij} > 0$).

(*d*) Determine the initial tunings for PI controllers for all allowable loop pairings.

20.11. Three CSTRs with the configuration of Example 3.10 and with the following design parameters are considered in this example; the common data is given below, and the unique data and steady states are given in Table Q20.11 for three cases.

$$F = 1 \text{ m}^3/\text{min}, \text{V} = 1 \text{ m}^3, C_{A0} = 2.0 \text{ kmole/m}^3, C_p = 1 \text{ cal/(gK)}, \rho = 10^6 \text{ g/m}^3,$$
$$k_0 = 1.0 \times 10^{10} \text{ min}^{-1}, E/R = 8330.1 \text{ K}^{-1}$$
$$(F_c)_s = 15 \text{ m}^3/\text{min}, C_{pc} = 1 \text{ cal/(g K)}, \rho_c = 10^6 \text{ g/m}^3, b = 0.5$$

The controlled variables are C_A and T, and the manipulated variables are C_{A0} and F_c. Answer the following questions for each chemical reactor and explain the differences among the designs. (Note that this question requires the linearized, steady-state model for each case.)

(*a*) Determine whether the input-output combination is controllable.

(*b*) Estimate whether the interaction changes the magnitude of the manipulated variable changes for a set point change between single-loop and multiloop control.

TABLE Q20.11

Case	I (Example 3.10)	II	III
$-\Delta H_{rxn} 10^6$ cal/(kmole)	130	13	−30
a (cal/min)/K	1.678×10^6	1.678×10^6	0.7746×10^6
T_0 K	323	370	370
T_{cin} K	365	365	420 (heating)
T_s K	394	368.3	392.7
C_{As}kmole/m^3	0.265	0.80	0.28

(*c*) Determine if either loop pairing can be eliminated based on the signs of the relative gains.

(*d*) Determine the initial tunings for PI controllers for all allowable loop pairings.

(*e*) Evaluate the transient responses for a concentration set point change of +0.02 kmole/m^3.

20.12. Discuss an empirical method for identifying the *inverse* of the process gain matrix directly from experimental data.

20.13. Determine whether $\mathbf{K}(\mathbf{K})^{-1}$ would give the same (correct) result as equation (20.27) for the elements of the relative gain array.

20.14. The process with two series chemical reactors in Example 3.3 is considered in this question. The process flexibility is increased by allowing the temperatures of the two reactors to be manipulated independently. The two controlled variables are the concentrations of reactant A in the two reactors. The rate constant can be expressed as $5.87 \times 10^5 e^{-5000/T}$ (with temperature in K).

(*a*) Determine whether the input-output combination is controllable.

(*b*) Determine if either loop pairing can be eliminated based on the signs of the relative gains ($\lambda_{ij} > 0$).

20.15. The following transfer functions were provided by Wood and Berry (1973) for a methanol-water separation in a distillation column similar to Figure 20.3. The products are expressed as mole % light key, and the reflux F_R and reboiler steam F_S are in lb/min. Time is in minutes.

$$\begin{bmatrix} X_D(s) \\ X_B(s) \end{bmatrix} = \begin{bmatrix} \dfrac{12.8e^{-s}}{16.7s+1} & \dfrac{-18.9e^{-3s}}{21s+1} \\ \dfrac{6.6e^{-7s}}{10.9s+1} & \dfrac{-19.4e^{-3s}}{14.4s+1} \end{bmatrix} \begin{bmatrix} F_R(s) \\ F_s(s) \end{bmatrix}$$

(*a*) Determine whether the input-output combination is controllable.

(*b*) Estimate whether the interaction changes the magnitude of the manipulated variable changes for a set point change between single-loop and multiloop control.

(*c*) Determine if either loop pairing can be eliminated based on the sign of the relative gains ($\lambda_{ij} > 0$).

(*d*) Determine the initial tuning for PI controllers for all allowable loop pairings.

(*e*) The model was determined empirically. Discuss the effects of likely model errors on the results in parts (*a*) to (*d*).

20.16. A series of nonisothermal CSTRs shown in Figure Q20.16 is analyzed in this question. The heat transfer is adjustable in each reactor, so that each reactor temperature can be considered a manipulated variable. The feed contains only a nonreacting solvent and

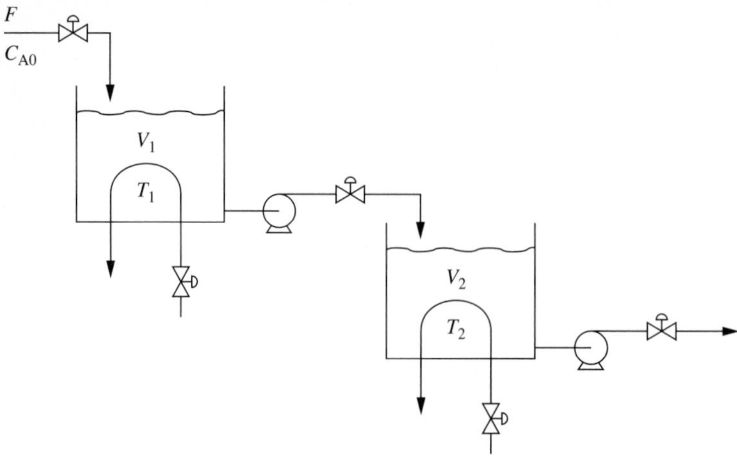

FIGURE Q20.16

component A. The potential manipulated variables are T_1, T_2, F, V_1, V_2, and C_{A0}. The variables to be controlled to independent steady-state values are the compositions of B and C in the effluent from the second reactor. For each of the sets of elementary reactions given below, determine (1) for which sets of two manipulated variables the system would be controllable and (2) for the variables selected in (1), whether either pairing of variables could be eliminated based on the relative gain.

(*a*) $A \xrightarrow{k_1} B \xrightarrow{k_2} C$

(*b*) $A \xrightarrow{k_1} B + C$

Assume that the rate constants can be expressed as Arrhenius functions of temperature and the heat of reaction is zero.

20.17. The mixing tank in Figure Q20.17 has two independent inlet streams of pure A and B that can be manipulated. The outlet flow cannot be manipulated by the unit; it is set by a unit of higher priority. The composition, the weight percent of B, and the level are to be controlled.

(*a*) Derive a linearized model of the system.

(*b*) Determine whether the system is controllable.

(*c*) Calculate the relative gain array for this process and make conclusions about the possible loop pairings for this system.

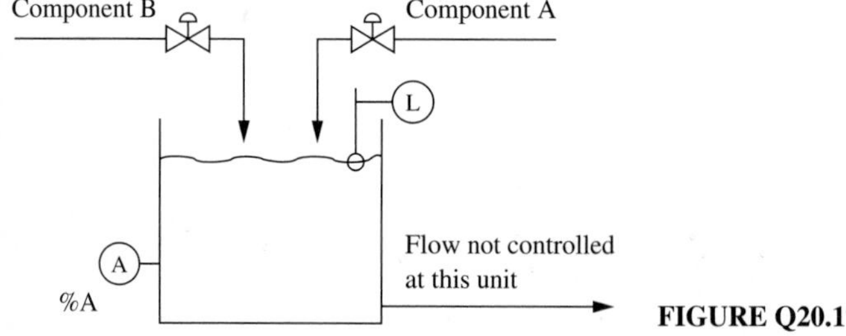

FIGURE Q20.17

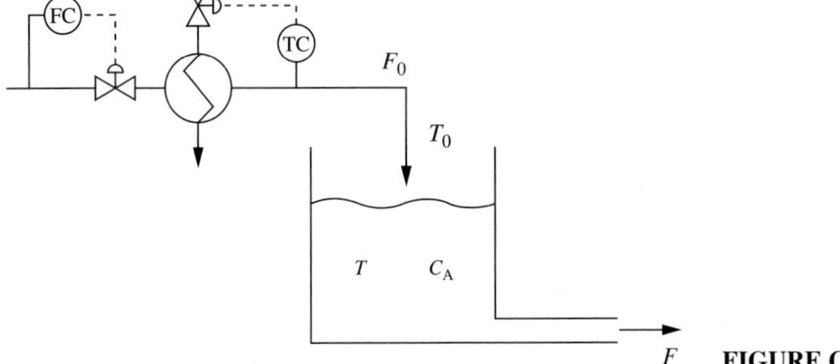

FIGURE Q20.18

20.18. A proposal is made to control the temperature (T) and composition (C_A) in the chemical reactor in Figure Q20.18 by manipulating the feed flow and the inlet temperature. The chemical reaction is A → B, with $r_A = -kC_A$ and no heat of reaction. The flow in the pipe is laminar, so that the flow out can be taken to be proportional to level, $F = KL$. The data for this system at the base case operation is the same as for Example 3.2; in addition, the temperature is 323 K and the reaction rate constant is $k = 2.11 \times 10^5 e^{-5000/T}$.

(*a*) Derive the linearized model for this system in deviation variables.

(*b*) Determine whether the system is controllable in the steady state.

(*c*) Derive the four individual single input-output transfer functions.

(*d*) Confirm the result in Figure 20.15 and explain the results physically.

(*e*) Evaluate the relative gain, both at steady state and as a function of frequency. Explain the differences.

(*f*) Select a feasible loop pairing and design a control system.

20.19. Evaluate the controllability and the interaction for the blending and distillation processes modelled in Section 20.2. Discuss the differences, if any, between the steady-state and frequency-dependent results.

20.20. The analysis of multiloop tuning summarized in Figure 20.9 considered only positive controller gains. Discuss the control performance when one or both of the controller gains is allowed to be negative.

CHAPTER

21

MULTILOOP CONTROL: PERFORMANCE ANALYSIS

21.1 INTRODUCTION

Multiloop process control systems were introduced in the previous chapter, where some important effects of interaction on steady-state and dynamic behavior were explained, and a quantitative measure of interaction—the relative gain—was presented. This understanding of interaction is now applied in the analysis of multiloop control performance and design. Three main facets of control performance analysis are presented and applied to the design of multiloop systems. The first is loop pairing: deciding the controlled and manipulated variables for each single-loop controller in a multiloop system. The second facet is controller tuning to achieve the desired performance, as well as to maintain stability. The third facet involves enhancements to the PID control calculations that can improve control performance while retaining the simplicity of the multiloop control strategy in selected applications.

As in the single-loop case, the first step is to define control objectives thoroughly. The main aspects of multivariable control performance are presented in the following list. Several are the same as for single-loop systems; however, items 2, 5, and 6 are new, and item 4 can assume even greater importance.

1. *Dynamic behavior of the controlled variables.* The control system should provide the desired control performance for expected disturbances and set point changes. The performance can be defined by any appropriate measures presented in Chapter 9 (e.g., IAE and decay ratio).
2. *Relative importance among controlled variables.* The multiloop control structure should be compatible with the relative importance of various controlled variables,

since some controlled variables may be very important and should be maintained close to their set points, while others may not be as important and can be allowed to experience larger short-term deviations.

3. *Dynamic behavior of the manipulated variables.* Feedback control reduces the variability in the controlled variables by adjusting manipulated variables; however, the variability in the manipulated variables should not be too large.
4. *Robustness to model errors.* The control system should be robust so that it performs well in spite of inevitable modelling errors. As with single-loop systems, this objective requires that feedback controllers be tuned to ensure stability and give the best feedback performance possible for the expected model errors. In addition, we shall see that some multivariable control systems are highly sensitive to model errors and can be applied only when models are very accurate.
5. *Integrity to controller status changes.* Each controller should retain reasonable performance for its basic objectives, even if performance is somewhat degraded, as changes occur in the automatic/manual status of interacting loops.
6. *Proper use of degrees of freedom.* The control system should be able to adapt itself to the degrees of freedom available in the process, which can change when a manipulated variable cannot be adjusted (e.g., because it reaches a physical limit). This topic is addressed in Chapter 22.

It would be possible to arrive at the best design by simulating all possible loop pairings and enhancements. However, simulating the numerous candidate designs would be a time-consuming task, especially since the controllers in every candidate would have to be tuned. In addition, such a "brute force" simulation technique would provide little insight into improving performance through changes in process equipment, operating conditions, or control structure.

The approaches presented here are selected because they address the most important issues and generally require less engineering effort than simulating all possibilities. Because these methods build on the results of the previous chapter, it will be assumed that all systems considered are controllable. The new analysis method for each major design decision is addressed in a separate section of the chapter; then, some advanced topics are introduced. Finally, a flowchart is provided to clarify the integration of major analysis steps in reducing potential candidate designs and making decisions for multiloop systems. The hierarchical analysis method eliminates candidates with a minimum of engineering effort and results in one or a few final designs. Because of assumptions in some of these methods, the final design selection may still require simulation, but of only a few candidates. Before the methods are covered, a few motivating examples are presented to highlight some important issues that distinguish multiloop from single-loop performance.

21.2 DEMONSTRATION OF KEY MULTILOOP ISSUES

In this section, four important multiloop issues are introduced through process examples that show the key effects of interaction on the dynamic performance of multiloop control systems. These issues were selected because they often influence control design for process units and they are unique to, or assume heightened significance for, multiloop systems. The analysis methods to address these issues are provided in subsequent sections of this chapter.

Example 21.1 Operating conditions. The first issue is the effect of operating conditions on multiloop control performance, which is introduced through consideration of the blending process in Figure 20.2. We begin by considering the same operating conditions previously considered in Table 20.5, which are repeated in Table 21.1 as the base case. For these operating conditions, the product is very dilute (5% A). Thus, changing the flow rate of component A by a small amount affects the product composition significantly while affecting the total product flow only slightly. This qualitative analysis was substantiated by the quantitative tuning analysis in Example 20.10, which leads to the recommendation of the pairing for the base case in Table 21.1.

Next, we investigate whether a different pairing is recommended for an alternative operating condition that involves a very concentrated product (95% A). In this operation, the product concentration is more sensitive to the flow of the solvent than to the flow of component A, as it was in the base case. The tuning for proportional-integral controllers is determined by the guidelines for 2 × 2 systems with one fast and one slow loop. For this alternative case the loop pairings A_1–F_1 and F_3–F_2 provide better control, because the tunings for the controllers in this configuration are not dependent on the automatic/manual status of the other controller. From this example, we can conclude:

> The proper control loop pairing depends on the operating conditions of the process.

Thus, it is not possible to specify a single control design for each unit operation, like blending or two-product distillation. Even though units may appear similar, at least with

TABLE 21.1
Effect of operating conditions on multiloop performance of the blending system

Operating condition	Set points A_1	F_3	Relative gain λ_{A1-F2}, λ_{F3-F1}	λ_{A1-F1}, λ_{F3-F2}	Pairing: A_1–F_2, F_3–F_1	Pairing: A_1–F_1, F_3–F_2
Base case	0.05	100	0.95	0.05	*Recommended* The controller tuning is essentially the same for single-loop and multiloop control.	*Not Recommended* The controller tuning depends strongly on the status of the interacting loop.
Alternative case	0.95	100	0.05	0.95	*Not Recommended* The controller tuning depends strongly on the status of the interacting loop.	*Recommended* The controller tuning is essentially the same for single-loop and multiloop control.

respect to equipment structure, their operating conditions and the resulting dynamic responses must be considered.

Example 21.2 Transmission interaction. The previous analysis selected the controller pairing that reduces transmission interaction. In fact, the best controller pairings for the two examples are consistent with selecting the multiloop pairings that yield relative gain values closest to 1.0, as verified by the relative gain values in Table 21.1. Given this result, it is tempting to assume that the multiloop control with relative gains closest to 1.0 always gives the best performance. This example demonstrates that this assumption is *not always valid* and that a more complete analysis is required.

This example consists of the two-product distillation tower separating a binary feed considered in Example 20.2. Both top and bottom product compositions are of equal importance, and the major disturbance is a change in feed composition. Two different regulatory loop pairings, which differ only in how the distillate and reflux flow rates are manipulated, are considered. The first, shown in Figure 21.1*a*, has the distillate manipulated to control the overhead drum level and the reflux manipulated to control the top product composition; this is called *energy balance* and was considered in Chapter 20. The second, shown in Figure 21.2*a*, has the distillate and reflux pairings interchanged; this is called *material balance* and is introduced here for the first time. It is important to recognize that the steady-state responses of these two systems are identical because the process equipment, controlled variables, and manipulated variables are the same. Only the transient behavior is different. The linear transfer functions, including 2 min analyzer dead times, for the two systems follow.

Energy balance.

$$\begin{bmatrix} X_D \\ X_B \end{bmatrix} = \begin{bmatrix} \dfrac{0.0747e^{-3s}}{12s+1} & \dfrac{-0.0667e^{-2s}}{15s+1} \\ \dfrac{0.1173e^{-3.3s}}{11.75s+1} & \dfrac{-0.1253e^{-2s}}{10.2s+1} \end{bmatrix} \begin{bmatrix} F_R \\ F_V \end{bmatrix} + \begin{bmatrix} \dfrac{0.70e^{-5s}}{14.4s+1} \\ \dfrac{1.3e^{-3s}}{12s+1} \end{bmatrix} X_F \tag{21.1}$$

Material balance.

$$\begin{bmatrix} X_D \\ X_B \end{bmatrix} = \begin{bmatrix} \dfrac{-0.0747e^{-2s}}{10s+1} & \dfrac{0.008e^{-2s}}{5s+1} \\ \dfrac{-0.1173e^{-2s}}{9s+1} & \dfrac{-0.008e^{-2s}}{3s+1} \end{bmatrix} \begin{bmatrix} F_D \\ F_V \end{bmatrix} + \begin{bmatrix} \dfrac{0.70e^{-5s}}{14.4s+1} \\ \dfrac{1.3e^{-3s}}{12s+1} \end{bmatrix} X_F \tag{21.2}$$

Tuning for these control systems can be determined by the methods in Chapter 20. The results are reported in Table 21.2.

The transient responses for well-tuned feedback control in response to a feed composition upset are given in Figures 21.1*b* and 21.2*b*, and the control performances are summarized in the IAE values in Table 21.2. Based on the total IAE values (0.52 for energy balance and 0.76 for material balance), the performance of the energy balance control design is better than the material balance controller for the feed composition disturbance—in spite of the fact that the interaction, as measured by the relative gain, is much further from 1.0 for the energy balance controller pairing. Thus, we conclude:

> The best-performing multiloop control system is not always the system with the least transmission interaction (i.e., with relative gain elements closest to 1.0).

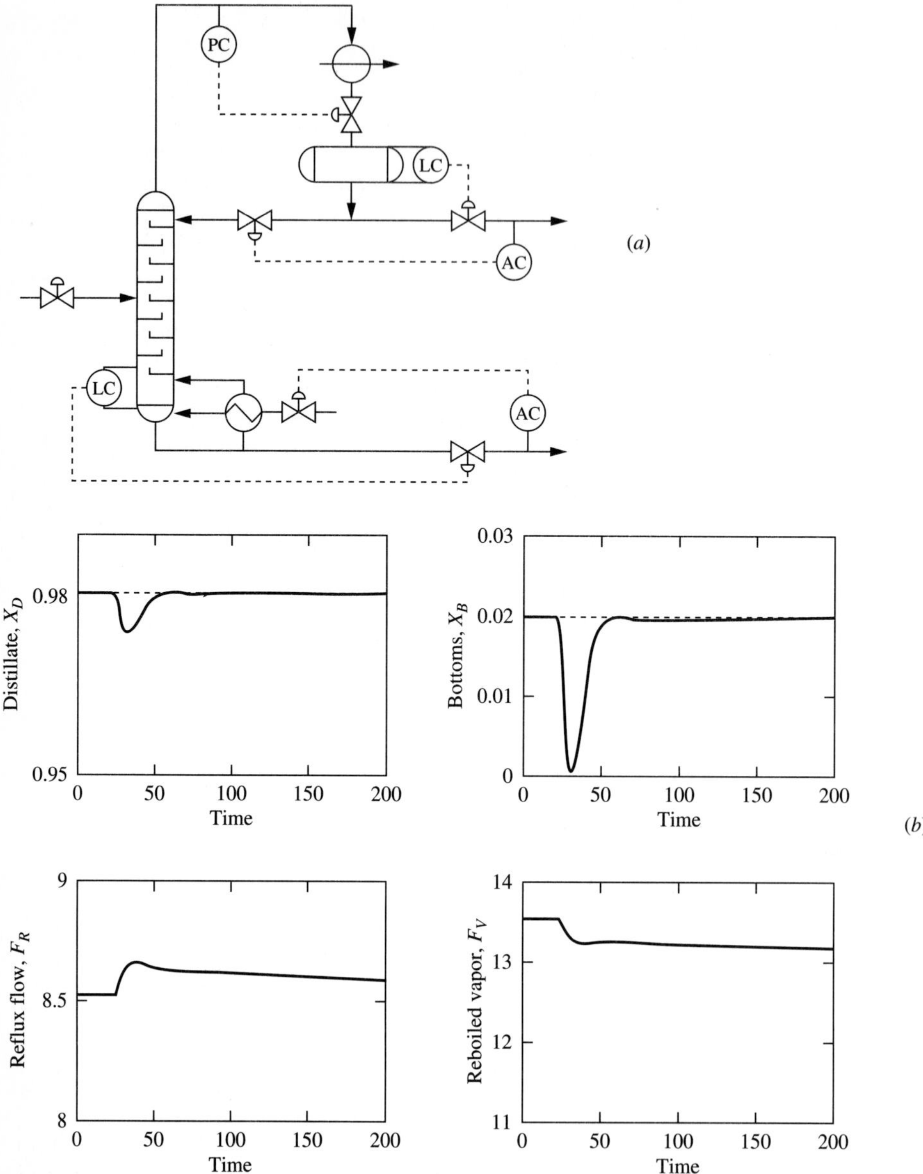

FIGURE 21.1
Energy balance distillation control: (*a*) schematic diagram; (*b*) transient response to a change in light key in feed of −0.04.

This result should not be surprising when one considers the closed-loop transfer function for a multiloop system, derived in Chapter 20 and repeated here.

$$\frac{\mathrm{CV}_1(s)}{D(s)} = \frac{\left(G_{d1}(s) - \dfrac{G_{d2}(s)G_{21}(s)G_{c2}(s)}{[1 + G_{c2}(s)G_{22}(s)]}\right)[1 + G_{c2}(s)G_{22}(s)]}{\mathrm{CE}(s)} \tag{21.3}$$

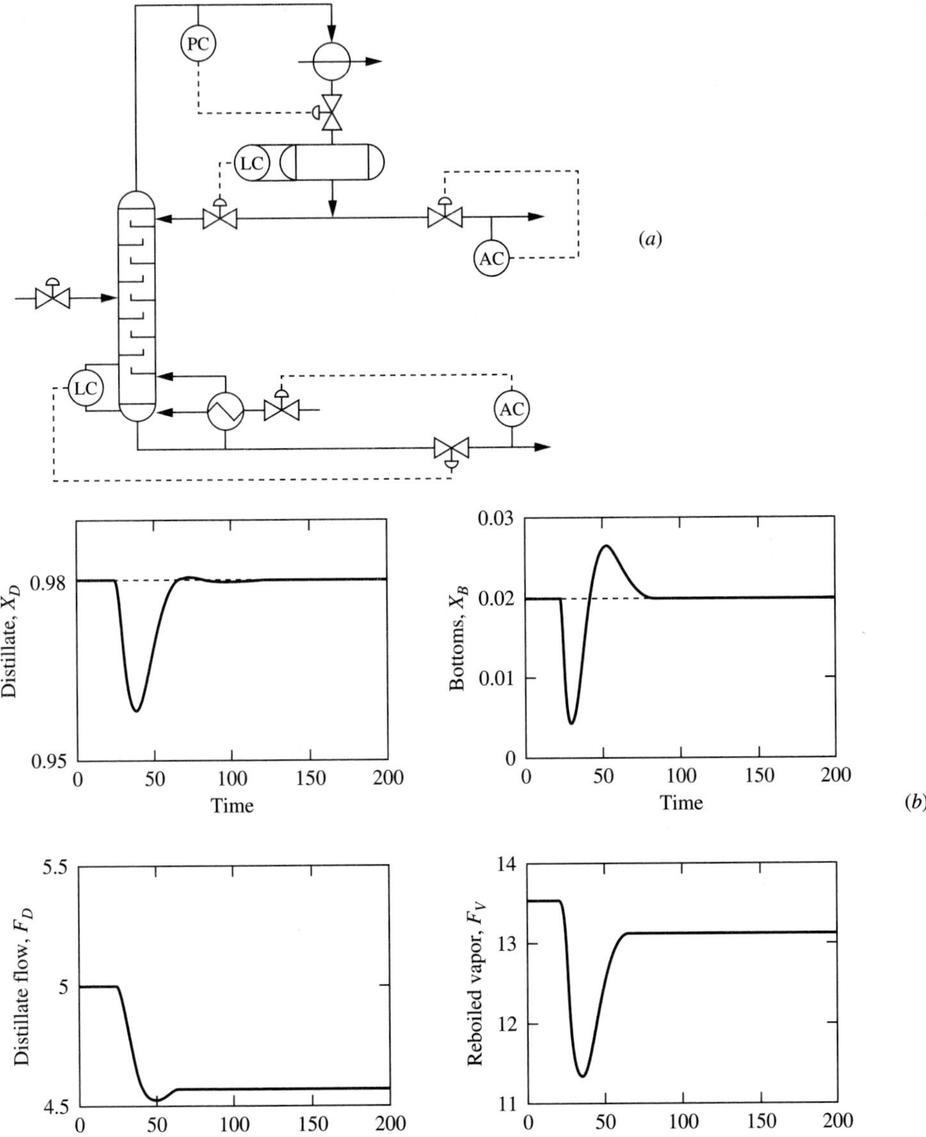

FIGURE 21.2
Material balance distillation control: (*a*) schematic diagram; (*b*) transient response to a change in light key in feed of −0.04.

with

$$\text{CE}(s) = 1 + G_{c1}(s)G_{11}(s) + G_{c2}(s)G_{22}(s) + \frac{G_{c1}(s)G_{11}(s)G_{c2}(s)G_{22}(s)}{\lambda_{11}(s)}$$

The dynamic response depends on all elements in the transfer function, so both numerator and denominator must be considered, especially in multivariable systems. However, the relative gain appears only in the denominator, whereas the disturbance transfer

TABLE 21.2
Tuning and performance data for distillation dynamics

		Energy balance	Material balance
λ_{XD-FB}		6.09	
λ_{XD-FD}			0.39
K_{CD}		10.4	−9.35
T_{ID}		9.0	10.0
K_{CB}		−6.8	−68.7
T_{IB}		6.1	6.7
Feed composition disturbance ($\Delta x_f = -0.04$)	IAE_{XD}	0.17	0.45
	IAE_{XB}	0.35	0.31
SP_{XD} disturbance ($\Delta\text{SP}_{XD} = 0.005$)	IAE_{XD}	0.35	0.0585
	IAE_{XB}	0.34	0.0456

function appears in the numerator. This result is a bit disappointing, since the design of multiloop systems would have been relatively easy if the pairing were determined completely by the relative gain. Transmission interaction is important and must be considered, but a simple pairing method based entirely on the relative gain is not always correct.

Example 21.3. Disturbance type. A further important question concerns the performance of candidate controls for different disturbances. Specifically, is it true that one candidate control pairing performs best for all disturbances? This issue is investigated by extending the study of the two distillation controller pairings for a different disturbance: a set point change to the distillate controller. The dynamic responses for a set point change in the top composition controller of +0.005 mole fraction, with the other set point and all disturbances constant, are given in Figures 21.3*a* and *b*. The results, summarized in Table 21.2, show that the total IAE values are 0.69 for energy balance and 0.104 for material balance. In this case, the material balance system performs better. Note that an attempt to "speed" the sluggish response of the energy balance system through tighter controller tuning will lead to instability.

From this example we conclude:

> The relative performance of control designs and the selection of the best design can depend on the specific disturbance(s) considered.

This result seems reasonable when considering the following closed-loop transfer function for the set point change:

$$\frac{\text{CV}_1(s)}{\text{SP}_1(s)} = \frac{G_{c1}(s)G_{11}(s) + G_{c1}(s)G_{c2}(s)[G_{11}(s)G_{22}(s) - G_{12}(s)G_{21}(s)]}{\text{CE}(s)} \tag{21.4}$$

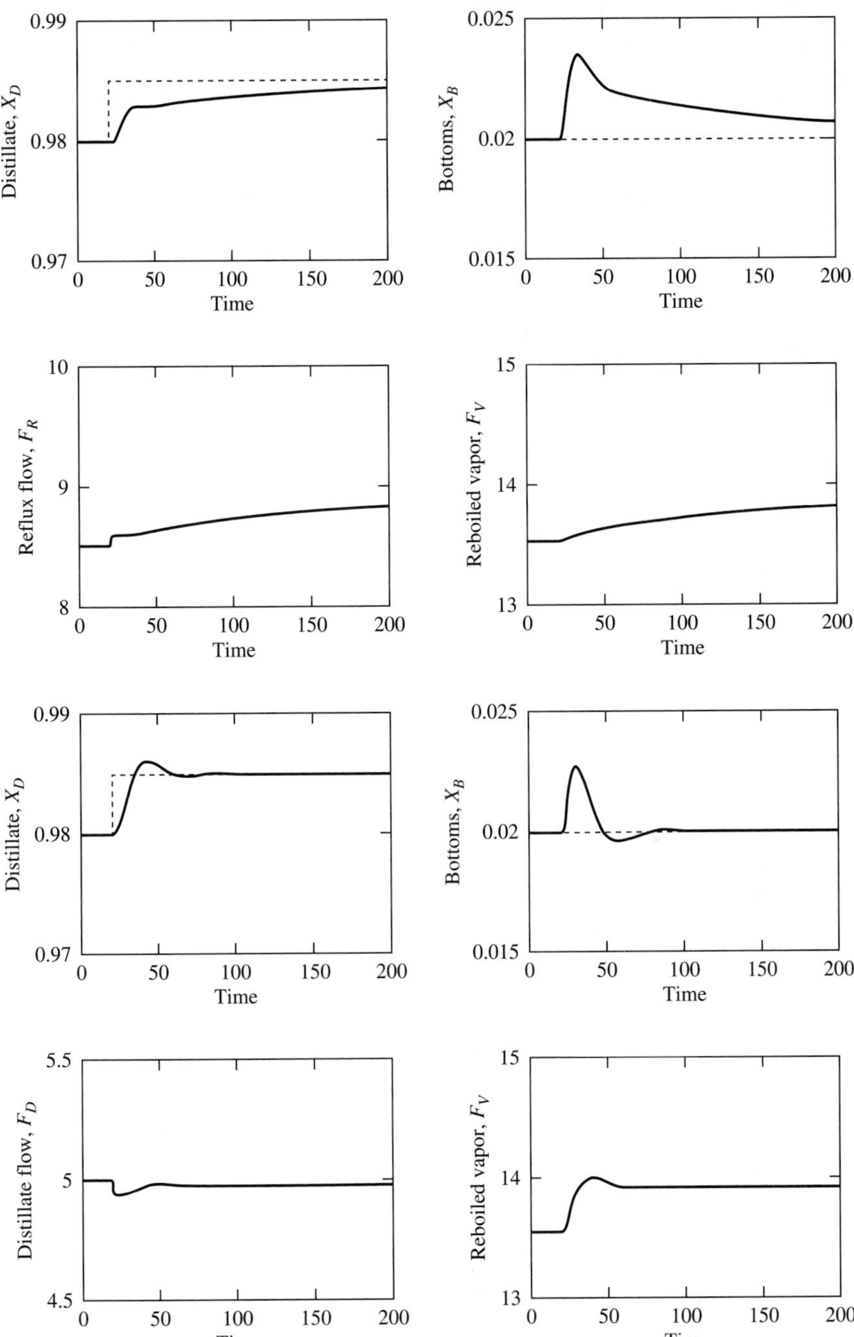

FIGURE 21.3
Transient response of distillation control to +0.005 distillate light key set point change: (*a*) energy balance design; (*b*) material balance design.

The characteristic equation is unchanged from equation (21.3), but the transfer function numerator is different for different disturbances, and thus the control performance could be different. The result again demonstrates the difficulty with having a single, standard design for a unit operation, because the types of disturbances a unit most often experiences depend on the entire plant design.

Example 21.4 Interactive dynamics. The examples covered to this point involved interactive systems in which the transmission interaction is not faster than the "direct" transfer function between the manipulated and controlled variables. Assuming that the controller is paired according to $CV_1(s)$–$MV_1(s)$, the systems studied to this point have had

$$G_{11}(s) \qquad \text{faster than} \qquad \frac{G_{21}(s)G_{c2}(s)G_{12}(s)}{1 + G_{c2}(s)G_{22}(s)}$$

A particularly difficult control challenge can occur when the transmission interaction is faster than the direct process response. As an example, two systems are considered; they have the same steady-state gains, and their steady-state interaction has been investigated as system B in Table 20.1 (page 651), but system B2 has fast transmission dynamics, whereas system B1 has similar dynamics for all transfer functions in the process model. In Example 20.9, system B1 has been shown to have "well-behaved" closed-loop dynamics and to be easily tuned.

System B1.

$$\begin{bmatrix} CV_1(s) \\ CV_2(s) \end{bmatrix} = \begin{bmatrix} \dfrac{1.0e^{-1.0s}}{1+2s} & \dfrac{0.75e^{-1.0s}}{1+2s} \\ \dfrac{0.75e^{-1.0s}}{1+2s} & \dfrac{1.0e^{-1.0s}}{1+2s} \end{bmatrix} \begin{bmatrix} MV_1(s) \\ MV_2(s) \end{bmatrix} \tag{21.5}$$

System B2.

$$\begin{bmatrix} CV_1(s) \\ CV_2(s) \end{bmatrix} = \begin{bmatrix} \dfrac{1.0e^{-3.0s}}{1+2s} & \dfrac{0.75e^{-0.1s}}{1+2s} \\ \dfrac{0.75e^{-0.1s}}{1+2s} & \dfrac{1.0e^{-0.1s}}{1+2s} \end{bmatrix} \begin{bmatrix} MV_1(s) \\ MV_2(s) \end{bmatrix} \tag{21.6}$$

System B2 has the same steady-state gains but very different dynamics. To first acquire some understanding of this system, the dynamic response is determined for a step change in $MV_1(t)$ with only the controller for variable 2 in automatic; this is the process reaction curve for the process $MV_1(t)$–$CV_1(t)$ with the other controller in automatic. The dynamic response in Figure 21.4*a* shows an inverse response, because the fast transmission effect produces an initial negative response before by the slower diagonal $[G_{11}(s)]$ effect produces a positive steady-state response.

A similar dynamic response was encountered in Section 5.4, where the complex dynamics were attributed to the effects of parallel paths between inputs and outputs, giving a right-half-plane (positive) numerator zero in the transfer function. It is important to recognize that the structure of a multiloop system with interaction ensures that parallel paths exist; the parallel paths include the direct transfer function and transmission interaction, as shown in Figures 20.7 and 20.8. These parallel paths do not always create complex feedback dynamics such as inverse response or initial overshoot, but

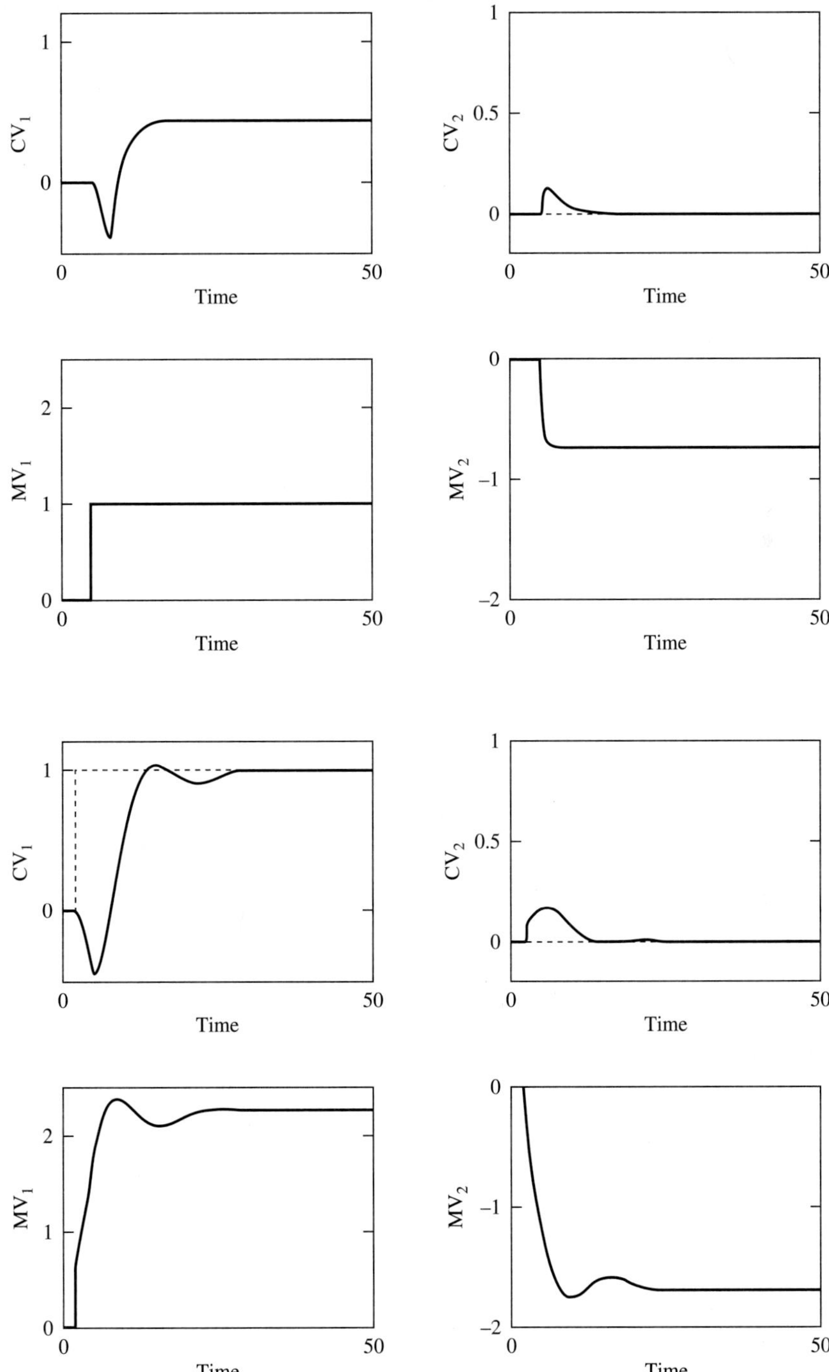

FIGURE 21.4
System B2. (*a*) Process reaction curve of MV_1–CV_1 with other loop closed; (*b*) multiloop transient response to set point change in CV_1.

TABLE 21.3
Effect of dynamics on multiloop performance

Case	K_{c1}	T_{I1}	K_{c2}	T_{I2}	IAE_1	IAE_2	$IAE_1 + IAE_2$
B1: Uniform interactive dynamics (Figure 20.11)	1.23	1.76	0.89	1.06	3.46	2.46	5.92
B2: Complex interactive dynamics (Figure 21.4*b*)	0.71	3.00	4.00	2.97	9.80	1.27	11.07

the possibility always exists. In system B2 the interactive path is faster and has an effect opposite to the direct effect, leading to the initial inverse response.

A process with an initial inverse response is usually difficult to control, as demonstrated in Section 13.5; thus, interaction with *fast transmission dynamics* can result in poor control performance. As an example, the control response of system B2 to a set point change in CV_1 with PI tunings that yield minimum ($IAE_1 + IAE_2$) is given in Figure 21.4*b*. (Again, this simple measure of control performance is selected for comparison purposes only.) The feedback controller cannot eliminate the initial inverse response, which results in a relatively long time during which $CV_1(t)$ is far from its set point.

The tuning and performance for systems B1 and B2 are compared in Table 21.3. This example clearly demonstrates the importance of interactive dynamics; recall that both systems B1 and B2 have the same steady-state interaction, but system B2 has poorer performance.

This example demonstrates:

> Multivariable systems with strong interaction and fast transmission dynamics can result in complex dynamic responses, involving inverse response or large overshoot, which can degrade control performance.

The examples considered in this section have demonstrated that the design of a multiloop control system is challenging task, involving more complex issues than single-loop systems, and that the process dynamic responses, operating conditions, disturbances, and extent of interaction must all be considered. The next three sections present methods for considering these issues when making the three main multiloop decisions: loop pairing, tuning, and enhancements.

21.3 MULTILOOP CONTROL PERFORMANCE THROUGH LOOP PAIRING

Loop pairing—the selection of controlled and manipulated variables to be linked through single-loop controllers—is an extremely important design decision. For the distillation examples in Figures 21.1*a*, the two possible pairings are (1) X_D–F_R and X_B–F_V and (2) X_D–F_V and X_B–F_R. However, for a system with more manipulated variables, the number of potential designs becomes very large; in fact, the number of initial candidates for a process with n manipulated and controlled variables is n factorial ($n!$). For example, there are 125 candidates for a five-controller, five-manipulated-variable distillation system in Figure 21.1a when the product compositions, pressure, and levels are considered! Clearly, the number of candidates must be reduced significantly, or the analysis task will require an enormous effort to evaluate all candidates. In this section, three separate analyses are described for eliminating clearly unacceptable pairing candidates and evaluating the remainder for likely performance. These analyses would be applied only to process designs that have been verified to be controllable and to have an adequate operating window. Also, the three analyses are employed sequentially, with only those candidates passing the prior steps evaluated at the next step.

Steady-State Process Gain

The first analysis builds on interpretations of the relative gain that were developed in the previous chapter. For 2×2 systems with the relative gain less than zero for a pairing, the sign of a controller gain in the system that provides stable performance depends on the statuses of other controllers (McAvoy, 1983; Grosdidier et al., 1985). In this situation, if a correct controller gain is positive when both controllers are in automatic, the controller gain has to be negative when the other controller is in manual. The same issue arises in higher-order systems and can be tested using the Niederlinski criterion given in Section 20.7.

Also, when the relative gain λ_{ij} is zero for a pairing, the steady-state gain of the pairing $\text{CV}_i(t)$–$\text{MV}_j(t)$ is zero when the other loops are open, and the controller cannot function unless the other controllers are in automatic to create a feedback loop via the transmission interaction path. The manner in which interaction forms the feedback loop can best be pictured through Figure 20.7 (page 649), where the transmission effect can be seen to provide a nonzero connection between the $\text{MV}_1(s)$ and $\text{CV}_1(s)$. The transmission relationship can be demonstrated with equation (20.13), which can have a nonzero relationship between $\text{MV}_1(s)$ and $\text{CV}_1(s)$ even if $G_{11}(s)$ is zero, as long as $G_{12}(s)G_{21}(s)G_{c2}(s)/[1 + G_{c2}(s)G_{22}(s)]$ is nonzero.

In both of these cases, proper functioning of a control loop requires that the adjustments from all other controllers be implemented at the final elements, which would not be satisfied if interactive control loops (1) were in manual mode or (2) were saturated at upper or lower bounds. It is not uncommon for these situations to occur, at least temporarily; thus, multiloop control designs with relative gains less than or equal to zero could often fail to provide stable feedback regulation. The only way to prevent these failures would be to prepare a computer program to monitor the control system continuously and change controller gains and automatic/manual statuses depending on the condition of all controllers in the multiloop system.

Therefore, the first loop-pairing guideline can be stated:

> Controller pairings with negative relative gains are normally rejected from further consideration, and those with zero relative gains are avoided.

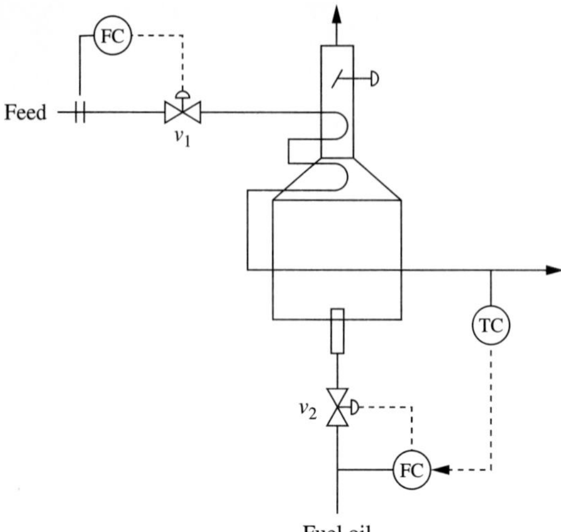

FIGURE 21.5
Furnace multiloop control pairing on variables with $\lambda > 0$.

As an example, consider the fired heater process in Figure 21.5. The process fluid flows through a pipe (termed a *coil*) and is heated by radiant and convective heat transfer from the combustion of fuel. The variables to be controlled are the process fluid flow rate and the process fluid outlet temperature, and the two manipulated valves are in the process fluid (v_1) and fuel (v_2) lines. When no feedback controllers are present, the process fluid flow rate is influenced directly only by v_1, and the outlet temperature is influenced by both v_1 and v_2. Thus, the 2×2 gain matrix has a zero, and as shown in Chapter 20, the relative gain array has ones in the diagonal elements and zeros in the off-diagonal elements. There is only one pairing with nonzero relative gain values, and this pairing is shown in Figure 21.5, which is the *common* loop pairing used in most industrial designs.

The guideline for eliminating pairings on nonpositive relative gains conforms to theory and common industrial practice; however, there are a *few cases* where the rule is violated and pairings with zero relative gains are used. These unconventional designs are employed, in spite of their recognized drawbacks, to achieve specific advantages—typically, very fast feedback dynamics for a particularly important controlled variable. An example of an exception is given in Figure 21.6. In this case, the tight control of the coil outlet temperature is very important, and the dynamic response between the process flow valve v_1, and the temperature can be very fast when the fluid residence time in the coils is short. Since the open-loop gain between valve v_2 and the process fluid flow is zero, the proper functioning of the flow controller in this case requires the operation of the temperature controller. This design is used industrially only when the temperature is of especially great importance, feed flow control need not be controlled tightly, and other steps to improve control performance are not possible or are extremely costly. Note that pairing on a relative gain of zero is allowed only when the remainder of the system, excluding the input-output pair, is controllable, so that the other feedback controllers can function properly.

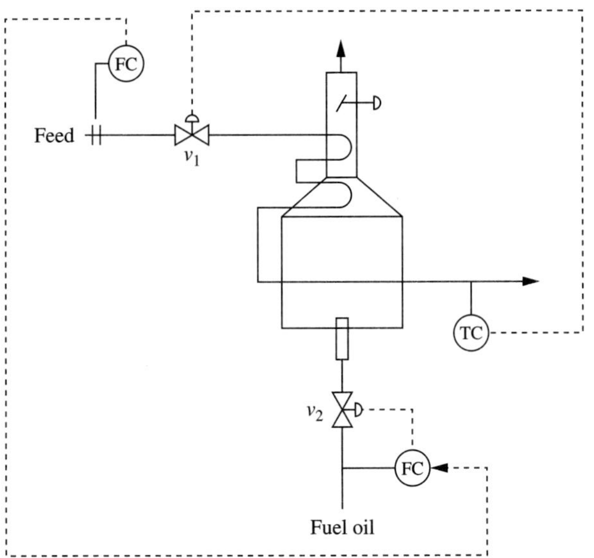

FIGURE 21.6
Furnace multiloop control pairing on variables with $\lambda = 0$.

Dynamics

Only the candidates with nonnegative relative gains are considered in the second analysis, which considers the feedback dynamics of the various pairings in conjunction with the control objectives. If one or a few controlled variables are much more important, the control loop pairing should be selected to give good performance for the most important variables. As demonstrated in discussions on single-loop control, control performance is much better when the feedback process dynamics involve a fast process with small fraction dead time. Thus, the second loop-pairing guideline is stated as follows:

> Very important controlled variables should be paired with manipulated variables that provide fast feedback dynamics with small dead times and time constants and negligible inverse response.

As an example of this guideline, consider the simplified system in Figure 21.7 in which two gases are mixed, as might occur where the heating value of the mixed gaseous fuel stream is to be controlled. The sources of the feeds are a gas stream L (lower heating value) and a vaporizer for the stream H (higher heating value). The controlled variables are the pressure and the composition in the pipe after mixing, and both manipulated variables affect both controlled variables. Generally, the pressure is of greatest importance, because variations could lead to unsafe conditions; short-term composition variations, while not desirable, can be more easily tolerated. Therefore, the pressure is controlled by manipulating the fast-responding gas feed, while the composition is controlled by manipulating the more slowly responding vaporization process. Since the pressure is most important, this pairing would be used

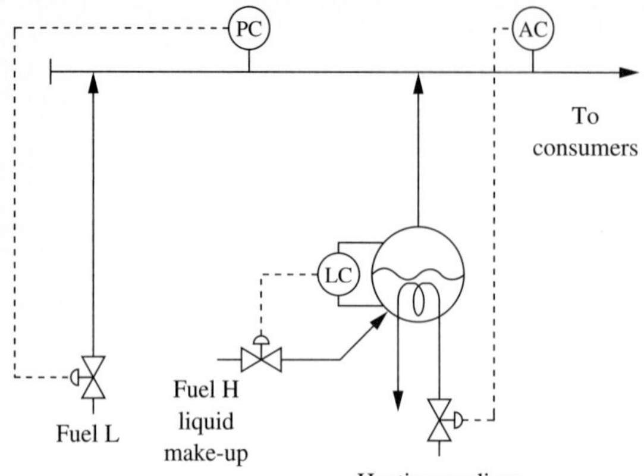

FIGURE 21.7
Fuel gas control system with key pressure variable paired with fast manipulated variable.

as long as the gas feed valve has the flexibility range to control pressure—in other words, as long as it does not go fully opened or closed in response to disturbances—regardless of the interaction effects on the composition.

Example 21.5. Evaluate the two possible loop pairings for the blending example process with base-case conditions in Table 21.1 according to the relative gain and dynamic responses.

The relative gain array for the blending process with dilute product (5% A) can be evaluated from the steady-state gains to be

Relative gain array:

	F_1	F_2
A_1	0.05	0.95
F_3	0.95	0.05

Since none of the elements is less than or equal to 0.0, both possible pairings are allowed based on the first guideline. Also, the data reported in Example 20.10 show the same dynamic responses for both pairings, since the dominant dynamics are due to the sensors. Therefore, neither pairing has an advantage regarding dynamics. Finally, since the two guidelines do not exclude either pairing, the results in Table 21.1 give strong evidence for preferring the A_1–F_2 and F_3–F_1 pairing, since the tuning of each controller does not depend on the automatic/manual status of the other.

Example 21.6. Evaluate the two possible loop pairings for the distillation example in Figure 20.3 according to the relative gain and dynamic responses.

The relative gain array can be evaluated from the steady-state gains in equation (20.24), giving

Relative gain array:

	F_R	F_V
X_D	6.09	−5.09
X_B	−5.09	6.09

Since only the pairing X_D–F_R and X_B–F_V has positive relative gains, only this pairing is allowed by the first guideline; this is the design in Figure 21.1*a*. The loop dynamics for the allowed pairing are not slower, and are even slightly faster, than the disallowed pairing, which indicates that there is no significant disadvantage to this design based on feedback dynamics.

Performance Measure

The third analysis addresses the remaining candidate pairings, involving controllable systems with positive relative gains, similar feedback dynamics, and controlled variables of equal importance, by investigating the control performance for specific disturbances. If only a few candidates remained at this point, one could simulate the systems for the important disturbance(s) to select the best design, as was done for the distillation tower in Examples 21.2 and 21.3. Here a shortcut method is outlined that provides a quick estimate of control performance and is useful in reducing the pairing candidates that can yield good control performance. Equally important, it provides insight into the effects of disturbances, specifically how interaction can be favorable or unfavorable in multiloop control (Stanley et al., 1986). The approach is introduced for 2×2 systems; however, it can be extended to higher-order systems (Skogestad and Morari, 1987). In spite of its advantages, the method does not provide a definitive recommendation, because of the assumptions required; thus, some care is required in its application, and the results may have to be verified through dynamic simulation.

The method takes advantage of a simple estimate of control performance that can be determined directly from the closed-loop transfer function. The control performance measure used here is integral error, which can be obtained directly by using the following relationship (see Appendix D):

$$\int_0^\infty E(t)dt = \lim_{s \to 0} \int_0^\infty E(t)e^{-st}dt = E(s)\big|_{s=0} \tag{21.7}$$

This relationship demonstrates that the integral of a variable, specifically the error, can be obtained from the transfer function of a stable system *without solving for the complete transient response* (Gibilaro and Lee, 1969). Naturally, much detailed information about the transient response is lost, but a useful single measure of control performance is easily obtained. A large integral error indicates poor performance and a pairing candidate that should be eliminated. A small integral error *can* result from good performance, and the pairing should be retained for further evaluation. However, large positive and negative errors occurring during the transient can cancel in this calculation (this is not the IAE!), so a small value of integral error does not definitely prove good control performance. Thus, the final selection requires further evaluation, such as a simulation, to determine the transient behavior.

The closed-loop disturbance response transfer function for a 2×2 system is given in equation (21.3). The relationship in equation (21.7) can be applied to equation (21.3) with $D(s) = 1/s$, resulting, after some rearrangement, in

$$\left[\int_0^\infty E_1(t)dt\right]_{\text{ML}} = \left[\int_0^\infty E_1(t)dt\right]_{\text{SL}} (f_{1,\text{tune}})(\text{RDG}_1) \tag{21.8}$$

with integral error under multiloop control $= \left[\int_0^\infty E_1(t)dt\right]_{\text{ML}}$

$$\text{integral error under single-loop control} = \left[\int_0^\infty E_1(t)dt\right]_{\text{SL}} = \frac{K_{d1}(T_{I1})_{\text{SL}}}{K_{11}(K_{c1})_{\text{SL}}} \quad (21.9)$$

$$\text{detuning for multiloop control} = f_{1,\text{tune}} = \frac{(K_{c1}/T_{I1})_{\text{SL}}}{(K_{c1}/T_{I1})_{\text{ML}}} \quad (21.10)$$

$$\textbf{Relative disturbance gain} = \text{RDG}_1 = \lambda_{11}\left(1 - \frac{K_{d2}K_{12}}{K_{d1}K_{22}}\right) \quad (21.11)$$

The multiloop control performance calculation in equation (21.8) is arranged to be the product of three factors so that separate facets of multiloop control are represented in each factor: (1) a factor for the single-loop performance, (2) a factor for tuning adjustment, (3) a factor accounting for interaction and disturbance. The first factor represents the single-loop performance that would be achieved if the other control loop were not in operation (e.g., in manual). This term again demonstrates that aspects of single-loop control performance, which are summarized in Chapter 13, also influence the controlled variables in a multiloop system. For example, fast feedback dynamics and small disturbance magnitudes are beneficial in multiloop systems.

The final two factors represent the change in control performance due to the multiloop structure. The tuning term f_{tune} represents the effects of detuning the PI controllers for multiloop control. The values of the multiloop tuning constants can be estimated using methods in Chapter 20 or alternative methods cited in the references. The multiloop tuning can be quickly estimated from Figures 20.12*a* and *b* (page 665) for systems which satisfy the appropriate assumptions. The results in those figures can be combined to provide the correlation between the relative gain and the detuning factor, given in Figure 21.8. Since the relative gain in most properly designed control systems is greater than about 0.7, the correlation shows that the detuning factor is usually bounded between 1.0 and 2.0 for 2×2 systems (Marino-Galarraga et al., 1987).

Thus, the effect of multivariable control is dominated by the third term, which is called the *relative disturbance gain,* RDG. The relative disturbance gain is the product of the relative gain and a disturbance factor. Recall that the relative gain is an inherent property of the feedback process, independent of the type of disturbance. In contrast, the disturbance term depends on the type of disturbance; for example, it has different values for feed composition and set point changes to a distillation tower.

The influence of the RDG is first analyzed from a mathematical, then a process point of view. The RDG is the product of two values, and its magnitude is small when control performance is good. The first factor is the relative gain; if the relative gain has a large value, its contribution will be to degrade control performance, because the integral error will tend to increase. The second factor represents the effect of the disturbance type, and because it is the difference of two values, it can have a magnitude ranging from zero to very large. A small magnitude of this factor indicates that

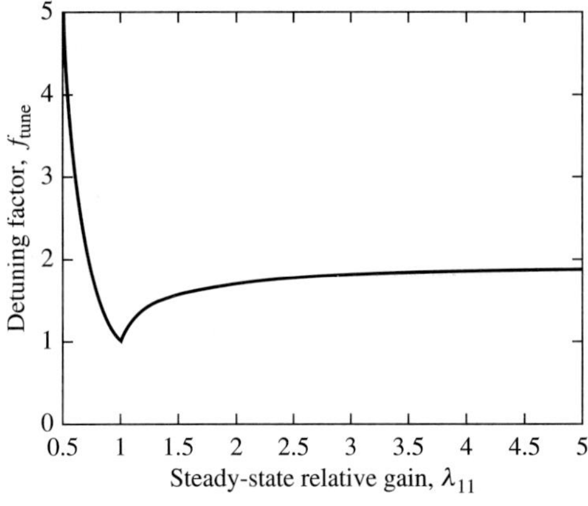

FIGURE 21.8
Correlation between detuning factor f_{tune} and relative gain for 2×2 system with equal input-output dynamics.

the multiloop performance could be much better than the single-loop performance. This situation would occur when the term $(1 - K_{d2}K_{12}/K_{d1}K_{22})$ has a value near zero, which is interpreted as favorable interaction. The other result, with a large disturbance contribution and much poorer multiloop performance, is also possible and is interpreted as unfavorable interaction.

> The combined effects of inherent process interaction and disturbance type determine the dominant difference between single-loop and multiloop control performance. These effects are reflected in the magnitude of the relative disturbance gain (RDG).

This clearly demonstrates that multiloop control performance can be better or worse than single-loop performance for some disturbances.

A key element in determining the effect of interaction in multiloop systems is the manner in which a disturbance affects both controlled variables, sometimes referred to as the "direction" of the disturbance. Thus, it is worthwhile considering the basis for favorable interaction. Favorable interaction occurs when controller 2, in correcting its own deviation from set point, makes an adjustment that improves the performance of controller 1, $CV_1(t)$. The net effect must consider the effects of the disturbances on both controlled variables (K_{d1} and K_{d2}), the manipulation taken to correct the $CV_2(t)$ deviation (characterized by $1/K_{22}$) and the interaction term (K_{12}). All of these parameters are in the interaction factor of the relative disturbance gain.

Example 21.7. For the distillation towers in Figures 21.1 and 21.2, evaluate the relative disturbance gain and provide an interpretation of the effect of interaction on the control performance of the distillate composition, X_D, for a disturbance in the feed composition.

The effect of interaction on control performance is predicted by equation (21.8), and the calculations are summarized in Table 21.4 for both distillation control designs. Recall that good (poor) control performance is predicted by a small (large) magnitude of the product of the detuning factor and the RDG. This analysis predicts that the energy balance performs better for feed composition disturbances, because its sum of values of

TABLE 21.4
Summary calculations of predicted control performance for the distillation tower in Examples 21.7 and 21.8

Data and calculated variable		Energy balance design in Figure 21.1*a* X_D	X_B	Material balance design in Figure 21.2*a* X_D	X_B
K_{FR}		0.0747	0.1173		
K_{FD}				−0.0747	−0.1173
K_{FV}		−0.0667	−0.1253	0.008	−0.008
λ		6.09		0.39	
f_{tune}		2.0		5.0	
Feed	K_d	0.70	1.3	0.70	1.3
composition	RDG	0.071	0.94	1.11	0.06
disturbance	$f_{\text{tune}} \cdot$ RDG	0.14	1.88	5.55	0.30
Set point	K_d	1.0	0.0	1.0	0.0
change (X_D)	RDG	6.09	*	0.39	*
	$f_{\text{tune}} \cdot$ RDG	12.2	*	1.53	*

* Predicted $\int E\,dt$ is finite, although RDG is infinite, due to cancellation of K_{d2} (which is zero) in numerator and denominator.

$f_{\text{tune}} \times \text{RDG}_i$ for the two compositions is smaller than for the material balance system. This conclusion is confirmed by the simulation results in Figures 21.1*b* and 21.2*b* and in Table 21.2.

The physical interpretation of the favorable interaction is considered here for the control design in Figure 21.1*a*. The initial effect of increased light key in the feed (before the effects of the analyzer controllers) results in the top and bottom products having too much light key. In response, the bottom controller increases the heating flow rate (i.e., reboiler duty). This adjustment by the bottom controller has the effect of decreasing the light key in the top product, exactly what the top controller is doing itself! The top controller must also take action by increasing the reflux; however, the (reinforcing) interaction from the bottoms controller improves the overall control performance. Therefore, the energy balance control pairing has favorable interaction and good multiloop performance for the top controller in response to a feed composition disturbance. The reader should repeat this thought experiment for the material balance system to confirm that the interaction is unfavorable for X_D.

Example 21.8. For the distillation towers in Figures 21.1 and 21.2, evaluate the relative disturbance gain for a change in the distillate composition controller set point and select the better design for X_D.

The analysis method, summarized in Table 21.4, correctly predicts that the material balance performs better for set point changes in the distillate controller, as was found by simulations in Figure 21.3*b*. Note that equation (21.3) can be used to represent a set point change by setting $G_{d1}(s) = 1.0$ and $G_{d2}(s) = 0.0$, and in this case the RDG_1 is equal to λ_{11}.

In summary, equation (21.8) provides the basis for estimating the major effect of multiloop control on the performance of each controlled variable. The information required to perform this calculation involves process gains in the feedback path K_{ij}

and the open-loop disturbance gains K_{di}, which can be easily determined from a steady-state analysis. One should consider the likely errors in the values of the gains, as well as in the simplifications in linearizing the process model, when interpreting the results. Small differences (10–20%) in predicted integral error should be considered within the accuracy of the information, and the candidate loop pairings should be considered indistinguishable.

This subsection introduced the consideration of disturbance type, which should be considered in all analyses of multiloop systems. However, it is necessary to repeat a caution concerning the use of the integral error, which can be small because of cancellations of large positive and negative errors. Thus, while large values of |RDG| definitely indicate poor control performance, small values do not necessarily indicate good performance. The best recourse to determine the effects of complex dynamics at this time is to perform a dynamic simulation. Note that the procedures described here are useful in substantially reducing the number of candidates for simulation, as well as providing insight into the importance of disturbance type (or "direction") on control performance.

Control Range

The method for determining controllability in Chapter 20 is valid for the linearized model at the point of linearization. For most processes that are not highly nonlinear, the results can be extended in a region about the point. However, there is no guarantee that the results can be extrapolated, especially when a manipulated variable encounters a constraint while attempting to make the change required by the controller. The method for identifying difficulties with range in achievable steady-state behavior is to determine the operating window of the process. Even if all steady states are feasible, manipulated variables may reach limits during transients; dynamic simulation would be required to determine the importance of a temporary saturation of a manipulated variable.

This section demonstrated a stepwise method for evaluating candidate multiloop control designs:

1. Use the relative gain to eliminate some pairings (which are not integral-controllable or lack integrity).
2. Use dynamic models to select pairings with fast dynamics for important variables.
3. Use approximate control performance analysis—the relative disturbance gain (RDG)—for specific disturbances to evaluate systems with controlled variables of equal importance.

Note that step 1 requires only steady-state information, which means that it is easy to perform with limited modelling information. Also, steps 2 and 3 require approximate dynamic information to identify where major differences in feedback dynamics are present. This approximate dynamic modelling information is also generally easy to obtain. If the effects of interactive dynamics are not easily predicted, so that the methods here cannot provide conclusive recommendations, the final design could be simulated to determine its performance.

21.4 MULTILOOP CONTROL PERFORMANCE THROUGH TUNING

The tuning of PID feedback controllers should be matched with the control objectives. Prior to tuning, the first two steps presented in the previous section should be applied, to eliminate inappropriate pairings by the use of the relative gain and Niederlinski criterion and to select pairings with fast feedback dynamics for the important controlled variables. In all cases, controllers for the most important controlled variables should be tuned tightly. The tuning of the controllers of lesser importance depends on the type of interaction present: favorable or unfavorable.

For systems with *unfavorable* interaction, as predicted by the relative disturbance gain, the effect of interaction degrades the performance of other loops; this degradation can be reduced through judicious controller detuning, consistent with the control objectives. Thus, the controllers for the important variable(s) would be tuned tightly, as close as possible to single-loop tuning. To ensure stability and prevent unfavorable interaction, the controllers for the less important variables would usually be detuned. However, if the interaction is *favorable,* as indicated by a small relative disturbance gain, interaction improves the performance of other loops and should be maintained by proper tuning. In this case, the interacting loop, even if not of great importance itself, should be tuned as tightly as possible to enhance the favorable interaction.

There are no exact guidelines for how the less important controllers should be tuned. When interaction degrades control performance, a starting approach is to tune the important loops close to their single-loop values and detune the less important loops by decreasing their controller gains. Normally, all feedback controllers would retain an integral mode to return the controlled variables to their set points (albeit very slowly for some variables) after disturbances. When both are to be tightly tuned, the method in Chapter 20 would give initial values. An example of how differences in control performance in the same process can be induced through different tuning is given in the results in Table 20.2.

Example 21.9. The effects of tuning on the control performance of the energy balance distillation control design in Figure 21.1*a* are investigated. For this example (only), the distillate product composition is assumed to be much more important than the bottoms composition, so the bottoms composition will be allowed to experience larger short-term variation about its set point. Since no strict guidelines exist for this tuning, the extent of detuning from the tight values in Table 20.7 used in this example represents exploratory results.

The effects of tuning, as determined by simulating the entire response, are given in Table 21.5. For a set point change in X_D, the interaction is unfavorable, as demonstrated by the large magnitude of RDG $\cdot$ f_{tune} (12.2) in Table 21.4. Therefore, tight tuning of the distillate composition controller, along with detuning the bottoms loop, reduces interaction and improves the performance of the distillate composition controller (reducing the IAE from 0.71 to 0.35). As expected, the variation in the bottoms composition (IAE) increased as the bottoms controller was detuned.

For the feed composition disturbance, the interaction is favorable, as demonstrated by the small magnitude of RDG $\cdot$ f_{tune} (0.14) in Table 21.4. Therefore, the control performance in the case with both controllers tightly tuned has better

TABLE 21.5
The effects of tuning on performance for Example 21.9

	Tuning				Performance	
Input change	K_{cXD}	T_{IXD}	K_{cXB}	T_{IXB}	IAE_{XD}	IAE_{XB}
Set point,	10.4	9.0	−6.8	6.1	0.71	0.68
($\Delta SP_{XD} = 0.01$)	20.75	9.0	−3.4	6.1	0.35	1.37
Feed composition,	10.4	9.0	−6.8	6.1	0.17	0.35
($\Delta X_B = -0.04$)	10.4	9.0	−2.0	6.1	0.36	1.18

distillate composition performance (IAE of 0.17) than the case with the bottoms controller detuned (IAE of 0.36), since detuning reduces the *favorable* interaction.

The discussion in this section and the results of Example 21.9 reinforce the importance of considering the effects of the disturbances in control design and tuning.

> Multiloop tuning should be chosen to retain favorable interaction and to reduce unfavorable interaction.

21.5 MULTILOOP CONTROL PERFORMANCE THROUGH ENHANCEMENTS: DECOUPLING

When the previous analyses are complete, it is possible to arrive at a design with two (or more) equally important controlled variables, which may not have the desired performance even with the best pairing and tuning. Often, the limiting factor is unfavorable interaction, which is indicated by a large magnitude of the relative disturbance gain (|RDG|). When poor control performance stems from unfavorable interaction, a potential solution involves reducing interaction through an approach called *decoupling,* which has the theoretical ability to improve performance in some loops without degrading performance in others.

Decoupling reduces interaction by transforming the closed-loop transfer function matrix into (an approximate) diagonal form, in which interaction is reduced or eliminated. There are at least three different decoupling approaches: (1) altering the manipulated variables, (2) altering the controlled variables, and (3) retaining the original variables but altering the feedback control calculation. Each is presented briefly in this section.

Manipulated Variables

The first decoupling approach involves changing the control structure to affect different manipulated secondary variables in a cascade structure, with the same final elements. This approach will be introduced by reconsidering the blending Example 20.1, in which both manipulated variables influence both controlled variables. The goal is to control the same variables (A_1 and F_3) with altered manipulated variables so that the altered system's gain matrix is diagonal or nearly diagonal. This goal is usually achieved through process insight. The restructured dynamic model can be developed from equations (20.1 and 20.2) without linearizing.

$$\tau_A \frac{dA_1(t)}{dt} = \left(\frac{F_2(t-\theta_A)}{F_1(t-\theta_A)+F_2(t-\theta_A)} \right) - A_1(t) = \text{MV}_1(t-\theta_A) - A_1(t) \quad (21.12)$$

$$\tau_F \frac{dF_3(t)}{dt} = F_1(t-\theta_F) + F_2(t-\theta_F) - F_3(t) = \text{MV}_2(t-\theta_F) - F_3(t) \quad (21.13)$$

From this model it becomes clear that the two controlled variables would be independent if the manipulated variables were defined as follows:

$$\text{Manipulated variable number 1} = \text{MV}_1 = F_2/(F_1+F_2)$$
$$\text{Manipulated variable number 2} = \text{MV}_2 = F_1+F_2$$

With this modification, the system in equations (21.12) and (21.13) has been altered to two independent input-output relationships, and as a side benefit the altered system is linear. Thus, standard single-loop control methods can be used to tune the controllers in this decoupled system.

The control strategy can be implemented using real-time calculations and cascade principles, as shown in Figure 21.9, because F_1 and F_2 are measured and respond essentially instantaneously to changes in the valve positions. For example, when the mixed flow (F_3) set point is increased, the initial response of controller F_3 is to increase the total flow (F_1+F_2) set point; this is achieved by adjusting v_1. This changes the flow ratio and is quickly followed by an adjustment by the flow ratio controller to increase v_2 to maintain the proper ratio $F_2/(F_1+F_2)$; this adjustment

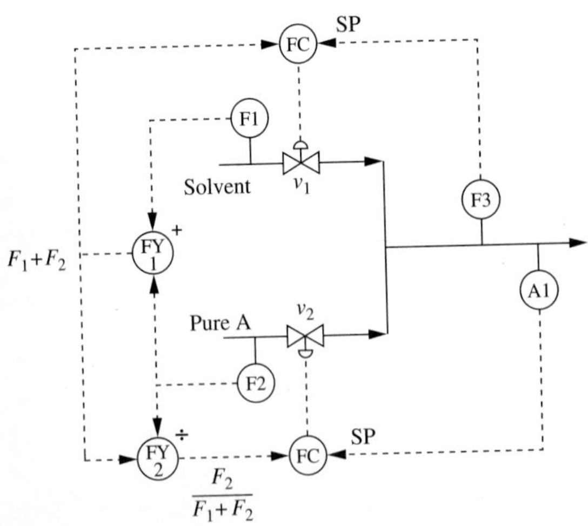

FIGURE 21.9
Manipulated-variable decoupled control of blending.

is made without feedback from the analyzer composition controller. These adjustments continue until the desired values of the total flow and ratio are achieved. By similar analysis, it can be shown that the analyzer controller output affects only the product composition, not the total flow. Thus, the interactions have been eliminated. As an added advantage, the decoupled control system is also easily understood by plant operating personnel. Naturally, the feedback controllers remain to account for small inaccuracies in the flow measurements, manipulated-variable calculations, and disturbances. Many similar strategies are used industrially to minimize unfavorable interactions and are the basis for the common water faucet design in which the total water flow and the ratio of hot to cold can be adjusted independently.

Controlled Variables

Another decoupling approach alters the controlled variables by replacing measured variables with calculated variables based on process output measurements. Again, the proper calculation is designed with knowledge of the process dynamics. As a simple example, the two-tank level control system in Figure 21.10 is considered; the levels are to be controlled by manipulating the set points of the flow controllers. If the goal were to design two decoupled controllers for maintaining the desired levels, calculated variables which yield independent equations would be sought in the basic linearized model of the process.

$$A\frac{dL_1'}{dt} = F_{1\text{in}}' - F_1' - K_{12}(L_1' - L_2') \tag{21.14}$$

$$A\frac{dL_2'}{dt} = F_{2\text{in}}' - F_2' + K_{12}(L_1' - L_2') \tag{21.15}$$

A decoupled system can be derived by noting that the sum of the levels depends on the sum of the manipulated variables, whereas the difference between the levels depends on the difference between the manipulated variables. This is easily shown by adding and subtracting equations (21.14) and (21.15) to give

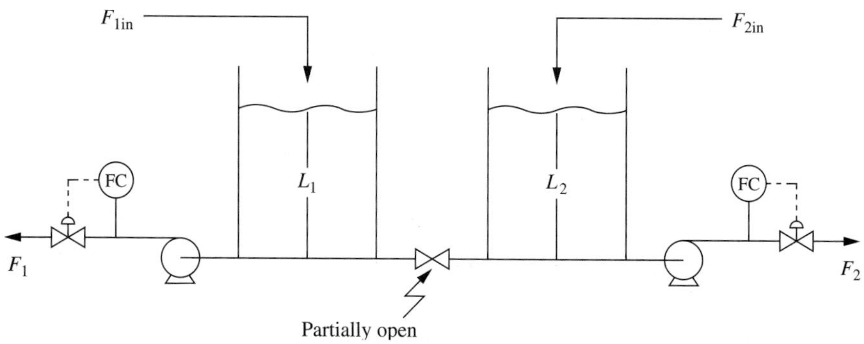

FIGURE 21.10
Level process.

$$A\frac{d(L_1' + L_2')}{dt} = (F_{1\text{in}}' + F_{2\text{in}}') - (F_1' + F_2') \tag{21.16}$$

$$A\frac{d(L_1' - L_2')}{dt} = (F_{1\text{in}}' - F_{2\text{in}}') - 2K_{12}(L_1' - L_2') - (F_1' - F_2') \tag{21.17}$$

Thus, a control design in which $(L_1 + L_2)$ and $(L_1 - L_2)$ are controlled by adjusting $(F_1 + F_2)$ and $(F_1 - F_2)$, respectively, is decoupled. Note that $(L_1 + L_2)$ is non–self-regulatory, whereas $(L_1 - L_2)$ is a first-order system. A process application of this principle to distillation reboiler level and composition control is given by Shinskey (1988).

This approach is not as widely applied as the approach based on manipulated variables, because it uses measured process output values in calculating the controlled variables. For this approach to function properly, all measured output variables should respond to adjustments in all manipulated variables with nearly the same dynamics so that the calculations are "synchronized." This criterion is easily satisfied for the example in Figure 21.10, because levels respond rapidly, but it is not commonly satisfied for complex units. Control designs for distillation composition using these concepts have been reported (Weber and Gaitonde, 1985; Waller and Finnerman, 1987).

Explicit Decoupling Calculations

The third approach to decoupling is to retain the original manipulated and controlled variables and alter the control calculation, while retaining the multiloop structure. There are two common implementations of this approach. The "ideal" decoupling compensates for interactions while leaving the input-output dynamic relationships for the feedback controllers unchanged from their single-loop behavior, $G_{ii}(s)$. While the concept is attractive, since controller tuning would not be affected by decoupling, experience has shown that the resulting system is very sensitive to modelling errors and generally does not perform well (Arkun et al., 1984; McAvoy 1979); thus, it is not considered further.

The "simplified" decoupling method presented here achieves a diagonal system by calculations that result in the interaction relationships between the controller outputs and controlled variables all being zero. Since it is not possible to eliminate the process interaction $G_{ij}(s)$, the decouplers are designed to provide compensating adjustments that cancel the process effects of manipulations in $\text{MV}_j(s)$ on $\text{CV}_i(s)$ for $i \neq j$ and thus yield independent, single-loop systems. The system is shown in Figure 21.11, with the decoupling transfer functions $D_{ij}(s)$ given by the following relationships:

$$D_{ij}(s) = -\frac{G_{ij}(s)}{G_{ii}(s)} \tag{21.18}$$

The reader may recognize the decoupler as similar to the feedforward controller, which compensates for measured disturbances; here the measured disturbance is the manipulated variable adjusted by an interacting feedback controller. The reader is referred to Chapter 15 on feedforward control for the derivation of this equation and a discussion of the possibility of the decoupler being unrealizable.

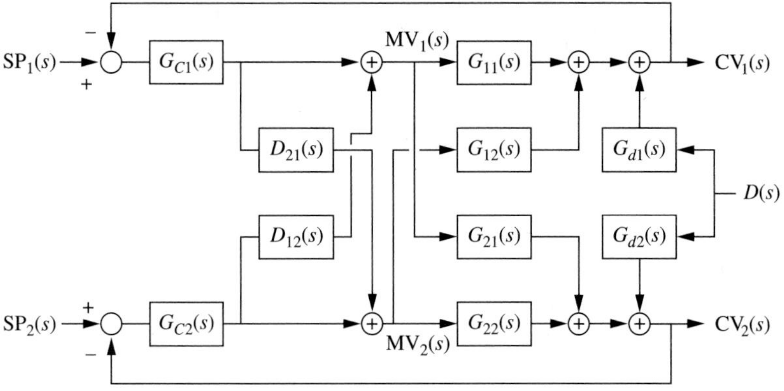

FIGURE 21.11
Block diagram of explicit decoupling.

When the process behavior can be modelled by first-order-with-dead-time transfer functions, the decoupler in equation (21.18) becomes

$$D_{ij}(s) = -\frac{K_{ij}}{K_{ii}}\frac{1+\tau_{ii}s}{1+\tau_{ij}s}e^{-(\theta_{ij}-\theta_{ii})s} \tag{21.19}$$

Again, this is the same form as feedforward controllers. The decoupling calculations in equation (21.19) can be implemented in digital form through the same procedures used with feedforward controllers in Chapter 15.

The explicit decoupler completely eliminates interaction only when the model is perfect. The resulting transfer function can be derived through block diagram manipulation assuming perfect decoupling, equation (21.18). The perfectly decoupled system is shown in Figure 21.12. Clearly, the "effective process" being controlled

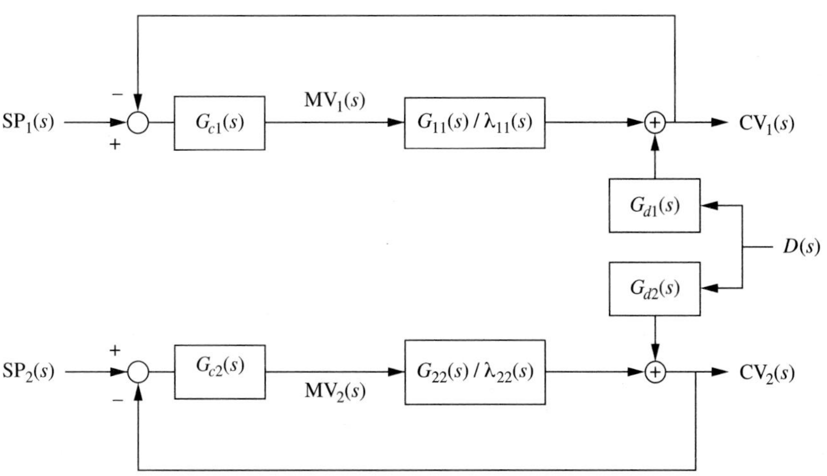

FIGURE 21.12
Consolidated block diagram explicit decoupling with perfect models. (Reprinted by permission. Copyright © 1983, Instrument Society of America. From *Interaction Analysis.*)

has changed because of the decoupling, and the controller tuning must be changed from single-loop values. Since the change in the "feedback process" transfer function is the inverse of the relative gain, the controller gain for the decoupled system should be taken as (approximately) the product of the single-loop controller gain, calculated using $G_{ii}(s)$, and the relative gain. This will maintain the $G_{OL}(s)$, product of the controller and the "process" $[\lambda_{11} G_c(s)][G_{11}(s)/\lambda_{11}]$, nearly constant, as a first approximation.

Errors in the models used in the decouplers affect the accuracy of the decoupling and, more seriously, affect the stability of the multiloop system. The sensitivity can be determined from an analytical expression of the performance as a function of the decoupler errors. The procedure to calculate the integral error in equation (21.7) can be applied to the closed-loop transfer function for the decoupled system with modelling errors. To simplify the analysis, only the decoupler gains have errors, with ϵ_i being a multiplicative error in the decoupler controller gain, K_{Dij}. The resulting expression for the performance is

$$\int_0^\infty E_1(t)dt = \lambda_{11}\lambda_{\epsilon 1 \epsilon 2}\left[\frac{K_{d1}T_{I1}}{K_{11}K_{c1}}\right]\left[\frac{1}{\lambda_{\epsilon 1}} + \frac{(\epsilon_1 - 1)K_{d2}K_{12}}{K_{d1}K_{22}}\right] \tag{21.20}$$

$$\text{with } \kappa = \frac{K_{12}K_{21}}{K_{11}K_{22}} \qquad \lambda_{11} = \frac{1}{1-\kappa} \qquad \lambda_{\epsilon i} = \frac{1}{1-\epsilon_i \kappa} \qquad \lambda_{\epsilon 1 \epsilon 2} = \frac{1}{1-\epsilon_1\epsilon_2\kappa}$$

$$D_{ij}(s) = -\epsilon_i \frac{G_{ij}(s)}{G_{ii}(s)} \qquad (\epsilon_i\text{=1 for perfect model})$$

Clearly, the *error relative gain,* $\lambda_{\epsilon_1\epsilon_2}$, plays a key role. As the decoupler errors increase, this factor and the integral error can become very large and the performance very poor. For processes with relative gains significantly greater than 1, even small decoupling errors can lead to very poor performance. For example, a small (5%) model error of $\epsilon_i = 1.05$ in a decoupler applied to the distillation example with energy balance control ($\lambda = 6.09$, $\kappa = 0.836$) would increase the integral error by about 100% over perfect decoupling! Thus, caution should be used when applying decoupling, since it requires model accuracies nearly impossible to achieve for real process systems with large relative gains. Similar results have been presented by McAvoy (1979), Shinskey (1988), and Skogestad and Morari (1987) using different analysis methods.

Several simplifications are possible in this decoupling approach. First, the dynamic decouplers in equation (21.18) can be approximated by the gains when this is sufficient for good control. Typically, the steady-state approximation is acceptable when $D_{ij}(s)$ has a small dead time and nearly equal lead (numerator) and lag (denominator) dynamics. Note that this simplification does not reduce the sensitivity to model gain errors shown in equation (21.20).

Also, decoupling can be simplified by using only one-way decoupling, with one $D_{ij}(s) = 0$. This approach would be applied to improve the performance of the more important controlled variable. Sensitivity analysis shows that one-way decoupling is much less sensitive to model gain errors than full decoupling, which presumably leads to its more frequent successful application in practice (McAvoy, 1979).

Example 21.10. Determine the performance with decoupling for the energy balance distillation control system in Figure 21.1. The disturbance is a set point change of +0.01 to the top composition controller.

The first question the engineer should ask is "Will error-free decoupling improve the control performance?" Recall that the magnitude of RDG $\cdot f_{\text{tune}}$ indicates the effects of interaction on multiloop controllers. Decoupling removes the effects of interaction, and the integral error will be the same as for a single-loop controller (i.e., with the other controllers in manual). Therefore, unfavorable interaction occurs when RDG $\cdot f_{\text{tune}} >$ 1.0, and decoupling can be used in such cases to remove the unfavorable interaction. The information required is given in Table 21.4, which gives the values of 12.2 for X_D and 0.0 for X_B. Since the value for X_D is so large, decoupling should be considered.

The values for the decoupler can be determined from the linear model of the energy balance system and are as follows:

$$D_{12}(s) = 0.893\frac{10.2s+1}{15s+1}e^{-(2-3.3)s} \qquad \text{(not realizable)}$$

$$\approx 0.893\frac{10.2s+1}{15s+1} \qquad \text{(physically realizable)}$$

$$D_{21}(s) = 0.930\frac{10.2s+1}{11.75s+1}e^{-1.3s}$$

A dynamic response for this decoupled system to a set point change of 0.01 in the top composition is given in Figure 21.13*a*, and the tuning values and performance are summarized in Table 21.6. This theoretically best decoupling performance is quite good, with a much lower IAE than the multiloop case reported in Table 21.2 (energy balance), although in this example the set point change has twice the magnitude. Note that both manipulated variables changed immediately when the set point was changed. The immediate change in MV_1 is from the controller G_{c11}, while the immediate change in MV_2 is from $G_{c11}D_{12}$, so that the decoupler acts before the controlled variable X_B is disturbed. Again, the similarity to feedforward is apparent, because the decoupler bases an adjustment in a process input on another process input.

However, the engineer must also consider the sensitivity to modelling errors. This decoupled system will become unstable for errors of about 10% in both decoupler gains; an example with 15% errors is given in Figure 21.13*b*, which shows the instability. *No amount of detuning* (short of $K_{c2} = 0$) in the feedback controllers will stabilize this response. Although the decoupler theoretically could improve performance, it is doubtful that sufficient model accuracy is generally available to use simplified (two-way) decoupling for processes with large relative gains.

With perfect decoupling, it is theoretically possible to improve control performance by reducing unfavorable interaction through decoupling as well as to

TABLE 21.6
Summary of decoupling Example 21.10

Case	K_{c1}	T_{I1}	K_{c2}	T_{I2}	K_{D12}	K_{D21}	IAE_1	IAE_2
Exact gains	60	9	−50	6.1	0.893	0.930	0.118	0.006
15% gain errors	60	9	−50	6.1	1.027	1.07	unstable	

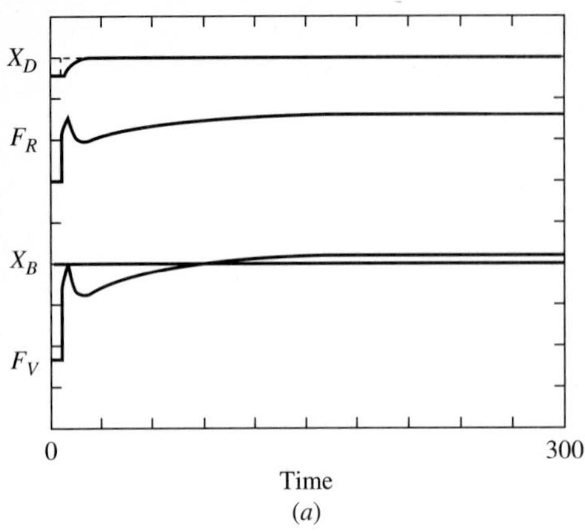

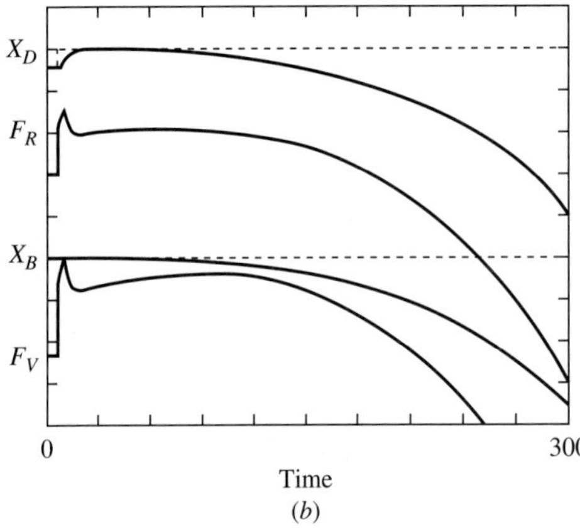

FIGURE 21.13
Explicit decoupling in distillation control, Example 21.6: (*a*) based on a perfect model; (*b*) with 15% gain errors in decouplers. (Scales: One tick = 0.02 for X_D and X_B, 0.50 for F_R, 0.30 for F_V)

degrade control performance by misapplying decoupling to a system that has favorable interaction. Decoupling should be considered only after an analysis of the relative disturbance gain has established that interaction is unfavorable for the expected disturbances and that performance with decoupling is not extremely sensitive to model errors.

- Decoupling improves control performance only when process interaction is *unfavorable,* so favorable interaction should not be reduced by decoupling.
- The stability and performance of full decoupling can be very sensitive to model errors when the relative gain is greater than 1. One-way decoupling has much lower sensitivity to model errors.

An important observation is that greater control system complexity does not always lead to better performance!

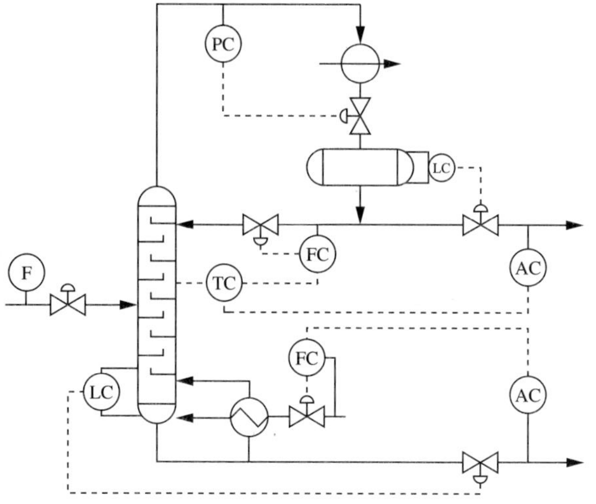

FIGURE 21.14
Multiloop distillation control with single-loop enhancements.

21.6 MULTILOOP CONTROL PERFORMANCE THROUGH ENHANCEMENTS: SINGLE-LOOP ENHANCEMENTS

Many enhancements were presented in Part IV to improve the performance of single-loop control systems. These methods are also widely applied to the control of multiloop systems, as will be covered in more depth in Part VI, but a brief example is presented here to complete the methods for achieving good multiloop performance. The distillation tower in Figure 21.14 has multiloop control of the two product compositions. In addition, the control performance is enhanced by inferential tray temperature control, which could provide a surrogate variable for control when the top analyzer provides an infrequent feedback measurement. Also, the reboiler utility and reflux flows have cascade control to reduce disturbances that result from changes in supply pressures. Other enhancements, such as feedforward, could be included as needed.

21.7 ADDITIONAL TOPICS IN MULTILOOP PERFORMANCE

The material in Chapters 20 and 21 presents only an introduction to the advances made in meeting the daunting challenges of multiloop control. The following subsections introduce a few selected additional topics.

Regulatory Control

Examples 21.2 and 21.3 on distillation control demonstrated that the regulatory control loops influence the composition control performance. An excellent control design objective is to select regulatory designs giving manipulated variables that simultaneously reduce transmission interaction (i.e., make the relative gain close to 1) and improve the disturbance rejection capability of the system (i.e., make the magnitude of the relative disturbance gain small). An example of such an approach is the simple distillation design developed by Rhyscamp (1980), which has proved remarkably successful on two-product distillation towers (Stanley et al., 1985, Waller et al., 1988). When simple regulatory loops do not provide these advantages, calculated variables can sometimes be derived that potentially improve multiloop performance (Haggblom and Waller et al., 1990; Johnston and Barton, 1987); however, the sensitivity of these approaches to model errors has not been fully evaluated.

Integrity

Another factor to be considered in multiloop control design is the integrity of the system; integrity means the ability of the control system to achieve a reduced set of its objectives when an individual element or elements are not functioning. Usually, the element that fails and cannot be replaced immediately is a sensor or final element, so the structure of the system changes. In such circumstances, interaction influences the stability and performance of such partially incapacitated systems. This issue has been studied by Grosdidier et al.(1985), Chiu and Arkun (1990), and Morari and Zafiriou (1989). The reader is cautioned that the large number of situations possible makes the analysis of integrity difficult for general multiloop systems, and a thorough analysis of many combinations of potential failures may be required.

One very simple and useful result is available based on the relative gain. If a control loop is paired using manipulated and controlled variables that have a *negative* relative gain element λ_{ij}, one of the following situations must exist:

1. The multiloop system is unstable with all controllers in automatic.
2. The single-loop system ij is unstable when all other controllers are in manual.
3. The multiloop system is unstable when the ijth controller is in manual and all other controllers are in automatic.

These results are general for any multiloop control design, as long as the feedback controllers have integral action; thus, they are applicable to multiple single-loop predictive controllers as well as PID controllers. Since all three situations are undesirable, the general conclusion is that single-loop designs should exclude pairings with negative relative gains.

Loop Pairing

Some alternative guidelines for loop pairings have been published by Yu and Luyben (1986), Economou and Morari (1986), Tzouanas et al. (1990). The selection of the final design, after many alternatives have been eliminated using methods in this chapter and references, relies on experience with similar units or dynamic simulation.

Robustness

The models used in control design never exactly match the true process behavior, and this factor would normally influence the performance of the system. While this issue could be addressed with simple assumptions and reasonable computation for single-loop systems, multiloop systems involve many more model parameters, all of which can be in error. Errors are introduced through empirical identification and as a result of changes in plant operation, such as flow rates and reactor conversions. Thus, the parameter errors in linearized models are not independent; that is, they have structure that must be considered in the analysis of robustness. The importance of robustness was discussed clearly by Doyle and Stein (1981) and is covered in Skogestad and Morari (1987) and extensively by Morari and Zafiriou (1989).

Dynamics

The results of Example 21.4 demonstrated the importance of considering interacting dynamics. The frequency-dependent relative gain was introduced in the previous chapter to evaluate interaction near the closed-loop critical frequency, and it has been shown that reliance only on steady-state analysis measures can result in good designs being improperly eliminated (e.g., Skogestad et al., 1990). Any predictions of control performance using the methods introduced in this chapter should be validated with a simulation of the closed-loop response. Since the design procedures usually result in a few candidates and simulation software is readily available, this final step should take little engineering effort.

21.8 CONCLUSION

The main result of Chapters 20 and 21 is the evaluation of the key effects of interaction on multiloop control. All of the factors that affect single-loop control affect multiloop control in similar ways; thus, the interpretations of these factors, presented in Chapters 9 through 13 and summarized in Table 13.2, remain essentially unchanged when considering multiloop control. Table 21.7 summarizes the effects of interaction on performance.

In this chapter, methods have been presented for achieving good control performance in multiloop systems through variable pairing, tuning, and simple enhancements. The methods have demonstrated that no single control performance predictor is available; for example, control strategies with relative gain values near 1.0 may not perform well for the disturbances of greatest importance. Even using the relative disturbance gain alone can lead to improper designs. For example, the pairing and tuning of a multiloop strategy can be selected to give better performance for a specific controlled variable (or variables) of particular importance over other variables of much less importance. Thus, the multiloop strategy must be selected with careful attention to the control objectives and process dynamic responses.

The flowchart in Figure 21.15 gives a procedure by which the analysis methods presented in this chapter can be applied to a 2×2 system analysis. Naturally, the control objectives must first be defined; then the necessary process information must be developed. The minimum information includes all steady-state gains as shown in Table 21.4 and some semiquantitative information on the relative dynamics between the manipulated and controlled variables is needed to select pairings based on dynamics and calculate the tuning factor. Finally, dynamic models, at least linear

TABLE 21.7
Effects of interaction on multiloop performance

Issue	Measure	Comments
Feasibility of feedback control	1. det $\mathbf{K} \neq 0$	1. Independent relationships exist between manipulated and controlled variables
	2. Specified set points can be achieved for expected disturbances	2. Manipulated variables have sufficient range; i.e., the process has sufficient capacity
Performance and Integrity	1. *for* $n \times n$, not integral-controllable if $\left(\frac{\det \mathbf{K}}{\prod_{i=1}^{n} K_{ii}}\right) < 0$ *for* 2×2, not integral-controllable if $\lambda_{ij} < 0$	1. Niederlinski criterion (or RGA for 2×2) used to evaluate whether controllers with integral modes can stabilize both single and multiloop systems without changing sign of controller gains
	2. $\lambda_{ij} > 0$	2. Usually, pairing selected that functions in single-loop and multiloop. ($\lambda_{ij} = 0$ sometimes acceptable)
Stability and tuning	for 2×2, λ_{11}	Interaction influences the characteristic equation, so it influences stability. Controller tuning must be modified for single-loop, usually detuned.
Performance	Relative disturbance gain (RDG)	Pairings are selected to reduce unfavorable interaction (|RDG| small) and provide fast feedback dynamics for important loops.
Enhancements		Designs, such as cascade and feedforward, that reduce the effects of disturbances are always beneficial. Decoupling can be used to reduce the effects of unfavorable interaction (|RDG| > 1) when the transmission interaction (RGA) is not too large

transfer functions and perhaps nonlinear models, are required if simulation verification is performed.

In the first step in the flowchart, the process is screened for the feasibility of multiloop control through evaluation of the controllability and operating window; if multivariable control is not possible, a different selection of variables or process

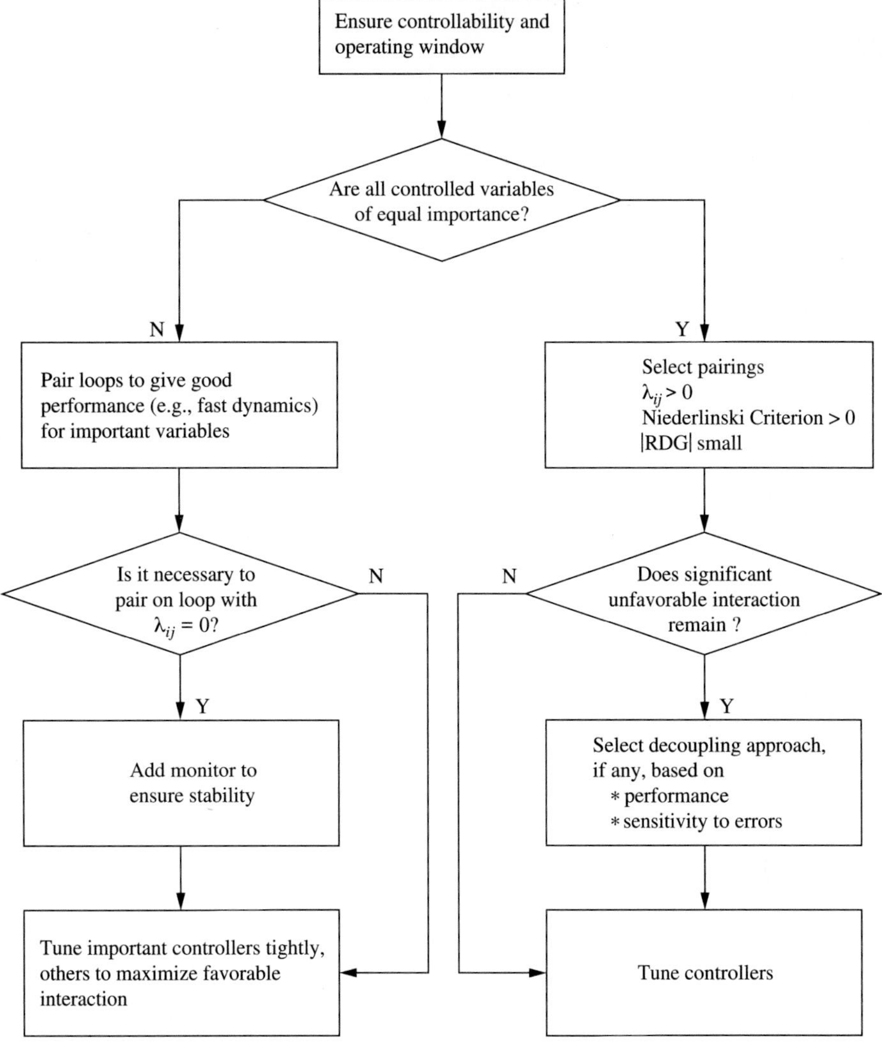

FIGURE 21.15
Flowchart for selecting 2 × 2 pairing and tuning.

modification is required. The first decision in the flowchart is whether both controlled variables are of equal importance. If one is of much greater importance, the left branch is taken. The important controlled variable is paired with the manipulated variable that provides the fastest feedback dynamics (along with satisfactory range), if a significant difference exists. A check is made to determine whether the controlled variable can be improved (through faster dynamics) by pairing it with a manipulated variable giving a zero relative gain; this step would be taken only in unusual situations in which the controlled variable is extremely important. After pairing has been selected, the control loops are tuned. Since the left-hand path is for unequal control priorities, the more important loops should be tuned to retain favorable interaction and reduce unfavorable interaction, and the less important loops should be tuned in a manner consistent with improving the overall performance and maintaining stability. Decoupling would probably not be considered, because detuning alone would reduce the effects of unfavorable interaction.

If the controlled variables are of equal importance, the pairings should be selected according to the analysis of the relative disturbance gain. If substantial unfavorable interaction remains, consideration would be given to decoupling, especially one-way decoupling to prevent the sensitivity problems encountered with two-way decoupling when the process has a large relative gain. Finally, the controllers would be tuned using methods described in Chapter 20. This procedure can lead to a good multiloop control strategy for the given process.

The concepts and methods presented in this chapter can be applied to a multiloop system of any order. However, the equations for the relative disturbance gain in this chapter are limited to a 2×2 system; they have been extended for higher-order systems by Skogestad and Morari (1987), who also introduce an alternative measure of multiloop performance.

Finally, this approach often, but not always, provides satisfactory performance. However, depending on factors such as the feedback dynamics and the disturbance type, magnitude, and frequency, situations exist in which no multiloop feedback design provides acceptable dynamic performance. Other steps for improving control performance include multivariable control; which is covered in Chapter 23, and process alterations, which are covered in Chapters 24 and 25.

REFERENCES

Arkun, Y., B. Monousiouthakis, and A. Palazuglu, "Robustness Analysis of Process Control Systems, A Case Study of Decoupling in Distillation," *IEC Proc. Des. Devel., 23,* 93–101 (1984).

Chiu, M. and Y. Arkun, "Decentralized Control Structure Selection Based on Integrity Considerations," *IEC Res., 29,* 369–373 (1990).

Doyle, J. and G. Stein, "Multivariable Feedback Design, Concepts for a Classical Modern Synthesis," *IEEE Trans. Auto. Cont. AC-26, 1,* 4–16 (1981).

Economou, C. and M. Morari, "Internal Model Control: 6, Multiloop Design," *IEC Proc. Des. Devel., 25,* 411–419 (1986).

Foss, A., J. Edmonds, and B. Kouvaritakis, "Multivariable Control for Two-Bed Reactors by the Characteristic Locus Method," *IEC Fund.,* 19, 109 (1980).

Gibilaro, L. and F. Lee, "The Reduction of Complex Transfer Function Models to Simple Models Using Method of Moments," *CES, 24,* 85–93 (1969).

Grosdidier, P., M. Morari, and B. Holt, "Closed-Loop Properties from Steady-State Gain Information," *IEC Fund., 24,* 221–235 (1985).

Haggblom, K. and K. Waller, "Control Structures for Disturbance Rejection and Decoupling of Distillation," *AIChE J., 36,* 1107–1111 (1990).

Johnston, R. and G. Barton, "Design and Performance Analysis of Control Systems Using Singular-Value Analysis," *IEC Res., 26,* 830–839 (1987).

Luyben, W., "Distillation Decoupling," *AIChE J., 16, 2,* 198–203 (1970).

Marino-Galarraga, M., T. McAvoy, and T. Marlin, "Short-Cut Operability Analysis: 2. Estimation of f_i Detuning Parameter for Classical Control Systems," *IEC Res., 26,* 511–521 (1987*a*).

Marino-Galarraga, M., T. McAvoy, and T. Marlin, "Short-Cut Operability Analysis: 3. Methodology for the Assessment of Process Control Designs," *IEC Res., 26,* 521–531 (1987*b*).

McAvoy, T., "Steady-State Decoupling of Distillation Columns," *IEC Fund., 18,* 269–273 (1979).

McAvoy, T., *Interaction Analysis,* ISA, Research Triangle Park, NC, 1983.

Morari, M. and E. Zafiriou, *Robust Process Control,* Prentice-Hall, Englewood Cliffs, NJ, 1989.
Rhyscamp, C., "New Strategy Improves Dual Composition Control," *Hydro. Proc., 60,* 51–59 (June 1980).
Skogestad, S., P. Lundstrom, and M. Morari, "Selecting the Best Distillation Control Configuration," *AIChE J., 36,* 753–764 (1990).
Skogestad, S. and M. Morari, "Effect of Disturbance Directions on Closed-Loop Performance," *IEC Res., 26,* 2029–2035 (1987).
Skogestad, S. and M. Morari, "Implications of Large RGA Elements on Control Performance," *IEC Res., 26,* 2323–2330 (1987).
Stanley, G., M. Marino-Galarraga, and T. McAvoy, "Short-Cut Operability Analysis: 1. The Relative Disturbance Gain," *IEC Proc. Des. Devel., 24,* 1181–1188 (1985).
Shinskey, F., *Process Control Systems* (3rd Ed.), McGraw-Hill, New York, 1988.
Tzouanas, V., W. Luyben, C. Georgakis, and L. Ungar, "Expert Multivariable Control. 1. Structure and Design Methodology," *IEC Res., 29,* 382–389 (1990).
Waller, K. and D. Finnerman, "On Using Sums and Differences to Control Distillation," *Chem. Eng. Commun, 56,* 253–268 (1987).
Waller, K., K. Haggblom, P. Sondelin, and D. Finnerman, "Disturbance Sensitivity of Distillation Control Structures," *AIChE J., 34,* 853–858 (1988).
Weber, R. and N.Y. Gaitonde, "Non-Interactive Distillation Tower Analyzer Control," *Proc. Amer. Cont. Conf., Boston, 1985,* 1072.
Yu, C. and W. Luyben, "Design of Multiloop SISO Controllers in Multivariable Processes," *IEC Proc. Des. Devel., 25,* 498–503 (1986).

ADDITIONAL RESOURCES

The "valve position" controllers introduced in the next chapter involve pairing on zero relative gain. In addition, a few examples of control designs with pairings on zero steady-state relative gains are given in McAvoy (1983) and in

Finco, M., W. Luyben, and R. Pollack, "Control of Distillation Columns with Low Relative Volatility," *IEC Res., 28,* 76–83 (1989).

In addition to Shinskey (1988), example multiloop control designs are presented in

Balchen, J. and K. Mumme, *Process Control: Structures and Applications,* Van Nostrand Reinhold, New York, 1988.

Additional results for systems with complex interactive dynamics (e.g., inverse responses), are given in

Holt, B. and M. Morari, "Design of Resilient Process Plants: VI The Effect of Right Half Plane Zeros on Dynamic Resilience," *CES, 40,* 59–74 (1985*a*).
Holt, B. and M. Morari, "Design of Resilient Process Plants: VII The Effect of Dead Time on Dynamic Resilience," *CES, 40,* 1229-1237 (1985*b*).

The methods in Chapters 20 and 21 can be applied in sequence, as shown in Figure 21.15, to eliminate poor alternatives, rank likely performance of feasible designs, and evaluate the appropriateness and sensitivity of decoupling. This analysis is based on quantitative analysis of the linearized system.

QUESTIONS

21.1. The following transfer functions were provided by Wood and Berry (1973) for a methanol-water separation in a distillation column similar to Figure 20.3. The products are expressed as mole % light key, and the reflux F_R, the reboiler steam F_S, and the disturbance feed flow F are in lb/min; time is in min.

$$\begin{bmatrix} X_D(s) \\ X_B(s) \end{bmatrix} = \begin{bmatrix} \dfrac{12.8e^{-s}}{16.7s+1} & \dfrac{-18.9e^{-3s}}{21s+1} \\ \dfrac{6.6e^{-7s}}{10.9s+1} & \dfrac{-19.4e^{-3s}}{14.4s+1} \end{bmatrix} \begin{bmatrix} F_R(s) \\ F_S(s) \end{bmatrix} + \begin{bmatrix} \dfrac{3.8e^{-8.1s}}{14.9s+1} \\ \dfrac{4.9e^{-3.4s}}{13.2s+1} \end{bmatrix} F(s)$$

Answer the following questions for the feed flow disturbance.

(*a*) Determine whether the input-output combination is controllable.

(*b*) Determine whether either loop pairing can be eliminated based on the sign of the relative gains ($\lambda_{ij} > 0$).

(*c*) Select the loop pairing based on an estimate of the control performance.

(*d*) Determine the initial tunings for PI controllers for the best loop pairing. Answer this question for (1) the two product compositions of equal importance and (2) the top product quality more important.

(*e*) Discuss whether decoupling is recommended and if so, design the decoupler.

(*f*) Discuss whether feedforward compensation would improve the control performance and if so, design the feedforward controller.

(*g*) The model was determined from empirical identification experiments. Discuss the likely errors in the model and the effects of these errors on the design conclusions.

For (*c*) through (*f*), compare the multiloop control performance for each controlled variable with its single-loop performance.

21.2. (*a*) Derive the expressions for the relative disturbance gain (RDG_1) and the integral error ($\int E_1 dt$) for the following inputs (1) ΔSP_1, (2) ΔSP_2, (3) a disturbance that has the same transfer function as MV_1, and (4) a disturbance that has the same transfer function as MV_2.

(*b*) Relate the value of the relative disturbance gain, RDG_1, to the ratio of changes in the manipulated variable for single-loop and multiloop control, $(\Delta\text{MV}_1)_{\text{ML}}/(\Delta\text{MV}_1)_{\text{SL}}$ to the same disturbance.

(*c*) Why is the magnitude, not the value, of the RDG used in evaluating performance?

(*d*) Is the RDG scale-dependent?

21.3. For a 2×2 control system with PID controllers and decoupling, write the equations for digital implementation of all control equations, or provide a sample computer program.

21.4. A linear transfer function model of a chemical reactor was determined by Foss et al. (1980) and simplified by Marino-Galarraga et al. (1987*a*). The reaction of oxygen and hydrogen over a catalyst occurs in two beds, with cold hydrogen quench added between the beds. The reactor is shown in Figure Q21.4, and the model is given below. The units are composition in mole%, temperatures in °C/167.4, flow in L/min/13.5, and time in sec/87.5. Assume that both controlled variables are of equal importance. Answer the following questions for two cases: (1) the input perturbation is a set point change to the composition controller and (2) the input perturbation is a change to the cooling medium temperature, so that the disturbance transfer function is the second

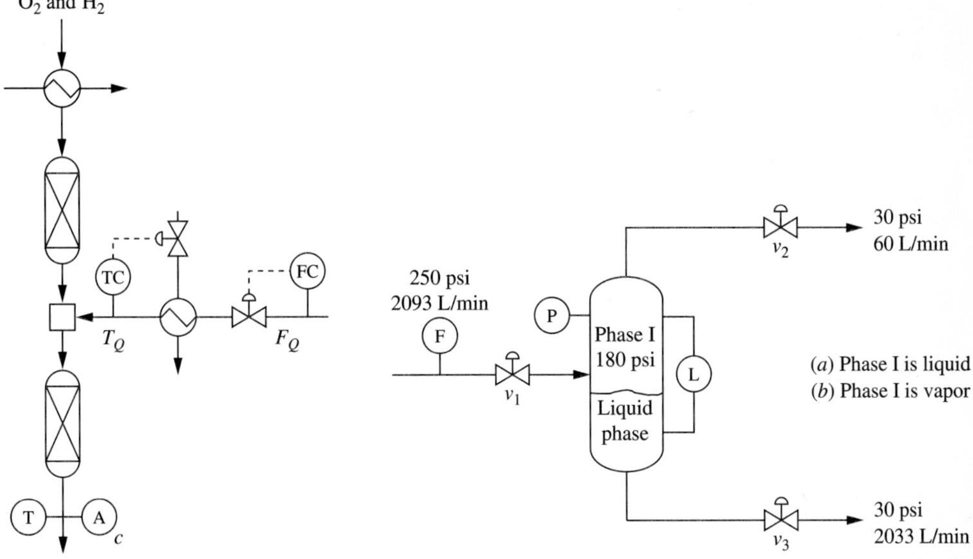

FIGURE Q21.4

FIGURE Q21.5

column of the following matrix (the same effect as a change in the manipulated quench temperature).

$$\begin{bmatrix} T(s) \\ C(s) \end{bmatrix} = \begin{bmatrix} \dfrac{-2.265e^{-1.326s}}{0.786s+1} & \dfrac{0.746e^{-2.538s}}{0.092s+1} \\ \dfrac{1.841e^{-0.445s}}{0.917s+1} & \dfrac{-0.654e^{-0.786s}}{0.870s+1} \end{bmatrix} \begin{bmatrix} F_Q(s) \\ T_Q(s) \end{bmatrix}$$

(*a*) Determine whether the input-output combination is controllable.

(*b*) Determine if either loop pairing can be eliminated based on the sign of the relative gains.

(*c*) Select the loop pairing based on an estimate of the control performance.

(*d*) Determine the initial tunings for PI controllers for the best loop pairing. Answer this question for (1) the temperature and product composition of equal importance and (2) the temperature more important.

(*e*) Discuss whether decoupling is recommended and if so, design the decoupler.

(*f*) Discuss whether feedforward compensation would improve the control performance and if so, design the feedforward controller.

For (*c*) through (*g*), compare the multiloop control performance for each controlled variable with the single-loop performance.

21.5. Two physical systems with exactly the same equipment structure, pressures, and flow rates in Figure Q21.5 are considered in this question. The only difference is that in system (*a*) phase I is a liquid (this is a decanter), whereas in system (*b*) phase I is a vapor (this is a flash drum). You may assume that the flows are proportional to the square root of the pressure drop and the valve % open; the valves are all 50% open at the base-case conditions. The three valves are available for manipulation, and three controlled variables are shown as sensors. The following additional information is provided about the variability of the process operation: the feed flow is 1400 to 2600 L/min, the percent overhead material in feed is 1 to 5%, and the external pressures are essentially constant. Select the best control loop pairing and discuss the differences, if any, between the results for systems (*a*) and (*b*).

21.6. Answer the following questions.

(*a*) Is there a feedback control system for system B2 in equation (21.6) that will prevent the inverse response?

(*b*) For System B1 in equation (21.5), can the multiloop feedback system experience an inverse response with two PID controllers?

(*c*) Values of the relative disturbance gain (RDG) can be related to the change in the manipulated variables under multiloop control. Determine the value of ΔMV_1 for a disturbance and relate this to RDG_1.

(*d*) Is it possible to have a relative gain $\approx$ 1.0 and a large RDG?

(*e*) Is it possible to have no interaction of any type (e.g., $K_{12} = K_{21} = 0$) and have a large RDG?

(*f*) Feedforward control can be applied on a multiloop system. Modify the calculation of the relative disturbance gain (RDG_1) and the integral error ($\int E_1 dt$) for various feedforward control designs (feedforward to MV_1 only, to MV_2 only, and to both) using the same disturbance.

(*g*) The relative disturbance gain provides the ratio of multiloop to single-loop performance. Discuss how to use this information when comparing the performance of two designs with different single-loop performances.

21.7. The outlet temperature of the process fluid and the oxygen in the flue gas can be controlled in the fired heater in Figure Q20.10 by adjusting the fuel pressure (flow) and the stack damper % open. A dynamic model for the fired heater in Figure Q20.10 was reported by Zhuang et al. (1987) and repeated here.

$$\begin{bmatrix} T(s) \\ A(s) \end{bmatrix} = \begin{bmatrix} \dfrac{0.6}{2400s^2 + 85s + 1} & \dfrac{-0.04}{3000s^2 + 90s + 1} \\ \dfrac{-1.1}{70s + 1} & \dfrac{0.30}{70s + 1} \end{bmatrix} \begin{bmatrix} P_{sp}(s) \\ V_2(s) \end{bmatrix}$$

The inputs and outputs are in percent of the range of each instrument, and the time is in sec.

(*a*) Determine whether the input-output combination is controllable.

(*b*) Determine whether either loop pairing can be eliminated based on the sign of the relative gains.

(*c*) Determine whether decoupling will improve the control performance.

(*d*) Determine the PI controller tuning for the best multiloop control, with or without decoupling.

21.8. The following transfer functions were provided by Waller et al. (1987) for a distillation column. System I was similar to Figure 20.3 except that the controlled product compositions were not measured directly; they were inferred from tray temperatures (°C) near the top, T_4, and near the bottom, T_{14}, trays. System II had the distillate/(distillate + reflux) as a manipulated variable rather than the reflux; this is designated as *R*. The flows are in kg/hr; time is in min. Answer the following questions for both systems (the same process with different regulatory control designs) and compare the results.

System I: Energy balance regulatory control:

$$\begin{bmatrix} T_4(s) \\ T_{14}(s) \end{bmatrix} = \begin{bmatrix} \dfrac{-0.045e^{-0.5s}}{8.1s+1} & \dfrac{0.048e^{-0.5s}}{11s+1} \\ \dfrac{-0.23e^{-1.5s}}{8.1s+1} & \dfrac{0.55e^{-0.5s}}{10s+1} \end{bmatrix} \begin{bmatrix} F_R(s) \\ F_S(s) \end{bmatrix} + \begin{bmatrix} \dfrac{0.004e^{-s}}{8.5s+1} \\ \dfrac{-0.65e^{-s}}{9.2s+1} \end{bmatrix} X_F(s)$$

System II: Modified regulatory control; $R = F_D/(F_D + F_R)$

$$\begin{bmatrix} T_4(s) \\ T_{14}(s) \end{bmatrix} = \begin{bmatrix} \dfrac{6.7e^{-0.5s}}{11s+1} & \dfrac{0.01e^{-0.5s}}{13s+1} \\ \dfrac{34e^{-1.3s}}{12s+1} & \dfrac{0.35e^{-0.5s}}{10s+1} \end{bmatrix} \begin{bmatrix} R(s) \\ F_S(s) \end{bmatrix} + \begin{bmatrix} \dfrac{-0.026e^{-2.5s}}{23s+1} \\ \dfrac{-0.81e^{-s}}{13s+1} \end{bmatrix} X_F(s)$$

(*a*) Determine whether the input-output combination is controllable.

(*b*) Determine whether either loop pairing can be eliminated based on the signs of the relative gains.

(*c*) Select the loop pairing based on an estimate of the control performance.

(*d*) Determine the initial tunings for PI controllers for the best loop pairing. Answer this question for (1) the two product compositions of equal importance and (2) the top product quality more important.

(*e*) Discuss whether decoupling is recommended and if so, design the decoupler.

(*f*) Discuss whether feedforward compensation would improve the control performance and if so, design the feedforward controller.

For (*c*) through (*f*), compare the multiloop control performance for each controlled variable with the single-loop performance.

21.9. (*a*) The limit for the integral error of a decoupled system in equation (21.20) as the gain errors approach zero is $\lambda_{11} K_{d1} T_{I1}/K_{c1} K_{11}$. Explain why this differs from equation (21.9).

(*b*) Explain why the gain decoupler errors in Example 21.10 lead to an unstable system. (Hint: consider the relative gain or Niederlinski criterion for the system with decouplers.)

(*c*) Derive the expression in equation (21.20) for the integral error for a 2×2 multiloop system with PI controllers and decouplers, with gain errors in the decouplers.

21.10. The process with two series chemical reactors in Example 3.3 is considered in this question. The process flexibility is increased by allowing the temperatures of the two reactors to be manipulated independently. The two controlled variables are the concentrations of reactant A in the two reactors. The rate constant can be expressed as $5.87 \times 10^5 e^{-5000/T}$ (with temperature in K), and the disturbance is feed composition, C_{A0}.

(*a*) Determine whether the input-output combination is controllable.

(*b*) Determine whether either loop pairing can be eliminated based on the signs of the relative gains.

(*c*) Determine whether decoupling could improve the dynamic performance, especially if the most important controlled variable is the concentration in the second reactor.

21.11. Doukas and Luyben (1978) reported the transfer function model for the distillation column with a side stream product, shown in Figure Q21.11. The feed contains

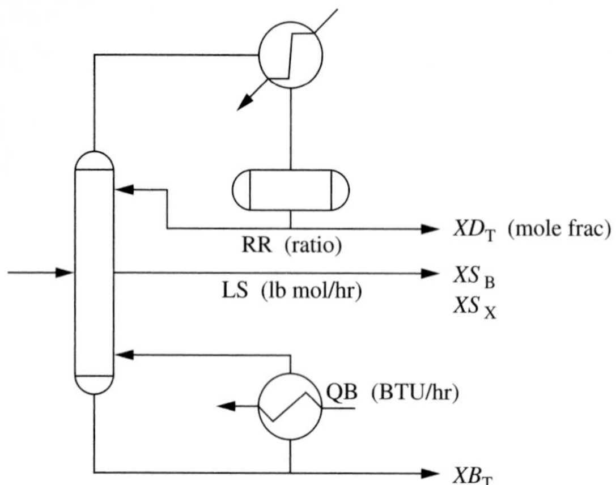

FIGURE Q21.11

benzene (B), toluene (T), and xylene (X). The controlled and manipulated variables are given in the figure, with the benzene in the side stream of much less importance than the other controlled variables. The linearized transfer function model is

$$\begin{bmatrix} XD_T(s) \\ XS_B(s) \\ XS_X(s) \\ XB_T(s) \end{bmatrix} = \begin{bmatrix} \dfrac{-1.986e^{-0.7s}}{66.7s+1} & \dfrac{5.24e^{-60s}}{400s+1} & \dfrac{5.984e^{-2.24s}}{14.3s+1} \\ \dfrac{0.002e^{-0.6s}}{(7.14s+1)^2} & \dfrac{-0.33e^{-0.7s}}{(2.4s+1)^2} & \dfrac{2.38e^{-0.42s}}{(1.43s+1)^2} \\ \dfrac{-0.176e^{-0.5s}}{(6.9s+1)^2} & \dfrac{4.48e^{-0.5s}}{11.1s+1} & \dfrac{-11.7e^{-1.9s}}{12.2s+1} \\ \dfrac{0.374e^{-7.75s}}{22.2s+1} & \dfrac{-11.3e^{-3.8s}}{(21.7s+1)^2} & \dfrac{-9.81e^{-1.6s}}{11.4s+1} \end{bmatrix} \begin{bmatrix} RR(s) \\ LS(s) \\ QB(s) \end{bmatrix}$$

For this system, determine the best loop pairing by following the method in Figure 21.15.

21.12. Design an improved control system to improve the dynamic performance of the composition in the fuel system in Figure 21.7 when

(*a*) A measurement of the total fuel flow to the consumers is available.

(*b*) A measurement of the gas fuel (L) is available.

21.13. Calculate the controller tuning for the blending system in Table 21.1 with $A_1 = 0.95$. Discuss which loop pairing would be preferred.

21.14 (*a*) Derive the closed-loop transfer function for a 2×2 system with decoupling.

(*b*) From the result in (*a*), determine whether one-way decoupling influences the stability of the closed-loop system.

21.15. The series of well-stirred chemical reactors with equal volumes shown in Figure Q21.15 is to be controlled. The controlled variables are the temperature and reactant concentration in the third reactor, and the manipulated variables are the inlet concentration set point and the cooling valve v_2. The chemical reaction is first-order, the rate constant has an Arrhenius relationship with temperature, and the heat of reaction is negligible. The heat exchanger dynamics are negligible. For this example, the concentration is much more important than the temperature, but both should have zero steady-state offset for a steplike disturbance. Design the loop pairings and tuning and discuss the rationale for the design.

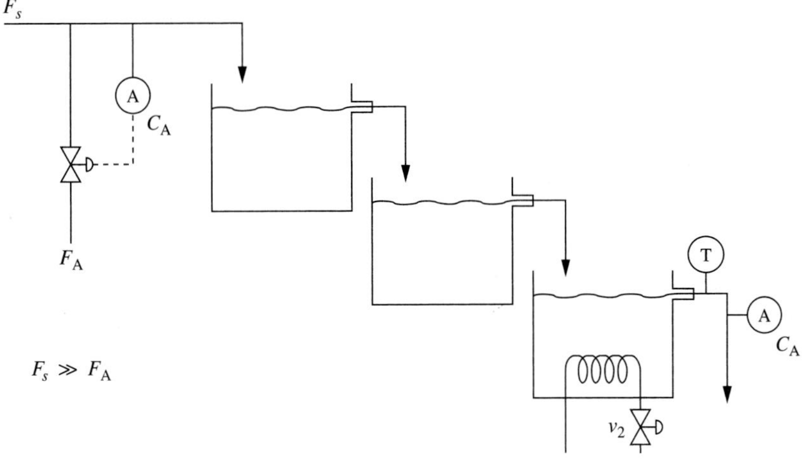

FIGURE Q21.15

21.16. Answer the following questions for two physical processes: (1) the chemical reactor described in Example 3.10, and (2) the same chemical reactor with no heat of reaction, $\Delta H_{rxn} = 0$. Both processes have two feedback PI controllers: $T \rightarrow F_c$ and $C_A \rightarrow C_{A0}$ (with the feed flow unchanged).

(*a*) Does process interaction influence the stability of the closed-loop system? Provide quantitative analysis to support your conclusion.

(*b*) Does process interaction influence the dynamic performance (behavior) of the closed-loop system? Explain your answer briefly.

21.17. Design feedforward controllers for the distillation column under energy balance control, described by equation (21.1), for a measured disturbance in feed composition. Design the feedforward controller for the two following situations, discuss the differences in the results, and discuss the implications for application of each.

(*a*) The distillate composition X_D is to be maintained constant, and the bottoms composition X_B is not controlled and may vary.

(*b*) The distillate composition X_D and the bottoms composition X_B are both to be maintained constant via the feedword controller.

CHAPTER 22

VARIABLE-STRUCTURE AND CONSTRAINT CONTROL

22.1 INTRODUCTION

To this point we have made the assumption that the multivariable process has the same number of manipulated and controlled variables. This situation is often referred to as a square or $n \times n$ system. Square systems are typical, because we consider dynamic behavior and control when designing plants and provide sufficient manipulated variables for at least the most important controlled variables. However, it is often the case that, due to process limitations and overriding control objectives, the number of manipulated and controlled variables are not always equal, and control approaches are needed to address these situations.

In this chapter, situations will be considered in which the number of manipulated variables is greater than or less than the number of controlled variables. When an excess of manipulated variables exists, the controlled variables can be returned to their set points at steady state by many combinations of the steady-state manipulated variables. Thus, the control system should operate the process in the most economical manner, in addition to providing good dynamic performance. When an excess of controlled variables exists, not all controlled variables can be maintained at their set points simultaneously. However, the control system can be designed to maintain the most important controlled variables at their set points.

The branch of process control that addresses these situations is known as *variable-structure* control. In this chapter, methods based on *single-loop control algorithms* are presented that provide the ability to change the input-output pairings of selected loops automatically. These methods are easy to design and simple to use and are therefore widely applied in practice. However, they are normally restricted to cases with limited dimensionality, such as one manipulated and several controlled variables or several manipulated variables and one controlled variable. A method that can address higher-dimensional structures, as well as square problems, is presented in the next chapter.

First, split range control systems are presented for processes with excess manipulated variables. Then, signal select control systems are presented for processes with excess controlled variables. In each section, examples demonstrate typical reasons for variable structure control, along with implementation guidelines. Finally, a few applications in the area of constraint control are provided; these demonstrate the combined application of split range and signal select, along with some frequently used extensions, such as multiple controllers with different set points and valve position controllers.

22.2 SPLIT RANGE CONTROL FOR PROCESSES WITH EXCESS MANIPULATED VARIABLES

The concept of split range control will be introduced through the example process in Figure 22.1. In this process, the flows of gaseous fuels from two sources are adjusted to control the pressure of a header, which is a pipe from which fuel is distributed to many consumers. The flow to the consumers is determined by many independent processes and cannot be adjusted to control the pressure. The following simple model of the (well-mixed) gas header system can be used to evaluate the degrees of freedom:

$$V\frac{dC_A}{dt} = F_A C_{A0} - F_{out} C_A \tag{22.1}$$

$$V\frac{dC_B}{dt} = F_B C_{B0} - F_{out} C_B \tag{22.2}$$

$$F_A = K_A v_A \sqrt{\frac{P_A - P}{\rho_A}} \tag{22.3}$$

$$F_B = K_A v_B \sqrt{\frac{P_B - P}{\rho_B}} \tag{22.4}$$

$$P = \frac{(VC_A + VC_B)RT}{V} \tag{22.5}$$

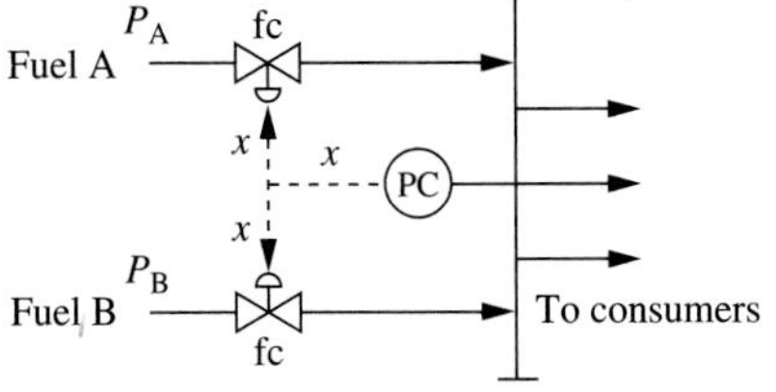

FIGURE 22.1
Split range pressure control.

Variables	External variables	Constants
P	F_{out}	R
F_A	T	V
F_B	P_A	K_A
C_A	P_B	K_B
C_B	C_{A0}	ρ_A
v_A	C_{B0}	ρ_B
v_B		

The model could be improved by including nonlinear valve characteristics and a nonideal gas law, but the model of this resolution is sufficient to demonstrate the degrees-of-freedom analysis. There are 5 equations and 7 variables; thus, the system is not specified. In this example, the prices of the two fuels are not equal; fuel A has a lower price than fuel B. Therefore, the control system should automatically adjust the valves so that as much of fuel A as possible is consumed before any fuel B is consumed, while providing good control of the pressure.

The split range control system in Figure 22.1 achieves the desired behavior in a simple manner. The pressure in the header is measured and used as the controlled variable to a standard PI feedback control algorithm, which has a single calculated output signal, x. This signal is sent to both control valves, but these valves are calibrated to open or close differently from the standard control valves. To achieve the desired behavior, the controller and valves obey the behavior defined in Table 22.1 and shown in Figure 22.2.

With this modification, the control equations become, for controller output $x < 50\%$,

$$\begin{aligned} v_A &= 2\left[K_c\left((P_{sp} - P) + \frac{1}{T_I}\int_0^t (P_{sp} - P)dt'\right) + I\right] \\ v_B &= 0.0 \end{aligned} \tag{22.6}$$

For controller output $x \geq 50\%$,

$$\begin{aligned} v_A &= 100 \\ v_A &= 100 - 2\left[K_c\left((P_{sp} - P) + \frac{1}{T_I}\int_0^t (P_{sp} - P)dt'\right) + I\right] \end{aligned} \tag{22.7}$$

TABLE 22.1
Typical valve adjustments for split range control

		Percent opening	
Controller output	**Pressure to valve**	**Valve A**	**Valve B**
0–50%	3–9 psig	0–100%	0%
50–100%	9–15 psig	100%	0–100%

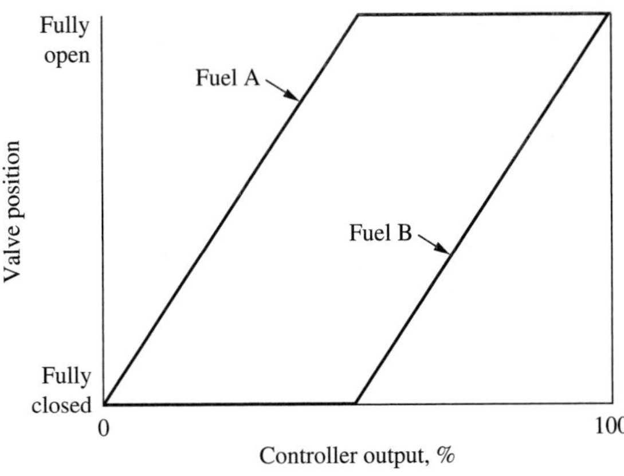

FIGURE 22.2
Fixed ranking of valve adjustments for split range.

Note that either set of two equations introduces no dependent variables and one external variable, P_{sp}, along with the controller tuning constants. The combination of the controller equations, either (22.6) or (22.7), with equations (22.1) through (22.5) results in a system with 7 equations and 7 variables. Thus, the process and control system is completely defined when the pressure set point has been specified.

Split range control is depicted in the process diagram in Figure 22.1. The labels x that indicate that the two signals have the same value after the split are optional. The fixed relationship between the controller output and the position of the two valves is shown in Figure 22.2. As the controller output initially begins to increase from 0%, the valve in the less expensive fuel line opens, while the valve in the more expensive fuel line remains closed. When the controller output reaches 50%, the fuel A valve is fully open, and the fuel B valve is closed. When the controller output continues to increase beyond 50%, the fuel A valve remains fully open, and the fuel B valve opens.

The behavior of the control system is given in Figure 22.3. The initial situation has a low total fuel demand, so that the pressure controller manipulates only the fuel A valve. At time 30, an increase in the fuel consumption occurs; the pressure in the header initially decreases; and the controller output increases. The fuel A valve is adjusted until the pressure is returned to its set point. At time 110 another increase in consumption occurs. The pressure controller responds by increasing its output. In this situation, valve A reaches its limit of 100%; then the fuel B valve is opened until the pressure is returned to its set point. This example demonstrates that the split range controller can smoothly adjust the two valves to maintain the controlled variable at the set point, while minimizing the cost of the fuel consumed.

Several important implementation issues arise in applying split range control. In principle, the concept of split range can be extended to any number of manipulated variables. However, there is a limit on how accurately a control valve can be adjusted. Therefore, split range is normally limited to two, or three at most, manipulated variables. Also, a feedback control system could tend to cycle if it had a "dead zone" in which neither valve is adjusted. To prevent this situation arising from inaccurate valve calibrations, the valves are normally calibrated to have an overlap (e.g., 0 to 55% and 45 to 100%).

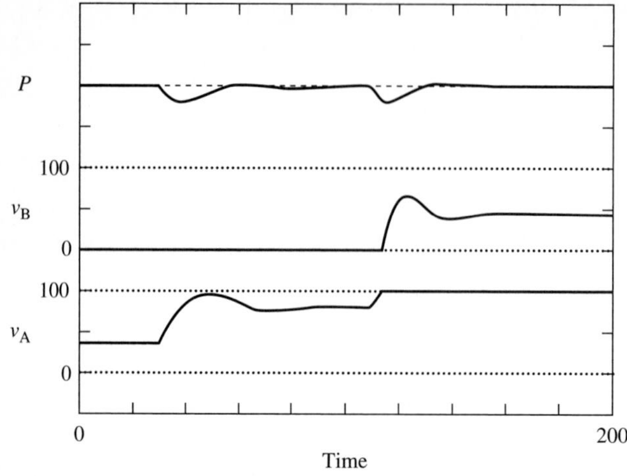

FIGURE 22.3
Dynamic response of split range control system.

Typical behavior can be easily implemented by a simple calibration of standard control valves, which has essentially no cost implication. Recall that the signal to the single-loop control valve is normally 3 to 15 psig, which relates to 0 to 100% of the controller output signal, respectively.

Another important issue is the stability and tuning of the control system. Note that the feedback process dynamics change when the controller output crosses the 50% value, as shown in Figure 22.4. Therefore, the controller tuning should remain constant only if the process dynamics for the two closed-loop paths are the same or similar; that is, if $G_{vA}(s)G_{pA}(s) \approx Gv_B(s)G_{pB}(s)$. If the closed-loop dynamics are significantly different, the controller tuning should be changed automatically by the control system. The controller tuning could be switched based on the value of the controller output, using one set of tuning constants for controller output values of 0 to 50% and another set of tuning for 50 to 100%. This retuning approach is another application of the adaptive tuning method referred to in Section 16.3 as *deterministic modification* of controller tuning.

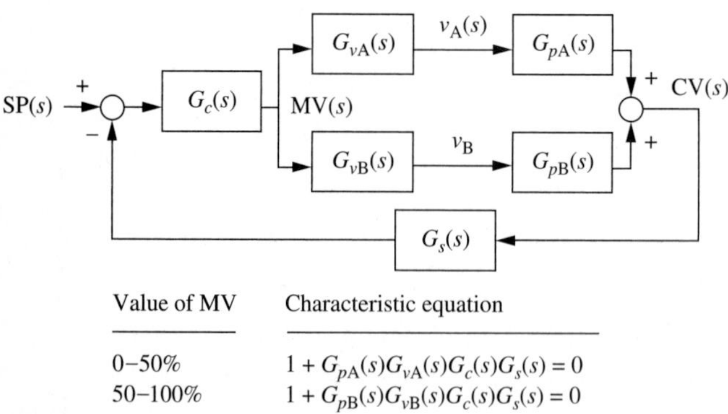

Value of MV	Characteristic equation
0–50%	$1 + G_{pA}(s)G_{vA}(s)G_c(s)G_s(s) = 0$
50–100%	$1 + G_{pB}(s)G_{vB}(s)G_c(s)G_s(s) = 0$

FIGURE 22.4
Schematic of split range control.

TABLE 22.2
Split range control criteria

Split range control is possible when
1. There is one controlled and more than one manipulated variable.
2. There is a causal relationship between each manipulated variable and the controlled variable.
3. The proper order of adjusting the manipulated variables adheres to a fixed priority ranking.

In conclusion, split range control is widely applied to processes with excess manipulated variables. The general criteria for split range control are summarized in Table 22.2. The feedback controller can use one tuning if the dynamics for all feedback paths are similar. If the dynamics are significantly different, the feedback controller must be (1) detuned to be stable without excessive oscillations for all situations or (2) retuned automatically via programmed modification.

22.3 SIGNAL SELECT CONTROL FOR PROCESSES WITH EXCESS CONTROLLED VARIABLES

Often, many control objectives exist for a process, and not all of these can be satisfied simultaneously. As an example, consider the chemical reactor shown in Figure 22.5, which has control objectives to maximize conversion while (1) maintaining the reactant composition in the effluent at or above a value $(C_A)_{min}$ and (2) preventing the reactor temperature from exceeding T_{max}. Each of these control objectives can be satisfied individually by adjusting the cooling medium flow rate. The engineer must determine the relative importance of the control objectives and design a control system that satisfies the priority ranking.

A signal select control strategy to implement this ranking is shown in Figure 22.5. An individual controller is implemented for each measured controlled variable, and the output signals from the two controllers are sent to a signal select element. The output of a signal select is either the minimum (low signal select) or maximum (high signal select) value of all inputs to the signal select. In the example, the proper element is a low signal select, since the largest flow of cooling medium is preferred and the valve is fail-open. (Recall that selecting the lowest signal to the

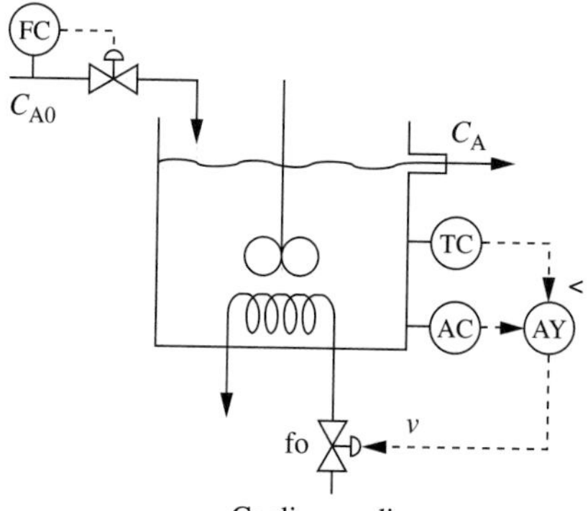

FIGURE 22.5
Example of signal select with two controllers.

cooling medium valve ensures the largest coolant flow.) The output of the signal select is sent to a control valve, as in this case, or can be sent to the set point of a secondary controller in a cascade system.

Again, the degrees of freedom of the control system should be analyzed. The equations that define the process and the control calculations for the example are as follows:

$$V\frac{dC_A}{dt} = F(C_{A0} - C_A) - Vk_0e^{-E/RT}C_A \tag{22.8}$$

$$V\rho C_p\frac{dT}{dt} = F\rho C_p(T_0 - T) - UA(T - T_c) + (-\Delta H_{rxn})Vk_0e^{-E/RT}C_A \tag{22.9}$$

$$MV_1 = K_{c1}\left((C_{Asp} - C_A) + \frac{1}{T_{I1}}\int_0^t (C_{Asp} - C_A)\,dt'\right) + I \tag{22.10}$$

$$MV_2 = K_{c2}\left((T_{sp} - T) + \frac{1}{T_{I2}}\int_0^t (T_{sp} - T)\,dt'\right) + I \tag{22.11}$$

$$UA = f(v) \tag{22.12}$$

$$v = \min(MV_1, MV_2) \tag{22.13}$$

Variables	**External variables**	**Constants**
C_A	C_{A0}	ρ
C_{Asp}	T_0	C_p
T	T_c	V
T_{sp}		
UA		R
MV_1		E
MV_2		k_0
v		$(-\Delta H_{rxn})$
		$K_{c1}, K_{c2}, T_{I1}, T_{I2}$

The system has 6 equations and 8 variables; thus, the system behavior is defined when two variables, C_{Asp} and T_{sp}, have been specified (i.e., are shifted to external variables). To achieve the objectives in this example, the set points must be set to the limiting values for these variables (i.e., $C_{Asp} = (C_A)_{min}$ and $T_{sp} = T_{max}$).

Depending on the operating conditions of the chemical reactor (e.g., feed composition, temperature, concentration of reaction inhibitors), either of the controllers could be selected to manipulate the cooling medium valve. Dynamic responses for two feedback systems are given to demonstrate the effect of the signal select. The initial steady-state conditions of the system result in the outlet composition being at its set point (minimum value) and the temperature below its set point (maximum value). After a short initial period of steady operation, reaction inhibitor is introduced with the feed, causing the reaction rate to decrease. The first response involves the reactor system with only composition control (and no temperature control), so no signal select exists. As shown in Figure 22.6*a*, the composition controller reduces the

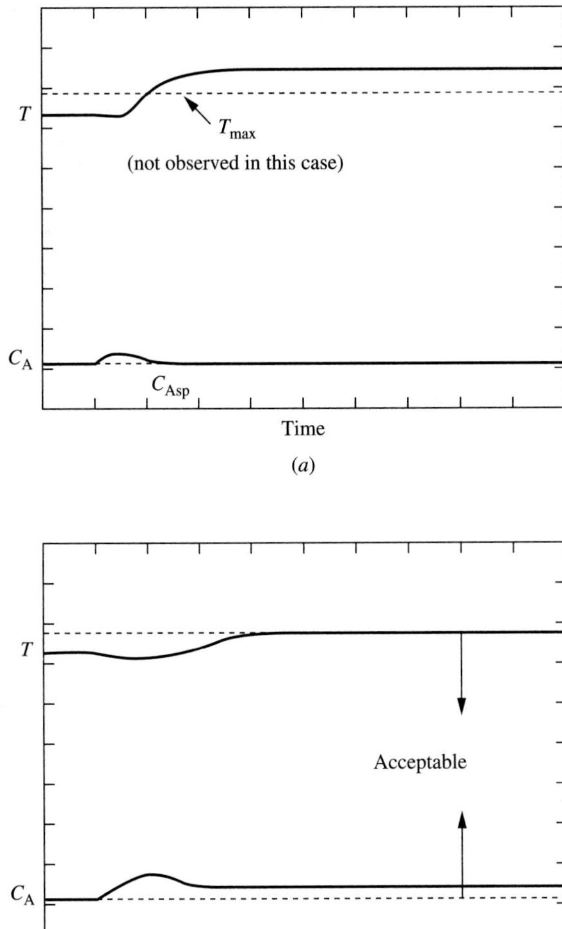

FIGURE 22.6
Reactor disturbance response (*a*) with only composition control; (*b*) with signal select design.

coolant flow to increase the reaction rate. The control system returns the composition to its set point, but it increases the reactor temperature above its maximum value.

The second response involves the same reactor and disturbance but with the signal select control design shown in Figure 22.5. The dynamic response is given in Figure 22.6*b*. Initially, the temperature is below its maximum limit, and the composition of product in the effluent is at its set point. Since the temperature controller is sending a higher signal (to increase the temperature) than the composition controller, the output of the composition controller is initially selected. In the initial response to the disturbance, the coolant is decreased by the concentration controller until the reactor temperature reaches its maximum value: the temperature controller set point. Then, the temperature controller output signal becomes smaller than the output from the composition controller. At the new steady-state conditions, the temperature is at its set point and the composition is above its set point. This situation may not be the

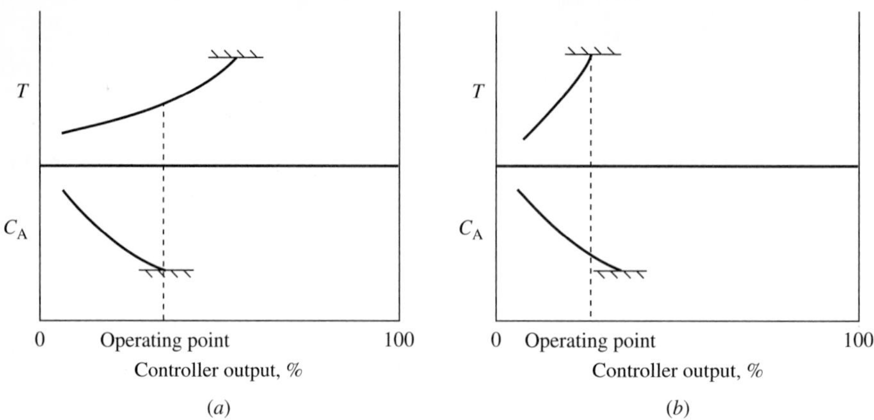

FIGURE 22.7
Determining the operating point for systems with signal select control.

most profitable in the short run, but it is the best operation, given the input variables, because it prevents damage to equipment due to extreme temperature. Improvement would require an elimination of the inhibitor in the feed.

The steady-state relationships between the split range manipulated and controlled variables are depicted in Figure 22.7 for the reactor example. Two cases show how the control objectives can be satisfied by adjusting the manipulated variable when the temperature or composition is the limiting factor. When sufficient range exists for the manipulated variable, the best steady-state operation can be achieved by opening the valve the least amount as constrained by the most limiting controlled-variable value. If the manipulated variable does not have sufficient range, the proper value of the output of the signal select is at either its minimum value (more cooling capacity required) or maximum value (zero cooling insufficient, heating required). No control system can do better; the equipment design or cooling medium temperature must be changed to satisfy the objectives. Thus, the simple signal select control system always achieves the best (unique) steady-state performance possible for the process design and control objectives.

However, signal selects are not appropriate for all cases of multiple controlled variables and one manipulated variable. An example where signal select is not appropriate is the same chemical reactor as in Figure 22.5 with different control objectives: Maintain (1) the effluent composition to be *no greater than* $(C_A)_{max}$ and (2) the temperature below T_{max}. This situation is depicted in Figures 22.8*a* and *b*. In Figure 22.8*a* both limits can be satisfied, but the control objectives are not defined completely enough to determine a unique value of the manipulated variable. In Figure 22.8*b*, no value of the manipulated variable satisfies the objectives. In either case, no unique operating point exists. Therefore, the control system must perform a task more complex than determining a limiting value. It must determine the best or "optimum" operation within acceptable limits (Figure 22.8*a*) or the operation that violates the important limits the least (Figure 22.8*b*). This task cannot be performed by signal selects but can be solved, using additional criteria entered by the engineer,

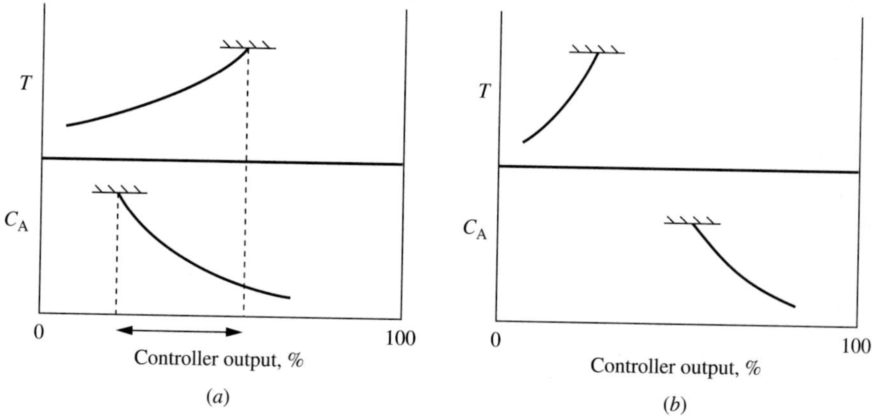

FIGURE 22.8
Systems for which signal select control is not appropriate.

with an optimization calculation. Control algorithms that are capable of performing optimization are introduced in Chapter 26.

The split range elements are designated by the symbols in Figure 22.9. As presented in Appendix A the designation "Y" is used for the second letter inside the symbol for a calculation and the less-than or greater-than sign to indicate low or high select. An older method that is still used frequently is to write LSS and HSS for low and high signal select, respectively.

The term *signal select* indicates that many different types of signals, not just controller outputs, can be used. As another example, the temperatures along a packed-bed reactor are monitored, and each temperature is to be maintained below its specified value. Two signal select control systems are shown in Figures 22.10*a* and *b*. In Figure 22.10*a* the measurements are input to a high signal select, and the output of the select is used as the *controlled* variable for a single controller, which adjusts the preheat. In Figure 22.10*b* each measurement goes to a separate

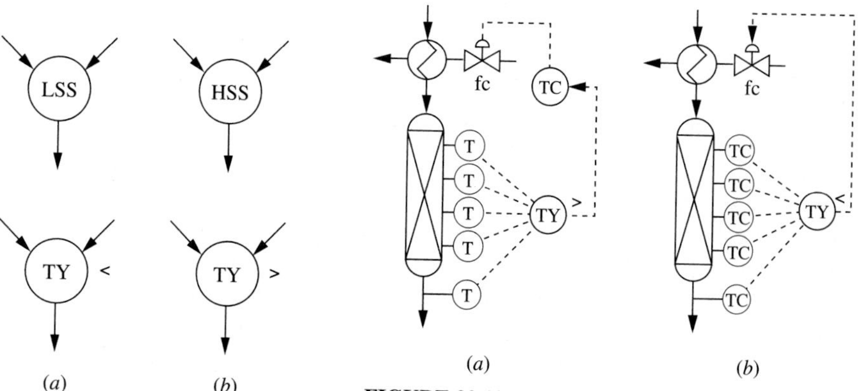

FIGURE 22.9
Symbols for signal selects.

FIGURE 22.10
Examples of signal select control on (*a*) measurements and (*b*) controller outputs.

controller, each controller output goes to the low signal select, and the output of the signal select goes to the *control valve.*

Both designs could succeed in maintaining the highest measured temperature at the set point. One difference is that the design in Figure 22.10*a* has one controller with one set of tuning constants, whereas the design in Figure 22.10*b* has separate tuning for each controller. The design in Figure 22.10*b* would be preferred if the feedback loop dynamics change with the measurement selected, as they might in this example. If the loop dynamics are essentially the same for all measurements, the design in Figure 22.10*a* would be preferred for its simplicity. Also, the design in Figure 22.10*a* enforces the same set point value for all measured variables, whereas the alternative design in Figure 22.10*b* allows different set points for different locations in the packed bed.

The designs in Figure 22.10 are similar, and often engineers have difficulty selecting between them. The proper selection is based on the recognition that the process dynamics in the feedback loop should be nearly constant (when the controller tuning constants are unchanged). The design in Figure 22.10*b* can have tuning tailored to each measurement and is thus a more general design. Three cases can occur:

1. When every closed-loop system has the same dynamics,

$$\frac{T_1(s)}{v(s)} = \frac{T_2(s)}{v(s)} = \cdots = G_p(s) \tag{22.14}$$

either design in Figure 22.10 can be used.

2. When the closed-loop systems have the same dynamics except for the steady-state gain,

$$\frac{T_1(s)}{v(s)} = K_1 G_p(s) \qquad \frac{T_2(s)}{v(s)} = K_2 G_p(s) \cdots \frac{T_i(s)}{v(s)} = K_i G_p(s) \tag{22.15}$$

the design in Figure 22.10*a* can be used if the controller gain is divided by a value K_i to compensate for the feedback process gain of the selected temperature, which would tune the single temperature controller to give the same stability margin:

$$G_{\mathrm{OL}}(s) = K_i G_p(s) \frac{K_c}{K_i} \left(1 + \frac{1}{T_I s}\right) \tag{22.16}$$

3. When the process dynamics are significantly different in each feedback loop, only the design in Figure 22.10*b* can be used for controllers with constant tuning values. (See Chapter 16 for evaluation of significant differences and methods for modifying controller tuning in real time.)

A very important implementation issue in the application of signal selects is the potential for reset (integral) windup in systems like the ones shown in Figures 22.5 and 22.10*b*. While all controller outputs are sent to the signal select, only one is used to determine the valve position; thus, there is only one feedback control system. The outputs from the other controllers do not influence the manipulated variable, and because of their controller integral modes, the outputs from the controllers not

TABLE 22.3
Signal select criteria

A signal select is possible when

1. There is one manipulated variable and several potential controlled variables.
2. There is a causal relationship between the manipulated variable and each controlled variable.
3. There is a unique, feasible operating point that satisfies all control objectives in the steady state (see Figure 22.7).

selected could wind up (i.e., increase or decrease without limit). Several possible solutions exist to prevent windup, with perhaps the clearest being the application of external feedback, which was introduced in Chapter 12. For signal select control, the value *after* the signal select is used as the external feedback variable for all controllers whose outputs go to the signal select. Such a system will not experience integral windup.

In conclusion, signal selects are widely applied to processes with excess controlled variables. The general approach for using a signal select is summarized in Table 22.3. Controlled variables can be used as inputs to a signal select if the feedback loop dynamics are similar for all controlled variables. The controller outputs should be used in the signal select if the feedback loop dynamics are different.

22.4 APPLICATIONS OF VARIABLE-STRUCTURE METHODS FOR CONSTRAINT CONTROL

The variable-structure control methods introduced in this chapter are based on a fixed ranking of controlled or manipulated variables. As a result, the operating conditions achieved through the control system maintain the process near a limiting or constraining value—for example, the minimum use of the more expensive fuel or the maximum reactor temperature. Control systems that result in process operation near a limit are generally termed *constraint controllers,* and since they often require variable-structure capability, constraint control is often implemented using split range and signal select methods. A few additional examples of constraint control are presented in this section.

Combined Variable-Structure Methods

This first example demonstrates how split range and signal select can be combined in control designs to achieve good control performance. Consider the situation in Figure 22.11, which shows two series processes. The product flow from unit 1 is usually not equal to the feed flow to unit 2; therefore, a large storage tank is located between the two units. One approach for dealing with the differences in flows would be to cool the entire production from unit 1, send it to the storage tank, and heat the feed to unit 2 as it flows at its desired rate from the storage tank. This approach would provide smooth and reliable flow control, but it would be very energy-inefficient.

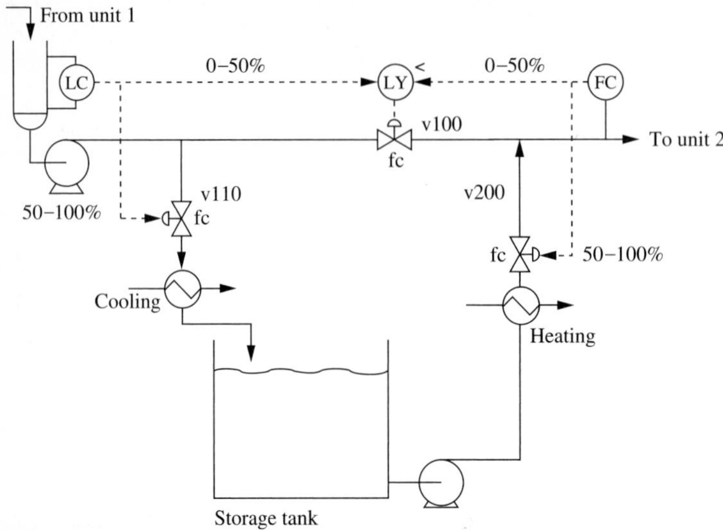

FIGURE 22.11
Example of combined split range and signal select.

A more efficient alternative approach would be to provide the maximum allowable direct flow from unit 1 to unit 2. The maximum direct flow between units would be determined by either the availability from unit 1 or the demand for unit 2, with the limiting condition changing as both unit operations change. A control system to maximize the direct flow automatically while always achieving proper level and flow control would be desirable. Such a system is shown in Figure 22.11, where both the level and flow controllers have split range outputs. Both controller outputs are sent to the low signal select, which determines the proper signal to manipulate the direct flow valve, which in this example is the smallest signal, which gives the smallest direct flow rate. Thus, one controller will adjust the direct flow valve, and the other controller will continue to increase its output until it adjusts the flow to the tank (for level control) or flow from the tank (for flow control), as appropriate. The resulting operations for the two situations are summarized in the following table.

	How each valve is adjusted			
Relative flows	**v100**	**v110**	**v200**	**Net flow**
Unit 1 flow > unit 2 flow	By FC	By LC	Closed	To storage
Unit 1 flow < unit 2 flow	By LC	Closed	By FC	From storage

With this control system the plant personnel need only input the set points to the level controller (normally 50% of range) and the flow controller (required flow to unit 2). The system automatically adjusts the valves as described to meet the level and flow requirements while minimizing flows to and from storage, thus minimizing energy use.

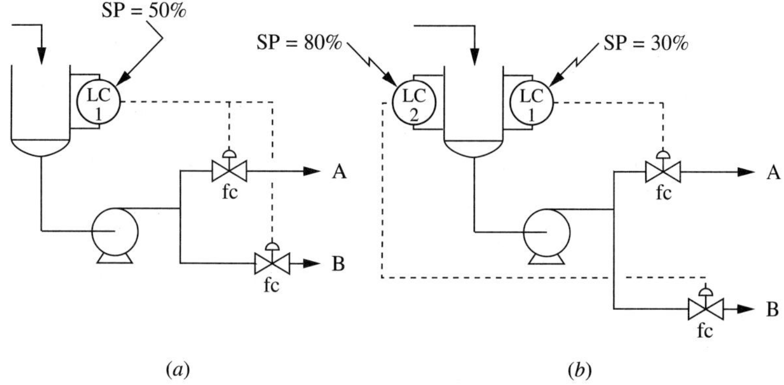

FIGURE 22.12
Alternative approaches to controlling inventory by *(a)* split range; *(b)* two controllers with different set points.

Multiple Controllers for One Variable

A design involving two separate controllers is sometimes used as an alternative to split range when the use of one manipulated variable is to be strictly minimized, even in short transients. For example, consider the level-flow system in Figure 22.12*a,* in which the level is normally controlled by manipulating the flow to a downstream unit but can be controlled by adjusting the flow to waste, if required. Naturally, the flow to waste is to be minimized. A split range controller is employed in the figure to achieve the control objective.

An alternative control system is given in Figure 22.12*b*, which employs two feedback controllers with *different set points.* Under normal conditions, the controller with the set point of 30% level (LC-1) adjusts the flow to the downstream unit, and the valve to waste is completely closed. If the flow in becomes large, the valve to the downstream unit is opened completely. If the flow in is still larger than the maximum flow to the downstream unit, the level then increases above 30%. If the large flow in remains for a long enough period of time, the level reaches the set point of the alternative controller with a higher set point (LC-2), here shown as 80%. When this level is reached, the alternative controller begins to increase the flow to waste.

The two-controller design in Figure 22.12*b* has three advantages. First, it uses the inventory in the vessel, so there is no flow to waste until the flow to downstream unit is at its maximum *and* the level increases to the upper set point. Thus, short-term disturbances in the inlet flow that can be accumulated in the vessel will not result in material being diverted to waste. Second, it has two sensors, valves, and controllers, so failures of elements in both control loops would have to occur before the level could overflow. This increase in reliability would be important if a large safety or economic penalty were incurred for an overflow. Finally, the use of two controllers allows separate tuning for the two feedback loops, although this probably would not be necessary in the example in Figure 22.12. The split range controller

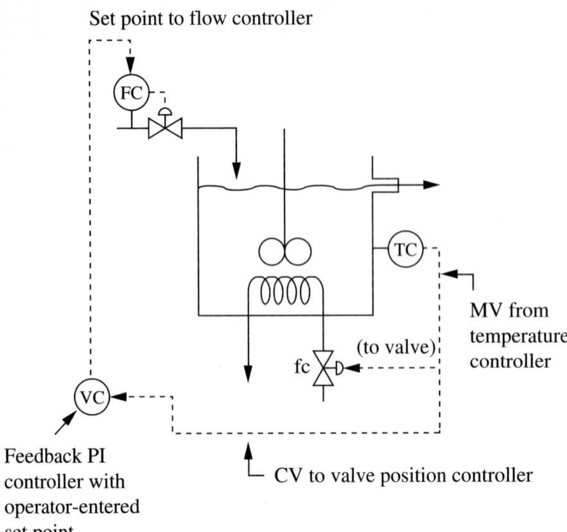

FIGURE 22.13
Example of valve position control.

has one set point and one set of tuning constants and is preferred, if it achieves the objectives, because of its simplicity.

Valve Position Control

Sometimes a limit is the result of equipment performance, and the approach to the limit is not easily inferred from measured process variables such as flow or temperature. This situation is demonstrated in Figure 22.13, in which the feed rate to a chemical reactor is to be maximized. The reactor temperature must be maintained constant, and the heat exchanger duty is the limiting factor in increasing the feed rate. There is no process variable that indicates how close the process operation is to the limit. One indication that the limit had been *exceeded* would be the reactor temperature remaining below its set point for a long time; however, this indication would be available only after the process had been upset. Thus, this measure of the limit is not normally acceptable.

Another potential indication of the limit is the temperature controller output, which is essentially the value of the heating medium valve position. When this value nears its maximum value of 100%, the limitation in heating duty is being approached. This analysis leads to the use of a *valve position controller* (VC), which uses the temperature *controller output* as its controlled variable and adjusts the feed flow controller set point. This is a feedback system and can use a standard proportional-integral algorithm; the set point of the valve position control system is chosen sufficiently far from the limiting value that the valve nearly never reaches a limit during a transient response to an upset. This approach ensures that the temperature control system has the range to respond to high-frequency disturbances and maintain the temperature at its set point. (A typical value might be 90%, but could be lower if the system experiences large temperature disturbances.)

The valve position controller feedback path involves the reactor and temperature control loop. Therefore, the valve position controller must be tuned loosely so as not to upset the temperature controller and to provide smooth, nonoscillatory approach to the constraint. Also, the feedback loop in the valve position controller includes the temperature controller; in other words, there is no causal feedback path without the temperature controller functioning. Thus, the valve position controller represents a loop pairing on a zero relative gain (see Chapter 20), and a monitoring program is recommended to determine whether the temperature controller is functioning and, if not, to switch the valve position controller into manual status.

Plantwide Variable-Structure Control

Sales demands and prices sometimes result in the pleasant circumstance that all of the plant's production can be sold. In this situation, the control system should be structured to result in the highest production rate possible, consistent with product quality and equipment performance limitations. Since most plants have several possible limiting factors, a variable-structure control system is normally used to monitor all likely limiting factors and adjust the feed rate so that the most restrictive factor does not exceed its limiting value.

The situation is shown in Figure 22.14 for a hypothetical plant in which three possible factors could limit the operation: heating medium availability for the reactor, maximum flow of vapor product from the flash drum, and maximum reboiler duty in the distillation tower. The control system monitors all three (two with valve position controllers), uses each in a feedback controller, and selects the lowest value of the

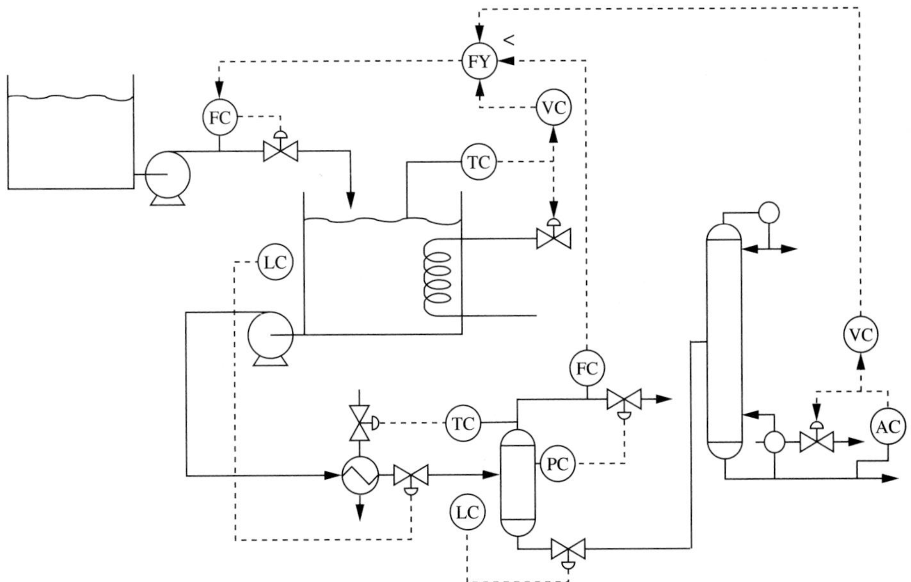

FIGURE 22.14
Example of maximum feed constraint control.

three controller outputs to adjust the feed flow set point. (Note that many important controllers are not shown in the simplified figure so that the feed maximization can be clearly shown.) Since the feedback processes are relatively slow for these constraint controllers, their set points should not be exactly the limiting values; the set points provide a safety margin from the limits, to account for the likely variability about the set point.

22.5 CONCLUSIONS

To achieve process objectives, engineers design equipment with appropriate capacities, provide measurements and manipulated variables, and design flexible control systems to respond to normal and upset conditions. Variable-structure control often enables the system to satisfy the operating objectives when the numbers of manipulated variables and controlled variables are not equal. Two methods have been presented in this chapter that are applicable to commonly occurring objectives. In split range control a single feedback controller output is sent to more than one final element, and the final elements are calibrated to operate over different ranges of the controller output signal (e.g., 0 to 50% and 50 to 100%). A signal select, on the other hand, is used when there are several controlled variables and one manipulated variable, and the signal select determines the most limiting control objective.

These methods are appropriate for situations in which the best operation resides on a constraint or "frame" of the steady-state operating window, as demonstrated in the following examples. First, consider the fuel pressure split range control in Figure 22.1. Since the flexibility in the system involves the manipulated variables, the operating window in Figure 22.15 has manipulated variables as the coordinates. Any point inside the steady-state window that satisfies $F_A + F_B = \sum F_{consumer}$ is a feasible plant operating point. Clearly, there are infinite combinations of fuel flows that can satisfy the total consumer demand. The best operation is designated by the dashed line, which shows the combination of flows of the two fuels that satisfies the consumer demand from zero to maximum while also minimizing fuel cost. The split range control system implements this strategy and therefore is appropriate for this example.

Then consider the chemical reactor signal select control system in Figure 22.5. Since the flexibility involves the controlled variables, the operating window in Figure 22.16 has controlled variables as the coordinates. Any point in the window

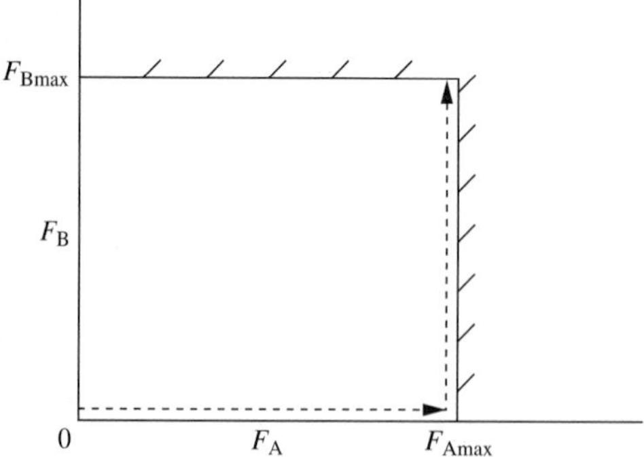

FIGURE 22.15
Operating window for fuel pressure split range control.

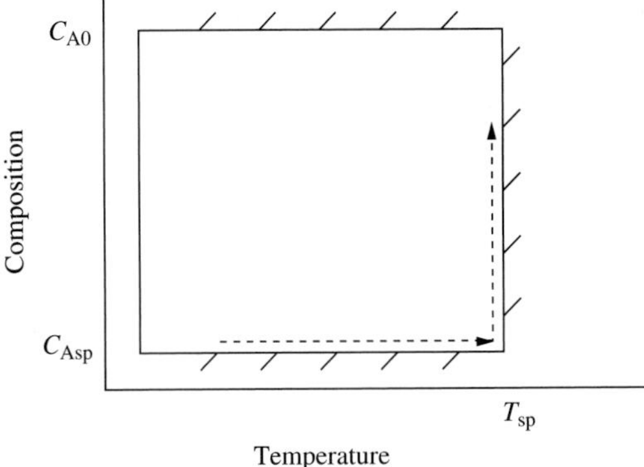

FIGURE 22.16
Operating window for reactor signal select control.

represents feasible plant operation, and there is an infinite number of these points. The best operation, which maximizes conversion subject to the limitations, is designated by the dashed line. The arrows on the dashed line represent the (quasi-steady-state) path followed as the inhibitor disturbance increases. The signal select implements this strategy and therefore is appropriate for this example.

Implicit in the use of the automated variable-structure methods in this chapter is the assumption that the change in structure must be made quickly, when required. If the required structural changes occur very infrequently and need not be made immediately, a simple switch could be used, and the position of the switch could be changed by a human operator. The simpler design using a switch is employed when the structure change is needed infrequently, such as during unit startups.

Variable-structure methods presented in this chapter employ single-loop controller algorithms. In this chapter, only PID controllers have been discussed; however, other algorithms, such as the model predictive controllers in Chapter 19, can be used.

The empirical model identification methods and controller tuning procedures for variable-structure systems are the same as presented previously for single-loop systems. Since the feedback path $G_{OL}(s)$ depends on the status of the variable structure, models and tuning for each feedback path must be determined and fine-tuned individually.

It is important to reiterate that the methods presented in this chapter, while simple and easy to apply, are limited to systems with low dimensionality. It is very difficult to implement a variable-structure system with many controlled *and* manipulated variables by the methods in this chapter. Fortunately, the multivariable control algorithm presented in the next chapter has the capability of practically solving high-dimension variable-structure control systems, as well as complex square control systems.

Finally, variable structure is applied widely in the process industries. It enables a process to operate with a specified (efficient) control pairing for normal operation and to maintain acceptable operation as large changes in input variables occur or unusual set points are entered. Thus, the integration of variable-structure control as a component of the control design is essential for proper operation of many process plants.

REFERENCES

MacGregor, J. and T. Harris, "Design of Multivariable Linear-Quadratic Controllers Using Transfer Functions," *AIChE J., 33, 9,* 1481–1495 (1987).

ADDITIONAL RESOURCES

High-frequency measurement noise can influence the performance of signal selects, causing a value that is not the lowest (highest) to be selected by a low (high) signal select. This behavior can prevent a constraint control design from achieving close approach to the best possible operation. This issue is discussed and improvements suggested in the following papers.

Weber, R. and R. Zumwalt, "The Effect of Measurement Noise on Feedback Controllers with Signal Selectors," *AIChE Natl. Meet.,* April 1979.

Giles, R. and L. Gaines, "Integral-Tracking Override Is Better than Output Tracking," *Cont. Eng.,* 63–65 (1978).

Constraint control has been applied to many unit operations. Examples in distillation control are presented in the following papers:

Maarleveld, A. and J. Rjinsdorp, "Constraint Control on Distillation Columns," *Automatica, 6,* 51–58 (1970).

Roffel, B. and H. Fontein, "Constraint Control of Distillation Processes," *Chem. Eng. Sci., 34,* 1007–1018 (1979).

A slightly different approach for valve position control is presented in

Love, B., "Advanced Technique Improves Two-Valve PRV Control," *Control, V,* 56–60 (1992).

Additional examples of variable-structure control are presented in

Shinskey, F.G., *Controlling Multivariable Processes,* Instrument Society of America, Research Triangle Park, NC, 1981.

> It is very helpful to describe the desired process operating policy in words first and then sketch the behavior in figures similar to Figures 22.2, 22.15, and 22.16. Follow this suggestion when designing controls for the questions in this chapter.

QUESTIONS

22.1. The three-tank mixer problem in Example 7.2 is considered here, with the slight modification that stream *B* is under flow control, with a sensor and valve added to the process. The goal is to maximize the production of material from the third tank. Limitations could be encountered in the flow rates of either stream *A* or *B*.

(*a*) Design at least one control system that would (1) control the product quality to 3% A and (2) maximize the production rate.

(*b*) Estimate the initial tuning for every controller in the design.

22.2. Prepare the detailed equations, with sequence of execution, or a sample digital control program for

(*a*) the split range controller in Figure 22.1

(*b*) the signal select system in Figure 22.5

22.3. Analyze the degrees of freedom based on models of the process and the control calculations for

(*a*) the system in Figure 22.11

(*b*) the system in Figure 22.12, both designs

22.4. Sketch the steady-state operating window and describe the path taken by the process under control for the following systems in response to selected disturbances: (*a*) Figure 22.10 (simplify this to two temperatures), (*b*) Figure 22.11, (*c*) Figures 22.12*a* and *b*, and (*d*) Figure 22.13.

22.5. For the stirred-tank heater system in Figure 22.13,

(*a*) Verify the degrees of freedom from models of the process and control.

(*b*) Specify the control calculations in analog or digital using external reset to prevent integral windup. Indicate the external variable clearly.

(*c*) Since the design includes a pairing on a zero relative gain, describe the monitoring program required. Clearly indicate the variables monitored and the actions taken when specific situations are encountered.

22.6. There are situations with excess manipulated variables in which neither variable should normally have a zero value. For example, consider the process in Figure 22.1, but with a contract that requires the plant to pay for a specified amount of fuel B (say 30% of valve opening), whether it is consumed or not. Design a control strategy that controls the pressure and provides good steady-state economic performance for this process.

22.7. Anti–reset windup is a very important aspect of successful control implementation. For the system in Figure 22.5,

(*a*) Sketch a block diagram in a continuous (analog) implementation of PI controllers with external feedback and clearly show the measurement used for the external variable.

(*b*) Provide the equations or sample program for digital control calculations, including all control elements, and an alternative method for anti–reset windup that does not have to use the external feedback principle.

22.8. Discuss how the two proportional-integral controllers in Figure 22.5 can be replaced with single-loop predictive controllers, either IMC or Smith predictors. In the discussion, provide all process and control equations, analyze the degrees of freedom, and explain how to prevent integral windup.

22.9. MacGregor and Harris (1987) describe a process that is shown schematically in Figure Q22.9. A moist film is dried using two sources of heat: an expensive electrical IR heater, which has a rapid effect on the moisture in the material, and a less costly steam heater, which has a slower response on the moisture in the material. Design a control system to provide tight control of the moisture and to minimize energy costs.

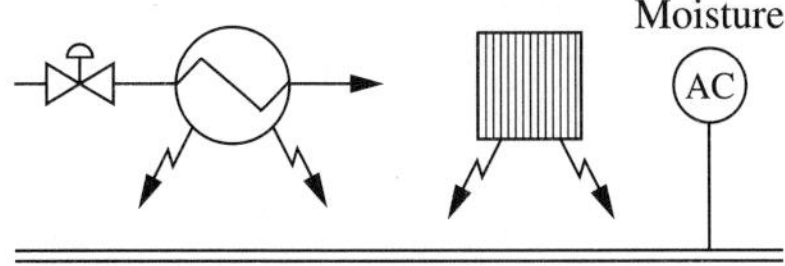

FIGURE Q22.9

22.10. Plantwide throughput maximization is certainly a good concept, but the slow dynamics between the downstream constraints and the manipulated feed flow rates could lead to extreme violations of the constraints as disturbances occur. An approach to prevent these violations is to include extra controllers that adjust manipulated variables that are "close" to the controlled variables (and have fast dynamics to the controlled variable). These *override controllers* prevent large, long constraint violations during the time required for the manipulation of the feed flow to affect the limiting plant variable. Apply this approach to the process in Figure 22.14 to prevent violations of the maximum vapor flow from the flash and excessive vapor boilup in the distillation column.

22.11. For a typical level process, as in Figure 18.1, design two control systems to ensure a minimum flow through the pump. (Some process equipment changes might be required.) Discuss the merits and demerits of each and recommend one for application.

22.12. Discuss the control objectives and control design in Figure 2.2.

22.13. Using methods in this chapter, design control systems to

(*a*) maximize the feed flow rate in Figure 15.5

(*b*) maximize the product rate of the valuable vapor product in Figure 13.19

(*c*) maximize the concentration of component B in Example 3.12

22.14. Discuss the steps necessary to identify linear dynamic models empirically, determine initial tuning, and fine-tune all controllers for the systems in Figures 22.1, 22.5, 22.10*a* and *b*, and 22.11.

22.15. The control design in Figure 22.11 has a deficiency, because the controllers experience a "gap" when switching between manipulated valves. Explain how this gap occurs and propose a design modification that eliminates this gap while retaining the good aspects of the original design.

22.16. For the following processes, design a variable-structure control system with a sketch, select feedback algorithms and modes, and estimate all initial tuning constants.

(*a*) The concentration of A in the effluent of the second reactor of the series chemical reactors in Example 5.4 is to be controlled. The preferred method is to adjust the flow of reactant A, F_A. When this variable saturates, the flow of solvent, F_s, is adjusted.

(*b*) The bottoms composition in the distillation tower in Example 20.2 is to be controlled. The preferred choice is to manipulate the reboiler duty, but if this saturates, the reflux can be adjusted. Consider two cases: (1) the distillate composition is free to vary: (2) the distillate composition is controlled by adjusting the reflux when possible, but the bottoms purity is of overriding importance.

(*c*) The temperature of the stirred-tank heat exchanger in Example 8.5 is to be controlled. The preferred choice is adjusting the coolant flow rate, but if this saturates, the feed flow rate can be adjusted.

(*d*) The reactor outlet temperature in the chemical reactor in question 21.4 is to be controlled. The preferred choice is the temperature of the quench, but if this saturates, the quench flow can be adjusted. With this approach, can the reactor outlet composition (in addition to the outlet temperature) be controlled by adjusting the quench flow when the quench temperature is not saturated?

CHAPTER 23

CENTRALIZED MULTIVARIABLE CONTROL

23.1 INTRODUCTION

The first three chapters in this section on multivariable control retained the proportional-integral-derivative (PID) control algorithm. This approach is generally preferred for its simplicity when it provides good performance, which is often the case. However, some especially challenging process control objectives are difficult or impossible to achieve using multiloop PID control. In this chapter, one centralized method for controlling multiple input-output processes is introduced. The term *centralized* denotes a control algorithm that uses all (process input and output) measurements simultaneously to determine the values of all manipulated variables. In contrast, multiloop control, also called *decentralized* control, involves many algorithms, with each using only one process output variable to determine the value of one manipulated variable. Further discussions on the need for centralized control are presented in Cutler and Perry (1983) and Prett and Garcia (1988).

In addition to all measurements, centralized controllers use a dynamic model of the process in the control calculation. The most common approach to using a model explicitly in the control calculation is the model predictive control structure described in Chapter 19. Since the discussions in this chapter are based on an understanding of the model predictive structure, the reader is advised to review Chapter 19 thoroughly before proceeding with this chapter.

This chapter begins with a straightforward extension of the model predictive controller to a multivariable system. This extension demonstrates the limitations in applying the analytical model inverse, which was easily determined for single-variable systems, to the multivariable case. Then, one approach to determining a controller design using numerical methods to obtain good dynamic performance is presented, first for single-variable and subsequently for multivariable systems. In this chapter, the digital algorithm is presented, because of the clarity and ease of implementation of this form. The presentation of the new control algorithm is concluded with discussions on implementation guidelines and extensions.

23.2 MULTIVARIABLE PREDICTIVE CONTROL

Model predictive control was introduced in Chapter 19, where some important properties were demonstrated for single-loop systems. The same principles can be applied to a multivariable system. For example, the following properties can be shown to hold for the general (open-loop stable) system in Figure 23.1.

1. The controlled variables will return to their set points for steplike inputs if

$$\mathbf{G}_{\mathrm{cp}}(0) = [\mathbf{G}_m(0)]^{-1} \tag{23.1}$$

Thus, the steady-state gain matrix of the controller must be the same as the inverse of the steady-state process model. Again, this can be achieved easily, because the engineer selects both of these elements in the control system. Note that the model does not have to match the plant exactly, although large model mismatch can degrade performance and lead to instability.

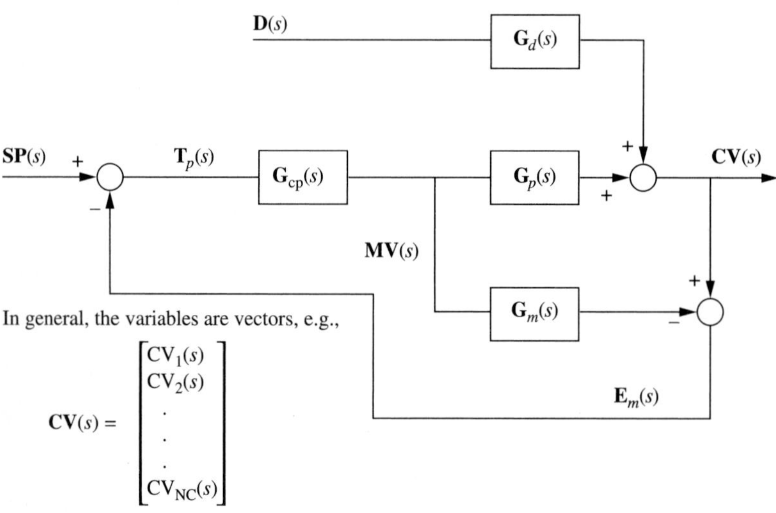

FIGURE 23.1
Model predictive control structure.

2. Perfect control (i.e., zero deviation from set point) is achieved when

$$\mathbf{G}_{cp}(s) = [\mathbf{G}_m(s)]^{-1} \tag{23.2}$$

Even if possible, this might involve excessive variability in the manipulated variable and thus not be desirable in practice.

3. If the model (and process) contains noninvertible elements, an *approximation* to equation (23.2) can be used to determine the controller, as follows.

$$\mathbf{G}_{cp}(s) = \left[\mathbf{G}_m^-(s)\right]^{-1} \tag{23.3}$$

$$\mathbf{G}_m(s) = \mathbf{G}_m^+(s)\,\mathbf{G}_m^-(s) \tag{23.4}$$

with $\mathbf{G}_m^+(s)$ The "noninvertible" factor has an inverse that is not causal or is unstable. The inverse of this term includes predictions, $e^{\theta s}$, and unstable poles, $1/(1 + \tau s)$, $\tau < 0$, appearing in $[\mathbf{G}_m(s)]^{-1}$. The steady-state gain of this factor must be the identity matrix.

$\mathbf{G}_m^-(s)$ The "invertible" factor has an inverse that is causal and stable, leading to a realizable, stable controller. The steady-state gain of this factor is the gain matrix of the process model, $\mathbf{K}_m$.

For single-variable systems, the design of the controller $G_{cp}(s)$ was relatively straightforward. However, the application of this analytical approach to multivariable systems encounters a significant barrier, as demonstrated in the following example.

Example 23.1. A multivariable predictive controller is to be applied to the binary distillation tower considered throughout the book (see Example 5.6 on page 180 and Chapters 20 and 21). The product compositions are to be controlled by adjusting the reflux and reboiler; thus, the energy balance regulatory control strategy provides the base control on which the composition control will be implemented. This approach, which shows the multivariable controller as an upper-level component in a cascade design, is given in Figure 23.2.

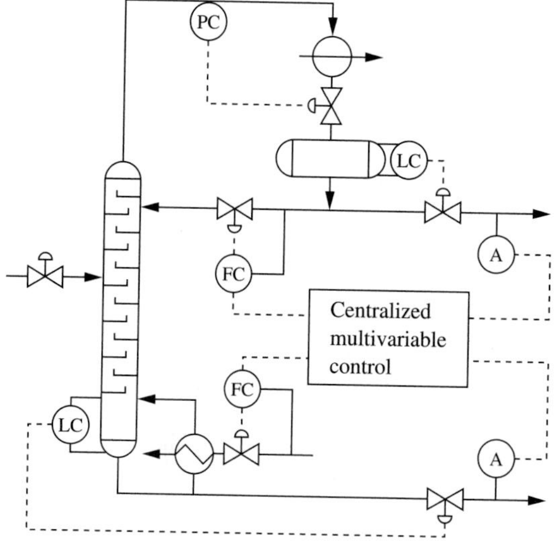

FIGURE 23.2
Centralized multivariable distillation control.

The model for the process is given in equation (21.1) and is repeated here:

$$\begin{bmatrix} X_D \\ X_B \end{bmatrix} = \begin{bmatrix} \dfrac{0.0747e^{-3s}}{12s+1} & \dfrac{-0.0667e^{-2s}}{15s+1} \\ \dfrac{0.1173e^{-3.3s}}{11.75s+1} & \dfrac{-0.1253e^{-2s}}{10.2s+1} \end{bmatrix} \begin{bmatrix} F_R \\ F_V \end{bmatrix} + \begin{bmatrix} \dfrac{0.70e^{-5s}}{14.4s+1} \\ \dfrac{1.3e^{-3s}}{12s+1} \end{bmatrix} X_F \qquad (23.5)$$

This two-variable system would be represented in the general symbols of Figure 23.1 as

$$\begin{bmatrix} \mathrm{CV}_1(s) \\ \mathrm{CV}_2(s) \end{bmatrix} = \begin{bmatrix} G_{11}(s) & G_{12}(s) \\ G_{21}(s) & G_{22}(s) \end{bmatrix} \begin{bmatrix} \mathrm{MV}_1(s) \\ \mathrm{MV}_2(s) \end{bmatrix} + \begin{bmatrix} G_{d1}(s) \\ G_{d2}(s) \end{bmatrix} D(s) \qquad (23.6)$$

By applying equation (23.1) the predictive controller is evaluated by determining the inverse of the feedback model.

$$[\mathbf{G}_m(s)]^{-1} = \frac{1}{\dfrac{0.00782e^{-5.3s}}{(15s+1)(11.75s+1)} - \dfrac{0.00936e^{-5s}}{(12s+1)(10.2s+1)}} \begin{bmatrix} \dfrac{-0.1253e^{-2s}}{10.2s+1} & \dfrac{0.0667e^{-2s}}{15s+1} \\ \dfrac{-0.1173e^{-3.3s}}{11.75s+1} & \dfrac{0.0747e^{-3s}}{12s+1} \end{bmatrix} \qquad (23.7)$$

The model in equation (23.5) cannot be factored uniquely into an invertible part. Also, the control performance of a control system that would satisfy equations (23.1) and (23.3) is not easily related to the analytical method of obtaining the invertible factor $\mathbf{G}_m^-(s)$.

> Thus, the analytical algorithm design method in equations (23.1 through 23.4) will not be used for multivariable systems in this chapter, although the model predictive structure will be retained.

The distillation example will be reconsidered after an alternative controller algorithm has been developed.

23.3 AN ALTERNATIVE DYNAMIC MODELLING APPROACH

The previous section demonstrated that a new approach to designing the model predictive algorithm is needed. Fortunately, several approaches have been developed, and one of these will be presented in the next section. However, the new method requires dynamic models in a format different from the standard transfer functions used to this point. The requisite modelling is described in this section using the symbols X for input and Y for output. This convention is used because these models can represent the input-output behavior for various variable combinations; for example, X could represent a disturbance or a manipulated variable.

Throughout the book, transfer function models have been determined from fundamental modelling and empirical identification. These transfer functions are very useful in representing the dynamic input-output behavior of linear (or linearized)

TABLE 23.1
Transfer function with its step response model

Transfer function			
$Y(s)/X(s) = K_p e^{-\theta s}/(\tau s + 1)$			
$= 1.0e^{-5s}/(5s + 1)$			
Step response			
Sample k	**Time** t	$X'(t)$	$Y'(t) = a_k$
0	0	1	0.
1	2.5	1	0.
2	5	1	0.
3	7.5	1	0.394
4	10	1	0.632
5	12.5	1	0.777
6	15	1	0.865
7	17.5	1	0.918
8	20	1	0.950
9	22.5	1	0.970
10	25	1	0.982
11	27.5	1	0.989

elements in a control system. They are parsimonious, in that the entire dynamic response can be represented by a small number of parameters. Also, their analytical structure enables the engineer to perform many transformations and calculations easily. However, alternative model structures are possible. For example, a dynamic model can be represented by the two forms in Table 23.1. This is the model of a single-tank mixing process with transportation delay shown in Figure 9.1 and used in Section 9.3, and it will be used in examples later in this chapter.

The example transfer function considered here is first-order with dead time, but more complex equations are common and can be modelled using this approach. An alternative model form is the step response, which is a set of discrete values representing the output response to a unit step input; these values are often referred to as the *step weights.* The transfer function gives a continuous model of the process, whereas the step response gives no information at times between the sampled points and has the same values as the continuous model at the sample points. The step response can be developed from the transfer function by solving for the output response of the continuous system to a unit (+1) step input at sample number 0. For the example first-order-with-dead-time system, the discrete form is

$$Y(s) = \frac{K_p e^{-\theta s}}{(\tau s + 1)} X(s) = \frac{K_p e^{-\theta s}}{s(\tau s + 1)} \qquad (\text{note } X(s) = \Delta X/s = 1/s) \qquad (23.8)$$

$$Y'(t) = (1.0)K_p(1 - e^{-(t-\theta)}) \qquad (t \geq \theta) \qquad (23.9)$$

This continuous model can be evaluated at sample points by setting time equal to multiples of the sample period, Δt. In the following equations the subscripts m

emphasize that the transfer function parameters refer to the model, which is only an approximate representation of the true plant.

Time	Sample no.	Input X'	Output Y'
0	0	1	0
Δt	1	1	0
$2\Delta t$	2	1	0
	{continues until the dead time}		
θ	$\theta_m/\Delta t$	1	0
$\theta_m + \Delta t$	$(\theta_m + \Delta t)/\Delta t$	1	$K_m(1 - e^{-\Delta t/\tau_m})$
$\theta_m + 2\Delta t$	$(\theta_m + 2\Delta t)/\Delta t$	1	$K_m(1 - e^{-2\Delta t/\tau_m})$
$\vdots$	$\vdots$	1	$\vdots$

The reader can verify that this method was used to develop the step weights from the transfer function in Table 23.1 by calculating the step response from an initial steady state of $Y' = 0$. The step response model can be used to calculate the value of the output Y' at any sample period k in response to a step of any size ΔX using the equation

$$Y'_k = a_k \Delta X_0 \tag{23.10}$$

Recall that the transfer function used in developing the step response can be derived from fundamental models using methods from Chapters 3 through 5, or it can be developed from empirical data using methods from Chapter 6.

Example 23.2. Determine two models for the data in Figure 23.3: a transfer function model and a discrete step model.

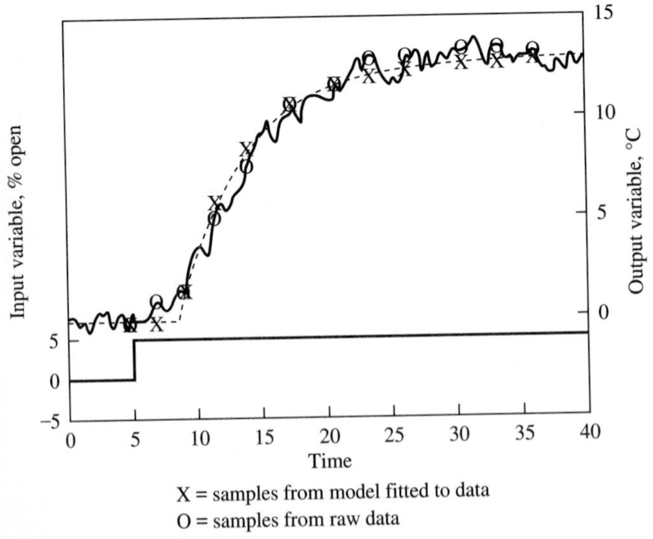

FIGURE 23.3
Process reaction data with continuous and discrete models.

The continuous transfer functions can be determined using the methods described in Chapter 6. This data was used in Examples 6.2 and 6.6, where it was concluded that a first-order-with-dead-time structure was adequate. For example, the parameters determined in Example 6.6 (page 215) using the statistical parameter method are $K_p = 2.56°C\%$ open, $\theta = 3.66$ min, and $\tau = 5.2$ min.

The discrete step response can also be determined from the data. One approach would be to use the measured values of the output variables as the step response; this approach would use the data indicated by the circles in Figure 23.3. While this represents the process behavior exactly for *this* experiment, the data includes noise, which would not be repeatable and should not be used for designing or tuning controllers. A better method for determining the step response would characterize the repeatable process response and ignore the higher-frequency noise. There are many methods for evaluating a step model from noisy data; one good method uses conventional modelling methods (for example, those in Chapter 6) to fit a transfer function model and subsequently evaluate the step response using the transfer function. This approach is demonstrated in Figure 23.3, where the dashed line is the continuous output from the transfer function model and the crosses are the step response from the estimated *model,* not the raw data. This modelling approach captures the dominant dynamic behavior while eliminating the effects of most of the noise. Further discussions of determining representative step response models from empirical data are given in MacGregor et al. (1991), Cutler and Yocum (1991), and Ricker (1988).

The step response model can be used to predict the dynamics of a system for any input function of time. This is achieved by sampling the input function and recognizing that it can be approximated by step changes at each sample point. The effect on the output of each input step is represented by the step response in equation (23.10). The overall effect of all of the input steps is the sum of each individual effect, assuming that the system is *linear.* This modelling method introduces potential errors, because the input may not be a perfect staircase function; however, the errors will be small if the sample period is short compared with the rate of change of the input and output variables. Assuming that the plant begins at a steady-state condition (Y_0), the step weights can be used to predict the output from the input values at the sample points as follows:

$$\begin{aligned} Y_1 &= Y_0 + a_1 \Delta X_0 \\ Y_2 &= Y_0 + a_2 \Delta X_0 + a_1 \Delta X_1 \\ Y_3 &= Y_0 + a_3 \Delta X_0 + a_2 \Delta X_1 + a_1 \Delta X_2 \end{aligned} \tag{23.11}$$

and so forth. This model can be expressed as an equation for any number of sample periods k for a single-input–single-output system as follows:

$$Y_{k+1} = Y_0 + \sum_{j=1}^{k+1} a_j \Delta X_{k-j+1} \tag{23.12}$$

Applying the model in equation (23.12) for a long time ($k \rightarrow$ large) would result in a sum over a very large number of samples, since every change in the past influences the current value of the output variable. We anticipate that such a large summation would cause difficulties for the controller calculation. However, the input changes have a constant effect as the time from the input step becomes large; that is, after the transient settles to the constant effect for a past ΔX. Thus, the model in equation (23.12) can be rewritten to give the following equation.

$$Y_{k+1} = \underset{\substack{\text{Initial} \\ \text{condition}}}{Y_0} + \underset{\substack{\text{Reached} \\ \text{steady state}}}{\sum_{j=\text{LL}+1}^{k+1} a_j \Delta X_{k-j+1}} + \underset{\substack{\text{Transient} \\ \text{response}}}{\sum_{j=1}^{\text{LL}} a_j \Delta X_{k-j+1}} \tag{23.13}$$

The last term on the right-hand side includes those past inputs whose effects have not yet reached their steady-state values. Thus, the number of samples multiplied by the period should be the settling time of the process; for example, LLΔt is approximately equal to the dead time plus four time constants for a first-order-with-dead-time process. The second term on the right-hand side involves the inputs whose effects have (essentially) reached their steady state, so that $a_k \approx K_p$ for $k >$ LL. It is not necessary to sum all of the values in the second term at each time step, because the summation only changes by one value each sample period: by 1 past ΔX. Thus, this can be calculated recursively using a new intermediate variable Y^* to include the initial value of Y and the effects of all ΔX values whose effects have reached steady state. (Recall that a recursive calculation uses only the past result and the new input to calculate the new result.)

$$\begin{aligned} Y_k^* &= Y_{k-1}^* + a_{\text{LL}+1} \Delta X_{k-\text{LL}} \\ Y_{k+1} &= Y_k^* + \sum_{j=1}^{\text{LL}} a_j \Delta X_{k-j+1} \end{aligned} \tag{23.14}$$

The approximation of the step response with its steady-state (or final) value introduces another potential error, which can be made small by proper choice of the number of steps (LL) to include in the summation in equation (23.13). Now the large sum of the steady-state effects has been eliminated by the recursive form of the model.

> The step response model in equation (23.14) does not require all past inputs to be stored and the large summation to be calculated each execution: the information about initial condition and inputs whose effects have reached steady state are accumulated in the Y^* term.

The modelling approach described in this section is simple and easily applied to any single input–single output process response. Because only discrete samples of the response are used, the step response model is not as complete a representation as a continuous transfer function model. However, the discrete step response model facilitates the design of centralized feedback controllers, as explained in the next section.

23.4 THE SINGLE-VARIABLE DYNAMIC MATRIX CONTROL (DMC) ALGORITHM

Several approaches can be used to develop a practical multivariable centralized controller. The method presented here is the Dynamic Matrix Controller, which was developed by Cutler (Cutler and Ramaker, 1979), was extended to include additional features (Prett and Gillette, 1979; Garcia and Morshedi, 1986), and has been applied successfully to complex processes (e.g., Kelly et al., 1988; Van Hoof et al., 1989).

The Dynamic Matrix Control algorithm can be implemented within the model predictive control structure, and the algorithm can be designed without determining the analytical inverse of the process model, so the extension to multivariable systems is straightforward. The DMC algorithm will be introduced here for the single-variable case and then will be extended to multivariable. This explanation will proceed in three steps, each introducing a key aspect of the overall algorithm.

I. Basic DMC Algorithm (Without Feedback)

The algorithm will be introduced by considering the situation encountered every time a model predictive feedback controller is executed. The dynamic response of a feedback control system is shown in Figure 23.4. The manipulated variable has been adjusted in the past, and the controlled variable has been influenced by these adjustments, as well as by disturbances. The prediction of the controlled variable, calculated using equations (23.14) and past values of the manipulated variable, is also shown in the figure. The task of the control algorithm is to determine *future* adjustments to the manipulated variable that will result in the predicted controlled variable returning quickly to the set point.

To determine the best controller moves, a measure of control performance must be selected. Here, the integral of the error squared, or the sum of the error squared at sample points, will be taken; we recognize that this measure is not complete, and

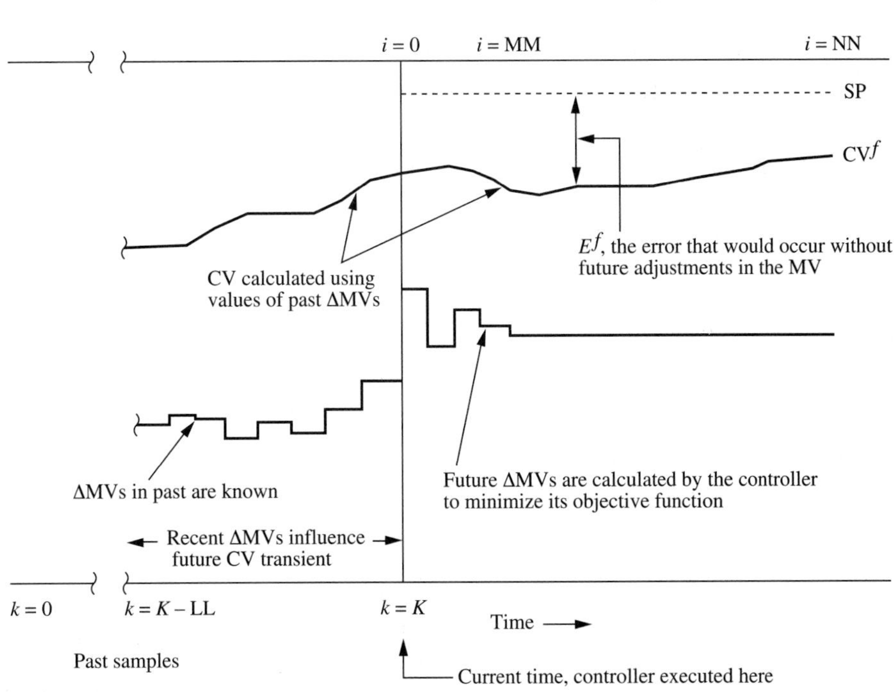

FIGURE 23.4
Dynamic response of variables for DMC control.

we will modify it later to consider robustness and the behavior of the manipulated variable. The error—deviation between set point and controlled variable—can be measured at the current time, but we know that it will change in the future because of recent adjustments to the manipulated variable. The behavior of the controlled variable *without adjustments in the future* should be used to determine the future error, which should be reduced by future adjustments. Thus, the DMC controller uses a dynamic model of the process to calculate the future behavior of the controlled variable that would occur without future control adjustments.

$$\text{CV}_i^f = \text{CV}_K + \sum_{j=1}^{\text{LL}} a_{j+i}\Delta\text{MV}_{K-j} \qquad \text{Note: Without feedback} \qquad (23.15)$$

with CV_i^f = the predicted (deviation) value of the controlled variable in the future as influenced by past changes in the manipulated variable

CV_K = the predicted value of the controlled variable at the current time based on all past inputs

i = sample periods in the future (i = 1 to NN)

The difference between the predicted values of the controlled variable and the set point are used to calculate the objective, the sum of the errors squared, which is to be minimized.

$$\text{OBJ}_{\text{DMC}} = \sum_{i=1}^{\text{NN}}\left[\text{SP}_i - (\text{CV}_i^f + \text{CV}_i^c)\right]^2 = \sum_{i=1}^{\text{NN}}\left[E_i^f - \text{CV}_i^c\right]^2 \qquad (23.16)$$

with SP_i = set point at each sample i in the future

CV_i^f = defined in equation (23.15) and cannot be influenced by the controller

CV_i^c = effect of future adjustments on the controlled variable at each sample i

E_i^f = ($\text{SP}_i - \text{CV}_i^f$), deviation from set point that would occur if no future control adjustment were made

NN = the future time over which the control performance is evaluated, termed the *output horizon*

In equation (23.16), the set point can remain constant at its current value in the future, but if it will vary in the future in a manner known when the controller is executed, a variable set point can be accommodated. Also, the future effects of *past* adjustments, CV_i^f, are calculated using equation (23.15). Thus, only the terms CV_i^c are influenced by the *future* adjustments determined by the controller algorithm. Finally, the output horizon (NN) should be long enough for the controlled variable to approach steady state under closed-loop control.

Now, the challenge is to determine the future adjustments in the manipulated variables to minimize the objective. This is an optimization problem that could be solved by many methods, including searching over a large grid of possible values of the manipulated adjustments, but that would involve wasteful, excessive calculations. A good method can be developed using the modelling approach introduced in the previous section. The step response model can be used to calculate the effects of *future* moves, by summing their effects.

$$\mathrm{CV}^c_{i+1} = \sum_{j=1}^{i+1} a_j \Delta \mathrm{MV}^c_{i-j+1} \qquad (23.17)$$

with CV^c = the effects of future adjustments in the manipulated variable on the controlled variable

$\Delta\mathrm{MV}^c$ = the future adjustments calculated by the controller

This model can be slightly rearranged to ease the optimization calculation. The same result can be obtained with the summation over all inputs at each sample i of the horizon by ensuring that the effects are zero for all adjustments after the sample at which the controlled variable is evaluated (i). This model can be expressed in matrix format as follows, using the step weights a_j that can be nonzero (where $i \geq j$) and 0.0 for the elements that must be zero (where $i < j$).

$$\begin{bmatrix} a_1 & 0 & 0 & \cdots & 0 \\ a_2 & a_1 & 0 & \cdots & 0 \\ a_3 & a_2 & a_1 & \cdots & 0 \\ \vdots & \vdots & \vdots & \ddots & \vdots \\ a_{\mathrm{NN}} & a_{\mathrm{NN}+1} & a_{\mathrm{NN}+2} & \cdots & a_{\mathrm{NN}-\mathrm{MM}+1} \end{bmatrix} \begin{bmatrix} \Delta\mathrm{MV}^c_0 \\ \Delta\mathrm{MV}^c_1 \\ \Delta\mathrm{MV}^c_2 \\ \vdots \\ \Delta\mathrm{MV}^c_{\mathrm{MM}-1} \end{bmatrix} = \begin{bmatrix} \mathrm{CV}^c_1 \\ \mathrm{CV}^c_2 \\ \mathrm{CV}^c_3 \\ \vdots \\ \mathrm{CV}^c_{\mathrm{NN}} \end{bmatrix} \qquad (23.18)$$

In this formulation, the adjustments in the manipulated variable could be allowed for all samples in the output horizon; however, experience indicates that this can lead to overly aggressive control action and oscillatory dynamic responses. Therefore, fewer manipulated-variable adjustments are allowed, and the number of adjustments is given by the *input horizon* MM, which must be less than the output horizon. Equations (23.17) and (23.18) are equivalent, and either one may be used to evaluate the effects of future adjustments on the control objective. Perhaps equation (23.18) provides a clearer picture of the calculation. The coefficient matrix in equation (23.18) is often designated by the symbol **A** and is referred to as the *dynamic matrix.* With this notation, equation (23.18) can be rewritten as

$$\mathbf{A}[\Delta\mathrm{MV}^c] = [\mathrm{CV}^c] \qquad (23.19)$$

The goal of perfect controlled-variable performance would be to have zero error for all samples in the future, which would be achieved if

$$\mathbf{E}^f = [\mathrm{CV}^c] \qquad \text{perfect control of } \mathbf{CV} \qquad (23.20)$$

However, this performance cannot be achieved in general, because of dead times, constraints on the manipulated variables, and right-half-plane zeros (see Sections 13.5 and 19.2). Another way of stating this conclusion is that an exact plant (model) inverse cannot be achieved because of limitations in the physical process. Therefore, the best control involves the manipulated-variable adjustments that minimize the sum of the error squared in equation (23.16), which in general is not zero. The solution to this problem is the least squares solution, which can be considered an approximate plant (model) inverse that has desirable properties for control performance. The solution to the optimization problem in equation (23.16) for the model in equation (23.18) is the well-known linear least squares result

$$\mathbf{K}_{\mathrm{DMC}} = (\mathbf{A}^{\mathrm{T}}\mathbf{A})^{-1}\mathbf{A}^{\mathrm{T}} \tag{23.21}$$

The Dynamic Matrix Controller $\mathbf{K}_{\mathrm{DMC}}$ can be used to calculate the future adjustments at each controller execution by

$$\mathbf{K}_{\mathrm{DMC}}\mathbf{E}^{f} = [\Delta \mathrm{MV}^{c}] \tag{23.22}$$

This equation shows that the model of the process in the feedback path, **A**, and the future errors are used to calculate the manipulated-variable adjustments. The calculated adjustment for the current time period, $\Delta \mathrm{MV}_0^c$, would be implemented after the controller calculation. The later adjustments would not be implemented, because they would be recalculated during later controller calculations.

Example 23.3. The process model in Table 23.1 describes the mixing process with dead time in Figure 9.1 (page 291). Feedback control using the proportional-integral-derivative (PID) algorithm has been evaluated for this process in Section 9.3. Assume that the process is initially at steady state, and its set point is changed by a 1% step. Design the DMC controller matrix and evaluate the closed-loop dynamic response, assuming that the model is perfect.

The following parameters must be chosen before the DMC design calculation can be performed.

Δt = the sample period

LL = number of sample periods required for the process model to reach steady state

NN = the controlled-variable (output) horizon

MM = the manipulated-variable (input) horizon

In this example, the analyzer update occurs only once every 2.5 min; thus, the controller execution is set by this limitation. The product (Δt)(LL) should be equal to or greater than the settling time of the open-loop process, and the product (Δt)(NN) should be equal to or greater than the settling time of the closed-loop process. The manipulated-variable horizon is usually selected to be greater than 1, to allow some manipulated-variable overshoot if desired, and to settle well before the end of the controlled-variable horizon; thus, $1 \leq \mathrm{MM} \leq \mathrm{NN}$. The values of the parameters for this example are summarized in Table 23.2. The horizons are somewhat shorter than usually used in practice, to enable the key matrices to be reported conveniently.

Since the system is initially at steady state, so that all past adjustments are zero, the future errors are equal to the current error. Some of the key values in the calculation of the future moves follow.

$$\mathbf{A}^{\mathrm{T}} = \begin{bmatrix} 0 & 0 & 0.394 & 0.632 & 0.777 & 0.865 & 0.918 & 0.950 & 0.970 & 0.982 & 0.989 \\ 0 & 0 & 0 & 0.394 & 0.632 & 0.777 & 0.865 & 0.918 & 0.950 & 0.970 & 0.982 \\ 0 & 0 & 0 & 0 & 0.394 & 0.632 & 0.777 & 0.865 & 0.918 & 0.950 & 0.970 \\ 0 & 0 & 0 & 0 & 0 & 0.394 & 0.632 & 0.777 & 0.865 & 0.918 & 0.950 \end{bmatrix}$$

TABLE 23.2
Summary of single-variable DMC simulation cases

	Algorithm parameters						Controlled-variable performance		MV performance
Case	Δt	**MM**	**NN**	**ww**	**qq**	**Controller model, difference from plant model****	**IAE**	**ISE**	$\sum(\Delta \mathbf{MV})^2$
Example 23.3 Figure 23.5	2.5	4	11	1	0	Same as process	6.0	5.6	8.8
Example 23.4 Figure 23.7	2.5	4	11	1	0	$K_m = 0.65$	12.9	7.7	29.5
Example 23.5 Figure 23.8*a*	2.5	4	11	1	0.2	$K_m = 0.65$	11.5	7.3	2.9
Example 23.5 Figure 23.8*b*	2.5	4	11	1	0.2	$K_m = 0.65$	8.9	4.2	0.8

*The process is represented by the model in Table 23.1.
**The model used in performing all model-based calculations for DMC
LL = large number, e.g., 2(NN)

$$\mathbf{K}_{\text{DMC}} = \begin{bmatrix} 0 & 0 & 2.54 & 0 & 0 & 0 & 0 & 0 & 0 & 0 & 0 \\ 0 & 0 & -4.08 & 2.54 & 0 & 0 & 0 & 0 & 0 & 0 & 0 \\ 0 & 0 & 1.53 & -4.08 & 1.86 & 0.95 & 0.40 & 0.065 & -0.14 & -0.26 & -0.34 \\ 0 & 0 & 0 & 1.54 & -2.04 & -0.93 & -0.26 & 0.14 & 0.39 & 0.54 & 0.62 \end{bmatrix}$$

$$(\mathbf{E}^f)^{\mathrm{T}} = [1 \quad 1 \quad 1 \quad 1 \quad 1 \quad 1 \quad 1 \quad 1 \quad 1 \quad 1 \quad 1]$$

These values can be used to calculate the future values of the manipulated-variable changes using equation (23.22). The changes can be summed to obtain the manipulated-variable values at each time in the future.

$$[\Delta \mathrm{MV}^c] = \begin{bmatrix} 2.54 \\ -1.54 \\ 0.00 \\ 0.00 \end{bmatrix} \qquad \mathrm{MV}^c = \begin{bmatrix} 2.54 \\ 1.00 \\ 1.00 \\ 1.00 \end{bmatrix}$$

Because the controller model is assumed perfect in this example, feedback does not change the results in later controller executions. The responses of the manipulated and controlled variables are given in Figure 23.5. The controlled variable cannot respond until after the process dead time, and for this system it can be changed to the set point in one sample period after the dead time. To achieve this performance, the manipulated variable must experience a rapid change of large magnitude, which may not be acceptable. However, the controller objectives, as stated to this point, have been achieved.

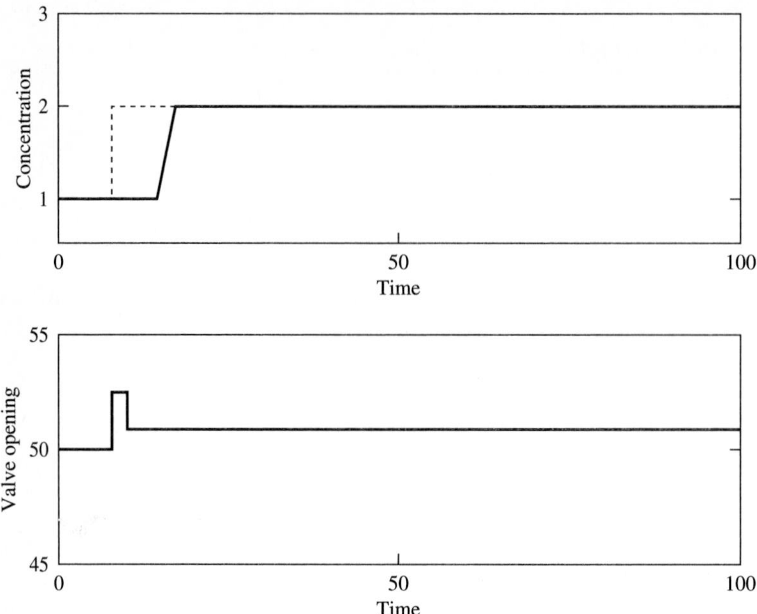

FIGURE 23.5
Dynamic response from Example 23.3.

Adding Feedback to the DMC Controller

To achieve acceptable feedback performance, the DMC controller must use the measured value of the controlled variable. The method for including the feedback is the same as employed in Chapter 19: the measured value is compared with a predicted value, and the difference, the feedback signal E_m, is added to the predicted value used by the controller. This scheme is shown in Figure 23.1; note that adding the feedback to the predicted controlled variable has the same effect on the sum of error squared as subtracting it from the set point, as seen by considering equation (23.16). This feedback approach is equivalent to adjusting a bias in the predictive model *without changing* the step weights a_j; thus, the feedback dynamics used by the controller to relate adjustments in the manipulated variable to the controlled variable are not influenced by the feedback. The result of the feedback, shown in Figure 23.6, is similar to that in the model predictive controllers in Chapter 19: zero steady-state offset for steplike disturbances, but no adaptation of dynamics for nonlinearities.

The model used to calculate the effects of future changes in the manipulated variables is similar to equation (23.15). However, the prediction of the future behavior without control is modified to combine the model with the feedback measurement signal as follows.

$$\mathrm{CV}_i^f = \mathrm{CV}_K + (E_m)_K + \sum_{j=1}^{\mathrm{LL}} a_{j+1}\Delta \mathrm{MV}_{K-j} \tag{23.23}$$

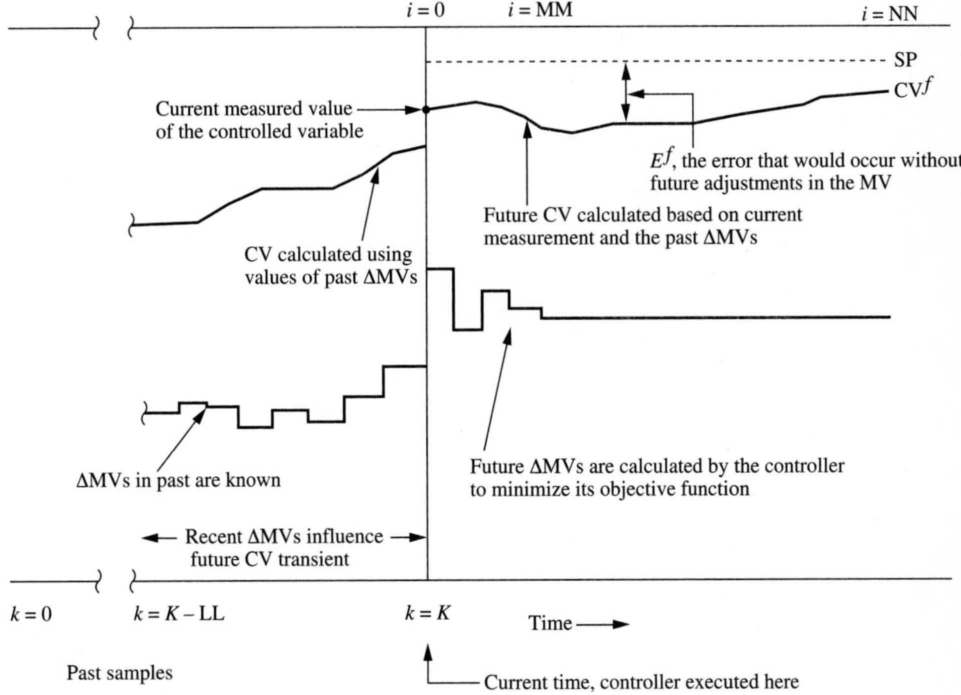

FIGURE 23.6
Dynamic response of variables with feedback.

The feedback signal is the difference between the measured and predicted values:

$$(E_m)_K = (\mathrm{CV}_{\mathrm{meas}})_K - \mathrm{CV}_K \tag{23.24}$$

Substituting equation (23.24) into equation (23.23) yields the model for the effects of future changes in the manipulated variables.

$$\mathrm{CV}_i^f = (\mathrm{CV}_{\mathrm{meas}})_K + \sum_{j=1}^{\mathrm{LL}} a_{j+i} \Delta \mathrm{MV}_{K-j} \tag{23.25}$$

> Thus, the feedback method is equivalent to setting the model prediction at the current time to the current measured value of the controlled variable.

The DMC controller $\mathbf{K}_{\mathrm{DMC}}$ can be designed with the same calculations, equation (23.21). Again only the manipulated-variable adjustment at the current sample period, $\Delta\mathrm{MV}_0$, is implemented. The entire controller calculation is repeated at the next sample period, because a new measured value of the controlled variable is available.

Some insight into the model predictive structure is gained by considering the meaning of the feedback signal when the controller model is perfect. In this situation, the effects of the manipulated variable on the true plant and the model are identical and cancel when E_m is calculated. Thus, the feedback signal is equal to the effect of the *disturbance* on the controlled variable. Since the same value of the feedback signal E_m is used to calculate all future values of the controlled variable without future adjustments, CV_i^f for all $i = 1$ to NN, the tacit assumption has been made that the disturbance will be the same in the future as it is currently. This is often a reasonable assumption, because we have no special information about the disturbance in this feedback control algorithm.

Example 23.4. The results in Example 23.3 were for the case when the controller model exactly represented the true process. In this example, the model differs from the plant; the model gain is 0.65%/% open, while the process gain remains 1.0%/% open. Determine the closed-loop performance for this system.

The step response for the model can be derived using the method in Table 23.1 with $K_m = 0.65$ (not 1.0), and this model can be used to derive the DMC controller from equation (23.21). The controller can then be employed with feedback, and the resulting dynamic response is shown in Figure 23.7 and summarized in Table 23.2. The model error led to considerable oscillation in this example, with increased ISE of the controlled variable and excessive manipulated-variable variation. However, the controlled variable ultimately returned to the set point, which was the goal of the feedback. Thus, feedback has improved the performance of the closed-loop system, but its dynamic behavior is not yet acceptable.

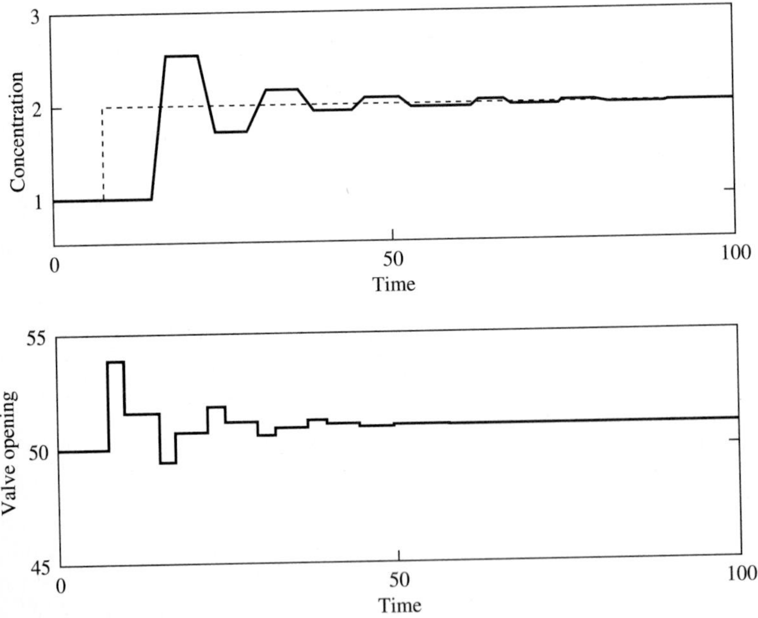

FIGURE 23.7
Dynamic response for Example 23.4.

Adding Tuning for Manipulated-Variable Behavior and Robustness

As with all controllers, adjustable parameters are needed to match the closed-loop performance to the particular needs (manipulated-variable variability) and circumstances (model mismatch) encountered in each application. In the DMC controller, the principal manner for addressing these needs is to expand the objective used in defining the control algorithm. This is done by adding a term that penalizes changes in the manipulated variable at each execution.

$$\begin{aligned} \text{OBJ}_{\text{DMC}} &= \sum_{i=1}^{\text{NN}} \left(\text{ww} \left[\text{SP}_i - (\text{CV}_i^f + \text{CV}_i^c) \right]^2 \right) + \sum_{i=1}^{\text{MM}} \left(\text{qq}(\Delta \text{MV}_i)^2 \right) \\ &= \sum_{i=1}^{\text{NN}} \left(\text{ww} \left[E_i^f - \text{CV}_i^c \right]^2 \right) + \sum_{i=1}^{\text{MM}} \left(\text{qq}(\Delta \text{MV}_i)^2 \right) \end{aligned} \tag{23.26}$$

with ww = (≥ 0) adjustable parameter weighting the controlled-variable deviations from set point, the ISE

qq = (≥ 0) adjustable parameter weighting the adjustments of the manipulated variable. This parameter is termed the *move suppression factor.*

The *relative* values of the two tuning parameters ww and qq determine how much importance is placed on the controlled variable ISE and on the variability of the manipulated variable; the original definition of the controller in equation (23.16) can be thought of as equation (23.26) with ww $=$ 1 and qq $=$ 0. Naturally, some variability in the manipulated variable must be tolerated to enable the control system to respond to disturbances and set point changes. However, the controller with qq $=$ 0 can be too aggressive, as seen in Figure 23.7. Also, because of model mismatch, the controller with qq $=$ 0 can lead to an unstable closed-loop system, and increasing the value of qq (more correctly, qq/ww) increases the range of model mismatch for which stable closed-loop performance is achieved. Finally, equation (23.26) contains no term for deviations of the manipulated variable from a target value, since the manipulated variable must be free to respond to disturbances of various magnitudes and directions; thus, the penalty is on the *adjustment* or change at each sample.

Again, the control algorithm determines the values of the future manipulated-variable changes that minimize the objective function. The result is

$$\mathbf{K}_{\text{DMC}} = (\mathbf{A}^{\text{T}}[\text{WW}]\mathbf{A} + [\text{QQ}])^{-1}\mathbf{A}^{\text{T}}[\text{WW}] \tag{23.27}$$

with $[\text{WW}]$ = diagonal matrix = $\text{ww}\mathbf{I}_{\text{NN}}$

$[\text{QQ}]$ = diagonal matrix = $\text{qq}\mathbf{I}_{\text{MM}}$

$\mathbf{I}_R$ = identity matrix of size $R \times R$

Again, only the current manipulated-variable adjustment is implemented at each controller execution. This is the form of the DMC control algorithm used in practice.

Example 23.5. Evaluate the control performance with model mismatch in Example 23.4, using the DMC algorithm in equation (23.27) with adjustable tuning.

The matrix algebra in equation (23.27) is slightly more complex, but the required model information (i.e., step weights) is the same. In this example, the number of parameters for the engineer to select is increased with the addition of ww and qq. For the single-variable DMC, ww can be set to 1.0 without loss of generality, which is not true for the extension to multivariable. The value for qq is selected to be 0.20 for this example, and the choice of this value is discussed in Section 23.6. The resulting transient response is shown in Figure 23.8*a*, and parameters and performance values are summarized in Table 23.2. The performance with qq = 0.2 is much more acceptable, with lower ISE of the controlled variable and about one-tenth the variability of the manipulated variable ($\sum \Delta MV^2$). An additional transient response has been evaluated for this system with the same feedback model mismatch and controller tuning parameters; this is a response to a unit step disturbance with a model $G_d(s) = 1/(5s + 1)$. The response in Figure 23.8*b* shows that DMC provides acceptable transient behavior and zero steady-state offset for this disturbance.

The addition of the variability of the manipulated variable to the controller objective with the associated tuning factor qq provides the engineer with the flexibility to tune the controller for a wide range of objectives and model mismatch.

23.5 MULTIVARIABLE DYNAMIC MATRIX CONTROL

It would be possible to employ the single-loop DMC as a replacement for the PID controller and to implement multiloop control with DMC using the approaches presented in Chapters 20 through 22. However, this approach would not realize the great power of Dynamic Matrix Control (or other similar centralized multivariable algorithms). Here, the goal is to achieve *centralized* multivariable control, in which the algorithm uses information from all controlled variables to calculate all manipulated-variable adjustments simultaneously each execution. Fortunately, the nature of the DMC algorithm makes its extension to multivariable control straightforward. In addition, the calculations performed at each controller execution remain relatively simple.

Again, the basis for the algorithm is the step response model. In the multivariable situation, one model exists for each input-output combination, and the form of each single input-output model remains as described in Section 23.3. The objective for the controller becomes

$$\mathrm{OBJ}_{\mathrm{DMC}} = \sum_{\mathrm{nc}=1}^{\mathrm{NC}} \mathrm{ww}_{\mathrm{nc}} \sum_{i=1}^{\mathrm{NN}} \left[E^{f}_{\mathrm{nc},i} - \mathrm{CV}^{c}_{\mathrm{nc},i} \right]^2 + \sum_{\mathrm{nm}=1}^{\mathrm{NM}} \mathrm{qq}_{\mathrm{nm}} \sum_{i=1}^{\mathrm{MM}} (\Delta \mathrm{MV}_{\mathrm{nm},i})^2 \tag{23.28}$$

with
- NC = the number of controlled variables; nc is the counter for the controlled variables (1 to NC)
- NM = the number of manipulated variables; nm is the counter for the manipulated variables (1 to NM)
- $\mathrm{ww}_{\mathrm{nc}}$ = adjustable parameter weighting the nc'th controlled variable's deviation from set point
- $\mathrm{qq}_{\mathrm{nm}}$ = adjustable parameter weighting the adjustments of the nm'th manipulated variable

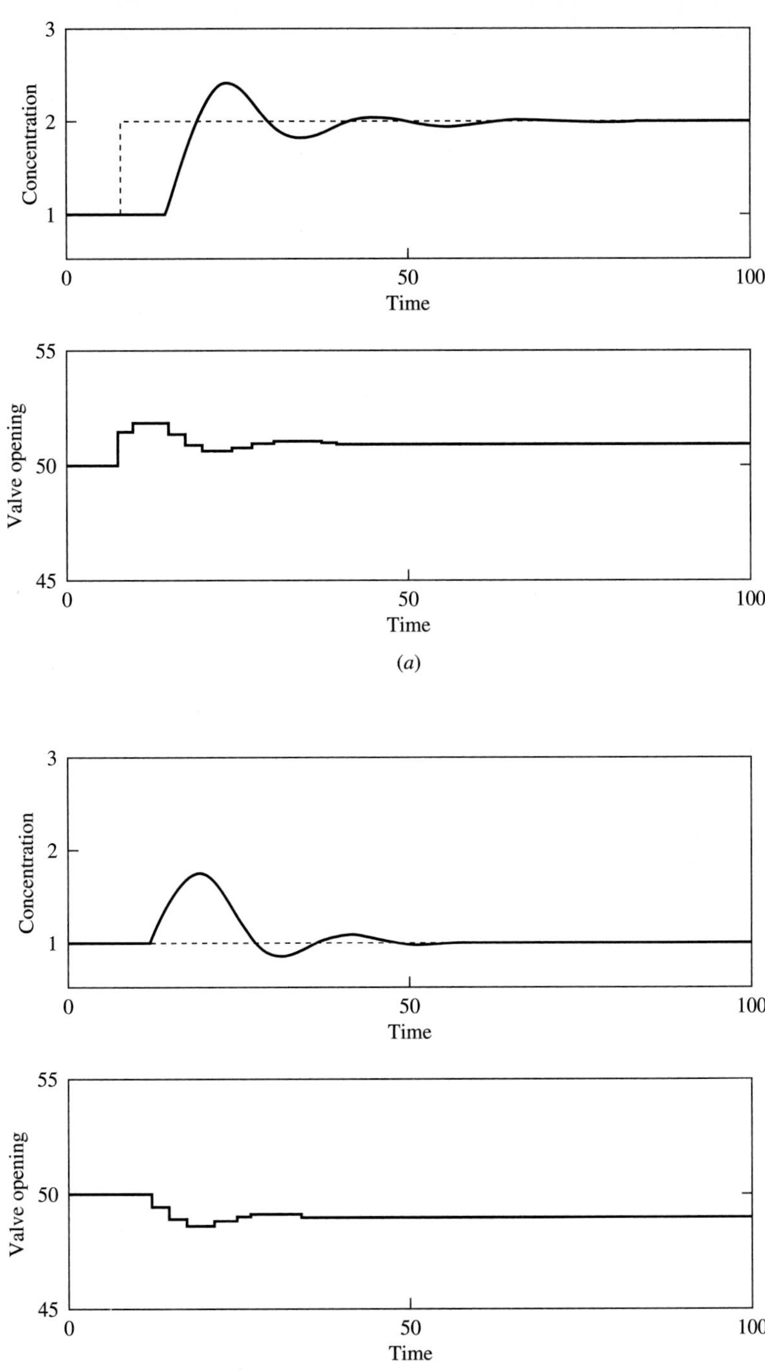

FIGURE 23.8
Dynamic response for Example 23.5: (*a*) set point change; (*b*) disturbance.

For multivariable control, a separate value for ww_{nc} is allocated to each controlled variable. The ratio of these values represent the relative importance of deviations from set point of the controlled variables, which can be used to tune the controller for different performance objectives. Also, each manipulated variable has an associated qq_{nm}, which gives the penalty for adjustments. A multivariable controller has many parameters that must be tuned in conjunction to obtain the desired performance.

The control algorithm that achieves this objective is

$$\mathbf{K}_{DMC} = (\mathbf{A}^T[WW]\mathbf{A} + [QQ])^{-1}\mathbf{A}^T[WW] \tag{23.29}$$

This is the same form as the result for single-loop control. However, the matrices in equation (23.29) are composed of individual blocks, with each block consisting of a single-variable matrix or zeros. For the two-variable control problem,

$$\mathbf{A} = \begin{bmatrix} \mathbf{A}_{11} & \mathbf{A}_{12} \\ \mathbf{A}_{21} & \mathbf{A}_{22} \end{bmatrix} \quad [WW] = \begin{bmatrix} [WW]_1 & \mathbf{0} \\ \mathbf{0} & [WW]_2 \end{bmatrix} \quad [QQ] = \begin{bmatrix} [QQ]_1 & \mathbf{0} \\ \mathbf{0} & [QQ]_2 \end{bmatrix} \tag{23.30}$$

where $\mathbf{A}_{nc,nm}$ = dynamic matrix shown in equation (23.18) for the controlled variable nc and the manipulated variable nm

$[WW]_{nc}$ = diagonal matrix = $ww_{nc}\mathbf{I}_{NN}$ for nc = 1 to NC

$[QQ]_{nm}$ = diagonal matrix = $qq_{nm}\mathbf{I}_{MM}$ for nm = 1 to NM

$\mathbf{0}$ = square matrix containing zeros

With this result, the errors of all controlled variables are considered simultaneously in determining the adjustments to all manipulated variables. The methods of modelling and feedback are identical to the design in Figures 23.1 and 23.6, with all models being multiple-input–multiple-output and the controlled and manipulated variables being vectors of values for each variable. As before, only the current manipulated-variable adjustments are implemented at each controller execution.

Example 23.6. Apply DMC control to the distillation tower considered in Example 23.1 and modelled in equation (23.5).

The first step is to develop the step response models using the method in Table 23.1 or, equivalently, equation (23.10) for each of the $\mathbf{A}_{nc,nm}$ matrices. In performing this modelling, the sample period and horizon lengths must be decided. Finally, the tuning parameters are selected. With this information, the DMC controller $\mathbf{K}_{DMC}$ can be calculated using equation (23.29). The controller design parameters used in this example are as follows:

$$\Delta t = 1 \quad MM = 5 \quad NN = 20 \quad ww_1 = ww_2 = 1 \quad qq_1 = qq_2 = 0.02$$

The transient response to a set point change in the distillate purity, with the bottoms set point unchanged, is given in Figure 23.9*a*. Also, the transient response to a −4% step change to the light key in the feed is given in Figure 23.9*b*. In both plots the plant is represented by the linear model in equation (23.5), so that these responses are for no model mismatch, although the tuning has been selected to give a reasonably moderate feedback response.

The performance measures for these dynamic responses are given in Table 23.3. These results can be compared with the performance achieved with two PI controllers,

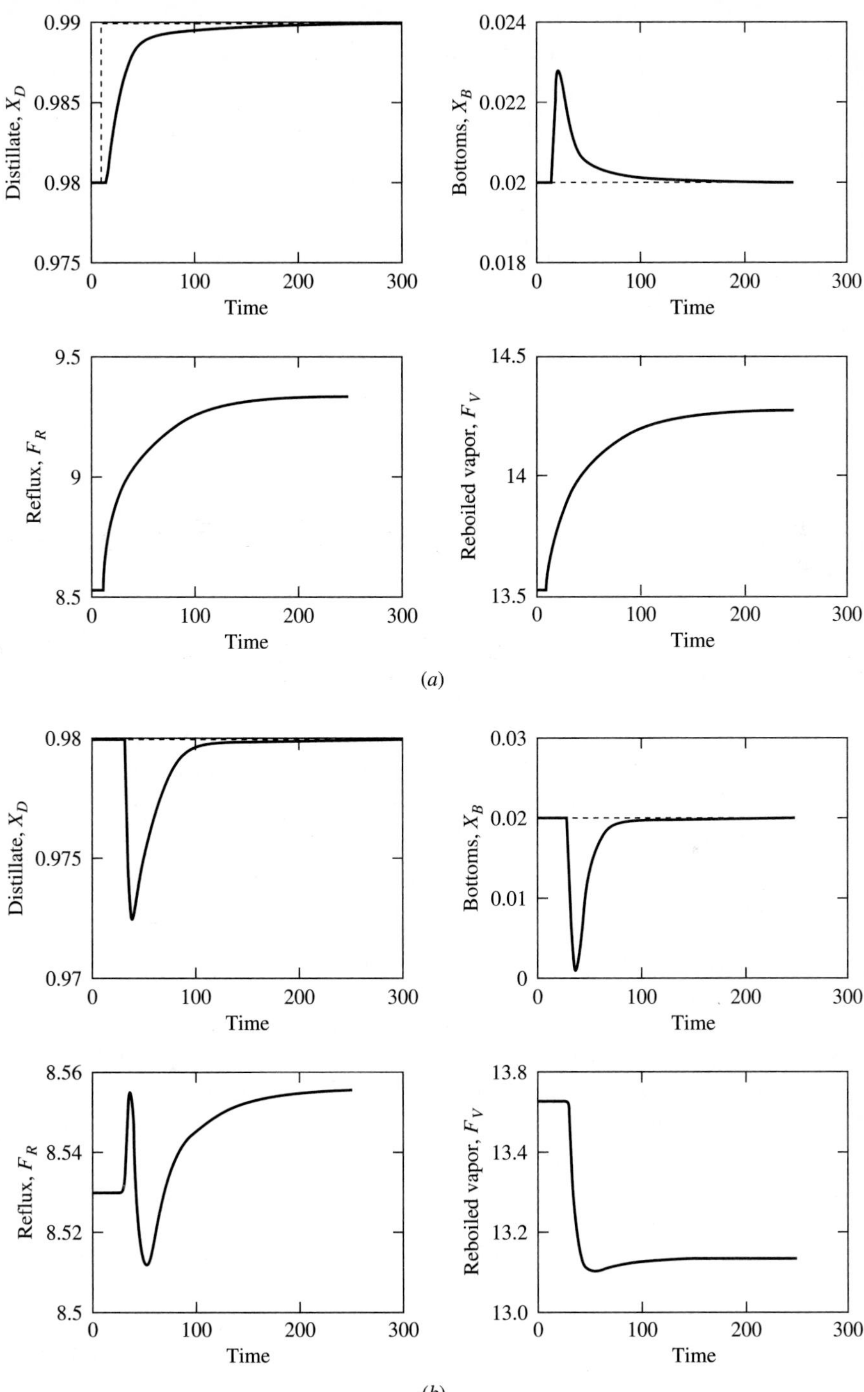

FIGURE 23.9
Responses for Example 23.6: (*a*) to set point change; (*b*) to disturbance.

TABLE 23.3
Summary of performance for Example 23.6

Case	IAE_{XD}	ISE_{XD}	IAE_{XB}	ISE_{XB}	$\sum(\Delta F_R)^2$	$\sum(\Delta F_V)^2$
Figure 23.9*a*	0.225	.00122	.073	.00010	.0141	.0097
Figure 23.9*b*	0.207	.00093	0.33	.00413	.00029	.0147

although no attempt was made to provide the best tuning to either control system. (Note that the set point change in Table 21.2 (page 688) was half the magnitude of that in Table 23.3.) In these examples, the DMC controller provided about the same performance for the disturbance and better performance for the set point change.

As demonstrated in this section, the multivariable Dynamic Matrix Controller is a straightforward extension of the single-variable controller. The controller algorithm can be calculated for any general (stable) process model, without regard for dead times or numerator dynamics. The dynamic responses in the example show that good performance can be achieved without excessive adjustments of the manipulated variables.

23.6 IMPLEMENTATION ISSUES IN DYNAMIC MATRIX CONTROL

While the design and implementation of centralized feedback control have been shown to be possible, a large number of design and implementation decisions must be made to achieve good performance. Some of the most important are discussed briefly in this section.

Real-Time Calculations

The distinction is important between the design calculations, which are performed once offline, and the control calculations, which are performed every control execution. Basically, the design calculation is given in equation (23.29). This calculation involves the inverse of a square matrix with dimensions (MM)(NM). This inverse could be computationally intensive, but it is calculated only during offline design. In contrast, the controller calculation requires the following calculations every execution:

1. Calculate the feedback signal $\mathbf{E}_m$, which requires advancing the prediction of the model $\mathbf{G}_m$ in the block diagram and equation (23.14) by one time step.
2. Calculate the future error that would occur without future adjustments, $E_i^f = ([\text{SP}]_i - [\text{CV}^f]_i)$ for $i = 1, \text{NN}$. This requires the model for $[\text{CV}^f]_i$ in equation (23.23) to be calculated for NN time steps.
3. Calculate the current adjustment to the manipulated variable. The basis for this calculation is equation (23.22), which will give the adjustments for the entire input horizon—more information than needed, because only the current change in manipulated variable is required. For example, the single-variable DMC needs

only ΔMV_0, which is the sum of the element products of the top row of $\mathbf{K}_{DMC}$ and the future error E^f. This vector-vector product requires fewer calculations.

Tuning

The Dynamic Matrix Controller has a large number of adjustable parameters, all of which influence the control performance. In addition, the best value of some parameters depend on the values of others. The following comments should help in selecting good initial values.

- Δt — Factors in selecting the execution period are the same as discussed in Chapter 11 on PID control. This should be a small fraction of the closed-loop dynamics (e.g., $\Delta t < 0.1(\theta + \tau)$).
- NN — The output horizon should be long enough for the closed-loop system to approach its steady-state in the time Δt(NN). Typical values for NN range from 20 to 50.
- MM — The input horizon is selected to be shorter than the output horizon. Typically, MM is about one-fourth to one-third of the output horizon.
- ww_{nc} — The weighting for each controlled variable represents the relative importance of each deviation from its set point. Increasing this number tends to reduce the deviation of this controlled variable, but the deviation of other controlled variables will increase. The engineer must recognize that the controller objective is calculated in engineering units, so that the weighting must reconcile the comparison of various variables, such as temperature and mole percent.
- qq_{nm} — The weighting for each manipulated variable represents the relative importance of the *adjustments* to each manipulated variable. Increasing this number will tend to slow the feedback adjustments, which would degrade the controlled-variable performance; however, increasing qq_{nm} also improves the robustness of the closed-loop system to model mismatch. Also, increasing qq_{nm} reduces the variability of the manipulated variable, which may be required in some circumstances. As a result, the parameter is often referred to as the *move suppression factor.*

Good values for ww_{nc} and qq_{nm} depend on their relative magnitudes, such as, ww_1/qq_2 and ww_1/ww_2. Thus, strong interactions exist among the effects of the many tuning parameters on the control performance, and often some simulation studies are required to determine good tuning.

The presentation in this chapter has assumed that the weighting matrices [WW] and [QQ] are diagonal. This assumption is valid when the desired behavior of one controlled variable does not depend on the behavior of other controlled variables. That condition might not be the case for some processes. For example, a high temperature and high reactant concentration might be a particularly bad condition; in such a case, a penalty could be introduced in the appropriate off-diagonal elements in [WW] for the deviations of both.

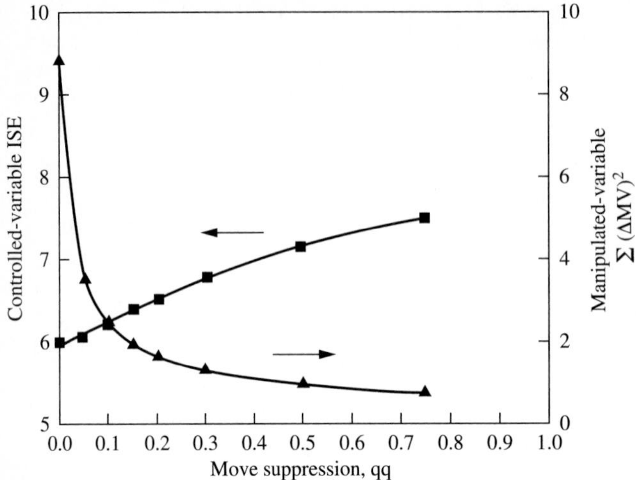

FIGURE 23.10
Effect of controller tuning on controlled and manipulated variables (ww = 1).

Example 23.7. Study the effects of tuning on the single-variable DMC controller in Examples 23.3 through 23.5. For this study, assume that no model mismatch exists and that the input forcing is a unit step set point change.

The common manner for presenting such a tuning study is to plot the performances of the manipulated and controlled variables against the tuning parameter, which for the single-loop case is qq/ww. This plot is given in Figure 23.10, with results that are typical of many systems. As the move suppression is increased from zero, the first effects are a rapid drop in the variability in the manipulated variable, with a small increase in the ISE of the controlled variable. After some value of qq, the effects on both variables is moderate. Often, the value of qq where the variability in the manipulated variable stops decreasing rapidly gives an acceptable initial tuning, with reasonable robustness to typical model mismatch and moderate variability in the manipulated variable. This study provided the basis for the value of qq, 0.2, used in Example 23.5.

Filtering

High-frequency noise in the controlled-variable measurement can be filtered for the reasons discussed in Section 12.3. The measurement can be filtered before calculating the feedback signal $\mathbf{E}_m$.

Cascade Implementation

Centralized multivariable controllers can output directly to final elements, but a more common design is to output to a single-loop system. As an example, the distillation control in Figure 23.2 and studied in Example 23.6 outputs to the set points of two flow controllers. The design of these lower-level loops follows the principles of single-loop enhancements (Part IV) and loop pairing (Chapters 20 through 22) already presented.

23.7 EXTENSIONS TO BASIC DYNAMIC MATRIX CONTROL

The method presented in detail in this chapter represents only the most basic form of the Dynamic Matrix Controller. Many extensions are possible, and some are essential for success in challenging applications. A few of the more important extensions are introduced briefly in this section.

Nonsquare Systems

Many control systems have an unequal number of controlled and manipulated variables. Methods for addressing these situations using single-loop (decentralized) control were presented in Chapter 22 on variable-structure control. The DMC controller can accommodate this situation, because no assumption has been made in developing the design for $\mathbf{K}_{\text{DMC}}$ in equation (23.29) regarding the number of process variables. If more controlled than manipulated variables exist, not all controlled variables can be maintained at their set points (at the steady state), and the DMC controller will achieve the minimum objective, as measured by equation (23.26). If more manipulated than controlled variables exist, all controlled variables can be maintained at their set points (in the steady state), and the manipulated variables can be adjusted to achieve additional benefits, such as low energy consumption. Methods are described in Cutler and Ramaker (1979) and Morshedi et al. (1985).

Feedforward

The centralized control method in this chapter addressed feedback control, but it can be extended to include feedforward compensation. If a measured disturbance satisfies the feedforward design criteria in Table 15.1 (page 509), it can be included by modelling its effect on the future controlled variable without feedback adjustment. Thus, the effect of the measured disturbance is simply another process input in calculating the values of $[\text{CV}^f]_i$ that are used in calculating $\mathbf{E}^f$ and the controller calculation in equation (23.22).

Constraints on Variables

Often, the behavior of the control system is limited by constraints. They can be physical limitations (e.g., a valve can only go to 100% open), or constraints can be imposed by the engineer (e.g., the temperature should not go above 300°C). The design of the DMC controller in equation (23.29) was based on least squares, which relied on the controlled and manipulated variables being continuous, not encountering constraints. The DMC approach can be extended to designs that minimize the same objective while observing constraints, by using a different optimization method. A common approach uses *quadratic programming*. A slight disadvantage of including constraints is a substantial increase in the calculations that must be performed each controller execution, which are on the order of the design equation (23.29). However, with powerful digital computers, this has not proved to be a barrier to practical application. Details are given in Garcia and Morshedi (1986), Morshedi et al. (1985), and Ricker (1985).

Non–Self-Regulating Processes

The step weight model described in Section 23.3 is limited to processes that are stable and self-regulating so that they attain a steady state after a step input. As discussed in Chapter 18, many inventory processes (levels) are not self-regulatory, because they are pure integrators. The step response modelling method has been extended to integrators, and details are provided by Cutler (1982).

23.8 CONCLUSIONS

A practical method for centralized process control has been presented in this chapter. The general model predictive structure provides the framework for the control method, but the analytical design approach proves a limit to direct extension of the methods from Chapter 19. The novel modelling and numerical calculations of the Dynamic Matrix Controller algorithm result in a method that can be applied to a wide range of processes. The addition of feedback and tuning parameters provides the basic centralized controller algorithm, with extensions possible for special situations. The performance of the Dynamic Matrix Controller has been demonstrated to be good for single- and multivariable systems.

REFERENCES

Caldwell, J. and J. Dearwater, "Model Predictive Control Applied to FCC Units," in Arkun, Y. and W.H. Ray (ed.), *Chemical Process Control—CPC IV,* CACHE, New York, 319–334 (1991).

Cutler, C., "Dynamic Matrix Control of Imbalanced Systems," *ISA Trans., 21, 1,* 1–6 (1982).

Cutler, C. and P. Perry, "Real-Time Optimization with Multivariable Control Is Required to Maximize Profit," *Comp. Chem. Eng., 7, 5,* 663–667 (1989).

Cutler, C. and B. Ramaker, "Dynamic Matrix Control—A Computer Control Algorithm," *AIChE Nat. Meet.,* April, 1979.

Cutler, C. and F. Yocum, "Experience with the DMC Inverse for Identification," in Arkun, Y. and W.H. Ray (ed.), *Chemical Process Control—CPC IV,* CACHE, New York, 297–318 (1991).

Garcia, C. and A. Morshedi, "Quadratic Programming Solution of Dynamic Matrix Control (QDMC)," *Chem. Eng. Comm., 46,* 73–87 (1986).

Kelly, S., M. Rogers, and D. Hoffman, "Quadratic Dynamic Matrix Control of Hydrocracking Reactors," *Proc. ACC Meet.,* 295–300 (1988).

MacGregor, J., D. Kourti, and J. Kresta, "Multivariate Identification: A Study of Several Methods," *Proc. ADCHEM '91, 14–16 Oct. 1991,* 369–375 (1991).

Morshedi, A., C. Cutler, and T. Skrovanek, "Optimal Solution of Dynamic Matrix Control with Quadratic Programming Techniques (QDMC)," *ISA Nat. Meet.,* paper no. 85-0732 (Oct. 1985).

Prett, D. and C. Garcia, *Fundamental Process Control,* Butterworths, Boston, MA, 1988.

Prett, D. and R. Gillette, "Optimization and Constrained Multivariable Control of a Catalytic Cracking Unit," *AIChE Nat. Meet.,* April 1979.

Ricker, L., "Use of Quadratic Programming for Constrained Internal Model Control," *IEC Proc. Des. Devel., 24,* 925 (1985).

Ricker, L., "The Use of Biased Least-Squares Estimators for Parameters in Discrete-Time Pulse Response Models," *IEC Res., 27,* 343 (1988).

Van Hoof, A., C. Cutler, and S. Finlayson, "Application of a Constrained Multi-Variable Controller to a Hydrogen Plant," *American Control Conference,* Pittsburgh, June 21–23, 1989.

ADDITIONAL RESOURCES

Centralized multivariable controllers make use of all elements in the feedback model, which can make these controllers sensitive to certain types of model mismatch. The causes of the sensitivity and experimental designs to improve model accuracy are discussed in

Kwong, C.W., and J. MacGregor, "Identification for Robust Multivariable Control: The Design of Experiments," *Automatica, 30,* 1541–1554 (1994).

Skogestad, S. and M. Morari, "Implications of Large RGA Elements on Control Performance," *IEC Res., 26,* 2323–2330 (1987).

A review of model predictive control that discusses Dynamic Matrix Control within this structure is in

Garcia, C., D. Prett, and M. Morari, "Model Predictive Control: Theory and Practice—a Survey," *Automatica, 25,* 335–348 (1989).

An alternative approach using linear programming is reported in

Brosilow, C. and G. Zhao, "A Linear Programming Approach to Constrained Multivariable Process Control," in C. Leondes (ed.) *Control and Dynamic Systems, 27,* 141 (1986).

Another approach to centralized model predictive control is described in

Mehra, R., R. Rouhani, and J. Etero, "Model Algorithmic Control: Basic Theoretical Properties," *Automatica, 18,* 401 (1982).

> The full power of centralized multivariable control becomes apparent through studies of closed-loop systems. These questions build understanding of the assumptions, theory, and preliminary calculations that can be performed without preparing a complete design and simulation package.

QUESTIONS

23.1. Determine step response models (i.e., the step weights) for the following systems based on the continuous models already developed. Select appropriate values for the sample period and the output horizon.

Single-variable:

(*a*) The three-tank mixing process, first-order-with-dead-time approximation (Example 6.4)

(*b*) The series chemical reactors in Example 5.4

(*c*) The nonisothermal chemical reactor in Examples 3.10 and 4.8; the input is the coolant flow rate, and the output is the temperature.

Two-variable:

(*d*) The blending process in Examples 20.6 and 20.10

(*e*) The two processes with simple and complex interactive dynamics, B1 and B2, in Example 21.4

(*f*) The distillation tower under material balance regulatory control in equation (21.2)

23.2. Calculate the dynamic matrix controller $\mathbf{K}_{DMC}$ for one of the single-loop processes already modelled in question 23.1. Select an appropriate input horizon and let ww = 1 for all controlled variables. The calculations can be performed on a spreadsheet or using a programming language. After the controller has been determined, evaluate the response of the controlled and manipulated variables to a step change in the set point without model error; this can be done by evaluating the product in equation (23.22) once. Begin with qq = 0, and increase it. Select an appropriate initial value for qq.

23.3. The step response model can be determined from empirical data.

(*a*) Discuss the advantages and disadvantages for using sampled values of the original data for the model.

(*b*) Discuss the procedure required and likely results of fitting the coefficients a_j in the following model to experimental data using linear least squares. Recall that this model will have between 20 and 50 coefficients.

$$Y_{k+1} = \sum_{j=1}^{k+1} a_j \Delta X_{k-j+1} + Y_0$$

(*c*) Are the dynamics of the sensor and the final element included in the models used in the design of the DMC controller?

23.4. The DMC objective function selected to be minimized is the ISE over the output horizon.

(*a*) What is the advantage of using the ISE rather than the IAE or $(\text{error})^4$?

(*b*) From a necessary condition for a minimum (the gradient is zero), derive the equation for the DMC controller in equation (23.21).

23.5. Derive the analytical model predictive controller for the following processes. For each, state whether the controller can be easily factored, and if so, select an IMC filter structure and time constant value(s) to give good dynamic performance.

(*a*) The systems B1 and B2 in Example 21.4.

(*b*) The blending process in Examples 20.6 and 20.10.

(*c*) The distillation tower with material balance regulatory control in equation (21.2).

23.6. Discuss the effect on the closed-loop performance of the following changes.

(*a*) Multiply every ww and qq by a positive constant.

(*b*) Add a constant to the DMC objective function.

(*c*) Change the units of one controlled variable, for example the bottoms composition in Example 23.5 from mole fraction to mole percent.

(*d*) Increase all qq by the same positive factor, maintaining all ww constant.

23.7. Develop the appropriate step response model for a pure integrating level process. Describe how this could be used to model the process over a long time, without involving a summation of infinite length.

23.8. Determine all calculations for adding feedforward control for a measured disturbance to the single-loop DMC control system in Example 23.5. The answer should include a block diagram, summary of controller execution calculations, and any new models and/or modifications to the controller $\mathbf{K}_{DMC}$. The model for the disturbance is $G_d(s) = 1.0e^{-2.5s}/(5s+1)$. Also, design a feedforward controller using methods in Chapter 15 and discuss the expected difference in performance.

23.9. Determine all calculations for adding feedforward control for a measured disturbance in the feed composition to the multivariable DMC control system in Example 23.6. The answer should include a block diagram, summary of controller execution calculations, and any new models or modifications to the controller, $\mathbf{K}_{DMC}$. The model for the disturbance is given in equation (23.5). Also, design a feedforward controller using methods in Chapter 15, and discuss the expected difference in performance.

23.10. Compare the DMC controller with the Smith predictor model predictive control structure. Based on this comparison, determine the requirement for the DMC system to achieve zero steady-state offset for steplike disturbances.

23.11. Compare the results in Figure 23.10 and in Figures 13.18*a* and *b*. What generalizations can you discern about feedback control, regardless of algorithm type?

23.12. Suppose that slower set point response was desired, but fast disturbance response was required. How could you modify the DMC control system design to accommodate this performance requirement?

23.13. The DMC controller was described in this chapter using step response models to calculate the model to compare with the feedback measurement, CV_K, and the future performance without control, CV_i^f.

(*a*) Describe how the discrete models derived in Appendix F could be used for these calculations.

(*b*) Could these models also be used to determine $\mathbf{K}_{DMC}$?

PART VI

PROCESS CONTROL DESIGN

In this final part, we complete the coverage of control engineering by addressing the design of process control systems. Design is perhaps the most challenging, yet enjoyable, subject in control engineering, because it enables us to use all of our analysis methods learned in the previous parts of the book. In fact, the entire point of the analysis methods is to enable us to design, and ultimately to build, equipment that functions according to requirements prescribed at the outset of the design procedure.

Before introducing some of the main concepts and methods in this part, the term *design* needs to be discussed. There have been many attempts to provide a general definition of the term, but no single definition has achieved wide acceptance. Here, we will simply describe the design function relevant to process control, without any claim to generality.

> *Design* is the procedure by which an engineer arrives at a complete control system specification that satisfies all performance objectives.

It is important to recognize that the performance specifications are determined by the engineer at the first stage of the design procedure based on physics, chemistry, and the marketplace, which defines product quality demands and economics. The initial objectives are specified independently of the solutions possible. The result of the design is a complete specification that satisfies the objectives, if possible. The simple flash process in Figure VI.1 is shown as a typical starting point for the design procedure. During the design the engineer must (1) clearly state the objectives, (2) determine needed process modifications, (3) specify instrumentation performance and alterations, (4) define control structure (loop pairing), (5) select algorithms, and (6) define special tuning requirements.

Five important features of design that distinguish it from previous topics are now discussed. One major feature of design that was not as prominent in previous material is the rich definition of the objectives or performance that the design is to satisfy. Thus, design involves considerable interaction between the objective statement and design results. The objective is usually stated to be the reduction of variability in the operation of a process plant. However, not all variability can be eliminated, and variability is much more important in some variables than in others. In fact, the plant is designed to provide specific variables and systems that can be easily adjusted with minimum effect on plant performance. For example, the cooling water, steam, electricity, and fuel systems are designed to be able to respond rapidly to demands in the plant. Thus, process control generally *moves variability* from important variables to less important variables. This is achieved by controlling the (important) controlled variables by adjusting the (less important) manipulated variables. Therefore, the control design must conform to the priority of variables indicated in the objective statement.

A second major feature of design is the large number of decisions that can be considered. For the purposes of this book, the following categories of design decisions are covered: (1) measurements and sensors, (2) final elements, (3) process design, (4) control structure, (5) control algorithms and tuning, and (6) performance monitoring. As has been seen many times in previous chapters, the process dynamics have a major effect on control performance. Thus, process design changes would be the preferred manner for achieving good control. When a plant is being designed initially, the engineers can make essentially any design changes, although equipment design changes to achieve good dynamic performance may be prohibitively

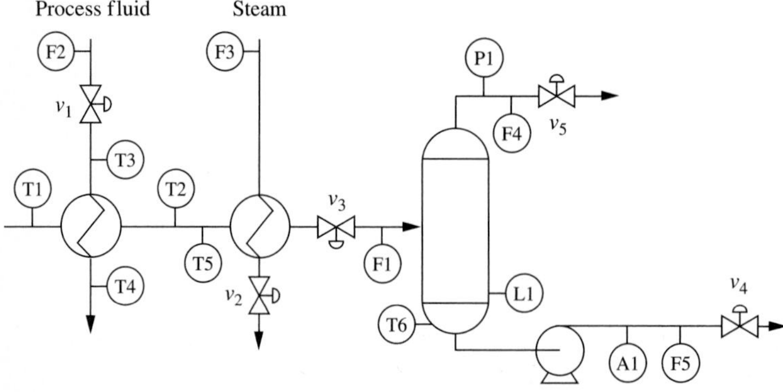

FIGURE VI.1.
Example flash process.

expensive when compared to the alternative of additional instrumentation and control algorithms. Major process equipment options require a thorough safety, reliability, and cost analysis of the alternatives, which is beyond the scope of an introductory process control book but should be included in a plant design project. Therefore, only "minor" process design changes are considered here; examples of minor changes are sizing inventory to attenuate variation and adding bypasses to add degrees of freedom and improve feedback dynamics. Typically, these are possible during initial design and as modifications to existing plants.

A third feature of design is the sequence in which the decisions can be considered. In previous chapters, relatively straightforward analysis methods were presented for, among other topics, controller tuning, cascade design, and multiloop pairing. Each procedure could be represented in a flowchart or table, with a fixed sequence of steps without iteration. This is not the case for control design, where iterations are frequently required. The order is an especially important issue, because the initial decisions will place limitations on future decisions, and the limitations may not be easily predicted when the initial decisions are made. Thus, the engineer must be ready to rethink previous decisions and be willing to *iterate* by changing some decisions and repeating the design.

A fourth feature is the ambiguity in determining the conclusion of the design procedure. One would have to evaluate most or all possible designs to be sure that the final design is the best. To respond quickly to market demands and limit total cost, the time for design is limited, and judgment must be used in deciding when the design is good enough. The typical procedure is to develop approximate bounds on the achievable performance and find a low-cost design that approaches the best performance.

Also, situations arise in which the initial objectives lead to unacceptable designs that are very costly and unreliable; in such cases, it is the engineer's task to alter the objective statement to meet the initial intent (e.g., make high-quality product safely), thereby preventing an unsatisfactory design. In fact, very restrictive objectives may not be achievable in the situation defined. For example, for a specified disturbances in the feed composition and flow rate and available sensors, it may not be possible to control the product quality of a chemical reactor. Clearly, a major change in the process design or performance specification is required.

A fifth feature of process control design is the concurrent application of process engineering and automatic control technologies. Automatic control principles may indicate that the feedback dynamics of a chemical reactor should be faster; then the chemical engineering principles can be used to select a process change (e.g., increasing the temperature or relocating the sensor). This tight coupling of process and control is the main reason why chemical engineers must learn control and why a "control specialist," without understanding of the process, cannot adequately perform the control design tasks.

The previous material in this book has prepared us well for the design task by providing an understanding of fundamental principles. For example, we can determine the relationship between the flow rate and the dynamic model of a process, and we can determine the manner in which the process dynamics affects feedback control system stability and performance. In addition to basic understanding, analysis provides methods for establishing quantitative relationships between adjustable factors (e.g., parameters and structures in control systems) and the behavior of the system. Thus, analysis directly provides methods for selecting cascade control or an inferential variable. The key point is that the fundamental analysis methods provide the *foundation for design* and thus are employed throughout these chapters.

It must be said at this point that control design—in fact all engineering design—is very challenging and requires considerable practice to master. Topics covered in previous chapters, such as single-loop controller stability analysis and tuning or feedforward controller design, can be learned quickly because they involve limited knowledge and a relatively straightforward analysis. As the previous discussion indicates, the design engineer has to master and apply all technologies concurrently. Adding to the challenge is the lack of a single, structured procedure for control design. This is to be expected, because design involves an element of creativity in adding process or control equipment, altering objectives and specifying control structures. As the reader has already experienced, procedures for stimulating creativity cannot be reduced to a flowchart. However, much can be presented and learned about the design procedure. Certainly, general procedures can be applied to the tasks of collecting information, defining objectives, and evaluating common checklists of potential decisions and outcomes. Also, typical sequences for considering control design decisions can be explained, although the best sequence is problem-dependent. Finally, examples demonstrating the interplay between process and control technology help the new engineer learn how to design. The chapters in Part VI provide guidance on performing the design procedure, by addressing its major features and supplying worked examples.

CHAPTER

24

PROCESS CONTROL DESIGN: DEFINITION AND DECISIONS

24.1 INTRODUCTION

Typically, the starting point for control system design and analysis is a preliminary process design, perhaps with some initial control loops, along with a specification of the desired process performance. This amount of initial information is realistic for existing plants, because the equipment is already in operation when an analysis to improve plant performance is carried out. It is also realistic for new plant designs, because a preliminary process structure (or alternative structures) must be available when dynamics and control are first analyzed.

The required information must be recorded concisely, and the Control Design Form described in the next section is proposed as a format for this record. A great advantage for using this form, in addition to giving excellent documentation, is that it provides a way to begin the design analysis. Often, the design problem seems so big and ill-defined that an engineer, especially one new to the technology, is unsure where to begin. By completing the thorough definition, the engineer begins the problem-solving process, and important issues and potential solutions become apparent.

Potential actions required to achieve the desired process performance include (1) defining the control strategy designs, (2) selecting measured variables and instrumentation (i.e., sensors and final elements), (3) specifying the process operating conditions, (4) making minor process changes such as adding a bypass, selecting an alternative manipulated variable, or changing the capacity of some equipment, or (5) making major process structure changes, such as changing from a packed- to fluid-bed reactor. The fifth possibility, involving major process alterations, is excluded from this discussion, because such a major decision would require an analysis of the steady-state and dynamic behavior of an integrated plant involving many units, which is beyond the scope of this book.

The six major categories of decisions made during the design procedure follow in the order covered in this chapter.

- *Measurements:* selecting measured variables and sensors
- *Final elements:* providing final elements with features contributing to good control performance
- *Process operability:* providing good steady-state and dynamic behavior that enables the control performance objectives to be achieved
- *Control structure:* providing the proper interconnection of measured and controlled variables via the control system
- *Control algorithms:* selecting and tuning the proper algorithms for feedback and feedforward control
- *Performance monitoring:* providing measurements and calculations for monitoring and diagnosing the process and control performance

The technology to design processes and control for dynamic performance has been introduced in previous chapters; here they are reinforced with design-related examples and placed in the context of a design procedure. At the completion of this chapter, the major tasks in defining the control problem and making individual decisions will have been covered. The next chapter presents a thought process for making the important decisions in a logical and systematic manner.

24.2 DEFINING THE DESIGN PROBLEM

The first step in the design task is the definition of the "problem," which perhaps should be referred to as an *opportunity* to apply our skills. We will retain the term *problem* because it is used commonly to describe the task of addressing complicated issues (e.g., "problem solving"). A complete definition of the design problem may be difficult in the beginning of the analysis, and the need for additional information may become apparent as the problem is analyzed. Therefore, the approach taken here is to provide a comprehensive form in which information can be recorded. The use of a form has several advantages. First, it serves as a convenient checklist so that the engineer is sure to address the important issues at the definition stage. Second, it provides a coherent, readable statement of goals, which can be reviewed by many members of a design team. Third, a form with topics concisely addressed under clear headings provides a structure that is easy to write and to use as a reference. Finally, additional information developed during the design analysis can be added at any time to the original form.

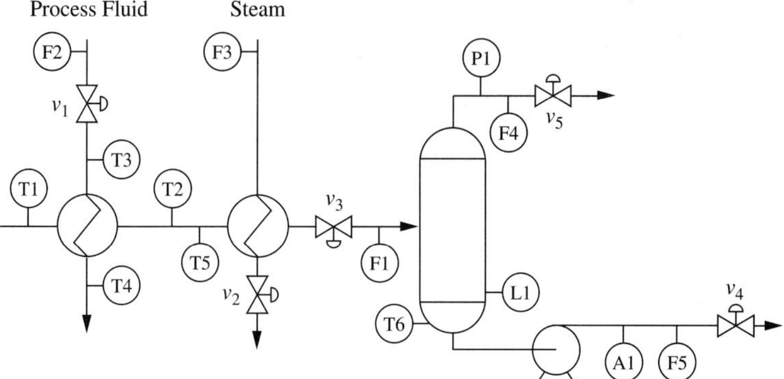

FIGURE 24.1
Preliminary process and instrumentation for the flash process.

The form used here is referred to as the Control Design Form (CDF). It will be introduced by discussing the initial draft in Table 24.1 for the proposed flash process shown in Figure 24.1; note that this initial design *is incomplete and may contain inconsistencies and errors.* The control design for this process will be addressed throughout the chapter as each of the design decisions is presented, and an improved process with control design will be presented at the end of the chapter.

As is typical in problem solving, we will start with a definition of the control objectives in the first major heading of the CDF. The control objectives are combined into the seven categories introduced in Chapter 2. The entries in each category must be concise but complete enough to provide the direction for the remaining design decisions. It is especially important to be as quantitative as possible regarding the performance, giving performance criteria for specific scenarios. This type of specification provides the basis for the design, along with a way to test the performance of the design against the objectives. Remember that the control performance should be specified for particular operating conditions and time periods; for example, (1) selected variables must remain within deviation limits from set point for a specified step disturbance; (2) the standard deviation for a variable must be no greater than specified over a day, week, or other interval; (3) a variable may not exceed its limits more than once per day; or (4) very undesirable conditions should not occur "under (essentially) any (conceivable) circumstances." Additional examples are given in Table 24.1 for the flash process.

The next heading contains information on the measurements provided for the control and monitoring system, which are crucial to the success of process control. The location of the sensor is shown in an accompanying drawing (i.e., Figure 24.1), and the physical principle of the sensor and range are given in the CDF. Special features of a sensor, such as the update frequency for a discrete sensor like a chromatograph, should also be recorded.

The final control elements are recorded under the next heading. The maximum capacity of the manipulated variable, typically the maximum flow through a valve, should be noted. Also, nonstandard features should be noted; for example, tight shutoff (i.e., the ability to prevent all flow); a valve that can open quickly; or a

TABLE 24.1
Preliminary Control Design Form for the flash process in Figure 24.1

TITLE: Flash drum PROCESS UNIT: Hamilton chemical plant DRAWING: Figure 24.1	ORGANIZATION: McMaster Chemical Engineering DESIGNER: I. M. Learning ORIGINAL DATE: January 1, 1994 REVISION No. 1

Control Objectives

1. Safety of personnel
   (*a*) The maximum pressure of 1200 kPa must not be exceeded under any (conceivable) circumstances.
2. Environmental protection
   (*a*) Material must not be vented to the atmosphere under any circumstances.
3. Equipment protection
   (*a*) The flow through the pump should always be greater than or equal to a minimum.
4. Smooth, easy operation
   (*a*) The feed flow should have small variability.
5. Product quality
   (*a*) The steady-state value of the ethane in the liquid product should be maintained at its target of 10 mole% for steady-state operating condition changes of
      (i) +20 to −25% feed flow
      (ii) 5 mole% changes in the ethane and propane in the feed
      (iii) −10 to +50°C in the feed temperature
   (*b*) The ethane in the liquid product should not deviate more than ±1 mole% from its set point during transient responses for the following disturbances:
      (i) The feed temperature experiences a step from 0 to 30°C.
      (ii) The feed composition experiences steps of +5 mole% ethane and −5 mole% or propane.
      (iii) The feed flow set point changes 5% in a step.
6. Efficiency and optimization
   (*a*) The heat transferred should be maximized from the process integration exchanger before using the more expensive steam utility exchanger.
7. Monitoring and diagnosis
   (*a*) Sensors and displays needed to monitor the normal and upset conditions of the unit must be provided to the plant operator.
   (*b*) Sensors and calculated variables required to monitor the product quality and thermal efficiency of the unit should be provided for longer-term monitoring.

Measurements

Variable	Sensor principle	Nominal value and sensor range	Special information
A1	Chromatograph	10, 0–15 mole%	Update every 2 minutes
F1	Orifice	100, 0–200	
F2	Orifice	120, 0–150	
F3	Orifice	100, 0–200	
F4	Orifice	45, 0–90	
F5	Orifice	55, 0–110	
L1	Δ pressure	Range is lower half of drum	
P1	Piezoelectric	5000–15000 kPa	

TABLE 24.1 (CONTINUED)

Measurements

Variable	Sensor principle	Nominal value and sensor range	Special information
T1	Thermocouple	0, (−)50–100 °C	
T2	Thermocouple	25, 0–100 °C	
T3	Thermocouple	90, 0–200 °C	
T4	Thermocouple	45, 0–200 °C	
T5	Thermocouple	25, 0–100 °C	
T6	Thermocouple	25, 0–100 °C	

Manipulated variables

I.D.	Maximum capacity (at design pressures)
	(%open, maximum flow)
v1	100%, 100
v2	53%, 189
v3	50%, 200
v4	14%, 340
v5	52%, 106

Constraints

Variable	Limit values	Measured/ inferred	Hard/ soft	Penalty for violation
Drum pressure	1200 kPa, high	P1, measured	Hard	Personnel injury
Drum level	15%, low	L1, measured	Hard	Pump damage
Ethane in F5 product	±1 mole%, (max deviation)	A1, measured, and T6, inferred	Soft	Reduced selectivity in downstream reactor

Disturbances

Source	Magnitude	Dynamics
Feed temperature (T_1)	−10 to 55°C	Infrequent step changes of 20°C magnitude
Feed rate (F_1)	70 to 180	Set point changes of 5% at one time
Feed composition	±5 mole% feed ethane	Frequent step changes (every 1 to 3 hr)

Dynamic responses
(Input = all manipulated variables and disturbances)
(Output = all controlled and constraint variables)

Input	Output	Gain	Dynamic model
v1	{see Example 24.7}		
v2			
v3			
v4			
v5			

Additional considerations

Liquid should not exit the drum via the vapor line.

final element that has a restricted range (e.g., cannot be closed). The failure mode of the final element is important but is not recorded here, because it is usually indicated on the drawing.

The next heading provides a place to document important limitations that could affect the control design. These are typically constraints on equipment and process variables. The limiting values and whether the constraint can be measured, along with the sensor type, should be recorded. The information should clearly indicate whether the constraint is soft or hard, along with the penalty for exceeding the constraint (e.g., yield loss, energy consumption, or equipment damage). A soft constraint can be violated for a short time and thus does not require the process to be shut down when the constraint is approached. An example would be a stream of material that, when not observing quality specifications, can be recycled or diverted to waste. Naturally, this is to be avoided but can be tolerated. The violation of a hard constraint causes severe safety or environmental hazards or costly equipment damage. Thus, a hard constraint must not be violated, and extreme measures, such as shutting down the process, are appropriate when a hard constraint is approached too closely.

Since the main reason for control is to respond to input changes (disturbances and set points), proper design depends on a good definition of these changes, which are recorded under the next heading. Recall that the importance of disturbances was recognized and included in methods presented in previous chapters, such as cascade, feedforward, gain scheduling, inferential control, and multiloop pairing. Therefore, each source of disturbance should be identified, along with its frequency of occurrence and magnitude; this information is useful in evaluating the potential need for and success of various design options. If the disturbance can be measured, that should be noted for possible feedforward and gain scheduling control.

The next-to-last heading covers dynamic responses between all process inputs (disturbances and manipulated variables), and all outputs (controlled variables and constraints). Naturally, this information is essential for control design. The models at the design stage might be very qualitative (fast, slow), semiquantitative (dominant time constants), or reasonably accurate (transfer function). The level of modelling performed should match the accuracy required for the decisions made during the control design; this design step might be less demanding than control implementation, which can be based on empirical models when the controllers are tuned.

The final heading provides a location for special information that does not fit under the other headings. For example, perhaps a particular flow should not be adjusted rapidly because of the sensitivity of product quality to flow rate. These special items, which require sound chemical engineering analysis of the process, must be considered in process control design.

This form may seem a bit pedantic, requiring excessive documentation for every decision; in fact, most control designs are performed in practice without such extensive documentation. The form is used here because it provides an excellent structure for beginning engineers who, after gaining proficiency, will often be able to perform the analysis without the form. However, even the most experienced engineers benefit from this type of documentation for complex designs. It is important to recognize:

> Experienced engineers can sometimes bypass the Control Design Form (CDF) documentation, but they must perform a thorough analysis involving all information and issues included in the CDF.

The subsequent sections of this chapter discuss issues related to the six major design decisions made in control engineering. These decisions are based on the information in the CDF. During the design, the engineer may find that the initial information is not complete and may have to return to enter additional information or enhance the measurements and final elements provided. In fact, the initial performance objectives might not be achievable with the initial process equipment and disturbances, in which case the engineer must reevaluate the objectives and either relax the specifications, alter the process design, or, if possible, reduce the disturbances. Such iterations are a natural part of the design process and do not necessarily indicate poor initial definition and analysis.

24.3 MEASUREMENTS

The success of automatic process control, real-time monitoring, and long-term performance tracking in improving plant performance depends crucially on measurements. The engineer must first determine the process variables to be measured and select a sensor for each. In this section, several important issues in selecting variables and sensors are discussed.

Measurement Feasibility

When the value of a variable is needed, it can be obtained from at least two real-time methods. First, it can be measured "directly" by a sensor; as an example, a temperature can be measured by a thermocouple, although the actual value sensed is the voltage generated for a bimetallic connection with nodes at two temperatures: the reference and process temperatures. This sensor is called *direct* because the physical principle underlying the measurement is independent of the process application, and the relationship between the sensor signal and process variable is reasonably accurate. Examples of variables that can usually be measured directly are level, pressure, temperature, and flows of many fluids. Also, the compositions and physical properties of some process streams can be determined in real time with on-stream analyzers.

In the second method, the variable cannot be measured, at least at reasonable cost, in real time, but it can be inferred using other measurements and a process-specific correlation. Inferential control is covered in detail in Chapter 17, so procedures for designing inferential variables will not be repeated here, except to emphasize that the acceptability of inferential control must be evaluated on a case-by-case basis. Examples of variables that are often inferred are composition of vapor-liquid equilibria (from temperature and pressure) and chemical reactor conversion (from temperature difference).

Not all variables can be measured or inferred in real time. These variables have to be determined through analysis of a sample of material in a laboratory. When the sample and analysis can be performed quickly, the laboratory measurement value can be used for feedback control. There are many industrial examples of controllers that use laboratory results and are executed every few hours, such as the one described by Roffel et al. (1989). While not providing control performance as good as would be possible with on-stream analysis, this approach usually gives much better performance than not using the laboratory value.

Accuracy

As covered in Chapter 12, the term *accuracy* refers to the error between the true process variable and the sensor signal. The error is a property of the sensor and, usually, its range; thus, the range should be maintained only as large as needed to measure the expected variation of the process variable about its normal operation. An associated property of the sensor is *reproducibility,* which indicates the differences in the sensor signal at different times for the same value of the true process variable. Often, sensors that provide good accuracy cost more than those that provide only good reproducibility; therefore, it is important to recognize which property is most important in a process control design and select the sensor accordingly.

For example, consider the process and control design in Figure 24.2, which includes cascade and feedforward. In determining whether accuracy or reproducibility is required, the key question is, "What is the purpose of the sensor?" For example, the objective of the feedforward controller is to adjust the manipulated variable for *changes* in the measured disturbance; therefore, it acts only on changes in the measured disturbance. In this situation, reproducibility of FI (i.e., reliable indications in the change of the disturbance variable), is more important than accuracy of the actual value. Similarly, the objective of the secondary controller in the cascade, FC, is to

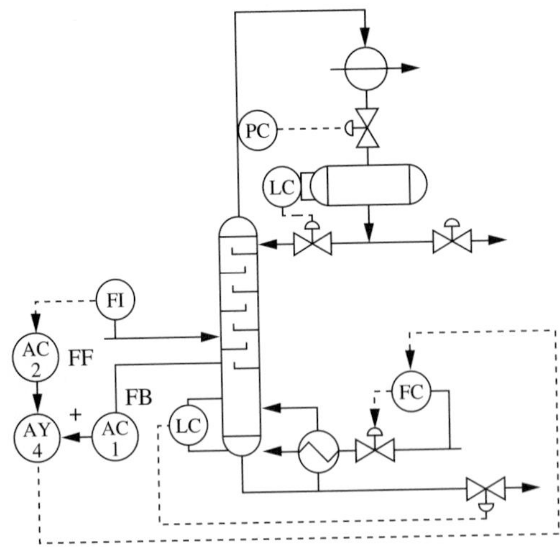

FIGURE 24.2
Example of feedforward-feedback control of a distillation tower product quality.

TABLE 24.2
Measurement objectives for various control structures

Control design	Measurement accuracy required	Only measurement reproducibility required*
Single-loop feedback	Product quality or other key variable	Tight control not important; proper set point can be adjusted infrequently by a person to attain the desired operating condition
Cascade	Primary controller	All secondary controllers
Feedforward-feedback	Measured variable for feedback controller	Measured disturbance for feedforward
Gain schedule	Measured variable used in correlation to determine tuning	
Inventory	Vapor: control must prevent violation of pressure limits for equipment	Liquid: reproducibility is acceptible if the inaccuracy is small with respect to the level range
Production rate	(1) The exact flow rate control is required or (2) The measurement is used to determine the sales volume	(1) Constant flow is important and (2) The goal is the proper average production over a day and (3) The production can be determined by accurate inventory measurement

*This is for control purposes; monitoring may require accuracy.

respond quickly to disturbances; therefore, reproducibility is again more important than accuracy. In contrast, the objective of the primary feedback controller in the cascade, A1, is to maintain the key output variable at the desired value; therefore, accuracy is required for this measurement. Analyses of sensor applications yield the summary of measurement objectives in Table 24.2, which readers should verify for themselves.

Some sensors have inherent inaccuracies that, if significant for a particular application, can be compensated in the input processing phase of the controller execution. As an example, the relationship between the pressure drop across an orifice and the volumetric flow rate is given by the equation

$$F = K\sqrt{\frac{\Delta P}{\rho}} \tag{24.1}$$

When the density of the fluid is not constant, both the density (ρ) and the pressure difference across the orifice plate (ΔP) could be measured and the appropriate calculation made in the control system to yield the corrected "measured" flow rate.

Dynamics

The dynamics of the process and sensors are present in the feedback loop and therefore influence control performance as discussed in Section 13.6. The first step to improve control is to select a location for the sensor that results in the fastest process dynamics in the feedback system. For example, the analyzer A1 in Figure 24.2 samples the vapor before the large first-order system that would have occurred if the analyzer had been downstream of the liquid inventory.

An estimate of the effect of sensor dynamics can be obtained by performing either a dynamic simulation or a frequency response analysis of the closed-loop system with and without the sensor dynamics. These analyses in Chapter 13 concluded that the sensor dynamics should be fast, certainly much faster than the process dynamics. For common flow, level, pressure, and temperature sensors, the dynamic response of the sensor is not usually a limiting factor in control performance, except for control of a few fast machinery systems. However, many analyzers are slow, because of (1) their sampling systems, which extract material from the process and transport it to a remote analyzer, and (2) the time for analysis. Thus, these sensors often contribute substantial dynamic delay to the closed-loop system and degrade the control performance. When this situation occurs, a common step to improve the control performance is to use a fast sensor as an inferential variable that can be reset in a cascade design by the slower analyzer controller (see Figure 17.12, page 569).

Reliability

Sensors used in control systems must be very reliable, because the failure of a sensor incapacitates the control loop and could lead to an unsafe situation. For example, a failure of the reboiler flow sensor in Figure 24.2, if not identified during input processing, could result in a zero value being used as the value of the controlled variable in the controller calculation. Since the measurement would be below the set point, the controller would rapidly open the reboiler valve completely, which could cause a pressure surge that might damage the trays. Some sensor characteristics that lead to lower reliability are (1) sensors contacting process fluids, (2) poorly designed sample systems that plug or extract an unrepresentative sample, and (3) complex chemical or physical analyses (Clevett, 1986). In many designs these characteristics cannot be eliminated, and the engineer should expect lower reliability.

Cost

The cost of a sensor is the total of equipment purchase, installation, maintenance, and operating costs. Most sensors have small operating costs, perhaps a small amount of electrical power for heating in cold weather; however, a sensor can occasionally contribute substantially to plant operating costs. An example is a flow sensor for gas in a pipe, where the standard orifice meter can be used to measure the flow, but the nonrecoverable pressure drop across the orifice can be large. If compression costs are significant, a sensor that has very low pressure losses (e.g., a venturi meter or pitot tube) could be used. The purchase and installation costs of the alternative meter would be greater than for the conventional orifice, but its total cost over several years would be lower.

Finally, the primary use of the measured value should be considered in selecting a sensor. Fast dynamics would be an important concern for sensors used in feedback control. However, measurements for monitoring, especially longer-term process performance, may be satisfactorily supplied by sensors that are slower or of lower cost.

Example 24.1. In this example, the sensors in the preliminary flash design in Figure 24.1 are considered. First, we notice that the sensors T2 and T5 are redundant and that redundancy is not needed, because this is not a critical measurement. Therefore, sensor T5 is removed. Second, it is noticed that the feed flow measurement F1 is located after the flash valve, where the material is composed of two phases. However, the pressure drop across an orifice meter, which is the sensor principle, does not accurately or reproducibly relate to the actual flow when the fluid has two phases. Therefore, the flow meter location is moved before the first heat exchanger, where the material is always one liquid phase in this example.

Third, the temperature indicating the flash, T6, is in the liquid inventory and will not rapidly respond to changes in the drum inlet temperature. Since this temperature will be used as an inference of composition, minimum feedback dynamics is desired. Therefore, T6 is relocated in the vapor space, which has little inventory. To provide a reliable indication regardless of the flow patterns in the drum, the sensor is located in the pipe leaving the top of the drum.

24.4 FINAL ELEMENTS

All final elements that are adjusted by an automatic controller or adjusted frequently by plant personnel must be automated. The automation of a final element requires a power source that changes the final element's value, usually the percentage valve opening, as determined by a signal transmitted from the control system. Many other final elements whose values change very infrequently are not automated and require a person to change their values manually at the equipment; thus, plants also contain many "hand valves." Some of the important features for an automated final element are discussed here.

Capacity and Precision

The final element should have the capacity to influence the manipulated process variable over the required range. As an initial guideline, a control valve should be 60 to 70% open at design conditions, so that the valve has considerable additional capacity to allow increased flow during disturbances or operation at increased production rates. However, each control system should be evaluated individually to ensure that the proper capacity exists.

Special designs are required when the range of the manipulated variable is large. For example, the feed to the flash drum in Figure 24.3 can vary from a small to a large amount of light, vaporized material. To accommodate the small, normal

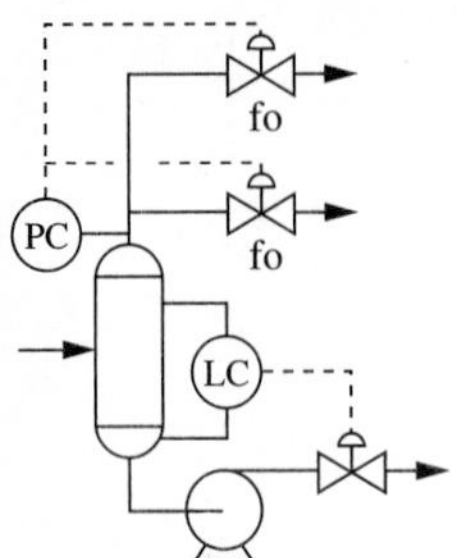

FIGURE 24.3
Example of use of final elements with small and large capacities to expand total range.

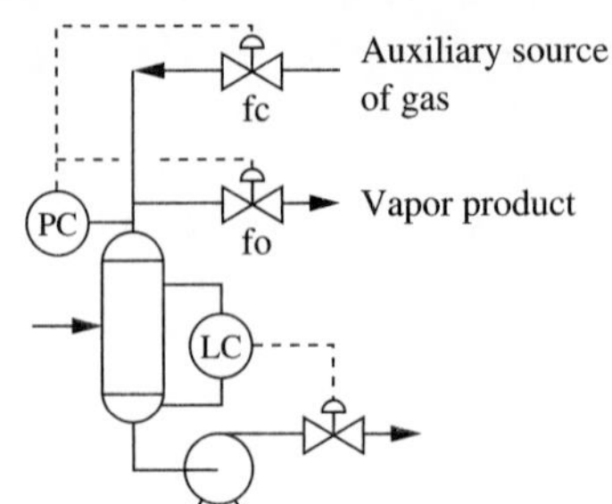

FIGURE 24.4
Example of final elements that allow in or out flow.

flow, a valve with a small capacity could be provided. However, a valve with a much larger capacity is provided to satisfy the infrequent, large vapor flow. The control design, using split range, is shown in the figure. Another example demonstrates the need to consider the sign of the manipulated variable as well as the magnitude. The drum in Figure 24.4 normally has a small vapor product; however, sometimes there is no vapor. To ensure that the pressure can be controlled for both cases, the pressure controller must be able to manipulate the outflow of product vapor or an inflow of a compatible gas. The control design, using split range, is shown in the figure.

A final element has a range over which it can accurately influence the manipulated process variable. For a typical control valve, the range of lowest to highest flows would be on the order of 1 : 20; thus, the range is quite large. In special cases, the control system might need to make quite small changes accurately when the total flow is relatively large. A two-valve arrangement that achieves this objective for strong acid–strong base pH control is shown in Figure 24.5. Normally, the larger valve is held constant, and the much smaller valve is adjusted by the controller. The larger valve is adjusted only when the smaller valve has reached a maximum or minimum limit. Finally, cascade principles can be employed to improve the valve performance by including a valve positioner, as explained in Chapter 14.

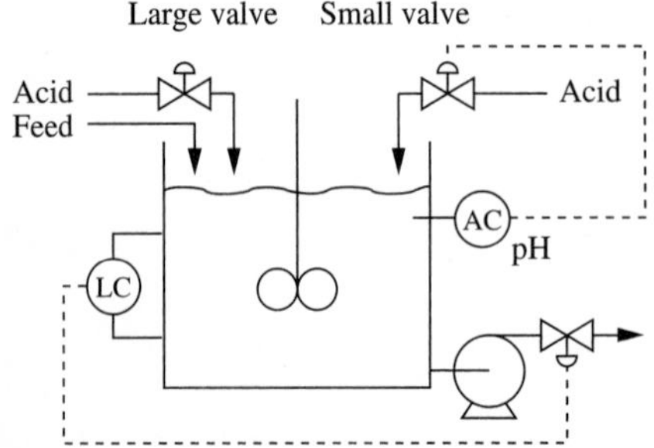

FIGURE 24.5
Example of the use of final elements with large and small capacities to improve accuracy in manipulations.

Dynamics

Again, dynamic elements in the feedback system degrade control performance. Therefore, the final element response should be much faster than that of other elements in the system. Most valve percent openings are achieved within a few seconds of a change in the signal to the valve, so that the valve dynamics are negligible for all but the fastest process control systems.

Failure Position

The failure position is selected to reduce the hazard to people and environment and damage to equipment when the signal to the final element is lost (i.e., when the signal to the valve attains its lowest value). Most valves are specified to go to either fully open or fully closed upon loss of signal. The proper failure position of a valve must be determined through an analysis of the integrated plant to determine the proper manner for relieving, storing, and venting material during an emergency. Naturally, the integrated plant must have the capacity to process (i.e., condense, combust, or store) material that cannot be vented to the environment.

Example 24.2. In this example, the final elements in the preliminary flash design in Figure 24.1 are considered. First, the valve in the liquid stream, v4, appears to be oversized, since its capacity is about seven times the design flow. Therefore, the valve specification should be changed so that the maximum flow through v4 is changed to 53% opened at design for a maximum flow of about twice its design value.

Second, the valve v2 is located in the condensate line, which means that the heat exchanger is behaving as shown in Figure 24.6*a*. In this design, the heat duty depends primarily on the area for condensation, which has a much higher heat transfer coefficient than the liquid-liquid film. As the valve is closed slightly, the liquid flow decreases, the area for condensation decreases, and the heat duty decreases. This is acceptable from a steady-state perspective; however, the dynamic response of the process depends on the direction of change. Increasing the duty is rapid because the liquid can flow quickly from the exchanger, but decreasing the duty is slow, because the liquid must condense and accumulate in the exchanger to reduce the area. A faster-responding design for both increasing and decreasing the duty is shown in Figure 24.6*b*, in which the steam flow is adjusted. This rapidly influences the steam pressure, and thus the

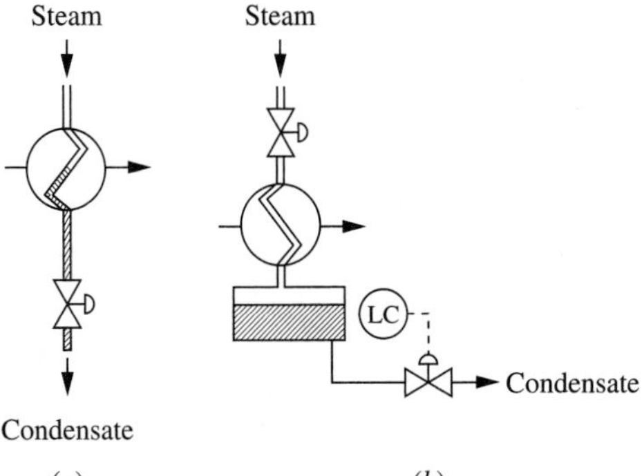

FIGURE 24.6
Alternative process designs for condensing heat transfer.

temperature difference for heat transfer, to provide the amount of condensation needed. To complete the water material balance, the liquid condensate is collected in an inventory outside of the exchanger (in a steam trap), from which it is returned to the steam generators. The design in Figure 24.6*b* is preferred and will be used for the flash drum example.

24.5 PROCESS OPERABILITY

One of the most important lessons in this book is that the process design and operating conditions have the most significant influence on control performance. Some processes are easily controlled; others require sophisticated algorithms to achieve satisfactory performance; and some processes cannot perform as required regardless of the type of control technology used. Thus, good control performance is one of the important goals of process design. Often, the ease with which a process is operated and controlled is referred to as *operability.* Some of the important factors that influence operability from the perspective of control performance are discussed in this section. The first topics address the possibility of control, and later topics address the quality of control performance.

Degrees of Freedom

The process must have sufficient manipulated external (independent) variables to control the specified (dependent) variables; if sufficient manipulated external variables are not provided, the desired control performance will not be achievable. Since the transient behavior is of interest, the degrees of freedom are determined by analyzing the *dynamic* model of the process. As presented in Chapter 3, the degrees of freedom of a system are

$$\mathrm{DOF} = \mathrm{NV} - \mathrm{NE} \tag{24.2}$$

with DOF = number of degrees of freedom, NV = number of variables, and NE = number of linearly independent equations. In modelling, we checked to ensure that the degrees of freedom were zero so that the model was consistent with the exactly defined problem statement. However, an essential part of the design task is to provide a process that can achieve the specified control objectives; therefore,

> The process without the controllers must have zero degrees of freedom when all external variables have been specified. To satisfy the control objectives, the number of manipulated external variables in the process alone must be equal to or greater than the number of dependent variables to be controlled.

The reason for the first requirement—zero degrees of freedom for the model—was presented in Chapter 3. The second requirement is a *minimum* requirement so that the process has the flexibility needed to satisfy the control objectives. If the number of manipulated variables is smaller than the number of controlled variables, the system is overspecified and cannot achieve all objectives. In other words, an attempt is being made to control more variables than is physically possible for a specific process design. Corrections include reducing the number of variables controlled or adding flexibility to the process by increasing the number of manipulated variables by, for example, adding heat exchangers, bypass flows, and so forth. When the number of manipulated variables is greater than the number of controlled variables, the system is underspecified; it is possible that the control objectives can be achieved by many combinations of manipulated-variable values, subject to further analysis of controllability and dynamic performance. Since the plant should have a unique operating policy, additional objectives, such as minimizing expensive fuel flows, can be added to the performance objectives.

When the controllers are added, the number of manipulated variables that are externally determined does not change, but the external variables change from the final element positions (for the open-loop system) to the set points (for the closed-loop system). Analyzing degrees of freedom is demonstrated through the following example.

Example 24.3. The design for the mixing process without reaction in Figure 24.7*a* is to be analyzed.

Additional information.

1. The inlet stream consists of two components, A and S, with equal densities.
2. The pressures P_1, P_2, and P_4 are determined externally.
3. Pressure P_3 is constant.
4. All pressure drop occurs across the control valves.

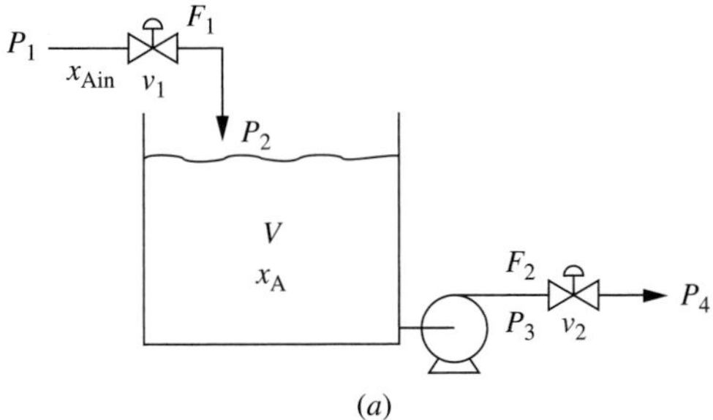

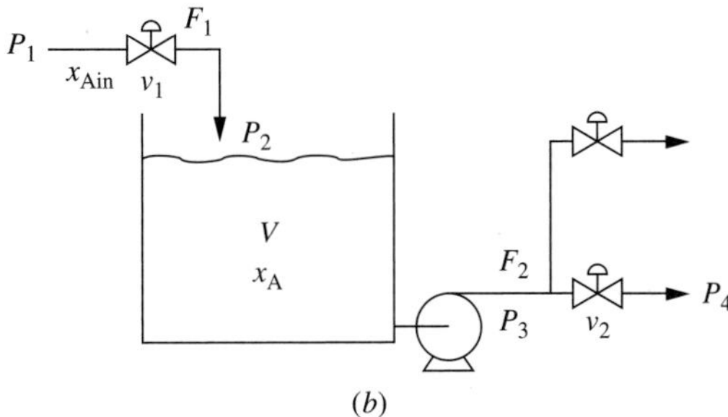

FIGURE 24.7
Level-concentration process: (*a*) example; (*b*) alternative.

5. The tank is well mixed.
6. The inherent characteristics of the valves are linear.

The dynamic model for this system is

$$F_1 = v_1 C_{v1}\sqrt{\frac{P_1 - P_2}{\rho}} \tag{24.3}$$

$$F_2 = v_2 C_{v2}\sqrt{\frac{P_3 - P_4}{\rho}} \tag{24.4}$$

$$\rho \frac{dV}{dt} = \rho\,(F_1 - F_2) \tag{24.5}$$

$$\rho \frac{d(x_A V)}{dt} = \rho\,(x_A \frac{dV}{dt} + V\frac{dx_A}{dt}) = \rho\,(x_{Ain}F_1 - x_A F_2) \tag{24.6}$$

Analysis. The analysis of degrees of freedom is summarized in Table 24.3. This analysis shows that the system has four equations and four dependent variables, which demonstrates that the DOF $= 4 - 4 = 0$ and the system is exactly specified. From our knowledge of cause and effect, we see that the valve positions are external variables that can be manipulated. The remaining four external variables (x_{Ain}, P_1, P_2, and P_4) depend on conditions of associated processes or the environment not included in this model and are usually referred to as *disturbance* variables. The conclusion is that only two of the external variables can be manipulated by a person or control system; therefore, at most two variables can be controlled. (In alternative process designs, x_{Ain} or even some of the pressures could be manipulated, but this is not the case in this example.)

The engineer has some options in model formulation. For example, it is possible to write a component balance on the solvent S, as follows.

$$\rho \frac{d(x_s V)}{dt} = \rho\,(x_S \frac{dV}{dt} + V\frac{dx_s}{dt}) = \rho\,(x_{sin}F_1 - x_s F_2) \tag{24.7}$$

The question arises concerning whether this equation affects the degrees of freedom. However, close examination of the model and recalling that $x_s + x_A = 1$ leads to the conclusion that equation (24.5) is the sum of equations (24.6 and 24.7); thus, these three equations are *not independent.* Since only linearly independent equations are considered in the analysis of degrees of freedom, (any) one of these three equations must be eliminated from the analysis; redundant (dependent) equations must not be included in the degrees-of-freedom analysis.

TABLE 24.3
Degrees-of-freedom analysis for Example 24.3

Dependent	Constants	Independent (specified)
F_1 = inlet flow	C_{vi} = valve constant	*External variables* (disturbances)
F_2 = outlet flow	ρ = density	x_{Ain} = inlet mass fraction of A (constant)
V = liquid volume	P_3 = pump outlet pressure	P_1, P_2, P_4
x_A = mass fraction A		
		External variables (manipulated)
		v_1 = percent valve opening
		v_2 = percent valve opening

Additional flexibility could be added to the process by including another pipe and valve at the outlet of the tank between the pump and the original valve, as shown in Figure 24.7*b*. This would result in an additional outflow and external manipulated variable, which means that three variables could be controlled. If only two controlled variables were specified, the system would be underspecified. The engineer would have to investigate the overall plant operating goals to determine the proper strategy for manipulating the two outlet valves, perhaps in a fixed ratio. Then, the number of external manipulated variables would again be equal to the number of controlled variables, which provides a proper statement for control design.

Controllability

A process design with the necessary degrees of freedom meets the important requirement that it be able to satisfy the proper *number* of objectives, but it is not sufficient to ensure that satisfactory control can be implemented. An additional requirement is that the process must be able to achieve the objectives for the specified controlled variables by adjusting the specified manipulated variables. The requirement to test this feature of the process is *controllability,* which was introduced in Chapter 20 for a multivariable process. The definition of controllability used in this book is repeated here:

> A system is **controllable** if the controlled variables can be maintained at their set points, in the steady state, in spite of disturbances entering the system.

Recall that the system is deemed controllable when the steady-state gain matrix relating the manipulated to controlled variables is nonsingular; that is, when its determinant is nonzero. (If the number of manipulated variables is greater than the number of controlled variables, the gain matrix must have a rank equal to or greater than the number of controlled variables. This means that a subset of the manipulated variables can be selected for which the square gain matrix including all controlled variables is nonsingular.)

The controllability criterion was derived using the final value theorem, which requires some limitations to be placed on the process transfer functions $G_{ij}(s)$, basically that each be stable. The use of the final value theorem precludes most liquid levels, which are pure integrators and have transfer functions of the form $G(s) = k/s$. Since most process plants have liquid levels, the method for determining controllability should be extended to levels. To include integrating processes and maintain a simple analysis, we choose to consider the *rate of change of the level* as the controlled variable for the controllability analysis. Thus, the controlled variable is $sL(s)$, and the transfer functions between the rate of change of level and the manipulated and disturbance variables are constants and thus stable. Then the final value theorem can be applied, and the test for controllability is valid. In this case, the definition of controllability is modified to include the rate of change of level being returned to its desired value of zero.

The controllability analysis can be applied to the system in Figure 24.7*a* for controlling the inlet flow F_1 and the volume V. Since the volume is non–self-regulatory, the rate of change of the volume is considered in this analysis. First, the defining equations are linearized and expressed in deviation variables.

$$F_1' = \left(C_{v1} \sqrt{\frac{P_1 - P_2}{\rho}} \right) v_1' \tag{24.8}$$

$$F_2' = \left(C_{v2} \sqrt{\frac{P_3 - P_4}{\rho}} \right) v_2' \tag{24.9}$$

$$\rho \frac{dV'}{dt} = \rho \left(F_1' - F_2' \right) \tag{24.10}$$

Then, the transfer functions are formed by taking the Laplace transforms and combining appropriate equations for $F_1(s)$ and $sV(s)$ as the controlled variables and $v_1(s)$ and $v_2(s)$ as the manipulated variables.

$$\begin{bmatrix} F_1(s) \\ sV(s) \end{bmatrix} = \begin{bmatrix} C_{v1}\sqrt{\dfrac{P_1 - P_2}{\rho}} & 0 \\ C_{v1}\sqrt{\dfrac{P_1 - P_2}{\rho}} & C_{v2}\sqrt{\dfrac{P_3 - P_4}{\rho}} \end{bmatrix} \begin{bmatrix} v_1(s) \\ v_2(s) \end{bmatrix} \tag{24.11}$$

Applying the final value theorem yields the gain matrix, and the determinant can be evaluated.

$$\det \begin{bmatrix} C_{v1}\sqrt{\dfrac{P_1 - P_2}{\rho}} & 0 \\ C_{v1}\sqrt{\dfrac{P_1 - P_2}{\rho}} & C_{v2}\sqrt{\dfrac{P_3 - P_4}{\rho}} \end{bmatrix} = C_{v1}\sqrt{\frac{P_1 - P_2}{\rho}}\, C_{v2}\sqrt{\frac{P_3 - P_4}{\rho}} \neq 0 \tag{24.12}$$

Since the determinant of the matrix is nonzero, the system is controllable, and we conclude that it is possible to control the inlet flow and the rate of change of the volume to zero deviation values. Note that this result does not address whether the level can be returned to its set point—only that the level can be stabilized.

Since the system is controllable, feedback control equations can be written that, when combined with the process model equations (24.3 to 24.6), describe the behavior of the closed-loop system. As an example, the controller equations for two proportional-integral feedback controllers follow, with the level measurement used for the volume variable.

$$v_1 = K_{c1} \left((F_{1\text{sp}} - F_1) + \frac{1}{T_{I1}} \int_0^t (F_{1\text{sp}} - F_1) dt' \right) + I_1 \tag{24.13}$$

$$v_2 = K_{c2} \left((L_{\text{SP}} - L) + \frac{1}{T_{I2}} \int_0^t (L_{\text{SP}} - L) dt' \right) + I_2 \tag{24.14}$$

For the degrees-of-freedom analysis, the number of equations increases from four to six, and since the controller parameters are constant, the number of dependent variables increases from four to six: the four original plus the two valve positions. The number of degrees of freedom remains zero, and the manipulated external variables are two: the controller set points. When the values for the two set points are defined, the system is completely specified and has no degrees of freedom. Remember that this system, which "has no degrees of freedom," has two external variables (the set points) that were determined by an engineer through an analysis of the plant.

Next, the controllability of the example process is evaluated for an alternative set of controlled variables: the concentration x_A and volume V. The nonlinear balance on component A can be simplified by multiplying equation (24.5) by x_A, subtracting the result from equation (24.6), and substituting equation (24.3) for F_1. The result is

$$\frac{dx_A}{dt} = \frac{C_{v1}\sqrt{\dfrac{P_1 - P_2}{\rho}}(x_{Ain} - x_A)}{V} v_1 \tag{24.15}$$

This equation can be linearized and expressed in deviation variables to give

$$\begin{aligned}\frac{dx'_A}{dt} &= \frac{C_{v1}\sqrt{\dfrac{P_1 - P_2}{\rho}}(x_{Ain} - x_{As})}{V_s} v'_1 \\ &\quad - \frac{C_{v1}\sqrt{\dfrac{P_1 - P_2}{\rho}}(x_{Ain} - x_{As})v_{1s}}{V_s^2} V' - \frac{C_{v1}v_{1s}\sqrt{\dfrac{P_1 - P_2}{\rho}}}{V_s} x'_A\end{aligned} \tag{24.16}$$

From the balance equation (24.6), the steady-state value of the concentration can be determined for this mixing process without reaction to be $x_{As} = x_{Ain}$. Therefore, the first two terms in equation (24.16) are identically zero, and the resulting model is

$$\begin{bmatrix} sV(s) \\ (\tau s + 1)x_A(s) \end{bmatrix} = \begin{bmatrix} C_{v1}\sqrt{\dfrac{P_1 - P_2}{\rho}} & C_{v2}\sqrt{\dfrac{P_3 - P_4}{\rho}} \\ 0 & 0 \end{bmatrix} \begin{bmatrix} v_1(s) \\ v_2(s) \end{bmatrix} \tag{24.17}$$

Since one row of the transfer function matrix contains all zeros, the determinant of the gain matrix, after the final value theorem is applied, is zero. Thus, although the system has two degrees of freedom, it is *not possible* to control the level and the composition in this process by manipulating the valves v_1 and v_2! The inability to control x_A may at first be unexpected, because the valve positions affect flows that appear in the material balance of component A, and one might (incorrectly) conclude that a desired value of x_A could be achieved by adjusting these valves. However, to control the level, the flows in and out must be equal at steady state, and the inlet concentration of A is not influenced by the valves. Thus, the outlet concentration is not controllable, in spite of the *(false) apparent cause-effect* relationship between the valves and the composition. This example highlights the need for a thorough controllability check for all designs that extends beyond a qualitative cause-effect analysis. It is worth noting that for a nonlinear process the determinant of the gain matrix may be nonzero for some operating conditions and zero for other operating conditions. Thus, the controllability should be evaluated for the operating condition(s) normally expected. In summary:

> The analysis of degrees of freedom and controllability evaluates whether the specified variables can be controlled by adjusting the specified manipulated variables in the region for which the linearized model is valid.

This analysis does not indicate the control structure required to achieve stable control or the range of disturbances that can be corrected; nor does it predict the control performance.

Operating Window

The degrees-of-freedom and controllability requirements ensure that for at least some disturbances of very small magnitude, the control system can return the controlled variables to their set points. For practical control performance, the process equipment must have the capacity or range to satisfy the control design objectives for disturbances of expected magnitudes. When analyzing the steady-state performance of a process, the capacity is often represented by an operating window, as presented in Chapters 3 and 20. The coordinates are important process variables, and the region of acceptable performance is indicated as a "window" that is surrounded by an "infeasible" region, which represents operation that is either undesirable or not possible. The boundary or "frame" of the window is defined by the constraints in the Control Design Form, and an important function of the control design is to maintain the process operation within the window. To achieve this goal, the manipulated variables must have sufficient capacity.

Equipment sizing is often determined by a steady-state analysis that chooses equipment designs (e.g., heat exchanger area, pump capacity, and distillation tower diameter) to maintain operation within the window for a defined set of expected operating conditions, including disturbances. However, a steady-state analysis is not always sufficient, because a process can exceed the steady-state limits of possible operation during transients, as demonstrated by the following example.

Example 24.4. Consider the nonisothermal, continuous stirred-tank chemical reactor described in detail in Example 3.10 (page 90). The nominal design operating conditions are the same as given in Example 3.10 except for the inlet concentration, which in this example has an initial value of 1.0 and experiences a step change to 2.0 kmole/m^3; thus, this exercise investigates a dynamic response returning to the initial conditions of Example 3.10.

The dynamic response of the system for the step in inlet concentration is evaluated through numerical solution of the differential equations, and the results are given in Figure 24.8, which shows the underdamped behavior consistent with Example 4.8. The same data is plotted in Figure 24.9 with concentration and temperature as the

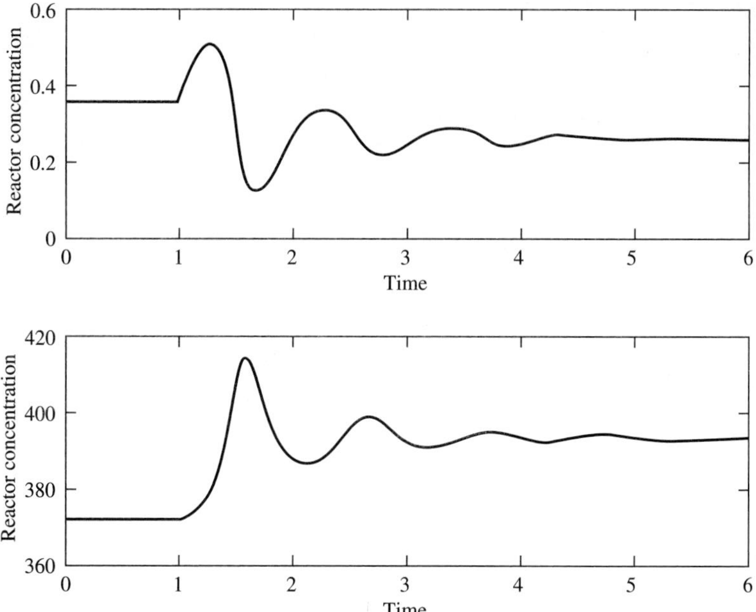

FIGURE 24.8
Dynamic response for Example 24.4.

coordinates, and the solid line defines the steady-state operating window: that is, the entire region of possible steady-state operation with $F_c = 0.5$ to $16.0\,\text{m}^3/\text{min}$ and $C_{\text{Ain}} = 1.0$ to $2.0\,\text{kmole/m}^3$. The trajectory in response to the step in C_{Ain} from 1.0 to 2.0 is shown as a dotted line with the arrows indicating the progression of time. Note that the transient begins and ends within the steady-state window, which it must, but that it violates the window by a considerable amount during the transient.

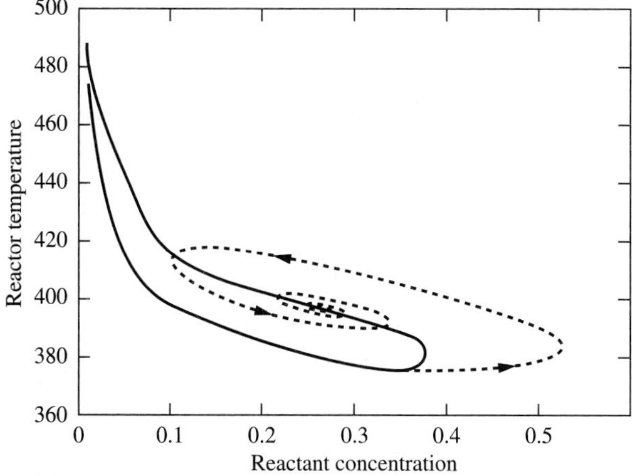

FIGURE 24.9
Operating window and trajectory for Example 24.4.

This example demonstrates that the dynamic behavior of the process must be analyzed when determining the possible operating conditions that can occur in a process.

Given the importance of maintaining the process variables within an acceptable region and the fact that designing for a steady-state region does not eliminate the possibility of violations during transients, some equipment may have to have a greater capacity than required to meet steady-state demands in order to maintain all variables inside the window during transients (Rinhard, 1981). Failure to consider dynamics could lead to process designs that cannot perform properly (e.g., they may explode or corrode) during dynamic operation.

After the feasibility of control has been determined from the steady-state analysis, the effect of process dynamics on control performance is evaluated. The dynamic performance of control systems has been addressed throughout the book; here a few of the major conclusions are reiterated. However, this is not a comprehensive summary of important prior results, which would be very lengthy. The highlights are separated into discussions of feedback and disturbance dynamics, and a few examples are given to demonstrate how minor process design changes can influence the control performance.

Feedback Dynamics

The first three items in this section addressed the possibility of control; now, the performance issues are addressed. The process typically contributes the dominant dynamics in the feedback system; therefore, improving the process dynamics is especially important in improving control performance, as presented thoroughly in Chapters 13 and 21. Feedback process characteristics that contribute to good control performance include the following:

1. The process should be self-regulatory and open-loop stable.
2. The process dynamics should be relatively constant as operating conditions change.
3. The process should have fast dynamics with a small dead time and no inverse response.
4. The multivariable process should have favorable interactions.

The first characteristics are not required for good closed-loop control performance; however, stable, self-regulating processes are easier to operate in open loop (i.e., manually). Since all processes are operated manually on some occasions, they are included as good characteristics. The second characteristic of unchanging dynamics allows a controller with constant tuning to provide good control performance. If the dynamics change significantly, methods in Chapter 16 may be applied to compensate partially, but if the process gain changes sign (see Example 3.12), feedback control using linear controllers is not possible.

Fast feedback dynamics can be achieved by reducing transportation delays through shortening pipes, reducing (numerous) time constants through decreasing inventories, and speeding thermal processes through lessening the accumulation terms

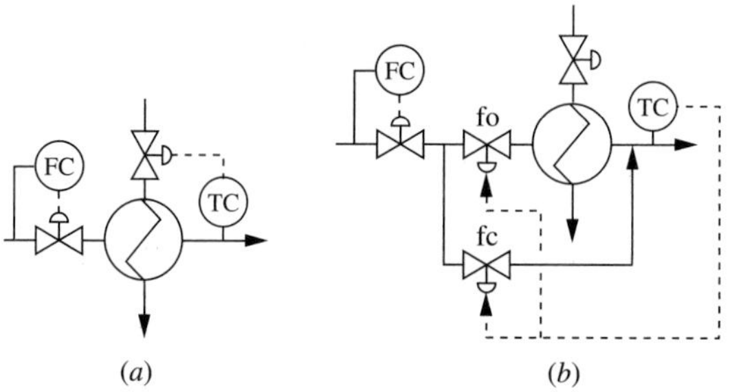

FIGURE 24.10
Alternative heat exchanger control designs.

associated with heat exchangers, tank walls, and so forth. These steps improve feedback dynamics and usually also reduce equipment size and cost. However, there is a limit beyond which process equipment cannot be modified, and other approaches are required to improve dynamics. For example, additional improvements can be achieved by selecting the proper manipulated variable from several available; an example of this approach is discussed here with respect to the two temperature control systems in Figure 24.10*a* and *b*. The dynamics between the cooling (or heating) fluid flow and the temperature in Figure 24.10*a* is slow, because the temperature of the fluid and metal in the heat exchanger must be changed to affect the controlled variable. The design in Figure 24.10*b* allows the ratio between the flow through the exchanger and the flow bypassing the exchanger to be adjusted to control the temperature. The dynamic behavior of a heat exchanger with bypass is analyzed in Example 5.3 (page 166), where the rapid response (with an overshoot) was ascertained. Thus, the design using the bypass would be preferred when good control performance is required, although the equipment cost would be slightly higher. Note that the engineer must be creative in adding flexibility in the equipment for improved control.

Disturbance Dynamics

The basic objective of process control is to compensate for disturbances; therefore, the process should be designed to reduce the occurrence and effects of disturbances. Previous analysis has established that feedback control is improved when disturbances have (1) small magnitude, ΔD, (2) small gain, K_d, and (3) favorable directions or interaction (small relative disturbance gain, |RDG|) and occur at frequencies much higher than the bandwidth of the disturbance process (where the open-loop amplitude ratio, $|G_d(j\omega)|$, is small) or much lower than the critical frequency of the closed-loop feedback system.

Many disturbances originate externally, such as from feed composition and cooling water temperature. However, the increased use of material and energy integration in process designs has increased the likelihood that variation in the process will negatively affect the dynamic performance of an associated process. As a simple example, consider the exothermic chemical reactor with a feed-effluent heat exchanger in Figure 24.11*a*, which was analyzed in Example 5.5 (page 173). (A more rigorous analysis, showing the possibility for multiple steady states, is given in Appendix C.) With no temperature control, an upset in the feed temperature affects the reactor inlet, which affects the reactor outlet, which again affects the reactor inlet.

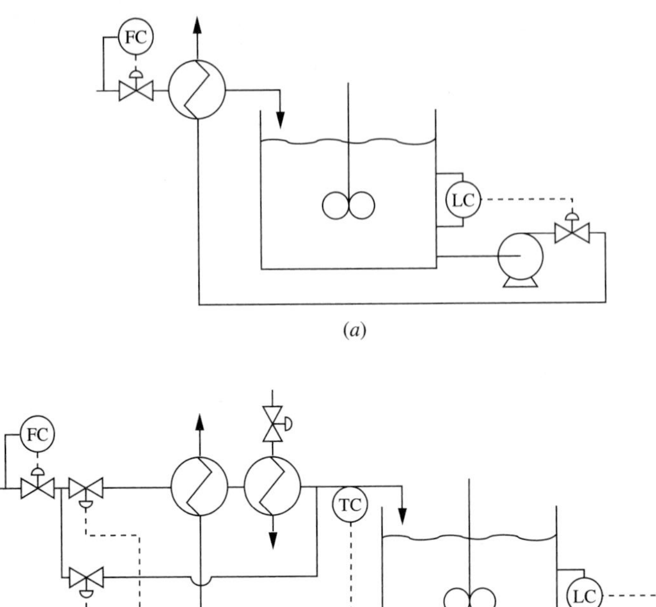

FIGURE 24.11
Example of control design to reduce the effects of process integration.

Thus, an *energy recycle structure* is created, which heightens the sensitivity to disturbances and could lead to instability for highly exothermic systems. Naturally, the recycle structure could be eliminated by using two exchangers with utility fluids: one to heat the feed and a second to cool the effluent. However, that design modification would lose the energy efficiency advantages of the design in Figure 24.11*a*.

An alternative way to improve the disturbance response and retain most of the energy savings is to control the inlet temperature so that it is nearly independent of the reactor outlet temperature. The approach requires an additional manipulated external variable, which can be supplied with a bypass placed around the feed-effluent heat exchanger. An additional heat exchanger—which would likely be needed for startup anyway—may be needed to provide the heat duty lost by to the bypass. As shown in Figure 24.11*b*, the reactor temperature could be controlled by adjusting the bypass around the feed-effluent exchanger, and the duty of the utility exchanger could be adjusted so that most of the feed preheat is supplied by the (inexpensive) heat integration.

Two general points demonstrated by this example can be applied to most material and energy recycle systems. First, feedback effects of disturbance propagation due to a recycle can be attenuated by adding an alternative path or source/sink where the recycle occurs. Second, the maximum steady-state benefit of process integration cannot always be achieved because of the poor dynamic behavior; however, most of the benefit can be realized by using the control methods demonstrated here while maintaining good control performance.

Inventory and Flow

Naturally, control of production rates and inventories is essential to good plant performance. The process should have sufficient inventories to ensure uninterrupted flows to pumps and smooth flow rate variations throughout the plant as shown in Figure 24.12*a*. Good performance depends on the proper combination of inventory size and level control, including a nonlinear feedback algorithm where warranted. A straightforward manner for reducing the effects of disturbances in stream properties, such as temperature and composition, is to locate an inventory between the disturbance source and the controlled variable, but not in the feedback path, as shown in Figure 24.12*b*. However, inventories have disadvantages such as cost and hazards, as discussed in Chapter 18, and large inventories are included sparingly—only when absolutely necessary to improve dynamic operation.

Some examples of the analysis of process design on dynamic performance are now presented. The first example demonstrates that small changes in the process design can improve both the feedback and disturbance dynamics. The subsequent examples continue the analysis of the flash process.

Example 24.5. A stirred-tank heat exchanger was modelled in Example 3.7, and the control performance of the linearized approximation was evaluated analytically in Example 8.5 (page 276). The results indicated reasonably good performance, because the feedback dynamics $G_p(s)$ were first-order. However, the question remains whether the performance could be improved by simple process modifications. A reasonable goal would be to change the process so that the feedback dynamic response is faster and the controlled variable is less sensitive to disturbances. This can be achieved by increasing the "self-regulatory" nature of the process without control. For the heat exchanger, the process will be more self-regulatory if the temperature driving force for exchange is small; then, a small increase in the exchanger fluid temperature due to a feed inlet temperature increase will substantially increase the cooling duty. Naturally, the heat exchanger area must be increased to achieve the same heat transfer rate as in the base case with a smaller temperature difference.

This concept is applied to the example heat exchanger by increasing the cooling temperature from the original value of 25°C to 65°C, with a commensurate increase in the heat exchanger area. The data for this example, which is the same as the original

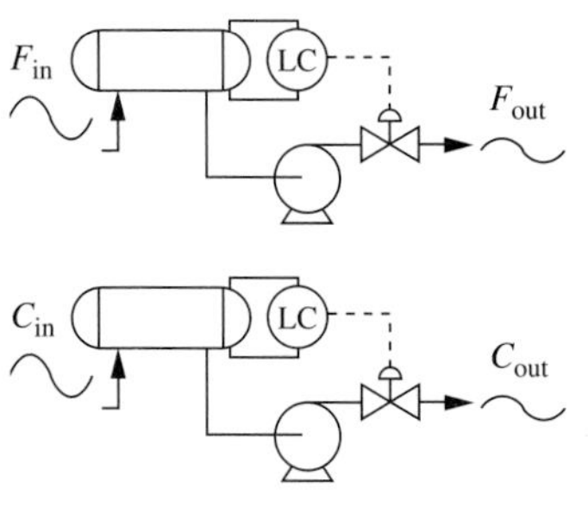

FIGURE 24.12
Use of inventory to improve control performance: (*a*) flow rate attenuation; (*b*) flow property attenuation.

TABLE 24.4
Data and selected results for Example 24.5

Parameter	Original value from Example 8.5	Modified value from Example 24.5	Comment
	Process data		
a (cal/min °C)	1.41×10^5	5.21×10^5	$UA = aF_c^b$
T_{cin} (°C)	25	65	
K_p (°C/(m^3/min))	−33.9	−19.6	
τ (min)	11.9	5.93	
K_d (°C/°C)	0.52	0.24	
τ_d (min)	11.9	5.93	
	Controller data		
K_c ((m^3/min)/°C)	−0.059	−0.10	K_cK_p the same
T_I (min)	0.95	0.47	T_I/τ the same
	Control system performance		
IAE (°C min)	5.31	1.27	due to smaller K_d and τ
Maximum deviation from set point (°C)	0.66	0.33	due to smaller K_d

process in Examples 3.7 and 8.5, are summarized below, and the modified data are summarized in Table 24.4.

$$F = 0.085\ \text{m}^3/\text{min}; \quad V = 2.1\ \text{m}^3; \quad T_s = 85.4°\text{C}; \quad \rho = 10^6\ \text{g/m}^3;$$

$$C_p = 1\ \text{cal/(g°C)}; \quad T_0 = 150°\text{C}$$

$$F_{cs} = 0.50\ \text{m}^3/\text{min}; \quad C_{pc} = 1\ \text{cal/(g°C)}; \quad \rho_c = 10^6\ \text{g/m}^3$$

The following linearized model can be derived for a disturbance in the inlet temperature using the model in Example 8.5.

$$\frac{T(s)}{T_0(s)} = \frac{G_d(s)}{1 + G_c(s)G_p(s)} = \frac{\left(\dfrac{K_d}{\tau_d s + 1}\right)}{1 + \left(\dfrac{K_pK_c}{\tau s + 1}\right)\left(1 + \dfrac{1}{T_I s}\right)} \tag{24.18}$$

$$\tau = \tau_d = \left(\frac{F}{V} + \frac{UA^*}{V\rho C_p}\right)_s^{-1}$$

$$K_p = \frac{\tau}{V\rho C_p}\left(\frac{-abF_c{}^b\left(F_c + \dfrac{a}{b}\dfrac{F_c{}^b}{2\rho_c C_{pc}}\right)(T - T_{cin})}{\left(F_c + \dfrac{aF_c{}^b}{2\rho_c C_{pc}}\right)^2}\right)_s$$

$$K_d = \left(\frac{F\rho C_p}{F\rho C_p + UA}\right)_s$$

$$UA^* = \left(aF_c^{b+1}/(F_c + aF_c^b/2\rho_c C_{pc})\right)_s$$

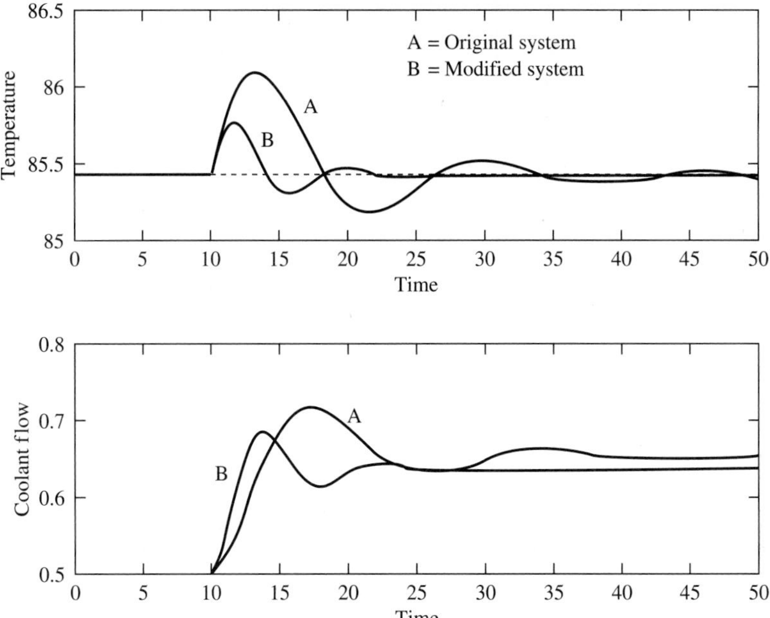

FIGURE 24.13
Transient responses for Example 24.5.

The data in Table 24.4 demonstrates that the approximate linear dynamic model has two significant improvements for the modified process. First, the feedback time constant is smaller, allowing better feedback performance. Second, the disturbance gain is smaller, meaning that the same feed inlet temperature disturbance has a smaller effect on the process without control because of the stronger self-regulation. The faster feedback dynamics and smaller disturbance gain would indicate that the feedback control performance should be better for the modified process. This analysis is confirmed by the results in Figure 24.13, which shows the temperature responses for the original and modified processes and by the results summarized in Table 24.4 on control performance. The tuning was similar for both systems, adjusted to have the same values for the key dimensionless parameters $K_c K_p$ and T_I/τ so that the manipulated-variable behavior is reasonable (and similar) for both transients. These responses were determined by numerically integrating the nonlinear differential equations for the process and controller.

The substantially improved performance for the inlet temperature disturbance has been accomplished with minor modification to the process. However, it is not without some negative impact. First, the heat exchanger area and cost have been increased. Second, the sensitivity of the process performance to disturbances in the coolant inlet temperature has increased. Thus, the best overall design and dynamic behavior must be tailored to each specific situation. This example demonstrates that strong self-regulation for key disturbances can improve control performance.

Example 24.6 Degrees of freedom. To perform the quantitative aspects of the design analysis in this chapter, a model of the flash process in Figure 24.1 is required. The goal of the model is to represent the dynamic input-output behavior of the system with

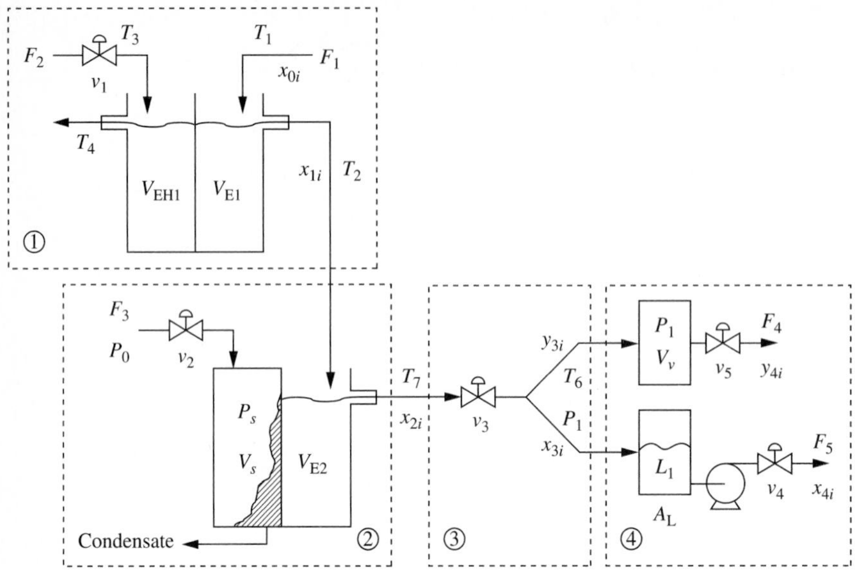

FIGURE 24.14
Approximate system used for modelling the flash process.

accuracy adequate to make the design decisions correctly within the mathematical methods consistent with this book. Therefore, the model presented here is simplified to involve algebraic and ordinary differential equations (not partial differential equations) and approximate physical property data. The model is reported in Appendix H, and the analysis of the model for control system design is presented in this example.

The physical system in this example, shown schematically in Figure 24.14, is the same as the flash process considered in Section 17.2 (page 557) on inferential control, except for the additional feed heat exchanger. Also, the changes in sensors and final elements proposed in previous examples have been included.

Assumptions. 1. All volumes are well mixed. 2. Densities, heat capacities, and heat transfer coefficients are constant. 3. Heat losses are negligible. These assumptions are common to all sections of the models.

Analysis. The analysis begins with a summary of the degrees-of-freedom analysis of the mathematical model, which is summarized in Table 24.5. The table presents the analysis of each section separately; however, the condition of zero degrees of freedom is required only for the complete process, not for any subsection. With all sections

TABLE 24.5
Degrees of freedom for the flash process

Section	1	2	3	4	Total
Number of equations	11	13	24	18	65
Number of dependent variables	12	13	24	17	65
Number of external manipulated variables	1	1	1	2	5

considered, the degrees of freedom for the entire system can be determined by summing the variables and equations to give DOF = 65 − 65 = 0; thus, the system is exactly specified. Also, the total number of manipulated external variables is 5; thus, no more than five dependent variables can be controlled.

Example 24.7 Controllability. Next, the controllability of the flash system is evaluated for some possible combinations of manipulated and controlled variables. Since five manipulated variables exist, the possibility of controlling five variables is investigated. Controlled variables are selected so that the control system achieves the specified objectives. Typical variables are the process feed flow (F_1) and the liquid product quality (A_1 measures the mole% ethane in the liquid product). The pressure of the flash drum (P_1) should be controlled for safety and product quality, and the liquid level (L_1) should be controlled for smooth operation and to prevent an overflow into the vapor line. Recall that the controllability of the rate of change of level, $sL_1(s)$, is determined, because the level process is an integrating process. Since this system is the same as the flash example in Chapter 17, which demonstrated that the flash temperature is a good indication of the liquid composition, the temperature (T_6) is provisionally selected as a fifth controlled variable.

The linear gains needed for the controllability check (see equation (20.11)) could be determined analytically for simple models. In this example they were determined numerically by introducing small changes in each manipulated variable and determining the steady-state value of the variables F_1, P_1, A_1, and T_6 and the rate of change of the level, L_1. The resulting equations are as follows:

$$\begin{bmatrix} F_1 \\ T_6 \\ A_1 \\ P \\ \dfrac{dL}{dt} \end{bmatrix} = \mathbf{K} \begin{bmatrix} v_1 \\ v_2 \\ v_3 \\ v_4 \\ v_5 \end{bmatrix} \qquad \text{with } \mathbf{K} = \begin{bmatrix} 0 & 0 & -2.00 & 0 & 0 \\ 0.0708 & 0.85 & -0.44 & 0 & -0.19 \\ -0.00917 & -0.11 & 0.132 & 0 & 0.043 \\ 0.567 & 6.80 & 1.39 & 0 & -5.86 \\ -0.0113 & -0.136 & 0.31 & -0.179 & -0.0265 \end{bmatrix} \tag{24.19}$$

$$\det \mathbf{K} = 6.7 \times 10^{-7} \approx 0.00 \tag{24.20}$$

The result indicates that this 5 × 5 system is *not* controllable. The reason becomes apparent when the values of the coefficients in the linearized models for v_1 and v_2 are compared. The first and second columns in the matrix in equation (24.19) are different by only a multiplicative constant, which indicates that these two manipulated variables have the same effect on all of the controlled variables. The lack of independence can be seen clearly in the block diagram of the effects of v_1 and v_2 on the controlled variables in Figure 24.15. Note that both manipulated variables affect the flash temperature, and it is only through the effect on flash temperature that they influence the other controlled variables. Therefore, it is not possible to achieve independent steady-state values for any two controlled variables in equation (21.19) by adjusting v_1 and v_2. As a result, it is concluded that it is not possible to control the five variables by adjusting the five manipulated variables in equation (24.19).

However, it is possible to control a different selection of five controlled variables in this process. For example, it is possible to control variables F_1, P_1, A_1, and T_2 and the rate of change of the level, sL_1, with the five valves v_1 through v_5. This can be seen in Figure 24.15 by the fact that T_2 is affected by v_1 but not by v_2, thus introducing an independent relationship.

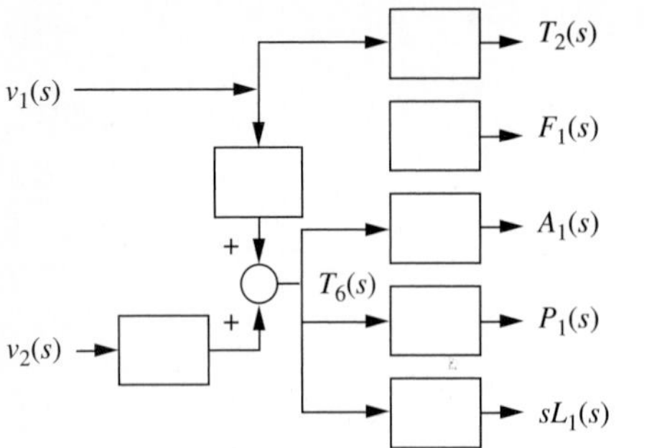

- Other manipulated variables and disturbances are not shown.
- v_1 and v_2 have no effect on F_1.

FIGURE 24.15
Block diagram of effects of v_1 and v_2.

Since T_2 is not related to the control objectives, the decision is to reduce the controlled variables to four and eliminate one manipulated variable. Since no control objective requires a specific behavior for T_6, it is eliminated; also, one of the two manipulated variables in Figure 24.15 must be eliminated: here, v_2 is retained and v_1 is eliminated. When this is done, the 4×4 system is controllable, as follows:

$$\begin{bmatrix} F_1 \\ A_1 \\ P_1 \\ \dfrac{dL}{dt} \end{bmatrix} = \mathbf{K} \begin{bmatrix} v_2 \\ v_3 \\ v_4 \\ v_5 \end{bmatrix} \qquad \text{with } \mathbf{K} = \begin{bmatrix} 0 & -2.0 & 0 & 0 \\ -0.11 & 0.132 & 0 & 0.043 \\ 6.8 & 1.39 & 0 & -5.86 \\ -0.136 & 0.31 & -0.179 & -0.265 \end{bmatrix} \tag{24.21}$$

$$\det \mathbf{K} = 0.126 \neq 0.00 \tag{24.22}$$

Example 24.8 Operating window. In addition to ensuring that the system is controllable, which is exact only in a small (differential) region about the steady state, the operating window should be analyzed to ensure that sufficient flexibility exists for expected changes in external disturbances and set point changes. A sample operating window is given in Figure 24.16 for the flash process with the product composition (A1) and pressure (P1) controlled at their set points and the design values for the other external variables, such as feed composition. In this example, the limits to the window are from

1. The minimum external feed temperature, $T_2 = -10$
2. The minimum feed flow, $F_1 = 60$
3. The maximum heating (v_2 fully opened)
4. The maximum flow of product (v_4 fully opened)
5. The minimum heating (v_1 fully closed)

In all of these cases, the frame of the window was selected so that all control valves are at least 5% from their limits of 0 through 100%; thus, all controlled variables can be regulated, at least for small disturbances, within the window and on the frame. Additional cases demonstrate that the process can satisfy the requirements specified in objective 5*a* in the CDF. If the linear model is valid over this range, the system is controllable for expected ranges of operating conditions. The large operating window involves the

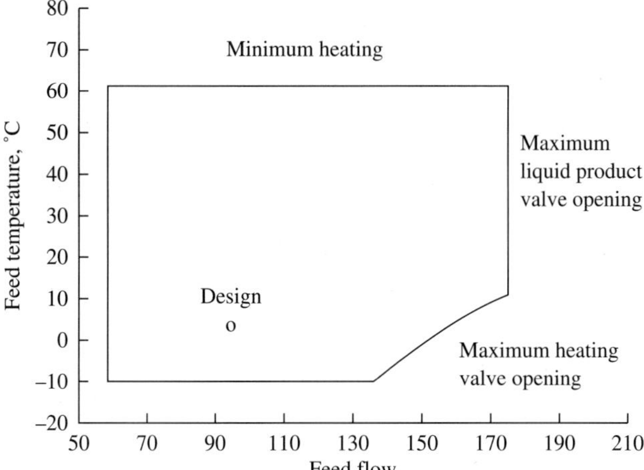

FIGURE 24.16
Operating window for flash process.

cost of purchasing larger equipment, and the capital costs must be balanced with the advantages of flexibility.

Further development of the control system, including the proper utilization of the inferential relationship and a strategy for adjusting the additional manipulated variable (v_2), is given in the next section on control structure.

In conclusion, the process design and operating conditions have important effects on control performance that should be carefully analyzed by the control engineer. First, the possibility of control is determined by evaluating the degrees of freedom, controllability, and operating window; if the results indicate that the control objectives cannot be achieved, equipment sizing and process structures would have to be modified. Second, those processes that satisfy the preliminary criteria are evaluated for control performance, which depends on the feedback and disturbance dynamic behavior. Quite simply, feedback dynamics should be fast, and disturbance dynamics should have a small gain and long time constants.

24.6 CONTROL STRUCTURE

The control system should be designed to give the best performance possible for the process. The comments here refer to multiloop control technology.

Controlled–Manipulated Variable Pairing

The variable pairing should yield a loop with a significant process gain. If the process gain is too small, the controller will not be able to return the controlled variable to its set point when disturbances occur. If the process gain is too large, the controller will be required to adjust the manipulated variable with great accuracy; since such accuracy is usually not possible (for example, because of valve sticking and hysteresis), oscillations will occur. The process gain can be expressed in dimensionless form (scaled), $(K_p)_s$, by relating the variables to their ranges (see Chapter 12).

$$(K_p)_s = K_p \frac{\text{Range of CV}}{\text{Range of MV}} \tag{24.23}$$

The typical range of values for this dimensionless process gain is 0.25 to 4.0. Values outside this range are possible but should be evaluated carefully so that satisfactory manipulated-variable capacity and sensor reproducibility are provided. Note that this evaluation requires an estimate of the expected disturbances. The control systems in Figures 24.3 to 24.5 demonstrate approaches to loop pairing with extreme demands on valve range.

The loop pairing should be selected with regard to the effects of interaction in multiloop systems. Analysis methods for multiloop systems were presented in Chapters 20 and 21 (which the reader should review at this point) and are briefly summarized as follows:

1. Automatic control should be provided for all non–self-regulatory or open-loop unstable variables, because if they are not controlled, they will drift out of the acceptable operating region. Manual regulation of such variables is difficult and time-consuming for plant personnel; reliable process operation requires automation.
2. Normally, variables are not paired when their relative gains are negative or zero. This will make the tuning process easier and will result in better performance when some control loops are not functioning (i.e., are in manual or have manipulated variables at their upper or lower limits).
3. The dynamics of the feedback loop pairings should be fast, with small dead times and little inverse response. The most important controlled variables should be paired to give fast feedback loops, even though this might somewhat degrade the performance of some variables of less importance.
4. The pairings should be selected to reduce unfavorable interaction and increase favorable interaction. The relative disturbance gain (RDG) can be used as an indication of how a pairing might affect the control system performance.

Finally, when the system has an unequal number of controlled and manipulated variables, the control structure should be able to alter the pairings to ensure that the objectives are attained. Methods for decentralized multiloop control are split range, signal select, and valve position controllers, which were presented in Chapter 22; a method for centralized multivariable control is Dynamic Matrix Control, presented in Chapter 23.

Disturbances

The effects of disturbances should be reduced through good control design. Two very effective designs that reduce the effects of disturbances are cascade and feedforward control, covered in Chapters 14 and 15.

Example 24.9. In this example, the control structure in the flash process is considered. First, the inventories should be controlled, and the natural pairings are the drum pressure with the vapor exit valve (v_5) and the drum liquid with the liquid exit valve (v_4). Also, the feed flow rate should be controlled, and valve v_3 should give fast control.

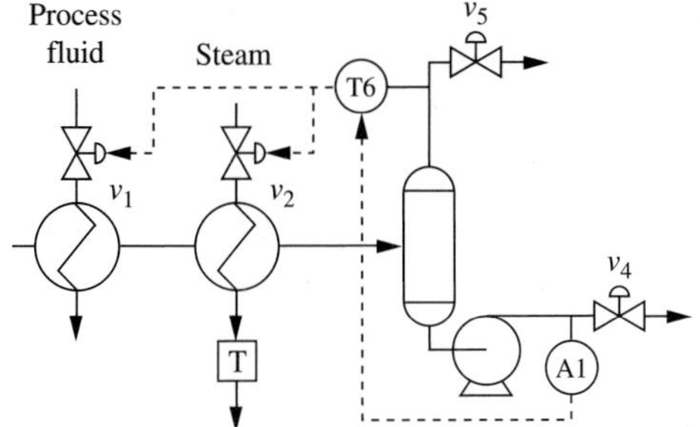

FIGURE 24.17
Control design to speed feedback disturbance response and optimize the use of heating sources.

Second, it is noticed that the analyzer is relatively slow, providing an update only every two minutes, whereas changes in the flow rate and feed composition disturbances can occur faster. Therefore, the continuously measured flash temperature T_6 is selected as an inferential variable for product composition. The selection process for temperatures applied to flash processes is thoroughly explained in Chapter 17, with calculations for this system, and is not repeated here. Since very tight control of the liquid product composition is required, the analyzer is retained to reset the temperature controller set point in a cascade design. This cascade observes the design rules introduced in Chapter 14: the secondary variable is measured, indicates important disturbances, depends in a causal manner on the manipulated variable (v_1) and has faster dynamics than the primary, because of the slow primary measurement. Recall that adding this controller does not change the degrees of freedom, because one external variable (the T_6 set point) becomes a dependent variable, one equation (the controller) is added, and one external variable (the analyzer set point) is added. Also, since the process equipment is unchanged, the operating window is not affected.

Third, the control objectives state that the process fluid flow to the first heat exchanger should be maximized before the steam is used to heat the feed. This is a system with one controlled variable and two manipulated variables with a fixed priority of adjustment. Therefore, a split range control design can be used. The resulting design for the product quality control is shown in Figure 24.17. Again, the split range controller does not violate degrees-of-freedom requirements, because, as discussed in Chapter 22, only one valve is adjusted at a time. The controllability of the system is ensured when either v_1 or v_2 is manipulated, as indicated in Figure 24.15 and as can be verified by evaluating the appropriate gain matrix.

The loop pairing can also be analyzed using methods introduced in Chapters 20 and 21. For example, the relative gain array (RGA) can be applied to ensure that the design does not violate guidelines such as not pairing on negative RGA elements. Following the suggestion of McAvoy (1983), the relative gain is calculated using the self-regulating variables and the rate of change of the integrating level. Thus, the steady-state gains for this 4×4 control system are those in equation (24.21). The relative gain array is

$$\text{RGA} = \begin{array}{c|cccc} & v_2 & v_3 & v_4 & v_5 \\ \hline F_1 & 0 & 1 & 0 & 0 \\ A_1 & 1.83 & 0 & 0 & -0.83 \\ P_1 & -0.83 & 0 & 0 & 1.83 \\ dL_1/dt & 0 & 0 & 1 & 0 \end{array} \qquad (24.24)$$

Based on selecting pairings with positive relative gains, the analysis recommends the pairings F1 → v_3, A1 → v_2 (which we have selected via T_6 as a cascade), P1 → v_5, and L1 → v_4. The analysis confirms the "common sense" selections based on semi-quantitative reasoning.

In conclusion, design of the proper control structure requires considerable knowledge of process dynamics, dominant disturbances, and equipment capacities. The control structure is tailored to satisfy the performance objectives for the process using the appropriate methods in Parts III through V.

24.7 CONTROL ALGORITHMS

After the control structure has been selected, the algorithms and tuning can be selected to give the best performance for that structure.

Feedback and Feedforward

Feedback control should be used extensively, because it corrects for all disturbances, even unmeasured disturbances, that influence the measured controlled variable. All of the feedback single-loop enhancements, such as cascade and gain scheduling, should be considered to improve the control performance of a feedback system. Feedforward control should be considered as an enhancement to feedback control when the feedback process is difficult to control because of long dead time and unfavorable interaction.

Algorithm

The control algorithm should be matched to the application. In particular, most feedback systems desire zero steady-state offset; therefore, this requirement should be satisfied by including the integral mode in the PID controller or by appropriate considerations in a model predictive controller. Based on its generally good performance and widespread acceptance, the PID controller should be used for most multiloop feedback control systems. Only when another algorithm provides demonstrably better performance should it be chosen over the PID. There are some cases, such as loops with inverse response or very long dead times (and large $\theta/(\theta + \tau)$), where a predictive controller might give better performance.

The feedback controller should be selected to be relatively insensitive to modelling errors, and the associated tuning errors, for the expected range of errors. Most single-loop feedback control algorithms satisfy this requirement. However, sensitivity analysis showed that some multivariable control designs (e.g., decoupling and centralized DMC control) are sensitive to certain model errors when the process has strong interactions (i.e., large elements in the relative gain array).

Tuning

Tuning parameters for all algorithms should be based on a careful analysis of the desired performance of all process variables. Typically, empirical methods are used for determining models for tuning. However, fundamental models are very useful for (1) verifying empirical results, (2) determining how model parameters depend on process operation (e.g., throughput), and (3) providing models for complex, nonlinear processes.

It is important to remember that the manipulated variable in a control system (e.g., steam flow) is another plant variable. The engineers involved with plant design and operations are responsible for ensuring the availability of appropriate utility systems that can be varied to control the process. However, extreme variation in the manipulated variable can cause disturbances in other units in the plant. The typical relationship in feedback systems was covered in Chapter 13, where it was shown that in the region of good tuning, the variability of the manipulated variable increases as the variability of the controlled variable decreases. In most cases, tuning can be selected to reduce the variability in the manipulated variable significantly, with only a small increase in the controlled-variable variability. For this reason, as well as for robustness, the controller is normally tuned to eliminate extreme variability in the manipulated variable.

Often, the tuning parameters do not have to be modified in response to moderate changes in process operation, because the dynamic responses do not change significantly over the range of operation. Recall that 10 to 20% errors in parameters are common. However, if the changes in process operating conditions are large or the process is highly nonlinear, the controller tuning should be adjusted in real time to maintain stability and acceptable performance. Approaches for adapting the tuning, having the goal of maintaining the same stability margin (and relative control performance), were explained in Chapter 16.

Finally, the tuning of multiloop controllers must be performed with consideration of the interaction among loops. This issue, along with tuning guidelines, was discussed in Chapters 20 and 21, where it was shown that the relative gain gives some indication of the extent that tuning must be adjusted to account for interaction. Also, the relative importance of the controlled variables is considered when tuning the controllers, with the tuning selected to reduce the deviation of the most important variables from their set points.

Example 24.10. In this example, a few issues related to tuning the controllers for the flash process are discussed. First, the order of the tuning is important. The level controller tuning can be determined without experimental modelling using the vessel size, and because the level is non–self-regulating, it should be tuned first. No specification is placed on the variability of the liquid leaving the drum, and a proportional-only controller with tight level tuning is selected because of the importance of not having liquid carry over. Also, the pressure in the drum could easily exceed its limits and should be tuned next using the standard methods. Then, the split range controller, T_6, will adjust v_1 and v_2. The dynamics between the valves and the temperature sensor can be expected to be different, and the gain matrix in equation (24.19) shows that the steady-state gains are different by a factor of about 12. Thus, the tuning of the T_6 PI controller should be adapted based on the condition of the split range. Finally, the analyzer measurement is updated only every two minutes; this long execution period for the feedback controller will require some detuning using the guidelines from Chapter 11.

24.8 CONTROL FOR SAFETY

Before completing the discussion on design decisions, safety must be discussed. Safety is addressed in the first control objective, and some control decisions, such as controlling the pressure in the flash process, have been made to satisfy safety requirements. However, special control system features are required, because of the importance of this objective. These features are often implemented in multiple layers, with every layer contributing to the safety of the system by taking actions only as aggressive as required for the particular situation (AIChE, 1993).

Basic Process Control System (BPCS)

The first layer involves the basic process control approaches discussed in prior sections, which employ standard sensors, final elements, and feedback control algorithms. This first layer maintains the process variables in a safe operating region through smooth adjustment of manipulated variables; this action does not interfere with, but rather usually enhances, the profitable production of high-quality material. However, the basic control system relies on sensors, signal transmission, computing, and final elements, which occasionally fail to function properly. In addition, the process equipment, such as pumps, can fail. Even if all elements are functioning properly, the control system may not maintain the system in the safe region in response to all disturbances; for example, a very large disturbance could cause a deviation of key variables into an unacceptable region.

The basic process control layer can employ standard techniques to improve its response to a fault. For example, the use of several sensors with a signal select reduces the effect of a sensor failure. An example is the temperature control system in Figure 22.10 (page 733), which can reduce the likelihood of a temperature excursion due to the failure of a single temperature sensor. Also, the use of split range control allows a controller to manipulate an additional (larger-capacity) valve in response to an unusual circumstance. An example of this approach is given in Figure 24.3. However, these techniques do not reduce the likelihood of injury or damage to an acceptably low probability; therefore, additional layers are implemented to increase safety.

Alarms

The second layer involves alarms, which are automatically initiated when variables exceed their specified limits. These alarms involve no automatic action in the process; their sole purpose is to draw the attention of the process operator to a specific variable and process unit. The person must review the available data and implement any actions required. A great advantage of involving operators is their ability to gather data not available to the computer. For example, an operator can determine the values of instruments that display values locally and can check the reliability of some sensors as part of the diagnosis. The operator usually takes action through the process control system; these actions could include placing a controller in manual status and adjusting the manipulated variable to a new value. Since the final element may not be functioning, the operator has the option of going directly to the process and adjusting the valve manually (or having this task performed by another person).

It is good practice for the alarm to be based on an independent sensor, because using the same sensor for alarm and control prevents the alarm from identifying the failure of the sensor to indicate the true value of the process variable. The alarms are often shown as a switch (e.g., PS for pressure switch) on a process drawing. The alarm is usually annunciated by activating a visual indicator (e.g., a blinking light) and an audio signal, beeping horn. These signals continue until the operator acknowledges the alarm; thereafter, the visual indicator remains active (e.g., a nonblinking light) until the variable returns within its acceptable limits. The blinking light indicates the variable involved, its current value, its alarm priority, and whether the variable has exceeded its high or low limit. Alarms can be arranged into three levels, depending on the severity of the potential consequences of the process fault or upset:

LEVEL 1 (HIGH). These alarms are designed to indicate conditions requiring prompt operator action to prevent hazards or equipment damage. Special color and visual displays and a distinct audio tone should be used to alert the operator. Examples of level 1 alarms are high pressure in a reactor; low water level in a boiler; and activation of a safety interlock system that has stopped operation of some processes (see next topic).

LEVEL 2 (MEDIUM). These alarms are designed to indicate conditions requiring close monitoring and operator action to prevent loss of production or other costly (but nonhazardous) situations. The operator typically has some time to analyze the alarm, along with other measurements, and make corrections that can maintain the process in an acceptable region of operation. These alarms should be annunciated in the same general manner as the level 1 alarms, although with distinct colors and tones.

LEVEL 3 (LOW). These alarms identify conditions that are not critical to the operation of the process and require no immediate action by the operator. These can be entered directly into a database for occasional review by the operators and engineers. These alarms should not be annunciated.

Some care must be taken in designing alarms. The major issue is the overuse of alarms. Kragt and Bonten (1983) report that an operator in an industrial processing plant experienced an average of 17 alarms per hour and that the operator took an action after only 8% of these alarms! Most of the alarms were not necessary and needlessly distracted the person. Such poorly designed alarm systems lead to lack of attention by the plant personnel to the occasional, but critical, *important* alarm.

Safety Interlock System (SIS)

The third layer involves automatic feedback control for situations when process variables approach "hard" constraints that should not be exceeded; these could cause injury to people or the environment or damage to expensive equipment. Because of the importance of preventing such situations, the actions taken are extreme and disrupt the process operation; usually, they stop all or part of the process operation by immediately closing (or opening) key valves to move the process to a safe condition. These control systems are termed *safety interlock systems* (SIS) or *emergency shutdown systems* (ESS).

As with alarms, this control layer should use a sensor independent of the basic control system; in addition, this automated system should use a final element independent of the basic control system. The equipment selected for this purpose must be of the highest reliability possible. Depending on the severity of the consequences, this layer may use several sensors and final elements. In some applications, three sensors are used, and the feedback control system bases its decision on the majority of the three; this approach prevents an occasional (individual) sensor failure from stopping process operation, while identifying an actual dangerous condition with high reliability. The control action taken is straightforward and simple to implement. Typically, a solenoid valve, which is normally closed to hold the air pressure to the pneumatic valve at a high value, opens and vents this pressure to atmospheric upon receiving a failure signal, allowing the pneumatic valve to attain its failure position. If this action is taken on a valve that is also used for basic process control, the solenoid valve is placed between the controller output and the valve; thus, under normal circumstances, the controller adjusts the control valve without alteration, whereas a failure signal disconnects the controller output from the valve, which goes to its failure position.

The valve selected for use in an SIS should have a capacity large enough to handle the largest expected flow. For example, a valve to vent a distillation tower may be based on a situation in which the condenser fails. Also, the manipulated variable should have a very fast effect on the key process variable and be able to maintain the process in the safe region; thus, dead times and time constants should be small. The limiting value for the initiation of the SIS is selected to be in the safe region and far enough from the undesired value that the largest expected disturbance will not cause an unsafe condition.

Safety Valves

The fourth layer involves feedback systems that are self-actuating; that is, which do not require electrical, pneumatic, or hydraulic power sources and have no significant distance of signal transmission. These features contribute to very high reliability. The major application at this layer is the safety valve, which is a valve normally held closed by a spring. When the pressure reaches the preset limit, the force due to the process pressure is high enough to overcome the force of the spring, and the valve begins to open. When the process pressure decreases, the safety valve is designed to close. The engineer must be sure that the material flowing through the safety valve can be either (1) released to the environment safely (e.g., steam), (2) processed to eliminate hazards (e.g., combusting hydrocarbons), or (3) retained in a containment vessel for later processing (e.g., wastewater storage and nuclear plant containment building).

These layers should be carefully designed, properly installed, and meticulously maintained. Through good or poor practices, the high level of safety may be enhanced or compromised. A few of these good practices are given in the following list:

Good Practices in Control for Safety

1. Never bypass the calculation (logic) for the SIS; that is, never turn it off.
2. Never mechanically block a control valve so that it cannot close.
3. Never open manual bypass valves around control and shutdown valves.
4. Never "fix" the alarm acknowledgment button so that new alarms will not require the action of an operator.
5. Avoid using the same sensor for control, alarm, and SIS.
6. Avoid combining high- and low-value alarms into one indication.
7. Evaluate the selection of alarms critically. Do not have too many alarms.
8. Use independent equipment for each layer, including computing equipment.
9. Select emergency manipulated variables with a fast effect on the key process variable.
10. Use redundant equipment for critical functions.
11. Provide capability for maintenance testing, because the systems are normally in standby.

Containment

The final layer involves containment, such as dikes, for major incidents. This layer may not prevent major hazards, but it can prevent their propagation to other sections of a plant and to the surrounding community. Other design issues, such as reliable electrical power supply, are also important for safety control; these are covered in the references.

Example 24.11. In this example, safety controls for the flash process are considered, and the results are shown in Figure 24.18.

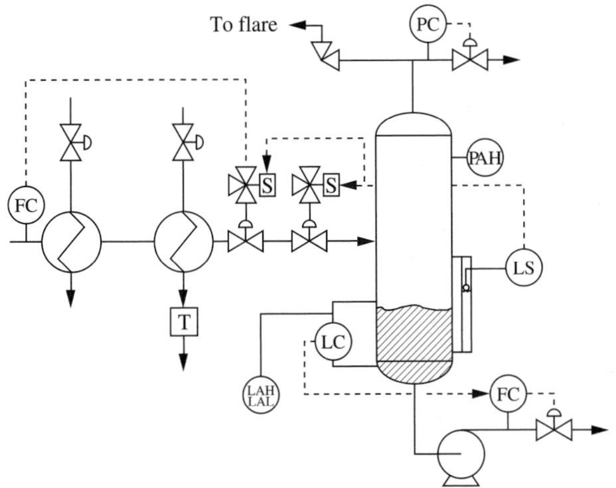

FIGURE 24.18
Safety-related controls for the example flash process.

There are several issues at the basic process control system layer. First, the pressure in the closed vessel should be controlled, and the valve in the overhead vapor line is a natural choice for the manipulated variable, because it has a very rapid effect on pressure. Second, the liquid level should be maintained within reasonable limits, and the valve in the bottom exit is a natural choice for manipulation. To prevent the liquid flow through the pump from falling below the minimum, the level controller could reset the flow controller set point, with the set point bounded to always be above the limit. Third, the use of a temperature cascade improves the reliability of the product quality control, because the analyzer would be much more likely to fail than the temperature sensor. Finally, the failure positions of the valves are selected to reduce the likelihood of high pressure, high temperature, and an overflow of liquid in the vapor line.

The alarm layer could conceivably include high and low alarms on every variable, but this would lead to excessive interruptions for the operator. Here, alarms will be placed on high pressure and high and low level. The analyzer measurement would normally not be alarmed unless composition variation led to unsafe conditions.

An SIS system would not normally be employed in this process. However, as an example, we will assume that the objective of preventing a liquid overflow in the vapor line from the drum is critical (see Kletz (1980) for an industrial example). A different type of sensor is used for the SIS; this sensor provides redundancy and diversity at the SIS level. This level sensor is used to determine when the level approaches the limiting value and to activate the emergency action to reduce the feed flow to zero. Both the control valve and an independent valve are used to enforce this SIS. When the safety interlock system activates, the process will experience a major disturbance, and the product will not observe the quality specifications.

The drum can be closed by the (improper) operation of the control valves; thus, a safety valve should be included, as shown in the figure. The combustible material must be contained or processed; typically, it would be diverted to a plant fuel system or combusted in a flare.

The multiple-layer approach described in this section provides excellent protection for most chemical processes. The reader must be aware of the importance of excellent detailed design and construction of equipment for safety control. This section simply presents some introductory concepts and is not meant to teach the practice of safety in design. The novice should refer to the many industrial standards and engage experienced consultants when designing safety control systems.

24.9 PERFORMANCE MONITORING

Monitoring should be considered at the design stage to ensure that the important performance measures are identified and that the sensors with required accuracy are provided. The most important purpose of short-term monitoring is to enable the plant operator to diagnose incipient problems, preferably before the problems worsen and cause major upsets. The purpose of longer-term monitoring is not only to record the performance but also to diagnose the reasons for good and poor performance. The results of this diagnosis can be used by the engineer as a basis for improving product quality, equipment performance, and profit through changes in operating conditions, control designs, and process equipment.

Real-Time Monitoring for Process Operation

The plant operators are part of the overall "control system"—that is, they are responsible for many feedback control tasks that are not automated, such as switching from one feed tank to another. Also, they are responsible for supervising the process equipment and automatic control system. To perform these tasks, operators require a thorough understanding of the process, along with rapid access to many measured values. The system designer must recognize that because the diagnosis of the control system, including sensors, is an important task, the operators need parallel information on key variables provided by independent sensors. The alarm feature of control systems, discussed in the previous section, can help the operator monitor hundreds of variables by drawing attention to variables that are outside of their normal operating ranges.

Variability of Key Process Variables

Individual measured variables can be analyzed as part of a longer-term monitoring program. The average values of most important variables provide a quick indication of the process performance, and when the average is not close to the desired value, improvement is clearly in order. However, good performance is not ensured when the average conforms to the desired value, as demonstrated in Chapter 2 (pages 32 to 38). The variability is important in determining the plant performance, because average process performance depends on the length of time each variable spends at values in the distribution. This concept is shown in Figure 24.19. The average performance can be calculated from the empirically measured distribution without making assumptions concerning the normality of the distribution, and a broad

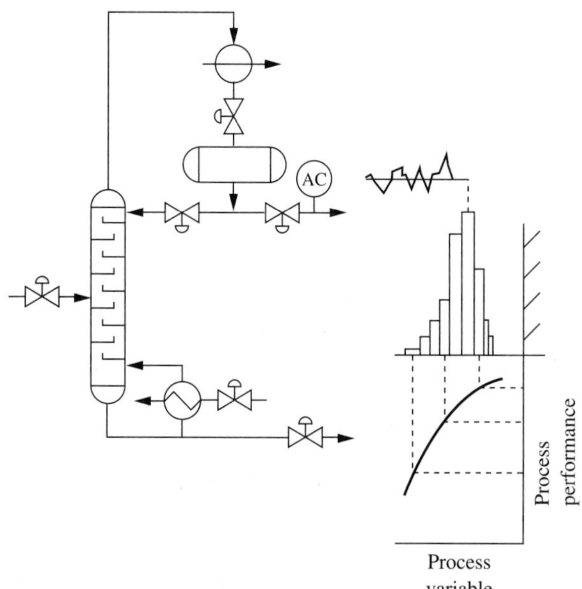

FIGURE 24.19
Schematic of the procedure relating a key process variable to performance.

distribution indicates considerable operating time far from the best conditions, even if the average conditions seem acceptable.

The total number of incidents also gives valuable insight into performance. One type of incident is the activation of alarms, with each important alarm monitored separately. Care should be taken in monitoring alarms, because one process disturbance can cause numerous alarms before the plant operation is returned to normal conditions. Other incidents include the number of times important constraints are violated, such as products outside specified quality limits, and the activation of safety interlock systems. Each major incident provides valuable information on the performance of the process and controls, which can be used in designing improvements.

Calculated Process Performance

In many cases, the performance of important process units can be estimated from measured variables. Some of these variables indicate the overall performance of the plant (e.g., energy consumption per kilogram of product sold). These are useful in indicating the overall performance but not usually complete enough to direct diagnosing and improving performance. However, monitoring the performance of individual units provides very useful information. For example, the efficiency of the boiler in Example 2.2 (page 36) gives insight into the performance of the excess-air control system, as well as other factors such as the heat transfer coefficient. Commonly monitored calculated variables are compressor and turbine efficiencies, heat transfer coefficients, fired-heater efficiencies, and the selectivity of chemical reactions to desired versus undesired products.

Utilization and Performance of Control

The fraction of time that each control system is in operation in automatic should be monitored. Although this information cannot be used to diagnose control performance, a low service factor (time used/time should be used) is a clear indication of unsatisfactory performance, at least in the opinion of the process operators. More information can be determined from dynamic plant data on the performance of the control system. Methods are available for estimating (1) the best possible feedback control, (2) the improvement possible with feedforward control, and (3) likely deficiencies in the existing control system (e.g., feedback controller tuning or feedforward disturbance model). These methods rely on mathematical analysis that is beyond the level of this book, but they require only simple interpretation of graphical results by plant personnel after they are implemented (see the Additional Resources in Chapter 9).

Example 24.12. In this example, monitoring for the flash process in Figure 24.1 is discussed. First, the averages and standard deviations of important process variables should be calculated from real-time data. Typically, the most important variables would be the flow rates, the flash temperature, and the liquid composition. The loss of heavy material in the vapor could be monitored through infrequent samples analyzed in a laboratory. Second, an appropriate sensor should be selected if an accurate measurement of the vapor flow rate is needed, perhaps for a record of sales. If an orifice meter is used in a stream with changing pressure, composition, or both, the density and the pressure drop are required to measure the mass flow rate accurately. Finally, the process performance would depend on the heat transfer coefficient in the process fluid heat exchanger. This could be monitored using measured temperatures and flows. A low value of the heat transfer coefficient, based on process data that satisfied material and energy balances (and thus is considered accurate), along with high steam use would indicate poor performance. Performance could be improved by taking the heat exchanger out of service for a short time to clean the surface.

24.10 THE FLASH EXAMPLE REVISITED

The original design in Figure 24.1 has been discussed in the examples in this chapter, where the improvements summarized in Table 24.6 were proposed. The final design, which incorporates all improvements except the SIS on high level, is shown in Figure 24.20. This design satisfies the objectives in the Control Design Form and is typical for industrial systems.

Example 24.13. In this example, the dynamic responses for the flash process in Figure 24.20 with recommended control is evaluated for a step change in feed composition,

TABLE 24.6
Summary of design decisions for the original flash process

Design Decision	
Measurements	F_1 moved to one-phase flow region T_5 redundant measurement removed T_6 moved to vapor space for faster response
Final elements	v_2 changed to steam flow v_4 reduced maximum flow
Process	Analyzed degrees of freedom; five manipulated variables exist Analyzed controllability to determine that only four (meaningful) variables can be controlled Operating window large enough to satisfy objectives
Control structure	Cascade from A_1 to T_6 inferential variable Split range to adjust both heating valves
Algorithms	Standard PI control except for P-only level controller Adaptive tuning for T_6 when changing the split range
Safety	Basic regulatory control of inventories with minimum liquid flow Valve failure modes Alarms on pressure and level Safety interlock system for high level
Monitoring	Correct vapor flow for density Monitor heat exchanger UA Monitor product quality (A_1) variance

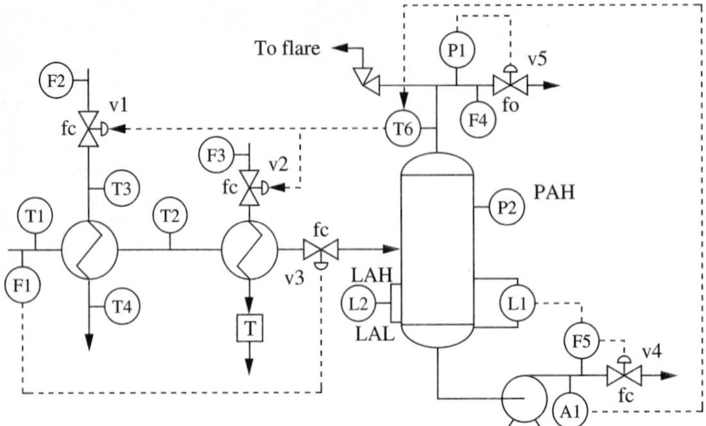

FIGURE 24.20
Modified design for the flash process incorporating improvements in Table 24.5.

with the ethane increasing 5 mole% and the propane decreasing by the same amount. The transient behavior is shown in Figure 24.21. The ethane in the liquid is maintained within the specified limits of 10 ± 1 mole% during the transient and returns to its set point; if the analyzer feedback were removed, the ethane concentration in the liquid product would exceed the acceptable limits during the transient and at the steady state. Other cases demonstrated that the design, including analyzer feedback, could maintain the ethane in the liquid within the limits for the other disturbances defined in the Control Design Form in Table 24.1. Thus, this process and control design is deemed acceptable.

24.11 CONCLUSION

Many issues must be considered in control design. Assuming that the principles in the previous parts of this book have been mastered, the challenges in design are to (1) recognize the issues important for control along with potential results and (2) develop a method for addressing the design task.

This chapter addressed the first challenge by presenting issues in the six categories of control design decisions: sensors, final elements, process design, control structure, control algorithms, and monitoring. Naturally, the issues discussed were not a complete listing of all possible items, but they included the most important issues in typical systems. The analysis of degrees of freedom and controllability were reviewed and their applications to control design demonstrated. Again, we see the importance of the process equipment design and operating conditions on process control, since these determine the operability of the system.

One means for defining and documenting the control problem (opportunity) is the Control Design Form. This helpful checklist partly addresses the second challenge by providing an organized manner for defining the design task. This is especially important, because the engineer often has the most difficulty in beginning the design task.

In addition, the engineer would benefit from a road map for the analysis and decision making during the design process. Is it best to start with the sensors, with the

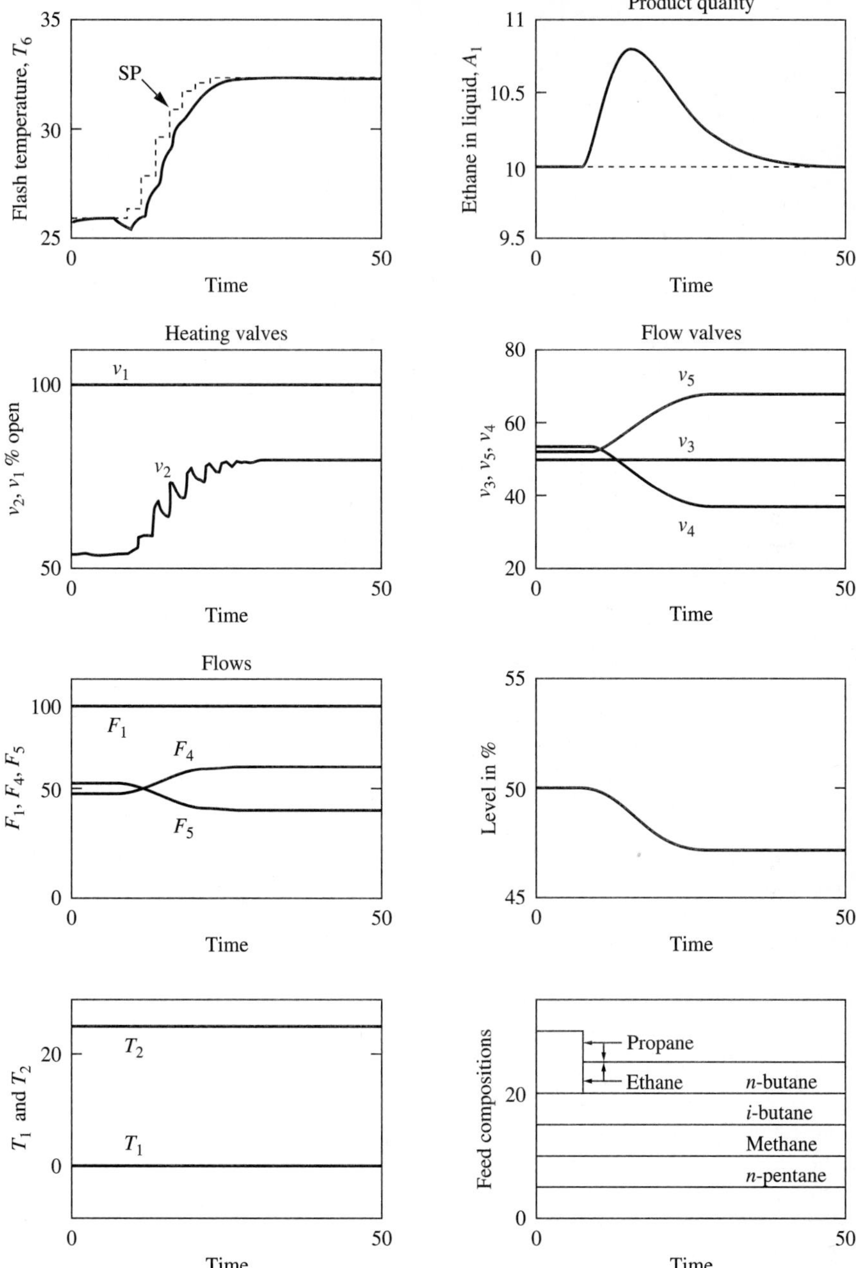

FIGURE 24.21
Transient response for final flash process and control design.

process, or with the algorithms? This important topic is covered in the next chapter, where a sequential design method, with checks for iterations, is presented along with some additional examples.

REFERENCES

AIChE, *Guidelines for Safe Automation of Chemical Processes,* American Institute of Chemical Engineers, New York, 1993.

Clevett, K., *Process Analyzer Technology,* Wiley-Interscience, New York, 1986.

Kletz, T., "Don't Let Control Loop Blunders Destroy Your Plant," *Instr. and Control Sys.,* 29–32 (1980).

Kragt, H., and J. Bonten, "Evaluation of a Conventional Process-Alarm System in a Fertilizer Plant," *IEEE Trans. Sys. Man. Cyber., SMC-13, 4,* 586–600 (1983).

McAvoy, T., *Interaction Analysis,* Instrument Society of America, Research Triangle Park, NC, 1983.

Rinhard, I., "Control Systems for Complete Plants—The Industrial View," in D. Seborg and T. Edgar, *Chemical Process Control II,* United Engineers Trustees, Sea Island, GA, 1982, p.541–546.

Roffel, J., J. MacGregor, and T. Hoffman, "The Design and Implementation of an Internal Model Controller for a Continuous Polybutadiene Polymerization Train," *Proceed. DYCORD-89, Maastricht, IFAC,* 9–15 (1989).

ADDITIONAL RESOURCES

Standards for programmable control equipment for use in safety applications are being completed by the Instrument Society of America.

Instrument Society of America, SP84—*Programmable Electronic Systems (PES) for Use in Safety Applications,* Research Triangle Park, NC, 1994.

An economic analysis of an SIS design is presented in

Roffel, B., and J. Rijnsdorp, *Process Dynamics, Control and Protection,* Ann Arbor Science, Ann Arbor, MI, 1982.

Quantitative reliability analysis of equipment and control strategy design is presented in the following.

Fisher, T. (ed.), "Control System Safety," *ISA Transactions, 30,* 1 (special edition), (1991).

Goble, W., *Evaluating Control System Reliability: Techniques and Applications,* Instrument Society of America, Research Triangle Park, NC, 1992.

Industrial experience and design recommendations for safety controls systems are given in the following.

Black, W., "Non-Self-Actuating Safety Systems for the Process Industries," *Hazards X. Process Safety in Fine and Specialty Chemical Plants,* Inst. Chem. Eng., London, 294–307 (1989).

Stewart, R., "High Integrity Protective Systems," *Inst. Chem. Eng. Symp. Ser. No. 34,* Inst. Chem. Eng., London, 99–104 (1971).

Wells, G., *Safety in Process Plant Design,* Wiley, New York, 1980.

The Proceedings from the series of meetings on "Hazards" organized by the Institute of Chemical Engineers, London, is a good source of up-to-date information.

Process and control designs are based on analysis of process behavior and control structure performance. This analysis may be analytical, numerical, or semi-quantitative, depending on the type of information available and the necessary accuracy of the results. The problems in this chapter give opportunities for all three.

QUESTIONS

24.1. (*a*) When pumping or compression costs are high, incentive exists for controlling flows at minimum cost. Suggest approaches for controlling flow rates with low pressure drops across sensors and valves.

(*b*) Discuss the response of systems when the sensor, rather than the valve, fails to function properly. How can safety be ensured in such situations?

(*c*) Discuss a quick method for determining the maximum number of variables that can be controlled for a completed process design. Assume that a detailed process schematic (piping and instrumentation drawing) is available, but a detailed mathematical model is not.

(*d*) Discuss why controllability is analyzed with a linear model whereas the operating window is determined based on a nonlinear model.

(*e*) If a system is controllable and has a sufficient operating window, will all possible loop pairings provide stable dynamic performance (assuming proper constant tuning)?

(*f*) If the process dynamics are overdamped, would variable values between two steady states within the operating window remain within the window during the transient response?

24.2. (*a*) Generalize the controllability test for non–self-regulating systems.

(*b*) The definition for controllability employed in this chapter is appropriate for many, but not all, processes. Discuss other definitions (e.g., for batch processing) and define appropriate tests.

(*c*) The test for controllability requires a "square" system (i.e., one with the same number of manipulated and controlled variables). What if the number of manipulated variables is greater than (or less than) the number of controlled variables?

24.3. Analysis of a heat exchanger is given in Example 24.5.

(*a*) Verify the relationships in the example.

(*b*) Compare the responses of the original and modified designs for a disturbance in the coolant inlet temperature.

(*c*) Compare the responses of the original and modified designs for a step set point change.

(*d*) Discuss the effect of process design on these results, along with the disturbances in inlet temperature, in determining the best process design.

24.4. A feedback control design is needed for the process in Example 3.12 (page 96). Critically evaluate the following proposals; in both, the reactor volumes are constant.

(*a*) The first proposal is to maximize the concentration of B by adjusting the temperature and control the conversion of A by adjusting the feed flow rate.

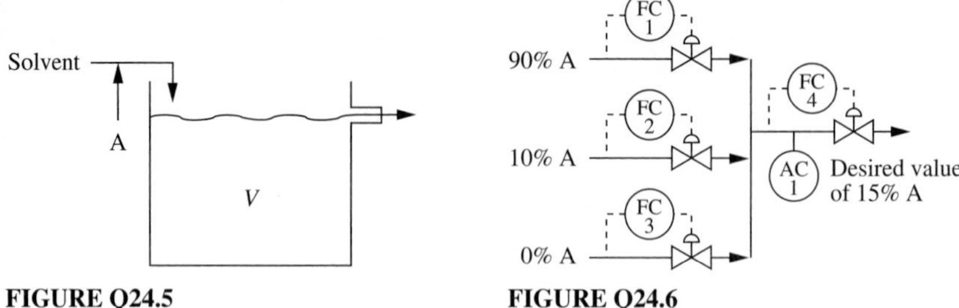

FIGURE Q24.5

FIGURE Q24.6

(*b*) The second proposal is to fix the flow rate and control the concentration of B at its maximum by manipulating the reactor temperature.

24.5. The chemical reactor in Figure Q24.5 has the following properties: well mixed, isothermal, constant volume, constant density. The chemical reaction occurring is A → B with the reaction rate $r_A = -kC_A$. The concentrations of the reactant and the product can be measured without delay.

(*a*) The total feed flow (F) and the feed concentration (C_{A0}) are the potential *manipulated* variables for the reactor effluent composition control. Construct a regulatory control scheme that will control these two variables (F, C_{A0}) simultaneously to independent set point values, and sketch it on the figure. You may place the sensors and final elements for these variables anywhere you think appropriate.

(*b*) Derive a dynamic model for $C_B(s)/C_{A0}(s)$. Analyze the model regarding (i) order, (ii) stability, (iii) periodicity, and (iv) step response characteristics.

(*c*) Derive a dynamic model for $C_B(s)/F(s)$. Analyze the model regarding (i) order, (ii) stability, (iii) periodicity, and (iv) step response characteristics.

(*d*) Based on the results in (*b*) and (*c*), which of these two manipulated variables would provide the best feedback control for C_B for a set point change using PI feedback control?

24.6. Part of a proposed control design for a blending process is given in Figure Q24.6. In addition, the composition of A is to be controlled by adjusting one or more flow set points. The objectives are to control the product flow tightly and the composition as tightly as possible; disturbances are in component compositions and changes to the product flow set point; no constraints are encountered. Critically evaluate the proposed control, make any changes to provide closed-loop flow, and design the composition feedback control.

24.7. Discuss the need for accuracy or reproducibility for the sensors in the control designs in the following Figures: 15.11, 15.13, 17.16, V.2, and 22.10.

24.8. The level process with control design in Figure Q24.8 is proposed to you. Evaluate whether the system can maintain the levels within their limits for changes in the flow from tank 2. Estimate the control performance and make changes, if required, to provide satisfactory performance.

24.9. The well-mixed, constant-volume chemical reactor with separator and recycle in Figure Q24.9 is considered in this question. The reactor has a single reaction A → B with $r_A = -kC_A$. The separator makes a perfect separation of the product and the pure reactant, which is recycled to the reactor feed, and the separator dynamics and transportation delays are very fast and will be assumed at quasi-steady state.

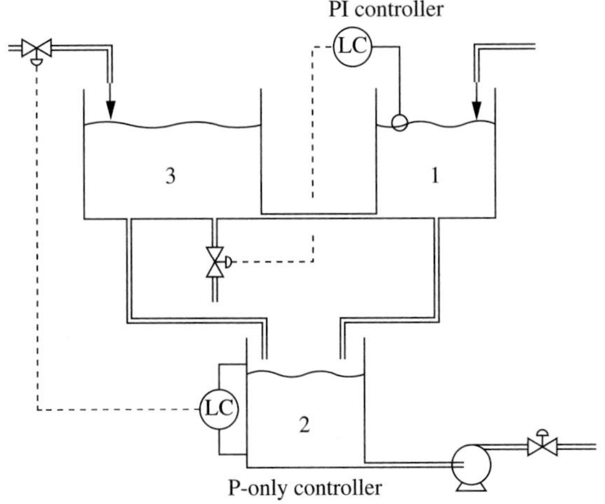

FIGURE Q24.8

(*a*) Assume that the fresh feed rate (F_1) is controlled constant. The major disturbance is temperature, which can be taken to be a change in the reaction rate constant. Based on a dynamic model of the process, determine an analytical relationship between the disturbance and (1) the recycle flow and (2) the reactor concentration. Determine how the dynamic behavior is affected by the steady-state conversion, $(C_{A0} - C_A)/C_{A0}$.

(*b*) Discuss the factors that would influence the choice of the best reactor conversion in a typical industrial process.

(*c*) Determine a simple change to the control design that substantially reduces the effect of the disturbance (without controlling temperature).

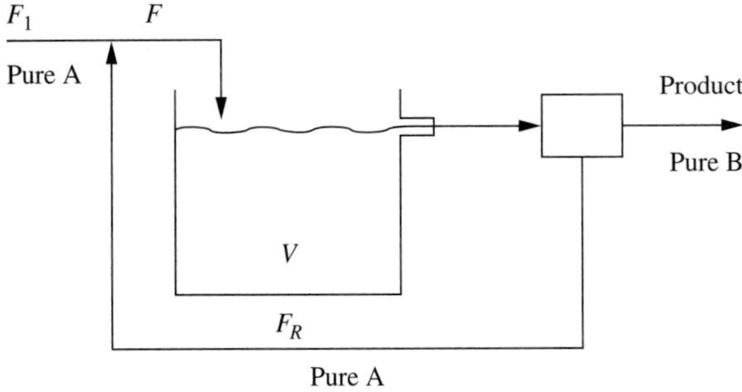

FIGURE Q24.9

24.10. If you have not completed questions 15.2 and 18.13, it would be worthwhile to do them now.

24.11. The design in Figure Q24.11 is proposed for an isothermal, well-mixed CSTR with a single reaction, A → B with $r_A = -kC_A$. The main disturbance is a steplike disturbance in the feed flow rate, and real-time measurement and control of the compositions is not possible. Evaluate the control performance (i.e., the deviation of the composition) for (*a*) perfect PI control of the level (the level exactly remains at its set point at all times) and (*b*) P-only control of the level. Which approach gives a smaller

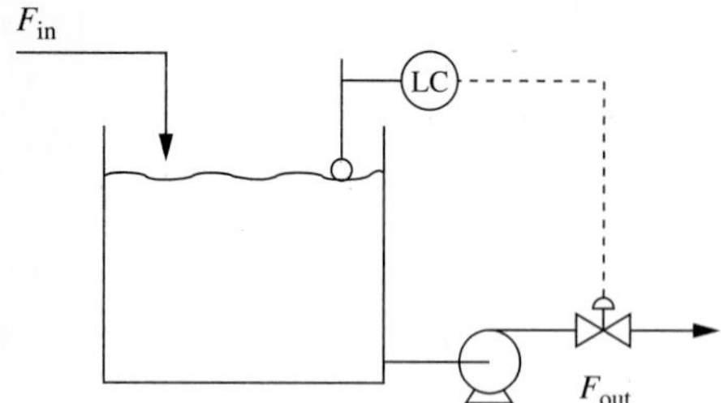

FIGURE Q24.11

deviation for the compositions from the initial conditions at the final steady state? Does your answer depend on the tuning of the P-only controller? If so, what is the best value of the controller gain?

24.12. Given the process schematic in Figure Q24.12 and the following data, determine the heat transfer coefficients for the three heat exchangers, and explain the assumptions you made in performing the calculations. If you performed this analysis over a long period of time, what useful information would you determine?

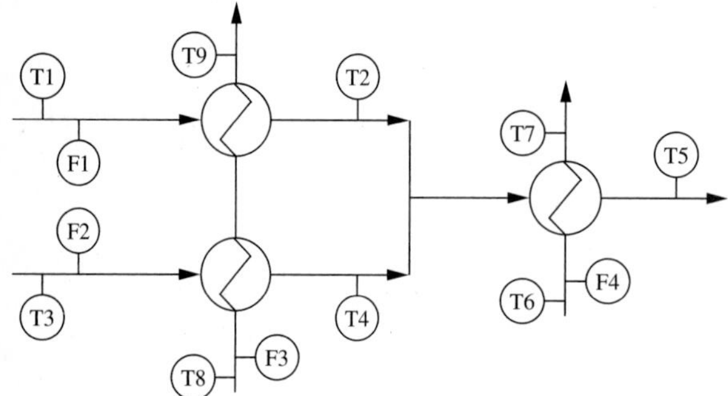

FIGURE Q24.12

$$T_1 = 20, T_2 = 42, T_3 = 45, T_4 = 68, T_5 = 76, T_6 = 88, T_7 = 71, T_8 = 75, T_9 = 31°\text{C}$$

$$F_1 = 50, F_2 = 50, F_3 = 56, F_4 = 150 \text{ m}^3/\text{hr}$$

$\rho = 0.8 \times 10^6 \text{g/m}^3$ and $C_p = 0.75$ cal/(g °C) for the streams measured by F_1 and F_2

$\rho = 1.0 \times 10^6 \text{g/m}^3$ and $C_p = 1.00$ cal/(g °C) for the stream measured by F_3

$\rho = 0.75 \times 10^6 \text{g/m}^3$ and $C_p = 0.71$ cal/(g °C) for the stream measured by F_4

24.13. Expand the following designs by adding (i) alarms, (ii) safety valves, (iii) final element failure directions, (iv) emergency safety controls, and (v) sensors for monitoring process performance. You may add sensors or final elements as necessary; also, specify whether each sensor should provide a signal that is highly accurate, or merely highly reproducible.

(*a*) The fired heater in Figure 17.17

(*b*) The distillation column in Figure 21.14

24.14. The CST mixing process in Figure Q24.14 is proposed. The goal is to control the effluent composition and temperature. Evaluate the design, suggest a proper loop pairing, and suggest process or control objective modifications, if necessary.

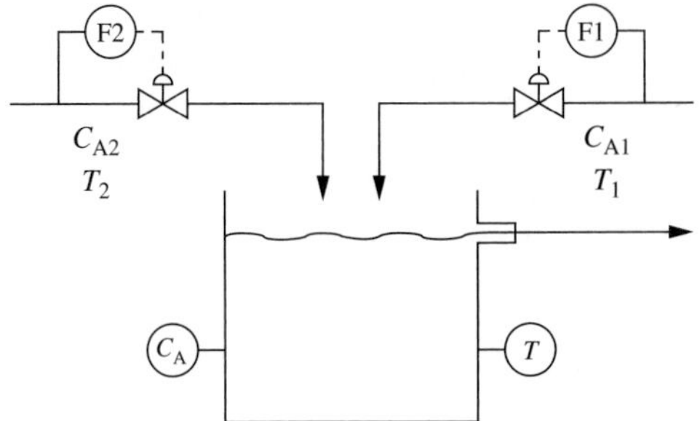

FIGURE Q24.14

24.15. A gas distribution system for a chemical plant is shown in Figure Q24.15. Several processes in the plant produce gas, and this control strategy is not allowed to interfere with these units. Also, several processes consume gas, and the rate of consumption of only one of the processes can be manipulated by the control system. The flows from producers and to consumers can change rapidly. Extra sources are provided by the purchase of fuel gas and vaporizer, and an extra consumer is provided by the flare. The relative dynamics, costs, and range of manipulation are summarized in Table Q24.15.

(*a*) Complete the blank entries in the Dynamics Column in Table Q24.15.

(*b*) Design a control strategy to satisfy the objectives of tight pressure control and minimum fuel cost. You may add sensors as required but make no other changes.

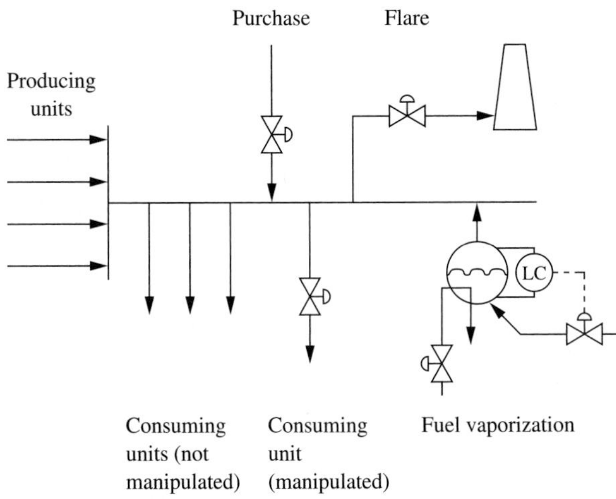

FIGURE Q24.15

TABLE Q24.15

Flow	Manipulated	Dynamics	Range (% of total flow)	Cost
Producing	No	Fast	0–100%	n/a
Consuming	Only one flow	Fast	0–20%	Very low
Generation	Yes		0–100%	Low
Purchase	Yes		0–100%	Medium
Disposal	Yes		0–100%	High

(*c*) Suggest process change(s) to improve the performance of the system.
(*Hint:* Before designing the controls, determine the correct response for all valves as the ratio of producing to consuming gas flows changes from much greater than 1.0 to much less than 1.0.)

24.16. The dynamics of an isothermal, constant-volume, constant-feed flow rate, well-mixed CSTR are to be evaluated for feedback control in this question. The feed consists entirely of component A, the chemical reaction is

$$\mathrm{A} \underset{k_\mathrm{B}}{\overset{k_\mathrm{A}}{\rightleftarrows}} \mathrm{B}$$

and the rates are first-order for both directions.

(*a*) Derive the dynamic model of the input-output system between C_{A0} and C_A. What conclusions can be determined regarding stability, periodicity, and either overshoot or inverse response for a step input? Describe the expected control performance for a step set point change. What tuning method could be used for a PID controller? Would you recommend feedforward control to improve the performance for a disturbance in temperature?

(*b*) Answer questions (*a*) for the input-output system C_{A0} and C_B.

(*c*) Would you expect that the control performance between $C_{A0} \rightarrow C_A$ would be better, the same, or worse than $C_{A0} \rightarrow C_B$, assuming that the feedback controllers were tuned on the same basis? Base your answer solely on the relative dynamics for the two possible systems. Consider a step set point change.

24.17. The process in Figure Q24.17 is a simplified head box for a paper-making process. The control objectives are to control the pressure at the bottom of the head box tightly and to control the slurry level within a range.

(*a*) Derive a model for the effects of the two inlet flows on the controlled variables.

(*b*) Design a control system by pairing the controlled and manipulated variables. Use the methods introduced in Chapters 20 and 21 as well as this chapter. Discuss the performance of your design and any special features that should be included in the implementation.

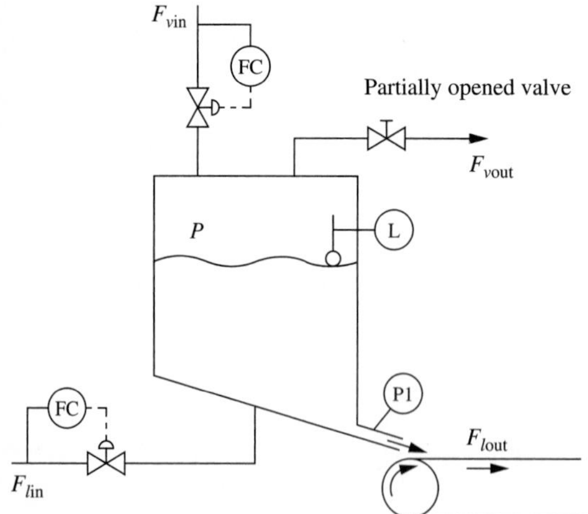

FIGURE Q24.17

24.18. The process in Figure Q24.18 includes a fired heater, chemical reactor, and heat exchangers to recover energy by heat transfer to other processes in the plant. The goals are to have tight flow control (F1), tight control of the reactor outlet temperature (T2), and good control of temperatures T3 and T4 in the integrated processes. The sensors and manipulated variables are shown in the figure. Disturbances are set point changes to the process flow and changes in the heating requirements of the heat-integrated processes.

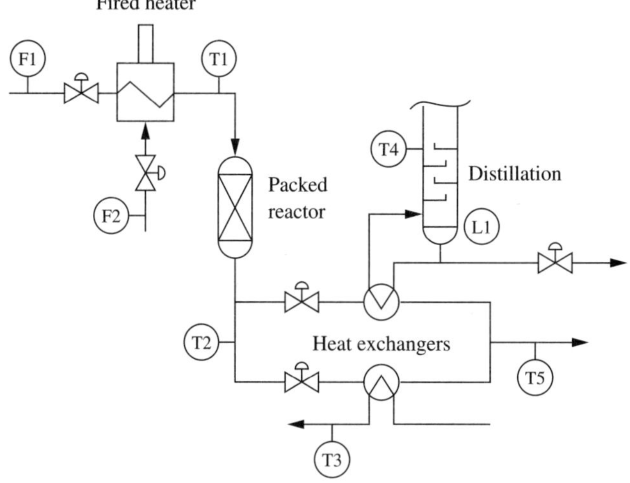

FIGURE Q24.18

(*a*) Without changing the instrumentation and process equipment, design a control system to achieve the objectives. Discuss whether all objectives can be achieved and if not, why.

(*b*) By making the minimum changes to the process equipment and instrumentation, design a system that improves the result in (*a*).

24.19. The mixing process in Figure Q24.19 involves a tank to mix components A and B without chemical reaction. The effluent from the mixing tank is blended with a stream of component C, and the flow of this stream is wild; that is, it cannot be adjusted by this control strategy. Note that waste is to be minimized.

(*a*) Using only the equipment shown in the figure, design a control system to tightly control the percentages of A, B, and C in *the blended product*. Can you achieve this and also control the total flow of blended product?

(*b*) Improve your result in (*a*) by adding an analyzer that can measure compositions in one stream. Decide the proper location and use it in the control system. Discuss why the analyzer would improve the performance.

24.20. A heat exchanger is shown in Figure Q24.20. The temperature measured by the sensor is to be controlled. Design four different control strategies to control this temperature and discuss the differences. Select the design that would give the best control performance, and discuss the reasons why.

24.21. Apply the Niederlinski criterion to the flash control system presented in this chapter. Discuss your results and the interpretation of the control system design.

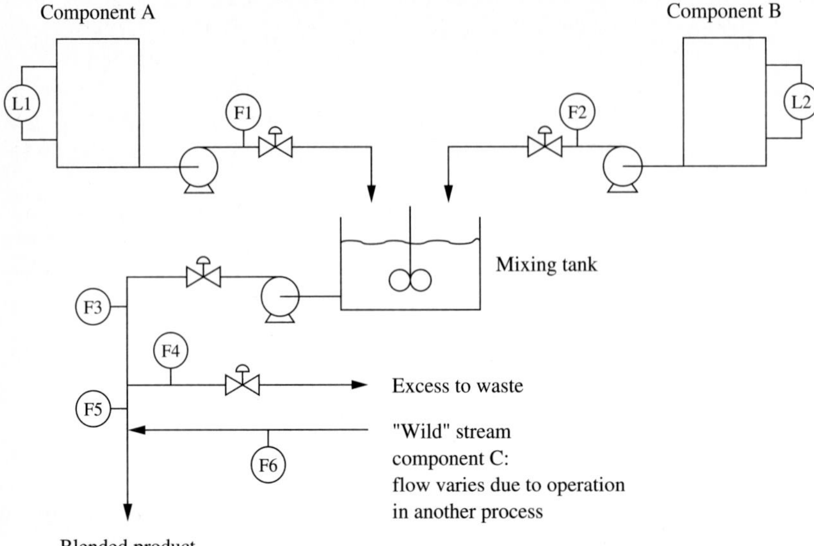

FIGURE Q24.19

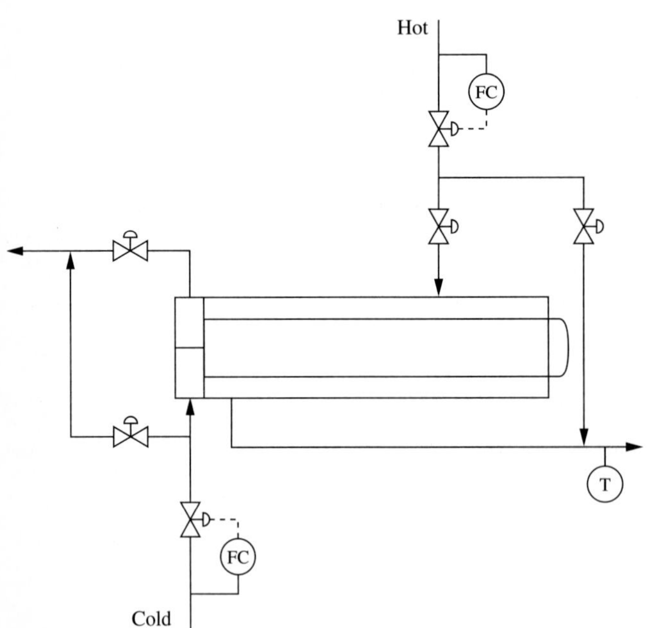

FIGURE Q24.20

24.22. Evaluate the controllability of the mixing process in Example 24.3 and Figure 24.7*a* for each of the following changes, which are to be considered individually, not cumulatively. The controlled variables are the liquid volume and the effluent concentration; valve v_2 is constant for this question.

(*a*) A pipe with valve is included from the feed to the effluent, bypassing the tank. The manipulated variables are v_1 and the new bypass valve.

(*b*) A new pipe and valve are added to inject a flow of pure component A into the feed stream. The manipulated variables are v_1 and the new valve in the pure A stream.

24.23. The dynamic responses of the heat transfer process in Figure Q24.23 are considered here. The medium in the coil and jackets is heating the fluid in the tanks. The tanks are well mixed, and all transportation delays are small compared with the time constants.

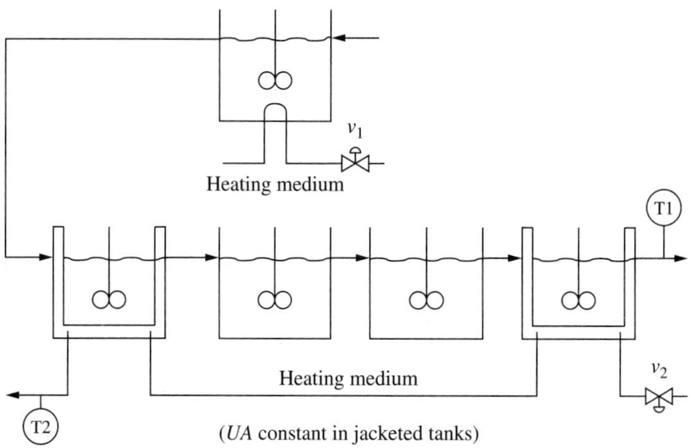

FIGURE Q24.23

(*a*) Describe the dynamic responses of each temperature to a step change in each valve.

(*b*) Discuss the likely control performances for each input-output pair for multiloop control.

(*c*) Assume that one feedback PI controller, with pairing T2 → v_2, is in operation and that a step change is made to v_1. Describe the dynamic response of both temperatures.

(*d*) Discuss the likely control performance if the other control loop, T1 → v_1, is closed.

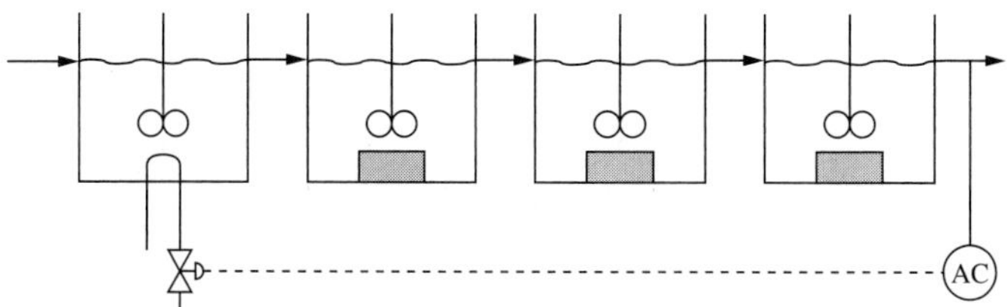

FIGURE Q24.24

24.24. The series of well-mixed stirred-tank chemical reactors for a first-order chemical reaction with negligible heat of reaction is shown in Figure Q24.24. Each reactor has a mass in the tank, which has the same temperature as the liquid and represents substantial energy accumulation. (This is a simplified representation of a packed bed, with the masses being the catalyst.) The concentration of the effluent from the last reactor is controlled by adjusting the heating medium valve.

(*a*) Discuss the effect of the masses on the control performance in response to feed stream temperature variations.

(*b*) Discuss the effect of the masses on the control performance in response to feed stream concentration variations.

(*c*) Draw general conclusions about the effects of the masses on the disturbance responses in (*a*) and (*b*) and on set point changes.

24.25. A two-product distillation tower with a single feed is considered in this question.

(*a*) Sketch two types of condensers (showing equipment and valves), describe their physical principles, and be sure to explain how the duty is adjusted for process control.

(*b*) Repeat (*a*) for reboilers.

(*c*) Discuss why distillation towers typically have an overhead liquid accumulator.

(*d*) Discuss why temperatures and pressures are measured on selected distillation trays.

(*e*) Where would you place safety valves and for what maximum flow should they be sized?

(*f*) Identify possible constraints (i.e., items that define the frame of the operating window).

(*g*) Identify potential disturbances.

(*h*) The composition sensors often provide new measurements only every two minutes. Give a reason why this might occur.

(*i*) Identify all inventories in the distillation process, determine which are non–self-regulatory, and describe potential control strategies for each.

(*j*) Discuss options for product quality control and the interaction between inventory and product quality controls.

24.26. The process in Figure Q24.26 involves the chemical reaction with the overall stoichiometry of 3A + B → C taking place in a packed-bed reactor. The inlet temperature

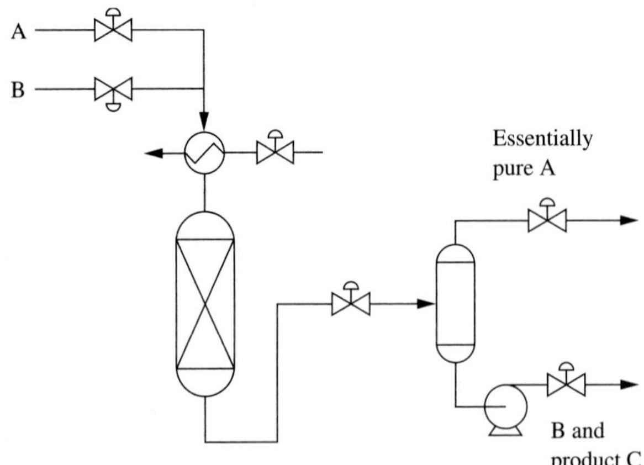

FIGURE Q24.26

has a strong effect on the rate of reaction, and there is no limit to any reasonable value for the bed temperature. The unreacted A is separated in a flash drum and sent to fuel at considerable cost because it cannot be recycled. Also, high temperatures tend to degrade the catalyst. The liquid product has a target of 80% product C. Design a control system to achieve the objectives just described, specify sensors, and sketch the design on the drawing.

24.27. For the fired heater in Figure Q24.27 (i) no change is allowed for the final elements, (ii) sensors must be added, and (iii) the feed rate and product outlet temperature must be controlled. Briefly state a reasonable set of control objectives and design a control strategy.

(*a*) Briefly give the algorithm and purpose for each controller.

(*b*) Sketch the strategy on the diagram.

(*c*) Give the failure positions for the final elements.

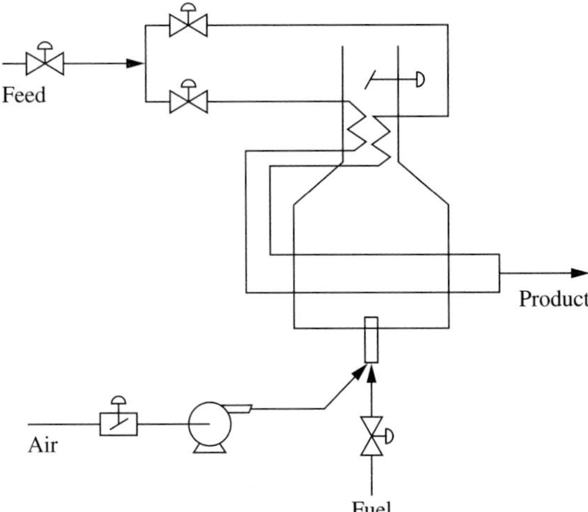

FIGURE Q24.27

24.28. The series of two chemical reactors described in Example 5.4 is the initial process on which this question is based. You may use all results from the modelling in Example 5.4 without proving; simply cite the source of the equations. The solvent flow and inlet composition are to be controlled by single-loop controllers. By adding sensors and final elements as required, describe briefly, and sketch a control system for this purpose. Given this strategy is functioning perfectly (maintaining C_{A0} *constant*), determine the model between the solvent flow and the concentration of the reactant in the second reactor, C_{A2}, and comment on the expected composition (C_{A2}) control performance using this manipulated–controlled variable pairing. Compare with the control performance in Example 13.8.

CHAPTER 25

PROCESS CONTROL DESIGN: MANAGING THE DESIGN PROCEDURE

25.1 INTRODUCTION

To this point, the control design problem has been defined, and the range of decisions has been presented. It becomes clear that tens to hundreds of decisions are made during the control design of an industrial process. One would expect, as is shown later in this chapter, that the sequence in which these decisions are made can influence the time required to complete the design and, perhaps, the quality of the control performance provided by the final design. Thus, the engineer is faced with the challenge of managing a large quantity of information and a large set of possible design decisions during the design procedure.

There is no single, correct way to manage this procedure. Different skilled engineers perform tasks in different sequences to reach equally good solutions, and different problems can be solved more easily by different sequences. However, the procedure presented here provides a structured problem-solving approach that is tailored to the control design task. The procedure represents, to the ability of the author to document such a fuzzy entity, the approach used by many practitioners.

There are several advantages to the novice engineer for using this procedure. Since the most difficult aspect of the design is often starting this ill-defined task, the first advantage is that a prescribed procedure provides a way to begin the design task. Second, the procedure provides a step-by-step approach that ensures that many important issues are addressed. Third, the procedure decomposes the problem in a manner that determines whether control is possible before continuing to detailed decisions on control strategies. Finally, the procedure provides some guidance on managing the interactions among the numerous design decisions. Interactions occur because some decisions made to satisfy specific control objectives affect the possible control performance with respect to other control objectives. Therefore, the engineer must try to make each decision with full recognition of its impact on the entire design and all control objectives. This thought process is demanding and not always possible, so the engineer often has to iterate by returning to initial decisions, changing some, and proceeding from these modified decisions to the completion of the design. The successful design engineer has the foresight to make (generally) good initial decisions, identify improper initial decisions early in the procedure, and minimize the iterations to the final design.

25.2 DEFINING THE DESIGN PROBLEM

We begin again with the definition of the problem provided in the Control Design Form (CDF) because of the crucial importance of this step to the quality of the design. In this section, some guidance is given on how an engineer goes about filling in a blank CDF. The CDF provides a useful checklist of the information needed in designing control systems and gives an organized manner for documenting the information.

Typically, people need some stimulation when defining problems; that is, they need some questions and issues to consider when beginning the design procedure. To stimulate the thought process, abbreviated tables of sample questions are presented here for the various control objectives. The first three objectives—safety, environmental protection, and equipment protection—are combined in Table 25.1 because they all address major deviations from normal operation, many of which could have common causes that influence all three objectives. Smooth operation, product quality, efficiency and optimization, and monitoring and diagnosis are addressed in Tables 25.2 through 25.5, respectively. The issues raised in the tables should be considered

TABLE 25.1
Checklist for safety, equipment, and environment

Limitations on operating conditions due to equipment, material, e.g.,
- Composition
- Flow
- pH
- Pressure
- Temperature

Explosion
- Fuel source
- Oxidizing source
- Energy source

Release of hazardous material
Failure of process equipment
Failure of control equipment
Human mistakes and their consequences

TABLE 25.2
Checklist for smooth operation

Unstable processes (do not reach steady state without control)
- Levels
- Chemical reactors

Processes that are very sensitive to disturbances
- Maximum rate of change of disturbance

Process integration that either propagates or attenuates disturbance
Manipulated variables that are easily interpreted by operating personnel
Disturbance sources

TABLE 25.3
Checklist for product quality

Target average value and variability
- One or multiple specifications
- Average value
- Variability
- ± deviation from target at which product is unacceptable

Variability in a property that affects future use by customer
- Standard deviation or other measure
- Nonlinearity between measurement and quality in future use

Disturbances that affect quality
- Magnitude
- Frequency

Factors affecting control performance
- Availability of on-stream measurement
- Degrees of freedom
- Controllability
- Feedback dynamics
- Modelling errors

for each design, and issues relevant to the plant should be noted in the Control Design Form, thereby developing a comprehensive statement of control objectives.

An additional way to identify control issues is to consider every stream or important location (e.g., the volume of a reactor or flash drum) in the process and to pose the following question:

What is the effect on A if the B in this stream or location C?

with A = each control objective (safety, environmental protection, equipment protection, smooth operation, product quality, efficiency, yield and profit, and monitoring and diagnosis)

B = property word indicating key operating variables (e.g., flow, temperature, pressure, composition, inventory, and so forth)

C = guide word indicating direction of changes in operation (e.g., increases, decreases) and rate of change (e.g., rapidly, slowly, periodically).

TABLE 25.4
Checklist for efficiency and optimization

External manipulated variables not used for control, potentially for optimization

Changes in targets or inputs (disturbances)
- Frequency
- Need for optimization
- Complexity of optimizing strategy

Parallel units
- Different product quality
- Different yields
- Energy consumption

Recycle flows
- Composition of recycle

Seperation units
- Energy-yield tradeoff

Chemical reactors
- Conversion
- Yield

Operating condition
- Internal optimum
- Operation at a constraint

TABLE 25.5
Checklist for monitoring and diagnosis

Performance that changes rapidly
- Alarms
- Emergency shutdowns
- Constraint violations
- Product quality
- Inventories

Performance that changes slowly
- Heat transfer coefficient
- Catalyst activity
- Corrosion
- Coking or fouling

Performance requiring complex calculations
- Fired heater efficiency
- Turbine and compressor efficiency

Utilization of control
- Percent of time in automatic

Temporal correlation of good or poor operation with external disturbances (feed type, equipment operation, and so forth)

The application of this question to the process will help the engineer identify the significant effects of the objectives. When a significant effect is identified, the engineer should determine the cause of the effect and how it can be retained (if the effect is beneficial) or prevented or compensated (if the effect degrades performance). This is a simplification of an approach that has been developed in much greater detail for hazards and operability (HAZOP) studies, which consider a broader range of issues influencing the safety of a process. Detailed descriptions of the procedures followed in HAZOP studies are available (AIChE, 1992).

The methods described in this section are intended to generate information on all major headings in the CDF, not just objectives, although the tables of questions are organized by objectives. When considering the objectives in such detail, information on the constraints and disturbances should also be identified and recorded in their proper locations. It is important to recognize that the CDF cannot be completed with only a cursory understanding of the process and quick review of a process sketch; a thorough understanding of the physics, chemistry, product quality, and economics is required.

At this preliminary design stage, the engineer should concentrate on determining the needs of the plant and not attempt to define the solutions. The control objectives and other critical issues should be clearly and quantitatively stated even when no solution is initially apparent, and the definition procedure should not be delayed by lengthy analysis of a particular issue, since too much attention to detail during the initial "brainstorming" activity tends to slow the flow of ideas. Also, it is important that the engineer not be overly concerned about the initial location for an element of information in a CDF. It is expected that the CDF will be reviewed and rationalized before the design procedure continues to the decision-making step.

25.3 SEQUENCE OF DESIGN STEPS

There is almost an infinite number of ways in which the numerous design decisions can be reached. There is no one best sequence for all control designs; in fact, various skilled practitioners use different sequences to arrive at equally good designs. However, there are certainly some sequences that are better than others, and some simple sequences can be used by novice engineers until they gain enough experience to modify the sequence to take advantage of their special insights. The sequence given in the flowchart in Figure 25.1 is recommended for control design and discussed further in this and the next sections.

Step 1: Definition

The first step involves the collection of information appearing in the Control Design Form and, for especially complex problems, the formal preparation of the Form. At this step, the objectives are translated to specific variables, either directly measured or calculated using measurements, which are to be controlled.

Step 2: Feasibility

The second step determines the feasibility of the control objectives for the equipment design, operating conditions, and disturbances given in the problem definition. An analysis of degrees of freedom and controllability determines whether it is possible to control the proposed controlled variables with the proposed manipulated variables. Since controllability rigorously addresses only the base-case operating point, the

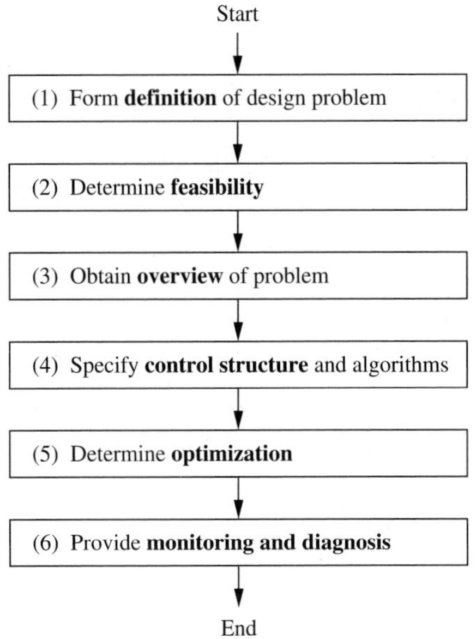

FIGURE 25.1
Overview of control design sequence.

operating window is determined to ensure that the process can be maintained within specified limits for the defined disturbance magnitudes. Thus, this step ensures that the system has sufficient capacity as well as degrees of freedom and controllability. As noted in Chapter 24, a dynamic analysis may have to be performed to evaluate the operating window fully. Also, the ability to measure or infer important variables is evaluated. If any of the results of these steps indicate that control is not possible, the design procedure must include an iteration in which the process is altered to make the control objectives possible.

Step 3: Overview

The third step establishes an integrated view of the plant operation, concentrating on the most important variables. The goal of this step is to obtain an *overview* of the feedback process dynamics, the disturbance dynamics, the interaction in the process, and the types of measurements and manipulated variables available for control. This overview is essential because the design engineer makes one decision at a time and needs this overview to be able to "look ahead" so that all decisions form a compatible design. Objectives that are easily achieved or likely to be difficult to achieve are noted. Also, potential changes to the instrumentation and process are identified for future use, if needed. However, no control designs are decided at this step.

Step 4: Control Structure

The fourth step involves specific decisions on control structure, algorithms, and tuning. Here, if single-loop control technology is used, the single-loop controlled and manipulated variables are paired, and the modes of the PID controllers are specified.

In addition, special requirements for the tuning are made in conjunction with the pairing. For example, level controllers are specified as tight or averaging. Also, tight and loose tuning of interacting loops is specified, to reduce the effects of unfavorable interaction while retaining the beneficial effects of favorable interaction, as required. The next sections of this chapter provide additional guidance on this step, discussing a hierarchy and decomposition for managing the design decisions.

Step 5: Optimization

The fifth step determines whether optimization opportunities are available after consistently high product quality has been achieved and, if so, whether additional manipulated variables, not used for control at previous steps, exist. It may be necessary to add sensors to provide information for optimization and to automate additional manipulated variables for optimization. If opportunities exist, an analysis is performed to determine the economic benefits which can be realized through optimization, as explained in Chapter 26. If significant benefits are available and can be realized through real-time control, the strategy is designed at this step.

Step 6: Monitoring and Diagnosis

The sixth and final step evaluates monitoring and diagnostics. At this step, the major analysis is the sensors required for this function. In addition, any calculations required for the monitoring are defined.

The sequence of steps is selected to maximize information gathering and understanding at the early steps and to reduce the need for iterations. The first two steps identify the capabilities of the process and instrumentation and the control objectives. Inconsistencies between process capability and objectives are identified so that they can be resolved soon in the design procedure, because inconsistencies should be resolved before further design steps are performed. Next, the overview of the process in the third step enables the engineer to understand the process responses before attempting to design controllers. The design of the controllers, up to and including product quality, is performed in the fourth step to give the best performance for the more important variables. Special controls for safety should be designed at this stage in an integrated manner. In the fifth step, the remaining degrees of freedom, which are not used at the previous stages (perhaps because they have the poorest dynamic responses for control of key variables), are used for profit maximization. Finally, the monitoring and diagnosis is designed.

25.4 TEMPORAL HIERARCHY OF CONTROL STRUCTURE

In this section the activities in the fourth step in the sequence, addressing control structure, are presented in greater detail. Proper design relies on an integrated analysis of the entire process or plant under consideration; however, the integrated design may involve too many variables and processes to be analyzed by currently available methods. Therefore, the engineer temporarily separates the design problem into smaller segments, and if the interactions among the segments are small, each can be analyzed individually to develop provisional control designs. Two approaches for selecting segments are discussed: *temporal hierarchy* in this section and *process decomposition* in the next section. It is important to recognize that these methods are used only when required by the large scope of the problem and that the methods employ approximations to simplify the analysis. It is essential that each decision contribute to the good performance when considering all factors in the integrated process.

A common approach for decomposing the design decisions is based on a temporal hierarchy, as originally suggested by Buckley (1964) and expanded here:

In hierarchical decomposition, the control decisions are usually made in the following order.

1. Flow and inventory
2. Process environment
3. Product quality (and safety)
4. Efficiency and profit
5. Monitoring and diagnosis

This hierarchy has the advantage of designing control loops in the order of the fastest to the slowest; the possible exception are the liquid and solid inventories, which may employ averaging controllers (see Chapter 18). In addition, the hierarchy is commonly used because it is difficult to design controllers for product quality without first defining how feed and product flows and process environments are controlled. Thus, the sequence makes sense from the viewpoint of control structure.

Flow and Inventory

At the first level, the flows and inventories considered are for the "process" materials, which are used to make the product. The flows of utility streams, such as fuel, cooling water, and steam, are not specified at this level, because they are manipulated to achieve other control objectives. The structures that control the process flows determine how the feed and production rates are specified and whether flow rates are nearly constant or are likely to vary significantly. Note that the inventories—liquid and solid levels and gas pressures—must be designed in conjunction with the flow controllers, to ensure that requirements for inventories and product deliveries are satisfied concurrently.

Process Environment

The second level addresses the process environment variables: pressure, temperature, feed ratios, catalyst addition, and so forth. These variables have a great influence on the product quality and are often manipulated, in a cascade structure, by the product quality controllers. Thus, this level provides tight control of the environment by compensating for many disturbances, and it can be adjusted by cascade feedback from higher levels.

Product Quality (and Safety)

The third level provides the essential product quality regulation. This is typically achieved by adjusting set points of controllers at the lower levels in a cascade structure, but it may adjust final elements directly. Control for safety should be addressed at this level of the decision hierarchy, because control strategies up to this level can influence the safe operation. As discussed in the previous chapter, the safety controllers will normally be implemented in a lower level of the implementation hierarchy.

Efficiency and Profit

The fourth level capitalizes on additional flexibility to improve profitability of the plant. These controllers perform their function slowly so that smooth operation and excellent product quality are not sacrificed. It is good practice for the optimizing controllers to influence the process through the lower levels in the implementation hierarchy; this ensures that higher-priority objectives such as safety and product quality are not compromised.

Monitoring and Diagnosis

The fifth level involves monitoring and diagnosis of process and control performance. This includes rapid monitoring and reporting to plant operating personnel, as well as longer-term monitoring for periodic analysis. Plant operations are influenced by decisions made at this level through actions of plant personnel, usually after detailed analysis of likely causes of unusual process performance. These decisions may not be implemented through the control strategies, because they may involve variables, such as feed purchases and reactor regeneration scheduling, that are outside of the purview of the continuous control system.

This analysis hierarchy conforms to the way many control systems are implemented. A typical implementation hierarchy is shown in Figure 25.2. The lowest level of the continuous control involves the flow and inventory loops and provides the basis for higher levels in the hierarchy. Note that the interaction between levels in the hierarchy is primarily through cascade control principles; this approach has several advantages:

1. It uses conventional technology.
2. It satisfies the requirements for relative dynamics so that good disturbance response is achieved.
3. It does not create conflicts in degrees of freedom (see Section 14.2).
4. The system is easily commissioned or decommissioned by changing controller cascade status between closed and open.

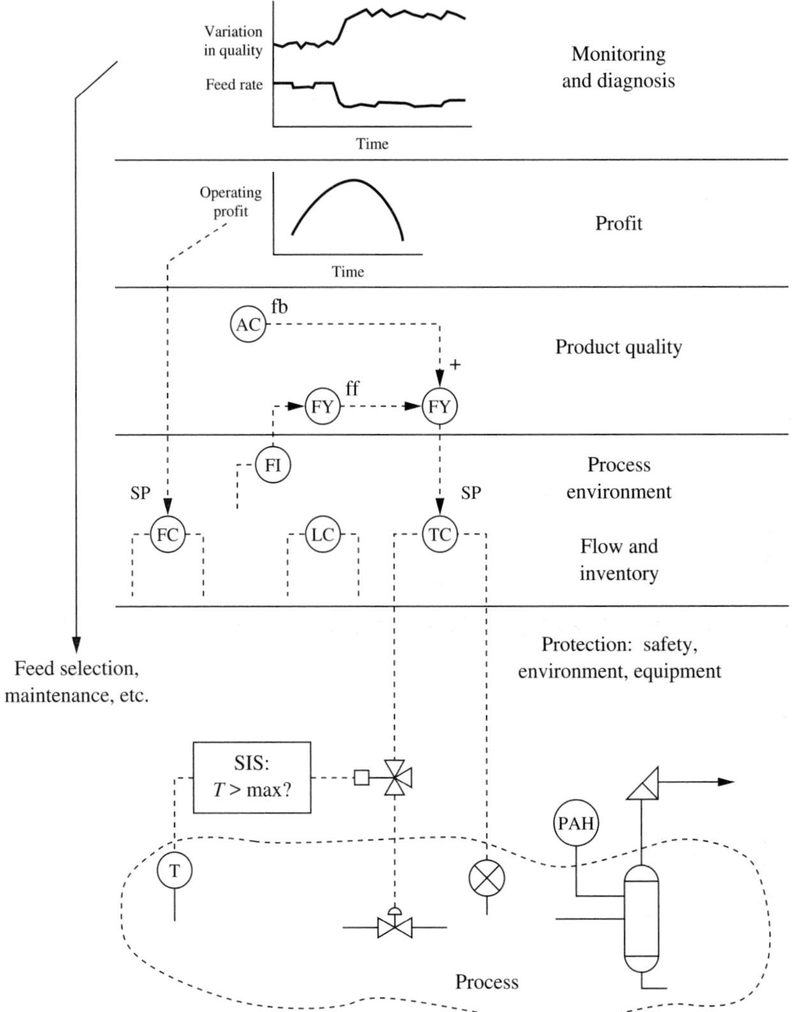

FIGURE 25.2
Schematic of the typical process control hierarchy.

Although this hierarchical approach has many advantages and has been found easy to apply by many engineers, it does not remove one of the most challenging features of the design procedure: the need for iteration. When making decisions at each level, the engineer attempts to look ahead to the completed design and determine the effects of the current decisions on the control performance. However, looking ahead is not always simple, or even possible, in complex plants; thus, the engineer may find that the final design is not satisfactory. When such a situation is encountered, the engineer should investigate whether the performance could be improved by another design that starts with different decisions at the previously designed, lower levels in the hierarchy.

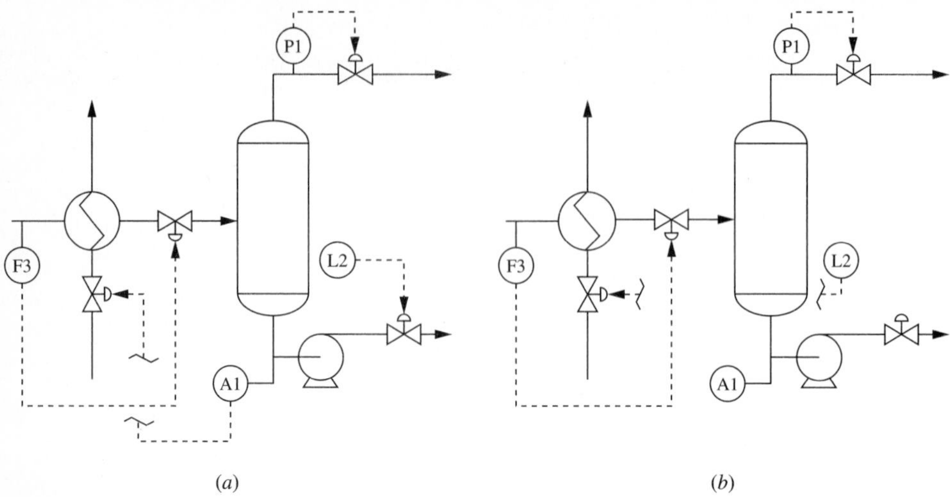

FIGURE 25.3
Two different flash control designs.

Example 25.1 Consider the flash process in Figure 25.3, which is similar to the process previously analyzed in Chapter 24. The case considered here involves two different initial flow and level control decisions, shown in Figure 25.3*a* and *b*. The first level of the hierarchy in Figure 25.3*b* has resulted in the control design in which the feed is on flow control, and the level is controlled by adjusting the heat transferred to the feed by adjusting the steam, which affects the amount of liquid vaporized. These initial decisions satisfy the relevant control objectives. However, given these flow and level decisions, the product quality controller has only one degree of freedom to adjust: the product flow rate. Therefore, the lower-level design decisions have dictated the higher-level control strategy.

To understand how the quality controller in Figure 25.3*b* would function, consider the case in which the light key in the liquid product component is too high. In response to the disturbance, the product quality controller would decrease the product flow rate, which would cause the level to increase; the level controller would increase the steam flow rate, which would increase the percentage vaporized; and the light key in the liquid product would decrease. Therefore, this quality control design is feasible, but it has slow dynamics, because the level control process and controller appear in the product quality feedback path. In fact, this design is another example of pairing single-loop controllers with a relative gain of zero. This can be verified using the following steady-state model, which has been extracted from equation (24.21) for the flash process in Chapter 24 (page 808):

$$\begin{bmatrix} A_1 \\ \dfrac{dL}{dt} \end{bmatrix} = \begin{bmatrix} -0.11 & 0.0 \\ -0.136 & -0.179 \end{bmatrix} \begin{bmatrix} v_2 \\ v_4 \end{bmatrix} \tag{25.1}$$

The relative gain for the $A_1 \rightarrow v_4$ pairing is zero, because the steady-state process gain between the product flow and the composition is zero, when all other loops are open. The relative gain for this system has ones on the diagonal and zeros on the off-diagonal elements. However, since the system is controllable for either pairing, a pairing on a zero relative gain would function, albeit with poor performance in this case.

Thus, the initial flow/inventory control design decisions have resulted in relatively poor product quality control. During the iteration, the engineer would be looking for a faster-responding manipulated variable for product control, because the cause of the poor performance is slow feedback dynamics. Another goal would be to find a pairing with a nonzero relative gain. After the iteration, the control design should be as shown in Figure 25.3*a*.

How does one properly perform the "look-ahead" to satisfy the control objective under consideration while preventing, as much as possible, an undesirable effect on other control objectives? The effect of "keeping in mind" is to ensure that the initial control design, in addition to meeting its control objectives,

1. Leaves unallocated some manipulated variables that can give good control performance for important controlled variables at higher levels in the hierarchy.
2. Attenuates disturbances and does not introduce unfavorable process control interactions.
3. Ensures that critical controllers can perform their tasks properly even if some other controllers are not functioning (e.g., are in manual).

This look-ahead requires an overview of all control objectives, which again reinforces the importance of a good problem definition and process overview in steps 1 to 3 of the sequence. Then the engineer must keep all of the key controlled variables in mind when designing the lower levels of the hierarchy.

> When performing the control design procedure, the engineer continually *looks ahead* to predict the effects of current decisions on later control objectives at higher levels in the hierarchy.

25.5 PROCESS DECOMPOSITION

Large plants may have hundreds or thousands of manipulated and controlled variables. Although the entire plant must be considered in designing controls, it is essentially impossible to analyze all aspects of the plant simultaneously while making each control decision. Therefore, the plant is often decomposed into several process units that have only weak interactions, if possible. The proper decomposition is particularly easy for the series process design structure of chemical process plants, shown in Figure 25.4*a*. For this process structure the upstream units affect the downstream units, but the downstream units do not affect the upstream units. Since the interaction among the units is in only one direction, upstream units are simply sources of disturbances to the downstream units. Thus, the general goal is to reduce the

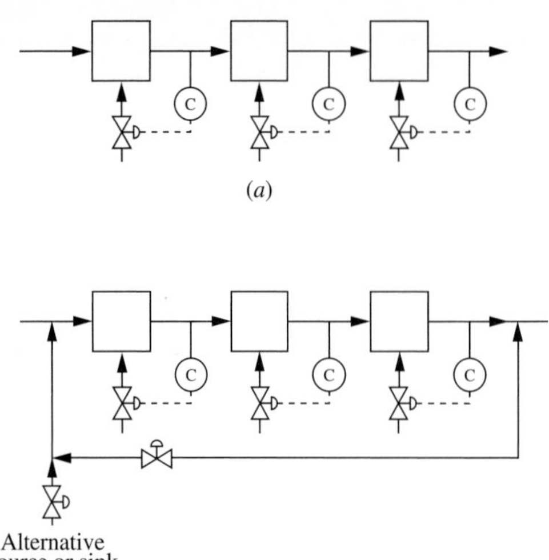

(b)

FIGURE 25.4
Typical structure of process plants: (*a*) series; (*b*) recycle without storage.

disturbances that leave one unit and propagate to downstream units, with special care to isolate units that are highly sensitive to disturbances. The controls within each process unit can then be designed using the standard procedures.

Process plants often have recycle streams, as shown in Figure 25.4*b*. These plants do not strictly allow such a simple decomposition, because two-way interaction occurs between processes. As demonstrated in Chapters 20 and 21, two-way interaction can significantly affect dynamic behavior and control performance. Usually, the control system is designed to reduce the effects of recycle on the overall plant dynamics. This is often achieved by providing alternative sources of the material or energy provided by the recycle, so that short-term variation in the recycle can be compensated by the alternative source. (This is the same concept used in Figure 24.11*a* and *b* for energy recycle.)

Two examples of material recycle are shown schematically in Figure 25.4*b* and 25.5. In the first, an alternative source of material is provided to ensure a steady recycle flow; in this design, the alternative must be available immediately to provide the total process flow required. In the second example in Figure 25.5, the recycle system includes an inventory so that the level in the inventory can vary while the material supplied to the beginning of the process remains undisturbed. Note that the level-flow pairing directs all recycled material to the storage tank, regardless of the current recycle flow returned to the process. Naturally, the storage tank must be large enough to remain within acceptable level limits during expected short-term transients, whereas the net flow in or out must compensate for longer-term accumulation. This concept is discussed by Buckley (1974), where he describes the principle that recycle level-flow systems should generally be paired in the manner shown in Figure 25.5.

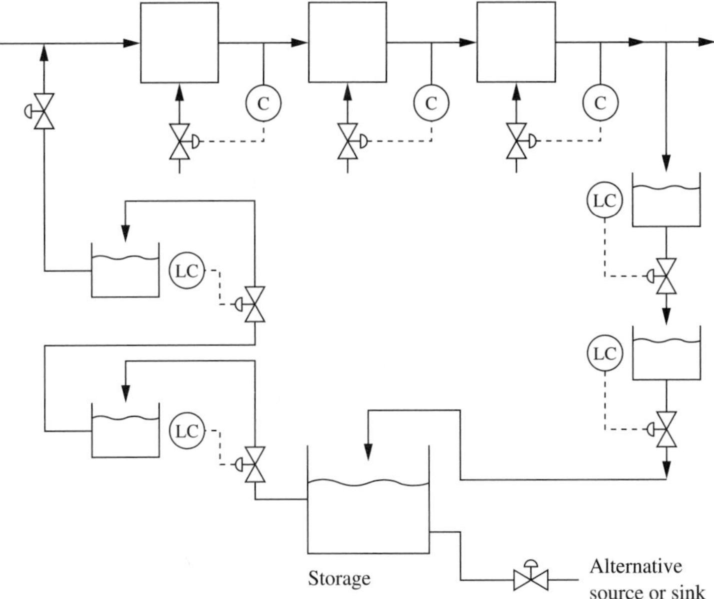

FIGURE 25.5
Typical recycle process with storage.

25.6 INTEGRATING THE CONTROL DESIGN METHODS

Several methods of organizing information and making design decisions have been presented in this and the previous chapter. In this section, the methods are combined into an integrated design thought process that demonstrates how the previously discussed methods can be combined to reach an adequate design. Novice engineers will most likely follow this integrated approach closely for their initial designs. As they gain more experience and learn to use their process understanding, they will adapt the approach to suit the problem at hand.

The integrated procedure is shown in Table 25.6, which combines the concepts of sequence, hierarchy, and design decisions. The procedure begins with a statement of the process design and plant requirements and ends with a complete control structure and algorithm specification. The major steps in the design sequence—(1) definition, (2) feasibility, (3) overview, (4) control structure, (5) optimization, and (6) monitoring and diagnosis—provide milestones for the procedure. Several quantitative design analyses are performed at each step in the design sequence, and the first three levels of the temporal hierarchy are performed for each process segment at the fourth step. The engineer will encounter all major design decisions presented in Chapter 24 in a logical order by using the procedure in Table 25.6.

Iterations are possible at several steps in the procedure. If the process does not have sufficient degrees of freedom, lacks independent input-output relationships to provide a controllable system, or lacks sufficient range, an iteration is required in step 2 to change the process. Also, if an analysis of the dynamics identifies poor

TABLE 25.6
Integrated control design procedure

START: Acquire information about the process

(a) Process equipment and flow structure — [Modify process and instrumentation] ◄
(b) Operating conditions
(c) Product quality and economics
(d) Preliminary location of sensors and final elements

1. DEFINITION: Complete the Control Design Form

(a) Use checklists
(b) Sample questions — [Modify objectives] ◄
(c) Prepare a preliminary set of controlled variables

2. FEASIBILITY: Determine whether objectives are possible

(a) Degrees of freedom — [Iterate] ►
(b) Controllability
(c) Operating window for key operating conditions

3. OVERVIEW: Develop understanding of entire process to enable "look-ahead" in decisions

(a) Key production rate variables
(b) Inventories for potential control
(c) Open-loop unstable processes
(d) Complex dynamics (long delays, inverse response, recycle, strong interactions)
(e) Key product qualities
(f) Key constraints
(g) Key disturbances
(h) Useful manner for decomposing the analysis (and control design), if necessary and appropriate

4. CONTROL STRUCTURE: Selection of controlled and manipulated variables, interconnections (pairings in decentralized control), and relevant tuning guidelines

(a) Preliminary decisions on overall process flows and inventories
(b) Process segment (Unit) 1
(c) Process segment (Unit) 2

Control Hierarchy (temporal decomposition) for every unit — [Iterate] ►

1. Flow and inventory
2. Process environment
3. Product quality
4. Safety

[Modify control] ◄

(d) Integrate control designs as needed for good overall performance

5. OPTIMIZATION: Strategy for excess manipulated variables

(a) Clear strategy for improved operation, or
(b) Measure of profit using real-time data
(c) Sensors and final elements — [Iterate] ►
(d) Minimize unfavorable interaction with product quality

6. MONITORING AND DIAGNOSIS

(a) Real-time operations monitoring
  1. Alarms 2. Graphic displays and trends — [Iterate] ►
(b) Process performance monitoring
  1. Variability of key variables (histogram and frequency range)
  2. Calculated process performances (efficiencies, recoveries, etc.)

FINISH: Completed specification, meeting objectives in step 1

(a) Process equipment and operating conditions
(b) Control equipment, sensors, and final elements
(c) Control structure and algorithms
(d) Tuning guidelines as needed, e.g., level control and interacting loops
(e) Safety controls and alarms
(f) Optimization
(g) Monitoring calculations

control performance, an iteration in step 4 is appropriate. Further iterations may be needed to provide all sensors necessary for optimization and monitoring in steps 5 and 6. During each iteration, the control objectives should also be reevaluated to be sure that the quantitative performance targets are proper and that the cost associated with achieving the demanding goals is justified.

As previously discussed, the engineer makes every effort to reduce or eliminate iterations by making the sequential design decisions with due consideration for future decisions. Information in steps 1 through 3 enable the engineer to identify the likely key elements of the design (i.e., the controlled variables requiring tight control). This enables the engineer to "set aside" manipulated variables that may be used for the control of the key variables. The integrated control design procedure is demonstrated in the following example.

25.7 EXAMPLE DESIGN: CHEMICAL REACTOR WITH RECYCLE

The integrated control design procedure will be applied to a simple chemical process in this section. The process, shown in Figure 25.6, involves feed of a raw material from storage to a chemical reactor. The reaction is A → B with first-order rate expression $-r_A = k_0 e^{-E/RT} C_A$ and negligible heat of reaction. The products of the reactor are heated and sent to a flash drum, from which the product is taken as a vapor flow which is predominantly component B, but contains some A. A liquid stream consisting of unreacted feed, along with some product B, is recycled to mix with the fresh feed and flows to the reactor. The base-case (initial) operating variables are given in Table 25.7.

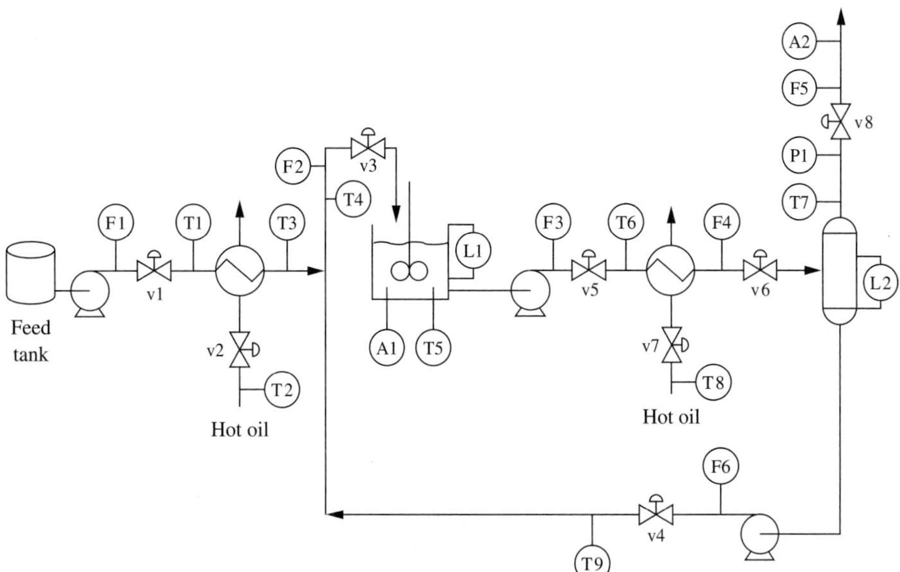

FIGURE 25.6
Chemical reactor and separator with recycle.

TABLE 25.7
Operating conditions for reactor with recycle

Variable	Symbol	Initial value	*Final value for Design I in Figure 24.9	*Final value for Design II in Fiqure 24.10
Fresh feed	F1	5	5	5.0
Reactor inlet flow	F2	20	34	20
Reactor outlet flow	F3	20	34	20
Vapor product	F5	5	5	5
Recycle flow	F6	15	29	15
Reactor level	L1, %	50	50	50
Flash level	L2, %	50	50	50
Fresh feed temperature	T3, °C	99	105	106.8
Reactor feed temperature	T4, °C	92	92	93.9
Reactor temperature	T5, °C	92	92	93.9
Flash temperature	T7, °C	90	90	90
Reactor concentration of A	A1, mole %	69.4	77.1	69.4
Vapor product concentration of A	A2, mole %	10	10	10

*After response to disturbance (1) in Table 25.8 on pages 854–855.

Definition Step

The control design will be developed through the procedure shown in Table 25.6. The first step in the sequence involves a complete definition of the problem, which is summarized in the Control Design Form in Table 25.8. (The reader should review the form on page 854 before proceeding.) This serves as the basis for all further design decisions.

Feasibility Step

The second step determines whether the control objectives are possible with the equipment available. This step involves the analysis of degrees of freedom and controllability. We assume that an analytical model of the process is not available; thus, the design is based on qualitative analysis from the process structure and on linear models identified empirically. There are eight manipulated external variables, so at most eight dependent variables can be controlled. A preliminary selection of controlled variables is made based on the CDF: (*a*) feed or production rate (1); (*b*) liquid and vapor inventories (3); and (*c*) product quality (1). Thus, at least five controlled variables exist. The number of external manipulated variables is greater than this minimum value. Therefore, it is concluded that the degrees of freedom do not preclude a possible design, and the design procedure can continue.

To extend the analysis further, the controllability of the system is evaluated. Controllability requires that linearly independent relationships must exist between the selected manipulated and controlled variables, or, in other words, the gain matrix must have a nonzero determinant. To perform this analysis, the model equations have to be linearized, and the matrix of gains evaluated at the base-case operation. Since inventory control is quite important, the level and pressure control loop pairings are decided first. The reactor level can be controlled with either v_5 or v_6, the flash drum level with v_4, and the flash drum pressure with v_8. The steady-state gains with these inventories under closed-loop control were determined by making small changes to the manipulated variable and determining the steady-state change in the potential controlled variables. The gain matrix for this example is

$$\begin{bmatrix} F_1 \\ F_2 \\ T_3 \\ T_5 \\ A_1 \\ A_2 \end{bmatrix} \begin{bmatrix} 0.020 & 0.0622 & 0.0106 \\ 0.45 & -0.38 & -0.127 \\ -0.13 & 2.14 & -0.13 \\ -0.02 & 0.44 & 1.105 \\ 0.0035 & -0.0063 & -0.0018 \\ 0.00025 & -0.0008 & 0.0006 \end{bmatrix} \begin{bmatrix} v_1 \\ v_2 \\ v_7 \end{bmatrix} \tag{25.2}$$

Note that the matrix is not square, so that control of all the potential controlled variables in equation (25.2) is not possible.

By the completion of the design procedure, there will be a strategy for every valve, and the system will be square, but at this point the goal is to determine whether the selected variables can be controlled. One way to answer this question is to select subsets of the manipulated variables until either (1) a subset results in a nonsingular gain matrix, in which case the system is controllable, or (2) all possibilities have been exhausted without finding a nonsingular system, in which case the system is not controllable. A more direct approach is to find the rank of the matrix, which gives the smallest square subset that is nonsingular. As subsets of the variables are selected, the controllability will be verified.

Overview Step

The third step of the control design sequence, which yields an overview of the process and control objectives, is now performed. The purpose of this step is to gather observations about the entire system that can be used when making sequential design decisions. The observations at this step are presented below by hierarchy level.

LEVEL 1: FLOW AND INVENTORY.

1. The feed tank has periodic deliveries of material and continuous outflow to the process. Therefore, it is not possible or necessary to control the level. The tank must be large enough so that it neither overflows nor goes empty for expected delivery and outflow policies.
2. The feed to the reactor is a combination of fresh feed and recycle. The flow and inventory design must consider this factor, to prevent oscillations caused by interactions. Also, there seem to be several possible ways to control the flow to the reactor, because there are valves in the fresh feed, recycle flow, and combined flow.
3. There is no option for the disposition of the reactor effluent; it must proceed directly to the flash drum.

TABLE 25.8

Preliminary Control Design Form for the chemical reactor and separator process in Figure 25.6

TITLE: Chemical reactor	ORGANIZATION: McMaster Chemical Engineering
PROCESS UNIT: Hamilton chemical plant	DESIGNER: I. M. Learning
DRAWING: Figure 25.6	ORIGINAL DATE: January 1, 1994
	REVISION No. 1

Control Objectives

1. Safety of personnel
   (*a*) The maximum pressure in the flash drum must not be exceeded under any circumstances.
   (*b*) No material should overflow the reactor vessel.
2. Environmental protection
   (*a*) None
3. Equipment protection
   (*a*) None
4. Smooth, easy operation
   (*a*) The production rate, F5, need not be controlled exactly constant; its instantaneous value may deviate by 1 unit from its desired value for periods of up to 20 minutes. Its hourly average should be close to its desired value, and the daily feed rate should be set to satisfy a daily total production target.
   (*b*) The interaction of fresh and recycle feed should be minimized.
5. Product quality
   (*a*) The vapor product should be controlled at 10 mole% A, with deviations of $\pm 0.7\%$ allowed for periods of up to 10 minutes.
6. Efficiency and optimization
   (*a*) The required equipment capacities should not be excessive.
7. Monitoring and diagnosis
   (*a*) Sensors and displays needed to monitor the normal and upset conditions of the unit must be provided to the plant operator.
   (*b*) Sensors and calculated variables required to monitor the product quality and thermal efficiency of the unit should be provided for longer-term monitoring.

Measurements

Variable	Sensor principle	Range	Special information
F1	Orifice	0–10	
F2	Orifice	0–40	
F3	Orifice	0–40	
F4	Orifice	0–40	
F5	Orifice	0–10	
F6	Orifice	0–40	
L1	Δ pressure		Reactor residence time is 5 minutes
L2	Δ pressure		Drum liquid hold-up time is 5 minutes
P1	Piezoelectric		
T1	Thermocouple	0–100°C	
T2	Thermocouple	100–200°C	
T3	Thermocouple	50–150°C	
T4	Thermocouple	50–150°C	
T5	Thermocouple	50–150°C	

TABLE 25.8 (CONTINUED)

Measurements

Variable	Sensor principle	Range	Special information
T6	Thermocouple	50–200°C	
T7	Thermocouple	50–150°C	
T8	Thermocouple	250–350°C	
A1	Continuous	0–100 mole%	mole% A in reactor
A2	Continuous	0–15 mole%	mole% A in product

Manipulated variables

I.D.	Capacity (at design pressures)	I.D.	Capacity (at design pressures)
	(% open, maximum flow)		(% open, maximum flow)
v1	50.6%, 10	v5	70.0%, 29
v2	9.6%, 100	v6	18.1%, 110
v3	50.0%, 40	v7	60.3%, 67
v4	26.9%, 58	v8	50.0%, 10

Constraints

Variable	Limit values	Measured/ inferred	Hard/ soft	Penalty for violation
Drum pressure	High	Measured	Hard	Personnel injury
Reactor level	Low	Measured	Hard	Pump damage
	High	Measured	Hard	Hazard
Light key A in product	High	Measured	Soft	Reduced selectivity in downstream reactor

Disturbances

Source	Magnitude	Period	Measured?
1. Impurity in feed (influences the reaction rate, basically affecting the frequency factor k_0)	10% rate reduction	Day	No
2. Hot oil temperature	±20°C	200+ min	Yes (T2)
3. Hot oil temperature	±20°C	200+ min	Yes (T8)
4. Feed rate	±1, step	shift-day	Yes (F1)

Dynamic responses
(Input = all manipulated variables and disturbances)
(Output = all controlled and constraint variables)

Input	Output	Gain	Dynamic model
	{see equation (25.2) for some steady-state gains}		

Additional considerations

None

4. The vapor product comes from a small drum inventory, and flow fluctuations can be expected. Since the control objectives allow for variability in the product rate, this is not likely to be a concern.
5. Two liquid levels are non–self-regulatory and should be controlled via feedback to prevent them from exceeding their limits. Also, one vapor space pressure, while theoretically self-regulating, can quickly exceed the acceptable pressure of the equipment; therefore, the pressure should also be controlled.

LEVEL 2: PROCESS ENVIRONMENT.

6. Several manipulated variables (v_1, v_2, v_3, v_4, v_7) and all disturbances affect the reactor temperature and thus, reaction rate.

LEVEL 3: PRODUCT QUALITY.

7. There appear to be several manipulated variables that affect the flash product quality, A_2.

LEVEL 4: PROFIT.

8. There are no objectives specified to increase profit beyond controlling product flow rate and quality. However, there appear to be extra manipulated variables, or at least extra valves in the process. This inconsistency must be resolved.

Control Structure Step

Since no severe difficulties were identified in the third step, we proceed to the fourth step, where we begin to design the control structure. Since we anticipate strong interaction among variables because of the process recycle, process decomposition is not applied. However, the control is designed according to the five-level temporal hierarchy. The overall structure is first selected; then, enhancements are added; finally, algorithms and modes are chosen.

LEVEL 1: FLOW AND INVENTORY. The first decision is usually the flow controller, which determines the throughput in the process. Usually, this controls either the feed rate or the production rate. The control objectives state that the production rate does not have to be maintained invariant, which is fortunate, because controlling the vapor flow from a flash drum would be difficult without allowing the pressure to vary excessively. For this process and objectives, the feed rate F_1 will be controlled. Any of three valves, v_1, v_3, or v_4, could be adjusted to control F_1. From the overview, it is realized that v_4 may be adjusted to control the liquid level control in the flash drum, so this is eliminated from consideration as a manipulated variable for controlling F_1. Either of the remaining valves may be adjusted to control F_2. Somewhat arbitrarily, we select v_1 as the manipulated variable; this selection has the minor advantage that the fresh feed can be reduced to zero and the system operated on total recycle for a short time. The remaining valve, v_3, is not needed and could be removed; in the example, we will simply maintain the valve position constant at its base-case value.

The reactor level must be controlled, because it is non–self-regulating, and the residence time affects the chemical reaction. The outlet flow is manipulated to control the level, because the inlet flow has already been selected as the feed flow controller. The outlet flow is affected by both valves v_5 and v_6; thus, there are one controlled and two manipulated variables. We select valve v_6, to maintain the highest pressure in the heat exchanger and so tend to prevent vaporization. The redundant valve, v_5, will not be adjusted.

The liquid level in the flash must also be controlled within limits, and no objective compels tight or averaging control. Tight level control is selected, because the level control is part of the recycle process, and the entire process would not attain steady-state operation until the level attains steady state. The valve v_4 was allocated to control the level when the feed flow was designed.

The final issue at this level of the hierarchy is the pressure control of the flash drum. The vapor valve v_8 is selected to give fast control of pressure.

In summary, the following allocation of controlled and manipulated variables has been made at this point.

Controlled	Manipulated
F_1	v_1
L_1	v_6
L_2	v_4
P_1	v_8

LEVEL 2: PROCESS ENVIRONMENT. The reactor environment is affected by the flow rates and temperatures of the incoming streams. The fresh feed flow rate was specified to satisfy objectives at level 1 of the hierarchy, but the total flow rate F_2 and the inlet temperature T_4 are available for reactor control unless we choose to iterate and change the earlier decision. Of these two variables, only the inlet temperature T_4 is directly influenced by the manipulated valve v_2, although the total flow may be (and is) influenced through the recycle. The valve v_7 would affect the reactor inlet temperature, but the dynamics would be slow because of the dynamics in the flash liquid inventory. Also, we can look ahead to the need to control the flash temperature, where v_7 would give fast dynamics. Thus, we choose to adjust v_2 to control reactor environment.

LEVEL 3: PRODUCT QUALITY. The flash composition is to be controlled, because it is the key measure of product quality; it is controlled directly, without a temperature cascade, because the composition sensor is continuous with fast dynamics. The proper choice for the manipulated variable would be the heating oil valve v_7, because it gives fast feedback dynamics over a large range of operation.

In summary, the following allocation of controlled and manipulated variables has been made at levels 2 and 3.

Controlled	Manipulated
Reactor	v_2
A_2	v_7

A reactor variable to be controlled has not yet been selected and could be temperature or concentration. Two alternative designs will be evaluated: temperature control and reactor concentration control.

Optimization Step

There are no optimization objectives in the Control Design Form. The control design to this point has allocated all manipulated variables, except for v_3 and v_6, which were found to be redundant for the previous control objectives. These valves provide no additional process flexibility, except that of controlling some intermediate pressures in liquid flow lines. There seems to be no reason to control these pressures, and ordinarily, these valves would be eliminated to save equipment and pumping costs. In this case, the valves will simply be retained at their base-case percent opening.

To complete this step, enhancements to the basic structure of controller pairings are considered. For this simple process, the enhancements will be restricted to cascade and feedforward, and each controlled variable is discussed individually.

F_1: The flow process is very fast, and the control design needs no enhancement. A PI controller is appropriate for this process, with nearly no dead time and significant high-frequency noise.

L_1: The process has little or no dead time, and the pump pressure is relatively constant. Thus, no cascade or feedforward is required, although a level-flow cascade may be used. The algorithm selected is a PI with tight tuning, because the level influences the residence time, and zero steady-state offset is desired.

L_2: The process has little or no dead time, and the pump pressure is relatively constant. Thus, no cascade or feedforward is required, although a level-flow cascade may be used. The algorithm is a PI with tight level tuning.

P_1: The process is fast, and the pressure should be maintained at its set point, because it affects the flash product composition. Therefore, a PI controller is selected.

A_2: The concentration of A in the product stream is the key product quality and is affected by the disturbance in T_8. Note that a cascade is not possible, because there is no causal relationship between the valve v_7 and the measured variable T_8. A feedforward controller is possible, because the criteria for feedforward would be satisfied. However, as a preliminary decision, no enhancement will be selected, because of the relatively fast feedback dynamics. This decision will be evaluated at the completion of this study. The feedback controller should have a PI or PID algorithm, depending on the dynamics, fraction dead time, and measurement noise.

Finally, the reactor environment control options are evaluated to determine the best control design. Each is discussed briefly as follows.

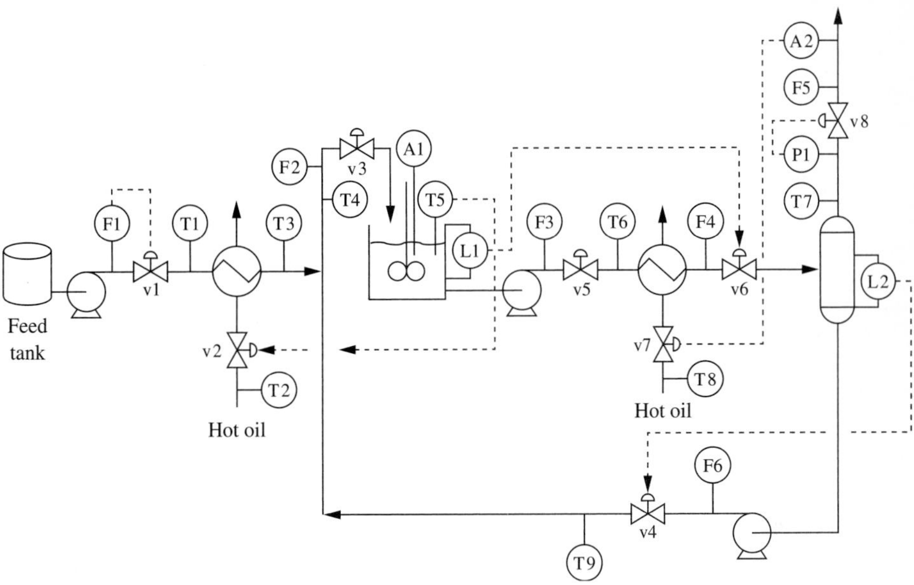

FIGURE 25.7
Control Design I.

1. Design I, shown in Figure 25.7, controls T_5. The reactor temperature is affected by several disturbances. These disturbances influence other measured variables before the reactor temperature measurement responds; thus, the potential for enhancements exists. For example, the measured fresh feed temperature T_1 could be a feedforward variable, and the feed temperature T_3 could be a secondary cascade variable. As a preliminary decision, the single-loop design $T_5 \rightarrow v_2$ is chosen with a PI algorithm. The resulting control of F_1, T_5, and A_2 is controllable, as can be verified using the gains in equation (25.2).
2. Design II, shown in Figure 25.8, controls the reactor composition A_1. A more direct measure of the reactor operation is the concentration of A, which can be controlled by adjusting valve v_2, although with slow dynamics. Therefore, the cascade design $A_1 \rightarrow T_4 \rightarrow v_2$ is selected, which gives good responses to temperature disturbances. The resulting control of F_1, A_1, and A_2 is controllable, as can be verified using the gains in equation (25.2).

Since no objectives have been stated for optimization, no further design decisions are needed at the fifth step in the sequence. Also, all manipulated variables have been allocated to control loops, except for v_3 and v_6, which will be held constant. Thus, no further degrees of freedom remain for adjustment.

Some control strategies would be required to ensure safe operation. The enclosed flash drum requires a reliable method for venting on high pressure, and a safety valve must be provided. Also, the objective of preventing an overflow from

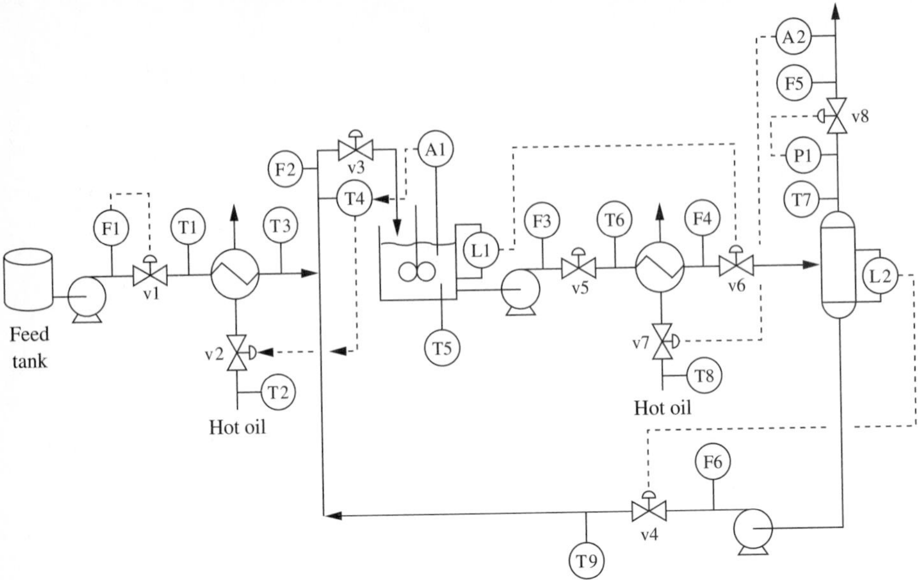

FIGURE 25.8
Control Design II.

the reactor could require a safety interlock system (SIS) to stop the feed flow if a high level is detected. If this feature is included, an alternative disposal for the liquid from the flash drum must be provided. The safety controls are not shown in Figures 25.7 and 25.8.

Monitoring and Diagnosis

All processes should be monitored for short-term operation and longer-term performance diagnostics. Shorter-term issues involve alarms for critical variables such as the liquid levels and the flash drum pressure. Some of the longer-term issues involve the reaction rate, which is influenced by impurities in the feed; recognition of poor feed characteristics would enable the engineer to trace the cause of the poor feed and take actions to prevent recurrence of such conditions. To monitor the product rate, the flow measurement F_5 should be accurate. If the density of the stream changes significantly, the conversion of sensor signal to the flow rate should be corrected based on a real-time sensor or on laboratory data of density. Another monitoring goal would involve the performance of the heat exchangers, which might foul over time. The measurements of the flows, temperatures, and valve positions enable some monitoring; for example, if the hot oil valve position increases over time at relatively constant production rate, the heat exchanger is most likely fouling. The lack of hot oil flow measurements prevents a complete check on the data; thus, the addition of flow and temperature sensors might be appropriate so that heat transfer coefficients can be calculated.

Evaluating the Design

Designs I and II are now complete. To evaluate their performances and select a final design, the dynamic performance of the process with each design was determined. In both cases, the process begins at the same initial steady state and is subjected to a change in feed impurity, which reduces the reaction rate (frequency factor) to 90% of its base-case value.

The response of Design I is shown in Figure 25.9. The operation of the process, especially the recycle flow rates, changes dramatically. The reason for this large change can be understood from process principles. The feed flow and the product purity remain unchanged. Therefore, the *rate* of production of B, $Vk_0e^{-E/RT}C_A$, must return to its initial value when steady state is attained. Because the reactor temperature and volume are maintained at their constant set points (in the steady state), the concentration of the reactant must increase to compensate for the impurity. Because of the low "single-pass" conversion in the reactor, a large recycle flow rate change accompanies the change in concentration. For successful operation the process equipment, pumps, pipes, and valves would have to have very large capacities, and thus the plant design would be costly.

The response of Design II is shown in Figure 25.10. After a transient, the process returns to nearly the same operating conditions, with the reactor concentration and volume at their initial values. To return the concentration to its set point, the A_1 controller increased the reactor temperature, thus maintaining the production rate of B constant. This response returns to steady state faster, satisfies all performance objectives for F_5 and A_2, and would not require excessive equipment capacity.

Control Design II should be evaluated for all disturbances in the CDF; these others are discussed briefly here but not plotted. Because of the T_4 temperature controller, it performs well for the +20°C disturbance in T_2, with only very small deviations in the compositions and product flow. The system experiences a rather large, but brief, disturbance when T_8 increases in a step of 20°C. The maximum allowable short-term variations in the product flow F_5 and the product composition A_2 are reached or slightly exceeded. If plant experience indicated that this disturbance occurred frequently, a feedforward compensation for changes in T_8, adjusting v_7 could be added to Design II. Finally, the response of a change in desired production rate, F_5, is rather sluggish, because the feed flow rate is manipulated manually, and the product increases slowly as the recycle system responds, finally attaining steady state. This is a direct result of the problem definition, because short-term variation in the product rate was stated to have negligible influence on the process performance in the CDF.

The IAE for the product quality variable (A_2) is 7.11 for Design I and 6.62 for Design II for the feed impurity disturbance. Since Design II has good performance for the key quality variable, has well-behaved dynamics for all variables, satisfies the control objectives, and requires equipment with smaller capacities, it is selected as the better control design for this process.

In performing this analysis, process decomposition was not employed, because of the strong integration, but temporal decomposition was helpful. The conclusion from this section is that the control design procedure was useful in ensuring that

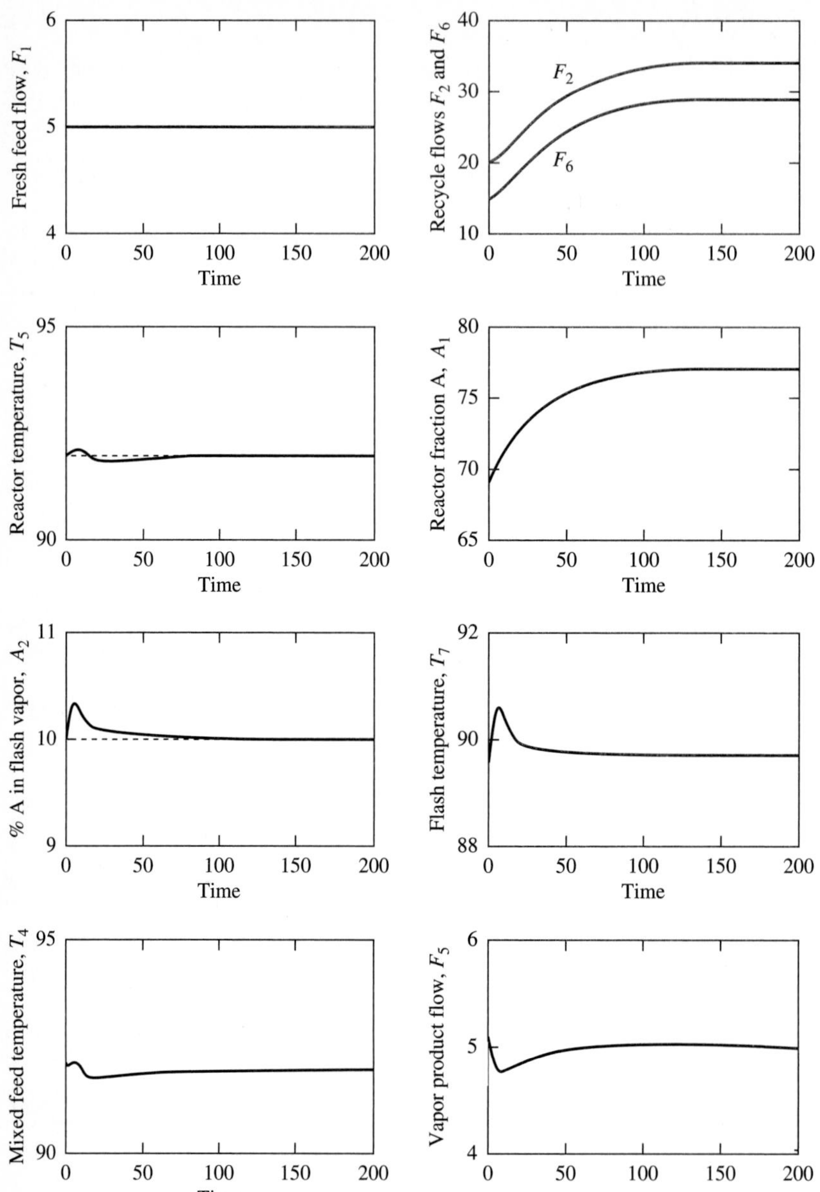

FIGURE 25.9
Transient response to feed impurity disturbance for Design I.

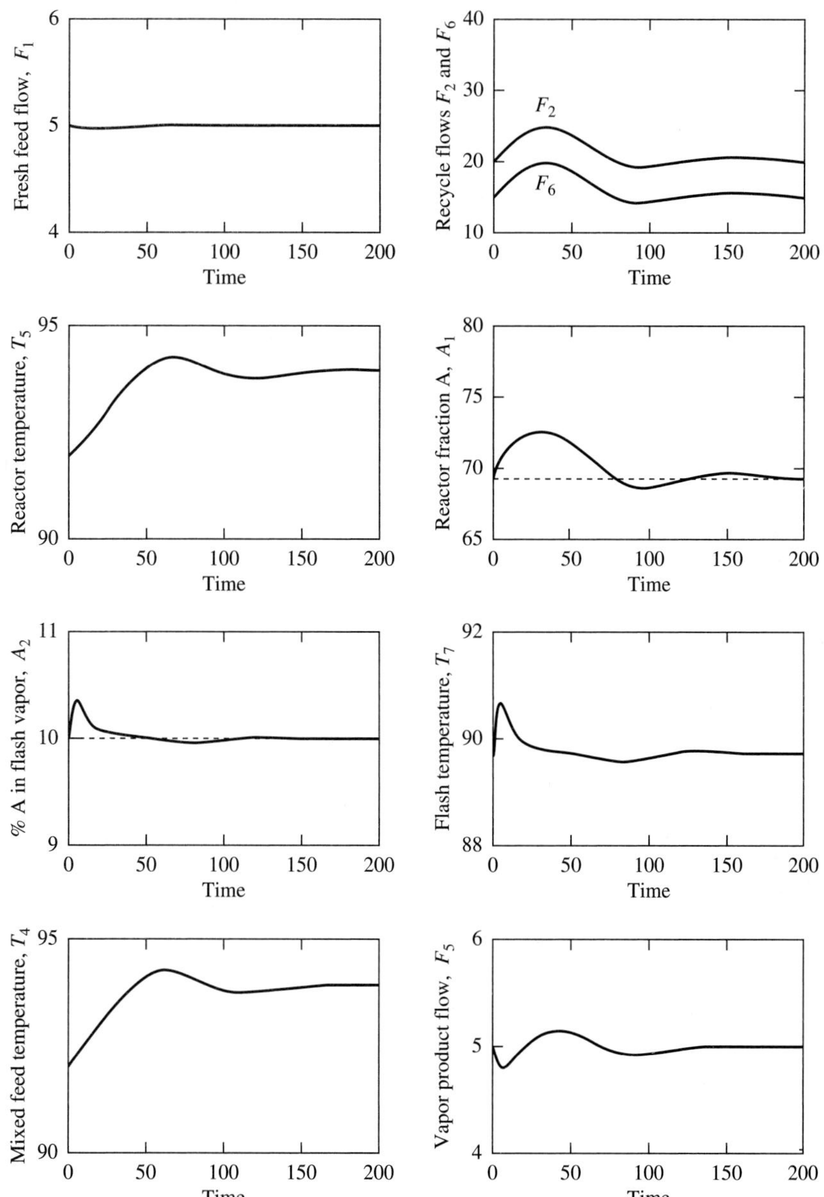

FIGURE 25.10
Transient response to feed impurity disturbance for Design II.

all important issues were considered, decisions were made in a reasonable order, and a good control design was completed. Other paths could have led to the same design, but proper shortcuts involve a very quick analysis of the factors covered in this procedure; shortcuts do not involve ignoring potentially important factors. Therefore, using the design procedure builds discipline and competence, enabling the engineer to make proper decisions quickly and, at times, in a less time-consuming manner.

25.8 SUMMARY OF KEY DESIGN GUIDELINES

Many useful guidelines have been developed in the preceding chapters for making control design decisions based on fundamental principles. Some of the more important and straightforward are summarized in this section. Before proceeding to the summary, the concept of control performance is reiterated. Here, control performance is defined with respect to the realistic situation of a nonlinear process with changing operating conditions; thus, a nominal linear model of the process used in analysis and tuning cannot be exact, and robustness under likely model uncertainty must be considered. The behavior of all process variables must be considered; this includes the controlled and manipulated variables and may include other "associated" variables, which may become limiting when they deviate too far from normal operation. Also, the possibility of noisy measurements must be considered in estimating performance. Finally, the performance must satisfy the requirements of the plant; thus, certain variables may have overriding influence on safety, product quality, and profit. Therefore, a simple summation of the IAE for all controlled variables often does not represent the process performance. Some controlled variables may be maintained close to their set points, at the expense of others varying far from their set points. This rich definition of control performance increases the difficulty of the design task, but it represents the realistic situation in most commercial enterprises.

The design procedure in Table 25.6 would generally encounter the decisions in the following order.

1. *Degrees of freedom.* A model of the system must have zero degrees of freedom when all external inputs are specified; this is simply requiring the model to be correctly formulated. The number of external manipulated variables (i.e., final elements) must be greater than or equal to the number of variables to be controlled. Recall that the degrees of freedom must be evaluated using the dynamic model of the process.
2. *Controllability.* The definition used in this book refers to the ability of the control system to maintain the controlled variables at their set points, in the steady-state conditions, as disturbances occur. For an $n \times n$ system, this requires that at least n linearly independent input-output process relationships exist. This can be tested by evaluating the determinant of the steady-state gain matrix; the system is controllable (by this definition) if the determinant is nonzero.

3. *Operating window*. This is the range of values of process variables for which the steady-state plant operation is acceptable (i.e., physically possible and within safety and product quality limits); it is also referred to as the *feasible operating region*. The window and operating points are typically evaluated using a nonlinear, steady-state model of the process. One or several operating points may be selected within the window to give good plant performance. If a process output variable appears at or near a constraint (frame) of the window, it should be controlled to prevent violations of the limit. If a manipulated variable appears at a constraint (frame), it should be held constant, if possible. Normally, the plant conditions have to be moved "inside" the window, or off the frame, to ensure that no violations occur during operation with disturbances. When important variables change from internal to on a constraint as conditions change, the engineer should anticipate the need for variable-structure control methods.
4. *Integral controllability*. The closed-loop system is integral-controllable if the sign of the controller gains, which stabilize each single-loop system with all other loops open, can result in stable control for the full $n \times n$ closed-loop system. This condition can be checked using the Niederlinski condition, which is sufficient but not necessary.
5. *Interaction*. The relative gain defined in equation (20.25) provides one measure of process interaction. It has limitations, because it represents only steady-state behavior and does not indicate strong one-way interaction, but when interpreted carefully, it gives useful information. Specifically, pairing control loops that represent negative relative gains results in *poor integrity* (specifically, systems whose stability depends on the manual/automatic status of the loops); thus, such pairings are judged to be unacceptable. Also, pairing on loops with zero relative gains results in systems whose proper functioning depends on the status of many loops; pairing on zero relative gains is to be avoided, but may be done if it provides a substantial improvement in control performance. Finally, control designs with loop pairings on relative gains near 1.0 are preferred, all other factors being equal.

   The performance of multiloop control systems depends on the type, or direction, of the disturbance. The relative disturbance gain, RDG, was introduced as an approximate indication of whether the interaction is favorable or unfavorable. Designs with relative disturbance gains of small magnitude for important controlled variables are generally favored, although evaluation of the dynamics is warranted before final selection.
6. *Feedback process dynamics*. Generally, feedback control performs well when the dynamics in the feedback path are fast, with a short dead time. Also, inverse responses were shown to degrade control performance, and, because multivariable control systems have a parallel structure, the closed-loop systems can experience inverse responses even though each individual input-output dynamic response does not. Improved control performance can be achieved in many cases by selecting from a suite of enhancements that improve dynamic performance, such as cascade, feedforward, adaptive tuning, inferential control, and process modifications that reduce the feedback dynamics.

Processes that are open-loop stable are preferred. Non–self-regulating levels and all pressures are noted for feedback control when reviewing a process. Also, processes that have significant inherent positive feedback should be evaluated to be sure that they are open-loop stable; if unstable, efforts should be made to modify the process design.

7. *Disturbance dynamics.* Additional steps can be taken to reduce the effect of the disturbance; the best action is to eliminate it at the source. Other steps include feedforward control, inventory sizing, and averaging level control, to modulate the rate of change in flow properties, and process operating condition changes, to reduce the sensitivity to a selected disturbance.
8. *Tuning guidance.* The control design and tuning should be selected concurrently. For example, certain levels may require averaging or tight level control, and interacting loops should be tuned to increase favorable interaction and minimize unfavorable interaction. These requirements should be documented as part of the control design; later implementation that does not adhere to the proper tuning is likely to be unsuccessful.

The methods used for the control design procedure involve a hierarchical analysis, in which the initial steps establish the feasibility of achieving the desired performance with the process and control designs. These initial evaluations are selected using "open-loop indicators" (Barton et al., 1991), which depend solely on the process and are independent of the control structure, algorithms, and tuning. The operating window, controllability, integral controllability, and relative gain are in this category. In these steps, many inappropriate design candidates are eliminated; also, many insights into the possible strengths and weaknesses in the remaining candidates are developed. Note that most of these evaluations can be based on steady-state models.

For final design of the process and selection of the best control design, the dynamic behavior of the closed-loop system *must* be considered. For example, Skogestad et al. (1990) demonstrate that reliance solely on steady-state analysis can result in the best control design being eliminated from consideration in distillation control. Further, a straightforward example of the importance of dynamics is the pairing of an important controlled variable with a manipulated variable that gives fast feedback dynamics; this can even lead to pairing on a zero relative gain, in extreme situations. In general, the behavior of multiloop systems can be quite complex, with poor designs yielding inverse response even when the process dynamics are well behaved (see Example 21.4). The frequency-dependent relative gain was briefly introduced to evaluate complex interactions, but the best approach is to simulate the final selection(s) to ensure good dynamic behavior. The use of nonlinear dynamic models for this final evaluation provides additional checks on the approximations inherent in the linear analysis methods used at earlier steps in the evaluation.

25.9 CONCLUSIONS

While no new technology was presented in this chapter, very important methods for managing the design procedure were presented. They enable the engineer to utilize information fully and effectively, to recognize when the problem is or is not fully defined, to apply the simplest decision methods at each stage, and to conclude the design procedure with high probability of success.

REFERENCES

AIChE, *Guidelines for Hazard Evaluation Procedures* (2nd Ed.), American Institute of Chemical Engineers, New York, 1992.

Barton, G., W. Chan, and J. Perkins, "Interaction Between Process Design and Process Control: The Role of Open-Loop Indicators," *J. Proc. Control, 1, 3,* 161–170 (1991).

Buckley, P., *Techniques in Process Control,* Wiley, New York, 1964.

Buckley, P., "Material Balance Control of Recycle Systems," *Instr. Tech.,* 29–34 (May 1974).

Skogestad, S., P. Lundstrom, and M. Morari, "Selecting the Best Distillation Control Configuration," *AIChE J. 36,* 753–764 (1990).

ADDITIONAL RESOURCES

Considerable material (including books, videos, and short courses) on HAZOP analysis is available from the Center for Chemical Process Safety of the American Institute of Chemical Engineering.

The concepts in the recycle example have been expanded upon in the following set of publications.

Luyben, W., "Dynamics and Control of Recycle Systems, 1. Simple Open-Loop and Closed-Loop Systems," *IEC Res. 32,* 466–475 (1993); "2. Comparison of Alternative Process Designs," *32,* 476–486 (1993); "3. Alternative Process Designs in a Ternary System," *32,* 1142–1153 (1993); Tyreus, B. and W. Luyben, "4. Ternary Systems with One or Two Recycle Streams," *32,* 1154–1162 (1993).

Additional checklists that can be useful in developing ideas for the Control Design Form can be found in

Marlin, T., J. Perkins, G. Barton, and M. Brisk, *Advanced Process Control Applications, Opportunities and Benefits,* Instrument Society of America, Research Triangle Park, NC, 1987.

The analysis approaches in the last few chapters are complemented by references giving the practice of control design for specific process units. A few examples are cited in Chapter 1 and below.

Balchen, J. and K. Mumme, *Process Control Structures and Applications,* Van Nostrand Reinhold, New York, 1988.

Baur, P., "Combustion Control and Burner Management," *Power, 126,* S-1 to S-16 (1982).

Duckelow, S., *The Control of Boilers* (2nd Ed.), Instrument Society of America, Research Triangle Park, NC, 1991.

Liptak, B., "Optimizing Controls for Chillers and Heat Pumps," *Chem. Engr., 90,* 40–51 (October 1983).

Luyben, W. (ed.), *Practical Distillation Control,* Van Nostrand Reinhold, New York, 1992.

Shinskey, F., *Distillation Control* (2nd Ed.), McGraw-Hill, New York, 1984.

Shinskey, F., *Energy Conservation through Control,* Academic Press, New York, 1978.

Starolesky, N. and L. Ladin, "Improved Surge Control for Centrifugal Compressors," *Chem. Engr., 86,* 175–184 (May 1979).

Leading research and practice in process control relating to process and control system design is presented in many technical meetings, including the following, which are organized periodically.

Process Systems Engineering (PSE).
International Federation of Automatic Control, *Advanced Process Control in the Chemical Industries* (ADCHEM).
International Federation of Automatic Control, *Workshop on the Interaction between Process Design and Control.*
International Federation of Automatic Control, *Symposium Series on Dynamics and Control of Chemical Reactors, Distillation Columns, and Batch Processes* (DYCORD).

> The procedures introduced in this chapter are applied using technology presented throughout the book. Questions for testing your learning are located at the end of Chapter 24. The questions at the end of Chapters 13 and 21 should also provide useful exercises. A few questions are given here that relate to the methods and examples introduced in this chapter.

QUESTIONS

25.1. Answer the following questions on the reactor with separator process.

(*a*) Verify that selected controlled variables in Designs I and II can be controlled with the selected manipulated variables.

(*b*) Check each of the Designs (I and II) to determine whether it is integral-controllable.

(*c*) Evaluate the relative gains for the two designs and discuss the implications.

(*d*) Demonstrate that flows F_1 and F_6 can be controlled with v_1, v_3, and v_4. Discuss reasons for selecting two of these three valves.

25.2. Discuss the performance of Designs I and II and propose better alternative designs, if possible, for the following situations. Each situation is to be considered separately, not cumulatively.

(*a*) The reactor temperature, T_5, must be maintained constant to obtain the best product selectivity. Is there an alternative reactor environment variable that can be adjusted? If yes, design a control strategy to meet the objectives.

(*b*) The analyzer for the reactor concentration is quite expensive. Is there another variable that can be used in its place?

(*c*) The control objectives are changed to include tight control of the product flow rate F_5. The disturbances are unchanged. How should the control strategy be changed?

(*d*) The daily total production of product B must be satisfied as close to its target as possible. How can the design be modified to satisfy this requirement?

(*e*) The recycle pump has been replaced with a spare pump of smaller capacity. Modify the control design to produce as much product as possible.

25.3. Using the checklists in Section 25.2, prepare Control Design Forms for the following processes. You should note information that you would need to determine from the plant personnel to complete the form.

(*a*) The distillation process in Examples 5.8, 20.2, 20.4, 20.5, and many examples in Chapter 21.

(*b*) The fired heater in Figure 15.17.

(*c*) The boiler in Figure 2.6.

(*d*) The gas distribution network in Question 24.15.

25.4. A series of processes is represented by the simplified system of flows and inventories in Figure Q25.4. Design a variable structure control system that will maximize the throughput while maintaining all levels within their maximum and minimum limits. The constraint that determines the maximum throughput could be the maximum feed target, the maximum product flow target, or any pump-valve combination in the system. (The targets are specified by the plant personnel.)

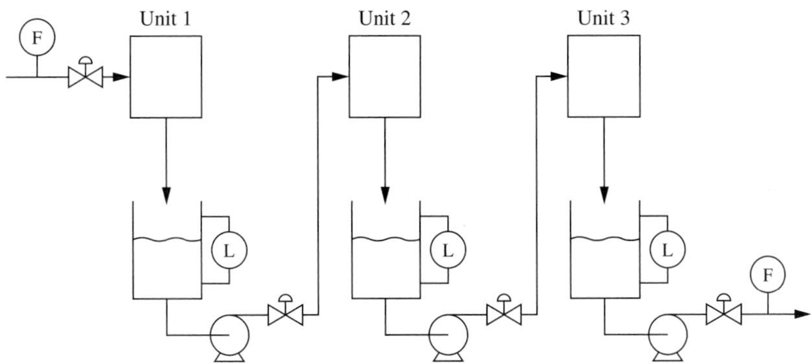

FIGURE Q25.4

25.5. An inverse response (right-half-plane zero) in the feedback process dynamics in a *single-loop* control system was analyzed in Examples 5.4 and 13.8. Assume that a two-input–two-output process has monotonic step responses for each input-output relationship. Discuss whether the 2 × 2 closed-loop control system can have an inverse response in one controlled variable, and if so, under what conditions. If yes, discuss how this situation may affect the control performance of the system.

25.6. Discuss the following issues in control design.

(*a*) What is the proper design for systems with more manipulated than controlled variables?

(*b*) How does the design engineer decide at which point(s) the process should be operated within the operating window?

(*c*) Is it impossible to implement feedback control for a system that is not integral-controllable, as determined by the Niederlinski index?

(*d*) If a nonzero operating window exists, is the process guaranteed to be controllable within the window?

(*e*) Is it appropriate to design a multiloop control system without giving guidance on tuning the controllers?

25.7. Discuss the following issues in control for safety.

(*a*) Give examples of how control strategies for temporal levels 1 through 3 (flow to product quality) contribute to safe operation.

(*b*) Give examples of how control strategies for temporal levels 1 through 3 can negatively influence the safety of the system. For each example, give a control design decision that would ameliorate the hazard.

25.8. For each of the processes in Question 25.3, determine process performance characteristics that should be monitored using real-time data. For each characteristic, define the calculations and sensors required and how the results would be interpreted, and discuss the actions taken when the process performance becomes unsatisfactory.

25.9. A major process design change is being evaluated for the reactor-with-recycle process. The stirred tank reactor can be replaced with a packed-bed reactor, as shown in Figure Q25.9. A new liquid byproduct, component C, is also produced, and it is separated from the recycle A (and B) in a liquid-liquid separator. Sketch a control system design for this process in the figure. You may add valves and sensors as needed.

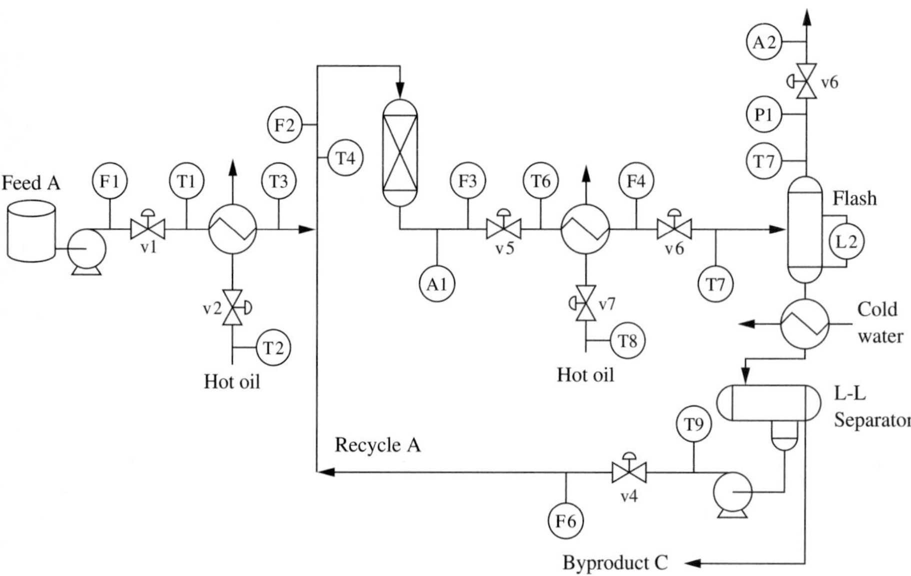

FIGURE Q25.9

CHAPTER 26

CONTINUAL IMPROVEMENT

26.1 INTRODUCTION

Decades of industrial experience have demonstrated the success of process control in maintaining selected variables near their desired values. Essentially all process plants apply automation, using feedback and feedforward principles to achieve safe and profitable production of consistently high-quality product. In general, process control is very effective when the desired operation has been determined from prior analysis and the control system has sufficient time to respond to disturbances (i.e., the feedback dynamics are fast compared with the disturbance frequency).

While process control, using the methods presented in this book, is required for regulating some process variables, the application of these methods may not be appropriate for all important variables. In some situations the best operating conditions change, and a fixed control design may not respond properly to these changes. In other situations, continuous feedback compensation can be too aggressive, leading to excessive variation in the controlled variables. Two approaches for continually improving plant operation are introduced in this chapter to address these situations. Both use the basic principle of feedback control: using outputs of a system to influence inputs to the system. However, these approaches involve very different technologies to address unique objectives. The approaches introduced in this chapter enhance the good performance achieved through process control.

Optimization. Optimization methods find the *extremum*—maximum or minimum—of an objective. Generally, the objective function will be profit, which we aim to maximize. When control objectives were discussed in Chapters 2 and 24, profit optimization was given less importance than safety, environmental and equipment protection, smooth operation, and product quality. Thus, these shorter-term objectives must be satisfied before we can turn our attention to profit, although the company will not survive in the long run without achieving profitable operation.

Statistical process control (SPC). The methods presented to this point in the book can be referred to as *automatic process control* (APC), because the control calculation is executed and the final element adjusted "automatically" as a result of the control calculation. In *statistical process control* (SPC) the process data is analyzed for opportunities for improvement, and when an opportunity exists, the data is diagnosed to ascertain an appropriate action. Thus, SPC involves statistical analysis of the real-time data, but not necessarily an action, at each execution. This additional analysis generally results in less frequent feedback actions, which can improve performance in some processes, as described in a subsequent section.

Both of these methods appear in the process control implementation hierarchy in Figure 25.2 (page 845), which shows them as higher levels in a cascade structure. Their decisions can be implemented through lower-level process control loops. For example, optimization systems can adjust the controller set points that regulate operating conditions such as temperatures and production rates. Alternatively, the highest-level decisions may involve complex manual intervention; in these cases, the results are provided in an advisory manner to plant personnel. Examples of such decisions are a change in feed material type and decisions on regenerating catalyst. Also, some diagnostic results indicate only that a significant change in process equipment performance has occurred, and further investigation by plant personnel is required to ascertain the cause and corrective actions.

Each of these topics is quite large, and entire books have been dedicated to their coverage. This chapter introduces some basic concepts for each approach and demonstrates how each relates to process control. It is important to recognize that most plants require excellent process control, to achieve safe and smooth operation, before the approaches in this chapter can be implemented and that opportunities for optimization and monitoring often exist. Thus, the engineer is not confronted with an "either/or" decision: all approaches in the hierarchy can be implemented concurrently.

26.2 OPTIMIZATION

The control design procedure in Chapters 24 and 25 allocates manipulated variables to achieve good dynamic performance, which is measured by the (hopefully, small) variability in key variables. Often, the number of manipulated variables exceeds the number of controlled variables. In these situations, safe operation and good product qualities can be achieved by manipulating selected process inputs that give the best control performance, and some manipulated variables can be maintained at arbitrary, constant values within an acceptable range. Alternatively, the excess manipulated variables can be adjusted to increase profit; these excess manipulated variables will be referred to as *optimization variables*. Some approaches for achieving high profit with excess manipulated variables have already been introduced; for example, the variable-structure controls in Chapter 22 provide the means for utilizing manipulated variables in a specified order, with the proper order based on the process economics. This approach has been employed in the examples of variable-structure control in Chapter 22 and Example 24.9. In this chapter, additional optimization approaches are introduced that address more complex situations, where a strategy for adjusting the excess variables is not as straightforward to determine and may change frequently. Three methods for optimizing process economics through adjusting optimization variables are discussed below and demonstrated with process examples.

I. Process Control Design

The first step in designing optimizing controls, after the regulatory controls have been designed, is an analysis to determine the proper strategy for the optimization variables. This analysis uses models of the process or plant data to answer two important questions:

1. *Do incentives exist for optimization?* In some situations the profit will not vary significantly as the values of the excess manipulated variables—the optimization variables—change. When the profit does not change, the optimization variables can be maintained at constant values selected for convenient operation. When the profit is significantly different for various values of the optimization variables, the next question is evaluated.
2. *Is the optimal strategy constant and simple?* When the profit is sensitive to the optimization variables, the response of these variables to external changes, disturbances, and set point changes (if relevant) should be evaluated. In some cases, the optimal response to these external changes (i) is nearly the same for all expected operating conditions and economics and (ii) can be implemented via straightforward real-time calculations as part of the control strategy.

When the answers to both questions are yes, a control strategy can be designed to approximate the best performance. Examples of this approach that have already been presented include the valve position controller in Figure 22.13 (page 738) and the production maximization in Figure 22.14 (page 739); in these examples, the best operating conditions were close to limiting values of key variables (i.e., they were "pushing constraints"). The method for process control design introduced in this subsection may not result in as simple a policy as operating near a constraint, but the concept is the same: implementing an operating policy that has been determined to be close to the best possible. The following example demonstrates the approach for answering the two questions above and, when appropriate, building the strategy to maximize profit via control calculations that do not explicitly involve economics.

Example 26.1. Steam is used in most process plants for power, driving turbines, and heat transfer. To satisfy the large and variable plant demands, many process plants have their own boilers and steam distribution networks. Typically, the boilers are arranged as shown in Figure 26.1, with all boilers providing steam to a single pipe, termed

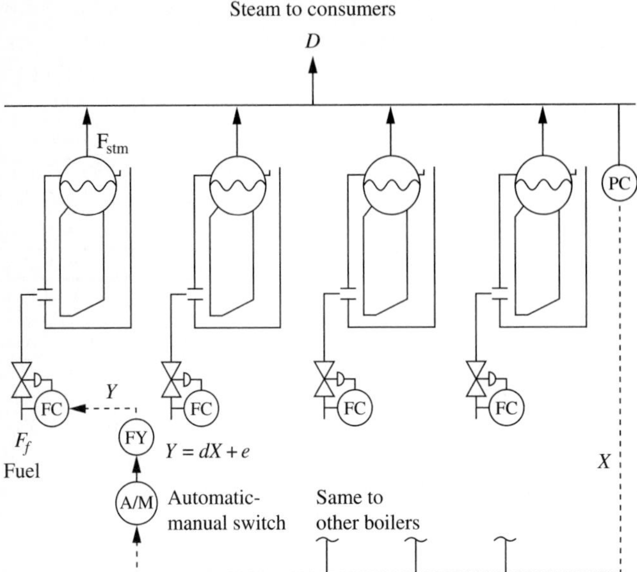

FIGURE 26.1
Multiple boiler and steam header.

a *header,* from which all consumers are supplied. The total steam demand can be provided by any combination of individual boiler productions that sums to the total demand. The boiler productions are often termed *loads,* expressed in units of fraction of the maximum from one boiler. This convention is used in the example, with all boilers having the same maximum and the total consumer steam demand expressed as a multiple of the maximum possible steam production from one boiler.

The basic requirement for process control is to ensure that the steam required by the consumers is produced by the boilers; in other words, the consumers and producers of steam are "in balance" at all times. This is achieved by controlling the

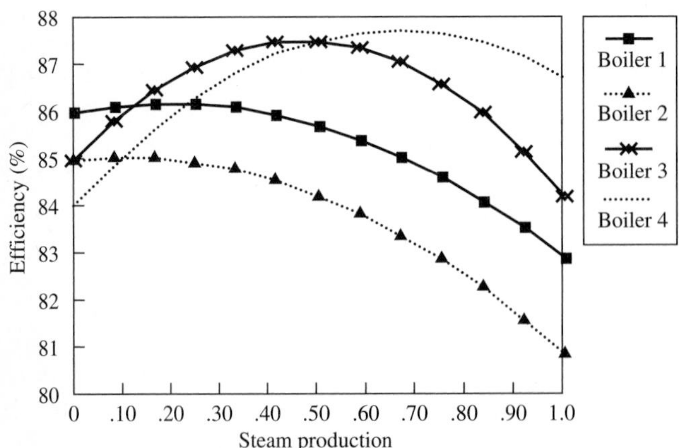

FIGURE 26.2
Boiler efficiencies for Example 26.1.

header pressure by adjusting the fuel to the boilers; any combination of steam productions from the four boilers that sums to the required total satisfies the basic objective. The percent efficiency for a boiler is defined as 100 × (energy transferred to the water)/(total heat of combustion); note that the energy to the water includes preheating the water, vaporization, and superheating the steam. Since the efficiencies vary as the demand changes and are different for different boilers, opportunity exists for influencing profit by using the minimum fuel, while satisfying the total demand from the steam consumers. In this example, the boiler efficiencies, from Cho (1978), are given in Figure 26.2.

Using this data, the process performance can be determined for any distribution of boiler loads at any steam production, D, which is the consumer demand. As explained in Chapter 2, the additional information required to calculate the benefits for automation is the distribution of plant operating conditions, which is here defined by the variability of the consumer demand. For this example, the demand is assumed to be uniform over the range of 0.8 to 2.5, as shown in Figure 26.3; for a real situation, this distribution would be determined based on process data.

The fuel requirements for any steam demand, which is the direct measure of economics, can be determined by the application of equation (26.1).

$$F_f = \sum_{i=1}^{N} \frac{(F_{\text{stm}})_i (H_0 - H_{\text{stm}})}{\Delta H_c (\eta_i / 100)} \tag{26.1}$$

with

F_f = total flow of fuel to all boilers

$(F_{\text{stm}})_i$ = flow of steam from boiler i

H_0 = specific enthalpy of water to boiler

H_{stm} = specific enthalpy of steam to the header

ΔH_c = heat of combustion of the fuel

N = number of boilers (in this example, 4)

η_i = efficiency of boiler i (see Figure 26.2)

The total steam demand D is determined by the consuming process units and is variable. The best boiler operation satisfies the steam demand and minimizes the total

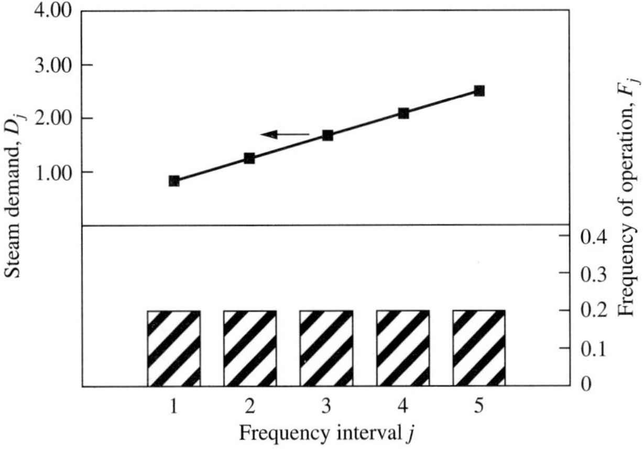

FIGURE 26.3
Data to calculate the average boiler performance.

fuel or, equivalently, maximizes the average efficiency. Also, the best operation is the average of the operations at the different demands weighted by the fuel at each. The maximization is defined mathematically in equations (26.2).

$$\max_{[(F_{\text{stm}})_i]_j} \eta_{\text{ave}} \tag{26.2a}$$

subject to

$$D_j = \left[\sum_{i=1}^{N} (F_{\text{stm}})_i\right]_j \tag{26.2b}$$

$$[\eta_i]_j = [a_i(F_{\text{stm}})_i^2 + b_i(F_{\text{stm}})_i + c_i]_j \tag{26.2c}$$

$$\eta_{\text{ave}} = \frac{\sum_{j=1}^{M} F_j \left[\sum_{i=1}^{N} (F_{\text{stm}})_i(\eta_i)\right]_j}{\sum_{j=1}^{M} F_j \sum_{i=1}^{N} [(F_{\text{stm}})_i]_j} \tag{26.2d}$$

with F_j = frequency at interval j (0.20 for all j in this example)
M = the total number of intervals (in the example, 5)
$F_{\text{stm}} \geq 0.0$

The solution to this nonlinear mathematical problem requires optimization mathematics, which are not central to this introductory coverage; this topic is explained elsewhere (Edgar and Himmelblau, 1988), and good software exists, such as GAMS (Brooke et al., 1992) and SPEEDUP (Aspen Technology, 1994). Thus, the results of the numerical solution of problem (26.2) are given in Figure 26.4 without details on the optimization method used. The best operation generates the required total steam by adjusting the steam produced from all boilers in response to a change in the demand. The approach gives the highest average efficiency, 87%. This complete optimization

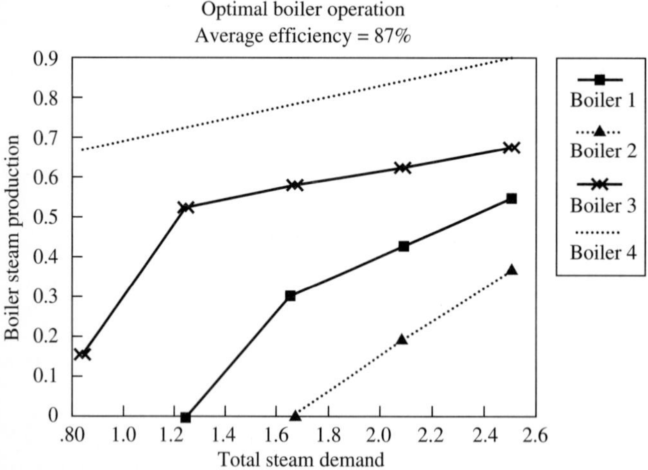

FIGURE 26.4
Optimal boiler load allocation for Example 26.1.

could be implemented as part of a control strategy but would require an optimization problem (26.2) to be solved frequently in real time.

The values of the optimization variables change as disturbances occur. In Section 26.2 two questions were posed in evaluating operations optimization. First, do incentives exist for optimization? This can be answered by determining the plant performance (here the average boiler efficiency) under the standard type of control. This base case is taken to be a load distribution for all boilers, so that the load of each boiler at any steam demand D would be D/N. The average efficiency for this example under the "equal loading" base case would be 86%, which is 1% lower than the optimal operation. Since this could represent a substantial increase in fuel consumption, incentives exist, and the second question will be considered.

The second question involved a simple control strategy that could, at least partially, replace the complex optimization calculations for real-time implementation. Since simplicity is always a goal—although not at the expense of poor product quality or significant loss in profit—an alternative approach to achieve partial optimization is evaluated. The simple alternative is to maintain the boiler loads at constant ratios, with the values of the constant ratios selected to give good (but suboptimal) economic performance. This design problem—which is solved only once, during design, to give parameters to be used in the real-time calculations—is the same as equations (26.2*a*) through (26.2*d*), but with the addition of equations (26.2*e*) for boilers $i = 2, N$ and interval $j = 1, M$:

$$[(F_{\text{stm}})_i]_j = R_i\,[(F_{\text{stm}})_1]_j \qquad (26.2e)$$

The solution of equations (26.2*a*) through (26.2*e*) determines the best values for the load ratios at each demand D_j. Note that R_i is the ratio of the steam from the *i*th boiler to the steam from the first boiler and that, once determined, this ratio does *not change* when the total steam demand changes. Thus, the resulting control strategy involves very simple calculations.

The solution of this problem is given in Figure 26.5. As expected from the optimal results, the ratios are selected to have a high steam production from the more

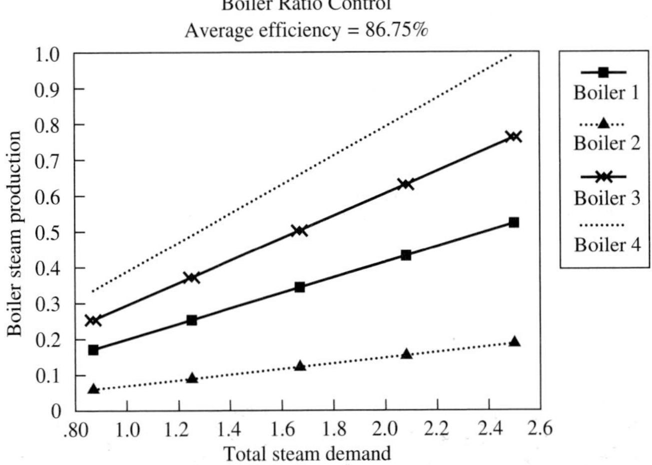

FIGURE 26.5
The best ratio boiler load allocation for Example 26.1.

efficient boilers. The average efficiency from this much simpler approach is only 0.25% less than the exact optimum for the wide range of operating conditions in Figure 26.3. Considering the likely accuracy of the boiler efficiency curves, this difference does not seem to be significant, and the simpler ratio control design would usually be selected. The ratio control could be implemented in a manner that would not influence good performance of the very important pressure controller. As shown in Figure 26.1, the pressure controller output influences every boiler fuel flow directly, and the controller output is modified to allow a ratio to be adjusted. The coefficients in the ratio calculation, d_i and e_i, would be determined from Figure 26.5.

An important lesson from this example is that tracking the best operating conditions does not always require extensive real-time calculations. The proper control calculations can be ratio control (this example), constraint pushing using signal selects (Figure 22.14), split range (Figure 22.1), or valve position controller (Figure 22.13). The correct design often requires careful process analysis to give a structure that closely follows the best operation in real-time control calculations.

This first approach, using a control design to approximate optimal operation, is appropriate when the control calculation need not change with time. For example, the ratios in Example 26.1 do not change as long as the efficiency curves for the boilers do not change with time. The result is a simple method that does not calculate or estimate the profit as part of the control calculation. In contrast, the next two approaches can respond to changes in plant performance by using process measurements in the calculation of profit, at the cost of much greater complexity.

II. Model-Based Optimizing Control

This second approach can be used when incentives exist for adjusting the optimization variables but the method for optimization cannot be implemented in a straightforward strategy such as constraint pushing or ratio control. In this approach a mathematical model of the process is used to calculate the best operating conditions for the current situation, and inevitable model errors are corrected (at least partially) using feedback measurements. Many technologies are available for real-time, model-based optimization. One of the simpler and frequently employed model-based approaches is introduced in the next example; it uses a linear model and a simple feedback updating method. When linear models are adequate, the model-based optimization can use the highly reliable linear programming solution of the optimization problem. When the feedback is introduced by adjusting the "bias" term in the linear model, the optimizing controller can be formulated in the model-predictive structure.

Example 26.2. In some cases, linear models can represent a process with satisfactory accuracy for the purpose of optimization of single process units. An industrially important control problem is the blending of several materials into a product mixture, with the control objectives to achieve the specified production rate and to maintain the product qualities within their limits. In this example, several hydrocarbon components are blended to produce gasoline. The product qualities, octane number (OCT) and vapor pressure (RVP), are important for the performance of the gasoline in an

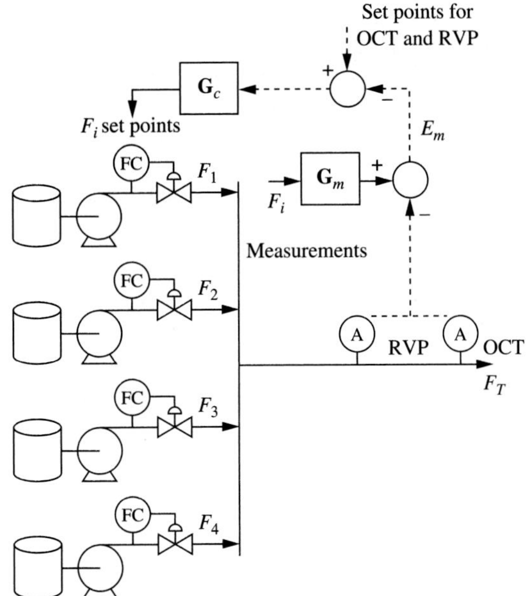

FIGURE 26.6
Blending process with optimizing, model-predictive controller.

internal-combustion engine (Stadnicki and Lawler, 1985). The component flows can take any values from zero to the maximum amount available.

This process is shown in Figure 26.6, and the linearized model is

$$(\text{RVP})F_T = \sum_{i=1}^{L} r_i F_i \tag{26.3a}$$

$$(\text{OCT})F_T = \sum_{i=1}^{L} o_i F_i \tag{26.3b}$$

$$F_T = \sum_{i=1}^{L} F_i \tag{26.3c}$$

with OCT = product octane
o_i = component octane
RVP = product vapor pressure
r_i = component vapor pressure
F_T = product flow
F_i = component flow
L = number of component flows (4 in this example)

In this example, the same model structure is used to represent both the true plant and the model used for control (i.e., $\mathbf{G}_m$ in Figure 26.6). The parameters in the controller model are *not* identical to those of the plant; these differences always occur in practice due to model error.

Note that the dynamics are so fast that the process is essentially at steady state, so the controller model is algebraic ($\mathbf{G}_m(s) = \mathbf{K}_m$). The controller in the model predictive structure involves an inverse of the process model, which for this case would be $\mathbf{G}_c(s) = \mathbf{K}_m^{-1}$. However, the process and the process model have more manipulated than controlled variables; four manipulated flows and only two controlled product qualities. In this situation many combinations of the manipulated-variable values can satisfy the controlled-variable values. This flexibility can be capitalized upon not only to satisfy the controlled-variable bounds in equation (26.3), but also to maximize profit by using the lowest-cost components. This flexibility is advantageous, but it leads to a mathematical problem that offers more challenge than taking the inverse of a square matrix. The statement of the problem to be solved by the *controller* $\mathbf{G}_c$ is

$$\max_{F_i} \text{ Profit} = V_T F_T - \sum_{i=1}^{L} V_i F_i \tag{26.4a}$$

subject to

$$(\text{RVP}_{\min})F_T \le \sum_{i=1}^{L} r_i F_i + F_T(E_m)_{\text{RVP}} \le (\text{RVP}_{\max})F_T \tag{26.4b}$$

$$(\text{OCT}_{\min})F_T \le \sum_{i=1}^{L} o_i F_i + F_T(E_m)_{\text{OCT}} \le (\text{OCT}_{\max})F_T \tag{26.4c}$$

$$F_T = \sum_{i=1}^{L} F_i \tag{26.4d}$$

$$0 \le F_i \le (F_i)_{\max} \tag{26.4e}$$

with V_T = value of the product

V_i = value of each component

E_m = feedback correction defined in equations (26.5), which would be zero if no feedback were implemented

Mathematical problems of this structure—linear equations that include both equalities and inequalities—are well known in applied mathematics as *linear programming* (Best and Ritter, 1985). The solution to this problem gives the values of the four manipulated variables (flows) that satisfy all equations under "subject to" and also maximizes the profit. The number of equations that are equalities at the solution is the number of original, strict equalities (26.4*d*) and the number of inequalities ($\le$ or $\ge$) that are at their limits at the solution. This forms the set of equations to be solved by adjusting the same number of manipulated variables. In this case, the solution contains one equality (26.4*d*) and two inequalities due to limits on the product quality (26.4*b* and 26.4*c*). Thus, three manipulated flows must be adjusted to values that satisfy the equalities. Since four flows exist, one flow is not specified, and linear programming theory demonstrates that this "excess" optimization variable must be at either its upper or lower limit, depending on which limit results in the highest profit.

Efficient computer programs are available to solve the linear program in equation (26.4), which is shown as $\mathbf{G}_c$ in Figure 26.6. If no feedback were included, the model would be used in a feedforward prediction of the correct flows to optimize the operation. The feedback control system in Figure 26.6 uses measurements of the product qualities to correct the model. Many possible methods can be used to correct the model, and in principle, all coefficients (o_i and r_i) could be adjusted when sufficient data is available. In this example, only the simplest feedback is considered, in which the difference between the measured value of the product quality and the model

TABLE 26.1
Table of data for plant behavior and controller model

System	Property	Component				Product	
		F_1	F_2	F_3	F_4	High	Low
Model	Octane	88	64.5	92.5	98	—	88.5
	Vapor pressure (psi)	5	14	138	5	10.5	—
Plant	Octane	91.8	64.5	92.5	96.5	—	88.5
	Vapor pressure (psi)	4	12	138	7	10.5	—
Model	Value ($bbl)	34	26	10.3	37	33	
Model and Plant	Maximum flow bbl /d	12000	6500	3000	7000	7000 (fixed)	

1 bbl (barrel) = 0.159 m^3; psi = 6.89 kPa

prediction is used to correct the model "bias" term. This is essentially the same type of feedback used in the model predictive controllers in Chapters 19 and 23. Thus, the E_m terms in equations (26.4) are

$$\begin{aligned}(E_m)_{RVP} &= RVP_{meas} - RVP_{model} \\ (E_m)_{OCT} &= OCT_{meas} - OCT_{model}\end{aligned} \tag{26.5}$$

with the subscript "meas" indicating the measured values in the product stream. While this type of feedback was shown to provide zero steady-state offset for steplike disturbances for the controllers in Chapter 19, it is *not guaranteed* to provide exact tracking of the true plant optimum for all situations of model errors, although it may under some conditions. Conditions for its success are given by Forbes and Marlin (1994).

In this example, the model used in the controller differs from the true plant performance, as would essentially always occur. The component qualities are given for the true plant and the controller model in Table 26.1. The dynamic response for this case under closed-loop control, with the model update equations (26.5) and the optimization problem (26.4) solved every controller execution, is given in Figures 26.7 through 26.9. In Figure 26.7, the component flow rates are shown for each controller iteration. The first iteration was performed without feedback ($E_m = 0$), so these results are a feedforward prediction of the best operation. After each controller execution, feedback measurements were taken and used to calculate the corrected biases E_m to be used by the controller for the next iteration. By the completion of the eighth iteration, the control system, using the feedback model corrections, achieved operating conditions that maximize profit in the true plant. The actual measured product qualities are shown in Figures 26.8 and 26.9. Both qualities should be within their upper and lower limits and, at the optimum, arrive at a limit—the upper limit for vapor pressure and the lower limit for octane, because this operation maximizes profit. Note that the qualities violate their limits during the transient responses in spite of the controller containing explicit equations for these limits, because the model errors are large enough to lead to significant, although temporary, violations of product quality limits in this example.

In general, many decisions must be made in designing and implementing a model-based real-time optimizer; some of these are model structure, parameters to be updated, measurements used for updating, and the updating calculation (e.g., least squares). Some guidance on these decisions is provided by Forbes and Marlin (1994)

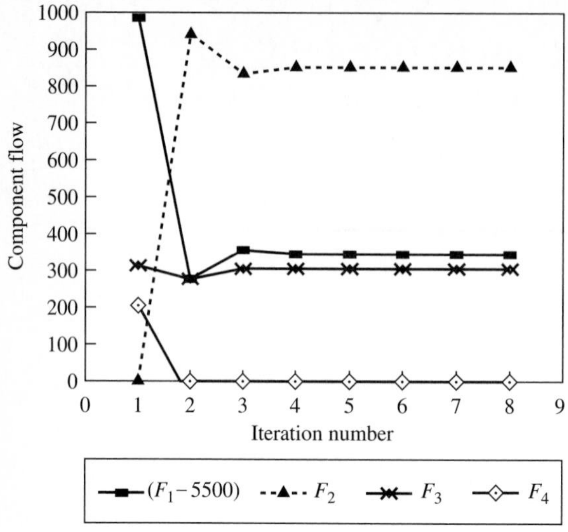

FIGURE 26.7
Component flow rates during feedback control.

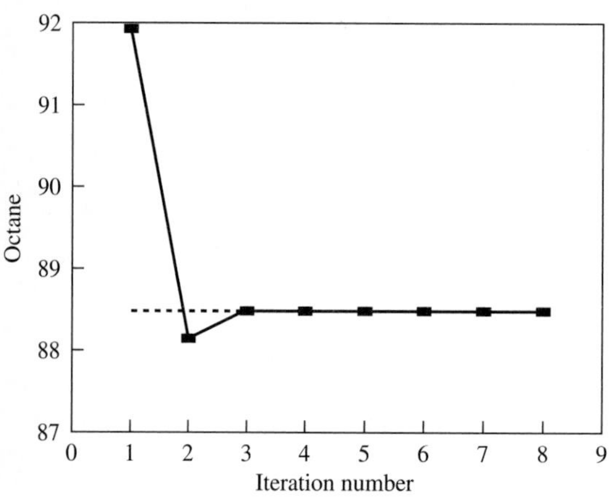

FIGURE 26.8
Response of octane under closed-loop optimizing control.

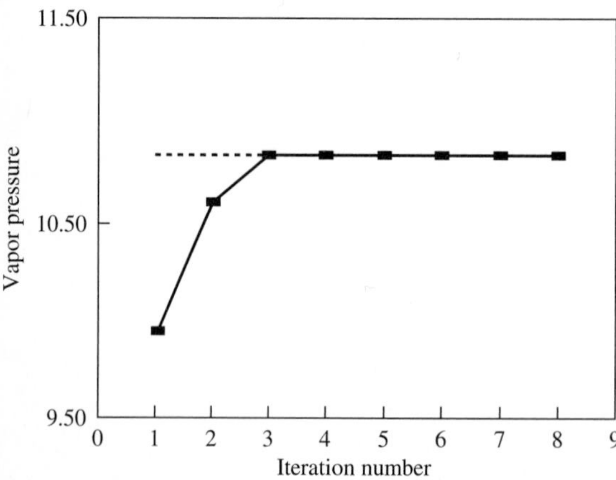

FIGURE 26.9
Response of vapor pressure under optimizing control.

and Krishnan et al. (1993). Industrial experience indicates great benefit for real-time optimization (e.g., Fatora et al., 1992; Larmon, 1977; Yang and Waldman, 1982). The best experiences are reported for plants with accurate models and good measurements, so that the feedback model updating leads to accurate representations. Also, substantial improvements occur more often in complex plants with many variables and changing conditions, where control structures, such as operating close to the same constraint, are not likely to yield the highest profit.

III. Direct Search

This third approach can be used when incentives exist for adjusting the optimization variables, but the strategy for optimization cannot be implemented in a straightforward strategy such as constraint pushing or ratio control, and accurate models do not exist. In these situations, a very simple, locally accurate model of the process is determined *empirically* from plant data. This model is used to determine the direction in which changes in the manipulated variables will increase profit. The plant operating conditions are then changed a small amount in this direction, and a new, updated model is evaluated. The direction for optimization is determined again from plant data, and another step is taken.

This iterative approach has been used for many years to study plant behavior and determine improved operating conditions. When the experiments are time-consuming and expensive, effort must be made to reduce the duration of the study; then, only a few experiments are performed and careful statistical evaluations are used to determine whether further improvement is likely and, if so, which direction is the best. Infrequent application of this concept in studies or "campaigns" is usually termed *evolutionary operation*, a term coined by Box and Draper (1969), who provided procedures, guidelines, and statistical tests for this periodic approach. While the periodic approach minimizes disturbances to the plant resulting from its designed experiments, it cannot track the best operation when it changes frequently.

The concept of building a locally accurate model for determining the direction of optimization can be extended to real-time, feedback control. Many algorithms are possible, and one of the simplest is discussed here (Bozenhart, 1986). The concept is shown in Figure 26.10, where the last few values of the optimization variable and calculated profit are plotted. Recall that the true plant profit is never known exactly; thus, an estimate of profit must be calculated from plant measurements. The direction in which the optimization variable should be changed to increase the calculated profit can be determined from the data in the plot. One method for determining the direction is to fit some of the most recent data with a straight line by least squares (Box et al., 1985). The slope of the line gives the correct direction (i.e., whether the optimization variable should be increased or decreased). The expression for the slope when there is one optimization variable is

$$S = \frac{\sum_{i=1}^{N_p} (P_i - P_{\text{ave}})(X_i - X_{\text{ave}})}{\sum_{i=1}^{N_p} (X_i - X_{\text{ave}})^2} \tag{26.6}$$

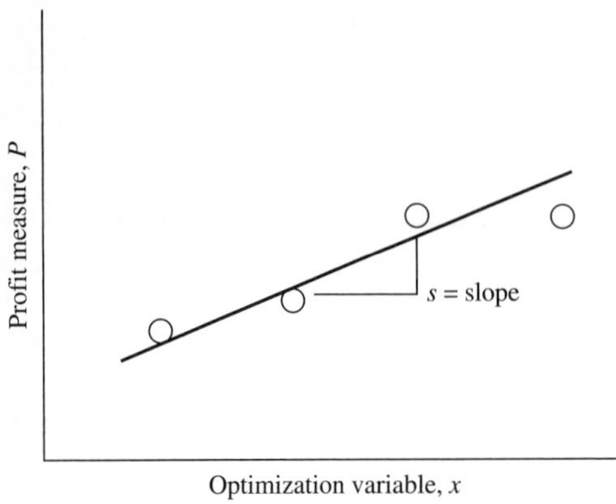

FIGURE 26.10
Using past data to determine the search direction.

with
N_p = number of points used in calculating the slope
P_i = profit at point i
P_{ave} = average profit (in N_p data points)
S = slope
X_i = optimization variable value at point i
X_{ave} = average value of the optimization variable (in N_p data points)

In this method, the optimizing controller makes a change in the optimization variable equal to $\Delta X[\text{sign}(S)]$, with the step size ΔX a fixed value independent of the magnitude of the slope. Note that this algorithm can be extended to more manipulated variables by modifying the expression for the slope. The following parameters appear in this algorithm and their selection and tuning are demonstrated in the next example.

CALCULATED PROFIT. The calculated variable should be directly related to plant profit and should be relatively insensitive to measurement noise and process disturbances.

OPTIMIZATION VARIABLE(S). The manipulated variables that yield excellent feedback control of safety-related variables and product quality should be allocated to these higher-priority tasks. The additional manipulated variable(s) that influence profit can be adjusted slowly to improve profit.

NUMBER OF PAST DATA POINTS. Past data provides a filter that makes the slope less sensitive to measurement noise; for this purpose, a large number would be good. However, too long a memory has two disadvantages. First, long memory gives importance to very old data that might not represent the current plant performance. Second, long memory requires many points on the "other side" of the maximum before the slope changes sign, which leads to large oscillations about the optimum operating point.

STEP SIZE. The step size should be small so that the change does not significantly influence important controlled variables, such as product quality. However, the step size should be large enough to cause a measurable change in the profit calculated from noisy plant measurements.

EXECUTION PERIOD. The approach to direct search described in this section requires the plant to achieve steady state between executions for measured values to represent the profit properly. Thus, the minimum execution period must be long enough for the process to achieve steady state: approximately the dead time plus four time constants for a first-order-with-dead-time process. Other approaches have been investigated that estimate parameters in a dynamic model and use the steady-state gain to determine the best direction (e.g., Bamberger and Isermann, 1978; Garcia and Morari, 1981).

CALCULATED DIRECTION. This method bases the direction on the slope. It would be possible to fit a higher-order curve to the data; however, the use of process measurements in calculating the profit estimate introduces noise into the method, which usually leads to unreliable estimates of coefficients of the higher-order terms.

Example 26.3. The steady-state behavior of a chemical reactor, shown in Figure 26.11, with series reactions A → B → C, was modelled in Example 3.12 (page 96). For constant temperature and flow rate, the composition of the intermediate component B experiences a maximum as the reactor volume, and residence time (space time), is changed; therefore, the optimum reactor volume depends on the reaction rate expression. Suppose that an inhibitor in the feed influences the rate constant for the first reaction without affecting the rate of the second reaction, and that the inhibitor concentration changes with time. If the inhibitor concentration is not measured but the level and the concentration of B in the product are, the direct-search optimization sketched in Figure 26.11 is possible.

In this case, all data are the same as in Example 3.12 for operation at 330 K except for the disturbance, when the parameter k_{01} deviates from its normal value of

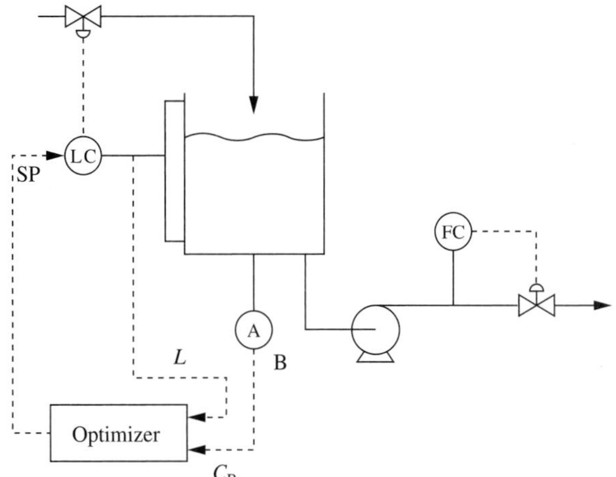

FIGURE 26.11
Stirred-tank reactor with direct-search optimizing control.

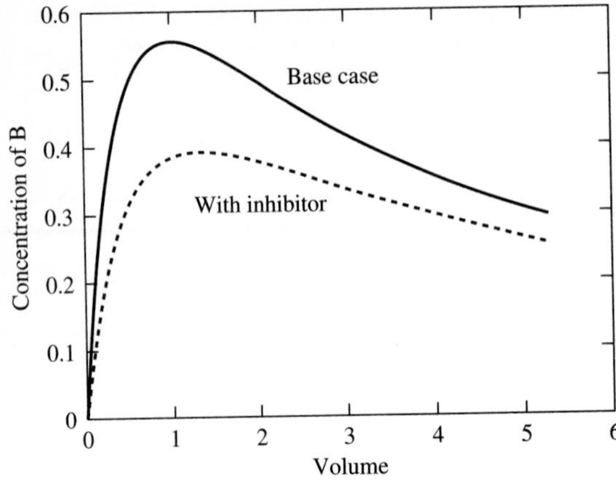

FIGURE 26.12
Concentration of B for two situations in Example 26.3.

17748.5 to 10000.0 for a period of time because of a change in inhibitor concentration; also, the flow rate is 2.65. The steady-state behavior of the system is summarized in Figure 26.12 for the two situations. The optimizing control algorithm just described is employed in this example with the following parameters:

Profit measure	= C_B
Optimization variable	= V (with T, C_{A0}, and F constant)
Number of points in memory	= 3
Step size	= 0.05 volume units
Execution period	= To achieve steady state
Calculated direction	= Slope, equation (26.6)

The performance of the direct-search optimization for the ideal situation, a plant without measurement noise, is shown in Figure 26.13. At controller iteration 20 the inhibitor in the feed increases in a step, and at iteration 50 it returns to its original value of 0.0. As a result of this disturbance, the concentration C_B decreases; then the search method adjusts the reactor volume V to achieve the maximum concentration of B for the current situation. Note that the optimum volume is shown in the figure only to aid in evaluating the performance of the optimizing controller; the optimum volume would normally not be known and was *not* used by the search algorithm.

The performance of the direct search for a realistic situation, in which the measurement of C_B includes noise, is given in Figure 26.14. The same scenario is involved in this data. As expected, the optimization performance is not as good, with some "wandering" around the optimum, but the algorithm was successful in changing the optimization variable in the proper direction and about the correct magnitude. Again, the true optimal value of the volume was *not* used by the direct-search controller.

26.3 STATISTICAL PROCESS CONTROL (SPC)

Automatic process control (APC) using feedforward and feedback principles identifies a deviation from desired operation (i.e., from the set point or points) and makes

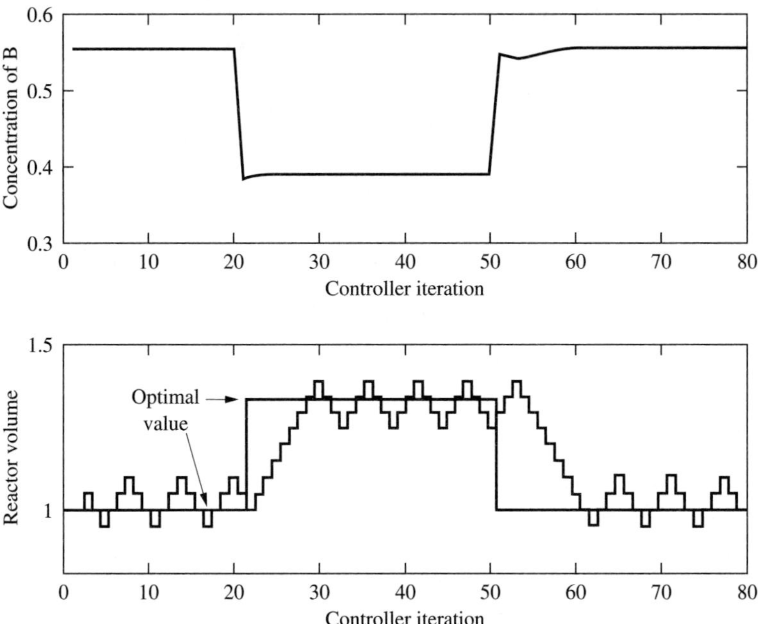

FIGURE 26.13
Direct-search optimization without measurement noise.

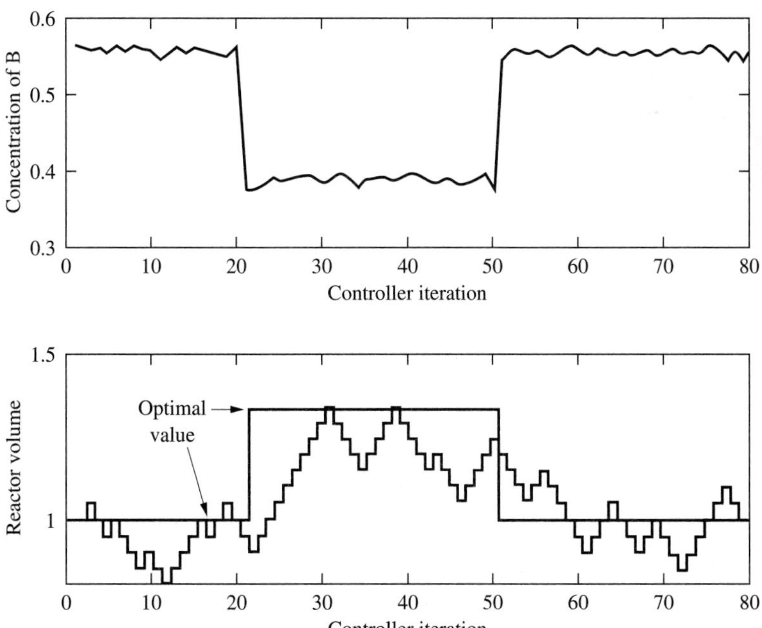

FIGURE 26.14
Direct-search optimization with measurement noise.

an immediate adjustment in a manipulated variable. Thus, automatic process control does *not eliminate* the cause of poor operation (i.e., the disturbance); the adjustment is selected to *compensate for the effects of* the disturbance and maintain the controlled variable at its desired value. Since the sources of disturbances have not been affected, the APC approach leaves the process susceptible to future disturbances from the same source. In contrast, statistical process control (SPC) has as a goal the *identification and elimination* of disturbances. By this approach of removing the source of disturbances, the long-term effect of SPC is to reduce variability in process operation and improve product quality. Since some variability in process operations is inevitable, statistical process control alone cannot adequately control most process operations. Fortunately, SPC and APC can provide complementary improvements and can be applied to the same process to improve the overall performance.

Statistical process control identifies deviations in process performance using real-time measurements. The base-case performance is established, not from a fundamental mathematical model, but rather from experience; thus, empirical data is used in establishing the typical variability in process variables. This variability results from many (small) disturbances and sensor noise, which are considered to be unavoidable. This typical variability is referred to as *common-cause,* which results in consistent variability over time. As each new set of process data is collected, it is evaluated by comparison with the common-cause variability, and the possibility of a significant change in process operation is evaluated. Significant deviation from the common-cause variability would then result from a disturbance that is not typical; this is referred to as a *special* (or *assignable*) *cause* of variability. If a change has occurred, the process is diagnosed to determine the proper corrective action. The corrective action may be as simple as adjusting a final control element, or it might be as involved as changing the source of feed material or catalyst to prevent the cause of the disturbance.

> Automatic process control compensates for deviations from set point. In contrast, statistical process control has the goal of identifying and eliminating causes of variability in key process variables.

Statistical process control is now demonstrated by way of its best-known method.

Shewhart Chart

The analysis of the process data to quickly and easily recognize changes in process performance is facilitated by the *Shewhart chart*, shown in Figure 26.15. The Shewhart chart provides a visual display of recent process data of a single measurement along with limits representing the typical, common-cause variability. The limits are determined empirically from "good" process operation and are typically set to include 99.7% of the data; if the data is normally distributed about its mean, the limits are located $\pm$ three standard deviations from the mean. The limits are referred to as

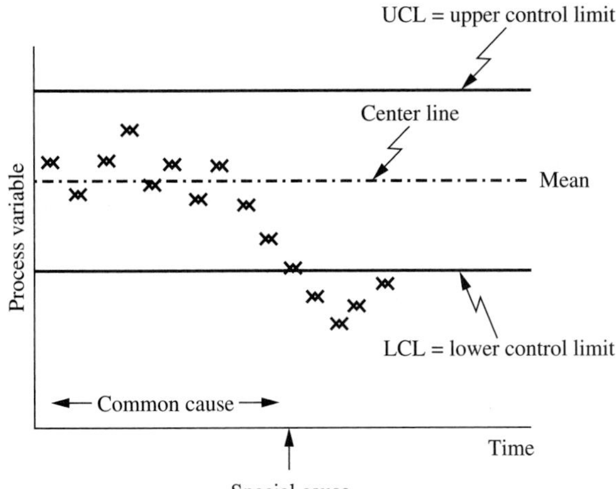

FIGURE 26.15
Shewhart chart.

the *upper* and *lower control limits* (UCL and LCL). Comparing a measured value with these limits is essentially a statistical hypothesis test on whether the mean of the variable has changed; this test could be calculated in a standard manner, although the clarity provided by the visual display of data with the limits increases the appreciation of the effects of variability (Montgomery, 1985). Also, modifications are available for variables with nonnormal distributions (Jacobs, 1990).

When the process is experiencing typical variability, often referred to as "in the state of statistical control," most data will be within the limits. Although there is variation of the measurement within these limits, this variation is accepted as inevitable and no action is taken, whereas automatic process control makes a feedback compensation for any nonzero error. If the measured value exceeds the limits, the SPC approach requires a diagnosis to determine the special or assignable cause and implementation of the appropriate corrective action.

Example 26.4. Reconsider the chemical reactor in Example 26.3 without the optimizer. The liquid level is controlled and the concentration of component B is measured online. Describe how the process could be monitored using a Shewhart chart.

The concentration of B is the key indicator of process performance and can be plotted on a Shewhart chart. Historical data, not shown, has been used to establish the common-cause variability and the control limits for the concentration. Some data are plotted in Figure 26.15 for this example. In the initial data, the concentration remains within the action limits, although it varies due to the common-cause disturbances: small changes in the level, flow, reactor temperature, and feed concentration. At a time indicated by an arrow, the concentration of B deviates from its usual range and remains outside this range for an extended time, which indicates a special-cause disturbance has occurred. In this example, the source of the disturbance is not obvious from the data, so additional diagnosis would be required. For example, the measures of the key process variables could be checked for errors, the reactor temperature could be determined, and the feed composition could be measured. As noted in Example 26.3, the inhibitor concentration is an important factor in the process performance and could be determined by laboratory analysis. If the inhibitor concentration has caused this deviation, as is likely for such a large disturbance, the underlying source of the disturbance should be determined; for example, the reason could be contamination in storage or poor quality from a supplier. Whatever the cause, the corrective action should not only eliminate the current disturbance but also prevent future occurrences. Notice that the optimization results in Example 26.3 can only give the best performance with a given level of inhibitor, which can represent a substantially lower concentration of B; only eliminating the disturbance can restore this process to its desired high concentration of B.

Reducing Variability

The distinction between APC and SPC can be clarified and the strengths of each can be demonstrated by considering two examples which could involve the same process, but experiencing different disturbances. Consider the packed-bed chemical reactor in Figure 14.11 (page 491), which is similar to the systems discussed in Examples 14.1 and 15.1. The objective is to maintain the concentration in the effluent measured by the sensor at a desired value, and concentration can be influenced by adjusting the heating medium valve in the reactor preheat exchanger. The performances of this process with and without feedback control are considered for two different scenarios.

SCENARIO I. For this scenario, the initial data is given without any feedback action in Figure 26.16. The cause of the variation for Scenario I is essentially random, uncorrelated noise about the constant mean value. For example, this could occur if (1) no (significant) disturbances occur in the reactor operating variables and (2) the sensor experiences a random error each time a sample is analyzed. In this situation, the proper operating policy for this common-cause variability is to make no adjustment to the valve, since the current error cannot be corrected by the adjustment and the current deviation does not provide an indication of the future deviations. As shown in Figure 26.16, implementing a standard proportional-integral feedback control calculation will *increase* the variability in the product quality. Thus, the SPC approach provides better performance for regulating the reactor in Scenario I.

SCENARIO II. In this scenario, the initial data is given without any feedback action in Figure 26.17. The variation for Scenario II is due not only to random sensor error but also to slower-changing disturbances in some process input variables such as feed composition and heating medium temperature. We can observe that the variability of the product composition appears correlated in time; that is, the composition includes a slow drift along with some random noise. In this situation, the current deviation provides an indication of the likely future deviations, and the feedback dynamics are fast enough that adjustments in the valve can compensate for the slowly changing disturbances. Thus, automatic process control is appropriate, as shown in Figure 26.17 which shows a *decrease* in the variability when a proportional-integral feedback controller is implemented. Thus, the APC approach provides better performance for regulating the reactor in Scenario II.

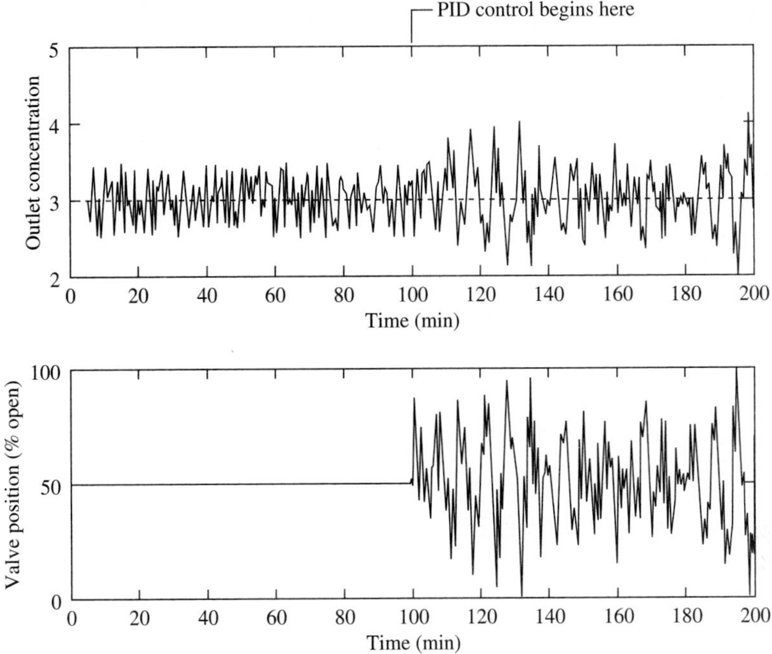

FIGURE 26.16
Dynamic response for Scenario I, in which feedback degrades performance.

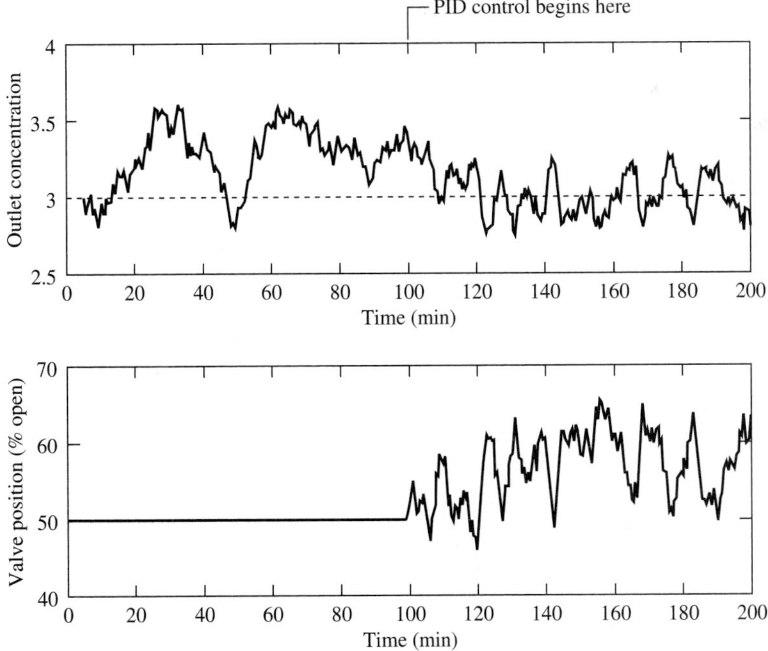

FIGURE 26.17
Dynamic response for Scenario II, in which feedback improves performance.

The comparison of the performance of SPC and APC for these two scenarios demonstrates that both have many applications. When the variability without feedback compensation is nearly random, so that feedback corrections cannot compensate for the deviations, an SPC approach is appropriate. When the variability without feedback compensation is due to slowly varying disturbances, APC can be effective in reducing the variability. For further discussion on this point, see MacGregor (1990).

Variability of the Manipulated Variable

Another distinction between APC and SPC stems from the frequency of corrective actions taken. APC involves an action every time the controller is executed; thus, it must be possible to adjust the final element without disrupting the process operation, which is possible with standard control valves. As a result, APC reduces the variance of the controlled variable while increasing the variance of the manipulated variable. This situation is sometimes described as "moving" the variability from the important controlled variable to the less important manipulated variable, as demonstrated in Figure 26.16. This situation has been discussed previously and has been shown in Figures 7.8, 7.9, 13.18, and 23.10.

In contrast, SPC involves infrequent adjustments—only when the measurement exceeds the control limits. This is advantageous for systems in which the cost of the control action is considerable. Examples of costly adjustments are changing the reactor catalyst, changing the feed material, and stopping and adjusting machinery. Since the special-cause disturbances occur infrequently and the action limits are set to result in few "false alarms" (only 3 in 1000), the SPC approach, when applied to appropriate scenarios, reduces the adjustments in the manipulated variable required to maintain the controlled variable within the upper and lower control limits.

This perspective suggests an approach for diagnosing process performance for variables that are under PID feedback control. In situations with effective feedback, the controlled variable may not deviate greatly from its set point, although significant disturbances occur. However, the occurrence of these disturbances can be determined by monitoring the manipulated variable, because it must be adjusted to compensate for disturbances.

Process Capability

The discussion to this point has addressed the variability of key process variables; now, the requirements of the market are added to the considerations. In particular, the comparison of the variability (here, assumed normally distributed) with the required minimum variability is an important factor in evaluating the success of the process operation. The comparison of actual with required variability is termed the *process capability*, defined as follows:

$$\text{Capability index} = C_p = \frac{\text{USL} - \text{LSL}}{6\sigma} \tag{26.7}$$

$$C_{pk} = \min\left[\frac{\text{USL} - X_m}{3\sigma}, \frac{X_m - \text{LSL}}{3\sigma}\right] \tag{26.8}$$

with C_p = process capability index
C_{pk} = process capability index
USL = the upper specification limit on acceptable variation in product variable
LSL = the lower specification limit on acceptable variation in product variable
X_m = mean value of the variable
σ = the standard deviation of the actual variability of the product quality

The variable C_p is meaningful when the target for the product specification is the mean of the range. The C_{pk} is meaningful when the target is not the mean of the range. The best situation occurs when the variability of the process is small compared with the variability allowed in the market:

$C_{pk} \ll 1$ Considerable "off- specification" material is produced

$C_{pk} \approx 1$ Most production satisfies specifications

$C_{pk} \gg 1$ Nearly all production well within the specifications

The capability index is a useful measure for evaluating the current process performance against the market needs. However, continual improvement efforts should not cease when C_p and C_{pk} are greater than 1.0.

> The reduction of variability should be a continual effort. The goals include the reduction in number of times special causes occur and the reduction of the common-cause variability.

The producer of the highest-quality product often can increase total sales or profit margins, and experience has shown that the lower-quality producers often cannot sell their products.

26.4 CONCLUSIONS

Two approaches for continual process improvement have been introduced in this chapter. Optimization is appropriate when the operating profit changes significantly because of frequent disturbances and there are available manipulated variables that can be adjusted to increase the profit without degrading the product quality. These variables tend to be set points of the underlying regulatory process controls. Thus, optimization generally functions as the highest level in a cascade control structure.

Statistical process control has as its goal the reduction of variability, primarily in the key product qualities. In contrast to automatic process control, statistical process control involves actions that address the root cause of the disturbance. By diagnosing and eliminating these causes, the number and severity of future disturbances are reduced, and the process performance is improved.

These approaches have merely been introduced in this chapter. The reader is encouraged to refer to the References and Additional Resources for further information. These methods can provide substantial improvement when applied continually to a process that is operating under excellent automatic process control.

REFERENCES

Aspen Technology, *SPEEDUP User Manual*, Boston, MA, 1994.

Bamberger, W. and R. Isermann, "Adaptive, Online Steady-State Optimization of Slow Dynamic Processes," *Automatica, 14,* 223–230 (1978).

Best, M. and K. Ritter, *Linear Programming*, Prentice-Hall, Englewood Cliffs, NJ, 1985.

Biles, W. and J. Swain, *Optimization and Industrial Experimentation*, John Wiley, New York, 1980.

Box, G. and N. Draper, *Evolutionary Operation: A Statistical Method for Process Improvement,* Wiley, New York, 1969.

Box, G., W. Hunter, and J. Hunter, *Statistics for Experimenters,* Wiley, New York, 1987.

Bozenhardt, H., "Hyperplane: A Case History," *Proc. Fifth Annual Control Engineering Conference,* May 1986.

Brooke, A., D. Kendrick, and A. Meeraus, *GAMS; A User's Guide (Release 2.25),* The World Bank, 1992.

Cho, C., "Optimal Boiler Load Allocation," *Inst. Tech.,* 55–58 (1978).

Edgar, T. and D. Himmelblau, *Optimization of Chemical Processes,* Wiley, New York, 1988.

Fatora, F., D. Kelly, and S. Davenport, "Closed-Loop Real-Time Optimization and Control of a World-Scale Olefins Plant," *AIChE Spring Meet.,* New Orleans, March 1992.

Forbes, F. and T. Marlin, "Model Accuracy for Economic Optimizing Controllers: The Bias Update Case," *IEC Res.,* accepted for publication (1994).

Garcia, C. and M. Morari, "Optimal Operation of Integrated Industrial Systems: Part I," *AIChE J., 27,* 960–968 (1981).

Jacobs, D., "Watch Out For Non-Normal Distributions," *Chem. Eng. Prog., 86,* 19–27 (1990).

Krishnan, S., G. Barton, and J. Perkins, "Robust Parameter Estimation in On-Line Optimization: Part I. Methodology and Simulated Case Study," *Comp. Chem. Eng., 16,* 545–562 (1992); "Part II. Application to an Industrial Process," *Comp. Chem. Eng., 17,* 663–669 (1993).

Larmon, F., "On-Line Optimization of an Ethylene Oxide Unit," in Van Nauta Lemke (ed.), *Digital Computing Applications to Process Control,* IFAC and North-Holland Publishing Company, 59–63 (1977).

MacGregor, J., "A Different View of the Funnel Experiment," *J. Qual. Control, 22,* 255–259 (1990).

Montgomery, D., *Introduction to Statistical Quality Control,* Wiley, New York, 1985.

Stadnicki, S. and M. Lawler, "An Integrated Planning and Control Package for Refinery Product Blending," *Control Engineering Conference,* 315–322 (1985).

Yang, C. and B. Waldman, "On-Line Optimization Boosts Ethylene Profits," *Oil and Gas J., 80,* 104–109 (September 1982).

ADDITIONAL RESOURCES

Many examples of process control designs that improve the economic performance of processes are given in

Liptak, B., *Optimization of Unit Operations,* Chilton Books, Radnor, PA, 1987.

A complete discussion of empirical methods for estimating response surfaces is given in the following reference:

Box, G. and N. Draper, *Empirical Model Building and Response Surfaces,* Wiley, New York, 1987.

An introduction to the standard methods in statistical process control is provided in the following references, which tend to provide methods for the parts manufacturing industries. They also give some guidance on managing the process diagnosis procedures.

Grant, E. and R. Leavenworth, *Statistical Process Control,* McGraw-Hill, New York, 1988.
Oakland, J., *Statistical Process Control, A Practical Guide,* Wiley, New York, 1986.

The process industries involve fast sampling of processes with relatively slow dynamics. In this situation, each data point can be correlated with points in the recent past. Although the general approaches in classical SPC are directly applicable, many of the specific methods and quantitative approaches must be modified. The following references provide insight into the special needs of the process industries.

Box, G. and T. Kramer, "Statistical Process Monitoring and Feedback Adjustment—A Discussion," *Technometrics, 34,* 251–267 (see also the extensive discussions on pages 268–285)(1992).
Harris, T. and W. Ross, "Statistical Process Control Procedures for Correlated Observations," *Can. J. Chem. Eng., 69,* 139–148 (1991).
Hunter, S., "The Exponentially Weighted Moving Average," *J. Qual. Techn., 18,* 203–210 (1986).
MacGregor, J., "On-Line Statistical Process Control," *Chem. Eng. Prog., 84,* 21–31 (1988).

The concepts of SPC can be extended to multivariable processes, although the direct, independent monitoring of many independent variables via Shewhart charts would be tedious and difficult to interpret. An alternative method is described in

Kresta, J., J. MacGregor, and T. Marlin, "Multivariate Statistical Monitoring of Process Operating Performance," *Can. J. Chem. Eng., 69,* 35–47 (1991).

QUESTIONS

26.1. Design an optimizing control strategy for the process in Figure Q26.1 to satisfy the following objectives.
(1) Tight control of the flow rate leaving the furnace via the coil
(2) The coil outlet temperature (TC) maintained close to its set point
(3) The total fuel consumption minimized

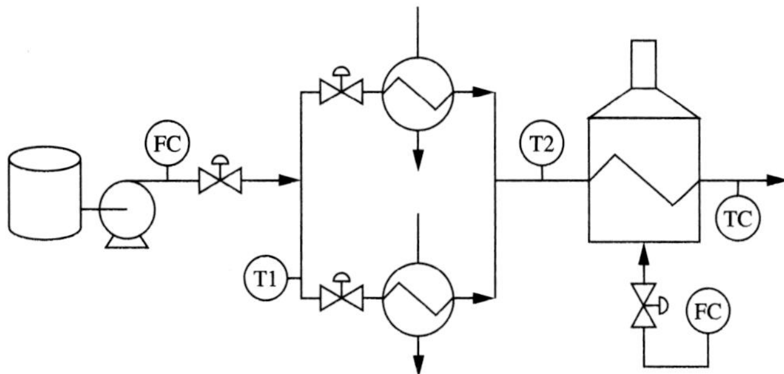

FIGURE Q26.1

Design a control strategy that achieves these objectives. Clearly define the measurements, calculation of the performance function, and the control algorithm and explain how interactions among the strategies will be considered.

26.2. Discuss the key elements of the single-stage refrigeration circuit in Figure Q26.2.

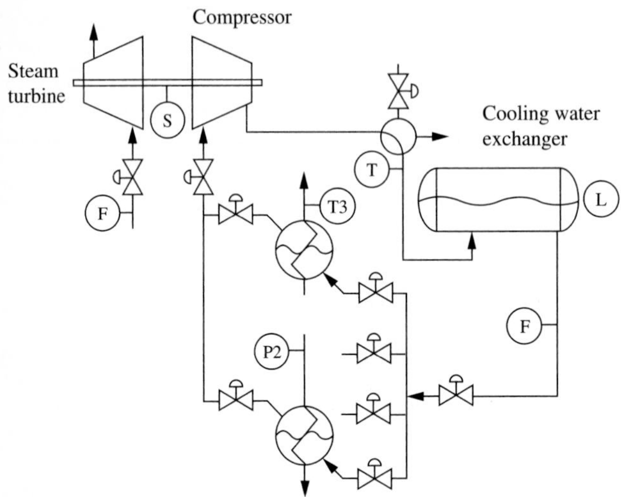

FIGURE Q26.2

(*a*) Design regulatory controls for this system that satisfy the demands of the consumers. Two consumers are shown as a heat exchanger (temperature controller) and a condenser (pressure controller).

(*b*) Add necessary controls to minimize the energy consumption (i.e., minimize the steam consumption) while satisfying the consumers' demands.
You may add sensors and add and delete values.

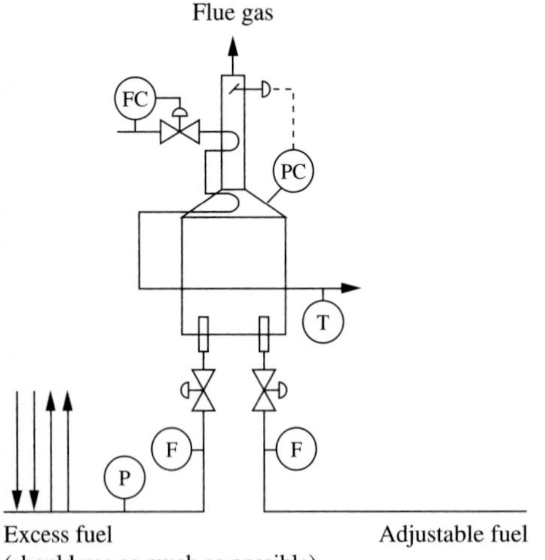

FIGURE Q26.3

26.3. The plant has byproducts that can be used as fuel or must be discarded with no value. Thus, all excess fuel should be consumed, if possible. Design a control strategy that provides good coil outlet temperature control and that consumes all possible excess fuel for the fired heater in Figure Q26.3. Note that (1) the two fuels have different compositions and (2) the excess fuel availability can change quickly and by large magnitudes.

26.4. In some plants, incentives exist to supply heat to the process via one (or a few) large, efficient fired heaters. The energy is transferred to consumers throughout the plant via an oil stream with good heat transfer and thermal stability properties. Design a control strategy for the process in Figure Q26.4 that satisfies the following objectives, listed in order of decreasing importance.

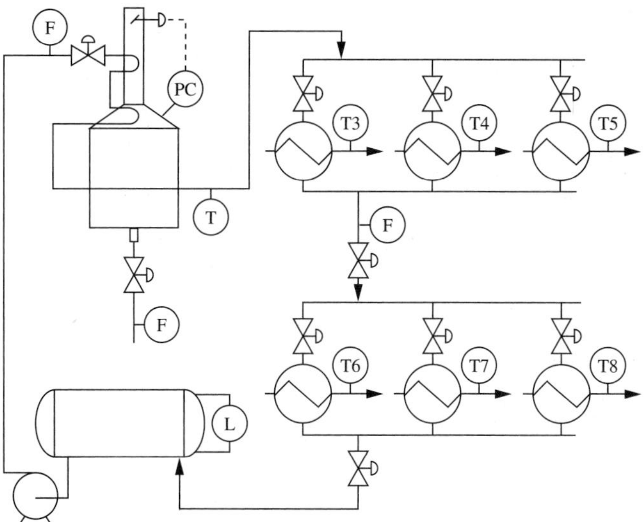

FIGURE Q26.4

(*a*) Control T3 and T4.

(*b*) Control T6 and T7.

(*c*) Determine the best value for the fired-heater outlet temperature, i.e., the value that satisfies (*a*) and (*b*) at minimum fuel.

(*d*) Recover as much energy as possible at the highest temperature.

You may add sensors and add or delete piping and valves.

26.5. The control design in Figure Q26.5 is proposed for maximizing the production rate in a chemical plant. The likely equipment limitations are the maximum reactor heating, the maximum flow of vapor from the flash, and the maximum reboiler duty in the distillation tower. The proposed design may not function well because of the long dynamics. Suggest enhancements that would ensure that (*a*) the maximum vapor flow from the flash is not exceeded and (*b*) the product quality in the distillation tower would be controlled close to its set point.

26.6. Derive the general equation for the direct search algorithm in Section 26.2 for any number of manipulated variables. Also, discuss potential drawbacks with the proposed method when applied to processes with more than one optimization variable.

26.7. The dynamic plots in Figure 26.13 have the iteration numbers as the abscissa. Determine an appropriate time between iterations for this process.

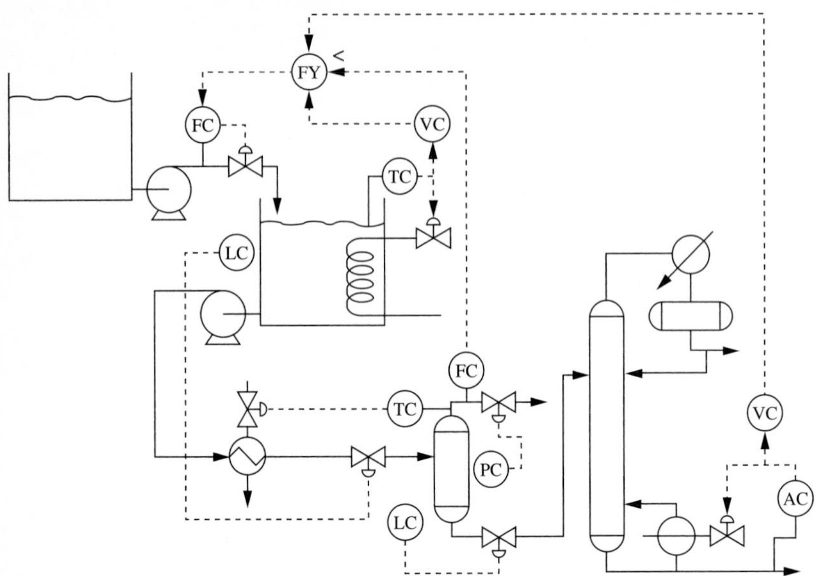

FIGURE Q26.5

26.8. Discuss additional considerations that should be included in a real-time boiler optimization as presented in Section 26.2. How could each consideration be integrated into the mathematical statement of the optimization?

26.9. Some Shewhart charts include warning limits, which are between the mean and the control limits. Discuss (*a*) the interpretation one could place on a single violation (or several sequential violations) of the warning limits, (*b*) reasonable values for the warning limits, and (*c*) types of actions which could be based on these limits.

26.10. The equations for the process capability used in Section 26.3 were based on normally distributed data. Describe a test of a data set to decide whether the data is normally distributed.

26.11. The Shewhart chart detects changes in mean via deviations beyond the control limits.

(*a*) Discuss the interpretation of several simultaneous data points above (or below) the mean, but within the control limits.

(*b*) Devise additional rules that could be used in conjunction with the standard Shewhart chart.

(*c*) Specify all assumptions required for the rules in (*b*) to be appropriate and when these assumptions are likely to be satisfied.

26.12. The Shewhart chart uses the data to identify a change in mean. Propose a different chart that could identify a change in the variability, as measured by the standard deviation or variance.

26.13. Often, the variable used in the Shewhart chart is an average of several samples taken at the same time from the process. Discuss the advantages and disadvantages of using the average of several samples rather than a single measurement.

26.14. A mixing tank after a process can, in some cases, reduce the effect of process variability prior to providing the product to the customer. Discuss the effects of product mixing on the following processes. Specifically, would the mixing reduce the variability important to the customer when the mean of the production is correct but the variance is too large?

(*a*) The bottoms product of a benzene-toluene distillation tower is mixed in a tank. The customer is interested in the percent benzene impurity.

(*b*) The ball bearings for a manufacturing plant are mixed in a bin. The customer is interested in the diameter of each ball bearing.

26.15. Discuss the differences between the control limits (UCL and LCL) and the specification limits (USL and LSL).

APPENDIX A

PROCESS CONTROL DRAWINGS

Drawings provide a simple visual representation of process designs and automation approaches. Since so many people are involved in the design, building, and operation of a process plant, drawing standards are essential, and the Instrument Society of America has prepared standards that are recognized in most countries and companies (ISA, 1986). The many design decisions lead to several typical levels of drawings; three common categories are

1 *Simplified*, which represents the use of measurements and calculations
2 *Conceptual*, which provides details on most calculations
3 *Detailed*, which specifies the computing resource in which each calculation is performed

Generally, simplified drawings are used in this book, and therefore, the simplified methods are presented in this appendix.

A.1 IDENTIFICATION LETTERS

Abbreviations of a few letters are used to identify the measurement types and calculations performed using measured values. Each abbreviation is located in a circle or "bubble," which indicates the location of the sensor in the process. The abbreviations usually consist of two to three letters, with the first letter indicating the variable type and the subsequent letter(s) giving some information about the function performed.

TABLE A.1
Identification letters

	First letter	Succeeding letters	
A	Analyzer	Alarm	
F	Flow	Ratio (fraction)	
H	Hand (manual operation)		High
L	Level		Low
P	Pressure		
S	Speed	Switch	
T	Temperature	Transmitter	
Y		General computation	

Some of the more common abbreviations are presented in Table A.1. Examples of typical abbreviations are

FC Flow control calculation

PIC Pressure measurement indicated (displayed) for the operating personnel and used in control calculation

LAH Level measurement used for signalling an alarm to operating personnel when the level exceeds a high limit

TS Temperature measurement used to open/close a switch that could shut down plant operation on a dangerous condition

AC Analyzer control calculation; the specific analysis is usually indicated just outside the bubble (e.g., ρ for density)

The symbol does not give much detail; for example, the flow measurement sensor could be an orifice plate, venturi meter, or pitot tube. Many additional details must be provided before the equipment can be selected and installed. These details are typically provided in tables, which complement drawings and are not discussed in this book.

Signals

The values of measurements or results of calculations must be transmitted between elements in a control system and ultimately to the final control element to influence the process. Many types of signals are used in a plant, and the three most common are shown in Figure A.1. The electric signal is represented by a dashed line in this book and is implemented with a 4–20 mA signal to represent measurements (e.g.,

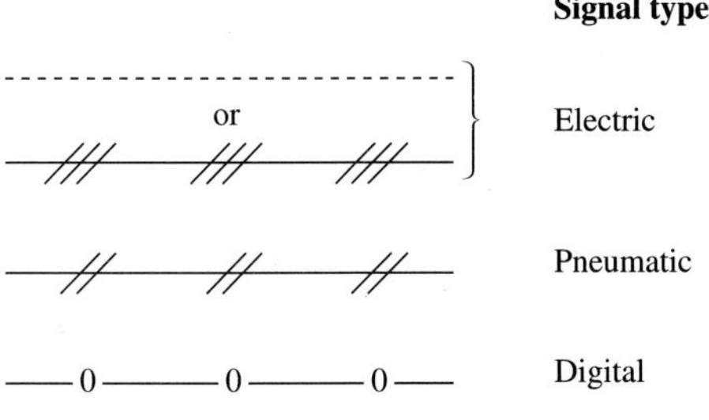

FIGURE A.1
Control signals. Reprinted by permission. Copyright ©1986, Instrument Society of America. From *Instrumentation Symbols and Identification,* ISA, 5.1-1984.

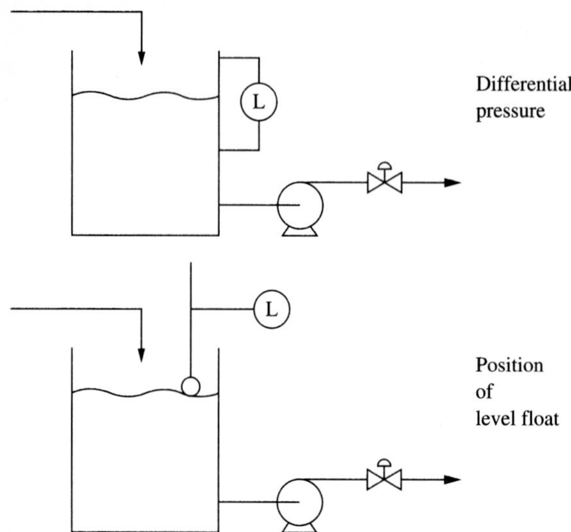

FIGURE A.2
Two-level sensors. Reprinted by permission. Copyright ©1986, Instrument Society of America. From *Instrumentation Symbols and Identification,* ISA, 5.1-1984.

300–400 K) and final element values (e.g., 0–100% valve opening). The pneumatic signal can be used for the same purposes, and it was the dominant signal type until the 1960s. It remains in wide use today, because the power source for most valves remains air pressure; typically, the electric signal is converted to pneumatic at the valve. Finally, many signals between calculations are internal to a digital computer, and these can be represented by the symbol in Figure A.1. Since most of the methods in this book can be implemented in a variety of computing equipment, analog or digital, the electric signal transmission is used throughout. This is common practice for simplified drawings and does not preclude digital implementation of the designs.

The general structure for signal transmission is shown in Figure 7.1 (page 234). The reader should be aware that the technology for signal transmission is changing rapidly. Soon, digital computation will be available at every sensor and final element, and most signal transmission will use digital principles. This revolution in signal transmission will not change the technology presented in this book, but it will open the door to advances in sensor diagnosis and improvements in process reliability.

Sensors

The drawing indicates the type of process variable measured and the location of the sensor. For the most part, details of the sensor physics and chemistry are not addressed in this book, because information is available in most books on fluid mechanics and heat transfer. In a few cases, some information on the sensor type is indicated in the drawing. For example, the drawings in Figure A.2 show two different types of level sensors: a differential pressure and a float.

A.2 FINAL ELEMENT

The predominant final element in the process industries is the diaphragm-actuated control valve, and this is essentially the only final element considered in this book.

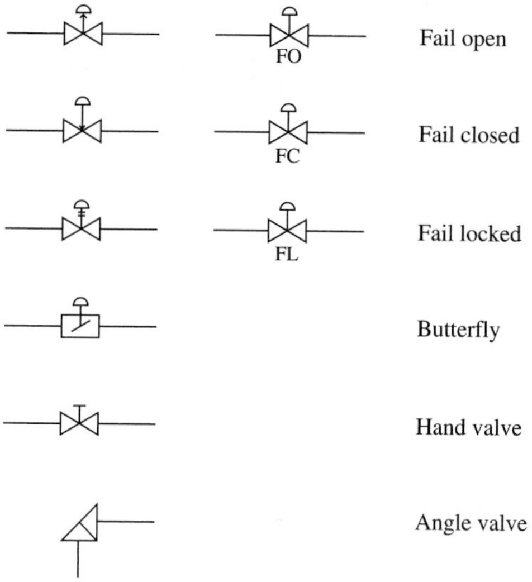

FO

Fail open

FC

Fail closed

FL

Fail locked

Butterfly

Hand valve

Angle valve

FIGURE A.3
Valve symbols. Reprinted by permission. Copyright ©1986, Instrument Society of America. From *Instrumentation Symbols and Identification,* ISA, 5.1-1984.

The valve is sketched as shown in Figure A.3. When the power (air pressure) is removed from the valve, it assumes its *failure position.* Three failure positions are shown; fail open, closed, and locked (unchanged). These positions are selected for safety, as discussed in Chapter 12. The typical control valve has a relatively large unrecoverable pressure drop; thus, a butterfly valve or damper is sometimes used for control. Many valves in a process design are not automated and must be opened or closed manually; an example of such a "hand" valve is shown in the figure. Finally, an angle valve is used in this book to represent safety valves, which open without an external power source when the process pressure exceeds a specified limit.

A.3 PROCESS EQUIPMENT

Control design drawings also include a simplified sketch of the process, which is included to clarify the control strategy but not to provide sufficient detail to build the process equipment. Some of the process schematic symbols are given in Figure A.4.

These elements are combined in the process drawing. An example is given in Figure A.5. The feed flow is maintained at the desired value using flow control, and the liquid level is controlled by adjusting the flow leaving the drum. The reactor effluent concentration is controlled by adjusting the heating medium valve opening. The bed temperature is measured and used to provide an alarm to the operating personnel when the temperature exceeds a specified value. Finally, the other sensors are used for display to operating personnel. Each sensor is numbered to allow unambiguous reference.

REFERENCES

Instrument Society of America, *Instrumentation Symbols and Identification*, ISA 5.1-1984, Research Triangle Park, NC, 1986.

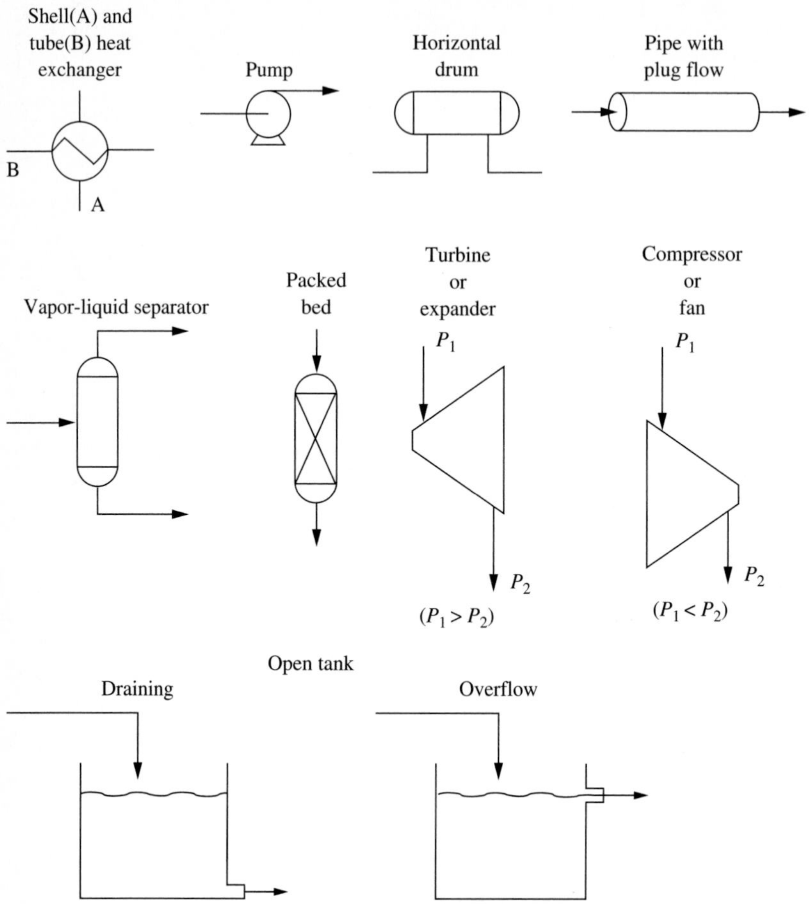

Shell(A) and tube(B) heat exchanger
B
A
Pump
Horizontal drum
Pipe with plug flow
Vapor-liquid separator
Packed bed
Turbine or expander
P_1
P_2
$(P_1 > P_2)$
Compressor or fan
P_1
P_2
$(P_1 < P_2)$
Open tank
Draining
Overflow

FIGURE A.4
Process schematics. Reprinted by permission. Copyright ©1986, Instrument Society of America. From *Instrumentation Symbols and Identification,* ISA, 5.1-1984.

F2
L1
T2
T3
T1
A2
F1
Feed
AC 1
Product

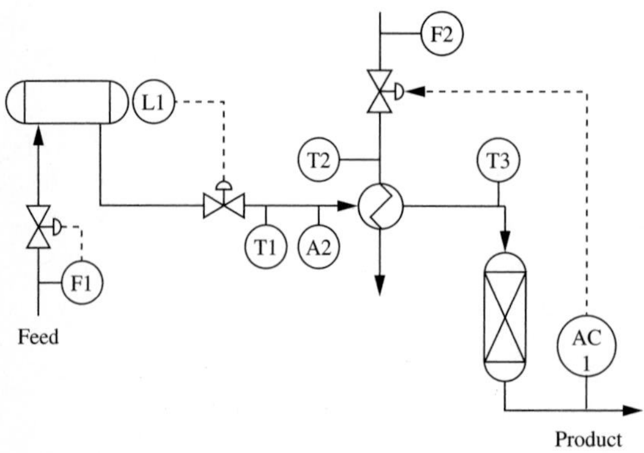

FIGURE A.5
Example process drawing with instrumentation.

APPENDIX B

INTEGRATING FACTOR

A single energy or material balance on a well-mixed system results in a first-order ordinary differential equation. Since this equation is often linearized in dynamic analysis, a linear first-order differential equation results. The differential equation is useful because it provides analytical relationships between the process equipment and operating parameters and key dynamic parameters such as time constants and gains. Often, an analytical solution is desired for the (open-loop) system output in response to one or more relatively simple input forcing functions. The integrating factor can be used to evaluate the analytical solution.

The general linearized model will be of the form

$$a(t)\frac{dY}{dt} + b(t)Y = c(t) \tag{B.1}$$

The functions $a(t)$, $b(t)$, and $c(t)$ are known functions of time, t. When the function $a(t) \neq 0$ during the time considered in the solution, equation (B.1) can be rearranged to give

$$\frac{dY}{dt} + f(t)Y = g(t) \tag{B.2}$$

with $g(t)$ the forcing function. This ordinary differential equation is linear and first-order but not separable. However, it can be modified to be separable, and directly solvable, by multiplying by a term called the *integrating factor,* IF.

The **integrating factor** is defined as

$$\text{IF} = \exp\left(\int f(t)\,dt\right) \tag{B.3}$$

Now, the standard equation (B.2) is multiplied by the integrating factor to give

$$\exp\left(\int f(t)\,dt\right)\frac{dY}{dt} + f(t)\exp\left(\int f(t)\,dt\right)Y = g(t)\exp\left(\int f(t)\,dt\right) \tag{B.4}$$

The left-hand side of equation (B.4) can be recognized to be the expansion of the derivative of a product:

$$\exp\left(\int f(t)\,dt\right)\frac{dY}{dt} + f(t)\exp\left(\int f(t)\,dt\right)Y = \frac{d}{dt}\left(Y\exp\left(\int f(t)\,dt\right)\right) \tag{B.5}$$

This can be substituted to yield a separable differential equation:

$$\frac{d}{dt}\left(Y\exp\left(\int f(t)\,dt\right)\right) = g(t)\exp\left(\int f(t)\,dt\right) \tag{B.6}$$

Equation (B.6) can be separated and integrated to give the final expression for the dependent variable.

$$Y = \exp\left(-\int f(t)\,dt\right)\int g(t)\exp\left(\int f(t)\,dt\right)dt + I\exp\left(-\int f(t)\,dt\right) \tag{B.7}$$

where I is a constant of integration to be evaluated from the initial condition. This method is successful when the integral in equation (B.7) can be evaluated analytically, which is possible for some simple functions $g(t)$ such as an impulse, step, and sine, which are useful in understanding the dynamic behavior of process systems.

This integrating factor method is applied to many first-order systems in Chapter 3. Also, it can be applied to a system of higher-order equations in which the equations can be solved sequentially; this type of system is referred to as a *noninteracting series* of first-order systems in Chapter 5. More complex systems, requiring simultaneous solution of equations, are addressed with Laplace transforms, as presented in Chapter 4.

APPENDIX C

CHEMICAL REACTOR MODELLING AND ANALYSIS

The chemical reactor is one of the most important unit operations considered by chemical engineers; thus, proper modelling and analysis are essential. The engineer should be able to derive the basic balances for typical reactor designs and to anticipate the range of likely dynamic behavior. This appendix is provided to complement and extend the coverage in Chapters 3 through 5 by deriving the energy balance and addressing more complex dynamic behavior.

C.1 ENERGY BALANCE

Material balances for reacting systems were derived in Chapter 3 and applied throughout the book. The energy balance for a continuous-flow chemical reactor is used, but not derived, in Section 3.5. The reactor energy balance is derived here, beginning with the general energy balance in equation (3.5), with the following assumptions:

1. The system volume is constant.
2. The heat capacity and density are constant.
3. $\Delta PE = \Delta KE = 0$.
4. The tank is well mixed.
5. One chemical reaction is occurring.

In this derivation the partial molar enthalpy of component i in a stream of n components, h_i, is assumed to be a function of temperature only.

$$\frac{\partial H}{\partial C_i} = h_i(T) = \text{partial molar enthalpy} \tag{C.1}$$

The symbol C_i is (moles/volume) of component i. The individual terms in equation (3.5) can be expressed as

$$\text{Accumulation:} \quad \frac{dE}{dt} \approx \frac{dH}{dt} = V\sum_{i=1}^{n} \frac{d(C_i h_i(T))}{dt} \tag{C.2}$$

$$\text{Flow in:} \quad FH_0 = F\sum_{i=1}^{n} C_{i0} h_i(T_0) \tag{C.3}$$

$$\text{Flow out:} \quad FH = F\sum_{i=1}^{n} C_i h_i(T) \tag{C.4}$$

The accumulation term can be expanded to give

$$V\sum_{i-1}^{n} \frac{d[C_i h_i(T)]}{dt} = V\left[\sum_{i=1}^{n} h_i \frac{dC_i}{dt} + \sum_{i=1}^{n} C_i \left(\frac{\partial h_i}{\partial T}\frac{dT}{dt}\right)\right] \tag{C.5}$$

The second term of the right-hand side of equation (C.5) can be simplified by noting that $\sum C_i(\partial h_i/\partial T) = \rho c_p$ (cal/[volume K]). Also, the first term on the right-hand side can be expanded by substituting the dynamic component material balance from equation (3.75) for dC_i/dt to give

$$\begin{aligned} \sum_{i=1}^{n} h_i \left[V\frac{dC_i}{dt}\right] &= \sum_{i=1}^{n} h_i[FC_{i0} - FC_i + V\mu_i r] \qquad \text{(C.6)} \\ &= F\sum_{i=1}^{n} C_{i0} h_i(T) - F\sum_{i=1}^{n} C_i h_i(T) + V\Delta H_{\text{rxn}} r \end{aligned}$$

The coefficients μ_i represent the amount of the component i generated from the extent of reaction r; for the example of a single reaction A $\rightarrow$ B, the coefficients are -1 for component A and $+1$ for component B. The sum of the products of these coefficients times their component enthalpies is commonly called the *heat of reaction* and is available in references. Combining the results gives

$$\begin{aligned} \rho V C_p \frac{dT}{dt} &= F\sum_{i=1}^{n} C_{i0}\,[h_i(T_0) - h_i(T)] \qquad \text{(C.7)} \\ &+ F\sum_{i=1}^{n} C_i[h_i(T) - h_i(T)] + V(-\Delta H_{\text{rxn}})r + Q - W_s \end{aligned}$$

Clearly, the second term on the right is zero. Also, the first term can be simplified, because partial molar enthalpy is assumed independent of composition, by expressing the total enthalpy of a stream as a function of its temperature to give

$$V\rho C_p \frac{dT}{dt} = F\rho C_p(T_0 - T) + V(-\Delta H_{\text{rxn}})r + Q - W_s \tag{C.8}$$

Equation (C.8) is the basic energy balance for a well-mixed, continuous-flow, liquid-phase chemical reactor. The second term on the right-hand side can be thought of as "generation due to reaction," but it is important to recognize that no generation term exists in the basic energy balance in equation (3.5). Also, it is important to recognize that many approximations have been employed that are not general. This equation is usually valid for liquid-phase systems but contains assumptions often not valid for gas-phase reactors. For alternative presentations and cogent discussion of reactor modelling, see Aris (1989) and Denn (1986).

C.2 MULTIPLE STEADY STATES

Some physical processes exhibit multiple steady states, a behavior that is not obvious without careful analysis. Recall that a steady state is defined as a condition in which all relevant balances are satisfied when the accumulation terms are zero. For linear equations, this situation would occur at only one (if any) set of operating conditions. However, the equations describing most chemical processes are nonlinear, and multiple solutions are possible, although they do not always, or even often, occur.

The steady-state balances for the system with a single reaction A → B were given in Section 3.6 and are repeated here as

$$C_A = \left[\frac{F}{F + Vk_0 e^{-E/RT}}\right] C_{A0} \tag{C.9}$$

The second equation, the steady-state energy balance, can be separated into two terms: Q_T for energy transfer and Q_R for release due to reaction, which sum to zero at steady state.

$$0 = Q_T + Q_R \tag{C.10}$$

with $$Q_T = F\rho C_P(T_0 - T) - \frac{aF_c^{b+1}}{F_c + \dfrac{aF_c^b}{2\rho_c C_{pc}}}(T - T_{cin})$$

$$Q_R = (-\Delta H_{rxn})Vk_0 e^{-E/RT} C_A$$

The steady-state solution is achieved when the two terms, $-Q_T$ and Q_R, are equal, and a simple calculation procedure is described in Section 3.6 to determine a single steady state. However, more than one solution can exist for this system. To check for multiple solutions, it is convenient to graph the two terms versus temperature, remembering that the concentration value used at each temperature is determined from equation (C.9) at the appropriate temperature.

This procedure has been carried out for three cases of reactor designs, which are described in Table C.1. Note that most parameters, including the chemical reaction, are the same in all cases; they differ in only the feed temperature and the coolant system. In addition to the design input variable values, the table presents the steady-state output variables, C_{As} and T_s. Also, the linearized stability analysis, in the form of the poles of the system (without control) at each steady state, is presented; the poles are the roots of the denominator of equation (4.60) (page 121).

TABLE C.1
Data for continuous-flow stirred-tank reactors

Variables	Case I	Case II			Case III
T_0(K)	323	343			323
T_{cin}(K)	365	310			340
a (cal/min K)/(m^3/min)	1.678×10^6	0.516×10^6			1.291×10^6
Steady-state C_A(kmole/m^3)	0.26	1.79	1.37	0.16	1.06
Steady-state T(K)	393.9	330.9	350	404.7	360
Poles (min^{-1})	$-0.89 \pm 5.92j$	$-0.96 \pm 0.47j$	1.94, −0.71	$-1.6 \pm 4.6j$	$0.34 \pm 1.41j$

Data:

$$F = 1 \text{ m}^3/\text{min},\ V = 1 \text{ m}^3,\ C_{A0} = 2.0 \text{ kmole/m}^3,\ C_p = 1 \text{ cal/(g°C)},\ \rho = 10^6 \text{g/m}^3,$$

$$k_o = 1.0 \times 10^{10} \text{ min}^{-1},\ E/R = 8330.1\text{K}^{-1},\ -\Delta H_{\text{rxn}} = 130 \times 10^6 \text{ cal/(kmole)}$$

$$(F_c)_s = 15 \text{ m}^3/\text{min},\ C_{pc} = 1 \text{ cal/(g K)},\ \rho_c = 10^6 \text{ g/m}^3,\ b = 0.5$$

Case I is identical to the reactor introduced in Example 3.10 (page 90) and considered often thereafter. The system has a single steady state, because a graph of the terms $-Q_T$ and Q_R in equation (C.10) has only one intersection. The local stability of this system can be determined by evaluating the poles (roots of the characteristic polynomial) of equation (4.60). This steady state is stable, because the real parts of the poles are negative, and the behavior is underdamped, because the poles are complex.

Case II has multiple steady states, as is demonstrated in Figure C.1, where the (negative of the) "energy transfer" and "release due to reaction" terms are equal at three temperatures! Thus, this chemical reactor can operate at three distinct sets of concentration and temperature for the *same values of all input variables and parameters.* Next, the stability in a small region about each steady state is evaluated, using linearized models about each steady state to determine whether the reactor would operate at the conditions without feedback control. The results in

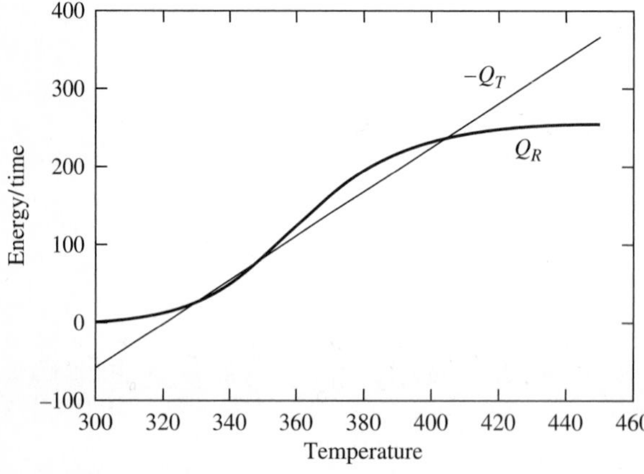

FIGURE C.1
Steady-state analysis of Case II.

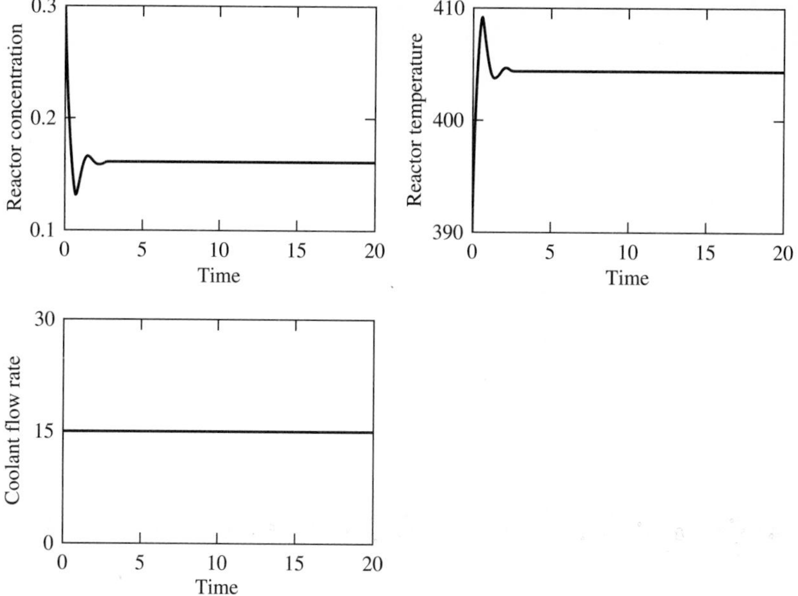

FIGURE C.2
Dynamic response of Case II without control.

Table C.1 indicate that two steady states are stable, whereas the steady state with the intermediate temperature is (locally) unstable, because it has a pole that is real and positive. This result indicates that only the two stable steady states can be achieved in practice without control. Any slight deviation from the *exact* values in the inputs in the table would result in the reactor dynamic response moving away from the unstable steady state toward one of the stable steady states. The final steady state achieved depends on the initial conditions of the reactor. For example, if the initial conditions are taken (arbitrarily) as 393.9 K and 0.26 kmole/m^3 (the values from Case I), the Case II reactor does not approach the unstable steady state, but rather approaches the steady state at the higher temperature, as shown in Figure C.2.

The instability of the intermediate temperature in Case II can be understood from steady-state arguments. It can be determined from Figure C.1 that as the temperature increases slightly from the intermediate steady state, the magnitude of the heat release increases faster than the magnitude of the heat transfer; that is, $d(-Q_T)/dT < d(Q_R)/dT$. Thus, any small positive deviation from the intermediate temperature will create a tendency to increase the temperature further. A similar conclusion can be determined for a small negative deviation in temperature. Thus, the intermediate temperature is unstable in the region about the intermediate steady state, as confirmed by the linearized stability analysis.

The previous analysis demonstrated that the intermediate steady state satisfies the steady-state balances but is not locally stable. However, these operating conditions can be achieved by stabilizing the system through feedback control. Thus, if a feedback PI controller is implemented to maintain temperature at 350 K by adjusting the coolant flow rate, the system reaches the intermediate steady state stabilized

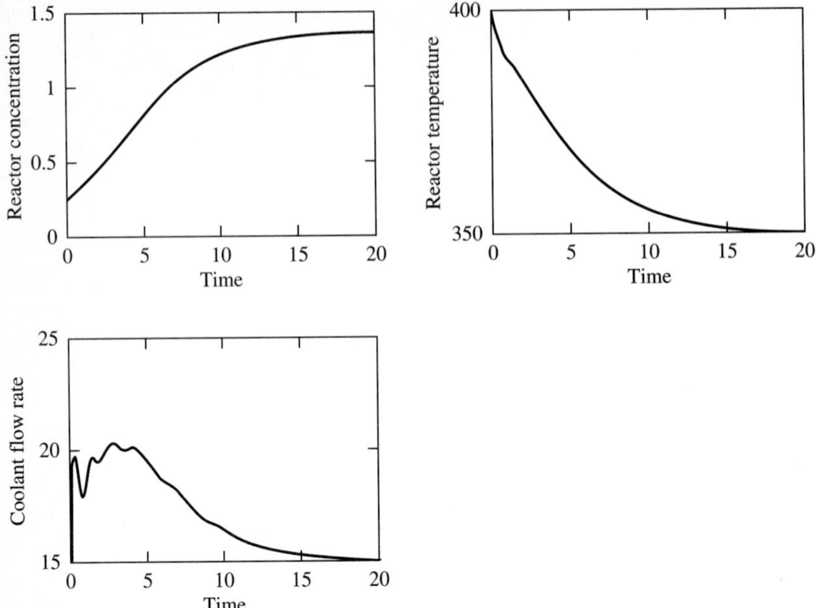

FIGURE C.3
Closed-loop dynamic response of Case II with PI feedback control (SP = 350 K, K_c = $-1(\text{m}^3/\text{min})/\text{K}$, T_I = 5 min).

by feedback control at exactly the operating conditions given in Table C.1; the dynamic response is given in Figure C.3 for this example. The occurrence of multiple steady states and the stabilization of an open-loop unstable steady state via feedback has been verified empirically for a stirred-tank reactor (Chang and Schmitz, 1975*b*).

Most models of processes used in design and analysis in chemical engineering do not exhibit multiple steady states. Typically, systems that are known to have multiple steady states are analyzed by ad hoc methods such as the one employed in this section, although some general correlations are available for CSTRs with simple kinetics (Perlmutter, 1972).

C.3 CONTINUOUS OSCILLATIONS DUE TO LIMIT CYCLES

Some strongly nonlinear systems can exhibit dynamic behavior that is quite surprising when first encountered: continuous oscillations in the output variables although the input variables are absolutely constant! Case III in Table C.1 is an example of a process with this behavior, which is termed a *limit cycle.* Notice that this system has a single steady state that is locally unstable, as demonstrated by the positive real part of its poles. This is a puzzle, because the only conditions for which the steady-state balances are satisfied cannot be approached stably; thus, how does the reactor behave? The answer is given in Figure C.4, which gives the results from the dynamic simulation of Case III. Clearly, the concentration and temperature never

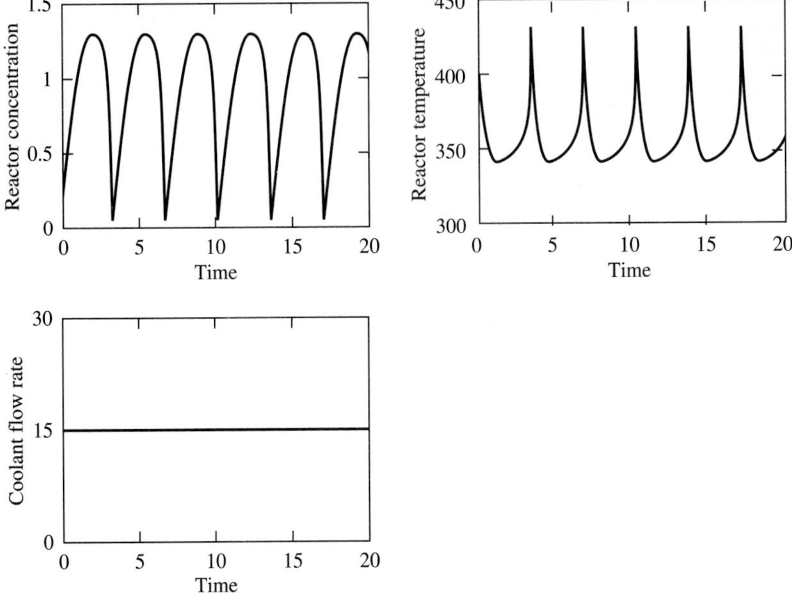

FIGURE C.4
Dynamic response of Case III without control, showing a limit cycle.

achieve their steady-state values, because they have periodic behavior that continues indefinitely without damping. This is "stable" periodic behavior, because the system will return to the same limit cycle after a pulse perturbation.

This behavior is not common but has occurred, to the surprise and consternation of practicing engineers in commercial situations (Bush, 1972). The behavior has also been analyzed mathematically (Aris and Amundson, 1958) and produced experimentally (Chang and Schmitz, 1975*a*). Systems that experience limit cycles can be stabilized through feedback control (e.g., Chang and Schmitz, 1975*a*), but sometimes process design changes are required to obtain acceptable performance (e.g., Penlidis et al., 1989).

C.4 THE EFFECT OF RECYCLE

The unique behaviors in the previous two sections resulted from the behavior of the chemical reactor without integration with other process units. Complex steady-state and dynamic behavior can also occur due to interactions among units. The effect of energy recycle was introduced in Example 5.5 (page 173), where the time constant and steady-state gain of the process were affected. Here, the concept of feed-effluent heat exchange is extended with a slightly more complex process, the system in Figure C.5 with reactor cooling and feed-effluent exchange. The reactor inlet temperature is a linear function of the reactor outlet temperature; although this model is extremely simple, it captures the essential interaction between the exchanger and reactor. Thus, the system is defined by the two basic dynamic balances on the

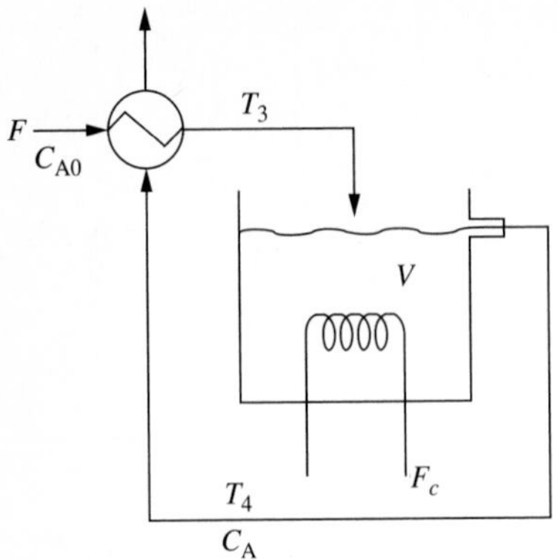

FIGURE C.5
Chemical reactor with feed-effluent exchanger.

reactor, equations (3.75) and (3.76), with the notation changed so that the reactor inlet temperature is T_3 and the reactor outlet temperature is T_4. The exchanger model is

$$\tau_{ex} \frac{dT_3}{dt} = K_{ex}(T_4 - 326.5) - (T_3 - 315) \tag{C.11}$$

In this example, $\tau_{ex} = 1$ minute and $K_{ex} = 0.60$. The data for the system without and with the exchanger are the same as in Section C.2 except for the variables given in Cases IV and V in Table C.2. Clearly, the system without the feed-effluent exchanger has a single, stable steady state. The addition of the exchanger results in a system with three steady states!

The local stability of the system about a steady state can be determined from the linearized model. For this system, the equations have the form

$$\frac{dC'_A}{dt} = a_{11}C'_A + a_{12}T'_4 + a_{15}T'_3 \tag{C.12}$$

$$\frac{dT'_4}{dt} = a_{21}C'_A + a_{22}T'_4 + a_{25}T'_3 \tag{C.13}$$

$$\frac{dT'_3}{dt} = a_{31}C'_A + a_{32}T'_4 + a_{35}T'_3 \tag{C.14}$$

TABLE C.2
Chemical reactor with feed-effluent heat exchange

Variables	Case IV without exchanger	Case V with feed-effluent exchanger		
T_3(K)	315	Depends on reactor steady state 315, 331, 362		
T_{cin}(K)	323	323		
a (cal/min K)/(m^3/min)	0.671×10^6	0.671×10^6		
Steady-state C_A(kmole/m^3)	1.85	1.85	1.28	0.16
Steady-state T(K)	326.5	326.5	353	405.2
Poles (min^{-1})	−1.55, −1.36	−2.21, −1.1, −.6	2.12, $-0.90 \pm 0.31j$	−0.75, $-2.15 \pm 5.19j$

The coefficients have the same meaning as in Example 3.10 (page 90), except for the newly introduced coefficients, $a_{15} = 0$, $a_{25} = F/V$, $a_{31} = 0$, $a_{32} = K_{ex}/\tau_{ex}$, and $a_{35} = -1/\tau_{ex}$. After taking the Laplace transform and combining the equations algebraically, the stability of the system can be determined from the roots of the denominator of the transfer function, which is the following third-order polynomial. (The same result can be obtained by evaluating the eigenvalues of the coefficient matrix in equations (C.12) to (C.14), which is more straightforward as the number of equations increases.)

$$s^3 - (a_{11} + a_{22} + a_{35})s^2 + (a_{11}a_{22} - a_{12}a_{21} - a_{25}a_{32} + a_{35}(a_{11} + a_{22}))s + (a_{11}a_{32}a_{25} - a_{35}(a_{11}a_{22} - a_{12}a_{21})) = 0 \quad \text{(C.15)}$$

The stability analysis in Table C.2 shows that the intermediate steady state is locally unstable! The qualitative explanation is that the exchanger results in an increase (decrease) in the inlet temperature as the reactor outlet increases (decreases). Thus, the exchanger introduces a positive, or destabilizing, feedback in the system, which in this case also leads to multiple steady states, with the intermediate one being locally unstable. The response of this system can be evaluated by simulation beginning at the lower-temperature steady state subject to a temporary decrease in cooling; no feedback control is present in this example. In Figure C.6, the system returned to its original conditions after the pulse disturbance which was shorter (3 min) in duration. In Figure C.7, the system moves to the higher-temperature steady state after the pulse disturbance of longer (6 min) duration. Note that the system will not move to the intermediate steady state, because it is locally unstable without

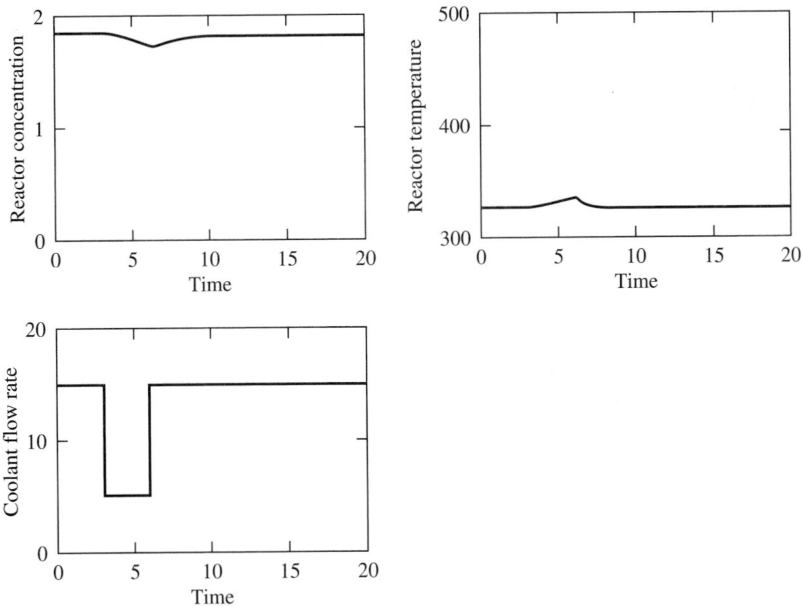

FIGURE C.6
Dynamic response of Case V reactor with exchanger, shorter disturbance.

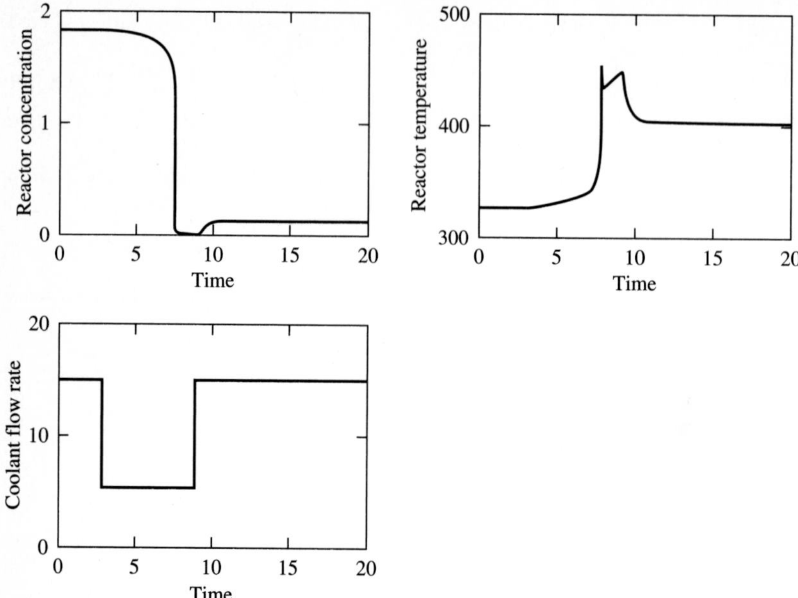

FIGURE C.7
Dynamic response for Case V reactor with exchanger, longer disturbance.

control. Although this example is simple, it captures the essential characteristics of realistic industrial designs (e.g., Silverstein and Shinnar, 1982).

C.5 CONCLUSION

This appendix has provided the derivation of the energy balance for chemically reacting systems and samples of complex behavior that can be exhibited by such nonisothermal reactors. Note that all of the examples in this section involved the same chemical kinetics; thus, a wide array of behaviors can be achieved by changing the process design parameters. Generally, the occurrence of multiple steady states and unstable steady states results from some type of positive feedback in the system. In the examples in this appendix, the positive feedback is provided by the exothermic chemical reaction and the feed-effluent heat exchanger. The analysis of steady-state multiplicity and stability is covered in greater detail in Perlmutter (1972), and the influence of these phenomena on design and control is reviewed by Seider et al. (1991).

REFERENCES

Aris, R., *Elementary Chemical Reactor Analysis,* Butterworth, Boston, 1989.

Aris, R. and N. Amundson, "An Analysis of Chemical Reactor Stability and Control," *CES, 7, 3,* 121–131 (1958).

Bush, S., *Proc. Royal Soc., 309A,* 1–26 (1969).

Chang, M. and R. Schmitz, "An Experimental Study of Oscillatory States in a Stirred Reactor," *CES, 30,* 21–34 (1975*a*).

Chang, M. and R. Schmitz, "Feedback Control of Unstable States in a Laboratory Reactor," *CES, 30,* 837–846 (1975*b*).

Denn, M., *Process Modeling,* Pitman Publishing, Marshfield, MA, 1986.

Penlidis, A., J. MacGregor, and A. Hamielec, "Continuous Emulsion Polymerization: Design and Control Considerations for CSTR Trains," *CES, 44,* 273–281 (1989).

Perlmutter, D., *Stability of Chemical Reactors,* Prentice-Hall, Englewood Cliffs, NJ, 1972.

Seider, W., D. Brengel, and S. Widagdo, "Nonlinear Analysis in Process Control," *AIChE J., 37,* 1–38 (1991).

Silverstein, J. and R. Shinnar, "Effect of Design on Stability and Control of Fixed Bed Catalytic Reactors with Heat Feedback," *IEC Proc. Des. Devel., 21,* 241–256 (1982).

APPENDIX D

METHODS OF MOMENTS

Real processes have complex dynamic responses and require models with many parameters to be characterized accurately. However, the engineer often seeks a simple model with few parameters to describe the main aspects of the dynamic behavior. Examples throughout this book demonstrate that the first-order-with-dead-time model is adequate for the process control analysis of many, but not all, processes. In this appendix a method is developed for determining a few parameters that can be used to fit a model to the expected dynamic behavior; this is the *method of moments*. The application of the method of moments described in this appendix was demonstrated by Paynter and Takahashi (1956) and Gibilaro and Lee (1969).

The basic approach is to evaluate several *moments* of the output behavior and use these to characterize the dynamic behavior. Thus, the first step is to define a moment.

> The nth **moment** of a variable $Y(t)$ is
>
> $$M_n = \frac{\int_0^\infty t^n Y(t)\,dt}{\int_0^\infty Y(t)\,dt} \tag{D.1}$$

Further moments are usually defined with respect to the first moment, which is the *mean*; thus, the moments of the variable $Y(t)$ about its mean are

$$T_n = \frac{\int_0^\infty (t - M_1)^n Y(t)\,dt}{\int_0^\infty Y(t)\,dt} \tag{D.2}$$

Given a function $Y(t)$ or a set of data Y, the integrals in equations (D.1) and (D.2) can be evaluated as long as they are bounded.

The moments can also be evaluated from the Laplace transform of a variable in a particularly simple manner, which is the application of moments in this book. The development begins with the input-output model of a single-variable system in transfer function form.

$$Y(s) = G(s)X(s) \tag{D.3}$$

with $X(s)$ being the input, $Y(s)$ the output, and $G(s)$ the transfer function, as defined in equation (4.67). The moment of the output variable will be evaluated for a unit impulse input, for which $X(s) = 1$ and all integrals in the moment equations are bounded. From the definition of the Laplace transform and equation (D.3),

$$\int_0^\infty e^{-st} Y(t)\,dt = Y(s) = G(s)X(s) = G(s) \tag{D.4}$$

Now, it is shown that any moment of an output in response to a unit impulse can be evaluated directly from the transfer function, using the result in equation (D.4) to evaluate the numerator and denominator of equation (D.1).

$$\int_0^\infty Y(t)\,dt = G(s)|_{s=0} \tag{D.5}$$

$$\int_0^\infty t^n Y(t)\,dt = (-1)^n \left(\frac{d^n}{ds^n} G(s)\right)_{s=0} \tag{D.6}$$

Equation (D.6) is verified using the results from equation (D.4).

$$(-1)^n \left(\frac{d^n}{ds^n} G(s)\right)_{s=0} = \left((-1)^n \int_0^\infty (-t)^n e^{-st} Y(t)\,dt\right)_{s=0} \tag{D.7}$$

$$= \int_0^\infty t^n Y(t)\,dt$$

The method of moments is used in this book for one important application: determining the *characteristic time* of a process. The first moment is used as the characteristic time to "time-scale" the dynamic responses in the dimensional analysis presented in the tuning correlations in Chapter 9 and the control performance correlations in Chapter 13. For example, the first moment is evaluated for a first-order-with-dead-time process model to be

$$\int_0^\infty Y(t)\,dt = \left(\frac{K_p e^{-\theta s}}{\tau s + 1}\right)_{s=0} = K_p \tag{D.8}$$

$$\begin{aligned} \int_0^\infty tY(t)\,dt &= (-1)\left(\frac{d}{ds}\frac{K_p e^{-\theta s}}{\tau s + 1}\right)_{s=0} \\ &= (-1)\left(\frac{-\theta K_p e^{-\theta s}}{\tau s + 1} + \frac{-\tau K_p e^{-\theta s}}{(\tau s + 1)^2}\right)_{s=0} = K_p(\theta + \tau) \end{aligned} \tag{D.9}$$

$$M_1 = \frac{K_p(\theta + \tau)}{K_p} = \theta + \tau \tag{D.10}$$

This result was used by Jeffreson (1981) in performance correlations.

The sum of the dead time and time constant is also the time at which the output response for a step in the manipulated variable reaches 63% of its final value ($t_{63\%}$) for the first-order-with-dead-time model. As a rough approximation, the first moment of many common transfer functions in the book can be used as an estimate of $t_{63\%}$. The first moment for a transfer function with dead time, multiple first-order numerator terms, and multiple first-order denominator terms is evaluated as follows:

$$\int_0^\infty tY(t)\,dt = (-1)\left(\frac{d}{ds}\frac{K_p e^{-\theta s}\prod_{j=0}^{m}(\tau_{\text{ld}j}s+1)}{\prod_{k=0}^{q}(\tau_k s+1)}\right)_{s=0}$$

$$= \theta\left(\frac{K_p e^{-\theta s}\prod_{j}(\tau_{\text{ld}j}s+1)}{\prod_{k}(\tau_k s+1)}\right)_{s=0} + \sum_{r=1}^{m} -\tau_{\text{ld}r}\left(\frac{K_p e^{-\theta s}\prod_{j}^{j\neq r}(\tau_{\text{ld}j}s+1)}{\prod_{k}(\tau_k s+1)}\right)_{s=0}$$

$$+ \sum_{\text{kk}=0}^{q}\left[\tau_{\text{kk}}\prod_{k=1}^{k\neq \text{kk}}(\tau_k s+1)\right]_{s=0}\left(\frac{K_p e^{-\theta s}\prod_{j}(\tau_{\text{ld}j}s+1)}{\left(\prod_{k}(\tau_k s+1)\right)^2}\right)_{s=0} \tag{D.11}$$

$$= K_p\left(\theta + \sum_{k=0}^{q}\tau_k - \sum_{j=0}^{m}\tau_{\text{ld}j}\right)$$

$$M_1 = \frac{K_p}{K_p}\left(\theta + \sum_{k=0}^{q}\tau_k - \sum_{j=0}^{m}\tau_{\text{ld}j}\right) \tag{D.12}$$

This is the basis for the approximation given in Chapter 5 that $t_{63\%}$ is approximately equal to the sum of the dead times and time constants for a series of noninteracting first-order-with-dead-time systems. This approximation is useful for estimating the general time for a complex series system to respond, but it does not give sufficient information in itself to design or tune controllers.

An additional application for the method of moments is in estimating the parameters in a simple model based on the parameters in a more complex model. In this approach, several moments of the simple and more complex models are determined analytically, and the unknown parameters are determined for the simple model. Naturally, one linearly independent moment equation is required for each parameter. This is demonstrated as follows by determining the parameters for a first-order-with-dead-time model based on a *known* second-order-with-dead-time model.

Second-order model:

$$G(s) = \frac{K_{p2}e^{-\theta_2 s}}{(\tau_{21}s+1)(\tau_{22}s+1)}$$

unit impulse $\int_0^\infty Y(t)\,dt$: K_{p2}

First moment: $(\theta_2 + \tau_{21} + \tau_{22})$

Second moment: $\theta_2^2 + 2\theta(\tau_{21} + \tau_{22}) - 2\tau_{21}\tau_{22} + 2(\tau_{21} + \tau_{22})^2$

First-order model:

$$\frac{K_{p1}e^{-\theta_1 s}}{\tau_1 s + 1}$$

unit impulse $\int_0^\infty Y(t)\,dt$: K_{p1}

First moment: $(\theta_1 + \tau_1)$

Second moment: $\theta_1^2 + 2\theta_1\tau_1 + 2\tau_1^2$

These equations can be applied to the second-order-with-dead-time model in question 6.5 to answer part (*b*) of the question: what is an approximate first-order-with-dead-time model? The results are summarized as follows.

Second-order:

$$\frac{T_4(s)}{T_2(s)} = \frac{1.87e^{-2.6s}}{(2s + 1)(2.7s + 1)}$$

Equating the moments gives

$$K_{p1} = 1.87$$
$$M_1 = \theta_1 + \tau_1 = 2.6 + 2.0 + 2.7 = 7.3$$
$$M_2 = \theta_1^2 + 2\theta_1\tau_1 + 2\tau_1^2 = 64.98$$

giving

$$\theta_1 = 3.3 \qquad \tau_1 = 4.0$$

Approximate first-order:

$$\frac{T_4(s)}{T_2(s)} = \frac{1.87e^{-3.3s}}{4s + 1}$$

REFERENCES

Paynter, H. and Y. Takahashi, "A New Method for Evaluating the Dynamic Response of Counterflow and Parallel-Flow Heat Exchangers," *Trans. ASME, 78,* 749–757 (1956).

Gibilaro, L. and F. Lee, "The Reduction of Complex Transfer Function Models to Simple Models Using the Methods of Moments," *CES, 24,* 84–93 (1969).

Jeffreson, C., "Controllability of Process Systems," *IEC Fund, 15, 3,* 171–179 (1976).

APPENDIX E

DETERMINING CONTROLLER CONSTANTS TO SATISFY PERFORMANCE SPECIFICATIONS

This appendix presents a procedure for determining the tuning constants for feedback controllers that satisfy robust, time-domain performance specifications. The specifications involve the behavior of the controlled and manipulated variables and include measurement noise and variable process dynamics, as defined in Table 9.1. Because the goals are formulated to minimize the controlled-variable IAE subject to limitation on the manipulated-variable values, the tuning constants are determined using optimization principles. It is not possible to derive analytical expressions relating the tuning constants to the IAE and manipulated-variable transient response; therefore, the control system performance is determined by numerical solution of the model, and the best tuning values are determined using an optimization method.

E.1 SIMULATION OF THE CONTROLLED SYSTEM TRANSIENT RESPONSE

The single-loop control system considered in this appendix is shown in Figure E.1. The real system consists of elements that are continuous and cannot be solved analytically. As an approximation, the closed-loop transient is determined by numerical solution of the equations that define the system. As discussed in Chapter 3, this approach can provide a set of points very close to the exact transient response—certainly accurate enough for use in the optimization approach.

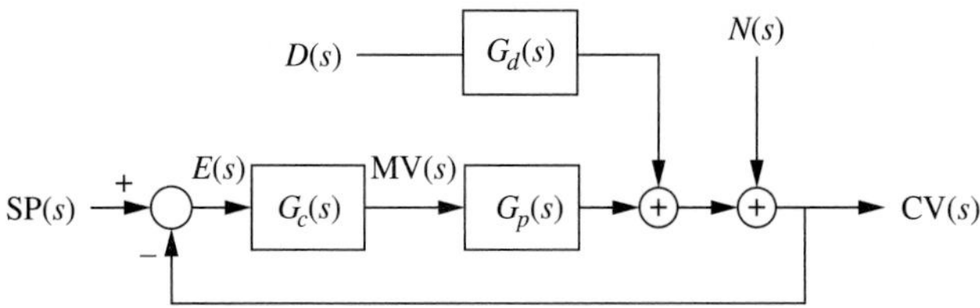

CV(s) = controlled variable
D(s) = disturbance
E(s) = error
MV(s) = manipulated variable
N(s) = noise
SP(s) = set point

$G_p(s)$ = process, valve, and sensor
$G_d(s)$ = disturbance
$G_c(s)$ = controller

FIGURE E.1
Block diagram of the feedback control system.

The model for the feedback process is assumed to be first-order with dead time. As discussed in Appendix F and Section 6.4, this model can be approximated by the following algebraic equation at each time step:

$$(\mathrm{CV_{fb}})_n = K_p(1 - e^{-\Delta t/\tau})\mathrm{MV}_{n-\Gamma-1} + e^{-\Delta t/\tau}(\mathrm{CV_{fb}})_{n-1} \tag{E.1}$$

The dead time is simulated by a delay, $\Gamma = \theta/\Delta t$. In equation (E.1) the dead time must be an integer multiple of Δt, but advanced modelling methods using modified z-transforms enable modelling of systems with noninteger dead times (Ogata, 1987). The model for the effect of the disturbance on the controlled variable is first-order.

$$(\mathrm{CV}_d)_n = K_d(1 - e^{-\Delta t/\tau_D})D_{n-1} + e^{-\Delta t/\tau_D}(\mathrm{CV}_d)_{n-1} \tag{E.2}$$

The noise, $(\mathrm{CV}_N)_n$, is based on a random perturbation passed through a dynamic process (Ciancone, 1990). It has a standard deviation, σ_N. The measured value of the controlled variable is the sum of the three effects.

$$\mathrm{CV}_n = (\mathrm{CV_{fb}})_n + (\mathrm{CV}_d)_n + (\mathrm{CV}_N)_n \tag{E.3}$$

These equations determine the behavior of the controlled variable given the manipulated and disturbance variables. The disturbance D is a step for the disturbance response cases and zero for the set point cases, and the set point is constant for disturbance response cases and a step for set point response cases.

The manipulated variable is determined by the feedback controller. The digital form of the PID controller is explained in Section 11.4 and repeated here:

$$\mathrm{MV}_n = \mathrm{MV}_{n-1} + K_c\left[(\mathrm{SP}_n - \mathrm{CV}_n) - (\mathrm{SP}_{n-1} - \mathrm{CV}_{n-1}) + \frac{\Delta t(\mathrm{SP}_n - \mathrm{CV}_n)}{T_I} + \frac{T_d}{\Delta t}(-\mathrm{CV}_n + 2\mathrm{CV}_{n-1} - \mathrm{CV}_{n-2})\right] \tag{E.4}$$

Equations (E.1) through (E.4) are solved at each time step from an initial steady state to a final time of about $6(\theta + \tau)$, which is sufficient to reach essentially the final steady state for a well-tuned system. The process equations and digital controller are executed at a frequency that gives $\Delta t/(\theta + \tau) = 0.1$, which is sufficient to approximate the continuous system closely although not exactly. By this method, the transient is evaluated for any set of tuning constants.

E.2 OPTIMIZATION OF THE TUNING CONSTANTS

The "best" values of the tuning constants are those that satisfy the performance goals. One goal requires that the integral of the controlled variable deviation, measured as IAE, be minimum. The IAE can be approximated using the discrete samples of the transient response as

$$\text{IAE} = \int_0^\infty |\text{SP} - \text{CV}|\,dt \approx \sum_{n=1}^{M} |\text{SP}_n - \text{CV}_n|\,\Delta t \tag{E.5}$$

with M the number of points in the transient. The second goal requires that model error be considered to ensure a reasonable amount of robustness. The approach used here is to evaluate the entire transient responses for three feedback control systems with *different* process models, each with the *same* controller tuning constants. Thus, the measure of the controlled-variable performance is modified to be

$$\sum_{i=1}^{3} \text{IAE}_i = \sum_{i=1}^{3} \left(\sum_{n=1}^{M} |\text{SP}_n - \text{CV}_n|\,\Delta t \right)_i \tag{E.6}$$

To include a range of process dynamics, the model parameters all change in a correlated manner as 75%, 100%, and +125% of their nominal values. This corresponds to changes in the feed flow rate in the example process in Figure 9.1 (page 291).

The third performance goal places a limitation on the variation of the manipulated variable. Here, the manipulated variable is restricted in the extent to which it may exceed its final steady-state value; the final value (with no measurement noise) would be $-\Delta D/K_p$ or $\Delta \text{SP}/K_p$ for disturbance or set point response, respectively. The region of allowable values for the manipulated variable is large during the initial part of the transient and becomes smaller as the final steady state is reached to prevent excessive oscillations. The final variability is *nonzero,* because higher-frequency noise in the controlled variable is propagated to cause (undesirable but unavoidable) variation in the manipulated variable. Thus, this third goal also includes a bound on the variability of the manipulated variable because of the measurement noise N, which is apparent at the end of the transient response. The equations for the manipulated-variable bound select the least limiting,

$$\begin{aligned}
(\text{MV}_n)_1 &\le \frac{-\Delta D}{K_p} + \left((\Delta \text{MV}_{\max}) \frac{-\Delta D}{K_p} - \frac{-\Delta D}{K_p} \right) \exp\left[\frac{-t_{\text{MV}}}{\beta_t(\theta + \tau)} \right] \\
(\text{MV}_n)_2 &\le \left(\frac{-\Delta D}{K_p} - \sigma_{\text{MV}} \right) \\
(\text{MV}_n)_{\min} &= \min\left[(\text{MV}_n)_1,\ (\text{MV}_n)_2 \right] \\
(\text{MV}_n)_{\min} &\le \text{MV}_n
\end{aligned} \tag{E.7}$$

TABLE E.1
Parameters used in tuning optimization

Factor	Symbol	Value	Comment
Measurement noise	σ_N	0.55% of scale	$\pm 4\sigma_N = \pm 2.2\%$ of scale
Maximum change in MV	$\Delta \text{MV}_{\text{max}}$	2.7	This allows 170% maximum overshoot at $t_{\text{MV}} = 0$ and decreases rapidly as time increases
Tune the time dependence for the allowable change in MV	β_t	1.5	This value reduces the allowable variation rapidly as time increases, damping the response
Allowable variation in MV at steady state, i.e., end of transient	σ_{MV}	2.5% of range	This is approximately $4\sigma_N$, the noise propagated via the proportional mode with $K_c K_p = 1.0$
Disturbance time constant	$\tau_D = \tau$	Depends on case	The disturbance time constant is the nominal feedback time constant
Input step magnitude	$-\Delta \text{SP}$ or ΔD ($K_d = 1$)	10% of scale	Should be larger than measurement noise and must be of the sign shown for the sign conventions in this appendix
Model error		25% of each parameter	The errors are due to a change in operating conditions in a nonlinear process and thus are correlated (all increase of decrease concurrently)
Execution period	Δt	$0.1(\theta + \tau)$	Relatively small compared with feedback process

with time t_{MV} measured from the initiation of the input step. The term ΔD is replaced with $-\Delta \text{SP}$ for a set point change. Several parameters in this equation are related to the dynamic response of the process. Other parameters are fixed at reasonable values selected by the author to suit the widest range of industrial process applications, as given in Table E.1. Naturally, this definition will not be appropriate for all systems, but it should provide good starting values for the tuning of many feedback systems.

Some values of the tuning constants will result in manipulated-variable values that violate the constraints defined in equation (E.7). These values will be considered invalid because of the violation and will not be acceptable, even if they result in a low value for $\sum$IAE. Only tuning constant values that result in the constraints in equation (E.7) being satisfied for the entire transient response will be considered when minimizing $\sum$IAE. This mathematical problem is of the general class of nonlinear, constrained optimization. Determining the best tuning consistent with the goals is conceptually straightforward; the engineer could perform many simulations and, by trial and error, eventually find the best values of the tuning constants. However, the trial-and-error approach would be very time-consuming and require excessive calculations. The approach taken here was to formulate equations (E.1) through (E.7) for all time steps and solve them simultaneously using a method which employs intermediate results to direct the search efficiently toward the best values of the tuning constants (Ciancone, 1990).

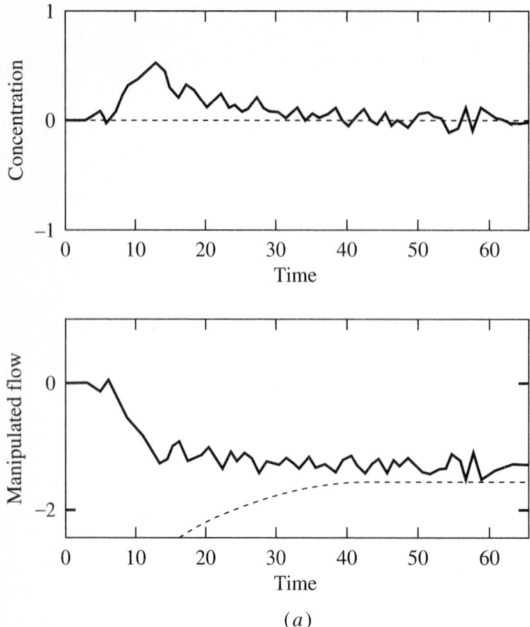

FIGURE E.2
Transient response for $\theta/(\theta+\tau) = 0.3$. (*a*) 75% of nominal feedback model parameters (high flow).

The transient responses in Figure E.2*a* through *c* show the results of the optimization for one value of the fraction dead time, the nominal $\theta/(\theta + \tau) = 0.3$. The "optimal" tuning for this case is $K_p K_c = 1.4$, $T_I/(\theta + \tau) = 0.7$, and $T_d/(\theta + \tau) = 0.02$, using the model parameters from the nominal case. The transient responses for the three cases, nominal (perfect) model and ±mismatch, demonstrate the importance of explicitly considering model error. Note that the feedback control is not too aggressive for the nominal case and is quite slow for the 75% case. However, the 125% case involving a slower process dynamics and a higher feedback process gain (i.e., smaller feed flow in the example process) is at the limit of the allowable manipulated-variable variation and exhibits oscillatory behavior. Thus, making the tuning more aggressive would result in unacceptable behavior for process dynamics for the 125% case, which is considered to occur often in this problem definition. Thus, all three goals are relevant in determining the best initial controller tuning. Finally, model errors larger than anticipated in the definition could cause closed-loop behavior deemed unacceptable; this situation would be rectified during fine tuning.

REFERENCES

Ciancone, R., "Selecting Appropriate Control Technology," undergraduate research project, McMaster University, 1990.

Ogata, K., *Discrete-Time Control Systems*, Prentice-Hall, Englewood Cliffs, NJ, 1987.

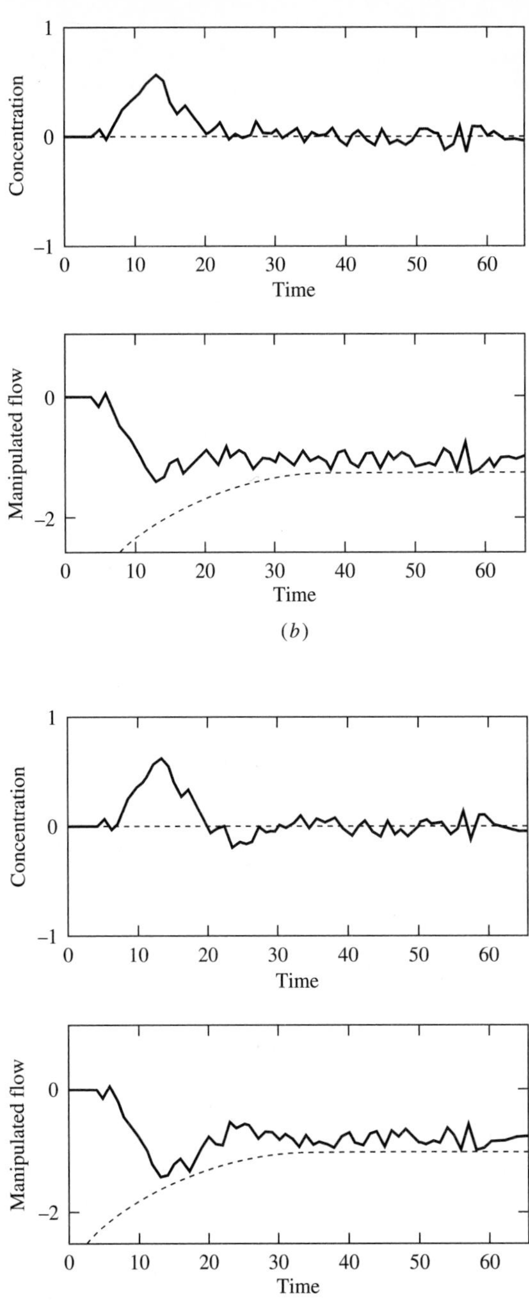

FIGURE E.2 ***(continued)***
Transient response for $\theta/(\theta + \tau) = 0.3$. (*b*) 100% of nominal feedback model parameters (nominal flow); (*c*) 125% of nominal feedback parameters (low flow).

APPENDIX F

DISCRETE MODELS FOR DIGITAL CONTROL

The chemical processes considered in this book involve continuous variables and can be modelled using algebraic and differential equations. Also, the control calculations have been introduced as equations involving continuous variables, which can be implemented using electronic or pneumatic analog calculating equipment. When all elements in the feedback loop are continuous, the system can be described using transfer functions involving Laplace transforms; this allows powerful analysis tools to be applied in determining the stability and performance of control systems. However, most control calculations are now implemented using digital computers, which introduce discrete equations in the control system. If the digital calculations are executed rapidly compared with the process dynamics, the analysis of continuous systems provides an accurate approximation of the dynamic behavior.

Because the controller is implemented in digital form, it is important that the engineer understand the digital forms of the models and control calculations used in this book. The major applications of digital calculations are summarized below.

Chapter 3: numerical solutions of differential equations using Euler or Runge–Kutta methods

Chapter 6: least squares fitting of parameters in dynamic models

Chapter 11: digital formulation of the PID controller

Chapter 12: implementation issues for digital control

Chapter 15: lead/lag elements for feedforward control

Chapter 19: digital formulation of the model predictive controller
Chapter 21: decoupling of multiple PID controllers
Chapter 23: dynamic matrix control

The material in Chapters 11, 12, and 23 is self-contained and will not be repeated here. The topics covered in this appendix involve the discrete forms of simple models used in process control. The process models will be represented using X as the input (cause) and Y as the output (effect). The current sampled value of a variable will be designated by the subscript n, the previous value by $n - 1$, and so forth, with the sample period being constant at Δt.

F.1 GAIN

The output of a gain is simply calculated as a proportion K of the input:

$$Y_n = KX_n \tag{F.1}$$

F.2 DEAD TIME

Dead time is simulated as a delay table of length Γ, which is an integer equal to $\theta/\Delta t$. At each time step, the model is executed by moving the past values to the location representing the next oldest value; the oldest value is discarded, and the previous value is placed in the table in the location of the most recent value. This calculation is summarized in Table F.1 with a delay table of length 4 (e.g., a dead time of 2 units of time and a sample period of 0.5), for eight time steps. The input is a pulse with a duration of two time steps.

This approach is simple to program and prevents the need to store all past data, because the table needs to store data for only the length of the dead time. More computationally efficient implementations move only one data point each execution and use an additional variable (pointer) to indicate the position of the oldest data in the table.

TABLE F.1
Example delay table for simulating dead time

Sample number, n	1	2	3	4	5	6	7	8
Input, X_n	0	1	1	0	0	0	0	0
Table entry 1, X_{n-1} (most recent)	0	0	1	1	0	0	0	0
Table entry 2, X_{n-2}	0	0	0	1	1	0	0	0
Table entry 3, X_{n-3}	0	0	0	0	1	1	0	0
Table entry 4, X_{n-4} (oldest)	0	0	0	0	0	1	1	0
Output, $Y_n = X_{n-4}$	0	0	0	0	0	1	1	0

F.3 FIRST-ORDER SYSTEM

Material and energy balances yield first-order differential equations, and the most common model is first-order with dead time. Thus, the first-order models are used frequently.

$$\tau \frac{dY}{dt} = KX - Y \tag{F.2}$$

The continuous model can be expressed as a discrete model by assuming that the input is constant at the value of X_{n-1} over the period t_{n-1} to t_n (or 0.0 to Δt). Then, integration of the differential equation from the initial condition Y_{n-1} can be performed to determine the value at Y_n. The solution can be determined using the integrating factor or Laplace transform; here the Laplace transform is demonstrated.

$$\tau s Y(s) - \tau Y_{n-1} = KX(s) - Y(s) \tag{F.3}$$

$$Y(s) = \frac{KX(s)}{\tau s + 1} + \frac{\tau Y_{n-1}}{\tau s + 1} \tag{F.4}$$

Note that $Y(t_{n-1}) = Y_{n-1}$. The input is evaluated as X_{n-1}/s, and the inverse transform can be taken to give

$$Y_n = K(1 - e^{-\Delta t/\tau})X_{n-1} + e^{-\Delta t/\tau} Y_{n-1} \tag{F.5}$$

Equation (F.5) gives exact sampled values if the process is truly first-order and the input is constant over the period. If the input changes during the period, then the use of X_{n-1} as a constant results in an approximation. An alternative, approximate model can be derived by approximating the derivative as a difference, $dY/dt \approx (Y_{n+1} - Y_n)/\Delta t$. This results in

$$Y_n = K\left(\frac{\Delta t}{\tau}\right) X_{n-1} + \left(1 - \frac{\Delta t}{\tau}\right) Y_{n-1} \tag{F.6}$$

Equations (F.5) and (F.6) give very similar results when the sample period is small compared with the time constant. For example, when $\Delta t/\tau = 0.05$, $e^{-\Delta t/\tau} = 0.951$ and $(1 - \Delta t/\tau) = 0.95$.

These discrete models can be used to represent a process and to implement a first-order filter, as described in Chapter 12. Also, the gain, dead time, and first-order discrete models can be combined to give for first-order with dead time:

$$Y_n = e^{-\Delta t/\tau} Y_{n-1} + K(1 - e^{-\Delta t/\tau}) X_{n-\Gamma-1} \tag{F.7}$$

Equation (F.7) is employed when using least squares to determine the values of the model parameters from discrete (sampled), empirical input-output data; it is also used as the prediction model in the IMC and Smith predictor model predictive control systems.

F.4 LEAD/LAG

The final discrete control calculation in this appendix is the lead/lag algorithm, which is as follows for a continuous system:

$$Y(s) = \frac{T_{\text{ld}}s + 1}{T_{\text{lg}}s + 1}X(s) \tag{F.8}$$

A straightforward manner for developing an approximate discrete lead/lag is to replace each "derivative," which is the product of the Laplace variable s and a variable, with its finite difference approximation. This gives

$$T_{\text{lg}}\left(\frac{Y_n - Y_{n-1}}{\Delta t}\right) + Y_n = T_{\text{ld}}\left(\frac{X_n - X_{n-1}}{\Delta t}\right) + X_n \tag{F.9}$$

This can be rearranged to give

$$Y_n = \left(\frac{\dfrac{T_{\text{lg}}}{\Delta t}}{\dfrac{T_{\text{lg}}}{\Delta t} + 1}\right)Y_{n-1} + \left(\frac{\dfrac{T_{\text{ld}}}{\Delta t} + 1}{\dfrac{T_{\text{lg}}}{\Delta t} + 1}\right)X_n - \left(\frac{\dfrac{T_{\text{ld}}}{\Delta t}}{\dfrac{T_{\text{lg}}}{\Delta t} + 1}\right)X_{n-1} \tag{F.10}$$

Equation (F.10) is used in feedforward controllers, as described in Chapter 15, and decouplers (another form of feedforward), as described in Chapter 21; it is also used for the combined IMC filter and controller, $G_f(s)G_{\text{cp}}(s)$, for a controller whose model has an invertible process factor that is first-order and with a filter that is first-order.

The discrete models of dynamic systems are in the form of difference equations, in which the current values of a variable can be expressed as a function of the last few values of the output and the input(s). In this appendix the difference equations have been formulated to calculate the nth sampled value. Any equation of this form can be modified to calculate, for example, the $(n+1)$th value. This can be done by substituting $n - 1 = m$ in the expressions; the result for the first-order system is

$$Y_{m+1} = K(1 - e^{-\Delta t/\tau})X_m + e^{-\Delta t/\tau}Y_m \tag{F.11}$$

Equations (F.5) and (F.11) are equivalent, and both formulations are commonly used, so the reader should be acquainted with both.

APPENDIX G

GUIDE TO SELECTED PROCESS EXAMPLES

Because of the strong interplay between process dynamics and control performance, examples should begin with process equipment and operating conditions. To this end, several process examples are introduced in the beginning chapters and used in many subsequent worked examples and questions. This approach has three advantages. First, the performance of different control approaches (e.g., tuning or control algorithm) can be evaluated on the same processes, allowing clear comparisons of competing methods. Second, the reader can concentrate on the learning objective applied to a familiar process. A final advantage is the reduction in the size of the book, since each example takes considerable space to introduce completely. Here we start as often as possible with a process and generalize to the behavior of similar process model structures.

Since the reader may want to review the control approaches applied to a process, this guide is provided. Major worked examples and questions involving the most important processes are summarized in the tables. The symbols used in the tables are Ex for a worked example, Q for a question at the end of a chapter, S for a chapter section, F for a figure, and T for a table; as elsewhere, the numbers before the period indicate the chapter.

G.1 HEAT EXCHANGER

This is a simple model of a heat exchanger. Since the process fluid side is well mixed and the utility side is at quasi-steady state, the basic model is first-order, which allows some analytical solutions to be determined. See Table G.1.

TABLE G.1
Heat exchanger

	Key issue addressed		Key issue addressed
Q 1.9	Possibility for feedback control	F 14.2	Cascade control
F 3.9	Process schematic	F 15.5	Feedforward control
F 3.10	Linearization	Q 15.2	Cascade and feedforward control
Ex 3.7	Derive balances and linearized approximation	F 16.10	Valve characteristic
		Q 19.6	IMC controller design
Ex 5.3	Exchanger with bypass	Q 19.12	IMC closed-loop frequency response
Q 5.1	Multiple input changes		
Q 5.2	Jacketed heat exhanger	Ex 24.5	Effect of operating conditions
Ex 8.5	Analytical solution for proportional-integral feedback system	Q 24.3	Further changes in operating conditions

G.2 THREE-TANK MIXING PROCESS

The most often used process example is the three-tank mixing process. An important aspect of the process is its simplicity, allowing the reader to easily relate the design and operating parameters to its dynamic behavior. However, the process has been selected to elucidate many important factors in process control systems. This process is third-order and can be made unstable with a proportional-only controller; is mildly nonlinear and can show the acceptable range of linearization; does not conform to the first-order-with-dead-time model and can show the effects of structural errors in a model; and has dynamics that depend on operating conditions and can demonstrate the use of adaptive retuning.

In addition to those listed in Table G.2, the following topics address closely associated series of tanks: Q 15.2 on multitank heat transfer and Q 21.13 on loop pairing.

G.3 NONISOTHERMAL STIRRED-TANK CHEMICAL REACTOR (CSTR)

The nonisothermal CSTR is an important industrial process that introduces the opportunity for a diverse range of process dynamics. This example involves only a single, exothermic chemical reaction and can have stable over- and underdamped steady states as well as a locally unstable steady state(s). Also, important in the presentation control technology is the opportunity to investigate different pairings of manipulated and controlled variables in a multiloop control system.

The final sections in Table G.3 refer to Appendix C, which introduces some advanced topics in reactor dynamics and control.

G.4 TWO-PRODUCT DISTILLATION COLUMN

The previous processes were of low order, so they could be represented by a few differential equations. In addition to being an important industrial process, distillation

TABLE G.2
Three-tank mixing process

	Key issue addressed		Key issue addressed
Ex 6.3	Process reaction curve	Q 11.9*a*	Execution period for digital control
Ex 7.2	Introduce the process model	S 13.5	Effect of model mismatch on closed-loop frequency response
S 8.3	Evaluate zero offset for P-only control	Q 13.1	Effect of process dynamics on performance and tuning
S 8.4	Evaluate zero offset for I-only control	Q 13.2	Rank process designs for performance
S 8.5	Evaluate zero offset for D-only control	Q 13.13	Repeat stability, tuning, and performance analysis after process change
Q 8.2	Dynamic simulation	S 16.2	Effect of flow rate on tuning
Q 8.12	Alternative process structure	S 16.3	Gain (tuning) scheduling
Ex 9.2	Tuning and performance	Q 16.1	Effect of set point on tuning
Ex 9.3	Effect of disturbance time constant on closed-loop performance	Q 16.2	Ziegler-Nichols tuning
Q 9.8	Dimensional analysis	Ex 19.3	IMC on third-order process
Ex 10.5	Roots of characteristic equation, root locus	Ex 19.4	IMC on approximate first-order-with-dead-time model
Ex 10.10	Ziegler-Nichols tuning	Ex 19.6	IMC digital implementation and simulation
Ex 10.18	Effect of model mismatch on stability analysis using $G_{OL}(s)$	Ex 19.7	IMC tuning correlations
Q 10.1	Effect on tuning of changing tank volume	Ex 19.8	IMC robustness
Q 10.11	The effect of process and control structures on possible dynamic responses	Ex 19.9	Smith predictor tuning and simulation
Q 10.12	Perfect feedback controller	Q 19.2	IMC tuning schedule
Q 10.17	Effect of adding dead time	Q 22.1	Variable-structure

is a high-order system whose linearized fundamental models are not normally analyzed. Also, the dynamic model formulation using the generalized tray concept is a worthwhile reinforcement to similar approaches covered in steady-state modelling. With two controlled compositions, the process offers a challenging two input–two output control system, when the control of pressure and levels is assumed. With no prior assumptions, the control design of a five input–five output system is a good control design case. To maintain simplicity, the case considered involves only binary distillation with constant relative volatility. In Table G.4, the cases not conforming to the exact parameters in Example 5.6 are marked with an asterisk (*).

G.5 TWO SERIES ISOTHERMAL CONTINUOUS STIRRED-TANK REACTORS (CSTR)

Additional low-order process examples are useful to reinforce principles. A series of two isothermal CSTRs is used throughout the book (see Table G.5) to provide many of these examples. Only one reaction occurs in each reactor, and the reactions

TABLE G.3
Nonisothermal CSTR

	Key issue addressed		Key issue addressed
Ex 3.10	Model, linearize and simulate	Q 8.17	General behavior under P-only and PD control
Ex 3.11	Steady-state model and operating window	Q 10.11	Effect of process and control structure on possible dynamic responses
Q 3.7	Effect of changing ΔH_{rxn}	Q 12.5	Failure modes
Ex 4.8	Solve step change using Laplace transforms	Q 19.9	IMC and Smith predictor
Ex 4.10	Transfer function	Q 20.2	The effect of ΔH_{rxn} on multiloop stability and dynamic response
Ex 4.17	Frequency response	Q 20.11	Integral controllability, loop pairing, and tuning for various sets of design parameters
Q 4.1	Emergency response	F 22.5, F 22.6	Variable-structure control, signal select
Q 4.5	Dynamics with $\Delta H_{rxn} = 0$	Ex 24.4	Dynamic transient exceeding steady-state operating window
Q 4.7	Transfer function with different inputs and outputs	S C.1	Derivation of energy balance
Ex 5.7	Multiple input-output block diagram and transfer functions	S C.2	Possibility of multiple steady states and their stability
Ex 5.8	Dynamic response for changes in two input variables	S C.3	Possibility of limit cycles
Q 6.17	Empirical model building		
Q 7.9	Evaluate proposed single-loop feedback control structures		
Q 7.11*b*	Operating window		
Q 8.15	Proportional-only control		

TABLE G.4
Two-product distillation column

	Key issue addressed		Key issue addressed
Q 2.8*	Effect of distribution on profit	Ex 21.2	Effect of control structure on multiloop control performance
Q 2.9*	Effect of distribution on profit	Ex 21.3	Effect of disturbance type on multiloop control performance
S 5.6	Model development	Ex 21.6	Relative gain and loop pairings
Ex 5.6	Simulated dynamic response	Ex 21.9	Match tuning with performance goals
Q 6.10	Process reaction curve	Ex 21.10	Decoupling, perfect and with model errors
Q 14.6*	Cascade control	Q 21.1*	Tuning, loop pairing, performance and decoupling
Ex 15.7*	Feedforward control	Q 21.8*	Tuning, loop pairing, performance and decoupling
S 17.5*	Inferential tray temperature	Q 21.11*	Control loop pairing
Ex 20.2	Linearized model	Ex 23.1	Complexity of analytical inverse
Ex 20.4	Operating window	Ex 23.6	DMC control
Ex 20.5	Evaluation of controllability		
Ex 20.6	Effect of interaction on the changes in manipulated variable		
Q 20.9*	Controllability, interaction, tuning		
Q 20.15*	Controllability, interaction, tuning		

TABLE G.5
Two isothermal CSTRs

	Key issue addressed		Key issue addressed
Ex 3.3	Derive process model and evaluate a step response	Q 5.4	Investigate system in Ex 5.4 with different input variable
Q 3.15	Impulse response	Q 7.11*c*	Operating window
Ex 4.6	Solve step response using Laplace transforms (slightly modified model)	Q 9.10	Effect of changing temperature on tuning
Ex 4.7	Solve step response using Laplace transforms	Ex 10.4	Roots of closed-loop characteristic equation (modified process)
Ex 4.9	Derive the transfer function	Ex 10.8	Repeat Ex 10.4 with additional dead time
Ex 4.12	Block diagram	Q 10.11	Effect of process and control structure on possible dynamic responses
Ex 4.16	Frequency response	Ex 13.8	Effect of inverse response on control performance
Q 4.1	Emergency response	Q 13.19	Effect of an alternative manipulated variable on control performance
Q 4.16	Derive model and dynamic response for a different input variable	Q 15.11	Effect of dynamics on feedforward-feedback control
Ex 5.4	Dynamic response for input yielding parallel dynamic structure and inverse response	Q 21.10	Multiloop control

are first-order. The model for this process is simple enough to enable the engineer to determine the effects of changes in equipment and operating parameters on the dynamics of the process and performance of the feedback control system.

G.6 HEAT EXCHANGE AND FLASH DRUM

A flash drum at controlled pressure and temperature is a simple method for effecting a physical separation of components with different vapor pressures. This process provides the opportunity to evaluate inferential control and pair loops for dynamic performance. See Table G.6.

TABLE G.6
Heat exchange and flash drum

	Key issue addressed		Key issue addressed
Q 1.6	Control system components	Ex 24.7	Controllability
S 2.2	Control objectives	Ex 24.8	Operating window
S 17.2	Inferential variable evaluation	Ex 24.9	Loop pairing
Q 21.5	Loop pairing	Ex 24.10	Algorithm selection and tuning
T 24.1	Control Design Form (CDF)	Ex 24.11	Control for safety
Ex 24.1	Sensors	Ex 24.12	Process monitoring
Ex 24.2	Final elements	Ex 24.13	Dynamic performance
Ex 24.6	Degrees of freedom		

APPENDIX H

MODEL FOR FLASH PROCESS

The flash process in Figures 24.1 and 24.20 is analyzed thoroughly in Chapter 24 as an example of process and control design. To perform several key steps of this analysis, a model of the process is required. The development of a simple model, involving algebraic and ordinary differential equations, is described in this appendix. The base-case conditions for the process are listed in Table H.1. This physical system

TABLE H.1
Base-case operating data for flash process

	Feed	After first exchanger	After second exchanger	After flash*
Liquid flow, moles/time	100	100	100	54.62
Temperature, °C	0	25	66	25.7
Liquid composition‡	0.10, 0.20, 0.30, 0.15, 0.20, 0.05	Same as the feed	Same as the feed	0.013, 0.10, 0.302, 0.201, 0.30, 0.084
Vapor flow, moles/time				46.38
Pressure, kPa				1000
Vapor composition‡				0.198, 0.312, 0.298, 0.088, 0.093, 0.011

*The same conditions as Table 17.1 for inferential control example (page 558).

‡Components are methane, ethane, propane, *i*-butane, *n*-butane, and *n*-pentane.

is the same as the flash process considered in Section 17.2 on inferential control, except for the additional feed heat exchanger. Also, the changes in sensors and final elements proposed in the examples in Sections 24.3 and 24.4 have been included.

The actual physical system in Figure 24.20 is represented by the approximate process in Figure 24.14 (page 806), which has only lumped parameter elements. To clarify the procedure, the models for each section of the process are presented individually, with relevant assumptions. The degrees of freedom of each section are also determined as the models are developed, although the degrees-of-freedom and controllability analysis must be performed on the integrated system. Assumptions that are common to all sections of the models are the following:

1. All volumes are well-mixed.
2. Densities, heat capacities, and heat transfer coefficients are constant.
3. Heat losses are negligible.
4. Valve characteristics are linear.

H.1 SECTION 1

The first section is the heat exchanger using the utility fluid. It is modelled as two stirred tanks exchanging heat, with the following assumptions:

1. The tank volumes are constant.
2. Heats of mixing are negligible.
3. $C_v \approx C_p$.

The energy balances are

$$\rho_H V_{\mathrm{EH1}} C_{pH} \frac{dT_4}{dt} = \rho_H C_{pH} F_2 (T_3 - T_4) - Q_H \tag{H.1}$$

$$\rho V_{\mathrm{E1}} C_p \frac{dT_2}{dt} = \rho C_p F_1 (T_2 - T_1) + Q_H \tag{H.2}$$

$$Q_H = U A_H (T_4 - T_2) \tag{H.3}$$

The utility flow is given by the following equation, and the process flow, F_1, cannot be determined by information in this section (but is modelled later).

$$F_2 = K_{v1} v_1 \tag{H.4}$$

The material balances for each component i, with X designating mole fraction, are

$$\frac{d}{dt}\left(\frac{V_{\mathrm{E1}}\rho}{(\mathrm{MW}_{\mathrm{ave}})_1} X_{1i}\right) = F_1 (X_{0i} - X_{1i}) \qquad i = 1, 6 \tag{H.5}$$

$$(\mathrm{MW}_{\mathrm{ave}})_1 = \sum_{i=1}^{6} \mathrm{MW}_i X_{1i} \tag{H.6}$$

The degrees-of-freedom analysis for the first section is

11 equations: (H.1) to (H.6)

12 dependent variables: $T_2, T_4, F_1, F_2, Q_H, X_{1i}$, and $(\text{MW}_{\text{ave}})_1$

7 specified external variables (disturbances): T_1, T_3, X_{0i} $(i = 1, 5 \text{ since } \sum X_i = 1)$

1 external manipulated: v_1

Note that the number of variables exceeds the number of equations for this section; however, there is no requirement that these values match for a section of an integrated process. The reason for the difference in this example is the process flow, F_1, which is affected by a valve in the flash section.

H.2 SECTION 2

The second section includes the heat exchanger in which steam is condensed, which is again modelled as two tanks. The dynamic energy balance for the process and the quasi-steady-state model for the steam are

$$\rho C_p V_{E2} \frac{dT_7}{dt} = \rho C_p F_1 (T_2 - T_7) + Q_s \tag{H.7}$$

$$Q_s = UA_s (T_s - T_7) \tag{H.8}$$

$$T_s = f(P_s) \tag{H.9}$$

The material balance on the steam is given by

$$\left(\frac{V_s \text{MW}_s}{RT_s} \right) \frac{dP_s}{dt} = \rho_s F_3 - F_{\text{cond}} \tag{H.10}$$

$$F_3 = K_{v2} v_2 \sqrt{\frac{P_0 - P_s}{\rho_s}} \tag{H.11}$$

$$F_{\text{cond}} = \frac{Q_s}{\Delta H_{\text{cond}}} \tag{H.12}$$

The component balances on the process liquid are

$$\frac{d}{dt} \left(\frac{V_{E2} \rho}{(\text{MW}_{\text{ave}})_2} X_{2i} \right) = F_1 (X_{1i} - X_{2i}) \qquad i = 1, 6 \tag{H.13}$$

$$(\text{MW}_{\text{ave}})_2 = \sum_{i=1}^{6} \text{MW}_i X_{2i} \tag{H.14}$$

The degrees-of-freedom analysis for the second section is

13 equations: (H.7) to (H.14)

13 dependent variables: $T_s, T_7, F_{\text{cond}}, F_3, P_s, Q_s, X_{2i}, \text{MW}_2$

1 specified external variable (disturbance): P_0

1 external manipulated variable: v_2

7 external variables from other sections: T_2, F_1, X_{1i}

Note that variables that are defined in a previous section are "external" variables to this section and do not contribute to the number of dependent variables in this system.

H.3 SECTION 3

The third section uses the standard model for a constant-enthalpy flash process (Smith and Van Ness, 1987). The quasi-steady-state material balance and equilibrium constants are

$$K_i = (a_i + b_i T_6)\frac{P_b}{P_1} \qquad i = 1, 6 \tag{H.15}$$

$$X_{3i} = \frac{Y_{3i}}{K_i} \qquad i = 1, 6 \tag{H.16}$$

$$Y_{3i} = \frac{\left(\dfrac{F_1 \rho}{(\mathrm{MW}_{\mathrm{ave}})_2}\right)}{\mathrm{FM}_{V3}} \left(\frac{K_i X_{2i}}{\dfrac{\mathrm{FM}_{\mathrm{L}3}}{\mathrm{FM}_{V3}} + K_i} \right) \qquad i = 1, 6 \tag{H.17}$$

$$\sum_{i=1}^{6} Y_{3i} = 1.0 \tag{H.18}$$

$$\rho F_1 = (\mathrm{MW}_{\mathrm{ave}})_{V3}\mathrm{FM}_{V3} + (\mathrm{MW}_{\mathrm{ave}})_{\mathrm{L}3}\mathrm{FM}_{\mathrm{L}3} \tag{H.19}$$

$$(\mathrm{MW}_{\mathrm{ave}})_{V3} = \sum_{i=1}^{6} \mathrm{MW}_i Y_{3i} \tag{H.20}$$

$$(\mathrm{MW}_{\mathrm{ave}})_{L3} = \sum_{i=1}^{6} \mathrm{MW}_i X_{3i} \tag{H.21}$$

with P_b the base-case pressure and FM designating molar flow. The quasi-steady-state energy balance involves no accumulation term:

$$\begin{aligned} \rho F_1 C_p (T_7 - T_{\mathrm{ref}}) = {} & (\mathrm{MW}_{\mathrm{ave}})_{V3}\mathrm{FM}_{V3} C_{pV}(T_6 - T_{\mathrm{ref}}) \\ & + (\mathrm{MW}_{\mathrm{ave}})_{\mathrm{L}3}\mathrm{FM}_{\mathrm{L}3} C_{pL}(T_6 - T_{\mathrm{ref}}) \\ & + \mathrm{FM}_{V3}\sum_{i=1}^{6} Y_{3i}\Delta H_{\mathrm{vap}i} \end{aligned} \tag{H.22}$$

Finally, the feed flow is determined by

$$F_1 = K_{v3} v_3 \tag{H.23}$$

The degrees-of-freedom analysis for the third section is

24 equations: (H.15) through (H.23)

24 dependent variables: T_6, P_1, FM_{L3}, FM_{V3}, K_i, X_{3i}, Y_{3i}, $(\text{MW}_{\text{ave}})_{V3}$, $(\text{MW}_{\text{ave}})_{\text{L3}}$

0 specified external variables (disturbances)

1 external manipulated variable: v_3

7 external variables from other sections: T_7, F_1, X_{2i}

H.4 SECTION 4

The fourth and final section includes the flash drum. An important assumption is that the vapor and liquid phases, which were in equilibrium at the flash point, are not in contact in the drum. This assumption is not entirely valid; changes in pressure, temperature, and composition would lead to some exchange between the phases in the drum. However, the exchange is complex to model, because the systems may not be in equilibrium; thus, the simplified model is used here.

The vapor space is assumed to be of constant volume. The overall and component material balances on the vapor space are

$$\frac{V_V(\text{MW}_{\text{ave}})_{V4}}{RT_6}\frac{dP_1}{dt} = (\text{MW}_{\text{ave}})_{V3}\text{FM}_{V3} - F_4\rho_V \tag{H.24}$$

$$F_4 = K_{v5}v_5\sqrt{\frac{P_1 - P_2}{\rho_V}} \tag{H.25}$$

$$\frac{d}{dt}\left(\frac{\rho_V V_V}{(\text{MW}_{\text{ave}})_{V4}}Y_{4i}\right) = \text{FM}_{V3}Y_{3i} - \left(\frac{F_4 P_{\text{st}}}{RT_6}\right)Y_{4i} \qquad i = 1, 6 \tag{H.26}$$

$$(\text{MW}_{\text{ave}})_{V4} = \sum_{i=1}^{6} \text{MW}_i Y_{4i} \tag{H.27}$$

with P_{st} the standard pressure for which the volumetric flow F_4 is expressed. The liquid inventory is modelled by the following overall and component balances:

$$\rho A_c \frac{dL_1}{dt} = (\text{MW}_{\text{ave}})_{\text{L3}}\text{FM}_{\text{L3}} - \rho_v F_5 \tag{H.28}$$

$$\frac{d}{dt}\left(\frac{\rho(A_c L_1)}{(\text{MW}_{\text{ave}})_{\text{L4}}}X_{4i}\right) = \text{FM}_{\text{L3}}X_{3i} - \frac{\rho F_5}{(\text{MW}_{\text{ave}})_{\text{L4}}}X_{4i} \qquad i = 1, 6 \tag{H.29}$$

$$F_5 = K_{v4}v_4 \tag{H.30}$$

$$(\text{MW}_{\text{ave}})_{\text{L4}} = \sum_{i=1}^{6} \text{MW}_i X_{4i} \tag{H.31}$$

The degrees-of-freedom analysis for the fourth section is

18 equations: (H.24) to (H.31)
17 dependent variables: P_1, L_1, F_4, F_5, $(\mathrm{MW}_{\mathrm{ave}})_{\mathrm{L4}}$, X_{4i}, Y_{4i}
1 specified external variable (disturbance): P_2
2 external manipulated variables: v_4, v_5
14 variables from other sections: $\mathrm{FM}_{\mathrm{L3}}$, FM_{v3}, T_6, X_{3i}, Y_{3i}

REFERENCES

Smith, J. and H. Van Ness, *Introduction to Chemical Engineering Thermodynamics* (4th Ed.), McGraw-Hill, New York, 1987.

INDEX

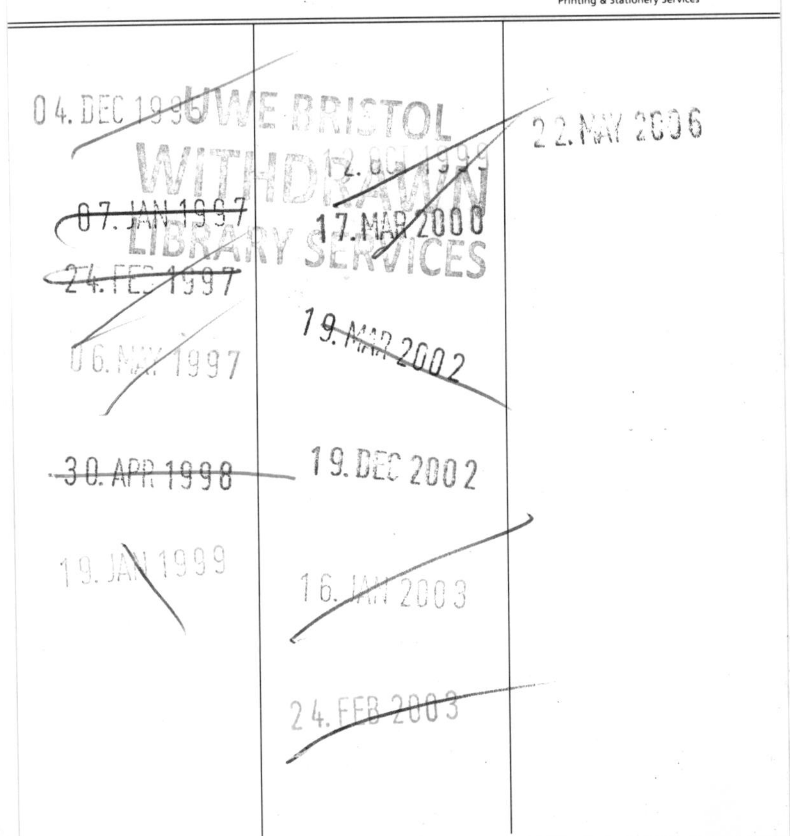

04. DEC 1996
07. JAN 1997
24. FEB 1997
06. MAY 1997
30. APR 1998
19. JAN 1999
12. OCT 1999
17. MAR 2000
19. MAR 2002
19. DEC 2002
16. JAN 2003
24. FEB 2003
22. MAY 2006
UWE BRISTOL
WITHDRAWN
LIBRARY SERVICES